The Dictionary of Substances and their Effects

ROYAL SOCIETY OF CHEMISTRY

Production team

Janet Crombie (Staff Editor)
Rebecca Allen
Barbara Edwards
Susan Knight
Alan Skull
Zelda Smith
Ghislaine Tibbs
Colin Newham
Carol Fletcher
Jayne Jackson

James Butler (Design)

The information contained in this book has been compiled with the greatest
care from sources believed to be reliable and to represent the best opinion
on the subject as of 1992. However, no warranty, guarantee, or
representation is made by the Royal Society of Chemistry as to the
correctness of any information herein, and the publisher assumes no
responsibility in connection therewith: nor can it be assumed that all
necessary warnings and precautionary measures are contained in this
publication, or that other or additional information or measures may not be
required or desirable because of particular or exceptional circumstances, or
because of new or changed legislation.

First published 1992
Reprinted 1995

ISBN 0-85186-331-0

Photocomposed by
Land & Unwin (Data Services) Ltd.
Bugbrooke, Northamptonshire
Printed in Great Britain
by Antony Rowe Ltd

CONTENTS

INTRODUCTION

The Royal Society of Chemistry's Dictionary of Substances and their Effects (DOSE) is a compilation of data on over 5000 chemicals which have some impact on the environment.

Assessment of the risks associated with chemicals entering the environment cannot be undertaken without an adequate knowledge of the chemistry of the substances involved. However, risk assessment is more than simply a matter of chemistry; it is a multidisciplinary subject in which knowledge of the widely differing biological effects of a chemical on various species is equally important. The Editorial Board was very aware of the need for a single work containing data on physico-chemical properties, human and other mammalian toxicology and adverse effects to aquatic life, birds and other non-target species.

The question of which chemicals to cover was a difficult one. Obviously DOSE could neither include all the chemicals in Chemical Abstracts - now over ten million - nor even the hundred thousand listed in the European Inventory of Existing Chemical Substances (EINECS). It was therefore decided to concentrate on those considered to have potentially the greatest impact on organisms in whatever environment, whether on land, in water or in the air. Chemicals were selected from sources such as the Authorised and Approved List from the EC's Classification, Packaging and Labelling Regulations; the EC's 'Black' and 'Grey' lists of dangerous substances; the 'Red' list drawn up by the UK Department of the Environment; the Priority Pollutant Lists from the USA and Canada; the Federal German Republic's Pollutant List; a list of chemicals identified by advanced analytical techniques as being present in water samples; and a variety of scientific papers listing chemicals with adverse effects. In choosing the data elements to include, particular note was taken of the legislative requirements of the European Community, the USA and Japan.

The probability that a particular level of a chemical in the environment will be hazardous to the exposed organisms can be extremely difficult to predict, and it is essential that rigorous scientific methods are used. DOSE is not a source of ready-made risk assessments, but by providing referenced data on the effects of chemicals in both the workplace and the environment, it will enable the health, safety or environmental professional to make a rapid, integrated assessment of risk for a given situation. For a number of the chemicals included, particularly those detected at low concentration in water samples, very few data are available. In these cases a precise risk assessment is not possible, but DOSE does provide background data from which to work.

Chemicals, whether natural or manmade, are essential to today's economy and lifestyle, and risks associated with exposure to them are inevitable. The way in which we assess these risks is important not only to ourselves but to the environment and to the species with which we share it. In a democratic society we all bear some responsibility for the decisions made on our behalf, decisions which may have worldwide consequences. Poison gases or pollutants in rivers do not recognise national boundaries.

DOSE will be of special value to scientists and others who are involved in making these complex decisions, and represents an important resource for anyone concerned with the impact of chemicals on human health or the environment.

HOW TO USE DOSE

Introduction

The data in DOSE is organised under the data headings summarised below:

Identifiers
Chemical name
Structure
CAS Registry Number
Synonyms
Molecular formula
Molecular weight
Uses
Occurrence

Physical properties
Melting point
Boiling point
Flash point
Specific gravity
Partition coefficient
Volatility
Solubility

Occupational exposure
Limit values
UN number
HAZCHEM number
Conveyance classification
Supply classification
Risk phrases
Safety phrases

Ecotoxicity
Fish toxicity
Invertebrate toxicity
Bioaccumulation
Effects to non-target species

Environmental fate
Nitrification inhibition
Carbonaceous inhibition
Anaerobic effects
Degradation studies
Abiotic removal
Absorption

Mammalian and avian toxicity
Acute data
Sub-acute data
Carcinogenicity and long-term effects
Teratogenicity and reproductive effects
Metabolism and pharmacokinetics
Irritancy
Sensitisation
Genotoxicity
Any other adverse effects to man
Any other adverse effects

Legislation

Any other comments

References

These headings only appear in an item when data have been identified for that heading. The reader can, therefore, assume that the absence of a heading means that no data has been found.

Identifiers and basic chemistry

The identifiers section of the item provides data on synonyms and approved codes used to define the precise chemical structure of the compound.

Chemical Abstracts Registry Number

The Chemical Abstracts Registry Number is a number sequence adopted by the Chemical Abstracts Service (American Chemical Society, Columbus, Ohio, USA) to uniquely identify specific chemical substances. The number contains no information relating to the chemical structure of a substance and is, in effect, a catalogue number relating to one of the 10 million or so unique chemical substances recorded in the Chemical Abstracts Registry: new numbers are assigned sequentially to each new compound identified by Chemical Abstracts Service. This information is also provided in the full index of CAS Registry Numbers available at the end of each volume.

Synonyms

For common chemicals, several chemical names and numerous trade names may be applied to describe the chemical in question. In the synonyms field, many of these names are identified to aid users on the range of names which have been used to describe each substance. This information is also provided in the full index of alternative names available at the end of each volume.

Molecular formula

This is the elemental composition of the compound. The elements appear alphabetically for inorganic compounds, ie. Ag_2CO_3, Cl_2Cr, etc, but for organic compounds, carbon and hydrogen content are shown first followed by the other elements in alphabetical order, ie. C_6H_5Br. This information is also provided in the molecular formula index at the end of each volume.

Molecular weight

This is directly calculated from the molecular formula. No molecular weights are given for polymers.

Uses

Principal uses of the substances are given, with information on other significant uses in industrial processes.

Occurrence

Natural occurrences, whether in plants, animals or fungi are reported. Natural emissions, combustion and/or pyrolysis processes are also given.

Physical properties

Melting/Boiling point

These data are derived from various sources.

Flash point

The flash point is the lowest temperature at which the vapours of a volatile combustible substance will sustain combustion in air when exposed to a flame. The flash-point information is derived from various sources. Where possible the method of determination of the flash point is given.

Specific gravity (density)
The specific gravity of each substance has been derived from a variety of sources. Where possible the data have been standardised.

Partition coefficient
Partition coefficients, important for structure-activity relationship considerations, particularly in the aquatic environment, are indicated. Ideally the n-octanol/water partition coefficient is quoted. The major data source for this measurement is:

Sangster, J. *J. Phys. Chem. Ref. Data* 1989, **18**(3), 1111-1229

Where no reference is quoted, it can be assumed that the information was derived from this source.

Volatility
The vapour pressure and vapour density are quoted were available. Where possible, the data have been standardised.

Solubility
Solubility data derived from several sources are quoted for both water and organic solvents.

Occupational exposure

Threshold limit values
The airborne limits of permitted concentrations of hazardous chemicals represent conditions under which it is believed that nearly all workers may be repeatedly exposed day after day without adverse effect. These limits are subject to periodic revision and vary between different countries. The term *threshold limit* relates primarily to the USA, but equivalent terms are available in most industrialised countries. In DOSE items, comparable values are available for the USA and United Kingdom. The data relates to concentrations of substances expressed in *parts per million (ppm)* and *milligrams per cubic meter (mg m^{-3})*.

USA threshold limits The threshold limit values for the USA have been taken from the *Threshold Limit Values and Biological Exposure Indices, 1991-1992* produced by the American Conference of Governmental Industrial Hygienists, Cincinnati, USA. The limits relate to *Threshold Limit - Time Weighted Average, Threshold Limit - Short Term Exposure Limit* and *Threshold Limit - Ceiling Limit*. The Threshold Limit Value - Time Weighted Average (TLV-TWA) allows a time-weighted average concentration for a normal 8-hour working day and a 40-hour working week, to which nearly all workers may be repeatedly exposed day after day, without adverse effect. The Threshold Limit Value - Short Term Exposure Limit (TLV-STEL) is defined as a 15-minute, time-weighted average which should not be exceeded at any time during a work day, even if the 8-hour time-weighted average is within the TLV. It is designed to protect workers from chemicals which may cause irritancy, chronic or irreversible tissue damage, or narcosis likely to cause the likelihood of accidental injury. Many STELs have been deleted pending further toxicological assessment. With Threshold Limit - Ceiling Values (TLV-C) the concentration should not be exceeded during any part of the working day.

UK exposure limits The occupational limits relating to airborne substances hazardous to health are published by the Health and Safety Executive annually in Guidance Note EH 40. The values in the data sheets have been taken from the 1992 edition.

In the United Kingdom, there are Maximum Exposure Levels (MEL) which are subject to regulation and which should not normally be exceeded. They derive from Regulations, Approved Codes of Practice, European Community Directives, or from the Health and Safety Commission. In addition, there are Occupational Exposure Standards (OES) which are considered to represent good practice and realistic criteria for the control of exposure. In an analogous fashion to the USA Threshold Limits, there are long-term limits, expressed as time-weighted average concentrations over an 8-hour working day, designed to protect workers against the effects of long-term exposure. The short-term exposure limit is for a time-weighted average of 10 minutes. For those substances for which no short-term limit is listed, it is recommended that a figure of three times the long-term exposure limit averaged over a 10-minute period be used as a guideline for controlling exposure to short-term excursions.

UN number
The United Nations Number is a four-figure code used to identify hazardous chemicals and is used for identification of chemicals transported internationally by road, rail and by air.

HAZCHEM code
The Hazchem Code is a code used by United Kingdom emergency services to classify hazards of chemicals transported by road. It is administered by the Chemical Industries Association.

Conveyance classification
The information presented for the transportation of substances dangerous for conveyance by road is derived from Information Approved for the Classification, Packaging and Labelling of Dangerous Substances for Supply and Conveyance by Road, HSC, 1990.

Supply classification
The information presented for the supply of substances dangerous for conveyance by road is derived from Information Approved for the Classification, Packaging and Labelling of Dangerous Substances for Supply and Conveyance by Road, HSC, 1990, and Approved Substance Identification Numbers, Emergency Action Codes and Classification for Dangerous Substances in Road Tankers and Tank Containers, HSC, 1989.

Risk and safety precautions
Risk and safety precaution phrases used in connection with DOSE items are approved phrases for describing the risks involved in the use of hazardous chemicals and have validity in the United Kingdom and throughout the countries of the European Community. The approved texts have designated R (Risk) and S (Safety) numbers from which it is possible to provide translations for all approved languages adopted by the European Community. The risk and safety precaution phrases relate to the Classification, Packaging and Labelling Regulations, 1984 and The Road Traffic (Carriage of Dangerous Substances in Packages etc.) Regulations, 1986. The information is provided in The Authorised and Approved List. Information approved for the classification, packaging and labelling of dangerous substances for supply and

conveyance by road (third edition) HSC, HMSO, 1990. The risk and safety phrases should be used to describe the hazards of chemicals on data sheets for use and supply; for labelling of containers, storage drums, tanks etc., and for labelling of articles specified as dangerous for conveyance by road. The law specifies supply labels as those required to warn/inform the user of acute and chronic exposure hazards. They are to be placed on receptacles from which a dangerous substance is to be dispensed. The law also specifies conveyance labels used to inform the transporter, emergency services and the public, which must be placed on the outer packaging and which must take into account the acute risks posed by the substance. Further details on the use and compilation of labels may be found in the HSE publication HS(R)22 A Guide to the Classification, Packaging and Labelling of Dangerous Substances Regulations, 1984.

Ecotoxicity

Information is presented on the effects of chemicals on various ecosystems. Results of studies carried out on aquatic species, primarily fish and invertebrates, but also fresh water and marine microorganisms and plants are reported. Persistence and potential for accumulation in the environment and any available information on the harmful effects to non-target species, i.e. the unintentional exposure of terrestrial and/or aquatic species to a toxic substance is given.

Fish toxicity
LC_{50}, with duration of exposure, are quoted for two species of freshwater and one marine species if available. Any additional information on bioassay type (static or flow through) and water condition (pH and temperature) is reported.

Invertebrate toxicity
$LC_{50,}$ with duration of exposure, are quoted for molluscs and crustaceans. EC_{50} values, i.e. concentrations which will immobilise 50% of the exposed population and/or the effective concentration (EC) which will inhibit microbial or algae growth are reported with the duration of the exposure if available.

Bioaccumulation
Bioaccumulation, biomagnification and bioconcentration data are quoted primarily for fish, invertebrates, bacteria and algae.

Effects to non-target species
Harmful effects arising from accidental exposure of terrestrial, avian, aquatic and insect populations to toxic chemicals are reported.

Environmental Fate

Degradation data is used to assess the persistence of a chemical substance in the environment, in water, soil and air. If the substance does not persist, information on the degradation products is also desirable. Intermediates may be either harmless or toxic substances which will themselves persist. Degradation occurs via two major routes, microbial degradation utilising microorganisms from a variety of habitats and decomposition by chemical methods. Microbial degradation is associated with the production of elemental carbon, nitrogen and sulfur from complex molecules. Standard biodegradation tests estimate the importance of microbial biodegradation as a persistence factor. Most tests use relatively dense microbial populations adapted to the compound being studied. Rapid degradation results in these tests implies that the compound will degrade under most environmental conditions, although specialised environments where degradation would not occur can exist. Compounds which are not readily degradable are likely to persist over a wide range of environmental situations. Chemical degradation processes include photolysis, hydrolysis, oxidation and removal by reversible/irreversible binding to sediment. Factors which influence degradation rates: duration of exposure; temperature; pH; salinity; concentrations of test substance, microbial populations, and other nutrients; must also be taken into account.

Nitrification inhibition

The nitrogen cycle is the major biogeochemical process in the production of nitrogen, an essential element contained in amino acids and proteins. Nitrogen is an essential element in microorganisms, higher plants and animals. Interference in the production of nitrogen from more complex molecules can be determined by standard tests using nitrogen-fixing bacteria. The degree of inhibition can be used to estimate the environmental impact of the test chemical.

Carbonaceous inhibition

Another major biogeochemical process is the recycling of carbon via the decomposition of complex organic matter by bacteria and fungi. In nature the process is important in the cycling of elements and nutrients in ecosystems. The degradation sequence occurs in stages, cellulose cellobiose glucose organic acids and carbon dioxide. Chemical inhibition of microbial processes at all or any of these stages is reported here.

Anaerobic effects

Anaerobic microbial degradation of organic compounds occurs in the absence of oxygen and is an important degradation process in both the natural environment and in waste treatment plants. Methods to assess the degree of importance of anaerobic processes in the degradation of a particular substance are reported here. These include laboratory simulations of the natural processes occurring in flooded soils using flasks containing samples of the medium and radiolabelled substrate. The other important method uses anaerobic digestion tests which compare the production of methane and carbon dioxide by anaerobic microbes in a sludge sample with and without added test material. Methane production is at the end of the food chain process used by a wide range of anaerobic microorganisms.

Degradation studies

This section focuses on microbial degradation in both soil and water. The half-life of the chemical substance in the environment is reported with its degradation products where possible, giving an indication of the degree of its persistence. Water pollution factors: BOD, biochemical oxygen demand; COD, chemical oxygen demand; and ThOD, theoretical oxygen demand are stated, where available. BOD estimates the extent of natural purification which would occur if a substance is discharged into rivers, lakes or the sea. COD is a quicker chemical method for this determination which uses potassium dichromate or permanganate to establish the extent of oxidation likely to occur. ThOD measures the amount of oxygen needed to oxidise hydrocarbons to carbon dioxide and water. When organic molecules contain other elements nitrogen, sulfur or phosphorus, the ThOD depends on the final oxidation stage of these elements.

Abiotic removal
Information on chemical decomposition processes is contained in this section. The energy from the sun is able to break carbon-carbon, and carbon-hydrogen bonds, cause photodissociation of nitrogen dioxide to nitric oxide and atomic oxygen and photolytically produce significant amounts of hydroxyl radicals. Hydrolysis occurs when a substance present in water is able to react with the hydrogen or hydroxyl ions of the water. Therefore the extent of photolytic and oxidative reactions occurring in the atmosphere and hydrolysis in water can be used as a measure of environmental pollution likely to arise from exposure to a substance.

Absorption
The environmental impact of a chemical substance is determined by its ability to move through the environment. This movement depends on the affinity of the chemical toward particulate matter; soil and sediment. Chemicals which have a high affinity for absorption are less readily transported in the gaseous phase or in solution, and therefore can accumulate in a particular medium. Chemical substances which are not readily absorbed are transported through soil, air and aquatic systems.

Mammalian and avian toxicity
Studies on mammalian and avian species are carried out to determine the potential toxicity of substances to humans. Procedures involve undertaking a series of established exposure studies on a particular substance using specific routes, oral, inhalation, dermal or injection for variable durations. Exposure durations include acute or single exposure to a given concentration of substance. Sub-acute or sub-chronic exposure, i.e. repeat doses increasing in concentration over an intermediate time period, up to 4 weeks for sub-acute and 90 day/13 week (in rodents) or 1 year (in dogs) for sub-chronic studies. Finally chronic/long term studies involve exposure to specific concentrations of chemical for a duration of 18 month-2 years. A variety of species are used in toxicity testing, most commonly rodents, (rats, mice, hamsters) and rabbits, but tests can also be carried out on monkeys, domestic animals and birds.

Acute Data
Single exposure studies quoting LD_{50}'s from at least two species.

Sub-acute data
Results of repeat doses, intermediate duration studies are quoted. Priority is given to reporting the adverse effects on the gastrointestinal, hepatic, circulatory, cardiopulmonary, immune, reproductive, renal and central nervous systems.

Carcinogenicity and long-term effects
Information on the carcinogenicity of substances unequivocally proven to cause cancer in humans and laboratory animals, together with equivocal data from carcinogenicity assays in laboratory animals are reported. Additionally treatment related chronic adverse effects are reported. Criteria for inclusion required the study to report the species, duration of exposure, concentration and target organ(s), sex is also given where available.

Teratogenicity and reproductive effects
The results of studies carried out in intact animal and *in vitro* systems to determine the potential for teratogenic, foetotoxic and reproductive damage are reported here.

Criteria for inclusion required the species, duration of exposure, concentration and details of the effect in relation to fertility to be stated. Adverse effects reported in this section include sexual organ dysfunction, developmental changes (to embryos and foetuses), malformations, increases in spontaneous abortions or stillbirths, impotence, menstrual disorders and neurotoxic effects on offspring.

Metabolism and pharmacokinetics

Data are quoted on the metabolic fate of the substance in mammals, and includes adsorption, distribution, storage and excretion. Mechanisms of anabolic or catabolic metabolism, enzyme activation and half-lives within the body are reported when available. Additionally findings from *in vitro* studies are reported.

Irritancy

Chemical substances which cause irritation (itching, inflammation) to skin, eye and mucous membranes on immediate contact in either humans or experimental animals is reported here. Exposure can be intentional in human or animal experiments, or unintentional via exposure at work or accident to humans.

Sensitisation

Sensitisation occurs where an initial accidental or intentional exposure to a large or small concentration of substance causes no reaction or irritant effects. However, repeat or prolonged exposure to even minute amounts of a sensitising chemical causes increasingly acute allergic reactions.

Genotoxicity

Genotoxicity testing is carried out to determine the mutagenic and/or carcinogenic potential of a chemical substance. A standard series of tests are carried out under controlled laboratory conditions on an established set of test organisms. A hierarchial system using bacteria, yeasts, cultured human and mammalian cells, *in vivo* cytogenetic tests in mammals and plant genetics is used to assess the genotoxic potential of the substance under study. Bacteria, unlike mammals, lack the necessary oxidative enzyme systems for metabolising foreign compounds to the electrophilic metabolites capable of reacting with DNA. Therefore, bacteria are treated with the substance under study in the presence of a post-mitochondial supernatant (S9) prepared from the livers of mammals (usually rats). This fraction is supplemented with essential co-factors to form the S9 mix necessary for activation. DOSE reports published studies: giving the test organisms, whether metabolic activation (S9) was required, and the result, positive or negative.

Any other adverse effects to man

Adverse effects to humans from single or repeat exposures to a substance are given. The section includes results of epidemiological studies, smaller less comprehensive studies of people exposed through their work environment and accidental exposure of a single, few or many individuals.

Any other adverse effects

Adverse effects to organisms or animals other than man are reported here.

Legislation

Any form of legislation, medical (food and drugs) or environmental from European, American and worldwide sources is reported.

Any other comments

All other relevant information including chemical instability and incompatibility is contained in this section. This includes reviews, phytotoxicity and toxic effects associated with impurities.

References

Contains references to all data from above sections.

Indexes

The most convenient means of accessing a chemical in DOSE is via one of the indexes at the back of the publication. DOSE contains three indexes: names and synonyms, CAS Registry Numbers and molecular formulae.

Index of names and synonyms

Contains the name of the chemical used in DOSE together with a number of synonyms for that chemical. All names are arranged alphabetically.

Index of CAS Registry Numbers

Contains a list of the CAS Registry Numbers of the chemicals appearing within the volume of DOSE in ascending order. This number is linked to the preferred DOSE name for that chemical and its DOSE number.

Index of molecular formulae

Contains a list of the molecular formulae of the chemicals appearing within the volume of DOSE in alphabetical order for inorganic compounds, i.e. Ag_2CO_3, Cl_2Cr, etc., but for organic compounds, carbon and hydrogen content are shown first followed by the other elements in alphabetical order, i.e. C_6H_5Br. This number is linked to the preferred DOSE name for that chemical and its DOSE number.

A1   Acenaphthene

CAS Registry No. 83-32-9
Synonyms 1,2-dihydroacenaphthylene; 1,8-ethylenenaphthalene;
periethylenenaphthalene
Mol. Formula $C_{12}H_{10}$ **Mol. Wt.** 154.21
Uses A dyestuff intermediate, insecticide and fungicide. In the manufacture of plastics.
Occurrence A product of coal combustion, in coal tar and diesel fuel emissions.
Volatile component of cassava and nectarines.

Physical properties

M. Pt. 93-95°C; **B. Pt.** 279°C; **Specific gravity** d_4^{90} 1.0242; **Partition coefficient** log
P_{ow} 3.92; **Volatility** v.p. 10 mmHg at 131°C.

Solubility
Organic solvent: ethanol, benzene

Ecotoxicity

Fish toxicity
LC_{50} (96 hr) fathead minnow, channel catfish, rainbow trout, brown trout 600-1700
$\mu g\ l^{-1}$ flow through bioassay, pH 7.5-7.6 (1).
LC_{50} (exposure unspecified) himedaka killifish 6.3 mg l^{-1} (2).
LC_{50} (96 hr) bluegill sunfish 1700 $\mu g\ l^{-1}$ static bioassay (3).

Invertebrate toxicity
LC_{50} (96 hr) snail 2040 $\mu g\ l^{-1}$ flow through bioassay (1).
LC_{50} (96 hr) mysid shrimp 970 $\mu g\ l^{-1}$ static bioassay (3).

Bioaccumulation
Mussels, scallops and snails have no detectable aryl-hydrocarbon hydroxylase
enzyme system and therefore accumulate acenaphthene (4).
Bluegill sunfish bioconcentration factor 387-389 (5,6).

Environmental fate

Nitrification inhibition
The microbial degradation of acenaphthene under denitrification conditions at
soil-to-water ratios of 1.25 with soil containing 10^5 denitrifying organisms g^{-1} was
investigated. Under excess nitrite conditions, acenaphthene was degraded to
undetectable levels in <9 wk. Acclimation periods of 12-38 days were observed in
tests with soil not previously exposed to polycyclic aromatic hydrocarbon (PAH)
compounds. The acclimation period resulted from the time required for a small
population of organisms capable of PAH degradation to attain sufficient densities to
exhibit detectable PAH reduction. Under nitrite limiting conditions the PAH
compounds were stable (7).

Anaerobic effects

Under anaerobic conditions no significant degradation was observed over a period of 70 days (7).

Degradation studies

Microbial degradation from initial aqueous phase concentration of 1 mg l^{-1} to non-detectable levels occurred within 10 days under aerobic conditions (7). Concentrations of 25-150 μg l^{-1} were degraded at ambient temperatures within 3 days in groundwater (8).

Abiotic removal

Acenaphthene, in common with other polycyclic aromatic hydrocarbons and cyclic alkanes, resists environmental hydrolysis (9).

The photolytic $t_{1/2}$ of acenaphthene in water at 20°C exposed to 100W Hg lamp is reported to be 3 hr. (10).

Acenaphthene adsorbed on coal fly ash resists photo-oxidation (11).

Mammalian and avian toxicity

Carcinogenicity and long-term effects

Mammalian studies have yielded equivocal findings (12).

Metabolism and pharmacokinetics

Acenaphthene, in common with other PAHs, accumulates in adipose tissue, but its transport into cells and between intracellular membranes is not well understood. Molecular volume is considered to be a rate-determining factor (13).

In mammals, PAHs are oxidised by aryl hydrocarbon hydroxylose activity and excreted as glucuronide conjugates (14).

Genotoxicity

8-azaguanine resistance was induced in *Salmonella typhimurium* with metabolic activation (15).

Salmonella typhimurium TA1538, TA1537 with and without metabolic activation negative. Although non- mutagenic, all the 1, 2-ring fused acenaphthenes were found to be indirect frameshift mutagens in strain TA1537 (16).

Any other adverse effects to man

A case-control study was undertaken in Montreal to investigate the possible associations between occupational exposures and cancers of oesophagus, stomach, colorectum, liver, pancreas, lung, prostrate, bladder, kidney, melanoma and lymphoid tissue. In total, 3726 cancer patients were interviewed between 1979-1985, to obtain detailed lifetime job histories, which were translated into a history of occupational exposures to polycyclic aromatic hydrocarbons (PAHs) (17).

Any other comments

In a survey of UK drinking water treatment facilities acenaphthene was detected in treated samples at 2 of 14 facilities (18).

Detected in water bottom sediments and fish in Japan (19).

Experimental toxicology, human health and environmental effects reviewed (12,20).

References

1. Holcombe G. W; et al. *Ecotoxicol. Environ. Safety* 1983, **7**(4), 400-409

2. Niinomi, J; et al. *Mie-ken Kankyo Kagaku Senta Kenkyu Hokoku* 1989, **9**, 53-60, (Jap.) (*Chem. Abstr.* **111**, 189142x)
3. *Ambient Water Quality Criteria Document* 1980, EPA B-4 Draft
4. Malins, D. C. *Ann. N.Y. Acad. Sci.* 1977, **298**, 482-496
5. Sittig, M. Ed. *Priority Toxic Pollutants* 1980, 46-47
6. Veith, G. D; et al. *ASTM STP 707 Aquatic Toxicology* 1980, 116-129, Easton, J. G. (Ed.), Am. Soc. Test Materials, US
7. Mihelcic J. R; et al. *Appl. Environ. Microbiol.* 1988, **54**(5), 1182-1198
8. Owaga I; et al. *Talanta* 1981, **28**(10), 25-729
9. Lyman, W. J; et al. *Handbook of Chemical Property Estimation Methods* 1982, 7-14, McGraw-Hill, New York
10. Fukuda, K; et al. *Chemosphere* 1988, **17**, 651-659
11. Korfmacher, W. A; et al. *Science* 1980, **207**, 763-765
12. *Health Effects Assessment for Acenaphthene* 1987, EPA 600/8-88/010
13. Plant, A. L; et al. *Chem-biol. Interact.* 1983, **44**(3), 237-246
14. Sittig, M. *Handbook of Toxic and Hazardous Chemicals and Carcinogens* 2nd ed., 1985, 741, Noyes Data Corp., Park Ridge, NJ
15. Krishnan, S; et al. *Environ. Sci. Technol.* 1979, **13**(12), 1532
16. Gatehouse, D. *Mutat. Res.* 1980, **78**, 121-135
17. Krewski, D; et al. *Environ. Sci. Res.* 1991, **39**, 343-352
18. Fielding, M; et al. *Organic Pollutants in Drinking Water* 1984, 49, TR-159, Water Res. Centre, UK
19. *JETOC Newsletters* 1988, **6**, 20
20. *ECETOC Technical Report No. 30(4)* 1991, European Chemical Industry Ecology and Toxicology Centre, B-1160 Brussels

A2 Acenaphthylene

CAS Registry No. 208-96-8
Synonyms cyclopenta[*de*]naphthalene
Mol. Formula $C_{12}H_8$ **Mol. Wt.** 152.20
Occurrence In cigarette smoke and in soots generated by the combustion of aromatic hydrocarbon fuels containing pyridine (1).

Physical properties

M. Pt. 92-93°C; **B. Pt.** 265-275°C; **Specific gravity** d_2^{16} 0.899; **Volatility** v.p. 9.12 × 10^{-4} mmHg at 25°C.

Solubility
Water: 16 mg l^{-1} at 25°C. Organic solvent: ethanol, diethyl ether, benzene

Ecotoxicity

Bioaccumulation
Bullhead catfish (Black River, Ohio) and striped bass (Potomac River, Maryland)

contained 270 and 43 ppb acenaphthylene, respectively. Oysters and clams contained 36 and 130 ppb, respectively (2,3).

Environmental fate

Degradation studies
Concentrations of 25-150 $\mu g\ l^{-1}$ were almost totally degraded within 3 days at ambient temperatures in groundwater (4).
Microbial degradation of acenaphthylene in water samples was low (5).

Abiotic removal
Acenaphthylene, in common with other polycyclic aromatic hydrocarbons (PAHs), is unlikely to undergo environmental hydrolysis (6).
Aqueous photolysis data for acenaphthylene indicate it is likely to undergo direct photolysis in the environment (7,8).
Acenaphthylene was completely degraded after 16 months incubation in the dark at 20°C, which suggests volatilisation is more important than biodegradation in the removal of this compound (9).

Mammalian and avian toxicity

Metabolism and pharmacokinetics
In mammals, acenaphthylene and other PAHs are oxidised by aryl hydrocarbon hydroxylase activity and excreted as glucuronide conjugates (10).

Genotoxicity
Salmonella typhimurium TA1538, TA1537 with and without metabolic activation negative. Although non-mutagenic, all 1,2-ring fused acenaphenes were found to be indirect frameshift mutagens in strain TA1537 (11).

Any other adverse effects to man
A case-control study was undertaken in Montreal to investigate the possible associations between occupational exposures and cancers of oesophagus, stomach, colorectum, liver, pancreas, lung, prostrate, bladder, kidney, melanoma and lymphoid tissue. In total, 3726 cancer patients were interviewed between 1979-1985 to obtain detailed lifetime job histories, which were translated into a history of occupational exposures to polycylic aromatic hydrocarbons (PAHs) (12).

Any other comments
Found in Canadian drinking water 0.1-20 ng l^{-1} (13).
Seafoods and agricultural produce can contain traces of acenaphthylene and other PAHs absorbed from the atmosphere and from contaminated water supplies (14).
Toxicity and hazards reviewed (15).

References
1. *Health Effects Assessment for Acenaphthylene* 1987, EPA 600/8-88/011
2. Vassilaros, D. L; et al. *Anal. Chem.* 1982, **54**, 106-112
3. McFall, J. A; et al. *Chemosphere* 1985, **14**, 1561-1569
4. Ogawa, I; et al. *Talanta* 1981, **28**(10), 725-729
5. Niinomi J; et al. *Mie-ken Kankagu Senta Kenkyu Hokoku* 1989, **9**, 53-60 (Jap.) (*Chem. Abstr.* **111** 189142x)

6. Lyman, W. J; et al. *Handbook of Chemical Property Estimation Methods* 1982, 7-14, McGraw Hill, New York
7. Fukada, K; et al. *Chemosphere* 1988, **1**(7), 651-659
8. Behymer, T. D; et al. *Environ. Sci. Technol.* 1985, **19**, 1004-1006
9. Bossert, I. D; et al. *Bull. Environ. Contam. Toxicol.* 1986**37**, 490-495
10. Sittig, M. *Handbook of Toxic and Hazardous Chemicals and Carcinogens* 2nd ed., 1985, 741, Noyes Data Corp., Park Ridge, NJ
11. Donnelly, K. C; et al. *Mutat. Res.* 1987, **180**(1), 31-42
12. Krewski, D; et al. *Environ. Sci. Res.* 1990, **39**, 343-352
13. Benoit, F. M; et al. *Bull. Environ. Contam. Toxicol.* 1979, **23**, 774
14. Ellenham, M. J; et al. *Medical Toxicology – Diagnosis and Treatment of Human Poisoning* 1988, 953, Elsevier, New York
15. Izmerov, N. F; *Scientific Reviews of Soviet Literature on Toxicity and Hazards of Chemicals* 1991, Eng. Trans. Richardson, M. L. (Ed.), UNEP/IRPTC, Geneva

A3   Acephate

CAS Registry No. 30560-19-1
Synonyms Acetamidophos; acetylphosphoramidothioic acid *O,S*-dimethyl ester; *O,S*-dimethyl acetylphosphoramidothioate; Orthene
Mol. Formula $C_4H_{10}NO_3PS$ **Mol. Wt.** 183.17
Uses Contact and systemic insecticide.

Physical properties

M. Pt. 82-93°C; **Specific gravity** 1.35 (temperature unspecified);
Volatility v.p. 1.72×10^{-6} mmHg.

Solubility
Water: 790 g l^{-1} at 20°C. Organic solvent: acetone, ethanol, ethyl acetate, benzene, hexane

Occupational exposure

Supply classification harmful.

Risk phrases
Harmful by inhalation, in contact with skin and if swallowed (R20/21/22)

Safety phrases
Keep out of reach of children – Keep away from food, drink and animal feeding stuffs (S2, S13)

Ecotoxicity

Fish toxicity
LC$_{50}$ (96 hr) bluegill sunfish, channel cat fish 2050-2230 mg l^{-1}. LC$_{50}$ (96 hr)

goldfish 9550 mg l^{-1} (1). Exposure (24 hr) to 400 mg l^{-1} depressed brain cholinesterase levels in rainbow trout for 15 days (2).

Invertebrate toxicity
LC_{50} (96 hr) pink shrimp, mysid shrimp 3.8-7.3 mg l^{-1} (3).

Effects to non-target species
LD_{50} oral mallard duck, ringneck pheasant 140-350 mg kg^{-1} (1,4).
LD_{50} (24 hr) oral little brown bats 197-1500 mg kg^{-1}, prevented 9 of 30 surviving bats from righting themselves when placed on their backs 24 hr after dosing, initial sample size 50. From this information the calculated toxicity was ED_{50} 687 mg kg^{-1} (5).
LD_{50} topically *Choristoneura occidentalis* and *Anagasta kuehniella* larvae 23-48 µg g^{-1} (6).
Administration of 0.25, 0.5, and 1 ppm acephate to bees for 14 days, caused a dose-dependent decrease in their numbers (7).
LD_{50} oral bee 1.2 µg adult bee^{-1}. Acephate classified as a poison without amorphogenic effects (8).
Field application of acephate did not have any adverse effect on a population of meadow vole, although brain acetylcholinesterase and plasma cholinesterase activities were depressed (9,10).

Environmental fate

Nitrification inhibition
Studies on the effect of acephate on growth and nitrogen fixation by *Westiellopsin protifica* and *Anabaena sp.* showed that low concentrations of 1.0-50 µg ml^{-1} enhanced growth, while higher concentrations were lethal to organisms. Concentrations >5 µg ml^{-1} decreased total nitrogen content (11).

Degradation studies
$t_{1/2}$ in soil 7-10 days, methamidophos was identified as a metabolite. In plants residual activity lasted for 10-15 days (1).

Abiotic removal
50% hydrolysis occured in 60 hr at pH9 and 40°C and in 710 hr at pH3 and 40°C (1). Hydrolytic products formed at 37°C and varying pH, included methamidophos, O,S-dimethyl phosphorothiolate, and O-methyl acetylphosphoramidothiolate (12).

Mammalian and avian toxicity

Acute data
LD_{50} oral rat 866-945 mg kg^{-1}.
LD_{50} oral dog >681 mg kg^{-1} (1).
LD_{50} oral mouse 351 mg kg^{-1} (13).
LD_{50} percutaneous rabbit >2000 mg kg^{-1} (1).

Sub-acute data
Mouse and meadow voles were fed 0-400 ppm acephate for 5 days. Brain and plasma cholinesterase activities were reduced in a dose-dependent manner. Body, liver weight, plasma enzyme activities and cytochrome content were not affected (13). The exposure of rats to 1 or 10 mg kg^{-1} of acephate for 15 wk, caused altered activity of the noncholinergic system without altering the cholinergic activity. The authors

suggest that low level chronic exposure to organophosphones cannot be predicted by measuring cholinesterase or acetylcholinesterase enzyme activities (14).

Carcinogenicity and long-term effects
In a 2 yr feeding trial, rats receiving 30 mg kg^{-1} diet and dogs receiving 100 mg kg^{-1} diet showed depression of cholinesterase but no other significant side effects (1).

Teratogenicity and reproductive effects
50 and 100 mg kg^{-1} acephate administered orally to white-footed mice inhibited brain acetyl cholinesterase activity 45% and 56% and reduced basal luteinising hormone concentration 29% and 25%, respectively, after 4 hr. Dietary exposure to 25, 100 and 400 ppm inhibited brain acetyl cholinesterase activity but did not affect plasma basal luteinising hormone. Reproductive function effects could be possible (13).

Metabolism and pharmacokinetics
Following an oral dose of acephate to mice, metabolic products detected in the liver up to 30 hr were methamidophos, O,S-dimethyl phosphoramidothiolate and S-methyl acetylphosphoramidothiolate. Acephate and methamidophos had inhibitory cholinesterase effects on mouse erythrocyte enzyme (12).

Genotoxicity

Acephate produced gene conversion and mitotic recombination in *Saccharomyces cerevisae* and unscheduled DNA synthesis in human fibroblasts in culture (15). *In vivo* tests in the mouse showed significant enhancement in chromosomal aberrations, differences in micronuclei and sperm abnormalities in acephate treated animals. In a dominant lethal assay in mice, dead implants were significantly higher at wk 3 in treated animals (16).
At dose levels limited by toxicity, negative results were observed for induction of sex-linked, recessive lethality in *Drosophila melanogaster* (17).
Acephate was positive in an assay for clastogenicity in *Vicia faba* (18).

Any other adverse effects to man

Inhalation and skin exposure to acephate was evaluated in 4 workers engaged in the formulation of 97% pure technical product. Urine content, erythrocyte and plasma cholinesterase levels were monitored. High correlation was found between skin exposure level and urine elimination. One subject with urinary excretion levels between 3-8 mg l^{-1} had slightly decreased values of plasma cholinesterase levels and erythrocytes (19).

Any other adverse effects

Investigation of inhibition *in vitro* of human erythrocyte, rat brain and insect acetylcholinesterase levels indicated that acephate inhibits acetyl cholinesterase *in vitro* in proportion to its toxicity *in vivo* (5,20).

Legislation

Limited under EC Directive on Drinking Water Quality 80/778/EEC. Pesticide: organophosphorus compounds maximum admissible concentration 0.1 µg l^{-1} (21).

Any other comments

Anticholinesterase properties of methamidophos and acephate in insects and mammals reviewed (16).
Acephate showed toxic effects on carbohydrate metabolism in rats and inhibited the

electron transfer of respiration in isolated mitochondria. The activity of cytochrome c oxidase was severely inhibited at alkaline pH (22).

References

1. *The Agrochemicals Handbook* 3rd ed. 1991, RSC, London
2. Zmkl, J. G; et al. *Bull Environ. Contam. Toxicol.* 1987, **38**(1), 22-28
3. *USEPA Report* 1981, EPA 600/4-81-041
4. *The Pesticide Manual* 9th ed., 1991, British Crop Protection Council, Farnham
5. Clark, D. R; et al. *Environ. Toxicol. Chem.* 1987, **6**(9), 705-708
6. Rosli bin Mohamad; et al. *Pertanika* 1987, **10**(1), 75-80
7. Fiedler, L. *Bull. Environ. Contam. Toxicol.* 1987, **38**(4), 594-601
8. Atkins, E. L; et al. *J. Apic. Res.* 1986, **25**(4), 242-255
9. Jett, D. A. *Environ. Toxicol. Chem.* 1986, **5**(3), 255-260
10. Sykes, P. W. Jr. *Condor* 1985, **87**(3), 438
11. Maharana, R. K. et al. *Geobios* 1986, **13**(5), 185-188
12. Chukwudebe, A. C; et al. *J. Environ. Sci. Health Part B* 1984, **198**(6), 501-522
13. Rattner, B. A; et al. *Toxicol. Lett.* 1985, **24**(1), 65-69
14. Singh, A. K; et al. *Environ. Res.* 1987, **43**(2), 342-349
15. *NTIS Report* 1980, PB 80-133226, National Technical Information Service, Springfield, VA
16. Behera, B. C; et al. *Mutat. Res.* 1989, **223**(3), 287-293
17. Carver, J. H; et al. *Toxicol.* 1985, **35**(2), 125-142
18. De Kergommeaux, J; et al. *Mutat. Res.* 1983, **124**(1), 69-84
19. Maroni, M; et al. *Arch. Environ. Contam. Toxicol.* 1990, **19**(5), 782-788
20. Hussain, M. A. *Bull. Environ. Contam. Toxicol.* 1987, **38**(1), 131-138
21. *EC Directive Relating to the Quality of Water Intended for Human Consumption* 1982, 80/778/EEC, Office for Official Publications of the European Communities, 2 rue Mercier, L-2985 Luxembourg
22. Ando, M. *Arch. Environ. Contam. Toxicol.* 1985, **14**(5) 535-540

A4   Acetal

$CH_3CH(OCH_2CH_3)_2$

CAS Registry No. 105-57-7

Synonyms acetaldehyde diethyl acetal; 1,1-diethoxyethane; diethyl acetal; ethylidene diethyl ether

Mol. Formula $C_6H_{14}O_2$ **Mol. Wt.** 118.18

Uses Solvent in synthetic perfumes such as jasmine. Used in organic synthesis, flavours. Formerly used as a hypnotic.

Physical properties

M. Pt. −100°C; **B. Pt.** 102.7°C; **Flash point** −20.5°C;

Specific gravity d_4^{20} 0.8254; **Volatility** v.p. 10 mmHg at 8°C; v. den 4.10.

Solubility

Water: 50 g l^{-1}. Organic solvent: ethanol, diethyl ether, heptane, ethyl acetate

Occupational exposure

UN No. 1088; **HAZCHEM Code** 3᎙E; **Conveyance classification** flammable liquid; **Supply classification** highly flammable and irritant.

Risk phrases
Highly flammable – Irritating to eyes and skin (R11, R36/38)

Safety phrases
Keep container in a well ventilated place – Keep away from sources of ignition – No
Smoking – Take precautionary measures against static discharges (S9, S16, S33)

Mammalian and avian toxicity

Acute data
LD_{50} oral rat, rabbit, mouse 3500-4600 mg kg^{-1} (1-3).
LC_{50} (4 hr) inhalation rat 4000 ppm (4).

Irritancy
Dermal rabbit (24 hr) 10 mg caused irritation and 500 mg instilled into rabbit eye
caused irritation (4).

Any other adverse effects
Central nervous system narcotic in high doses (3).

Any other comments
Data on the physico-chemical properties, experimental toxicology and human health
effects have been reviewed (5).

References
1. *Proc. Soc. Exp. Biol. Med.* 1903/1904, **1**, Academic Press, New York
2. *Med. v. Ernaehrung* 1967, **8**, 244
3. Bhzina, A. Z; et al. *Gig. Sanit.* 1977, **3**, 12-15 (*Chem Abstr.* **88**, 32649b)
4. Smyth, H. F; et al. *J. Ind. Hyg. Toxicol.* 1949, **31**, 60
5. *ECETOC Technical Report No. 30(4)* 1991, European Chemical Industry Ecology and
   Toxicology Centre, B-1160, Brussels

A5   Acetaldehyde

CH3CHO

CAS Registry No. 75-07-0
Synonyms acetic aldehyde; ethanal; ethylaldehyde
Mol. Formula C_2H_4O **Mol. Wt.** 44.05
Uses In the manufacture of aniline dyestuffs, perfumes, flavours, plastics, synthetic
rubbers and for silvering mirrors and hardening gelatin fibres.

Physical properties
M. Pt. –123.5°C; **B. Pt.** 20.2°C; **Flash point** 38°C; **Specific gravity** d^{20} 0.783;
Partition coefficient log P_{ow} –0.40 (calc.) (1).; **Volatility** v.p. 740 mmHg.

Solubility
Water: miscible. Organic solvent: ethanol, diethyl ether

Occupational exposure

US TLV (TWA) 100 ppm (180 mg m^{-3}); US TLV (STEL) 150ppm (270 mg m^{-3}); UK Long-term limit 100 ppm (180 mg m^{-3}) under review; UK Short-term limit 150 ppm (270 mg m^{-3}) under review; UN No. 1089; HAZCHEM Code 2YE; Conveyance classification flammable liquid; Supply classification extremely flammable and harmful.

Risk phrases

Extremely flammable – Irritating to eyes and respiratory system – Possible risk of irreversible effects (R12, R36/37, R40)

Safety phrases

Keep away from sources of ignition – No Smoking – Take precautionary measures against static discharges – Wear protective clothing and gloves (S16, S33, S36/37)

Ecotoxicity

Fish toxicity

LC$_{50}$ (24 hr) pinperch 70 mg l^{-1} (2).
LC$_{50}$ (96 hr) bluegill sunfish 53 mg l^{-1} (3).

Invertebrate toxicity

Cell multiplication inhibition test *Uronema parduczi* 57 mg l^{-1} (4).
EC$_{50}$ (48 hr) *Daphnia magna* 9-14 g l^{-1} (5).

Environmental fate

Anaerobic effects

67-97% degradation occured in an anaerobic system (6,7).

Degradation studies

Biodegradable (8).
A number of studies confirm the degradability of acetaldehyde by acclimated sludge. Some loss may be attributed to volatilisation (9-13).

Abiotic removal

Photolytic t$_{1/2}$ 8-16 hr (calc.) (14).

Mammalian and avian toxicity

Acute data

LD$_{50}$ oral rat 1930 mg kg^{-1} (15).
LD$_{50}$ intravenous mouse 212 mg kg^{-1} (16).

Carcinogenicity and long-term effects

Inadequate evidence for carcinogenicity to humans, sufficient evidence for carcinogenicity to animals, IARC classification group 2B (17).
Inhalation ♂/♀ rat (≤28 month) 0, 1500 or 3000 ppm (6 hr day^{-1} 5 day wk^{-1}) gradually reduced to 1000 ppm during the first 52 wk. Major compound related effects include increased mortality, growth retardation, nasal tumours, and non-neoplastic nasal changes in each of the test groups. The treatment related nasal changes comprised: degeneration, hyperplasia, metaplasia, and adenocarcinomas of the olfactory epithelium at all exposure levels; squamous metaplasia accompanied by slight to severe keratinisation and squamous cell carcinomas of the respiratory

epithelium at the two highest exposure levels; and slight to severe rhinitis and sinusitis in the highest concentration group of rats (18,19).

Long- term inhalation and intratracheal instillation studies of acetaldehyde were carried out in Syrian hamsters. Exposure to acetaldyhyde vapour at a concentration of 1500 ppm resulted in epithelial hyperplasia and metaplasia accompanied by inflammation in the nasal cavity and trachea. Extensive peribronchiolar adenomatoid lesions often accompanied by inflammatory changes occured in the lungs after intratracheal instillation of acetaldhyde. There was no evidence of acetaldehyde possessing carcinogenic activity (20).

Teratogenicity and reproductive effects
In mice caused decreased weight and abnormal closure of neural tube (16).
Rat malformations included microcephaly, micromelia and digital anomalies (21).

Genotoxicity
Acetaldehyde produced chromosomal aberrations including chromosomal fragments, achromatic lesions and chromatid breaks in metaphases at 12 hr and 24 hr in primary cultures of rat skin fibroblasts. Dose-related increases in aneuploidy were also observed (22).

Any other adverse effects to man
General narcotic. In humans large doses cause death by respiratory paralysis (23).

Any other comments
Toxicity and hazards reviewed (24).
Exposure data, experimental toxicology and human health effects reviewed (25,26).

References
1. Hansch, C; et al. *Med. Chem. Proj.* 1981, No. 19, Pomona College, Claremont, CA
2. Garrett, J. T; et al. *Texas J. Sci.,* 1951, 3, 391
3. McKee, J. E; et al. *Water Quality Criteria* 1963, California State Water Quality Board
4. Bringmann, G; et al. *Z. Wasser/Abwasser Forsch.* 1980, **1**, 26
5. Takahashi, I. T; et al. *Bull. Environ. Contamin. Toxicol.* 1987, **39**, 229-236
6. Chou, W. L; et al. *Biotech. Bioeng. Symp.* 1979, **8**, 391-414
7. *Selected Biodegradation Techniques for Treatment/Ultimate Disposal of Organic Materials* 1979, EPA 600/2-79-006
8. *MITI Report* 1984, Ministry of International Trade and Industry, Tokyo
9. Gerhold, R. M; et al. *J. Water Pollut. Control Fed.* 1957, **38**, 562-579
10. Hatfield, R. *Ind. Eng. Chem.* 1957, **49** 192-196
11. Heukelekian, H; et al. *J. Water Pollut. Control Fed.* 1955, **29** 1040-1053
12. Rogovskaya, T; et al. *Inst. Vodosnakzh Kanaliz Gidrotekhn Sooruzhenii Inzh Gidrogeol* 1962, 178-213 (Russ.) (*Chem. Abstr.* **57**, 11659b)
13. Ludzach, J. R; et al. *J. Water Pollut. Control Fed.* 1960, **32** 1173-2000
14. Weaver, J. *Diss. Abstract. Int. B 1977* 1976, **37** 3427
15. *Arch. Ind. Hyg. Occup. Med.* 1951, **4**, 199
16. O 'Shea, K. S; et al. *J. Anat.* 1979, **128**(1), 65
17. *IARC Monograph* 1987, **Suppl. 7**, 77
18. Wouterson, R. A; et al. *Toxicology* 1986, **41**(2), 213-231
19. *Eur. J. Cancer Clin. Oncol.* 1982, **18** 13
20. Feron, V. J. *Prog. Exp. Tumour Res.* 1977, 162
21. Padmanabhan, R; et al. *Indian J. Pharmacol.* 1982, **14**(3), 247
22. Bird, R. P; et al. *Mutat. Res.* 1982, **101**(3), 237-246
23. *The Merck Index* 11th ed. 1989, Merck & Co. Inc, Rahway, NJ

24. Izmerov, N. F. *Scientific Reviews of Soviet Literature of Toxicity & Hazards of Chemicals* 1991, **111**, Eng. Trans. Richardson, M. L. (Ed.), UNEP/IRPTC, Geneva
25. *BIBRA Toxicity Profile* 1989, British Industrial Biological Research Asociation, Carshalton
26. *ECETOC Technical Report No. 30(4)* 1991, European Chemical Industry Ecology and Toxicology Centre, B-1160 Brussels

A6   Acetaldehyde formylmethylhydrazone

H3CCH=NNCH3CHO

CAS Registry No. 16568-02-8
Synonyms acetaldehyde *N*-methyl-*N*-formylhydrazone; ethylidene gyromitrin; ethylidene methyl hydrazine carboxaldehyde; gyromitrin
Mol. Formula $C_4H_8N_2O$ **Mol. Wt.** 100.12
Occurrence Fungal toxin from false morels, *Gyromitra sp.*

Physical properties
M. Pt. 5°C.

Mammalian and avian toxicity

Acute data
LD_{50} oral rat, mouse 320-340 mg kg^{-1} (1, 2).
LD_{50} oral rabbit 50 mg kg^{-1} (3).
LD_{Lo} unspecified route human 10-20 mg kg^{-1} (4).

Sub-acute data
TD_{Lo} (90 day) oral rabbit 5 mg kg^{-1} day^{-1} liver degeneration observed.
TD_{Lo} (90 day) oral chicken 0.5 mg kg^{-1} day^{-1} degeneration of liver, kidneys and heart observed (5).

Carcinogenicity and long-term effects
No adequate evidence for carcinogenicity to humans, limited evidence for carcinogenicity to animals, IARC classification group 3 (6). Acetaldehyde methylformylhydrazone administered 100 µg g^{-1} wk^{-1} by intragastric instillations for 52 wk in mice, induced tumours of the lungs, preputial glands, forestomach and clitoral glands (6,7).

Metabolism and pharmacokinetics
After oral administration of acetaldehyde formylmethylhydrazone to rabbits, rats and chickens, some of the compound was excreted unchanged in the urine of rabbits (2). At 37°C under acidic conditions (pH 1 to 3), acetaldehyde formylmethylhydrazone is converted to methylhydrazine a known tumour inducer in mice and hamsters via an intermediate, *N*-methyl-*N*-formylhydrazine (8).

Any other adverse effects to man
Poisoning of a family of 4 at an unspecified exposure level has been reported, toxic effects included liver injury, seizures and haemolysis. Recovery occurred within 4-8 days (9).

References

1. *Mutat. Res.* 1978, **54**, 167
2. Makinen, S. M; et al. *Food Cosmet. Toxicol.* 1977, **15**(6), 575-578
3. Pyysalo, H. *Naturwissenschaften* 1975, **62**, 395
4. Schmidlin-Meszaros, J. *Mitt. Geb. Lebensmittelunters. Hyg.* 1974, **65**, 453
5. Niskanen, A; et al. *Food Cosmet. Toxicol.* 1976, **14**(5), 409-415
6. *IARC Monograph* 1987, **Suppl. 7**, 64, 391
7. Toth, B; et al. *J. Natl. Cancer Inst.* 1981, **67**(4), 881-887
8. Nagel, D; et al. *Cancer Res.* 1977, **37**(9), 3458-3460
9. Garnier, R; et al. *Toxicol. Eur. Res.* 1978, **1**(6), 359-364

A7   Acetaldoxime

$CH_3CH=NOH$

CAS Registry No. 107-29-9
Synonyms acetaldehyde oxime; aldoxime; ethanal oxime; ethylidine hydroxylamine
Mol. Formula C_2H_5NO **Mol. Wt.** 59.07
Uses Chemical intermediate. Corrosion inhibitor.

Physical properties

M. Pt. 47°C; **B. Pt.** 115°C; **Volatility** v.p. 18.9 mmHg at 25°C.

Solubility
Water: 185 g l^{-1} at 25°C. Organic solvent: ethanol, diethyl ether

Occupational exposure

UN No. 2332; **HAZCHEM Code** 2SE; **Conveyance classification** flammable liquid.

Ecotoxicity

Bioaccumulation
The calculated bioconcentration factor of 0.5 indicated that environmental accumulation is unlikely (1).

Environmental fate

Carbonaceous inhibition
Acetaldoxime was utilised as a sole carbon source by one bacteria isolate and one fungi isolate obtained from a silty clay soil (2).

Degradation studies
Reduced to NO_2- by *Pseudomonas aeruginosa* (3).
Soil mobility and leaching potential was predicted to be high (4,5).

Abiotic removal
The photolytic $t_{1/2}$ of acetaldoxime was calculated to be 7.3 days (5).

Mammalian and avian toxicity

Acute data
LD_{50} intraperitoneal mouse 100 mg kg^{-1} (6).

LD$_{50}$ unspecified route mouse 115 mg kg^{-1} (7).

References

1. Lyman, W. J; et al. *Handbook of Chemical Property Estimation Methods* 1982, 5-14, McGraw-Hill, New York
2. Doxtader, K. G; et al. *Soil Sci. Soc. Am. Proc.* 1966, **30**, 351-355
3. Amarger, N; et al. *J. Bacteriol.* 1968, **95**(5) 1651-1657
4. Swann, R. L; et al. *Res. Rev.* 1983, **85**, 16-28
5. Hansch, C; et al. *Medchem Project No. 26* 1985, Pomona College, Claremont, CA
6. *NTIS Report* AD 277-689 Natl. Tech. Inf. Ser., Springfield, VA.
7. Freidman, A. L. *Khim.-Farm. Zh.* 1978 **12**(2), 88-92 (Russ.) (*Chem. Abstr.* **88**, 182874k)

A8   Acetamide

CH3CONH2

CAS Registry No. 60-35-5
Synonyms acetic acid amide; amide C$_2$; ethanamide; methane carboxamide
Mol. Formula C$_2$H$_5$NO   **Mol. Wt.** 59.07
Uses Solvent, plasticiser and stabiliser. Alcohol denaturant.

Physical properties

M. Pt. 79-81°C; **B. Pt.** 222°C; **Specific gravity** d$_4^{20}$ 1.159; **Partition coefficient** log P$_{ow}$ −1.26.

Solubility
Water: 2 kg l^{-1}. Organic solvent: ethanol, pyridine, chloroform, glycerol, hot benzene

Ecotoxicity

Fish toxicity
LC$_{50}$ (24-96hr) mosquito fish 26-13 g l^{-1} (1).

Invertebrate toxicity
Cell multiplication inhibition test, *Microcystis aeruginosa* 6200 mg l^{-1}, *Entosiphon sulcatum* 99 mg l^{-1}, *Pseudomonas putida* >10,000 mg kg^{-1} (2).

Environmental fate

Nitrification inhibition
At 100 mg l^{-1} no inhibition of NH$_3$ oxidation by *Nitrosomonas sp.* (3).

Mammalian and avian toxicity

Acute data
LD$_{50}$ intraperitoneal and subcutaneous rat 10 g kg^{-1} (4).

Carcinogenicity and long-term effects
No adequate data for evidence of carcinogenicity to humans, sufficient evidence for carcinogenicity to animals, IARC classification group 2B (5).

Administration of 2.5% acetamide (1 yr) diet rats induced malignant liver tumours, hyperplastic nodules and precancerous lesions (6).

Oral administration of acetamide to rats induced benign and malignant liver tumours and an increased incidence of malignant lymphomas in ♂ mice (7-9).

Genotoxicity

Salmonella typhimurium TA98, TA100, TA1535, TA1537 with metabolic activation negative (4).

Acetamide was inactive in morphological transformation assays in mouse embryo cells (10).

Any other adverse effects

Acetamide, a known animal carcinogen, is discussed in relation to humans receiving metronidazole therapy (11).

Any other comments

Human health effects and experimental toxicity reviewed (12).

References

1. Meinck, F; et al. *Les Eaux Residuaires Industrielles* 1970
2. Bringmann, G; et al. *Water Res.* 1980, **14**, 231-241
3. Hockenbury, M. R; et al. *J. Water Pollut. Control Fed.* 1977, **49**(5), 768-777
4. McCann, J; et al. *Proc. Nat. Acad. Sci. USA* 1975, **72**(12), 5135-5139
5. *IARC Monograph* 1987, **Suppl. 7**, 56, 389-390
6. Weisburger, J. H; et al. *Toxicol. Appl. Pharmacol.* 1969, **14**, 163-175
7. *IARC Monograph* 1974, **7**, 197-202
8. Flaks, B; et al. *Carcinogenesis* 1983, **4**, 1117-1125
9. Fleischman, R. W; et al. *J. Environ. Path. Toxicol.* 1980, **3**(5-6), 149-170
10. Patierno, S. R; et al. *Cancer Res.* 1989, **49**(4), 1038-1044
11. Koch, R. L; et al. *Science* 1981, **211**(4488), 398-400
12. *BIBRA Toxicity Profile* 1989, British Industrial Biological Research Association, Carshalton

A9   Acetamide, *N*-(aminothiooxomethyl)

CH₃CONHCSNH₂

CAS Registry No. 591-08-2

Synonyms acetyl thiourea; *N*-acetyl thiourea; 1-acetyl-2-thiourea; *N*-(aminothioxomethyl)acetamide

Mol. Formula $C_3H_6N_2OS$ **Mol. Wt.** 118.16

Uses Reagent for the synthesis of thiols.

Physical properties

M. Pt. 165-169°C.

Solubility

Organic solvent: ethanol, diethyl ether

Ecotoxicity

Bioaccumulation

The calculated bioconcentration factor of 0.9 indicated that environmental accumulation is unlikely (1).

Abiotic removal

The hydrolytic $t_{1/2}$ is 2.7 hr at pH 9.6. Hydrolysis products include acetic acid and thiourea. Acetamide, N-(aminothiooxomethyl) may undergo direct photolysis in the environment (2,3).

Mammalian and avian toxicity

Acute data

LD_{Lo} oral rat and mouse 50-94 mg kg^{-1} (4,5).
LD_{50} intraperitoneal mouse 100 mg kg^{-1} (6).

References

1. Lyman, W. J; et al. *Handbook of Chemical Property Estimation Methods* 13th ed., 1985, 7-90, McGraw-Hill, New York
2. Congdon, W. I; *Can. J. Chem.* 1974, **52**, 697-701
3. Sadtler, *11381 UV* 1966, Sadtler Res. Lab., Philadelphia, PA
4. Schaefer, E. W; et al. *Arch. Environ. Contam. Toxicol.* 1985, **14**(1), 111-129
5. Diecke, S. H; et al. *J. Pharmacol. Exp. Therap.* 1947, **90**, 260-270
6. *NTIS Report* AD277-689, Natl. Tech. Inf Ser., Springfield, VA

A10   Acetaminophen

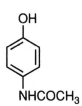

CAS Registry No. 103-90-2

Synonyms acetamide, N-(4-hydroxyphenol); p-acetamidophenol; N-acetyl-p-aminophenol; 4-hydroxyacetanilide; Paracetamol

Mol. Formula $C_8H_9NO_2$ **Mol. Wt.** 151.17

Uses Analgesic and antipyretic. In the manufacture of azodyestuffs and photographic chemicals.

Physical properties

M. Pt. 167-169°C.

Solubility

Organic solvent: methanol, ethanol, acetone, ethyl acetate

Mammalian and avian toxicity

Acute data

LD_{50} oral rat 2400 mg kg^{1} (1).

LD_{Lo} (unspecified exposure) oral human 150-800 mg kg^{-1} central nervous system, gastrointestinal tract, and liver effects (2-4).

A single oral dose was administered to fasted adult ♂ mice, which were subsequently sacrificed 30 min to 48 hr after treatment. Damage to liver, kidney, necrosis of bronchiolar epithelium, lymphoid necrosis and testicular changes were observed. Spermatid degeneration with early development of spermatid multinucleated giant cells was a characteristic feature (5).

Sub-acute data

Single or multiple intraperitoneal doses of 300-1000 mg kg^{-1} with pretreatment (P-448 inducer) in C57BL/6 and DBA/2 mice and 500-1500 mg kg^{-1} with pretreatment (P-450 inducer) in New Zealand white and chinchilla rabbits showed that susceptibility to acetaminophen cataractogenesis can be genetically predetermined and may involve enzymic bioactivation possibly independent of hepatic biotransformation and toxicity (6).

Carcinogenicity and long-term effects

No adequate evidence for carcinogenicity to humans, IARC classification group 3 (7).
Acetaminophen promoted kidney tumour development in rats (8).
Acetaminophen, 1000 or 5000 ppm for 40 wk, produced transient chronic hepatic hyperplasia without evidence of carcinogenicity in B6C3F1 (6 wk old) mice (9).
Acetaminophen did not reveal tumour initiating potential in rats with preexisting fatty liver induced by a choline-deficient diet (10).
Administration of 1.1% or 1.25% acetaminophen respectively to B6C3F1 and NIH mice for 41 wk caused severe liver injury characterised by centrilobular necrosis. The results are discussed in terms of their importance to the interpretation of carcinogenicity studies (11).

Teratogenicity and reproductive effects

Low birth rate, foetal hepatoxicity and neonatal death occured when ♀ mice (16-19 days pregnant) were administered 42-84 mg kg^{-1} acetaminophen simultaneously with a fatty diet (12).
Acetaminophen administered orally at 14 mg kg^{-1} to near term rats resulted in constriction of the foetal ductum (13).
Addition of acetaminophen to rat embryos in culture produced an increased incidence of morphologically abnormal anterior neuropores. The data suggest that the visceral yolk sac plays a vital role in the metabolic transformation of acetaminophen to catechol and quinone-imine reactive metabolites (14).
Acetaminophen inhibited Na^+, K^+ and Mg^{2+}-ATPase activities *in vitro* in human foetal cerebrum and cerebellum in a dose-dependent manner (15).

Metabolism and pharmacokinetics

Acetaminophen is metabolised in the liver via two routes, one pathway leads to the formation of *N*- acetyl-*p*-benzoquinone imine (NAPQI) mediated by cytochrome P-450, the other detoxification pathway leads to the formation of glucuronide, sulfate and glutathione conjugates which are excreted in the urine, $t_{1/2}$ 1-4 hr (16,17).
The toxic pathway metabolites predominate in the bile, whereas non-toxic metabolites are excreted in the urine (18).
Acetaminophen passes rapidly into the breast milk of nursing mothers.
Acetaminophen and its glucuronide, sulfate, cysteine and mercapturate conjugates were found in the urine of neonates (19).

Following a single intravenous administration of acetaminophen (50 mg kg^{-1}) to pregnant guinea pigs (60-65 days gestation), the major detoxification metabolites in the plasma and urine of dams were in the form of gluconurides, the sulfate metabolites predominated in the foetal and neonatal animals (20).

Acetaminophen clearance in healthy human ♀ volunteers was unaffected by chronic conjugated oestrogen treatment (21).

The analgesic effect of acetaminophen was directly related to the circulation level of the compound (22).

Irritancy

Skin rashes have been reported (23).

Genotoxicity

In vitro testing, acetaminophen was cytotoxic to Chinese hamster ovary V79 cells, inhibited DNA synthesis and increased sister chromatid exchanges (24,25).

Effects on hydroxyurea-resistant mouse mammary tumour cell line (TA3H2) showed that acetaminophen reduced DNA synthesis by inhibiting ribonucleotide reductase activity. In wild-type cells acetaminophen produced a concentration-dependent induction of chromosomal aberrations and sister chromatid exchanges (26).

In vivo testing in the mouse micronucleus test, an increased occurrence of micronuclei in polychromatic bone marrow erthrocytes was seen (27).

In human volunteers, following the oral administration of acetaminophen (1000 mg x 3 doses during 8 hr), blood and buccal mucosal cell samples were taken at time 0, 24, 72 and 168 hr after the first dose. Treatment reduced the level of unscheduled DNA synthesis in 1-methyl- 3-nitro-1-nitrosoguanidine (MNNG) treated lymphocytes, and increased the frequency of micronucleated cells in the buccal mucosa at 72 hr (28).

In ouabain-resistant mouse embryo cells (3H10T1/2 clone 8), acetaminophen produced negative mutation, positive induction of non-neoplastic cells and morphological transformation (29)

Any other comments

The pharmacokinetics, toxicity and clinical side-effects, including allergic and skin reactions, of acetaminophen have been extensively reviewed (30-37).

References

1. *Toxicol. Appl. Pharmacol.* 1971, **18**, 185
2. *Lancet* 1973, **1**, 66
3. *Br. Med. J.* 1981, **282**, 191
4. *Pediatrics* 1978, **61**, 68
5. Placke, M. E; et al. *Toxicol. Pathol.* 1987, **15**(4), 381-387
6. Lubek, B. M; et al. *Fundam. Appl. Toxicol.* 1988, **10**(4), 596-606
7. *IARC Monograph* 1990, **50**, 307-332
8. Kurata, Y. *Nagoya-shiritsu Daigaku Igakkai Zasshi* 1989, **40**(2), 429-448 (*Chem. Abstr.* **111**, 210387t)
9. Ward, J. M; et al. *Toxicol. Appl. Pharmacol.* 1988, **96**(3), 494-506
10. Takashima, Y; et al. *J. Toxicol. Pathol.* **3**(1), 57-63
11. Maruyama, H; et al. *Arch Toxicol.* 1988, **62**(6), 465-469
12. Miranda, G. A; et al. *Proc. Ann. Meet. Electron. Micros. Soc. Am.* 1987, 718-719
13. Momma, J; et al. *Am. J. Obster. Gynaecol.* 1990, **162**(5), 1304-1310
14. Stark, K. L; et al. *J. Pharmacol. Exp. Ther.* 1990, **255**(1), 74-82
15. Sarker, A. K; et al. *Indian J. Exp. Biol.* 1989, **27**(9), 802-804

16. *Martindale: The Extra Pharmacopoeia*, 29th ed., 1989, The Pharmaceutical Press, London
17. Savides, M. C. *J. Appl. Toxicol.* 1983, **3**(2), 96-111
18. Gregus, Z; et al. *J. Pharmacol. Exp. Ther.* 1988, **244**(1), 91-99
19. Notarianni, L. J; et al. *Br. J. Clin. Pharmacol.* 1987, **24**(1), 63-67
20. Ecobichon, D. J; et al. *J. Biochem. Toxicol.* 1989, **4**(4), 235-240
21. Scavone, J. M; et al. *Eur. J. Clin. Pharmacol.* 1990, **38**(1), 97-98
22. Levy, G. *J. Pharm. Sci.* 1987, **76**(1), 88-89
23. Tsuca, H; et al. *Carcinogenesis (London)* 1988, **9**(4), 547-554
24. Hongslo, J. K; et al. *Mutat. Res.* 1988, **204**(2), 333-341
25. Horner, S. A; et al. *Toxicol. In Vitro* 1987, **1**(3), 133-138
26. Hongslo, J. K; et al. *Mutagenesis.* 1990, **5**(5), 475-480
27. Sicardi, S. M; et al. *Acta Farm Bonaerensa* 1987, **6**(2), 71-75, (Span.) (*Chem. Abstr.* **108**, 143055f)
28. Topinka, J; et al. *Mutat. Res.* 1989, **227**(3), 147-152
29. Patierno, S. R; et al. *Cancer Res.* 1989, **49**(4), 1038-1044
30. Thesen, R; et al. *Pharm. Ztg.* 1988, **133**(23), 9-12, (Ger.) (*Chem. Abstr.* **109**, 47708w)
31. McCay, P. B; et al. *Coenzymes Cofactors* 1989, **3**, 111-157
32. Potter, D. W; et al. *Drug Metab. Rev.* 1989, **20**(2-4), 341-358
33. Wendel, A. *Reakt. Sauersotffspezies Med.* 1986, 99-107, (Ger.) (*Chem. Abstr.* **108**, 92350p)
34. Vial, T; et al. *Lyon Pharm.* 1988, **39**(3), 187-191
35. Brine, K. *Agents Actions Suppl.* 1988, **25**, 9-19
36. Prescott, L. F. *Int. Congr. Ser. – Excerpta Med.* 1987, **734**, 161-174
37. Fowler, P. O. *Med. Toxicol.* 1987, **2**(5), 338-366

A11   Acetanilide

NHCOCH$_3$

CAS Registry No. 103-84-4
Synonyms acetylaminobenzene; acetylaniline; antifebrin; *N*-phenylacetamide
Mol. Formula C$_8$H$_9$NO   **Mol. Wt.** 135.17
Uses Manufacture of medical and dyestuff compounds. An antipyretic and analgesic. Used as a stabiliser for peroxide solutions and as an additive to cellulose ester varnishes.

Physical properties
M. Pt. 114-115°C; **B. Pt.** 304-305°C; **Flash point** 173°C (open cup); **Specific gravity** d$_4^4$ 1.2105; **Partition coefficient** log P$_{ow}$ 1.16; **Volatility** v.p. 1 mmHg at 114°C; v. den. 4.65.

Solubility
Water: 5 g l^{-1}. Organic solvent: ethanol, chloroform, diethyl ether, benzene

Ecotoxicity

Fish toxicity

LC_{50} (96 hr) bluegill sunfish 100 mg l^{-1} (40% survival in static bioassay at 23°C).
LC_{50} (96 hr) inland silverside 115 mg l^{-1} (100-20% survival in static bioassy at 23°C)
(1).

Bioaccumulation

Bioconcentration factor in goldfish 1.2 (2).

Environmental fate

Degradation studies

Confirmed biodegradable (3).
BOD_{10} 1.20 mg l^{-1} oxygen using standard dilute sewage (4).
94% COD, 14.7 mg COD g dry $inoculum^{-1}$ hr^{-1} at 20°C in an activated sludge system
using the substance as sole carbon source. With influent of 50 mg l^{-1} and 365+ days
acclimation 50% at 20°C was recorded in a observation period of 10 days, at a
concentration of 600-1000 mg l^{-1} and under similar conditions inhibition was reported
(5).
Aryl acylamine amidohydrolase (EC3.5.1), isolated from *Aspergillus nidulans* shows
activity to acetanilide. Enzyme activitiy occurs over a range of pH values between 7.8
and 10.2 (6).

Mammalian and avian toxicity

Acute data

LD_{50} oral rat 800 mg kg^{-1} (7).
LD_{50} oral dog, intraperitoneal mouse 500 mg kg^{-1} (8,9).

Sub-acute data

LD_{Lo} (1 hr intermittently) oral human 56 mg kg^{-1} central nervous system and
gastrointestinal effects (10).

Metabolism and pharmacokinetics

Acetanilide is readily excreted in the urine as sulfate and glucuronide conjugates (11).
>99.9% of acetanilide remains unionised at body pH, facilitating absorption from
blood to cerebrospinal fluid (12).
Readily absorbed from the gastrointestinal tract (13).
In healthy subjects given acetanilide orally at concentrations of 50 mg kg^{-1}, plasma
clearance values varied from 12-25 ml hr^{-1} (14).

Any other comments

Acetanilide has been replaced by safer analgesics (15).

References

1. Dawson, G; et al. *J. Hazardous Materials* 1975/77, **1**, 303-318
2. Nakatsugawa, T; et al. *Environ. Toxicol. Pesticides* 1972, Matsumura, F. (Ed.), Academic Press
3. *MITI Report* 1984, Ministry of International Trade and Industry, Tokyo
4. Mills, E. J; et al. *Proc. 8th Purdue Ind. Waste Conf.* 1953, 492
5. Ludzack, F. J; et al. *J. Hazardous Materials* 1975/77, **1**
6. Pelsy, F; et al. *Pestic. Biochem. Physiol.* 1987, **27**(2), 182-188
7. *J. Pharm. Exp. Therp.* 1935, **54**, 159

8. *NTIS Report* AD 277-689, National Technical Information Service, Springfield, VA
9. *Abdernelden's Handbuch der Biologischen Arbeitsmethoden* 1935, **4**, 1290
10. *Am. J. Med. Sci.* 1901, **122**, 770
11. Patty, F. A. *Industrial Hygiene and Toxicology* 2nd ed., 1963, 1835, Interscience Publishers, New York
12. LaDu, B. N; et al. *Fundamentals of Drug Metabolism and Disposition* 1971, 82, Williams & Wilkins, Baltimore
13. Thienes C; et al. *Clinical Toxicology* 5th ed., 1972, Lea & Febiger, Philadelphia
14. Cunningham, J. L; et al. *Eur. J. Clin. Pharmacol.* 1974, 7(6), 461
15. *Martindale: The Extra Pharmacopoeia* 29th ed., 1989, The Pharmaceutical Press, London

A12   Acetic acid

CH_3COOH

CAS Registry No. 64-19-7
Synonyms ethanoic acid; ethylic acid; glacial acetic acid; methanecarboxylic acid; vinegar acid
Mol. Formula $C_2H_4O_2$ **Mol. Wt.** 60.05
Uses In the manufacture of various acetates and acetyl compounds. In plastics, rubber, tanning, printing and dyeing silks. An acidulant and preservative in foods and a solvent for gums, resins, volatile oils.

Physical properties

M. Pt. 17°C; **B. Pt.** 118°C; **Flash point** 39°C (closed cup);
Specific gravity d_{25}^{25} 1.049; **Volatility** v.p. 11.4 mmHg at 20°C.

Solubility
Water: miscible. Organic solvent: ethanol, diethyl ether, acetone, benzene

Occupational exposure

US TLV (TWA) 10 ppm (25 mg m^{-3}); **US TLV (STEL)** 15 ppm (37 mg m^{-3});
UK Long-term limit 10 ppm (25 mg m^{-3}); **UK Short-term limit** 15 ppm (37 mg m^{-3}); **UN No.** 2789; 2790; **HAZCHEM Code** 2 P (glacial or >80% by weight); 2R (solution >10% but ≤80%); **Conveyance classification** corrosive substance; **Supply classification** (>90%) flammable and corrosive; (>25% but <90%) corrosive.

Risk phrases
>90% – Flammable – Causes severe burns (R10, R35); >25% but <90% – Causes burns (R34)

Safety phrases
Keep out of reach of children – Do not breathe vapour – In case of contact with eyes, rinse immediately with plenty of water and seek medical advice (S2, S23, S26)

Ecotoxicity

Fish toxicity
LC$_{50}$ (96 hr) bluegill sunfish 75 mg l^{-1} (1).

LC$_{50}$ (96 hr) fathead minnow 88 mg l^{-1} static bioassay 18-22°C (2).
LD$_{50}$ (24 hr) goldfish 423 mg l^{-1}. Period of survival (48-96 hr) 100 mg l^{-1} at pH 6.8; period of survival (96 hr) 10 mg l^{-1} at pH 7.3 (3).

Invertebrate toxicity

EC$_{50}$ (24-48 hr) *Daphnia magna*, brine shrimp 47-32 mg l^{-1} (1,4).
Cell multiplication inhibition test *Microcystis aeruginosa* 90 mg l^{-1}, *Scenedesmus quadricauda* 4000 mg l^{-1}, *Entosiphon sulcatum* 78 mg l^{-1}, *Uronema parduczi* 1350 mg l^{-1} (2,5).

Bioaccumulation

Acetic acid shows no potential for biological accumulation or food chain contamination (6).

Environmental fate

Carbonaceous inhibition

Cell multiplication inhibition test, *Pseudomonas putida* 2850 mg l^{-1} (7).

Degradation studies

Biodegradable (8).
BOD$_{10}$ 82% reduction dissolved oxygen in fresh water and 88% reduction dissolved oxygen in seawater at 20°C (1).
ThOD$_5$ 40% 24 hr incubation activated sludge (4).

Mammalian and avian toxicity

Acute data

LD$_{50}$ oral rat 3310 mg kg^{-1} (9).
LC$_{50}$ (1 hr) inhalation guinea pig 5000 ppm (10).
A single 50 µl intratesticular injection of 36% acetic acid produced sterility in male rats (11).
LD$_{50}$ dermal rabbit 1060 mg kg^{-1} (12).

Sub-acute data

Suckling rats were given 0.3 g l^{-1} acetic acid in drinking water from parturition until the pups were 18 days old. Offspring exposed to acetic acid were less active and showed significant weight gain compared to controls (13).

Carcinogenicity and long-term effects

Acetic acid applied repeatedly to the skin of papilloma-bearing mice resulted in an increased incidence of skin cancer (14).

Irritancy

Exposure can cause burns to skin and eye irritation (15).

Genotoxicity

Salmonella typhimurium TA98, TA100, TA1535, TA1537 with metabolic activation negative (16).
A single application of acetic acid to mouse epidermis induced a sustained stimulation of DNA, RNA and protein synthesis, indicating that acetic acid acted as a promoting agent (17).

Any other adverse effects

Chronic exposure may cause erosion of dental enamel and bronchitis. Ingestion may cause severe corrosion of mouth and gastrointestinal tract with vomiting, haematemesis, diarrhoea, circulatory collapse, uremia and death (15).

Any other comments

The toxicity of acetic acid has been reviewed (18).
Incompatible with carbonates, hydroxides, oxides, phosphates. Corrosive.

References

1. Price, K. S. *J. Water Pollut. Control Fed.* 1974, **46**, 1
2. Mattson, V. R; et al. *Acute Toxicity of Selected Organic Compounds to Fathead Minnows* 1976, EPA 600/3-76-097
3. Ellis, M. M. *Detection and Measurement of Stream Pollution* 1937, US Dept of Commerce, Bureau of Fisheries
4. Elkins, H. F; et al. *Sewage Ind. Wastes* 1956, **28**(12), 1475
5. Bringmann, G; et al. *Water Research* 1980, **14**, 231-241
6. *Environment Canada, Tech. Info. for Problem Spills on Acetic Acid Draft* 1981, 1, 76
7. Bringmann, G; et al. *Z. Wasser/Abwasser Forsch.* 1980, (1), 26-31
8. *MITI Report* 1984, Ministry of International Trade and Industry, Tokyo
9. *Union Carbide Data Sheet* 1963, Union Carbide Corp., New York
10. Verschuern, K. *Handbook of Environmental Data on Organic Chemicals* 1983, Van Nostrand Reinhold, New York
11. Freeman, C; et al. *Fertility Sterility* 1973, **24**, 884-890
12. *Am. Ind. Hyg. Assoc. J.* 1972, **33**, 624
13. Barrett, J; et al. *Neurobehavioral Toxicol. Teratol.* 1982, **4**, 105-108
14. Rotstein, J. B; et al. *Cancer Lett.* 1988, **42**(1-2), 87-90
15. *The Merck Index* 11th ed. 1989, Merck & Co. Inc, Rahway, NJ
16. McCann, J; et al. *Proc. Nat. Acad. Sci.* 1975, **72**(12), 5135-5139
17. Slaga, T. J; et al. *J. Natl. Cancer Instit.* 1975, **55**(4), 983-987
18. *BIBRA Toxicity Profile* 1990, British Industrial Biological Research Association, Carshalton

A13   Acetic acid, butyl ester

$CH_3CO_2(CH_2)_3CH_3$

CAS Registry No. 123-86-4
Synonyms butyl acetate; butyl ethanoate
Mol. Formula $C_6H_{12}O_2$ **Mol. Wt.** 116.16
Uses Used in lacquers, artificial leathers, photographic films, plastics and safety glass.

Physical properties

M. Pt. −77°C; **B. Pt.** 125-126°C; **Flash point** 38°C;
Specific gravity d_{20}^{20} 0.88; **Volatility** v.p. 15 mmHg at 25°C.

Solubility

Water: 8 ml l^{-1} at 25°C. Organic solvent: miscible in ethanol, diethyl ether, hydrocarbons

---

Occupational exposure

US TLV (TWA) 150 ppm (713 mg m^{-3}); US TLV (STEL) 200 ppm (950 mg m^{-3}); UK Long-term limit 150 ppm (710 mg m^{-3}); UK Short-term limit 200 ppm (950 mg m^{-3}); UN No. 1123; HAZCHEM Code 3啞E; Conveyance classification flammable liquid; Supply classification flammable.

Risk phrases
Flammable (R10)

Ecotoxicity

Fish toxicity
LC$_{50}$ (96 hr) fathead minnow 18 mg l^{-1} (1).
LC$_{50}$ (96 hr) bluegill sunfish 100 ppm, freshwater static bioassay at 23°C.
LC$_{50}$ (96 hr) inland silverside 185 ppm, seawater static bioassay at 23°C (2).
In fish, butyl acetate is metabolised by *in vivo* hydrolysis of the carboxylic acid esters. Lethality properties of this class of compounds cannot be compared unless relative carboxylase esterase activities for the species are known (3).

Invertebrate toxicity
EC$_{50}$ (48 hr) *Daphnia magna* 44 ppm at 23°C conditions of bioassay unspecified (3). Cell multiplication inhibition test *Pseudomonas putida* 78 mg l^{-1}, *Microcystis aeruginosa* 420 mg l^{-1}, *Scenedesmus quadricauda* 3700 mg^{-1}, *Entosiphon sulcatum* 970 mg l^{-1} (4,5).

Bioaccumulation
Calculated bioconcentration factors range 4- 14, indicated that environmental accumulation is unlikely (6).

Environmental fate

Degradation studies
Hydrolytic calculated t$_{1/2}$ 3 yr at pH 7 (7).
Estimated evaporation rate from solid surfaces, t$_{1/2}$ 50 min (8).

Mammalian and avian toxicity

Acute data
LD$_{50}$ oral mouse, rabbit 7100-7400 mg kg^{-1} (9).
Acute oral rat 8-16 g kg^{-1} affected kidney and liver, caused granular dystrophy of the cytoplasm of hepatocytes and dystrophy of ganglion cells in the brain (10).
TC$_{Lo}$ (exposure unspecified) inhalation human 200 ppm – effects to nose, eye and pulmonary system (11).

Sub-acute data
Sub-acute intoxication of rats (route unspecified) with 0.8-1.6 g kg^{-1} day^{-1} for 1 month caused glomerulonephritis. Administration of 0.5 mg kg^{-1} for 6 months caused no change in organs (10).

Irritancy
Reported to cause irritation and conjunctivitis (species unspecified) (12).
A pharmaceutical worker developed dermatitis from butyl acetate, it was due to its attack on PVC gloves (13).

Any other adverse effects

Reported to have mild irritant and central nervous system depressant effects (species unspecified) (14).

Increased levels of serum bile acids, indicative of early signs of liver failure, were found in a group of workers exposed to a mixture of organic solvents including butyl acetate (15).

Any other comments

Narcotic in high concentrations (16).

Considered safe for use as a cosmetic ingredient (17).

Experimental toxicology, epidemiology and human health effects reviewed (18,19).

References

1. Brooke, L. T; et al. *Acute Toxicity of Organic Chemicals to Fathead Minnows* 1984, 175, Lake Superior Environmental Centre, Univ. Wisconsin
2. Verschueren, K. *Handbook of Environmental Data on Organic Chemicals* 2nd ed., 1983, Van Nostrand Rheinhold, New York
3. Kamlet, M. J; et al. *Environ. Sci. Technol.* 1987, **21**, 149-155
4. Bringmann, G; et al. *Water Res.* 1980, **14**, 231
5. Bringmann, G; et al. *Gwf-Wasser/Abwasser* 1976, **117**(9), 119
6. Lyman, W. J; et al. *Handbook of Chemical Property Estimation Methods* 1982, 4-9, McGraw-Hill, New York
7. Mabey, W; et al. *J. Phys. Chem. Ref. Data* 1978, **7**, 383
8. Park, J. G; et al. *Ind. Eng. Chem.* 1932, **24**, 132
9. Updyke, D. C. *J. Food Cosmet. Toxicol.* 1979, **17**, 509
10. Petrovskaya, O. G; et al. *Gig. NaselenNew Yorkkh Mest.* 1969, **8**, 120-124, (Russ.) (*Chem. Abstr.* **73**, 64466w)
11. *J. Ind. Hyg. Toxic.* 1943, **25**, 282
12. *The Merck Index* 11th ed. 1989, Merck & Co. Inc, Rahway, NJ
13. Roed-Peterson, J. *Contact Dermatitis* 1980, **6**(1), 55
14. *Martindale: The Extra Pharmacopoeia* 29th ed., 1989, Pharmaceutical Press, London
15. Franco, G. *Occup. Environ. Chem. Hazard* 1986, 173-177
16. Browning, E. *Toxicity Metabolism Industrial Solvents* 1965, 591-593
17. *J. Am. Coll. Toxicol.* 1989, **8**(4), 681-705
18. *Gov. Rep. Announce. Index US* 1981, **81**(11), 2171, PB81-147993
19. *ECETOC Technical Report No. 30(4)* 1991, European Chemical Industry Ecology and Toxicology Centre, B-1160 Brussels

A14   Acetic acid, ethyl ester

CH3CO2CH2CH3

CAS Registry No. 141-78-6

Synonyms ethyl acetate; acetic ether; ethyl ethanoate; vinegar naphtha; acetoxyethane

Mol. Formula $C_4H_8O_2$ **Mol. Wt.** 88.11

Uses Used in artificial fruit essences. As a solvent for nitrocellulose, varnishes, lacquers and in cleaning textiles. In the manufacture of photographic film, artificial silk and perfumes.

Physical properties

M. Pt. −83°C; **B. Pt.** 76-77°C; **Flash point** −4.4°C (closed cup); **Specific gravity** d_4^{20} 0.900; **Partition coefficient** log P_{ow} 0.73; **Volatility** v.p. 100 mmHg at 27°C; v. den. 3.04.

Solubility

Water: 100 ml l^{-1} at 25°C. Organic solvent: acetone, diethyl ether, ethanol

Occupational exposure

US TLV (TWA) 400 ppm (1400 mg m^{-3}); **UK Long-term limit** 400 ppm (1400 mg m^{-3}); **UN No.** 1173; **HAZCHEM Code** 3☒E; **Conveyance classification** flammable liquid; **Supply classification** highly flammable.

Risk phrases
Highly flammable (R11)

Safety phrases
Keep away from sources of ignition – No Smoking – Do not breathe vapour – Do not empty into drains – Take precautionary measures against static discharges (S16, S23, S29, S33)

Ecotoxicity

Fish toxicity
LC_{50} (96 hr) common Indian catfish 212 mg l^{-1}. Lowering of hepatic glycogen, hyperglycemic and hyperlacticemia reported (1).
LC_{50} (96 hr) fathead minnow 230 mg l^{-1} (2).
In fish, ethyl acetate is metabolised by *in vivo* hydrolysis of carboxylic acid esters. Lethality properties of this class of compounds cannot be compared unless relative carboxylase esterase activities for the species are known (3).

Invertebrate toxicity
Cell multiplication inhibition test, *Pseudomonas putida* 650 mg l^{-1}, *Entosiphon sulcatum* 202 mg l^{-1} (4).

Effects to non-target species
LC_{50} (48 hr) Mexican axolotl 150 mg l^{-1} (5).

Environmental fate

Nitrification inhibition
Nitrosomonas growth inhibited by 50% at 18 g l^{-1} (6).

Anaerobic effects
96% utilisation occurred in an anaerobic reactor with a 20 day retention time (7).

Degradation studies
BOD_5 36-68% reduction in dissolved oxygen using sewage inoculum (8-11).
100% degradation occurred in 20 hr using activated sludge (12).

Abiotic removal
Photochemical reactivity $t_{1/2}$ measured 2-8 days, variance depended on atmospheric conditions (13).

Mammalian and avian toxicity

Acute data
LD_{50} oral rat 11.3 mg kg^{-1} (14).
LC_{50} (8 hr) inhalation rat 1600 ppm (15).
LC_{Lo} (1 hr) inhalation mouse 31 mg m^{-3} (16).
LD_{50} intraperitoneal mouse 709 mg kg^{-1} (17).

Sub-acute data
Inhalation rats, mice 10 and 43 mg m^3 (unnecessary) for 90 days, increased numbers of leukocytes, decreased blood cholinesterase activity and liver mixed-function oxidase activity reported (18).

Metabolism and pharmacokinetics
The extent of metabolism of ethyl acetate inspired into the upper respiratory tract was measured as 40-65% in rats and 63-90% in hamsters (19).

Irritancy
Irritant to mucous membranes, especially eyes, buccal cavity and respiratory passage (20).
The skin on hands of workers exposed to ethyl acetate measured pH 6.6- 8.4, compared to the pH 3-6 of controls. Dermatitis is reported to have developed on the hands of workers exposed to ethyl acetate (21).

Genotoxicity
Saccharomyces cerevisiae increased aneuploidy frequency (22).
Chinese hamster V79 cells negative for micronuclei induction (23).
Vicia faba induced stimulation of mitotic index and chromosome aberrations in anaphase cells and micronuclei in interphase cells (24).

Any other adverse effects to man
Increased levels of serum bile acids, indicative of early signs of liver failure, were found in a group of workers exposed to a mixture of organic solvents including ethyl acetate (25).

Any other adverse effects
Inhalation rat (4 hr) unspecified concentration caused leukopenia without any change in differential or red blood cell count (25).

Any other comments
Human odour perception 0.6 mg m^{-3} (16).
Considered safe for use as a cosmetic ingredient (26).
Experimental toxicology, exposure and human health effects reviewed (27).

References
1. Gupta, A. B; et al. *Ecotoxicol. Environ. Saf.* 1982, **6**(2), 166-170
2. Brooke, L. T; et al. *Acute Toxicities of Organic Chemicals to Fathead Minnows* 1984, 27, Center for Lake Superior Environmental Studies, Univ. of Wisconsin
3. Kamlet, M. J; et al. *Environ. Sci. Technol.* 1987, **21**, 149-155
4. Bringmann, G; et al. *Water Res.* 1980, **14**, 231
5. Sloof, W; et al. *Bull. Environ. Contam. Toxicol.* 1980, **24**, 439
6. Clark, C; et al. *J. Bact.* 1967, **93**, 1309

7. Chou, W. L; et al. *Biotech. Bioeng. Symp.* 1979, **8**, 391-414
8. Price, K. S; et al. *J. Water Pollut. Control Fed.* 1974, **46**, 63-77
9. Ettinger, M. B. *Ind. Eng. Chem.* 1956, **48**, 256-259
10. Young, R. H. F; et al. *J. Water Pollut. Control Fed.* 1968, **40**, 354-368
11. Heukelekian, H; et al. *J. Water Pollut. Control Assoc.* 1955, **29**, 1040-1053
12. Slave, T; et al. *Rev. Chim.* 1974, **25**, 666-670
13. Campbell, I. M; et al. *Chem. Phys. Lett.* 1978, **53**, 385-389
14. Smyth; et al. *Am. Ind. Hyg. Assoc. J.* 1962, **23**, 95
15. *Patty's Industrial Hygiene and Toxicology* 2nd ed., 1958-1963, Interscience Publishers, New York
16. *Comm. Eur. Commun. Rep. Ethyl Acetate* 1988, European Atomic Energy Community, EUR11553
17. Sax N. I. *Dangerous Properties of Industrial Materials* 7th ed., 1989, Van Nostrand Reinhold, New York
18. Solomin, G. I; et al. *Kosm. Biol. Aviakosmicheskaya Med.* 1975, **9**(1), 40-44
19. Morris, J. B. *Toxicol. Appl. Pharmacol.* 1990, **102**(2), 331-345
20. Gu, Z; et al. *Weisheng Dulixue Zazhi* 1988, **2**(1), 1-4 (*Chem. Abstr.* **112**, 93613w)
21. Mirzoyan, I. M; *Vop. Gig. Tr. Profzabol. Mater. Nauch Konf.* 1971, 244-246, A. P. Filin, (Ed.) (*Chem. Abstr.* **116**, 10900w)
22. Mayer, V. M; et al. *Mutat. Res.* 1987, **187**(1), 31-35
23. Franco, G; et al. *Occup. Environ. Chem. Hazard* 1986, 173-177
24. Gomez-Arroyo, S; et al. *Cytologia (Tokyo)* 1986, **51**(1), 133-142
25. Brondeau, M. T; et al. *J. Appl. Toxicol.* 1990, **10**(2), 83-86
26. *J. Am. Coll. Toxicol.* 1989, **8**(4), 681-705
27. *ECETOC Technical Report No. 30(4)* 1991, European Chemical Industry Ecology and Toxicology Centre, B-1160 Brussels

A15 Acetic acid, fluoro-

FCH₂COOH

CAS Registry No. 144-49-0
Synonyms fluoroethanoic acid; monofluoroacetic acid; 2-fluoroacetic acid; Cymonic acid; Gibflaar poison
Mol. Formula $C_2H_3FO_2$ **Mol. Wt.** 78.04
Uses Chemical warfare agent. Catalyst for coatings.
Occurrence Occurs in *Dichapetalum cymosum* and *Palicourea maregiavii* leaves (1).

Physical properties

M. Pt. 33°C; **B. Pt.** 165°C.

Occupational exposure

UN No. 2642; **Conveyance classification** toxic substance; **Supply classification** toxic.

Risk phrases

Very toxic if swallowed (R28)

Safety phrases

Keep locked up and out of reach of children – When using do not eat or drink – Do not breathe dust – In case of contact with eyes, rinse immediately with plenty of water and seek medical advice – In case of accident or if you feel unwell, seek medical advice immediately (show label where possible) (S1/2, S20, S22, S26, S45)

Ecotoxicity

Invertebrate toxicity
Cell multiplication inhibition test *Microcystis aeruginosa* 0.4 µg l^{-1}, *Entosiphon sulcatum* 31 mg l^{-1} (2,3).

Environmental fate

Carbonaceous inhibition
Pseudomonas cepacia is capable of growing in fluoracetate enriched solutions without any reduction in growth rate. Fluoroacetate was degraded to carbon dioxide at a rate of 23.5 ng 10^9 cells^{-1} hr^{-1} (4).

Mammalian and avian toxicity

Acute data
LD_{50} oral rat 5 mg kg^{-1} (5).
LD_{50} intraperitoneal mouse 7 mg kg^{-1} (6).
LD_{50} intravenous mouse 13 mg kg^{-1} (7).

Sub-acute data
In man, the symptoms of sub-acute fluoroacetate poisoning began between 30 min and several hr after exposure. Death occurred rapidly preceeded by convulsions and arrhythmia (8).

Metabolism and pharmacokinetics
Fluoroacetate administered by intraperitoneal injection to rats and mice was defluorinated to yield the fluoride ion which was detected in urine and kidney. Urinary metabolites included an *S*-(carboxymethyl) conjugate complex in rats and mice and sulfoxidation products in rats. Bile metabolites included *S*-(carboxymethyl)glutathione or a related conjugate and an *O*-conjugate of fluoroacetate. Metabolic defluorination of fluoroacetate was attributed to conjugation of fluoroacetate with reduced glutathione and conversion to (-)-*erythro*- fluorocitrate. Urine of rats and mice poisoned with fluoroacetate show elevated citrate and glucose levels and diminished urea consistent with disruptions in the TCA cycle and ammonia metabolism (9).
After intraperitoneal administration to rats, fluoroacetic acid was detected in the liver, gastrointestinal tract, lung, kidneys and brain (10).

Any other adverse effects to man
Affects the human central nervous system causing convulsions and ventricular fibrillation (11).

Any other adverse effects
Neurotoxin (1,12).
In chicken liver gluconeogenesis, fluoroacetate intoxication blocked the tricarboxylic acid (TCA) cycle in both fed and fasted birds. Inhibition of TCA cycle caused near maximum vasodilation in respiratory muscles and myocardium without affecting resting skeletal muscle (13,14).

References

1. Eckschmidt, M; et al. *Braz. J. Med. Biol. Res.* 1989, **22**(8), 975-977
2. Bringmann, G; et al. *GWF-Wasser/Abwasser* 1976, **117**(9) 119
3. Bringmann, G; et al. *Water Res.* 1980, **14**, 231-241
4. Meyer, J. J; et. al; *Appl. Environ. Microbiol* 1990, **56**(7), 2152-2155
5. *Toxicol. Appl. Pharmacol.* 1968, **13** 189
6. Patterson, F. L. M; et al. *J. Org. Chem.* 1956, **21**, 883
7. *U.S. Army, Arm. Res. and Ser. Command* NX 03014
8. Peters, R. A; et al. *J. Occup. Med.* 1981, **23**(2), 112-113
9. Tecle, B; et al. *Chem. Res. Toxicol.* 1989, **2**(6), 429-435
10. *Foreign Compound Metabolism in Mammals* 1970, **1**, 93, The Chemical Society, London
11. Gribble, G. *J. Chem. Educ.* 1973, **50**(7), 460-462
12. Hornfeldt, C. S; et al. *Eur. J. Pharmacol.* **179**(3), 307-313
13. Bobyleva-Guarrievo, V; et al. *Fluoride* 1984, **17**(2), 94-104
14. Johnson, R. L. *J. Appl. Physiol.* 1988, **64**(1), 174-180

A16   Acetic acid, isopropyl ester

$$CH_3CO_2CH(CH_3)_2$$

CAS Registry No. 108-21-4

Synonyms isopropyl acetate; 1-methylethyl acetate; 1-methylethyl ethanoate; acetic acid, 1-methylethyl ester

Mol. Formula $C_5H_{10}O_2$ **Mol. Wt.** 102.13

Uses Used in perfumery, and as a solvent for cellulose derivatives, plastics, oils and fats.

Physical properties

M. Pt. –73°C; **B. Pt.** 87-89°C; **Flash point** 4°C (closed cup);

Specific gravity d_4^{20} 0.871; **Volatility** v.p. 40 mmHg at 17°C; v. den. 3.52.

Solubility

Water: 40 ml l^{-1} at 27°C. Organic solvent: miscible in diethyl ether, ethanol

Occupational exposure

US TLV (TWA) 250 ppm (1040 mg m^{-3}); **US TLV (STEL)** 310 ppm (1290 mg m^{-3}); **UK Short-term limit** 200 ppm (840 mg m^{-3}); **UN No.** 1220;
HAZCHEM Code 3☒E; **Conveyance classification** flammable liquid;
Supply classification highly flammable.

Risk phrases

Highly flammable (R11)

Safety phrases

Keep away from sources of ignition – No Smoking – Do not breathe vapour – Do not empty into drains – Take precautionary measures against static discharges (S16, S23, S29, S33)

Ecotoxicity

Bioaccumulation
The calculated bioconcentration factor of 1.8 indicated that environmental accumulation is unlikely (1).

Environmental fate

Degradation studies
BOD_5 61% reduction in dissolved oxygen using a settled domestic wastewater seed (2).
The calculated soil adsorption coefficient of 14.8 indicates that isopropyl acetate will be highly mobile in soil (3).
Evaporation was significant in both dry and wet soil (4).

Abiotic removal
Low atmospheric reactivity (5).
Resistant to hydrolysis under acidic conditions (6).

Mammalian and avian toxicity

Acute data
LD_{50} oral rat 3-7 g kg^{-1} (7,8).
LC_{Lo} (4 hr) inhalation rat 32000 ppm (7).
LD_{50} gavage rat, mouse (duration unspecified) 10.9 g kg^{-1} and 6.6 g kg^{-1}, respectively. Increased motor activity, interrupted respiration and death within 1-3 days reported (9).

Any other adverse effects to man
12 human subjects of both sexes were exposed to isopropyl acetate for 15 min. Majority of subjects experienced some degree of eye irritation at 200 ppm. Highest concentration which majority of subjects estimated satisfactory for 8 hr exposure (10).

Any other comments
Experimental toxicology, epidemiology and human health effects reviewed (11).

References

1. Stephan, H; et al. *Solubilities of Inorganic and Organic Compounds in Binary Systems* 1963, **1**, 1-79
2. Price, K. S; et al. *J. Water Pollut. Contr. Fed.* 1974, **46**, 63-77
3. Swann, R. L; et al. *Res. Rev.* 1983, **85**, 17-28
4. Ambrose, D; et al. *J. Chem. Therm.* 1981, **13**, 795-802
5. Farley, F. F. *Int. Conf. Photochem. Oxid. Pollut. Control Proc.* 1977, **2**, 713-726
6. Elam, E. U. *Kirk-Othmer Enc. Chem. Tech.* 3rd ed., 1978, **9**, 311-337
7. Smyth H. F; et al. *Arch. Ind. Hyg. Occup. Med.* 1954, **10**, 61
8. *Patty's Industrial Hygiene and Toxicology* 3rd ed., 1981, **2** Clayton, G. D.; et al. (Eds.), John Wiley, New York
9. Guseinov, V. G; et al. *Azerb. Med. Zh.* 1988, **63**(5), 41-5, (Russ.) (*Chem. Abstr.* **106**, 79939b)
10. Silverman, L; et al. *J. Ind. Hyg. Toxicol.* 1946, **28**, 262
11. *ECETOC Technical Report No. 30(4)* 1991, European Chemical Industry Ecology and Toxicology Centre, B-1160 Brussels

A17   Acetic acid, lead salt

(CH3CO2)2Pb

CAS Registry No. 301-04-2

Synonyms lead acetate; lead methanoate; neutral lead acetate; normal lead acetate; salt of saturn; sugar of lead

Mol. Formula $C_4H_6O_4Pb$ **Mol. Wt.** 325.28

Uses In the manufacture of lead salts. In the dyeing and printing of cotton.

Physical properties

M. Pt. 200°C (decomp.); **Specific gravity** d_4^{20} 3.25.

Solubility
Water: 6 g l^{-1} at 20°C. Organic solvent: glycerol, ethanol

Occupational exposure

US TLV (TWA) 0.15 mg m^{-3}; **UK Long-term limit** 0.15 mg m^{-3} of air (as Pb); **UN No.** 1616; **HAZCHEM Code** 2Z; **Conveyance classification** harmful substance; **Supply classification** harmful.

Risk phrases
Harmful by inhalation and if swallowed – Danger of cumulative effects (R20/22, R33)

Safety phrases
Keep away from food, drink and animal feeding stuffs – When using do not eat, drink or smoke (S13, S20/21)

Ecotoxicity

Fish toxicity
LC$_{50}$ (96 hr) bluegill sunfish 400 ppm. No lethal effects up to 300 ppm (1).

Invertebrate toxicity
Cell multiplication inhibition test *Uronema parduczi* 0.07 mg l^{-1}, *Pseudomonas putida* 1.8 mg l^{-1} (1,2).

Microcystis aeruginosa lowest effective concentration observed to cause reproductive effects 0.45 mg l^{-1} (2).

LC$_{50}$ (96 hr) freshwater crabs 20 ppm. On exposure the haemolymph, pH, lactate and carbon dioxide levels increased while the oxygen level decreased (3).

EC$_{50}$ (24-48 hr) *Daphnia magna* 5-2.7 mg l^{-1} (4).

Environmental fate

Nitrification inhibition
Threshold inhibition at 0.5 mg l^{-1} (5).

Mammalian and avian toxicity

Acute data
LD$_{50}$ intraperitoneal rat 200 mg kg^{-1} (6).

Sub-acute data

Rats receiving 10 mg kg^{-1} daily (unspecified route) for 3 months exhibited reticulocytosis within 36 days (6).
An oral dose of lead acetate at 0.2 mg kg^{-1} to rats for 21 days caused growth retardation, blood haemoglobin decrease and metabolic disorders (7).

Carcinogenicity and long-term effects

Inadequate evidence for carcinogenicity to humans, sufficient evidence for carcinogenicity to animals for inorganic lead compounds, IARC classification group 2B (8).
The carcinogenic potential of lead in humans is not well established, however a number of cases of renal adenocarcinoma in lead workers have been reported (9).
The level of human exposure equivalent to the level of lead acetate producing renal tumours in rats is 810 mg day^{-1}. This level exceeds the maximum tolerated dose in humans (10).

Teratogenicity and reproductive effects

Injection of lead acetate into yolk sac of chick embryos caused malformations at a rate of 50% at the 50-75 µg egg^{-1} level (11). Survival rate of 100 µg egg^{-1} dosed group reduced from 100% (untreated control) to 62.5% (12).
Lead salts are reported to cause morphological sperm abnormalities in mice (13).
In ♂ humans, concentrations of lead (unspecified dose) were associated with a high frequency of altered spermatogenesis. Relatively low levels of lead absorption have a direct toxic action on the male gonads (14).

Metabolism and pharmacokinetics

Lead is readily absorbed from gastrointestinal tract and accumulates in various tissues, in particular the bone (14).
Increased serum arginase activity may indicate liver damage, while decreased kallikrein activity may indicate kidney damage in workers exposed to lead (15).

Genotoxicity

Escherichia coli SOS chromatest negative (16).
Lead acetate significantly induced sister chromatid exchange in a dose-dependent manner in Chinese hamster ovary cells (17).
Lead acetate caused dose-related transformation of cultured Syrian hamster embryo cells. Isolated cells from the transformed colonies produced fibrosarcomas when injected subcutaneously into either Syrian hamster or nude mice. Results suggest that lead acetate-induced transformation may be a result of decreased accuracy of DNA synthesis (18).

Any other adverse effects to man

Clinical studies have found increased chromosomal defects in workers with blood lead levels above 600 µg l^{-1} (19).
Lead acetate was formerly applied in aqueous solution to the eye for astringent effect, but induced opacities and lead encrustation of the cornea and conjuctiva. Opacities occured in patients when the cornea epithelium had been injured allowing the lead solution to penetrate. The epithelium normally acts as a barrier to lead salts penetrating the cornea (20).

Any other adverse effects

Symptoms of lead poisoning include thirst, persistent metallic taste, nausea, abdominal pain, vomiting, diarrhoea, and constipation. Central nervous system symptoms include paresthesia, pain and muscle weakness, other symptoms include anaemia, haemoglobinemia, oliguria and urinary changes. Severe exposure can cause death in 1-2 days. Degenerative changes in motorneurons and axons have been reported. Can cause visage and ocular disturbances (9).

Organic lead poisoning produces mainly central nervous system symptoms, but can have gastrointestinal and cardiovascular effects, and cause renal and hepatic damage (14).

Recent clinical research has focused on neuropsychiatric, reproductive and renal effects of chronic low dose lead exposure (21).

Legislation

Control limits for lead are set out in the Health & Safety Commission approved code of practice supporting the Control of Lead at Work Regulations 1980 (22).

Any other comments

Lead acetate intoxication, epidemiology studies and genotoxicity testing are reported (16,23-27).

References

1. Bringmann, G. B; et al. *Z. Wasser/Abwasser Forsch.* 1980, (1), 26-31
2. Bringmann, G. B; et al. *G. w.f. Wasser/Abwasser* 1976, **117**(9)
3. Tulasi, S. J; et al. *Bull. Environ. Contam. Toxicol.* 1988, **40**(2), 198-203
4. Richardson, M. L. *Nitrification Inhibition in the Treatment of Sewage* 1985, RSC, London
5. Klangarot, B.S; et al. *Bull. Environ. Contam. Toxicol.* 1987, **38**, 722-726
6. Sudakova, A. I; et al. *Tsitol Genet.* 1983, **17**(4), 3-7 (*Chem. Abstr.* **99**, 117430a)
7. Ivanitskii, A. M; et al. *Vopr. Pitan.* 1985, (2), 63-66 (*Chem. Abstr.* **103**, 36304m)
8. *IARC Monograph* 1987, **Suppl. 7**, 230-232
9. Gilman, A. G; et al. *The Pharmacological Basis of Therapeutics* 7th ed., 1985, Macmillan Publishing, New York
10. *IARC Monograph* 1972, **1**, 48
11. Matsuo, T; et al. *Kanyo Kagaku Kenkyusho Kenkyu (Kiuki Diagaku)* 1984, (12), 179-185 (*Chem. Abstr.* **103**, 82952)
12. Matsuo, T; et al. *Kanyo Kagaku Kenkyusho Kenkyu (Kiuki Diagaku)* 1983, (11), 191-195 (*Chem. Abstr.* **100**, 46550r)
13. Lancranjan, I; et al. *Arch. Environ. Health* 1975, **30**, 396-401
14. *Martindale: The Extra Pharmacopoeia* 29th ed., 1989, The Pharmaceutical Press, London
15. Chmielnicka, J; et al. *Br. J. Ind. Med.* 1981, **38**(2), 175-178
16. Rozalski, M; et al. *Bromatol. Chem. Toksykol.* 1982, **15**(1-2), 129-130
17. DiPaolo, J. A; et al. *Br. J. Cancer* 1978, **38**(3), 452-455
18. Willems, M. I; et al. *Arch Toxicol.* 1982, **50**, 149-157
19. Doull, J; et al. *Casarett and Doull's Toxicology* 3rd Ed., 1986, Macmillan, New York
20. Grant, W. M. *Toxicology of the Eye* 3rd ed., 1986, 550, C. C. Thomas, Springfield, IL
21. Ellenham, M. J; et al. *Medical Toxicology, Diagnosis and Treatment of Human Poisoning* 1988, Elsevier, New York
22. *Food Surveillance Paper No. 10: Survey of Lead in Food: Second Supplementary Report* 1982, Min. Agric. Fisheries Food, HMSO, London
23. Malcolm, D; et al. *Br. J. Ind. Med.* 1982, **39**, 404-410
24. Selevan, S. G; et al. *Am. J. Epidemiol.* 1985, **122**, 673-783

25. Marzin, D. *IARC Med.* 1986, **6**(5), 252-254 (*Chem. Abstr.* **105**, 129225t)
26. Leonard, A. *Metal Ions in Biological Systems* 1986, **20**, 229-258
27. Tai, E. C; et al. *Bull. Inst. Zool. Acad. Sin.* 1990, **29**(2), 124-125

A18   Acetic acid, methyl ester

$CH_3CO_2CH_3$

CAS Registry No. 79-20-9
Synonyms methyl acetate; methyl methanoate
Mol. Formula $C_3H_6O_2$ **Mol. Wt.** 74.08
Uses Solvent for resins and oils. Used in the manufacture of artificial leather.

Physical properties

M. Pt. –98°C; **B. Pt.** 56.9°C; **Flash point** –10°C (closed cup); **Specific gravity** d_4^{25} 0.93; **Volatility** v.p. 100 mmHg at 9°C; v. den. 2.55.

Solubility
Organic solvent: miscible in ethanol, diethyl ether

Occupational exposure

US TLV (TWA) 200 ppm (606 mg m^{-3}); **US TLV (STEL)** 250 ppm (757 mg m^{-3}); **UK Long-term limit** 200 ppm (610 mg m^{-3}); **UK Short-term limit** 250 ppm (760 mg m^{-3}); **UN No.** 1231; **HAZCHEM Code** 2⬛E; **Conveyance classification** flammable liquid; **Supply classification** highly flammable.

Risk phrases
Highly flammable (R11)

Safety phrases
Keep away from sources of ignition – No Smoking – Do not breathe vapour – Do not empty into drains – Take precautionary measures against static discharges (S16, S23, S29, S33)

Ecotoxicity

Fish toxicity
LC$_{50}$ (96 hr) fathead minnow 320-399 mg l^{-1} (1,2).

Bioaccumulation
Calculated bioconcentration factors of 0.57-0.81 indicate that environmental accumulation is unlikely (3).

Environmental fate

Carbonaceous inhibition
0.2% (v/v) methyl acetate was hydrolysed to methanol and acetate within 24-36 hr by *Pseudomonas sp.* strains which possess non-specific inducible carboxyl esterase activity (4).

Anaerobic effects

>66% of methyl acetate was degraded to methane in 90 days using an anaerobic digester seed acclimated to acetic acid (5).

Degradation studies

Oxidised by *Alcaligenes faecalis* isolated from activated sludge (6).

BOD_5 26% reduction in dissolved oxygen using a sewage seed inoculum (7).

Calculated soil adsorption coefficients indicated that methyl acetate will be highly mobile in soil (8).

Abiotic removal

Generally alkyl esters were resistant to hydrolysis under acidic or neutral conditions typically found in the environment (9).

Evaporation was significant in the removal of methyl acetate from the environment (10).

Photochemical reactivity, $t_{1/2}$ 47-94 days at 19-30°C (11).

Mammalian and avian toxicity

Acute data

LD_{50} oral rabbit 3700 mg kg^{-1} (12).
LC_{50} (1 hr) inhalation cat 67 g m^{-3}.
LC_{Lo} (4 hr) inhalation mouse 34 g m^{-3}.
LD_{Lo} subcutaneous cat 3000 mg kg^{-1} (13).

Metabolism and pharmacokinetics

2 hr exposure twice daily (species unspecified) at concentration of 200 ppm for 3-4 days caused an increase in urinary excretion of methanol (14).

Irritancy

Dermal rabbit (24 hr) 500 mg caused mild irritation, 100 mg instilled into rabbit eye in 24 hr caused severe irritation (15,16).

Genotoxicity

Saccharomyces cerevisiae induced sex chromosome loss and non-disjunction (17).

Any other adverse effects to man

A 17 yr old girl, who had sniffed a lacquer thinner for 3 months, suffered acute blindness followed by optic atrophy. Brain scan revealed symmetrical low attentuation areas in the bilateral putamen. Major components of the thinner in the vapourised state were methanol and methyl acetate (18).

Any other adverse effects

Narcotic (19).

Any other comments

Experimental toxicology, epidemiology and human health effects reviewed (20).

References

1. Greiger, D. L; et al. *Acute Toxicity of Organic Chemicals to Fathead Minnows* 1985, 50, Lake Superior Environmental Studies Centre, Univ. of Wisconsin
2. Brooke, L. T; et al. *Acute Toxicity Of Organic Chemicals to Fathead Minnows* 1984, 57-58, Lake Superior Environmental Studies Centre, Univ. of Wisconsin

3. Lyman, W. J; et al. *Handbook of Chemical Property Estimation Methods* 1982, **4**, 1-33, McGraw-Hill, New York
4. Rakov, D. Y; et al. *Microbiol. Lett.* 67 1990, 67-72
5. Chou, W. L; et al. *Biotech. Bioeng. Symp.* 1979, **8**, 391-414
6. Marion, C. V; et al. *J. Water Control Fed.* 1963, **35**, 1269-1284
7. Heukelekian, H; et al. *J. Water Pollut. Control Assoc.* 1955, **29**, 1040-1053
8. Swann, R. L; et al. *Res. Rev.* 1983, **85**, 17-28
9. Drossman, H; et al. *Chemosphere* 1987, **17**, 1509-1530
10. Ambrose, D; et al. *J. Chem. Therm.* 1981, **13**, 795-802
11. Atkinson, R. *Chem. Rev.* 1985, **85**, 69-201
12. *Ind. Med. Surg.* 1972, **41**, 31
13. *Arch. Gerwebenpathologie Hyg.* 1933, **1** 5, Int. Arch. Environ. Health, Springer Verlag, Berlin
14. Tada, O; et al. *Rodo Kagaku* 1974, **50**(4), 239-248 (*Chem. Abstr.* **82**, 814m)
15. Opdyke, D. L. J. *Food Cosmet. Toxicol.* 1979, **17**(Suppl.), 695
16. Marhold, J. V. *Sbornik Vysledku Toxixologickeho Vystreni Latek A Pripravku* 1972, 44 Prague
17. *Mutat. Res.* 1985, **149**, 339
18. Kohriyama, K; *J. Univ. Occup. Environ. Health* 1989, **11**(4), 449-453 (Jap.) (*Chem. Abstr.* **113**, 146807s)
19. *The Merck Index* 11th ed. 1989, Merck & Co. Inc, Rahway, NJ
20. *ECETOC Technical Report No. 30(4)* 1991, European Chemical Industry Ecology and Toxicology Centre, B-1160 Brussels

A19   Acetic acid, propyl ester

$CH_3CO_2CH_2CH_2CH_3$

CAS Registry No. 109-60-4
Synonyms *n*-propyl acetate; *n*-propyl methanoate
Mol. Formula $C_5H_{10}O_2$ **Mol. Wt.** 102.13
Uses Used in the manufacture of flavours and perfumes. As a solvent for resins, cellulose derivatives and plastics.

Physical properties

M. Pt. –92°C; **B. Pt.** 99-102°C; **Flash point** 14°C (closed cup); **Specific gravity** d_4^{20} 0.887; **Partition coefficient** log P_{ow} 1.23; **Volatility** v.p. 40 mmHg at 29°C; v. den. 3.52.

Solubility
Water: 16 ml l^{-1} at 16°C. Organic solvent: miscible in ethanol, diethyl ether

Occupational exposure
US TLV (TWA) 200 ppm (835 mg m^{-3}); **US TLV (STEL)** 250 ppm (1040 mg m^{-3}); **UK Long-term limit** 200 ppm (840 mg m^{-3}); **UK Short-term limit** 250 ppm (1050 mg m^{-3}); **UN No.** 1276; **HAZCHEM Code** 3￼E;
Conveyance classification flammable liquid;corrosive substance; **Supply classification** highly flammable.

Risk phrases
Highly flammable (R11)

Safety phrases
Keep away from sources of ignition – No Smoking – Do not breathe vapour – Do not empty into drains – Take precautionary measures against static discharges (S16, S23, S29, S33)

Ecotoxicity

Fish toxicity
LC_{50} (96 hr) fathead minnow 60 mg l^{-1} (1).
In fish, n-propyl acetate was metabolised by *in vivo* hydrolysis of carboxylic acid esters. Lethality properties of this class of compounds cannot be compared unless carboxylase esterase activities for the species are known (2).

Invertebrate toxicity
Cell multiplication inhibition test, *Pseudomonas putida* 1700 mg l^{-1}, *Entosiphon sulcatum* 97 mg l^{-1} (3).

Bioaccumulation
Calculated bioconcentration factors of 2-5 indicate that environmental bioaccumulation will be insignificant (4).

Environmental fate

Degradation studies
BOD_5 62 % reduction in dissolved oxygen using a settled domestic wastewater seed (5).
High mobility in soil reported (6).
Evaporation was significant in the removal of n-propyl acetate from dry and wet soil (7).

Abiotic removal
Hydrolysed rapidly to alcohol and acid (7).

Absorption
Activated sludge 0.149 g g^{-1} carbon (8).

Mammalian and avian toxicity

Acute data
LD_{50} oral rat, rabbit 6630-9370 mg kg^{-1} (9).
LC_{Lo} (4 hr) inhalation rat 8000 ppm (9).

Metabolism and pharmacokinetics
Esters, used as industrial solvents and of interest as occupational exposure hazards, were incubated for 1-8 hr in water or in blood. No hydrolysis occurred in the pure aqueous solution but progressive hydrolysis occurred in blood (10).

Any other adverse effects to man

Reported to be narcotic in high concentrations in humans (11).

Any other comments

Experimental toxicology, epidemiology and human health effects reviewed (12).

References

1. Brooke, L. T; et al. *Acute Toxicity of Organic Chemicals to Fathead Minnows* 1984, 125, Lake Superior Environ. Studies Centre., Univ. of Wisconsin
2. Kamlet, M. J; et al. *Environ. Sci. Tech.* 1987, **21**, 149-155
3. Bringman, G; et al. *Water Res.* 1980, **14**, 231
4. Lyman, W. J; et al. *Handbook of Chemical Property Estimation Methods* 1982, **4**, 1-33, McGraw-Hill, New York
5. Price, K. S; et al. *J. Water Pollut. Control Fed.* 1974, **46**, 63-77
6. Swann, R. L; et al. *Res. Rev.* 1983, **85**, 17-28
7. Ambrose, D; et al. *J. Chem. Therm.* 1981, **13** 795-802
8. *Ind. Med. Surg.* 1972, **41**, 31
9. *Am. Ind. Hyg. Assoc. J.* 1969, **30**, 470
10. Ghittori, S; et al. *Boll. Soc. Ital. Biol. Sper.* 1984, **60**(11), 2207-2214
11. *The Merck Index* 11th ed. 1989, Merck & Co. Inc, Rahway, NJ
12. *ECETOC Technical Report No. 30(4)* 1991, European Chemical Industry Ecology and Toxicology Centre, B-1160 Brussels

A20   Acetic acid, thallium(I) salt

CH3CO2Tl

CAS Registry No. 563-68-8
Synonyms thallium(I) acetate; thallium(I) methanoate; thallous acetate
Mol. Formula $C_2H_3O_2Tl$ **Mol. Wt.** 263.42
Uses In ore flotation. Previously used as a medicine for the treatment of ringworm and as an ingredient of depilatory creams.

Physical properties

M. Pt. 126-128°C; **Specific gravity** d^{137} 3.76.

Solubility

Organic solvent: ethanol, chloroform

Occupational exposure

US TLV (TWA) 0.1 mg m^{-3} (as Tl); **UK Long-term limit** 0.1 mg m^{-3} (as Tl); **UN No.** 1707; **Conveyance classification** toxic substance;
Supply classification toxic.

Risk phrases

Very toxic by inhalation and if swallowed – Danger of cumulative effects (R26/28, R33)

Safety phrases

Keep out of reach of children – Keep away from food, drink and animal feeding stuffs – After contact with skin, wash immediately with plenty of water – In case of accident or if you feel unwell, seek medical advice immediately (show label where possible) (S2, S13, S28, S45)

Ecotoxicity

Fish toxicity
LC_{50} (96 hr) Atlantic salmon 0.03 ppm as thallium (1).

LC_{50} (96 hr) bluegill sunfish 170 ppm static bioassay at 23°C.
LC_{50} (96 hr) inland silverside 31 ppm static bioassay at 23°C (2).

Invertebrate toxicity
LC_{50} (96 hr) brown shrimp 10 ppm (1).
EC_{50} (48 hr) *Daphnia magna* (Strauss test) <1 mg l^{-1} (3).

Bioaccumulation
The bioconcentration factor in marine invertebrates was 150,000, in freshwater and marine fish and plants 100,000. The bioconcentration factor in clams was 17-18, in mussels 11-12, and Atlantic salmon 27-1430 (4).

Environmental fate

Degradation studies
No evidence was found for the formation of volatile thallium compounds in the environment (4).
Natural thallium levels in plants are reported as 0.01-3800 ppm by weight, 0.5 ppm was the average value in most species (5).

Absorption
Thallium salts are strongly absorbed by montmorillonite clays (4).

Mammalian and avian toxicity

Acute data
LD_{50} oral mouse 35 mg kg^{-1}.
LD_{50} intraperitoneal rat 30 mg kg^{-1}.
LD_{Lo} intravenous rabbit 26 mg kg^{-1} (6).
Thallium salts are toxic when inhaled, ingested or absorbed through the skin. Symptoms of poisoning may appear within 12-24 hr of a single toxic dose and include severe abdominal pain, vomiting, diarrhoea, gastrointestinal haemorrage, tremors, delirium, convulsions, paralysis and coma leading to death in 1-2 days. The acute reaction may subside to be followed in 10 days by the development of polyneuritis, psychosis, delirium encephalopathy, lachycardia, hypertension, skin eruptions and hepatorenal injury. Alopecia occurs within 15-20 days. Death may result from respiratory failure (7).

Sub-acute data
In a 90 day feeding study in rats, 0.003% caused growth depression and depilation. Atrophy of hair follicles and sebaceous glands were observed in the skin. The kidney was the principal site of deposition in rats followed by bone, liver, lung, spleen and brain (8).

Teratogenicity and reproductive effects
In humans (and laboratory animals), following thallium poisoning, the testis had some of the highest thallium levels of any organ. The evidence suggests that reproductive systems are highly susceptible to thallium toxicity (5).

Metabolism and pharmacokinetics
In humans, thallium was detected in urine, faeces and hair 5 months after a single exposure. Excretion takes place via the kidney, gut and salivary glands (5).
Excretory routes differ in importance for different species. Following exposure to thallium by unspecified route(s), (3 day) in humans faecal and urinary excretion was measured at 0.4% and 11.3%, respectively. In comparison, (2 day) rat faecal

excretion was 9% and urinary 6%, and dog (36 day) 60% was excreted in the urine (8).

Genotoxicity

Escherichia coli SOS chromatest negative (5).

Any other adverse effects

In acute thallium poisoning, thallium accumulates in the grey matter of the brain (5). In mouse myelinated cord-ganglia-muscle 10 μg ml^{-1} caused enlarged mitochondria in the axons of peripheral nerve fibres within 2 hr (9).

Following inhalation of thallium and its salts, thallium was rapidly absorbed from mucous membranes of respiratory tract, mouth and lung. Distribution to the tissues was rapid via the blood, part of the thallium was absorbed into erythrocytes. Thallium crosses the blood/brain and placental barriers (10).

Any other comments

UK legislation prohibits the use of thallium compounds in cosmetic products (7).
The toxicity of thallium and its salts are discussed in detail (8).
Thallium acetate as a hazardous material is reviewed (11).
Cases of human poisoning by thallium and its salts are reviewed (12-14).

References

1. Zitko, V. *The Science of the Total Environment* 1973, **4**, 185
2. Dawson, G. W; et al. *J. Haz. Mater.* 1975/77, **1**, 303-318
3. Bringmann, G; et al. *Z. Wasser Abwasser Forsch.* 1982, **15**(1), 1-6
4. Callahan, M. A; et al. *Water-Related Environmental Fate of 129 Priority Pollutants* 1979, 1, EPA 440/4 79-029a
5. Seiler, H. G; et al. *Handbook on the Toxicity of Inorganic Compounds* 1988, 678, 686-681, Marcel Dekker, New York
6. Spector, W. S. (Ed.), *Handbook of Toxicology* 1956, **1**, 294-295, Saunders, Philadelphia
7. *Martindale: The Extra Pharmacopoeia* 29th ed., 1989, The Pharmaceutical Press, London
8. *Patty's Industrial Hygiene and Toxicology* 3rd ed., 1981, **2**, Clayton, G. D; et al. (Eds.), John Wiley, New York
9. Spencer, P. S; et al. *J. Cell Biol.* 1973, **58**, 79
10. Venugopal, B; et al. *Metal Toxicity in Mammals* 1978, (2), Plenum Press, London
11. *Dangerous Prop. Ind. Mater. Rep.* 1987, **7**(2), 92-94
12. Conley, B. E. *J. Am. Med. Assoc.* 1957, **165** 1566
13. Munch, J. C. *J. Am. Med. Assoc.* 1934, **102** 1929
14. Poliakova, M. M. *Gig. Tr. Prof. Zabol.* 1977, **2**, 14

A21   Acetic anhydride

$(CH_3CO)_2O$

CAS Registry No. 108-24-7
Synonyms acetic acid, anhydride; acetic oxide; acetyl anhydride; acetyl ether; acetyl oxide; ethanoic anhydrate
Mol. Formula $C_4H_6O_3$ **Mol. Wt.** 102.09

Uses Manufacture of acetyl compounds and cellulose acetate fibres and plastics. An acetylating agent and solvent in examining wool, fat, glycerol, fatty and volatile oils, resins. Widely used in organic synthesis. A dehydrating agent and acetylating agent in the production of pharmaceuticals, dyestuffs, perfumes, and explosives.

Physical properties

M. Pt. $-73°C$; **B. Pt.** $139°C$; **Flash point** $49°C$ (closed cup); **Specific gravity** d_4^{20} 1.082; **Volatility** v.p. 3.5 mmHg at 20°C; v. den. 3.52.

Solubility
Organic solvent: miscible in ethanol, acetone, diethyl ether, soluble in chloroform, benzene, dimethyl sulfoxide

Occupational exposure

US TLV (TWA) Ceiling limit 5 ppm (21 mg m^{-3}); **UK Short-term limit** 5 ppm (20 mg m^{-3}); **UN No.** 1715; **HAZCHEM Code** 2P; **Conveyance classification** corrosive substance; **Supply classification** flammable and corrosive.

Risk phrases
Flammable – Causes burns (R10, R34)

Safety phrases
In case of contact with eyes, rinse immediately with plenty of water and seek medical advice (S26)

Ecotoxicity

Fish toxicity
Aquatic toxicity rating, designated non-toxic to trout, bluegill sunfish and goldfish (1).

Invertebrate toxicity
Cell multiplication inhibition test, *Pseudomonas putida* 1150 mg l^{-1}, *Scenedesmus quadricauda* 3400 mg l^{-1}, *Chlorella pyrenoidosa* 360 mg l^{-1}, *Entosiphon sulcatum* 30 mg l^{-1} (2,3).

Mammalian and avian toxicity

Acute data
LD$_{50}$ oral rat 1780 mg kg^{-1} (4).
LD$_{50}$ dermal rabbit 4000 mg kg l^{-1} (5).
LC$_{50}$ (4 hr) inhalation rat 1000 ppm (6).

Irritancy
10 mg applied to rabbit skin for 24 hr caused mild irritation, 250 μg instilled in rabbit eye caused severe irritation (4).
Skin, eye and upper respiratory tract irritant (7).
May cause dermatitis and occasional sensitisation (8).

Genotoxicity
Salmonella typhimurium TA98, TA100, TA1535, TA1537 with and without metabolic activation negative (9).

Any other adverse effects to man

Workers exposed to (undetermined) high vapour concentrations reported burning sensations in nose and throat and dyspnoea (10).

Any other adverse effects

Can cause bronchial and lung injury (11).
Lachrymator and may cause conjunctival oedema and corneal burns. Temporary or permanent interstitial keratitis with corneal opacity and loss of vision have been reported (12).

Any other comments

Physical and chemical properties, hazards and current French legislation on acetic anhydride reviewed (13).
Experimental toxicology and human health effects of acetic anhydride reviewed (14).
Reacts with water to form acetic acid. Explosion risk.

References

1. Wood, E. M. *Toxicity of 3400 Chemicals to Fish* 1987, EPA 560/6-87-002, PB 87-200-275
2. Bringmann, G; et al. *Water Res.* 1980, **14**, 231-241
3. Jones, H. R. *Environmental Control in the Organic and Petrochemical Industries* 1971, Noyes Data Corporation, New York
4. *Arch. Ind. Hyg. Occup. Med.* 1951, **4**, 119
5. *Union Carbide Data Sheet* 1963, Union Carbide Corp., New York
6. Deichmann, W. B. *Toxicology of Drugs and Chemicals* 1969, Academic Press, New York
7. *Chemical Safety Data Sheets* 1990, **3**, 11-15, RSC, London
8. Fassett, D. W; et al. *Toxicology* 1963, **2**, Wiley-Interscience, New York
9. Mortelmans, K; et al. *Environ. Mutagen.* 1986, **8**(7), 1-119
10. Hygienic Guide Series: Acetic Anhydride *Am. Ind. Hyg. Assoc. J.* 1971, **32**, 64
11. Henderson, Y; et al. *Noxious Gases* 1943, 129, Van Nostrand Reinhold, New York
12. Colman, J. P; et al. *Inorg. Synth.* 1963, **7**, 205-207
13. *Cah. Notes Doc.* 1986, **125**, 593-596
14. *ECETOC Technical Report No. 30(4)* 1991, European Chemical Industry Ecology and Toxicology Centre, B-1160 Brussels

A22   Acetochlor

$CH_3CH_2OCH_2$, $COCH_2Cl$

CH_3CH_2, CH_2CH_3

CAS Registry No. 34256-82-1
Synonyms 2-chloro-*N*-(ethoxymethyl)-*N*-(2-ethyl-6-methyl phenyl)acetamide; 2-chloro-*N*-ethoxymethyl-6'-ethylacet-*o*-toluidide
Mol. Formula $C_{14}H_{20}ClNO_2$ **Mol. Wt.** 269.77
Uses Herbicide.

Physical properties

M. Pt. 82-93°C.

Solubility
Water: 223 mg l^{-1} at 25°C. Organic solvent: acetone, diethyl ether, benzene, ethanol, ethyl acetate, toluene

Ecotoxicity

Fish toxicity
LC_{50} (96 hr) rainbow trout, bluegill sunfish 0.5-1.3 mg l^{-1} (1).

Effects to non-target species
LD_{50} 1.715 mg bee^{-1} (1).

Environmental fate

Degradation studies
Microbial degradation accounts for most loss from soil (2).

Abiotic removal
Strongly absorbed by soil (1).

Mammalian and avian toxicity

Acute data
LD_{50} oral bobwhite quail 1260 mg l^{-1} (1).
LD_{50} oral rat 1063-2183 mg kg^{-1} (2,3).
LD_{50} percutaneous rabbit 4166 mg kg^{-1} (2).

Sub-acute data
Oral rat (42 day) 10-50 mg kg^{-1} administered 5 day wk^{-1} caused changes in enzyme activity, including cytochrome oxidase, lactate dehydrogenase and glucose-6-phosphate dehydrogenase, suggesting adverse effects on mitochondrial metabolic function (3).
Oral rabbit (12 month) 0.3-30 mg kg^{-1} induced atherosclerotic changes in aorta. Simultaneous administration of cholesterol and acetochlor induced more severe changes than single administration of either compound (4).

Irritancy
Dermal guinea pig (10 day) 0.1 mg of 50% acetochlor caused irritation, the severity lessened with dilution (5).

Sensitisation
Intracutaneous guinea pig 50-200 μg l^{-1} followed by skin test showed no positive reaction in response to stimulation.
Intracutaneous guinea pig 250 μg with subsequent 7-fold application onto guinea pig skin caused hyperaemic reaction in a 24 hr spot test.
In a guinea pig epidermal sensitisation test, 0.5 % acetochlor was negative (5).

Legislation
Limited under EC Directive on Drinking Water Quality 80/778/EEC. Herbicide: maximum admissible concentration 0.5 μg l^{-1} (6).

Any other comments

Occupational health hazard from exposure to acetochlor discussed (4).

References

1. *The Agrochemicals Handbook* 3rd ed., 1991, RSC, London
2. Breaux, E. J. *Weed Sci.* 1987, **35**(4), 463-468
3. Khalkova, Z; et al. *Probl. Khig.* 1990, **15**, 96-102, (Bulg.) (*Chem. Abstr.* **116**, 122936u)
4. Ivanov, Y. V. *Gig. Sanit.* 1988, **5**, 84-85, (Russ.) (*Chem. Abstr.* **109**, 18531c)
5. Gadalina, I. D; et al. *Gig. Sanit.* 1984, (8), 85-86, (Russ.) (*Chem. Abstr.* **101**, 185630a)
6. *EC Directive Relating to the Quality of Water Intended for Human Consumption* 1982, 80/778/EEC, Office for Official Publications of the European Communities, 2 rue Mercier, L-2985 Luxembourg

A23   Acetohexamide

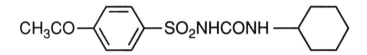

CAS Registry No. 968-81-0

Synonyms benzenesulfonamide, 4-acetyl-*N*-[(cyclohexylamino)carbonyl];
1-(*p*-acetylbenzenesulfonyl)-3-cyclohexylurea;
3-cyclohexyl-1-(*p*-acetylphhenylsulfonyl)urea; Dimelor; Ordimel

Mol. Formula $C_{15}H_{20}N_2O_4S$ **Mol. Wt.** 324.40

Uses Antidiabetic.

Physical properties

M. Pt. 188-190°C.

Solubility
Organic solvent: pyridine

Mammalian and avian toxicity

Acute data
LD$_{50}$ oral rat 5000 mg kg^{-1} (1).

Carcinogenicity and long-term effects
A Computer-Automated Structure Evaluation (CASE) prediction of the carcinogenicity of acetohexamide was negative, actual experimental evidence was also negative (2). B6C3F mice, test routes feed, water, oral gavage and skin painting reported non carcinogenic (3).

Metabolism and pharmacokinetics
Acetohexamide was readily absorbed from the gastrointestinal tract and bound to plasma proteins. Maximum hypoglycaemic activity absorbed 3 hr after ingestion. Total duration of action was 12-24 hr. Major metabolite hydroxyhexamide had a plasma t$_{1/2}$ of 6 hr. Acetohexamide had plasma t$_{1/2}$ 1.3 hr (4).

Genotoxicity

Salmonella tymphimurium TA98, TA100, TA1535, TA1537 with and without metabolic activation negative (3).

Any other comments

Metabolic reduction of acetohexamide reviewed (5).

References

1. Goldstein, M. J; et al. *New Engl. J. Med.* 1966, **275**, 97
2. Rosenkranz, H. S; et al. *Mutat. Res.* 1990, **228**, 105-124
3. Zeiger, E. *Cancer Res.* 1987, **47**, 1287-1296
4. Goodman, L. S; et al. *The Pharmacological Basis of Therapeutics* 5th ed., 1975, 1521, Macmillan Publishing, New York
5. Imamura, Y. *Yakubutsu Dotai* 1990, **5**(2), 263-271, (Jap.) (*Chem. Abstr.* **113**, 204291g)

A24 Acetone

CH_3COCH_3

CAS Registry No. 67-64-1
Synonyms 2-propanone; dimethyl ketone; dimethyl formaldehyde; dimethylketal; β-ketopropane; pyroacetic ether
Mol. Formula C_3H_6O **Mol. Wt.** 58.08
Uses Solvent for fats, oils, waxes, resins, rubber, plastics, rubber cements and pharmaceuticals. Intermediate in chemical synthesis of rayon and photographic film. Used in paints and varnishes. In extraction processes to obtain various principles from animal and plant substances.
Occurrence May be released into the environment as stack emissions, fugitive emissions and in wastewater. Naturally occurring volatile metabolite in vegetation and insects (1).

Physical properties

M. Pt. –94°C; **B. Pt.** 56.2°C; **Flash point** –18°C (closed cup); **Specific gravity** d_{25}^{25} 0.788; **Partition coefficient** log P_{ow} –0.24; **Volatility** v.p. 231 mmHg at 25°C; v. den. 2.00.

Solubility

Water: miscible. Organic solvent: miscible in ethanol, dimethylformamide, chloroform, diethyl ether

Occupational exposure

US TLV (TWA) 750 ppm (1780 mg m^{-3}); **US TLV (STEL)** 1000 ppm (2380 mg m^{-3}); **UK Long-term limit** 750 ppm (1780 mg m^{-3}) proposed change; **UK Short-term limit** 1500 ppm (3560 mg m^{-3}) proposed change; **UN No.** 1090; **HAZCHEM Code 2YE**; **Conveyance classification** flammable liquid; **Supply classification** highly flammable.

Risk phrases
Highly flammable (R11)

Safety phrases
Keep container in a well ventilated place – Keep away from sources of ignition – No Smoking – Do not breathe vapour – Take precautionary measures against static discharges (S9, S16, S23, S33)

Ecotoxicity

Fish toxicity
LC_{50} (96 hr) rainbow trout 5540 mg l^{-1} static bioassay (2).
LC_{50} (24 hr, 96 hr) goldfish, bluegill sunfish, mosquito fish 5000-13000 mg l^{-1} (2,3).
LC_{50} (24 hr) harlequin fish 5700 ppm (4).

Invertebrate toxicity
Asellus aquaticus (3 day) 3 ml l^{-1} caused 100% mortality (5).
ED_{50} (24, 48 hr) *Daphnia magna* 10 mg l^{-1} (6).
LD_{50} (24, 48 hr) brine shrimp 2100 mg l^{-1} at 24°C (7).
Cell multiplication inhibition test, *Entosiphon sulcatum* 28 mg l^{-1} (2).

Bioaccumulation
Potential for acetone bioconcentration is negligible (1).

Environmental fate

Nitrification inhibition
EC_{50} 8100 mg l^{-1} caused nitrification inhibition (8).
0.1%-0.4% (5 hr) blue green algae caused an unmeasured increase of nitrogen fixation at 22°C (9).

Anaerobic effects
The physiology of anaerobic acetone degradation was studied with a strain of gram-negative denitrifying bacteria. The results provided evidence that acetone was channelled into the intermediary metabolism of the strain via carboxylation to acetoacetate (10).
100% degradation of acetone in 4 day, after a 5 day time lag, under anaerobic conditions was observed (11).

Degradation studies
In soil acetone can volatilise and leach into the ground where it is readily biodegradable (1).
Theoretical BOD_5 averaged 37-55% using results from a range of sewage inocula (12-15).
Activated sludge inocula, after 20 hr time lag, acetone degraded at a rate of 0.016 l hr^{-1}, theoretical BOD was 42% after 155 hr (16).
EC_{50} >1000 mg l^{-1} using Organisation for Economic Cooperation and Development (OECD) 209 Test closed system (17).

Abiotic removal
Volatilisation is likely in natural water (1).
Acetone reacts photochemically in atmosphere, annual average $t_{1/2}$ 22 day (18).
No photodegradation occurred when acetone was exposed to sunlight for 15 and 23 hr in stream and distilled water, respectively (19).

Mammalian and avian toxicity

Acute data
LD$_{50}$ oral rat, mouse, rabbit, dog 4-11 g kg^{-1} (20).

Sub-acute data
LC$_{50}$ (5 day diet) oral Japanese quail >40,000 ppm (21).

Teratogenicity and reproductive effects
No developmental toxic effects were seen in rats and mice exposed to atmospheric concentrations of acetone vapour up to 11,000 ppm and 6000 ppm respectively 6 hr day^{-1} for 7 day (22).

Metabolism and pharmacokinetics
In the hamster, 8% acetone administered in the drinking water, induced cytochrome P-450 in the liver, kidneys and lung, and produced differential changes in mixed function oxidase enzyme activities in these tissues (23). In perfused livers of rats, acetone was found to be metabolised to cholesterol and glycogen (20,24). Urinary elimination of acetone in workers occupationally exposed to the compound showed that there was a linear relationship between the amount of acetone absorbed and that excreted in the urine (25).

Irritancy
Rabbit eye 100% acetone caused severe irritation (26).

Genotoxicity
Salmonella typhimurium TA92, TA94, TA98, TA100, TA1535, TA1537 with metabolic activation negative. Chinese hamster fibroblast (24 hr) without metabolic activation induced chromosomal aberrations (27).

Any other adverse effects to man
Screen-printing workers exposed to acetone reported significantly greater frequencies of neurasthenic symptoms and other symptoms of peripheral neuropathy and autonomic dysfunction than a control group (28).
Women occupationally exposed to solvents including acetone in chemical plants showed deviations in haemoglobin and haematocrit values, erythrocyte, leukocyte, lymphocyte, monocyte and granulocyte counts, coagulation and bleeding times and in methanol absorption (29).

Any other adverse effects
Prolonged or repeated topical use may cause erythema and dryness, while inhalation can cause headache, excitement, bronchial irritation and narcosis (30).

Any other comments
Experimental toxicology, human health effects and metabolic fate reviewed (31-35).

References
1. Howard, P. H. (Ed.) *Fate and Exposure Data for Organic Chemicals* 1990, **2**, 1-18
2. Verschueren, K. *Handbook of Environmental Data on Organic Chemicals* 2nd ed., 1983, 150, Van Nostrand Reinhold, New York
3. Bridie, A. L; et al. *Water Res.* 1979, **13**, 623-626
4. Alabaster, J. S. *Int. Pest. Control* 1969, **3/4**, 29-35

5. Hodson, P. V; et al. *Environ. Toxicol. Chem.* 1988, **7**(6), 443-454
6. Elkins, H. F; et al. *Sewage Ind. Wastes* 1956, **12**(12), 1475
7. Price, K. S; et al. *J. Water Pollut. Control Fed.* 1974, **46**, 1
8. Chudoba, J; et al. *Vodni. Hospod.* 1986, **36**, 180, (*Chem. Abstr.* **69**, 89571g)
9. Stratton, G. W; et al. *Bull. Environ. Contam. Toxicol.* 1980, **24**(4), 562-569
10. Platen, H; et al. *J. Gen. Microbiol.* 1989, **135**(4), 883-891
11. Chou, W. L; et al. *Bioeng. Symp.* 1979, **8**, 391-414
12. Young, R. H. F; et al. *J. Water Pollut. Control Fed.* 1968, **40**, 354-368
13. Bridie, A; et al. *Water Res.* 1979, **13**, 627-630
14. Vaishnav, D. D; et al. *Chemosphere* 1987, **16**, 695-703
15. Price, K. S; et al. *J. Water Pollut. Control Fed.* 1974, **346**, 63-77
16. Urano, K; et al. *Haz. Mater.* 1986, **13**, 147-159
17. Kwan, K. K. *Toxicity Assess.* 1988, **3**, 93
18. Meyrahn, H; et al. *J. Atmos. Chem.* 1986, **4**, 227-291
19. Rathburn, R. E; et al. *Chemosphere* 1982, **11**, 1097-1114
20. Commission of the European Communities, *Solvents in Common Use: Health Risks to Workers* 1988, RSC, London
21. Hill, E. F; et al. *Lethal Dietary Toxicities of Environmental Contaminants and Pesticides to Coturnix* 1986, 22, Fish and Wildlife Technical Report No. 2, US Dept. of Int. Fish. Wildlife
22. Mast, T. J; et al. *Energy Res. Abstr.* 1989, **14**(7), Abstr. No. 13672
23. Veng, T. H. *Arch. Toxicol.* 1991, **65**(1), 45-51
24. Gavino, V. C; et al. *J. Biol. Chem.* 1987, **262**(14), 6735-6740
25. Pezzagno, G; et al. *Scand. J. Work, Environ. Health* 1986, **12**, 603-608
26. Kennah, H. E; et al. *Fundam. Appl. Toxicol.* 1989, **12**, 281-290
27. Ishidate, M; et al. *Food Chem. Toxicol.* 1984, **22**(8), 623-636
28. Ng, T. P; et al. *Proc. ICMR Seminar* 1988, **8**, 371-377
29. Lembas-Bogaczyk, J; et al. *Med. Pr.* 1989, **40**(2), 111-118, (Pol.) (*Chem. Abstr.* **113**, 102608y)
30. *The Merck Index* 11th ed. 1989, Merck & Co. Inc, Rahway, NJ
31. *ECETOC Technical Report No. 30(4)* 1991, European Chemical Industry Ecology and Toxicology Centre, B-1160 Brussels
32. Lopez-Sariano, J; et al. *Speculation Sci. Technol.* 1986, **9**(3), 219-220
33. *Comm. Eur. Communities, [Rep.] EUR,EUR 11553, Solvents Common Use* 1988, 13-38, European Atomic Energy Community
35. Izmerov, N. F. *Scientific Reviews of Soviet Literature on Toxicity & Hazards of Chemicals* 1991, **97**, Eng. Trans. Richardson, M. L., (Ed.) UNEP/IRPTC, Geneva

A25   Acetone cyanohydrin

$(CH_3)_2C(OH)CN$

CAS Registry No. 75-86-5
Synonyms propanenitrile, 2-hydroxy-2-methyl-; lactonitrile, 2-methyl-; α-hydroxyisobutyronitrile; 2-hydroxyisobutyronitrile; methyl lactonitrile; 2-methyllactonitrile
Mol. Formula C_4H_7NO **Mol. Wt.** 85.11
Uses Used in the manufacture of insecticides, pharmaceuticals, foaming agents and polymerisation inhibitors. Intermediate for organic synthesis, especially methyl methacrylate.

Physical properties

M. Pt. –19°C; **B. Pt.** 95°C; **Flash point** 74°C (closed cup); **Specific gravity** d_4^{25} 0.9267; **Volatility** v. den. 2.93.

Solubility
Water: freely soluble. Organic solvent: diethyl ether, ethanol

Occupational exposure

UN No. 1541; **HAZCHEM Code** 2XE; **Conveyance classification** toxic substance; **Supply classification** toxic.

Risk phrases
Very toxic by inhalation, in contact with skin and if swallowed (R26/27/28)

Safety phrases
Keep container tightly closed and in a well ventilated place – Take off immediately all contaminated clothing – In case of accident or if you feel unwell, seek medical advice immediately (show label where possible) (S7/9, S27, S45)

Ecotoxicity

Fish toxicity
LC_{50} (96 hr) bluegill sunfish 0.57 ppm in fresh water at 23°C with mild aeration at 24 hr. LC_{50} (96 hr) inland silverside 0.50 ppm in synthetic sea water at 23°C with mild aeration at 4 hr (1).

Environmental fate

Degradation studies
1 mg l^{-1} degraded completely to acetone and hydrogen cyanide in river water. Removal of the cyanide ion occurs by evaporation, therefore cyanide pollution is a danger in closed systems (2).

Absorption
Adsorption on activated carbon: initial concentration 1000, 200 and 100 ppm, carbon dosage ×10. Residual concentrations 400, 110, 70 ppm, respectively, corresponding to removal of 60%, 45% and 30% acetone cyanohydrin (3).

Mammalian and avian toxicity

Acute data
LD_{50} oral rat, mouse, rabbit 13-17 mg kg^{-1} (4,5). LC_{Lo} (4 hr) inhalation rat 63 ppm (6).

Sub-acute data
Repeated administration (duration unspecified) of 5 mg twice weekly, orally to rats caused kidney and liver damage (7).

Irritancy
In humans can cause burns and irritation to skin, mucous membranes and eye (8,9).

Genotoxicity
Salmonella typhimurium TA97, TA98, TA100, TA1535, TA1537 with and without metabolic activation negative (10).

Any other adverse effects

Very toxic by inhalation. Symptomatic effects include giddiness, headache, weakness, confusion, unconsciousness, convulsions, nausea and vomiting. Breathing may stop in severe cases due to hydrogen cyanide poisoning (9).

Any other comments

A structure-activity relationship comparing the acute and sub-acute data, teratogenicity and biochemical mechanism of the toxicity of a series of structurally related aliphatic nitriles has been published (11).
Experimental toxicology, human health and environmental effects reviewed (12).

References

1. Dawson, G. W; et al. *J. Haz. Mater.* 1975, **8**(13)
2. Kampars, V; et al. *Latv. Kim. Z.* 1991, **4**, 473-476 (Russ.) (*Chem. Abstr.* **116**, 10900w)
3. Dahm, D. B; et al. *Environ. Sci. Technol.* 1974, **8**(13)
4. Marhold, J. V. *Inst. Pro. Vychovu Veloucin Pracovniku Chem. Prum. Praha* 1972, 161
5. *NTIS Report* PB 214-270, National Technical Information Service, Springfield, VA
6. Smyth, H. F; et al. *Am. Ind. Hyg. Assoc. J.* 1962, **23**, 95
7. Motoc, F; et al. *Arch. Mel. Prof. Med. Trav. Secur. Soc.* 1971, **32**(10-11), 653-658 (Fr.) (*Chem. Abstr.* **76**, 122561y)
8. *The Merck Index* 11th ed. 1989, Merck & Co. Inc, Rahway, NJ
9. *Chemical Safety Data Sheets* 1991, **4a**, 1-3, RSC, London
10. Zeiger, E; et al. *Environ. Mol. Mutagen.* 1988, **11**(Suppl 12), 1-157
11. Johannsen, F. R; et al. *Fundam. Appl. Toxicol.* 1986, **7**(4), 690-697
12. *ECETOC Technical Report No. 30(4)* 1991, European Chemical Industry Ecology and Toxicology Centre, B-1160 Brussels

A26 Acetone thiosemicarbazide

(CH3)2C=NNHC(S)NH2

CAS Registry No. 1752-30-3
Synonyms thiosemicarbazone acetone; 2-(1-methylethylidene)hydrazine carbothioamide
Mol. Formula $C_4H_9N_3S$ **Mol. Wt.** 131.20
Uses Chemical intermediate. Fungicide.

Physical properties

M. Pt. 178-179°C.

Solubility

Organic solvent: acetone

Mammalian and avian toxicity

Acute data

LD_{Lo} oral rat 10 mg kg^{-1} (1).
LD_{50} intraperitoneal mouse 23 mg kg^{-1} (2).

Any other comments

Experimental toxicology, human health and environmental effects of acetone thiosemicarbazide reviewed (3).

References

1. *Natl. Acad. Sci. Natl. Res. Counc. Chem-Biol. Coord. Cent. Rev.* 1953, **5**, 43
2. Jenny, E. H; et al. *J. Pharmacol. Exptl. Ther.* 1958, **122**, 110-123
3. *ECETOC Technical Report No. 30(4)* 1991, European Chemical Industry Ecology and Toxicology Centre, B-1160 Brussels

A27   Acetonitrile

CH3CN

CAS Registry No. 75-05-8

Synonyms methyl cyanide; cyanomethane; ethanenitrile; ethylnitrile; methylcarbonitrile

Mol. Formula C_2H_3N **Mol. Wt.** 41.05

Uses A solvent for organic and inorganic compounds including polymers and gases. Used in plastics casting and moulding, and in dyeing textiles. Increasingly used in its highly purified form in electrochemistry and high- performance liquid chromatography. Used as a catalyst and stabiliser for chlorinated solvents.

Physical properties

M. Pt. –45°C; **B. Pt.** 81.1°C; **Flash point** 6°C (open cup); **Specific gravity** d_4^{15} 0.7868; **Partition coefficient** log P_{ow} 0.34; **Volatility** v.p. 74 mmHg at 20°C; v. den. 1.42.

Solubility

Water: miscible. Organic solvent: miscible in ethanol, methanol, methyl acetate, ethyl acetate, acetone, diethyl ether

Occupational exposure

US TLV (TWA) 40 ppm (67 mg m^{-3}); **US TLV (STEL)** 60 ppm (101 mg m^{-3}); **UK Long-term limit** 40 ppm (70 mg m^{-3}); **UK Short-term limit** 60 ppm (105 mg m^{-3}); **UN No.** 1648; **HAZCHEM Code** 2WE; **Conveyance classification** flammable liquid; toxic substance; **Supply classification** highly flammable and toxic.

Risk phrases

Highly flammable – Toxic by inhalation, in contact with skin and if swallowed (R11, R23/24/25)

Safety phrases

Keep away from sources of ignition – No Smoking – Take off immediately all contaminated clothing – If you feel unwell, seek medical advice (show label where possible) (S16, S27, S44)

Ecotoxicity

Fish toxicity
LC_{50} (96 hr) fathead minnow, guppy, bluegill sunfish 1000-1850 mg l^{-1} (1-3).

Invertebrate toxicity
Cell multiplication inhibition test, *Pseudomonas putida* 680 mg l^{-1}, *Entosiphon sulcatum* 1810 mg l^{-1}, *Scenedesmus quadricauda* 7300 mg l^{-1} (4). *Microcystis aeruginosa* 520 mg l^{-1} (4,5).

Environmental fate

Nitrification inhibition
100 mg l^{-1} did not inhibit NH_3 oxidation by *Nitrosomonas sp.* (6).

Carbonaceous inhibition
Chromobacterium sp. and *Pseudomonas aeruginosa* both utilised high concentrations of acetonitrile as their sole carbon source. Maximum growth attained after 96 hr of incubation. Concentration >25 g l^{-1} inhibited growth and oxygen uptake (7).

Degradation studies
Confirmed biodegradable (8).
Wastewater treatment, biodegradation by mutant microorganisms 500 mg l^{-1} at 20°C. 100% disruption in parent strain in 9 hr, 100% in mutant strain in 1.5 hr (9).
In soil *Pseudomonas putida* was capable of utilising high concentration of acetonitrile as the sole source of carbon and nitrogen. Metabolites included acetic acid and ammonia (10).
BOD_5 17% reduction in dissolved oxygen (11).
Acinetobacter sp. 51-2 degraded 25 g l^{-1} acetonitrile in 48 hr (12).

Mammalian and avian toxicity

Acute data
LD_{50} oral rat 3800 mg kg^{-1} (13).
LC_{50} (8 hr) inhalation rat 7500 ppm (14).
LD_{50} intraperitoneal mouse 500 mg kg^{-1}.
LD_{50} subcutaneous rabbit 130 mg kg^{-1} (13).

Teratogenicity and reproductive effects
Some toxicity in high dose animal experiments (species unspecified), but no overt evidence of teratogenesis (15)

Metabolism and pharmacokinetics
Acetonitrile was metabolised anaerobically by gastrointestinal microflora to produce cyanide and thiocyanate (16).
Elevated concentrations of cyanide and thiocyanate were detected in all tissues studied at 2.5 hr after an oral or intraperitoneal dose of acetonitrile (100- 400 mg kg^{-1}) in Syrian golden hamsters (14).

Irritancy
In humans, may cause skin irritation (17).
Reported to cause dermatitis in rabbits at concentrations of 3900-7850 mg kg^{-1} (18).

Genotoxicity

Salmonella typhimurium TA98, TA100, TA1535, TA1537 with and without metabolic activation negative (19).

Sacchromyces cerevisiae D.61M positive mitotic chromosomal, malsegregation in diploid strain (20).

Chinese hamster ovary, sister chromatid exchange without metabolic activation weakly positive, with metabolic activation negative (21).

Inhalation exposure adult ♀ *Drosphila melanogaster* induced aneuploidy, both chromosome loss and gain (22).

Any other adverse effects to man

Symptomatic effects in humans include flushing of face, tightness of chest, increased salivation, eye and nose irritation. Exposure to high concentrations may cause headache, dizziness, rapid respiration and pulse rate, nausea, vomiting, unconsciousness, convulsions and death. Onset of symptoms may be delayed for several hr, perhaps due to the low release of cyanide (23-25).

Any other comments

Glutathione in brain and liver tissue and the activity of cytochrome oxidase in the brain of rats were studied. Acetonitrile reduced tissue glutathione content (26). Physical and chemical properties, experimental toxicology, epidemiology and human health effects reviewed (27-29).

References

1. Brooke, L. T; et al. *Acute Toxicities of Organic Chemicals to Fathead Minnows* 1984, 27, Center for Lake Superior Environmental Studies, Univ. of Wisconsin
2. Jones, H. R. *Environmental Control in the Organic and Petrochemical Industries* 1971, Noyes Data Corporation, New York
3. Henderson, C; et al. *The Effect of Some Organic Cyanides (Nitriles) on Fish* 1960, 120-130, Purdue Univ. Eng. Bull. Ext. Ser. 106
4. Bringmann, G; et al. *Water Res.* 1980, **14**, 231
5. Bringmann, G; et al. *GWF-Wasser/Abswassser* 1976, **117**, (9), 119
6. Verschueren, K. *Handbook of Environmental Data on Organic Chemicals* 2nd ed., 1983, Van Nostrand Reinhold, New York
7. Chapatwala, K. D; et al. *J. Ind. Microbiol.* 1990, **5** (2-3), 65-69
8. *MITI Report* 1984, Ministry of International Trade and Industry, Tokyo
9. Worne, E. *The Activity of Mutant Microorganisms in the Biological Treatment of Industrial Wastes* Tijdschrift Van Het, BECWA, Liege
10. Nawoz, M. S; et al. *Appl. Environ. Microbiol.* 1989, **55**(9), 2267-2274
11. *Hazardous Chemical Data* 1984-1985, **2**, US Dept. of Transportation, Washington, DC
12. Xie, S; et al. *Weishengwu Xuebao* 1990, **30** (6), 471-474, (Ch.) (*Chem. Abstr.* **114**, 181840q)
13. Spector, S. Ed. *Handbook of Toxicology* 1956, Vol. 1, Saunders, Philadelphia
14. Willhite, C. C. *Teratology* 1983, **27**, 313-325
15. Johannsen, F. R. *Fundam. Appl. Toxicol.* 1986, **7**(1), 33-40
16. Koch, R. L. *Pharmacol. Exp. Ther.* 1990, **254**(2), 612-615
17. *The Merck Index* 11th ed. 1989, Merck & Co. Inc, Rahway, NJ
18. *Chemical Safety Data Sheets* 1989, **1**, 5-7, RSC, London
19. Mortelmans, K; et al. *Environ. Mutagen.* 1986, **8**(7), 1-119
20. Zimmermann, F. K; et al. *Mutat. Res.* 1985, **149**, 339
21. Galloway, S. M; et al. *Environ. Mol. Mutagen.* 1987, **10**(Suppl. 10), 1-175

22. Osgood, C; et al. *Mutat. Res.* 1991, **259**(2), 147-163
23. Amdur, M. L. *J. Occup. Med.* 1959, **1**, 625
24. Dequidt, J; et al. *Bull. Soc. Pharm. Lille* 1972, **4**, 143
25. Dequidt, J; et al. *Eur. J. Toxicol.* 1974, **7**, 91
26. Kutsenko, S. A; et al. *Vopr. Med. Khim.* 1991, **37**(2), 72-74, (Russ.), (*Chem. Abstr.* **114**, 223135w)
27. *ECETOC Technical Report No. 30(4)* 1991, European Chemical Industry Ecology and Toxicology Centre, B-1160 Brussels
28. *Dang. Prop. Ind. Mater. Rep.* 1989, **9**(6), 46-60
29. *Cah. Notes Doc.* 1989, **137**, 717-721, (Fr.) (*Chem. Abstr.* **112**, 124222z)

A28   Acetophenone

CAS Registry No. 98-86-2
Synonyms methyl phenyl ketone; acetylbenzene; Hypnone; benzoyl methide; 1-phenylethanone; phenyl methyl ketone; Dymex
Mol. Formula C_8H_8O **Mol. Wt.** 120.15
Uses Used in the perfume industry. Catalyst for the polymerisation of olefins. Photosensitiser in organic synthesis. Hypnotic.
Occurrence In oils of castoreum and labdanum resin, and buds of balsam poplar. In heavy oil fractions of coal tar.

Physical properties

M. Pt. 19-20°C; **B. Pt.** 202°C; **Flash point** 82°C (closed cup); **Specific gravity** d_4^{20} 1.03; **Partition coefficient** log P_{ow} 1.58; **Volatility** v.p. 1 mmHg at 15°C; v. den 4.14.

Solubility
Organic solvent: ethanol, chloroform, diethyl ether, glycerol

Ecotoxicity

Fish toxicity
LC_{50} (96 hr) fathead minnow 155 mg l^{-1}, static bioassay at 18-22°C (1).

Environmental fate

Carbonaceous inhibition
Pseudomonas putida can oxidise acetophenone, the metabolite identified was a monohydroxy product 3-acetyl-2,4-cyclohexadien-1-ol (2).

Degradation studies
Confirmed biodegradable (3).

Water pollution factors, BOD5 0.518 mg l^{-1} oxygen consumed, estimated ThOD 32 mg l^{-1} (4).

Acetophenone was reported to be 14% biodegradable (5).

Microbial metabolism of acetophenone was reported to occur via the formation of phenyl acetate with further metabolism occurring by hydrolysis (6).

Absorption
Adsorbed by activated carbon 0.194 g g^{-1} carbon, 97% reduction, influent 1000 mg l^{-1}, effluent 28 mg l^{-1} (7,8).

Mammalian and avian toxicity

Acute data
LD50 oral rat 815 mg kg^{-1} (9).
LD50 intraperitoneal mouse 200 mg kg^{-1} (10).
LDLo subcutaneous mouse 330 mg kg^{-1} (11).

Irritancy
Dermal rabbit (24 hr) 515 mg caused mild irritation and 771 μg instilled into rabbit eye for 24 hr caused severe irritation (12,13).

Genotoxicity
Chinese hamster lung cells with metabolic activation induced chromosome aberrations (14).

Any other adverse effects
May be narcotic at high concentrations (15).

Any other comments
USSR maximum permissible concentrations in drinking water 1 μg l^{-1} (16).
Experimental toxicology and human health effects reviewed (17).

References

1. Mattson, V; et al. *Acute Toxicity of Selected Organic Compounds to Fathead Minnows* 1976, EPA 600/2-76-097
2. Howard, P. W; et al. *J. Chem. Soc. Chem. Comm.* 1990, **17**, 1182-1184
3. *MITI Report* 1984, Ministry of International Trade and Industry, Tokyo
4. Donaghue, N. A; et al. *The Microbial Metabolism of Cyclohexane and Related Compounds* 1976, Univ. Coll. of Wales, Aberstwyth
5. Ludzack, F. J; et al. *J. Water Pullut. Control Fed.* 1960, **32**, 1173
6. Cripps, R. E. *Biochem. J.* 1975, **152**(2), 233-241
7. Guisti, D. M; et al. *J. Water Pollut. Control. Fed.* 1974, **46**(5), 947-965
8. Meinck, F; et al. *Les Eaux Residuaires Industrielles* 1970
9. *Gig. Tr. Prof. Zabol.* 1982, **26**, 63
10. *NTIS Report* AD 277-6891, National Technical Information Service, Springfield, VA
11. Herrlen, G. *Pharmak Prufuny v. Analgetiku* 1933
12. *Union Carbide Material Safety Data Sheet* 27 Dec. 1971, Union Carbide Corp., New York
13. *Am. J. Opthalmol.* 1946, **29**, 136
14. Sofuni, T; et al. *Eisei Shikensho Hokoku* 1985, **103**, 64-75, (Jap.) (*Chem. Abstr.* **104**, 181539s)
15. Sax, N. I; et al. *Dangerous Properties of Industrial Materials* 7th ed., 1989, Van Nostrand Reinhold, New York

16. *Russian Toxicological Data for Chemicals in Sources of Drinking Water* 1978, Central Water Planning Unit, Technical Note No. 20
17. *ECETOC Technical Report No. 30(4)* 1991, European Chemical Industry Ecology and Toxicology Centre, B-1160 Brussels

A29   Acetyl acetone

CH₃COCH₂COCH₃

CAS Registry No. 123-54-6
Synonyms 2,4-pentanedione; diacetylmethane
Mol. Formula $C_5H_8O_2$ **Mol. Wt.** 100.12
Uses Forms organometallic complexes which are used as gasoline additives, lubricant additives, driers for varnishes, printing inks, fungicides and insecticides.

Physical properties

M. Pt. –23°C; **B. Pt.** 140°C; **Flash point** 40°C; **Specific gravity** d_4^{25} 0.9721.

Solubility
Water: 125 g l^{-1} (temperature unspecified). Organic solvent: ethanol, benzene, chloroform, diethyl ether, acetone, glacial acetic acid

Occupational exposure

UN No. 2310; **HAZCHEM Code** 2S; **Conveyance classification** flammable liquid; **Supply classification** flammable and harmful.

Risk phrases
Flammable – Harmful if swallowed (R10, R22)

Safety phrases
When using do not smoke – Do not breathe vapour – Avoid contact with skin and eyes (S21, S23, S24/25)

Ecotoxicity

Fish toxicity
LC$_{50}$ (96 hr) bluegill sunfish 29-201 mg l^{-1}. Ventilatory patterns and time to death suggested that acetyl acetone was lethal by mechanisms other than narcosis (1).

Invertebrate toxicity
Cell multiplication inhibition test, *Pseudomonas putida* 67 mg l^{-1}, *Entosiphon sulcatum* 11 mg l^{-1}, *Scendesmus quadricauda* 2.7 mg l^{-1}, *Microcystis aeruginoisa* 8.5 mg l^{-1} (1-3).

Mammalian and avian toxicity

Acute data
LD$_{50}$ oral rat 1000 mg kg^{-1} (4).
LC$_{50}$ (4 hr) inhalation rat 1224 ppm (5).
LD$_{50}$ intraperitoneal mouse 750 mg kg^{-1} (6).

Sub-acute data

Fischer-344 rats were exposed to 0-650 ppm acetyl acetone vapour 6 hr day^{-1}, 5 days week^{-1} for 14 weeks. Damage to thymus and brain was reported in animals exposed to 650 ppm (7).

Carcinogenicity and long-term effects

Subcutaneous rat (40 weeks) 200 mg kg^{-1} 5 day week^{-1} neurotoxic effects were reported (8).

Teratogenicity and reproductive effects

At concentrations of 398 ppm, acetyl acetone showed maternal toxicity (reduced body weight) and foetotoxicity (reduced body weight and ossification), and at 202 ppm foetotoxicity (reduced body weight) was observed. Embryotoxicity or teratogenicity were not observed at any concentration. The no observable effect concentration was 53 ppm for both maternal and developmental toxicity (9).

Irritancy

In rabbits, contact with 0.5 ml produced mild local erythema and oedema. Instilling 0.1 ml into the conjunctival sac caused mild conjunctivitis which lasted 24 hr. No corneal injury reported (6).

Genotoxicity

Salmonella typhimurium TA104 without metabolic activation positive. Induced sister chromatid exchanges in Chinese hamster ovary cells without metabolic activation. Genotoxic activity can be related to the reactivity of acetyl acetone with DNA and its lipophilic character (10).
Bacillus subtilis direct or indirect DNA damaging potential in microsome recombination assay (11).

Any other adverse effects

Inhalation ♂ Fischer 344 rats (5 day) 0-694 ppm 6 hr day^{-1}. No histopathological changes observed in brain, testes, or thymus from ♂ rats exposed to high concentrations, sacrificed after 8 wk of mating. Minor transient reproductive and gestational effects were present and the authors report, although not statistically significant, the results are dose related and compatible with a transient slight dominant lethal effect at the spermatid stage of spermatogenesis (12).

Any other comments

Experimental toxicology and human health effects reviewed (13).

References

1. Carlson, R. W. *Comp. Biochem. Physiol. C: Comp. Pharmacol. Toxicol.* 1990, **95C**(2), 181-196
2. Bringmann, G; et al. *Water Res.* 1980, **14**, 231
3. Bringmann, G; et al. *GWF-Wasser Abwasser* 1976, **117**(9), 119
4. *J. Ind. Hyg. Toxicol.* 1944, **26**, 269
5. *NTIS Report* AD 691-490, Natl. Tech. Inf. Ser., Springfield, VA
6. Ballantyne, B; et al. *Drug Chem. Toxicol.* 1977, **9**(2), 133-146
7. Dodd, D. E; et al. *Fundam. Appl. Toxicol.* 1986, **7**(2), 329-339
8. Nagano, M; et al. *Sangyo Igaku* 1983, **25**(6), 471-482, (Jap.) (*Chem Abstr* **101**, 49780x)
9. Tyl, R. W; et al. *Toxicol. Ind. Health* 1990, **6**(3-4), 461-474
10. Gava, C; et al. *Toxicol. Environ. Chem.* 1989, **22**(1-4), 149-157

11. Matsui, S; et al. *Water Sci. Technol.* 1989, **21**(8-9), 875-887
12. Tyl, R. W; et al. *Toxicol. Ind. Health* 1989, **5**(3), 463-477
13. *ECETOC Technical Report No. 30(4)* 1991, European Chemical Industry Ecology and Toxicology Centre, B-1160 Brussels

A30   4-Acetyl aminobiphenyl

CH₃CONH —⟨ ⟩—⟨ ⟩

CAS Registry No. 4075-79-0
Synonyms *p*-phenylacetanilide; (4-phenyl)acetanilide; (1,1′-biphenyl-4-yl)acetamide; 4-acetamidobiphenyl
Mol. Formula $C_{14}H_{13}NO$ **Mol. Wt.** 211.27
Uses In genotoxicity and cytotoxicity testing.
Occurrence Metabolite of 4-aminobiphenyl.

Physical properties
M. Pt. 170-172°C.

Mammalian and avian toxicity

Carcinogenicity and long-term effects
A single subcutaneous injection of 25-100 µg 4-acetylaminoliphenyl induced hepatoma, pulmonary adenoma and lymphoma in neonatal ICR/Ha mice, duration of study 1 yr (1).
In ♀ rats, 0.025% dietary dose as a supplement for 12-15 month induced distant tumours at one or more unspecified sites (2).
8-10 month feeding trials in rats (concentrations unspecified) induced a high number of adenocarcinomas of the mammary glands in ♀ rats and a small number of adenocarcinomas of the small intestine and squamous cell carcinomas and sebaceous gland carcinomas of the earduct in both ♂ and ♀ mice (3).

Metabolism and pharmacokinetics
In rabbits 4-acetyl aminobiphenyl administered orally and intraperitoneally was excreted in urine as the glucuronide and *N*-hydroxy derivatives (4).
Intraperitoneal rat 186 mg kg^{-1}, 4-acetyl aminobiphenyl appeared slowly in the blood, accompanied by the appearance of aminobiphenyl and 4-hydroxy-acetyl aminobiphenyl. After 4 hr the concentration of 4-acetyl aminobiphenyl and aminobiphenyl were in dynamic equilibrium with the N-deacetylation of 4-acetyl aminobiphenyl and the acetylation of aminobiphenyl (5).
Single dose oral rat 100 mg kg^{-1} metabolised to arylamine, *N*- arylformamide and *N*-aryl acetamide. The compounds were excreted in the urine and faeces. Percentages of arylamine, *N*-arylformamide and *N*-arylacetamide excreted in urine were 0.25%, 0.01% and 0.44% respectively. Percentages excreted in the faeces were 0.29%, 0.04% and 8.21% respectively (6).

Genotoxicity

Salmonella typhimurium TA98, TA100 with metabolic activation induced frame-shift and base-pair mutations (7).

Acetylaminobiphenyl reduced the survival of *Escherichia coli Uvr* endonuclease-deficient strains and reduced the ability of PBR322 plasmid DNA to transform the *Escherichia coli* strains (8).

Inactive in cytotoxicity tests using *Xeroderma pigmentosum* human cells (9).

References

1. Fujii, K; et al. *Oncology* 1979, **36**(3), 105-112
2. Morris, H. P; et al. *J. Natl. Cancer Inst.* 1957, **18**, 101-111
3. Miller, E. C; et al. *Cancer Res.* 1956, **16**, 525-534
4. Fries, W; et al. *Xenobiotica* 1973, 3(8), 525-540
5. Karreth, S; et al. *Xenobiotica* 1991 **21**(3), 417-428
6. Tatsumi, K; et al. *Cancer Res.* 1989, **49** 2059-2064
7. Connor, T. H; et al. *J. Mol. Toxicol.* 1989, **2**, 53-65
8. Tamura, N; et al. *Carcinogenesis (London)* 1990, **11**(4), 535-540
9. Maher, V. M; et al. *Proc. Am. Assoc. Canc. Res. Conf. Proc.* 1974, 140, No. 560

A31   2-Acetylaminofluorene

CAS Registry No. 53-96-3

Synonyms *N*-2-fluorenylacetamide; *N*-9*H*-fluoren-2-ylaceamide; *N*-acetyl-2-aminofluorene; 2-acetamidofluorene; 2-acetaminofluorene; 2-fluorenylacetamide; AAF

Mol. Formula $C_{15}H_{13}NO$ **Mol. Wt.** 223.28

Physical properties

M. Pt. 194°C; **Partition coefficient** $\log P_{ow}$ 3.22 (1)..

Solubility
Organic solvent: ethanol, glycols, diethyl ether

Ecotoxicity

Bioaccumulation
Calculated bioconcentration factor 165 (2).

Environmental fate

Degradation studies
Incubation of 500 ppm acetylaminofluorene at 20°C in activated sludge for 6 days was toxic to microorganisms and some degradation occurred. Theoretical BOD 7-12% (3).

Abiotic removal

Chemical hydrolysis and oxidation are unlikely to be environmentally significant (4). Vapour phase photolytic reaction in water and air with hydroxyl radicals $t_{1/2}$ 6 hr at 25°C (1).

Absorption

Estimated soil adsorption coefficient suggests low mobility of acetylaminofluorene in soil and strong adsorption to suspended solids and sediments in water (5).

Mammalian and avian toxicity

Acute data

LD_{50} oral mouse 1020 mg kg^{-1} (6).
LD_{50} intraperitoneal mouse 2200 mg kg^{-1} (7).

Sub-acute data

Oral mouse (28 day) 30-150 mg kg^{-1} induced N-deoxyguanosin-8-yl-2-aminofluorene in liver and bladder DNA. In liver, correlating incidence of tumours was observed (8).

Carcinogenicity and long-term effects

Oral BALB/C mouse (6 month) 500 ppm. The mice were 1, 7 or 13 months old and were killed immediately, 3 or 6 months after treatment terminated. Urinary bladder hyperplasia was less severe in young mice, mammary tumourigenesis increased with age, old mice were sensitive to liver karyomegaly, and cytoplasmic and nuclear inclusions. Bladder tumourigenesis was not influenced by age (9).

Metabolism and pharmacokinetics

Repeated exposure can increase metabolic rate and increase excretion of N-hydroxy-N-2- fluorenylacetylacetylamine in urine as glucuronide conjugate in rabbits. The major enzyme involved in the bioactivation of 2-acetylaminofluorene in the human liver is P-450$_{PA}$ (phenacetin O-deethylase) (10).

Genotoxicity

Salmonella typhimurium TA1535 with metabolic activation positive (11).
In vitro mouse bone marrow micronucleus cytochrome test positive (12).
In vivo/in vitro assay, ♂ rats were given single oral dose (concentration unspecified) of 2-acetylaminofluorene. Slight increase in the frequency of chromosomal aberrations and chromosome breaks. Sister chromatid exchange equivocal results (13).

Any other comments

Genotoxic and nongenotoxic effects reviewed (14-16).

References

1. *GEMS: Graphical Exposure Modeling System* 1986, FAP. Fate of Atmos. Pollut.
2. Hansch, C; et al. *Medchem. Project Issue No. 19* 1985
3. Melaney, G. W; et al. *J. Water Pollut. Control Fed.* 1967, **39**, 2020-2029
4. Tabak, H. H; et al. *Test Protocols for Environ. Fate and Movement of Toxicant* 1981, Proc. Sym. Assoc. of Official Anal. Chem. 94th Ann. Meeting, Washington, DC
5. Swann, R. L; et al. *Res. Rev.* 1983, **85**, 17-28
6. *Toxicol. Appl. Pharmacol.* 1973, **25**, 447
7. *Prog. in Mutat. Res.* 1981, **1**, 682
8. Beland, F. A; et al. *Prog. Clin. Biol. Res.* 1990, **331**, 121-129
9. Greenman, D. L; et al. *J. Toxicol. Environ. Health* 1987, **22**(2), 113-129

10. Searle, C. E. *Chemical Carcinogens. ACS Monograph 173* 1976, 372, American Chemical Society, Washington, DC
11. Shimada, T; et al. *Cancer Res.* 1989, **49**(12), 3218-3228
12. Pot-Deprun, J; et al. *Eval. Short-term Tests Carinogen.* 1988, **1**, 1.158-1.161, J. Ashby (Ed.), Cambridge Univ. Press, Cambridge
13. Clare, M. G; et al. *Eval. Short-term Tests Carcinog.* 1988, **1**, 1.393-1.399, J. Ashby (Ed.) Cambridge Univ. Press, Cambridge
14. Neumann, H. G; et al. *Environ. Health Perspect.* 1990, **88**, 207-211
15. Coombes, R. D; *Eval. Short-Term Tests Carcinog.* 1988, **1**, 1.315-1.320, J. Ashby (Ed.), Cambridge Univ. Press, Cambridge
16. Ashby, J. (Ed.) *Eval. Short-Term Tests Carcinog.* 1988, **1**, 57-77, Cambridge University Press, Cambridge

A32   Acetyl bromide

CH3COBr

CAS Registry No. 506-96-7
Synonyms acetic bromide; ethanoyl bromide; acetic acid bromide
Mol. Formula C_2H_3BrO **Mol. Wt.** 122.95
Uses Paper manufacturing.

Physical properties

M. Pt. –96°C; **B. Pt.** 74-77°C; **Specific gravity** d_4^{16} 1.6625.

Solubility
Organic solvent: miscible in benzene, diethyl ether, chloroform

Occupational exposure

UN No. 1716; **HAZCHEM Code** 4WE; **Conveyance classification** corrosive substance.

Ecotoxicity

Fish toxicity
LC_{50} (96 hr) fathead minnow 40 mg l^{-1} (1).

Mammalian and avian toxicity

Acute data
LC_{50} inhalation rat, mouse 48 g m^{-3} (2).
LD_{50} intraperitoneal rat, mouse 250 mg kg^{-1} (2).

Irritancy
Reported to be irritating to eyes (species unspecified) (3).

Any other comments

Decomposes violently in water and ethanol (3).

References

1. Curtis, M. W; et al. *J. Hydrology* 1981, **51**, 359-367
2. Lyublina, E. I. *Gig. Tr. Prof. Zabol.* 1974, (4), 55-57, (Russ.) (*Chem. Abstr.* **81**, 115438p)
3. *The Merck Index* 11th ed. 1989, Merck & Co. Inc, Rahway, NJ

A33   Acetyl chloride

CH3COCl

CAS Registry No. 75-36-5
Synonyms ethanoyl chloride; acetic acid chloride; acetic chloride
Mol. Formula C_2H_3ClO **Mol. Wt.** 78.50
Uses Used in the synthesis of pharmaceuticals and dyestuffs, in the esterification of cellulose and in improving the wetfastness of yarns. Used as a rubber antiscorch agent.

Physical properties

M. Pt. –112°C; **B. Pt.** 52°C; **Specific gravity** d_4^{20} 1.1051; **Volatility** v.p. 352.5 mmHg at 24°C.

Solubility
Organic solvent: miscible in acetone, glacial acetic acid, diethyl ether, benzene, toluene, chloroform, carbon disulfide

Occupational exposure

UN No. 1717; **HAZCHEM Code** 4WE; **Conveyance classification** flammable liquid; corrosive substance; **Supply classification** highly flammable and corrosive.

Risk phrases
Highly flammable – Reacts violently with water – Causes burns (R11, R14, R34)

Safety phrases
Keep container in a well ventilated place – Keep away from sources of ignition – No Smoking – In case of contact with eyes, rinse immediately with plenty of water and seek medical advice (S9, S16, S26)

Ecotoxicity

Fish toxicity
LC_{50} (96 hr) fathead minnow 42 mg l^{-1} (1).

Environmental fate

Degradation studies
In view of the instability of the compound in aqueous media, it is unlikely to persist in an aquatic environment and soil (2).

Abiotic removal
Estimated $t_{1/2}$ for atmospheric photochemical reaction 3 months (3).

Mammalian and avian toxicity

Acute data
TC_{Lo} (1 min) inhalation human 2 ppm (4).

Irritancy
Extremely irritating to eyes. Corrosive, causes severe burns (5).

Genotoxicity

Salmonella typhimurium TA97, TA98, TA100, TA1535, TA1537 with and without metabolic activation negative (6).

Any other comments

Physico-chemical properties, experimental toxicology and human health effects reviewed (7).

When heated to decomposition it emits highly toxic fumes of phosgene. Reacts violently with water and other hydrogen active compounds like amines, phenols, and alcohols (2).

References

1. Curtis, M. W; et al. *J. Hydrol.* 1981, **51**(1-4), 359-367
2. Howard P. H. (Ed.) *Fate and Exposure Data For Organic Chemicals* 1990, **1**, 1-4, Lewis Publi., MI
3. *GEMS: Graphical Exposure Modeling System* 1986, Fate of Atmospheric Pollutants
4. *Handbook of Organic Industrial Solvents* 2nd ed., 1961, National Association of Mutual Casualty Companies, Chicago, IL
5. *The Merck Index* 11th ed. 1989, Merck & Co. Inc, Rahway, NJ
6. Zeiger, E; et al. *Environ. Mol. Mutagen.* 1988, **11**(12), 1-158
7. *ECETOC Technical Report No. 30(4)* 1991, European Chemical Industry Ecology and Toxicology Centre, B-1160 Brussels

A34 Acetylene

$$HC \equiv CH$$

CAS Registry No. 74-86-2
Synonyms ethine; ethyne; acetylen; narcylen
Mol. Formula C_2H_2 **Mol. Wt.** 26.04
Uses Used in soldering, in oxyacetylene welding and cutting.
Used in the chemical synthesis of acetaldehyde and acetic acid.
Occurrence Occurs in emission from vehicle exhausts.

Physical properties

M. Pt. −81°C (sublimes); **Flash point** −18°C; **Specific gravity** 0.62 (liquefied).

Solubility

Organic solvent: benzene, acetone, chloroform

Occupational exposure

UN No. 1001; **Conveyance classification** flammable gas; **Supply classification** highly flammable.

Risk phrases

Heating may cause an explosion – Explosive with or without contact with air – Extremely flammable (R5, R6, R12)

Safety phrases

Keep container in a well ventilated place – Keep away from sources of ignition – No Smoking – Take precautionary measures against static discharges (S9, S16, S33)

Ecotoxicity

Fish toxicity

LC_{50} (33 hr) river trout 200 mg l^{-1} (1).

Environmental fate

Nitrification inhibition

Effective in inhibiting the reduction of N_2O in denitrification, even for prolonged incubation periods of up to 96 hr under moist or saturated soil/water conditions (2). Inhibits biodegradation processes in *Nitrosomonas europaea* and *Nitrosococcus oceanus* by inhibiting ammonia monooxygenase enzyme activity (3).

Carbonaceous inhibition

Acetylene vapour 1.5-6.5% completely inhibited carbon dioxide fixation by nitrifying and methanotrophic microorganisms, while 24-45% acetylene inhibited carbon dioxide fixation by thiomic and heterotrophic bacteria (4).

Any other adverse effects to man

Concentrations of 20-40% acetylene reported to cause dyspnoea, headache in humans, higher concentrations can cause asphyxiation and narcosis (5). Inhibits cytochrome P-450 catalysed reactions by suicide inactivation (6).

Any other comments

Not explosive at atmospheric pressure but at ≥ 2 atmospheres is explosive by spark or decomposition.
Forms insoluble explosive compounds with copper and silver, therefore copper and brass containers must not be used.
Mixtures with air between 3% and 65% gas are explosive (5).
Physico-chemical properties, exposure, experimental toxicology and human health effects reviewed (7).
The effectiveness of acetylenic structures for the inactivation of enzymes in cancer cell metabolism reviewed (8).

References

1. Meinck, F; et al. *Les Eaux Residuares Industrielles*, 1970
2. Aulakh,M. S; et al. *Commun. Soil Sci. Plant Anal.* 1990, **21**(19-20), 2233-2243
3. Rasche, M. E; et al. *Appl. Environ. Microbiol.* 1990, **5608**, 2568-2571
4. *Tabulae Biol.* 1933, **3**, 231
5. *The Merck Index* 11th ed. 1989, Merck & Co. Inc, Rahway, NJ
6. *Casarett & Doull's Toxicology*, 3rd ed., 1989, 87, Macmillan, New York
7. *ECETOC Technical Report No. 30(4)* 1991, European Chemical Industry Ecology and Toxicology Centre, B-1160 Brussels
8. Bador, P; et al. *Pharm. Acta. Helv.* 1990, **65**(11), 305-310, (Fr.) (*Chem. Abstr.*, **114**, 16967d)

A35   Acetyl iodide

CH_3COI

CAS Registry No. 507-02-8
Mol. Formula C_2H_3IO **Mol. Wt.** 169.95

Physical properties

B. Pt. 108°C; **Specific gravity** d_4^{20} 2.0674; **Partition coefficient** log P_{ow} 1.5491.

Solubility
Organic solvent: benzene, diethyl ether

Occupational exposure

UN No. 1898; **HAZCHEM Code** 2R; **Conveyance classification** corrosive substance.

Mammalian and avian toxicity

Irritancy
Reported to be a strong irritant (species and concentrations unspecified) (1).

Any other adverse effects

Acetyl iodide vapours reported to cause pulmonary oedema (species and concentrations unspecified) (1).

Any other comments

Decomposes in water and ethanol.

References

1.   *The Merck Index* 11th ed. 1989, Merck & Co. Inc, Rahway, NJ.

A36   Acetyl ketene

CAS Registry No. 674-82-8
Synonyms 3-buteno-β-lactone; 2-oxetanone, 4-methylene; diketen; diketene
Mol. Formula $C_4H_4O_2$ **Mol. Wt.** 84.08
Uses Used in the manufacture of pigments and in the synthesis of acetoacetic acid esters. Bactericide.

Physical properties

M. Pt. –6.5°C; **B. Pt.** 127.4°C; **Flash point** 45°C; **Specific gravity** d_{20}^{20} 1.0897; **Volatility** v.p. 8 mmHg temperature unspecified; v. den. 2.9 at 20°C.

Solubility
Water: decomp. in water.

Occupational exposure

US TLV (TWA) 250 ppm; **US TLV (STEL)** 310 ppm; **UN No.** 2521; **HAZCHEM Code** 2S; **Conveyance classification** flammable liquid; **Supply classification** Flammable and harmful.

Risk phrases
Flammable – Harmful by inhalation (R10, R20)

Safety phrases
Keep in a cool place (S3)

Mammalian and avian toxicity

Acute data
LD_{50} oral rat 560 mg kg^{-1} (1).
LC_{50} (24 hr) inhalation guinea pig 3 mg m^{-3} (2).
LC_{50} (1 hr) inhalation rat 2000 ppm (3).
LD_{50} dermal rabbit 2830 mg kg^{-1} (1).

Any other comments

Polymerisation occurs slowly on standing. Violent polymerisation is catalysed by acids and bases. Storage hazard (4).
Physico-chemical properties, experimental toxicology and human health effects reported (5).

References

1. Carpenter, C. I; et al. *Toxicol. Appl. Pharmacol.* 1974, **28** 313
2. Demel, I; et al. *Wochenbl. Papierfabr.* 1983, **111**(3), 95-102
3. *Ind. Hyg. Found. Am. Chem. Toxicol. Ser. Bull.* 1961, **26**, 1
4. *Union Carbide Data Sheet* 1963, Union Carbide Corp. New York
5. *ECETOC Technical Report No. 30(4)* 1991, European Chemical Industry Ecology and Toxicology Centre, B-1160 Brussels

A37   Acetylsalicylic acid

CAS Registry No. 50-78-2
Synonyms 2-acetoxybenzoic acid; 2-acetyloxybenzoic acid; salicylic acid acetate; aspirin; *acidum acetylsalicylicum*; acetol

Mol. Formula $C_9H_8O_4$ **Mol. Wt.** 180.16

Uses Analgesic, antipyretic, anti-inflammatory agent.

Physical properties

M. Pt. 135°C (rapid heating); **B. Pt.** 140°C.

Solubility

Water: 3.3 g l^{-1} at 25°C. Organic solvent: ethanol, chloroform, diethyl ether

Occupational exposure

US TLV (TWA) 5 mg m^{-3}; **UK Long-term limit** 5 mg m^{-3}.

Ecotoxicity

Fish toxicity

No effect on stickleback, steelhead trout or sockeye salmon at 30 mg l^{-1} (1).

Mammalian and avian toxicity

Acute data

LD_{50} oral rat, mouse, rabbit 1010-1800 mg kg^{-1} (2-5).
LD_{50} intraperitoneal mouse, rabbit 280-500 mg kg^{-1} (4,6).

Sub-acute data

TD_{Lo} (16 hr) oral child 81 mg kg^{-1} pulmonary, central nervous system and systemic effects (7).
TD_{Lo} (8 wk) oral human 2880 mg kg^{-1} ear and gastrointestinal tract effects (8).

Teratogenicity and reproductive effects

Salicylates readily cross the placenta and have been shown to be teratogenic in animals. Although some studies and anecdotal reports have implicated acetylsalicylic acid in the formation of congenital abnormalities, most large studies have failed to find any significant risk or evidence of teratogenicity (9-11).

Metabolism and pharmacokinetics

Absorption of non-ionised acetyl salicylic acid occurs in the stomach.
Acetylsalicylates and salicylates are also readily absorbed from the intestine.
Hydrolysis to salicyclic acid occurs rapidly in the intestine and in the circulation.
Salicylates are extensively bound to plasma proteins, aspirin to a lesser degree.
Aspirin and salicylates are rapidly distributed to all body tissues, they appear in milk and cross the placenta. Major excretory products in the urine are free salicylic acid and glucuronide conjugates of salicylic and gensitic acids, the rate of urinary excretion increasing with pH, and is greatest at pH 7.5 and above (12,13).

Sensitisation

Some adults exhibit hypersensitivity to aspirin, symptoms include rhinitus, sinusitis, asthma and urticaria (12,14,15).
In a retrospective study in 86 patients with urticaria, confirmed by a positive challenge test in 84, aspirin was the responsible agent in one case (16).

Genotoxicity

Salmonella typhimurium TA98, TA100, TA1535, TA1537, TA1538 with and without metabolic activation negative (17).

In vitro mouse C3H/10T1/2 0.5-2 mg ml^{-1} cytotoxic effects demonstrated by decreased plating efficiency (18).

Any other adverse effects to man

In humans symptomatic effects include nausea, dyspepsia, vomiting, dizziness, tinnitus, deafness, sweating and headache. Symptoms of severe intoxication including overdosing include hyperventilation, fever, restlessness, ketosis, respiratory alkalosis and metabolic acidosis. Depression of the central nervous system may lead to coma, cardiovascular collapse and respiratory failure. In children drowsiness and metabolic acidosis commonly occur. Has been implicated in Reye's syndrome in children (12). Transient myopia occurred in a patient following the ingestion of 2.7 g acetylsaliclic acid (19).

The Boston Collaborative Drug Surveillence Program monitored consecutively 32,812 medical inpatients. Drug induced deafness occurred in a dose related manner in 32 of 2974 patients given acetylsalicylic acid (20).

Of 15,438 patients hospitalised between 1975 and 1982 none of the allergic skin reactions detected were attributed to acetylsalicylic acid among 984 recipients of the drug (21).

Thirteen cases of toxic epidermal necrolysis have been reported with the use of acetylsalicylic acid or methyl salicylate (22).

Any other adverse effects

Oral administration of 600 mg acetylsalicylic acid to 5 adult Sprague-Dawley rats, followed by fasting for 16 hr caused urinary excretion of γ-glutamyl transpeptidase, N-acetyl-β-δ- glucosaminidase proteins, glucose and blood urea nitrogen after 24 hr indicative of nephrotoxic effect (23).

Any other comments

Caution required in administration to patients with renal or hepatic problems. Should not be given to patients suffering from haemophilia or other haemorrhagic disorders. Stable in dry air, but gradually hydrolyses in contact with moisture to acetic and salicylic acids (12).

The properties of acetylsalicylic acid including workplace experience, experimental toxicology, pharmacokinetics and human health effects have been extensively reviewed (24-30).

References

1. *Fish Toxicity Screening Data*, 1989, EPA 560/6-89-001, PB 89-156715
2. Gnezdilova, A. *Gigiena Truda i Professional'nye Zabolevaniia* 1980, **24**(3), 43-45, (*Chem. Abstr.*, **93**, 37079d)
3. *The Merck Index* 11th ed. 1989, Merck & Co. Inc, Rahway, NJ
4. Sunshine, I. *CRC Handbook of Analytical Toxicology* 1969, 3, The Chemical Rubber Co., Cleveland
5. Hayes, W. J. Jr. *Toxicology of Pesticides* 1975, 68, Williams & Wilkins, Baltimore
6. *J. Pharm. Pharmacol.* 1952, **4**, 872
7. *J. Am. Med. Assoc.* 1944, **126**, 806
8. *Arzneimittel-Forschung* 1983, **33**, 631
9. Slone, D; et al. *Lancet* 1976, **1**, 1373
10. Shapiro, S; et al. *Lancet* 1976, **1**, 1375
11. Winship, K. A; et al. *Arch. Dis. Child.* 1984, **59**, 1052

12. *Martindale: The Extra Pharmacopoeia* 29th ed., 1989, The Pharmaceutical Press, London
13. Goodman, L. S; et al. *The Pharmacological Basis of Therapeutics* 5th ed., 1975, 334, Macmillan, New York
14. Graham, D. Y; et al. *Ann. Intern. Med.* 1986, **104**, 390
15. Asad, S. I; et al. *Br. Med. J.* 1984, **288**, 745
16. Kauppinen, K; et al. *Br. J. Derm.* 1985, **112**, 575
17. Jasiewicz, M. L; et al. *Mutat. Res.* 1987, **190**(2), 95-100
18. Patierno, S. R; et al. *Cancer Res.* 1989, **49** (4), 1038-1044
19. Sandford-Smith, J. H; *Brit. J. Ophthalmol.* 1974, **58**, 698
20. Porter, J; et al. *Lancet* 1977, **1**, 587
21. Bigby, M; et al. *J. Am. Med. Assoc.* 1986, **256**, 3358
22. Lowney, E. O; et al. *Arch. Derm.* 1967, **95**, 359
23. Chakrabarti, S; et al. *Res. Commun. Subst. Abuse* 1987, **7**(3-4), 153-160
24. *ECETOC Technical Report No. 30(4)* 1991, European Chemical Industry Ecology and Toxicology Centre, Brussels
25. Burnat, P; et al. *Lyon Pharm.* 1988, **39** (6), 361-374
26. Fowler, P. D. *Med. Toxicol.* 1987, **2**(5), 338-366
27. Beaufils, M; et al. *Lancet* 1985, **1**, 840
28. Lubbe, W. F; et al. *Am. J. Obstet. Gynecol.* 1985, **132**, 322
29. Ocak, F; et al. *FABAD Farm. Bilimzer. Derg.* 1989, **14**(1), 51-63
30. Szczeklik, A. *Eur. Respir. J.* 1990, **3**(5), 588-593

A38   Acifluorfen

CAS Registry No. 50594-66-6
Synonyms 5-(2-chloro-α,α,α-trifluoro-*p*-tolyloxy)-2-nitrobenzoic acid;
5-(2-chloro-4-(trifluoromethyl)phenoxy)-2-nitrobenzoic acid; Blazer
Mol. Formula $C_{14}H_7ClF_3NO_5$ **Mol. Wt.** 361.66
Uses Post-emergence herbicide.

Physical properties

M. Pt. 151-157°C; **Partition coefficient** log P_{ow} –4.85 (calc.) (1);
Volatility v.p. 7.52×10^{-8} mmHg at 20°C.

Solubility
Water: >120 mg l^{-1} at 25°C. Organic solvent: acetone, ethanol, dichloromethane, xylene

Ecotoxicity

Fish toxicity
LC$_{50}$ (96 hr) rainbow trout, bluegill sunfish 17-62 mg l^{-1} (2).

Effects to non-target species
LD$_{50}$ oral mallard duck 281 mg kg^{-1} (2).

Environmental fate

Anaerobic effects
Under anaerobic conditions in soil the acetamide, amino and denitro derivatives were formed, $t_{1/2}$ 1 month (1).

Degradation studies
In soil, photodecomposes to non-herbicidal products, $t_{1/2}$ 30-60 days (2).
In an aerobically incubated soil, $t_{1/2}$ 170 days. Dominant residue compounds after 6 months were unchanged acifluorfen and bound materials (1).

Abiotic removal
Stable to hydrolysis. No degradation observed in solutions at pH3, pH6 or pH9 within a 28 day interval, temperature range 18-40°C. Under laboratory conditions with continuous exposure to light approximating natural sunlight, $t_{1/2}$ 92 hr. Primary degradation product detected in solution was the decarboxy derivative.
In a 179 day study of silt loam soil removal by leaching was negligible (1).

Mammalian and avian toxicity

Acute data
LD_{50} oral rat 1370-2025 mg kg^{-1}.
LD_{50} oral bobwhite quail 325 mg kg^{-1}.
LD_{50} (4 hr) inhalation rat >6.9 mg l^{-1}.
LD_{50} dermal rabbit >2000 mg kg^{-1} (2).

Sub-acute data
LC_{50} (8 day) oral bobwhite quail, mallard duck >10 g kg^{-1} in diet (2).

Carcinogenicity and long-term effects
Oral ♂, ♀ B6C3F1 (18 month) 0, 625, 1250, 2500 ppm in diet induced benign and malignant liver tumours in both sexes and stomach papillomas in ♀ mice (3).
Oral ♂, ♀ CR-CD 1 mice (2 yr) 0, 7.5, 45, 270 ppm induced benign and malignant liver tumours in both sexes (4).

Metabolism and pharmacokinetics
Oral rat acifluorfen was rapidly and completely absorbed from the gastrointestinal tract (70-90%) within 96 hr. No unusual organ or tissue accumulation of acifluorfen or its metabolites occurred. 46-82% excreted via the urine as unchanged acifluorfen (minor amounts of amine and glucuronide moities detected) and 5-41% via the faeces, majority as amine metabolite with minor amounts of unchanged acifluorfen and acetamide (3).

Genotoxicity
Saccharomyces cerevisiae D5 DNA damage (mitotic recombination) positive (4).
Salmonella typhimurium TA1535, TA1537, TA1538, TA98, TA100 with and without metabolic activation negative.
Saccharomyces cerevisiae D4 with and without metabolic activation negative (1).
Drosophila melanogaster chromosomal aberrations positive.
In vitro rodent hepatocytes unscheduled DNA synthesis negative.
In vivo rat bone marrow and mouse lymphoma chromosome aberrations and gene mutation negative (4).

Any other adverse effects

Oral F344 rats (2 yr) 0, 25, 150, 500, 2500, 5000 ppm in diet. High dose animals suffered increased mortality and reduced body weight gain, increases in liver weight and liver and renal enzyme activities (alkaline phosphatase, blood urea nitrogen and creatine, renal changes (nephritis, pyelonephritis and glomerulonephritis), a reduction in testes size and stomach ulcers. Lowest no effect dose 500 ppm (4).

Legislation

Limited under EC Directive on Drinking Water Quality 80/778/EEC. Pesticides and related products: maximum admissible concentration 0.5 μg l^{-1} (5).

Any other comments

Extensive animal toxicity studies have been carried out using acifluorfen sodium 62476-59-9 (1).

References

1. *Drinking Water Health Advisory: Pesticides* 1989, 1-22, Lewis Publishers, Chelsea, MI
2. *The Agrochemicals Handbook* 3rd ed., 1991, RSC, London
3. Quest, J. A; et al. *Regul. Toxicol. Pharmacol.* 1989, **10**(2), 149-159
4. *Peer Review Acifluorfen* 1987, USEPA Pesticide Toxicology Program, Washington, DC
5. *EC Directive Relating to the Quality of Water Intended for Human Consumption* 1982, 80/778/EEC, Office for Official Publications of the European Communities, 2 rue Mercier, L-2985 Luxembourg

A39   Acinitrazole

CAS Registry No. 140-40-9
Synonyms aminitrozole; N-(5-nitro-2- thiazolyl)acetamide; 2-acetamido-5-nitrothiazole; 2- acetylamino-5-nitrothiazole; Tritheon; Trichorad; acetyl enheptin; Trichloral; Gynofon; Pleocide
Mol. Formula $C_5H_5N_3O_3S$ **Mol. Wt.** 187.18
Uses Used as an antibacterial agent and in veterinary medicine.

Physical properties

M. Pt. 264-265°C.

Mammalian and avian toxicity

Acute data
LD_{50} oral mouse 1000 mg kg^{-1} (1).
LD_{50} oral dog 125 mg kg^{-1} (2).

Genotoxicity

Salmonella typhimurium TA100 and *Klebsiella pneumoniae* with and without metabolic activation positive (3).

In vivo oral administration for mouse bone marrow micronucleus test, bone marrow depression observed. Acinitrazole was reported to be extremely toxic (4).

Any other adverse effects

At toxic concentration in unspecified species caused blood cell count changes and weight loss (1,2).

References

1. *Farmaco, Ed. Sci* 1964, **19**, 301
2. *Antibiotics Chemotherapy* 1955, **5**, 540
3. Voogd, C. E; et al. *Mutat. Res.* 1983, **118**, 153
4. Paik, S. G. *Environ. Mutagen. Carcinogen.* 1985, **5**(2), 61-72

A40   Aclonifen

CAS Registry No. 74070-46-5
Synonyms 2-chloro-6-nitro-3-phenoxyaniline;
2-chloro-6-nitro-3-phenoxybenzenamine; Aclofen
Mol. Formula $C_{12}H_9ClN_2O_3$ **Mol. Wt.** 264.67
Uses Pre-emergence herbicide.

Physical properties

M. Pt. 81-82°C; **Volatility** v.p. 6.8×10^{-6} mmHg at 20°C.

Solubility
Water: 2.5 mg l^{-1} at 20°C. Organic solvent: methanol, hexane, toluene

Ecotoxicity

Fish toxicity
LC_{50} (96 hr) rainbow trout, carp 1-3.3 mg l^{-1} (1,2).

Effects to non-target species
LD_{50} oral canary >15,000 mg kg^{-1}.
LD_{50} oral >26 mg bee^{-1} (1,2).

Mammalian and avian toxicity

Acute data
LD_{50} oral Japanese quail >15,000 mg kg^{-1}.
LD_{50} oral rat, mouse >5000 mg kg^{-1}.
LD_{50} percutaneous rat >5000 mg kg^{-1} (1,2).

Sub-acute data
In a 90 day feeding study in rats and a 180 day study in dogs, the no observed effect level were 28 mg kg^{-1} day^{-1} and 3 mg kg^{-1}, respectively (1).

Metabolism and pharmacokinetics
In rats, following oral administration, 70% was excreted in urine within 24 hr, principally as polar compounds (1).

Legislation

Limited under EC Directive on Drinking Water Quality 80/778/EEC. Herbicide: maximum admissible concentration 0.5 μg l^{-1} (3).

References

1. *The Agrochemicals Handbook* 3rd ed., 1991, RSC, London
2. *The Pesticide Manual* 9th ed., 1991, British Crop Protection Council, Farnham
3. *EC Directive Relating to the Quality of Water Intended for Human Consumption* 1982, 80/778/EEC, Office for Official Publications of the European Communities, 2 rue Mercier, L-2985 Luxembourg

A41   Aconitine

CAS Registry No. 302-27-2
Synonyms
16-ethyl-1,16,19-trimethoxy-4-(methoxymethyl)aconitane-3,8,10,11,18-pentol 8-acetate 10-benzoate; acetyl benzoyl aconine; aconitin cristallisat; aconitane
Mol. Formula $C_{34}H_{47}NO_{11}$ **Mol. Wt.** 645.75
Uses An antipyretic drug.

Physical properties

M. Pt. 204°C.

Solubility
Water: 300 mg l^{-1}. Organic solvent: chloroform, benzene, absolute ethanol

Occupational exposure

Supply classification toxic.

Risk phrases
Very toxic by inhalation and if swallowed (R26/28)

Safety phrases
Keep locked up – Avoid contact with skin – In case of accident or if you feel unwell, seek medical advice immediately (show label where possible) (S1, S24, S45)

Mammalian and avian toxicity

Acute data
LD_{50} oral mouse 1 mg kg^{-1} (1).
LD_{50} intravenous rat 80 µg kg^{-1} (2).
LD_{Lo} intraperitoneal rat 125 µg kg^{-1} (3).
LD_{Lo} subcutaneous rabbit 131 µg kg^{-1} (4).

Sub-acute data
LD_{Lo} (duration unspecified) oral human 28 mg kg^{-1} central nervous system and gastrointestinal tract effects (5).

Any other adverse effects to man
Moderately toxic doses produce a tingling of the tongue, mouth, stomach and skin followed by numbness and anaesthesia (6).
Aconitine has variable effects on the heart leading to heart failure. It also affects the central nervous system. Symptoms of aconitine poisoning may appear almost immediately and are rarely delayed beyond 1 hr. In fatal poisoning death usually occurs within 6 hr, although with large doses it may be instantaneous (6).
Symptomatic effects include gastrointestinal disturbances, irregular pulse, difficult respiration, cold, clammy and livid skin, muscular weakness, incoordination and vertigo (6).

Any other comments
Toxicology and arrhythmogenic properties reviewed (7).

References

1. *CRC Handbook of Antibiotic Compounds* 1982, **8**(1), 159
2. *Arzneimittel-Forschung* 1955, **5**, 324
3. *Proceedings of the Society for Experimental Biology and Medicine* 1928, **26**, 221
5. Deichmann, W. B. *Toxicology of Drugs and Chemicals* 1969, 72, Academic Press, New York
6. *Martindale: The Extra Pharmacopeia* 29th ed., 1989, The Pharmaceutical Press, London
7. Boyadzhiev, T. *Nauchni Tr. Vissh. Med. Inst. Sofia* 1966, **45**(5), 69-74, (Bulg.) (*Chem. Abstr.* **67**, 31049k)

A42   Acridine

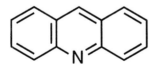

CAS Registry No. 260-94-6
Synonyms 10-azaanthracene; 2,3,5,6-dibenzopyridine; 2,3-benzoquinoline; benzo(b)quinoline; dibenzo(b,e)pyridine
Mol. Formula $C_{13}H_9N$ **Mol. Wt.** 179.22
Uses Intermediate in manufacture of dyestuffs, alkaloids and antibacterials.
Occurrence In coal tar, wood preservative sludge and coke oven emissions.

Physical properties

M. Pt. 110.5°C; **B. Pt.** 346°C; **Specific gravity** d$^{19.7}$ 1.005;
Partition coefficient log P_{ow} 3.40; **Volatility** v.p. 1 mmHg at 129.4°C.

Solubility

Organic solvent: ethanol, diethyl ether, hydrocarbons, carbon disulfide, benzene

Occupational exposure

UN No. 2713; **Conveyance classification** toxic substance.

Ecotoxicity

Fish toxicity

Trout, bluegill sunfish, goldfish 5 ppm death occurred in 4-6 hr (1).

Invertebrate toxicity

LC_{50} (24 hr) *Daphnia pulex* 2.9 mg l^{-1} (2).
EC_{50} (5-30 min) *Photobacterium phosphoreum* 7.47 mg l^{-1} Microtox test (3).

Bioaccumulation

Daphnia pulex bioconcentration factor 29.6 (2).
Fathead minnow bioconcentration factors 125-874. Depuration was rapid, possible
pathways for uptake include interaction, ingestion of contaminated zooplankton and
invertebrates and direct uptake (4).

Environmental fate

Anaerobic effects

Using 3 different inocula, degradation was investigated under methanogenic,
denitrifying and sulfate- reducing conditions at 1-6 µg ml^{-1} in 1-3 wk (5).

Mammalian and avian toxicity

Acute data

LD_{50} subcutaneous mouse 400 mg kg^{-1}.
LD_{50} intravenous rabbit 100 mg kg^{-1} (6).

Irritancy

Severe irritant to eyes, skin and respiratory tract (7,8).

Sensitisation

It is a skin photosensitiser (8).

Genotoxicity

Salmonella typhimurium TA98, TA100, TA1535, TA1537, TA1538 with and without
metabolic activation positive.
Saccharomyces cerevisiae forward mutation studies non-definite.
In vitro mouse BALB/c-3T3 cell transformation studies negative (9).

Any other comments

US EPA suggested ambient air limit of 162 µg m^{-3} and an ambient level in water of 800 µg
l^{-1} (8).
Toxicology and genotoxicity reviewed (9,10).

References

1. Hollis, E. H; et al. *The Toxicity of 1085 Chemicals to Fish* 1987, 111, EPA560/6-87-002, US Fish and Wildlife Service, WV
3. Southworth, G. R; et al. *Environ. Sci. Techn.* 1978, **12**(9), 1062-1066
4. Southworth, G. R; et al. *Water, Air and Soil Poll.* 1979, **12**, 331-341
5. Knezovich, J. P; et al. *Environ. Toxicol. Chem.* 1990, **9**(10), 1235-1243
6. Rubbo, S. D. *Br. J. Exper. Pathol.* 1947, **28**, 1
7. *Dangerous Prop. Ind. Mater. Rep.* 1981, **1**(8)
8. Sittig, M. *Handbook of Toxic and Hazardous Chemicals and Carcinogens* 2nd ed., 1985, Noyes Publications
9. Ferguson, L. R; et al. *Mutat. Res.* 1991, **258**, 123-160
10. *Dangerous Prop. Ind. Mater. Rep.* 1988, **8**(5), 49-55

A43   Acrolein

CH2=CHCHO

CAS Registry No. 107-02-8

Synonyms acrylaldehyde; allyl aldehyde; 2-propenal; prop-2-enal; ethylene aldehyde; acrylic aldehyde

Mol. Formula C_3H_4O **Mol. Wt.** 56.06

Uses In the manufacture of glycerol and glutaraldehyde. Aquatic herbicide and fungicide used in water and wastewater treatment. Used in military poison gas mixtures, manufacture of plastics perfumes and warning agent in methyl chloride refrigerant.

Occurrence Occurs in exhaust gas emissions, tobacco smoke, wastewater effluents and as a photooxidation product of hydrocarbon pollutants (1). Volatile component of essential oil extracted from the wood of oak trees. Has been detected in emissions from plants manufacturing acrylic acid and coffee roasting (2).

Acrolein is produced in the pyrolysis of animal fats.

Physical properties

M. Pt. -87°C; **B. Pt.** 52°C; **Flash point** -26°C; **Specific gravity** d^{20} 0.843; **Partition coefficient** $\log P_{ow}$ 0.101 (3).; **Volatility** v.p. 210 mmHg at 20°C; v. den. 1.9.

Solubility

Water: 20.6% at 20°C. Organic solvent: ethanol, diethyl ether, acetone

Occupational exposure

US TLV (TWA) 0.1 ppm (0.23 mg m^{-3}); **US TLV (STEL)** 0.3 ppm (0.69 mg m^{-3}); **UK Long-term limit** 0.1 ppm (0.25 mg m^{-3}); **UK Short-term limit** 0.3 ppm (0.8 mg m^{-3}); **UN No.** 1092; **HAZCHEM Code** 2WE; **Conveyance classification** flammable liquid; **Supply classification** Highly flammable and toxic.

Risk phrases

Highly flammable – Toxic by inhalation – Irritating to eyes, respiratory system and skin (R11, R23, R36/37/38)

Safety phrases

Do not empty into drains – Take precautionary measures against static discharges – If you feel unwell, seek medical advice (show label where possible) (S29, S33, S44)

Ecotoxicity

Fish toxicity

LC_{50} (96 hr) bluegill sunfish, goldfish 0.04-0.31 mg l^{-1} (1-4).

Death occurred at 30 mg l^{-1} for stickleback and steelhead trout in 1-2 hr and 0-1 hr respectively (5).

5 ppm of acrolein administered to trout, bluegill sunfish, perch and goldfish caused death in 2-6 hr (6).

LC_{50} (4 day) fathead minnow 20 μg l^{-1} (7).

LC_{50} (24, 48 hr) harlequin fish 0.14- 0.06 mg acrolein l^{-1}. Estimated lethal threshold 0.03 mg acrolein l^{-1} (8).

Invertebrate toxicity

EC_{50} (24-48 hr) *Daphnia magna* 0.23-0.083 mg l^{-1} (conditions unspecified) (9).

Cell multiplication inhibition test, *Pseudomonas putida* 0.21 mg l^{-1}, *Microcystis aeruginosa* 0.04 mg l^{-1}, *Uronema parduczi* 0.44 mg l^{-1} (10).

After a 24 hr exposure to 10 mg l^{-1} acrolein, 98% of adult snails and 100% of snail embryos died (11).

EC_{50} (5 min) *Photobacterium phosphoreum* 0.674 mg l^{-1} Microtox test (12).

Bioaccumulation

Bioconcentration factor has been measured as 344 in bluegill sunfish (13).

However, the calculated bioconcentration factor 0.6 suggests that bioaccumulation in aquatic organisms is not significant (14).

Environmental fate

Degradation studies

Nonbiodegradable/qualified (15).

Under aerobic conditions, acrolein undergoes reversible hydration to β-hydroxypropionaldehyde $t_{1/2}$ 21 days (16).

BOD_5 no oxygen removal (17).

No evidence of degradation was observed for 50 mg l^{-1} of acrolein as organic carbon when incubated for 8 wk in a 10% anaerobic sludge inoculum (18).

Abiotic removal

Evaporates rapidly from dry soil surfaces. In the atmosphere, acrolein reacts photochemically with hydroxyl radicals ($t_{1/2}$ 10-13 hr) to produce carbon dioxide, formaldehyde and glycoaldehyde. When nitrogen oxides are present, products include peroxynitrate and nitric acid (1).

Estimated $t_{1/2}$ 18 days with ozone (19).

Absorption

The low soil adsorption coefficient of 24 suggests that acrolein will be unlikely to adsorb to suspended solids and sediments in water and will be highly mobile in soil (2,20).

Mammalian and avian toxicity

Acute data
LD_{50} oral rabbit, mouse 7-40 mg kg^{-1} (21,22).
LD_{Lo} (6 hr) inhalation rabbit 24 mg m^3 (23).
LC_{Lo} (10 min) inhalation human 153 ppm (24).
LD_{50} dermal rabbit 562 mg kg^{-1} (25).
LD_{50} intraperitoneal rat, mouse 4-9 mg kg^{-1} (26,27).

Sub-acute data
Inhalation rat (3 wk) 0-3 ppm 6 hr day^{-1} 5 day wk^{-1} induced exfoliation, erosion and necrosis of the respiratory epithelium and squamous metaplasia. Body and spleen weights decreased for 3 ppm exposed rats (28).

Carcinogenicity and long-term effects
No adequate evidence for carcinogenicity to humans, limited evidence for carcinogencity to animals, IARC classification group 3 (29).
Oral rats, hamsters (2 yr) (unspecified dose) in drinking water. The authors report a nonsignificant increase in the incidence of neoplasms, target tissues included liver, uterus and thyroid gland (30).

Teratogenicity and reproductive effects
Acrolein is a toxic and reactive metabolite of the widely used anticancer drug and known teratogen cyclophosphamide. To determine if acrolein was responsible for the production of limb malformations, excised limbs were exposed to 10 or 50 μg ml^{-1} acrolein for the first 20 hr of a 6 day culture period. Exposure to acrolein produced malformed limbs, had little effect on growth parameters, and did not effect alkaline phosphatase and creatine phosphatase activity in excised limbs (31).
In vitro rat embryo caused 25% and 48% deaths at 50 and 100 μg, respectively and 46% incidence of malformations and growth retardation at 100 μg (32).

Irritancy
Dermal rabbit (24 hr) 500 mg caused severe irritation, while 50 μg instilled into rabbit eye for 24 hr caused severe irritation (33).

Sensitisation
The alleged dermal sensitisation properties of acrolein were evaluated by its ability to induce allergic contact dermatitis. Challenge treatment with acrolein (species unspecified) failed to produce positive skin reactions (34).

Genotoxicity
In vitro Chinese hamster V79 cells with and without metabolic activation positive (35).
Chinese hamster ovary HGPRT assay with and without metabolic activation, results showed acrolein was very cytotoxic but no significant mutagenic response observed (36).

Legislation
Limited under EEC Directive on Drinking Water Quality 80/778/EEC. Pesticides and related products: maximum admissible concentration 0.5 μg l^{-1} (37).

Any other comments

Acrolein residues have been detected in some foodstuffs (2).
Human health and environmental effects, experimental toxicology, ecotoxicology and
exposure levels reviewed (38-42).

References

1. Howard, P. H. *Fate and Exposure Data for Organic Chemicals* 1990, **1**, 5-12
2. *IARC Monograph* 1985, **36**, 133-161
3. Lipnick, R. C; et al. *Xenobiotica* 1987, **17**(8), 1011-1025
4. Carlson, R. W. *Comp. Biochem. Physiol.* 1990, **95**C(2), 181-196
5. MacPhee, C; et al. *Lethal Effects of 2014 Chemicals upon Sockeye Salmon, Steelhead Trout and Threespine Stickleback* 1989, 7, EPA 560/6-89-001
6. Wood, E. M. *The Toxicity of 3400 Chemicals to Fish* 1953, 100, US Fish and Wildlife Service, WV
7. Geiger, D. L; et al. *Acute Toxicities of Organic Chemicals to Fathead Minnows* 1988, **4**, 355
8. Alabaster, J. S. *International Pest Control* 1969, 29-35
9. Le Blanc, G. A. *Bull. Environ. Contam. Toxicol.* 1980, **24**, 684-691
10. Verschueren, K. *Handbook of Environmental Data of Organic Chemicals* 2nd ed., 1983, 158, Van Nostrand Reinhold, New York
11. Ferguson, F. F; et al. *Ambient Water Quality Criteria Document: Acrolein* 1980, B-3, EPA 440/5-80-016
12. Kaiser, K. L. E; et al. *Water Pollut. Res. J. Canada* 1991, **26**(3), 361-431
13. Barrows, M. E; et al. *Dynamics, Exposure Hazard. Assess. Toxic Chem.* 1980, 279-292, Ann Arbor Science, Ann Arbor, MI
14. Lyman, W. J; et al. *Handbook of Chemical Property Estimation Methods* 1982, McGraw-Hill, New York
15. *MITI Report* 1984, Ministry of International Trade and Industry, Tokyo
16. *Herbicide Handbook* 5th ed., 1983, 8-12, Weed Science Society of America, Champaign, IL
17. Bridie, A. L; et al. *Water Res.* 1979, **13**, 627-630
18. Shelton, D. R; et al. *Development of Tests for Determining Anaerobic Biodegradation Potential* 1981, EPA 560/5-81-013, NTIS PB84-166 495
19. Atkinson, R; et al. *Chem. Rev.* 1984, **84**, 437-470
20. Swann, R. L; et al. *Res. Rev.* 1984, **85**, 17-28
21. *Pesticide Chemicals Official Compendium* 1966, 1
22. *Biochem. J.* 1940, **34** 1196
23. *Br. Med. J.* 1956, **2**, 913
24. *NTIS Report* PB 214-270, National Technical Information Service, Springfield, VA
25. *World Rev. Pest Control* 1970, **9**, 119
26. Astry, C. L. *Toxicol. Appl. Pharmacol.* 1983, **71**, 84
27. *National Cancer Inst. Screening Program Data Summary* 1986, Developmental Therapeutics Program, Bethesda, MD
28. Leach, C. L; et al. *Toxicol. Lett.* 1987, **39** (2-3), 189-198
29. *IARC Monograph* 1987, **Suppl. 7**, 77
30. Lijinsky, W. *Ann. New York Acad. Sci.* 1988, **534**, 246-254
31. Hales, B. F. *Teratology* 1989, **40**(1), 11-20
32. Slott, V. L; et al. *Biochem. Pharmacol.* 1987, **36**(12), 2019-2025
33. Marhold, J. V; et al. *Sbornik Vysledku Toxixologickeno Vysetreni Latek A Pripravku* 1972, Prague
34. Susten, A. S; et al. *Contact Dermatitis* 1990, **22**(5), 299-300
35. Smith, R. A; et al. *Carcinogenesis (London)* 1990, **11**(3), 497-498
36. Parent, R. A; et al. *J. Appl. Toxicol.* 1991, **11** (2), 91-95
37. *EC Directive Relating to the Quality of Water Intended for Human Consumption* 1982, 80/778/EEC, Office for Official Publications of the European Communities, 2 rue Mercier, L-2985 Luxembourg

38. *USEPA Report* 1987, EPA 600/8-88/013, PB88-179494/GAR
39. Izmerov, N. F; *Scientific Reviews of Soviet Literature on Toxicity & Hazards of Chemicals* 1991, **50**, Eng. Trans. Richardson, M. L. (Ed.), UNEP/IRPTC, Geneva
40. *ECETOC Technical Report No. 30(4)* 1991, The Chemical Industry Ecology and Toxicology Centre, B-1160 Brussels
41. Wesenberg, J; et al. *Lebensmittelindustrie* 1990, **37**(4), 159-162, (Ger.) (*Chem. Abstr.* **113**, 209934c)
42. *Dangerous Prop. Ind. Mater. Rep.* 1990, **10**(3), 9-27

A44   Acrolein dimer

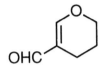

CAS Registry No. 100-73-2
Synonyms 2*H*-pyrancarboxaldehyde, 3,4-dihydro-; pyran aldehyde; 5-hexanal, 2,6-epoxy; 2-formyl-3, 4-dihydro-2*H*-pyran; 3,4-dihydro-2*H*-pyran-2-carboxaldehyde
Mol. Formula $C_6H_8O_2$ **Mol. Wt.** 112.13
Uses Intermediate for pharmaceuticals and dyestuffs.

Physical properties

B. Pt. 153°C; **Flash point** 48°C (open cup); **Specific gravity** d_4^{20} 1.0775; **Partition coefficient** $\log P_{ow}$ 1.0775 at 20°C.

Occupational exposure

UN No. 2607; **HAZCHEM Code** 2S; **Conveyance classification** flammable liquid.

Mammalian and avian toxicity

Acute data
LD_{50} oral rat 4920 mg kg^{-1} (1).

Carcinogenicity and long-term effects
Inadequate evidence for carcinogenicity to humans and animals, IARC classification group 3 (2).

Irritancy
Dermal rabbit 500 mg caused mild irritation, while 750 μg instilled into rabbit eye caused severe irritation (1,3).

References

1. *Arch. Ind. Hyg. Occ. Med.* 1954, **10**, 61
2. *IARC Monograph* 1987, **Suppl. 7**, 56
3. *Union Carbide Data Sheet* 1965, Union Carbide Corp., New York

A45   Acronycine

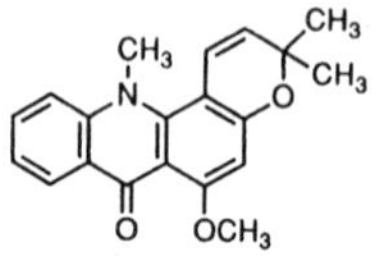

CAS Registry No. 7008-42-6

Synonyms 3,12-dihydro-6-methoxy-3,3,12-trimethyl-7H-pyrano(2,3-c)acridin-7-one; acronine; acromycine

Mol. Formula $C_{20}H_{19}NO_3$ **Mol. Wt.** 321.38

Uses Antineoplastic agent with broad-spectrum antitumour activity.

Occurrence Alkaloid from *Acroynchia baueri* and *Melicope leptococca*.

Physical properties

M. Pt. 175-176°C.

Ecotoxicity

Fish toxicity

LC_{Lo} (24-96 hr) harlequin fish 850-200 mg l^{-1} (1).

Mammalian and avian toxicity

Acute data

LD_{50} oral, intraperitoneal mouse 522-613 mg kg^{-1} (2).

Metabolism and pharmacokinetics

Acronycine is metabolised by humans and excreted in urine mainly as the 9-hydroxy- and 11-hydroxyacronycine compounds (3).

References

1. Svoboda, G. H; et al. *J. Pharm. Sci.* 1966, **55**, 758-768
2. *National Cancer Inst. Screening Program Data Summary* 1978, NCI-CG-TR, 49
3. Sullivan, H. R; et al. *J. Med. Chem.* 1970, **13**(5), 904-909

A46   Acrylamide

$$H_2C=CHCONH_2$$

CAS Registry No. 79-06-1

Synonyms acrylic amide; ethylene carboxamide; propenamide; 2-propenamide

Mol. Formula C_3H_5NO **Mol. Wt.** 71.08

Uses Intermediate in polymer manufacture, especially polyacrylamides which are used in the treatment of municipal drinking water and wastewater. Flocculant. In the pulp and paper, textile, and paints and coatings industries. Has been used as thickening agent in cosmetics and toiletries.

Occurrence Occurs in discharges from water treatment plants which use polyacrylamides and flocculants.

Physical properties

M. Pt. 84.5°C; **B. Pt.** 125°C at 25 mmHg; **Flash point** 138°C; **Specific gravity** d_4^{30} 1.122; **Volatility** v.p. 1.6 mmHg at 84.5°C; v. den. 2.45.

Solubility
Water: 2155 g l^{-1} at 30°C. Organic solvent: methanol, ethanol, acetone, ethyl acetate, chloroform, benzene

Occupational exposure
US TLV (TWA) 0.03 mg m^{-3}; **UK Long-term limit** (MEL) 0.3 mg m^{-3}; **UN No.** 2074; **HAZCHEM Code** 2PE; **Conveyance classification** harmful substance; **Supply classification** toxic.

Risk phrases
Toxic by inhalation, in contact with skin and if swallowed – Danger of cumulative effects (R23/24/25, R33)

Safety phrases
Take off immediately all contaminated clothing – If you feel unwell, seek medical advice (show label where possible) (S27, S44)

Ecotoxicity

Fish toxicity
LC_{50} (24-96 hr) brown trout, harlequin fish 130-460 mg l^{-1} (1,2).
Rainbow trout exposed to waterborne acrylamide monomer 0-50 mg l^{-1} for 14 days, reversible dose-related histological lesions detected in the liver, characterised by necrosis around the central vein (3).

Bioaccumulation
Fingerling trout bioconcentration factor in carcass and viscera were 0.86 and 1.12, respectively at 12°C for 72 hr under static conditions indicating that no appreciable bioaccumulation occurred. In freshwater, acrylamide levels declined to 75% after 96 hr (4).

Environmental fate

Anaerobic effects
In 14 days under waterlogged conditions 64-89% degradation occurred (5).
In 56 days, no degradation occurred when acrylamide was incubated with digester sludge (6).

Degradation studies
Complete degradation occurred in filtered river water in 10-12 days. In soil, 100% degradation occurred in 6 days with 60% converted to carbon dioxide (7).
BOD_5 standard dilution technique, acclimated sludge oxygen (0.97 mg l^{-1}) consumed at 20°C (1).
BOD_5 69-75% theoretical BOD obtained using acclimated sewage seed (8,9).
In river and estuary water with and without added sediment, 0.5 and 10 ppm acrylamide was completely degraded within 8 days (10).

Abiotic removal
Acrylamide does not absorb light at >250 nm, therefore direct photolysis is not expected (11).
Vapour phase acrylamide reacts with photochemically produced hydroxyl radicals $t_{1/2}$ 6.6 hr (12).

Absorption
Activated carbon 8 mg l^{-1}, pH 5.0, 0.5 hr 13% removed (1).
No significant adsorption occurred in short term expermiments on natural sediments, industrial and sewage sludges, clays, peat or synthetic resins (13).

Mammalian and avian toxicity

Acute data
LD_{50} oral mouse, rat 107-170 mg kg^{-1} (14,15).
LD_{50} intraperitoneal mouse 170 mg kg^{-1} (16).

Carcinogenicity and long-term effects
No adequate evidence for carcinogenicity to humans, sufficient evidence for carcinogenicity to animals, IARC classification group 2B (17).
Fischer 344 rats (2 yr) 0, 0.01, 0.1, 0.5 or 2 mg kg^{-1} in drinking water. Increased incidence of adrenal pheochromocytomas, mesotheliomas of the tunica of the testes and follicular adenomas of the thyroid were observed in ♂. In ♀, increased incidences of pituitary adenomas, thyroid follicular tumours, mammary adenomas and adenocarcinomas, oral cavity papillomas, uterine adenocarcinomas and clitoral gland adenomas were reported. In ♀ rats tumour of the central nervous system of glial origin or glial proliferation suggestive of early tumour were also increased (18).

Teratogenicity and reproductive effects
Gavage mice (6-17 day gestation) 0, 3, 15 or 45 mg kg^{-1} day^{-1}, gavage rats (6-20 day gestation) 0, 2.5, 7.5, 15 mg kg^{-1} day^{-1}. Maternal toxicity effects including reduced body weight gain were observed in both species, while hindlimb splaying occurred in mice only. Embryo/foetal toxicity was not observed in rats but foetal weight was reduced in mice given 45 mg kg^{-1} day^{-1}. No increased in incidence of malformations was observed but the incidence of variations, predominantly extra rib, increased with the dose (19).

Metabolism and pharmacokinetics
Intravenous rat 10 mg kg^{-1} distributed rapidly to tissues. Initial absorption in liver, kidney, testes, plasma and fat cells. Concentration in blood remained constant at 12% of the initial dose for 7 day, while plasma concentrations were eliminated readily. Less that 2% of initial dose was excreted unchanged in urine and bile (20).
Wholebody autoradiographs of mice administered orally 120 mg kg^{-1} acrylamide revealed accumulated radioactivity in the gastrointestinal tract, liver pancreas, testis, gall bladder, brain, epithelia of oral cavity, oesophagus and bronchi (21).
Pregnant intravenous rat, rabbit, dog, pig 5 or 10 mg kg^{-1} acrylamide crossed the placenca in all 4 species within 1-2 hr (22).

Irritancy
Dermal rabbit (3 day) 50 mg caused mild irritation and 10 mg instilled in rabbit eye for 30 sec caused mild irritation (15).
100 mg installed into rabbit eye for 24 hr caused moderate irritation (23).

Genotoxicity

Salmonella typhimurium TA98, TA100 with metabolic activation weakly positive (24).

In vitro mouse lymphoma L51178Y tk+/tk- without metabolic activation positive (25). Mouse embryo spot test (18 day foetus) examined. Single treatment of 50-75 mg kg^{-1} to mice induced increased number of offspring with colour spots, indicating positive mutagen and 3 treatments with 75 mg kg^{-1} caused tail kink demonstrating teratogenic potential (26).

In vivo mouse bone marrow cells chromosomal aberrations and micronucleus assay positive (27).

Any other adverse effects to man

In industry main route of contamination occurs via skin, though poisoning via inhalation and ingestion is possible (16,28).

Risk of irreversible nervous system damage. Time delay of up to 8 yr before onset of systems (15).

A study was undertaken of 71 acrylamide workers and 51 unexposed controls. Early symptoms of poisoning included weak legs, numb hands and feet, skin peeling from hands, impairment of vibration sensation in toes and loss of ankle reflexes. Three cases had cerebellar involvement followed by polyneuropathy due to heavy exposure. Total prevalence of acrylamide poisoning was 73.2% (29,30).

The mortality 371 employees in acrylamide monomer and polymer production was studied for deaths from cancers of central nervous system, thyroid gland, other endocrine glands and mesotheliomas. Until 1982 a total of 29 deaths observed versus 38 expected. No statistically significant excesses were noted in the total cohort and no deaths were found for hypothesised sites of cancer. Total observed deaths from all cancers was 11 versus 7.9 expected; however, this was due entirely to cancers of respiratory system and digestive tract in workers with previous exposure to organic dyestuffs. The study does not support a relationship between exposure to acrylamide and overall mortality, total malignant neoplasms or any specific cancers (30).

Any other adverse effects

Acute oral mouse 100 or 150 mg kg^{-1}, effects to testis were studied for 10 days after treatment. Severe initial testicular damage was repaired 7-10 days after treatment. Vulnerable stage identified as Golgi phase I-III (31). Causes peripheral neuropathy, skin peeling. Variety of symptoms mainly occuring in skin, hands and feet (15,16). Can be absorbed through intact skin, mucous membranes, lung and gastrointestinal tract. Can cause central nervous system paralysis (species unspecified) (32,33).

In vitro rats, acrylamide inhibited creatine kinase activity from brain and sciatic nerve. *In vivo* rats (8 day) 50 mg kg^{-1} day^{-1} caused paralysis of hind limbs and suppressed creatine kinase activity in the cerebrum, cerebellum, spinal cord and muscle (34).

Exposure of 4 monkeys, *Macaca fascicularis*, to prolonged intoxication with low levels of acrylamide caused a slow response across the spinal-medullary function. Associated morphological changes were preterminal accumulation of axonal neurofilaments without synaptic disruption in the nucleus. Shortlatency somatosensory response and axon morphology were reversible after 7 month of recovery. However, the extreme delay in onset of neurological dysfunction 940 days after administration of a presumed safe level of acrylamide exposure prompts the authors to conclude that permissible levels of human exposure should be reassessed (35).

Any other comments

Human health effects, experimental toxicology, environmental effects, ecotoxicology and exposure levels reviewed (36,37).

Neurotoxicity, metabolism, developmental and reproductive effects, genotoxicity and carcinogenicity reviewed (37-39).

Rapid and exothermic polymerisation occurs at melting point.

References

1. Verschueren, K. *Handbook of Environmental Data on Organic Chemicals* 2nd ed., 1983, Van Nostrand Reinhold, New York
2. Tooloy, T. E; et al. *Chem. Ind. (London)* 1975, 523-526
3. Petersen, D. W; et al. *Toxicol. Appl. Pharmacol.* 1987, **89**(2), 249-255
4. Petersen, D. W; et al. *Toxicol. Appl. Pharmacol.* 1985, **80**, 58-65
5. Abdelmagid, H. M; et al. *J. Environ. Qual.* 1982, **11**, 701-704
6. Shelton, D. R; et al. *Development of Tests for Determining Anaerobic Biodegradation Potential* 1981, 92, EPA 560/5-81-013
7. *Drinking Water and Health* 1986, **6**, 297, National Research Council, Washington, DC
8. Bridie, A. L; et al. *Water Res.* 1979, **13**, 627-630
9. Cherry, A. B; et al. *Sewage Indust. Wastes* 1956, **28**, 137-146
10. Brown, L; et al. *Water Res.* 1980, **14**, 779-781
11. Carpenter, E. L; et al. *J. Appl. Chem.* 1957, **7**, 671-676
12. *GEMS: Graphical Exposure Modeling System* 1986, FAP Fate of Atmos. Pollut.
13. Rhead, M. M; et al. *Water Res.* 1980, **14**(7), 775-778
14. *Archives of Toxicology* 1981, **47**, 179
15. Marhold, J. V. *Sbornik Vysledka Toxixologickeho Vysetreni Latek A Pripravku* 1972, 54, Prague
16. Petersen, R. E; et al. *Toxicol. Appl. Pharmacol.* 1975, **33**, 142
17. *IARC Monograph* 1987, **Suppl. 7**, 56
18. Johnson, K. A; et al. *Toxicol. Appl. Pharmacol.* 1986, **85**(2), 154-168
19. Field, E. A; et al. *Fundam. Appl. Toxicol.* 1990, **14**(3), 502-512
20. *IARC Monograph* 1982, **39**, 41
21. Miller, M. J; et al. *Toxicol. Appl. Pharmacol.* 1982, **63**(1), 36-44
22. Marlowe, C; et al. *Toxicol. Appl. Pharmacol.* 1986, **86**(3), 457-465
23. *Toxicol. Appl. Pharmacol.* 1964, **6**, 172
24. Zeiger, E; et al. *Environ. Mutagen.* 1987, 9(Suppl. 9), 1-109
25. Moore, M. M; et al. *Environ. Mutagen.* 1987, **9**(3), 261-267
26. Neuhaeuser-Klaus, A; et al. *Mutat. Res.* 1989, **226**(3), 157-162
27. Adler, I. D; et al. *Mutat. Res.* 1988, **206**(3), 379-385
28. *Kirk-Othmer Encyclopedia of Chemical Technology* 3rd ed., Vol. 1, 298-311, John Wiley, New York
29. He, F; et al. *Scand. J. Work Environ. Health* 1989, **15**(2), 125-129
30. Sobel, W; et al. *Br. J. Ind. Med.* 1986, **43**(11), 785-788
31. Sakamoto, J; et al. *Exp. Mol. Pathol.* 1988, **48**(3), 324-334
32. *The Merck Index* 11th ed. 1989, Merck & Co. Inc, Rahway, NJ
33. Grant, W. M. *Toxicology of the Eye* 3rd ed., 1986, 50, C. C. Thomas Publisher, Springfield, IL
34. Matsuoka, M; et al. *Brain Res.* 1990, **507**(2), 351-353
35. Ostro, B. D; et al. *JAPCA. J. Air Waste Manag. Assoc.* 1989, **39**(10), 1313-1318
36. *ECETOC Technical Report No. 30(4)* 1991, The Chemical Industry Ecology and Toxicology Centre, B-1160 Brussels
37. Izmerov, N. F; *Scientific Reviews of Soviet Literature on Toxicity & Hazards of Chemicals* 1991, **104**, Eng. Trans. Richardson, M. L. (Ed.), UNEP/IRPTC, Geneva
38. Spencer, P. S; et al. *Recent Adv. Nerv. Syst. Toxicol.* 1988, **100**, 163-176, NATO ASI Ser., Ser. A. New York
39. Dearfield, K. L; et al. *Mutat. Res.* 1988, **193**(1), 45-77

A47   Acrylic acid

$$H_2C=CHCO_2H$$

CAS Registry No. 79-10-7
Synonyms acroleic acid; ethylenecarboxylic acid; propene acid; 2-propenoic acid; vinylformic acid
Mol. Formula $C_3H_4O_2$ **Mol. Wt.** 72.06
Uses Intermediate in polymer manufacture. Also used as a tackifier and flocculant.
Occurrence Occurs in wastewater effluents as a byproduct of propylene oxidation (1).

Physical properties

M. Pt. 13°C; **B. Pt.** 141°C (polymerises); **Flash point** 68°C (open cup); **Specific gravity** 1.062 at 16°C; **Partition coefficient** log P_{ow} 0.31/0.43 (calc.) (1).; **Volatility** v.p. 10 mmHg at 39.9°C; v. den. 2.45.

Solubility
Water: miscible. Organic solvent: ethanol, diethyl ether, acetone, benzene

Occupational exposure

US TLV (TWA) 2 ppm (5.9 mg m^{-3}); **UK Long-term limit** 10 ppm (30 mg m^{-3}); **UK Short-term limit** 20 ppm (60 mg m^{-3}); **UN No.** 2218; **HAZCHEM Code** 2PE; **Conveyance classification** corrosive substance; **Supply classification** flammable and corrosive.

Risk phrases
Flammable – Causes burns (R10, R34)

Safety phrases
In case of contact with eyes, rinse immediately with plenty of water and seek medical advice – Wear suitable protective clothing (S26, S36)

Ecotoxicity

Fish toxicity
LC_{50} (24-96 hr) harlequin fish, brown trout 130-460 mg l^{-1} (2-4).

Invertebrate toxicity
Cell multiplication inhibition test, *Pseudomonas putida* 41 mg l^{-1}, *Microcystis aeruginosa* 0.15 mg l^{-1}, *Scenedesmus quadricauda* 18 mg l^{-1}, *Entosiphon quadricauda* 20 mg l^{-1}, *Uronema parduczi Chatton-Lwoff* 11 mg l^{-1} (2,3).

Bioaccumulation
Calculated bioconcentration factor 0.78 indicates environmental accumulation will be unlikely (4).

Environmental fate

Nitrification inhibition
No inhibition of nitrifying bacteria at 10 mg l^{-1} (4).

Anaerobic effects
8 wk incubation using 10% activated sludge caused >75% theoretical methane

production while another study showed acrylic acid was toxic and poorly utilised by unacclimated anaerobic acetate-enriched cultures. A possible explanation suggested is that acetate cultures have to exhaust acetic acid as a carbon and energy source before using alternative compounds (5).

Degradation studies
Confirmed biodegradable (6).
Rhodococcus converted acrylic acid to ammonia and carbon dioxide (7).
In a 42 day screening study using a sewage sludge inculum, 71% of acrylic acid was mineralised (8).

Abiotic removal
Vapour phase photochemical reaction with atmospheric hydroxyl radicals $t_{1/2}$ 6.5 days (9).

Absorption
Activated carbon adsorbability 0.129 g g^{-1} carbon 64.5% reduction: influent 1000 mg l^{-1} (2).

Mammalian and avian toxicity

Acute data
LD_{50} oral rat 250 mg kg^{-1} (10).
LD_{50} oral mouse 2400 mg kg^{-1} (11).
LD_{50} dermal rabbit 280 mg kg^{-1} (12).

Carcinogenicity and long-term effects
No adequate data for carcinogenicity to humans and animals, IARC classification group 3 (13).

Irritancy
Dermal rabbit 500 mg and 1 mg instilled in rabbit eye caused severe irritation (duration unspecified) (14).

Genotoxicity
Salmonella typhimurium TA98, TA100, TA1535, TA1537 with and without metabolic activation negative (15).
In vitro mouse lymphoma L5178Y tk+/tk- without metabolic activation positive (16).
Syrian hamster embryo fibroblast cells with and without metabolic activation negative (17).

Legislation
No US criteria for drinking water.
USSR limit 0.5 mg l^{-1} for drinking water (18).

Any other comments
Corrosive, exothermic polymerisation at room temperature may become explosive if confined. Fire hazard when exposed to heat or flame.
Acrylic acid is rapidly oxidised in water therefore wastewater containing the compound can deplete reservoirs of oxygen (19).
Human health effects, experimental toxicology and environmental effects reviewed (20).

References

1. Wise, H. E; et al. *Occurence and Predictability of Priority Pollutants in Wastewater of the Organic Chemicals and Plastics Synthetic Fibers Industrial Categories* 1981, Amer. Chem. Soc. Ind. Div.
2. Verschueren, K. *Handbook of Environmental Data on Organic Chemicals* 2nd ed., 1983, Van Nostrand Reinhold, New York
3. Bringmann, G; et al. *Water Res.* 1980, **14**, 231-241
4. Richardson, M. L. *Nitrification Inhibition in the Treatment of Sewage* 1985, 4, RSC, London
5. Shelton, D. R; et al. *Appl. Environ. Microbiol.* 1984, **47**, 850-857
6. *MITI Report* 1984, Ministry of International Trade and Industry, Tokyo
7. Arai, T; et al. *Zentrabl. Bakteriol. Mikobiol. Hyg.* 1981, **11**, 297-307
8. Chou, W. L; et al. *Biotech. Bioeng. Symp.* 1978, **8**, 391-414
9. Atkinson, R. *Int. J. Chem. Kinetics* 1987, **19**, 799-828
10. Deichman, W. B. *Toxicology of Drugs and Chemicals* 1969, 76, Academic Press, New York
11. *Biochem. J.* 1940, **34**, 1196
12. *Toxicol. Appl. Pharmacol.* 1974, **28**, 313
13. *IARC Monograph* 1987, **Suppl. 7**, 56
14. *Union Carbide Data Sheet* 1965, Union Carbide Corp., New York
15. Zeiger, E; et al. *Environ. Mutagen* 1987, **9**(Suppl. 9), 1-109
16. Moore, M. M; et al. *Environ. Mol. Mutagen* 1988, **11**(1), 49-63
17. Wiegand, H; et al. *Arch. Toxicol.* 1989, **63**(3), 250-251
18. Sittig, M. *Handbook of Toxic and Hazardous Chemicals and Caracinogens* 2nd ed., 1985, 43, Noyes Publications, NJ
19. Ekhina, R. S; et al. *Combined Effect of Six Acrylates on the Sanitary Status of a Reservoir Vodemov* 1977, 157-164
20. *ECETOC Technical Report No. 30(4)* 1991, The Chemical Industry Ecology and Toxicology Centre, B-1160 Brussels

A48   Acrylic chloride

$H_2C=CHCOCl$

CAS Registry No. 814-68-6
Synonyms acrylic acid chloride; acrylyl chloride; 2-propenoyl chloride
Mol. Formula C_3H_3ClO **Mol. Wt.** 90.51
Uses Chemical intermediate.

Physical properties

B. Pt. 72-76°C; **Flash point** 16°C; **Specific gravity** 1.114.

Solubility
Organic solvent: chloroform

Occupational exposure

UK Long-term limit 10 ppm; **UK Short-term limit** 20 ppm.

Mammalian and avian toxicity

Acute data
LC_{Lo} (4 hr) inhalation rat 25 ppm (1).
LC_{50} (2 hr) inhalation mouse 92 mg m^{-3} (2).
LD_{50} intravenous mouse 180 mg kg^{-1} (3).

Sub-acute data
Inhalation rat (3 wk) <25 ppm in air caused respiratory damage (1).

Any other comments
Decomposes on contact with water.

References
1. Gage, J. C. *Br. J. Ind. Med.* 1970, **27**, 1-18
2. Izmerov, N. F; et al. *Toxicometric Parameters of Industrial Toxic Chemicals Under Single Exposure* Centre of International Projects, GKNT, Moscow, 1982, 17
3. US Army Armament Research and Development Command, Chemical Systems Laboratory, Report NX 03367

A49   Acrylonitrile

CH2=CHCN

CAS Registry No. 107-13-1
Synonyms 2-propenenitrile; vinyl cyanide; cyanoethylene
Mol. Formula C_3H_3N **Mol. Wt.** 53.06
Uses In the production of acrylic and modacrylic fibres, nitrile rubbers, and various plastics. Also used in fumigants.
Occurrence In auto exhaust and cigarette smoke emissions and in wastewater discharges associated with its production.

Physical properties

M. Pt. $-83.5°C$; **B. Pt.** 77-79°C; **Flash point** $-5°C$; **Specific gravity** d_4^{20} 0.806; **Partition coefficient** log P_{ow} -0.92.

Solubility
Water: 73 g l^{-1} at 20°C. Organic solvent: acetone, ethanol, benzene, diethyl ether

Occupational exposure

US TLV (TWA) 2 ppm (4.3 mg m^{-3}); **UK Long-term limit** MEL 2 ppm (4 mg m^{-3}); **UN No.** 1093; **HAZCHEM Code** 3WE; **Conveyance classification** flammable liquid; toxic substance; **Supply classification** highly flammable and toxic.

Risk phrases
May cause cancer – Highly flammable – Toxic by inhalation, in contact with skin and if swallowed – Irritating to skin (R45, R11, R23/24/25, R38)

Safety phrases

Avoid exposure-obtain special instructions before use – Keep away from sources of ignition – No Smoking – Take off immediately all contaminated clothing – If you feel unwell, seek medical advice (show label where possible) (S53, S16, S27, S44)

Ecotoxicity

Fish toxicity

LC_{50} (96 hr) bluegill sunfish, fathead minnow, pinfish, guppy 10-33 mg l^{-1} flowthrough and static bioassays (1,2).

Invertebrate toxicity

LC_{50} (24 hr) *Asellus aqualticus* 50 mg l^{-1}.
LC_{50} (50 hr) *Crammorus gassorium* 32 mg l^{-1} (3).
LC_{50} (24(hr) brown shrimp 10-33
mg l^{-1} (4).
EC_{50} (24, 48 hr) *Daphnia magna* 13-7.6 mg l^{-1} (5).
EC_{50} (30 min) *Photobacterium phosphoreum* 254 mg l^{-1} Microtox test (6).

Bioaccumulation

The bioconcentration factor for bluegill sunfish exposed to unspecified concentrations of acrylonitrile for 28 days or until equilibrium was obtained in a flowing water system was 48 (7).
Bioaccumulation of acrylonitrile in aquatic organisms is unlikely, but cyanoethylation of proteins in the aquatic environment is probable (8).

Environmental fate

Nitrification inhibition

NH_3 oxidation, not inhibitory to *Nitrosomonas sp.* (9).
Saprophytic microorganisms can utilise 150 mg l^{-1} acrylonitrile. Concentrations of >50 ppm may inhibit bacterial nitrification, adversely affecting activated sludge processes (10).
Inhibition cell multiplication *Pseudomonas putida* commences at 53 mg l^{-1} (11).

Degradation studies

Concentrations of <20 mg l^{-1} acrylonitrile was readily biodegraded during anaerobic digestion processes operating in municipal sewage treatment facilities.
Pseudomonas putida in sludge degraded up to 35% acrylonitrile at concentrations of 500 mg l^{-1}.
BOD_5 70% reduction in dissolved oxygen (8).
Numerous studies indicate that acrylonitrile was degraded in aerobic systems after acclimation. Using activated sludge inocula >95% degradation and 100% degradation occured within 7 days in screening studies with sewage seed (12,13).

Abiotic removal

Acrylonitrile does not adsorb strongly to soil and is very volatile (14).
Acrylonitrile does not absorb light >290 nm, therefore does not undergo direct photolysis (15).
Can react photochemically with hydroxyl radicals in the atmosphere, $t_{1/2}$ 3.5 day (16).
Hydrolysis is negligible, 10 ppm acrylonitrile was stable in aqueous solution pH 4-10 for 23 day (17).

Mammalian and avian toxicity

Acute data
LD_{50} oral rat, rabbit 93 mg kg^{-1} (18).
LD_{50} (4 hr) inhalation guinea pig 0.99 mg kg^{-1} (19).
LD_{50} intraperitoneal mouse 44 mg kg^{-1} (20).
LD_{50} intravenous rabbit 69 mg kg^{-1} (21).

Carcinogenicity and long-term effects
Limited evidence for carcinogenicity to humans, sufficient evidence for carcinogenicity to animals, IARC classification 2A (22).
Inhalation rat ($\leq$104 wk) 5-60 ppm increased the incidence of encephalic glioma in both sexes, mammary gland neoplasm in ♀ and Zymbal gland neoplasm in ♂ (23).
♂ Sprague-Dawley rats (2 yr) 0, 20, 500 ppm in drinking water. High dose animals exhibited early mortality, retarded weight gain and tumours of Zymbal gland. No increases in tumours of organ systems detected but a trend toward the development of forestomach papillomas was noted (24).

Teratogenicity and reproductive effects
Administration of 63 mg kg^{-1} (repeat doses) on days 8-15 gestation were toxic to ♀ Sprague-Dawley rats, produced malformations and embryotoxicity (25,26).
In rats treated with 5 mg kg^{-1} from day 5-21 gestation (route unspecified), pups showed no adverse effects on functional or postnatal morphological development. However, altered brain levels of 5-hydroxytryptamine, norepinephrine and monoamine oxidase were detected (27).

Metabolism and pharmacokinetics
Metabolised to cyanide and then to thiocyanate which in turn is eliminated in urine (28).
Inhalation rat (8 hr) 1, 5, 10, 50 or 100 ppm, urine samples were collected up to 32 hr after initial exposure, metabolites included hydroxyethyl mercapuric acid, S-carboxymethyl cysteine, cyanoethyl mercapturic acid and at higher concentrations unmetabolised acrylonitrile (29).
Oral and intravenous (unspecified concentrations) of distribution of ^{14}C- acrylonitrile in rat and monkey was studied by wholebody autoradiography. Radioactivity was observed in blood, liver, kidney, lung, adrenal cortex and stomach mucosa (30).

Genotoxicity
Salmonella typhimurium TA1535, pSK1002 with and without metabolic activation negative (31).
In vitro human KB cells (72 hr incubation) 5.7 μg ml^{-1} showed cytoplasmic projections and altered morphology on cell surface. After 72 hr, 30 μg ml^{-1} completely inhibited cell growth (32).
In vitro human bronchial epithelial cells induced of sister chromatid exchange and DNA single-strand breaks at concentrations of 150-500 μg ml^{-1} (33).

Any other adverse effects to man
Effects in humans included significant excess of prostate cancers observed in workers exposed to acrylonitrile (34).
Fatalities have been reported following the use of fumigant mixtures containing

acrylonitrile, carbon tetrachloride and methylene chloride. Exact exposure conditions not known (4).

There were no statistically significant excesses in overall cancer incidence or mortality among ♂ employees occupationally exposed to acrylonitrile at a textile fibre plant from 1944-1970 and followed through 1981 for mortality and through 1983 for cancer incidence. However, a statistically significant excess of prostate cancer was found (5 cases versus 1.9 expected). Mortality from all causes was significantly lower than expected based on US rates (35).

Any other adverse effects

Acrylonitrile can be readily absorbed by mouth, through intact skin or by inhalation (species unspecified) (36).

In humans, symptoms of over exposure include headache, sleeplessness, nausea, vomiting, diarrhoea, fatigue, mild jaundice, irritation and inflammation of the eyes and respiratory tract, including nose and throat. In severe cases unconsciousness and convulsions can occur (4).

Oral mice (60 day) 10 mg kg^{-1} day^{-1} caused a decrease in testicular sorbitol dehydrogenase and acid phosphatase and increases in lactate dehydrogenase and β-glucuronidase activities. Decrease in sperm count and degeneration of seminferous tubules detected. The authors conclude that acrylonitrile may affect the ♂ reproductive function by causing testicular injury (37).

Acrylonitrile interacts with rat blood by binding to cytoplasmic and membrane proteins and may cause damage in red cells by mechanisms other than the release of cyanide (38).

Any other comments

Physico-chemical properties including acute and chronic toxicity, metabolic fate, teratogenicity, genotoxicity and carcinogenicity extensively reviewed (39- 47).

References

1. Henderson, C; et al *Eng. Bull. Ext. Ser. Purdue Univ.* 1961, **106**, 130
2. Buccafusco, R. J; et al. *Bull. Environ. Contam. Toxicol.* 1980, **24**, 684-691
3. Tonagau, Y. I. *Ecological Chem (Seitai Kayaku)* 1984, **7**(1), 17
4. *Environ. Health Criteria No. 28. Acrylonitrile* 1983, **7**(1), World Health Organisation, Geneva
5. Le Blanc, G. A. *Bull. Environ. Contam. Toxicol.* 1990, **24**, 684-691
6. Kaiser, K. L. E; et al. *Water Poll. Res. J. Canada* 1991, **26**(3), 361-431
7. Barrows, M. E; et al. *Am. Chem. Soc. Div. Environ. Chem.* 1978, **18**, 345-346
8. Callahan, M. A; et al. *Environmental Fate of 129 Priority Pollutants* 1979, **1**, EPA 440/4 79-029a
9. Hockenberry, M. R; et al. *J. Water Pollut. Control Fed.* 1977, **49**(5), 768-777
10. Chekhovskaya, N; et al. *Ambient Water Quality Criteria: Acrylonitrile* 1980, C-10-15, USEPA, Washington, DC
11. *Toxic and Hazardous Industrial Chemicals Safety Manual* 1972, International Technical Information Institute, Tokyo
12. Ludzack, F. J; et al. *J. Water Pollut. Control Fed.* 1961, **33**, 492-505
13. Cherry, A. B; et al. *Sewage Ind. Wastes* 1956, **28**, 1137-1146
14. Norris, M. V. *Encycl. Ind. Chem. Anal.* 1967, **4**, 368-381
15. Grasselli, J; et al. *Atlas of Spectral Data and Physical Constrants for Organic Compounds* 2nd ed., 1975, **6**
16. Edney, E; et al. *Atmospheric Chemistry of Several Toxic Compounds* 1983, EPA 600 53-82-092

17. Going, J; et al. *Environmental Monitoring Near Industrial Sites: Acrylonitrile* 1979, EPA 560/6-79-003
18. Shackelford, W. W; et al. *USEPA Report* 1976, EPA 600/4-76-062
19. Hradec, J; et al. *Carcinogenisis (London)* 1988, **9**(5), 843-846
20. Paulet, G; et al. *Arch. Sat. Pharmacyn.* 1961, **131**, 54
21. Knoblach, K; et al. *Med. Pract.* 1972, **23**, 243
22. *IARC Monograph* 1987, **Suppl. 7**, 57
23. Maltoni, C; et al. *Ann. New York Acad. Sci.* 1988, **534**, 179-202
24. Gallagher, G. T; et al. *J. Am. Coll. Toxicol.* 1988, **7**95), 603-615
25. *Fed. Rep.* 1977, **42**, 44170, 44178
26. Farooqui, M. Y. U; et al. *Chem-biol. Interact.* 1982, **38**(2), 145-160
27. Mehrotra, J; et al. *Ind. Health* 1988, **26**(4), 251-255
28. Muller, G; et al. *Arch. Toxicol.* 1987, **60**(6), 464-466
29. Carson, P. A; et al. *Safe Handling of Chemicals in Industry* 1988, Longman, Harlow
30. Ashby, J; et al. *J. Occup. Med.* 1988, **30**(10)
31. Nakamura, S; et al. *Mutat. Res.* 1987, **192**(4), 239-246
32. Mochida, K; et al. *Bull. Environ. Contam. Toxicol.* 1989, **42**(3), 424-426
33. Chang, C. M; et al. *Mutat. Res.* 1990, **241**(4), 355-360
34. *IARC Monograph* 1987, **Suppl. 6**, 813-818
35. Chen, J. L; et al. *Am. J. Ind. Med.* 1987, **11**(2), 157-163
36. Clayton, G. D; et al. *Patty's Industrial Hygiene and Toxicology* 3rd ed., 1981-1982, 4865, John Wiley, New York
37. Tandon, R; et al. *Toxicol. Lett.* 1988, **42**(1), 55-63
38. Sandberg, E. C; et al. *Toxicol. Lett.* 1980, **6**(3), 187-192
39. *ECETOC Technical Report No. 30(4)* 1991 European Chemical Industry Ecology and Toxicology Centre, B-1160 Brussels
40. Johannsen, F. R; et al. *Fundam. Appl. Toxicol.* 1986, **7**, 690-697
41. *IARC Monograph* 1982, **Suppl. 4**, 25-27
42. Derland, D. E; et al. *Drug Metab. Review* 1989, **20**(2-4), 233-246
43. *Dangerous Prop. Ind. Mater. Rep.* 1989, **9**(3), 2-8
44. Bolt, H. M; et al. *Adv. Exp. Med. Biol.* 1986, **197**(3), 1013-1016
45. Guengerich, F. P; et al. *IARC Sci. Publ.* 1986, **70**, 255-260
46. Radeva, M; et al. *Khig. Zdraveopaz* 1988, **31**(3), 95-101, (Bulg.) (*Chem. Abstr.* **109**, 105913x)
47. Strother, D. E; et al. *Ann. New York Acad. Sci.* 1988, **534**, 169-178

A50 Actinomycin C

CAS Registry No. 8052-16-2

Synonyms 1H-pyrrolo[2,1-i][1,4,7,10,13]oxatetraazacyclohexadecine, cyclic peptide deriv.; 3H-phenoxazine, actinomycin C deriv.; Cactinomycin; Sanamycin; HBF 386; NSC 8722

Mol. Formula $C_{62}H_{86}N_{12}O_{16}$ **Mol. Wt.** 1255.45

Uses Antineoplastic antibiotic.

Occurrence Obtained from cultures of *Streptomyces chrysomallus, S. antibiotious* and *S. parvulus*.

Physical properties

M. Pt. 246-247°C (trihydrate).

Solubility

Organic solvent: ethanol, propylene glycol (trihydrate)

Mammalian and avian toxicity

Acute data

LD_{50} oral rat, mouse 7-13 mg kg^{-1} (1,2).

LD_{50} intraperitoneal rat 100 µg kg^{-1} (1).

Sub-acute data

Rats treated with 7% of the LD_{50} dose for 14 wk (unspecified route) caused immunosuppressive effects (3).

Carcinogenicity and long-term effects

Inadequate evidence for carcinogenicity to humans, limited evidence for carcinogenicity to animals, IARC classification group 3 (4).

Irritancy

Systemic skin irritant by intravenous route (5).

Genotoxicity

Drosophila melanogaster wing mosaic assay displayed low level of genotoxicity (6).

Any other adverse effects to man

In humans can cause bone marrow depression, oral and gastrointestinal effects and tissue damage (2,7).

In children hepatoxic effects observed after a single dose of 60 µg kg^{-1} in 5 out of 40 patients (7).

Any other adverse effects

Symptons include nausea, vomiting and alopecia (7).

References

1. Sax, N. I; et al. *Dangerous Properties of Industrial Materials* 7th ed., 1989, Van Nostrand Reinhold, New York
2. *Ann. N. Y. Acad. Sci* 1960, **89**, 285-485
3. Scherf, H. R; et al. *Recent Results Cancer Res.* 1975, **52**, 76-87
4. *IARC Monograph* 1987, **Suppl. 7**, 56
5. D'Angio, G. J. *Lancet* 1987, **2**, 104
6. Szabad, J; et al. *Mutat. Res.* 1986, **173**(3), 197-200
7. *Martindale: The Extra Pharmacopoeia* 29th ed., 1989, The Pharmaceutical Press, London

A51   Actinomycin D

MeGly   MeGly
L-Pro Me-L-Val   L-Pro Me-L-Val

D-Val   O   D-Val   O

L-Thr   L-Thr

CAS Registry No. 50-76-0

Synonyms 1*H*-pyrrolo[2,1- *i*][1,4,7,10,13]oxatetraazacyclohexadecine, cyclic peptide deriv.; 3*H*-phenoxazine, actinomycin D deriv.; 3*H*-phenoxazine-1,9-dicarboxamide, 2-amino-N,N′- bis(hexadecahydro-6,13-dissopropyl-2,5, 9-trimethyl-1,4,7,11,14-pentaoxo-1H-pyrrolo[2,1- *i*][1,4,7,10,13]oxatetraazacyclohexadecin-10-yl)-4, 6- dimethyl-2-oxo-; Actinomycin C1; Dactinomycin; Meractinomycin; Oncostatin K

Mol. Formula $C_{62}H_{86}N_{12}O_{16}$ **Mol. Wt.** 1255.45

Uses Antineoplastic.

Occurrence Antibiotic substance belonging to the actinomycin complex, produced by several *Streptomyces spp.*

Physical properties

M. Pt. 241.5-243°C (decomp.).

Solubility
Organic solvent: ethanol, propylene glycol

Mammalian and avian toxicity

Acute data
LD_{50} oral mouse, rat 7.2-13 mg kg^{-1} (1).

Carcinogenicity and long-term effects
Inadequate evidence for carcinogenicy to humans, limited evidence for carcinogenicity to animals, IARC classification group 3 (2).
Mice and rats given 3 intraperitoneal injections wk^{-1} of the maximum tolerated dose (duration unspecified) induced cancer in the peritoneal cavity (3).

Teratogenicity and reproductive effects
Intraperitoneal albino rats (8 day pregnant) 0.2 µg g^{-1} inhibited collagen formation by chondrocytes and the formation of matrix in the epiphysical phase of newborn rats (4).
Gavage (8-12 day gestation) ICR/SIM mice, Chernoff/Kavlock development toxicity screen positive teratogen, significant reduction in live-born litter size (5).

Metabolism and pharmacokinetics
A 1 mg kg^{-1} dose of actinomycin D administered intravenously to rabbits resulted in <10% of initial concentration detected in blood after 2 hr, rising slightly after 10 hr. Actinomycin D was found in kidney, heart, spleen, liver, bile and urine (6).

Genotoxicity
Saccharomyces cerevisiae induction of diploidy and aneuploidy negative (7).

Salmonella tymphimurium TA1538, TA1537, TA1535 with and without metabolic activation positive (8).

Mouse lymphoma tk+/tk- forward mutation assay positive (9).

Mouse leukaemia L1210 DNA topoisomerases I and II, Actonomycin D induced cleavage by topoisomerases I and II (10).

In vitro human lymphocyte sister chromatid exchange positive. *Vicia faba* sister chromatid exchange positive (11).

Any other adverse effects

Studies with human neutrophils indicated that actinomycin D impaired both intracellular killing by human neutrophils through an effect on respiratory activity and directed cell migration of human neutrophils (12).

Direct injection of 8 µg ml^{-1} into encephalon of 42 hr-old chick embryos and incubated for 6-8 hr induced lesions of the ocular structures within 8 hr. Damage to nerve cell layers was observed (13).

Any other comments

Historical background of actinomycin D and chemistry, carcinogenicity, toxicology, pharmacology of actinomycin reviewed (1,6).

References

1. Philips, F.S; et al. *Ann. New York Acad. Sci.* 1960, **89**, 285-485
2. *IARC Monograph* 1987, **Suppl. 7**, 80-82
3. Weisburger, J. H; et al. *Recent Results Cancer Res.* 1975, **52**, 1-17
4. Yun, Y. J; et al. *HaNew Yorkang Uidae Haksulchi* 1987, **7**(1), 95-107, (Kor.) (*Chem. Abstr.* **107**, 190458d)
5. Seidenberg, J. M; et. al. *Terat. Carcinog. Mutagen.* 1986, **6**, 361-374
6. *IARC Monograph* 1976, **10**, 35
7. Sora, S; et al. *Mutat. Res.* 1988, **210**, 375-384
8. Beljanski, M; et al. *Exp. Cell Biol.* 1982, **50**(5), 271-280
9. Wangenheim, J; et al. *Mutagenesis* 1988, **3**(3), 193-205
10. Wasserman, K; et al. *Mol. Pharmacol.* 1990, **38**(1), 38-45
11. Warijin, Xing; et al. *Mutat. Res.* 1990, **241**(2), 109-113
12. Chang, F. Y; et al. *Microbiol. Immunol.* 1990, **34**(3), 311-321
13. Gendre, L; et al. *Arch. Biol.* 1988, **99**(1), 55-65, (Fr.) (*Chem. Abstr.*, **110**, 50785v)

A52   Adamantane

CAS Registry No. 281-23-2
Synonyms tricyclo[3.3.1.1^{3,7}]decane
Mol. Formula $C_{10}H_{16}$ **Mol. Wt.** 136.24
Uses A chemical intermediate used in polymers, lubricating oils, pharmaceuticals and sublimation carriers.
Occurrence In petrochemical fractions.

Physical properties

M. Pt. 268°C (sealed tube), 205-210°C (sublimes); **Specific gravity** 1.07.

Solubility

Organic solvent: benzene, acetone

Any other comments

Review of adamantane in medicinal chemistry (1).

Reference

1. Gerzon, K; et al. *Am. Chem. Soc., Div. Petrol. Chem., Prepr.* 1970, **15**(2), B80-B84 (*Chem. Abstr.* **76**, 10106x).

A53   Adipic acid

$$HO_2C(CH_2)_4CO_2H$$

CAS Registry No. 124-04-9
Synonyms hexanedioic acid; 1,4-butanedicarboxylic acid
Mol. Formula $C_6H_{10}O_4$ **Mol. Wt.** 146.14
Uses Artificial resins, nylon, urethane foams, intermediate in oil additive lubricants, sometimes substituted for tartaric acid in baking powder (due to lower hygroscopicity). pH control in food and drugs.
Occurrence Found in beetroot juice.

Physical properties

M. Pt. 152°C; **B. Pt.** 337.5°C; **Flash point** 196°C; **Specific gravity** d_4^{25} 1.36; **Partition coefficient** log P_{ow} 0.08 (1).

Solubility

Water: 14.4 g l^{-1}. Organic solvent: ethanol, diethyl ether, chloroform

Occupational exposure

Supply classification irritant.

Risk phrases

Irritating to eyes (R36)

Ecotoxicity

Fish toxicity
LC_{50} (24-96 hr) bluegill sunfish, fathead minnow 330-97 mg l^{-1} static bioassay at 18-22°C (2).

Bioaccumulation
The low partition coefficient indicates that bioconcentration in fish will be negligible (3).

Environmental fate

Anaerobic effects

After 10 days acclimation 82% adipic acid was degraded under anaerobic conditions (4).

Degradation studies

BOD_5 0.598 mg l^{-1} oxygen consumed at 20°C, 36% ThOD (2).
In river water 50-90% degradation occurred in 3.5 and 7 days respectively, initial concentrations 700 mg l^{-1} (5).

Abiotic removal

Apidic acid exists mainly in its dissociated form in the environment, forming salts with cations (6).
Photochemical reactions occur with hydroxyl radicals in the atmosphere, $t_{1/2}$, 4.4 days (7).

Mammalian and avian toxicity

Acute data

LD_{50} oral mouse 1900 mg kg^{-1} (8).
LD_{50} intravenous, intraperitoneal mouse 275-680 mg kg^{-1} (8,9).

Sub-acute data

Fifteen repeat exposures inhalation rat (6 hr), total exposure 126 g l^{-1} no adverse effects reported (10).

Metabolism and pharmacokinetics

Only partially metabolised by humans, remainder eliminated unchanged in urine (11).

Irritancy

20 mg instilled in rabbit eye for 24 hr caused severe irritation (12).

References

1. Hansch, C; et al. *Medchem. Project Issue No. 19* 1981, Pomona College, Claremont, CA
2. Verschueren, K. *Handbook of Environmental Data on Organic Chemicals* 2nd ed., 1983, 165, Van Nostrand Rheinhold, New York
3. Lyman, W. J; et al. *Handbook of Chem. Property Estimation Methods: Environ. Behav. of Organic Compounds* 1982, **5**, 1-30, McGraw-Hill, New York
4. Chou, W. C; et al. *Bioeng. Symp.* 1979, **8**, 391-414
5. Zahn, R; et al. *Wasser Abwasser Forsch* 1980, **13**, 1-7
6. Serjeant, E. P; et al. *Ionisation Constants of Organic Acids in Aqueous Solution* 1979, 989, IUPAC Chemical Data Series No. 23, Pergamon Press, New York
7. *Gems: Graphcal Exposure Modeling System: Fate of Atmospheric Pollutants*, 1986, GEMS.FAP
8. *J. Agric. Food Chem.* 1957, **5**, 759
9. *Toxicol. Appl. Pharmacol.* 1975, **32**, 566
10. Cage, J. C. *Br. J. Ind. Med.* 1970, **27**
11. Furia, T. E. *Handbook of Food Additives* 1968, Chem. Rubber Co., Cleveland, OH
12. Marhold, J. V. *Sbornik Vysledku Toxixologickeho Vystreni Latek A Pripravku* 1972, 51, Prague

A54 Adiponitrile

NC(CH2)4CN

CAS Registry No. 111-69-3
Synonyms 1,4-dicyanobutane; hexanedinitrile; adipic acid dinitrile; hexanedioic acid dinitrile; tetramethylene cyanide
Mol. Formula $C_6H_8N_2$ **Mol. Wt.** 108.14
Uses Organic syntheses. Intermediate in the manufacture of nylon.

Physical properties

M. Pt. 1-3°C; **B. Pt.** 295°C; **Flash point** >110°C;
Specific gravity 0.951 (temperature unspecified); **Volatility** v. den. 3.73.

Solubility
Organic solvent: ethanol, acetone, chloroform

Occupational exposure

UN No. 2205; **HAZCHEM Code** 3X; **Conveyance classification** toxic substance.

Ecotoxicity

Bioaccumulation
Calculated bioconcentration factor <1, indicated environmental accumulation is unlikely (1).

Environmental fate

Degradation studies
Confirmed biodegradable (2).
Unacclimated river water treated with initial concentration 0.5-10 mg l^{-1} stored sewage seed, negligible degradation in 2 days, 40% theoretical BOD in 5 days and 100% theoretical BOD in 12 days (3).
Activated sludge exposed to initial substrate concentration 500 mg l^{-1} theoretical BOD 2.2-2.8% in 72 hr (4).
Bench scale activated sludge unit, influent concentration equivalent to 275-350 mg l^{-1}, mean aeration detention time 7-13 hr, 93-98% BOD removal achieved (5).
Aeromonas sp. BN 7013 isolated from soil used adiponitrile as its sole carbon source (6).
Adiponitrile vapour reacts photochemically with hydroxyl radicals $t_{1/2}$ 11 days (7).
Brevibacterium sp. can utilise adiponitrile as a carbon and nitrogen source, suggesting these strains could be used to degrade nitriles in waste water (8).

Abiotic removal
Calculated soil adsorption coefficients 9-16 were estimated for adiponitrile using linear regression equation based on a measured log k_{ow} of –0.32. These results suggest adiponitrile will be highly mobile in soil with no significant adsorptions to suspended solids and sediments in water (9).

Mammalian and avian toxicity

Acute data
LD_{50} oral rat, mouse, rabbit 172-300 mg kg^{-1} (1,10).
LC_{50} (4 hr) inhalation rat 1710 mg m^3 (11).
LD_{50} subcutaneous guinea pig 50 mg kg^{-1} (12).
LD_{50} intraperitoneal mouse 40 mg kg^{-1} (13).

Sub-acute data
Inhalation rat (4 or 13 wk) 493 or 99 mg m^{-3}, respectively, 6 hr day^{-1}, 5 day wk^{-1} increased mortality, reduced weight gain and slight anaemia for high dose animals. No histopathological evidence of organ toxicity or abnormal reproductive effects in animals exposed to 99 mg m^{-3} (14).

Teratogenicity and reproductive effects
Gavage Sprague-Dawley rats (gestation day 6- 19) 0, 20, 40 and 80 mg kg^{-1} induced maternal toxicity in high doses, maternal effects at middle dosages, slight foetotoxicity at the highest dose and no teratogenic effects (15).

Metabolism and pharmacokinetics
Toxicity due in part to its action as a cyanide when absorbed or ingested (1).
Of a single 50 mg kg^{-1} dose administered to guinea pigs 79% was excreted as thiocyanate in the urine (16).

Genotoxicity
Salmonella typhimurium TA97, TA98, TA100, TA1535, TA1537 with and without metabolic activation negative (13).

Any other adverse effects
Causes disturbances of respiration and circulation, irritation of stomach and intestines, and weight loss (17).

Any other comments
Toxicological properties including acute and subcronic toxicity, teratogenicity and biochemical mechanism studies discussed (18,19).

References
1. Tanii, H; et al. *Arch. Toxicol.* 1985, **57**(2), 88-93
2. *MITI Report* 1984, Ministry of International Trade and Industry, Tokyo
3. Ludzack, L. J; et al. *Sewage and Ind. Wastes* 1959, **31**, 33-44
4. Lutin, P. A. *J. Water Poll. Contr. Fed.* 1970, **42**, 1632-1642
5. Ludzack, F. J; et al. *Proc. of the 14th Industrial Waste Conf., Eng. Ext. Ser. 7* 1959, **104**, 547-565
6. Kuwahara, M; et al. *Hakkokogaku Kaishi* 1980, **58**, 441-448
7. Atkinson, R. *Intern. J. Chem. Kinet.* 1987, **19**, 799-828
8. Thiery, A; et al. *Zentralbl. Mikrobiol.* 1986, **141**(8), 575-582
9. Swann, R. L; et al. *Res. Rev.* 1983, **85**, 17-28
10. Hahn, W; et al. *Water Quality Characteristics of Hazardous Materials* 1974, 3, Texas A&M University
11. Smith, W. L; et al. *Toxicol. Appl. Pharmacol.* 1982, **65**(2), 257-263
12. *Med. Lav.* 1955, **46**, 221, Industrial Medicine, Milan
13. Patty, F. A. *Industrial Hygiene and Toxicology* 2nd ed., 1963, Interscience Publishers, New York

14. Short, R. D; et al. *J. Toxicol. Environ. Health* 1990, **30**(3), 199-207
15. Johannsen, F. R; et al. *Fundam. Appl. Toxicol.* 1986, **7**(1), 33-40
16. Zeiger, E; et al. *Environ. Mol. Mutagen.* 1988, **11**(Suppl. 12), 1-158
17. *NTIS Report* AD 691-490, Nat. Tech. Inf. Ser., Springfield, VA
18. Johannsen, F. R; et al. *Fundam. Appl. Toxicol.* 1986, **7**, 690-697
19. *Dangerous Prop. Ind. Mater.* 1987, **7**(6), 35-40

A55   Adrenaline-D

CAS Registry No. 150-05-0
Synonyms 1-(+)-adrenaline; d-epinephrine
Mol. Formula $C_9H_{13}NO_3$ **Mol. Wt.** 183.21
Uses Cardiac stimulant, vasoconstrictor and bronchodilator.
Occurrence The principal sympathomimetic hormone produced by the adrenal medulla in most species. Occurs as the *l*- form in animals and humans.

Physical properties

M. Pt. 211-212°C.

Solubility
Organic solvent: ethanol, acetic acid

Ecotoxicity

Effects to non-target species
LD_{Lo} subcutaneous frog 5000 mg kg^{-1} (1).

Mammalian and avian toxicity

Acute data
LD_{Lo} subcutaneous rat, mouse, rabbit 4- 80 mg kg^{-1} (1-3).
LD_{50} intravenous rat, rabbit 50-800 μg kg^{-1} (1,3).
LD_{50} intravenous mouse 50 mg kg^{-1} (2).

Metabolism and pharmacokinetics
Adrenaline is methylated to metanephrine by catechol-*O*-methyltransferase (COMT) followed by oxidative deamination by the mixed function oxidase system to 4-amino-3-methoxymandelic acid, or first oxidatively deaminated by the mixed function oxidase system to 3, 4- dihydroxymandelic acid which is then is methylated by COMT, and finally converted to 4-hydroxy-3-methoxymandelic acid. The metabolities are excreted in the urine mainly as their glucuronide and ether sulfate conjugates (4).

Any other comments

It is important to note that endogenous adrenaline is the laevo isomer (4).
Adrenaline crosses the placenta to enter foetal circulation (4).

References

1. Abdernalden's *Handbuck der Biologischen Arbeitsmethoden*, 1935, **4**, 1294
2. Sax, N. I; et al. *Dangerous Properties of Industrial Materials* 7th ed., 1989, Van Nostrand Reinhold, New York
3. Hoppe, J. O. *J. Pharmacol. Exp. Therap.* 1949, **95**, 502-505
4. *Martindale: The Extra Pharmacopoeia* 29th ed., 1989, The Pharmaceutical Press, London

A56   Adrenaline-L

CAS Registry No. 51-43-4

Synonyms Adrenaline-L; Epinephrine; 1-(3,4-dihydroxyphenyl)-2-(methylamino)ethanol; 3, 4-dihyroxy-1-[1-hydroxy-2-(methylamino)-ethyl]benzene; methylaminoethanolcatechol; Epifrin; Glaucon; Simplene

Mol. Formula $C_9H_{13}NO_3$ **Mol. Wt.** 183.21

Uses Cardiac stimulant, vasoconstrictor and bronchodilator.

Occurrence The principal sympathomimetic hormone produced by the adrenal medulla in most species. Occurs as the *l*-form in animals and humans.

Physical properties

M. Pt. 211-212°C.

Solubility
Organic solvent: acetic acid

Mammalian and avian toxicity

Teratogenicity and reproductive effects
Oral administration of 500 µg kg^{-1} on day 7-10 of pregnancy caused preimplantation wastage in hamsters (1).

Metabolism and pharmacokinetics
In the rat major metabolites included 3,4- dihydroxyphenylacetic acid, homovanillic acid, 3-methoxy, 4- hydroxyphenylethylene glycol and 5-hydroxyindoleacetic acid (2).

♀ Wistar rat intraperitoneal (unspecified dose) reduced cytochrome P450 and inhibited hexobarbital biotransformation. Glycogen and cytochrome P450 loss and functional impairment of the mixed function monooxygenase system were dose related (3).

Genotoxicity
Salmonella typhimurium TA98, TA100, TA102, TA104 without metabolic activation positive.
In vitro mouse lymphoma L5178V tk+/tk- without metabolic activation positive (4).

Any other adverse effects to man

Adrenaline was infused for 8.5 hr into normal, healthy adult males on 4 separate occasions at concentrations of 0, 0.5, 1 and 2 μg min^{-1} to elevate circulating adrenaline into the high physiological range seen in stress and trauma. Adrenaline produced long-term elevation of the metabolic rate with minimal effect on protein metabolism beyond acute changes affecting amino acid levels (5).

In trabecular meshwork explants from human eyes, adrenaline at 1.8 mg l^{-1} caused abnormal cytokinesis and cell retraction, inhibited mitosis and phagocytosis and induced a 4 to 5 fold increase in cAMP. After 7-10 days degenerative changes observed (6).

Any other adverse effects

Neck implant rat 15-30 mg adrenaline slow release control caused damage to kidneys and impaired their function (7).

Any other comments

Effects of adrenaline on fish uterus function are discussed (8).
Effects on central nervous system and neurological disorders reviewed (9,10).
Physical properties, biosynthetic pathways and human health effects of adrenaline reviewed (11-13).

References

1. Hirsch, K. S; Fritz, H. I. *Teratology* 1981, **23**(3), 287-291
2. Herregodts, P; et al. *Biog. Amines* 1990, **7**(1), 71-80
3. Kulcsar, A; et al. *Exp. Pathol.* 1990, **38**(3), 171-175
4. McGregor, D. B; et al. *Environ. Mol. Mutagen.* 1988, **11**, 523-544
5. Nakahari, T; et al. *Am. J. Physiol.* 1990, **258**(6*I*), G878-G886
6. Tripathi, B. J; et al. *Lens Eye Toxic Res.* 1989, **6**(1-2), 141-156
7. Sadjak, A; et al. *Prog. Clin. Biol. Res.* 1987, **242**, 83-88
8. Shimada, K; et al. *Nippon Kakin Gakkaishi* 1990, **27**(4), 257-265, (Jap.) (*Chem. Abstr.* **114** 39457b)
9. Stokes, P. E; et al. *Neurol. Neurobiol.* 1988, **42C**, 237-242
10. Chalmers, J. P; et al. *Neurol. Neurobiol.* 1988, **42B**, 291-295
11. Brown, M. J. *Pharmacol. Toxicol. (Copenhagen)* 1987, **63**(Suppl. 1), 16-20
12. Gava, R; et al. *G. Clin. Med. (Bologna)* 1988, **69**(4), 309-311, (Ital.) (*Chem. Abstr.* **110**, 899q)
13. Mefford, I. N. *Prog. Neuro-Psychopharmacol. Biol. Psychiatry* 1988, **12**(4), 365-368

A57   Adriamycin

CAS Registry No. 23214-92-8
Synonyms Doxorubicin; (8S-cis)-10-[(3-amino-2,3,6-

trideoxy-α-L-lyxo-hexopyranosyl)oxy]-7,8,9,10-tetrahydro-6,8,11-trihydroxy-
8-(hydroxyacetyl)-1-methyoxy- 5,12-naphthacenedione; 14-hydroxydaunomycin;
Adriablastina

Mol. Formula $C_{27}H_{29}NO_{11}$ **Mol. Wt.** 543.53

Uses Antineoplastic agent, forms stable complex with DNA.

Occurrence Isolated from *Streptomyces peucetius* var. *caesius*.

Physical properties

M. Pt. 204-205 °C.

Solubility

Organic solvent: methanol

Mammalian and avian toxicity

Acute data

LD_{Lo} intravenous human 15 mg kg^{-1} (1).

A single dose 9 mg kg^{-1} (route unspecified) caused 67% mortality in rats, whereas 3
repeat doses of 3 mg kg^{-1} administered every third day caused 7% mortality (2).

Sub-acute data

LD_{Lo} (31 wk) intravenous human 380 mg kg^{-1} (1).

Intravenous dog 0.5 mg kg^{-1} day^{-1} was lethal after 5-10 doses, while 0.125-0.25 mg
kg^{-1} day^{-1} was toxic to rats and dogs but not lethal. Inhibition of haemopoiesis was
observed which persisted 2-3 days after discontinuation of treatment (3).

Carcinogenicity and long-term effects

Inadequate evidence for carcinogenicity to humans, sufficient evidence for
carcinogenicity to animals, IARC classification group 2A (4).

Single intravenous injection of 5 or 10 mg adriamycin to rats produced mammary
tumours and single or repeat subcutaneous injection produced local sarcomas and
mammary tumours (5).

Intravesicular instillation of adriamycin in rats resulted in a low incidence of bladder
papillomas (4).

Teratogenicity and reproductive effects

Embryonic chicks (4-5 day) 2.5-10 μg g egg $^{-1}$ (route unspecified). High doses
increased the mortality rate and frequencies of ventricular septal defect,
dextroposition of the aorta and aortic arch anomalies (6).

Metabolism and pharmacokinetics

In humans, following intravenous injection, it is rapidly cleared from blood, and
distributed into lungs, liver, heart, spleen and kidneys. It undergoes rapid metabolism
in the liver to the metabolite doxorubicinol (7).

In rats, adriamycin was excreted in urine at 5.7% of initial intravenous injection after
5 days. After initial decrease, plasma levels remained constant for at least 7 days (8).

Intravenous mice 5 mg kg^{-1}, adriamycin was readily bound to tissues and after 30 min
tissue concentrations were 10 times greater than in blood, 50% of dose was excreted
within 32 hr. Intravenous rabbit 5 mg kg^{-1} 17% of dose excreted intact in bile and 2%
in urine in 8 hr (9).

Sprague-Dawley intraperitoneal ♀ rats 2-8 mg kg^{-1} decreased levels of α_c actin
mRNA, β actin mRNA, glyceraldehyde-3-phosphate dehydrogenase mRNA in the

heart and α_{sk} actin mRNA, glyceraldehyde-3- phosphate dehydrogenase mRNA in gastrocnemius muscle correlating with induced muscle disease (10).

A pharmacokinetic study showed anthracycline drugs are rapidly transferred to the tissues and then slowly released. Adriamycin showed highest levels in liver and kidney (11).

Genotoxicity

Salmonella typhimurium TA1535, pSK1002 without metabolic activation positive (12).
Urine samples from cancer patients tested for mutagenicity over a 14-day period in *Salmonella typhimurium* TA98, TA100, UTH 8413, UTH 8414 without metabolic activation positive (13).

In vitro mouse lymphoma L5178 tk+/tk- without metabolic activation induced chromosome aberrations, were mutagenic and clastogenic to a dose of 5 ng ml^{-1} (14).

In vitro HeLa S3 cells 0.01 µg ml^{-1} retarded cell cycle and DNA formation, 0.1 µg ml^{-1} cells accumulated in G2 phase, 1.0 µg ml^{-1} DNA formation slow and cells accumulated in S phase (15).

Human lymphocytes ≤100 ng ml^{-1} induced chromosome aberrations and micronuclei (16).

In mice treated with 2-6 mg kg^{-1} 0.6% spermatocytes had chromosomal translocations. No increase in dominant lethality was observed with doses up to 6 mg kg^{-1} (17).

Drosophila melanogaster Cross N and Cross S, positive induction of mosaic light spots and twin spots but the increase of twin spots were only significant after treatment of 46 hr larvae (18).

Drosophila melanogaster white-ivory reversion test positive (19).

Any other adverse effects to man

In a study of 9170 patients, 2 or more yr survivors of childhood cancer. The influence of therapy on subsequent leukaemic risk was detected by a case control study conducted on 25 cases and 90 matched controls. Adriamycin was identified as a possible risk factor although treatment with alkylating agents was identified as primay cause (20).

In vitro chemosensitivity was evaluated in 28 patients with head and neck squamous cell carcinomas (including 12 pharyngeal cancers, 7 oral cavity cancers, 4 laryngeal cancers, 4 maxillary sinus cancers, 1 eosophageal cancer and 19 thyroid cancers. Tumour fragments obtained at biopsy or surgery were exposed to anticancer drugs and assayed for succinate dehydrogenase activity. The average of succinate dehydrogenase activity in squamous cell carcinomas was 41% and for thyroid cancer 38.3% (21).

Any other adverse effects

♂ Wistar rats (1-4 wk) 1 mg kg^{-1} three times per wk (route unspecified) produced persistant dose-related reductions in testis, epididymis and seminal vesicle weight but did not alter ventral prostrate weight. Serum leutenising hormone was increased after treatment, while binding of iodanoted hCG to testicular leutenising hormone was reduced. Adriamycin produces significant and persistent damage to the endocrine and spermatogenic compartments of the testis (22).

Caused pronounced bone-marrow depression with leucopenia 10-15 days after administration (23).

Any other comments

Pharmacokinetics of adriamycin reviewed (24).

Incompatible with heparin sodium and possibly with aluminium salts, aminophylline, cephalothin sodium, methasone, fluorouracil and hydrocortisone (25).

Effects of cardiomyopathy, tumour radiotherapy and adriamycin reviewed (26,27).

In vitro and *in vivo* experimental results on the effects of adriamycin on interleukin 2 formation is reviewed (28).

Antimitotic efficiency and resistance in chronic lymphocytic leukaemia reviewed (29,30).

References

1. *Cancer* 1974, **34**, 518
2. Monti, E; et al. *Med. Sci. Res.* 1988, **16**(6), 283-284
3. Shepelevtseva, N. G; et al. *Antibiot. Med. Biotekhol.* 1986, **31**(10), 768-774, (Russ.), (*Chem. Abstr.* **106**, 346z)
4. *IARC Monograph* 1987, **Suppl 7**, 82
5. Bucciarelli, E; *J. Natl. Cancer Inst.* 1981, **66**(1), 81-84
6. Takagi, Y; et al. *Teratology* 1989, **39**(6), 573-580
7. *Martindale: The Extra Pharmacopoeia* 29th Ed., 1989, The Pharmaceutical Press, London
8. *Foreign Compound Metabolism in Mammals* 1975, **3**, 192, The Chemical Society, London
9. *IARC Monograph* 1976, **10**, 46
10. Papoain, T; et al. *Am. J. Pathol.* 1990, **136**(6), 1201-1207
11. Fujita, H; et al. *Jpn. J. Antibiot.* 1986, **39**(5), 1321-1336
12. Nakamura, S; et al. *Mutat. Res.* 1987, **192**(4), 239-246
13. Monteith, D. K; et al. *Environ. Mol. Mutagenesis* 1987, **10**(4), 341-356
14. Moore, M. M; et al. *Mutat. Res.* 1987, **191**(3-4), 183-188
15. Tenjin, T; *J. Jpn. Soc. Cancer Therapy* 1988, **23**(8), 1623-1632, (Jap.) (*Chem. Abstr.* **110**, 33451r)
16. Capomazza, C; et al. *Med. Sci. Res.* 1990, **18**(9), 351-352
17. Meistrich, M. L; et al. *Cancer Res.* 1990, **50**(2), 370-374
18. Clements, J; et al. *Mutat. Res.* 1988, **209**(1-2), 1-5
19. MacKay, D. *Trends Pharmacol. Sci.* 1990, **11**(1), 17-22
20. Tucker, M. A; et al. *J. Natl. Cancer. Inst.* 1987, **78**(3), 459-464
21. Nakashima, T; et al. *Ganto Kagaku Ryohu* 1990, **17**(1), 79-84, (Jap.) (*Chem. Abstr.* **112**, 229378h)
22. Ward, J. A; et al. *Reprod. Toxicol.* 1988, **2**(2), 117-126
23. Eskborg, S; et al. *Eur. J. Clin. Pharmacol.* 1985, **28**, 205
24. Lampidis, T. J; et al. *Colloq. INSERM Anticancer Drugs* 1989, **191**, 29-38
25. D'Arcy, P. F. *Drug Intell. and Clin. Pharmacol.* 1983, **17**, 532
26. Gindrey, B; et al. *Pathol. Biol.* 1987, **35**(1), 54-60, (Fr.), (*Chem. Abstr.* **106**, 152154y)
27. Singal, P. K; et al. *J. Mol. Cell. Cardiol.* 1987, **19**(8), 817-828
28. Turk, J. L; et al. *Agents Actions* 1989, **26**(1-2), 156-157
29. Selber, R; et al. *Adv. Enzyme Reg.* 1989, **29**, 267-276
30. Goormaghtigh, E; et al. *Biophys. Chem.* 1990, **35**(2-3), 247-257

A58 Aflatoxicol

CAS Registry No. 29611-03-8

Synonyms aflatoxin R_0; [(1S- (1α,6aβ,9a,β)]-2,3,6a, 9a-tetrahydro-1-hydroxy-4-methoxycyclopenta[c]furo[3′,2′:4,5]furo[2,3-h][1]benzopyran- 11(1H)-one

Mol. Formula $C_{17}H_{14}O_6$ **Mol. Wt.** 314.30

Occurrence Metabolic product of Aflatoxin B_1.

Physical properties

M. Pt. 224-226°C.

Solubility
Organic solvent: methanol, ethanol

Mammalian and avian toxicity

Sub-acute data
TD_{Lo} (1 yr) oral rat 1092 mg kg^{-1} (1).

Carcinogenicity and long-term effects
Sufficient evidence for carcinogenicity to humans and animals, IARC classification group 1 (2).
Weanling rat (1 yr) 50 and 200 ppb in feed induced 20% and 70% hepatocellular carcinoma (1). Rainbow trout (1 yr) in feed induced hepatocellular carcinoma (3-5).

Metabolism and pharmacokinetics
Oral pig 0.1 mg kg^{-1} distributed to kidney, liver and muscle (6). Liver postmitochondrial and microsomal fractions from humans and 8 other species were compared. Major metabolic pathway was the dehydrogenation of aflatoxicol yielding aflatoxin B1. The aflatoxicol dehydrogenase activity was associated with the microsomal fraction and required a hydrogen ion acceptor but was inhibited by carbon monoxide, indicating that it was not dependent on the haem containing microsomal drug metabolising system. Postmitochondrial liver fractions oxidised aflatoxicol to at least 5 other metabolites including, aflatoxin Q1, P1, H1, M1 and B2, none of which were formed in the presence of carbon monoxide (7).

Genotoxicity

Salmonella typhimurium TA98, TA100 without metabolic activation positive. Direct mutagenesis occurred with stereo-specificity shown to be important with unnatural aflatoxicol having greater mutagenicity potential than its natural epimer (8).
Escherichia coli sfiA::lacZ SOS spot test some inhibition of bacterial growth (9).
Escherichia coli K12 PQ37 SOS chromotest with metabolic activation positive (10).
Intraperitoneal rat (2 hr) inhibition of total liver nuclear RNA synthesis 68.2% and activity of total RNA polymerase I, II and III reduced 51.7% (11).

Any other comments

Toxicity and hazards reviewed (12).

References

1. Nixon, J. E; et al. *J. Natl. Cancer. Instit.* 1981, **66**(6), 1159-1163
2. *IARC Monograph* 1987, **Suppl. 7**, 83
3. Hendricks, J. D. *Proc. Int. Symp. Princess Takamatsu Cancer Res. Fund* 1980-1981, 227-240
4. Schoenhard, G. L; et al. *Fed. Proc.* 1974, **33**(254), 247
5. Schoenhard, G. L; et al. *Cancer Res.* 1981, **41**(3), 1011-1014
6. Trucksess, M. W; et al. *J. Assoc. Off. Anal. Chem.* 1982, **65**(4), 884
7. Salhab, A. S; et al. *Cancer Res.* 1977, **37** (4), 1016-1021
8. Yourtree, P. M; et al. *Res. Commun. Chem. Pathol. Pharmacol.* 1987, **57**(1), 55-76
9. Auffray, Y; et al. *Mutat. Res.* 1986, **171** (2-3), 79-82
10. Auffray, Y; et al. *Mycopathologie* 1987, **100**(1), 49-53
11. Chen, S. C. G; et al. *Proc. Natl. Sci. Counc. Repub. China, Part B* 1983, **7**(3), 379-386
12. Izmerov, N. F; *Scientific Reviews of Soviet Literature on Toxicity & Hazards of Chemicals* 1991, **101**, Eng. Trans. Richardson, M. L. (Ed.), UNEP/IRPTC, Geneva

A59   Aflatoxin B₁

CAS Registry No. 1162-65-8
Synonyms AFBI; aflatoxin B;
2,3,6a,9a-tetrahydro-4-methoxycyclopenta[*c*]furo[3,2′:4,5]furo[2, 3-*h*]-[1]benzopyran-1,11-dione
Mol. Formula $C_{17}H_{12}O_6$ **Mol. Wt.** 312.28
Occurrence Mycotoxin isolated from *Aspergillus flavus* and *Aspergillus parasiticus*.

Physical properties

M. Pt. 268-269°C.

Solubility
Water: <1 mg ml^{-1}. Organic solvent: ethanol, acetone

Ecotoxicity

Fish toxicity
Exposure to aflatoxin B1 caused inhibition of larval development in zebra fish and hepatocarcinogenicity in rainbow trout, concentration and duration unspecified (1).

Invertebrate toxicity
EC$_{50}$ (5-15 min) *Photobacterium phosphoreum* 19.3-23.2 mg l^{-1} Microtox text (2).

Mammalian and avian toxicity

Acute data
LD_{50} oral hamster, rat, monkey 2-10 mg kg^{-1} (3,4).
LD_{50} oral duckling 730 μg kg^{-1} (5).

Sub-acute data
Chickens fed dietary 1.5 mg kg^{-1} (duration unspecified) developed liver lesions (6).
Macaca fascicularis, dietary level of 1.8 mg kg^{-1} (duration unspecified) produced liver damage characterised by centrilobular necrosis, bile ducts proliferation and fibrosis (7).

Carcinogenicity and long-term effects
Sufficient evidence for carcinogenicity to humans and animals, IARC classification group 1 (8).
Intraperitoneal rat (2 hr) 3 mg kg^{-1} inhibited RNA polymerase II activity only in the target tissue, liver, and not in the non-target tissues *e.g.* lung and brain (9).
♂ Fischer rats (7 wk – 19 month) 50 μg kg^{-1} in diet induced hepatocellular carcinomas (10).
Fed rainbow trout (8 month) 56% incidence of carcinoma (11).

Teratogenicity and reproductive effects
Aflatoxin B1 is unable to pass the blood-brain barrier of pregnant rats but it passed the uteroplacental junction to foetuses and subsequently reached liver, brain and other organs. No significant change observed in plasma consitutents of offspring of ♀ rats treated with a 2 mg kg^{-1} intraperitoneal dose during day 8-10 or 15-17 of gestation. The only significant increase detected was liver triglyceride content in the offspring of rats exposed to B1 during 8-10 day of gestation (12).

Metabolism and pharmacokinetics
Mammalian (including human) metabolism involves the conversion of aflatoxin B1 to hydroxylated metabolites prior to excretion in (mainly) bile and urine.
In the milk of farm animals fed a diet containing aflatoxin B1, a significant amount of the toxic metabolite M1 has been found.
This metabolite has also been seen in the urine of human subjects who have ingested aflatoxin contaminated foods (13).
Intraperitoneal monkey 5.6% of initial unspecified dose was retained by the liver principally bound to liver proteins after 4 days.
Oral monkey (unspecified dose) during day 1-4 excreted in urine and faeces as unchanged parent compound, aflatoxin M1, aflatoxin B1 β-glucuronide and sulfate conjugate (14).
Acid hydrolysis of aflatoxin B1 adducts yielded 2, 3-dihydro-2,3-dihydroxyaflatoxin B1 (15).
Oral pig 0.1 mg kg^{-1} distributed to kidney, liver and muscle (16).
Aflatoxin B1 and its metabolites exist as protein conujugates in systemic blood which is specific to plasma albumin, proceeding irreversibly enzymically by liver and kidney cells (17).
Biotransformation potential investigated using hepatic microsomes from rat, mouse, monkey and human. At low substrate concentrations representative of environmental exposure striking differences were observed in ratios of metabolites between species.
Exposure to 38 mg l^{-1} primate liver formed large amounts of aflatoxin Q1 but failed

to produced detectable P1 levels and the proportion of B1 converted to B1 8, 9 expoxide increased in rat and human microsomes but not in mouse or monkey (18). The metabolism of aflatoxin B1 was investigated in tracheal cultures and purified tracheal microsomes from rabbit, hamster and rat. Metabolic pathways involved cytochrome P450 enzymes and cytosolic GSH transferase activities but differed between the species. In the upper airway epithelium in rabbit metabolic activity involved aflatoxin B1 activation whereas detoxification pathways predominated in the hamster (19).

Genotoxicity

Salmonella typhimurium TA100, TA98, TA1538 with metabolic activation positive (20).
Salmonella typhimurium, Streptococcus sanguis, Mycobacterium tuberculosis all tests positive (1).
Escherichia coli PA37 with and without metabolic activation positive (21).
Early cultures of LNRL (untransformed rat liver cell line) were co-cultivated with Chinese hamster ovary cells, sister chromatid exchange positive (22).
Chinese hamster V79 cells with metabolic activation induced ouabain-resistant mutants of V79 cells (23).

Any other adverse effects to man

Numerous studies have been carried out on the possible impact of aflatoxin contamination on human groups. A study has also implicated aflatoxins in the development of colonic cancer in 2 research workers.
Aflatoxins have been implicated in the deaths of patients who were suffering from Reye's syndrome, as well as a variety of liver diseases (24).
Values of aflatoxin B1 of >0.15 μg ml^{-1} were obtained in sera of 3 patients with primary liver cell carcinoma. The authors conclude, although it is inconclusive as to the carcinogenic potential of aflatoxins in adult man, foodstuffs should be protected from excessive aflatoxin contamination (25).
In a study of primary liver cancer patients aflatoxin B1 was found in 51.3% of urine samples compared to 38.4% for controls (26).

Any other adverse effects

Toxin causing turkey 'X' disease.

Any other comments

Carcinogenicity, genotoxicity and adduct formations with DNA and oncogene mutation reviewed (27- 29).
Interactions between nutrition, toxicology and pharmacology reviewed (30).
Toxicity and hazards reviewed (31).

References

1. Lafont, P; et al. *Microbiol. Aliments. Nutr.* 1989, **7**(1), 1-8, (Fr.) (*Chem. Abstr.* **112**, 93687y)
2. Kaiser, K. L. E; et al. *Water Pollut. Res. J. Canada* 1991, **26**(3), 361-431
3. *Toxicol. Appl. Pharmacol.* 1971, **19**, 169
4. Newberne, P. M; et al. *Cancer Res.* 1969, **29**, 236-250
5. Wogan, G. N; et al. *Cancer Res.* 1971, **31**, 1936-1942
6. Carnaghan, R. B. A; et al. *Pathol. Vet.* 1966, **3**, 601-615
7. Cuthbertson, W. F. J; et al. *Br. J. Nutr.* 1967, **21**, 893-908

8. *IARC Monograph* 1987, **Suppl. 7**, 83
9. Yu, F. L.; et al. *Carcinogenesis (London)* 1982, **3**(9), 1005-1009
10. Cullen, J. M; et al. *Cancer Res.* 1987, **47**(7), 1913-1917
11. Schoenhard, G. L; et al. *Cancer Res.* 1981, **41**(3), 1011-1014
12. Chentanez, T; et al. *Nutr. Rep. Int.* 1986, **34**(3), 379-386
13. Nixon, J. E; et al. *J. Natl. Cancer Inst.* 1981, **6**(6), 1159-1163
14. *Foreign Compound Metabolism in Mammals* 1975, **3**, 304, The Chemical Society, London
15. *Foreign Compound Metabolism in Mammals* 1979, **5**, 232, TheChemical Society, London
16. Trucksess, M. W; et al. *J. Assoc. Off. Anal. Chem.* 1982, **65**(4), 884
17. Nassar, A. Y; et al. *Mycopathologia* 1982, **79**(1), 35
18. Ramsdell, H. S; et al. *Cancer Res.* 1990, **50**(3), 615-620
19. Ball, R. W; et al. *Carcinogenesis (London)* 1991, **12**(2), 305-310
20. Engel, G; et al. *Kiel. Milchwirtsch. Forschungsber.* 1976, **28**(3), 359-365, (Ger.) (*Chem. Abstr.*, **86**, 84291g)
21. Krivobok, S; et al. *Mutagenesis* 1987, **2**(6), 433-439
22. Ray-Chaudhuri, R; et al. *Carcinogenesis (New York)* 1980, **1**(9), 779-786
23. Langenbach, R; et al. *Proc. Natl. Acad. Sci. USA* 1978, **75**(6), 2864-2867
24. *Environmental Health Criteria, No. 11: Mycotoxins* 1979, World Health Organisation, Geneva
25. ONew Yorkemelukwe, G. C; et al. *Toxicol. Lett.* 1982, **10**(2-3), 309-312
26. Salamat, L. A; et al. *ICMR Ann.* 1985, **5**, 131-138
27. Nagao, M. *Rinsho Eryo* 1990, **77**(7), 879-884, (Jap.) (*Chem. Abstr.* **114**, 241406q)
28. Wogan, G. N. *Pure Appl. Chem.* 1989, **61**(1), 1-6
29. Lotlikar, P. D. *Toxicol. Toxin Rev.* 1989, **8**(1-2), 97-109
30. Hart, R. W; et al. *J. Nutr. Biochem.* 1990, **1**(8), 396-409
31. Izmerov, N. F; *Scientific Reviews of Soviet Literature on Toxicity & Hazards of Chemicals* 1991, **101**, Eng. Trans. Richardson, M. L. (Ed.), UNEP/IRPTC, Geneva

A60   Aflatoxin B$_2$

CAS Registry No. 7220-81-7
Synonyms dihydroaflatoxin B$_1$; 2,3,6aα,8,9,9aα-hexahydro-4-methoxycyclopenta[*c*]furo[3',2':4,5]furo[2, 3- *h*][1]benzopyran-1,11-dione
Mol. Formula C$_{17}$H$_{14}$O$_6$ **Mol. Wt.** 314.30
Occurrence Fungal toxin from *Aspergillus flavus* and *Aspergillus parasiticus*.

Physical properties

M. Pt. 286-289°C (decomp.).

Solubility
Organic solvent: ethanol, chloroform

Ecotoxicity

Invertebrate toxicity
EC_{50} (5-30 min) *Photobacterium phosphoreum* 62.7 mg l^{-1} Microtox test (1).

Effects to non-target species
LD_{50} oral duck 1700 µg kg^{-1} (2).

Mammalian and avian toxicity

Acute data
LD_{50} oral mouse 570 mg kg^{-1} (3).
Oral rat 116 mg kg^{-1} (duration of exposure unspecified) non toxic (4).
LD_{50} intravenous rat 10.5 mg kg^{-1} (5).

Carcinogenicity and long-term effects
Sufficient evidence for carcinogenicity to humans and animals, IARC classification group 1 (6). Intraperitoneal rat (2 hr) 3 mg kg^{-1} inhibited RNA polymerase II activity only in the target tissue, liver, and not in the non-target tissues, *e.g.* lung and brain (7). Unspecified levels were detected in sera of 20 patients with primary cell carcinoma. The authors conclude, although it is inconclusive as to the carcinogenic potential of aflatoxins in man, that foodstuffs should be protected from excessive aflatoxin contamination (8).

Teratogenicity and reproductive effects
Pregnant ♀ rats administered 25% aflatoxin B2 0.7-7.0 mg kg^{-1} in 8 or 16 day gestation no foetal malformation observed (9).

Metabolism and pharmacokinetics
After administration of aflatoxin B2 to rats, adducts were found in hepatic DNA and ribosomal RNA. Levels of hepatic protein aflatoxin adducts were 35-70% as great for aflatoxin B2-treated rats as for aflatoxin B1-treated rats (10).

Genotoxicity
Salmonella typhimurium TA100, TA98, TA1538 with and without metabolic activation negative (11).
Escherichia coli PQ37 with and without metabolic activation positive (12).

Any other comments
Investigations in the Sudan, Ghana, Kenya and Nigeria confirmed aflatoxins cross the human placental membrane (13).
Toxicity and hazards reviewed (14).

References
1. Kaiser, K. L. E; et al. *Water Pollut. Res. J. Canada* 1991,**26**(3), 361-431
2. Carnaghan, R. B. A; et al. *Nature (London)* 1963,**200**(4911), 1101
3. *Antipologiki (Moscow)* 1983, **28**, 298
4. *Toxicol. Appl. Pharmacol.* 1985, **79**, 412
5. Wogan, G. N; et al. *Cancer Res.* 1971, **31**(12), 1936-1942
6. *IARC Monograph* 1987, **Suppl. 7**, 83
7. Yu, F. L.; et al. *Carcinogenesis (London)* 1982, **3**(9), 1005-1009
8. Onyemelukure, G. C; et al. *Toxicol. Lett.* 1982, **10**(2-3), 309-312
9. Sharma, A; et al. *Indian J. Anim. Res.* 1987, **21**(1), 35-40

10. Swenson, D. H; et al. *Cancer Res.* 1977, **37**(1), 172
11. Engel, G; et al. *Kiel. Milchwirtsch. Forschungsber.* 1976, **28**(3), 359-365, (Ger.) (*Chem. Abstr.*, **86**(13), 84291g)
12. Krivobok, S; et al. *Mutagenesis* 1987, **2**(6), 433-439
13. Maxwell, S. M; et al. *J. Toxicol. Toxin Rev.* 1989, **8**(1-2), 19-29
14. Izmerov, N. F; *Scientific Reviews of Soviet Literature on Toxicity & Hazards of Chemicals* 1991, **101**, Eng. Trans. Richardson, M. L. (Ed.), UNEP/IRPTC, Geneva

A61   Aflatoxin G1

CAS Registry No. 1165-39-5
Synonyms
3,4,7aα,10aα-tetrahydro-5-methoxy-1H,12H-furo[3′,2′:4,5]furo[2,3-h]pyranol[3,4-c][1]benzopyran-1,12-dione
Mol. Formula $C_{17}H_{12}O_7$ **Mol. Wt.** 328.28
Occurrence Fungal toxin from *Aspergillus flavus* and *A. parasiticus*.

Physical properties
M. Pt. 244-246°C.

Solubility
Organic solvent: ethanol, chloroform

Ecotoxicity

Fish toxicity
TD_{Lo} (12 min) rainbow trout 20 ppb (1).
LC_{50} (24 hr) zebra fish 0.75 μg ml^{-1} (2).
LC_{50} (96 hr) rainbow trout 1.9 mg kg^{-1} (3).

Invertebrate toxicity
EC_{50} (5-30 min) *Photobacterium phosphoreum* 36 mg l^{-1} Microtox test (4).

Effects to non-target species
LD_{50} oral mallard duck 784 μg kg^{-1} (5).

Environmental fate

Degradation studies
Streptomyces sp. exposed to 5 μg aflatoxin G1 showed a decrease in proteolytic, amylolytic, denitrification and/or antibiotic activity (6).

Mammalian and avian toxicity

Acute data
LD_{50} oral rat 2-4 mg kg^{-1} (7).
LD_{50} intraperitoneal rat 14.9 mg kg^{-1} (8).

Carcinogenicity and long-term effects

Sufficient evidence for carcinogenicity to humans and animals, IARC classification group 1 (9).

Intramuscular injection and oral intubation ♀ rhesus monkey (5 yr) dose unspecified developed a metastasising intrahepatic bile-duct carcinoma which was detected at autopsy at death, 25 yr after toxin administration was discontinued (10).

Intraperitoneal rat (2 hr) 3 mg kg^{-1} RNA polymerase II activity only in the target tissue, liver, and not in non-target tissues, *e.g.* lung and brain (11).

Gavage rat (8 wk) 40 equal doses 1.4-2.0 mg rat^{-1} induced hepatocellular carcinoma and adenocarcinomas of the kidney (7).

Unspecified levels detected in sera of 20 patients with primary liver cell carcinoma. The authors conclude, although it is inconclusive as to the carcinogenic potential of aflatoxins in man, that foodstuffs should be protected from excessive aflatoxin contamination (12).

Metabolism and pharmacokinetics

Bioactivation occurs via the mixed function oxidase system (cytochrome P-450$_{NF}$) (13).

Genotoxicity

Salmonella typhimurium TA100, TA98, TA1538 with metabolic activation positive (14- 16).

Escherichia coli PQ37 with metabolic activation positive (17).

Any other adverse effects

A study of 35 women from Songkhla, Thailand, has samples of human cord sera and sera immediately after birth analysed for aflatoxin contamination. Aflatoxin at concentrations 0.064-13.6 nmol ml^{-1} were obtained from 48% of participants. Results confirmed that transplacental transfer occurs (18).

Any other comments

Toxicity and hazards reviewed (19).

References

1. Ayres, J. L; et al. *J. Natl. Cancer Inst.* 1971, **46**, 561-564
2. Abedi, Z. H; et al. *J. Assoc. Off. Anal. Chem.* 1966, **52**, 962-969
3. Bayer, D. H; et al. *Toxicol. Appl. Pharmacol.* 1969, **15**(2), 415
4. Kaiser, K. L. E; et al. *Water Pollut. Res. J. Canada* 1991, **26**(3), 361-431
5. Carnaghan, R. B. A; et al. *Nature (London)* 1963, **200**(4911), 1101
6. Marcinowska, K; et al; *Acta Agrar. Silvestria, Ser. Agrar.* 1987, **27**, 69-80 (Pol.) (*Chem. Abstr.* **110**, 35084x)
7. Wogan, G. N; et al. *Cancer Res.* 1971, **31**(12), 1936-1942
8. Butler, W. H; et al. *J. Pathol.* 1970, **102**, 209-212
9. *IARC Monograph* 1987, **Suppl. 7**, 56
10. Tilak, T. B. G. *Food Cosmet. Toxicol.* 1975, **13**(2), 247-244
11. Yu, F. L; et al. *Carcinogenesis (London)* 1982, **3**(9), 1005-1009
12. Onyemelukwe, G. C; et al. *Toxicol. Lett.* 1982, **10**(2-3), 309-312
13. Shimada, T; et al. *Cancer Res.* 1989, **49**(12), 3218-3228
14. Yourtree, D. M; et al. *Res. Commun. Chem. Pathol. Pharmacol.* 1987, **57**(1), 55-76
15. Engel, G; et al. *Kiel. Milchwirtsch. Forschungsber.* 1976, **28**(3), 359-365, (Ger.), (*Chem. Abstr.* **86**, 84291g)

16. Wong, J. J; et al. *Proc. Natl. Acad. Sci. USA* 1976, **73**(7), 2241-2244
17. Krivobok, S; et al. *Mutagenesis* 1987, **2**(6), 433-439
18. Denning, D. W; et al. *Carcinogenesis (London)* 1990, **11**(6), 1033-1035
19. Izmerov, N. F; *Scientific Reviews of Soviet Literature on Toxicity and Hazards of Chemicals* 1991, **101**, Eng. Trans. Richardson, M. L. (Ed.), UNEP/IRPTC, Geneva

A62   Aflatoxin G$_2$

CAS Registry No. 7241-98-7
Synonyms
3,4,7aα,9,10,10aα-hexahydro-5-methoxy-1H,12H-furo[3′,2′:4,5]furo[2,3-h]pyrano[3,4-c][1]benzopyran-1,12-dione
Mol. Formula C$_{17}$H$_{14}$O$_7$ **Mol. Wt.** 330.30
Occurrence Fungal toxin from *Aspergillus flavus* and *Aspergillus parasiticus*.

Physical properties
M. Pt. 237-240°C.

Solubility
Organic solvent: ethanol, chloroform

Ecotoxicity

Fish toxicity
LC$_{50}$ (24 hr) zebra fish larvae 4.2 μg ml^{-1} (1).

Invertebrate toxicity
EC^{50} (30 min) *Photobacterium phosphoreum* 69 mg l^{-1} Microtox test (2).

Effects to non-target species
LD$_{50}$ oral duckling 3450 μg kg^{-1} (1 day old) (3).

Mammalian and avian toxicity

Acute data
Oral rat 232 mg kg^{-1} non toxic (4).

Carcinogenicity and long-term effects
Sufficient evidence for carcinogenicity to humans and animals, IARC classification group 1 (5).
Intraperitoneal rat (2 hr) 3 mg kg^{-1} inhibited RNA polymerase II activity only in the target tissue, liver, and not in the non target tissues, e.g. lung and brain (6).
Unspecified levels were detected in the sera of 20 patients with primary liver cell carcinoma. The authors conclude, although its carcinogenic potential in adult humans is inconclusive, foodstuffs should be protected from excessive aflatoxin contamination (7).

Genotoxicity

Salmonella typhimurium TA100, TA98, TA1538 with metabolic activation negative (8). *Salmonella typhimurium* strain unspecified with metabolic activation weakly positive (9). *Escherichia coli* PQ37, PQ35 with and without metabolic activation negative (10). Early cultures of LNRL (untransformed rat liver cell line) were cocultivated with Chinese hamster ovary cells, sister chromatid exchange positive (11). Chinese hamster V79 cells with and without metabolic activation negative (12).

Any other comments

Toxicity and hazards reviewed (13).

References

1. Abedi, Z. M; et al. *J. Assoc. Off. Anal. Chem.* 1966, **52**, 962-969
2. Kaiser, K. L. E. *Water Pollut. Res. J. Canada* 1991, **26**(3), 361-431
3. Carnaghan, R. B. A; et al. *Nature (London)* 1963,**200**(4911), 1101
4. Wogan, G. N; et al. *Cancer Res.* 1971, **31**(12), 1936-1942
5. *IARC Monograph* 1987, **Suppl. 7**, 83
6. Yu, F. L; et al. *Carcinogenicity (London)* 1982, 3(9), 1005-1009
7. ONew Yorkemelukwe, G. C; et al. *Toxicol. Lett.* 1982, **10**(2-3), 309-312
8. Engel, G; et al. *Kiel. Milchwirtsch. Forschungsber.* 1976, **28**(3), 359-365, (Ger.) (*Chem. Abstr.* **86**(13), 84291g)
9. Wong, J. J; et al. *Proc. Natl. Acad. Sci. USA* 1976, **73**(7), 2241-2244
10. Krivobok, S; et al. *Mutagenesis* 1987, **2**(6), 433-439
11. Ray-Chaudhuri, R; et al. *Carcinogenesis (New York)* 1980, **1**(9), 779-786
12. Langenbach, R; et al. *Proc. Natl. Acad. Sci. USA* 1978, **75**(6), 2864-2867
13. Izmerov, N. F; *Scientific Reviews of Soviet Literature on Toxicity & Hazards of Chemicals* 1991, **101**, Eng. Trans. Richardson, M. L. (Ed.), UNEP/IRPTC, Geneva

A63   Aflatoxin M₁

CAS Registry No. 6795-23-9

Synonyms 4-hydroxyaflatoxin B_1;
2,3,6a,9a-tetrahydro-9a-hydroxy-4-methoxycyclopenta[c]furo[3',2':4,5]furo[2,3-h]benzopyran-1,11-dione

Mol. Formula $C_{17}H_{12}O_7$ **Mol. Wt.** 328.28

Occurrence Metabolic product of aflatoxin B_1. A common source is animal milk, *e.g.* cattle fed plant material containing aflatoxin B_1.

Physical properties

M. Pt. 299°C (decomp.).

Solubility

Organic solvent: methanol, ethanol, dimethylformamide

Ecotoxicity

Fish toxicity
Hepatocarcinogenic to rainbow trout (1,2).

Effects to non-target species
LD_{50} oral duckling 16 µg kg^{-1} (3).

Mammalian and avian toxicity

Acute data
LD_{Lo} oral rat 1500 µg kg^{-1} (4).

Sub-acute data
TD_{Lo} (8 wk) oral rat 8 mg kg^{-1} (5).

Carcinogenicity and long-term effects
Sufficient evidence for carcinogencity to humans and animals, IARC classification group 1 (6). ♂ Fischer rats (7 wk-21 month) 0-50 µg kg^{-1} in diet induces hepatocellular carcinomas and neoplastic nodules (7).
Rainbow trout (1 yr) in food (concentrations unspecified) induced hepatocellular carcinoma (8).

Genotoxicity

Salmonella typhimurium TA100, TA98, TA1535 with metabolic activation positive (9,10). *Salmonella typhimurium, Streptococcus sanguis, Mycobacterium tuberculosis*, inhibition of larval development in zebra fish and heptacarcinogenicity in rainbow trout, all tests positive (11).

Any other adverse effects to man

In a study of primary cancer liver patients aflatoxin M1 was found in 51.3% of urine samples compared to 38.4% for controls (12).

Any other comments

Contamination of dairy produce reviewed (13).
Toxicity and hazards reviewed (14).

References

1. Sinnhuber, R. O; et al. *Fed. Proc.* 1970, **29**, 568
2. Hendricks, J. D. *Proc. Int. Symp. Princess Takamatsu Cancer Res. Fund* 1980 (pub. 1981) 227-240
3. Purchase, I. F. H. *Food Cosmet. Toxicol.* 1967, **5**, 339-342
4. Pong, R. S; et al. *J. Natl. Cancer Inst.* 1971, **47**, 585
5. Wogan, G. N; et al. *Food Cosmet. Toxicol.* 1974, **12**, 381-384
6. *IARC Monograph* 1987, (Suppl. 7), 83
7. Cullen, J. M; et al. *Cancer Res.* 1987, **47**(7), 1913-1917
8. Schoenhard, G. L; et al. *Cancer Res.* 1981, **41**(3), 1011-1014
9. Engel, G; et al. *Kiel. Milchwirtsch. Forschungsber.* 1976, **28**(3), 359-365, (Ger.) (*Chem. Abstr.* **86**(13), 84291g)
10. Wong, J. J; et al. *Proc. Natl. Acad. Sci. USA* 1976, **73**(7), 2241-2244
11. Lafont, P; et al. *Microbiol. Aliments. Nutr.* 1989, **7**(1), 1-8, (Fr.) (*Chem. Abstr.* **112**, 93687y)
12. Salamat, L. A; et al. *ICMR Ann.* 1985, **5**, 131-138

13. Luis Blanco, J; et al. *Alimentaria (Madrid)* 1989, **26**(205), 35-46, (Span.) (*Chem. Abstr.* **112**, 137606c)
14. Izmerov, N. F; *Scientific Reviews of Soviet Literature on Toxicity & Hazards of Chemicals* 1991, **101**, Eng. Trans. Richardson, M. L. (Ed.), UNEP/IRPTC, Geneva

A64   Agar

CAS Registry No. 9002-18-0

Synonyms agar-agar; Bengal isinglass; Ceylon isinglass; Chinese isinglass; Japan isinglass; Japan agar

Uses Suspending or thickening agent in pharmaceutical and food products used as a substitute for gelatin. Corrosion inhibitor. Sizing agent for paper and textiles. In nutrient media for bacterial cultures. In human and veterinary medicine used as a laxative.

Occurrence Polysaccharide complex obtained from various species of Rhodophyceae algae (British Pharmacopoeia specifies the genus *Gelidum*) (1,2).

Ecotoxicity

Invertebrate toxicity

Cytophaga HK-5, a marine bacterial strain was capable of degrading a number of marine plant polysaccharides, including agar (3).

Mammalian and avian toxicity

Acute data

LD$_{50}$ oral rat, mouse, rabbit, hamster 6-16 g kg^{-1} (1).

Carcinogenicity and long-term effects

US National Toxicology Program ♂ and ♀ F344 rats and B6C3F1 mice (2 yr) oral in food (concentration unspecified) negative (4).

Oral rat (103 wk) 25,000 to 50,000 ppm increased adrenal cortex adenoma for higher dose in ♀ and increased liver adenoma for both doses in ♂ (5).

Any other comments

The use of agar in food is limited only by good manufacturing practice. Toxicological information, properties and applications of agar in food products reviewed (6,7).

References

1. Sax, N. I; et al. *Dangerous Properties of Industrial Materials.*, 7th ed., 1989, Van Nostrand Reinhold, New York
2. *Martindale: The Extra Pharmacopoeia* 29th ed., 1989, 1432, The Pharmacentical Press, London
3. Kondrat'eva, L. M; et al. *Mikrobiologiya* 1989, **58**(6), 990-994, (Russ) (*Chem. Abstr.*, **112**, 73610q)
4. *National Toxicology Program Research & Testing Division* 1992, Report No.TR-230, NIEHS, Research Triangle Park, NC
5. *NIH Publ. NIH-82-1786* 1982, US Dept. of Health and Human Services

6. *Food Additives Service No. 5* 1974, World Health Organisation, Geneva
7. Yang, X; et al. *Shipin Kexue (Beijing)* 1988, **100**, 27-32, (Ch.) (*Chem. Abstr.* **109**, 148025b)

A65   Alachlor

$$CH_3OCH_2 - N - COCH_2Cl$$

CH$_3$CH$_2$ — CH$_2$CH$_3$

CAS Registry No. 15972-60-8
Synonyms 2-chloro-2′,6′-diethyl-*N*- methoxymethylacetanilide; 2-chloro-*N*-(2, 6-diethylphenyl)-*N*-(methoxymethyl)acetamide; *N*- (methoxymethyl)-2, 6-diethy-l-chloroacetanilide; 2-chloro- 2′, 6′-diethyl-*N*-(methoxymethyl)acetanilide; α-chloro-2′,6′-diethyl-*N*-methoxymethl acetanilide
Mol. Formula $C_{14}H_{20}ClNO_2$ **Mol. Wt.** 269.77
Uses Pre- or early post-emergence herbicide.

Physical properties

M. Pt. 39.5-41.5°C; **B. Pt.** $_{0.02}$ 100°C; **Specific gravity** 1.133 at 25°C; **Volatility** v.p. 2.1×10^{-5} mmHg at 25°C.

Solubility

Organic solvent: diethyl ether, acetone, benzene, chloroform, ethanol, ethyl acetate

Ecotoxicity

Fish toxicity
LC_{50} (96 hr) guppy, rainbow trout, bluegill sunfish 0.75-2.8 mg l^{-1} (1-3).

Invertebrate toxicity
EC_{50} (96 hr) *Daphnia magna* 0.05 mg l^{-1} (3).
LC_{50} (96 hr) crayfish 19.5 ppm (4).
Mud crab at salinity >24 ppt were hypoosmotic while at lower salinities they were hyperosmotic. Salinity decreases from 10-0 ppt elevated the oxygen consumption rate and critical oxygen temperature but this response was unaffected by alachlor concentrations as high as 25 ppm (5).
LC_{50} (24-96 hr) mud crab larvae 27-10 mg l^{-1} (6).

Environmental fate

Nitrification inhibition
Nitrogen fixation was adversely affected by concentrations of 20-80 μg l^{-1} in *Nostoc muscorum* (7).

Anaerobic effects
Anaerobic degradation was less rapid in surface soils $t_{1/2}$ 100 day, in subsurface soil (0.5- 2.4 m) $t_{1/2}$ 144 day and in aquifer samples $t_{1/2}$ 337-553 days (8).

Degradation studies
Loss from soil primarily by microbial degradation. Persists in soil for 6-10 wk depending on conditions (1,2).

Under aerobic conditions, $t_{1/2}$ 23 day in surface soil, $t_{1/2}$ 73-284 day in the vadose zone and $t_{1/2}$ 320-324 day in aquifers. Addition of organic nutrients enhanced aerobic degradation in subsurface soils and one aquifer sample (8).

Rapidly biodegraded by soil fungi to release the chloride ion. Other metabolites detected include 2-chloro-2′,6′-diethylacetanilide, 2, 6-diethyl-N-(methoxymethyl)aniline, 2,6-diethylaniline and 1-chloroacetyl-2,3-dihydro-7-ethylindole. Soil incubation studies using alkaline hydrolysis suggest that metabolites were bound to soil organic matter (9).

Alachlor incubated under upland soil conditions for 80 days yielded 4 major degradation products 8-ethyl-2- hydroxy-N-(methoxymethyl)-1,2,3,4-tetrahydroquinoline, N-hydroxyacetyl-2,3-dihydro-7-ethylindole, 2-hydroxy-2′, 6′-diethyl-N-(methoxymethyl)acetanilide and 9- ethyl-1, 5-dihydro-1-(methoxymethyl)-5-methyl-4,1-benzoxazeprin-2($3H$)-one (10).

Mammalian and avian toxicity

Acute data
LD_{50} oral mouse, rat 462-1200 mg kg^{-1} (11,12).
LD_{50} oral mallard >2000 mg kg^{-1} (13).
LD_{50} dermal rabbit 3500 mg kg^{-1} (14).

Sub-acute data
LC_{50} (8 day feeding trial) oral pheasant, bobwhite quail >5000-10,000 ppm (4).
Ninety day feeding trials, no effect observed in rats or dogs ≤200 mg kg^{-1} diet (2).

Metabolism and pharmacokinetics
In vitro incubation with microsomal fractions prepared from liver and nasal turbinates of rats and mice. Biotransformation to 3, 5-diethylbenzoquinone-4- imine occurred via oxidation of 2,6-diethylaniline and 4- amino-3,5-diethyl phenol intermediates (15). Intraperitoneal ♂ rat (7 day) 1 or 100 mg kg^{-1} metabolised via the monooxygenase system (16).

Incubation of alachlor in the presence of GSH with the cytosolic fraction from rat, mouse and monkey livers produced the GSH conjugate of alachlor as the initial metabolite; further degradation occurred via the mercapturic acid pathway to yield cysteinylglycine, cysteine and N- acetylcysteine conjugates of alachlor. Species and gender differences were observed (17).

Following intravenous administration to rhesus monkeys, an average of 88% and 10% of radiolabelled alachlor was recovered in the urine and faeces respectively within 48 hr (18).

Irritancy
Reported to be a mild irritant to rabbit skin and non-irritating to rabbit eyes (1,2).

Genotoxicity
Salmonella typhimurium TA1535, TA1537, TA1538, TA98, TA100 with and without metabolic activation negative.
Saccharomyces cerevisiae D4 with metabolic activation positive for technical grade alachlor (19).
Tradescantia paludosa (18-24 hr) 0.8% alachlor increased the incidence of chromosomal aberrations (20).
In vitro human lymphocytes and *in vivo* rat bone marrow cells dose dependent increase in chromosomal aberrations and clastogenic effects (21).

Any other adverse effects to man

Passive exposure studies carried out by the US EPA to predict the dosage range received by US farmers during use established a range of 0.0054-0.54 μg kg^{-1} lb^{-1} of active ingredient during open pour mixing and loading and a range of 0.0034-0.34 μg kg^{-1} lb^{-1} of active ingredients during mechanical mixing and loading (22).

Legislation

Limited under EEC Directive on Drinking Water Quality 80/778/EEC. Pesticides and related products: maximum admissible concentration 0.1 μg l^{-1} (23).

Any other comments

Tolerable daily intake for humans 0.1 mg kg^{-1} day^{-1} (24).

The properties, toxicology, pharmacokinetics, environmental fate and health effects in humans and animals reviewed (25-27).

References

1. *The Agrochemicals Handbook* 1991, RSC, London
2. Worthing, C. R; et al. *The Pesticide Manual* 9th ed., 1991, British Crop Protection Council, Farnham
3. Strateva, A; et al. *Probl. Khig.* 1986, **11**, 32-37, (Bulg.) (*Chem. Abstr.* **107**, 72436)
4. *Herbicide Handbook* 4th ed., 1979, 9, Weed Science Society of America, Champaign, IL
5. Diamond, D. W; et al. *Comp. Biochem. Physiol. A: Comp. Physiol.* 1989, **93A**(2), 313-318
6. Tukacs, R. L; et al. *Estuaries* 1988, **11**(2), 79-82
7. Davies, H. M; et al. *Weed Sci.* 1990, **38**(3), 206-214
8. Pothuluri, J. V; et al. *J. Environ. Qual.* 1990, **19**(3), 525-530
9. Kearney, P. C; et al. *Herbicides: Chemistry, Degradation and Mode of Action* 2nd ed., 1975, **1,2**, 369, Marcel Dekker, New York
10. Lee, J. K; et al. *Han' guk Nonghwa Hakhoechi* 1986, **29**(2), 182-189, (*Chem. Abstr.* **106**, 97928b)
11. Evans, D. M. *Chem. Ind. (London)* 1969, 615-616
12. Pan'shina, T. N. *Gig. Tr. Prof. Zabol.* 1977, **21**(12), 30, (Russ.) (*Chem. Abstr.* **88**, 125762)
13. *Handbook of Toxicity of Pesticides to Wildlife* 1984, 9, US Dept. of Interior, Fish and Wildlife Service, Resource Publication 153, Washington, DC
14. *Guide to the Chemicals Used in Crop Protection* 1973, **6**, 3, Information Canada, Ottawa
15. Feng, P. C. C; et al. *Drug Metab. Dispos.* 1990, **18**(3), 373-377
16. Leslei, C; et al. *Arch. Environ. Contam. Toxicol.* 1989, **18**(6), 876-880
17. Feng, P. C. C; et al. *Pest. Biochem. Physiol.* 1988, **31**(1), 84-90
18. Kronenberg, J. M; et al. *Fundam. Appl. Toxicol.* 1988, **10**(4), 664-671
19. Plewa, M. J; et al. *Mutat. Res.* 1984, **136** (3), 233-245
20. Dryanovska, O. *Dokl. Bolg. Akad. Nauk.* 1987, **40**(6), 73-75
21. Georgian, L; et al. *Mutat. Res.* 1983, **116** (3-4), 341-348
22. Maddy, K. T; et al. *ACS Symp. Ser.* 1989, 382, 338-353
23. *EC Directive Relating to the Quality of Water Intended for Human Consumption* 1982, 80/778/EEC, Office for Official Publications of the European Communities, 2 rue Mercier, L-2985 Luxembourg
24. Sittig, M. *Handbook of Toxic and Hazardous Chemicals and Carcinogens* 2nd ed., 1985, Noyes Publications, New York
25. *Dang. Prop. Ind. Mater. Rep.* 1990, **10**(2), 23-30
26. Chesters, G; et al. *Rev. Environ. Contam. Toxicol.* 1989, **110**, 1-74
27. *Gov. Rep. Announc. Index US* 1987, **87**(18), Order No. PB87-200176, 1-264, USEPA, Washington, DC

A66   Aldicarb

$$CH_3SC(CH_3)_2CH=NOCONHCH_3$$

CAS Registry No. 116-06-3
Synonyms Ambush; Carbanalate; 2-methyl-2-(methylthio)propionaldehyde;
O-(methylcarbamoyl)oxime; 2-methyl-2-(methylthio)propanol;
O-[(methylamino)carbonyl]oxime; Temik; UC21149
Mol. Formula $C_7H_{14}N_2O_2S$ **Mol. Wt.** 190.27
Uses An insecticide, acaricide and nematocide.
Occurrence Aldicarb residues have been found in potable water sources from wells
in the US and some fruit and vegetables (1,2).

Physical properties

M. Pt. 99-100°C; **Specific gravity** d_{25}^{20} 1.1950; **Partition coefficient** log P_{ow} 1.359
(3).; **Volatility** 9.8×10^{-6} mmHg at 20°C.

Solubility
Water: 6 g l^{-1} at 25°C. Organic solvent: acetone, benzene, xylene, dichloromethane

Occupational exposure

Supply classification toxic.

Risk phrases
Very toxic by inhalation, in contact with skin and if swallowed (R26/27/28)

Safety phrases
Keep locked up – Keep away from food, drink and animal feeding stuffs – After
contact with skin, wash immediately with plenty of water – In case of accident or if
you feel unwell, seek medical advice immediately (show label where possible) (S1,
S13, S28, S45)

Ecotoxicity

Fish toxicity
LC_{50} (96 hr) rainbow trout, bluegill sunfish, fathead minnow 0.88-13.4 mg l^{-1} (4- 6).
Puntius conchonius chronic exposure to sublethal concentrations 0.8 ppm caused
hepatic lesions, including hypertrophy, vacuolisation, nuclear pycnosis and karyolysis
(7).
Exposure of *Barbus conchonius* (15 and 30 day) 48 µg l^{-1} caused
hypercholesterolaemia, moderate polycythemia, a rise in haemoglobin content and
decrease in blood glucose levels. Main target organ liver (8).
LC_{50} value varies at different temperatures and water hardness (3).

Invertebrate toxicity
LC_{50} (24 hr) *Paramecium multimicronucleatum* static bioassay 93 ppm (9).

Effects to non-target species
Toxic to bees (4).
LC_{50} (14 day) *Lumbricus terrestris* 530 mg kg^{-1} dry soil substrate.
LC_{50} (14 day) *Eisenia foetida* 65 mg kg^{-1} dry soil substrate (10).

Environmental fate

Nitrification inhibition

Aldicarb at concentrations of 5 ppm caused 100% inhibition of *Nitrosomonas europaea* (11).

Severe damage was observed to a soil nitrifying population for the first 16 wk after application of 2.5 g m^{-2} aldicarb. Incubations in the field and under laboratory conditions without substrate addition showed negligible effects whereas incubation with ammonium sulfate led to an reduction in nitrification (12).

Anaerobic effects

No degradation was observed in sterile or unsterile groundwater in 60-65 day at pH 5.2 and 6.0. Its metabolite sulfone or its hydrolysis products were not detected (13,14).

Degradation studies

Surface soils up to 75 cm deep $t_{1/2}$ 20- 361 days and in subsurface soils up to 183 cm $t_{1/2}$ 131- 233 days. Metabolites detected included aldicarb sulfoxide and sulfone and their oximes and sulfoxide nitrile (14).

Under aerobic conditions, aldicarb was metabolised rapidly to aldicarb sulfoxide which was slowly oxidised to aldicarb sulfone. These reactions were more rapid in surface and shallow subsurface soils than in deeper subsurface soils. Additional metabolites detected were oxime, nitrile and acid derivatives of aldicarb sulfoxide, $t_{1/2}$ 29-78 days for total toxic residues (15).

Acclimated *Pseudomonas sp.* degraded 50% of an unspecified initial concentration of aldicarb in 24-32 hr (16).

In a model ecosystem aldicarb demonstrates a high degree of persistence and a low biodegradability potential (17).

Degraded rapidly in soils depending on soil type loamy, sandy and clay soils were 5, 6, 10 days, respectively. Main degradation products were sulfoxide and sulfone derivatives which were further degraded by soil microorganisms, including *Bacillus* (18).

In loamy sand soil reported $t_{1/2}$ 9 days with no residues present 4 months after application (19).

Abiotic removal

Hydrolytic $t_{1/2}$ at 20°C in water were 6 and 131 days at pH 8.85 and 3.95 respectively, while hydrolytic $t_{1/2}$ at 15°C in soil were 10 and 990 days at pH7 and 5.4, respectively (20-22).

Mammalian and avian toxicity

Acute data

LD$_{50}$ oral rat 0.93 mg kg^{-1} (23).
LD$_{50}$ (5 min) inhalation rat, mouse, guinea pig 200 mg m^{-3} (3).
LD$_{50}$ dermal rabbit 5 mg kg^{-1} (23).
LD$_{50}$ dermal rat 850 mg kg^{-1} (3).

Sub-acute data

LC$_{50}$ 7 day dietary bobwhite quail 2400 mg kg^{-1} (4).
Adult ♀ mice (34 day) 1, 10, 100 ppb day^{-1} in water. Effects to T-cells, T- suppressor, T-helper and B-cells were evaluated and the authors conclude that aldicarb does not result in adverse effects to the immune system in mice (24).

Carcinogenicity and long-term effects
A National Toxicology Program 2 yr study of ♂ and ♀ F344 rats and B6C3F1 mice given aldicarb (dose unspecified) in feed non- carcinogenic and nongenotoxic (25).

Teratogenicity and reproductive effects
Pregnant ♀ rats (18th day gestation) were given a single 0.1 mg kg^{-1} dose aldicarb by gastric intubation. Rats were sacrificed at 1, 6, 12 and 24 hr after administration. Significant reduction of acetylcholinesterase activity detected in blood, brain and liver tissues of both dams and foetuses (26).
Pregnant rats were fed 0-1.0 mg kg^{-1} aldicarb throughout pregnancy until pups were weaned. No significant effects on fertility, viability of offspring, lactation or other parameters observed (3).

Metabolism and pharmacokinetics
In numerous animal studies the principal excretion route for aldicarb and its metabolites, which include aldicarb sulfoxide, aldicarb sulfone, oxime sulfoxide, oxime sulfone, nitrile sulfoxide and nitrile sulfone, is via the urine > 90%. Small amounts are also excreted via the faeces and exhaled as carbon dioxide (27).

Irritancy
Four cases of contact dermatitis and one case of eye irritation (chemical conjunctivitis) have been reported after contact with Temik (28).

Genotoxicity
Salmonella typhimurium TA97, TA98, TA100, TA1535, TA1537 with and without metabolic activation negative (29).
In vitro mouse lymphoma L5178Y tk+/tk- with metabolic activation positive (30).

Any other adverse effects to man
A study of 1500 subjects who had consumed water from wells contaminated with 8-66 μg l^{-1} aldicarb during 1981. The rate of spontaneous abortions was high in women consuming water from wells contaminated with >66 μg l^{-1}+ of aldicarb (3).
In 1985, a study linked ingestion of aldicarb-contaminated drinking water with altered T-cell distribution in humans. A follow-up study in 1987, 45 of the 50 initial participants and a further 27 women took part. From this group only 5 were found to be exposed currently. This group of 5 women compared to 39 unexposed controls had an increased percentage of lymphocytes and an increased number of CD2 and T-cells. No identified water contaminant apart from aldicarb could explain these findings (31). However, it is reported that the T-lymphocyte data fell within the normal range (3).

Any other adverse effects
The toxicity of aldicarb is based on its transient inhibition of acetylcholinesterase. Carbamates form unstable complexes with cholinesterase by carbamolyation of the active site of the enzymes. The process is quickly reversible (32,33).
Symptomatic effects include headache, dizziness, anxiety, excessive sweating, salivation, lachrymation, increased bronchial secretions, vomiting, diarrhoea, abdominal cramps, muscle fasciculations and pinpoint pupils (3).

Legislation
In 1982 the FAO/WHO set the acceptable daily intake in food of 5 mg kg^{-1} body weight (34).

The Office of Drinking Water of the US Environmental Protection Agency established a Health Advisory Level of 10 μg l^{-1} for residues of aldicarb in drinking water in 1987, with a proposed revision of 3 μg l^{-1} in 1991 (35).
Limits under EEC Directive on Drinking Water Quality 80/778/EEC. Pesticides and related products: maximum admissible concentration 0.1 μg l^{-1} (36).

Any other comments

Mint can absorb aldicarb in sufficient concentrations rendering the plant toxic to pests and highly dangerous when consumed by humans in small amounts (37).
Biochemical properties, toxicology, mutagenicity, teratogenicity, carcinogenicity and environmental effects reviewed (38-42).
Toxicity and hazards reviewed, to be published in 1992-1993 (43).

References

1. Miller, W. L; et al. *Water Resour. Bull.* 1989, **25**(1), 79-86
2. Duggan, R. E; et al. *Pest. Res. Levels in Foods in the United States* 1983, FDA Div. Chem. Technol., Washington, DC
3. *Environmental Health Criteria No. 121. Aldicarb* 1991, **121**, 1-130, World Health Organisation, Geneva
4. *The Agrochemicals Handbook* 3rd ed., 1991, RSC, London
5. *American Hospital Formulary Service – Drug Information 85* 1985, American Society Hospital Pharmacists, Bethesda
6. Pickering, Q. H; et al. *Arch. Environ. Contam. Toxicol.* 1982, **11**(6), 699-702
7. Gill, T. S; et al. *J. Environ. Sci. Health Part A* 1990, **A25**(6), 653-663
8. Pant, J; et al. *Bull. Environ. Contam. Toxicol.* 1987, **38**(1), 36-41
9. Edmiston, C. E. Jr; et al. *Environ. Res.* 1985, **36**(2), 338-350
10. Hague, A; et al. *Z. Pflanzenkr. Pflanzenschutz* 1983, **90**, 395-408
11. Kuseske, D. W; et al. *Plant Soil* 1974, **41**, 255-269
12. Mathes, K; et al. *Toxic. Assess.* 1988, **3**(3), 271-286
13. Delfino, J. J; et al. *Soil Crop Sci. Soc. Florida Proc.* 1985, **44**, 9-14
14. Ou, L. T; et al. *J. Agric. Food Chem.* 1985, **33**, 72-78
15. Ou, L. T; et al. *J. Environ. Qual.* 1986, **15**(4), 356-363
16. Chaudhry, G. R; et al. *Water Sci. Technol.* 1988, **20**(11-12), 89-94
17. *USEPA Initial Scientific and Minieconomic Review of Aldicarb* 1975, 66, EPA 540/1-75-013
18. Shi, G; et al. *Zhongguo Huanjing Kexue* 1987, **7**(1), 38-43, (Ch.) (*Chem. Abstr.* **107**, 111070v)
19. Hegg, R. O; et al. *Agric. Ecosyst. Environ.* 1988, **20**(4), 303-315
20. Given, C. J; et al. *Bull. Environ. Contam. Toxicol.* 1985, **34**, 627-633
21. Bromilow, R. H. *Pest. Sci.* 1980, **11**, 389-395
22. Smelt, J. H; et al. *Pest. Sci.* 1978, **9**, 293-300
23. Weiden, M. H. J; et al. *J. Econ. Entomol.* 1965, **58**, 154
24. Thomas, P; et al. *Fundam. Appl. Toxicol.* 1990, **15**(2), 221-230
25. *National Toxicology Program, Research and Testing Division* 1992, Report No. TR-136, NIEHS, Research Triangle Park, NC 27709
26. El-Elaimy, I. *Proc. Zool. Soc. A. R. Egypt* 1986, **10**, 41-49
27. *National Research Council Drinking Water and Health* 1977, **1**, National Academy Press, Washington, DC
28. Peoples, S. A; et al. *Vet. Hum. Toxicol.* 1978, **20**(5), 321-324
29. Zeiger, E; et al. *Environ. Mol. Mutagen* 1988, **11**(Suppl. 12), 1-157
30. Myhr, B. C; et al. *Environ. Mol. Mutagen* 1988, **12**, 103-194
31. Mirkin, I. R; et al. *Environ. Res.* 1990, **51**(1), 35-50
32. Done, A. K. *Emerg. Med.* 1979, **11**, 167-175

33. Mortensen, M. L. *Pediatric Clin. North Am.* 1986, **33**, 421-425
34. *Pesticide Residues in Food* 1982, FAO Plant Production and Protection Paper 46, Rome
35. *National Primary Drinking Water Regulations- Fed. Reg. 56, 3600-3614* 1991, US Environmental Protection Agency
36. *EC Directive Relating to the Quality of Water Intended for Human Consumption* 1982, 80/778/EEC, Office for Official Publications of the European Communities, 2 rue Mercier, L-2985 Luxembourg
37. Hayes, W. J. Jr. *Toxicology of Pesticides Baltimore* 1975, 271, Williams & Wilkins, Baltimore, MD
38. *Gov. Rep. Announce. Index US* 1987, **87**(18), Order No. PB-87200176
39. Mink, F. L; et al. *Environ. Pollution* 1989, **24**(4), 241-251
40. Baron, R. L; et al. *Rev. Environ. Contam. Toxicol.* 1988, **105**, 1-70
41. Risher, J. F; et al. *Environ. Health Perspect.* 1987, **72**, 267-281
42. *IARC Monograph* 1991, **53**, 93-113/539
43. Izmerov, N. F; *Scientific Reviews of Soviet Literature on Toxicity and Hazards of Chemicals* 1991, **114**, Eng. Trans, Richardson, M. L. (Ed.), UNEP/IRPTC, Geneva

A67 Aldol

CH₃CH(OH)CH₂CHO

CAS Registry No. 107-89-1
Synonyms 3-hydroxybutanal; 3-hydroxybutyraldehyde; acetaldol; β-hydroxybutyraldehyde; oxybutyric aldehyde
Mol. Formula $C_4H_8O_2$ **Mol. Wt.** 88.11
Uses In the manufacturing of rubber vulcanisers, accelerators and age resisters. Ore flotation. An hypnotic and sedative.

Physical properties

M. Pt. 0°C; **B. Pt.** $_{20}$ 83°C; **Flash point** 65.5°C (open cup); **Specific gravity** d_4^{16} 1.109; **Volatility** v.p. 21 mmHg at 20°C.

Solubility
Water: miscible. Organic solvent: miscible with ethanol, diethyl ether

Occupational exposure

UN No. 2839; **HAZCHEM Code** 2R; **Conveyance classification** toxic substance.

Environmental fate

Degradation studies
BOD_{10} 0.9 using standard dilute sewage (1).

Mammalian and avian toxicity

Acute data
LD_{50} oral rat 2180 mg kg^{-1} (2).
LD_{50} dermal rabbit 140 mg kg^{-1} (3).

Carcinogenicity and long-term effects
The length of exposure prior to the appearance of cancer was 26 ± 4 yr for 4 patients

55-59 yr old and 4 patients >65 yr, all smoked 5-10 cigarettes day^{-1}. One patient developed cancer at 58 yr after 13 yr exposure and smoking >30 cigarettes day^{-1}. Of the cancers, 5 affected the bronchi, 2 the mouth, 1 the stomach and 1 the caecum. Syncarcinogenic effects of the aliphatic aldehyde mixture and the possible carcinogenicity of acetaldehydes are discussed (4)

Irritancy
100 mg instilled in rabbit eye (duration unspecified) caused mild irritant effects (2).

Any other comments
Decomposes to crotonaldehyde and water when heated.

References
1. Mills, E. J. Jr; et al; *Biological Oxidation of Synthetic Organic Chemicals Proc.* 1953, 8th Purdue Ind. Waste Conf.
2. Smyth, H. F; et al. *J. Ind. Hyg. Toxicol.* 1949, **31**, 60
3. *Union Carbide Data Sheet* 1967, Union Carbide Corp., New York
4. Bittersohl, G. *Arch. Geschwulstforsch.* 1974, **43**(2), 172-176

A68   Aldoxycarb

CH3SO2C(CH3)2CH=NOCONHCH3

CAS Registry No. 1646-88-4
Synonyms 2-methyl-2-methylpropionaldehyde *O*-methylcarbamoyloxime;
2-methyl-2-methylsulfonylpropionaldehyde *O*-methylcarbamoyloxime;
2-methyl-2-methyl sulfonyl propionaldehyde *O*-methylcarbamoyloxime;
2-methyl-2-(methlsulfonyl)propanal *O*-(methylcarbamoyl)oxime
Mol. Formula $C_7H_{14}N_2O_4S$ **Mol. Wt.** 222.26
Uses Systemic insecticide and nematocide.

Physical properties
M. Pt. 140-142°C; **Volatility** v.p. 9.0×10^{-5} mmHg at 25°C.

Solubility
Organic solvent: acetone, methanol, acetone, acetonitrile, chloroform

Ecotoxicity

Fish toxicity
LC_{50} (96 hr) trout, bluegill sunfish 40-55 mg l^{-1} (1,2).

Effects to non-target species
LD_{50} oral mallard duck 33.5 mg kg^{-1} (1,2).
Low toxicity to bees (1).

Environmental fate

Anaerobic effects
In anaerobic reduced subsoil yielded corresponding nitriles and aldehydes as degradation products (3).

Degradation studies

Residual activity in soil for 4-8 wk (1). Major metabolite of aldicarb (4).

Mammalian and avian toxicity

Acute data

LD_{50} oral rat 26.8 mg kg^{-1} (technical material in corn oil) (1,2, 5).
LC_{50} (8 day) inhalation rat 120 mg m^{-3} (1, 2).
LD_{50} (4 hr) inhalation rat 0.14 mg l^{-1} air (2).
LD_{50} dermal rabbit 1000mg mg kg^{-1} (6).
LD_{50} percutaneous rabbit 200 mg kg^{-1} (in corn oil) (1).

Sub-acute data

LC_{50} (8 day dietary) bobwhite quail, mallard duck 5706->10,000 mg kg^{-1} (2).
Life span feeding trials (2 yr) no effect levels were mouse 9.6 mg kg^{-1} day^{-1} and rat
2.4 mg kg^{-1} day^{-1} (1,2)

Metabolism and pharmacokinetics

Aldoxycarb is degraded (species unspecified) through the hydrolysis of the carbamate
ester to aldoxycarb oxime, and elimination of the methylcarbamate group to give
aldoxycarb nitrile (1).

Legislation

Limited under EEC Directive on Drinking Water Quality 80/778/EEC. Pesticides and
related products: maximum admissible concentration 0.1 μg l^{-1} (7).

References

1. *The Agrochemicals Handbook* 3rd ed., 1991, RSC, London
2. *The Pesticide Manual* 9th ed., 1991, British Crop Protection Council, Farnham
3. Bromilow, R. H; et al. *Pest. Sci.* 1986, **1**(5), 535-547
4. Meher, H. C; et al. *Indian. J. Agric. Sci.* 1989, **59**, 771-777
5. *Farm Chemicals Handbook* 1980, **D**, 287
6. *Special Publication of the Entomological Society of America* 1978, **78**(1), 61
7. *EC Directive Relating to the Quality of Water Intended for Human Consumption* 1982,
    80/778/EEC, Office for Official Publications of the European Communities, 2 rue Mercier,
    L-2985 Luxembourg

A69   Aldrin

CAS Registry No. 309-00-2
Synonyms
(1*R*,4*S*,4a*S*,5*S*,8*R*,8a*R*)-1,2,3,4,10,10-hexachloro-1,4,4a,5,8,8a-hexahydro-1,4:5,
8-dimethanonaphalene; 1,2,3,4,10,10-hexachloro- 1,4,4a,5,8,8a-hexahydro-exo-

1,4-endo-5,8-dimethanonaphthalene; HHDN; 1,2,3,4,10,10-hexachloro-
1α,4α,4aβ,5α,8α,8aβ-hexahydro-1,4:5,8-dimethanonaphthalene
Mol. Formula $C_{12}H_8Cl_6$ **Mol. Wt.** 364.92
Uses Insecticide.
Occurrence Has been detected but not quantified in drinking water in the
Netherlands, Canada and the US (1-3).

Physical properties

M. Pt. 104-104.5°C; **B. Pt.** $_2$ 145°; **Specific gravity** 1.70 (20°C); **Volatility** v.p. 6.45
$\times 10^{-5}$ mmHg at 20°C.

Solubility
Organic solvent: acetone, benzene, xylene

Occupational exposure

US TLV (TWA) 0.25 mg m^{-3}; **UK Long-term limit** 0.25 mg m^{-3}; **UK Short-term
limit** 0.75 mg m^{-3}; **Supply classification** toxic.

Risk phrases
Toxic in contact with skin and if swallowed – Possible risk of irreversible effects –
Danger of serious damage to health by prolonged exposure (R24/25, R40, R48)

Safety phrases
Do not breathe dust – Wear protective clothing and gloves – If you feel unwell, seek
medical advice (show label where possible) (S22, S36/37, S44)

Ecotoxicity

Fish toxicity
LC_{50} (96 hr) chinook salmon, rainbow trout, fathead minnow, black bullhead, channel
catfish, bluegill sunfish, largemouth bass 2.6-53 µg l^{-1} in a static bioassay at 13-24°C
(4).
LC_{50} (96 hr) threespine stickleback 27.4 ppb in a static bioassay (5).
Subacute dose 0.14 ppm aldrin induced hyperchloremia in catfish within 4 days,
while sublethal doses 0.035 ppm caused hypochloremia at 15, 25, 35, 50 and 70 days
after exposure. The decrease in chlorine concentration was not significant in treated
fish at 25 and 35 days (6).
In long-term static bioassays ≤4 months *Puntius conchonius* 0.0466 µg l^{-1} only stage
III oocytes were totally resistant to damage (7).

Invertebrate toxicity
EC_{50} (48 hr) *Daphnia magna* 28 mg l^{-1}.
Mercenaria mercenaria 10 day eggs treated with 1000 ppb 0% survival (5).
LC_{50} (96 hr) *Pteronarcys california, Acroneuria pacifica* 180- 200 µg l^{-1} (8).
LC_{50} (96 hr) scud, glass shrimp, stonefly 1.3-4300 µg l^{-1} in a static bioassay at
15-21°C (4).
EC_{50} (48 hr) daphnid, seed shrimp 18-32 µg l^{-1} in a static bioassay at 15-21°C (4).
In vitro administration of aldrin (concentration and duration unspecified) *Panaeid*
prawn caused inhibition of acid and alkali phosphatase activity in stomach, muscle,
gill and brain in a dose-dependent manner (9).

Bioaccumulation

Confirmed to be accumulated on a high level (10).

Bioconcentration factors in molluscs 4571, golden orfe 3890 and *Chlorella fusca* 12,260 (11- 13).

Anabaena sp. and *Aulosira fertilissima* bioconcentration ranges were 3.9-247.5 μg g^{-1} and 6.3-302.3 μg g^{-1}, respectively. Maximum concentration of aldrin was reached in 8.16 hr and metabolism to dieldrin occurred (14).

Effects to non-target species

Toxic to bees (15).

Environmental fate

Degradation studies

Biodegradable (10).

75%-100% disappearance from soil in 1-6 yr (5).

No biodegradation of aldrin at 5 and 10 mg l^{-1} was observed using a mixed culture inoculum from sewage (16).

Dunaliella sp. degraded 23.3% initial aldrin (concentration unspecified) to dieldrin and 5.2% to the diol (17).

Calculated soil sorption coefficient suggest minimal leaching to groundwater (18).

Abiotic removal

Degradation products of 5 mg saturated aldrin vapour treated with a sunlamp for 45 hr were dieldrin (50-60 μg) and photoaldrin (20-30 μg) (19).

Mammalian and avian toxicity

Acute data

LD$_{Lo}$ oral child 1250 μg kg^{-1} (20).

LD$_{50}$ oral rat, rabbit, hamster, guinea pig 33-100 mg kg^{-1} (21).

LD$_{50}$ oral bobwhite quail (3-4 month $\female$) 7 mg kg^{-1} (4).

LD$_{50}$ dermal rat, rabbit 15-98 mg kg^{-1} (21).

Carcinogenicity and long-term effects

Inadequate evidence for carcinogenicity to humans, IARC classification group 3 (22). Feed F344 rats, B6C3F1 mice (74-80 wk) 0.0008-0.006% caused liver carcinomas only in mice, US National Toxicology Program classification D, single sex of a single species with a single tissue affected (23).

Metabolism and pharmacokinetics

Converted into the major metabolite dieldrin in soil, water and living organisms. Dieldrin is oxidised in the liver to a hydrophilic hydroxy compound and is subsequently eliminated in the urine as the glucuronic acid conjugate (15,24-28). Intravenous rat (concentration unspecified) detected in liver, duodenum, intestine and faeces (29).

Genotoxicity

Escherichia coli PQ37 SOS-Chromotest with or without metabolic activation negative (30).

Oral mouse 13, 19.5 and 39 mg kg^{-1} caused dose-dependent increases in the frequency of autosomal univalents and sex univalents. Aneuploidy and translocations were reported only for highest dose while polyploidy was significantly increased at 19.5 and 39 mg kg^{-1} (31).

Any other adverse effects to man

In a study of 100 women, aldrin was detected in maternal blood, placenta and umbilical cord blood indicating placental transfer (32).

Any other adverse effects

Central nervous system disturbances including nausea, vomiting, tremors, ataxia, muscular incoordination, epileptic convulsions, renal damage, albuminuria, haematuria and respiratory failure (25).

Oral rat (13 or 26 day) unspecified concentration of aldrin caused extensive degeneration of all varieties of germ cells at stage VII, reduction in sperm count, luteinising hormone and testosterone. It was concluded that aldrin may have a direct inhibitory influence on gonadotropin release, but the possibility of direct action on the testes is discussed (33).

Legislation

Cited as a prescribed substance for release into water under Schedule 5, S.I. No. 472, 1991 (34).

Limits under EEC Directive on Drinking Water Quality 80/778/EEC. Pesticides and related products: maximum admissible concentration 0.1 mg l^{-1} (35).

Any other comments

Toxicology, human exposure and health effects reviewed (36-39).

The environmental fate and occurrence of aldrin in human food and human tissue has been extensively reviewed (39,40).

Toxicity and hazards reviewed, to be published in 1992-1993 (26).

The response of fish to toxicants of this type is heavily dependant upon the manner in which they are formulated, because this effects the distribution of the compound in the water.

References

1. Kraybill, H. F. *NY Acad. Sci. Ann.* 1977, **298**, 80-89
2. Kopfler, F. C; et al. *Adv. Environ. Sci. Technol.* 1977, **8**, 419-433
3. Kool, H. J; et al. *CRC Crit. Rev. Environ. Cont.* 1982, **12**, 307-357
4. *Handbook of Acute Toxicity of Chemicals to Fish and Aquatic Invertebrates* 1980, 137, US Dept. of the Interior, Fish and Wildlife Service, US Govt. Print. Off., Washington, DC
5. Verschueren, K. *Handbook of Environmental Data on Organic Chemicals* 2nd ed., 1983, Van Nostrand Reinhold, New York
6. Srivastava, A. K; et al. *J. Environ. Biol.* 1988, **9** (Suppl 1), 91-95
7. Kumar, S; et al. *Bull. Environ. Contam. Toxicol.* 1988, **41**(2), 227-232
8. Jensen, L.D; et al. *J. Water Pollut. Control Fed.* 1966, **38**(8), 1273-1286
9. Reddy, M. S; et al. *Biochem. Int.* 1990, **22**(6), 1033-1040
10. *MITI Report* 1984, Ministry of International Trade and Industry, Tokyo
11. Hawker, D. W; et al. *Ecotoxicol. Environ. Safety* 1986, **11**, 184-197
12. Freitag, D; et al. *Ecotoxicol. Environ. Saf.* 1982, **6**, 60-81
13. Geyer, H; et al. *Chemosphere* 1984, **13**, 269-284
14. Dhanaraj, P. S; et al. *Agric. Ecosyst. Environ.* 1989, **25**(2-3), 187-193
15. *The Agrochemicals Handbook* 3rd ed., 1991, RSC, London
16. Tabak, H. H; et al. *Proc. Sym. Assoc. Off. Anal. Chem. 94th Ann. Meeting* 1981, 267-328

17. Patil, K. C; et al. *Environ. Sci. Technol.* 1972, **6**, 629-632
18. Kenaga, E. E. *Ecotoxicol. Environ. Saf.* 1980, **4**, 26-38
19. Crosby, D. G; et al. *Arch. Environ. Contam. Toxicol.* 1974, **2**, 62-74
20. Deichmann, W. *Toxicology of Drugs and Chemicals* Academic Press,
21. Sax, N. I; et al. *Dangerous Properties of Industrial Materials* 7th ed., 1989, Van Nostrand Reinhold, New York
22. *IARC Monograph*, 1987, **Suppl. 7**, 88
23. Ashby, J; et al. *Mutat. Res.* 1988, **204**, 17-115
24. Manno, M. *Chemistry, Agriculture and the Environment* 1991, Richardson, M. L. (Ed.), RSC, London
25. Keith, L. H; et al. (Ed.) *Compendium of Safety Data Sheets for Research and Industial Chemicals* 1985, VCH Publishers
26. Izmerov, N. F; *Scientific Reviews of Soviet Literature on Toxicity and Hazards of Chemicals* 1992-1993, **114**, Eng. Trans. Richardson, M. L. (Ed.) UNEP/IRPTC, Geneva
27. Hayes, W. J. *Pesticides Studied in Man* 1982, 234, Williams and Wilkins, Baltimore
28. Menzie, C. M. *Metabolism of Pesticides* 1969, 25, US Dept. of the Interior, Fish and Wildlife, Publication 127, Washington DC
29. Saxena, M. C; et al. *Arch. Toxicol.* 1981, **48**(2-3), 127-134
30. Mersch-Sundermann, V; et al. *Zentralbl. Hyg. Umweltmed.* 1989, **189**(2), 135-146, (Ger.) (*Chem. Abstr.* **112**. 113966u)
31. Usha Rani, M. V; et al. *IRCS Med. Sci.* 1986, **14**(11), 1125-1126
32. Gaines, T. B. *Toxicol. Appl. Pharmacol.* 1969, **14**, 515-534
33. Chatterjee, S; et al. *J. Endocrinol.* 1988, **119**(1), 75-81
34. *S.I. 1991, No. 472 The Environmental Protection (Prescribed Processes and Substances) Regulations* 1991, HMSO, London
35. *EC Directive Relating to the Quality of Water Intended for Human Consumption* 1982, 80/778/EEC, Office for Official Publications of the European Communities, 2 rue Mercier, L-2985 Luxembourg
36. *Gov. Rep. Announce Index US* 1989, **89**(22), Order No. PB89-214514
37. *Gov. Rep. Announce Index US* 1988, **88**(13), Order No. PB88-179403
38. *Dangerous Prop. Ind. Mater. Rep.* 1988, **8**(2), 45-50
39. Scheunert, I. *SCOPE Ecotoxicol. Clim.* 1989, **38**, 299-316, Muenchen Inst. Bodenoekol. Gesellsch. Strahlen-Umwelstforsch, Neuherberg
40. *Environmental Health Criteria 91, Aldrin and Dieldrin* 1989, World Health Organisation, Geneva

A70 Alkyl(C$_{14}$-C$_{16}$)dimethylbenzylammonium chloride

$$\left[\text{C}_6\text{H}_5-\text{CH}_2\overset{\overset{\displaystyle CH_3}{|}}{\underset{\underset{\displaystyle CH_3}{|}}{N}}{}^+-(CH_2)_nCH_3 \right] Cl^- \quad n = 13\text{-}15$$

CAS Registry No. 63449-41-2
Synonyms benzyldimethylalkyl(C$_{14}$-C$_{16}$)ammonium chloride; Roccal; Tret-o-lite XC511
Uses Bactericide. Fungicide.

Ecotoxicity

Fish toxicity
LC_{50} (24-96 hr) harlequin fish 2.45-0.62 mg l^{-1} (1).

Mammalian and avian toxicity

Acute data
LD_{50} oral rat, mouse 150-300 mg kg^{-1} (2,3).
LD_{50} dermal rat 1420 mg kg^{-1} (2).

Irritancy
2 mg instilled into eye of rat, mouse, dog, guinea pig and hamster, caused severe irritation (4,5).

Any other adverse effects to man

Human fatalities have been reported for alkyldimethylbenzylammonium chloride where the alkyl group ranges from C_8-C_{18} and C_{15}-C_{18} after intramuscular or intravenous administration as well as intrauterine instillation (6).

References

1. Toobey, E; et al. *Chem. Ind.* 1975, **6**, 52
2. *Pharm. Chem. J.* 1978, **12**, 1593
3. *Pesticide Index* 1976, College Sci. Publ., PA, **4**, 38
4. *Food and Cosmetics Toxicology* 1977, **15**, 131
5. *Am. J. Ophthal.* 1974, **78**, 98
6. Gosselin, R. E; et al. *Clinical Toxicology of Commercial Products* 5th ed., 1984, **III**, 63-66, Williams & Wilkins, Baltimore, MD

A71   *N*-Alkyl(C8-C18)dimethylbenzylammonium chloride

$$\left[\text{C}_6\text{H}_5 - \text{CH}_2 - \overset{\overset{\displaystyle CH_3}{|}}{\underset{\underset{\displaystyle CH_3}{|}}{N}}{}^+ - CH_2(CH_2)_{6\text{-}12}CH_3 \right] Cl^-$$

CAS Registry No. 8001-54-5
Synonyms cetalkonium chloride; benzalkonium chloride
Uses Cationic surfactant, germicide and fungicide. Used in leather processing and textile dyeing industries. As a general antibacterial agent.

Physical properties
Specific gravity d^{20} 0.988.

Solubility
Organic solvent: ethanol, acetone, diethyl ether, carbon tetrachloride

Ecotoxicity

Fish toxicity

Threespine stickleback exposed to 2 mg l^{-1} died within 4-6 hr and steelhead trout died within 2-4 hr (1).

Mammalian and avian toxicity

Sensitisation

An allergic reaction was reported in one patient to N-alkyl(C_8- C_{18}) dimethylbenzylammonium chloride used as a preservative in nose drops and confirmed by a challenge which produced nasal congestion and irritation of the eyes and throat lasting 48 hr (2).

At a 0.01-0.1% concentration no allergenic activity was observed in rabbits, guinea pigs and dogs and skin irritating action was dependent on the nature, concentration and number of applications (3).

Any other adverse effects to man

Reported to be toxic in humans (4). Inflammation of the eye and deterioration of vision occurred 3 days after change of soaking solution for a soft contact lens to one containing N-alkyl(C_8- C_{18})dimethylbenzylammonium chloride (5).

In addition to inhibiting sperm motility there is evidence that benzalkonium chloride disturbs the electrolyte balance in the aqueous phase of cervical mucous, making it hostile to sperm (6).

A woman occupationally exposed to disinfectant containing N-alkyl(C_8-C_{18})dimethylbenzylammonium chloride developed asthmatic symptoms and a skin test and bronchial provocation test proved her sensitivity which could be blocked by sodium chromoglycate (7).

Any other adverse effects

Intramuscular rat induced myofascial edematous swelling with increased succinate oxidation and Na^+, K^+, Mg^{2+} ATPase activity 30 min after administration (8).

Dermal application of 13% and 15% N-alkyl(C_8-C_{18})dimethylbenzylammonium chloride caused death in 9 of 48 and 20 of 48 mice, respectively. Survivors developed skin lesions and lost weight. Applications of 0.08% or 3% had minor effects (8).

Any other comments

Incompatible with soaps and other anionic surfactants, citrates, iodides, nitrates, permanganates, salicylates, silver salts and tartrates. Incompatible with some commercial rubber mixes or plastics, aluminium, cotton dressings, fluorescein sodium, hydrogen peroxide, kaolin, hydrous wool fat and some sulfonamides (9). Exposure, experimental and environmental toxicology, and human health effects reviewed (10).

References

1. *Lethal Effects of 2014 Chemicals Upon Sockeye Salmon, Steelhead Trout and Threespine Stickleback* 1989, EPA 560/6-89-001
2. Hillerdal, G. *J. Otorhinolayngol Borderl.* 1985, **47**, 278
3. Krivoshein, Yu. S; et al. *Tr. Krym. Gos. Med. Inst.* 1977, **72**, 98-100

4. Honigman, J. L. *Pharm. J.* 1975, **2**, 523
5. Gasset, A. R. *Am. J. Ophthal.* 1977, **48**, 169
6. Pearson, R. M. *Pharm. J.* 1985, **1**, 686
7. Innocenti, A. *Med. Lav.* 1978, **69**(6), 713-715
8. Futami, T; et al. *Jpn. J. Pharmacol.* 1977, **27**(4), 523-529
9. *Martindale: The Extra Pharmacopoeia* 29th ed., 1989, The Pharmaceutical Press, London
10. *ECETOC Technical Report No. 30(4)* 1991, European Chemical Industry Ecology and Toxicology Centre, B-1160 Brussels

A72 *N*-Alkyl(C$_8$-C$_{18}$)dimethyl-3,4-dichlorobenzylammonium chloride

$$\left[\text{Cl} - \text{(ring)} - \text{CH}_2\overset{\displaystyle \text{CH}_3}{\underset{\displaystyle \text{CH}_3}{\text{N}}}{}^+ - (\text{CH}_2)_{7\text{-}17}\text{CH}_3 \right] \text{Cl}^-$$

CAS Registry No. 8023-53-8

Synonyms Tetrosan 3,4D; 3,4-dichlorobenzylammonium chloride, *N*-alkyldimethyl-
Uses Pesticide. Antiseptic, germicide, algicide, sanitiser and deodorant.

Mammalian and avian toxicity

Acute data
LD$_{50}$ oral guinea pig, rat, mouse 316-2000 mg kg^{-1} (1).
LD$_{50}$ intravenous mouse 50 mg kg^{-1} (2).

Irritancy
1% instilled in rabbit eye (unspecified duration) caused severe irritant effects (2).

Any other comments
Toxicity reviewed (3).

References
1. *Soap Sani. Chem.* (SSCHAH), 1949, **25**, 125
2. *J. Am. Pharm. Assoc. Sci. Ed.* 1949, **38**, 428
3. *Arch. Toxicol.* 1974, **32**, 245

A73   N-Alkyl-1,3-propanediamine

Tallow-NH(CH₂)₃NH-Tallow

CAS Registry No. 61791-55-7
Synonyms *N*-tallow-1,3-propanediamine
Uses In water treatment (1).

Ecotoxicity

Effects to non-target species
LC_{50} larvae and pupae Southern house mosquito 0.9 mg kg^{-1} (2,3).

Any other comments
When a series of long-chain aliphatic diamines of the general formation $RR_1N(CH_2)_nNR_2C_2H_{2n+1}$ were tested *in vitro* for their activity against *Streptococus mutans*, all compounds with n=14-18 showed high activity (4).

References
1. Moran, F. *Fr. Demande* No 2359076, 17 Feb 1978 (*Chem. Abstr.* **90**, 141902r)
2. Mulla, M. S; et al. *J. Econ. Entomol.* 1970, **63**, 1972
3. Mulla, M. S; et al. *J. Econ. Entomol.* 1970, **60**, 115
4. Bass, G. E; et al. *J. Dent. Res.* 1975, **54**, 972

A74   Allethrin

CAS Registry No. 584-79-2
Synonyms cyclopropanecarboxylic acid, 2,2-dimethyl-3- (2-methyl-1-propenyl)-, 2-methyl-4-oxo-3-(2-propenyl)-2- cyclopenten-1-yl ester; 2-cyclopenten-1-one, 2-allyl-4- hydroxy-3-methyl, 2, 2-dimethyl-3-(2-methylpropenyl)cyclopropanecarboxylate; chrysanthemummonocarboxylic acid, 3-allyl-2-methyl-4-oxo- cyclopentene-1one; cyclopropanecarboxylic acid, 2,2-dimethyl-3-(2-methylpropenyl)-, ester with 2-allyl-4-hydroxy-3-methyl-2-cyclopenten-1-one; Allyl cinerin I; d- Allethrin; Pynamin; Pyresyn
Mol. Formula $C_{19}H_{26}O_3$ **Mol. Wt.** 302.42
Uses Synthetic pyrethroid, non-systemic with contact and respiratory action, used against household pests.

Physical properties

M. Pt. ≈ 4°C (*dl-trans-* allethrin 51°C); **B. Pt.** 140°C at 0.1 mmHg; **Specific gravity** d_4^{25} 1.005; **Volatility** v.p. 1.2×10^{-4} mmHg at 30°C.

Solubility
Organic solvent: miscible ethanol, carbon tetrachloride, 1,2-dichloroethane, hexane, xylene, petroleum ether

Occupational exposure
Supply classification harmful.

Risk phrases
Harmful by inhalation, in contact with skin and if swallowed (R20/21/22)

Safety phrases
Keep out of reach of children – Keep away from food, drink and animal feeding stuffs (S2, S13)

Ecotoxicity
Fish toxicity
LC_{50} (96 hr) channel catfish, bluegill sunfish, rainbow trout, steelhead, salmon 10-56 $\mu g\ l^{-1}$ (1,2).

Invertebrate toxicity
LC_{50} (96 hr) freshwater shrimp 8-11 $\mu g\ l^{-1}$.
EC_{50} (48 hr) *Daphniidae* 21-56 $\mu g\ l^{-1}$.
LC_{50} (96 hr) *Pteronarcys california* 2.1 $\mu g\ l^{-1}$ (1).

Effects to non-target species
LD_{50} oral honey bee 3.4 $\mu g\ bee^{-1}$ at 26-27°C (3).
LD_{50} contact honey bee 9.1 $\mu g\ bee^{-1}$ at 26-27°C (3).

Abiotic removal
Photochemical changes occurring in the acid moiety involve stepwise oxidation of the *trans*-methyl group to the alcohol, aldehyde, carboxyl derivatives and oxidation of the double bond to a keto group with subsequent formation of *trans*-carbonic acid esters (4).

Mammalian and avian toxicity
Acute data
LD_{50} oral rat, mouse 480-1100 mg kg^{-1} (5,6).
LD_{50} oral rabbit, bobwhite quail 2030- 4290 mg kg^{-1} (5-7).
LD_{50} percutaneous rat >2500 mg kg^{-1} (5,6).
LD_{Lo} intravenous rat, intracerebral mouse 4 mg kg^{-1} (8,9).

Sub-acute data
Wistar rats in diet (12 wk) 5000-15,000 mg kg^{-1} caused decrease in body-weight gain and an increase in liver to kidney weight ratio (10).

Metabolism and pharmacokinetics
In mammals, following oral administration one of the two terminal methyl groups of the chrysanthemic acid moiety is oxidised in the liver to an alcohol group, and further to a carboxyl group (6).
Allethrin administered orally to mammals is absorbed from the intestinal tract and distributed to tissues. A 500 mg kg^{-1} dose of the *trans*-isomer was excreted via the urine and faeces within 20 days (10).

Oral rat 1-5 mg kg^{-1} 60% eliminated in urine and faeces within 48 hr. Major metabolic reactions include ester hydrolysis, oxidation at *trans*-methyl of the isobutenyl group, *gem*- dimethyl of cyclopropane ring, and the methylene of the allyl group and 2,3-diol formation at the allylic group. Major urinary metabolites were chrysanthemum dicarboxylic acid, and allethrolone (11).

Irritancy

Application of 10% olive oil solution to rabbit eye, slight hyperaemia of the conjunctiva and eye discharge observed 10 min and 2 hr after application, respectively. A 5% olive oil solution applied to guinea pig skin on alternate days for 20 days, animals were challenged with an intradermal injection 2 wk later, no sensitisation observed but slight lymphocytic and monocytic infiltration of the dermis was noted (11).

Genotoxicity

In vitro Chinese hamster ovary cells sister chromatid induction negative (12).

Any other adverse effects

Symptomatic effects in a range of species include hyperactivity, tremors, convulsion and paralysis (11).

Legislation

Limited under EC Directive on Drinking Water Quality 80/778/EEC. Pesticide: maximum admissible concentration 0.1 μg l^{-1} (13).

Any other comments

Comprehensive review of toxicology and environmental fate (11).
The toxicology of pyrethroids including epidemiology studies and neurotoxicological potential have been reviewed (14,15).
Toxicity and hazards reviewed, to be published in 1992-1993 (16).
Allethrin refers to a mixture of *cis* and *trans* allethrins.

References

1. Verschueren *Handbook of Environmental Data on Organic Chemicals* 2nd ed., 1983, Van Nostrand Reinhold, New York
2. Mauck, W. L; et al. *Arch. Environ. Contam. Toxicol.* 1976, **4**, 18-29
3. Stevenson, J. H; et al. *Poisoning of Honey-Bee by Pesticides* 1978, **2**, 55-72
4. Menzie, C. M. *Metabolism of Pesticides. An Update. US Dept. of the Interior, Fish, Wildlife Service, Special Scientific Report* 1974, 307, US Govt. Print. Off., Washington, DC
5. *The Pesticide Manual* 9th ed., 1991, British Crop Protection Council, Farnham
6. *The Agrochemicals Handbook* 3rd ed., 1991, RSC, London
7. *Special Publication of the Entomological Society of America* 1978, **78**(1), 7
8. Verschoyle, R. D; et al. *Pesticide Biochem. Physiol.* 1972, **2**, 308-311
9. Gammon, D. W; et al. *Toxicol. Appl. Pharmacol.* 1982, **66**, 290
10. Miyamoto, J. *Environ. Health Perspect.* 1976, **14**, 15-28
11. *Environ. Health Criteria 87: Allethrin* 1989, WHO, Geneva
12. Wang, T. C; et al. *Bull. Inst. Zool. Acad. Sin.* 1988, **27**(2), 111-117
13. *EC Directive Relating to the Quality of Water Intended for Human Consumption* 1982, 80/778/EEC, Office for Official Publications of the European Communities, 2 rue Mercier, L-2985 Luxembourg
14. Aldridge, W. N. *CRC Critical Reviews in Toxicology* 1990, **21**(2), 89-104

15. Vijverberg, H. P. M; et al. *CRC Critical Reviews in Toxicology* 1990, **21**(2), 105-126
16. Izmerov, N. F; *Scientific Reviews of the Society Literature on Toxicity Hazards of Chemicals* 1991, **119**, Eng. Trans. Richardson, M. L. (Ed.), UNEP/IRPTC, Geneva

A75   Allidochlor

ClCH2CON(CH2CH=CH2)2

CAS Registry No. 93-71-0
Synonyms *N,N*-diallylchloroacetamide; 2-chloro-*N,N*-di-2-propenylacetamide; CDAA; Randox; α-chloro-*N, N*-diallylacetamide
Mol. Formula $C_8H_{12}ClNO$ **Mol. Wt.** 173.64
Uses Herbicide.

Physical properties

B. Pt. 125°C (decomp.); **Specific gravity** d^{25} 1.09; **Volatility** 0.0094 mmHg at 20°C.

Solubility
Water: 2 g l^{-1}. Organic solvent: ethanol, hexane

Occupational exposure

Supply classification harmful.

Risk phrases
Harmful by inhalation, in contact with skin and if swallowed – Irritating to eyes and skin (R20/21/22, R36/38)

Safety phrases
Keep out of reach of children – Keep away from food, drink and animal feeding stuffs (S2, S13)

Ecotoxicity

Fish toxicity
LC_{50} (24 hr) carp, rainbow trout 2-8 mg l^{-1} at 15°C (1-3).

Effects to non-target species
Toxic to bees (4).
LC_{50} (duration unspecified) tadpole 3.3 ppm at pH 7-9 (5).

Environmental fate

Degradation studies
Major route of decomposition is via microbial degradation (6).

Abiotic removal
Hydrolysis is secondary to microbial breakdown when considering degradation (6).

Mammalian and avian toxicity

Acute data
LD_{50} oral rat 700 mg kg^{-1} (7).
LD_{50} percutaneous rabbit 1350 mg kg^{-1} (4).

Metabolism and pharmacokinetics

Allidochlor is readily absorbed through skin, rapid hydrolytic action leads to the removal of the chlorine atom. Metabolite detected was glycol acid (7).

Irritancy

Skin irritant in humans. Can be absorbed directly via the skin (4, 8).

Legislation

Limited under EC Directive on Drinking Water Quality 80/778/EEC. Pesticides and related products: maximum admissible concentration 0.5 μg l^{-1} (9).

Any other comments

Experimental toxicology and human health effects reviewed (10).

References

1. Gosselin, E. E; et al. *Clinical Toxicology of Commercial Products* 5th ed., 1981, Williams and Wilkins, London
2. Nishiuchi, Y. *Suisan Zoshoko* 1977, **24**(4), 140-145
3. Nishiuchi, Y. *Suisan Zoshoko* 1980, **28**(2), 107-112
4. Hamm, P. C; et al. *J. Agric. Food Chem.* 1956, **4**, 518
5. Miles, J. R. W; et al. *J. Environ. Sci. Health* 1983, **B18**(3), 305-315
6. *Herbicide Handbook* 4th ed., 1979, Weed Society of America, Champaign, IL
7. Bailey, G. W; et al. *Residue Reviews*, 1965, **10**, 97
8. *The Merck Index* 11th ed. 1989, Merck & Co. Inc., Rahway, NJ
9. *EC Directive Relating to the Quality of Water Intended for Human Consumption* 1982, 80/778/EEC, Office for Official Publications of the European Communities, 2 rue Mercier, L-2985 Luxembourg
10. *ECETOC Technical Report No. 30(4)* 1991, European Chemical Industry Ecology and Toxicology Centre, B-1160 Brussels

A76   Allopurinol

CAS Registry No. 315-30-0
Synonyms 1,5-dihydro-4*H*-pyrazolo[3,4-d]pyrimidin-4-one; 1*H*-pyrazolo[3,4-d]pyrimidin-4-ol; 4-hydroxypyrazolo[3,4-d]pyrimidine; Zyloric; Zyloprim
Mol. Formula $C_5H_4N_4O$ **Mol. Wt.** 136.11
Uses Treatment of hyperuricaemia and gout; inhibitor of xanthine oxidase.

Physical properties

M. Pt. >350°C.

Solubility

Water: 0.5 g l^{-1} at 25°C. Organic solvent: dimethyl formamide

Mammalian and avian toxicity

Acute data

LD_{50} intraperitoneal rat 900 mg kg^{-1} (1).

Sub-acute data

LD_{Lo} (22 day) oral woman 88 mg kg^{-1} intermittent blood effects.
LD_{Lo} oral man 22.4 mg kg^{-1} 5 day intermittent central nervous system, blood and liver effects (2).

Metabolism and pharmacokinetics

Converted in the liver to oxypurinol, an inhibitor of xanthine oxidase activity, clearance $t_{1/2}$ 18-30 days (1,3).

Oral 20% of inital unspecified dose was excreted in faeces in 48-72 hr. Peak plasma concentration was reached in 2-6 hr. Allopurinol and its metabolite alloxanthine are distributed in total tissue water (except the brain) without binding to plasma proteins (4).

Genotoxicity

Subcutaneous rat 10-100 mg kg^{-1} (duration unspecified) negative results for mitotic index and chromosomal aberations (5).

Any other adverse effects to man

Can cause cataractogenesis if given as a treatment for gout in human diabetic patients (6).
A 15 yr old girl who swallowed 22.5 g allopurinol did not suffer any ill effects and after 84 hr, $t_{1/2}$ 3.6 hr (7).

Any other adverse effects

Symptoms of allergy include fever chills, leucopenia, leucocytosis, eosinophilia, arthelgia and vasculitis leading to renal and hepatic damage. These reactions can be severe or fatal. Other side effects include peripheral neuritis, nausea, vomiting, diarrhoea, headache, drowsiness and vertigo (3).

Used for treating heart or stroke victims to increase blood flow and reduce tissue damage to the injured ischemic tissue (8).

Any other comments

Use of allopurinol in providing organ protection following hepatic transplantation reviewed (9).

Pharmacokinetics in humans reviewed (10).

References

1. *Advances in Teratology* 1968, **3**, 181
2. *Ann. Rheum. Dis.* 1981, **40**, 245
3. *Martindale: The Extra Pharmacopoeia* 29th ed., 1989, The Pharmaceutical Press, London
4. Goodman, L. S; et al. *The Pharmacological Basis of Therapeutics* 5th ed., 1975, 353-354, MacMillan Publishing Co., New York
5. Rakhmatullin, E. K; et al. *Veterinariya (Moscow)* 1989, (9), 60-61, (*Chem. Abstr.*, no, **111**, 208736n)
6. Ansari, N. H. *Biochem. Biophys. Res. Commun.* 1990, **168**(3), 939-943
7. Ferner, R. E; et al. *Hum. Toxicol.* 1988, **7**(3), 293-294
8. Babbs, C. F; et al. *Purdue Research Foundation* 1990, US-4978668A
9. Toledo-Pereyra, L.H. *Basic Life Sci.* 1988, **49**, (Oxygen Radicals Biol. Med.) 1047-1052
10. Murrell, G; et al. *Clin. Pharmacokinet.* 1986, **11**(5), 343-353

A77   Alloxydim-sodium

CAS Registry No. 55635-13-7
Synonyms Cyclohexanecarboxylic acid,
2,2-dimethyl-4,6-dioxo-5-[1-[(2-propenyloxy)imino]butyl]-, methyl ester, ion(1-),
sodium; alloxydimedon sodium; BAS 9021; Fervin; Kusgard; NP 48; Tritex
Mol. Formula $C_{17}H_{24}NNaO_5$ **Mol. Wt.** 345.37
Uses Post-emergence herbicide.

Physical properties

M. Pt. 185.5°C (decomp.); **Volatility** v.p. $<1.0 \times 10^{-6}$ mmHg at 25°C.

Solubility
Organic solvent: dimethyl formamide, methanol, ethanol

Ecotoxicity

Fish toxicity
LC_{50} (96 hr) carp, trout 2000-2600 mg l^{-1} [1,2].

Effects to non-target species
Non-toxic to bees.
LD_{50} oral Japanese quail 2970 mg kg^{-1} [1,2].

Environmental fate

Degradation studies
$t_{1/2}$ in soil 2-10 days, variable with time of year [1,3].

Mammalian and avian toxicity

Acute data
LD_{50} oral rat 2260-2322 mg kg^{-1} [1,2].
LD_{50} oral mice 3000-4600 mg kg^{-1} [1,2].
LD_{50} dermal rabbit, rat 2000->5000 mg kg^{-1} [1,2].
LD_{50} intraperitoreal rat 1700 mg kg^{-1} [1,2].

Carcinogenicity and long-term effects
In 2 yr feeding studies mice and rats given 100 mg kg^{-1} in diet showed no adverse
effects [2].

Metabolism and pharmacokinetics
Oral rat (7 day) 1,5-^{14}C labelled compound was excreted within 48 hr in the urine [4].

References

1. *The Agrochemicals Handbook* 3rd ed., 1991, RSC, London
2. *The Pesticide Manual* 9th ed., 1991, British Crop Protection Council, Farnham
3. Shigeo, O; et al. *Nippon Noyaku Gakkaishi* 1984, **9**(3), 471-480 (Jap.) (*Chem. Abstr.* **103**, 18379K)
4. Tanoue, T; et al. *Nippon Noyaku Gakkaishi* **4**(3), 315-322 (Jap.) (*Chem. Abstr.* **92**, 5306060a)

A78 Allyl acetate

$CH_3CO_2CH_2CH=CH_2$

CAS Registry No. 591-87-7
Synonyms acetic acid, allyl ester; acetic acid, 2-propenyl ester; 3-acetoxypropene
Mol. Formula $C_5H_8O_2$ **Mol. Wt.** 100.12
Uses Intermediate in polymer manufacture.

Physical properties

B. Pt. 104°C; **Flash point** 22°C; **Specific gravity** 0.928;
Volatility v. den. 3.45.

Solubility
Organic solvent: ethanol, diethyl ether, acetone

Occupational exposure

UN No. 2333; **HAZCHEM Code** 3WE; **Conveyance classification** flammable
liquid;toxic substance.

Mammalian and avian toxicity

Acute data
LD_{50} oral rat, mouse 130-170 mg kg^{-1} (1-3).
LC_{50} (1 hr) inhalation rat 1000 ppm (1).
LD_{50} dermal rabbit 1021 mg kg^{-1} (2).

Irritancy
Dermal rabbit (24 hr) 10 mg and 100 mg instilled in rabbit eye caused mild irritation
(2,4).

Any other adverse effects

Allyl acetate is absorbed through intact skin and rapidly hydrolysed in the body to
allyl alcohol and acetic acid (4).

References

1. *J. Ind. Hyg. Toxicol.* 1949, **31**, 60
2. Smyth, H. F; et al. *J. Ind. Hyg. Toxicol.* 1949, **31**, 60-62
3. Jenner, P. M; et al. *Food Cosmet. Toxicol.* 1964, **2**, 327-343
4. *Patty's Industrial Hygiene and Toxicology* 3rd ed., 1981, **2**, Clayton, G. D; et al. (Eds.)
   John Wiley, New York

A79 Allyl alcohol

$CH_2=CHCH_2OH$

CAS Registry No. 107-18-6
Synonyms 2-propen-1-ol; 3-hydroxypropene; vinyl carbinol
Mol. Formula C_3H_6O **Mol. Wt.** 58.08

Uses Used in the manufacture of resins and plasticisers. As an intermediate in the manufacture of pharmaceuticals, flavourings and other organic chemicals. In the manufacture of glycerol, acrolein, military poisons. Herbicide.

Physical properties

M. Pt. –50°C; **B. Pt.** 96-97°C; **Flash point** 21°C; **Specific gravity** d^{15} 0.8573; **Partition coefficient** log P_{ow} –0.25; **Volatility** v.p. 10 mmHg 10.5°C; v. den. 2.0.

Solubility

Water: miscible. Organic solvent: miscible in ethanol, chloroform, diethyl ether, petroleum ether

Occupational exposure

US TLV (TWA) 2 ppm (4.8 mg m^{-3}); **US TLV (STEL)** 4 ppm (9.5 mg m^{-3}); **UK Long-term limit** 2 ppm (5 mg m^{-3}); **UK Short-term limit** 4 ppm (10 mg m^{-3}); **UN No.** 1098; **HAZCHEM Code** 2PE; **Conveyance classification** flammable liquid; **Supply classification** highly flammable and toxic.

Risk phrases

Highly flammable – Very toxic by inhalation – Irritating to eyes, respiratory system and skin (R11, R26, R36/37/38)

Safety phrases

Keep away from sources of ignition – No Smoking – Wear eye/face protection – In case of accident or if you feel unwell, seek medical advice immediately (show label where possible) (S16, S39, S45)

Ecotoxicity

Fish toxicity

LC$_{50}$ (24 hr) goldfish 1 mg l^{-1} (1).

Environmental fate

Nitrification inhibition

Ammonia oxidation inhibition in activated sludge was 75% at concentrations of 19.5 mg l^{-1} (2).

Degradation studies

Confirmed biodegradable (3).
BOD$_5$ 9.1% reduction of dissolved oxygen concentration using settled sewage seed at 20°C increasing to 81.8% in 20 days (4).

Abiotic removal

13.9% degradation to carbon dioxide after 24 hr following photo-oxidation in aqueous medium at 50°C (5).

Mammalian and avian toxicity

Acute data

LD$_{50}$ oral rat, mouse, rabbit 71-105 mg kg^{-1} (6).
LC$_{50}$ (1-4 hr) inhalation rat 76-165 ppm (7).
LD$_{50}$ percutaneous rabbit 89 mg kg^{-1} (8).
LD$_{50}$ intraperitoneal mouse 60 mg kg^{-1} (9).

Carcinogenicity and long-term effects

Gavage ♂ Syrian hamsters (lifetime study) 2 mg wk^{-1} induced tumours of the pancreatic ducts and forestomach. The authors reported that the incidence was insignificant (9).

Teratogenicity and reproductive effects

Chronic treatment with allyl alcohol (dose and duration unspecified) to ♂ rats caused no adverse reproductive effects (10).

Metabolism and pharmacokinetics

Converted by alcohol dehydrogenase in the liver to acrolein resulting in liver toxicity (11).

Allyl alcohol is readily oxidised in min following an intravenous injection of 30 mg kg^{-1} to rats and within 1 hr the alcohol had almost disappeared from blood (12). Oral rat 42 mg kg^{-1} was oxidised to acrolein within 10-15 min. Allyl alcohol elevated alanine aminotransferase, α-glutamyl transpepsidase and glutamate dehydrogenase activity in the plasma and induced lesions in the periportal regions of the liver (13).

Irritancy

Vapour and liquid are intensely irritating to skin and mucous membrane. Produces lachrymation and corneal burns (14).

Sensitisation

Inhalation mice 3.9 ppm depressed the respiratory rate by 50% due to sensory irritation (15).

Genotoxicity

In vitro Chinese hamster V79 cells with and without metabolic activation positive (16).

Any other adverse effects

Can be absorbed through intact skin in both toxic and lethal concentrations. Exposure to toxic concentrations can lead to deep muscle pain, presumably due to spasm (17,18).

Any other comments

Experimental toxicology, epidemiology and human health effects reviewed (19-21).

References

1. Lipnick, R. L; et al. *Xenobiotica* 1987, **17**(8), 1011-1025
2. Tomlinson, T. G; et al. *J. Appl. Bacteriol.* 1966, **29**(2), 266-291
3. *MITI Report* 1984, Ministry of International Trade and Industry, Tokyo
4. Lamb, C. B; et al. *Proc. 8th Indust. Waste Conf. Purdue Univ.* 1952, 329
5. Verschueren, K. *Handbook of Environmental Data of Organic Chemicals* 2nd ed., 1983, 177, Van Nostrand Reinhold, New York
6. Dunlap, M. K; et al. *AMA Arch. Ind. Health* 1958, **18**, 303-311
7. *Health and Environmental Effects Profile for Allyl Alcohol* 1985, 36, USEPA, EC AO-CIN-121
8. *Farm Chemicals Handbook* 1989, C-16, Meister Publishing Co, Willoughby, OH
9. Lijinsky, W; et al. *Toxicol. Ind. Health* 1987, **3**(3), 337-345
10. Jenkinson, P. C; et al. *Mutat. Res.* 1990, **229**(2), 173-184

11. Miccadei, S; et al. *Arch. Biochem. Biophys.* 1988, **265**(2), 302-310
12. Clayton, G. D; et al. *Patty's Industrial Hygiene and Toxicology* 3rd ed., 1981-1982, **2A, 2B**, 2C(Toxicology), 4668, John Wiley Sons, New York
13. Penttila, K. E; et al. *Pharmacol. Toxicol. (Copenhagen)* 1987, **60**(5), 340-344
14. Gosselin, R. E; et al. *J. Clinical Toxicology of Commercial Products* 5th ed., 1981, Williams & Wilkins, London
15. Nielson, G. D; et al. *Acta Pharmacol. Toxicol.* 1984, **54**(4), 292-298
16. Smith, R. A; et al. *Carcinogenesis (London)* 1990, **11**(3), 497-498
17. Arena, J. M; et al. *Poisoning* 5th ed., 1986, 275, Charles C. Thomas, Springfield, IL
18. Hamilton, A; et al. *Industrial Toxicology* 3rd ed., 1974, 299, Publishing Sciences Group, Acton, MA
19. *ECETOC Technical Report No. 30(4)* 1991, European Chemical Industry Ecology and Toxicology Centre, Brussels
20. *BIBRA Toxicity Profile* 1988, British Industrial Biological Research Association, Carshalton
21. Izmerov, N. F; *Scientific Reviews of Soviet Literature on Toxicity & Hazards of Chemicals* 1991, **73**, Eng. Trans. Richardson, M. L. (Ed.), UNEP/IRPTC, Geneva

A80   Allylamine

$CH_2=CHCH_2NH_2$

CAS Registry No. 107-11-9

Synonyms 2-propen-1-amine; 3-aminopropene; 3-aminopropylene; 2-propenamine; monoallylamine; 2-propenylamine

Mol. Formula C_3H_7N **Mol. Wt.** 57.10

Uses Used in the manufacture of pharmaceuticals especially mercurial diuretics. Used to improve dyeability of acrylic fibres.

Physical properties

M. Pt. −88.2°C; **B. Pt.** 56.5°C; **Flash point** −29°C (open cup); **Specific gravity** 0.762 at 20°C; **Volatility** v. den. 2.0.

Solubility

Water: miscible. Organic solvent: miscible in ethanol, diethyl ether

Occupational exposure

UN No. 2334; **HAZCHEM Code** 2WE; **Conveyance classification** toxic substance;flammable liquid; **Supply classification** highly flammable and toxic.

Risk phrases

Highly flammable – Toxic by inhalation, in contact with skin and if swallowed (R11, R23/24/25)

Safety phrases

Keep container in a well ventilated place – Keep away from sources of ignition – No Smoking – Avoid contact with skin and eyes – If you feel unwell, seek medical advice (show label where possible) (S9, S16, S24/25, S44)

Ecotoxicity

Invertebrate toxicity
Cell multiplication inhibition test, *Microcystis aeruginosa* 0.35 mg l^{-1}, *Scenedesmus quadricauda* 2.2 mg l^{-1}, *Pseudomonas putida* 700 mg l^{-1}, *Entosiphon sulcatum* 23 mg l^{-1} (1,2).

Effects to non-target species
LC_{50} (48 hr) clawed toad 5.0 mg l^{-1} (3).

Environmental fate

Degradation studies
Nonbiodegradable (4).
Degradation by *Aerobacter sp.* 200 mg l^{-1} at 30°C (5).

Mammalian and avian toxicity

Acute data
LD_{50} oral rat, mouse 57-106 mg kg^{-1}.
LC_{50} (4 hr) inhalation rat 286 ppm.
LD_{50} percutaneous rabbit 35 mg kg^{-1} (6).

Sub-acute data
TC_{Lo} inhalation man 5 ppm min^{-1} adverse pulmonary effects (7).
Gavage Sprague-Dawley rat 100 mg kg^{-1} day^{-1} 10 doses in 11 days caused myocardial necrosis (8).
Oral rat 100 mg kg^{-1} two doses on successive days and sacrificed 24 hr after final dose, endothelial cell profileration and interstitial cell activiation reported (9).

Metabolism and pharmacokinetics
Allylamine was metabolised to acrolein which caused generalised tissue destruction (10).

Irritancy
Dermal rabbit (24 hr) 500 mg and 50 mg instilled in rabbit eye for 20 sec caused severe irritation (6).

Any other adverse effects to man

Mucous membrane irritation and chest discomfort in some humans at 2.5 ppm, intolerable to most at 14 ppm (6).

Any other adverse effects

Chick myocardial myocyte aggregates treated with 0.5 or 5 mg l^{-1} allylamine showed myocyte necrosis and reduction in tissue compactness (10).

Any other comments

Experimental and chemical use and general toxicity and human health effects reviewed (11,12).

References

1. Bringmann, G; et al *Gwf-Wasser/Abwasser* 1976, **117**(9), 119
3. Bringmann, G; et al. *Water Res.* 1980, **14**, 231
3. Warne, H. E. *Tijdschrift van het BECEWA* Liege

4. *MITI Report* 1984, Ministry of International Trade and Industry, Tokyo
5. Verschueren, K. *Handbook of Environmental Data on Organic Chemicals* 177
6. Hine, C. H; et al. *Arch. Environ. Health* 1960, **1**, 345-352
7. Sittig, M. *Handbook of Hazardous Chemicals and Carcinogens* 1985, Noyes Data Corp., Park Ridge, NJ
8. Kumar, D; et al. *Toxicol. Appl. Pharmacol.* 1990, **103**(2), 288-302
9. Boor, P. J; et al. *Res. Commun. Chem. Pathol. Pharmacol.* 1990, **68**(1), 3-11
10. Kesingland, K; et al. *Toxicol. In Vitro* 1991, **5**(2), 145-156
11. Boor, P. J; et al. *Toxicology* 1987, **44**(2), 129-145
12. *ECETOC Technical Report No. 30(4)* 1991, European Chemical Industry Ecology and Toxicology Centre, B-1160 Brussels

A81   4-Allylanisole

CAS Registry No. 104-46-1
Synonyms Esdragol; Chavicol methyl ether; *p*-propenylanisole; 4-methoxypropenylbenzene; Oil of Aniseed; *p*-allylanisole
Mol. Formula $C_{10}H_{12}O$ **Mol. Wt.** 148.21
Uses Used in the manufacture of anisaldehyde. Used in the perfume and flavouring industries. A sensitiser in bleaching colours in colour photography. An imbedding material in microscopy.
Occurrence Oil from *Artemisia dracunculus* (tarragon).

Physical properties

B. Pt. $_{25}$ 108-114°C; **Specific gravity** d_4^{20} 0.9882.

Solubility
Organic solvent: ethanol, chloroform, benzene, diethyl ether

Ecotoxicity

Effects to non-target species
LD_{50} oral redwing blackbird 316 mg kg^{-1} (1).

Mammalian and avian toxicity

Acute data
LD_{50} oral rat, mouse, guinea pig 2090-3050 mg kg^{-1} (2).

Metabolism and pharmacokinetics
In rabbit and rat studies, the *trans*- isomer was found concentrated in liver, lungs and brain after intravenous administration whereas after oral dosage, it remained in the stomach (3).
Is absorbed from the digestive tract by passive diffusion (4).

Any other comments

Forms azeotropic mixtures in water.

References

1. Schafer, E. W; et al. *Arch. Environ. Contam. Toxicol.* 1983, **12** 355-382
2. Opdyke, D. L. J. *Food Cosmet. Toxicol.* 1964, **2**, 327
3. Opdyke, D. L. J. *Monographs on Fragrance Raw Materials* 1972, 92, Pergamon Press, New York
4. Fritsch, P; et al. *Food Cosmet. Toxicol.* 1975, **13**(3), 359

A82   Allyl bromide

$H_2C{=}CHCH_2Br$

CAS Registry No. 106-95-6
Synonyms 3-bromo-1-propene; 3-bromopropylene; bromallylene; 1-bromo-2-propene
Mol. Formula C_3H_5Br **Mol. Wt.** 120.98
Uses Used in the manufacture of synthetic perfumes and the synthesis of other alkyl compounds.

Physical properties

M. Pt. −119°C; **B. Pt.** 71°C; **Flash point** −1°C; **Specific gravity** d_4^{20} 1.398; **Volatility** v. den. 4.17.

Solubility

Organic solvent: miscible with ethanol, diethyl ether, carbon disulfide, carbon tetrachloride

Occupational exposure

UN No. 1099; **HAZCHEM Code** 2WE; **Conveyance classification** flammable liquid; toxic substance.

Ecotoxicity

Invertebrate toxicity

EC_{50} (48 hr) *Daphnia magna* 100-10 ppm (1).

Mammalian and avian toxicity

Acute data

LD_{50} oral guinea pig 30 mg kg^{-1} (1).
LD_{50} intraperitoneal mouse 108 mg kg^{-1} (2).

Metabolism and pharmacokinetics

Gavage rat 120 mg kg^{-1} in water, 3-hydroxypropylmercapturic acid an intermediate in the formation of acrolein was detected at levels of 3% in the urine (3).

Genotoxicity

Salmonella typhimurium TA100 without metabolic activation positive (4).

Any other adverse effects

May injure liver and kidneys and can lead to fatalities (5).
Toxic amount may be absorbed through the skin (6).

References

1. Hann, W; et al. *Water Quality Characteristics of Hazardous Materials*, 1974, **4**, A & M Univ., Texas
2. Fischer, G. W; et al. *J. Prakt. Chem.* 1978, **320**, 133
3. Sanduja, R; et al. *J. Appl. Toxicol.* 1989, **9**(4), 235-238
4. Eder, E; et al. *Biochem. Pharmacol.* 1980, **29**, 993
5. *Handling Chemicals Safely* 1980, Dutch Chemical Industry Association and Dutch Safety Institute, Amsterdam
6. *National Fire Protection Association. Fire Protection Guide on Hazardous Materials* 7th ed., 1978, 49, National Fire Protection Association, Boston, MA

A83   Allyl butyrate

$$CH_3(CH_2)_2CO_2CH=CH_2$$

CAS Registry No. 2051-78-7
Synonyms allyl butanoate; vinyl carbinyl butyrate
Mol. Formula $C_7H_{12}O_2$ **Mol. Wt.** 128.17
Uses Used in perfumery.

Physical properties

B. Pt. 142°C.

Solubility
Organic solvent: miscible with ethanol

Mammalian and avian toxicity

Acute data
LD_{50} oral rat 250 mg kg^{-1} (1).
LD_{50} dermal rabbit 530 mg kg^{-1} (2).

Irritancy
Human skin irritant (1,2).

Any other comments

Experimental toxicology and human health effects reviewed (3).

References

1. Taylor, J. M; et al. *Toxicol. Appl. Pharmacol.* 1964, **6**, 378
2. Opdyke, D. L. J. *Food Cosmet. Toxicol.* 1977, **15**, 611
3. *ECETOC Technical Report No. 30(4)* 1991, European Chemical Industry Ecology and Toxicology Centre, B-1160 Brussels

A84 Allyl caprylate

$$CH_3(CH_2)_6CO_2CH_2CH=CH_2$$

CAS Registry No. 4230-97-1
Synonyms allyl octanoate; allyl octylate; octanoic acid, 2-propenyl ester
Mol. Formula $C_{11}H_2O_2$ **Mol. Wt.** 166.14
Uses Food flavouring material (synthetic pineapple flavour).

Physical properties
B. Pt. $_{5.5}$ 87-88°C; **Specific gravity** d^{30} 0.8729.

Solubility
Organic solvent: ethanol, diethyl ether

Mammalian and avian toxicity

Acute data
LD$_{50}$ oral rat 570 mg kg^{-1} (1).

Irritancy
Dermal rabbit (24 hr) 310 mg caused moderate irritancy (1).

Any other comments

Experimental toxicology and human health effects reviewed (2).

References
1. Opdyke, D. C. J. *Food Cosmet. Toxicol.* 1978, **16**, 637
2. *ECETOC Technical Report No. 30(4)* 1991, European Chemical Industry Ecology and Toxicology Centre, B-1160 Brussels

A85 Allyl chloride

$$CH_2=CHCH_2Cl$$

CAS Registry No. 107-05-1
Synonyms 1-propene, 3-chloro-; 3-chloropropene; 3-chloropropylene; 2-propenyl chloride; chlorallylene; chloroallylene; 1-chloroprop-2-ene; 1-chloro-2-propene; α-chloropropylene; 3-chloro-1-propylene; NCI-CO4615
Mol. Formula C_3H_5Cl **Mol. Wt.** 76.53
Uses Used as a chemical intermediate in epichlorohydrin manufacture. A polymerisation monomer in the manufacture of resins, polymers, varnishes and adhesives, and in the synthesis of medicinal derivatives, such as barbituates, diuretics and cyclopropane.

Physical properties
M. Pt. −134.5°C; **B. Pt.** 44.6°C; **Flash point** −31.7°C (closed cup); **Specific gravity** d_4^{20} 0.9376; **Partition coefficient** log P$_{ow}$ 1.450.

Solubility
Organic solvent: miscible with ethanol, chloroform, diethyl ether, petroleum ether

Occupational exposure
US TLV (TWA) 1 ppm (3 mg m^{-3}); **US TLV (STEL)** 2 ppm (6 mg m^{-3}); **UK Long-term limit** 1 ppm (3 mg m^{-3}) under review; **UK Short-term limit** 2 ppm (6 mg m^{-3}) under review; **UN No.** 1100; **HAZCHEM Code** 3WE; **Conveyance classification** flammable liquid; toxic substance; **Supply classification** highly flammable and toxic.

Risk phrases
Highly flammable – Very toxic by inhalation (R11, R26)

Safety phrases
Keep away from sources of ignition – No Smoking – Do not empty into drains – Take precautionary measures against static discharges – In case of accident or if you feel unwell, seek medical advice immediately (show label where possible) (S16, S29, S33, S45)

Ecotoxicity

Fish toxicity
LC_{50} (96 hr) fathead minnow 10 mg l^{-1} (1).
LC_{50} (24 hr) goldfish 19.8 mg l^{-1} (2).
LC_{50} (2 wk) guppy 1.2 mg l^{-1} (3).

Invertebrate toxicity
Cell multiplication inhibition test, *Scenedesmus quadricauda* 6.3 mg l^{-1}, *Entosiphon sulcatum* 8.4 mg l^{-1}, *Uronema parduczi* >240 mg l^{-1}, *Pseudomonas putida* 115 mg l^{-1} (4).

Environmental fate

Nitrification inhibition
75% Inhibition of NH_3 oxidation by activated sludge 180 mg l^{-1} (5).

Degradation studies
BOD_5 14% and 25% reduction of dissolved oxygen concentration using nonacclimated and acclimated seed, respectively (2).
Using activated sludge, allyl chloride was readily degradable (6).

Abiotic removal
Photochemical degradation occurs via reaction with hydroxyl radicals in the atmosphere, calculated daily loss of 91.1% in 12 hr (7).
Ozonolytic $t_{1/2}$ 9 hr (6).
Hydrolytic $t_{1/2}$ 7.2 days in water at 25°C produces allyl alcohol and hydrochloric acid (8).

Mammalian and avian toxicity

Acute data
LD_{50} oral rat 700 mg kg^{-1} (9).
LD_{50} (4 hr) inhalation guinea pig 2200 mg kg^{-1} (10).
LD_{50} dermal rabbit 2200 mg kg^{-1} (11).

Sub-acute data
Inhalation rat, guinea pig, rabbit (1 month) 7 hr day^{-1}, 5 day wk^{-1} 8 ppm induced liver injury (12).

Inhalation rat (34 wk) 10, 50 and 100 ppm 8 hr day^{-1} 5 day wk^{-1} reduction of motor and sensory nerve conduction velocities and nerve action potentials after 28 wk for higher concentration and retarded motor distal latency for the last period of exposure. Depressed amplitude of nerve action potentials was evident in rats exposed to 50 ppm (13).

Carcinogenicity and long-term effects
No adequate evidence for carcinogenicity to humans, limited evidence for carcinogenicity to animals, IARC classification group 3 (14).
TD_{Lo} oral mouse (78 wk intermittent) 50 g kg^{-1} equivocal tumourigenic effects (15).
TD_{Lo} intraperitoneal mouse (8 wk intermittent) 5880 mg kg^{-1} equivocal tumourigenic effects (16).

Teratogenicity and reproductive effects
Inhalation rat (7 hr day^{-1} during major organogenesis) 30 or 300 ppm caused maternal toxicity at 300 ppm. Slight delay in foetal development at 300 ppm may have been associated with observed effects in maternal animals (17).

Irritancy
Upper respiratory tract irritant and concentrations of 50-100 ppm very irritating to eyes in humans (12).

Genotoxicity
Salmonella typhimurium TA100, TA1535 without metabolic activation positive (18).
Salmonella typhimurium TA98, TA100, TA1535, TA1538 with and without metabolic activation negative. *Escherichia coli* WP2, WP2$uv1A$ with and without metabolic activation positive. *Sacchromyces cerevisiae* JD1 with and without metabolic activation positive (19).

Any other adverse effects
Acute exposure can cause unconsciousness while chronic exposure can cause injury to liver and kidneys (species unspecified) (20).

Any other comments
Hardness of water decreases toxicity (1).
Hazardous potential reviewed (21).
Experimental toxicology, human health effects, epidemiology and workplace experience reviewed (22).
A dangerous fire hazard when exposed to heat, flame or oxidisers, and releases hydrogen chloride on combustion. It is incompatible with nitric acid, ethyleneimine, ethylenediamine, chlorosulfonic acid, oleum and sodium hydroxide. Violent exothermic polymerisation may occur on contact with aluminium chloride, boron trifluoride or sulfuric acid.

References
1. Pickering, Q. H; et al. *J. Water Pollut. Control Feder.* 1966, **38**, 1419-1429
2. Bridie, A. L; et al. *Water Res.* 1979, **13**, 627-630
3. Hermens, J; et al. *Ecotox. Environ Saf.* 1985, **9**, 321-326
4. Bringmann, G; et al. *Water Res.* 1980, **14**, 231-241
5. Tomlinson, T. G; et al. *J. Appl. Bacteriol.* 1966, **29**(2), 266-291
6. Brown, S. L; et al. *Research Program on Hazardous Priority Ranking of Manufactured Chemicals* 1975, 70B1-70C6, NTIS PB-263 164

7. Singh, H. B; et al. *Environ. Sci. Technol.* 1982, **16**, 872-880
8. *Kirk Othmer Encyclopedia of Chemical Technology* 3rd ed., 1978, **5**, 563-573, John Wiley, New York
9. Fengsheng, H. E; et al. *Scand. J. Work Environ. Health* 1985, **11**(Suppl. 4), 43-45
10. *J. Ind. Hyg. Toxicol.* 1940, **22**, 79
11. *J. Ind. Hyg. Toxicol.* 1948, **30**, 63
12. *Patty's Industrial Hygiene and Toxicology* 3rd ed., 1981, **2**, Clayton, G. D; et al. (Eds.), John Wiley, New York
13. Nagano, M; et al. *Sangyo Igaku* 1991, **33**(2), 73-80, (Jap.) (*Chem. Abstr.* 1991, **115**, 86985c)
14. *IARC Monograph* 1987, **Suppl. 7**, 56
15. *Cancer Res.* 1979, **39**, 391
16. *National Cancer Institute Carcinogenesis Technical Report Series* 1979, **39**, 391
17. John, J. A; et al. *Fundam. Appl. Toxicol.* 1983, **3**, 437-442
18. Eder, E; et al. *Biochem. Pharmacol.* 1980, **29**, 993
19. Dean, B. J; et al. *Mutat. Res.* 1985, **153**, 57-77
20. *The Merck Index* 11th ed. 1989, Merck & Co. Inc, Rahway, NJ
21. *Dang. Prop. Ind. Mater. Rep.* 1988, **8**(1), 20-28
22. *ECETOC Technical Report No. 30(4)* 1991, European Chemical Industry Ecology and Toxicology Centre, B1160 Brussels

A86   Allyl chloroformate

$CH_2=CHCH_2CO_2Cl$

CAS Registry No. 2937-50-0
Synonyms allyl chlorocarbonate; formic acid, chloro-, allyl ester; chloroformic acid, allyl ester
Mol. Formula $C_4H_5ClO_2$ **Mol. Wt.** 120.54

Physical properties

M. Pt. −80°C; **B. Pt.** 106-114°C; **Flash point** 31°C (closed cup); **Specific gravity** 1.1 at 25°C; **Volatility** v. den. 4.2.

Occupational exposure

UN No. 1722; **HAZCHEM Code** 2WE; **Conveyance classification** corrosive substance.

Mammalian and avian toxicity

Acute data

LD_{50} oral mouse, rat 210-244 mg kg^{-1} (1).
LC_{Lo} (duration unspecified) inhalation mouse 2000 mg m^{-3} (2).

Any other adverse effects

Affects the respiratory system and causes dyspnoea (1,3).

References

1. Oskerko, E. F; et al. *Gig. Tr. Prof. Zabol.* 1984, **5**, 51-52 (*Chem. Abstr.* **101**, 49741k)
2. *Nat. Defence Res. Com. Progress Rep.* 1943
3. Lenga, R. E. *The Sigma-Aldrich Library of Chemical Safety Data* 2nd ed, 1988, **2**

A87   Allyl cinnamate

$$CH=CHCO_2CH_2CH=CH_2$$

CAS Registry No. 1866-31-5
Synonyms allyl-3-phenylpropenoate; allyl-β-phenylacrylate; propenyl cinnamate; vinyl carbinylcinnamate
Mol. Formula $C_{12}H_{12}O_2$ **Mol. Wt.** 188.23
Uses Flavouring ingredient in foods and cosmetics.

Physical properties

B. Pt. $_{15}$ 150-152°C; **Specific gravity** d^{23} 1.048.

Solubility
Organic solvent: ethanol, diethyl ether

Mammalian and avian toxicity

Acute data
LD_{50} oral rat 1520 mg kg^{-1} (1).
LD_{50} dermal rabbit <5 g kg^{-1} (2).

Irritancy
Human skin irritant (2).

Sensitisation
Non-sensitiser in humans (2).

Any other comments

Experimental toxicology and health effects reviewed (3).

References

1.   Jenner, P. M; et al *Food Cosmet. Toxicol.* 1964, **2**, 327-343
2.   Opdyke, D. C. J. *Food Cosmet. Toxicol.* 1977, **15**(6), 615-616
3.   *ECECTOC Technical Report No. 30(4)* 1991, European Chemical Industry Ecology and Toxicology Centre, B-1160 Brussels

A88   Allyl cyclohexylpropionate

$$(CH_2)_2CO_2CHCH=CH_2$$

CAS Registry No. 2705-87-5
Synonyms allyl cyclohexane propionate; 3-allylcyclohexyl propionate; allyl

hexahydrophenyl propionate; allyl-3-cyclohexyl propionate; allyl β-cyclohexyl propionate

Mol. Formula $C_{12}H_{20}O_2$ **Mol. Wt.** 196.29

Uses Food flavouring material.

Physical properties

M. Pt. 196.3°C; **B. Pt.** 91°C at 1 mm kg; **Specific gravity** d_{25}^{25} 0.95.

Solubility

Organic solvent: ethanol

Mammalian and avian toxicity

Acute data

LD$_{50}$ oral rat, guinea pig 380-585 mg kg^{-1} (1).

Any other comments

Experimental toxicology and human health effects reviewed (2).

References

1. Jenner, P. M; et al. *Food Cosmet. Toxicol.* 1964, **2**, 327-343
2. *ECETOC Technical Report No. 30(4)* 1991, European Chemical Industry Ecology and Toxicolgy Centre, B-1160 Brussels

A89   Allyl ethyl ether

CH2=CHCH2OCH2CH3

CAS Registry No. 557-31-3

Synonyms 3-ethoxy-1-propene; ethyl allyl ether

Mol. Formula $C_5H_{10}O$ **Mol. Wt.** 86.13

Physical properties

M. Pt. 66-67°C; **B. Pt.** <21°C; **Flash point** −20°C;

Specific gravity d_4^{20} 0.765; **Partition coefficient** log P_{ow} 1.388.

Solubility

Organic solvent: miscible with ethanol, diethyl ether

Occupational exposure

UN No. 2335; **HAZCHEM Code** 3WE; **Conveyance classification** flammable liquid;toxic substance.

Mammalian and avian toxicity

Acute data

LD$_{50}$ oral rat 320 mg kg^{-1} (1).

Reference

1. Smyth, H. F; et al. *J. Ind. Hyg. Toxicol.* 1949, **31**, 60

A90　Allyl formate

$$HCO_2CH_2CH=CH_2$$

CAS Registry No. 1838-59-1

Synonyms formic acid, 2-propenyl ester; formic acid, allyl ester; 3-propenyl methanoate

Mol. Formula $C_4H_6O_2$　　　　　　　　　　　**Mol. Wt.** 86.09

Physical properties

B. Pt. 83°C; **Flash point** <22°C (closed cup); **Specific gravity** d_4^{18} 0.948.

Solubility

Organic solvent: ethanol

Occupational exposure

UN No. 2336; **HAZCHEM Code** 3WE; **Conveyance classification** flammable liquid;toxic substance.

Ecotoxicity

Fish toxicity

Oral trout (24 hr) 100 μl kg^{-1} induced liver necrosis (1).

Environmental fate

Degradation studies

Biodegrades to allyl alcohol and formic acid which are ultimately biodegradable (2).

Mammalian and avian toxicity

Acute data

LD_{50} oral rat, mouse 124-136 mg kg^{-1} (2).

LC_{50} (3 hr) inhalation mouse 14 g m^{-3} (3).

References

1.　Droy, B. F; et al. *Mar. Environ. Res.* 1988, **24**(1-4), 259-264
2.　*Food Cosmet. Toxicol.* 1964, **2**, 237
3.　Izmerov, N. F; et al. *Parameters Ind. Toxic Chemicals* 1982, 18, Centre for International Projects, Moscow

A91　Allyl glycidyl ether

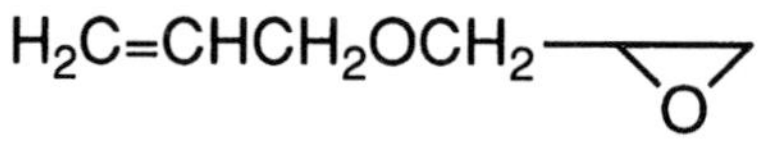

$$H_2C=CHCH_2OCH_2 \underset{O}{\triangle}$$

CAS Registry No. 106-92-3

Synonyms allyl-2,3-epoxypropyl ether; 1-(allyloxy)-2,3-epoxpropane; ((2-propenyloxy)methyl)oxirane; 1, 2-epoxy-3-allyloxypropane

Mol. Formula $C_6H_{10}O_2$ **Mol. Wt.** 114.15

Uses Used as an additive for epoxy resins and as a co-monomer in polyglycols and polyolefins synthesis. Stabiliser for chlorinated compounds.

Physical properties

M. Pt. −100°C; **B. Pt.** 153.9°C; **Flash point** 56°C (open cup);

Specific gravity d_4^{20} 0.9698; **Volatility** v.p. 4.7 mmHg at 25°C; v. den. 3.94.

Solubility
Organic solvent: ethanol, acetone, diethyl ether, benzene

Occupational exposure

US TLV (TWA) 5 ppm (23 mg m^{-3})); **US TLV (STEL)** 10 ppm (47 mg m^{-3});
UK Long-term limit 5 ppm (22 mg m^{-3}); **UK Short-term limit** 10 ppm (44 mg m^{-3});
UN No. 2219; **HAZCHEM Code** 2S; **Conveyance classification**
flammable liquid;harmful substance; **Supply classification** harmful.

Risk phrases
Harmful by inhalation − May cause sensitisation by skin contact (R20, R43)

Safety phrases
Avoid contact with skin and eyes (S24/25)

Ecotoxicity

Fish toxicity
LD$_{50}$ (96 hr) goldfish 30 mg l^{-1} (1).

Mammalian and avian toxicity

Acute data
LD$_{50}$ oral rat, mouse 390-922 mg kg^{-1} (2,3).
LC$_{Lo}$ (4 hr) inhalation rat, mouse 270-860 ppm (2).
LD$_{50}$ dermal rabbit 2550 mg kg^{-1} (3).

Sub-acute data
Inhalation mouse (4-14 day) 7.1-2.5 ppm 6 hr day^{-1} induced lesions in nasal cavity, necrosis of respiratory epithelium and erosion of olfactory epithelium (4).

Carcinogenicity and long-term effects
Inhalation (2 yr) rat, mouse 0, 5 or 10 ppm for 6 hr day^{-1} 5 day wk^{-1} induced papillary adenoma, squamous cell carcinoma and adenocarcinomas of the nasal passage (5).

Irritancy
Dermal rabbit (24 hr) 500 mg and 250 µg instilled in rabbit eyes for 24 hr caused severe irritation (2).
In rats exposed to 260 ppm slight eye irritation was reported, while concentrations of 600 and 900 ppm caused severe irritation and corneal opacities. Severe but reversible conjunctivitis, iritis and corneal opacity were reported in rabbits (6).

Sensitisation
Dermatitis, with itching, swelling and blistering, and sensitisation have been reported in humans. Contact allergies and reaction to allyl glycidyl ether subsequent to patch tests have been reported in resin workers (7,8).

Genotoxicity

Escherichia coli PQ37 SOS chromotest with and without metabolic activation positive (9).
Salmonella typhimurium TA100, TA1535 with and without metabolic activation positive (10).

Any other adverse effects

Exposure to allyl glycidyl ether (concentration and duration unspecified) has been reported to cause testicular atrophy in rats, rabbits and dogs (11).
Inhalation mice 1.9-6.8 ppm (duration unspecified) caused irritation of the nasal mucosa and decreased respiratory rate. Exposure to 7.1 ppm for 4 day caused nasal cavity lesions but no lung injury, indicating it primarily affects the upper airways (12).
Rats exposed to levels above 260 ppm (route and duration unspecified) were found to have bronchopneumonia, emphysema, bronchiectasis, pneumonitis, haemorrhage, discoloured liver and adrenals at autopsy (13).

Any other comments

Exposure data, human health effects and experimental toxicology reviewed (14).

References

1. Verschueren, K. *Handbook of Environmental Data on Organic Chemicals* 2nd ed., 1983, Van Nostrand Reinhold, New York
2. Sax, N. I; et al. *Dangerous Properties of Industrial Materials* 7th ed., 1989, Van Nostrand Reinhold, New York
3. Hine, C. H; et al. *Arch. Ind. Health* 1956, **14**, 250
4. Gagnaire, F; et al. *Toxicol. Lett.* 1987, **39**(2-3), 139-145
5. Boorman, G. *Gov. Rep. Announce. Index US* 1990, **90**(21), No. 005,294
6. Hine, C. H; et al. *Arch. Ind. Health* 1956, **13**, 250-264
7. Jolanki, R; et al *Contact Dermatitis* 1987, **16**, 87-92
8. Fregert, S; et al. *Acta Allergol.* 1964, **19**, 269-299
9. Hude von der, W; et al. *Mutat. Res.* 1990, 231(2), 205-218
10. Canter, D. A; et al *Mutat. Res.* 1986, **172**(2), 105-138
11. Lane, J. M; *Vet. Hum. Toxicol.* 1980, **22** (2), 99-101
12. Gagnaire, F; et al. *Toxicol. Lett.* 1987, **39**(2-3), 139-45
13. *Chemical Safety Data Sheets* 1992, **5**, RSC, London
14. *ECETOC Technical Report No. 30(4)* 1991, European Chemical Industry Ecology and Toxicology Centre, B-1160 Brussels

A92   Allyl iodide

CH2=CHCH2I

CAS Registry No. 556-56-9
Synonyms 3-iodo-1-propene; 3-iodopropylene
Mol. Formula C_3H_5I **Mol. Wt.** 167.98
Uses In the synthesis of allyl compounds.

Physical properties

M. Pt. –99°C; **B. Pt.** 101-103°C; **Flash point** < 21°C;
Specific gravity d^{12} 1.848.

Solubility

Organic solvent: miscible with ethanol, chloroform, diethyl ether

Occupational exposure

UN No. 1723; **HAZCHEM Code** 2WE; **Conveyance classification** flammable
liquid; corrosive substance; **Supply classification** Flammable and corrosive.

Risk phrases

Flammable – Causes burns (R10, R34)

Safety phrases

Keep container tightly closed – In case of contact with eyes, rinse immediately with
plenty of water and seek medical advice (S7, S26)

Mammalian and avian toxicity

Irritancy

Produces irritation of eyes and respiratory passages. Readily absorbed through skin (1).

Any other adverse effects to man

Acute exposure can cause unconsciousness while chronic exposure can cause injury
to liver and kidney (1).

Any other comments

Health effects and safety precautions are similar to those for allyl bromide and allyl chloride (1).
Experimental toxicology and human health effects reviewed (2).

References

1. *The Merck Index* 11th ed. 1989, Merck & Co. Inc, Rahway, NJ
2. *ECETOC Technical Report No. 30(4)* 1991, European Chemical Industry Ecology and
   Toxicology Centre, B-1160 Brussels

A93   Allylisopropylacetamide

$$CH_2=CHCH_2CH(CONH_2)CH(CH_3)_2$$

CAS Registry No. 299-78-5
Synonyms 2-(1-methylethyl)-4-pentenamide; 2-isopropyl-4-pentenamide
Mol. Formula $C_8H_{15}NO$ **Mol. Wt.** 141.21

Physical properties

M. Pt. 107°C.

Mammalian and avian toxicity

Sub-acute data

A single dose of 225 mg kg^{-1} to rabbits or doses of 35 mg kg^{-1} day^{-1} for 7 days

---

caused an increase in liver microsomal enzymes (1).

Oral administration of 2 g to rabbits caused a short-term rise in liver and bile porphyrin levels (2).

Subcutaneous administration of 600 mg kg^{-1} over 24 hr induced porphyrin in rats (3). Allyl isopropylacetamide was found to stimulate an overproduction of porphyrins, causing porphyria in rabbits, fowls and rats. These were found to excrete relatively high levels of porphobilinogen and porphyrins (4).

Any other adverse effects

In vivo administration (dose and duration unspecified) to phenobarbital-pretreated rats results in a marked loss of hepatic cytochrome P450 content in liver and kidney (5). A dose of 3 mg increased the level of δ-aminovulinic acid synthetase in a culture of liver parenchyma cells from a chick embryo (6).

References

1. Ivanov, E; et al. *Eksp. Med. Morfol.* 1982, **21**(3), 110-115, (Russ.) (*Chem. Abstr.* **98**, 192842u)
2. Benard, M; et al. *Compt. Rend. Soc. Biol.* 1954, **148**, 838-840 (Fr.) (*Chem. Abstr.* **49**, 1952i)
3. Labbe, R. F; et al. *Arch. Biochem. Biophys.* 1961, **92**, 373-374
4. Goldberg, A; et al. *Proc. Royal Soc.* 1955, **B143**, 257-279
5. Bornheim L. M; et al. *Mol. Pharmacol.* 1987, **32**(2), 299-308
6. Granick, S. *J. Biol. Chem.* 1963, **238**(6), 2247-2249

A94   Allyl isothiocyanate

H$_2$C=CHCH$_2$NCS

CAS Registry No. 57-06-7

Synonyms 2-propenyl isothiocyanate; 3-isothiocyanato-1-propene; allyl thiocarbonimide; allyl isosulfocyanate; allyl isorhodanide; synthetic mustard seed; mustard oil

Mol. Formula C$_4$H$_5$NS   **Mol. Wt.** 99.16

Uses Used in the manufacture of ointments and counterirritants. Fumigant. Manufacture of military poison gas.

Occurrence Isolated from black mustard seed *Brassica nigra*.

Physical properties

M. Pt. −80°C; **B. Pt.** 151°C; **Flash point** 46°C; **Specific gravity** d$_4^{15}$ 1.015; **Partition coefficient** log P$_{ow}$ 2.11 (calc.) (1).; **Volatility** v.p. 10 mmHg at 38.3°C; v. den. 3.41.

Solubility

Organic solvent: ethanol, acetone, diethyl ether, benzene

Occupational exposure

UN No. 1545; **HAZCHEM Code** 3WE; **Conveyance classification** toxic substance.

Environmental fate

Nitrification inhibition
75% inhibition of nitrification process in activated sludge at 1.9 mg l^{-1} (1).

Mammalian and avian toxicity

Acute data
LD_{50} oral rat, mouse 108-339 mg kg^{-1} (2-4).
LD_{50} (duration unspecified) inhalation mouse 69 mg kg^{-1} (2).
LD_{50} subcutaneous rat, mouse 80-92 mg kg^{-1} (5).
LD_{50} dermal rabbit 88 mg kg^{-1} (3).

Sub-acute data
Oral rat (6 wk) 0-40 mg kg^{-1} 5 day wk^{-1}. High doses caused histopathological changes in liver and kidneys (6).

Carcinogenicity and long-term effects
Inadequate evidence for carcinogenicity to humans, limited evidence for carcinogenicity to animals, IARC classification group 3 (7).
Gavage rat (2 yr) 25 mg kg^{-1}, 5 day wk^{-1} induced urinary bladder transitional cell papilloma and subcutaneous tissue fibrosarcoma (8).
Gavage F344 rats, B6C3F1 mice (2 yr) 25 mg kg^{-1}, US National Toxicology Program classification D, single sex of a single species with single tissue affected, genotoxic carcinogen (9).

Irritancy
Powerful vesicant and irritant and should not be inhaled or tasted undiluted (10).

Genotoxicity
In vitro Chinese hamster ovary cells with metabolic activation induced sister chromatid exchange, with and without metabolic activation chromosome abberation weakly positive (11).
Mouse lymphoma L5178Y tk+/tk- cells without metabolic activation positive (12).

Any other comments
Experimental toxicology and human health effects reviewed (13,14).

References
1. Verschueren, K. *Handbook of Environmental Data on Organic Chemicals* 2nd ed., 1983, Van Nostrand Reinhold, New York
2. Poloz, D. D; et al. *Veterinariya (Moscow)* 1980, (8), 54-56 (*Chem. Abstr.* **93**, 198676m)
3. Sax, N. I; et al. *Dangerous Properties of Industrial Materials* 7th ed, 1989, Van Nostrand Reinhold, New York
4. Jenner, P. M; et al. *Food Cosmet. Toxicol.* 1964, **2**, 327
5. Nishie, K; et al. *Food Cosmet. Toxicol.* 1980, **18**(2), 159-172
6. Lewerenz, H. J; et al. *Nahrung* 1988, **32**(8), 723-728
7. *IARC Monograph* 1987, **Suppl. 7**, 57
8. Haseman, J. K; et al. *Environ. Mol. Mutagen.* 1990, **16**(18), 15-31
9. Ashby, J; et al. *Mutat. Res.* 1988, **204**, 17-115
11. Gallaway, S. M; et al. *Environ. Mol. Mutagen.* 1987, **10**(Suppl. 10), 1-175
12. McGregor, D. B; et al. *Environ. Mol. Mutagen.* 1988, **12**(1), 85-154

13. *ECETOC Technical Report No. 30(4)* 1991, European Chemical Industry Ecology and Toxicology Centre, B-1160 Brussels
14. *BIBRA Toxicity Profile* 1990, British Industrial Biological Research Association, Carshalton

A95   Allyl isovalerate

$(CH_3)_2CHCH_2CO_2CH_2CH=CH_2$

CAS Registry No. 2835-39-4
Synonyms 3-methylbutanoic acid, 2-propenyl ester; allyl isovalerianate;
3-methylbutanoic acid, allyl ester; 2-propenyl isovalerate; isovaleric acid, allyl ester
Mol. Formula $C_8H_{14}O_2$ **Mol. Wt.** 142.20
Uses Flavouring material for foods and cosmetics.

Physical properties

B. Pt. 89-90°C.

Solubility
Organic solvent: ethanol, acetone

Mammalian and avian toxicity

Acute data
LD_{50} oral rat 230 mg kg^{-1} (1).
LD_{50} dermal rabbit 560 mg kg^{-1} (1).

Sub-acute data
Rats fed at 31-62 mg kg^{-1}, 5 times per wk over 103 wk, showed cholangiofibrosis, nodular degeneration, cirrhosis, focal necrosis, fatty metamorphosis and cytoplasmic vacuolation of the liver (2).

Carcinogenicity and long-term effects
Limited evidence for carcinogenicity to experimental animals, in the absence of epidemiological data no evaluation was made for its carcinogenicity to humans, IARC classification group 3 (3).
Gavage rat, mouse (2 yr) 62 mg kg^{-1} 5 day wk^{-1} induced haematopoietic system leukaemia and malignant lymphoma (4).
US National Toxicology Program gavage F344 rats and B6C3F1 mice (2 yr) 62 mg kg^{-1} positive in ♂ rat and ♀ mice (5).

Irritancy
Dermal rabbit (24 hr) 500 mg caused moderate irritancy (1).

References
1. Opdyke, D. C. J. *Food Cosmet. Toxicol.* 1979, **17**, 703
2. *National Toxicology Program Technical Report Series* 1983, 253
3. *IARC Monograph* 1987, **Suppl. 7**, 56
4. Haseman, J. K; et al. *Environ. Mol. Mutagen.* 1990, **16**(Suppl. 18), 15-31
5. *National Toxicology Program, Research & Testing Division* 1992, Report No. TR-253, NIEHS, Research Triangle Park, NC

A96 Allyl phenoxyacetate

$$OCH_2CO_2CH_2CH=CH_2$$

CAS Registry No. 7493-74-5
Synonyms acetate PA
Mol. Formula $C_{11}H_{12}O_3$ Mol. Wt. 192.22
Uses Flavouring in foods and cosmetics.

Physical properties

B. Pt. 100-102°C.

Solubility
Organic solvent: ethanol, diethyl ether

Mammalian and avian toxicity

Acute data
LD_{50} oral rat 475 mg kg^{-1} (1).
LD_{50} dermal rabbit 820 mg kg^{-1} (1).

Irritancy
Non-irritant in humans (1).

Sensitisation
Non-sensitiser in humans (1).

Any other comments

Experimental toxicology and human health effects reviewed (2).

References

1. Opdyke, D. C. J. *Food Cosmet. Toxicol.* 1975, **13**(Suppl.), 681
2. *ECETOC Technical Report No. 30(4)* 1991, European Chemical Industry Ecology and Toxicology Centre, B-1160 Brussels

A97 Allyl phenylacetate

$$CH_2CO_2CH_2CH=CH_2$$

CAS Registry No. 1797-74-6
Synonyms allyl α-toluate; 2-propenyl phenylacetate; benzeneacetic acid, 2-propenyl ester; phenylacetic acid, allyl ester
Mol. Formula $C_{11}H_{12}O_2$ Mol. Wt. 176.22

Uses Flavouring in food and cosmetics.

Physical properties
B. Pt. $_3$ 89-93°C.

Solubility
Organic solvent: ethanol, diethyl ether

Mammalian and avian toxicity

Acute data
LD_{50} oral rat 650 mg kg^{-1} (1).

Irritancy
Dermal human (48 hr) 30 mg caused irritation, while dermal rabbit (24 hr) 310 mg caused moderate irritation (1).

Sensitisation
Sensitisation reported in humans but thought to be due to free allyl alcohol (1).

Any other comments

Experimental toxicology and human health effects reviewed (2).

References
1. Opdyke, D. C. J. *Food Cosmet. Toxicol.* 1977, **15**(6), 621
2. *ECETOC Technical Report No. 30(4)* 1991, European Chemical Industry Ecology and Toxicology Centre, B-1160 Brussels

A98   Allyl propyl disulfide

$$H_2C=CHCH_2SSCH_2CH_2CH_3$$

CAS Registry No. 2179-59-1
Synonyms onion oil; 2-propenyl propyl disulfide
Mol. Formula $C_6H_{12}S_2$ **Mol. Wt.** 148.29
Occurrence Chief volatile constituent of onion oil.

Physical properties
M. Pt. −15°C; **B. Pt.** 56.5°C.

Solubility
Organic solvent: diethyl ether, benzene, ethanol, acetone

Occupational exposure
US TLV (TWA) 2 ppm (12 mg m^{-3}); **US TLV (STEL)** 3 ppm (18 mg m^{-3}).

Mammalian and avian toxicity

Sub-acute data
Rats fed with allyl propyl disulfide (dose and duration unspecified) exhibited lower blood sugar, lipid and albumin levels, compared to controls, together with increased levels of adipose tissue lipase (1).

Oral administration of 100 mg kg^{-1} day^{-1} for 15 days improved glucose tolerance of alloxan diabetic rabbits (2).

Sensitisation
In garlic sensitive humans, allyl propyl disulfide provokes allergic reactions including dermatitis (3).

Genotoxicity
Salmonella typhimurium TA97, TA98, TA100, TA1535, TA1537 with and without metabolic activation negative (4).

Any other adverse effects to man
Oral administration to healthy, fasted humans caused a significant drop in blood glucose levels and a rise in serum insulin (5).

Any other comments
Experimental toxicology and human health effects reviewed (6).

References
1. Wilcox, B. F; et al. *Indian J. Biochem. Biophys.* 1984, **21**(3), 214-216
2. Augusti, K. T; et al. *Experientia* 1974, **30**(10), 1119-1120
3. Papageorgiou, C; et al. *Arch. Dermatol. Res.* 1983, **275**(4), 229-234
4. Zeiger, E; et al. *Environ. Mol. Mutagen.* 1988, **11**(Suppl. 12), 1-158
5. Augusti, K. T; et al. *Clin. Chim. Acta* 1975, **60**(1), 121-123 (*Chem. Abstr.* **83**, 22593m)
6. *ECETOC Technical Report No. 30(4)* 1991, European Chemical Industry Ecology and Toxicology Centre, B-1160 Brussels

A99 Allyl thiourea

H$_2$C=CHCH$_2$NHCSNH$_2$

CAS Registry No. 109-57-9
Synonyms allyl thiocarbamide; thiosinamine; allyl sulfocarbamide; 2-propenyl thiourea; Rhodalline; 1-allyl-2-thiourea; Aminosin
Mol. Formula C$_4$H$_8$N$_2$S   **Mol. Wt.** 116.19
Uses Corrosion inhibitor. In veterinary medicine has been used to minimise scar tissue. As a laboratory reagent for BOD testing.

Physical properties
M. Pt. 78°C; **Specific gravity** 1.22.

Solubility
Organic solvent: ethanol, diethyl ether

Environmental fate

Nitrification inhibition
No inhibition occurred at 0.58 mg l^{-1}, 16% inhibition at 1.1 mg l^{-1} (oxidised N production), 38% inhibition at 1.1 mg l^{-1} (inhibition of ammonia loss) using activated sludge inocula (1).

Approximately 50% inhibition of ammonia oxidation by *Nitrosomonas sp.* at 1.2 mg l^{-1} (2).

Mammalian and avian toxicity

Acute data
LD_{50} oral rat 200 mg kg^{-1} (3).
LD_{50} intraperitoneal rat 500 mg kg^{-1} (4).
LD_{Lo} intravenous dog 110 mg kg^{-1} (5).

Sensitisation
Causes eczema in sensitised humans (6).

Genotoxicity
Salmonella typhimurium TA97, TA98, TA100, TA1535, TA1537 with and without metabolic activation negative (7).

Any other adverse effects
Neuroteratogenic to cultured brain cells (3).

Any other comments
Neurotoxic potential reviewed (6).

References
1. Wood, L. B; et al. *Water Res.* 1981, **15**, 543-551 in *Nitrification Inhibition in the Treatment of Sewage* 1985, Richardson, M. L., (Ed.), RSC, London
2. Hooper, A; et al. *J. Bacteriol.* 1973, **115**, 480
3. Dieke, S. H; et al. *J. Pharm. Exp. Ther.* 1947, **90**, 260
4. Dubois, K. P; et al. *J. Pharm. Exp. Ther.* 1947, **89**, 186
5. *Abdernaldeu's Handbuch der Biologiochem Arbeitsmethoden* 1935, **45**, 1289
6. Khera, K. S; et al. *Toxicol. in Vitro* 1988, **2**(4), 257-273, (*Chem. Abstr.* **110**, 90357p)
7. Zeiger, E; et al. *Environ. Mol. Mutagen.* 1988, **11**(Suppl 12), 1-158

A100   Allyltrichlorosilane

$CH_2=CHCH_2SiCl_3$

CAS Registry No. 107-37-9
Synonyms trichloro-2-propenylsilane; silane, trichloro-2-propenyl-
Mol. Formula $C_3H_5Cl_3Si$ **Mol. Wt.** 175.52
Uses Chemical synthesis in cycloaddition reactions. In heat-resisting adhesives.

Physical properties
B. Pt. 117.5°C; **Flash point** 35°C (open cup); **Specific gravity** d^{27} 1.217; **Volatility** v. den. 6.0.

Occupational exposure
UN No. 1724; **HAZCHEM Code** 4WE; **Conveyance classification** corrosive substance.

Mammalian and avian toxicity

Acute data
LD_{50} intravenous mouse 56 mg kg^{-1} (1).

Any other comments
Only minimal toxicity information is reported to be available on chlorosilanes (1).
Corrosive.

References

1. *U.S. Army Armament R&D Command* NX 04219

A101   Aloe Emodin

CAS Registry No. 481-72-1
Synonyms 9,10-anthracenedione, 1,8-dihydroxy-3-(hydroxymethyl)-
Mol. Formula $C_{15}H_{10}O_5$ **Mol. Wt.** 270.24

Physical properties
M. Pt. 223-224°C.

Genotoxicity
Salmonella typhimurium TA1537, TA98, TA1538 without metabolic activation
positive (1).
In vivo primary rat hepatocytes caused a 2- to 3-fold increase in DNA synthesis (2).

Any other adverse effects
Intraperitoneal mice with P388 leukemia (7 day) 40 mg kg^{-1} increased the survival
time of the mice by 36%, also reduced the ascites volume and tumour cell number.
Results indicated inhibition of biosynthesis of DNA, RNA and P388 leukemia cell
protein (3).
Intraperitoneal mice (7 day) 44 mg kg^{-1} day^{-1} decreased cytochrome P450 in hepatic
microsomes by 66.3% (4).

References

1. Westendorf, J; et al. *Mutat. Res.* 1990, **240**(1), 1-12
2. Woelfe, D; et al. *Cancer Res.* 1990, **50**(20), 6540-6544
3. Lu, M; et al. *Zhongguo Yaoke Daxue Xuebao* 1989, **20**(3), 155-157, (Ch.) (*Chem. Abstr.*
    1989, **111**, 126567u)
4. Sun, Y; et al. *Zhongguo Yaoke Daxue Xuebao* 1988, **19**(2), 110-112, (Ch.) (*Chem. Abstr.*
    1988, **109**, 104177y)

A102 Aluminium

Al

CAS Registry No. 7429-90-5

Synonyms aluminum; aluminium fibre; C.I. 77000; aluminium flake; aluminium powder

Mol. Formula Al   **Mol. Wt.** 26.98

Uses Used as the pure metal or as alloys for aircraft, utensils, apparatus and electrical conductors. Flashlight powder in photography, explosives, fireworks and in paints.

Occurrence The toxicity of aluminium is dependent on the ability of the organism to absorb it. Therefore toxicity data refer to bioavailable forms, such as the ion in solution or particulate matter.

Does not occur in free state in nature, but is found combined with oxygen, fluorine and silicon. Bauxite is the principle ore, and the richest source of aluminium. The solution chemistry of aluminium is complex, and the response of the biota to the metal is dependant upon the chemical form of the toxicant. pH values and water hardness are highly influential.

Physical properties

M. Pt. 660°C; **B. Pt.** 2450°C; **Specific gravity** d^{25} 2.70; **Volatility** 1 mmHg at 1284°C.

Occupational exposure

US TLV (TWA) 10 mg m^{-3} (metal dust), 5 mg m^{-3} (pyro powders, welding fumes); **UK Long-term limit** 10 mg m^{-3} (total inhalable dust), 5 mg m^{-3} (respirable dust) proposed change; **UN No.** 1396 (powder uncoated); 1309 (powder coated); **HAZCHEM Code** 4Y (powder uncoated); 4⊠ (powder coated); **Conveyance classification** substance which in contact with water emit flammable gas (powder uncoated); flammable solid (powder coated); **Supply classification** highly flammable.

Risk phrases

Al powder (pyrophoric) – Contact with water liberates highly flammable gases – Spontaneously flammable in air (R15, R17) Al powder (stabilised) – Flammable – Contact with water liberates highly flammable gases (R10, R15)

Safety phrases

Keep container tightly closed and dry – In case of fire, use dry chemical powder extinguisher (S7/8, S43)

Ecotoxicity

Fish toxicity

LC$_{50}$ (4 wk) rainbow trout 0.56 mg l^{-1} (1).

Lowest observed effective concentration (3 wk) longnose sucker, brook trout 0.1 and 0.2 mg l^{-1}, respectively (2).

Yolk fry of Atlantic salmon exposed to 135 μg l^{-1} at pH 5 and 1°C for ≈30 day caused ≈6% mortality. Acidification increased aluminium accumulation but ≈60% of aluminium was absorbed by the fish body surface and was lost early during depuration (3).

The effect on brown trout of lime being administered to an acid stream rich in aluminium was studied from 0-100 metres of the mixing zone. Within 15 metres of dosing, filterable aluminium content fell from 580 to 230 μg l^{-1} and to 120 μg l^{-1} within 30 metres although total aluminium was unchanged. After 24 hr, fish mortality was 100% at untreated acidic sites, 80% at <30 metres downstream of loading and 0% at 100 metres. Mortalities correlated with aluminium concentration in gill tissues and filterable aluminium in the water (4).

Invertebrate toxicity

LC$_{50}$ (48-96 hr) *Asellus aquaticus* 6.57-4.37 mg l^{-1} specific toxicological reaction to mobility (5).

EC$_{50}$ (48 hr) *Daphnia magna* 1.4 mg l^{-1} (6).

EC$_{50}$ (3 wk) *Daphnia magna* 0.68 mg l^{-1} reproductive effects (6).

Lowest observed effective concentration (3 wk) *Daphnia magna* 0.32 mg l^{-1} reproductive effects (6).

Freshwater clams *Anodonta anatina* and *Unio pictorium* (3 wk continous or 24 day fluctuating) 300 and 900 μg l^{-1} caused accumulation in kidney; midgut gland, rest gill and mantle in decreasing order of concentration. During the 3 wk exposure, aluminium in the gills and kidney increased linearly and saturation was not reached. Ambient pH had a significant effect on accumulation in the gills whereas water hardness did not (7).

Effects to non-target species

Elevated aluminium levels in common frog and moor frog larvae increased the rate of defects including spinal curvature and vesicles on the head and thorax, and altered the feeding behaviour (8).

Mammalian and avian toxicity

Sub-acute data

Intravenous Japanese quail (18 hr) accumulation of 51% of aluminium ion in liver in ♂. Cumulative maximum in egg components (10 day) 38% in yolks and 0.54% in shells (9).

Metabolism and pharmacokinetics

Poorly absorbed through the gastrointestinal tract, the absorption which does occur has been suggested to be mediated, at least in part, by an active transport process controlled by parathyroid hormone (PTH) (10).

70- 90% of total aluminium was bound to plasma proteins (60-70% to a high molecular weight protein and 10-20% to albumin while only 10-30% was unbound). This high affinity of aluminium for plasma proteins strongly suggests high levels of binding of aluminium to a variety of tissue proteins (11).

Any other adverse effects to man

Workers exposed to aluminium had urinary concentrations 80-90 times higher than those occupationally non-exposed workers t$_{1/2}$ 5-6 wk. Among retired workers t$_{1/2}$ <1-8 yr and were related to the number of yr since retirement, suggesting aluminium is retained and stored in several compartments of the body and eliminated at different rates (12).

Exposure of 65 welders to aluminium for a long period revealed neuropsychiatric symptoms (13).

Any other adverse effects

Aluminium affects the central nervous system (14).

Aluminium content in blood serum, aqueous humour and lenses of humans with senile cataracts gradually increased and was optimum during the period of the mature cataract (15).

Chronic inhalation of fumes and dusts containing high concentrations of aluminium may cause dyspnoea, cough, weakness, emphysema, and non-nodular pulmonary fibrosis (aluminosis). Shavers disease is the only industrial disease attributable to aluminium exposure, and is characterised by pulmonary fibrosis and emphysema (16). Inhibits both the cytosolic and mitochondrial hexokinase activities in the rat brain. IC_{50} 4-9 mg kg^{-1} (17).

Any other comments

Nutritional aspects of aluminium toxicity reviewed in relation to its possible role in Alzheimer's disease and other neurodegenerative diseases (18).

Absorption, plasma $t_{1/2}$ and excretion reviewed (19).

Mutagenic and carcinogenic potential reviewed (20,21).

Physico-chemical properties, human health effects, experimental toxicology, environmental effects and workplace experience reviewed (22).

Toxicity and hazards reviewed, to be published in 1992-1993 (23).

References

1. Birge, W. J; et al. *USEPA Report* 1980, EPA 600/9-80-022, NTIS Dept. of Commerce, Springfield, VA
2. Baker, J. P; et al. *Water Air Soil Pollut.* 1982, **18**, 289-309
3. Parent, L; et al. *Water Pollut. Res. J. Canada* 1988, **23**(2), 227-242, (Fr.) (*Chem. Abstr.* 1989, **110**, 226849k)
4. Weatherley, N. S; et al. *Water, Air, Soil Pollut.* 1991, **55**(3-4), 345-353
5. Martin, T. R; et al. *Water Res.* 1986, **20**(9), 1137-1147
6. Biesinger, K. E; et al. *J. Fish. Res. Board Canada* 1972, **29**, 1691-1700
7. Pynnoenen, K. *Comp. Biochem. Physiol., C: Comp. Pharmacol. Toxicol.* 1990, **97**C(1), 111-117
8. Olsson, M; et al. *Bull. Environ. Contam. Toxicol.* 1987, **39**(1), 37-44
9. Robinson, G. A; et al. *Poultry Sci.* 1990, **69**(2), 300-306
10. Berlyne, G. M; et al. *Lancet* 1972, **1**, 564
11. Elliott, H. L; et al. *Lancet* 1978, **2**, 1255
12. Ljunggren, K. G; et al. *Br. J. Ind. Med.* 1991, **48**(2), 106-109
13. Sjoegren, B; et al. *Br. J. Ind. Med.* 1990, **47**(10), 704-707
14. Wicniewski, H. M. *J. Environ. Pathol. Toxicol. Oncol.* 1985, **6**(1), 1
15. Wu, X; et al. *Shandong Yike Daxue Xuebao* 1989, **27**(3), 59-64, (Ch.) (*Chem. Abstr.* 1990, **112**, 74812u)
16. *Chemical Safety Data Sheets* 1989, **2**, 1-5, RSC, London
17. Lai, J. C. K. *J. Neurochem.* 1984, **42**(2), 438
18. Klein, G. L. *Nutr. Res. Rev.* 1990, **3**, 117-141
19. Wilhelm, M; et al. *Pharmacol. Toxicol. (Copenhagen)* 1990, **66**(1), 4-9
20. Leonard, A; et al. *Toxicol. Environ. Chem.* 1989, **23**(1-4), 27-31
21. Leonard, A; Gerber, G. B. *Mutat. Res.* 1988, **196**(3), 247-257
22. *ECETOC Technical Report No. 30(4)* 1991, European Chemical Industry Ecology and Toxicology Centre, B-1160 Brussels
23. Izmerov, N. F; *Scientific Reviews of Soviet Literature on Toxicity & Hazards of Chemicals* 1992-1993, **127**, Eng. Trans, Richardson, M. L. (Ed.), UNEP/IRPTC, Geneva

A103   Aluminium bromide

AlBr₃

CAS Registry No. 7727-15-3
Mol. Formula AlBr₃   Mol. Wt. 266.71
Uses Catalyst in organic synthesis.
Occurrence The solution chemistry of aluminium is complex, and the response of the biota to the metal is dependant upon the chemical form of the toxicant. pH values and water hardness are highly influential. The aquatic toxicity of aluminium is dependant upon the chemical state of the metal, and/or the associated hydroxide complexes.

Physical properties

M. Pt. 94-98°C; **B. Pt.** 250-270°C; **Specific gravity** d_4^{18} 3.205; **Volatility** v.p. 1 mmHg at 81°C.

Solubility
Organic solvent: ethanol, acetone, carbon disulfide, benzene, nitrobenzene, toluene, xylene, simple hydrocarbons

Occupational exposure

US TLV (TWA) 2 mg m^{-3}; **UK Long-term limit** 2 mg m^{-3} (as Al); **UN No.** 1725 (anhydrous); 2580 (solution); **HAZCHEM Code** 4X (anhydrous);2R (sol.)ion); **Conveyance classification** corrosive substance.

Mammalian and avian toxicity

Irritancy
Inhalation of powders and vapours can cause severe irritation, Aluminium bromide has a high affinity for water which can cause severe tissue burns on contact with skin (1).

Any other adverse effects to man

In humans, symptoms of mild poisoning include shortness of breath, coughing, wheezing, nausea, aching muscles and slight fever, while severe poisoning can cause convulsions, liver and kidney damage. Inhalation may prove fatal as a result of spasm, inflammation and oedema of the larynx and bronchi, chemical pneumonitis and pulmonary oedema. Prolonged exposure may lead to inorganic bromide poisoning, the symptoms of which include depression, emaciation and, in severe cases, psychoses and mental deterioration (1).

Legislation

Limited under EEC Directive on Drinking Water Quality 80/778/EEC. Aluminium: guide level 0.05 mg l^{-1}, maximum admissible concentration 0.2 mg l^{-1} (2).

Any other comments

Aluminium and its compounds have been implicated in Alzheimers disease (3-6).
Toxicity and hazards reviewed, to be publshed in 1992-1993 (7).
Highly corrosive to metals. Reacts violently with water, alcohols and acids.
Incompatible with strong oxidising agents and mixtures with sodium or potassium.

References

1. *Chemical Safety Data Sheets* 1989, **2**, 6-8, RSC, London
2. *EC Directive Relating to the Quality of Water Intended for Human Consumption* 1982, 80/778/EEC, Office for Official Publications of the European Communities, 2 rue Mercier, L-2985 Luxembourg
3. *Lancet* 1985, **1**, 616
4. Wheater, R. H. *J. Am. Med. Assoc.* 1985, **253**, 2288
5. King, R. G; et al. *Med. J. Aust.* 1985, **142**, 352
6. Candy, J. M; et al. *Lancet* 1986, **1**, 354
7. Izmerov, N. F; *Scientific Reviews of Soviet Literature on Toxicity & Hazards of Chemicals* 1992-1993, **127**, Eng. Trans. Richardson, M. L. (Ed.), UNEP/IRPTC, Geneva

A104   Aluminium carbide

Al_4C_3

CAS Registry No. 1299-86-1
Mol. Formula Al_4C_3 **Mol. Wt.** 143.96
Uses Reduction of metal oxides. Manufacture of aluminium nitride. Ceramic manufacture. Generation of methane.
Occurrence The solution chemistry of aluminium is complex, and the response of the biota to the metal is dependant upon the chemical form of the toxicant. pH values and water hardness are highly influential. The aquatic toxicity of aluminium is dependant upon the state of the metal, and/or the associated hydroxide complexes.

Physical properties

M. Pt. 2100°C; **B. Pt.** 2200°C (decomp.).

Occupational exposure

US TLV (TWA) 2 mg m^{-3} (as Al); **UK Long-term limit** 2 mg m^{-3}; **UN No.** 1394;
Conveyance classification substance which in contact with water emit flammable gas.

Any other comments

Aluminium has been implicated in Alzheimers disease (1-4).
Toxicity and hazards reviewed, to be published in 1992-1993 (5).
For other toxic properties see aluminium and its compounds.
Reacts violently with acids and is incompatible with strong oxidisers.

References

1. *Lancet* 1985, **1**, 616
2. Wheater, R. H. *J. Am. Med. Ass.* 1985, **253**, 2288
3. King, R. G; et al. *Med. J. Aust.* 1985, **142**, 352
4. Candy, J. M; et al. *Lancet* 1986, **1**, 354
5. Izmerov, N. F; *Scientific Reviews of Soviet Literature on Toxicity & Hazards of Chemicals* 1992-1993, **127**, Eng. Trans. Richardson, M. L. (Ed.), UNEP/IRPTC, Geneva

A105   Aluminium chloride

AlCl₃

CAS Registry No. 7446-70-0

Synonyms trichloroaluminium; aluminium trichloride

Mol. Formula AlCl₃   **Mol. Wt.** 133.34

Uses Acid catalyst especially in Friedal-Crafts type reactions. In cracking of petroleum and in the manufacture of rubbers, lubricants and antiperspirants, and treatment of wastewaters.

Occurrence The aquatic toxicity of aluminium is dependant upon the chemical state of the metal, and/or the associated hydroxide complexes.

Physical properties

M. Pt. 194°C (5.2 atm); **B. Pt.** 181°C (sublimes), 262°C (decomp.); **Specific gravity** d^{25} 2.44; **Volatility** v.p. 1 mmHg at 100°C.

Solubility

Organic solvent: ethanol, diethyl ether, benzene, carbon tetrachloride, chloroform

Occupational exposure

US TLV (TWA) 2 mg m^{-3} (as Al); **UK Long-term limit** 2 mg m^{-3} (as Al); **UN No.** 1726; **HAZCHEM Code** 4X; **Conveyance classification** corrosive substance; **Supply classification** Corrosive.

Risk phrases

Causes burns (R34)

Safety phrases

Keep container tightly closed and dry – After contact with skin, wash immediately with plenty of water (S7/8, S28)

Ecotoxicity

Fish toxicity

LC_{50} (12-96 hr) goldfish 100 mg l^{-1} (1).

Invertebrate toxicity

EC_{50} (48 hr) *Daphnia magna* 3.9 mg l^{-1} (2).

Mammalian and avian toxicity

Acute data

LD_{50} oral rat, mouse 770-3730 mg kg^{-1} (3,4).

Teratogenicity and reproductive effects

LD_{50} 1.1 µg l^{-1} injected egg^{-1} chick embryo air sac of egg on day 3 of incubation. Minimal lethal dose was 0.3 µg l^{-1} egg^{-1}. Development of the embryos was retarded and teratogenic effects were apparent (5).

Rats administered 0.025-5 mg l^{-1} in drinking water for 6 months before and during pregnancy embryotoxicity and neurotoxicity reported (6).

Metabolism and pharmacokinetics

Aluminium chloride (concentration unspecified) added to the drinking water of mice resulted in the accumulation of aluminium in organs, especially the brain and bones, which interfered with calcium and phosphorus metabolism, increased brain acetylcholinesterase activity and damaged kidney, bone and brain and reduced growth and development (7).

Irritancy

Acute biological hazards of aluminium chloride are mostly due to the extremely acid products of its reaction with water which it takes from the tissues. Aluminium chloride dust and vapours are irritants and can cause severe burns or allergic skin reactions. In its hydrated form, inhibits sweating and causes clinical irritation after prolonged exposure (8).

Genotoxicity

Bacillus subtilis H17 (rec +), M45 (rec –) negative DNA damage (9).
Escherichia coli SOS chromotest negative (10).
Intraperitoneal mouse 1.3-13 g l^{-1} induced chromosomal aberrations in bone marrow cells (11).

Any other adverse effects

Inhalation can be fatal as a result of spasm, inflammation of the larynx and bronchi, chemical pneumonitis and pulmonary oedema. Symptoms of exposure may take several hr to appear and include a burning sensation, coughing and wheezing, headache, nausea and vomiting. Prolonged exposure may result in lung damage. Increased lung:body weight ratio, bronchiolitis and secondary effects on liver and kidney weights have been reported in chronic inhalation studies (12).
Can cause irritation, especially if applied to damp skin, attributed to the formation of hydrochloric acid (13).

Legislation

Limit under EEC Drinking Water Quality Directive 80/778/EEC. Aluminium: guide level 0.05 mg l^{-1}, maximum admissible concentration 0.2 mg l^{-1}. Chloride guide level 25 mg l^{-1} (14).

Any other comments

Physico-chemical properties, experimental toxicology, human health and environmental effects reviewed (15).
Aluminium and its compounds have been implicated in Alzheimers disease (16-19).
Toxicity and hazards reviewed, to be published in 1992-1993 (20).
Reacts violently with water to release heat and toxic fumes of hydrogen chloride, aluminium oxide and aluminium chlorate.

References

1. McGautrey, P. H. *Eng. Management of Water Quality* 1968, McGraw-Hill, New York
2. Biesinger, K. E; et al. *J. Fish. Res. Board Can.* 1972, **29**, 1691-1700
3. *Teratology, A Journal of Abnormal Development* 1974, **9**, A14
4. *Environmental Quality and Safety* 1975, **1**(Suppl.), 1
5. Sun, Y; et al. *Jiepou Xuebao* 1990, **21**(1), 102-106, (*Chem. Abstr.* **113**, 167212j)
6. Pestova, L. V; et al. *Gig. Sanit.* 1990, (9), 23-25, (Russ.), (*Chem. Abstr.* 1991, **114**, 19259k)
7. Yu, D; et al. *Weisheng Dulixue Zazhi* 1990, **4**(4), 227-229, 249, (Ch.), (*Chem. Abstr.* 1991, **115**, 2720b)
8. Serban, G. P; et al. *J. Soc. Cosmet. Chem.* 1984, **35**(8), 391-410
9. Kanematsu, M; et al. *Mutat. Res.* 1980, **77**, 109-116

10. Olivier, P; et al. *Mutat. Res.* 1987, **189**(3), 263-269
11. Berlyne, G. M; et al. *Lancet* 1972, **I**(7750), 564-567
12. Stone, C. J; et al. *Toxicol. Appl. Pharmacol.* 1979, **49**(1), 71-76
13. *Martindale. The Extra Pharmacopoeia* 29th ed., 1989, 777, The Pharmaceutical Press, London
14. *EC Directive Relating to the Quality of Water Intended for Human Consumption* 1982, 80/778/EEC, Office for Official Publications of the European Communities, 2 rue Mercier, L-2985 Luxembourg
15. *ECETOC Technical Report No. 30(4)* 1991, European Chemical Industry Ecology and Toxicology Centre, B-1160 Brussels
16. *Lancet* 1985, **1**, 616
17. Wheater, R. H. *J. Am. Med. Assoc.* 1985, **253**, 2288
18. King, R. G; et al. *Med. J. Aust.* 1985, **142**, 352
19. Candy, J. M; et al. *Lancet* 1986, **1**, 354
20. Izmerov, N. F. *Scientific Reviews of Soviet Literature on Toxicity & Hazards of Chemicals* 1992-1993, **127**, Eng. Trans. Richardson, M. L. (Ed.), UNEP/IRPTC, Geneva

A106   Aluminium hydride

AlH$_3$

CAS Registry No. 7784-21-6
Synonyms alane; aluminium trihydride; α-aluminium trihydride
Mol. Formula AlH$_3$ **Mol. Wt.** 30.01
Uses Catalyst for polymerisation. Reducing agent. In photoimaging. Former interest as a high energy additive to solid rocket propellants.
Occurrence The solution chemistry of aluminium is complex, and the response of the biota to the metal is dependant upon the chemical form of the toxicant. pH values and water hardness are highly influential. The aquatic toxicity of aluminium is dependant upon the chemical state of the metal, and/or the associated hydroxide complexes.

Occupational exposure

US TLV (TWA) 2 mg m^{-3} (as Al); **UK Long-term limit** 2 mg m^{-3}; **UN No.** 2463;
Conveyance classification substance which in contact with water emit flammable gas.

Any other adverse effects

Alkaline reaction on hydrolysis causing chemical burns on skin and other tissues.

Any other comments

Environmental, human health effects and experimental toxicology reviewed (1).
Aluminium and its compounds have been implicated in Alzheimers disease (2- 5).
Toxicity and hazards reviewed, to be published in 1992-1993 (6).

References

1. *ECETOC Technical Report No. 30(4)* 1991, European Chemical Industry Ecology and Toxicology Centre, B-1160 Brussels
2. *Lancet* 1985, **1**, 616

3. Wheater, R. H. *J. Am. Med. Ass.* 1985, **253**, 2288
4. King, R. G; et al. *Med. J. Aust.* 1985, **142**, 352
5. Candy, J. M; et al. *Lancet* 1986, **1**, 354
6. Izmerov, N. F; *Scientific Reviews of Soviet Literature on Toxicity & Hazards of Chemicals* 1992-1993, **127**, Eng. Trans., Richardson, M. L. (Ed.), UNEP/IRPTC, Geneva

A107   Aluminium lithium hydride

AlH4Li

CAS Registry No. 16853-85-3
Synonyms lithium tetrahydroaluminate; lithium alanate; lithium aluminohydride; lithium aluminium hydride; lithium tetrahydroaluminate
Mol. Formula AlH$_4$Li   **Mol. Wt.** 37.95
Uses Reducing agent in preparation of ether hydrides. Used to identify structure of drugs.
Occurrence The solution chemistry of aluminium is complex, and the response of the biota to the metal is dependant upon the chemical form of the toxicant. pH values and water hardness are highly influential. The aquatic toxicity of aluminium is dependant upoon the chemical state of the metal, and/or the associated hydroxide complexes.

Physical properties

M. Pt. 125°C (decomp.); **Volatility** Stable in dry air at 20°C (1).
Decomposes above 125°C (1).

Solubility
Organic solvent: ether, tetrahydrofuran, dibutyl ether, 1,4- dioxane

Occupational exposure

UK Long-term limit 2 mg m^{-3} (as Al); **UN No.** 1410; **Conveyance classification** substance which in contact with water emit flammable gas; **Supply classification** Highly flammable.

Risk phrases
Contact with water liberates highly flammable gases (R15)

Safety phrases
Keep container tightly closed and dry – Avoid contact with skin and eyes – In case of fire, use specially manufactured dry powder extinguishers (S7/8, S24/25, S43)

Ecotoxicity

Fish toxicity
Aluminium lithium hydride would release ionic lithium on contact with water.
Lithium is lethal to rainbow trout in 35 day at a concentration of 1.4 mg l^{-1} (2).

Any other adverse effects

Inhibits adenylate cyclase activity and therefore hormone function in human cells (3).

Legislation

Limited under EEC Directive on Drinking Water Quality 80/778/EEC. Aluminium: guide level 0.05 mg l^{-1}, maximum admissible concentration 0.2 mg l^{-1} (4).

Any other comments

The main hazards associated with it relate to its highly caustic reaction on inhalation, injection or contact with skin, and the formation of lithium hydroxide on contact with moisture (5).

Aluminium and its compounds have been implicated in Alzheimers disease (6-9).

Toxicity and hazards reviewed, to be published in 1992-1993 (10).

References

1. *The Merck Index* 10th ed., 1983, Merck & Co. Inc., Rahway, NJ
2. *Report of the Director for Water Pollut. Res.* 1970, 61, HMSO, London
3. Ebstein, R. P; et al. *Prog. Neuro- Psych. Biol. Psych.* 1986, **10**, 323
4. *EC Directive Relating to the Quality of Water Intended for Human Consumption* 1982, 80/778/EEC, Office for Official Publications of the European Communities, 2 rue Mercier, L-2985 Luxembourg
5. *Chemical Safety Data Sheets* 1989, **2**, 248-250, RSC, London
6. *Lancet* 1985, **1**, 616
7. Wheater, R. H. *J. Am. Med. Ass.* 1985, **253**, 2288
8. King, R. G; et al. *Med. J. Aust.* 1985, **142**, 352
9. Candy, J. M; et al. *Lancet* 1986, **1**, 354
10. Izmerov, N. F. *Scientific Reviews of Soviet Literature on Toxicity & Hazards of Chemicals* 1992-1993, **127**, Eng. Trans. Richardson, M. L. (Ed.), UNEP/IRPTC, Geneva

A108   Aluminium nitrate nonahydrate

Al(NO$_3$)$_3$.9H$_2$O

CAS Registry No. 7784-27-2

Mol. Formula AlH$_{18}$N$_3$O$_{18}$ **Mol. Wt.** 375.13

Uses Tanning leather, anti-perspirant, corrosion inhibitor, salting out agent in extraction of actinides.

Occurrence Occurs in several states of hydration of which the nonahydrate is the most stable. The solution chemistry of aluminium is complex, and the response of the biota to the metal is dependant upon the chemical form of the toxicant. pH values and water hardness are highly influential. The aquatic toxicity of aluminium is dependant upon the chemical state of the metal, and/or the associated hydroxide complexes.

Physical properties

M. Pt. 73°C; **B. Pt.** 135°C (decomp.).

Solubility

Organic solvent: ethanol, acetone

Occupational exposure

UK Long-term limit 2 mg m^{-3} (as Al); **UN No.** 1438; **HAZCHEM Code** 1S;

Conveyance classification oxidising substance.

Ecotoxicity

Fish toxicity
LC_{50} (10 day) stickleback 0.07 mg l^{-1} as aluminium metal (1).

Mammalian and avian toxicity

Acute data
LD_{50} oral rat 4.28 g kg^{-1} (2).

Legislation

Limited under EEC Directive on Drinking Water Quality 80/778/EEC. Aluminium: guide level 0.05 mg l^{-1}, maximum admissible concentration 0.2 mg l^{-1}. Nitrate: guide level 25 mg l^{-1}, maximum admissible concentration 50 mg l^{-1} (3).

Any other comments

Aluminium and its compounds have been implicated in Alzheimers disease (4-7).

References

1. Doudoroft, P; et al. *Sewage Ind. Wastes* 1953, **25**(7), 802
2. Smyth, H. F; et al. *Am. Ind. Hyg. Assoc. J.* 1969, **30**, 470
3. *EC Directive Relating to the Quality of Water Intended for Human Consumption* 1982, 80/778/EEC, Office for Official Publications of the European Communities, 2 rue Mercier, L-2985 Luxembourg
4. *Lancet* 1985, **1**, 616
5. Wheater, R. H. *J. Am. Med. Assoc.* 1985, **253**, 2288
6. King, R. G; et al. *Med. J. Aust.* 1985, **142**, 352
7. Candy, J. M; et al. *Lancet* 1986, **1**, 354

A109   Aluminium oxide

Al_2O_3

CAS Registry No. 1344-28-1
Synonyms alumina; activated aluminium oxide; aluminium sesquioxide; α-alumina; β-alumina; γ-alumina
Mol. Formula Al_2O_3 **Mol. Wt.** 101.96
Uses Used as an adsorbent, desiccant, a filler for paints and varnishes. Manufacture of alloys, ceramics, electrical insulators and resistors. Catalyst for organic reactions.
Occurrence As the minerals bauxite, bayerite, boehmite, corundum, disapore, gibbsite. The aquatic toxicity of aluminium is dependant upon the chemical state of the metal, and/or the associated hydroxide complexes.

Physical properties

M. Pt. 2072°C; **B. Pt.** 2977°C; **Specific gravity** d_4^{20} 4.0.

Occupational exposure

US TLV (TWA) 10 mg m^{-3}; UK Long-term limit 10 mg m^{-3} (as total inhalable dust), 5 mg m^{-3} (as respirable dust).

Mammalian and avian toxicity

Acute data

5 mg intratracheal rat caused a mild increase in the number of polymorphonuclear leukocytes in the lung 7 days after administration (1).

Carcinogenicity and long-term effects

Cancer morbidity and total morbidity pattern were studied among workers manufacturing abrasive materials who had been exposed to aluminium oxide (total dust levels 0.1-1 mg mg^{-3}) from 1958 to 1983. No significant increase was found in mortality cases or incidence of nonmalignant respiratory diseases (2).

Occasional mesotheliomata reported in rats after intrapleural injection of 0.5-20 mg aluminium oxide (3).

Irritancy

Skin irritation and acronesthesia, a congestive, anaesthetic condition, can result from prolonged exposure (4).

Genotoxicity

Bacillus subtilis H17 (rec$^+$), M45 (rec$^-$) negative DNA damage (5).

Any other adverse effects to man

Thirty-three foundry workers were exposed to aluminium oxide dust inhalation of <1 mg m^{-3}. Aluminium in serum was significantly raised but not in urine, suggesting incomplete excretion of aluminium (6).

Exposure to inhaled aluminium oxide in 38 potroom workers with no airway symptoms and 20 healthy office workers (all non-smokers) was low, 15-20% of the Swedish exposure limits (7).

Any other adverse effects

Intratracheal hamster induced dose-related increase in the incidence and severity of alveolar septal fibrosis (duration of exposure unspecified) (8).

Inhalation of particles of aluminium oxide has been implicated in Shavers disease, an often fatal and rapidly progressive interstitial fibrosis of the lung, although it is believed that silica fume may be partially responsible (9).

Any other comments

Toxicity to lungs reviewed (10).

Aluminium and its compounds have been implicated in Alzheimers disease (11-14).

Human health effects, experimental toxicology, epidemiology and environmental effects reviewed (15).

Toxicity and hazards reviewed, to be published 1992-1993 (16).

References

1.   White, L. R; et al. *Environ. Res.* 1987, **42**(2), 534-545
2.   Edling, C; et al. *Br. J. Ind. Med.* 1987, **44**(1), 57-59
3.   Wagner, J. C; et al. *Br. J. Cancer* 1973, **28**(2), 173-185

4. *Chemical Safety Data Sheets* 1989, **2**, 28-30, RSC, London
5. Kanematsu, M; et al. *Mutat. Res.* 1980, **77**, 109-116
6. Rollin, H. B; et al. *Br. J. Ind. Med.* 1991, **48**(4), 243-246
7. Larsson, K; et al. *Scand. J. Work, Environ. Health* 1989, **15**(4), 296-301
8. Renne, R. A; et al. *Gov. Rep. Announce. Index US* 1983, **83**(24), 6055, Report Order No. PB83-24430
9. Shaver, C. G; et al. *J. Ind. Hyg. Tox.* 1947, **29**, 145
10. Dinman, B. D. *J. Occup. Med.* 1988, **30**(4), 328
11. *Lancet* 1985, **1**, 616
12. Wheater, R. H. *J. Am. Med. Assoc.* 1985, **253**, 2288
13. King, R. G; et al. *Med. J. Aust.* 1985, **142**, 352
14. Candy, J. M; et al. *Lancet* 1986, **1**, 354
15. *ECETOC Technical Report No. 30(4)* 1991, European Chemical Industry Ecology and Toxicology Centre, B-1160 Brussels
16. Izmerov, N. F; *Scientific Reviews of Soviet Literature on Toxicity & Hazards of Chemicals* 1992-1993, **127**, Eng. Trans. Richardson, M. L. (Ed.), UNEP/IRPTC, Geneva

A110   Aluminium phosphide

AlP

CAS Registry No. 20859-73-8
Synonyms Phostoxin; Celphos
Mol. Formula AlP   **Mol. Wt.** 57.96
Uses Fumigant for killing insects in stored feed, grain, seeds, nuts. Source of phosphine in semiconductor research. Acute rodenticide.
Occurrence The solution chemistry of aluminium is complex, and the response of the biota to the metal is dependant upon the chemical form of the toxicant. pH values and water hardness are highly influential. The aquatic toxicity of aluminium is dependant upon the chemical state of the metal, and/or the associated hydroxide complexes.

Physical properties

M. Pt. Does not melt below 1000°C; **Specific gravity** d_4^{25} 2.85.

Occupational exposure

UK Long-term limit 2 mg m^{-3} (as Al); **UN No.** 1397; **Conveyance classification** substance which in contact with water emit flammable gas; **Supply classification** highly flammable and toxic.

Risk phrases

Contact with water liberates toxic, highly flammable gas – Very toxic if swallowed (R15/29, R28)

Safety phrases

Keep locked up and out of reach of children – Do not breathe dust – In case of fire, use dry chemical powder extinguisher – In case of accident or if you feel unwell, seek medical advice immediately (show label where possible) (S1/2, S22, S43, S45)

Mammalian and avian toxicity

Acute data
LC_{Lo} inhalation rat 1 ppm (1).

Any other adverse effects to man
Phosphine detected post mortem in stomach and contents, blood and liver specimens of a man who had ingested tablets containing aluminium phosphide (2).
Two incidents were reported of children exposed to fumigated grain on ship, symptoms included headache, nausea, vomiting, dyspnoea, fatigue and jaundice. Fatalities reported, myocardial infiltration with necrosis and pulmonary oedema were found on autopsy (3,4).

Any other adverse effects
Aluminium phosphide releases phosphine on contact with moisture. Symptomatic effects of phosphine exposure include weakness, vertigo, pains around the diaphragm, dyspnoea, bronchitis, oedema and other lung damage, convulsions and coma (5).

Any other comments
Reacts with water to produce phosphine. Hazardous potential reviewed (6).
Physico-chemical properties, human health and environmental effects, experimental toxicology, ecotoxicology and exposure levels reviewed (7).
Aluminium and its compounds have been implicated in Alzheimers disease (8-11).
Toxicity and hazards reviewed, to be published in 1992-1993 (12).

References
1. *Pesticide Chemicals Official Compendium* 1966, 25, Association of the American Pesticide Control Officials, Inc.
2. Chan, L. T. F; et al. *J. Anal. Toxicol.* 1983, **7**, 165-167
3. Heyndrickx, A; et al. *Eur. J. Toxicol. Environ. Hyg.* 1976, **9**, 113-118
4. Wilson, R; et al. *J. Am. Med. Assoc.* 1980, **244**, 148-150
5. *Chemical Safety Data Sheets* 1989, **2**, 31-33, RSC, London
6. *Dangerous Prop. Ind. Mater. Rep.* 1990, **10** (4), 39-46
7. *ECETOC Technical Report No. 30(4)* 1991, European Chemical Industry and Toxicology Centre, B-1160 Brussels
8. *Lancet* 1985, **1**, 616
9. Wheater, R. H. *J. Am. Med. Assoc.* 1985, **253**, 2288
10. King, R. G; et al. *Med. J. Aust.* 1985, **142**, 352
11. Candy, J. M; et al. *Lancet* 1986, **1**, 354
12. Izmerov, N. F; *Scientific Reviews of Soviet Literature on Toxicity & Hazards of Chemicals* 1992-1993, **127**, Eng. Trans. Richardson, M. L. (Ed.), UNEP/IRPTC, Geneva

A111   Aluminium potassium sulfate

$KAl(SO_4)_2$

CAS Registry No. 10043-67-1
Synonyms *Anhydrous:* sulfuric acid, aluminium potassium salt; burnt alum; exsiccated alum; *Dodecahydrate:*alum; potassium alum; kalinite; alum flour
Mol. Formula $AlKO_8S_2$ **Mol. Wt.** 258.21

Uses Used in dyeing, printing fabrics. Manufacture of dyestuffs, paper, vegetable glue, cement and explosives. Used in the tanning, hardening and electrolytic copperplating industries.

Occurrence The aquatic toxicity of aluminium is dependant upon the chemical state of the metal, and/or the associated hydroxide complexes.

Physical properties

M. Pt. 92.5°C.

Solubility
Water: 1 g in 20 ml (cold).

Occupational exposure

US TLV (TWA) 2 mg m^{-3} (as Al); **UK Long-term limit** 2 mg m^{-3} (as Al).

Environmental fate

Anaerobic effects
Alum, iron and lime treated fluid and air- dried anaerobically digested sewage sludges were added to soil. The sludge supplied twice the total nitrogen than for common fertilisers (1).

Degradation studies
Nitrification took place in the reactors treating both primary and mixed primary (alum added) chemical sludge.
The high content of aluminum did not inhibit the nitrifiers (2).

Genotoxicity

Escherichia coli SOS/umu test negative (3).

Legislation

Limited under EEC Directive on Drinking Water Quality 80/778/EEC. Aluminium: guide level 0.05 mg l^{-1}, maximum admissible concentration 0.2 mg l^{-1}. Sulfate: Guide level 25 mg l^{-1}, maximum admissible concentration 250 mg l^{-1} (4).

Any other comments

Aluminium and its compounds have been implicated in Alzheimers disease (5-8).
Toxicity and hazards reviewed, to be published in 1992-1993 (9).

References

1. Eikum, A. S. *Water Res.* 1974, **8**(11), 927
2. Cohen, *Res. Rep. – Res. Program Abatement Munic. Pollut. Provis. Can. – Ont. Agreement Great Lakes Water Qual.* 1978, **79**, 128
3. Kosaka, H; Nakamura, S. *Osaka- furitsu Koshu Eisei Kenkyusho Kenkyu Hokoku, Rodo Eisei Hen* 1991, **29**, 33-37
4. *EC Directive Relating to the Quality of Water Intended for Human Consumption* 1982, 80/778/EEC, Office for Official Publications of the European Communities, 2 rue Mercier, L-2985 Luxembourg
5. *Lancet* 1985, **1**, 616
6. Wheater, R. H. *J. Am. Med. Assoc.* 1985, **253**, 2288
7. King, R. G; et al. *Med. J. Aust.* 1985, **142**, 352
8. Candy, J. M; et al. *Lancet* 1986, **1**, 354
9. Izmerov, N. F. *Scientific Reviews of Soviet Literature on Toxicity & Hazards of Chemicals* 1992-1993, **127**, Eng. Trans., Richardson, M. L. (Ed.), UNEP/IRPTC, Geneva

A112 Aluminium resinate

Al(C$_{44}$H$_{63}$O$_5$)$_3$

CAS Registry No. 61789-65-9
Synonyms Resin acids and rosin acids, aluminium salts; Size precipitate
Mol. Formula C$_{132}$H$_{189}$AlO$_{15}$ **Mol. Wt.** 2042.95
Uses In paper size.
Occurrence The solution chemistry of aluminium is complex, and the response of the biota to the metal is dependant upon the chemical form of the toxicant. pH values and water hardness are highly influential. The aquatic toxicity of aluminium is dependant upon the chemical state of the metal, and/or the associated hydroxide complexes.

Occupational exposure

UK Long-term limit 2 mg m^{-3} (as Al); **UN No.** 2715; **Conveyance** classification flammable solid.

Any other comments

Aluminium and its compounds have been implicated in Alzheimers disease (1-4). Toxicity and hazards reviewed, to be published in 1992-1993 (5).

References

1. *Lancet* 1985, **1**, 616
2. Wheater, R. H. *J. Am. Med. Assoc.* 1985, **253**, 2288
3. King, R. G; et al. *Med. J. Aust.* 1985, **142**, 352
4. Candy, J. M; et al. *Lancet* 1986, **1**, 354
5. Izmerov, N. F; *Scientific Reviews of Soviet Literature on the Toxicity & Hazards of Chemicals* 1992-1993, **127**, Eng. Trans. Richardson, M. L. (Ed.), UNEP/IRPTC, Geneva

A113 Aluminium sulfate

Al$_2$(SO$_4$)$_3$

CAS Registry No. 10043-01-3
Synonyms anhydrous aluminium sulfate
Mol. Formula Al$_2$O$_{12}$S$_3$ **Mol. Wt.** 342.15
Uses Tanning leather, sizing paper, mordant. Water purification. Fireproofing and waterproofing cloth. Antiperspirant.
Occurrence Occurs in nature as the mineral alunogenite. The solution chemistry of aluminium is complex, and the response of the biota to the metal is depandant upon the chemical form of the toxicant. pH values and water hardness are highly influential. The aquatic toxicity of aluminium is dependant upon the chemical state of the metal, and/or the associated hydroxide complexes.

Physical properties

M. Pt. 770°C (decomp.); **Specific gravity** 1.61.

Solubility
Water: 31.3 g in 100 g H_2O at 0°C and 89.1 g at 100°C.

Occupational exposure
US TLV (TWA) 2 mg m^{-3} (as Al); **UK Long-term limit** 2 mg m^{-3} (as Al).

Ecotoxicity
Fish toxicity
LC_{50} goldfish (12-96 hr) 100 mg l^{-1} (1).

Invertebrate toxicity
EC_{50} (48-72 hr) *Asellus aquaticus* 6.57-4.37 mg l^{-1}. EC_{50} (48-96 hr) *Crangonyx pseudogracilis* 12.80-9.19 mg l^{-1} (2).

Bioaccumulation
Bioconcentration of aluminium in rainbow trout tissue and plankton was studied from aluminium sulfate- contaminated water. Statistical comparison of experimental and control tissues revealed no significant differences between exposed and non-exposed organisms (3).

Environmental fate
Nitrification inhibition
Threshold for sulfate is 500 mg l^{-1} (4).

Degradation studies
Rapid biodegradation occurred in 3 sludges which contained aluminium sulfate. Degradation rates of the sludge increased as pH increased (5).

Mammalian and avian toxicity
Acute data
LD_{50} oral rat, mouse 1930-6207 mg kg^{-1} (6,7).

Sub-acute data
Oral rat 0.25-2 g kg^{-1} (duration unspecified) caused significant reduction of relative liver weight but no lethal effect (8).

Teratogenicity and reproductive effects
TD_{Lo} (30 day) subcutaneous mouse 27.4 mg kg^{-1} reproductive effects (9).

Metabolism and pharmacokinetics
Gastrointestinal absorption of ingested aluminium is poor due to transformation of salts into insoluble aluminium phosphate in the digestive tract, brought about by pH changes and presence of phosphate in the diet. In rats only 10% of 200 mg kg^{-1} aluminium administered orally as the sulfate was absorbed. Distributed to all tissues, including bone, liver, testes and brain. Main route of excretion was via faeces (7).

Genotoxicity
Bacillus subtilis H17 (rec$^+$), M45 (rec$^-$) negative DNA damage (10).
In vitro human lymphocyte cells (72 hr) 20 µg ml^{-1} induced chromosomal aberrations in cells from ♂ and ♀ subjects, while the frequency of translocations and dicentrics was low (11).

Oral rat (prolonged exposure) induced dose-dependent inhibition of dividing cells and increased chromosomal aberrations, uninfluenced by duration of exposure (12).

Any other adverse effects to man

In 1988 a substantial quantity of aluminium sulfate was accidentally released into the drinking water supply of 20,000 people in the vicinity of Camelford, UK. Exposures were very variable but it is likely that for up to 3 days consumers were supplied with water of pH 3.9-5.0. The maximum aluminium concentration recorded in this water was 620 mg l^{-1}, but it is estimated that most consumers received concentrations of 10-50 mg l^{-1}. A wide range of short term symptoms were reported immediately following the incident including gastrointestinal disturbances, rashes and mouth ulcers. The existence of longer term effects is still under study (13).

A study was undertaken to determine blood and urine levels of workers employed in the production of aluminium sulfate. All workers had significantly higher blood and urine concentrations than unexposed control group (14).

Any other adverse effects

Ingestion may result in ulceration and necrosis of the mucosa of the mouth, throat and oesophagus. Systemic effects include epigastric pain, nausea, vomiting, diarrhoea, thirst, haemorrhagic gastroenteritis and circulatory collapse (15).

Legislation

Limited under EEC Directive on Drinking Water Quality 80/778/EEC. Aluminium: guide level 0.05 mg l^{-1}, maximum admissible concentration 0.2 mg l^{-1}.
Sulfate: Guide level 25 mg l^{-1}, maximum admissible concentration 250 mg l^{-1} (16).

Any other comments

Aluminium and its compounds have been implicated in Alzheimers disease (17-20). Toxicity and hazards reviewed, to be published in 1992-1993 (21).

References

1. McGauhey, P. H. *Engin. Management of Water Quality* 1968, McGraw-Hill, New York
2. Martin, T. R; et al. *Water Res.* 1986, **20**(9), 1137-1147
3. Buergel, P. M; et al. *J. Freshwater Ecol.* 1983, **2**(1), 37-44
4. Richardson, M. L. *Nitrification Inhibition in the Treatment of Sewage* 1985, RSC, London, cites R. Vismala, *Ingegn Ambient* 1982, **11**, 634- 643
5. Gaynor, J. D. *Env. Pollution* 1979, **20**(1), 57
6. *British J. Indust. Med.* 1966, **23**, 305
7. Venugopal, B; et al. *Metal Toxicity in Mammals* 1978, **2**, Plenum Press, New York
8. Kanoh, S; et al. *Oyo Yakuri Pharmacokinetics* 1982, **24**, 1
9. *J. Reproductive Med.* 1964, **7**, 21
10. Kanematsu, M; et al. *Mutat. Res.* 1980, **77**, 109-116
11. Roy, A. K; et al. *Mutat. Res.* 1990, **244**(2), 179-183
12. Roy, A. K; et al. *Cytobios* 1991, **66**(265), 105-111
13. Clayton, B. E; et al. *Report of the Lowermoor Incident Advisory Group* 1989, DoE, London
14. Sjoegren, B; et al. *Br. J. Ind. Med.* 1983, **40**(3), 301-304
15. *Chemical Safety Data Sheets* 1989, **2**, 34-36, RSC, London
16. *EC Directive Relating to the Quality of Water Intended for Human Consumption* 1982, 80/778/EEC, Office for Official Publications of the European Communities, 2 rue Mercier, L-2985 Luxembourg
17. *Lancet* 1985, **1**, 616

18. Wheater, R. H. *J. Am. Med. Ass.* 1985, **253**, 2288
19. King, R. G; et al. *Med. J. Aust.* 1985, **142**, 352
20. Candy, J. M; et al. *Lancet* 1986, **1**, 354
21. Izmerov, N. F; *Scientific Reviews of Soviet Literature on Toxicity & Hazards of Chemicals* 1992-1993, **127**, Eng. Trans. Richardson, M. L. (Ed.), UNEP/IRPTC, Geneva

A114   Aluminium triisopropoxide

Al[OCH(CH3)2]3

CAS Registry No. 555-31-7

Synonyms Aluminium isopropoxide; Aluminium isopropylate

Mol. Formula $C_9H_{21}AlO_3$ **Mol. Wt.** 204.25

Uses Formulation of paints, waterproofing textiles, formation of aluminium soaps. Chemical synthesis of alkoxides, chelates, acylates. In ester exchange and Meervein Pondorf reaction.

Occurrence The solution chemistry of aluminium is complex, and the response of the biota to the metal is dependant upon the chemical form of the toxicant. pH values and water hardness are highly influential. The aquatic toxicity of aluminium is dependant upon the chemical state of the metal, and/or the associated hydroxide complexes.

Physical properties

M. Pt. 134-138°C; **B. Pt.** $_{10}$135°C.

Solubility

Organic solvent: ethanol, isopropanol, benzene, toluene, chloroform

Occupational exposure

US TLV (TWA) 2 mg m^{-3} (as Al); **UK Long-term limit** 2 mg m^{-3} (as Al); **Supply classification** highly flammable.

Risk phrases

Highly flammable (R11)

Safety phrases

Keep container dry – Keep away from sources of ignition – No Smoking (S8, S16)

Mammalian and avian toxicity

Acute data

LD$_{50}$ oral rat 11.3 g kg^{-1} (1).

Any other comments

Physico-chemical properties, experimental toxicology, human health and environmental effects reviewed (2).

Aluminium and its compounds have been implicated in Alzheimers disease (3-6).

Toxicity and hazards reviewed, to be published in 1992-1993 (7).

References

1. Smyth; et al. *Am. Ind. Hyg. Ass. J.* 1969, **30**, 470

2. *ECETOC Technical Report No. 30(4)* 1991, European Chemical Industry Ecology and Toxicology Centre, B-1160 Brussels
3. *Lancet* 1985, **1**, 616
4. Wheater, R. H. *J. Am. Med. Assoc.* 1985, **253**, 2288
5. King, R. G; et al. *Med. J. Aust.* 1985, **142**, 352
6. Candy, J. M; et al. *Lancet* 1986, **1**, 354
7. Izmerov, N. F; *Scientific Reviews of Soviet Literature on Toxicity & Hazards of Chemicals* 1992-1993, **127**, Eng. Trans. Richardson. M. L. (Ed.), UNEP/IRPTC, Geneva

A115   Ametryn

CH₃S⟍N⟍NHCH₂CH₃

CAS Registry No. 834-12-8

Synonyms 1,3,5-triazine-2,4-diamine, *N*-ethyl- *N'*-(1-methylethyl)-6-(methylthio)-; *N*-ethyl- *N'*-(1-methylethyl)-6-(methylthio)-1,3,5-triazine-2,4-diamine; 2-(ethylamino)-4-(isopropylamino)-6-(methylthio)-*s*-triazine; 2-ethylamino-4-isopropylamino-6- methylmercapto-*s*-triazine; Ametrex; Gesapax

Mol. Formula $C_9H_{17}N_5S$ **Mol. Wt.** 227.33

Uses Herbicide.

Physical properties

M. Pt. 84-85°C; **Specific gravity** d_4^{20} 1.19; **Partition coefficient** log P_{ow} −1.72 (calc.) (1); **Volatility** v.p. 8.4×10^{-7} mmHg at 20°C.

Solubility
Water: 185 mg l^{-1} at 20°C. Organic solvent: acetone, hexane, methanol

Risk phrases
Harmful by inhalation and if swallowed (R20/22)

Safety phrases
Keep out of reach of children – Keep away from food, drink and animal feeding stuffs (S2, S13)

Ecotoxicity

Fish toxicity
LC_{50} (96 hr) bluegill sunfish, rainbow trout, goldfish 4.1-14.1 mg l^{-1} (2).
No acute mortality to mosquito fish within 48 hr at concns. up to 10 ppm in laboratory experiments, or within 5 days at 5 lb/acre active ingredient in field ponds. At larvicidal rates (1-5 ppm), most formulations were highly toxic in the laboratory, but no acute toxicity was observed at the larvicidal rate (0.25-1.0 lb/acre) in the field. Surface breathing insects and water skimming spiders, however, were adversely affected (3).

Invertebrate toxicity
Chlorococcum sp. (technical solution) 20 ppb 50% decrease in oxygen evolution.

Chlorococcum sp. 10 ppb 50% decrease in growth after 10 days. *Phaeodactylum tricornutum* 20 ppb 50% decrease in growth after 10 days (4).

LC_{50} (96 hr) oyster >1.0 ppm (conditions of bioassay unspecified) (5).

Ametryn, *Scenedesmus sp.* growth rate, chlorophyll a content, and ratio of chlorophylls decreased as the concentration. increased. Microscopical examination in all tested cultures showed marked morphological changes (6).

Ametryn was a strong photosynthesis poison to *Ankistrodesmus falcatus* (7).

Bioaccumulation
Estimated bioconcentration factor is 33 using water solubility of 185 mg l^{-1} which suggests that bioaccumulation in aquatic organisms will be insignificant (8).

Effects to non-target species
LC_{50} (8 day) oral mallard ducks 23,000 mg kg^{-1} (2).

Low toxicity to bees (2).

Environmental fate

Nitrification inhibition
Toxic to *Nitrosomonas* in soil. No inhibition at 5 ppm.

Toxic to *Nitrobacter sp.* in soil, inhibitory concentration 100 ppm (9).

Anaerobic effects
Rate of metabolism decreased $t_{1/2}$ 122 days (1).

A concentration of 5 and 10 mg l^{-1}. Ametryn in feed reduced methane gas production by 12.5%. The removal value was 37% at the lower concentration investigated (10).

Degradation studies
Under aerobic conditions $t_{1/2}$ 2-3 wk in soil. Metabolites, included 2-amino-4-isopropylamino-6-methylthio-5-triazine;2-amino-4-ethylamino-6-methylthio-5-triazine and 2, 4-diamino-6- methylthiotriazine (1).

Ametryn exerted a depressive effect on the total count of cellulose decomposing fungi after 1 and 3 wk of treatment with a high dose (54 mg active ingredient/kg dry soil), and 5 wk after treatment with medium (27 mg) and low doses. This inhibitory effect was alleviated after 8 wk, whereas after 12 wk ametryn had a promoting effect on decomposing fungi at the low dose (5.4 mg) (11).

When incorporated in the agar medium, this herbicide was toxic to the total count and to the counts of almost all fungal genera and species at the 3 doses (25, 125, 250 ppm). The growth and sporulation of test fungal species were partially or completely inhibited by the 3 doses, except, *Asperigillus niger*, *Chaetomium globosum*, and *Gliocladium roseum* which were not affected by the low dose (11).

Abiotic removal
Under aqueous conditions, ametryn is stable in natural sunlight $t_{1/2}$ >1 wk. When exposed to artificial sunlight for 6 hr, 75% remained, with 2-ethylamino-4-hydroxy-6-isopropylamino-5-triazine as photolysis product (1).

Direct photolysis can occur when exposed to sunlight. On the surface of three sandy loam soils <10-30% photolytic loss was observed in 7 days thought to be a result of light-induced free radical oxidation (12,13).

Volatilisation from soil is an important route of removal from the environment despite its very low vapour pressure (14).

At 20°C hydrolysis occurs to herbicidally inactive 6-hydroxy analogue, with 50% loss in 32 days at pH 1, and over 200 days at pH 13 (15).

Absorption
Ametryn has a pKa of 3.12, indicating that it is almost entirely undissociated at environmental pHs. Ametryn and humic acid form stable complexes and ionic, hydrogen bonding, donor-acceptor and covalent forces contribute to the binding (16,17).

Mammalian and avian toxicity

Acute data
LD_{50} oral rat, mouse 965-1110 mg kg^{-1} (2,6).
LD_{50} dermal rabbit >8160 mg kg^{-1} (17).
LC_{50} (8 day) oral bobwhite quail 30g kg^{-1} (2).

Sub-acute data
Oral rats (90 day) at 100 mg kg^{-1} day^{-1} in feed, animals were comparable to controls except for slight histological changes in the liver (18).

Metabolism and pharmacokinetics
In 24 hr, following oral administration to rat (dose unspecified) 52% of ametryn was excreted in urine and 18% in faeces. Within 72 hr, elimination was almost complete, a further 6% had been excreted in urine, 14% in faeces, leaving <2% in carcass. After 6 hr, ametryn levels were maximal in stomach, liver, kidneys, spleen and lung, decreasing with time, although blood levels remained constant for 72 hr (19).

Irritancy
76 mg instilled into rabbit eye caused mild irritation (20).

Genotoxicity
Salmonella typhimurium TA98, TA100, TA1535, TA1537, TA1538 with and without metabolic activation negative (1).
Bacillus subtilis (strains unspecified) rec-assay negative. *Escherichia coli* WP2 utilizing auxotrophic strains in reversion assays negative (21).

Legislation
Limited under EEC Directive on Drinking Water Quality 80/778/EEC. Pesticides and related products: maximum admissible concentration 0.5 μg l^{-1} (22).

Any other comments
Ametryn was severely phytotoxic when foliage (windbreak trees) was sprayed (23).

References
1. *Drinking Water Health Advisory: Pesticides* 1989, 23-34, Lewis Publishers, Chelsea, MI
2. *The Agrochemicals Handbook* 3rd ed., 1991, RSC, London
3. Darwazeh, H. A; et al. *Mosq. News* 1974, **34**(2), 214-219
4. Walsh, G. E. *Hyacinth Control J.* 1972, **10**, 45-48
5. *Weed Science Society of America. Herbicide Handbook* 5th ed., 1983, Weed Science Society of America, Champaign, IL
6. El-Dib, M. A; et al. *Water, Air, Soil, Pollut.* 1989, **48**(3-4), 307-316
7. Tscheu-Schlueter, M. *Acta Hydrochim. Hydrobiol.* 1976, **4**(2), 153-170
8. Kenega, E. E. *Ecotoxicol. Environ. Saf.* 1980, **4**, 26-38
9. Parr, J. F. *Pest. Soil Water* 1974, 321-340

10. El-Gohary, F. A; et al. *Biogas Technol., Transfer Diffus., [Proc. Int. Conf.]* El- Halwagi, M. M. (Ed.), Elsevier, 1986, 454-462
11. Abdel-Mallek, A. Y; et al. *Folia Microbiol. (Prague)* 1986, **31**(5), 375-381
12. Jordan, L. S; et al. *Res. Rev.* 1970, **32**, 267-286
13. Miller, G. C; et al. *Amer. Chem. Soc. Div. Environ. Chem. 193rd Natl. Meeting* 1987, **27**, 463-465
14. Spencer, W. F; et al. *J. Environ. Qual.* 1988, **17**, 504-509
15. *The Pesticide Manual* 9th ed., 1991, British Crop Protection Council, Farnham
16. Senesi, N; et al. *Geoderma* 1982, **28**, 129-146
17. Rahman, A; et al. *Weed Sci.* 1979, **27**, 158-161
18. *Pesticide Dictionary* 1975, Farm Chemicals, Meister Publishing Co, Willoughby, OH
19. *The Chemical Society. Foreign Compound Metabolism in Mammals* 1970, **1**, 82, The Chemical Society, London
20. *Ciba-Geigy Toxicology Data/Indexes* 1977
21. Shirasu, Y; et al. *Mutat. Res.* 1976, **40**, 19-30
22. *EC Directive Relating to the Quality of Water Intended for Human Consumption* 1982, 80/778/EEC, Office for Official Publications of the European Communities, 2 rue Mercier, L-2985 Luxembourg
23. Takahara, T. *Kaju Shikenjo Hokoku D* 1983, **5**, 27-46, (Jap.) (*Chem. Abstr.* **9**, 117444h)

A116   Amidithion

$(CH_3O)_2PS_2CH_2CONHCH_2CH_2OCH_3$

CAS Registry No. 919-76-6
Synonyms S-(N-2- methoxyethylcarbamoylmethyl)-O,O-dimethyl dithiophosphate; O, O-dimethyl-S-(2- methoxyethylcarbamoyl methyl)dithiophosphate; S-[2-[(2-methoxyethyl)amino]-2-oxoethyl] O, O-dimethyl phosphorodithioic acid ester; Thiocron
Mol. Formula $C_7H_{16}NO_4PS_2$ 　　　　　　　　　　**Mol. Wt.** 273.31
Uses Acaricide and insecticide.

Occupational exposure

Supply classification harmful.

Risk phrases
Harmful by inhalation, in contact with skin and if swallowed (R20/21/22)

Safety phrases
Keep out of reach of children – Keep away from food, drink and animal feeding stuffs (S2, S13)

Mammalian and avian toxicity

Acute data
LD_{50} oral rat 600 mg kg^{-1} (1).
LD_{50} dermal rat 1600 mg kg^{-1} (2).

Sub-acute data
Oral rat (3-4 month) 8 mg kg^{-1} day^{-1} decreased activity of cholinesterase in the blood, brain, liver and kidneys and that of alkaline phosphatase in the serum (3).

Legislation

Limited under EC Directive on Drinking Water Quality 80/778/EEC. Pesticide: maximum admissible concentration 0.1 µg l^{-1} (4).

Any other comments

Use as an insecticide discontinued.

References

1. *The Pesticide Manual* 8th ed., 1987 The British Crop Protection Council, Farnham
2. *Pesticide Index* 1976, **5**, 9
3. Sasinovich, L. M; et al. *Farmakol. Toksikol. (Kiev)* 1973, (8), 120-121, (Russ.) (*Chem. Abstr.* **80**, 116881n)
4. *EC Directive Relating to the Quality of Water Intended for Human Consumption* 1982, 80/778/EEC, Office for Official Publications of the European Communities, 2 rue Mercier, L-2985 Luxembourg

A117   2-Aminoanthracene

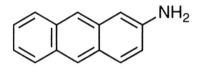

CAS Registry No. 613-13-8
Synonyms 2-aminoanthracenamine; β- aminoanthracene; 2-anthracylamine; 2-anthrylamine; 2- anthramine
Mol. Formula $C_{14}H_{11}N$ **Mol. Wt.** 193.25
Uses Herbicide, dyestuffs intermediates.

Physical properties

M. Pt. 238°C; **B. Pt.** sublimes at 293°C; **Partition coefficient** log P_{ow} 3.4.

Solubility
Organic solvent: ethanol, diethyl ether

Ecotoxicity

Fish toxicity
Exposure of rainbow trout to 122 mg l^{-1} for 6 days initiated biochemical effects including glycogen uptake, changes to cholesterol levels and lipids (1).

Bioaccumulation
Using the calculated bioconcentration factor of 3140 based on a measured water solubility, it is concluded that accumulation in aquatic organisms is likely (2).

Environmental fate

Abiotic removal
2-Aminoanthracene contains no hydrolyzable functional groups and therefore, is not expected to undergo environmental hydrolysis (2).
Photochemical reaction with atmospheric hydroxyl radicals estimated $t_{1/2}$ 1.8 hr (3).

Absorption

Extremely strong adsorption to soil and suspended solids and sediments in water is reported (4).

Mammalian and avian toxicity

Metabolism and pharmacokinetics

Human peripheral blood lymphocytes were exposed at 37°C for 18 hr and showed 2 or 3 adducts from 8-1500 mol μg^{-1} DNA (5).

Genotoxicity

Salmonella typhimurium TA1535 with and without metabolic activation caused SOS-induction (6).

Salmonella typhimurium TA98, TA100 with metabolic activation positive, mutagenic effects decreased as preincubation temperature increased, optimal at 15°C (7).

SOS Chromotest on *Escherichia coli* PQ37 was used to detect DNA damage induced by 2-anthracenamine with metabolic activation (8).

Any other comments

All reasonable efforts have been taken to find information on isomers of this compound, but no relevant data are available.

References

1. Miyauchi, M; et al. *Bull. Environ. Contam. Toxicol.* 1987, **39**, 175
2. Lyman, W. J; et al. *Handbook of Chemical Property Estimation Methods* 1982, McGraw-Hill, New York
3. Atkinson, R. *Intern. J. Chem. Kinet.* 1987, **19**, 799-828
4. Swann, R. L; et al. *Res. Rev.* 1983, **85**, 117-128
5. Gupta, R. C; et al. *Proc. Natl. Acad. Sci. USA* 1988, **85**(10), 3513-3517
6. Egorov, I. A; et al. *Dokl. Akad. Nauk SSSR* 1991, **318**(5), 1230-1232
7. Johnson, B. T. *Environ. Toxicol.* 1990, **9**(9), 1183-1192
8. Venier, P; et al. *Mutagenesis* 1989, **4**(1), 51-57

A118   2-Aminoanthraquinone

CAS Registry No. 117-79-3

Synonyms 9,10-anthracenedione, 2-amino-; 2-amino-9,10- anthracenedione; β-aminoanthraquinone; NCI- CO1876

Mol. Formula $C_{14}H_9NO_2$ **Mol. Wt.** 223.23

Uses Important intermediate in the preparation of indanthrene colourants. Used in the manufacture of pharmaceuticals.

Physical properties

M. Pt. 292-295°C; **B. Pt.** sublimes.

Solubility

Organic solvent: ethanol, benzene, acetone, chloroform

Mammalian and avian toxicity

Acute data

LD_{50} intraperitoneal rat 1500 mg kg^{-1} (1).

Sub-acute data

Oral Fischer 344 rat (1 wk) 2% in feed led to nephrotoxicity in ♀ caused by deposits of crystalline materials in the kidney tubules (2).

Carcinogenicity and long-term effects

No adequate evidence for carcinogenicity to humans, limited evidence for carcinogenicity to animals, IARC classification group 3 (3,4).
F344 rats and B6C3F1 mice (78 wk) 0.69 and 1% (daily) in feed, respectively, induced haematopoietic system tumours, liver carcinomas and adenomas in ♂ rats, ♀ and ♂ mice (5).

Genotoxicity

Salmonella typhimurium TA98, TA100, TA1535, TA97, TA1537 with and without metabolic activation positive (6).

Any other comments

Experimental toxicology, human health and environmental effects reviewed (7).
All reasonable efforts have been taken to find information on isomers of this compound, but no relevant data are available.
It is known that some quinonoid compounds cause blindness in fish by retinal detachment.

References

1. *Gig. Tr. Prof. Professional' naya v Estenskoi SSR* 1977, **21**(12), 27
2. Gothoskar, S. V. *Xenobiotica* 1979, **9**(9), 533
3. *IARC Monograph* 1987, **Suppl 7.**, 56
4. *IARC Monograph* 1982, **27**, 199
5. Ashby, J; et al. *Mutat. Res.* 1988, **204**(1), 17-115
6. Zeiger, E; et al. *Cancer Res.* 1987, **47**, 1287-1296
7. *ECETOC Technical Report No. 30(4)* 1991, European Chemical Industry Ecology and Toxicology Centre, B-1160 Brussels

A119   4-Aminoantipyrene

CAS Registry No. 83-07-8
Synonyms 4-amino-2,3-dimethyl-1-phenyl-3-pyrazolin-5- one; 4-amino-phenazone; ampyrone; 3*H*-pyrazol-3-one, 4-amino-1,2-dihydro-1,5-dimethyl-2-phenyl-; ampyrone

Mol. Formula $C_{11}H_{13}N_3O$ **Mol. Wt.** 203.25

Uses Used as an analgesic and antipyretic stabiliser. Production of azo dyestuffs. Reagent for glucose and for detection of alkylphenols.

Physical properties

M. Pt. 107-109°C.

Solubility

Organic solvent: ethanol, benzene, diethyl ether

Mammalian and avian toxicity

Acute data

LD_{50} intraperitoneal mouse, rat 270-1700 mg kg^{-1} (1-3).

Metabolism and pharmacokinetics

Observed dose-dependent disposition of 4- aminoantipyrene in rabbits is a result of reduced renal and hepatic blood flow caused by the drug itself (4).

Vitamin B deficiency increased urinary excretion of total and acetylated 4-aminoantipyrine following intraperitoneal administration of 30 mg kg^{-1} aminopyrine (5).

1 g single oral dose of 4-aminoantipyrene to healthy human volunteers showed peak plasma concentrations 2.7 μg ml^{-1} and 1.6 μg ml^{-1} $t_{1/2}$ 5.5 hr and 3.8 hr in slow and rapid acetylations, respectively (6).

Genotoxicity

Salmonella typhimurium TA97 with metabolic activation positive. Possible long-term hazards are discussed in view of their pluripotent direct genotoxicity (7).

Any other comments

All reasonable efforts have been taken to find information on isomers of this compound, but no relevant data are available.

References

1. *Boll. Chim. Farm.* 1978, **117**, 638
2. *Collect. Czech. Chem. Commun.* 1982, **47**, 636
3. *Arzneim-Forsch.* 1960, **10**, 820
4. Mitsuyama, S. *J. Pharmacobio-dyn.* 1985, **8** (5), 365 (*Chem. Abstr.* **103**, 47818h)
5. Lychko, A. P; et al. *Vopr. Med. Khim.* 1988, **34**(3), 45-48, (Russ.) (*Chem. Abstr.* **109**, 53698j)
6. Levy, M. *Eur. J. Clin. Pharmacol.* 1984, **27**(4), 453 (*Chem. Abstr.* **102**, 55628v)
7. Parisis, D; et al. *Mutat. Res.* 1988, **206**(3), 317-326

A120   4-Aminoazobenzene

CAS Registry No. 60-09-3

Synonyms benzeneamine, 4-(phenylazo)-; aniline yellow; *p*-aminodiphenylimide; C.I. Solvent Yellow 1; *p*-(phenylazo)aniline; C.I.11000

Mol. Formula $C_{12}H_{11}N_3$ **Mol. Wt.** 197.24

Uses Used in the manufacture of dyestuffs. Insecticide.

Physical properties

M. Pt. 123-126°C; **B. Pt.** >360°C; **Partition coefficient** log P_{ow} 2.98 (calc.) (1).

Solubility
Organic solvent: ethanol, benzene, chloroform, diethyl ether

Ecotoxicity

Invertebrate toxicity
EC_{50} (5-30 min) *Photobacterium phosphoreum* 2.66 mg l^{-1} Microtox test (2).

Bioaccumulation
Using the calculated bioconcentration factor of 58 based on the estimated log P_{ow} it is concluded that accumulation in aquatic organisms will be minimal (3).

Effects to non-target species
Frogs (*Rana pipiens*) administered 0.3- 0.5 mg in olive oil directly below the kidney capsule developed kidney nodules, which induced adenocarcinomas (4).

Environmental fate

Nitrification inhibition
Ammonia oxidation by *Nitrosomonas sp.* at 100 mg l^{-1} 54% inhibition, at 50 mg l^{-1} 47% inhibition, at 10 mg l^{-1} 0% inhibition (5).

Degradation studies
Aeromonas hydrophila 24B is able to degrade 4-aminoazobenzene to aniline and can be applied to wastewater treatment (6).
Pseudomonas cepacia 13NA degraded 4-aminoazobenzene to yield metabolites, including aniline, *p*-phenylenediamine, acetanilide, *p*- aminoacetanilide and *p*-phenylenediacetamide (7).
Readily degradable using an activated sludge inoculum with 89% degradation occurring in 13 days, including a 7-day lag (8).
No BOD consumption with sewage and activated sludge inocula after 5 and 6 day incubations, respectively (9,10).
For static cultures, the lag period increased with concentration and with 100 ppm of 4-aminoazobenzene, 46% degradation occurred after 24 hr and 59% degradation after 48 hr in shaking and static cultures, respectively, at 37°C (11).
Strongly inhibited the growth of activated sludge microorganisms. The partition coefficients of azobenzenes between octanol and water correlated inversely with growth inhibition (12).

Abiotic removal
The photochemical $t_{1/2}$ produced from hydroxyl radicals was estimated at 5.8 hr (13).

Mammalian and avian toxicity

Acute data
LD_{Lo} intraperitoneal mouse 3.3 mg kg^{-1} (14).

Carcinogenicity and long-term effects

No adequate evidence for carcinogenicity to humans, sufficient evidence for carcinogenicity to animals, IARC classification group 2B (14).

TD_{Lo} (2 yr intermittent) dermal rat 1965 mg kg^{-1} neoplastic liver tumours (15).

Single intraperitoneal injection to mice (concentration unspecified) induced hepatomas in 46-93% of animals tested (16).

Subcutaneous injection to pregnant ♀ and newborn ♂ mice (concentration unspecified) increased the incidence of liver tumour and tumours of haematopoietic and lymphoid tissues (17).

Genotoxicity

Salmonella typhimurium TA98 and TA100 without metabolic activation positive (18).

Escherichia coli 700 μg $well^{-1}$ gene conversion and mitotic recombination (19).

Intravenous rat 0.00098 mg l^{-1} caused unscheduled DNA synthesis (20).

Any other comments

Physico-chemical properties, human health effects and experimental toxicology reviewed (21).

All reasonable efforts have been taken to find information on isomers of this compound, but no relevant data are available.

References

1. Verschueren, K. *Handbook of Environmental Data on Organic Chemicals* 2nd ed., 1983, Van Nostrand Reinhold, New York
2. Kaiser, K. L. E; et al. *Water Pollut. Res. J. Canada* 1991, **26**(3), 361-431
3. Lyman, W. J; et al. *Handbook of Chemical Property Estimation Methods* 1982, 5.1-5.30, McGraw-Hill, New York
4. Strauss, E; et al. *Cancer Res.* 1964, **24**, 1969
5. Hockenbury, M. R. *J. Water Pollut. Control Fed.* 1977, **49**(5), 768-777
6. Yatome, C; et al. *J. Soc. Dyers Colour.* 1987, **103**(11), 395-398, (Jap.), (*Chem. Abstr.* 1988, **108**, 118274u)
7. Idaka, E; et al. *Bull. Environ. Contam. Toxicol.* 1987, **39**(1), 108-113
8. Urushigawa, Y; et al. *Bull. Environ. Contam. Toxicol.* 1977, **17**, 214-218
9. Heukelekian, H; et al. *J. Water Pollut. Control Fed.* 1955, **29**, 1040-1053
10. Lutin, P. A; et al. *Purdue Univ. Eng. Bull. Ext. Series* 1965, **118**, 131-145
11. Idaka, E; et al. *J. Soc. Dyers Colour* 1978, **94**, 91-94
12. Yonezawa, Y; et al. *Bull. Environ. Contam. Toxicol.* 1977, **17**(2), 208-213
13. Atkinson, R. *Int. J. Chem. Kinet.* 1987, **19**, 799-828
14. *IARC Monograph* 1987, **Suppl. 7**, 390
15. Delclos, K. B; et al. *Cancer Res.* 1966, **26**, 2406
16. Delclos, K. B; et al. *Cancer Res.* 1984, **44**, 2540-2550
17. Fujii, K. *Cancer Lett.* 1983, **17**, 321
18. Degawa, M; et al. *Carcinogenesis* 1982, **3**, 1113
19. *Mutat. Res.* 1977, **46**, 53
20. *Environmental Mutagenesis* 1981, **3**, 11
21. *ECETOC Technical Report No. 30(4)* 1991, European Chemical Industry Ecology and Toxicology Centre, B-1160 Brussels

A121   2-Aminobenzenesulfonic acid

CAS Registry No. 88-21-1
Synonyms orthanilic acid; aniline-2-sulfonic acid; *o*-sulfanilic acid;
o-aminophenylsulfonic acid; benzenesulfonic acid, 2-amino; *o*-aniline sulfonic acid;
anilino-*o*-sulfonic acid
Mol. Formula $C_6H_7NO_3S$ **Mol. Wt.** 173.19
Uses Manufacture of azo dyestuffs. Component of water- based hydraulic
fluids.

Physical properties
M. Pt. >300°C.

Environmental fate

Degradation studies
Decomposition by a soil microflora in >64 day (1).

Mammalian and avian toxicity

Metabolism and pharmacokinetics
Orthanilic acid interacts with rat liver glutathione S-transferase (GST) by direct
binding (3).

Genotoxicity
Salmonella typhimurium TA98 without metabolic activation weakly positive (4).

References

1. Alexander, M; et al. *J. Agric. Food Chem.* 1966, **14**, 410
2. Yoshioka K. *Nara Igaku Zasshi* 1957, **8**, 427, (Jap.) (*Chem. Abstr.* **52**, 13009b)
3. Dierickx, P. J. *Res. Commun. Chem. Pathol. Pharmacol.* 1982, **37**(3), 385
4. Zeiger, E; et al. *Environ. Mol. Mutagen.* 1988, **11**(Suppl. 12), 1-158

A122   3-Aminobenzenesulfonic acid

CAS Registry No. 121-47-1
Synonyms 1-aminobenzene-3-sulfonic acid; *m*- sulfanilic acid; aniline-*m*-sulfonic
acid; benzenesulfonic acid, 3-amino-metanilic acid

Mol. Formula $C_6H_7NO_3S$ **Mol. Wt.** 173.19

Uses Synthesis of azo dyes and certain sulfa drugs.

Physical properties

M. Pt. Decomposes without melting at ≈288°C; **Specific gravity** 1.69.

Solubility
Water: 10.8 g l^{-1} at 20°C. Organic solvent: hot methanol

Occupational exposure

Supply classification harmful.

Risk phrases
Harmful by inhalation, in contact with skin and if swallowed (R20/21/22)

Safety phrases
Avoid contact with eyes – After contact with skin, wash immediately with plenty of water (S25, S28)

Ecotoxicity

Bioaccumulation
Non-accumulative or low accumulative (1).

Environmental fate

Degradation studies
Decomposition period by soil microflora in >64 days (2).
An adapted activated sludge utilises 3- aminobenzenesulfonic acid as sole carbon source at 20°C.
95% COD, 4 mg COD g^{-1} dry inoculum hr^{-1} (3).
BOD_5 1.1 mg l^{-1} standard diluting sewage (4).

Mammalian and avian toxicity

Acute data
LD_{50} oral rat 12 g kg^{-1} (5).

Metabolism and pharmacokinetics
3-Aminobenzenesulfonic acid interacts with rat liver glutathione-S-transferase (GST) by direct binding (6).

Irritancy
500 mg instilled in rabbit eye for 24 hr caused mild irritation (5).

Genotoxicity

Salmonella typhimurium TA100, TA1535, TA97, TA98 with and without metabolic activation negative (7).

Any other comments

Physico-chemical properties, experimental toxicology and human health effects reviewed (8).

References

1. *MITI Report* 1984, Ministry of International Trade and Industry, Tokyo

2. Alexander, M. *J. Agric. Food. Chem.* 1966, **14**, 410
3. Pitter, P. *Water Res.* 1976, **10**, 231-235
4. Meissner, B. *Wasserwirtschaft- Wassertechnik* 1954, **4**, 166
5. Marhold, J. V. *Sbornik Vysledku Tox. Vysetreni Latek A Primpravku* 1972, 179, Prague
6. Dierickx, P. *J. Res. Commun. Chem. Pathol. Pharmacol.* 1982, **37**(3), 385-394
7. Zeiger, E; et al. *Environ. Mol. Mutagen.* 1988, **11**(Suppl. 12), 1-158
8. *ECETOC Technical Report No. 30(4)* 1991, European Chemical Industry and Toxicology Centre, B- 1160 Brussels

A123   4-Aminobenzenesulfonic acid

CAS Registry No. 121-57-3
Synonyms aniline-4-sulfonic acid; sulfanilic acid; *p*-aminophenylsulfonic acid; benzenesulfonic acid, 4- amino-
Mol. Formula $C_6H_7NO_3S$ **Mol. Wt.** 173.19
Uses Antibacterial uses. Manufacture of dyestuffs and organic chemicals. Reagent for nitrate.

Physical properties

M. Pt. 288°C (decomp.); **Specific gravity** d_4^{25} 1.485.

Solubility
Water: 10.8 g l^{-1} at 20°C, 66.7 g l^{-1} at 100°C. Organic solvent: hot methanol

Occupational exposure
Supply classification harmful.

Risk phrases
Harmful by inhalation, in contact with skin and if swallowed (R20/21/22)

Safety phrases
Avoid contact with eyes – After contact with skin, wash immediately with plenty of water (S25, S28)

Ecotoxicity

Fish toxicity
LC50 (96 hr) fathead minnow 100.4 mg l^{-1} static bioassay (1).

Invertebrate toxicity
EC50 (5-30 min) *Photobacterium phosphoreum* 114 mg l^{-1} Microtox test (2).

Environmental fate

Degradation studies
Microbial decomposition in >64 days (3).
4-Aminobenzenesulfonic acid was utilised as the sole carbon source by adapted

activated sludge at 20°C. COD 95%, 4 mg COD g^{-1} dry inoculum hr^{-1} (4).
BOD_5 1.1 mg l^{-1} oxygen consumed using standard dilute sewage (5).

Mammalian and avian toxicity

Acute data
LD_{50} oral mouse >3.2 g kg^{-1} (6).
LD_{50} intravenous rat 6 g kg^{-1} (7).

Metabolism and pharmacokinetics
Following oral administration to rats 53% of the dose was excreted in the urine (8).

Irritancy
Dermal rabbit (24 hr) 500 mg caused mild irritation, 100 mg instilled in rabbit eye caused moderate irritation (9).

Genotoxicity
Salmonella typhimurium TA100, TA1535, TA97, TA98 with and without metabolic activation negative (10).

Any other adverse effects
Variable allergic response according to administration route in rat (11).

Any other comments
Physico-chemical properties, experimental toxicology and human health effects reviewed (12).

References

1. Curtis, M. W; ert al. *J. Hydrol.* 1981, **51**, 359-367
2. Kaiser, K. L. E; et al. *Water Pollut. Res. J. Canada* 1991, **26**(3), 361-431
3. Alexander, M. *J. Agric. Food Chem.* 1966, **14**, 410
4. Pitter, P. *Water Res.* 1976, **10**, 231
5. Meissner, B. *Wasserwirtschaft- Wassertechnik* 1954, **4**, 166
6. Patty, F. A. *Industrial Hygiene and Toxicology* 1967, **2**, Interscience Publishers, New York
7. Marhold, J. V. *Sb. Vysledku Toxixologiekeho Vysetreni Latek A Prinpravku* 1972, 180, Prague
8. Scheline, R. R; et al. *Acta Pharmacol. Toxicol.* 1965, **23**(1)
9. *Naunyn-Schmiedebergs Arch. Exp. Pathol. Pharmakol.* 1950, **211**, 367
10. Zeiger, E; et al. *Environ. Mol. Mut.* 1988, **11**(Suppl. 12), 1-158
11. Guerin, B. *Immunotoxicol. Proc. Int. Symp.* 1982, 457
12. *ECETOC Technial Report No. 30(4)* 1991, European Chemical Industry Ecology and Toxicology Centre, B-1160 Brussels

A124   2-Aminobenzoic acid

CAS Registry No. 118-92-3
Synonyms *o*-anthranilic acid; Vitamin L; *o*- aminobenzoic acid

Mol. Formula $C_7H_7NO_2$ **Mol. Wt.** 137.14

Uses Acaricide. Dyestuffs, pharmaceuticals, perfume.

Physical properties

M. Pt. 144-146°C; **Specific gravity** d^{20} 1.412; **Partition coefficient** log P_{ow} 1.21.

Solubility

Organic solvent: ethanol, diethyl ether

Ecotoxicity

Fish toxicity

Designated non-toxic to trout, bluegill sunfish, yellow perch and goldfish (1).

Environmental fate

Nitrification inhibition

At 100 mg l^{-1} no inhibition of NH_3 oxidation by *Nitrosomonas sp.* (2).

Anaerobic effects

Under anaerobic conditions in the presence of nitrate, *o*-aminobenzoate was oxidised to carbon dioxide by *Pseudomonas sp.* which involved 2- aminobenzoyl-CoA reductase. Aerobic degradation was via gentisic acid (3).

Degradation studies

Biodegradable (4).

97.5% COD, activated sludge at 20°C (5).

Decomposition by soil microflora 2 days (6).

Four strains of the two actinomycete species *Streptomyces violaceoruber* and *Amycolata autotrophica* degraded *o*-aminobenzoic acid, *A. autotrophica* 43093 being the most active strain (7).

o-Aminobenzoate was degraded as sole source of carbon and energy in methanogenic enrichment cultures obtained from anoxic sediments and sewage sludge, to acetate, carbon monoxide, methane and ammonia (8).

Mammalian and avian toxicity

Acute data

LD_{50} oral rat 4550 mg kg^{-1} (9).

LD_{50} intraperitoneal mouse 2500 mg kg^{-1} (10).

Carcinogenicity and long-term effects

No adequate evidence for carcinogenicity to humans, limited evidence for carcinogencity to animals, IARC classification group 3 (11).

National Toxicology Program Evaluation of 2-aminobenzoic acid in rats and mice in feed negative (12).

Metabolism and pharmacokinetics

Intraperitoneal administration to rat resulted in <10% of dose (18 or 101 mg kg^{-1} body weight) excreted in bile (13).

Genotoxicity

Salmonella typhimurium TA98, TA100, TA1535, TA1537 with and without metabolic activation negative (13).

In vitro L5178Y tk+/tk- mouse lymphoma cells without metabolic activation weakly positive (15).

In vivo B6C3FI mice carcinogenic/noncarcinogenic pair was *o*-toluidine hydrochloride/*o*-aminobenzoic acid, respectively. Intraperitoneal administration of the pair up to the maximum tolerated dose to mice bone marrow cells did not increase the frequency of chromosomal aberrations or micronuclei. *o*-Aminobenzoic acid had a positive effect on sister chromatid exchange (16).

Any other comments
Human health effects and experimental toxicology reviewed (17).

References
1. Wood, E. M. *The Toxicity of 3400 Chemicals to Fish* 1987, EPA560/6-87-002, US Fish and Wildlife Service, WV
2. Hockenbury, M. R; et al. *J. Water. Pollut. Control. Fed.* 1977, May
3. Ziegler, K; et al. *Arch. Microbiol.* 1987, **149**(1), 62-69
4. *MITI Report* 1984, Ministry of International Trade and Industry, Tokyo
5. Haller, H. D. *J. Water. Pollut. Control. Fed.* 1978, 2771
6. Verschueren, K. *Handbook of Environmental Data of Organic Chemicals* 1983, Van Nostrand Reinhold Co., New York
7. Taha, K. M; et al. *DECHEMA Biotechnol. Conf.* 1989, **3**(B), 793-796
8. Tschech, A; et al. *Syst. Appl. Microbiol.* 1988, **11**(1), 9-12
9. Sax, N. I; et al. *Dangerous Properties of Industrial Materials* 7th ed., 1989, Van Nostrand Reinhold, New York
10. *Russ. Pharmacol. Toxicol.* 1974, **37**, 105
11. *IARC Monograph* 1987, **Suppl. 7**, 57
12. *National Toxicology Program Research and Testing Div.* 1992, Report No. TR-036, NIEHS, Research Triangle Park, NC27709
13. *IARC Monograph* 1972, **16**, 43
14. Zeiger, E; et al. *Environ. Mol. Mutagen.* 1987, **9**(Suppl. 9), 1-110
15. Mitchell, A. D; et al. *Environ. Mol. Mutagen.* 1988, **12**(Suppl. 13), 37-101
16. McFee, A. F; et al. *Environ. Mol. Mutagen.* 1989, **14**(4), 207-220
17. *ECETOC Technical Report No. 30(4)* 1991, European Chemical Industry Ecology and Toxicology Centre, B-1160 Brussels

A125 3-Aminobenzoic acid

CAS Registry No. 99-05-8
Synonyms *m*-aminobenzoic acid
Mol. Formula $C_7H_7NO_2$ **Mol. Wt.** 137.14

Physical properties

M. Pt. 174°C; **Specific gravity** d_4^{20} 1.51; **Partition coefficient** log P_{ow} 0.14.

Solubility
Organic solvent: ethanol, diethyl ether

Ecotoxicity

Fish toxicity
Designated non-toxic to trout, bluegill sunfish, yellow perch and goldfish (1).

Environmental fate

Degradation studies
Biodegradable (2).
97% removal in adapted activated sludge at 20°C (3).
Decomposition by soil microflora in >64 days (4).

Mammalian and avian toxicity

Acute data
LD_{50} intraperitoneal mouse 500 mg kg^{-1} (5).

References

1. Wood, E. M. *The Toxicity of 3400 Chemicals to Fish* 1953, U.S. Fish and Wildlife Service, WV
2. *MITI Report* 1984, Ministry of International Trade and Industry, Tokyo
3. Pitter, P. *Water Res.* 1976, **10**, 231
4. Verschueren, K. *Handbook of Environmental Data of Organic Chemicals* 1983, Van Nostrand Reinhold, New York
5. *Summary Tables of Biological Tests* 1954, **6**, 53, National Research Councils, Washington, DC

A126   4-Aminobenzoic acid

CAS Registry No. 150-13-0
Synonyms *p*-aminobenzoic acid; PABA; AMBEN; paraminol; Vitamin B_x; Chromotrichia factor
Mol. Formula $C_7H_7NO_2$ **Mol. Wt.** 137.14
Uses Veterinary products. Manufacture of esters for local anaesthetics. Production of azo dyestuffs. Sunburn prevention treatments.

Physical properties

M. Pt. 187°C; **Specific gravity** d_4^{20} 1.474; **Volatility** log P_{ow} 0.68.

Solubility
Organic solvent: ethanol, ethyl acetate

Ecotoxicity

Invertebrate toxicity
EC_{50} (30 min) *Photobacterium phosphoreum* 27.4 mg l^{-1} Microtox test (1).

Environmental fate

Nitrification inhibition
At 100 mg l^{-1} no inhibition of NH_3 oxidation by *Nitrosomonas sp.* (2).

Anaerobic effects
Complete degradation was achieved at a concentration of 50 mg carbon l^{-1} of *p*-aminobenzoic acid using anaerobic digesting sludge under methanogenic conditions (3).

Degradation studies
Biodegradable (4).
Decomposition by soil microflora in >8 days (2).
Adapted activated sludge removal 96% at 20°C (5).

Mammalian and avian toxicity

Acute data
LD_{50} oral rabbit, mouse, rat 1.8-6 g kg^{-1} (6,7).
LD_{50} intravenous rabbit 2000 mg kg^{-1} (7).

Sub-acute data
Intramuscular rabbits (2 wk) 500 mg kg^{-1} day^{-1} showed 1-2 large nucleoli/oocyte nucleus with sizes 3-4-fold greater than those in controls. Increases corresponded to periods of accelerated oogenesis (8).

Carcinogenicity and long-term effects
No adequate evidence for carcinogenicity to humans, limited evidence of carcinogencity in experimental animals, IARC classification group 3 (9,10).
Exposure of 4-aminobenzoic acid to UV radiation and application to the backs of hairless light-pigmented mice prior to daily UV irradiation for 30 wk retarded the induction time of tumours and reduced the number of squamous cell carcinomas (11).

Metabolism and pharmacokinetics
Percutaneous absorption and metabolism through hairless guinea pig skin was greater through nonviable skin. 4-Aminobenzoic acid was extensively N-acetylated during dermal absorption (12).
4-Aminobenzoic acid, a metabolite of procaine, appeared in the intestinal mucosa of rat ileum, duodenum and jejunum and increased with parent compound concentration (13).
When administered to humans orally, 4-aminobenzoic acid is absorbed from the gastrointestinal tract, metabolised in the liver and excreted in the urine as the unchanged drug and metabolites (14).

Irritancy
Contact and photocontact allergic dermatitis reported following the topical administration of aminobenzoate sunscreen agents (14).

Any other adverse effects

Subcutaneous, intraperitoneal, intravenous application to mice caused no toxicity with single or repeated doses. It did not stimulate immediate or delayed allergic responses after repeated administration of antigens in sensitised animals (15).

Aminobenzoate sunscreen agents should not be used by patients with previous experience of photosensitive or allergic reactions to chemically related drugs, such as sulfonamides, thiazide diuretics and certain local anaesthetics, particularly benzocaine (14).

Any other comments

Human health effects and experimental toxicology reviewed (16).

References

1. Kaiser, K. L. E; et al. *Water Pollut. Res. J. Canada* 1991, **26**(3), 361-431
2. Verschueren, K. *Handbook of Environmental Data of Organic Chemicals* 1983, Van Nostrand Reinhold, New York
3. Battersby, N. S; et al. *Appl. Environ. Microbiol.* 1989, **55**(2), 433-439
4. *MITI Report* 1984, Ministry of International Trade and Industry, Tokyo
5. Pitter, P. *Water Res.* 1976, **10**, 231
6. Scott, C. C; et al. *Proc. Soc. Expt. Biol. Med.* 1942, **49**, 184
7. *Fed. Am. Soc. Exp. Biol.* 1942, **1**, 71
8. Khvoles, A. G; et al. *Dokl. Akad. Nauk. SSSR* 1988, **301**(4), 985-988, (Russ.) (*Chem. Abstr.*, 1988, **109**, 184702d)
9. *IARC Monograph* 1987, **Suppl. 7**, 56
10. *IARC Monograph* 1978, **Suppl. 6**, 249
11. Flindt-Hansen, H; et al. *Photodermatology* 1989, **6**(6), 263-267
12. Nathan, D; et al. *Pharm. Res.* 1990, **7**(11), 1147-1151
13. Rummel, J; et al. *Naunyn- Schmiedeberg's Arch. Pharmacol.* 1990, **342**(2), 228- 233
14. *Martindale The Extra Pharmacopoeia* 29th ed., 1989, 1450, The Pharmaceutical Press, London
15. Grigor'eva, L. V; et al. *Izr. Akad. Nauk. SSSR, Ser. Biol.* 1988, (6), 923-926, (Russ.) (*Chem. Abstr.* 1989, **110**, 52569b)
16. *ECETOC Technical Report No. 30(4)* 1991, European Chemical Industry Ecology and Toxicology Centre, B-1160 Brussels

A127   2-Aminobiphenyl

CAS Registry No. 90-41-5
Synonyms 2-biphenylamine; *o*-aminobiphenyl; *o*-biphenylamine; *o*-phenylaniline; 2-phenylaniline; (1, 1'-biphenyl)-2-amine
Mol. Formula $C_{12}H_{11}N$ **Mol. Wt.** 169.23
Uses Intermediate in organic synthesis of carbazoles, resins and synthetic rubbers.

Physical properties

M. Pt. 51-53°C; **B. Pt.** 299°C; **Flash point** 110°C;
Partition coefficient log P_{ow} 2.84; **Volatility** v. den. 5.8.

Solubility
Organic solvent: ethanol, diethyl ether

Ecotoxicity

Fish toxicity
Lethal concentration (4 hr) trout, bluegill sunfish and goldfish 5 ppm (1).

Invertebrate toxicity
EC_{50} (5-30 min) *Photobacterium phosphoreum* 6.74 mg l^{-1} Microtox test (2).

Mammalian and avian toxicity

Acute data
LD_{50} oral rabbit, rat 1020-2340 mg kg^{-1} (3).

Carcinogenicity and long-term effects
♂ mouse 3000 ppm in feed day^{-1} equivocal evidence of carcinogenicity.
♀ mouse 3000 ppm feed day^{-1} positive carcinogenic effects, circulatory system
haemangiosarcoma (4).

Metabolism and pharmacokinetics
No specific data found but some aromatic amines are metabolised in humans and
dogs to the proximate carcinogen, while failing to produce tumours when implanted
directly into experimental animals (5).
2-Aminobiphenyl is metabolised at *o* and *p* positions with respect to the amino group
but does not form *N* oxidation products *in vitro* (6).
Predominantly metabolised *in vivo* to 3- and 5-hydroxy conjugated derivatives in
mice, rats, hamsters and guinea pigs. In some species, 2- aminobiphenyl is also
excreted as N-conjugated derivatives. During 24 hr, renal excretion accounts for
≈30-40% of the administered dose. The 5- *o*-sulfate and 5-*o*-glucuronide of
2-amino-5- hydroxybiphenyl are major metabolites in all species, and 2-
amino-3-hydroxybiphenyl-*o*-sulfate to a lesser extent (7).

Genotoxicity

Salmonella typhimurium TA98, TA100 with metabolic activation negative (8).
Escherichia coli with metabolic activation phage inhibition capacity (9).
Chinese hamster ovary cells without metabolic activation induced chromosomal
aberrations (9).
Drosophila melanogaster wing spot test negative (10).

References

1. Wood, E. M. *Toxicity of 3400 Chemicals to Fish* 1987, EPA 560/6-87-002, PB 87-200-275 1987
2. Kaiser, K. L. E; et al. *Water Pollut. Res. J. Canada* 1991, **26**(3), 361-431
3. *J. Ind. Hyg. Toxicol.* 1947, **29**, 1
4. Haseman, J. K; et al. *Environ. Mol. Mutagen.* 1990, **16**(Suppl. 18), 15-31
5. Thompson, C. Z. *Environ. Mutagen.* 1983, **5**(6), 803-811
6. Ioannides, C; et al. *Carcinogenesis* 1989, **10**(8), 1403-1407
7. Kajbaf, M; et al. *Eur. J. Drug Metab. Pharmacokinet.* 1987, **12**(4), 285-290

8. Yuk, L. Ho; et al. *Cancer Res.* 1981, **41**, 532-536
9. Tennant, R. W. *Science* 1987, **236**, 933-941
10. Tripathy, N. K; et al. *Mutat. Res.* 1990, **242**(3), 169-180

A128   3-Aminobiphenyl

CAS Registry No. 2243-47-2
Synonyms (1,1'-biphenyl)-3-amine; *m*- aminobiphenyl; *m*-phenylaniline;
3-phenylaniline; 3- biphenylamine
Mol. Formula $C_{12}H_{11}N$ **Mol. Wt.** 169.23

Physical properties

M. Pt. 36°C; **B. Pt.** 254°C.

Solubility
Organic solvent: ethanol, diethyl ether, acetone, benzene

Mammalian and avian toxicity

Metabolism and pharmacokinetics
Metabolised by rat liver microsomal preparations to hydroxylamines and the nitro compounds of 3- nitrosobiphenyl and 3-nitrobiphenyl (1).

Genotoxicity

Salmonella typhimurium TA100 without metabolic activation weakly positive (2).
Salmonella typhimurium TA98 with and without metabolic activation positive.
In vitro F344 rat hepatocytes unscheduled DNA synthesis negative (3).

References

1. Bayraktar, N; et al. *Arch. Toxicol.* 1987, **60**(1-3), 91-2
2. El-Bayonny, K; et al. *Mut. Res.* 1983, **5**, 803
3. Thompson C. Z; et al. *Environ. Mutagen.* 1983, **5**, 803-811

A129   4-Aminobiphenyl

CAS Registry No. 92-67-1
Synonyms 4-biphenylamine; *p*-biphenylamine; *p*-aminobiphenyl; *p*-aminodiphenyl;

4- aminodiphenyl; *p*-phenylaniline; (1,1'-biphenyl)-2- amine; biphenyl-4-amine; xenylamine

Mol. Formula $C_{12}H_{11}N$ **Mol. Wt.** 169.23

Uses In chemical analysis to detect sulfate ion. As a carcinogen in research. Formerly used as a rubber antioxidant.

Occurrence Found in tobacco smoke.

Physical properties

M. Pt. 53°C; **B. Pt.** 302°C; **Specific gravity** d^{20} 1.16; **Partition coefficient** log P_{ow} 2.86.

Solubility

Organic solvent: ethanol, chloroform

Occupational exposure

Supply classification Toxic.

Risk phrases

May cause cancer – Harmful if swallowed (R45, R22)

Safety phrases

Avoid exposure-obtain special instructions before use – If you feel unwell, seek medical advice (show label where possible) (S53, S44)

Ecotoxicity

Invertebrate toxicity

EC_{50} (5-30 min) *Photobacterium phosphoreum* 6.74 mg l^{-1} Microtox test (1).

Bioaccumulation

Using the estimated log P_{ow} of 2.80, bioconcentration factor is 79 (2).

Environmental fate

Degradation studies

In a static biodegradability test in which 2 mg l^{-1} of 4-aminobiphenyl was seeded with sludge, 50% degradation occurred after 7 days (3).

Abiotic removal

Sensitive to oxidation in air and darkens on standing (4).

Photochemical reaction with hydroxy radicals in the atmosphere, estimated $t_{1/2}$ 6.9 hr which suggests that hydrolysis will not be significant (5).

Absorption

Estimated soil adsorption coefficient of 417 indicates moderate adsorption to soil (2).

Mammalian and avian toxicity

Acute data

LD_{50} oral rat, rabbit, mouse 205-690 mg kg^{-1} (6,7).
LD_{Lo} intraperitoneal mouse 250 mg kg^{-1} (8).

Carcinogenicity and long-term effects

Sufficient evidence for carcinogenicity in humans and experimental animals, IARC classification group 1 (7).

Dose-related neoplasms angiosarcomas, bladder urothelial carcinomas, and heptocellular neoplasms were found in ♂ and ♀ BALB/cStCr1fC3Hf/Nctr mice given up to 300 ppm in their drinking water (duration unspecified). Non neoplasmic dose-related lesions were left atrical thrombosis, bladder urothelical hyperplasia, splenic heamosiderosis and spleneic erythropoeisis. The incidences of bladder carcinomas and atrial thrombosis were higher in ♂ and the incidences of hepatocellular neoplasms and angiosarcomas were higher in the ♀ (8).

Intraperitoneal B6C3FI mouse administered maximum tolerated dose on day 1, 8, 15 and 22 after birth, killed after 9 and 12 months revealed multiple hepatocellular adenomas and carcinomas (9).

Teratogenicity and reproductive effects
Oral pregnant ICR mice (day 18 of gestation) killed 24 hr after treatment with 135 mg kg^{-1} 4- aminobiphenyl dissolved in trioactamin revealed binding of 4-aminobiphenyl to the DNA of maternal and foetal liver, lung, kidney, heart, brain, intestine, skin, maternal uterus and placenta. 4-Aminobiphenyl bound preferentially to DNA of maternal liver and kidney but showed no preference among foetal tissues (10).

Metabolism and pharmacokinetics
A single intraperitoneal dose of 5 mg to rat, had a $t_{1/2}$ 15.6, 17 and 17 hr, respectively for urinary, faecal and total elimination (11).

Catalysed in humans by cytochrome P-450$_{PA}$ and in rats by cytochrome P-450$_{ISF-G}$ (12).

Thin-layer chromatography of the 24-48 hr urine of rats dosed with 4-aminobiphenyl (route and concentration unspecified), metabolites included 4- acetylaminobiphenyl; 4'-hydroxy-4-aminobiphenyl, 2'-hydroxy- 4-acetylaminobiphenyl; 4'-hydroxy-4-acetylaminobiphenyl, 3'- hydroxy-4'-methoxy-4-acetylaminobiphenyl; 4'-hydroxy-3'- methoxy-4-acetylaminobiphenyl and 3', 4'-dihydroxy-4-acetylaminophenyl (13).

The metabolic pathway appears to be via N-hydroxylation and N-glucuronide in the liver, N-glucuronides are transported to the bladder where hydrolysis to highly reactive electrophillic amyl nitrenium ions occurs (14).

Genotoxicity
Salmonella typhimurium TA98, TA100 with metabolic activation positive (15-17).
Escherichia coli with metabolic activation included prophage λ dts 857 appreciably, prophage λ negligibly (18).
Chinese hamster bone marrow positive induction of sister chromatid exchange (19).
♂ B6C3FI mice bone marrow and peripheral blood micronuclei test positive (20, 21).
In vitro primary rat hepatocytes weakly induced unscheduled DNA synthesis (22).
No significant heterogenicity in the survival of human epithelial cells from 5 donors after exposure to 4-aminobiphenyl. Cultures of normal fibroblasts from 41 donors showed an unexpected heterogenous response to the cytotoxic effects of 4-aminobiphenyl (23).

Any other adverse effects to man
Human peripheral lung tissue samples obtained from 17 workers of known occupational and smoking histories revealed 4-aminobiphenyl-DNA adducts detected by ELISA (24).

Any other adverse effects

In experimental animals convulsions, ataxia, dyspnoea, methaemoglobinemia, carbohaemoglobinemia have been reported (9,25).

Any other comments

The potent tobacco carcinogen 4-aminobiphenyl can cross the human placenta and bind to foetal haemoglobin in concentrations that are significantly higher in smokers than in non-smokers (26).

Human health effects, epidemiology, workplace experience and experimental toxicology reviewed (27).

References

1. Kaiser, K. L. E; et al. *Water Pollut. Res. J. Canada* 1991, **26**(3), 361-431
2. Lyman, W. J; et al. *Handbook of Chemical Property Estimation Methods* 1982, McGraw-Hill, New York
3. Tabak, H. H; et al. *Test Protocols for Environmental Fate and Movement of Toxicants.* *AOAC 94th Mtg.* 1981, 267-328
4. *IARC Monograph* 1972, **17**, 74-79
5. *GEMS: Graphical Exposure Modeling System. Fate of Atmospheric Pollutants* Office of Toxic Substances, USEPA
6. *J. Ind. Hyg. Toxicol.* 1947, **29**, 1
7. *IARC Monograph* 1987, **Suppl. 7**, 91
8. Schiererstein, G. J; et al. *Eur. J. Cancer Clin. Oncol.* 1985, **21**, 865
9. *Summary Tables of Biological Tests* 1954, **6**, 54, NRC, Chem. –Biol. Cord. Centre, Washington, DC
10. Lu, L. J. W; et al. *Cancer Res.* 1986, **46**, 3046-3054
11. Karreth, S; et al. *Xenobiotica* 1991, **21**(3), 417-428
12. Butler, M. A; et al. *Proc. Natl. Acad. Sci. USA* 1989, **86**(20), 7696-7700
13. Karreth, S; et al. *Xenobiotica* 1991, **21**(6), 709-724
14. *IARC Monograph* 1981, **40**, 13- 30
15. McCann, J; et al. *Proc. Natl. Acad. Sci. (USA)* 1975, **72**, 5135
16. Bos, R. P; et al. *Mutat. Res.* 1982, **93**, 317
17. Haworth, S; et al. *Environ. Mutagen.* 1983, **5**(Suppl. 1), 3
18. Yuk, L. Mo; et al. *Cancer Res.* 1971, **41**, 532-536
19. Neal, S. B; et al. *Mutat. Res.* 1983, **113**, 33
20. Gulati, D. K; et al. *Mutat. Res.* 1990, **234**(3-4), 135-139
21. Shelby, M. D; et al. *Environ. Mol. Mutagen.* 1989, **13**(4), 339-342
22. Harvach, P. R; et al. *Mutat. Res.* 1981, **90**, 345-354
23. Reznikoff, C. A; et al. *Carcinogenesis (London)* 1986, **7**(10), 1625-1632
24. Wilson, V. L; et al. *Carcinogenesis (London)* 1989, **10**(11), 2149-2153
25. *Science* 1970, **167**, 992
26. Coghlin, J; et al. *J. Natl. Cancer Inst.* 1991, **83**(4), 274-280
27. *ECETOC Technical Report No. 30(4)* 1991, European Chemical Industry Ecology and Toxicology Centre, B-1160 Brussels

A130   Aminocarb

OCONHCH$_3$

CH$_3$

N(CH$_3$)$_2$

CAS Registry No. 2032-59-9
Synonyms 4-dimethylamino-3-methylphenyl methylcarbamate; methylcarbamic acid,
4-dimethylamino-*m*-tolyl ester; 4-dimethylamino-3-cresyl methylcarbamate; Metacil
Mol. Formula C$_{11}$H$_{16}$N$_2$O$_2$ **Mol. Wt.** 208.26
Uses Insecticide and molluscicide.

Physical properties
M. Pt. 93-94°C; **Volatility** v.p. 1.3×10^{-5} mmHg at 20°C.

Occupational exposure
Supply classification toxic.

Risk phrases
Toxic by inhalation, in contact with skin and if swallowed (R23/24/25)

Safety phrases
Keep out of reach of children – Keep away from food, drink and animal feeding stuffs
– If you feel unwell, seek medical advice (show label where possible) (S2, S13, S44)

Ecotoxicity

Fish toxicity
LC$_{50}$ (96 hr) walleye 880 µg l^{-1} at 18°C.
LC$_{50}$ (96 hr) bluegill sunfish, largemouth bass, yellow perch 3.1-6.4 mg l^{-1} in hard
water 12-20°C.
LC$_{50}$ (96 hr) Atlantic salmon, fathead minnow, channel catfish, rainbow trout, brown
trout, brook trout 7.6-16 mg l^{-1}.
LC$_{50}$ (96 hr) cutthroat trout 31 mg l^{-1} in hard water at 10°C (1).

Invertebrate toxicity
LC$_{50}$ (96 hr) *Gammarus pseudolimnaeus* >50 µg l^{-1} (1).
EC$_{50}$ (48 hr) *Daphnia magna*, *Chironomus* 10-270 µg l^{-1} (1).
LC$_{50}$ (96 hr) *Gammarus lacustris* 12 µg l^{-1} (2).

Bioaccumulation
Bioconcentration factor *Mytilus edulis* 3.8-4.9 (3).
Aminocarb and its metabolite, 4-methylamino-*m*-tolyl *N*-methylcarbamate were
detected in the tissue of fingerling rainbow trout 96 hr after exposure to 21.3, 29.1 or
0.36 mg l^{-1} aminocarb, and >50% of the total residue was the parent compound.
Bioaccumulation factor was 1.70-3.32. Both compounds were eliminated rapidly after
transfer of the fish to clean water (4).

Effects to non-target species
LD$_{50}$ oral redwing blackbird, starling 50 mg kg^{-1} (5).
Toxic to bees (6).

Environmental fate

Degradation studies
Chlamydomonas variabilis and *Selenastrum capricornutum* degraded 50% and 100% aminocarb, within 6 and 14 days of incubation, respectively (7).

Abiotic removal
An initial concentration of 0.01 mg l^{-1} in river water was completely degraded in sunlight and artificial fluorescent light in 4 wk (8).

Mammalian and avian toxicity

Acute data
LD_{50} oral rat 50 mg kg^{-1}.
LD_{50} dermal rat 275 mg kg^{-1} (9).
LD_{50} intraperitoneal rat, mouse 7-21 mg kg^{-1} (10,11).

Sub-acute data
Rat 2 yr feeding study 200 mg kg^{-1} in the diet suffered no ill effects (6).
Gavage mice (exposure unspecified) sublethal doses and bone marrow was assessed by marrow transplantation to normal mice. Exposure of 0.08-5.0 mg kg^{-1} to donor animals did not affect regenerating bone marrow in the recipient mice. At 0.08 and 0.32 mg kg^{-1} a marked shift in surface IgM density on marrow B cells was noted (12).

Metabolism and pharmacokinetics
Metabolised in liver to 4-amino-3-cresyl methylcarbamate, 4-methylamino-3-cresyl methylcarbamate, and 4-dimethylamino-3-cresyl *N*-hydroxymethylcarbamate (13). Aminocarb was hydrolysed to 4-(dimethylamino)-3- methylphenol which in turn was converted to 2-methyl-1, 4- benzoquinone by a direct means or via 2-methyl-1,4-dihydroquinone (species unspecified) (14).
In rhesus monkeys, 74% of aminocarb ($t_{1/2}$ 25 hr) was absorbed from the forehead whereas 37% ($t_{1/2}$ 31 hr) was absorbed from ventral forearm. In rats, 88% of aminocarb ($t_{1/2}$ 17 hr) was absorbed from the middorsal region (15).

Genotoxicity
Topical administration to mouse of aminocarb over 24 hr induced a dose-dependent increase in the frequency of hair follicle nuclear aberrations (16).

Any other adverse effects
Carbamate insecticide which acts as a cholinesterase inhibitor. Symptomatic effects may include respiratory discomfort, nausea, vomiting, diarrhoea, headache, blurred vision, salivation, sweating and confusion. Central nervous system effects include ataxia, slurred speech and paralysis. Severe exposure may result in hypotension, pulmonary oedema, convulsions, coma and death from respiratory failure or cardiac arrest (17,18).

Legislation
Limited under EEC Directive on Drinking Water Quality 80/778/EEC. Pesticides and related products: maximum admissible concentration 0.1 μg l^{-1} (19).

Any other comments
Probable human lethal dose is 5-50 mg kg^{-1} (17).

References

1. *U.S. Dept. of Interior, Fish and Wildlife Service. Handbook of Acute Toxicity of Chemicals to Fish and Aquatic Invertebrates* 1980, (137), 11, U.S. Govt. Print. Off., Washington, DC
2. Saunders, H. O. *Toxicity of Pesticides to the Crustacean Gammarus lacustus*, 1969, Tech. paper No. 25, US Govt. Print. Off., Washington DC
3. McLesse, D. W., et al. *Bull. Environ. Contam. Toxicol.* 1980, **24**, 575
4. Szeto, S. Y; et al. *J. Environ. Sci. Health Part B* 1982, **B17**(1), 51-61
5. Schafer, E. W; et al. *Arch. Environ. Contam. Toxicol.* 1983, **12**, 355-382
6. *The Agrochemicals Handbook* 2nd ed., 1987, RSC, London
7. Menzie, C. M. *Metabolism of Pesticides. U.S. Dept. of the Interior, Bureau of Sport Fisheries and Wildlife* 1969, 237, US Govt. Print. Off., Washington, DC
8. Eichelberger, J. W; et al. *Environ. Sci. Technol.* 1971, **5**(6), 541-544
9. Ames, B. N; et al. *Proc. Nat. Acad. Sci. USA* 1973, **70**, 2281-2285
10. Baron, R. L; et al. *Toxicol. Appl. Pharmacol.* 1964, **6**, 402
11. *Pesticide Manual* 8th ed., 1987, 270, British Crop Protection Council, Farnham
12. Bernier, J; et al. *Pest. Biochem. Physiol.* 1990, **36**(1), 35-45
13. Krishna, J. G; et al. *J. Agric. Food Chem.* 1966, **14**, 98
14. Leger, D. A; et al. *J. Agric. Food Chem.* 1988, **36**(1), 185-189
15. Moody, R. P; et al. *J. Toxicol. Environ. Health* 1987, **20**, 209
16. Schop, R. N; et al. *Fundam. Appl. Toxicol.* 1990, **15**(4), 666-675
17. Gosselin, R. E; et al. *Clinical Toxicology of Commercial Products*, 5th ed., 1984, Williams & Wilkins, Baltimore
18. *Pestline* 1991, **2**, 1455, Occupational Health Services Inc., Van Nostrand Reinhold, New York
19. *EC Directive Relating to the Quality of Water Intended for Human Consumption* 1982, 80/778/EEC, Office for Official Publications of the European Communities, 2 rue Mercier, L-2985 Luxembourg

A131   2-Amino-4-chlorophenol

CAS Registry No. 95-85-2

Synonyms *p*-chloro-*o*-aminophenol; C.I. Oxidation base 18; C.I. 76525; Fouramine PY

Mol. Formula C_6H_6ClNO **Mol. Wt.** 143.57

Uses Dyestuff intermediate.

Physical properties

M. Pt. 139°C.

Occupational exposure

UN No. 2673; **HAZCHEM Code** 2X; **Conveyance classification** toxic substance.

Environmental fate

Degradation studies

Degraded aerobically by *Alcaligenes* TK-2 (1).

Mammalian and avian toxicity

Acute data
LD$_{50}$ oral mouse, rat 690-1030 mg kg^{-1} (2).

Genotoxicity
Salmonella typhimurium TA1535 with metabolic activation, weakly positive, negative results reported in other strains (3).

Any other comments
All reasonable efforts have been taken to find information on isomers of this compound, but no relevant data are available.

References

1. Bellnink, J; et al. *Appl. Microbiol. Biotechnol.* 1990, **34**(1), 108-115
2. Vasilenko, N. M. *Gig. Tr. Prof. Zabol.* 1981, **25**(8), 50-52
3. Zeiger, K; et al. *Environ. Mol. Mutagen.* 1988, **11**(Suppl. 12), 1-157

A132   1-Amino-2,4-dibromoanthraquinone

CAS Registry No. 81-49-2
Synonyms 2,4-dibromo-1-anthraquinololylamine; 1-amino- 2, 4-dibromo-9,10-anthracenedione
Mol. Formula C$_{14}$H$_7$Br$_2$NO$_2$ **Mol. Wt.** 381.04
Uses Dyestuff synthesis.

Physical properties
M. Pt. 226-227°C.

Mammalian and avian toxicity

Carcinogenicity and long-term effects
Retrospective characterisation of morphological and stereological features of altered hepatocellular foci in haematoxylin and eosin stained sections was performed in 2 yr carcinogenicity studies in Fischer 344 rats by the National Toxicological Program. In the 1-amino-2,4-dibromoanthraquinone study there was clear evidence of hepatocarcinogenicity (1).
Oral rats, mice in feed 2 yr study currently being carried out by the National Toxicology Program Pathological Working Group (2).

Irritancy
500 mg instilled into rabbit eye (24 hr) caused mild irritation effects (3).

Genotoxicity

Salmonella typhimurium TA1537 with and without metabolic activation positive (4). *In vitro* Chinese hamster ovary cells with and without metabolic activation induced sister chromatid exchange, with metabolic activation did not induce chromosomal aberrations (5).

Any other comments

Human health effects and experimental toxicology reviewed (6).
All reasonable efforts have been taken to find information on isomers of this compound, but no relevant data are available.

References

1. Harada, T; et al. *Toxicol. Pathol.* 1989, **17**(4, Part 1), 690-708
2. *National Toxicology Program Research & Testing Division* 1992, Report No. 383, NIEHS, Research Triangle Park, NC 27709
3. Marhold, J. V. *Sbornik Vysledku Tox. Vys. Lat. A Priprarku* 1972, 88, Prague
4. Haworth, S; et al. *Environ. Mutat.* 1983, **5**(Suppl. 1), 3
5. Loveday, K. S; et al. *Environ. Mol. Mutagen.* 1990, **16**(4), 272-303
6. *ECETOC Technical Report No. 30(4)* 1991, European Chemical Industry Ecology and Toxicology Centre, B-1160 Brussels

A133   4-Amino-*N*,*N*-diethylaniline

CAS Registry No. 93-05-0
Synonyms *N*,*N*-diethyl-*p*-phenylenediamine; *N*,*N*′-diethyl-1,4-benzenediamine; *p*-aminodiethylaniline; *N, N*-diethyl-1,4-diaminobenzene
Mol. Formula $C_{10}H_{16}N_2$ **Mol. Wt.** 164.25
Uses Photography. Analytical reagent to detect chlorine residues in water. As a dyestuff intermediate and as a source of diazonium compounds in diazo copying processes.

Physical properties

M. Pt. 19-21°C; **B. Pt.** 260-262°C; **Flash point** >107°C;
Specific gravity 1.00.

Occupational exposure

Supply classification toxic.

Risk phrases

Toxic if swallowed – Causes burns (R25, R34)

Safety phrases

In case of contact with eyes, rinse immediately with plenty of water and seek medical advice – Wear suitable protective clothing – If you feel unwell, seek medical advice (show label where possible) (S26, S36, S44)

Mammalian and avian toxicity

Acute data

LD_{Lo} oral rabbit 450 mg kg^{-1}.

LD_{Lo} dermal rabbit 125 mg kg^{-1}.

LD_{Lo} subcutaneous rat, rabbit, guinea pig 100- 250 mg kg^{-1} (1).

Genotoxicity

Salmonella typhimurium TA97, TA98, TA100 with metabolic activation positive, TA97 without metabolic activation weakly positive (2).

In vitro Chinese hamster lung and ovary cells chromosome aberrations with metabolic activation negative, without metabolic activation weakly positive for lung cells and distinctly positive for ovary cells (3).

Any other adverse effects

Toxic effects are expected to be similar to those of *p*-phenylenediamine, causing eye and skin irritation and dermatitis (4).

Following *in vivo* application to guinea pig skin, binding to epidermal proteins was demonstrated, although it was relatively labile and no oligopeptide could be isolated (4).

Any other comments

Physico-chemical properties, human health effects and experimental toxicology reviewed (5,6).

All reasonable efforts have been taken to find information on isomers of this compound, but no relevant data are available. See also Bardrowski's base.

References

1. *J. Ind. Hyg.* 1923, **4**, 386
2. Zeiger, E; et al. *Environ. Mol. Mutagen.* 1988, **11**(Suppl. 12), 1-157
3. Sofuni, T; et al. *Mutat. Res.* 1990, **241**(2), 175-213
4. Reynolds; et al. *Food Cosmet. Toxicol.* 1970, **8**(6), 635
5. *EPA Chemical Profiles* 1985, 86- 1603, 3
6. *ECETOC Technical Report No. 30(4)* 1991, European Chemical Industry Ecology and Toxicology Centre, B-1160 Brussels

A134   2-Amino-3,4-dimethylimidazo[4,5-*f*]quinoline

CAS Registry No. 77094-11-2

Synonyms Me-IQ

Mol. Formula $C_{12}H_{12}N_4$ **Mol. Wt.** 212.26

Occurrence Isolated from broiled fish and meat and beef extract.

Abiotic removal
Sensitive to activation by sunlight and fluorescent light (1).

Mammalian and avian toxicity

Carcinogenicity and long-term effects
No adequate data for evidence of carcinogenicity to humans, insufficient evidence for carcinogenicity to animals, IARC classification group 3 (2).
Oral rat (58 wk) 14 doses of 10 mg kg^{-1} were administered in drinking water, a small number of Zymbal's gland tumours were detected. Short exposures to low doses produced persistent procarcinogenic lesions. Secondary factors, promoters or high cell turnover may, over time, develop these lesions into cancer (3).
Gavage ♀ CFI mice, Sprague-Dawley rats (unspecified dose) administered twice with a 4-day interval induced colon cancer (4).
Fed CDF$_1$ mice (91 wk) 0.04 or 0.01% induced hepatocellular carcinomas and hepatocellular adenomas in ♀. Incidence of squamous cell carcinomas and papillomas were higher in both sexes (5).

Genotoxicity
Salmonella typhimurium TA98, pYG 121 with metabolic activation positive (6,7).
Intraperitoneal rat (2 hr) 80 mg kg^{-1} no DNA damage observed in stomach, small and large intestine, liver, kidney or testis. Intraperitoneal ♂ F344 rat 80 mg kg^{-1} or fed 0.03% 2-amino-3, 4-dimethylimidazo[4, 5-*f*]quinoline (13 day) induced DNA damage in large intestine, liver and kidney (8).
In vitro Chinese hamster V79 cells with metabolic activation induced sister chromatid exchange (9).

Any other comments
In vivo and *in vitro* mice bacterial mutation assays demonstrated that bran in the diet reduced the genotoxicity potential by restricting uptake of 2-amino-3,4-dimethylimidazo[4,5-*f*]quinoline from the gut lumen. Feeding mice a high-fat diet led to its hepatic conversion to an active genotoxin (10).

References
1. De Flora, S; et al. *Carcinogenesis (London)* 1989, **10**(6), 1089-1097
2. *IARC Monograph* 1987, **Suppl. 7**, 65
3. Kristiansen, E; et al. *Pharmacol. Toxicol. (Copenhagen)* 1989, **65**(5), 332-335
4. Tudek, B; et al. *Cancer Res.* 1989, **49**(5), 1236-1240
5. Ohgaki, H; et al. *Carcinogenesis (London)* 1986, **7**(11), 1889-1893
6. Petry, T. W; et al. *Carcinogenesis (London)* 1989, **10**(12), 2201-2207
7. Nagao, M; et al. *Carcinogenesis (London)* 1981, **2**(11), 1147-1149
8. Holme, J. A; et al. *Mutat. Res.* 1991, **251**(1), 1-6
9. Brunborg, G; et al. *Mutagenesis* 1988, **3**(4), 303-309
10. Brennan-Craddock, W. E; et al. *Food Addit. Contam.* 1990, **7**(Suppl. 1), 553-554

A135   2-Amino-3,8-dimethylimidazo[4,5-*f*]quinoxaline

CAS Registry No. 77500-04-0
Synonyms Me-IQx
Mol. Formula $C_{12}H_{11}N_5$ **Mol. Wt.** 225.25
Occurrence Isolated from fried beef.

Mammalian and avian toxicity

Carcinogenicity and long-term effects
No adequate data for evidence of carcinogenicity to humans, insufficient evidence for carcinogenicity to animals, IARC classification group 3 (1).

Metabolism and pharmacokinetics
Oral, intravenous, intraperitoneal mouse (duration unspecified) 7-16% excreted unchanged, with 15-35% in the urine and 30-55% in the faeces (2).
In rabbit and human liver, activation of 2-amino-3,8-dimethylimidazo[4, 5-*f*]quinoxaline was catalysed by cytochrome P450IA2 (2).
Metabolised to 2-hydroxyamino-3,8- dimethylimidazo[4,5-*f*]quinoxaline by microsomal cytochrome P450 and is further activated to an acetylated form (3).
Incubation with mixed human faecal microflora under anaerobic conditions yielded 2-amino-3, 6-dihydro-3,8- dimethylimidazo[4,5-*f*]quinoxalin-7-one as the major metabolite, but with low overall conversion (4).
In rat hepatocytes, 10 metabolites were identified, including 2-(hydroxyamino)-3,8-dimethylimidazo[4,5*f*]quinoxaline and its *N*-hydroxy-*N*-glucuromide (5).

Genotoxicity
Salmonella typhimurium TA98 with metabolic activation positive (6).
In vitro human lymphocyte cells with metabolic activation induced a low frequency of sister chromatid exchange (7).
Chronic and acute treatment of mice with 2-amino-3,8-dimethylimidazo[4,5-*f*]quinoxaline did not induce *DL6-1* mutations (8).
In vitro hamster hepatocytes unscheduled DNA synthesis weak dose-related response observed (9).

Any other adverse effects to man
Excreted in the urine of men after consumption of fried ground beef patties with 1.8-4.9% of the amount consumed excreted unchanged within 12 hr (1).

Any other comments
The binding of mutagens known to occur in fried or broiled food, including MeIQx, was investigated *in vitro*. Results showed water insoluble fibre components were responsible for most of the binding capacity. There was a significant correlation

between Klason lignin content and the binding of mutagens. Dietary fibre from sorghum had the highest binding capacity (10).

References

1. *IARC Monograph* 1987, **Suppl. 7**, 65
2. Gooderham, N. J; et al. *Prog. Clin. Biol. Res.* 1990, **340**, 67-76
3. Negishi, C; et al. *Mutat. Res.* 1989, **210**(1), 127-134
4. Bashir, M; et al. *Heterocycles* 1990, **31**(7), 1333-1338
5. Turesky, R. J; et al. *Chem. Res. Toxicol.* 1990, **3**(6), 524-535
6. Kasai, H; et al. *Chem. Lett.* 1981, 485
7. Aeschbacher, H. U; et al. *Carcinogenesis (London)* 1989, **10**(3), 429-433
8. Winton, D. J; et al. *Cancer Res.* 1990, **50**(24), 7992-7996
9. Howes, A. J; et al. *Food Chem. Toxicol.* 1986, **24**(5), 383-387
10. Sjoedin, P. B; et al. *J. Food Sci.* 1985, **50**(6), 1680-1684

A136   3-Amino-1,4-dimethyl-5*H*-pyrido(4,3-*b*)indole

CAS Registry No. 62450-06-0

Synonyms 3-amino-1,4-dimethyl-γ-carboline; TRP-P-1; Trp-P-1; tryptophan-P-1

Mol. Formula $C_{13}H_{13}N_3$ **Mol. Wt.** 211.27

Occurrence Pyrolysis product of cooked food, especially sardines.

Physical properties

M. Pt. 252-260°C.

Solubility

Organic solvent: methanol

Abiotic removal

After a 30 min irradiation by a mercury vapour lamp, mutagenic activity was reduced to less than half (1).

Mammalian and avian toxicity

Acute data

LD_{50} oral mouse, hamster 200-380 mg kg^{-1} (2).

Carcinogenicity and long-term effects

No adequate data for evidence of carcinogenicity to humans, sufficient evidence for carcinogenicity in experimental animals, IARC classification group 2B (3,4).

Subcutaneous neonatal mice, observed for 1 yr, induced liver tumours in 45% of ♂, malignant lymphomas in 13% of ♂ and in 24% of ♀ (5).

ICR mice (1 yr) 50 µg g^{-1} induced lung, liver, lymphoma tumours and leukaemia in both ♂ and ♀ animals (6).

Hepatocellular carcinomas reported in rats and mice given
3-amino-1,4-dimethyl-5*H*-pyrido(4, 3- b)indole in the diet at 0.02-0.08% (7,8).

Genotoxicity

Salmonella typhimurium TA98, TA100 with metabolic activation positive (8).
SOS Chromotest using *Escherichia coli* K12, PG37 and PQ35 with metabolic
activation positive (10).
Induced sister chromatid exchanges in cultured human lymphocytes with metabolic
activation (11).
Intraperitoneal *in vivo* rat bone marrow cells (5 daily injections) 0.210-10.5 ng kg^{-1}
dose-response relationship obtained (12).

Any other adverse effects

Taken up into PC12h cells by the transport system specific for dopamine,
accumulated in cells and reduced the enzyme activity of tyrosine hydroxylase and
aromatic L-amino acid decarboxylase (13).

Any other comments

Identified in human, cataractous lenses, but not in young bovine lenses, concentrated
in the insoluable lens proteins (14).

References

1. Yoo, Y. S; et al. *Environ. Mutagen. Carcinog.* 1989, **8**(2), 99-104
2. Miller, E. C; et al. (Ed.) *Naturally Occurring Carcinogens – Mutagens and Modulators of Carcinogenesis* 1979, 167, Japan. Sec. Soc., Tokyo
3. *IARC Monograph* 1983, **31**, 247
4. *IARC Monograph* 1987, **Suppl. 7**, 73
5. Fujii, K; et al. *Carcinogenesis (London)* 1987, **8**(11), 1721-1723
6. Fujii, K; *Carcinogenesis (London)* 1991, **12**(8), 1409-1415
7. Ohyaki, H; et al. *Environ. Health Perspect.* 1986, **67**, 129-134
8. Takayama, S; et al. *Jpn. J. Cancer Res.* 1985, **76**(9), 815-817
9. Nagao, M; et al. *Carcinogenesis (London)* 1980, **1**, 451
10. Thybaud-Lambay, V; et al. *Mutat. Res.* 1986, **173**(3), 177-180
11. Inoue, K; et al. *Mutat. Res.* 1983, **117**(3-4), 301-309
12. Fujii, K; et al. *Kobe Daigaku Igakubu Kiyo* 1987, **48**(4), 247-252, (Jap.) (*Chem. Abstr.* **109**, 793e)
13. Takahasi, T; et al. *Adv. Behav. Biol.* 1990, **38A**, 345-348
14. Manabe, S; et al. *Exp. Eye Res.* 1989, **48**(3), 351-363

A137   2-Amino-4,6-dinitrophenol

CAS Registry No. 96-91-3
Synonyms picramic acid; dinitroaminophenol; 4,6- dinitro-2-aminophenol

Mol. Formula $C_6H_5N_3O_5$ **Mol. Wt.** 199.12

Uses Manufacture of azo dyestuffs. Reagent for albumin. Rarely used as indicator.

Physical properties

M. Pt. 168°C; **Flash point** 210°C.

Solubility

Water: 0.65 g l^{-1} at 22-25°C. Organic solvent: benzene, aniline, ethanol, acetic acid

Occupational exposure

Supply classification explosive and harmful.

Risk phrases

Explosive when dry – Harmful by inhalation, in contact with skin and if swallowed (R1, R20/21/22)

Safety phrases

This material and its container must be disposed of in a safe way (S35)

Ecotoxicity

Fish toxicity

No inhibition in growth of rainbow trout exposed to 0.02 mg l^{-1} but petechial haemorrhages along abdomen wall developed with over 80% having lesions in 42 days (1).
LC$_{50}$ (96 hr) rainbow trout 46.2 mg l^{-1} (2).

Invertebrate toxicity

American oysters exposed to 0.02 mg l^{-1} for 42 day showed significant inhibition of shell deposit. Also discolouration of nacre layers and body mass (1).
LC$_{50}$ (14 hr) American oyster 70 mg l^{-1} (2).

Bioaccumulation

Rainbow trout exposed for 42 days to picramic acid showed no bioconcentration in epaxial muscle tissues.
American oyster bioconcentration factor after 42 day exposure to 0.02 mg l^{-1} was 49.3 (3).
$t_{1/2}$ for elimination in trout, measured 9-9.5 days (4).

Genotoxicity

Salmonella typhimurium TA97, TA98, TA100, TA1535 with and without metabolic activation positive (5).

Any other comments

Toxicity is reported similar to 2,4- dinitrophenol, for which LD$_{50}$ oral rat is 30 mg kg^{-1}, and which is readily absorbed through the intact skin causing rise in metabolic rate and temperature, even collapse and death. It may cause dermatitis, cataracts and weight loss (3).
Comments on drinking water guidelines have been published (6).
Physico-chemical properties, human health effects and experimental toxicology reviewed (7).
All reasonable efforts have been taken to find information on isomers of this compound, but no relevant data are available.

References

1. Goodfellow, W. L; et al. *Chemosphere* 1983, **12**, 1259-1268
2. Goodfellow, W. L; et al. *Water Resour. Bull.* 1983, **19**, 641-648
3. Burton, D. T; et al. *Report 1983, JHU/APL/CPE-8303 (Chem. Abstr.* **99**, 189293f)
4. Cooper, K. R; et al. *J. Toxicol. Environ. Health* 1984, **14**, 731-747
5. Zeiger, E; et al. *Environ. Mol. Mutagen.* 1988, **11**(Suppl. 12), 1-157
6. *Gov. Rep. Announc. Index (US)*, 1983, **83**, 2147, Report No. NRC-TOX-P897 (*Chem. Abstr.* **99**, 58498d)
7. *ECETOC Technical Report No. 30(4)* 1991, European Chemical Industry Ecology and Toxicology Centre, B-1160 Brussels

A138   2-(2-Aminoethoxy)ethanol

$NH_2CH_2CH_2OCH_2CH_2OH$

CAS Registry No. 929-06-6
Synonyms diglycolamine
Mol. Formula $C_4H_{11}NO_2$ **Mol. Wt.** 105.14
Uses Rust-proofing steels. Gas sweetening. Gas purification (especially removal of COS). In zeolite preparation.

Physical properties

B. Pt. 218-224°C; **Specific gravity** d_4^{25} 1.048.

Occupational exposure

UN No. 3055; **HAZCHEM Code** 2R; **Conveyance classification** corrosive substance.

Mammalian and avian toxicity

Acute data
LD_{50} oral rat 5660 mg kg^{-1}.
LD_{50} dermal rabbit 1190 mg kg^{-1} (1).

Irritancy
Dermal rabbit (24 hr) 10 mg caused severe irritation, and 250 μg instilled into rabbit eye caused severe irritation (1).

Any other comments

N-nitrosodiethanolamine and *N*- nitrosomorpholine have been found in cutting oils containing diglycolamine after the fluid has been heated to 100°C for 48 hr (2).

References

1. *AMA Arch. Ind. Hyg. Occup. Med.* 1951, **4**, 119
2. Loeppky, R. N; et al. *Food Chem. Toxicol.* 1983, **21**, 607-6139

A139   3-Amino-9-ethylcarbazole

CAS Registry No. 132-32-1
Synonyms 3-amino-N-ethylcarbazole; 9-ethyl- 9H-carbazol-3-amine
Mol. Formula $C_{14}H_{14}N_2$ Mol. Wt. 210.28
Uses Pigment synthesis. Colourimetric enzyme assay, peroxidase enzyme activity assay.

Physical properties

M. Pt. 98-100°C.

Mammalian and avian toxicity

Acute data
LD_{50} oral rat 144 mg kg^{-1} (1).
LD_{50} intraperitoneal mouse 150 mg kg^{-1} (2).

Any other comments
Human health and environmental effects, epidemiology, workplace experience and experimental toxicology reviewed (3).
All reasonable efforts have been taken to find information on isomers of this compound, but no relevant data are available.

References

1. *Nat. Inst. Health* 1971, E-2144, Bethesda, MD
2. *NTIS Report.* AD 691-490, Natl. Tech. Inf. Ser., Springfield, VA
3. *ECETOC Technical Report No. 30(4)* 1991, European Chemical Industry Ecology and Toxicology Centre, B-1160 Brussels

A140   N-Aminoethylpiperazine

CAS Registry No. 140-31-8
Synonyms N-(β-aminoethyl)piperazene
Mol. Formula $C_6H_{15}N_3$ Mol. Wt. 129.21
Uses Epoxy curing agent.

Physical properties
M. Pt. –19°C; B. Pt. 220°C; Flash point 93°C (open cup);

Specific gravity d_{20}^{20} 0.98; **Volatility** v. den. 4.4.

Occupational exposure
UN No. 2815; **HAZCHEM Code** 2R; **Conveyance classification** corrosive substance.

Mammalian and avian toxicity

Acute data
LD_{50} oral rat 2140 mg kg^{-1} (1).
LD_{50} dermal rabbit 880 mg kg^{-1} (2).
LD_{50} intraperitoneal mouse 250 mg kg^{-1} (3).

Irritancy
100 mg instilled into rabbit eye caused moderate irritation (duration unspecified) (4).

Any other comments
Evaluation of morpholine piperazine and analogues in the mouse lymphoma L-5178Y and BALB-3T3 transformation assays have been carried out. Results available as conference proceedings. (5).

References
1. *Am. Ind. Hyg. Assoc. J.* 1962, **23**, 95
2. *Union Carbide Data Sheet* 13 June 1969, Union Carbide Corp, New York
3. *NTIS Report* AD 277-689, Natl. Tech. Inf. Ser., Springfield, VA
4. Deichmana, W. B. (Ed.) *Toxicology of Drugs and Chemicals* 1989, Academic Press, New York
5. Conaway, C. C; et al. *Environ. Mol. Mutagen.* 1982, **4**(3), 390

A141   Aminomethane

CH3NH2

CAS Registry No. 74-89-5
Synonyms methylamine; methanamine; monomethylamine; carbinamine; mercurialin
Mol. Formula CH$_5$N   **Mol. Wt.** 31.06
Uses Used in tanning, organic synthesis, dyeing of acetate textiles, paint removers. Intermediate for accelerators, dyestuffs, pharmaceuticals, insecticides, fungicides and surface active agents.
Occurrence Occurs in certain plants such as *Mentha aquatica*.

Physical properties

M. Pt. −93.5°C; **B. Pt.** −6.3°C; **Flash point** 0°C (closed cup); **Specific gravity** d_4^{20} 0.662; **Partition coefficient** log P_{ow} 1.77; **Volatility** v.p. 2650 mmHg at 25°C; v. den. 1.07.

Solubility
Organic solvent: ethanol, benzene, ethanol, acetone miscible with diethyl ether

---

Occupational exposure

US TLV (TWA) 12 mg m^{-3}; **UK Long-term limit** 10 ppm (12 mg m^{-3});
UN No. 1061 (anhydrous);1235 (aqueous solution); **HAZCHEM Code** 2PE;
Conveyance classification flammable gas; **Supply classification** Highly flammable and irritant.

Risk phrases

Extremely flammable liquefied gas – Irritating to eyes and respiratory system (R13, R36/37)

Safety phrases

Keep away from sources of ignition – No Smoking – In case of contact with eyes, rinse immediately with plenty of water and seek medical advice – Do not empty into drains (S16, S26, S29)

Ecotoxicity

Fish toxicity
LC$_{50}$ (24 hr) creek chub 10-30 mg l^{-1} (1).

Invertebrate toxicity
LD$_{Lo}$ *Daphnia magna* 480 mg l^{-1} (2).
EC$_{50}$ (5 min) *Photobacterium phosphoreum* 34.8 mg l^{-1} Microtox test (3).

Bioaccumulation
The estimated bioconcentration factor of 0.22 using a measured log K$_{ow}$ of –0.57 indicates that environmental accumulation is unlikely (4).

Environmental fate

Nitrification inhibition
50% inhibition of NH$_3$ oxidation in *Nitrosomonas* at 310 mg l^{-1} (5).

Anaerobic effects
Anaerobic degradation occurred using mixed cultures from marine sediments and pure cultures of *Methanosarcina barkeri* (6,7).

Degradation studies
Pseudomonas sp. MA utilises aminomethane as a sole source of carbon and nitrogen (8).
In the OECD screening test and closed bottle test, degradation was 96% and 107%, respectively (9).
BOD$_{13}$ 67.8% reduction of dissolved oxygen concentration (10).

Abiotic removal
A rate constant for aqueous hydroxy radical reaction with aminomethane in water is 1.1×10^7 molecules sec^{-1}. At a typical aqueous hydroxyl radical concentration of 10^{-7} 3 µg l^{-1}, the aminomethane t$_{1\backslash2}$ would be 199 yr suggesting aqueous hydrolytic degradation is insignificant (9,11,12).
Atmospheric t$_{1\backslash2}$ from hydroxyl radicals is 3-22 hr (13).

Absorption
Estimated soil adsorption coefficient of 12 indicates that aminomethane will not strongly adsorb to organic matter in soil or sediment and is expected to leach readily through most soils (4).

Mammalian and avian toxicity

Acute data
LD_{50} oral rat 0.1-0.2 g kg^{-1} (10% solution) (14).
LC_{50} (2 hr) inhalation mouse 2.4 g m^{-3} (15).
LD_{50} subcutaneous mouse 2.5 g kg^{-1} (16).

Sub-acute data
Inhalation rat (2 wk) 6 hr day^{-1} 5 day wk^{-1} 0, 75, 250 or 750 ppm aminomethane. The highest dose caused mortality or severe body weight losses, clinical pathological changes suggestive of liver damage, nasal degenerative changes and haematopoietic changes were observed during the recovery period of 14 days. Exposure to 250 ppm produced damage of the respiratory mucosa of the nasal turbinates whilst mild irritation of the nasal turbinate occurred after 75 ppm (17).

Teratogenicity and reproductive effects
Intraperitoneal pregnant CD-1 mice (1-17 day gestation) 75 and 155 mg kg^{-1} day^{-1} no obvious maternal or foetal effects. When administered to embryos in culture, caused dose-dependent decreases in size, DNA, RNA and protein content as well as survival suggesting aminomethane acts as an endogenous teratogen under certain conditions (18).

Metabolism and pharmacokinetics
After ingesting fish containing aminomethane, the major excretory route in humans was in the urine. A 4- fold increase in excretion of dimethylamine and >8-fold increase in excretion of trimethylamine was reported. There is potential for *in vivo* conversion of dimethylamine to nitrosodimethylamine, a carcinogen, although no studies have determined that the ingestion of fish increases the risk of cancer (19).

Any other adverse effects to man
Workers at a plant processing dimethylamine where <36.7 mg m^{-3} of aminomethane was present in ambient air from 0600 to 1800 had urinary concentrations of 1.3-2.48 mg l^{-1} aminomethane during a 24 hr period (20).

Any other comments
Physico-chemical properties, human health effects, epidemiology, workplace experience and experimental toxicology reviewed (21).
Fire hazard, moderately explosive with nitromethane.

References
1. McKee, J. E; et al. *Water Quality Criteria* 1963, The Resources Agency of California, State Water Quality Control Board
2. Meinck, F; et al. *Les Eaux Residuaires Industrielles* 1970
3. Kaiser, K. L. E; et al., *Water Pollut. Res. J. Canada* 1991, **26**(3), 361-431
4. Lyman, W. J; et al., *Handbook of Chem. Property Estimation Methods* 1982, McGraw-Hill, New York
5. Hooper, A; et al. *J. Bacteriol.* 1973, **115**, 480
6. King, G. M; et al. *Appl. Environ. Microbiol.* 1983, **45**, 1848-1853
7. Hippe, H; et al., *Proc. Natl. Acad. Sci. USA* 1979, **76**, 494-498
8. Jones, J. G; et al., *J. Biol. Chem.* 1991, **266**(18), 11705-11713
9. Schmidt-Bleek, F; et al., *Chemosphere* 1982, **11**, 383-415
10. *US Coastguard. Dept. of Transportation. CHRIS – Hazardous Chemical Data* 1984- 1985, **II**, US Govt. Print. Off., Washington, DC

11. Anbar, M; et al. *Int. J. Appl. Radiation Isotopes* 1967, **18**, 493-523
12. Mill, T; et al. *Science* 1980, **270**, 886-887
13. Atkinson, R; et al. *J. Chem. Phys.* 1978, **68**, 1850-1853
14. Patty, F. A. *Industrial Hygiene and Toxicology* 1967, **2**, Interscience Publishers, New York
15. Izmerov, N. F; *Toxicometric Parameters of Industrial Toxic Chemicals Under Single Exposure* 1982, Moscow
16. *The Merck Index* 11th ed. 1989, Merck & Co. Inc, Rahway, NJ
17. Kinney, L. A; et al. *Inhalation Toxicol.* 1990, **2**(1), 29-35
18. Guest, I; et al. *J. Toxicol. Environ. Health* 1991, **32**(3), 319-330
19. Zeisel, S. H; et al. *Cancer Res.* 1986, **46**(12, Pt. 1), 6136-6138
20. Bittersohl, G; et al., *Z. Gesamte Hyg. Ihre Grenzgeb* 1980, **26**(4), 258-259, (*Chem. Abstr.* **95**, 11968j)
21. *ECETOC Technical Report No. 30(4)* 1991, European Chemical Industry Ecology and Toxicology Centre, B-1160 Brussels

A142   1-Amino-2-methoxy-5-methylbenzene

CAS Registry No. 120-71-8
Synonyms 5-methyl-*o*-anisidine; 3-amino-*p*- cresol, methyl ether;
3-amino-4-methoxytoluene; 2-amino-4- methylanisole; 2-methoxy-5-methylaniline;
2-methoxy-5- methylbenzenamine; Azoic Red 36; *p*-cresidine
Mol. Formula $C_8H_{11}NO$ **Mol. Wt.** 137.18
Uses Dyestuff intermediate for cotton, chemical intermediate.

Physical properties

M. Pt. 51.5°C; **B. Pt.** 235°C; **Flash point** >112°C.

Solubility
Organic solvent: ethanol, diethyl ether, benzene, hot petroleum ether

Ecotoxicity

Bioaccumulation
Non-accumulative or low accumulative (1).

Mammalian and avian toxicity

Acute data
LD_{50} oral rat 1450 mg kg^{-1} (2,3).

Carcinogenicity and long-term effects
No adequate data for evidence of carcinogenicity to humans, sufficient evidence for carcinogenicity in experimental animals, IARC classification group 2B (4,5).
Oral ♂ ♀ rat (104 wk) 0.5% or 1% in feed induced adenomas, adenocarcinomas and squamous cell carcinomas of the nasal cavity. Poorly differentiated adenocarcinomas infiltrating the skull and brain were observed in 49% of ♂ animals (6).

Genotoxicity

Salmonella typhimurium TA1538, TA98, TA100 with and without metabolic activation positive (7).

Any other adverse effects

Induced methaemoglobinaemia in rodents indicating that the carcinogen is absorbed and metabolically oxidised (8).

Any other comments

Human health effects and experimental toxicology reviewed (9).
All reasonable efforts have been taken to find information on isomers of this compound, but no relevant data are available.

References

1. *MITI Report* 1984, Ministry of International Trade and Industry, Tokyo
2. *Huntingdon Research Centre Report* (Brocklandville, MD), 1972
3. Lewis, R. J, Sr; et al. (Eds.) *Registry of Toxic Effects of Chemicals* 1979, 116
4. *IARC Monograph* 1982, **27**, 92
5. *IARC Monograph* 1987, **Suppl. 7**, 61
6. Reznik, G; et al. *Anticancer Res.* 1981, **1**(5), 279-286
7. Rosenkranz, H. S; et al. *Contract No I-CP-65855, National Cancer Institute* 1981, 1-10, 871-9
8. Ashby, J; et al. *Mutat. Res.* 1991, **250**(1-2), 115-133
9. *ECETOC Technical Report No.30(4)* 1991, European Chemical Industry Ecology and Toxicology Centre, B-1160 Brussels

A143　3-Amino-1-methyl-5*h*-pyrido[4,3- *b*]indole

CAS Registry No. 62450-07-1
Synonyms 3-amino-1-methyl-γ-carboline; 1-methyl-5*H*-pyrido[4,3-*b*]indol-3-amine; Trp-P-2; tryptophan P2
Mol. Formula $C_{12}H_{11}N_3$　　　　　　　　　　　**Mol. Wt.** 197.24
Occurrence Broiled fish and meat.

Physical properties

M. Pt. 248-250°C.

Solubility

Organic solvent: methanol

Mammalian and avian toxicity

Carcinogenicity and long-term effects

No adequate evidence for carcinogenicity to humans, sufficient evidence for

carcinogenicity to experimental animals, IARC classification group 2B (1).
ICR mice (1 yr) 25 μg g^{-1} induced tumours of lung and liver in σ mice and tumours of the lymphatic system or leukaemia in $\female$ mice (2). Dermal application of 2 mg to $\female$ mice, twice weekly for 5 wk followed by 2.5 μg TPA (CAS RN. 16561-29-8) caused skin squamous cell papillomas and carcinomas, alone did not cause skin tumours, indicating it acts mainly as an initiator rather than a complete carcinogen (3).

Teratogenicity and reproductive effects
SOS Chromotest *Escherichia coli* K12 PG37 and PQ35 with metabolic activation positive (4).
Intraperitoneal administration of 4.2 mg kg^{-1} day^{-1} to mice on day 8-9 pregnancy; mammalian mutagenicity spot test positive (5).

Genotoxicity
Salmonella typhimurium TA98, TA100 with metabolic activation positive (6).
Chinese hamster lung fibroblasts induced diptheria toxin resistant mutations (7).

Any other comments
All reasonable efforts have been taken to find information on isomers of this compound, but no relevant data are available.

References
1. *IARC Monograph* 1987, **Suppl. 7**, 73
2. Fujii, K; *Carcinogenesis (London)* 1991, **12**(8), 1409-1415
3. Takahashi, M; et al. *Jpn. J. Cancer Res.* 1986, **77**(6), 509-513
4. Thybaud-Lambay, V; et al. *Mutat. Res.* 1986, **173**(3), 177-180
5. Jensen, N. J. *Cancer Lett.* 1983, **20**(2), 241-244
6. Nagao, M; et al. *Carcinogenesis (London)* 1980, **1**, 451
7. Tenada, M; et al. (Ed.) *International Conference on Environmental Mutagens*, 1981, Tokyo

A144   3-Aminomethyl-3,5,5-trimethylcyclohexylamine

CAS Registry No. 2855-13-2
Synonyms cyclohexanemethylamine, 5-amino-1,3,3-trimethyl-; isopherone diamine
Mol. Formula $C_{10}H_{22}N_2$ **Mol. Wt.** 170.30
Uses Synthesis of diisocyanate plastics.

Physical properties
B. Pt. 247°C.

Solubility
Organic solvent: hydrocarbons

Occupational exposure

UN No. 2289; HAZCHEM Code 2R; **Conveyance classification** corrosive substance; **Supply classification** corrosive.

Risk phrases

Harmful in contact with skin and if swallowed – Causes burns – May cause sensitisation by skin contact (R21/22, R34, R43)

Safety phrases

In case of contact with eyes, rinse immediately with plenty of water and seek medical advice – Wear suitable protective clothing, gloves and eye/face protection (S26, S36/37/39)

Any other adverse effects to man

A man handling epoxy resins and wood in a shipyard experienced chronic itching fissured dermatitis of the fingertips and palms. Similarly, a man employed in an electronics company using epoxy resins, metals, plastics and varnishes suffered from chronic itching and scaly contact dermatitis on the hands and fingers. Patch tests were positive for 3-aminomethyl-3,4,5-trimethylcyclohexylamine (1).

Any other comments

Physico-chemical properties, human health effects and experimental toxicology reviewed (2).

All reasonable efforts have been taken to find information on isomers of this compound, but no relevant data are available.

References

1. Camarasa, J. G; et al. *Contact Dermatitis* 1989, **20**(5), 382-383
2. *ECETOC Technical Report No. 30(4)* 1991, Eurpoean Chemical Industry Ecology and Toxicology Centre, B-1160 Brussels

A145   1-Aminonaphthalene-4-sulfonic acid

CAS Registry No. 84-86-6
Synonyms 1-naphthylamine-4-sulfonic acid; Piria's acid; 1-amino-4-sulfonaphthalene; 1,4-naphthionic acid; USAF M-5
Mol. Formula $C_{10}H_9NO_3S$ **Mol. Wt.** 223.25
Uses The sodium salt is an important dyestuff intermediate in the manufacture of Congo red, Fast Red A, azo rubine and similar azo dyestuffs.

Physical properties

M. Pt. Decomposes on heating without melting; **Specific gravity** d_4^{25} 1.673.

Solubility
Water: 0.31 g l^{-1} at 20°C.

Environmental fate

Degradation studies
Nonbiodegradable/qualified (1).

Mammalian and avian toxicity

Acute data
LD$_{50}$ intraperitoneal mouse 300 mg kg^{-1} (2).

Carcinogenicity and long-term effects
Tested in inbred A/St ♂ and ♀ mice by the pulmonary adenoma bioassay. No tumourigenic activity observed (3).

Metabolism and pharmacokinetics
Identified as a metabolite in the faeces of rat, mouse and guinea pig following administration of carmoisine, amaranth and Brown HT (4).

Any other comments

The sodium salt is an haemostatic agent (5).

References

1. *MITI Report* 1984, Ministry of International Trade and Industry, Tokyo
2. *NTIS Report* AD 277-689, National Technical Information Service, Springfield, VA
3. Theiss, J. C; et al. *J. Natl. Cancer Inst.* 1981, **67**(6), 1299-1302
4. Phillips, J. C; et al., *Food Chem. Toxicol.* 1987, **25**(12), 927-925, 947-954, 1013-1019
5. *The Merck Index* 11th ed. 1989, Merck & Co. Inc, Rahway, NJ

A146   2-Aminonaphthalene-1-sulfonic acid

CAS Registry No. 81-16-3
Synonyms 2-amino-1-naphthalenesulfonic acid; 2-naphthylamine-1-sulfonic acid; Tobias acid
Mol. Formula $C_{10}H_9NO_3S$ **Mol. Wt.** 223.25
Uses Dyestuff intermediate.

Physical properties

M. Pt. 180°C (diethylammonium salt).

Solubility
Organic solvent: ethanol, diethyl ether

Environmental fate

Degradation studies

Pseudomonas sp. TA-2 degraded 2- aminonaphthalene-1-sulfonic acid to produce intermediate metabolites 1-naphthalenesulfonate and 2-naphthol-1- sulfonate, and the release of ammonia, sulfate and sulfite. It is suggested that degradation occurrs as a result of 1,2- dioxygenation in the initial process (1).

In an aqueous screening test, 95-97% loss was observed in 108-118 days at a concentration of 1 mg l^{-1} (2).

Abiotic removal

Reported as generally resistant to hydrolysis (3).

Atmospheric $t_{1/2}$ 2 hr based on photochemically produced hydroxyl radicals (4).

Absorption

High mobility of 2-aminonaphthalene-1-sulfonic acid in soil is suggested (5).

Mammalian and avian toxicity

Acute data

LD_{50} oral rat 19.4 g kg^{-1} (6).

Carcinogenicity and long-term effects

Oral BALB/c mice (66 wk) 2500, 5000 and 10,000 ppm in diet. The animals were kept under observation until 140 wk of age when the expt. was terminated. No tumour type at any site was related to treatment (7).

Metabolism and pharmacokinetics

Intravenous and oral administration of 1 mg kg^{-1} to rats resulted in almost exclusive urine elimination and equal elimination in urine and faeces, respectively. There was significant absorption from the gastrointestinal tract (8).

Irritancy

500 mg instilled into rabbit eye for 24 hr caused mild irritation (6).

Genotoxicity

Salmonella typhimurium TA98, TA100 with and without metabolic activation negative (9).

Any other adverse effects to man

The lymphocytotoxicity of workers exposed to 2-naphthylamine-1-sulfonic acid was investigated. Workers using tobias acid have lymphocytes with a normal range of reactivity towards bladder cancer cells, which is in keeping with the suggestion that tobias acid is noncarcinogenic (10).

References

1. Ohe, T; et al. *Agric. Biol. Chem.* 1990, **54**(3), 669-675
2. Bosch, F. M; et al. *Doc. Eur. Sewage Refuse Symp. Eas. 4th* 1978, 272-286
3. Lyman, W. J; et al. *Handbook of Chemical Property Estimation Methods* 1982, McGraw-Hill, New York
4. Atkinson, R. *Intern. J. Chem. Kin.* 1987, **19**, 799-828
5. Swann, R. L; et al. *Res. Rev.* 1983, **85**, 16-28
6. Marhold, J. V. *Sbovnik Vystedku Toxixologickeho Vysetreni Latek A Pripravku* 1972, 187, Prague

7. Dela Porta, G; et al. *Carcinogenesis (London)* 1982, **3**(6), 647-649
8. Marchisio, M. A; et al. *Br. J. Ind. Med.* 1976, **33**(4), 269-271
9. Pogodina, O. N; et al. *Eksp. Onkol.* 1984, **6**(4), 23-25, (Russ.) (*Chem. Abstr.* **101**, 185749w)
10. Kumar, S; et al. *Br. J. Ind. Med.* 1981, **38**(2), 167-169

A147   5-Aminonaphthalene-2-sulfonic acid

CAS Registry No. 119-79-9
Synonyms 5-amino-2-naphthalenesulfonic acid; 1-naphthylamine-6-sulfonic acid;
1,6-Cleve's acid
Mol. Formula $C_{10}H_9NO_3S$ **Mol. Wt.** 223.25
Uses Dyestuff intermediate.

Physical properties

M. Pt. 180-190°C.

Solubility
Water: 0.1 g in 100 ml.

Environmental fate

Degradation studies
Pseudomonas sp. BN6 degraded 5-amino-2-naphthalenesulfonic acid (1).

Mammalian and avian toxicity

Acute data
LD_{50} oral rat 14200 mg kg^{-1} (2).

Irritancy
500 mg instilled in rabbit eye for 24 hr caused mild irritant effects (2).

References

1. Noertemann, B; et al. *Appl. Environ. Microbiol.* 1986, **52**(5), 1195-1202
2. Le, J; et al. *Food. Chem. Toxicol.* 1985, **23**, 695

A148   6-Aminonaphthalene-2-sulfonic acid

CAS Registry No. 93-00-5
Synonyms 6-amino-2-naphthalenesulfonic acid; 2-naphthylamine-6-sulfonic acid;
Broenners acid
Mol. Formula $C_{10}H_9NO_3S$ **Mol. Wt.** 223.25
Uses Manufacture of azo dyestuffs, e.g. CI Direct Red 4.

Environmental fate

Degradation studies
The degradation of 6-aminonaphthalene-2- sulfonic acid is inhibited by simultaneous
oxidation of α-aminonaphthalene-2-sulfonic acid which in submerged cultures leads
to accumulation of inhibiting fermentation products (1).
Degradation by a mixed bacterial community in river water induced the action by
Pseudomonas BN6 converting 6-aminonaphthalene-2- sulfonic acid into
5-aminosalicylate (2).

References

1. Hattendorf, C; et al. *DECHEMA Biotechnol. Conf. 7 1990*, **4**(4), 581-584
2. Noertemann, B; et al. *Appl. Environ. Microbiol.*, 1986, **52**(5), 1195-1202

A149   7-Aminonaphthalene-2-sulfonic acid

CAS Registry No. 494-44-0
Synonyms 7-amino-2-naphthalenesulfonic acid; 2-naphthylamine-7-sulfonic acid;
β-naphthylamine-8-sulfonic acid; Cassella's acid F; Bayer's acid; Amido-F acid
Mol. Formula $C_{10}H_9NO_3S$ **Mol. Wt.** 223.25
Uses Manufacture of azo dyestuffs.

Solubility
Water: 0.3 g in 100 g at 100°C.

Any other comments

There is evidence of low chronic toxicity and low carcinogenic potential in these
sulfonated derivatives (1).

References

1. Anliker, R. in *Toxic Hazard Assessment of Chemicals* 1986, M. L. Richardson (Ed.), RSC, London.

A150   8-Aminonaphthalene-2-sulfonic acid

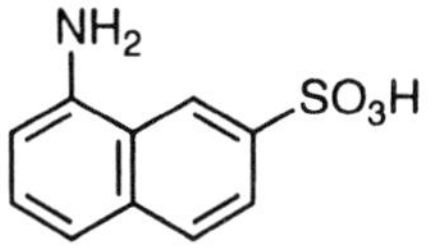

CAS Registry No. 119-28-8

Synonyms 8-amino-2-naphthalenesulfonic acid; 1-naphthylamine-7-sulfonic acid; Cleve's acid 1,7

Mol. Formula $C_{10}H_9NO_3S$ **Mol. Wt.** 223.25

Uses Manufacture of azo dyestuffs, e.g. CI Direct Green 51.

Physical properties

M. Pt. >300°C.

Solubility

Water: 0.45 g in 100 g. Organic solvent: ethanol

Any other comments

The sulfonated aromatic amines of this family appear to have low chronic toxicity and low carcinogenic potential (1).

Reference

1. Anliker, R. in *Toxic Hazard Assessment of Chemicals*, M. L. Richardson (Ed.), 1986, 166-187, RSC, London.

A151   2-Amino-5-(5-nitro-2-furyl)-1,3,4-thiadiazole

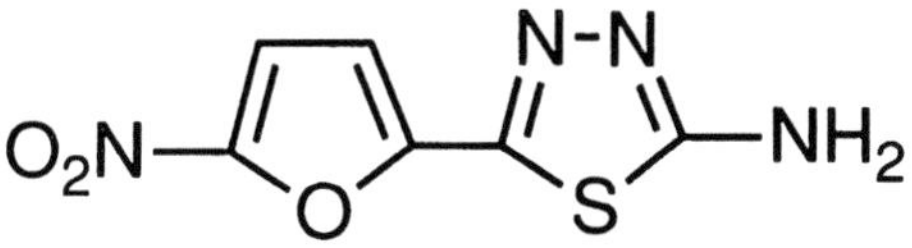

CAS Registry No. 712-68-5

Synonyms 2-(5-nitro-2-furyl)-5-amino-1,3,4- thiadiazole;
5-amino-2-(5-nitro-2-furyl)-1,3,4-thiadiazole;
5-(5-nitro-2-furanyl)-1,3,4-thiadiazol-2-amine; 5-(5-nitro-
2-furyl)-2-amino-1,3,4-thiadiazole; furidiazine; Triafu r

Mol. Formula $C_6H_4N_4O_3S$ **Mol. Wt.** 212.19

Uses Used as antimicrobial agent in human and veterinary medicine, as conservation chemical, and as ingredient in cosmetic preparations and synthetic textile materials. As stimulator of livestock production.

Physical properties

M. Pt. 280°C.

Solubility

Organic solvent: acetic acid, dimethyl formamide

Mammalian and avian toxicity

Sub-acute data

TD_{Lo} (46 wk intermittent) oral rat 6 g kg^{-1} (1).

TD_{Lo} (32 wk continuously) oral rat 2240 mg kg^{-1} (2).

Carcinogenicity and long-term effects

No adequate evidence for carcinogenicity to humans, sufficient evidence for carcinogenicity to animals, IARC classification 2B (3).

Sprague Dawley ♀ rats (75 wk) 4 g (cumulative dose rat^{-1}) via feed induced tumours of mammary glands, fibroadenomas and adenocarcinomas, tumours of forestomach, lung and kidney were also observed. Initial dose levels were reduced due to growth retardation (4).

Genotoxicity

Salmonella typhimurium TA100 with and without metabolic activation positive.
Escherichia coli WP2 or WP2uvrA with and without metabolic activation negative.
DNA repair test positive (5).

Any other comments

Toxic to *Salmonella, Shigella, Escherichia coli* and *Staphlococcus aureus* (6).
Human health effects and experimental toxicology reviewed (7).
All reasonable efforts have been taken to find relevant information on isomers of this compound, but no data are available.

References

1. Cohen, S. M; et al. *J. Natl. Cancer Inst.* 1975, **54**(4), 841-850
2. Erturk, E; et al. *Fed. Proc.* 1970, **29**, 817
3. *IARC Monograph* 1987, **Suppl. 7**, 57
4. Cohen, S. M; et al. *J. Natl. Cancer Inst.* 1975, **54**(4), 841-850
5. Ebringer, L; et al. *Acta Fac. Rerum. Nat. Univ. Comenianae, Microbiol.* 1984, **11**, 49-56 (*Chem. Abstr.* **101**, 224470v)
6. Skagius, K. *Antibiotics and Chemotherapy* 1961, **10**, 37-45
7. *ECETOC Technical Report No. 30(4)* 1991, European Chemical Industry Ecology and Toxicology Centre, B-1160 Brussels

A152   2-Amino-4-nitrophenol

CAS Registry No. 99-57-0

Synonyms 3-amino-4-hydroxynitrobenzene; *p*-nitro-*o*-aminophenol;
2-hydroxy-5-nitroaniline

Mol. Formula $C_6H_6N_2O_3$ **Mol. Wt.** 154.13

Uses In dyestuff synthesis. Organic synthesis intermediate.

Physical properties
M. Pt. 143-145°C.

Solubility
Organic solvent: ethanol, diethyl ether, methanol, acetic acid, hot benzene

Mammalian and avian toxicity

Acute data
LD_{50} oral rat 1030 mg kg^{-1} (1).
LD_{50} intraperitoneal rat 246 mg kg^{-1} (2).

Carcinogenicity and long-term effects
Gavage F344/N rats and B6C3F1 mice (2 yrs) 0, 125 or 250 mg kg^{-1}, 5 day wk^{-1} increased incidence of renal cortical ademonas in ♂ rats. No evidence of carcinogenic activity in ♀ rats or ♂ and ♀ mice (3).

Irritancy
100 mg instilled into rabbit eye caused moderate irritation in 24 hr (1).

Genotoxicity
Salmonella typhimurium TA98, TA100 with metabolic activation positive (4).
Bacteriophage T4D 22.7 μg ml^{-1} induced rapid lysis mutants (5).
In vitro Chinese hamster ovary cells with and without metabolic activation positive for chromosomal aberration and sister chromatid exchange (6).
In vitro L5178Y mouse lymphoma cells without metabolic activation positive (7).

References
1. Marhold, J. V; *Sbvornik Vysledku Toxicologickeho Vysetreni Latek A Pripravku* 1972, 107, Prague
2. Bennett, C; et al. *J. Toxicol. Environ. Health* 1977, **2**(3), 657-662
3. *National Toxicology Program Research & Testing Division* 1988, Report No. TR-339, NIEHS, Research Triangle Park, NC 27709
4. Zeiger, E; et al., *Environ. Mol. Mutagen.* 1987, **9**(Suppl. 9), 1-110
5. Kuelland, I. *Hereditas* 1985, **102**(1), 151-154
6. Anderson, B. E; et al. *Environ. Mol. Mutagen.* 1990, **16**(Suppl. 18), 55-137
7. Myhr, B; et al. *Environ. Mol. Mutagen.* 1990, **16**(Suppl. 18), 138-167

A153   2-Amino-5-nitrophenol

CAS Registry No. 121-88-0
Synonyms 2-hydroxy-4-nitroaniline; 4-amino-3-hydroxynitrobenzene; 5-nitro-2-aminophenol

Mol. Formula $C_6H_6N_2O_3$ **Mol. Wt.** 154.13

Uses Organic synthesis intermediate. Ingredient in hair dyes.

Physical properties

M. Pt. 207-208°C.

Solubility

Organic solvent: ethanol, diethyl ether, benzene

Mammalian and avian toxicity

Carcinogenicity and long-term effects

Gavage F344/N rats and B6C3F1 mice (2 yr) 100- 200 mg kg^{-1} 5 day week^{-1}.
Increased incidence of adenomas of the pancreas in ♂ rats given 100 mg kg^{-1}.
Marginally increased incidences of preputial or clitoral gland adenomas or
carcinomas in ♂ and ♀ rats administerd 200 mg kg^{-1}. A 13 wk study of mice given
400 mg kg^{-1} showed no evidence of carcinogenic activity. Reduced survival in a 2 wk
study of mice given 800 mg kg^{-1} (1).

Genotoxicity

Salmonella typhimurium TA98, TA100, TA1537 with and without metabolic
activation positive (2).
In vitro L5178Y mouse lymphoma cells without metabolic activation positive (3).

Any other comments

Non-toxic to rabbits by percutaneous administration of hair dyes containing
2-amino-5-nitrophenol (4).

References

1. *National Toxicology Program Research & Testing Division* 1992, Report No. TR-334,
    NIEHS, Research Triangle Park, NC 27709
2. Zeiger, E; et al., *Environ. Mol. Mutagen.* 1987, **9**(9), 1-110
3. Myhr, B; et al. *Environ. Mol. Mutagen.* 1990, **16**(Suppl. 18), 138-167
4. Burnett, C; et al. *J. Toxicol. Environ. Health* 1976, **1**(6), 1027-1040

A154   4-Amino-2-nitrophenol

CAS Registry No. 119-34-6

Synonyms 4-hydroxy-3-nitroaniline; *o*-nitro-*p*-aminophenol; 2-nitro-4-aminophenol;
Oxidation base 25

Mol. Formula $C_6H_6N_2O_3$ **Mol. Wt.** 154.13

Physical properties

M. Pt. 131°C.

Solubility
Organic solvent: diethyl ether, ethanol, acetone

Ecotoxicity

Fish toxicity
LC_{50} (96 hr) fathead minnow 34 mg l^{-1} (1).

Invertebrate toxicity
ED_{50} (5-30 min) *Photobacterium phosphoreum* 38.7 mg l^{-1} Microtox test (2).

Mammalian and avian toxicity

Acute data
LD_{50} oral rat, mouse 1470 mg kg^{-1} (3).
LD_{50} intraperitoneal rat 302 mg kg^{-1} (4).

Carcinogenicity and long-term effects
No adequate evidence for carcinogenicity to humans, limited evidence for carcinogenicity to animals, IARC classification group 3 (5).
National Toxicology Program evaluation of 4-amino-2-nitrophenol in rats and mice in feed. ♂ rat positive; ♂ and ♀ mice negative (6).

Irritancy
100 mg instilled into rabbit eye caused severe irritation in 24 hr (7).

Genotoxicity
Salmonella typhimurium TA98 without metabolic activation positive (8).
Salmonella typhimurium TA1535, TA100, TA1537, TA1538, TA98 with or without metabolic activation negative (9).

Any other comments
Human health effects, epidemiology, workplace occupation and experimental toxicology reviewed (10).
All reasonable efforts have been taken to find information on isomers of this compound, but no relevant data are available.

References
1. Holcombe, G. W; et al. *Environ. Pollut. Ser. A.* 1983, **35**(4), 367-381
2. Kaiser, K. L. E; et al. *Water Poll. Res. J. Canada* 1991, **26**(3), 361-431
3. *Natl. Cancer Inst. Progress Report* 1973, Contract No. NIH-NCI-E-C-72-3252
4. Burnett, C; et al. *J. Toxicol. Environ. Health* 1977, **2**(3), 657-662
5. *IARC Monograph*, 1987, **Suppl. 7**, 57
6. *National Toxicology Program Research and Testing Div.* 1992, Report No. TR-094, NIEHS, Research Triangle Park, NC27709
7. Marhold, J. V; *Sbornik Vysledku Toxixologickeho Vysetreni Latek A Pripravku* 1972, 107, Prague
8. Zeiger, E; et al. *Environ. Mol. Mutagen.* 1987, **9**(9), 1-110
9. Shahin, M. M. *Int. J. Cosmet. Sci.* 1985, **7**(6), 277-289
10. *ECETOC Technical Report No. 30(4)* 1991, European Chemical Industry Ecology and Toxicology Centre, B-1160 Brussels

A155   2-Amino-5-nitrothiazole

CAS Registry No. 121-66-4

Synonyms aminonitrothiazole; aminonitrothiazolium; 5-nitro-2-aminothiazole; 5-nitro-2-thiazolylamine; enheptin

Mol. Formula $C_4H_4N_3O_2S$ **Mol. Wt.** 158.16

Uses Antiprotozoal agent.

Physical properties

M. Pt. 195-196°C.

Solubility
Organic solvent: ethanol, diethyl ether

Mammalian and avian toxicity

Acute data
LD_{50} intraperitoneal mouse 200 mg kg^{-1} (1).

Sub-acute data
TD_{Lo} (10 day) oral ♂ rat 600 mg kg^{-1} caused temporary sterility and decreased prostate weight (2).

Carcinogenicity and long-term effects
No adequate evidence for carcinogenicity to humans, limited evidence for carcinogenicity to experimental animals, IARC classification group 3 (3).

Genotoxicity

Salmonella typhimurium TA98, TA100, TA1535, TA1537, TA1538 with and without metabolic activation positive (4).
Escherichia coli PQ37 SOS chromotest without metabolic activation positive (5).
Mouse lymphoma L5178Y tk+/tk- without metabolic activation positive (6).

Any other comments

Experimental toxicology and human health effects reviewed (7).
All reasonable efforts have been taken to find information on isomers of this compound, but no relevant data are available.

References

1. Sax, N. I; et al. *Dangerous Properties of Industrial Materials*, 7th ed., 1989, Van Nostrand Reinhold, New York
2. Snair, D. W; et al. *Toxicol. Appl. Pharmacol.* 1960, **2**, 418-429
3. *IARC Monograph* 1987, **Suppl. 7**, 57
4. Ashby, J; et al. *Mutat. Res.* 1988, **204** (1), 17-115
5. Van der Hude, W; et al. *Mutat. Res.* 1988, **203** (2), 81-94
6. Caspary, W. J; et al. *Environ. Mol. Mutagen.* 1988, **12**(Suppl. 13), 19-36
7. *ECETOC Technical Report No. 30(4)* 1991 European Chemical Industry Ecology and Toxicology, B-1160 Brussels

A156   4-Aminophenazone

CAS Registry No. 58-15-1

Synonyms aminopyrine; 4-dimethylamino-2,3-dimethyl-1- phenyl-3-pyrazolin-5-one; dimethylaminophenyldimethylpyrazdone; 4- dimethylaminoantipyrine

Mol. Formula $C_{13}H_{17}N_3O$ **Mol. Wt.** 231.30

Uses Antipyretic, analgesic.

Physical properties

M. Pt. 107-109°C.

Solubility
Organic solvent: ethanol, benzene, chloroform, diethyl ether

Ecotoxicity

Effects to non-target species
LD_{Lo} parenteral frog 900 mg kg^{-1} (1).

Mammalian and avian toxicity

Acute data
LD_{50} oral rat 1.7 g kg^{-1} (2).
LD_{50} oral rabbit 600 mg kg^{-1} (3).

Carcinogenicity and long-term effects
Syrian golden hamsters (8-10 wk) combined administration of 0.1% nitrite and 0.1% aminopyrine in drinking water resulted in subsequent development of hepatocellular nodules and cholangiofibrotic lesions and cholangiocellular carcinomas (4).
B6C3F1 mice (104 wk) 0, 0.04, 0.08% in drinking water no significant incidence in tumour frequency between treated and control groups (5).

Teratogenicity and reproductive effects
Subcutaneous injection C57BL/6N and DBA/2N mice (7, 8 and 9 day gestation) 200 mg kg^{-1} induced malformations, including omphalocele, club foot and kinky tail (6).

Metabolism and pharmacokinetics
Injection rats (concentration unspecified), after 30 min uniform distribution was detected throughout the body, preferential localisation was found in the nasal mucosa and liver (7).

Sensitisation
A life-threatening asthmatic attack has been reported due to administration of 4-aminophenazone (8).

Genotoxicity

Salmonella typhimurium TA100, TA98, TA97, TA102, TA104 with and without metabolic activation negative (9).

Any other adverse effects to man

Agranulocytosis has been reported in humans after administration of 4-aminophenazone (10,11).

The Boston Collaborative Drug Surveillance Program monitored consecutively 32,812 medical inpatients. Drug-induced anaphylaxis occurred in 1 out of 1992 patients given 4- aminophenazone (12).

Implicated as a causative agent in immune haemolytic anaemia (13).

Any other comments

Metabolism, pharmacokinetics and human health effects reviewed (14-16).

References

1. *Naunyn-Schmedeberg's Arch. Exp. Path. Pharm.* 1932, **166**, 437
2. Hart, J. *J. Pharm. Exp. Ther.* 1947, **89**, 205
3. Oyo, Y. *Pharmacometrics* 1978, **16**, 1011
4. Thamavit, W; et al. *Jpn. J. Cancer Res.* 1988, **79**(8), 909-916
5. Inai, K; et al. *Jpn. J. Cancer Res.* 1990, **81**(2), 122-128
6. Takeno, S; et al. *Res. Commun. Chem. Pathol. Pharmacol.* 1987, **57**(3), 409-419
7. Brittebo, E. B; *Acta Pharmacol. Toxicol.* 1982, **51**(3), 227
8. Bartoli, E; et al. *Lancet* 1976, **1**, 1357
9. Parisis, D. M; et al. *Mutat. Res.* 1988, **206**(3), 317-326
10. Urbach, E; et al. *J. Am. Med. Assoc.* 1946, **131**, 893
11. *Martindale: The Extra Pharmacopoeia* 29th ed., 1989, The Pharmaceutical Press, London
12. Porter, J; et al. *Lancet* 1977, **1**, 587
13. Beutler, E; *Pharmac. Rev.* 1969, **21**, 73
14. Baba, S; *Radioisotopes* 1986, **35**(4), 201-205
15. Roth, H. J; *Agents Actions Suppl.* 1986, **19**, 205-221
16. Forth, W; *Agents Actions Suppl.* 1986, **19**, 169-175

A157   2-Aminophenol

CAS Registry No. 95-55-6

Synonyms 2-amino-1-hydroxybenzene; 2-hydroxyaniline; *o*-aminophenol

Mol. Formula C_6H_7NO **Mol. Wt.** 109.13

Uses Manufacture of azo- and sulfur dyestuffs. Used in the dyeing of fur and hair.

Physical properties

M. Pt. 170-174°C; **Partition coefficient** log P_{ow} 0.52-0.62.

Solubility

Water: 17 g l^{-1} at 0°C. Organic solvent: diethyl ether, ethanol

Occupational exposure

UN No. 2512; **HAZCHEM Code** 2X; **Conveyance classification** harmful substance; **Supply classification** harmful.

Risk phrases
Harmful by inhalation, in contact with skin and if swallowed (R20/21/22)

Safety phrases
After contact with skin, wash immediately with plenty of water (S28)

Ecotoxicity

Fish toxicity
LD_{100} (48 hr) goldfish 20 mg l^{-1} (1).

Invertebrate toxicity
Cell multiplication inhibition test *Chlorella pyrenoidosa* 47 mg l^{-1} (2).
EC_{50} (5 and 30 min) *Photobacterium phosphoreum* 134 mg l^{-1} Microtox test (3).

Environmental fate

Carbonaceous inhibition
Adapted activated sludge at 20 °C used product as sole carbon source 21.1 mg COD g^{-1} dry innoculum hr^{-1} (4).

Anaerobic effects
Concentrations of 50 mg C^{-1} l^{-1} >80% degradation occurred using anaerobic digesting sludge (5).

Degradation studies
Biodegradable (6).
95% COD, 21 mg COD g^{-1} dry innoculum l^{-1} (7).
Reported decomposition period by soil microflora was 4 days (8).

Mammalian and avian toxicity

Acute data
LD_{50} oral quail 320 mg kg^{-1} (9).
LD_{50} oral mouse 1250 mg kg^{-1} (10).
LD_{50} intraperitoneal mouse 200 mg kg^{-1} (10).
LD_{50} subcutaneous cat 37 mg kg^{-1} (11).

Teratogenicity and reproductive effects
TD_{Lo} (8 day pregnant) intraperitoneal hamster 150 mg kg^{-1} caused body wall and musculoskeletal system abnormalities (12).

Irritancy
100 mg instilled in rabbit eye caused mild irritant effects (13).
The position of the substituted amino or hydroxyl group influences the potency of primary skin irritation, sensitivity and cytotoxicity. The *o*- position was marginally less potent than the *p*isomer (14).

Genotoxicity

Salmonella typhimurium TA98 without metabolic activation positive (15,16).
Vicia faba induced chromosome aberrations and sister chromatid exchanges positive (17).

Any other adverse effects

May produce dermatitis, methaemoglobinia, bronchial asthma and restlessness (18).

Any other comments

Experimental toxicology and human health effects reviewed (19).

References

1. McKee J. E; et al. *Water Quality Criteria* 1963, California State Water Board
2. Jones, H. R. *Environmental Control in Organic and Petrochemical Industries* 1971, Noyes Data Corp., Park Ridge, NJ
3. Kaiser, K. L. E; et al. *Water Pollut. Res. J. Canada* 1991, **26**(3), 361-431
4. Alexander, M; et al *J Agric. Food Chem.* 1966, **14**, 410
5. Battersby, N. S; et al. *Appl. Environ. Microbiol.* 1989, **55**(2), 433-439
6. *MITI Report* 1984, Ministry of International Trade and Industry, Tokyo
7. Pitter, P. *Water Res.* 1976, **10**, 231
8. *NTIS Report* AD 691-490, National Technical Information Service, Springfield, VA
9. *Arch. Environ. Contam. Toxicol.* 1983, **12**, 355
10. Vasilenko, N. M; et al. *Gig. Tr. Prof. Zabol.* 1981, **25**(8), 50, (Russ.) (*Chem. Abstr.* **95**, 163298)
11. Yakkyoka, K. *Pharmacy* 1981, **32**, 1093
12. Putkowski, J. V; et al *Toxicol. Appl. Pharmacol.* 1982, **63**, 264
13. Updyke, D. C. J. *Food Chem. Toxicol.* 1982, **20**, 573
14. Masamoto, Y; et al. *Shinshu Igaku Zasshi* 1987, **35**(2), 185-193, (Jap.) (*Chem. Abstr.* **107**, 128652j)
15. *Environ. Mutagen.* 1983, **5**(Suppl 1), 3
16. De Flora, S; et al. *Mutat. Res* 1984 **133** 161
17. Kanaya, N; *Mutat. Res.* 1990, **245**(4), 311-315
18. Plunckell, E. R. *Handbook of Industrial Toxicology* 1976, Chemical Publishing Co. Inc., New York
19. *ECETOC Technical Report No. 30(4)* 1991, European Chemical Industry Ecology and Toxicology Centre, B-1160 Brussels

A158   3-Aminophenol

CAS Registry No. 591-27-5
Synonyms 3-amino-1-hydroxybenzene; 3-hydroxyaniline; *m*-aminophenol
Mol. Formula C_6H_7NO **Mol. Wt.** 109.13
Uses Dyestuff intermediate.

Physical properties

M. Pt. 122-123°C; **Partition coefficient** log P_{ow} 0.15-0.17.

Solubility
Water: 26 g l^{-1}. Organic solvent: ethanol, diethyl ether, amyl alcohol

Occupational exposure

UN No. 2512; **HAZCHEM Code** 2X; **Conveyance classification** harmful substance; **Supply classification** harmful.

Risk phrases
Harmful by inhalation, in contact with skin and if swallowed (R20/21/22)

Safety phrases
After contact with skin, wash immediately with plenty of water (S28)

Ecotoxicity

Invertebrate toxicity
Cell multiplication inhibition test *Chlorella pyrenoidosa* 100 mg l^{-1} (1).

Bioaccumulation
Non-accumulative or low accumulative (2).

Effects to non-target species
LD_{50} oral redwing blackbird, starling 240 mg kg^{-1} (3).

Environmental fate

Carbonaceous inhibition
Inhibition of glucose degradation *Pseudomonas fluorescens* 0.6 mg l^{-1} (4).

Degradation studies
Biodegradable (2).
Arthrobacter sp. mA3 was capable of utilising 3-aminophenol as sole source of carbon and nitrogen (5).
Adapted activated sludge 20°C as sole carbon source 90% COD removal (6).
Reported decomposition by soil microflora >64 days (7).

Mammalian and avian toxicity

Acute data
LD_{50} dermal rat 1000 mg kg^{-1} (8).
LD_{50} intraperitoneal mouse 150 mg kg^{-1} (9).

Irritancy
Dermal rabbit (24 hr) 12.5 mg caused mild irritant effects (8).

Sensitisation
Guinea pig maximisation test (GPMT) positive sensitiser (10).
CBA/Ca mice 25 μl^{-1} of a 5% solution applied to the dorsum of both ears for 3 consecutive days. 3-6 days after initial topical application, mice were challenged, results confirm positive sensitisation potential (11).

Genotoxicity

Salmonella typhimurium TA97, TA98, TA100, TA1535, TA1537 with and without metabolic activation positive (12).
In vivo mouse bone marrow (4 month study) increased the rate of chromosomal aberrations at toxic concentrations (13).
Vicia faba sister chromatid exchange negative, chromosome aberrations negative (14).
Neurospora crassa 220 mg l^{-1} induced sex chromosome loss and nondisjunction (15).

Any other adverse effects

ICR mice (concentration and duration unspecified) blood GSH levels decreased but no difference in tissue GSH observed. Slight nephrotic effect observed (16).

Any other comments

Experimental toxicology and human health effects reviewed (17).

References

1. Jones, H. R. *Environmental Control in Organic and Petrochemical Industries* 1971, Noyes Data Corp., Park Ridge, NJ
2. *MITI Report* 1984, Ministry of International Trade and Industry, Tokyo
3. *Arch. Environ. Contam. Toxicol.* 1983, **12**, 355
4. Bringmann, G; et al. *GWF-Wasser/Abwasser* 1960, **81**, 337
5. Lechner, U; et al. *J. Basic Microbiol.* 1988, **28**(9-10), 629-637
6. Pitter, P. *Water Res.* 1976, **10**, 231
7. Alexander, M; et al. *J. Agric. Food Chem.* 1966, **14**, 410
8. *Food Cosmet. Toxicol.* 1977, **15**, 607
9. *NTIS Report* AD 691-490, National Technical Information Service, Springfield, VA
10. Magnusson, B; et al. *J. Invert. Dermatol.* 1969, **52**, 268-276
11. Basketter, D. A; et al. *Toxicology Methods* 1991, **1**(1), 30-43
12. Zeiger, E; et al. *Environ. Mol. Mutagen.* 1988, **11**(Suppl. 12), 1-157
13. Babayan, E. A; et al. *Biol. Zn. Arm.* 1987, **40**(1), 62-67, (Russ.) (*Chem. Abstr.* **107**, 2398n)
14. Kanaya, N; *Mutat. Res.* 1990, **245**(4), 311-315
15. *Mutat. Res.* 1986, **167**, 35
16. Ito, I. *Tokyo Joshi Ika Daigaku Zasshi* 1987, **57**(12), 1655-1666, (*Chem. Abstr.* **108**, 181763)
17. *ECETOC Technical Report No. 30(4)* 1991, European Chemical Industry Ecology and Toxicology Centre, B-1160 Brussels

A159   4-Aminophenol

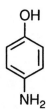

CAS Registry No. 123-30-8

Synonyms 4-amino-1-hydroxybenzene; *p*-hydroxyaniline; azol; paranol; Unal

Mol. Formula C_6H_7NO　　　　　　　　　　**Mol. Wt.** 109.13

Uses Photography. Intermediate in azo and sulfur dyestuff manufacture.

Physical properties

M. Pt. 189-190°C; **B. Pt.** 284°C (decomp.); **Partition coefficient** log P_{ow} 0.04.

Solubility

Water: 11 g l^{-1} at 0°C. Organic solvent: methyl ethyl ketone

---

Occupational exposure

UN No. 2512; **HAZCHEM Code** 2X; **Conveyance classification** harmful substance; **Supply classification** harmful.

Risk phrases

Harmful by inhalation, in contact with skin and if swallowed (R20/21/22)

Safety phrases

After contact with skin, wash immediately with plenty of water (S28)

Ecotoxicity

Fish toxicity

LC_{50} (48 hr) goldfish 2 mg l^{-1} (1).
Exposure of trout, bluegill sunfish and goldfish to 5 ppm caused death in 22 hr (2).
LC_{50} (96 hr) fathead minnow 24 mg l^{-1} static bioassay at 18-22°C (3).

Invertebrate toxicity

Cell multiplication inhibition test *Chlorella pyrenoidosa* 140 mg l^{-1}, *Scenedesmus quadricauda* 6 mg l^{-1} (4).
EC_{50} (5 and 30 min) *Photobacterium phosphoreum* 3.3 mg l^{-1} Microtox test (5).

Effects to non-target species

LD_{50} oral redwing blackbird, starling 56 mg $^{-1}$ (6).

Environmental fate

Degradation studies

Adapted activated sludge 87% COD (7).
Nostoc linckia and *Nostoc muscorum* >2 μg ml^{-1} decreased cell numbers, chlorophyll and total carbohydrate and inhibited carbon dioxide uptake, and nitrate reductase and nitrogenase (8).

Mammalian and avian toxicity

Acute data

LD_{50} oral rat, mouse 375-420 mg kg^{-1} (9,10).

Sub-acute data

Intraperitoneal mice (5 day) 500 mg kg^{-1} caused sperm morphology effects (11).

Metabolism and pharmacokinetics

In the presence of oxyhaemoglobin 4- aminophenol forms numerous adducts with GSH and is converted to thioethers within the erythrocytes (12).

Irritancy

100 mg instilled in rabbit eye caused mild effects (duration unspecified) (9).
The position of the substituted amino or hydroxyl groups in aminophenol influences the potency of primary skin irritation, delayed contact sensitivity and cell toxicity. The *p*-isomer exhibited the greatest irritation, sensitivity and cytotoxicity in guinea pigs (13).

Genotoxicity

Salmonella typhimurium TA97, TA98, TA100, TA1535, TA1537 with and without metabolic activation negative (14).
In vitro mouse lymphoma with metabolic activation positive (15).

Chinese hamster ovary hypoxanthine-guanosine phosphoribosyl transferase assay with and without metabolic activation negative (16).

In vivo mouse hepatocytes increased the incidence of micronuclei (17).

Drosphila melanogaster wing somatic mutation and recombination test positive (18).

Vicia faba chromosome aberrations and sister chromatid exchange positive (19).

Any other adverse effects

In a study of mutagenicity, teratology, haemotology and histopathological changes in rats dosed with 0.07, 0.2 or 0.7% in diet for 6 months, no significant haemotological changes observed, nephrosis was observed in high dose animals. The compound was considered non- teratogenic although an increase in developmental variations associated with maternal toxicity was noted at the mid- and high-dose levels. Mutagenic activity was not detected in the urine (20).

Intraperitoneal rat (concentration and duration unspecified) caused necrosis of renal tubular epthelial cells and elevated urinary N-acetyl- β-D-glucosaminidase and γ-glutamyltranspeptidase activities (21).

Any other comments

Experimental toxicology and human health effects reviewed (22).

References

1. McKee, J. E.; et al. *Watwr Quality Criteria* 1963, California
2. Wood, E. M. *Toxicity of 3400 Chemicals to Fish* 1987, US Fish and Wildlife Service, EPA 560/6-87-002
3. Vincent, R. M; et al. *Acute Toxicity of Selected Organic Compounds to Fathead Minnows* 1976, EPA 600/3-76-097
4. Jones, H. R. *Environmental Control in Organic and Petrochemical Industries* 1971, Noyes Data Corp
5. Kaiser, K. L. E; et al. *Water Pollut. Res. J. Canada* 1991, **26**(3), 361-431
6. Schafer, E. W; et al. *Arch. Environ. Contam. Toxicol.* 1983, **12**, 355
7. Pitter, P. *Water Res.* 1976, **10**, 231
8. Megharaj, M; et al. *Pestic. Biochem. Physiol.* 1991, **10**(3), 266-273
9. BIOFAX *Industrial Biotext Laboratories* 29 April 1973
10. Zholdakova, Z. I. *Gig. Sanit.* 1985, **50**, 4, (Russ.) (*Chem. Abstr.* **102**, 216505)
11. McCann, J; et al. *Proc. Nat. Acad. Sci. USA* 1975, **72**, 5135
12. Eckert, K. G; *Xenobiotica* 1988, **18**(11), 1319-1326
13. Masamoto, Y; et al. *Sinshu Igaku Zasshi* 1987, **35**(2), 185-193, (Jap.) (*Chem. Abstr.* **107**, 128652j)
14. Zeiger, E; et al. *Environ. Mol. Mutagen.* 1988, **11**(Suppl. 12), 1-157
15. Garberg, P; et al. *Mutat. Res.* 1988, **203**(3), 155-176
16. Oberly, T. J; et al. *Environ. Mol. Mutagen.* 1990, **16**(4), 260-271
17. Cliet, I; et al. *Mutat. Res.* 1989, **216**(6), 321-326
18. Eiche, A; et al. *Mutat. Res.* 1990, **240**(2), 87-92
19. Kanaya, N; *Mutat. Res.* 1990, **245**(4), 311-315
20. Burnett, C. M; et al. *Food Chem. Toxicol.* 1989, **27**(10), 691-698
21. Yoshida, M; et al. *J. Toxicol. Sci.* 1989, **14**(4), 257-269
22. *ECETOC Technical Report No. 30(4)* 1991, European Chemical Industry Ecology and Toxicology Centre, B-1160 Brussels **16**(4), 260-271

A160   2-(4-Aminophenyl)-6-methyl-7-benzothazolylsulfonic acid

CAS Registry No. 130-17-6
Synonyms 2-(4-aminophenyl)-6-methyl-7-benzothiazolesulfonic acid;
6-methyl-2-(p-aminophenyl)-7-benzothiazolesulfonic acid
Mol. Formula $C_{14}H_{12}N_2O_3S_2$ **Mol. Wt.** 320.39
Uses Manufacture of dyestuffs.

Mammalian and avian toxicity

Acute data
LD_{50} intravenous mouse 178 mg kg^{-1} (1).

Any other comments

Based on data for benzothiazole, significant toxicity may be expected.

References

1. *U.S. Army Armament R.&D Report. NX00718.* Aberdeen Proving Ground, MD 21010

A161   Aminophylline

CAS Registry No. 317-34-0
Synonyms 3,5-dihydro-1-methyl-1H-purine-2,6-dione with 1,2-ethanediamine (2:1);
ethylenediamine, compound with theophylline; Cardiomin; Diuxanthine; Euphyllin;
Stenovasan
Uses Pharmaceutical applications including being a smooth muscle relaxant, and a
bronchodilator.

Physical properties

Solubility
Water: 200 g l^{-1} at 25°C.

Mammalian and avian toxicity

Acute data
LD_{50} oral mouse 540 mg l^{-1} (1).

LD_{50} intraperitoneal rat, guinea pig 200-250 mg kg^{-1} (2,3).

Intravenous single dose Wistar rat 5, 10, 20, 40 or 80 mg kg^{-1} hypoventilation with corresponding hypoxemia and hypercapnia observed in phrenicotomised rats (4).

Sub-acute data

Oral NJH mice (8-10 day) 25, 50 or 100 mg kg^{-1} day^{-1} immune system effects included increased haemolytic ability of plaque-forming cells and antibody concentration and decreased delayed type hypersensitivity reaction and peripheral white blood cell phagocytosis (5).

Intravenous newborn New Zealand rabbits 0.6 mg kg^{-1} (duration unspecified) increased diuresis, renal vascular resistance and filtration fraction (6).

Metabolism and pharmacokinetics

Pulmonary absorption of aminophylline 500 µg rat^{-1} was 96% within 1 min (7). At least 7 hr are required to clear blood completely in normal adult subjects after intravenous administration of 300-500 mg. >10 hr are required before an oral dose disappears entirely from the blood (8).

Any other adverse effects to man

In human poisoning symptomatic effects include restlessness, anorexia, nausea, fever, vomiting and dehydration. Can result in cardiovascular and respiratory collapse, shock, cyanosis and death (8).

Any other comments

Incompatible with a range of materials, including acids, bleomycin sulfate, chlorpromazine hydrochloride, corticotrophin, doxorubicin and erythromycin. Commonly administered as the hydrate, human health effects reviewed (9). Mixture of theophylline and ethylenediamine.

References

1. Thompson, C. R; et al. *J. Lab. Clin. Med.* 1946, **31**, 1337
2. *J. Am. Pharm. Assoc.* 1947, **36**, 248
3. *J. Pharmacol. Exp. Therap.* 1945, **83**, 120
4. Nachazel, J; et al. *Eur. Respir. J.* 1990, **3**(3), 311-317
5. Lui, F; et al. *Zhongguo Yaoli Xuebao* 1989, **10**(5), 457-460, (Ch.) (*Chem. Abstr.* **111**, 187103m)
6. Gouyon, J. B; et al. *Life Sci.* 1988, **42**(13), 1271-1278
7. Arakawa, E; et al. *Chem. Pharm. Bull.* 1987, **35**(5), 2038-2044
8. Gosselin, R. E; et al. *Clinical Toxicology of Commercial Products* 4th ed., 1976, 16-20, William & Wilkins, Baltimore
9. *Martindale: The Extra Pharmacopoeia* 29th ed., 1989, The Pharmaceutical Press, London

A162   Aminopropane

$$CH_3CH_2CH_2NH_2$$

CAS Registry No. 107-10-8

Synonyms *n*-propylamine; 1-aminopropane; 1-propanamine; mono-*n*-propylamine

Mol. Formula C_3H_9N **Mol. Wt.** 59.11

Uses Chemical intermediate.

Occurrence Occurs naturally in various species of marine algae (1).

Physical properties

M. Pt. –83°C; **B. Pt.** 48-49°C; **Flash point** –12°C (closed cup); **Specific gravity** d_{20}^{20} 0.7191; **Partition coefficient** log P_{ow} 0.48; **Volatility** v.p. 248 mmHg at 20°C.

Solubility
Water: miscible. Organic solvent: miscible ethanol, diethyl ether

Ecotoxicity

Fish toxicity
LC_{50} (24 hr) creek chub 40-60 mg l^{-1} (2).

Environmental fate

Nitrification inhibition
No inhibition of NH_3 oxidation *Nitrosomonas sp.* at 100 mg l^{-1} (3).

Degradation studies
Degradation by *Aerobacter sp.* 200 mg l^{-1} 30°C. Parent 100% degradation in 31 hr while mutant strains 100% degradation in 9 hr (2).
Arthrobacter P1 metabolised aminopropane via primary amine oxidase activity and utilised both carbon and nitrogen as sources of growth (4).
BOD_{13} of 102% was measured using a non-activated sludge inoculum (5).

Abiotic removal
Photochemical reaction with hydroxyl radicals in the atmosphere (6).

Mammalian and avian toxicity

Acute data
LD_{Lo} oral rat 250-570 mg kg^{-1} (7,8).
LC_{50} (2 hr) inhalation mouse 2500 mg m^{-3} (7).
LC_{50} (4 hr) inhalation rat 2310 ppm (9).
LD_{50} dermal rabbit 560 mg kg^{-1} (10).

Irritancy
Dermal rabbit (24 hr) 100 µg caused irritation, while 720 µg instilled into rabbit eye caused severe irritation (10,11).

References

1. Steiner, M; et al. *Planta* 1968, **79**, 113-121
2. Verschueren, K. *Handbook of Environmental Data on Organic Chemicals* 2nd ed., 1983, Van Nostrand Reinhold, New York
3. Hockenbury, M. R; et al. *J. Water. Pollut. Control Fed.* 1977, **49**(5), 768-777
4. De Boer, L; et al. *Antony van Leeuwenhoek* 1989, **56**(3), 221-232
5. Chudoba, J; et al. *Chem. Prum.* 1969, 1969, **19**, 76-80
6. Atkinson, R. *Inter. J. Chem. Kinet.* 1987, **19**, 799-828
7. Izermov, N. F; et al. *Toxicometeric Parameters of Industrial Toxic Chemicals Under Single Exposure*, 1982, 102, Centre of International Projects, Moscow
8. Colosi-Esca, D; et al. *Rev. Chim. (Bucharest)* 1987, **38**(10), 933-937 (Rum.) (*Chem. Abstr.* **109**, 18308k)
9. *Arch. Environ. Health* 1960, **1**, 343
10. Smyth, H. F; et al. *Am. Ind. Hyg. Assoc. J.* 1962, **23**, 9
11. *Union Carbide Material Safety Data Sheet* 1968, Union Carbide Corportation, New York

A163 2-Aminopropane

CH3CH(NH2)CH3

CAS Registry No. 75-31-0
Synonyms isopropylamine; 1-methylethylamine; 2-propylamine; 2-propanamine; *sec*-propylamine
Mol. Formula C_3H_9N **Mol. Wt.** 59.11
Uses Solvent and chemical intermediate.
Occurrence Detected in tobacco leaves and cigarette smoke. Emitted to the air from decomposing manure in animal feed lots (1).

Physical properties

M. Pt. –101.2°C; **B. Pt.** 33-34°C; **Flash point** –2°C (open cup); **Specific gravity** d_4^{15} 0.694; **Partition coefficient** log P_{ow} 0.26; **Volatility** v.p. 579.6 mmHg at 25°C; v. den. 2.03.

Solubility
Water: miscible. Organic solvent: miscible ethanol, diethyl ether, soluble in acetone, benzene, chloroform

Occupational exposure

UN No. 1221; **HAZCHEM Code** 2WE; **Conveyance classification** flammable liquid; **Supply classification** highly flammable and irritant.

Risk phrases
Extremely flammable – Irritating to eyes, respiratory system and skin (R12, R36/37/38)

Safety phrases
Keep away from sources of ignition – No Smoking – In case of contact with eyes, rinse immediately with plenty of water and seek medical advice – Do not empty into drains (S16, S26, S29)

Ecotoxicity

Invertebrate toxicity
EC_{50} (48 hr) *Daphnia magna* 91.5 mg l^{-1}.
IC_{50} (96 hr) *Selenastrum capricornutum* 120 mg l^{-1} (buffered with 10% HCl) (2).

Bioaccumulation
Calculated bioconcentration factor 0.43 (3).

Abiotic removal
Very mobile in soil (4).

Mammalian and avian toxicity

Acute data
LD_{50} oral rat 820 mg kg^{-1} (5).
LC_{50} (4 hr) inhalation rat 4000 ppm (6,7).
LD_{50} dermal rabbit 380-550 mg kg^{-1} (7).

Irritancy

Dermal rabbit (24 hr) 10 mg caused severe irritant effects (2), while 50 μg instilled into rabbit eye (24 hr) caused severe irritant effects (8).

Genotoxicity

Salmonella typhimurium TA98, TA100, TA1537 with and without metabolic activation negative (9).

Any other comments

Physical, chemical properties, experimental toxicology, epidemiology, human health effects and hazards reviewed (10).

Experimental toxicology, epidemiology and human health effects reviewed (11).

References

1. Mosier, A. R; et al. *Environ. Sci. Technol.* 1973, **7**, 642-644
2. Haley, M. V; et al. *Gov. Rep. Announce. Index. US* 1989, **89**(18), 1-14, Report CRDEC-TR-052 Order No. AD-A208165
3. Lyman, W. J; et al. *Handbook of Chemical Property Estimation Methods* 1982, **5**, 4, McGraw-Hill, New York
4. Swann, R. L; et al. *Res. Rev.* 1983, **85**, 16-28
5. Smyth, H. F; et al. *Arch. Ind. Hyg. Occup. Med.* 1951, **4**, 119-122
6. *Interagency Collaborative Group on Environmental Carcinogenesis* 17 Jun 1974
7. Sax, N. I; et al. *Dangerous Properties of Industrial Materials* 7th ed., 1989, Van Nostrand Reinhold, New York
8. Marhold, J. V. *Sbornik Vysleaku Toxixogickeho Bystreni Latek A Priprauku* 1972, 62, Prague
9. Zeiger, E; et al. *Environ. Mutagen.* 1987, **9**(Suppl. 9), 1-109
10. *Cah. Notes Doc.* 1990, **141**, 875-878, (Fr.) (*Chem. Abstr.* **114**, 233948c)
11. *ECETOC Technical Report No. 30(4)* 1991, European Chemical Industry Ecology and Toxicology Centre, B-1160 Brussels

A164   1-Aminopropan-2-ol

CH₃CHOHCH₂NH₂

CAS Registry No. 78-96-6

Synonyms 2-hydroxypropylamine; 1-amino-2-propanol; MIPA; α-aminoisopropyl alcohol; isopropanolamine; monoisopropanolamine

Mol. Formula C_3H_9NO **Mol. Wt.** 75.11

Uses Emulsifying agent, dry cleaning, soaps, soluble textile oils, wax removers, metal cutting oils, cosmetics, emulsion paints, plasticisers, insecticides.

Physical properties

M. Pt. 1.4°C; **B. Pt.** 160°C; **Flash point** 77.2°C; **Specific gravity** 0.9619; **Partition coefficient** log P_{ow} –0.96; **Volatility** v.p. <1 mmHg at 20°C; v. den. 2.6.

Occupational exposure

Supply classification corrosive.

Risk phrases
Causes burns (R34)

Safety phrases
Do not breathe vapour – In case of contact with eyes, rinse immediately with plenty of water and seek medical advice – Wear suitable protective clothing (S23, S26, S36)

Ecotoxicity

Fish toxicity
LC_{50} (24 hr) goldfish >5000 mg l^{-1} (1).

Invertebrate toxicity
EC_{50} (30 min) *Photobacterium phosphoreum* 27.3 mg l^{-1} Microtox test (2).

Environmental fate

Carbonaceous inhibition
BOD_{10} 34% ThOD 2.5 mg l^{-1} in mineralised dilution water with settled sewage seed at 20°C (3).

Degradation studies
BOD_5 4% using non-acclimated sewage inocula and a standard BOD dilution method. Adaptation of the sewage inocula result theoretical BOD_5 of 43% (4).

Absorption
Activated carbon: adsorbability 40 mg g^{-1}; 20% reduction, influent 1 g l^{-1}, effluent 800 mg l^{-1} (5).

Mammalian and avian toxicity

Acute data
LD_{50} oral rat 2100 mg kg^{-1} (3).
LD_{50} dermal rabbit 1640 mg kg^{-1} (6).

Irritancy
Dermal rabbit 485 mg caused irritant effects and 970 µg instilled in rabbit eye caused irritant effects (duration unspecified) (6).

Genotoxicity

Salmonella typhimurium TA98, TA100, TA1535, TA1537 with and without metabolic activation equivocal (7).

Any other adverse effects

At toxic concentrations causes somnolence and gastrointestinal hypermotility and diarrhoea (6).

Any other comments

Corrosive and moderately flammable. Incompatible with oxidizing materials (6). Based on available data from a study of the chemical, metabolic and toxicological properties, it was concluded that isopropanolamine is safe as a cosmetic ingredient in the present practices of use and concentration (8).
All reasonable efforts have been taken to find information on isomers of this compound, but no relevant data are available.

References

1. Bridie, A. L; et al. *Water Res.* 1979, **13**, 623
2. Kaiser, K. L. E; et al. *Water Poll. Res. J. Canada* 1991, **26**(3), 361-431
3. Ettinger, M. B. *Ind. Eng. Chem.* 1956, **48**, 256
4. Bridie, A. L; et al. *Water Res.* 1979, **13**, 627-630
5. Guisti, D. M; et al. *J. Water Pollut. Control. Fed.* 1974, **46**, 947
6. *Union Carbide Data Sheet*, 1971, Union Carbide Corporation, New York
7. Zeiger, E; et al. *Environ. Mutagen.* 1987, **9**(Suppl. 9), 1-109
8. Beyer, K. H; et al. *J. Am. Coll. Toxicol.* 1987, **6**(1), 53-76

A165   4'-Aminopropiophenone

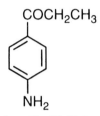

CAS Registry No. 70-69-9

Synonyms 1-(4-aminophenyl)-1-propanone; ethyl *p*-aminophenyl ketone; *p*-aminopropiophenone; paraminopropiophenone; *p*-aminophenylpropanone

Mol. Formula $C_9H_{11}NO$ **Mol. Wt.** 149.19

Uses Chemical intermediate. Cyanide antidote.

Physical properties

M. Pt. 140°C.

Solubility
Organic solvent: ethanol, chloroform

Ecotoxicity

Effects to non-target species
LD_{50} oral starling 133 mg kg^{-1} (1).

Environmental fate

Nitrification inhibition
Nitrosomonas sp. 75-100% inhibition of ammonia oxidation at 100 mg l^{-1} (2).

Mammalian and avian toxicity

Acute data
LD_{50} oral mouse, rat 168-177 mg kg^{-1} (3,4).
LD_{50} oral cat, mouse 5600 μg kg^{-1} (3).
LD_{50} oral guinea pig 1020 mg kg^{-1} (5).
LD_{Lo} intraperitoneal rat 525 mg kg^{-1} (6).
LD_{50} intraperitoneal mouse 80 mg kg^{-1} (7).

Any other adverse effects

Oral dog 0.5 mg kg^{-1} protected against cyanide poisoning induced by intravenous HCN dose of 0.67 or 1.34 mg kg^{-1}. Protection caused sequestration of cyanide inside red cells (8).

Oral rat (unspecified concentration) brought about peak methaemoglobin levels at 15-40 min while intravenous administration caused peak methaemoglobin levels 15-25 min after dosing (9).

Any other comments

All reasonable efforts have been taken to find information on isomers of this compound, but no relevant data are available.

References

1. Schafer, E. W; et al. *Arch. Environ. Contam. Toxicol.* 1983, **12**(3), 355-382
2. Hockenbury, M. R; et al. *J. Water Pollut. Contr. Fed.* 1977, **49**(5), 768-777
3. Savaric, P. J; et al. *Bull. Environ. Contam. Toxicol.* 1983, **30**(1), 122-126
4. Pan, H. P; et al. *Gen. Pharmacol.* 1983, **14**(4), 465-467
5. Scawin, J. W; et al. *Toxicol. Lett.* 1984, **23**(3), 359-365
6. Coleman, I. W; et al. *Can. J. Biochem. Physiol.* 1960, **38**, 667-672
7. *NTIS Report* AD 277-689, National Technical Information Service, Springfield, VA
8. Marrs, T. C; et al. *Hum. Toxicol.* 1987, **6**(2), 139-145
9. Marrs, T. C; et al. *Hum. Exp. Toxicol.* 1991, **10**(3), 183-188

A166   3-Aminopropyldiethylamine

$(CH_3CH_2)_2N(CH_2)_3NH_2$

CAS Registry No. 104-78-9
Synonyms *N,N*-diethyl-1,3-propanediamine; *N,N*-diethyl-1,3-diaminopropane; 3-diethylamino-1-propylamine
Mol. Formula $C_7H_{18}N_2$ **Mol. Wt.** 130.23
Uses Used as a hardening agent in epoxy resins.

Physical properties

B. Pt. 169-171°C; **Flash point** 60°C; **Specific gravity** 0.8264; **Volatility** v. den. 4.48.

Occupational exposure

Supply classification flammable and corrosive.

Risk phrases

Flammable – Harmful in contact with skin and if swallowed – Causes burns – May cause sensitisation by skin contact (R10, R21/22, R34, R43)

Safety phrases

In case of contact with eyes, rinse immediately with plenty of water and seek medical advice – Wear suitable protective clothing, gloves and eye/face protection (S26, S36/37/39)

Mammalian and avian toxicity

Acute data
LD_{50} oral rat 1410 mg kg^{-1} (1).
LD_{50} dermal rabbit 750 mg kg^{-1} (2).

Irritancy
Dermal rabbit (24 hr) 100 µg caused irritant effects (2).
Dermatitis can occur either by primary irritation or allergic sensitisation. Symptoms of skin contact include skin rash, tenderness and eczema. Exposure to vapours can cause erythema of the face, oedema and pruritus (3).

Any other adverse effects to man
Significant cross reactions to aliphatic polyamines were observed in patients allergic to topical ethylenediamine. Antihistamines given topically or orally failed to inhibit ethylenediamine-induced allergic dermatitis (4).

Any other comments
Experimental toxicology and human health effects reviewed (5).
Incompatible with acids, acid chlorides, acid anhydrides and strong oxidising materials.
All reasonable efforts have been taken to find information on isomers of this compound, but no relevant data are available.

References
1. *Union Carbide Safety Datasheet* 27 Feb 1967, Union Carbide Corporation, New York
2. *Am. Ind. Hyg. Assoc. J.* 1962, **95**, 23
3. *Chemical Safety Data Sheets* 1990, **3**, 79-81, RSC, London
4. Balato, N; et al. *Contact Dermatitis* 1986, **15**(5), 263-265
5. *ECETOC Technical Report No. 30(4)* 1991, European Chemical Industry Ecology and Toxicology Centre, B-1160 Brussels

A167   3-Aminopropyldimethylamine

$(CH_3)_2N(CH_2)_3NH_2$

CAS Registry No. 109-55-7
Synonyms *N,N*-dimethyl-1,3-propanediamine; *N,N*-dimethyl-1,3-diaminopropane; 3-dimethylamino-1-propylamine
Mol. Formula $C_5H_{14}N_2$ **Mol. Wt.** 102.18
Uses As a hardener for epoxy resins, a dispersant- detergent for high quality motor oils and in antistatic agents for synthetic fibres.

Physical properties
M. Pt. < -70°C; **B. Pt.** 123°C; **Flash point** 38°C; **Specific gravity** 0.817 at 30°C; **Volatility** v.p. 10 mmHg at 30°C; v. den. 3.52.

Occupational exposure

Supply classification flammable and corrosive.

Risk phrases
Flammable – Harmful if swallowed – Causes burns – May cause sensitisation by skin contact (R10, R22, R34, R43)

Safety phrases
In case of contact with eyes, rinse immediately with plenty of water and seek medical advice – Wear suitable protective clothing, gloves and eye/face protection (S26, S36/37/39)

Mammalian and avian toxicity

Acute data
LD_{Lo} oral rat 1870 mg kg^{-1}.
LD_{Lo} (24 hr) dermal rabbit 100 μg (1).

Irritancy
5 mg instilled into rabbit eye (duration unspecified) caused moderate irritant effects (2).
Dermal rabbit (duration unspecified) 0.5 ml of a 1% solution caused severe burns (3).

Genotoxicity

Salmonella typhimurium TA98, TA100, TA1535, TA1537 with and without metabolic activation negative (4).

Any other adverse effects to man

Significant cross reactions to aliphatic polyamines were observed in patients allergic to topical ethylenediamine. Antihistamines given topically or orally, failed to inhibit ethylenediamine-induced allergic dermatitis (5).

Any other adverse effects

Corrosive to mucous membranes and upper respiratory tract. Inhalation can cause bronchial spasm, inflammation, oedema and death (3).

Any other comments

Manufacture and industrial applications, experimental toxicology and human health effects reviewed (6,7).
Lachrymatory. Reacts with 1,2-dichloromethane to form acetylene gas. Incompatible with acids, acid chlorides and acid anhydrides. Ignites spontaneously in contact with cellulose nitrate.
All reasonable efforts have been taken to find information on isomers of this compound, but no relevant data are available.

References

1. *Am. Ind. Hyg. Assoc. J* 1962, **23**, 95
2. *Union Carbide Datasheet* 1971, Union Carbide Corporation, New York
3. *Chemical Safety Data Sheets* 1990, **3**, 83-85, RSC, London
4. Zeiger, E; et al. *Environ. Mutagen.* 1987, **9**(Suppl. 9), 1-109
5. Balato, N; et al. *Contact Dermatitis* 1986, **15**(5), 263-265
6. Lapper, P; et al. *Chem-Ztg.* 1987, **111**(4), 117-125, (Ger.) (*Chem. Abstr.* **109**, 22504p)
7. *ECETOC Technical Report No. 30(4)* 1991, European Chemical Industry Ecology and Toxicology Centre, B-1160 Brussels

A168 Aminopterin

CAS Registry No. 54-62-6
Synonyms aminopteridine; N-(p-[(2, 4-diamino-6-pteridylmethyl)amino]benzoyl)glutamic acid; ((((diaminopteridinyl)methyl)amino)benzoylglutamic acid
Mol. Formula $C_{19}H_{20}N_8O_5$ Mol. Wt. 440.42
Uses Antineoplastic and antileukaemic. Antagonist of folic acid.

Physical properties

M. Pt. (L-form) 260-265°C (decomp.).

Mammalian and avian toxicity

Acute data
LD_{Lo} oral rat 2500 µg kg^{-1} (1).
LD_{50} intraperitoneal rat, mouse 1900-3400 µg kg^{-1} (1,2).

Sub-acute data
Intraperitoneal mice (5 day) 2 mg kg^{-1} caused sperm morphology changes (2).

Teratogenicity and reproductive effects
TD_{Lo} (40 day pregnant) oral woman 200 µg kg^{-1} caused spontaneous abortion (3).
Injection mouse (meiosis 1 or 3 hr prior to meiosis I) 2-4 mg kg^{-1} no effect on ovulation, rate of fertilisation, cleavage or implantation (4).
In vitro microassay of human embryonic palatal mesenchymal (HEPM) cells with metabolic activation to determine teratogenic potential. IC_{50} 38.7 µg l^{-1} positive teratogenic potential (5).

Genotoxicity

Drosophila melanogaster mutant *vg* strains yielded more offspring on a medium containing aminopterin than on normal medium (6).

Any other comments

Physiology, metabolism and pharmacology reviewed (7,8).

References

1. Philips, F. S; et al. *J. Pharmacol. Exptl. Therap.* 1949, **95**(3), 303-311
2. Sax, N. I; et al. *Dangerous Properties of Industrial Materials* 7th ed., 1989, Van Nostrand Reinhold, New York
3. Thiersch, J. B. *Am. J. Obstet. Gynecol.* 1952, **63**, 1298-1304
4. Hashimoto, T; et al. *J. Toxicol. Sci.* 1986, **11**(4), 279-291
5. Tsuchiya, T; et al. *Teratogen. Carcinogen. Mutagen.* 1988, **8**, 265-272
6. Silber, J; et al. *Mol. Gen. Genet.* 1989, **218**(3), 475-480
7. Sirotnak, F. M; et al. *Chem. Biol. Pteridines. 1989 Proc. Int. Symp. Pteridines Folic Acid Deriv. 9th* 1990, 1185-1191, Curtis, H. C. (Ed.), Kettering Cancer Centre, New York
8. Matherly, L. H; et al. *Pharmacol. Ther.* 1987, **35**(1-2), 27-56

A169   2-Aminopyridine

CAS Registry No. 504-29-0
Synonyms α-aminopyridine; *o*-aminopyridine; amino-2-pyridine; α-pyridylamine; α-pyridinamine; 2-pyridinamine
Mol. Formula $C_5H_6N_2$ **Mol. Wt.** 94.12
Uses Organic synthetic intermediate. Used in pharmaceutical manufacture, particularly antihistamines.

Physical properties

M. Pt. 57-58°C; **B. Pt.** 204-210°C; **Flash point** 92°C; **Partition coefficient** log P_{ow} –0.22 (1).; **Volatility** v. den. 3.25.

Solubility
Organic solvent: diethyl ether, ethanol, acetone

Occupational exposure

US TLV (TWA) 0.5 ppm (1.9 mg m^{-3}); **UK Long- term limit** 0.5 ppm (2 mg m^{-3}); **UK Short-term limit** 2 ppm (8 mg m^{-3}); **UN No.** 2671; **Conveyance classification** toxic substance.

Ecotoxicity

Invertebrate toxicity
EC_{50} (5-30 min) *Photobacterium phosphoreum* 284 mg l^{-1} Microtox test (2).
EC_{50} (60 hr) *Tetrahymena pyriformis* 390 mg l^{-1} (3).

Bioaccumulation
Non-accumulative or low accumulative (4).
Calculated bioconcentration factor 0.14 (5).

Effects to non-target species
LD_{50} oral redwing blackbird, starling 32 mg kg^{-1} (6).

Environmental fate

Degradation studies
Adapted activated sludge at 20°C used product as sole carbon source, 97.3% COD 41.0 mg COD g^{-1} dry inoculum hr^{-1} (7).
Of 17 ppm incubated in soil at pH 7 and 28°C <1% degraded within 30 days, as evidenced via the release of inorganic nitrogen (8).

Mammalian and avian toxicity

Acute data
LD_{50} oral quail 130 mg kg^{-1} (6).
LD_{50} intraperitoneal mouse 35 mg kg^{-1} (9).
LD_{50} subcutaneous mouse 70 mg kg^{-1}.
LD_{50} intravenous mouse 23 mg kg^{-1} (10).

Genotoxicity

Salmonella typhimurium TA98, TA100, TA1535, TA1537 with and without metabolic activation negative (11).

Any other comments

Experimental toxicology and human health effects reviewed (12).

References

1. Verschueren, K; *Handbook of Environmental Data on Organic Chemicals* 2nd ed., 1983, 192, Van Nostrand Reinhold, New York
2. Kaiser, K. L. E; et al. *Water Pollut. Res. J. Canada* 1991, **26**(3), 361-431
3. Shultz, T. W; et al. *Ecotox. Environ. Safety* 1985, **10**, 75-85
4. *MITI Report* 1984, Ministry of International Trade and Industry, Tokyo
5. Lyman, W. J; et al. *Handbook of Chemical Property Estimation Methods* 1984, **5**, 4, McGraw-Hill, New York
6. Schafer, E. W; et al. *Arch. Environ. Contam. Toxicol.* 1983, **12**(3), 355-382
7. Pitter, P. *Water Res.* 1976, **10**, 231-235
8. Sims, G. K; et al. *Appl. Environ. Microbiol.* 1986, **51**, 963-968
9. Vohra, M. M; et al. *J. Med. Chem.* 1965, **8**(3), 296-304
10. Watrous, R. M; et al. *Ind. Med. Surg.* 1950, **19**, 317
11. Zeiger, E; et al. *Environ. Mutagen.* 1987, 9(Suppl. 9), 1-109
12. *ECETOC Technical Report No. 30(4)* 1991, European Chemical Industry Ecology and Toxicology Centre, B-1160 Brussels

A170   3-Aminopyridine

CAS Registry No. 462-08-8
Synonyms amino-3-pyridine; *m*-aminopyridine; 3-pyridinamine; 3-pyridylamine
Mol. Formula $C_5H_6N_2$ **Mol. Wt.** 94.12
Uses Organic synthetic intermediate. In drug and dyestuff manufacture.

Physical properties

M. Pt. 64°C; **B. Pt.** 251°C.

Solubility

Organic solvent: ethanol, diethyl ether, benzene

Occupational exposure

UN No. 2671; **Conveyance classification** toxic substance.

Ecotoxicity

Invertebrate toxicity

EC_{50} (5-30 min) *Photobacterium phosphoreum* 682 mg l^{-1} Microtox test (1).
EC_{50} (60 hr) *Tetrahymena pyriformis* 283 mg l^{-1} (2).

Bioaccumulation
Non-accumulative or low accumulative (3).

Effects to non-target species
LD_{50} oral redwing blackbird, starling 13.3 mg kg^{-1} (4).

Mammalian and avian toxicity

Acute data
LD_{50} oral quail 178 mg kg^{-1} (1).
LD_{50} intraperitoneal, intravenous mouse 24-30 mg kg^{-1} (5-7).

References

1. Kaiser, K. L. E; et al. *Water Pollut. Res. J. Canada* 1991, **26**(3), 361-431
2. Shultz, T. W; et al. *Ecotoxicol. Environ. Safety* 1987, **13**, 76
3. *MITI Report* 1984, Ministry of International Trade and Industry, Tokyo
4. Schafer, E. W; et al. *Arch. Environ. Contam. Toxicol.* 1983, **12**(3), 355-382
5. Vohra, M. M; et al. *J. Med. Chem.* 1965, **8**(3), 296-304
6. Fastier, F. N; et al. *Aust. J. Exp. Biol. Med. Sci.* 1958, **36**, 365
7. Lechat, P; et al. *Ann. Pharm. Fr.* 1968, **26**(5), 345-349 (Fr.) (*Chem. Abstr.* 1968, **69**, 65970k)

A171   4-Aminopyridine

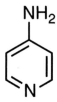

CAS Registry No. 504-24-5
Synonyms amino-4-pyridine; γ-aminopyridine; 4-pyridinamine; 4-pyridylamine; *p*-aminopyridine
Mol. Formula $C_5H_6N_2$ **Mol. Wt.** 94.12
Uses Organic synthetic intermediate. Bird repellent.

Physical properties
M. Pt. 155-158°C; **B. Pt.** 273°C; **Partition coefficient** log P_{ow} 0.28.

Solubility
Organic solvent: benzene, diethyl ether

Occupational exposure
UN No. 2671; **Conveyance classification** toxic substance.

Ecotoxicity

Fish toxicity
LC_{50} (96 hr) bluegill sunfish 2.8-7.5 mg l^{-1} static bioassay at 12-22°C. LC_{50} (96 hr) channel catfish 2.4-5.8 mg l^{-1} static bioassay at 12- 22°C (1).

Invertebrate toxicity

EC_{50} (5-30 min) *Photobacterium phosphoreum* 284 mg l^{-1} Microtox test (2).
EC_{50} (60 hr) *Tetrahymena pyriformis* 260 mg l^{-1} (3).

Bioaccumulation

Non-accumulative or low accumulative (4).

Environmental fate

Anaerobic effects

Soil degradation under anaerobic conditions was negligible for up to 2 months (5).

Degradation studies

Under aerobic conditions, using ^{14}C- labelled 4-aminopyridine, at 30°C and 50% moisture, $^{14}CO_2$ was evolved at 0.4% to >50% depending upon the soil type (5). In a 90 day period, 54.6, 10.74 and 4.88% of original radio-labelled concentration of 10 ppm were mineralised to CO_2 from a loamy sand and sandy clay loam at 30°C and pHs of 7.8, 7.7 and 7.6, respectively, which correspond to $t_{1/2}$ of 90, 330 and 960 days. For a loam with pHs of 5.8, 5.6 and 4.1 losses of 5.95, 16.52 and 0.35% corresponding to $t_{1/2}$ 600, 240 and >660 days (5).

Abiotic removal

Oxidation with ozone in aqueous solution increases with increasing pH, complete oxidation occurs at pH 9.3 in 50 min (6).
<1% of an initial 14.6 ppm was mineralised in >30 days as evidenced via the release of inorganic nitrogen (7).
Photochemical reaction with hydroxyl radicals in the atmosphere $t_{1/2}$ 8 hr (8).

Mammalian and avian toxicity

Acute data

LD_{50} oral rat, mouse 20-42 mg kg^{-1} (4,9).
LD_{50} intraperitoneal rat, mouse 6.5-11.5 mg kg^{-1} (9-12).
LD_{50} subcutaneous rat, mouse 5- 18.5 mg kg^{-1} (11,12).

Metabolism and pharmacokinetics

Nine healthy subjects (7 men and 2 women) received a single intravenous injection of 20 mg 4- aminopyridine. Five of the subjects received the same dose in the form of enteric-coated tablets and 4 the same dose in uncoated tablets, treatments were 2 wk apart. Saliva concentrations were higher than those in serum after 5 min. The $t_{1/2}$ and volume distribution calculated form serum and saliva concentrations were of the same order. The total urinary excretion of unchanged drug was 90.6% after intravenous doses and 88.5% after oral doses of enteric- coated tablets (13).

Genotoxicity

Salmonella typhimurium TA1537, TA2637 without metabolic activation negative (14).

Any other adverse effects

The effect of 4-aminopyridine on [^{3}H]acetylcholine release was studied in rat cerebral cortical synaptosomes. Results suggested that 4- aminopyridine blocks potassium channels involved in regulating membrane potential in isolated cholinergic terminals but that changes in these channels are not important in the nerve terminal's response to depolarisation (15).

Reported to reverse the effects of non-depolarising muscle relaxants and to have analeptic effects. Improvement of myasthenia gravis has been reported (16).

References

1. Schafer, E. W; et al. *J. Wildlife Management* 1975, **39**, 807-811
2. Kaiser, K. L. E; et al. *Water Pollut. Res. J. Canada* 1991, **26**(3), 361-431
3. Shultz, T. W; et al. *Ecotox. Environ. Safety* 1985, **10**, 97-111
4. *MITI Report* 1984, Ministry of International Trade and Industry, Tokyo
5. Schafer, E. W; et al. *Arch. Environ. Contam. Toxicol.* 1985, **14**(1), 111-129
6. Starr, R. I; et al. *Arch. Environ. Contam. Toxicol.* 1975, **3**(1), 72-83
7. Matsneu, A. I; et al. *Kozh.-Obovn. Prom-st.* 1988, **10**, 57-60, (Russ.) (*Chem. Abstr.* **110**, 13040c)
8. Sims, G. K; et al. *Appl. Environ. Microbiol.* 1986, **51**, 963-968
9. Atkinson, R. *Intern. J. Chem. Kin.* 1987, **19**, 799-828
10. Schafer, E. W; et al. *Toxicol. Appl. Pharmacol.* 1973, **26**(4), 532-538
11. Vohra, M. M; et al. *J. Med. Chem.* 1965, **8**(3), 296-304
12. Lechat, P; et al. *Ann. Pharm. Fr.* 1968, **26**(5), 345-349 (Fr.) (*Chem. Abstr.* 1968, **69**, 65970k)
13. Mitson, V; et al. *Eksp. Med. Morfol.* 1972, **11**(3), 162-165 (*Chem. Abstr.* 1973, **79**, 939u)
14. Uges, A; et al. *Clin. Pharmacol. Ther.* 1982, **31**(5), 587
15. Ogawa, H. I; et al. *Mutat. Res.* 1986, **172**(2), 97-104
16. Meyer, E. M; et al. *Neurochem. Res.* 1989, **14**(2), 157-160

A172   11-Aminoundecanoic acid

$$NH_2(CH_2)_{10}CO_2H$$

CAS Registry No. 2432-99-7
Mol. Formula $C_{11}H_{23}NO_2$ **Mol. Wt.** 201.31
Uses Polymer intermediate.

Physical properties
M. Pt. 190-192°C.

Mammalian and avian toxicity

Sub-acute data
Intravenous cat (duration unspecified) 2 mg kg^{-1} caused transient lowering of blood pressure (1).

Carcinogenicity and long-term effects
No adequate evidence for carcinogenicity in humans, limited evidence for carcinogenicity in experimental animals, IARC classification group 3 (2,3).
F344 rats and B6C3F1 mice (103 wk) 1.5% (daily) in food induced liver adenomas and urinary/bladder tumours (4).
Fischer 344 rats, B6C3F1 mice (2 yr) gavage in corn oil benign liver tumours observed (5).

Metabolism and pharmacokinetics
Oral rat (concentration unspecified) minor incorporation in DNA nucleosides of the liver within 24 hr. No DNA alkylation observed (6).

Genotoxicity

Salmonella typhimurium TA98, TA100, TA1535, TA1537 with and without metabolic activation negative (2).

Chinese hamster ovary cells without metabolic activation sister chromatid exchange positive (7).

Mouse lymphoma L-5178Y tk+/tk- with and without metabolic activation negative (8).

Any other comments

Experimental toxicology and human health effects reviewed (9).

References

1. Gofman, S. M; et al. *Tr. Volgograd. Gos. Med. Inst.* 1968, **21**(2), 152-153 (Russ.) (*Chem. Abstr.* **74**, 41021b)
2. *IARC Monograph*, 1987, **Suppl. 7**, 57
3. *IARC Monograph* 1986, **39**, 239
4. Ashby, J; et al. *Mutat. Res.* 1988, **204**(1), 17-115
5. Mirsalis, J. C; et al. *Environ. Mol. Mutagen.* 1989, **14**(3), 155-164
6. Peter, H; et al. *Arch. Toxicol.* 1987, **61**(1), 86-87
7. Galloway, S. M; et al. *Environ. Mol. Mutagen.* 1987, **10**(Suppl. 10), 1-175
8. McGregor, D. B; et al. *Environ. Mol. Mutagen.* 1988, **12**(1), 85-154
9. *ECETOC Technical Report No. 30(4)* 1991, European Chemical Industry Ecology and Toxicology Centre, B-1160 Brussels

A173 Amiton

$$CH_3CH_2O \underset{CH_3CH_2O}{\overset{O}{\underset{\diagup}{\overset{\diagdown}{P}}}}-SCH_2CH_2N(CH_2CH_3)_2$$

CAS Registry No. 78-53-5

Synonyms Tetram; Inferno; Metramac; S-[2-(diethylamino)ethyl]phosphorothioic acid *O,O*-diethyl ester; *O, O*-diethyl S-(β-diethylamino)ethyl phosphorothiolate

Mol. Formula $C_{10}H_{24}NO_3PS$ **Mol. Wt.** 269.35

Uses Insecticide, acaricide.

Physical properties

B. Pt. 88°C.

Mammalian and avian toxicity

Acute data

LD_{50} oral rat 5.4 mg kg^{-1} (1).

LD_{50} subcutaneous rat, mouse, hamster, guinea pig 80-120 µg kg^{-1} (2).

Any other adverse effects

Organophosphate pesticide, symptomatic effects of this type of compound include respiratory discomfort, nausea, vomiting, diarrhoea, headache and blurred vision. Can cause central nervous system disorders. Death is primarily due to respiratory failure although cardiac arrest can be implicated (3).

Legislation

Limited under EC Directive on Drinking Water Quality 80/778/EEC. Pesticide: maximum admissible concentration 0.1 $\mu g\,l^{-1}$ (4).

Any other comments

Insecticidal and acaricidal properties reported (5,6).
Contact insecticide, acaricide, no longer in widespread use.

References

1. Frawley, et al. *Toxicol. Appl. Pharmacol.* 1963, **5**, 605
2. Coleman, I. W; et al. *Can. J. Phy. Pharmacol.* 1968, **46**, 109
3. *Pestline* 1991, **2**, 1465-1467, Occupational Health Services Inc., Van Nostrand Reinhold, New York
4. *EC Directive on Drinking Water Quality Intended for Human Consumption* 1982, 80/778/EEC, Office for Official Publications of the European Communities, 2 rue Mercier, L-2985 Luxembourg
5. Ghosh, R. *Chem. Ind. (London)* 1955, 118
6. Baldit, G. L. *J. Sci. Food Agric.* 1958, **9**, 516

A174   Amiton oxalate

$$\left[\begin{array}{l} CH_3CH_2O \\ CH_3CH_2O \end{array} \!\!\! P \!\!\! \begin{array}{l} O \\ S \end{array} \!\!\!- (CH_2)_2N(CH_2CH_3)_2 \right]$$

$$\left[HO_2CCO_2H \right]$$

CAS Registry No. 3734-97-2
Synonyms phosphorothioic acid, *S*-[2-(diethylamino)ethyl]-,*O,O*-diethyl ester, oxalate; amiton hydrogen oxalate
Mol. Formula $C_{12}H_{28}NO_7PS$ **Mol. Wt.** 361.40
Uses Contact insecticide and acaricide.

Physical properties

M. Pt. 98-99°C.

Mammalian and avian toxicity

Acute data
LD_{50} oral rat 3-9 mg kg^{-1} (1,2).
LD_{50} intraperitoneal mouse 500 $\mu g\,kg^{-1}$ (3).

Any other adverse effects

Irreversible anticholinesterase agent (4).
Organophosphate pesticide; symptomatic effects of this type of compound include

---

respiratory discomfort, nausea, vomiting, diarrhoea, headache and blurred vision. Can cause central nervous system disorders. Death is primarily due to respiratory failure, although cardiac arrest can be implicated (5).

Legislation
Limited under EC Directive on Drinking Water Quality 80/778/EEC. Pesticide: maximum admissible concentration 0.1 μg l^{-1} (6).

Any other comments
Insecticide and acaricide properties reported (7,8).

References
1. Shafer, E. W; et al. *Toxicol. Appl. Pharmacol.* 1960, **2**, 1
2. *Pesticide Index* 1969, **4**, 162
3. *Pharmacol. Rev.* 1959, **11**, 636
4. Henderson, E. G; et al. *J. Pharmacol. Exp. Ther. 1989,* **251**(3), 810-816
5. *Pestline* 1991, **2**, 1465-1467, Occupational Health Services, Inc., Van Nostrand Reinhold, New York
6. *EC Directive Relating to the Quality of Water Intended for Human Consumption* 1982, 80/778/EEC, Official Publications of the European Communities, 2 rue Mercier, L-2985 Luxembourg
7. Ghosh, R. *Chem. Ind. (London)* 1955, 118
8. Baldit, G. L. *J. Sci. Food Agric.* 1958, **9**, 516

A175   Amitraz

$$\left[H_3C{-}\!\!\overset{\displaystyle CH_3}{\underset{}{\bigcirc}}\!\!{-}N{=}CH \right]_2 \!\!{-}N{-}CH_3$$

CAS Registry No. 33089-61-1
Synonyms ATA; Mitac; Triatox; N'-(2,4- dimethylphenyl)-N-[[2, 4-dimethylphenyl)imino]methyl]- N-ethylmethanimidamide; N-methyl-N'-2,4-xylyl-N-(N-2,4-xylylformimidoyl)formamidine; N,N'-[(methylimino)dimethylidyne]di-2,4-xylidine; 2-methyl-1,3-di(2,4-xylylimino)-2-azapropane
Mol. Formula $C_{19}H_{23}N_3$ **Mol. Wt.** 293.42
Uses Veterinary acaricide used in the treatment of demodecosis in dogs. Insecticide.

Physical properties
M. Pt. 86-87°C; **Specific gravity** d^{25} 1.128; **Volatility** v.p. 3.8×10^{-5} mmHg.

Solubility
Water: 1 mg l^{-1} at 25°C. Organic solvent: acetone, toluene, xylene

Ecotoxicity

Fish toxicity
LC_{50} (48 hr) rainbow trout 2.7-4.0 mg l^{-1}.

LC$_{50}$ (48 hr) Japanese carp 1.17 mg l^{-1}.
LC$_{50}$ (96 hr) bluegill sunfish, harlequin fish 1.3-4.2 mg l^{-1} (1).

Effects to non-target species
LD$_{50}$ (8 day) oral Japanese quail, mallard duck 1800-7000 mg kg^{-1} in diet (2).
LD$_{50}$ (exposure unspecified) *Apis mellifera* 12 μg active ingredient bee^{-1} ingested,
LC$_{50}$ *Apis mellifera* by direct spray 3.6 g l^{-1} (20% emulsifiable concentrate) while
residual contact/fumigant >10g l^{-1} active ingredient (as 20% emulsifiable concentrate)
(2).
LD$_{50}$ 4.42 μg bee^{-1} in feed (3).

Environmental fate

Degradation studies
Amitraz has been investigated in laboratory microcosms using 3 different sediments
(acidic, loamy sand and clay) and their associated water. Amitraz was applied to the
water surface and the microcosms incubated at 25°C in a stream of moist air. Rapidly
dissipated from the water column via hydrolysis and adsorption to sediment. Times
for 90% decline of amitraz in the microcosms ranged from 1.3-8 days. Degradation
was slower in acidic compared to alkaline medium (4).

Abiotic removal
Amitraz was rapidly dissipated from water via hydrolysis and adsorption to sediment (4).

Mammalian and avian toxicity

Acute data
LD$_{50}$ oral mouse, rat 800->1600 mg kg^{-1} (2).
LD$_{50}$ percutaneous rat >1600 mg kg^{-1} (5).
A single oral, intraperitoneal dose rat 20, 60 and 100 mg kg^{-1} caused a depressant
effect on the central nervous system (6).
Amitraz 555 ppm was administered to fasting dogs 4 hr prior to a single intravenous
injection 0.6 g kg^{-1} of glucose. Plasma glucose concentrations increased but the
increase in plasma insulin concentration which usually follows intravenous
administration of glucose was suppressed. It was concluded that amitraz induces
hyperglycemia

Sub-acute data
No adverse effect level in rats 50 mg kg^{-1} in diet. ≥100 mg kg^{-1} inhibited monoamine
oxidase in rats, negligible inhibition of acetylcholine esterase (8).
Repeat doses intravenous dog 1, 2 and 5 mg kg^{-1} caused transient increase in blood
pressure, decrease in heart rate and depressed respiratory rate. High dosage caused
hyperventilation (9).

Metabolism and pharmacokinetics
In the rat, the principal metabolites included 2,4-dimethylformanilide,
2,4-dimethylaniline, 4-formamido-3- methylbenzoic acid and
4-amino-3-methylbenzoic acid (10).

Genotoxicity
Salmonella typhimurium TA98, TA100, TA97, and TA102 with and without
metabolic activation negative.
Escherichia coli PQ243 SOS chromotest negative (11).

Any other adverse effects

In Sprague-Dawley, Long-Evans and Fischer 344 rats symptoms of amitraz exposure included excitability, hyperreactivity and physiological and autonomic changes. All effects increased with repeated dosing (12).

Intraperitoneal Long-Evans rat (1-4 hr, 1-8 day) 10, 25, 50, 100 or 200 mg kg^{-1} symptoms included depressed arousal and rearing activity, hypothermia, body weight loss, ptosis, chromodacryorrhea resulting in facial crustiness, loss of pupil reflex and decreased defaecation. Altered gait and decreased foot splay landing ability were also observed (13).

Legislation

Limited under EC Directive on Drinking Water Quality 80/778/EEC. Pesticide: maximum admissible concentration 0.1 μg l^{-1} (14).

Any other comments

Unstable at pH >7. Slow deterioration of moist compound on prolonged standing.

References

1. *The Agrochemicals Handbook* 3rd ed., 1991, RSC, London
2. *Pesticide Manual, A World Compendium* 9th ed., 1991, Worthing, C. R. (Ed.), British Crop Protection Council, Farnham
3. Iranov, Y. A; et al. *Vestn. S-kh. Nauki. Kaz* 1987, **4**, 68-70, (Russ.) (*Chem. Abstr.* **107**, 190409p)
4. Allen, R; et al; *Brighton Crop Prot. Conf. Pests Dis.* 1990, **3**, 1023-1028
5. Harrison, I. R; et al. *Pest. Sci.* 1972, **3**, 679
6. Florio, J. C; et al. *Brazil. J. Med. Biol. Res.* 1989, **22**(10), 1291-1303
7. Hsu, W. H; et al. *Am. J. Vet. Res.* 1988, **49**(1), 130-131
8. Moses, V. C; et al. *Fundam. Appl. Toxicol.* 1989, **12**(1), 12-22
9. Cullen, L. K; et al. *J. Vet. Pharmacol. Ther.* 1987, **10**(2), 134-143
10. Knowles, C. O; et al. *J. Environ. Sci. Health Part B* 1981, **B16**, 547-556
11. Tudek, B; et al. *Mutat. Res.* 1988, **204**(4), 585-591
12. Moser, V. C; et al. *Toxicol. Appl. Pharmacol.* 1991, **108**(2), 267-283
13. Moser, V. C. *Fundam. Appl. Toxicol.* 1991, **17**(1), 7-16
14. *EC Directive Relating to the Quality of Water Intended for Human Consumption* 1982, 80/778/EEC, Office for Official Publications of the European Communities, 2 rue Mercier, L-2985 Luxembourg

A176   Amitriptyline

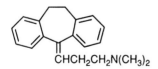

CHCH$_2$CH$_2$N(CH$_3$)$_2$

CAS Registry No. 50-48-6
Synonyms 1-propanamine, 3-(10,11-dihydro-5*H*-dibenzo-(a,d)cyclohepten-5-ylidene, *N,N*-dimethyl-; Triptisol
Mol. Formula C$_{20}$H$_{23}$N   **Mol. Wt.** 277.41

Uses Anti-depressant.

Physical properties
M. Pt. 196-197°C.

Solubility
Organic solvent: ethanol, acetone, chloroform, methanol

Mammalian and avian toxicity

Acute data
LD$_{50}$ oral rat, mouse 140-320 mg kg^{-1} (1,2).
LD$_{50}$ intraperitoneal, intravenous rat, mouse 16-72 mg kg^{-1} (3).

Sub-acute data
Oral rat (21 day, twice daily) 10 mg kg^{-1} reduced the concentration of 5-α-dihydrotestosterone in serum, cerebral cortex and hypothalamus, a decrease in testosterone level was also observed in the hypothalamus (4).

Metabolism and pharmacokinetics
Mostly hydroxylated at position 10, N- demethylation and glucuronide conjugation. About 1% is converted to the N-oxide. It is excreted in the urine, mainly in the form of its metabolites, either free or in a conjugated form (5,6).
Intraperitoneal rat (6-9 hr) 20 mg kg^{-1} major metabolites included E-10-hydroxyamitriptyline, 10,11-dihydroxyamitriptyline, and the phenol E-2-hydroxyamitriptyline. Minor metabolites included 2,10- and 2,11-dihydroxyamitriptyline, 2,10,11- trihydroxyamitriptyline and 2-hydroxy-3-methoxyamitriptyline and the dehydration product 3-hydroxy-10,11-dehydroxyamitriptyline (7).

Genotoxicity
Salmonella typhimurium TA98, TA100, TA1535 with and without metabolic activation negative (8).
In vitro human lymphocyte exposed to plasma levels showed no evidence of chromosome aberrations, mitotic index and sister chromatid exchange. High doses 1 and 10 μg ml^{-1} slightly increased frequencies of chromosome aberrations and sister chromatid exchange (9).
Drosophila melanogaster wing spot somatic mutation and recombination test negative (10).

Any other adverse effects to man
Six drug-free, healthy elderly subjects received a single oral dose 50 mg of amitriptyline. Observed effects included reduced salivary volume, drowsiness and impaired psychomotor performance (11).
In a 13 day study of 12 healthy men, 150 mg day^{-1} symptomatic effects included delay in intracardiac conduction, increased heart rate, decreased salivation, constipation and sedation (12).

Any other adverse effects
In perfused rat heart amitriptyline caused protein degradation and structural damage (13).

Any other comments

Often administered as hydrochloride salt. Human health effects reviewed (14).

References

1. *Clin. Toxicol.* 1977, **10**, 327
2. *Therapie* 1971, **26**, 459
3. *Pol. J. Pharmacol. Pharm.* 1975, **27**, 503
4. Przegalinski, E; et al. *Pol. J. Pharmacol. Pharm.* 1987, **39**(6), 683-689
5. Garland, W. A; et al. *Clin. Pharm. Ther.* 1979, **25**(6), 844
6. Florey, K. *Analytical Properties of Drug Substances* 1974, **3**, 127, Academic Press Inc., New York
7. Breyer-Ptaff, U; et al. *Drug Metals. Dispos.* 1987, **15**(6), 882-889
8. Ferguson, L. R; et al. *Mutat. Res.* 1988, **209**(1-2), 57-62
9. Saxena, R; et al. *Environ. Mol. Mutagen.* 1988, **12**(4), 421-430
10. Van Schaik, N; et al. *Mutat. Res.* 1991, **260**(1), 99-104
11. Ghose, K; et al. *Eur. J. Clin. Pharmacol.* 1987, **33**(5), 505-509
12. Warrington, S. J; et al. *Br. J. Clin. Pharmacol.* 1989, **27**(3), 343-351
13. Hull, B. E; et al. *Toxicol. Appl. Pharmacol.* 1986, **86**(2), 308-324
14. *Martindale: The Extra Pharmacopoeia* 29th ed., 1989, The Pharmaceutical Press, London

A177   Amitrole

CAS Registry No. 61-82-5

Synonyms aminotriazole; $1H$-1,2,4-triazol-3-ylamine; 3-amino-$1H$-1,2,4-triazole; 3-amino-s-triazole; amizol; cytrol; Weedazol

Mol. Formula $C_2H_4N_4$ **Mol. Wt.** 84.08

Uses Non-selective herbicide. Cotton defoliant. Photographic reagent. Use restricted to non-food crops.

Physical properties

M. Pt. 157-159°C; **Specific gravity** 1.138 at 20°C; **Volatility** 7.5×10^{-6} mmHg at 20°C.

Solubility

Water: 280 g l^{-1} at 25°C. Organic solvent: methanol, ethanol, chloroform, dichloromethane, acetonitrile

Occupational exposure

US TLV (TWA) 0.2 mg m^{-3}.

Risk phrases

Harmful if swallowed – Possible risk of irreversible effects – Danger of serious damage to health by prolonged exposure (R22, R40, R48)

Safety phrases
Wear suitable protective clothing – Wear suitable gloves (S36, S37)

Ecotoxicity

Fish toxicity
LC_{50} (96 hr) mosquito fish 2100 mg l^{-1} (1).
LC_{50} (48 hr) silver salmon 325 mg l^{-1} (2).
LC_{50} (48 hr) bluegill sunfish >180 mg l^{-1} (3).
LC_{50} (2 hr) whitebait 0.45% corresponding to active ingredient concentration of 26 mg l^{-1} (4).

Invertebrate toxicity
Mud crab (4 hr exposure) direct spray full strength mixture no mortality (4).
EC_{50} (5-30 min) *Photobacterium phosphoreum* 180-582 mg l^{-1} Microtox test (5).
No effect level (48 hr) for *Gammarus fasciatus, Asellus brevicaudas, Palaemonetes kadiakensis, Orconectes nais* all 100 mg l^{-1} (6).

Bioaccumulation
Calculated bioconcentration 0.5 (7).
Reported as having no or low bioaccumulation (8).

Effects to non-target species
Non-toxic to bees (9).
LC_{50} oral Japanese quail, ring-necked pheasant >5000 ppm in diet, no mortality in 10- or 12-day old birds (10).

Environmental fate

Nitrification inhibition
50% inhibition of ammonia oxidation in *Nitrosomas sp.* at 70 mg l^{-1} (11).

Degradation studies
Persists in soil for 2-4 wk, breakdown principally by microbial action (9).

Mammalian and avian toxicity

Acute data
LD_{50} oral mouse, rat 1.1-2.5 g kg^{-1} (7, 12).
LD_{50} intraperitoneal mouse 200 mg kg^{-1} (13).

Sub-acute data
Prolonged feeding study (68 wk) 50 mg kg^{-1} in diet caused enlarged thyroids in ♂ rats after 13 wk (9,14).
Intraperitoneal rat (8-10 wk 1 g kg^{-1} 3 times wk^{-1} liver effects included reduced activity of catalase and superoxide dismutase and increased GSH level. Heart catalase activity was also decreased (15).
Oral rabbits (43 day) 0.2% in drinking water produced catarctous changes and 50% reduction in iris and ciliary process catalase activity (16).

Carcinogenicity and long-term effects
Inadequate evidence for carcinogenicity to humans, sufficient evidence for carcinogenicity to animals, IARC classification 2B (17).
Oral, subcutaneous rat, mice (dose unspecified) induced thyroid and liver tumours (18).

Teratogenicity and reproductive effects

Gavage (8-12 day gestation) ICR/SIM mouse 10 g kg^{-1} total dose Chernoff/Kavlock development toxicity screen non-teratogenic, nonembryotoxic (19).

Genotoxicity

Salmonella typhimurium TA97, TA98, TA100, TA1535, TA1537 with and without metabolic activation negative (20).
Mouse C3H/10T1/2 embryonic fibroblasts without metabolic activation negative (21).
L5178Y mouse lymphoma cell forward mutation assay tk+/tk- negative (22).
Drosphila melanogaster wing spot test (48 and 72 hr) positive (23).
Asperigillus nidulans mutagenic effects negative (24).

Any other adverse effects to man

Studies on Swedish railway workers (1957-78) showed increased incidence of tumours but not corroborated (14).
In a study of 348 railroad workers exposed to amitrole 5 reported deaths resulted from cancer, including 2 lung cancers, 1 pancreatic cancer, 1 reticulum-cell sarcoma and 1 maxillary sinus cancer versus 3.3 expected (25).

Any other adverse effects

1 g kg^{-1} fed to ♂ rats produced moderate liver necrosis and increased serum glutamic-pyruvic transaminase activity (26).
Antithyroid agent, inhibits liver and kidney catalase (27).

Legislation

Limited under EC Directive on Drinking Water Quality, 80/778/EEC. Herbicide: maximum admissible concentration 0.5 μg l^{-1} (28).

Any other comments

Environmental fate, experimental toxicology and human health effects reviewed (29).

References

1. Johnson, C. R; *Proc. Roy. Soc. Queensland* 1978, **89**, 25
2. Bond, C. E; et al. *R. A. Saft San. Eng. Cen. Tech. Rep.* 1960, 96-101, Report No. W603
3. Robertson, E. B; *Govt. Report Announcements* 1975, **75**(14), Abstr. No. 75-058
4. Gillespie, P. A. *Environ. Toxicol. Chem.* 1989, **8**(9), 809-815
5. Kaiser, K. L. E; et al. *Water Pollut. Res. J. Canada* 1991, **26**(3), 361-431
6. Sanders, H. O; *J. Water Poll. Contr. Fed.* 1970, **42**(8), 1544-1550
7. Lyman, W. J; et al. *Handbook Chemical Property Estimation Methods* 1982, McGraw-Hill, New York
8. *Japan Chemical Industry Ecology-Toxicology & Information Center (JETOC)* 1991, **5**, 13
9. *The Agrochemicals Handbook* 3rd ed., 1991, RSC, London
10. Hill, E. F; *Lethal Dietary Toxicities of Environmental Pollutants to Birds* 1975, **191**, 9, US Dept. Inter. Fish. Wildlife, Washington, DC
11. Hooper, A; et al. *J. Bacteriol.* 1973, **115**, 480
12. *The Merck Index* 11th ed. 1989, Merck & Co. Inc, Rahway, NJ
13. *NTIS Report* AD 277-689, National Technical Information Service, Springfield, VA
14. Axelson, O; et al. *Scand. J. Work, Environ. Health* 1980, **6**(1), 73-79
15. Antonenkov, V. D; et al. *Vo'r. Med. Khim.* 1987, **33**(6), 59-64, (Russ.) (*Chem. Abstr.* **108**, 33463e)
16. Costarides, A; et al. *Lens Eye Toxic. Res.* 1989, **6**(1-2), 167-173

17. *IARC Monograph* 1987, **Suppl. 7**, 57
18. Tsuda, H; et al. *J. Natl. Cancer Inst.* 1976, **57**, 861-864
19. Seidenberg, T. M; et al. *Terat. Carcinog. Mutagen.* 1986, **6**, 361-374
20. Zeiger, E; et al. *Environ. Mol. Mutagen.* 1988, **11**(Suppl. 12), 1-157
21. Dunkel, V. C; et al. *Environ. Mol. Mutagen.* 1988, **12**(1), 21-31
22. McGregor, D. B; et al. *Environ. Mol. Mutagen.* 1987, **9**(2), 143-160
23. Tripathy, N. K; et al. *Mutat. Res.* 1990, **242**(3), 169-180
24. Crebelli, R; et al. *Mutat. Res.* 1986, **172**(2), 139-149
25. *IARC Monograph* 1982, **Suppl. 4**, 38-40
26. Butler, W. H; et al. *Toxicology* 1978, **9**(1/2), 103-107; (*Pollution Abstr.* 1978, 78-05673)
27. *Compendium of Safety Data Sheets for Research and Industrial Chemicals* 1985, L. H. Keith and D. B. Walters (Eds.), VCH Publishers, New York
28. *EC Directive Relating to the Quality of Water Intended for Human Consumption* 1982, 80/778/EEC, Office for Official Publications of the European Communities, 2 rue Mercier, L-2985 Luxembourg
29. *ECETOC Technical Report No. 30(4)* 1991, European Chemical Industry Ecology & Toxicology Centre, B-1160 Brussels

A178   Ammonia

NH₃

CAS Registry No. 7664-41-7
Synonyms ammonia gas; AM-FOC; NITRO-SIL; R717; Spirit of Hartshorn; anhydrous ammonia
Mol. Formula NH_3 **Mol. Wt.** 17.03
Uses In the manufacture of nitric acid, explosives, synthetic fibres and fertilisers. In refrigeration and as a chemical intermediate in the production of cyanides, amides, nitrates and dyestuffs.
Occurrence As a result of farming practices, crude sewage, breakdown of animal and vegetable waste. In fuel emissions-coal, oil, natural gas, wood (1-4).
At neutral pH ammonia exists as its ionic form which is less toxic than ammonia.

Physical properties

M. Pt. –77.7°C; **B. Pt.** –33.4°C (decomp.); **Specific gravity** 0.771 g l^{-1} at 760 mmHg (temperature unspecified); **Volatility** v.p. 7600 mmHg at 25.7°C; v. den. 0.6.

Solubility

Water: 531 g l^{-1} at 20°C. Organic solvent: diethyl ether, ethanol, methanol, chloroform

Occupational exposure

US TLV (TWA) 25 ppm (17 mg m^{-3}); **US TLV (STEL)** 35 ppm (24 mg m^{-3}); **UK Long-term limit** 25 ppm (17 mg m^{-3}); **UK Short-term limit** 35 ppm (24 mg m^{-3}); **UN No.** 1005 (anhydrous, liquefied or solns. s.g. <0.88 at 15°C in H_2O with >50% NH₃); 2073 (solns. s.g. <0.88 at 15°C with 35% ≤50% NH₃); 2673 (solns. s.g. 0.88-0.957 at 15°C with >10% but ≤35% NH₃); **HAZCHEM Code** 2PE;
Conveyance classification toxic gas (anhydrous, liquidified or solns. s.g. <0.88 at 15°C in H_2O with >35% NH₃); corrosive substance (solns. s.g. 0.88-0.957 at 15°C

with >10% but ≤35% NH₃); **Supply classification** corrosive (>35% NH₃); irritant (≥10% ≤35% NH₃).

Risk phrases
>35% NH₃ – Causes burns – Irritating to eyes, respiratory system and skin (R34, R36/37/38); ≥10%≤35% NH₃ – Irritating to eyes, respiratory system and skin (R36/37/38)

Safety phrases
>35% NH₃ – Keep container tightly closed – In case of contact with eyes, rinse immediately with plenty of water and seek medical advice (S7, S26); ≥10%≤35% NH₃ – Keep out of reach of children – In case of contact with eyes, rinse immediately with plenty of water and seek medical advice (S2, S26)

Ecotoxicity

Fish toxicity
LC_{50} (96 hr) fathead minnow, goldfish 2.5-8.2 mg l⁻¹ (5,6).
LC_{50} (96 hr) rainbow trout 0.53 mg l⁻¹ (7).
Goldfish were exposed to 213 mg l⁻¹ and 0.91 mg l⁻¹. Fish exposed to 213 mg l⁻¹ died within 96 hr. Exposure caused hyperexcitability, hyperventilation, postmortem investigation showed gill congestion and haemorrhage (8).
LC_{50} (96 hr) 1 and 4 months old spotted seatrout 0.98 and 1.72 mg l⁻¹ respectively (9).
Juvenile rainbow trout (4 wk) exposure to 0.25 mg l⁻¹ ammonia. Initial effects recorded were increased ventilation frequency, reduced food intake and decreased weight gain. Rapid adaption took place and no change from exposure to the pollutant could be detected in the number of mucous cells in the epidermis at the end of the experiment (10).
LC_{50} (96 hr) juvenile chinook salmon 0.45 mg l⁻¹ in water at 7°C. Ammonia concentration in aquatic sediment was 31 g l⁻¹ (11).

Invertebrate toxicity
Ammonia toxicity to *Gammarus pacustris* and *Asellus aquaticus* at room temperature increased with decreasing pH. Synergism of ammonium ion and hydrogen ion toxicity to sodium transport in gills of crustaceans is discussed in relation to water pollution with acids (12).
LC_{50} (96 hr) *Musculium transversum* 1.10 mg l⁻¹ (7).
LC_{50} (24, 48, 96, 120 hr) juvenile *Penaeus chinensis* 3.29, 2.10, 1.53 and 1.44 mg l⁻¹ for unionised ammonia as nitrogen by a static renewal method in 33% seawater at pH 7.94 and 26°C. The threshold was found at 120 and 192 hr for ammonia and nitrate respectively (13).
LC_{50} (48 hr) juvenile grass shrimp 1.2 mg l⁻¹.
LC_{50} (48 hr) juvenile killifish 1.6 mg l⁻¹ (14).
EC_{50} (5 min) *Photobacterium phosphoreum* 2 mg l⁻¹ Microtox test (15).

Bioaccumulation
32 μg l⁻¹ accumulation in channel catfish (3).

Environmental fate

Nitrification inhibition
Synechoccoccus sp. SF1 isolated from *Sargassum fruitans* was capable of autotrophic growth using ammonia as the sole source of nitrogen; carbonate supplied the carbon source (16).

Nitrobacter sp. ammonia tolerance limit 0.1-1.0 mg l^{-1} dependent on concentration, pH, time of exposure and biomass concentration (17).

Arthrobacter P1 ammonia assimilation involved NADP-dependent glutamate dehydrogenase, NAD-dependent alanine dehydrogenase and glutamine synthetase as key enzymes (18).

Degradation studies
In soil, major sources of ammonia are the aerobic degradation of organic matter and the application and atmospheric deposition of synthetic fertilisers (19).

Abiotic removal
In the atmosphere ammonium ions are oxidised to nitrous oxides and the nitrate ion which represent a significant contribution to acid rain (20).

Absorption
Ammonia strongly adsorbs to soil and sediment particles and colloids in water (20).

Mammalian and avian toxicity

Acute data
LD_{50} oral rat 350 mg kg^{-1} (21).
LC_{Lo} (5 min) guinea pig 3500 mg m^{-3}.
LC_{50} (1 hr) inhalation rat, mouse, rabbit 3360-7050 mg m^{-3} (20,22-23).
Inhalation human (5 min) 134 ppm caused irritation to eyes nose and throat (24).

Sub-acute data
Inhalation rat (3-42 day) 30-150 ppm decreased ability to clear bacteria from lungs and caused encephalopathy (25,26).
Inhalation rat (5-15 day) 25 or 300 ppm 6 hr day^{-1} showed dose-dependent increases of blood ammonia, blood and brain glutamine levels and hepatic citrulline synthesis. High dose animals exhibited slight acidosis (27).

Metabolism and pharmacokinetics
Foetal uptakes of amino acids and ammonia via umbilical circulation were measured in single pregnant ewes at mid-gestation 66-81 days. Significant net fluxes from placenta to foetus of ammonia and 12 amino acids and net fluxes from foetus to placenta of glutamate and serine took place (28). Intravenous rat (6 hr) exposure to ammonia increased brain glutamine, decreased brain serotonin (29).
Inhalation rat (5-15 day) 25 ppm increased brain glutamine (30).
Ammonia is normally present in all tissues constituting a metabolic pool. Its distribution is pH dependent, since NH_3 diffuses more easily that NH_{4+}. Ammonia is taken up by glutamic acid in many tissues, and this will take part in a variety of transaminations and other reactions, the nitrogen being incorporated in non-essential amino acids. In the liver, ammonia is used in the synthesis of urea by the Krebs- Henseleit cycle. Mammals excrete urea and secrete ammonium salts in the kidney tubules as a means of hydrogen ion abstraction. Faecal and respiratory excretion are insignificant (19).

Irritancy
In humans is an eye irritant and causes lachrymation. Can cause conjunctival oedema, corneal damage and acute glaucoma. Late complications include closed-angle glaucoma, opaque corneal scars, atrophy of the iris and formation of cataracts (31).

Genotoxicity

Escherichia coli without metabolic activation positive (32).

Any other adverse effects to man

In 2 plants in the USSR, workers engaged in the manufacture of carbon fibres by pyrolysis of polyacrylonitrile were exposed to carbon dust and low concentrations of ammonia, hydrogen cyanide, acrylonitrile and carbon monoxide. Occupational diseases of the respiratory tract, skin and eyes reported (33).

Increased rates of skin, laryngeal, gastrointestinal and bronchopulmonary diseases were observed among workers engaged in the manufacture of enzyme preparations by microbial fermentation. The workers were exposed to airborne enzymes protease and pectinase in excess of permissible limits in addition to airborne ammonia and formaldehyde (34).

Eight human subjects were exposed to 2 mg m^{-3} ammonia for 42 days in a closed chamber. Exposure time related changes were observed in urinary concentrations of adrenaline, noradrenaline, DOPA, dopamine and urea, blood concentrations of histamine, serotonin, urea and activities of acetylcholinesterose, nonspecific cholinesterose and ammonia concentration in expired air (35).

Healthy, mature, non-starved brain incorporated 7.22 µg ammonia 100 g min^{-1}. In patient suffering from incipient early onset dementia of the Alzheimer type, the brain released 25 µg ammonia 100 g min^{-1}. Ammonia may be involved in the morphological changes in astrocytes and in the gliosis observed in early degeneration related to Alzheimer (36).

Five yr follow-up study of a patient whose persistent airflow obstruction was caused by accidental inhalation of concentrated ammonia fumes (37).

Any other adverse effects

Symptoms of exposure include a burning sensation in the eyes, nose and throat, respiratory distress, lachrymation, coughing and increased respiratory rate. Severe exposure can result in laryngeal and pulmonary pneumonia and bronchopneumonia. Symptoms of exposure are usually reversible but chronic bronchitis and bronchiectasis have been reported (38).

Legislation

Limits under EEC Directive on Drinking Water Quality 80/778/EEC. Ammonium ion guide level 0.05 mg l^{-1}, maximum admissible concentration 0.5 mg l^{-1} (39).
In the UK the use of ammonia in cosmetics is prohibited by law (31).

Any other comments

Physical properties, neurotoxicology, metabolism, toxicolgy, fish toxicity, human health effects, environmental fate and storage reviewed (40-58).

Dilute solutions of ammonia have been used as reflex stimulants either as smelling salts, solutions or oral administration (31).

Ammonia does not present a direct threat to humans except as a result of accidental exposure, particularly in industry.

Decomposes to hydrogen and nitrogen at high temperatures. Explosion risk.

References

1. Weiss, W. P; et al. *J. Animal Sci.* 1986, **63**, 525

2. *Code of Federal Regulations* 1984, **49**, 172.102
3. *Code of Federal Regulations* 1986, **49**, 172.101
4. *Morbid. Mortal. Weekly Rep.* 1985, Centre for Disease Control, Atlanta, GA
5. Jones, H. R. *Environmental Control in the Organic and Petrochemical Industries* 1971, Noyes Data Corp., Park Ridge, NJ
6. McGauhey, P. H. *Engineering Management of Water Quality* 1968, McGraw-Hill, New York
7. Arthur, J. W; et al. *Bull. Environ. Contam. Toxicol.* 1987, **38**(2), 324-331
8. Tarazona, J. V; et al. *Aquacult. Fish Manage.* 1987, **18**(2), 167-172
9. Daniels, H. V; et al. *Prog. Fish. Cult.* 1987, **49**(4), 260-263
10. Lang, T; et al. *Dis. Aquat. Org.* 1987, **3**(3), 159-165
11. Servizi, J. A; et al. *Bull. Environ. Contam. Toxicol.* 1990, **44**(4), 650-656
12. Vinogradov, G. A; et al. *Eksp. Vodn. Toksikol.* 1985, **10**, 35-40, (Russ.) (*Chem. Abstr.* **106**, 79887h)
13. Chen, J. C; et al. *Mar. Biol. (Berlin)* **107**(3), 427-431
14. Burton, D. T; et al. *Bull. Environ. Contam. Toxicol.* 1990, **44**(5), 776-783
15. Kaiser, K. L. E; et al. *Water Pollut. Res. J. Canada* 1991, **26**(3), 361-431
16. Spiller, H.: et al. *J. Bacteriol.* 1987, **169**(12), 5379-5384
17. Sutharsan, S; et al. *Water Pollut. Res. J. Can.* 1986, **21**(2), 257-266
18. De Boer, L; et al. *Antonie van Leevwenhoek* 1989, **56**(3), 221-232
19. *IRPTC Bulletin* 1988, **9**(1), 20-24, United Nations Environment Programme, Geneva
20. *Technical Information for Problem Spills: Ammonia* 1981, 99-103, Environment, Canada
21. *CRC. Crit. Rev. Toxicol.* 1977, CRC Press, Boca Raton, FL
22. *NTIS Report* PB 214-270, National Technical Information Service, Springfield, VA
23. *J. Ind. Hyg. Toxicol.* 1944, **26**, 29
24. NIOSH, *Criteria for a Recommended Standard, Occupational Exposure to Ammonia*, 1974, HEW (NIOSH) Pub. No. 74-136, Washington, DC
25. Kastner, P; et al. *Archiv. Exp. Vet.* 1989, **43**, 191
26. Raabe, W. *Exptl. Neurol.* 1987, **96**, 601
27. Manninen, A; et al. *Pox. Soc. Exp. Biol. Merd.* 1988, **187**(3), 278-281
28. Bell, A. W; et al. *Q. J. Exp. Physiol.* 1989, **74**(5), 635-643
29. Bugge, M; et al. *Res. Exptl. Med.* 1989, **189**, 101
30. Manninen, A; et al. *Proc. Soc. Exptl. Biol. Med.* 1988, **187**, 278
31. *Martindale: The Extra Pharmacopoeia* 29th ed., 1989, The Pharmaceutical Press, London
32. *Am. Nat.* 1951, **85**, 119
33. Troitskaya, W. A. *Gig. Sanit.* 1988, **4**, 21-23, (Russ.) (*Chem. Abstr.* **108**, 209411c)
34. Kuchuk, A. A; et al. *Gig. Tr. Prof. Zabal.* 1989, **4**, 16-18, (Russ.) (*Chem. Abstr.* **110**, 236527t)
35. Saving, V. P; et al. *Kam. Biol. Aviakosm. Med.* 1988, **22**(5), 76-80, (Russ.) (*Chem. Abstr.* **110**, 19516j)
36. Hoyer, S; et al. *Neurosci. Lett.* 1990, **117**(3), 358-362
37. Flury, K. E; et al. *Mayo Clin. Proc.* 1983, **58**, 389
38. *Chemical Safety Data Sheets* 1990, **3**, 33-40, RSC, London
39. *EC Directive Relating to the Quality of Water Intended for Human Consumption* 1982, 80/778/EEC, Office for Official Publications of the European Communities, 2 rue Mercier, L-2985 Luxembourg
40. Jayaweera, G. R; et al. *Adv. Agron.* 1991, **45**, 3033-356
41. Felipo, V; et al. *J. Clin. Gastroenterol.* 1990, **5**(3), 165-169
42. Cooper, A. J. L. *Adv. Exp. Med. Biol.* 1990, **272**, 23-46
43. Raabe, W. *Brain Res. Ser.* 1986, **14**, 396-403
44. Reitzer, L. J; et al. *Escherichia coli Salmonella typhimurium* 1987, **1**, 302-320, Dept. Biol. Massachusetts Inst. Technol., Cambridge, MA
45. Dougall, D. K. *Plant. Biol.* 1987, **3**, 97-117
46. *Cah. Notes Doc.* 1987, **128**(3), 461-465, (Fr.) (*Chem. Abstr.* **107**, 160534h)

47. Randall, D. J. *Fish Physiol. Biochem.* 1987, **3**(3), 107-120
48. Hertz, L; et al. *Neurochem. Pathol.* 1987, **6**(1), 97-129
49. Sturzenegger, E. *Sulzer Tech. Rev.* 1987, **69**(1), 35-37
50. Souba, W. W. *J. Parenter. Enteral. Nutr.* 1987, **11**(6), 569-579
51. Halussinger, D; et al. *Int. Congr. Ser. Excerpta. Med.* 1988, 761, 26-36
52. Walter, J. H; et al. *J. Appl. Med.* 1988, **14**(5), 305-311
53. Moyana, F. J; et al. *Ars. Pharm.* 1988, **29**(2), 145-152, 163-172, (Span.) (*Chem. Abstr.* **112**, 93277g, 116008v)
54. Schilling, N; et al. *Muench. Beitr. Abwasser-. Fisch.-Flussbiol.* 1989, **43**, 122-147, (Ger.) (*Chem. Abstr.* **113**, 36439r)
55. Byrnes, B. H. *Fert. Res.* 1990, **26**(1-3), 209-215
56. Nagami, G. T. *Miner. Electrolyte Metab.* 1990, **16**(5), 259-263, 270-276
57. Singh, V. K. *Fert. News* 1990, **35**(11), 41-46
58. *ECETOC Technical Report No. 30(4)* 1991, European Chemical Industry Ecology and Toxicology Centre, B-1160 Brussels

A179   Ammonium acetate

CH3CO2NH4

CAS Registry No. 631-61-8
Synonyms acetic acid, monoammonium salt; ethanoic acid, ammonium salt
Mol. Formula $C_2H_7NO_2$ **Mol. Wt.** 77.08
Uses A diaphoretic and diuretic in pharmaceutical applications. A mordant in dyeing wool. A reagent in analytical chemistry. In preserving meats. In the manufacture of foam rubbers and vinyl plastics.

Physical properties

M. Pt. 114°C; **B. Pt.** decomp.; **Specific gravity** d_4^{20} 1.17.

Solubility
Water: 148 g in 100 g at 4°C. Organic solvent: ethanol, acetone

Ecotoxicity

Fish toxicity
LC_{50} (24 hr) mosquito fish 238 mg l^{-1} in fresh water, conditions of bioassay not specified (1).
LC_{50} (48 hr) carp for unionised ammonia 1.15- 1.06 mg l^{-1} (2).
In a study of *Tilapia mossambica* it was concluded that ammonium acetate harmed fish because of its neurotoxic effect. Target organ muscle (3).

Environmental fate

Nitrification inhibition
NH_3 inhibition of nitrification 436 and 1000 mg l^{-1} (4).

Degradation studies
Treatment of wastewater by activated sludge was facilitated by the presence of 1-50 mg l^{-1} ammonium acetate which prevented sludge bulking when the carbon:nitrogen ratio was low (5).

Mammalian and avian toxicity

Acute data
LD$_{50}$ intravenous mouse 1.8 mg [NH$_{4+}$] 20g-1 (6).
LD$_{50}$ intraperitoneal rat 632 mg kg^{-1} (7).

Legislation
Limited under EEC Directive on Drinking Water Quality 80/778/EEC. Ammonium guide level 0.05 mg l^{-1}, maximum admissible concentration 0.5 mg l^{-1} (8).

Any other comments
Short-term regulation of the urea cycle, the mechanism of ammonia toxicity and clinincal implications are discussed (9).

References
1. *Hazardous Chemical Data* 1984/1985, **2**, US Dept. of Transport, Coast Guard
2. Dabrowska, H; et al. *Pol. Arch. Hydrobiol.* 1986, **33**, 121-128, (Pol.) (*Chem. Abstr.* **105**, 185467k)
3. Prakash, J; et al. *Indian J. Environ. Health* 1990, **32**(4), 416-419
4. Vismara, R. *Ingegn Ambient.* 1982, **11**, 634
5. Shozo, H; et al. *Jpn. Kokai Tokkyo Koho JP 63 07,899* 1988, Appl. 86/151925, (*Chem. Abstr.* **100**, 173016m)
6. Welch, et al. *J. Lab. Clin. Med.* 1944, **29**, 809
7. *Arch. Biochem. Biophys.* 1956, **64**, 342
8. *EC Directive Relating to the Quality of Water Intended for Human Consumption* 1982, 80/778/EEC, Office for Official Publications of the European Communities, 2 rue Mercier, L-2985 Luxembourg
9. Costell, M; et al. *Biochem. Biophys. Res. Commun.* 1990, **167**(3), 1263-1270

A180   Ammonium arsenate

(NH$_4$)$_2$HAsO$_4$

CAS Registry No. 7784-44-3
Synonyms diammonium monohydrogen arsenate; ammonium acid arsenate
Mol. Formula AsH$_9$N$_2$O$_4$ **Mol. Wt.** 176.00

Occupational exposure
US TLV (TWA) 0.2 mg m^{-3}; **UK Long-term limit** 0.1 mg m^{-3}; **UN No.** 1546; **HAZCHEM Code** 2X; **Conveyance classification** toxic substance; **Supply classification** toxic.

Risk phrases
Toxic by inhalation and if swallowed (R23/25)

Safety phrases
Keep locked up and out of reach of children – When using do not eat, drink or smoke – After contact with skin, wash immediately with plenty of water – If you feel unwell, seek medical advice (show label where possible) (S1/2, S20/21, S28, S44)

Ecotoxicity

Fish toxicity
Designated non-toxic to trout, bluegill sunfish and goldfish (1).

Mammalian and avian toxicity

Carcinogenicity and long-term effects
Sufficient evidence for carcinogenicity to humans, limited evidence for carcinogenicity to animals, IARC classification group 1. Applies to arsenic and its compounds as a group and not necessarily the individual compounds (2).

Legislation
Limited under EEC Directive on Drinking Water Quality 80/778/EEC. Ammonium ion guide level 0.05 mg l^{-1}, maximum admissible concentration 0.5 mg l^{-1}. As arsenic ion maximum admissible concentration 50 μg l^{-1} (3).

Any other comments
Genetic toxicology of arsenic compounds has been reviewed (4).

References
1. Wood, E. M. *The Toxicity of 3400 Chemicals to Fish* 1987, US Fish and Wildlife Service, EPA 560/6-87-002
2. *IARC Monograph*, 1987, **Suppl. 7**, 100-106
3. *EC Directive Relating to the Quality of Water Intended for Human Consumption* 1982, 80/778/EEC, Office for Official Publications of the European Communities, 2 rue Mercier, L-2985 Luxembourg
4. *IARC Monograph* 1987, **Suppl. 6**, 71

A181   Ammonium benzoate

CAS Registry No. 1863-63-4
Synonyms benzoic acid, ammonium salt; Vulnoc AB
Mol. Formula $C_7H_9NO_2$ **Mol. Wt.** 139.16
Uses Component in certain rubber formulations.
Used to preserve glue and latex. Urinary anti-infective.

Physical properties
M. Pt. 198°C (decomp.); **Specific gravity** d^{25} 1.26.

Solubility
Water: 19.6 g in 100 ml at 14°C. Organic solvent: glycerol, diethyl ether, ethanol

Legislation

Limited under EEC Directive on Drinking Water Quality 80/778/EEC as ammonia. Ammonium ion guide level 0.05 mg l^{-1}, maximum admissible concentration 0.5 mg l^{-1} (1).

Any other comments

See also benzoic acid. Gradually loses ammonia on exposure to air. Incompatible with ferric salts, acids, alkali hydroxides or carbonates.

Reference

1. *EC Directive Relating to the Quality of Water Intended for Human Consumption* 1982, 80/778/EEC, Office for Official Publications of the European Communities, 2 rue Mercier, L-2985 Luxembourg

A182   Ammonium bicarbonate

NH4HCO3

CAS Registry No. 1066-33-7
Synonyms acid ammonium carbonate; ammonium hydrogen carbonate
Mol. Formula CH_5NO_3 **Mol. Wt.** 79.06

Physical properties

M. Pt. 60°C (decomp.); **Specific gravity** 1.586.

Solubility

Water: 174 g l^{-1} at 20°C. Organic solvent: glycerol

Environmental fate

Degradation studies

Antifungal activity attributed to ammonium bicarbonate results from concentrations of disassociated free ammonia (1).

Legislation

Limited under EEC Directive on Drinking Water Quality 80/778/EEC. Ammonium ion guide level 0.05 mg l^{-1}, maximum admissible concentration 0.5 mg l^{-1} (3).

Any other comments

Pharmaceutical incompatibility with acids and caustic alkalis (4).
See also ammonia, ammonium hydroxide.

References

1. DePasquale, D. A; et al. *Appl. Environ. Microbiol.* 1990, **56**(12), 3711-3717
2. Wilson, R. P; et al. *Am. J. Vet. Res.* 1968, **29**, 897
3. *EC Directive Relating to the Quality of Water Intended for Human Consumption* 1982, 80/778/EEC, Office for Official Publications of the Euroepean Communities, 2 rue Mercier, L-2985 Luxembourg
4. *The Merck Index* 11th ed. 1989, Merck & Co. Inc, Rahway, NJ

A183 Ammonium bisulfite

NH4HSO3

CAS Registry No. 10192-30-0
Synonyms ammonium bisulfite
Mol. Formula H_5NO_3S **Mol. Wt.** 99.11
Uses Preservative.

Physical properties
M. Pt. (decomp.).

Solubility
Water: 2670 g l^{-1} at 10°C.

Legislation
Limited under EC Directive on Drinking Water Quality 80/778/EEC. Ammonia: maximum admissible concentration 0.5 mg l^{-1} (1).

References
1. *EC Directive Relating to the Quality of Water Intended for Human Consumption* 1982, 80/778/EEC, Office for Official Publications of the European Communities, 2 rue Mercier, L-2985 Luxembourg

A184 Ammonium carbamate

NH2CO2NH4

CAS Registry No. 1111-78-0
Synonyms ammonium carbaminate; ammonium aminoformate; ''Anhydride'' of ammonium carbonate
Mol. Formula $CH_6N_2O_2$ **Mol. Wt.** 78.07
Uses Ammoniating agent. Pesticide.

Physical properties
M. Pt. 60°C volatises.

Solubility
Water: miscible. Organic solvent: ethanol

Mammalian and avian toxicity

Acute data
LD_{50} intravenous rat, mouse 39-77 mg kg^{-1} (1).

Legislation
Limited under EC Directive on Drinking Water Quality 80/778/EEC. Maximum admissible concentration 0.1 µg l^{-1}. Ammonia: guide level 0.05 mg l^{-1}, maximum admissible concentration 0.5 mg l^{-1} (2).

References

1. *Am. J. Vet. Res.* 1968, **29**, 897
2. *EC Directive Relating to the Quality of Water Intended for Human Consumption* 1982, 80/778/EEC, Office for Official Publications of the European Communities, 2 rue Mercier, L-2985 Luxembourg

A185   Ammonium carbonate

$(NH_4)_2CO_3$

CAS Registry No. 506-87-6

Synonyms Sal volatile; Hartshorn; carbonic acid ammonium salt; carbonic acid diammonium salt

Mol. Formula $CH_8N_2O_3$ **Mol. Wt.** 96.09

Uses Baking powders, washing and defatting woollens, dyeing, manufacture of rubber articles.

Physical properties

M. Pt. 58°C (decomp.).

Solubility

Water: 250 g l^{-1}. Organic solvent: aqueous methanol

Ecotoxicity

Fish toxicity

LC_{50} (96 hr) fathead minnow 37 mg l^{-1} (1).

Invertebrate toxicity

Ammonium carbonate at 5, 10, 25 or 50 mg NH_{4+}. Decreased the fertility of *Daphnia magna* at 50 mg NH_{4+}. Disrupted embryonic development, and at 10, 25 or 50 mg NH4+. Impaired postembryonic growth of the crustacea. Industrial sewage entering fishing waters apparently should contain ≤ 1 mg NH4+l-1 (2).

Mammalian and avian toxicity

Acute data

LD_{50} intravenous mouse 96 mg kg^{-1} (3).

Any other comments

See ammonia and ammonium hydroxide.

References

1. Curtis, M. W; et al. *J. Hydral.* 1981, **51**, 359
2. Dyga, A. K; et al. *Eksp. Vod. Toksikol.* 1972, **3**, 51-8, (Russ.) (*Chem. Abstr.* **78** 67813w)
3. Wilson, R. P; et al. *Am. J. Vet. Res.* 1968, **29**(4), 897

A186   Ammonium chloride

NH4Cl

CAS Registry No. 12125-02-9

Synonyms sal ammoniac; Salmiac; Amchlor; Darammon; ammonium muriate

Mol. Formula ClH_4N **Mol. Wt.** 53.49

Uses Flux for galvanising and tinning. In dry and Leclanche batteries. Used in the dyeing, tanning and electroplating industries. Detergents. Veterinary expectorant, diaphosetic and acidifying diuretic.

Physical properties

M. Pt. 340°C (sublimes); **B. Pt.** 520°C; **Specific gravity** d^{25} 1.5275; **Volatility** v.p. 3.6×10^{-2} mmHg.

Solubility

Water: 28.3% at 26°. Organic solvent: ethanol, methanol

Occupational exposure

US TLV (TWA) 10 mg m^3 (fume); **US TLV (STEL)** 20 mg m^3 (fume); **UK Long-term limit** 10 mg m^{-3} (fume); **UK Short-term limit** 20 mg m^{-3} (fume); **Supply classification** harmful.

Risk phrases

Harmful if swallowed – Irritating to eyes (R22, R36)

Safety phrases

Do not breathe dust (S22)

Ecotoxicity

Fish toxicity

LC_{50} (48 hr) carp 109 mg l^{-1} (total ammonia), 1.6 mg l^{-1} (unionised ammonia) static bioassay (1).

Invertebrate toxicity

Ammonium chloride at 5, 10, 25 or 50 mg NH_{4+} l-1 decreased the fertility of *Daphnia magna* at 50 mg NH_{4+} l-1, disrupted embryonic development at 10, 25 or 50 mg NH_{4+} l-1, and impaired postembryonic growth of the crustacea. Industrial sewage entering fishing waters apparently should contain $\leq$1 mg NH_{4+} l-1 (2).

Mammalian and avian toxicity

Acute data

LD_{50} oral rat 1650 mg kg^{-1} (3).

LD_{50} oral rabbit 1000 mg kg^{-1} (4).

LD_{Lo} intravenous guinea pig 240 mg kg^{-1} (5).

Metabolism and pharmacokinetics

Following oral administration to ewes, ammonium chloride is rapidly absorbed from the gastrointestinal tract, complete absorption occurring within 3-6 hr (6).

Any other adverse effects

Ewes fed ammonium chloride unspecified dose once a day with a mineral-trace element supplement (50 g) for 7-40 days around the time of fertilisation were 100% fertile as compared with 18% barren for controls, and had 17.2% (av.) more lambs. Blood calcium, phosphorous and total protein levels were increased by the supplements and blood haemoglobin levels were decreased (7).

Any other comments

Sublimes without melting. Strongly endothermic in water. Incompatible with alkalis and their carbonates, lead and silver salts.
Human health effects and experimental toxicology reviewed (8).

References

1. Dabrowska, H; et al. *Pol. Arch. Hydrobiol.* 1986, **33**(1), 121-128
2. Dyga, A. K; et al. *Eksp. Vod. Toksikol.* 1972, **3**, 51-8, (Russ.) (*Chem. Abstr.* **78**, 67813w)
3. Marhold, J. V. *Sbnorik Vysledku Toxixologickeho Vysetreni Latek A Pripravku* 1972, 15, Prague
4. *Abernalden's Handbuch der Biologischen Arbeitsmethoden* 1935, **4**, 1289
5. *J. Pharmacol. Exp. Therap.* 1915, **6**, 695
6. *Am. Hosp. Form. Service-Drug Inform. 88* 1988, 1384, Am. Soc. Hosp. Pharm., Bethesda, MD
7. Gorev, E. L; et al. *Tr. Nauchno-Issled. Vet. Inst.* 1977, **7**, 120-125, (Russ.) (*Chem. Abstr.* **91**, 55022t)
8. *ECETOC Technical Report No. 30(4)* 1991, European Chemical Industry Ecology and Toxicology Centre, B-1160 Brussels

A187   Ammonium chloroplatinate

$(NH_4)_2PtCl_6$

CAS Registry No. 16919-58-7
Synonyms ammonium hexachloroplatinate (IV); ammonium platinic chloride; platinic ammonium chloride
Mol. Formula $Cl_6H_8N_2Pt$ **Mol. Wt.** 443.89
Uses Platinum plating. Manufacture of spongy platinum.

Physical properties

M. Pt. decomp.; **Specific gravity** 3.07.

Occupational exposure

US TLV (TWA) 0.002 mg m^{-3} (as Pt).

Mammalian and avian toxicity

Acute data
LD$_{50}$ oral rat 1 mg kg^{-1} (1).

Any other adverse effects to man

Acute rhinitis, conjunctivitis and bronchial asthma in platinum refinery workers (2,3).

References

1. Veselov, V. G. *Gig. Tr. Prof. Zabol.*, 1977, (7), 55, (Russ.) (*Chem. Abstr.* **87**, 112601z
2. Freedman, S. D; et al. *J. Allergy* 1968, **42**, 233
3. Pepys, J; et al. *Clin. Allergy* 1972, **2**(4), 39

A188 Ammonium chromate

(NH4)2CrO4

CAS Registry No. 7788-98-9

Synonyms ammonium chromate(VI)

Mol. Formula CrH$_8$N$_2$O$_4$ **Mol. Wt.** 152.07

Uses Used in textile printing pastes. Sensitising gelatin in photography. Fixer for chromate dyes on wool. Reagent in analytical chemistry.

Occurrence Chromium (VI) detected in industrial effluent from chromate manufacturing processes and landfill sites (1).

Background ambient air concentrations of total chromium have ranged from as low as 0.005 ng m^{-3} (at the South Pole) to 1.1 ng m^{-3} in other remote areas of the world. Because Cr(III) is highly stable and Cr(VI) reacts over time to form Cr(III), it is assumed that most chromium in ambient air occurs in the trivalent state (1). Detected in drinking water, ground water in US and Canada (2).

Physical properties

M. Pt. 185°C (decomp.); **Specific gravity** 1.8.

Solubility

Water: 198 g l^{-1} at 0°C. Organic solvent: methanol, acetone

Occupational exposure

US TLV (TWA) 0.05 mg m^{-3} (as Cr); **UK Long-term limit** MEL 0.05 mg m^{-3}.

Ecotoxicity

Fish toxicity

LC$_{50}$ (48-96 hr) mosquito fish 212-136 ppm, respectively (3).

Bioaccumulation

Trout can accumulate hexavalent chromium even at levels of 0.001 ppm. Bioconcentration factor for chromium in marine plants 2000, freshwater and brown algae 100-500, marine invertebrates 2000, marine fish 400 and freshwater fish 200 (3). Bioconcentration factors for chromium(6+) range from 125 to 236 for bivalve molluscs and polychaetes (1).

Effects to non-target species

LD$_{50}$ silkworm lava <10 ppm (4).

Abiotic removal

Up to 300 ppm were removed by all types of soil in column studies, soil pH is a major factor for uptake. Crop damage may result from levels above 0.5 ppm free chromate in soil (3).

Chromium is usually present as Cr(III) in the soil and is characterised by its lack of mobility, except in cases where Cr(VI) is involved. Chromium(VI) of natural origin is rarely found (2).

Mammalian and avian toxicity

Carcinogenicity and long-term effects

Sufficient evidence for carcinogenicity to humans and experimental animals for chromium(VI) compounds, IARC classification group 1 (5).

Chromium salts are human and experimental carcinogens of the lungs, nasal cavity and paranasal sinus and are also experimental carcinogens of the stomach and larynx (6). Cancer causing chromium exposure has been attributed to industrial processes involving ammonium chromate (7).

Metabolism and pharmacokinetics

In the airways and in the gastrointestinal tract, soluble Cr(VI) compounds are apparently taken up by epithelial cells, by simple diffusion, through the plasma membrane. After entry, Cr(VI) reduction occurs by enzymatically mobilised electrons, available from GSH, NADPH and NADH. The reducing capacity inside the cell is limited, so that Cr(VI) and Cr(III) exist simultaneously inside the cytoplasm; Cr(VI) is then released from the cell by simple diffusion into the bloodstream and taken up into blood cells (1).

Any other comments

Unlike the trivalent compounds, those of Cr(VI) tend to cross biological membranes fairly easily and are somewhat more readily absorbed through the gut or the skin (1).

References

1. *USEPA Health Assessment Document: Chromium* 1984, 2-27, EPA 600/8-83-014F
2. *Effects of Chromium in the Canadian Environment* 1976, 33-40, NRCC No. 15017
3. Fujii, M; et al. *Nippon Sanshigaka Zassh* 1972, **41**(2), 104
4. *Toxic and Haz. Ind. Chem. Saf. Man.* 1982, The International Techniocal Information Institute, Tokyo
5. *IARC Monograph* 1990, **49**, 49-256
6. Maltoni, C. *Occupational Chemical Carcinogenesis: New Facts, Priorities and Perspectives, Intern. Symposia Series*, 1976, **52**, 127
7. Higgens, I. T. *Epidemiological Evidence on the Carcinogenic Risk of Air Pollution, Intern Symposia Series*, IARC, 1976, **52**, 13

A189   Ammonium dichromate

$(NH_4)_2Cr_2O_7$

CAS Registry No. 7789-09-5
Synonyms Ammonium bichromate (VI)
Mol. Formula $Cr_2H_8N_2O_7$ **Mol. Wt.** 252.06

Uses Used in pyrotechnics, lithography and photoengraving. Mordant. Catalyst.Used in porcelain finishes. Intermediate in the manufacture of pigments. Magnetic recording materials. Source of pure nitrogen.

Occurrence Detected in drinking water, ground water in US and Canada (1-2). Chromium (VI) detected in industrial effluent from chromate manufacturing processes and landfill sites (1).

Background ambient air concentrations of total chromium have ranged from as low as 0.005 ng m^{-3} (at the South Pole) to 1.1 ng m^{-3} in other remote areas of the world. Because Cr(III) is highly stable and Cr(VI) reacts over time to form Cr(III), it is assumed that most chromium in ambient air occurs in the trivalent state (1).

Physical properties

M. Pt. 180°C (decomp.); **Specific gravity** 2.15.

Solubility
Water: 27% at 20 °C. Organic solvent: ethanol

Occupational exposure

US TLV (TWA) 0.05 mg m^{-3} (as Cr); **UN No.** 1439; **HAZCHEM Code** 2X; **Conveyance classification** oxidising substance; **Supply classification** explosive and irritant.

Risk phrases
Explosive when dry – Contact with combustible material may cause fire – Irritating to eyes, respiratory system and skin – May cause sensitisation by skin contact (R1, R8, R36/37/38, R43)

Safety phrases
After contact with skin, wash immediately with plenty of water – This material and its container must be disposed of in a safe way (S28, S35)

Ecotoxicity

Fish toxicity
LC$_{50}$ (48-96 hr) mosquito fish 212-136 ppm, respectively (3).

Bioaccumulation
Trout can accumulate hexavalent chromium even at levels of 0.001 ppm. Bioconcentration factors for chromium in marine plants 2000, freshwater and brown algae 100-500, marine invertebrates 2000, marine fish 400, freshwater fish 200 (3). Bioconcentration factors for chromium (VI) range from 125 to 236 for bivalve molluscs and polychaetes (4).

Abiotic removal
Up to 300 ppm were removed by all types of soil in column studies, soil pH is a major factor for uptake. Crop damage may result from levels above 0.5 ppm free chromate in soil (3).

Chromium is present usually as Cr(III) in the soil and is characterised by its lack of mobility, except in cases where Cr(VI) is involved. Chromium(VI) of natural origin is rarely found (5).

Mammalian and avian toxicity

Acute data
LD_{50} intravenous rat 30 mg kg^{-1} (1,6).
LD_{Lo} subcutaneous guinea pig 25 mg kg^{-1} (1).

Sub-acute data
Inhalation rat (1-6 month) dust at 1 mg m^{-3} 2 hr 3 × wk^{-1} caused disturbances in the blood circulation in lungs and emphysema in the perivascular and peribronchial tissues (7).
Inhalation rat (6 month) (0.05 mg m^{-3}) 2 hr every other day papillary growths in the bronchi, and in some cases pneumonia (8).

Carcinogenicity and long-term effects
Precancerous changes were reported in rats pretreated with a noncarcinogenic dose of benzoapyrene, following intrapleural administration of 0.5-1.0 μg m^{-3} ammonium dichromate, 2 hr day^{-1} 3 x wk^{-1} for 18 months (9).

Metabolism and pharmacokinetics
In the airways and in the gastrointestinal tract, soluble Cr(VI) compounds are apparently taken up by epithelial cells, by simple diffusion, through the plasma membrane. After entry, Cr(VI) reduction occurs from the action of enzymatically mobilised electrons, available from GSH, NADPH and NADH. The reducing capacity inside the cell is limited, so that Cr(VI) and Cr(III) exist simultaneously inside the cytoplasm; Cr(VI) is then released from the cell by simple diffusion into the bloodstream and taken up into blood cells (1).

Irritancy
Causes skin irritation, ulceration (''chrome sores''), perforation of nasal septum, pulmonary irritation (1).

Any other adverse effects

If swallowed prompts vomiting, but if retained leads to kidney injury and stomach ulceration. Unlike the trivalent compounds, those of Cr(VI) tend to cross biological membranes fairly easily and are more readily absorbed through the gut or the skin (1).

Any other comments

Flammable self-sustaining decomposition at 225°C with swelling and evolution of heat and nitrogen, leaving chromic (III) oxide (10).
Human health effects and experimental toxicology reviewed (11).

References

1. *USEPA Health Assessment Document: Chromium* 1984, 2-27, EPA 600/8-83-014F
2. *Effects of Chromium in the Canadian Environment* 1976, 35-37, NRCC No. 15017, National Research Council, Canada
3. *NIH/USEPA Report* 1985, OHM/TADS
4. *USEPA Ambient Water Quality Criteria: Chromium* 1984, 18, EPA 440/5-84-029
5. *Environ. Qual. Saf. Suppl.* 1975, **1**, 1
6. *Toxic and Hazardous Industrial Chemicals Safety Manual* 1982, International Technical Information Institute, Tokyo
7. Inerbaeva, G. S; et al. *Gig. Tr. Prof. Zabol.*, 1974, (Russ.) **6**, 48-9, (*Chem. Abstr.* **81** 100316y)
8. Kovalenko, V. R; et al. *Zdravookhr. Kaz.* 1972, **7**, 50-1, (Russ.) (*Chem. Abstr.* **78** 67952r)

9. Zikeev, V. V; et al. *Zdravookhr. Kaz.* 1973, **8**, 83-4, (Russ.) (*Chem. Abstr.* **80** 56305x)
10. *The Merck Index* 11th ed. 1989, Merck & Co. Inc, Rahway, NJ
11. *ECETOC Technical Report No. 30(4)* 1991, European Chemical Industry Ecology and Toxicology Centre, B-1160 Brussels

A190   Ammonium fluoride

NH4F

CAS Registry No. 12125-01-8
Synonyms neutral ammonium fluoride
Mol. Formula FH$_4$N   **Mol. Wt.** 37.04
Uses Laboratory reagent. Etching and frosting glass. Preserving wood. Printing and dyeing.

Physical properties

M. Pt. 125.6°C; **Specific gravity** d^{25} 1.009.

Solubility
Water: 453 g l^{-1} at 25°C. Organic solvent: ethanol

Occupational exposure

US TLV (TWA) 2.5 mg m^{-3} (as F); **UK Long-term limit** 2.5 mg m^{-3} (as F); **UN No.** 2505; **HAZCHEM Code** 2X; **Conveyance classification** harmful substance; **Supply classification** toxic.

Risk phrases
Toxic by inhalation, in contact with skin and if swallowed (R23/24/25)

Safety phrases
Keep locked up and out of reach of children – In case of contact with eyes, rinse immediately with plenty of water and seek medical advice – If you feel unwell, seek medical advice (show label where possible) (S1/2, S26, S44)

Ecotoxicity

Fish toxicity
LC$_{50}$ (96 hr) silver carp 1.6 ppm. 100 ppm caused 70% inhibition of cholinesterase activity. Teratogenicity reported in white amur grass carp exposed to 15 ppm fluoride (1).
LC$_{50}$ (96 hr) silver carp, white amur grass carp, carp were 1.6, 9.3 and 11.8 ppm, respectively. NH$_4$F at 70 ppm induced the formation of micronuclei erythocytes in carp induced amitosis in the cell. Inhibitory effects on the activity of cholinesterase in fish. NH$_4$F at 100 ppm caused a 70% inhibition, and at 10 ppm caused a 20% inhibition of the enzyme activity. Teratogenicity was found when eggs were hatched and young fish were kept in water containing 15 ppm fluoride. In teratomatous fish, maximum bone fluoride contents were found to be 9608 ppm, which was 100-300-fold higher than the values found in the bones of normal fish (1).
LC$_{50}$ (96 hr) fathead minnow 364 mg l^{-1} (2).

Invertebrate toxicity
LC_{50} (96 hr) grass shrimp 75 mg l^{-1} (2).

Effects to non-target species
LD_{50} silkworm larva <15 ppm (3).

Mammalian and avian toxicity

Acute data
LD_{50} intraperitoneal rat 32 mg kg^{-1} (4).

Metabolism and pharmacokinetics
Fluorides are absorbed from gastrointestinal tract, lung, and skin. Gastrointestinal tract is major site of absorption. Fluoride is preponderantly deposited in the skeleton and teeth and the degree of skeletal storage is related to intake and age. Major route of excretion is via kidneys. Also excreted in small amounts by sweat glands, breast milk and gastrointestinal tract. About 90% of fluoride ion filtered by glomerulus is reabsorbed by renal tubules (5,6).

Following ingestion soluble fluorides are rapidly absorbed from the gastrointestinal tract at least to the extent of 97%. Absorbed fluoride is distributed throughout the tissues of the body by the blood. Fluoride concentrations in soft tissues fall to pre-exposure levels within a few hours of exposure. Fluoride exchange with hydroxyl groups of hydoxyapatite (the inorganic constituent of bone) to form fluorohydroxyapatite. Fluoride that is not retained is excreted rapidly in urine (6).

Any other adverse effects to man

Ingestion causes nausea, vomiting, diarrhoea and abdominal pains. Chronic effects include shortness of breath, cough, elevated temperature and cyanosis (4).

Any other comments

Human health effects and experimental toxicology reviewed (7).
Toxicity and human health effects reviewed (8).
Corrodes glass. On heating decomposes to ammonia and hydrogen fluoride.

References
1. Zhang, R; et al. *Huanjing Kexue* 1982, 3(4), 1-5, (Ch.) (*Chem. Abstr.* 97, 176345k)
2. Curtis, M. W; et al. *Water Res.* 1979, 13, 137-141
3. Fujii, M; et al. *Nippon Sanshigaka Zasshi* 1972, 41(2), 104, (Jap.) (*Chem. Abstr.* 77, 122710s)
4. *Atomic Energy Commission, University of Rochester, R & D Reports* 1951, UR-154, Rochester, New York
5. Gilman, A G; et al. (Ed.) *Goodman and Gilman's The Pharmacological Basis of Therapeutics* (6th ed.) 1980, 1546, Macmillan Publ., New York
6. *USEPA Office of Drinking Water; Criteria Document (Draft): Fluoride* 1985, (III), 19
7. *ECETOC Technical Report No. 30(4)* 1991, European Chemical Industry Ecology and Toxicology Centre, B-1160 Brussels
8. *Martindale: The Extra Pharmacopaeia* 29th ed., 1989, The Pharmaceutical Press, London

A191   Ammonium fluoroborate

NH₄BF₄

CAS Registry No. 13826-83-0
Synonyms ammonium tetrafluoroborate; ammonium borofluoride
Mol. Formula BF₄H₄N   Mol. Wt. 104.84
Uses Flame retardant. Flux for inert atmosphere soldering.

Physical properties
M. Pt. 230°C; Specific gravity 1.87.

Solubility
Water: 258 g l^{-1} at 25°C.

Occupational exposure
US TLV (TWA) 2.5 mg m^{-3} (as F); UK Long-term limit 2.5 mg m^{-3} (as F); UN No. 2811.

Mammalian and avian toxicity

Metabolism and pharmacokinetics
Following ingestion, soluble fluorides are rapidly absorbed from the gastrointestinal tract at least to the extent of 97%. Absorbed fluoride is distributed throughout the tissues of the body by the blood. Fluoride concentrations in soft tissues fall to pre-exposure levels within a few hours of exposure. Fluoride exchange with hydroxyl groups of hydroxyapatite (the inorganic constituent of bone) to form fluorohydroxyapatite. Fluoride that is not retained is excreted rapidly in urine (1). Borates are rapidly absorbed from mucous membranes and abraded skin, but not from intact or unbroken skin. Borate excretion occurs mainly through kidneys; ≈ 50% is excreted in first 12 hr and remainder is eliminated over a period of 5 to 7 days (2).

References
1. *USEPA Office of Drinking Water Criteria Document (Draft): Fluoride* 1985, (III), 19
2. Gosselin, R. E; et al. *Clinical Toxicology of Commercial Products* 5th ed., 1984, Williams & Wilkins, Baltimore

A192   Ammonium fluorosilicate

(NH₄)₂SiF₆

CAS Registry No. 16919-19-0;001309-32-6
Synonyms ammonium silicofluoride; ammonium hexafluorosilicate; silicate(2-), hexafluoro-, diammonium; diammonium hexafluorosilicate; cryptohalite; ammonium fluosilicate; diammonium fluosilicate
Mol. Formula F₆H₈N₂Si   Mol. Wt. 178.15
Uses Used in the manufacture of pesticides. Soldering flux. Etching glass.
Occurrence In nature as the mineral cryptohalite.

Physical properties

M. Pt. 120°C (decomp.); **Specific gravity** 2.01.

Solubility
Water: 18.1 g in 100 ml at 17°C.

Occupational exposure

US TLV (TWA) 2.5 mg m^{-3} (as F); **UK Long-term limit** 2.5 mg m^{-3} (as F); **UN No.** 2854; **HAZCHEM Code** 1Z; **Conveyance classification** harmful substance; **Supply classification** toxic.

Risk phrases
Toxic by inhalation, in contact with skin and if swallowed (R23/24/25)

Safety phrases
Keep locked up and out of reach of children – In case of contact with eyes, rinse immediately with plenty of water and seek medical advice – If you feel unwell, seek medical advice (show label where possible) (S1/2, S26, S44)

Ecotoxicity

Fish toxicity
Exposure of steelhead trout and bridgelip sucker to 10 mg l^{-1} in a static 24 hr bioassay resulted in neither death nor loss of equilibrium in either species (1).
Exposure of threespine stickleback to 10 mg l^{-1} in a 24 hr bioassay resulted in loss of equilibrium within 1-2 hr and death in 6-24 hr (2).

Effects to non-target species
LD_{Lo} subcutaneous frog 224 mg kg^{-1} (3).
LD_{50} oral silkworm larvae > 10 ppm (4).

Mammalian and avian toxicity

Acute data
LD_{Lo} oral rat 100 mg kg^{-1} (5).
LD_{50} intragastric rat, mouse 45-64 mg kg^{-1} (6).

Metabolism and pharmacokinetics
Fluorides are absorbed from the gastrointestinal tract, approximately 97% absorption. Absorbed fluoride is distributed throughout the tissues of the body by the blood. Fluoride exchanges with hydroxyl groups of hydroxyapatite (the inorganic constituent of bone) to form fluorohydroxyapatite. Fluoride that is not retained is excreted rapidly in urine (7).

Irritancy
50 mg instilled into rabbit eye caused severe corneal damage after 3 hr, while weak hyperaemia was observed after skin contact in rabbits (6).

Any other adverse effects to man

In the manufacture of ammonium fluorosilicate workers are exposed to air pollution by hydrogen fluoride, fluorosilicate and aerosols containing ammonium fluorosilicate. The urine of exposed workers showed the presence of fluoride. Exposure to ammonium fluorosilicate was associated with disorders of the nervous system and liver function (8).

The dust is irritating to the respiratory tract and inhalation may be fatal due to spasm. Symptoms of acute exposure include inflammation and oedema of the larynx and bronchi, chemical pneumonitis and pulmonary oedema. Chronic effects include coughing, sore throat, dyspnoea, headache, nausea and vomiting (9).

Any other adverse effects

Minimum toxic concentration inhalation (4 hr) rat 7.4-9.6 mg m^{-3}, non-toxic concentration 0.8 mg m^{-3}. Main toxic effects were decreases in the number of blood cells and decreased activities of cholinesterase and lactate dehydrogenase in blood serum (6).

Legislation

Limited under EEC Directive on Drinking Water Quality 80/778/EEC. Ammonium: Guide Level 0.05 mg l^{-1}, maximum admissible concentration 0.5 mg l^{-1}. Fluoride: maximum admissible concentration 1500 μg l^{-1} (8-12°C), 700 μg l^{-1} (25-30°C) (10).

Any other comments

Serves as preservation in impregnation of prefabricants used for house building (11).

References

1. *Lethal Effects of 964 Chemicals Upon Steelhead Trout and Bridgelip Sucker* 1989, EPA 560/6-89-001
2. *Lethal Effects of 2014 Chemicals Upon Sockeye Salmon, Steelhead Trout and Threespine Stickleback* 1989, EPA 560/6-89-001
3. *Compt. Rend. Serv. Soc. Biol.* 1937, **124**, 133
4. Fujii, M; et al. *Nippon Sanshigaki Zasshi* 1972, **4**(2), 104-110, (Jap.) (*Chem. Abstr.* **77**, 122710
5. *NRC Chemical Biological Coordination Centre* 1953, **5**, 27, Nat. Acad. Sci., Washington DC
6. Rumanaryantsev, G.I; et al. *Gig. Sanit.* 1988, **11**, 80-82, (Russ.) (*Chem. Abstr.* **110**, 34951r)
7. *Office of Drinking Water: Criteria Document Fluoride (Draft)* 1985, **3**, 19, US Environmental Protection Agency
8. Levchenko, N. I. *Gig. Tr. Prof. Zabol.* 1987, **9**, 52-54, (Russ.) (*Chem. Abstr.* **108**, 43173)
9. *Dangerous Prop. Ind. Mater. Rep.* 1984, 4(3), 36-38
10. *EC Directive Relating to the Quality of Water Intended for Human Consumption* 1982, 80/778/EEC, Office for Official Publications of the European Communities, 2 rue Mercier, L-2985 Luxembourg
11. Gorshin, S. N; et al. *Derevoobrab Prom-St* 1986, **12**, 3-5, (Russ.) (*Chem. Abstr.* **107**, 60884d)

A193   Ammonium hydrogen difluoride

(NH4)HF2

CAS Registry No. 1341-49-7
Synonyms ammonium bifluoride; ammonium acid difluoride; ammonium hydrogen fluoride
Mol. Formula F$_2$H$_5$N   **Mol. Wt.** 57.04

Uses Aluminium anodising. Corrosion resisting treatment of magnesium and its alloys. Sterilising dairy and other food equipment. Solubilising silica.Textiles. Wood preservative.

Occurrence Manufactured by gas phase reaction of anhydrous ammonia and anhydrous hydrogen fluoride.

Physical properties

M. Pt. 125°C; **B. Pt.** 239.5°C; **Specific gravity** 1.51.

Solubility
Water: 415 g l^{-1} at 25°C. Organic solvent: ethanol

Occupational exposure

US TLV (TWA) 2.5 mg m^{-3} (as F); **UK Long-term limit** 2.5 mg m^{-3} (as F); **UN No.** 1727 (solid); 2817 (solution); **HAZCHEM Code** 2X (solid); 2R (solution); **Conveyance classification** corrosive substance (solid); toxic substance (solution); **Supply classification** corrosive.

Risk phrases
Toxic if swallowed – Causes burns (R25, R34)

Safety phrases
Do not breathe dust – In case of contact with eyes, rinse immediately with plenty of water and seek medical advice – Wear suitable gloves (S22, S26, S37)

Environmental fate

Nitrification inhibition
Fluoride ions caused 10% inhibition of nitrification on biological film reactor at 135 mg F l^{-1} (1).

Mammalian and avian toxicity

Acute data
LD$_{50}$ oral guinea pig 150 mg kg^{-1} (2).

Metabolism and pharmacokinetics
Fluorides are absorbed from the gastrointestinal tract, lung and skin. Gastrointestinal tract is major site of absorption. Fluoride is preponderantly deposited in the skeleton and teeth and the degree of skeletal storage is related to intake and age. Major route of excretion is via kidneys. Also excreted in small amount by sweat glands, breast milk and gastrointestinal tract. About 90% of fluoride ion filtered by glomerulus is reabsorbed by renal tubules (3).

Following ingestion, soluble fluorides are rapidly absorbed from the gastrointestinal tract at least to the extent of 97%. Absorbed fluoride is distributed throughout the tissues of the body by the blood. Fluoride concentrations in soft tissues fall to pre-exposure levels within a few hours of exposure. Fluoride exchanges with hydroxyl groups of hydroxyapatite (the inorganic constituent of bone) to form fluorohydroxyapatite (4).

Any other adverse effects to man
0.15 g F l^{-1} caused poisoning of 48 subjects (5).

Legislation

Limited under EC Directive on Drinking Water Quality. 80/778/EEC.
Maximum admissible concentration 0.5 mg l^{-1} as NH_{4+}, 1500 μg l^{-1} as fluoride at
8-12°C, 700 μg l^{-1} fluoride at 25-30°C (6).

References

1. Beg, S. A; et al. *35th Ind. Waste Conf. Perdue Univ.* 1980, **35**, 826, (*Chem. Abstr.* **95**, 2975u) cited in *Nitrification Inhibition in the Treatment of Sewage* 1985, Richardson, M. L. (Ed.), RSC, London
2. Hodge, H. C; et al. in Simon, J. H. ed., *Fluorine Chemistry* 1965 Vol. 4, p192 Academic Press, New York
3. *Goodman and Gilman's The Pharmacological Basis of Therapeutics* (6th ed.) 1980, 1546, Gilman, A.G. (Ed.), Macmillan Publ. Co., New York
4. *USEPA Office of Drinking Water; Criteria Document (Draft): Fluoride* 1985, (III), 19-20
5. Werner, U; et al. *Z. Gesamte Hyg. Ihre Grenzgeb.* 1985, **31**(10), 568, (Ger.) (*Chem. Abstr.* **104**, 831836)
6. *EC Directive Relating to the Quality of Water Intended for Human Consumption* 1982, 80/778/EEC, Office for Official Publications of the European Communities, 2 rue Mercier, L-2985 Luxembourg

A194   Ammonium hydroxide

NH₄OH

CAS Registry No. 1336-21-6
Synonyms ammonia water; aqua ammonia; "Spirits of Hartshorn" (29%)
Mol. Formula H_5NO **Mol. Wt.** 35.05
Uses Detergent. Stain remover. Bleaching agent in calico printing. Manufacture of ammonium salts and aniline dyestuffs.

Physical properties

M. Pt. –77°C; **Specific gravity** d^{15} 0.947 (aqueous ammonia).

Occupational exposure

UN No. 2073 (35-50% NH_3); 2672 (10-35% NH_3; 1006 ≥50% for NH_3)

Ecotoxicity

Fish toxicity

LC_{50} (48 hr) bluegill sunfish 0.024- 0.093 mg l^{-1} (1).
LC_{50} (7 day) channel catfish 0.974-1.97 mg l^{-1}. Temperature range 21.1-22.8°C, pH range 7.7-8.0 (2).
LC_{50} (96 hr) fathead minnow 8.2 mg l^{-1} (3).
LC_{50} (24 hr) Atlantic salmon smolt 5-8 mg l^{-1} (4).
LC_{50} (24 hr) Atlantic salmon smolt 0.08 mg NH_3 l^{-1} (dissolved O2 of 3.2 mg l^{-1} fresh water) (5).
LC_{50} (24 hr) chinook salmon 2.2 mg NH_3 l^{-1} (9.6% salinity) (6).

Invertebrate toxicity

EC_{50} (48 hr) *Daphnia magna* 0.66 mg l^{-1} at 22°C. LC_{50} (120 hr) diatom 420 mg l^{-1} 50% growth reduction hard and soft water at 22°C. LC_{50} (96 hr) snail 90 mg l^{-1} in soft water at 20°C (7).

Mammalian and avian toxicity

Acute data

LD_{50} oral rat 350 mg kg^{-1} (8).
LC_{50} (1 hr) inhalation rat 7338 ppm (9).

Legislation

Limited under EC Directive on Drinking Water Quality 80/778/EEC. Ammonia: guide level 0.05 mg l^{-1}, maximum admissible concentration 0.5 mg l^{-1} (10).

Any other comments

Human health effects and experimental toxicology reviewed (11).
Toxicity and human health effects reviewed (12).

References

1. Turnbull, H; et al. *Ind. Eng. Chem.* 1954, **46**(2), 324-333
2. Knepp, G.L; et al. *Prog. Fish-Cult* 1973, **35**(4), 221-224
3. Veschueren, K. *Handbook of Environmental Data of Organic Chemicals* 2nd ed. 1983, 195, Van Nostrand Reinhold, New York
4. *Tech. Info. for Problem Spills:Ammonia (Draft) 1983, 85*
5. Alabaster, J. S; et al. *J. Fish Biol.* 1983, **22**(2), 215-222
6. Harader, R. R; et al. *Trans. Am. Fish Soc.* 1983, **112**(6), 834-837
7. *Tech. Info. for Problem Spills: Ammonia (Draft)* 1983, 86, Environment Canada
8. *J. Ind. Hyg. Toxicol.* 1941, **23**, 259
9. Vernot, E. H; *Toxicol. Appl. Pharm.* 1977, **42**, 417-423
10. *EC Directive Relating to the Quality of Water Intended for Human Consumption* 1982, 80/778/EEC, Office for Official Publications of the European Communities, 2 rue Mercier, L-2985 Luxembourg
11. *ECETOC Technical Report No. 30(4)* 1991, European Chemical Industry Ecology and Toxicology Centre, B-1160 Brussels
12. *Martindale: The Extra Pharmacopaeia* 29th ed., 1989, The Pharmaceutical Press, London

A195   Ammonium metavanadate

NH4VO3

CAS Registry No. 7803-55-6
Synonyms ammonium vanadate; vanadic acid, ammonium salt
Mol. Formula H_4NO_3V **Mol. Wt.** 116.98
Uses Catalyst. Dyeing and printing. Photographic developer. Producing vanadium lustre on pottery. Reagent in analytical chemistry.

Physical properties

M. Pt. 200°C (decomp.); **Specific gravity** 2.33.

Solubility
Water: 6 g l^{-1}.

Occupational exposure
UN No. 2859; HAZCHEM Code 1Z; Conveyance classification toxic substance.

Ecotoxicity
Fish toxicity
LC$_{50}$ (144 hr) goldfish, guppy 1.5-3.8 ppm (1).

Bioaccumulation
Studies of vanadium transfer and accumulation to molluscs revealed that the type of absorption of vanadium was dependent on species; work on marine bacteria, phytoplankton, invertebrates, crustaceans and fish showed that direct absorption of vanadium from water was more important than by feeding. Vanadium uptake rates in a mussel varied inversely with both salinity and vanadium concentration in water (2).

Mammalian and avian toxicity

Acute data
LD$_{50}$ oral rat 160 mg kg^{-1} (3).
LD$_{100}$ subcutaneous guinea pig 3 mg kg^{-1} (4).
LD$_{100}$ intravenous rabbit 3 mg kg^{-1} (4).

Teratogenicity and reproductive effects
The exposure of pregnant Syrian golden hamsters to ammonium metavanadate from days 5-10 of gestation resulted in a significant increase in skeletal anomalies and a decrease in the male:female foetal sex ratio. Skeletal anomalies included micrognathia, supernumerary ribs, and alterations in sternebral ossification. Although not significant, external anomalies included meningocoele, 1 foetus with multiple anomalies, and the presence of a molar pregnancy. Soft tissue anomalies did not differ significantly among groups but included hydronephrosis/hydroureter and kidney dysplasia. The numbers of malformed offspring were small, and there was no dose- response relationship (5).

Metabolism and pharmacokinetics
Absorbed vanadium is widely distributed in the body. In animals the highest values are found in bone, kidney, liver, spleen and, after intratracheal instillation, in lung. Bone maintains essentially unchanged levels for several weeks. The lowest values are found in the brain (6).

Genotoxicity
Bacillus subtilis gene conversion and mitotic recombination positive (7,8).
Saccharomyces cerevisiae D61M induced aneuploidy.
In vitro Chinese hamster V79 cells with and without metabolic activation negative. Potent cytotoxic agent (9).
In vitro Chinese hamster ovary cells with and without metabolic activation sister chromatid exchange positive (10).

Any other comments
Vanadium compounds when absorbed, are rapidly excreted and exhibit low degrees of toxicity, as indicated by minor irritation and lack of systemic effects (11).

References

1. Knudtson, B. K. *Bull. Environ. Contam. Toxicol.* 1979, **23**(1-2), 95
2. *Handbook on the Toxicity of Inorganic Compounds* 1988, Marcel Dekker, New York
3. Smyth, H. F; et al. *Am. Ind. Hyg. Assoc. J.* 1969, **30**, 470
4. Venugopal, B; et al. *Metal Toxicity in Mammals* 1978, Plenum Press, New York
5. Carlton, B. D; et al. *Environ. Res.* 1982, **29**(2), 256-262
6. Kanematsu, N; et al. *Mutat. Res.* 1980, **77**(2), 109
7. Arlauskas, A; et al. *Environ. Res.* 1985, **36**(2), 379
8. Friberg, L; et al. *Handbook of the Toxicolgy of Metals* 2nd ed., **I,II**, V2, Elsevier Science Publishers, Amsterdam
9. Galli, A; *Teratog. Carcinog. Mutagen.* 1991, **1**(4), 175-83
10. Owusu-Yaw, J; et al. *Toxicol. Lett.* 1990, **50**(2-3), 327-36
11. *NIOSH Criteria Document: Vanadium* 1977, 101, NIOSH 77-222

A196   Ammonium nitrate

NH_4NO_3

CAS Registry No. 6484-52-2
Synonyms ammonium saltpeter
Mol. Formula $H_4N_2O_3$ **Mol. Wt.** 80.04
Uses Oxidising agent. Used in the manufacture of nitrous oxide. Explosive. Fertiliser. Anaesthetic.

Physical properties

M. Pt. 210°C (decomp.); **Specific gravity** d_4^{20} 1.725.

Solubility
Water: 2000 g l^{-1}. Organic solvent: ethanol, methanol, acetone

Occupational exposure

UN No. 2067 (>90% but <0.2% combustible material); 2069 (>45% <70 but <0.4% combustible material); 2068 (>80% but <90% and <0.4% combustible material); 2070 (>70% <90% and <0.4% combustible material); 2426 (liquid <93%); 1942 (<0.2% combustible substances); **HAZCHEM Code** 1P; **Conveyance classification** oxidizing substance.

Ecotoxicity

Invertebrate toxicity
Ammonium nitrate at 5, 10, 25 or 50 mg NH_4^+ l^{-1} decreased the fertility of *Daphnia magna* at 50 mg NH_4^+ l^{-1}. Disrupted embryonic development, and at 10, 25 or 50 mg NH_4^+ l^{-1}. Impaired postembryonic growth of the crustacea. Industrial sewage entering fishing waters apparently should contain ≤1 mg NH_4^+ l^{-1} (1).

Effects to non-target species
The nematodes *Aphelenchus avenae*, *Meloidogyne incognita* and *Ditylenchus dipsaci* isolated from mushroom *Fusarium graminearum* mycelia, cucumber roots, and onions, respectively had decreased body dimensions when the plants were fertilised

with ammonium nitrate. Egg laying was also decreased in *Aphelenchus avenae*. Various morphophysiological changes in these neamtodes under the above fertiliser conditions are discussed (2).

Environmental fate

Degradation studies
LC_{50} 40 hr *Aspergillus niger* 15 mg l^{-1} 36°C (3).
Ammonium nitrate will be taken up by bacteria. Nitrate is more persistent in water than the ammonium ion. Nitrate degradation is fastest in anaerobic conditions (4).

Abiotic removal
The immediate loss of fertiliser nitrogen as nitrous oxide (by biochemical and microbiological action) into the atmosphere was determined by *in situ* measurements of the nitrous oxide evolution rates from uncultivated Eolian sand. The net loss was equivalent to 0.05% for ammonium nitrate. The total immediate loss of nitrous oxide-nitrogen after application of mineral fertiliser is estimated to be 0.004-1.2 teragram yr^{-1} (5).

Mammalian and avian toxicity

Acute data
LD_{50} oral rat 4820 mg kg^{-1} (6).

Metabolism and pharmacokinetics
12 healthy volunteers ingested orally 7-10.5 g ammonium nitrate in a single dose. Samples of blood, saliva and urine were collected just before a 24 hr period. Saliva and urine were analysed for volatile *N*-nitrosamines (NA), nitrate and nitrite. Blood was analysed for nitrate. Neither in urine nor in saliva were NA other than *N*-nitrosodiethylamine (NDMA) detected. Of the 188 urine samples, 13% contained greater than 0.1 mg NDMA kg^{-1} the highest level being 0.5µg kg^{-1}. In 92% of the 179 saliva samples, less than 0.5µg NDMA l^{-1} was found. Peak values for nitrate plus nitrate 2- 6 hr after intake 320-3440 mg l^{-1}. Nitrite was detected in 26% of the urine samples. An average of 75% of administered nitrate was excreted in urine in 24 hr. Nitrate contents in blood, urine and saliva after 24 hr were still higher than before the nitrate intake (7).

Legislation
Limited under EC Directive on Drinking Water Quality 80/778/EEC. Ammonia: guide level 0.05 mg l^{-1}, maximum admissible concentration 0.5 mg l^{-1}. Nitrates: guide level 25 mg l^{-1} (NO_3), maximum admissible concentration 50 mg l^{-1} (NO_3). Nitrites: maximum admissible concentration 0.1 mg l^{-1} (NO_2) (8).

References
1. Dyga, A. K; et al. *Eksp. Vod. Toksikol.* 1972, **3**, 51-8
2. Zinoveva, S. V; et al. *Tr. Gel'mintol. Lab. Akad. Nauk SSSR* 1980, **30**, 5-19
3. *Tech. Info. for Problem Spills: Ammonium Nitrate (Draft)* 1981, 55, Environment Canada
4. *Tech. Info. for Problem Spills: Ammonium Nitrate (Draft)* 1981, 57, Environment Canada
5. Conrad, R; et al. *Field measurements of the loss of fertilizer nitrogen into the atmosphere as nitrous oxide* 1980, **14**(5), 555-8
6. *Gig. Sanit.* 1981, **46**(9), 76
7. Ellen, G; et al. *Volatile N-nitrosamines, nitrate and nitrite in urine and saliva of healthy volunteers after administration of large amounts of nitrate* 1982, 365-78, IARC Sci. Publ. 41

8. *EC Directive Realting to the Quality of Water Intended for Human Consumption* 1982, 80/778/EEC, Office for Official Publications of the European Communities, 2 rue Mercier, L-2985 Luxembourg

A197   Ammonium oxalate

$$\begin{array}{c} CO_2NH_4 \\ | \\ CO_2H \end{array}$$

CAS Registry No. 5972-73-6
Synonyms ethanedioic acid, monoammonium salt monohydrate
Mol. Formula $C_2H_5NO_4$ **Mol. Wt.** 107.07

Physical properties

M. Pt. decomposes; **B. Pt.** 240-273°C; **Flash point** 118°C;
Specific gravity 1.5.

Solubility
Water: 25.4 g l^{-1} at 0°C.

Environmental fate

Degradation studies
Using ammonium oxalate as the sole carbon source 92.5% was degraded at 9.3 mg COD g^{-1} dry inoculum hr^{-1} by adapted activated sludge (1).

Legislation

Limited under EEC Directive on Drinking Water Quality 80/778/EEC. Ammonium: guide level 0.05 mg l^{-1}, maximum admissible concentration 0.5 mg l^{-1} (2).

References

1. Pitter, P. *Water Res.* 1976, **10**, 231-235
2. *EC Directive Relating to the Quality of Water Intended for Human Consumption* 1982, 80/778/EEC, Office for Official Publications of the European Communities, 2 rue Mercier, L-2985 Luxembourg

A198   Ammonium perfluorooctanoate

$$F_3C(CF_2)_6CO_2NH_4$$

CAS Registry No. 3825-26-1
Synonyms pentadecafluorooctanoic acid, ammonium salt
Mol. Formula $C_8H_4F_{15}O_2N$ **Mol. Wt.** 431.10
Uses Surfactant. Dispersant for manufacture of powdered poly(tetrafluoroethylene).

Occupational exposure
US TLV (TWA) 0.1 mg m^{-3}.

Mammalian and avian toxicity

Acute data
LD$_{50}$ oral rat 430 mg kg^{-1} (1).
LC$_{50}$ inhalation (4 hr) rat 980 mg m^{-3}. This concentration produced both an increase in liver size and corneal opacity (2).
LD$_{50}$ dermal rabbit 4300 mg kg^{-1} (3).

Sub-acute data
Inhalation (2 wk) rat 0, 1, 8 or 84 mg m^{-3} exposure for 6 hr day^{-1} 5 days week^{-1}. Suppressed body weight gain observed at 84 mg m^{-3}. Reversible liver weight increases, reversible increases in serum enzyme activities and microscopic liver pathology including necrosis occurred at exposures of 8 and 84 mg m^{-3}. No ocular changes were produced. Concentrations of organofluoride in the blood showed a dose relation with initial levels of 108 ppm in rats treated with 84 mg m^{-3} falling to 0.84 ppm after 84 days with a blood t$_{1/2}$ of 5-7 days. The no observed effect level was 1 mg m^{-3} and a mean organofluoride blood level of 13 ppm was detected in rats immediately after 10th exposure to an atmospheric level of 1 mg ammonium perfluorooctanoate m^{-3} (2).
Mouse (14 or 21 days) ≥3 ppm in diet increased the weight of mouse liver in a dose-dependent manner (4).

Teratogenicity and reproductive effects
Inhalation rat (6-15 days gestation) 0, 0.1, 1, 10 and 25 mg m^{-3} 6 hr day^{-1} or by gavage 100 mg kg^{-1} day^{-1} in corn oil. Maternal deaths occurred in the groups given highest levels by each route and overt toxicity was evident among the surviving dams and among those of the 10 mg m^{-3} group. No teratogenic responses were demonstrated (5).

Irritancy
0.5 g applied to rabbit skin for 24 hr caused mild irritation (3).

Any other adverse effects
Total hepatic DNA content was similar in control and ammonium perfluorooctanoate treated rats, therefore hepatomegaly represented a hypertrophic rather than a hyperplastic response. Cytochrome P450 content and activity of benzphetamine N-demethylase increased in the livers of treated rats, indicating the proliferation of the smooth endoplasmic reticulum. Glutathione S-transferase and UDP-glucuronyltransferase were unaffected. Morphological studies confirmed the proliferation of the endoplasmic reticulum, mitochondria and peroxisomes in the livers of treated animals. Ammonium perfluorooctanoate does not possess hypolipidemic activity (6).

Legislation
Limited under EEC Directive on Drinking Water Quality 80/778/EEC. Ammonium: guide level 0.05 mg l^{-1}, maximum admissible concentration 0.5 mg l^{-1} (7).

Any other comments

Experimental toxicology and human health effects reviewed (8).

References

1. *Am. Ind. Hyg. Assoc. J.* 1980, **41**, 576
2. Kennedy, G. L; et al. *Food Chem. Toxicol.* 1986, **24**(12), 1325-1329
3. Kennedy, G. L. *Toxicol. Appl. Pharmacol.* 1985, **81**(2), 348-355
4. Kennedy, G. L. *Toxicol. Letters* 1987, **39**(2-3), 295-300
5. Staples, R. E; et al. *Fundam. Appl. Toxicol.* 1984, **4**(3/1), 429-440
6. Pastoor, T. P; et al. *Exp. Mol. Pathol.* 1987, **47**(1), 98-109
7. *EC Directive Relating to the Quality of Water Intended for Human Consumption* 1982, 80/778/EEC, Office for Official Publications of the European Communities, 2 rue Mercier, L-2985 Luxembourg
8. *ECETOC Technical Report No. 30(4)* 1991, European Chemical Industry Ecology and Toxicology Centre, Brussels

A199   Ammonium persulfate

$(NH_4)_2S_2O_8$

CAS Registry No. 7727-54-0

Synonyms ammonium peroxydisulfate; diammonium peroxydisulfate

Mol. Formula $H_8N_2O_8S_2$ **Mol. Wt.** 228.20

Uses Agent for bleaching foodstuffs. Water purification treatment. Reducer and retarder in photography. In the manufacture of aniline dyestuffs. In etching and metal cleaning. Maturing agent for wheat flour. For the detection and determination of manganese.

Physical properties

M. Pt. 120°C (decomp.); **Specific gravity** 1.982.

Occupational exposure

UN No. 1444; **Conveyance classification** oxidizing substance.

Mammalian and avian toxicity

Acute data

LD_{50} oral rat 820 mg kg^{-1} (1).

Carcinogenicity and long-term effects

Dermal ♀ Sencar mice (1 yr) 5 µg diluted in 0.2 ml acetone, administered twice weekly. Two of the 20 mice developed skin tumours. The authors conclude ammonium persulfate was inactive as a promoter or complete carcinogen (2).

Any other adverse effects

Ammonium persulfate inhibited intracellular uptake of calcium and accelerated calcium release, thus raising the cytosolic calcium concentration and causing cell contraction in isolated rat heart cells in a concentration- and time-dependent manner (3).

Legislation

Limited under EEC Directive on Drinking Water Quality 80/778/EEC. Ammonium: guide level 0.05 mg l^{-1}, maximum admissible concentration 0.5 mg l^{-1}. Sulfates: guide level 25 mg l^{-1}, maximum admissible concentration 250 mg l^{-1} (4).

Any other comments

Experimental toxicology and human health effects reviewed (5).
Decomposes in the presence of moisture (6).

References

1. Smith, H. F; et al. *Am Ind. Hyg. Assoc. J.* 1969, **30**, 470
2. Kurokawa, Y; et al. *Cancer Letters* 1984, **24**, 299-304
3. Kaminishi, K; et al. *Can. J. Cardiol.* 1989, **5**(3), 168-174
4. *EC Directive Relating to the Quality of Water Intended for Human Consumption* 1982, 80/778/EEC, Office for Official Publications of the European Communities, 2 rue Mercier, L-2985 Luxembourg
5. *BIBRA Toxicity Profile* 1989, The British Industrial Biological Research Association, Carshalton
6. *The Merck Index* 11th ed. 1989, Merck & Co. Inc, Rahway, NJ

A200   Ammonium picrate

CAS Registry No. 131-74-8

Synonyms 2,4,6-trinitrophenol, ammonium salt; ammonium picronitrate; ammonium carbazoate; Obeline picrate; Explosive D; picrate of ammonia

Mol. Formula $C_6H_6N_4O_7$ **Mol. Wt.** 246.14

Uses In explosives, fireworks and rocket propellants.

Physical properties

M. Pt. decomposes; **B. Pt.** 423°C (explodes); **Specific gravity** 1.72.

Solubility

Water: 1.1 g in 100 ml at 20°C. Organic solvent: ethanol.

Occupational exposure

UN No. 1310; **Conveyance classification** flammable solid.

Ecotoxicity

Fish toxicity

LC_{50} (96 hr) bluegill sunfish 220 ppm static bioassay in fresh water, mild aeration applied after 24 hr, at 23 °C.
LC_{50} (96 hr) inland silverside 66 ppm static bioassay in synthetic seawater, mild aeration applied after 24 hr, at 23°C (1).

Legislation

Limited under EEC Directive on Drinking Water Quality 80/778/EEC. Ammonium: guide level 0.05 mg l^{-1}, maximum admissible concentration 0.5 mg l^{-1} (2).

Any other comments

Physiolologic efects and protective measures in cases of exposure to ammonium picrate discussed (3).
Experimental toxicology and human health effects reviewed (4).
Explodes easily from heat or shock (5).

References

1. Dawson, G. W; et al. *J. Haz. Mater.* 1975/77, **1**, 303-318
2. *EC Directive Relating to the Quality of water Intended for Human Consumption* 1982, 80/778/EEC, Office for Official Publications of the European Communities, 2 rue Mercier, L-2985 Luxembourg
3. Foulger, *US Armed Forces Med. J.* 1953, **4**, 1425
4. *Dangerous Prop. Ind. Mat. Rep.* 1988, **8**(2), 42-44
5. *The Merck Index* 11th ed. 1989, Merck & Co. Inc, Rahway, NJ

A201 Ammonium polysulfide

$(NH_4)_2S_3$

CAS Registry No. 9080-17-5;012259-92-6
Synonyms ammonium trisulfide; AP-S; diammonium trisulfide
Mol. Formula $H_8N_2S_3$ **Mol. Wt.** 132.27
Uses Photography, chemical reagent.

Physical properties

Specific gravity 1.10.

Solubility
Water: miscible.

Occupational exposure

UN No. 2818; **HAZCHEM Code** 2X; **Conveyance classification** corrosive substance; **Supply classification** corrosive.

Risk phrases
Contact with acids liberates toxic gas – Causes burns (R31, R34)

Safety phrases
In case of contact with eyes, rinse immediately with plenty of water and seek medical advice (S26)

Mammalian and avian toxicity

Acute data
LD_{50} oral rat 150 mg kg^{-1}.
LD_{50} dermal rabbit 1790 mg kg^{-1} (1).

Legislation
Limited under EEC Directive on drinking Water Quality 80/778/EEC. Ammonium: guide level 0.05 mg l^{-1}, maximum admissible concentration 0.5 mg l^{-1} (2).

References
1. *NTIS Report* AD A062-138, National Technical Information Service, Springfield, VA
2. *EC Directive Relating to the Quality of Water Intended for Human Consumption* 1982, 80/778/EEC, Office for Official Publications of the European Communities, 2 rue Mercier, L-2985 Luxembourg

A202   Ammonium sulfamate

$NH_2S(O_2)ONH_4$

CAS Registry No. 7773-06-0

Synonyms AMS; ammonium sulfamidate; monoammonium sulfamate; Ammate; Amcide; ammonium amidosulfonate; ammonium amidosulfonate

Mol. Formula $H_6N_2O_3S$ **Mol. Wt.** 114.12

Uses Non-selective herbicide. Fire-retardant for flame-proofing textiles and paper products. In electroplating solutions.

Physical properties
M. Pt. 131°C; **B. Pt.** 160°C (decomp.).

Solubility
Water: 684 g l^{-1} at 25°C. Organic solvent: glycerol, formamide.

Occupational exposure
US TLV (TWA) 10 mg m^{-3}.

Ecotoxicity

Fish toxicity
Exposure of threespine stickleback to 10 mg l^{-1} in a 24 hr bioassay resulted in death within 1 hr (1).

Environmental fate

Degradation studies
Microbial degraded in soil to ammonium sulfate within 6-8 weeks (2).

Mammalian and avian toxicity

Acute data
LD_{50} oral rat 2000-3900 mg kg^{-1} (2,3).
LD_{Lo} intraperitoneal rat 800 mg kg^{-1} (4).

Sub-acute data
Oral rat (27-42 day study) 1.7 g kg^{-1} day^{-1}. No gross pathological changes reported. Microscopic pathology indicated superficial damage to stomach mucosa, slight

---

vacuolation of the cytoplasm of liver cells and moderate numbers of macrophages filled with hemosiderin in spleen (5).

Oral rat (90 days) 0, 100, 250 or 500 mg kg^{-1} 6 days wk^{-1}. Body weights in high dose animals were significantly increased compared to controls (6).

Irritancy

Repeated application of 50% aqueous solutions to shaved skins of rats caused no irritation or systemic toxicity (7).

Any other adverse effects to man

A cohort mortality study of 1225 workers who had worked ≥ 6 months during 1950-82 in the forestry trade at a public utility, using phenoxy acids, s- triazines, substituted urea and other herbicides including ammonium sulfamate showed no excess mortality relative to the reference population. There was a statistically significant increase in deaths due to suicide for the cohort as a whole. There were no deaths due to cancers such as soft-tissue sarcoma and non-Hodgkins lymphoma (8).

Legislation

Limited under EEC Directive on Drinking Water Quality 80/778/EEC. Ammonium: guide level 0.05 mg l^{-1}, maximum admissible concentration 0.5 mg l^{-1}. Sulfate: guide level 25 mg l^{-1}, maximum admissible concentration 250 mg l^{-1} (9).

Any other comments

Experimental toxicology and human health effects reviewed (10,11).

References

1. *Lethal Effects of 2014 Chemicals upon Sockeye Salmon, Steelhead Trout and Threespine Stickleback* 1989, EPA 560/6-89-001
2. *The Agrochemicals Handbook* 3rd ed., 1991, RSC, London
3. *Arch. Ind. Health* 1956, **14**, 178
4. *J. Ind. Hyg. Toxicol.* 1943, **25**, 26
5. *USEPA Drinking Water Health Advisory: Pesticides* 1989, 35-42, Lewis Publishers, Chelsea, MI
6. Gupta, B. N; et al. *Toxicology* 1979, **13**, 45-49
7. Ambrose, *J. Ind. Hyg. Toxicol.* 1943, **25**, 26
8. Green, L. M. *Br. J. Ind. Med.* 1991, **48**(4), 234-248
9. *EC Directive Relating ot the Quality of Water Intended for Human Consumption* 1982, 80/778/EEC, Office for Official Publications of the European Communities, 2 rue Mercier, L-2985 Luxembourg
10. *ECETOC Technical Report No. 30(4)* 1991, European ChemicalIndustry Ecology and Toxicology Centre, Brussels
11. *Dangerous Prop. Ind. Mater. Report* 1987, **7**(5), 95-99

A203   Ammonium sulfate

(NH4)2SO4

CAS Registry No. 7783-20-2

Synonyms sulfuric acid, diammonium salt; diammonium sulfate; Mascagnite; Dolamin

Mol. Formula $H_8N_2O_4S$ **Mol. Wt.** 132.14

Uses Manufacture of ammonium alum. In analysis. Freezing mixtures. For flameproofing fabrics and paper. In the manufature of viscose silk. In the tanning industry. In the fractionation of proteins. The commercial grade is used as a fertiliser. Food additive in fermenation processes.

Occurrence In nature as mascagnite.

Physical properties

M. Pt. 280°C (decomp.); **Specific gravity** d_4^{20} 1.77.

Solubility
Water: 1g in 100 g.

Ecotoxicity

Fish toxicity
LC_{50} (96 hr) bleak 310 mg l^{-1} (1).

A 6 month exposure of snakeheadfish to 100 ppm (considered a safe concentration) and 500 ppm (sublethal concentration) of ammonium sulfate caused hypertrophy, degranulation, nuclear pyknosis and focal necrosis in hepatocytes. Thyroid follicles exhibited various degrees of hypertrophy, hyperplasia, hyperaemia and a reduction in colloid content. The damage was more pronounced in high dose fish (2).

Snakeheadfish exposed to single concentrations of 100 and 500 ppm ammonium sulfate exhibited testicular abnormalities including disorganisation of lobules, degeneration of spermatogenic elements, necrosis of interstitial cells and proliferative fibrosis of lobule walls. The absence of intact germ cells in some lobules suggested irreversible damage in high dose fish. The inhibition of testicular development and deleterious changes in spermatogenic elements result from direct action on the testis itself or indirectly via the hypothalamic-pituitary- testicular axis (3).

Invertebrate toxicity
LC_{50} (25-100 hr) *Daphnia magna* 423-292 mg l^{-1} (4).

Environmental fate

Nitrification inhibition
Inhibition of nitrification-threshold concentration 500 mg l^{-1} (5).

Mammalian and avian toxicity

Acute data
TD_{Lo} oral man 1500 mg kg^{-1} (6).
LD_{50} oral rat 3000 mg kg^{-1} (7).
LD_{50} intraperitoneal mouse 610 mg kg^{-1} (8).

Sub-acute data
Intratracheal (4 or 8 month) rat 0.5 mg m^{-3} 5 hr day $^{-1}$, 5 days wk^{-1}. At 4 months cellular immunological responsiveness was not impaired but physiological changes were detected. Effects included bronchiolar epithelial hyperplasia and changes in alveolar mean chord length (9).

The effects of ammonium sulfate aerosols on asthmatic dyspnoea (immediate type), induced in guinea pigs by repeated inhalation of a mixture of bovine serum and egg albumin, was investigated. Inhalation (38 wk) guinea pig 0.2, 0.4, and 2.0 mg m^{-3}

ammonium sulfate aerosol for 2 hr day $^{-1}$ 5 days wk^{-1}. Results show asthmatic dyspnoea was increased by exposure to aerosol (10).

Genotoxicity

Chinese hamster V-79 cells treated with a hypotonic solution of ammonium sulfate resulted in an increase of chromosomal aberrations. These can be attributed to direct DNA damage due to the hypotonic treatment or changes to internal pH or damage of chromosomal proteins (11).

Legislation

Limited under EEC Directive on Drinking WAter Quality 80/778/EEC. Ammonium: guide level 0.05 mg l^{-1}, maximum admissible concentration 0.5 mg l^{-1}. Sulfate: guide level 25 mg l^{-1}, maximum admissible concentration 250 mg l^{-1} (6).

Any other comments

A small scale ecosystem simulating hydrological isolated lentic soft waters was exposed to artificial rain pH 5.6 containing ammonium sulfate for 2 yrs. Remarkable changes in water quality and flora were observed. These included pH decrease to 3.5, accumulation of both ammonium and sulfate. Nitrification of ammonium appeared to be the dominant acidifying process resulting in increases in the concentrations of aluminium, cadmium, calcium, iron, magnesium, manganese, and zinc. In acidified systems exposed to high depositions of ammonium sulfate typical soft water plants such as *Littorella uniflora* disappeared and a luxuriant growth of *Sphagnum cuspidatum* and *Juncus bulbosas* occurred. The filamentous green algae *Oedogonium sp.* and *Mougeotic sp.* became dominant species (13).

Experimental toxicology and human health effects reviewed (14).

References

1. Linden, E; et al. *Chemosphere* 1979, **11/12**, 843-851
2. Ram, R.N; et al. *Ecotoxicol. Environ. Saf.* 1987, **13**(2), 185-190
3. Ram, R.N; et al. *Indian J. Exp. Biol.* 1987, **25**(10), 667-670
4. Verschueren, K. *Handbook of Environmental Chemicals* 1983, Van Nostrand Reinhold, New York
5. Vismara, R. *Ingegn. Ambien.* 1982, **11**(8), 634-643
6. *Gig. Sanit.* 1977, **42**(2), 100
7. *Can. J. Comp. Med. Vet. Sci.* 1948, **12**, 216
8. *Publications in Pharmacology* 1941, **2**, 1, Univ. California
9. Smith, L. G; et al. *Environ. Res.* 1989, **49**(1), 60-78
10. Kitabatake, M; et al. *J. Toxicol. Environ. Health* 1991, **33**(2), 157-170
11. Nowak, C. *Teratogen. Carcinogen. Mutagen.* 1987, **7**(6), 515-525
12. *EC Directive Relating to the Quality of Water Intended for Human Consumption* 1982, 80/778/EEC, Office for Official Publications of the European Communities, 2 rue Mercier, L-2985 Luxembourg
13. Schuurkes, J. A. A. R; et al. *Aquat. Bot.* 1987, **28**(3-4), 199-226
14. *ECETOC Technical Report No. 30(4)* 1991, European Chemical Industry Ecology and Toxicology Centre, Brussels

A204   Ammonium sulfide

$(NH_4)_2S$

CAS Registry No. 12135-76-1
Synonyms diammonium sulfide; ammonium monosulfide
Mol. Formula H_8N_2S Mol. Wt. 68.14
Uses To apply patina to bronze. In photography development. In textile manufacture.
In chemical analysis. Fungicide.

Physical properties

M. Pt. –18°C; B. Pt. Decomposes at high temperatures to $(NH_4)_2S$, ammonia and
polysulfides.

Solubility
Organic solvent: ethanol.

Occupational exposure

UN No. 2683; HAZCHEM Code 2X; Conveyance classification corrosive
substance; flammable liquid; toxic substance.

Ecotoxicity

Fish toxicity
LC_{50} (48 hr) carp 6.6-109 mg l^{-1} total ammonia concentration in static bioassay
ammonium sulfide solution replaced every 24 hr.
LC_{50} (48 hr) carp 1.15-1.96 mg l^{-1} unionised ammonia in static bioassay (1).
LC_{50} (48 hr) mosquito fish 248 ppm in fresh water (2).

Environmental fate

Nitrification inhibition
Inhibition of NH_3 oxidation-activated sludge when 1 mg l^{-1} 28%, 5 mg l^{-1} 67%, 3.2
mg l^{-1} 100% (3).

Mammalian and avian toxicity

Acute data
LD_{Lo} oral mouse 80 mg kg^{-1} (3).
LD_{Lo} dermal mouse, rabbit 119-2460 mg kg^{-1} (4).

Metabolism and pharmacokinetics
Alkali sulfides are rapidly absorbed from the intestine. Excretion occurs via 3 routes:
the sulfide radical is excreted partially via the lung and kidney, is oxidised partially to
sulfate and thiosulfate and excreted by the kidneys and it can also be incorporated into
metallic sulfides, *e.g.* iron sulfide which is excreted in the faeces (5).

Genotoxicity

Escherichia coli PQ37, SOS chromotest without metabolic activation negative (6).

Legislation

Limited under EEC Directive on Drinking Water Quality 80/778/EEC. Ammonium: guide level 0.05 mg l^{-1}, maximimum admissible concentration 0.5 mg l^{-1} (7).

References

1. Dabrowska, H; et al. *Pol. Arch. Hydrobiol.* 1986, **33**(1), 121-128, (Pol.) (*Chem. Abstr.* **105**, 185467k)
2. *CHRIS Hazardous Chemical Data* 1984-1985, **2**, US Coast Guard, Dept. of Transportation, Washington, DC
3. Beccari, M; et al. *Environ. Technol. Lett.* 1980, **1**, 245-252
4. *Kirk-Othmer Encyclopedia of Chemical Technology* 1982, John Wiley & Sons, New York
5. Thienes, C; et al. *Clinical Toxicology* 5th ed., 1972, 59, Lea & Feliger, Philadelphia
6. Olivier, P; et al. *Mutat. Res.* 1987, **189**(3), 263-269
7. *EC Directive Relating to the Quality of Water Intended for Human Consumption* 1982, 80/778/EEC, Office for Official Publications of the European Communities, 2 rue Mercier, L-2985 Luxembourg

A205 Ammonium sulfite

$(NH_4)_2SO_3$

CAS Registry No. 10196-04-0
Mol. Formula $H_8N_2O_3S$ **Mol. Wt.** 116.14
Uses In photography. As a reducing agent. In bricks for blast furnace linings. In lubricants for metal cold working.

Physical properties

M. Pt. 60-70°C (decomp.); **B. Pt.** 150°C (sublimes); **Specific gravity** d^{25} 1.41.

Solubility

Water: 32.4 g in 100 ml at 0°C. Organic solvent: ethanol.

Ecotoxicity

Fish toxicity
LC_{50} (48 hour) mosquito fish 240 ppm (1).

Invertebrate toxicity
LC_{50} (25-100 hr) *Daphnia magna* 300-200 mg l^{-1} (2).

Legislation

Limited under EEC Directive on Drinking Water Quality 80/778/EEC. Ammonium: guide level 0.05 mg l^{-1}, maximum admissible concentration 0.5 mg l^{-1}. Sulfate: guide level 25 mg l^{-1}, maximum admissible concentration 250 mg l^{-1} (3).

References

1. *CHRIS Hazardous Chemical Data* 1984-1985, **2** US Coast Guard, Dept. of Transportation, Washington, DC
2. Verschueren, K. *Handbook of Environmental Chemicals* 2nd ed., 1983, Van Nostrand Reinhold, New York

3. *EC Directive Relating to the Quality of Water Intended for Human Consumption* 1982, 80/778/EEC, Office for Official Publications of the European Communities, 2 rue Mercier, L-2985 Luxembourg

A206 Ammonium tartrate

$$CO_2NH_4$$
$$|$$
$$CHOH$$
$$|$$
$$CHOH$$
$$|$$
$$CO_2NH_4$$

CAS Registry No. 3164-29-2

Synonyms butanedioic acid, 2,3-dihydroxy-, diammonium salt; 2,3-dihydroxybutanedioic acid, diammonium salt; ammonium threonate; 2,3-dihydroxysuccinic acid, diammonium salt

Mol. Formula $C_4H_{12}N_2O_6$ **Mol. Wt.** 184.15

Physical properties

M. Pt. decomposes; **Specific gravity** 1.601.

Solubility
Water: 58 g in 100 ml at 15°C. Organic solvent: ethanol

Mammalian and avian toxicity

Acute data
LD_{50} intravenous rabbit 113 mg kg^{-1}.
LD_{50} subcutaneous rabbit 1130 mg kg^{-1} (1).

Legislation

Limited under EEC Directive on Drinking Water Quality 80/778/EEC. Ammonium: guide level 0.05 mg l^{-1}, maximum admissible concentration 0.5 mg l^{-1} (2).

References

1. *Abdernalden's Handbuch der Biologischen Arbeitsmethoden (Leipzig)* 1935, **4**, 35
2. *EC Directive Relating to the Quality of Water Intended for Human Consumption* 1982, 80/778/EEC, Office for Official Publications of the European Communities, 2 rue Mercier, L-2985 Luxembourg

A207 Ammonium thiocyanate

NH_4SCN

CAS Registry No. 1762-95-4

Synonyms thiocyanic acid, ammonium salt; ammonium rhodanate; ammonium rhodanide; ammonium sulfocyanate; ammonium sulfocyanide

Mol. Formula CH_4N_2S **Mol. Wt.** 76.12

Uses In matches. Dyeing of fabrics, photography and coatings on zinc. In the manufacture of artificial resins, thiourea and pesticides. In rustproofing compositions. Detection and determination of iron, silver, mercury.

Physical properties
M. Pt. 149°C; **B. Pt.** 170°C (decomp.); **Specific gravity** 1.305.

Solubility
Water: 128 g in 100 ml at 0°C. Organic solvent: ethanol, methanol, acetone

Occupational exposure
Supply classification harmful.

Risk phrases
Harmful by inhalation, in contact with skin and if swallowed – Contact with acids liberates very toxic gas (R20/21/22, R32)

Safety phrases
Keep out of reach of children – Keep away from food, drink and animal feeding stuffs (S2, S13)

Ecotoxicity
Fish toxicity
Bluegill sunfish exposed to 280-300 ppm died within 1 hr (1).

Environmental fate
Nitrification inhibition
Ammonia oxidation inhibited by activated sludge concentration 100 mg l^{-1} (2).

Mammalian and avian toxicity
Acute data
LD_{Lo} oral mouse, guinea pig, 330-600 mg kg^{-1} (3,4).
TD_{Lo} oral human 375 mg kg^{-1} (5).
LD_{50} intragastric rat, mouse 750 mg kg^{-1} (5,6).
LD_{50} intragastric rat, guinea pig 500 mg kg^{-1} (6).
LD_{Lo} intraperitoneal mouse 500 mg kg^{-1} (7).

Metabolism and pharmacokinetics
Thiocyanate readily diffuses into all tissues. It appears early in saliva and is excreted slowly via the urine. It is not decomposed to cyanide in appreciable quantities (8,9). Intragastric rat (duration unspecified) 375 mg kg^{-1} uniform distribution throughout the body, except some accumulation in the thyroid gland and kidney. Excreted mainly in urine $t_{1/2}$ 36 hr. During chronic aerosol inhalation by rats of maximum permissible concentration 5 mg m^{-3}, ammonium thiocyanate was rapidly excreted from the body within 24 hr (5).

Sensitisation
Subcutaneous guinea pig (concentration and duration unspecified) weak allergic reaction occurred (6).
In human subjects occupationally exposed the incidence of allergic dermatitis was

above normal and phagocytic activity of neutrophils was impaired and the concentration of immunoglobulin A (IgA) was below controls (10).

Any other adverse effects

In chronic inhalation studies in rats (duration unspecified) minimum toxic and maximum non-toxic concentrations were 20 and 5 mg m^{-3}, respectively (6).

Legislation

Limited under EEC Directive on Drinking Water Quality 80/778/EEC. Ammonium: guide level 0.05 mg l^{-1}, maximum admissible concentration 0.5 mg l^{-1} (11).

Any other comments

Herbicidal activity reviewed (12).

References

1. *CHRIS Hazardous Chemical Data* 1984-1985, US Coast Guard, Dept. of Transportation, Washington DC
2. Becconi M; et al. *Environ. Technol. Lett.* 1980, **1**, 245-252
3. *Deutsches Archiv fur Experimentelle Pathologie und Pharmakologie* 1966, **169**, 429-433
4. *J. Am. Med. Pharmac. Assoc.* 1940, **29**, 152
5. Ivanova, L. A; et al. *Gig. Trud. Prof. Zabol.* 1988, (10), 52-53, (Russ.) (*Chem. Abstr.* **109**, 224339z)
6. Talakin, Y. N; et al. *Gig. Tr. Prof. Zabol.* 1986, **10**, 51-52, (Russ.) (*Chem. Abstr.* **106**, 14311w)
7. *NTIS Report* AD277-689, National Technical Information Service, Springfield, VA
8. Gosselin, R. E; et al. *Clinical Toxicology of Commercial Products* 5th ed., 1985, **2**, 122, Williams & Wilkins, Baltimore
9. Thienes, C; et al. *Clinical Toxicology* 5th ed., 1972, Lea & Febiger, Philadelphia
10. Savchenko, M. V. *Gig. Sanit.* 1987, **11**, 29-32, (Russ.) (*Chem. Abstr.* **108**, 50965n)
11. *EC Directive Relating to the Quality of Water Intended for Human Consumption* 1982, 80/778/EEC, Office for Official Publications of the European Communities, 2 rue Mercier, L-2985 Luxembourg
12. Duncan, H. J; et al. *Adjuvants Agrochem.* 1989, **1**, 27-33

A208 Ammonium thiosulfate

(NH$_4$)$_2$S$_2$O$_3$

CAS Registry No. 7783-18-8

Synonyms thiosulfuric acid, diammonium salt; ammonium hyposulfite

Mol. Formula H$_8$N$_2$S$_2$O$_3$ **Mol. Wt.** 148.20

Uses Photographic chemicals. Metal refining. Fertilisers.

Physical properties

M. Pt. 150°C (decomp.); **Specific gravity** 1.679.

Solubility

Water: 103 g in 100 ml at 100°C. Organic solvent: acetone.

Environmental fate

Nitrification inhibition
Inhibits nitrite to nitrate conversion in soils (1).

Mammalian and avian toxicity

Acute data
Intraperitoneal or subcutaneous injections of 0.2 ml of an aqueous solution into 10 g mice caused convulsions and death (2).

Legislation
Limited under EEC Directive on Drinking Water Quality 80/778/EEC. Ammonium: guide level 0.05 mg l^{-1}, maximum admissible concentrations 0.5 mg l^{-1} (3).

Any other comments
The US Environmental Protection Agency has removed ammonium thiosulfate from its list of hazardous substances since the median lethal concentration is well above 500 mg l^{-1} for aquatic toxicity (4).
Experimental toxicology and human health effects reviewed (5).

References
1. Janzen, H. H; et al. *Soil Sci. Soc. Am. J.* 1986, **50**(3) 803-806
2. Iwahashi, N. *Kumamoto Med. J.* 1956, **9**, 97-107 (*Chem. Abstr.* **51**, 8281f)
3. *EC Directive Relating to the Quality of Water Intended for Human Consumption* 1982, 80/778/EEC, Office for Official Publications of the EC, 2 rue Mercier , L-2985 Luxembourg
4. *Fed. Registration* 1989, **54**(155) EPA 3342b-84
5. *ECETOC Technical Report No. 30(4)* 1991, European Chemical Industry Ecology and Toxicology Centre, Brussels

A209   Amoxycillin

CAS Registry No. 26787-78-0
Synonyms 4-thia-1-azabicyclo(3.2.0)heptane-2- carboxylic acid, 6-((amino(4-hydroxyphenyl)acetyl)amino)-3,3-dimethyl-7-oxo,(2S-(2α,5α,6β(S)+))-; BRL-2333; *D*-(-)-α-amino-*p*- hydroxybenzyl penicillin; Amoxil Histocillin; (6*R*)-6- [α-*D*-(14-hydroxyphenyl)glycylamino]p enicillanic acid; α-amino-*p*- hydroxyphenylacetamido]penicillanic acid; α-amino-*p*-hydroxybenzylpenicillin; *p*- hydroxyampicillin
Mol. Formula $C_{16}H_{19}N_3O_5S$ **Mol. Wt.** 365.41
Uses Antimicrobial. Antibacterial. Semi-synthetic antibiotic related to penicillin

Ecotoxicity

Fish toxicity

Young yellowtail fish were exposed to 80 and 400 mg kg^{-1} amoxicillin for 10 days. At 400 mg kg^{-1} the fish showed no abnormal appetite, movement or haematological effects except an increased erythrophagocytosis by splenic macrophages (1).

Mammalian and avian toxicity

Acute data

LD$_{50}$ intraperitoneal rat, mouse 2870- 3590 mg kg^{-1} (as trihydrate) (2).

Teratogenicity and reproductive effects

LD$_{Lo}$ (7-13 day pregnant) oral rat 2800 mg kg^{-1}.
LD$_{Lo}$ (7-13 day pregnant) oral mouse 9100 mg kg^{-1} teratogenicity reported (2).

Metabolism and pharmacokinetics

The bioavailability of amoxycillin was studied in healthy human volunteers. Single doses of 250, 500, 750 and 1000 mg were administered. The mean urinary recovery was 59.6, 55.9, 58.5 and 45.8%, respectively (3).

The serum t$_{1/2}$ of amoxycillin was 0.96 hr in humans given a single 500 mg oral dose of the drug (4).

Amoxycillin is more rapidly and completely absorbed from gastrointestinal tract than is ampicillin. After a single 250 mg dose (species unspecified) peak plasma concentrations are reached in 2 hr and average 4 μg ml^{-1}. Approximately 20% is bound to plasma proteins and 60% is excreted in active form via the urine (5). Demonstrated *in vitro* to use the dipeptide carrier-mediated system as a transport mechanism in rodent small intestine (6).

Amoxycillin (2 x 250 mg capsules was administered orally to healthy human ♂. Venous blood and urine were taken at intervals for 6 and 12 hr, respectively. Peak serum concentration measured 7.6 mg l^{-1} and was reached in 1.4 hr (7).

Irritancy

Analysis by the Boston Collaborative Drug Surveillance Program of data on 15438 patients hospitalised between 1975 and 1982 detected 63 allergic skin reactions attributed to amoxycillin among 1225 recipients of the drug. This was the highest incidence of skin reactions among the drugs studied (8).

Photosensitivity reactions have been reported with amoxycillin (9).

Any other adverse effects to man

Electron microscopy of a skin biopsy from an infant with neonatal adrenleukodystrophy detected numberous myelinic bodies within fibroblasts. These myelinic bodies were absent from skin fibroblasts from the same patient when examined approximately 1 yr later. The role of amoxycillin and gentamycin which had been administered for a urinary tract infection at the time of the first biopsy in inducing myelinic bodies was examined but the results were inconclusive (10).

Administration of amoxycillin (trihydrate) by mouth in 3 patients caused neutropenia 13-23 days after onset of treatment. It was considered to be immune-mediated (11).

Non-pseudomembranous colitis, presenting with abdominal pain and bloody diarrhoea was associated with the administration of amoxycillin by mouth in 4 patients (12).

Any other adverse effects

Allergic response in breast fed infants (13).
Chlamydia trachomatis minimum inhibitory concentration 0.5 μg l^{-1} (14).

Any other comments

Physiochemical properties, antibacterial pharmacology, pharmacokinetics and toxicity of amoxicillin for domestic animals reviewed (15).
Cows milk is rejected if it contains >0.02 i.u. ml^{-1} (12 μg l^{-1}) (16).
Often administered as the sodium and trihydrate salts to enhance bioavailability (13).

References

1. Nakauchi, R; et al. *Gyobyo Kenkyu* 1988, **23**(4), 251-255, (Jap.) (*Chem. Abstr.* **110**, 128143f)
2. *Kiso to Rinsho Clinical Report* 1973, **7**, 3040,3074,3113, Yubunsha Co Ltd, Tokyo
3. Arancibia, A; et al. *Int. J. Clin. Pharmacol. Ther. Toxicol.* 1988, **26**(6), 300-303
4. Lui, C. H; et al. *T'ai-wan Yao Hsueh Tsa Chih* 1986, **38**(4), 262-269
5. Goodman, L. S; et al. *The Pharmacological Basis of Therapeutics* 5th ed., 1975, 1145, MacMillan, New York
6. Westphal, J. F; et al. *J. Antimicrob. Chemother.* 1991, **27**(5), 647-654
7. Sum, Z. M; et al. *J. Antimicrob. Chemother.* 1989, **23**(6), 861-868
8. Bigby, M; et al. *J. Am. Med. Assoc.* 1986, **256**, 3358
9. Stone, K. *Aust. J. Pharma.* 1985, **66**, 415
10. Pampols, T; et al. *NATO ASI Ser. Ser. A* 1988, **150**, 435-441
11. Rouveix, B; et al. *Br. Med. J.* 1983, **287**, 1832
12. Iida, M; et al. *Endoscopy* 1985, **17**, 64
13. *Martindale: The Extra Pharmacopoeia* 29th ed., 1989, The Pharmacutical Press, London
14. Boersum, T; et al. *Chemotherapy (Basel)* 1990, **36**(6), 407-415
15. Matsubara, S. *Kachiku Kokinzai Kenkyukaiho* 1988, **9**, 51-62, (Jap.) (*Chem. Abstr.* **114**, 16959c)
16. Milk Marketing Board Information Services Leaflet 1981, No.5

A210   Amphetamine

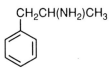

CH$_2$CH(NH$_2$)CH$_3$

CAS Registry No. 300-62-9
Synonyms (±)-α- methylbenzene ethanamine; *dl*-α- methylphenethylamine; 1-phenyl-2-aminopropane; (phenylisopropyl)amine; β-aminopropylbenzene; Phenedrine; Actedron; Simpatedrin; Mecodrin
Mol. Formula C$_9$H$_{13}$N   **Mol. Wt.** 135.21
Uses Central nervous system stimulant.

Physical properties

B. Pt. 200-203°C; **Flash point** <100°C (open cup); **Specific gravity** d$_4^{25}$ 0.913; **Partition coefficient** log P$_{ow}$ 1.76; **Volatility** v. den. 4.7.

Solubility
Organic solvent: ethanol, diethyl ether.

Ecotoxicity

Invertebrate toxicity
Tetrahymena pyriformis W 2 hr exposure to 5 μg l^{-1} caused increased phagocytic activity (1).

Mammalian and avian toxicity

Acute data
LD$_{50}$ oral rat 30 mg kg^{-1} (2).
LD$_{50}$ subcutaneous rat 180 mg kg^{-1} (3).
LD$_{50}$ intraperitoneal mouse 15 mg kg^{-1} (4).

Sub-acute data
Injection study mouse (duration unspecified) 0.4 mg kg^{-1} day^{-1} showed a reduction in thymus and spleen cellularity and in peripheral T lymphocyte population. Inhibition of T-cell proliferation and a reduction in the capacity of mice to development and passive transfer of immunity to *Listeria monocytogenes* was observed (5).

Carcinogenicity and long-term effects
National toxicology program investigated DL-amphetamine sulfate in rat, mouse. Designated non-carcinogen in rat and mouse (6).
A COMPACT computer-optimized molecular parametric analysis of chemical toxicity was used to determine genotoxic and carcinogenic potential. The potential of amphetamine was negative (7).

Metabolism and pharmacokinetics
Metabolised to 4-hydroxyamphetamine and 4- hydroxynorephedrine (8).
Amphetamine concentrates in the kidney, lungs and brain (9).
Considerable species difference in biotransformation exists, but not in the excretion of ^{14}C after administration of ^{14}C- amphetamine. After intraperitoneal administration to dogs and oral administration to other species excretory percentages in urine were rats 86%, rabbits 86% and dogs 78% (10).

Genotoxicity
Amphetamine produced marked clastogenic activity and affected the cell proliferation in the bone marrow of mice. In mouse somatic cells, a dose-dependent increase in micronucleated polychromatic erythrocytes was observed (11).

Any other adverse effects to man
A single dose 10-25 mg of amphetamine given to human volunteers produce peak plasma levels within 1-2 hr and was rapidly absorbed from gastrointestinal tract. Amphetamine absorption was usually complete within 4-6 hr (9).
An amphetamine-abusing mother, who had taken methamphetamine 5 hr prior to onset of labour, confirmed previous findings of premature delivery and retarded intrauterine development (12).

Any other adverse effects
In vitro horse liver alcohol dehydrogenase activity inhibited by amphetamine (13).

Respiration by rat brain homogenates was inhibited by 18% by 0.27 g l^{-1} amphetamine (14).

Behavioural effects include excitement, tremor, convulsions or effect on seizure threshold drug of abuse (15).

To assess the effects on independent feeding during development, preweanling rats were administered amphetamine and allowed to ingest milk through oral cannulas. In 1 hr milk-deprived pups, milk intake was stimulated at 3, 7 and 10 days of age and suppressed at 15 days. In 22 hr deprived pups, no effect observed at 3, 7 and 10 days but reliably suppressed in take at 15 days (16).

Any other comments

Neurological and psychosis effects, toxicology and pharmacology, behavioural effects, biotransformation and metabolism have been extensively reviewed (17-27).

References

1. Stefanidou, M; et al. *Toxicol. in Vitro* 1990, **4**(6), 779-781
2. *Arzneimittel-Forschung* 1973, **23**, 810
3. Warren, M. R; et al. *J. Pharmacol. Exp. Ther.* 1945, **85**, 119
4. *Toxicol. Appl. Pharmacol.* 1965, **7**, 566
5. Freire-Garabal, M; et al. *Life Sci.* 1991, **49**(16), PL107-PL112
6. Ashby, J; et al. *Mutat. Res.* 1991, **257**(3), 229-306, (NTP Report No. 387, 1990)
7. Lewis; D. F; et al. *Mutagenesis* 1990, **5**(5), 433-435
8. Arai, Y; et al. *Neurochem. Int.* 1990, **17**(4), 587-592
9. Ellenhorn, M. J; et al. *Medical Toxicology Diagnosis Treatment of Human Poisoning* 1988, 631, Elsevier, New York
10. *Foreign Compound Metabolism in Mammals* 1970, **1A**, 65, The Chemical Society, London
11. Tariq, M; et al. *Mutat. Res.* 1987, **190**, 153-157
12. Bost, R. O; et al. *J. Anal. Toxicol.* 1989, **13**(5), 300-302
13. Roig, M. G; et al. *J. Pharm. Sci.* 1991, **80**(3), 267-270
14. Nag, M; et al. *Biosci. Rep.* 1991, **11**(1), 11-14
15. *Martindale: The Extra Pharmacopoeia* 29th Ed., 1989, Pharmaceutical Press, London
16. Capuano, C. A; et al. *Pharmacol. Biochem. Behav.* 1989, **33**(3), 567-572
17. Dougan, D; et al. *Trends Pharmacol. Sci.* 1987, **8**(7), 277-280
18. Pitts, D. K; et al. *Life Sci.* 1988, **42**(9), 949-968
19. Segal, D. S; et al. *Psychopharmacol. Bull.* 1987, **23**(3), 417-424
20. Nichols, D. E; et al. *Cocaine, Marijuana, Des. Drugs: Chem. Pharmacol. Behav.* 1989, 175- 185, Redde, K. K; et al.(Eds.)
21. Middough, L. D. *Ann. N.Y. Acad. Sci.* 1989, **562**(Prenatal Abuse With Illicit Drugs) 308-318
22. Miczek, K. A; et al. *NIDA Res. Monograph* 1989, **94**, 68-100
23. Gold, L. H; et al. *NIDA Res. Monograph* 1989, **94**, 101-126
24. Groves, P. M; et al. *NIDA Res. Monograph* 1989, **94**, 127-145
25. Molliver, M. E; et al. *NIDA Res. Monograph* 1989, **94**, 270-305
26. Yamada, H; et al. *Eisei Kagaku* 1989, **35**(6), 383-396, (Jap.) (*Chem. Abstr.* **112**, 151110t)
27. Rendic, S; et al. *Farm. Glas.* 1989, **45**(10), 317-326, (Serb.-Croat.) (*Chem. Abstr.* **112**, 131686g)

A211　Ampicillin

CAS Registry No. 69-53-4

Synonyms 4-thia-1-azabicyclo(3.3.0)heptane-2- carboxylic acid,
6-((aminophenylacetyl)amino)-3, 3-dimethyl- 7-oxo-2,S- 12α,5α,6β7(S+)); Penbritin;
Penicline; amino benzylpenicillin; Omnipen; Britacil;
(6R)-6-(α-D- phenylglycylamino)penicillanic acid

Mol. Formula $C_{16}H_{19}N_3O_4S$ **Mol. Wt.** 349.41

Uses Broad spectrum antibiotic. Semi-synthetic derivative of penicillin.

Physical properties

M. Pt. 199-202°C (anhydrous).

Mammalian and avian toxicity

Acute data

TD_{Lo} oral man 400 mg kg^{-1} (1).

LD_{50} intraperitoneal, intravenous mouse 3250- 4990 mg kg^{-1} (2,3).

Sub-acute data

Subcutaneous (5 day) guinea pig 6, 8, 10 mg kg^{-1} 3 times day^{-1}. Over a period of 12 days, the lowest ampicillin dose appeared to be tolerated well. However, significant body weight reduction and mortality occurred with the 2 higher dose regimens. Caecal cultures of dead animals confirmed the presence of *Clostrichum difficile* associated with antibiotic-induced enterotoxaemia. Ampicillin accumulated in the urine and bile (4).

Carcinogenicity and long-term effects

National Toxicology Program evaluation of ampicillin trihydrate in rats and mice by gavage (2 yr) five times per week, maximum dose rats 1500 mg kg^{-1}, maximum dose mice 3000 mg kg^{-1}. Induced haemotopoietic system leukaemia and adrenal gland pheochromocytoma in ♂ rats (5-8).

Metabolism and pharmacokinetics

A penicillin amidase or acylase-type enzyme in fungi, yeasts, actinomycetes and bacteria causes hydrolysis to 6-aminopenicillanic acid (1).

Oral sheep, a single dose of 750 mg caused peak blood plasma concentration of 0.38 μg ml^{-1} within 1 hr. The biological $t_{1/2}$ was 110 min with an elimination rate constant of 0.006 min^{-1} (9).

Studies of ampicillin pharmacokinetics in the pancreas of dogs and rats following intravenous administration of 50 mg kg^{-1} revealed an elimination $t_{1/2}$ of 50 min in dogs, whereby levels attained approximately 6% of that of blood serum levels in both rats and dogs, and the degree of permeation into the pancreatic secretion of dogs was 2%. In neither case did concentrations reach minimum inhibitory concentration required for *Escherichia coli* (10).

In humans, peak plasma concentrations are obtained 1-2 hr, and following a dose of 500 mg are reported to range from 2-6 $\mu g\ ml^{-1}$ (11).

After oral administration of 500 mg to human serum $t_{1/2}$ was 0.92 hr (12).

Irritancy

Allergic reactions can occur in sensitised persons. Skin rashes are among the most common side effects and are either urticaria or maculopapular (12).

Genotoxicity

Salmonella typhimurium TA97, TA98, TA100, TA1535 and TA1537 with and without metabolic activation negative (13).

Escherichia coli PQ37 with and without metabolic activation negative (14).

Chinese hamster ovary with and without metabolic activation, sister chromatid exchange or chromosomal aberrations negative (15).

Mouse lymphoma L5178Y with and without metabolic activation negative (16).

The *in vitro* and *in vivo* clastogenic potential of ampicillin was investigated using cultured human lymphocytes and the rat micronucleus test. No increase in chromosome damage *in vitro* up to test concentrations of 10 mg ml^{-1}. Negative in *in vivo* rat micronucleus test to 5 g kg^{-1} (17).

Any other adverse effects to man

In humans symptomatic effects include diarrohea, nausea and vomiting. Pseudomembranous colitis and supra-infections have been reported (12).

The oral bioavailability of ampicillin when bound to sulbactam (sultamicillin) was compared with ampicillin alone in 16 healthy subjects using an open-label, multiple crossover study. It was demonstrated that the bioavailability of ampicillin was increased by sulbactam (18).

Ampicillin is poorly bound to proteins which results in higher concentrations in foetal tissues and amniotic fluid than would occur with highly protein-bound penicillins. High foetal-to-maternal peak serum concentration ratios of between 0.3 and 0.9 had been reported for ampicillin (19).

In a study of 42 women, administration of 500 mg every 6 hours by mouth resulted in concentrations of 0.4-5.1 $\mu g\ ml^{-1}$ in the amniotic fluid collected between 3.25 and 5.75 hours after the third dose and 0.24-2 $\mu g\ ml^{-1}$ in cord sera collected of delivery after 2-13 doses (20).

Neutropenia and thrombocytopenia in one patient has been reported following treatment with ampicillin (21).

Any other adverse effects

Oral (103 wk) rat 750 or 1500 mg kg^{-1} and oral mouse (103 wk) 1500 or 3000 mg kg^{-1}, administered as ampicillin trihydrate, clinical signs included diarrhoea, excessive urination and chromodacryorrhoea in rats and increased salivation and decreased activity in mice. In mice ampicillin administration was associated with an increased incidence of forestomach lesions, inflammation, hyperkeratosis, acanthosis and fungal infection (22). Intramuscular (7 days) rat 30 mg $kg^{-1}\ day^{-1}$ some nephrotoxic effects (unspecified) reported (23).

Any other comments

Pharmacokinetics, antibacterial activities and side-effects and antiobiotic activity in veterinary medicine (24,25).

Cows' milk is restricted for sale if it contains >0.02 I.U. l[1] (12 µg) penicillin (26).
Drug often administered as trihydrate or sodium salts to enhance bioavailability.

References

1. Claridge, C. A; et al. *Proc. Soc. Exp. Biol. Med.* 1963, **113**, 1008-1012
2. *Eksperimetnal'naya i Klinicheskaya Farmakoterapiya* 1980, **9**, 83
3. *Drugs in Japan: Ethical Drugs* 6th ed., 1982, **6**, 57
4. Young, J. D; et al. *Lab. Anim. Sci.* 1987, **37**(5), 652-656
5. Haseman, J. K; et al. *Environ. Mol. Mutagen.* 1990, **16**(Suppl. 18), 15-31
6. Haseman, J. K; et al. *J. Appl. Toxicol.* 1988, **8**(4), 267-273
7. *Gov. Rep. Announce. Index (U.S.)* 1987, **87**(20), Order No. PB87-204160, National Toxicology Program, Research Triangle Park, NC
8. Dunnick, J. K; et al. *Fundam. Appl. Toxicol.* 1989, **12**(2), 252-257
9. Ahmad, I; et al. *Arch. Exp. Veterinaermed.* 1990, **44**(3), 465-470
10. Drewelow, B; et al. *Wiss. Z. Wilhelm- Pieck- Univ. Rostock. Naturwiss. Reihe.* 1988, **37**(9), 85-89
11. *Martindale: The Extra Pharmacopoeia* 29th ed., 1989, The Pharmaceutical Press, London
12. Lui, C. H; et al. *T'ai-wan Yao Hsueh Tsa Chih* 1986, **38**(4), 262-269
13. Zeiger, E. *Environ. Mol. Mutagen.* 1990, **16**(Suppl. 18), 32-54
14. Venier, P; et al. *Mutagenesis* 1989, **4**(1), 51-57
15. Anderson, B. E; et al. *Environ. Mol. Mutagen.* 1990, **16**(Suppl. 18), 55-137
16. Myhr, B; et al. *Environ. Mol. Mutagen.* 1990, **16**(Suppl. 18), 138-167
17. Stemp, G; et al. *Mutagenesis* 1989, **4**(6), 439-445
18. Desager, J. P; et al. *J. Int. Med. Res.* 1989, **17**(6), 532-538
19. Chow, A. W; et al. *Rev. Infect. Dis.* 1985, **7**, 287
20. Blecher, T. E; et al. *Br. Med. J.* 1966, **1**, 137
21. Hughes, G. S. *Ann. Intern. Med.* 1983, **99**, 573
22. *Natl. Toxicol. Program Tech. Rep. Ser.* 1987, **318**, 1-190, National Toxicology Program, Research Triangle Park, NC
23. Deb, C; et al. *Indian J. Pharmacol.* 1986, **18**(2), 110-112
24. Flournoy, D. J. *Med. Actual.* 1988, **24**(3), 169-174
25. Bichin, S; et al. *Prax. Vet.* 1986, **34**(1-2), 53-61, (Serb. Croat.) (*Chem. Abstr.* **106**, 207046x)
26. *Milk Marketing Board Information Service Pamphlet* 1981, No. 5

A212   Amyl acetate

$$CH_3CO_2(CH_2)_4CH_3$$

CAS Registry No. 628-63-7
Synonyms pentyl acetate; acetic acid, amyl ester; amylacetic ester; 1-pentanol acetate; 1-pentyl acetate; primary amyl acetate; *n*-amyl acetate
Mol. Formula $C_7H_{14}O_2$ **Mol. Wt.** 130.19
Uses Solvent and chemical intermediate. Used in the production of acrylic resins, photographic films, glass, polishes. Food flavouring agent.
Occurrence Commercial amyl acetate is a mixture of isomers, the composition depending on its grade and derivation (1). Has been reported in an effluent of the explosives industry and of the porcelain/enamelling industry (2).

Physical properties

M. Pt. –78.5°C; **B. Pt.** 149°C; **Flash point** 23°C; **Specific gravity** d_4^{20} 0.879;

Partition coefficient log P_{ow} 2.258 (3); **Volatility** v.p. 5 mmHg at 25°C; v. den. 4.5.

Solubility
Water: 800 mg l^{-1} at 20°C. Organic solvent: miscible ethanol, diethyl ether.

Occupational exposure

US TLV (TWA) 100 ppm (532 mg m^{-3}); **UK Long- term limit** 100 ppm (530 mg m^{-3}); **UK Short-term limit** 150 ppm (800 mg m^{-3}); **UN No.** 1104; **HAZCHEM Code 3🔲**; **Conveyance classification** flammable liquid; **Supply classification** flammable.

Risk phrases
Flammable (R10)

Safety phrases
Do not breathe vapour (S23)

Ecotoxicity

Fish toxicity
LC_{50} (24-96 hr) mosquito fish 65 mg l^{-1} (4).
LD_0 (24 hr) creek chub 50 mg l^{-1} in Detroit river water.
LD_{100} (24 hr) creek chub 120 mg l^{-1} in Detroit river water (5).
LC_{50} (96 hr) bluegill sunfish 650 ppm static bioassay in fresh water, mild aeration applied after 24 hr, at 23°C.
LC_{50} (96 hr) inland silverside 180 ppm static bioassay in synthetic seawater, mild aeration after 24 hr, at 23°C (6).

Invertebrate toxicity
Toxic threshold effect (48 hr) *Daphnia magna* 440 ppm (7).
Cell multiplication inhibition test *Pseudomonas putida* 145 mg l^{-1}, *Microcystis aeruginosa* 63 mg l^{-1}, *Scenedesmus quadricauda* 80 mg l^{-1}, *Entosiphon sulcatum* 226 mg l^{-1} (8,9).

Bioaccumulation
Calculated bioconcentration factor 31 (10).

Effects to non-target species
Blocks muscle electrical activity in *Rana pipiens* (11).

Environmental fate

Degradation studies
Inhibition of glucose degradation by *Pseudomonas fluorescens* at 350 mg l^{-1}.
Inhibition of glucose degradation by *Escherichia coli* >1 g l^{-1} (12).
BOD_5 0.9 at 440 mg l^{-1} standard dilution (13).
BOD_5 64 and 35% theoretical reduction in dissolved oxygen in fresh and salt water respectively using non-acclimated sewage sludge. After 20 days the respective values were 72 and 87%, respectively (14).

Abiotic removal
The calculated hydrolysis rate for *n*-amyl acetate at pH >8 is 5.9×10^{-2} l^{-1} molecules sec^{-1} at 25°C, resulting in $t_{1/2}$ 13.5 days at pH 9 (15).

Absorption

Activated carbon absorbability 0.175 g g^{-1} carbon, 88% reduction. Influent 985 mg l^{-1}, effluent 119 mg l^{-1} (16).

Mammalian and avian toxicity

Acute data

LD$_{50}$ oral rat 6500 mg kg^{-1} (17).
LC$_{Lo}$ (8 hr) inhalation rat 5200 ppm (18).
LD$_{Lo}$ intraperitoneal guinea pig 1500 mg kg^{-1} (19).

Genotoxicity

Non-cytotoxic in Ehrlich-Landschuetz diploid ascites tumour cells (20,21).

Any other adverse effects to man

TC$_{Lo}$ inhalation human (30 min) 5000 mg m^{-3} central nervous system, eye and pulmonary effects (22).
Women exposed occupationally to amyl acetate (concentrations unspecified) showed deviations in haemoglobin and haematocrit values, erythrocyte, leukocyte, lymphocyte, monocyte, granulocyte counts, coagulation and bleeding times (23).
In humans, prolonged exposure may result in headache, fatigue and depression of the central nervous system. Irritation of the gastrointestinal tract, respiratory tract and eyes may also occur (24).
A 27 yr old man developed headache, nausea and vomiting after using a paint containing amyl acetate as the solvent in an unventilated room. Some days later chest pain and dyspnoea developed. He was admitted to hospital 2 wk after exposure with congestive heart failure which slowly responded to treatment (25).

Any other comments

Hydra sp. in vitro assay adult, embryo minimal effective ratio equals 1. This predicts that when subject to standard testing in pregnant mammals, amyl acetate would be equally toxic to mother and embryo/foetus. However, the concentrations necessary to produce adverse effects in *Hydra* do not reliably predict the levels required to produce adverse effects in the standard mammalian species (26).
Experimental toxicology and human health effects reviewed (27).
In a survey of drinking water in the UK, *n*-amyl acetate was detected in the drinking water in one of 14 treatment plants tested. The source of water for this plant was groundwater (28).

References

1. Brodzinsky, R; et al. *Volatile Organic Chemicals in the Atmosphere* 1982, 128-129, Stanford Research Institute Contract 68-02-3452, Menlo Park, CA
2. Shackleford, W. M; et al. *Anal. Chem. Acta* 1983, **146**, 15-27
3. Brass, H. J; et al. *Drinking Water Quality Enhancement Source Prot.* 1977, 393-416
4. James, H. R. *Environmental control in the organic and petrochemical industry* Noyes Data Corp., 1971
5. Gillette, L. A; et al. *Sewage and Industrial Waste* 1952, **24**(11), 1397-1401
6. Dawson, G. W; et al. *J. Haz. Mater.* 1975/77, **1**, 303-318
7. McKee, J. E; et al. *Water Quality Criterea* April 1971, US PHS Control Board Pub. –A
8. Bringmann, G.: et al. *Gwt- Wasser/Abwasser* 1976, **117**(9)
9. Bringmann, G; et al. *Wat. Res.* 1980, **14**, 231

10. Lyman, W. J; et al. *Handbook of Chemical Property Estimation Methods Environ. Behaviour of Organic Compounds* 1982, **4**, 1-33, McGraw-Hill, New York
11. Kamlet, M. J. *Quant. Struct. Act. Relat.* 1988, **7**, 71
12. Bringmann, G; et al. *Verg. Toxicol. Berfunde Wasser-Bakterien* 1960, **81** 337
13. Lund, H. F. *Industrial Pollution Control Handbook* 1971, 14-21, McGraw-Hill, New York
14. Price, K. S; et al. *J. Water Pollut. Contr. Fed.* 1974, **46**, 63-77
15. *GEMS: Graphical Exposure Modeling System: HYDRO* 1987
16. Guisti, D. M; et al. *J. Water. Pollut. Contr. Fed.* 1974, **46**(5), 947-965
17. *Raw Material Data Handbook* 1974, **1**, 3, Nat. Assoc. Print. Ink Res. Inst., Bethlehem, PA
18. *Documentation of Threshold Limit Values for Substances in Workrom Air* Am. Conf. Gov. Industrial Hygienists, 1971, 12
19. *Am. Ind. Hyg. Assoc. J.* 1974, **35**, 21
20. Holmberg, B; et al. *Environ. Res.* 1974, **7**(2), 183-192
21. *Chemical Safety Data Sheets Volume 1: Solvents* 1988, RSC, London
22. Lehmann, K. B. *Arch. Hyg.* 1913, **78**, 260
23. Lembas-Bogaczyk, J; et al. *Med. Pr.* 1989, **40**(2), 111-118, (Pol.) (*Chem. Abstr.* **113**, 102608y)
24. *Martindale: The Extra Pharmacopoeia* 29th ed., 1989, The Pharmaceutical Press, London
25. Weissberg, P. L; et al. *Br. Med. J.* 1979, **2**, 1113
26. Newman, L. M; et al. *J. Am. Coll. Toxicol.* 1990, **9**(3), 361-365
27. *ECETOC Technical Report No. 30(4)* 1991, The Chemical Industry Ecology Toxicology Centre, Brussels
28. Fielding, M; et al. *Organic Micropollutants in Drinking Water* 1981, Report No. TR- 159, Eng. Water Res. Centre, Medmenham

A213   *tert*-Amyl acetate

CH3CO2CH2C(CH3)3

CAS Registry No. 625-16-1
Synonyms 2-methyl-2-butanol acetate; *tert*-pentyl acetate
Mol. Formula $C_7H_{14}O_2$ **Mol. Wt.** 130.19
Uses Used as an industrial solvent.

Physical properties

M. Pt. −11.9°C; **B. Pt.** 124-125°C; **Flash point** d_4^{20} 0.87.

Solubility
Water: 140 g l^{-1} at 30°C.

Occupational exposure

UN No. 1104; **HAZCHEM Code** 3☑; **Conveyance classification** flammable liquid.

Any other comments

Experimental toxicology and human health effects reviewed (1).

References

1. *ECETOC Technical Report No. 30(4)* 1991, European Chemical Industry Ecology and Toxicology Centre, Brussels

A214 Amyl alcohol

CH3CH2CH2CH2CH2OH

CAS Registry No. 71-41-0
Synonyms 1-pentanol; pentyl alcohol; *n*-amyl alcohol; Amylol; Pentasol;
n-butylcarbinol
Mol. Formula $C_5H_{12}O$ **Mol. Wt.** 88.15
Uses An organic solvent and petroleum additive.
Occurrence Decomposition product of 15-hydroperoxyeicosatetranoic acid (1).

Physical properties

M. Pt. -79°C; **B. Pt.** 137.5°C; **Flash point** 38°C; **Specific gravity** d_2^{20} 0.824;
Partition coefficient log P_{ow} 1.51; **Volatility** v.p. 2 mmHg at 25°C; v. den. 3.0.

Solubility

Water: 2.7 g 100 ml^{-1} at 22°C. Organic solvent: miscible in ethanol, diethyl ether,
soluble in acetone

Occupational exposure

UN No. 1105; **HAZCHEM Code** 3ᵈ; **Conveyance classification** flammable liquid;
Supply classification flammable and harmful.

Risk phrases

Flammable – Harmful by inhalation (R10, R20)

Safety phrases

Avoid contact with skin and eyes (S24/25)

Ecotoxicity

Fish toxicity

LC_0 (24 hr) creek chub 350 mg l^{-1}.
LC_{100} (24 hr) creek chub 500 mg l^{-1} (2).
LC_{50} (96 hr) fathead minnow 470 mg l^{-1} (3).
LC_{50} (96 hr) zebra fish 530 mg l^{-1}.
LC_{50} (48 hr) ide 479-492 mg l^{-1} (4).

Invertebrate toxicity

LC_{50} (48 hr) *Daphnia magna* 440 mg l^{-1} (5).
EC_{50} (5 min, 30 min) *Photobacterium phosphoreum* 394 mg l^{-1} Microtox test (6).
Cell multiplication inhibition test, *Pseudomonas putida* 220 mg l^{-1}, *Microcystis
aeruginosa* 17 mg l^{-1} and *Scenedesmus sp.* toxic at 280 mg l^{-1} (7,8).
LC_{50} (96 hr) *Nitocra spinipes* 440 mg l^{-1} (9).

Environmental fate

Degradation studies

Aerobic heterotrophs, *Nitrosomas sp.* and methanogens were tested for toxicity to
amyl alcohol. LC_{50} *Nitrosomonas sp.* 520 mg l^{-1}, methanogens 4700 mg l^{-1} (3).
Desulfovirio vulgaris Marburg DSM 2119 metabolised amyl alcohol to the
corresponding acids in the presence of sulfate (10).

Abiotic removal
Air flotation, 89% removal occurred after chemical addition (11).

Absorption
Absorbability 0.155 g g^{-1} carbon, 71.8% reduction (12).

Mammalian and avian toxicity

Acute data
LD$_{50}$ oral rat 2700-3030 mg kg^{-1} (13,14).
LC$_{50}$ (6 hr) inhalation mouse, rat 14,000 mg m^{-3}.
LD$_{50}$ dermal rabbit 3600 mg kg^{-1} (15).
Probable human oral lethal dose 0.5-5 mg kg^{-1} (1).

Teratogenicity and reproductive effects
Inhalation rat (1-19 days gestation) 14,000 mg m^{-3} 7 hr day^{-1} reduced maternal feed intake and weight gain but there were no decreases in foetal weights and no teratogenic effects (16).

Genotoxicity
Single intragastric administration to rat bone marrow of (0.2 LD$_{50}$) caused chromosomal aberrations and polyploidy (17).

Any other adverse effects to man
Narcotic in humans. Symptomatic effects include headache, nausea, depression, diarrhoea, coughing and irritability (15).

Any other comments
Sedative effects (species unspecified) on the central nervous system reported (18).

References

1. Gosselin, R. E; et al. *Clinical Toxicology of Commercial Product*, 5th ed., 1984, Williams & Wilkins, Baltimore
2. *Code of Federal Regulations* 1984, **49**, 172.102
3. Blum, D. J. W; et al. *Res. J. Water. Pollut. Control Fed.* 1991, **63**(3), 198-207
4. Wellens, H. Z. *Wasser Abwasser Forsch.* 1982, **15**, 49
5. Verschueren, K. *Handbook of Environmental Data on Organic Chemicals* 2nd ed., 1983, Van Nostrand Reinhold, New York
6. Kaiser, K. L. E; et al. *Water Pollut. Res. J. Canada* 1991, **26**(3), 361-431
7. Bringmann, G; et al. *GWF. Wasser/Abwasser* 1976, **117**, H(9)
8. Bringmann, G; et al. *Water Res.* 1980, **14**, 231-241
9. Linden, E; et al. *Chemosphere* 1979, **11/12**, 1
10. Tanaka, K; et al. *Arch. Microbiol.* 1990, **155**(1), 18-21
11. Howe, R. H. L. *Hazardous Chemicals Handling and Disposal* 1971, Noyes Data Corp., Park Ridge, NJ
12. Ginsti, D. M; et al. *J. Water Pollut. Control Fed.* 1974, **46**, 949
13. Jenner, P. M; et al. *Food Cosmet. Toxicol.* 1964, **2**, 327
14. Scala, R. A; et al., *Am Ind. Hyg. Assoc. J.* 1973, **34**(11), 493-499
15. *Raw Materials Data Handbook* 1974, **1**, 4, Nat. Assoc. Print. Ink Res. Instit., Bethlehem, PA
16. Nelson, B. K; et al. *J. Am. Coll. Toxicol.* 1989, **8**(2), 405-410
17. Barilyak, I. R; et al. *Tsitol. Genet.* 1988, **22**(2), 49-52, (Russ.) (*Chem. Abstr.* **109**, 68620b)
18. Feller, D. J; et al. *J. Pharmacol. Exp. Ther.* 1991, **256**(3), 947-953

A215 Amyl butyrate

$$CH_3CH_2CH_2CO_2CH_2(CH_2)_3CH_3$$

CAS Registry No. 540-18-1
Synonyms butanoic acid, pentyl ester; butyric acid, pentyl ester; m-amyl butyrate; pentyl butyrate
Mol. Formula $C_9H_{18}O_2$ **Mol. Wt.** 158.24
Uses In flavours such as apricot, pineapple, pear, plum and sparingly in some perfume compositions.

Physical properties

M. Pt. $-73.2°C$; **B. Pt.** $185°C$; **Flash point** $57°C$; **Specific gravity** d_4^{15} 0.8713; **Volatility** v. den. 5.5.

Solubility
Water: 0.54 g l^{-1} at 50°C. Organic solvent: ethanol, diethyl ether

Occupational exposure
UN No. 2620; **HAZCHEM Code** 3☑; **Conveyance classification** flammable liquid.

Mammalian and avian toxicity

Acute data
LD_{50} oral rat, guinea pig $\approx$12 g kg^{-1} (1).

References
1. Jenner, P. M; et al. *Food Cosmet. Toxicol.* 1964, **2**, 327

A216 Amyl chloride

$$CH_3(CH_2)_4Cl$$

CAS Registry No. 543-59-9
Synonyms pentyl chloride; 1-chloropentane; n- butylcarbonyl chloride
Mol. Formula $C_5H_{11}Cl$ **Mol. Wt.** 106.60
Uses Solvent. Chemical intermediate.

Physical properties

M. Pt. $-99°C$; **B. Pt.** $107.8°C$; **Flash point** $12.2°C$; **Specific gravity** d_4^{20} 0.8818; **Volatility** v. den. 3.67.

Solubility
Organic solvent: miscible in ethanol, diethyl ether, soluble in benzene, chloroform

Occupational exposure
UN No. 1107; **HAZCHEM Code** 3☑E; **Conveyance classification** flammable liquid; **Supply classification** highly flammable and harmful.

Risk phrases
Highly flammable – Harmful by inhalation, in contact with skin and if swallowed
(R11, R20/21/22)

Safety phrases
Keep container in a well ventilated place – Do not empty into drains (S9, S29)

Ecotoxicity

Invertebrate toxicity
EC_{50} (5 min, 30 min) *Photobacterium phosphoreum* 244 mg l^{-1} Microtox test (1).

Environmental fate

Carbonaceous inhibition
A thermophilic obligate methane oxidizing bacterium H-2 (type 1) can degrade liquid
monochloro- and dichloro-*n*-alkanes (C5,C6) using the ribulose monophosphate and
serine pathways. Compounds are oxidised yielding their corresponding acids or
haloacids (2).

Degradation studies
Activated sludge from 3 operational waste treatment plants (Columbus, OH, Hilliard,
OH, and Linworth, OH) were used to determine the extent of oxidation for a series of
aliphatic compounds. ThOD (6, 12, 24 hr) were 1.5, 1.8 and 2.8%, respectively (3).
Aerobic heterotrophs, *Nitrosomonas sp.* and methanogens were tested for toxicity to
amyl chloride. IC_{50} *Nitrosomonas sp.* 99 mg l^{-1}, methanogens 150 mg l^{-1} and aerobic
heterotrophs 68 mg l^{-1} (4).
Acinetobacter sp. GJ70 isolated from activated study degraded amyl chloride, primary
step was the release of the halide (5).

Any other adverse effects
Exposure to amyl chloride changes the structure of the myelin sheath of rat nerve
tissue *in vitro* (6).
The effect of amyl chloride on hepatic triglyceride secretion was investigated *in vivo*
and *in vitro* (species unspecified). A dose-related decrease in hepatic triglyceride
secretion was reported (7).

References

1. Kaiser, K. L. E; et al. *Water Pollut. Res. J. Can.* 1991, **26**(3), 361-431
2. Imai, T; et al. *Appl. Environ. Microbiol.* 1986, **52**(6), 1403-1406
3. Gerhold, R. M; et al. *J. Water Pollut. Control Fed.* 1966, **38**, 562-579
4. Blum, D. J. W; et al. *Res. J. Water Pollut. Control Fed.* 1991, **63**(3), 198-207
5. Janssen, D. B; et al. *Appl. Environ. Microbiol.* 1987, **53**(3), 561-566
6. Rumsby, M. G; et al. *J. Neurochem.* 1967, **13**(12), 1513-1515
7. Selan, F. M; et al. *Res. Commun. Chem. Pathol. Pharmacol.* 1987, **35**(2), 249-269

A217 Amyl-3-cresol

$$CH_3(CH_2)_4 \text{ — } \begin{array}{c} OH \\ \end{array} \text{ — } CH_3$$

CAS Registry No. 1300-94-3
Synonyms 6-amyl-*m*-cresol; 6-*n*-amyl- *m*-cresol; 5-methyl-2-pentylphenol;
6-*n*-pentyl- *m*-cresol; amyl-*m*-cresol
Mol. Formula $C_{12}H_{18}O$ **Mol. Wt.** 178.28
Uses Prevention of moulds, bactericide, germicide, antiseptic (1).

Physical properties
M. Pt. 24°C; **B. Pt.** $_{15}$ 137°C; **Flash point** 115.6°C; **Specific gravity** 0.97.

Solubility
Organic solvent: acetone, ethanol

Occupational exposure
Supply classification flammable.

Risk phrases
Flammable (R10)

Safety phrases
Do not breathe vapour (S23)

Mammalian and avian toxicity

Acute data
LD_{50} oral rat 1500 mg kg^{-1} (2).

References
1. *Martindale: The Extra Pharmacopoeia* 29th ed., 1989, The Pharmaceutical Press, London
2. *Proc. Soc. Expt. Biol. Med.* 1935, **32**, 592

A218 Amylene

$$(CH_3)_2C=CHCH_3$$

CAS Registry No. 513-35-9
Synonyms 2-methyl-2-butene; trimethylethylene; β-isoamylene
Mol. Formula C_5H_{10} **Mol. Wt.** 70.14
Uses Organic synthesis. Polymeristion inhibition.

Physical properties

M. Pt. –124°C; **B. Pt.** 37-39°C; **Flash point** –18°C; **Specific gravity** d_4^{15} 0.66;
Volatility v. den. 2.3.

Solubility

Organic solvent: miscible ethanol, diethyl ether

Occupational exposure

UN No. 1108; **HAZCHEM Code** 3☑E; **Conveyance classification** flammable liquid.

Genotoxicity

Salmonella typhimurium TA98, TA100, TA1535, TA1537, TA1538 with and without metabolic activation negative.
Escherichia coli WP2, WP2 uvrA with and without metabolic activation negative.
Saccharomyces cerevisae JD1 with and without metabolic activation negative (1).

Any other adverse effects to man

A simple asphyxiant in humans (2).

Any other comments

Critical evaluation of solubility data in water (3).

References

1. Dean, B. J; et al. *Mutat. Res.* 1985, **153** (1-2), 57-77
2. *The Merck Index* 11th ed. 1989, Merck & Co. Inc, Rahway, NJ
3. Hefter, G. T. *Solubility Data Ser.* 1989, **37**, 22-26

A219   Amyl mercaptan

CH3(CH2)4SH

CAS Registry No. 110-66-7
Synonyms pentylmercaptan; 1-pentanethiol
Mol. Formula $C_5H_{12}S$ **Mol. Wt.** 104.22
Uses Chemical intermediate.

Physical properties

M. Pt. –75.7°C; **B. Pt.** 126°C; **Flash point** 18°C; **Specific gravity** d_4^{20} 0.857; **Volatility** v.p. 13.8 mmHg at 25°C; v. den. 3.59..

Solubility

Organic solvent: miscible ethanol, diethyl ether

Occupational exposure

UN No. 1111; **HAZCHEM Code** 3WE; **Conveyance classification** flammable liquid.

Mammalian and avian toxicity

Acute data

LC_{Lo} (4 hr) inhalation rat 2000 ppm (1).
LD_{50} intraperitoneal mouse 100 mg kg^{-1} (2).

Any other adverse effects

Long-chain 1-mercapto-*n*-alkanes showed potent inhibitory effects with horse liver alcohol dehydrogenase. Results suggest that thiols interact simultaneously with ≥2 sites of the enzyme; it is thought the primary interaction could be with the zinc atom in the active site and the other with the hydrophobic binding site for alkyl carbon atoms (3).

Any other comments

Experimental toxicology and human health effects reviewed (4).

References

1. Carpenter, C. P; et al. *J. Ind. Hyg. Toxicol.* 1949, **31**, 343-346
2. Howell, W. C; et al. *J. Am. Chem. Soc.* 1956, **78**, 3843-3846
3. Miwa, K; et al. *Biophys. Res. Commun.* 1987, **142**(3), 993-998
4. *ECETOC Technical Report No. 30(4)* 1991, The Chemical Industry Ecology and Toxicology Centre, Brussels

A220   Amyl nitrate

CH3(CH2)4NO3

CAS Registry No. 1002-16-0
Synonyms nitric acid, pentyl ester; *n*-pentyl nitrate
Mol. Formula $C_5H_{11}NO_3$ **Mol. Wt.** 133.15
Uses Additive for diesel fuel.

Physical properties

B. Pt. 153-157°C; **Flash point** 48°C(open cup); **Specific gravity** d_4^{20} 0.997.

Solubility

Organic solvent: alcohol, ether

Occupational exposure

UN No. 1112; **HAZCHEM Code** 3☒; **Conveyance classification** flammable liquid.

Mammalian and avian toxicity

Sub-acute data

LC_{Lo} (3 × 7 hr) inhalation rat exposures 1703 ppm (1). Human exposure during animal testing produced nausea and headaches (1).

References

1. Treon, J. F; et al. *AMA, Arch. Ind. Health* 1955, **II**, 290

A221   1-Amyl-1-nitrosourea

$$CH_3(CH_2)_4N(NO)CONH_2$$

CAS Registry No. 10589-74-9

Synonyms N-nitroso-N-pentylurea; N- amyl-N-nitrosourea; N-pentyl-N- nitrosourea; ANU

Mol. Formula $C_6H_{13}N_3O_2$ **Mol. Wt.** 159.19

Mammalian and avian toxicity

Acute data

LD_{50} oral rat 570 mg kg^{-1} [1].

TD_{Lo} subcutaneous rat 510 mg kg^{-1} (1).

Sub-acute data

LD_{50} (50 wk) dermal mouse 629 mg kg^{-1} (2).

Carcinogenicity and long-term effects

Gavage Fischer 344 rats (50 wk) administered as equimolar doses twice weekly, induced tumours of the liver, forestomach, thyroid and nervous system in both sexes, single sex tumours recorded ♀ mammary gland, uterus, colon, ♂ ileum, jejunum, duodenum, mesothelioma, Zymbal gland, bladder and skin (3).

Dermal ♀ Swiss mice 6.4 g l^{-1} painted on twice weekly induced skin carcinomas, lung adenomas and adenocarcinomas, and tumours of the forestomach, mammary gland, uterus, ovary and leukaemia (4).

Dermal application to ♀ mice (40-50 weeks) 25 µl of a 0.04 ml solution in acetone. Animals were observed until death or 100 weeks. 1-Amyl-1- nitrosourea induced skin tumours in 11 of 20 mice (2).

A single subcutaneous injection to rats (concentration unspecified) at 1 or 10 days of age. Tumours of the nervous system tissue and neurinomas of the heart were induced. Carcinogenic potential was reduced when administered to older animals (5).

Genotoxicity

Salmonella typhimurium TA1535 without metabolic activation positive and *Escherichia coli* BR339 λ positive (6).

In vitro Chinese hamster cells incubation (48 hr) 0.25 mg ml^{-1} induced >50% chromosome aberrations (7).

References

1. *Ann. New York. Acad. Sci.* 1982, **381**, 250
2. Lijinsky, W; et al. *J. Cancer. Res. Clin. Oncol.* 1981, **102**(1), 13-20
3. Lijinsky, W; et al. *Toxicol. Ind. Health* 1989, **5**(6), 925-935
4. Lijinsky, W; et al. *J. Cancer. Res. Clin. Oncol.* 1988, **114**, 245-249
5. Ivanovic, S; et al. *Arch. Geschwulstforsch* 1981, **51**(2), 187-203, (Ger.) (*Chem. Abstr.* **95**(5), 36806t)
6. Lijinsky, W; et al. *Mutat. Res.* 1987, **178**(2), 157-165
7. Ishidate, M. Jr. *Mutat. Res.* 1977, **48**(3-4), 337-353

A222 Amylobarbitone

CAS Registry No. 57-43-2

Synonyms amobarbitol; barbamic; penymal; somnal; dormythal; isomythal;
5-ethyl-5-isopentylbarbituric acid; 5- ethyl-5-(3-methylbutyl)-
2,4,6-(1H,3H,5H)-pyrimidinetrione

Mol. Formula $C_{11}H_8N_2O_3$ **Mol. Wt.** 216.20

Uses Hypnotic and sedative.

Physical properties

M. Pt. 156-158°C.

Solubility
Water: 0.77 g l^{-1}. Organic solvent: benzene, ethanol, chloroform, diethyl ether

Ecotoxicity

Invertebrate toxicity
EC_{50} (5 min) *Photobacterium phosphoreum* 1011 mg l^{-1} Microtox test (1).

Mammalian and avian toxicity

Acute data
LD_{50} oral mouse 345 mg kg^{-1} (2).
LD_{50} subcutaneous mouse 212 mg kg^{-1} (3).
LD_{50} intraperitoneal rat 115 mg kg^{-1} (4).

Metabolism and pharmacokinetics
Readily absorbed from gastrointestinal tract, 60% bound to plasma proteins.
$t_{1/2}$ 20-25 hr longer in neonates.
50% excreted in urine as 3'- hydroxyamylobarbitone and 30% as
N-hydroxyamylobarbitone (5).
Readily absorbed from the gastrointestinal tract and following absorption,
approximately 60% is bound to plasma proteins. $t_{1/2}$ 20-25 hr in adults. Metabolised
in the liver with 50% excreted in the urine as 3'-hydroxyamylobarbitone, and 30% as
N-hydroxyamylobarbitone. Less than 1% appears as unchanged parent compound.
Approximately 5% is excreted in the faeces (6).
Significant urinary metabolite reported to be 5-(3'-carboxybutyl)-5-ethylbarbituric acid (7).

Genotoxicity
In vivo Chinese hamster V-79 cells with and without metabolic activation positive (8).

Any other adverse effects to man
In humans, adverse effects are similar to those for barbiturates and include respiratory

depression, allergic skin rashes, hepatitis, cholestasis and photosensitivity. Erythema multiforme (Stevens-Johnson syndrome) and exfoliative dermatitis (sometimes fatal) have been reported. As with other sedatives, paradoxical excitement and irritability may occur. Nystagmus and ataxia may occur with excessive doses. Toxic effects of overdose result from profound central depression and include coma, respiratory and cardiovascular depression with hypotension and shock leading to renal failure (5). Intravenous injection 200 mg to 9 healthy young human adults (4 ♂, 5 ♀). Blood samples were taken 48 hr after infusion, the mean values for clearance and apparent volume of distribution were $0.032 \, l \, hr^{-1} \, kg^{-1}$ and $1.08 \, l \, kg^{-1}$, respectively (9). One human subject ingested amylobarbitone (concentration unspecified) 7 times over 3 yr, plasma clearances $32.1 \, ml \, min^{-1}$ exhibited constancy, while the distribution volume 73.6 l showed some fluctuation (10).

Legislation

Controlled substance in the US (11).

Any other comments

Often administered as the sodium salt to enhance bioavailability (7).
Properties, experimental toxicology and metabolism reviewed (12).

References

1. Kaiser, K. L. E; et al. *Water Pollut. Res. J. Can.* 1991, **26**(3), 361-431
2. *J. Am. Chem. Soc.* 1939, **61**, 96
3. Irrgang, K. *Arzneimittel-Forsch.* 1965, **15**, 688
4. Knoll, J; et al. *Arzneimittel. Forsch.* 1971, **21**, 719
5. *Martindale: The Extra Pharmacopoeia* 29th ed., 1989, The Pharmaceutical Press, London
6. Tang, B. K; et al. *Drug. Metab. Disposit.* 1975, **3**, 479
7. Baldeo, W; et al. *J. Pharm. Pharmac.* 1977, **29**, 254
8. Bohrman, J. S; et al. *Environ. Mol. Mutagen.* 1988, **12**(1), 33-51
9. Bachmann, K. *Br. J. Clin. Pharmacol.* 1987, **23**(1), 95-98
10. Inaba, T; et al. *Clin. Pharmacol. Therap.* 1976, 439-444
11. *US Code of Federal Regulations* 1987, **21**(329.1, 1308.12, 1308.13)
12. Mian, N; et al. *Anal. Profiles Drug. Subst.* 1990, **19**, 27-58

A223   4-*tert*-Amylphenol

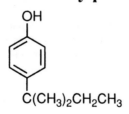

OH

C(CH₃)₂CH₂CH₃

CAS Registry No. 80-46-6
Synonyms *p-tert*-pentylphenol; 2-methyl-2- *p*-hydroxyphenylbutane; Pentaphen; *p-tert*- Amylphenol; Amilfenol; 4-(1,1-dimethylpropyl)phenol; 4- *tert*-Amylphenol; 4-*tert*-pentylphenol; *p*- (α, α-dimethylpropyl)phenol; *p*-1,1-dimethylpropyl)phenol
Mol. Formula $C_{11}H_{16}O$ **Mol. Wt.** 164.25
Uses Manufacture of oil-soluble resins. Intermediate for organic mercury germicides

and pesticides. Intermediate in the manufacture of chemicals used in the rubber and petroleum industries.

Physical properties

M. Pt. 94-95°C; **B. Pt.** 262°C; **Flash point** 112°C; **Specific gravity** d_4^{20} 0.96.

Solubility

Organic solvent: ethanol, diethyl ether, chloroform, benzene

Ecotoxicity

Fish toxicity
LC_{50} (96 hr) fathead minnow 2.5 mg l^{-1} (1).

Invertebrate toxicity
EC_{50} (48 hr) *Tetrahymena pyriformis* 9.6 mg l^{-1} (2).

Mammalian and avian toxicity

Acute data
LD_{50} oral rat 1830 mg kg^{-1} (3).
LD_{50} dermal rabbit 2000 mg kg^{-1} (4).

Irritancy
A 1% solution instilled into rabbit eye caused severe irritancy (4).
Application of 100 μg to rabbit skin for 24 hr caused irritancy (5).

Genotoxicity

Salmonella typhimurium TA97, TA98, TA100, TA1535 with and without metabolic activation negative (6).

Any other comments

Toxicology reviewed (7).

References

1. Holcombe, G. W; et al. *Environ. Pollut. Series A* 1984, **35**, 367-381
2. Schultz, T. W. *Ecotoxicol. Environ. Saf.* 1987, **14**, 178
3. *Ind. Hyg. Foundation of America, Chem. and Toxic. Ser. Bull.* 1967, **1**, 6
4. *Union Carbide Material Safety Data Sheet* 13 Aug 1964
5. *Am. Ind. Hyg. Assoc. J.* 1962, **23**, 95
6. Zeiger, E; et al. *Environ. Mol. Mutagen.* 1988, **11**(Suppl. 12), 1-157
7. *ECETOC Technical Report No. 30(4)* 1991, The European Chemical Industry Ecology and Toxicology Centre, Brussels

A224 Amyl propionate

$CH_3CH_2CO_2(CH_2)_4CH_3$

CAS Registry No. 624-54-4
Synonyms *n*-pentyl propionate; *n*-pentyl propanoate
Mol. Formula $C_8H_{16}O_2$ **Mol. Wt.** 144.22

Occurrence Found in flavour component of bananas.

Physical properties

M. Pt. −73.1°C; **B. Pt.** 168.6°C; **Flash point** 41°C (open cup); **Specific gravity** d_4^{20} 0.8761; **Volatility** v. den. 5.0.

Solubility

Organic solvent: benzene, diethyl ether, ethanol

Genotoxicity

Saccharomyces cerevisiae D61.M without activating system negative (1).

References

1. Zimmerman, F. K; et al. *Mutat. Res.* 1989, **224**(2), 287-303

A225 Amyltrichlorosilane

$CH_3(CH_2)_4SiCl_3$

CAS Registry No. 107-72-2
Synonyms trichloropentylsilane; pentylsilicon trichloride; pentyltrichlorosilane; trichloroamylsilane
Mol. Formula $C_5H_{11}Cl_3Si$ **Mol. Wt.** 205.59
Uses Intermediate for silicone synthesis.

Physical properties

B. Pt. 160°C; **Flash point** 62°C (open cup); **Specific gravity** 1.142.

Occupational exposure

UN No. 1728; **HAZCHEM Code** 4XE; **Conveyance classification** corrosive substance.

Mammalian and avian toxicity

Acute data

LD_{50} oral rat 2340 mg kg^{-1} (1).
LC_{50} (4 hr) inhalation rat 2000 ppm (2).
LD_{50} dermal rabbit 780 mg kg^{-1} (1).

Irritancy

100 µg applied to rabbit for 24 hr caused moderate irritancy (1).

Any other comments

Readily hydrolysed with liberation of hydrogen chloride.

References

1. Smyth, H. F; et al. *Am. Ind. Hyg. Assoc. J.* 1962, **23**, 95
2. Smyth, H. F; et al. *J. Ind. Hyg. Toxicol.* 1949, **31**, 343

A226   Ancymidol

CAS Registry No. 12771-68-5
Synonyms α-cyclopropyl-α-14- methoxyphenyl-5-pyrimidinemethanol; iso-F; EL-531; A-Rert; Reducymol
Mol. Formula $C_{15}H_{16}N_2O_2$ **Mol. Wt.** 256.31
Uses Plant growth regulator and retardant.

Physical properties

M. Pt. 110-111°C; **Volatility** v.p. 1.0×10^{-6} mmHg at 50°C.

Solubility
Water: 650 mg l^{-1}. Organic solvent: acetone, ethanol, methanol, chloroform, acetonitrile, hexane

Ecotoxicity

Fish toxicity
LC_{50} (duration unspecified) rainbow trout, bluegill sunfish, goldfish (fingerlings) 55 to >100 mg l^{-1} (1).
LC_{50} (duration unspecified) bluegill sunfish 146 mg l^{-1} (2).

Effects to non-target species
Non-toxic to bees (2).

Mammalian and avian toxicity

Acute data
LD_{50} oral rat, mouse 4500 to >5000 mg kg^{-1} (2,3).
LD_{50} oral dog >500 mg kg^{-1} (3).
LD_{50} percutaneous rabbit >200 mg kg^{-1} (2).

Sub-acute data
In 90 day feeding trials, rats and dogs receiving up to 8000 mg kg^{-1} diet showed no ill- effects (2).

References

1. *The Pesticides Manual* 9th ed., 1991, British Crop Protection Council, Farnham
2. *The Agrochemicals Handbook* 3rd ed., 1991, RSC, London
3. *European Directory of Agrochemical Products* 1990, RSC, London

A227 Anilazine

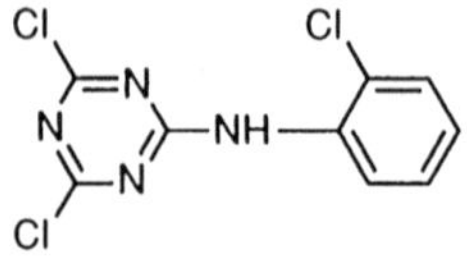

CAS Registry No. 101-05-3

Synonyms 4,6-dichloro-N-(2-chlorophenyl)-1,3,5- triazin-2-amine;
2, 4-dichloro-6-(o-chloroanilino)-s-triazine; (o-choroanilino) dichlorotriazine;
Dyrene; Boktrysan; Triazine

Mol. Formula $C_9H_5Cl_3N_4$ **Mol. Wt.** 275.53

Uses Fungicide.

Physical properties

M. Pt. 159°C; **Specific gravity** d^{20} 1.8.

Solubility
Organic solvent: toluene, xylene, acetone

Ecotoxicity

Fish toxicity
LC_{50} (96 hr) rainbow trout, channel catfish, bluegill sunfish 140 µg l^{-1} (1).

Invertebrate toxicity
EC_{50} (duration unspecified) *Daphnia pulex* 4.5 mg l^{-1} (2).

Effects to non-target species
LD_{50} oral redwing blackbird 100 mg kg^{-1} (3).
Harmful to bees (4).

Environmental fate

Degradation studies
The $t_{1/2}$ in damp soil measured 12 hr (5).

Absorption
Adsorption through exchange process to organic matter and clay minerals is
dependent on pH of solution and acidity of adsorbent surface (6).

Mammalian and avian toxicity

Acute data
LD_{50} oral rat, rabbit 400-2700 mg kg^{-1} (7,8).
LD_{50} intraperitoneal mouse, rat 25-50 mg kg^{-1} (9).
LD_{50} percutaneous rabbit, rat 5000 to >9400 mg kg^{-1} (5).

Sub-acute data
In 2 yr feeding trials, rats receiving up to 200 mg kg^{-1} diet showed no ill-effects (5).

Carcinogenicity and long-term effects
National Toxicology Program tested rat, mouse with anilazine via feed. Negative
results for ♂ and ♀ rats and mice (10).

Metabolism and pharmacokinetics

20% of a dose was excreted in the faeces, 64% in urine and 16% remained in the carcass after 3 day (11).

Irritancy

Dermal rabbit (duration unspecified) 500 mg caused severe irritancy (8).

Genotoxicity

Salmonella typhimurium TA97, TA98, TA100, TA1535, TA1537 with and without metabolic activation negative (12).

Legislation

Limited under EC Directive on Drinking Water Quality 80/778/EEC. Maximum admissible concentration 0.1 µg l^{-1} (13).

Any other comments

Toxicity studied (14).
Incompatible with oils and alkaline materials.

References

1. *Handbook of Acute Toxicity of Chemicals to Fish and Aquatic Invertebrates* 1980, 35, US Dept. of Interior, Fish and Wildlife Service, US Govt. Print. Off., Washington, DC
2. Nishiuchi, Y; et al. *Botyu-Kagaku* 1967, **32**, 5-11
3. Schafer, E. W; et al. *Arch. Environ. Contam. Toxicol.* 1983, **12**(3), 355-382
4. *European Directory of Agrochemical Products; Fungicides* 1990, RSC, London
5. *The Agrochemicals Handbook* 3rd ed., 1991, RSC, London
6. Doull, J; et al. *Casarett and Doull's Toxicology* 2nd ed., 1980, 640, Macmillan Publishing Co., New York
7. *Agric. Res. Serv.* 1966, **20**, 9
8. *Toxicology of Drugs and Chemicals* 1969, 235
9. Cohen, S. D; et al. *J. Agric. Food Chem.* 1973, **21**, 140
10. *National Toxicology Program Research and Testing Div.* 1992, Report No. TR-104, NIEHS, Research Triangle Park, NC 27709
11. Gosselin, R. E; et al. *Clinical Toxicology of Commercial Products* 5th ed., 1984, II-333, Williams and Wilkins, Baltimore
12. Zeiger, E; et al. *Environ. Mol. Mutagen.* 1988, **11**(Suppl. 12), 1-157
13. *EC Directive Relating to the Quality of Water Intended for Human Consumption* 1982, 80/778/EEC, Office for Offical Publications of the European Communities, 2 rue Mercier, L-2985 Luxembourg
14. Cohen, S. D; et al. *Agric. Food Chem.* 1973, **21**, 140

A228   Aniline

CAS Registry No. 62-53-3

Synonyms benzenamine; phenylamine; aminobenzene; aminophen; Kyanol; Anyvim; Blue Oil; C.I. 76000; C.I. Oxidation Base; aniline oil; NCI-C03736

Mol. Formula C_6H_7N **Mol. Wt.** 93.13

Uses In the manufacture of dyestuffs, medicinals, resins, varnishes and shoe blacks.

Physical properties

M. Pt. $-6.2°C$; **B. Pt.** 184.4°C; **Flash point** 70-76°C (closed cup); **Specific gravity** d_4^{20} 1.02173; **Partition coefficient** log P_{ow} 0.90; **Volatility** v.p. 0.3 mmHg at 20°C; v. den. 3.22.

Solubility
Organic solvent: benzene, diethyl ether, ethanol, acetone

Occupational exposure

US TLV (TWA) 2 ppm (7.6 mg m^{-3}); **UK Long-term limit** 2 ppm (10 mg m^{-3}) under review; **UK Short- term limit** 5 ppm (20 mg m^{-3}) under review; **UN No.** 1547; **HAZCHEM Code** 3X; **Conveyance classification** toxic substance; **Supply classification** toxic.

Risk phrases
Toxic by inhalation, in contact with skin and if swallowed – Danger of cumulative effects (R23/24/25, R33)

Safety phrases
After contact with skin, wash immediately with plenty of water – Wear protective clothing and gloves – If you feel unwell, seek medical advice (show label where possible) (S28, S36/37, S44)

Ecotoxicity

Fish toxicity
LC$_{50}$ (0-8 day) bass, goldfish, catfish 47.3-4.4 mg l^{-1} (1).
LC$_{50}$ (96 hr) fathead minnow 134 mg l^{-1} (2).
LC$_{50}$ (7 day) rainbow trout 8.2 mg l^{-1} (3).

Invertebrate toxicity
Inhibition of cell multiplication *Pseudomonas putida* 130 mg l^{-1} (4).
EC$_{50}$ (48 hr) *Daphnia magna, Daphnia pulex, Daphnia cucculata* 0.10-0.68 mg l^{-1} (5).
EC$_{50}$ (5-30 min) *Photobacterium phosphoreum* 70.6 mg l^{-1} Microtox test (6).

Bioaccumulation
Measured bioconcentration was ≤148, demonstrating that aniline does not accumulate in fish (7-9).

Effects to non-target species
LD$_{50}$ oral redwing blackbird 562 mg kg^{-1} (10).

Environmental fate

Nitrification inhibition
Ammonia oxidation *Nitrosomonas sp.* 76% inhibition 2.5 mg l^{-1} (11).

Degradation studies
Biodegradable (12).
Adsorption on Amberlite XAD 7 100% (13).
Decomposition by soil microflora 4 days (14).

Decomposition by *Aerobacter* 500 mg l^{-1} at 30°C, parent 100% ring disruption in 54 hr, mutant 100% ring disruption in 12 hr (15).

Degraded by many common species of bacteria and fungi found in soil and acetanilide, 2-hydroxyacetanilide, 4-hydroxyaniline and catechol are reported metabolites (16-19).

Desulfobacterium anilini completely degraded aniline to carbon dioxide and ammonia with stoichiometric reduction of sulfate to sulfide (19).

Abiotic removal

Oxidised on exposure to sunlight in air forming products including hydrazobenzene, 4- aminodiphenylamine, 2-aminodiphenylamine, benzidine and azobenzene (20).

Photolytic $t_{1/2}$ in the atmosphere has been estimated to be 2.1 days based on a measured reaction rate constant of 0.32 l day^{-1} (21).

Estimated $t_{1/2}$ for aniline vapour reacting with photochemically generated hydroxyl radicals in the atmosphere is 3.3 hr (22).

Soil-catalysed oxidation occurs in sterilised soil with 9 products being detected after 2 days, including azobenzene, azoxybenzene, phenazine, formanilide and acetanilide (23).

$t_{1/2}$ in estuarine water was 27 and 173 hr in the light and dark (microbial), respectively. Photolysis rates decreased in winter due to decreased surface irradiation and temperature (24).

Absorption

0.150 g g^{-1} carbon, 74.9% reduction (4).

Soil adsorption increases with the percent of organic carbon and decreases with the pH of the soil (25).

The soil adsorption coefficient in coleqoidal organic carbon from groundwater was 3900 which effectively increases aniline solubility and leaching into groundwater (26).

Mammalian and avian toxicity

Acute data

LD$_{50}$ oral rat 250 mg kg^{-1} (27).
LC$_{50}$ (7 hr) inhalation mouse 175 ppm (28).
LD$_{Lo}$ dermal rabbit 820 mg kg^{-1} (29).
LD$_{50}$ dermal guinea pig 290 mg kg^{-1} (30).
LD$_{50}$ intraperitoneal rat 420 mg kg^{-1} (31).

Sub-acute data

Inhalation rat (4 day) 150 ppm methaemaglobinaemia, decreased haemocrit reported (32).

Carcinogenicity and long-term effects

Inadequate evidence for carcinogenicity to humans or animals, IARC classification group 3 (33).

Induced tumours in spleen of ♂ and ♀ rats, but not mice (34).

Metabolism and pharmacokinetics

Converted via hepatic microsomal enzymes to aminophenols and *N*-hydroxylamines (35,36).

Following administration to rabbits, 80% of dose is excreted in urine as conjugates of 4-aminophenol (55%), 2-aminophenol (10%) and 3-aminophenol (0.1%), and as aniline (3.5%), aniline-N- glucuronide (6%), phenylsulfamic acid (8%) and acetanilide (0.2%). Traces of the metabolites (1%) were excreted in faeces and no aniline was exhaled (37).

In rats aniline is absorbed by simple diffusion from rectum at rates related to degree of ionisation and lipid-to-water partition coefficient of compound (38).

Irritancy
Dermal rabbit (24 hr) 500 mg caused moderate irritation, while 102 mg instilled into rabbit eye caused severe irritation (39,40).

Genotoxicity
Escherichia coli PQ37 SOS-Chromotest negative (41).
Escherichia coli WP2s (λ) negative induction of prophage in microscreen assay (42).
Mouse L5178Y tk+/tk- lymphoma cells with and without metabolic activation positive (43).
Positive response was obtained in the mouse bone marrow micronucleus test 24 hr after the second of two intraperitoneal injections of 80% of the median LD. Negative results were observed at lower dose levels and at earlier and later sampling times (44).

Any other adverse effects to man
Following exposure of two workers to aniline, methaemoglobin and urinary metabolites decreased while no significant change was observed for methaemoglobin reductase level. $t_{1/2}$ of methaemoglobin was 13 hr (45). Inhalation, ingestion or cutaneous absorption results in methaemoglobinaemia, with cyanosis, headache, weakness, stupor and coma. Skin sensitisation, nausea, liver and kidney damage and cardiac arrhythmias may occur and haemolysis has been reported (46).

Any other comments
The US government has alerted US companies using aniline that new evidence clearly associates exposure to aniline with an increased risk of bladder cancer, and that they should reduce workers' exposure to the chemical to the lowest feasible concentration. Other recommendations include: engineering controls and work practices to limit exposure, using personal protective equipment including special clothing, environmental monitoring and medical screening of workers (47).
Occupational exposure in organic synthesis and dyestuffs manufacturing, properties, risks, metabolism, toxicity, handling and storage, and first aid reviewed (48).
Analysis of air samples in school buildings containing self-levelling flooring material containing casein showed the presence of aniline, but at concentrations well below official threshold limits. The pollution was attributed to biodegradation of the casein by alkali-resistant *Clostridium sp*. People in these buildings reported sick building syndrome-type symptoms (49).
Physical and chemical properties, toxicity, hazards and recommendations for storage and handling, and first aid in case of poisoning reviewed (50).
Physico-chemical properties, human health effects, experimental toxicology, epidemiology, environmental effects and workplace experience reviewed (51,52).
Incompatible with oxidisers, albumin, solutions of iron, zinc, aluminium, acids and alkalis.

References
1. Marking, L. L; et al. *Aquatic Toxicol.* 1979, 131-147

2. Brooke, L. T; et al. *Acute Toxicities of Organic to Fathead Minnows (Pimephales promelas)* 1984, Center for Lake Superior Environmental Studies, Univ. of Wisconsin, Superior, WI
3. Abram, F. H. S. *Water Res.* 1982, **16**(8), 1309
4. Simpson, R. M. *Progress in Hazardous Chemicals Handling and Disposal* 1972, Noyes Data Corp., Park Ridge, NJ
5. Canton, J. H; et al. *Hydrobiol.* 1978, **59**, 135-140
6. Kaiser, K. L. E; et al. *Water Pollut. Res. J. Canada* 1991, **26**(3), 361-431
7. Lu, P. Y; et al. *Environ. Health Perspect.* 1975, **10**, 269-284
8. Freitag, D; et al. *Ecotox. Environ. Safety* 1982, **6**, 60-81
9. Freitag, D; et al. *Chemosphere* 1985, **14**, 1589-1616
10. Schafer, E. W. *Toxicol. Appl. Pharmacol.* 1972, **21**, 315
11. Hockenbury, M. R; et al. *J. Water Pollut. Control Fed.* 1977, **49**(5), 768-777
12. *MITI Report* 1984, Ministry of International Trade and Industry, Tokyo
13. Bringmann, G; et al. *GWF- Wasser/Abwasser* 1976, **117**, 119
14. Alexander, M; et al. *J. Agric. Food Chem.* 1966, **14**, 410
15. Worne, H. E. *Tijdschrift van het BECEWA* Liege, Belgium
16. Aoki, K; et al. *Agric. Biol. Chem.* 1982, **46**, 2563-2571
17. Smith, E. V; et al. *Arch. Biochem. Biophys.* 1974, **161**, 551-558
18. Subramanian, V; et al. *Indian Instit. Sci. J.* 1978, **60**, 143-178
19. Schnell, S; et al. *Arch. Microbiol.* 1989, **152** (6), 556-563
20. Zepp, R. G; et al. *Chemosphere* 1981, **10**, 109 –117
21. Mill, T; et al. *ACS Div. Environ. Chem. 192nd Natl. Meeting* 1986, **26**, 59-63
22. Atkinson, R. *Inter. J. Chem. Kinet.* 1987, **19**, 799-828
23. Pillai, P; et al. *Chemosphere* 1982, **11**, 299 –317
24. Hwang, H. M; et al. *Water Res.* 1987, **21**(3), 309-316
25. Moreale, A; et al. *J. Soil Sci.* 1976, **27**, 48-57
26. Means, J. C. *Amer. Chem. Soc. 186th Natl. Meet.* 1982, **23**, 250-251
27. *J. Pharmacol. Exper. Therap.* 1947, **90**, 260.
28. *NTIS Report* PB 214-270, National Technical Information Service, Springfield, VA
29. *Polish J. Pharmacol. Pharmacy* 1980, **32**, 223
30. *Toxic and Hazardous Industrial Chemicals Safety Manual* 1975, International Technical Information Service, Tokyo
31. *Arch. Gewerbepathol und Gewerbehygiene* 1957, **15**, 447
32. Gosselin, R. E; et al. *Clinical Toxicology of Commercial Products* 5th ed., 1984, Williams & Wilkins, Baltimore
33. *IARC Monograph* 1987, **Suppl. 7**, 99
34. Ashby, J. *Mutat. Res.* 1988, **204**, 17-115
35. Keise, M; et al. *Euro. J. Clin. Pharmacol.* 1972, **4**, 115
36. Mennes, W. C; et al. *Inst. Natl. Sante. Rech. Med.* 1988, **164**, 351
37. Parke, D. V. *The Biochemistry of Foreign Compounds* 1968, 224, Pergamon Press, Oxford
38. La Du, B. N; et al. *Fundamentals of Drug Metabolism and Disposition* 1971, 36, Williams & Wilkins, Baltimore
39. Miyajima, K; et al. *Sangyo Igaku* 1991, **33**(2), 106-107, (Jap.) (*Chem. Abstr.* 1991, **115**, 98376t)
40. Marhold, J. V. *Sbornik Vysledku Toxixologickeho Vysetreni Latek A Pripravku* 1972, 65, Prague
41. *BIOFAX Industrial Bio-Test Laboratories, Inc., Data Sheets* 1-5/69
42. Von der Hude, W; et al. *Mutat. Res.* 1988, **203**(2), 81-94
43. Rossman, T. G; et al. *Mutat. Res.* 1991, **260**(4), 349-367
44. McGregor, D. D; et al. *Environ. Mol. Mutagen.* 1991, **17**(3), 196-219
45. Ashby, J; et al. *Mutat. Res.* 1991, **263**(2), 115-117
46. *Martindale: The Extra Pharmacopoeia* 29th ed., 1989, The Pharmaceutical Press, London
47. *Chemical Marketing Reporter* 1991, **239**(14), 3,16

48. Bressa, G. D. A., Dif. Ambientale 1990, **14**(6-7), 29, (Ital.) (Chem. Abstr. 1991, **115**, 141412r)
49. Karlsson, S; et al. Mater. Struct. 1989, **22**(129), 163-169
50. Cah. Notes Doc. 1989, **134**, 161-165, (Fr.) (Chem. Abstr. 1989, **110**, 236410z)
51. ECETOC Technical Report No. 30(4) 1991, European Chemical Industry Ecology and Toxicology Centre, B-1160 Brussels
52. Izmerov, N. F; Scientific Reviews of Soviet Literature on Toxicity & Hazards of Chemicals 1992–1993, **53**, Eng. Trans, Richardson, M. L. (Ed.), UNEP/IRPTC, Geneva

A229   Aniline hydrochloride

CAS Registry No. 142-04-1
Synonyms aniline salt; benzenamine hydrochloride; anilinium chloride
Mol. Formula C_6H_8ClN **Mol. Wt.** 129.59
Uses See aniline.

Physical properties

M. Pt. 189°C; **B. Pt.** 245°C; **Flash point** 193°C (open cup); **Specific gravity** 1.2215; **Volatility** v. den. 4.5.

Solubility
Organic solvent: ethanol

Occupational exposure

UN No. 1548; **HAZCHEM Code** 2Z; **Conveyance classification** harmful substance; **Supply classification** toxic.

Risk phrases
Toxic by inhalation, in contact with skin and if swallowed – Danger of cumulative effects (R23/24/25, R33)

Safety phrases
After contact with skin, wash immediately with plenty of water – Wear protective clothing and gloves – If you feel unwell, seek medical advice (show label where possible) (S28, S36/37, S44)

Ecotoxicity

Fish toxicity
LC_{50} (48 hr) goldfish 5.5 mg l^{-1} (1).

Mammalian and avian toxicity

Acute data
LD_{50} oral mouse, rat 840-1070 mg kg^{-1} (2).
LD_{50} intraperitoneal mouse 300 mg kg^{-1} (3).

Carcinogenicity and long-term effects

After oral administration, no increase in tumour incidence was observed in mice, but fibrosarcomas, sarcomas and haemangiosarcomas of the spleen and peritoneal cavity occurred in rats (4).

National Toxicology Program tested rats and mice with aniline hydrochloride maximum tolerated dose via feed. Positive results for ♂ and ♀ rats (5).

Irritancy

Dermal rabbit (24 hr) 500 mg caused moderate irritation, while 20 mg instilled into rabbit eye for 24 hr caused moderate irritation (6).

Genotoxicity

Vicia faba root cells induced chromosomal aberrations (7).

Oral ♂ CRH mouse (24 hr) 1000 mg kg^{-1} induced micronuclei in the bone marrow (8).

References

1. McKee, J. W; et al. *Water Quality Criteria* 1963, Resources Agency of California, State Water Quality Control Board
2. *NTIS Report* PB 214-270, National Technical Information Service, Springfield, VA
3. *NTIS Report* AD 277-689, National Technical Information Service, Springfield, VA
4. *IARC Monograph* 1982, **27**, 39-61
5. *National Toxicology Program, Research and Testing Div.* 1992, Report No. TR-130, NIEHS, Research Triangle Park, NC 27709
6. Marhold, J. V; *Sbornik Vysledku Toxixologickeho Vysetreni Latek A Pripravku* 1972, 65, Prague
7. Kanaya, N. *Mutat. Res.* 1990, **245**(4), 311-315
8. Westmorland, C; et al. *Carcinogenesis (London)* 1991, **12**(6), 1057-1059

A230   Aniline, 3,5-dichloro-

CAS Registry No. 626-43-7
Synonyms 3,5-dichlorobenzenamine
Mol. Formula $C_6H_5Cl_2N$ **Mol. Wt.** 162.02

Physical properties

M. Pt. 51-53°C; **B. Pt.** $_{741}$ 259- 260°C; **Flash point** >110°C.

Occupational exposure

UN No. 1590; **HAZCHEM Code** 2X; **Conveyance classification** toxic substance; **Supply classification** toxic.

Risk phrases

Toxic by inhalation, in contact with skin and if swallowed – Danger of cumulative effects (R23/24/25, R33)

Safety phrases

After contact with skin, wash immediately with plenty of water – Wear protective clothing and gloves – If you feel unwell, seek medical advice (show label where possible) (S28, S36/37, S44)

Ecotoxicity

Fish toxicity

LC_{50} (14 day) guppy 3.9 mg l^{-1} (1).

Invertebrate toxicity

EC_{50} (5, 15, 30 min) *Photobacterium phosphoreum* 10.5 mg l^{-1} Microtox test (2).

Environmental fate

Degradation studies

BOD_{10} aniline, 3,5-dichloro- was highly resistant to biodegradation. In simulated activated-sludge process, approximately 97% is bioeliminated after 10-16 days adaptation time; approximately 15-20% dichloroaniline is removed by vaporisation, bioelimination and sorption by apparatus (3).

The effects on guelph loam were studied. Aniline, 3,5-dichloro- at concentrations of 5-100 µg g^{-1} soil was inhibitory against the oxidation of the nitrogen of ammonia to nitrite nitrogen, but not nitrite nitrogen to nitrate nitrogen. Dichloroanilines were more persistent than aniline or the monochloroanilines (4).

No degradation was observed in an anaerobic-water screening test over a 28 days incubation period using digester sludge inocula (5).

Calculated bioconcentration factor 7 (6).

Abiotic removal

Photochemical reaction with atmospheric hydroxyl radicals has been estimated to be 77.17×10^{-12} cm^3 mol-sec^{-1} at 25°C, which corresponds to an atmospheric $t_{1/2}$ of about 5 hr at an atmospheric concentration of 5×10^5 hydroxyl radicals per cm^3 (7).

Absorption

Aromatic amines have been observed to undergo rapid and reversible covalent binding with humic materials in aqueous solution; the initial binding reaction is followed by a slower and much less reversible reaction believed to represent the addition of the amine to quinoidal structures followed by oxidation of the product to give an amino-substituted quinone. These processes represent pathways by which aromatic amines may be converted to nascent forms in the biosphere (8).

Genotoxicity

Salmonella typhimurium TA100, TA98 with and without metabolic activation negative (9).

Any other adverse effects

In vivo (48 hr) ♂ Fischer rat 65-160 mg kg^{-1}. Renal effects induced include decreased urine volume, increased proteinuria, haematuria, modest elevations in blood urea nitrogen concentrations, decreased accumulation of *p*-aminohippurate by renal cortical slices and no change or slight decrease in kidney weight (10).

References

1. Hermans, J; et al. *Ecotox. Environ. Safety* 1985, **9**, 321-326
2. Kaiser, K. L. E; et al. *Water Pollut. Res. J. Canada* 1991, **26**(3), 361-431
3. Janicke, W; et al. *Gas-Wasserfach: Wasser/Abwasser* 1980, **121**(3), 131-135
4. Thompson, F. R; et al. *Can. J. Microbiol.* 1969, **15**(7), 791-796
5. Shelton, D. R; et al. *Development of Tests for Determining Anaerobic Biodegradation Potential* 1981, 39, EPA 560/5-81-013
6. Hansch, C; et al. *Medchem. Project Issue No. 26* 1985, Clarmont, CA
7. Atkinson, R. *J. Inter. Chem. Kinet.* 1987, **19**, 799-828
8. Parris, G. E; et al. *Environ. Sci. Technol.* 1980, **14**, 1099-1106
9. Rashid, K. A; et al. *J. Environ. Sci. Health Part B* 1987, **B22**(6), 721-729
10. Lo, H.H; et al. *Toxicology* 1990, **63**(2), 215-231

A231 Aniline, 2,4,5-trimethyl-

CAS Registry No. 137-17-7

Synonyms 2,4,5-trimethylaniline; 2,4,5- trimethylbenzenamine; ψ-cumidine; pseudocumidine; 1-amino-2,4, 5-trimethylbenzene; 1,2,4- trimethyl-5-aminobenzene

Mol. Formula $C_9H_{13}N$ **Mol. Wt.** 135.21

Uses Dyestuff synthesis.

Physical properties

M. Pt. 68°C; **B. Pt.** 234-245°C; **Specific gravity** 0.957; **Volatility** v.p. 1 mmHg at 68°C.

Solubility
Organic solvent: ethanol

Occupational exposure

Supply classification toxic.

Risk phrases
Toxic by inhalation, in contact with skin and if swallowed – Danger of cumulative effects (R23/24/25, R33)

Safety phrases
After contact with skin, wash immediately with plenty of water – Wear protective clothing and gloves – If you feel unwell, seek medical advice (show label where possible) (S28, S36/37, S44)

Mammalian and avian toxicity

Acute data
LD_{50} oral rat 1250 mg kg^{-1} (1).

Carcinogenicity and long-term effects

No adequate evidence for carcinogenicity to humans, limited evidence for carcinogenicity to animals, IARC classification group 3 (2).

National Toxicology Program tested rats and mice with 2,4,5-trimethylaniline maximum tolerated dose via feed. Positive results for ♂ and ♀ rats and ♀ mice, equivocal results for ♂ mice (3).

Metabolism and pharmacokinetics

In ♀ rats haemoglobin bound covalently to 2,4, 5-trimethylaniline at a binding index of 0.7. The haemoglobin adducts were hydrolysed under alkaline conditions (4).

Genotoxicity

Salmonella typhimurium TA98, TA100 with metabolic activation positive (5).
Drosophila melanogaster wing spot test positive for single small spots, negative for large single spots (6).
Chinese hamster ovary cells with and without metabolic activation induced sister chromatid exchange and without metabolic activation induced chromosomal aberrations (7).

Any other comments

Human health effects, experimental toxicology, epidemiology and workplace experience reviewed (8).

References

1. Sax, N. I; et al. *Dangerous Properties of Industrial Materials* 7th ed., 1989, Van Nostrand Reinhold, New York
2. *IARC Monograph* 1987, **Suppl. 7**, 73
3. *National Toxicology Program, Research and Testing Div.* 1992, Report No. TR-160, NIEHS, Research Triangle Park, NC 27709
4. Birner, G; et al. *Arch. Toxicol.* 1988, **62**(2-3), 110-115
5. Zimmer, D; et al. *Mutat. Res.* 1980, **77**, 317
6. Kugler-Steigmeier, M. E; et al. *Mutat. Res.* 1989, **211**(2), 279-289
7. Loveday, K. S; et al. *Environ. Mol. Mutagen.* 1990, **16**(4), 272-303
8. *ECETOC Technical Report No. 30(4)* 1991, European Chemical Industry Ecology and Toxicology Centre, B-1160 Brussels

A232   Aniline, 2,4,6-trimethyl-

CAS Registry No. 88-05-1
Synonyms 2-aminomesitylene; aminomesitylene; 2,4,6- trimethylaniline; 2,4,6-trimethylbenzenamine; mesidine; mesitylamine
Mol. Formula $C_9H_{13}N$ **Mol. Wt.** 135.21

Uses Dyestuff synthesis.

Physical properties

M. Pt. −5°C; **B. Pt.** 232-233°C; **Specific gravity** d_{20} 1.5495.

Occupational exposure

Supply classification toxic.

Risk phrases

Toxic by inhalation, in contact with skin and if swallowed − Danger of cumulative effects (R23/24/25, R33)

Safety phrases

After contact with skin, wash immediately with plenty of water − Wear protective clothing and gloves − If you feel unwell, seek medical advice (show label where possible) (S28, S36/37, S44)

Mammalian and avian toxicity

Acute data

LD_{50} oral mouse, rat 590-740 mg kg^{-1} (1,2).
LC_{50} (2 hr) inhalation mouse 290 mg m^{-3} (1).

Carcinogenicity and long-term effects

No adequate evidence for carcinogenicity to humans or animals, IARC classification group 3 (3).

Irritancy

Dermal rabbit (24 hr) 500 mg caused moderate irritation, while 20 mg instilled into rabbit eye for 24 hr caused severe irritation (4).

Genotoxicity

Drosophila melanogaster wing spot test positive for small single spots at higher concentrations, negative for large single spots (5).
Salmonella typhimurium TA98, TA100 with metabolic activation positive (6).

Any other comments

Human health effects and experimental toxicology reviewed (7).
All reasonable efforts have been taken to find information on isomers of this compound, but no relevant data are available.

References

1. Izmerov, N. F; et al. *Toxicometric Parameters of Industrial Toxic Chemicals under Single Exposure* 1982, 20, Moscow
2. Sax, N. I; et al. *Dangerous Properties of Industrial Materials* 1989, 7th ed., Van Nostrand Reinhold, New York
3. *IARC Monograph* 1987, **Suppl. 7**, 73
4. Marhold, J. V. *Sbornik Vysledku Toxixologickeho Vysetreni Latek A Pripravku* 1972, 66, Prague
5. Kugler-Steigmeier, M. E; et al. *Mutat. Res.* 1989, **211**(2), 279-289
6. *Mutat. Res.* 1980, **77**, 317
7. *ECETOC Technical Report No. 30(4)* 1991, European Chemical Industry Ecology and Toxicology Centre, B-1160 Brussels

A233   Anilofos

$$S$$
$$\|$$
$$(H_3CO)_2PSCH_2CONCH(CH_3)_2$$

CAS Registry No. 64249-01-0
Synonyms S-[2-[4-chlorophenyl(1-methylethyl)amino]-2-oxo-ethyl]-O,O-dimethyl phosphorodithioate; S-4-chloro-N- isopropylcarbaniloylmethyl-O,O-dimethyl phosphorodithioate
Mol. Formula $C_{13}H_{19}ClNO_3PS_2$ **Mol. Wt.** 367.86
Uses Herbicide.

Physical properties

M. Pt. 50.5-52.5°C; **B. Pt.** 150°C (decomp.); **Volatility** v.p. 1.65×10^{-5} mmHg at 60°C.

Solubility

Water: 13.6 mg l^{-1}. Organic solvent: acetone, chloroform, toluene, benzene, ethanol, dichloromethane, hexane

Ecotoxicity

Fish toxicity
LC_{50} (96 hr) goldfish, trout 2.8-4.6 mg l^{-1} (1).

Environmental fate

Degradation studies
$t_{1/2}$ in soil 30-45 days at 23°C (1).

Mammalian and avian toxicity

Acute data
LD_{50} oral ♀ rat 470 mg kg^{-1} (1,2).
LD_{50} oral Japanese quail 2300-3350 mg kg^{-1} (2).
LD_{50} percutaneous rat >2000 mg kg^{-1} (1).

Legislation

Limited under EC Directive on Drinking Water Quality 80/778/EEC. Pesticide: organophosporous compound maximum admissable concentration 0.1 µg l^{-1} (3).

References

1. *The Agrochemicals Handbook* 3rd ed., 1991, RSC, London
2. *The Pesticide Manual* 9th ed., 1991, British Crop Protection Council, Farnham
3. *EC Directive Relating to the Quality of Water Intended for Human Consumption* 1982, 80/778/EEC, Office for Official Publications of the European Communities, 2 rue Mercier, L-2985 Luxembourg

A234   2-Anisidine

NH$_2$
OCH$_3$

CAS Registry No. 90-04-0
Synonyms *o*-methoxyaniline; 2-methoxybenzenamine; *o*-aminoanisole; *o*-anisidine
Mol. Formula C$_7$H$_9$NO   **Mol. Wt.** 123.16
Uses Manufacture of azo dyestuffs.

Physical properties

M. Pt. 6.2°C; **B. Pt.** 224°C; **Flash point** 107°C (closed cup); **Specific gravity** d$_4^{20}$
1.0923; **Partition coefficient** log P$_{ow}$ 0.95; **Volatility** v.p. <0.1 mmHg at 30°C; v.
den. 4.25.

Solubility
Organic solvent: diethyl ether (miscible), ethanol, acetone, benzene

Occupational exposure

US TLV (TWA) 0.1 ppm (0.5 mg m^{-3}); **UK Long- term limit** 0.1 ppm (0.5 mg m^{-3});
UN No. 2431; **HAZCHEM Code** 3X; **Conveyance classification** harmful substance;
Supply classification toxic.

Risk phrases
Very toxic by inhalation, in contact with skin and if swallowed – Danger of
cumulative effects (R26/27/28, R33)

Safety phrases
After contact with skin, wash immediately with plenty of water – Wear protective
clothing and gloves – In case of accident or if you feel unwell, seek medical advice
immediately (show label where possible) (S28, S36/37, S45)

Ecotoxicity

Bioaccumulation
Based on water solubility of 14 g l^{-1} and log K$_{ow}$ of 1.18, the estimated
bioconcentration factor is 3-5 which suggests bioaccumulation in aquatic organisms
will be unlikely (1).

Effects to non-target species
LD$_{50}$ oral redwing blackbird 420 mg kg^{-1} (2).

Environmental fate

Degradation studies
Japanese MITI test, initial *o*-anisidine 100 ppm, initial activated sludge 30 mg l^{-1},
14-day incubation period, 69.1% theoretical BOD (nitrate end product), 81.7%
theoretical BOD (ammonia end product) (3).
20 µg l^{-1} inoculated with a mixed culture of soil microorganisms in an aqueous
mineral salts media persisted for >64 days (4).

Escherichia coli converted *o*-anisidine in the presence of nitrate to its corresponding (phenylazo)naphthol (5).

Abiotic removal
Atmospheric $t_{1/2}$ with photochemically generated hydroxyl radicals estimated as 3 hr (6).

Absorption
Relatively immobile in soil and strongly binds to humic material in suspended solids and sediments in water (7).

Mammalian and avian toxicity

Acute data
LD_{50} oral mouse, rat 1400-2000 mg kg^{-1}.
LD_{50} oral rabbit 870 mg kg^{-1} (8).

Carcinogenicity and long-term effects
No adequate evidence for carcinogenicity to humans, sufficient evidence for carcinogenicity in animals, IARC classification group 2B (8).

Metabolism and pharmacokinetics
Horseradish peroxidase metabolism yielded diimine, quinone imine, azo dimer, polymer metabolites (9).

Genotoxicity

Salmonella typhimurium TA98, TA100 with metabolic activation positive (10).

Any other comments

Physico-chemical properties, human health effects and experimental toxicology reviewed (11).

References

1. Lyman, W. J; et al. *Handbook of Chemical Property Estimation Methods* 1982, 5-5, McGraw- Hill, New York
2. Shafer, E. W; et al. *Arch. Environ. Contam.* 1983
3. Kitano, M. *OECD Tokyo Meeting* 1978, 8-13
4. Alexander, M; et al. *J. Agric. Food Chem.* 1966, **14**, 410-413
5. Lammerding, A. M; et al. *J. Agric. Food Chem.* 1982, **30**, 644-647
6. Atkinson, R. *Inter. J. Chem. Kinet.* 1987, **19**, 799-828
7. Berry, D. F. *Diss. Abst. Int. B* 1985, **45**, 3799
8. *IARC Monograph* 1987, **Suppl. 7**, 57
9. Thompson, D. C; et al. *Chem. Res. Toxicol.* 1991, **4**(4), 474-487
10. Zeiger, E; et al. *Environ. Mol. Mutagen.* 1992, **19**(Suppl. 21), 2-141
11. *ECETOC Technical Report No. 30(4)* 1991, European Chemical Industry Ecology and Toxicology Centre, B-1160 Brussels

A235　3-Anisidine

CAS Registry No. 536-90-3
Synonyms *m*-methoxyaniline; 3-methoxybenzenamine; *m*-aminophenol methyl ether; *m*-aminoanisole; *m*-anisylamine; *m*-anisidine
Mol. Formula C_7H_9NO　　　　　　　**Mol. Wt.** 123.16
Uses Manufacture of azo dyestuffs.

Physical properties

M. Pt. –10°C; **B. Pt.** 251°C; **Specific gravity** d_4^{20} 1.096; **Partition coefficient** log P_{ow} 0.93.

Solubility
Organic solvent: ethanol, diethyl ether, acetone, benzene

Occupational exposure

UN No. 2431; **HAZCHEM Code** 3X; **Conveyance classification** harmful substance.

Ecotoxicity

Effects to non-target species
LD_{50} wild bird 560 mg kg^{-1} (1).

Environmental fate

Degradation studies
Decomposition by a soil microflora >64 days (2).

Genotoxicity

Salmonella typhimurium TA98, TA100, TA1538 with metabolic activation positive (3).

References

1. Schafer, E. W; et al. *Arch. Environ. Contam.* 1983
2. Alexander, M; et al. *J. Agric. Food Chem.* 1966, **14**, 410
3. Zeiger, E; et al. *Environ. Mol. Mutagen.* 1992, **19**(Suppl. 21), 2-141

A236　4-Anisidine

CAS Registry No. 104-94-9
Synonyms 1-amino-4-methoxybenzene; benzenamine, 4- methoxy-;

p-aminoanisole; *p*-anisidine; *p*- anisylamine; methoxyaniline;
4-methoxybenzenamine; *p*- methoxyphenylamine
Mol. Formula C_7H_9NO **Mol. Wt.** 123.16
Uses Manufacture of azo dyestuffs.

Physical properties

M. Pt. 57°C; **B. Pt.** 246°C; **Flash point** 5°C (closed cup); **Specific gravity** d_4^{57} 1.071;
Partition coefficient log P_{ow} 0.95; **Volatility** v.p. 0.1 mmHg at 20°C; v.den. 4.25.

Solubility
Organic solvent: methanol, ethanol, diethyl ether, acetone, benzene

Occupational exposure

US TLV (TWA) 0.1 ppm (0.5 mg m^{-3}); **UK Long-term limit** 0.1 ppm (0.5 mg m^{-3});
UN No. 2341; **HAZCHEM Code** 3X; **Conveyance classification** harmful substance;
Supply classification toxic.

Risk phrases
Very toxic by inhalation, in contact with skin and if swallowed – Danger of
cumulative effects (R26/27/28, R33)

Safety phrases
After contact with skin, wash immediately with plenty of water – Wear protective
clothing and gloves – In case of accident or if you feel unwell, seek medical advice
immediately (show label where possible) (S28, S36/37, S45)

Ecotoxicity

Invertebrate toxicity
EC$_{50}$ (30 min) *Photobacterium phosphoreum* 14.5 mg l^{-1} Microtox test (1).

Bioaccumulation
Estimated bioconcentration factor of 3 which suggests bioaccumulation will be
unlikely in aquatic organisms (2).

Environmental fate

Degradation studies
Escherichia coli converted 4-anisidine in the presence of nitrate to its corresponding
(phenylazo)naphthol (3).
Japanese MITI test, initial 4- anisidine concentration 100 ppm, initial activated sludge
30 mg l^{-1} 14-day incubation period, 65.3% theoretical BOD (nitrogen dioxide end
product), 78.5% theoretical BOD (ammonia end product) (4).
25 µg l^{-1} of 4-anisidine inoculated with a mixed culture of soil microorganisms in an
aqueous mineral salts media underwent complete degradation in 64 days (5).

Abiotic removal
Estimated $t_{1/2}$ 3 hr reacting with photochemically generated hydroxyl radicals (6).

Mammalian and avian toxicity

Acute data
LD$_{50}$ oral rat 1400 mg kg^{-1}.
LD$_{50}$ dermal rat 3200 mg kg^{-1}.

LD$_{50}$ intraperitoneal rat 1400 mg kg^{-1} (7).

Intraperitoneal ♂ rats 120 mg kg^{-1} induced swelling of the tubular epithelial cells and a significant elevation in urinary N-acetyl-β- D-glucosaminidase activity (8).

Carcinogenicity and long-term effects

No adequate evidence for carcinogencity to humans or animals, IARC classification group 3 (9).

Genotoxicity

Salmonella typhimurium TA100 with metabolic activation positive (10).
In vitro study, horseradish peroxidase metabolism yielded a diimine metabolite which subsequently hydrolysed to form a quinone imine; also observed was a dimeric metabolite with an azo bond (11).

Any other comments

Experimental toxicology and human health effects reviewed (12).

References

1. Kaiser, K. L. E; et al. *Water Pollut. Res. J. Canada* 1991, **26**(3), 361-431
2. Lyman, W. J; et al. *Handbook of Chemical Property Estimation Methods* 1982, 5-5, McGraw- Hill, New York
3. Lammerding, A. M; et al. *J. Agric. Food Chem.* 1982, **30**, 644-647
4. Kitano, M. *OECD Tokyo Meeting* 1978, 8-13
5. Alexander, M; et al. *J. Agric. Food Chem.* 1966, **14**, 41-413
6. Atkinson, R. *Inter. J. Chem. Kinet.* 1987, **19**, 799-828
7. *Archiv fuer Gewerbepathol. und Gewerbehygiene* 1957, **15**, 447
8. Yoshida, M; et al. *J. Toxicol. Sci.* 1989, **14**(4), 257-268
9. *IARC Monograph* 1987, **Suppl. 7**, 57
10. Zeiger, E; et al. *Environ. Mol. Mutagen.* 1992, **19**(Suppl. 21), 2-141
11. Thompson, D. C; et al. *Chem. Res. Toxicol.* 1991, **4**(4), 474-481
12. *ECETOC Technical Report No. 30(4)* 1991, European Chemical Industry Ecology and Toxicology Centre, B-1160 Brussels

A237   Anisole

CAS Registry No. 100-66-3
Synonyms methoxybenzene; phenyl methyl ether; methyl phenyl ether
Mol. Formula C_7H_8O **Mol. Wt.** 108.14
Uses Used in perfumery.

Physical properties

M. Pt. −37.3°C; **B. Pt.** 153.8°C; **Flash point** 52°C; **Specific gravity** d_4^{20} 0.9961;
Partition coefficient log P_{ow} 2.11; **Volatility** v.p. 10 mmHg at 42.2°C; v. den. 3.72.

Solubility
Organic solvent: ethanol, diethyl ether, acetone, benzene

Occupational exposure
UN No. 2222; HAZCHEM Code 3Y; Conveyance classification flammable liquid.

Ecotoxicity

Invertebrate toxicity
EC_{50} (30 min) *Photobacterium phosphoreum* 18.8 mg l^{-1} Microtox test (1).

Environmental fate

Degradation studies
Confirmed biodegradable (2).
Decomposition by a soil microflora in 8 days (3).

Mammalian and avian toxicity

Acute data
LD_{50} oral rat 3700 mg kg^{-1} (4).
LD_{Lo} subcutaneous rat 3500 mg kg^{-1} (5).
Mice exposed to an average concentration of 1880 mg m^{-3} anisole for 2 hr, suffered damage to mucous membranes, exhibited slight excitation, then ataxia, and died within 2 days (6).

Metabolism and pharmacokinetics
A major urinary metabolite is 4-hydroxyphenyl methyl ether which was excreted unconjugated (2%) and conjugated with glucuronic acid (48%) and sulfuric acid (29%). Administration of anisole to dogs caused increased excretion of ethereal sulfate (5).

Irritancy
Dermal rabbit (24 hr) 500 mg caused moderate irritation (7).

References

1. Kaiser, K. L. E; et al. *Water Pollut. Res. J. Canada* 1991, **26**(3), 361-431
2. *MITI Report* 1984, Ministry of International Trade and Industry, Tokyo
3. Alexander, M; et al. *J. Agr. Food Chem.* 1966, **14**, 410
4. Jenner, P. M; et al. *Food Cosmet. Toxicol.* 1964, **2**(3), 327-343
5. Patty, F. A. *Industrial Hygiene and Toxicology* 2nd ed., 1963, 1690-1682, Interscience Publishers, New York
6. Uzhdowini, E. R; et al. *Gig. Tr. Prof. Zabol.* 1984, (6), 43-44 (Russ.) (*Chem. Abstr.* **101**, 67331f)
7. Opdyke, D. C. J. *Food Cosmet. Toxicol.* 1979, **17**(3), 243-245

A238 Anisoyl chloride

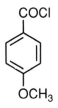

CAS Registry No. 100-07-2
Synonyms *p*-anisoyl chloride; *p*-anisyl chloride; *p*-methoxybenzoyl chloride;
4-methoxybenzoyl chloride
Mol. Formula $C_8H_7ClO_2$ **Mol. Wt.** 170.60
Uses Chemical intermediate.

Physical properties

M. Pt. 22°C; **B. Pt.** 262-263°C (decomp.); **Flash point** >107°C; **Specific gravity** d_4^{20}
1.261.

Solubility
Organic solvent: diethyl ether, acetone, benzene

Occupational exposure
UN No. 1729; **HAZCHEM Code** 2X; **Conveyance classification** corrosive substance.

Ecotoxicity

Invertebrate toxicity
EC_{50} (5-30 min) *Photobacterium phosphoreum* 1.91 mg l^{-1} Microtox test (1).

Any other adverse effects to man

In humans exposure to vapours can cause serious eye burns (2).

References

1. Kaiser, K. L. E; et al. *Water Pollut. Res. J. Canada* 1991, **26**(3), 361-431
2. *The Merck Index* 11th ed. 1989, 106-107, Merck & Co. Inc, Rahway, NJ

A239 Anthracene

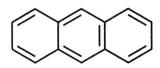

CAS Registry No. 120-12-7
Synonyms Antracin; Green oil; paranaphthalene; Tetra olive N2G
Mol. Formula $C_{14}H_{10}$ **Mol. Wt.** 178.24
Uses Intermediate for anthraquinone dyestuffs.

Occurrence Urban air, oil spills, incomplete combustion.
Obtained from coal tar.

Physical properties

M. Pt. 218°C; **B. Pt.** 342°C; **Flash point** 121°C; **Specific gravity** d_4^{27} 1.25; **Partition coefficient** log P_{ow} 4.45; **Volatility** v. den. 6.15.

Solubility
Water: 1.24 mg l^{-1}. Organic solvent: ethanol, benzene, carbon disulfide, toluene, carbon tetrachloride

Ecotoxicity

Fish toxicity
LC_{50} (96 hr) bluegill sunfish 11.9 μg l^{-1} (1).
Fathead minnow (6 wk) 0, 6 or 12 μg l^{-1} followed by increases to 12 and 20 μg l^{-1} for 3 wk. Eggs were collected daily and placed into clean water. Significant bioconcentration of anthracene was observed in the eggs laid, in the gonads and carcasses of the spawning fish. Decreased reproductive output was observed in all anthracene-exposed fish and maternal exposure in the absence of solar UV radiation caused reduction in percent hatch and percent survival to 96 hr post hatch. Teratogenic effects including internal haemorrhaging, oedema, eye and yolk deformities were observed in fry maternally exposed with subsequent solar UV radiation exposure (2).

Invertebrate toxicity
LC_{50} *Culicid* mosquito larvae 26.8 μg l^{-1} (1).

Bioaccumulation
Estimated bioconcentration factor in whole fish was 675, a lower value than that predicted by log P_{ow} 4.45 because of biotransformation (3).
Measured bioconcentration factor in goldfish exposed to 1 mg l^{-1} was 162 (4).
Rainbow trout were exposed for 72 hr to ^{14}C anthracene alone and in an oil shale retort water. Tissues were analysed at 24, 48, and 96 hr, measured bioconcentration factor 9000-9200 (5).
Daphnia pulex bioconcentration factor measured as 759-912 (6).
Accumulation of anthracene administered to young coho salmon in food and by intraperitoneal injection was in key organs, e.g. liver and brain. After intraperitoneal injection the highest percent of metabolites occurred in the gall bladder, but significant amounts were also found in the liver, brain, flesh and carcass (7).

Environmental fate

Degradation studies
The $t_{1/2}$ in soil was 108-175 days (8).
Theoretical BOD_5 using inoculum from 3 polluted surface waters 2% (9).
Significant degradation with gradual adaptation reported for 5 and 10 mg l^{-1} anthracene incubated with sewage seed, 43% and 26% degradation after 7 days, 92% and 51% degradation after 28 days and 3 weekly subcultures (10).
Phanerochaete chrysoporium degraded anthracene to form the metabolite anthraquinone (11).

Alcaligenes denitrificans WW1 utilised anthracene as sole carbon source (12).
Slight degradation was reported with benzene acclimated sludge in 8 hr at 20°C (13).
The $t_{1/2}$ were 57-210 days in unacclimatised sediments and 5-7 days in oil-treated sediments (14).
Proposed pathway for bacterial catabolism is via metabolites 1,2-dihydroxyanthracene, 2-hydroxy-3- naphthaldehyde, 2-hydroxy-3-naphthioic acid and 2,3- dihydroxynaphthalene salicylic acid (14).

Abiotic removal
The $t_{1/2}$ was 35 min in distilled water exposed to midday sunlight (15).
Estimated atmospheric $t_{1/2}$ is 1.67 days after reaction with photochemically produced hydroxyl radicals (16).

Absorption
Soil adsorption coefficient of 26,000 indicates strong adsorption to soil and anthracene may degrade before it reaches ground water (14).

Mammalian and avian toxicity

Carcinogenicity and long-term effects
No adequate evidence for carcinogenecity to humans, limited evidence for carcinogenicity to animals, IARC classification group 3 (17).

Any other comments
Anthracene and related compounds were found in French alcoholic drinks from 1-10 ppb (18).
Experimental toxicology, human health effects, ecotoxicology and environmental effects reviewed (19).

References

1. Oris, J. T; et al. *Stud. Environ. Sci.* 1984, **25**, 639-658
2. Hall, A. T; et al. *Aquat. Toxicol.* 1991, **19**(3), 249-264
3. Spacie, A; et al. *Ecotoxicol. Environ. Saf.* 1983, **7**(3), 330-341
4. Ogata, M; et al. *Bull. Environ. Contam. Toxicol.* 1984, **33**, 561-567
5. Linder, G; et al. *Environ. Toxicol. Chem.* 1985, **4**, 549-558
6. Southworth, G. R; et al. *Water Res.* 1978, **12**, 973-977
7. Roubal, W. T; et al. *Arch. Environ. Contam. Toxicol.* 1977, **5**(4), 513-529
8. Sims, R. C; et al. *Res. Rev.* 1983, **88**, 1-68
9. Dore, M; et al. *Trib. Cebedeau* 1975, **28**, 3-11, (Fr.) (*Chem. Abstr.* **83**, 32670y)
10. Tabak, H. H; et al. *Proc. Symp. AOAC 94th Ann. Meeting* 1981, 267-328, Washington, DC
11. Hammel, K. E; et al. *USEPA Report* 1990, EPA 600/9-90/041, 55-56
12. Weissenfels, W. D; et al. *Appl. Microbiol. Biotechnol.* 1991, **34**(4), 528-535
13. Malaney, D. C; et al. *Water Sew. Works* 1966, **113**, 302-309
14. *USEPA Health and Environmental Effects Profile for Anthracene* 1987, ECAO-CIN-P230
15. Southworth, G. R. *Aquatic Toxicology* 1979, 359-380, ASTM STP-667, Philadelphia, PA
16. *GEMS: Graphical Exposure Modeling System. Fate of Atmospheric Pollutants* 1986, Database Office of Toxic Substances, USEPA
17. *IARC Monograph* 1987, **Suppl. 7**, 57
18. Toussaint, G; et al. *J. Chromatog.* 1979, **171**, 448-452
19. *ECETOC Technical Report No. 30(4)* 1991, European Chemical Industry Ecology and Toxicology Centre, B-1160 Brussels

A240   Anthraflavic acid

CAS Registry No. 84-60-6

Synonyms 2,6-dihydroxyanthraquinone; 2,6-dihydroxy- 9,10-anthracenedione; anthraflavin

Mol. Formula $C_{15}H_8O_4$ Mol. Wt. 252.23

Uses Dyestuffs.

Physical properties

M. Pt. >330°C.

Solubility

Organic solvent: ethanol

Mammalian and avian toxicity

Acute data

LD_{50} intravenous mouse 180 mg kg^{-1} (1,2).

Genotoxicity

Salmonella typhimurium TA102, TA1537 with and without metabolic activation negative (3).

In vitro V79-HGPRT mutagenicity assay negative (3).

Any other comments

Administration of the antimutagen anthraflavic acid to rats gave rise to significant increases in the hepatic microsomal O-deethylations of ethoxyresorufin and ethoxycoumarin, but not in the O-dealkylation of pentoxyresorufin nor in cytosolic glutathione S-transfer activity. Immunoblot studies of solubilised microsomes from anthraflavic acid-treated rats revealed that anthraflavic acid induced the apoproteins P450 I, A1 and A2 but not P450 B1 and B2.

Pretreatment with anthraflavic acid resulted in a marked increase in the *in vitro* bioactivation of 2- amino-6-methyldipyrido[1,2-a:3′, 2′-d]imidazole and IQ to mutagenic intermediate(s). IQ is a carcinogen against which anthraflavic acid has desplayed strong antimutagenic effect in the Ames test when incorporated into the metabolic activation system. The increase in mutagenicity of IQ was the result of enhancement of both the microsomal and cytosolic activation steps. Thus, antraflavic acid is a specific inducer of P450 I proteins in the rat and this compound is not only unlikely to exhibit any anticarcinogenic effect *in vitro* but may act as a cocarcinogen (4). Caused a reduction in the binding of polycyclic aromatic hydrocarbons to DNA and protein following dermal application to mice (5).

Inhibited the mutagenicity of the cooked food mutagen IQ, by virtue of its ability to inhibit both the microsomal and cytosolic activation pathways (6).

References

1. Sax, N. I; et al. *Dangerous Properties of Industrial Materials* 7th ed., 1989, Van Nostrand Reinhold, New York
2. *US Army Armament R&D Command Report* Chemical Systems Lab, NX-06773
3. Westendorf, J; et al. *Mutat. Res.* 1990, **240**(1), 1-12
4. Ayrton, A. D; et al. *Food Chem. Toxicol.* 1988, **26**(11-12), 909-915
5. Das, M; et al. *Cancer Res.* 1987, **47**(3), 767-773
6. Ayrton, A. D; et al. *Mutat. Res.* 1988, **207**(3-4), 121-125

A241   Anthraquinone

CAS Registry No. 84-65-1
Synonyms 9,10-anthraquinone; 9,10-anthracenedione; Cordit
Mol. Formula $C_{14}H_8O_2$ Mol. Wt. 208.22
Uses Intermediate for dyestuffs manuafacture. Bird repellant.

Physical properties

M. Pt. 286°C (sublimes); B. Pt. 377°C; Flash point 185°C; Specific gravity d_4^{20} 1.42-1.44; Volatility v. den. 7.2.

Solubility
Organic solvent: ethanol, diethyl ether, benzene, toluene

Environmental fate

Degradation studies
Confirmed biodegradable (1).

Mammalian and avian toxicity

Acute data
LD_{50} oral mouse >5000 mg kg^{-1} (2).
LC_{50} (4 hr) inhalation rat >1.3 mg l^{-1} (3).
LD_{50} intraperitoneal rat 3500 mg kg^{-1} (4).

Sub-acute data
In 90 day feeding trials, rats receiving up to 15 mg kg^{-1} diet showed no ill-effects (3).

Any other comments
Experimental toxicology reviewed (5).
Selected for general toxicology study by National Toxicology Program (6).

References

1. *MITI Report* 1984, Ministry of International Trade and Industry, Tokyo

2. *The Pesticide Manual* 9th ed., 1991, British Crop Protection Council, Farnham
3. *The Agrochemicals Handbook* 3rd ed., 1991, RSC, London
4. Volodchenko, V. A. *Gig. Tr. Prof. Zabol.* 1977, **21**, 27-30, (Russ.) (*Chem. Abstr.* **88**, 109859e)
5. *ECETOC Technical Report No. 30(4)* 1991, European Chemical Industry Ecology and Toxicology Centre, B-1160 Brussels
6. *National Toxicology Program, Research and Testing Div.* 1992, NIEHS, Research Triangle Park, NC 27709

A242   Antimony

Sb

CAS Registry No. 7440-36-0
Synonyms antimony black; antimony regulus
Mol. Formula Sb   **Mol. Wt.** 121.75
Uses Manufacture of alloys, fireworks, thermoelectric piles, coating metals, paints, rubber, ceramics, medicines and semiconductors.
Occurrence The toxicity of antimony is dependent on the ability of the organism to absorb it. Therefore toxicity data refers to bioavailable forms, such as the ion in solution or particulate matter.
In China, Mexico, Bolivia it is mined as stibnite Sb_2S_3.

Physical properties

M. Pt. 630°C; **B. Pt.** 1635°C; **Specific gravity** d^{25} 6.68; **Volatility** v.p. 1 mmHg at 886°C.

Occupational exposure

US TLV (TWA) 0.5 mg m^{-3}; **UK Long-term limit** 0.5 mg m^{-3}; **UN No.** 2871; **HAZCHEM Code** 2Z; **Conveyance classification** harmful substance.

Ecotoxicity

Fish toxicity
LC_{50} (28 day) rainbow trout 0.66 mg l^{-1} (1).

Invertebrate toxicity
EC_{50} (48 hr) *Daphnia magna* 423 mg l^{-1} for trivalent antimony (2).

Mammalian and avian toxicity

Acute data
LD_{Lo} intraperitoneal rat, guinea pig 100-150 mg kg^{-1} (3).

Metabolism and pharmacokinetics
Inorganic trivalent antimony is excreted in the bile after conjugation with glutathione and also in urine. A significant proportion of that excreted in bile undergoes an enterohepatic circulation. In workers exposed to pentavalent antimony, urinary antimony excretion is related to the intensity of exposure. After an 8 hr exposure to 500 µg m^{-3} the increase in urinary antimony at the end of the shift is on average 25 µg g^{-1} creatinine (4).

Rats were fed 40 mg kg^{-1} day^{-1} antimony for 7.5 months *ad libitum*; another group were fed similarly by increasing the dose to 1 g kg^{-1} day^{-1} or 40 mg kg^{-1} day^{-1} for 4 months. An average of 1 mg of antimony was found in the carcasses of exposed rats regardless of the dose indicating that accumulation in animals is insignificant (5). Antimony may enter body through lung where it can then be absorbed and taken up by blood and tissues (6).

Following 6 intraperitoneal injections of 50 mg kg^{-1} each to rats with grafted sarcoma 45 tumour, antimony was present in higher concentrations in the blood of sarcoma 45-bearing rats than in controls. Antimony did not selectively accumulate in the tumour tissue; highest levels were observed in muscles, lung and skin of the tumour-bearing rats and in the lung and skin of normal rats (7).

Irritancy

Following repeated antimony application to skin the intensity of skin responses and the number of animals affected were reduced gradually. Responsive animals may suffer from acute attacks of interstitial pneumonia and die after inhalation of antimony compound dust (8).

Any other adverse effects to man

Lung tissue samples (200) taken from women aged over 40 yr with lung cancer and from urban areas were reported to have traces of antimony due to air pollution (9). Women working in an antimony metallurgical plant (unspecified amounts of antimony) had blood antimony levels 10-times greater than in controls. Urine levels of antimony ranged from 2.1-2.9 mg 100 ml^{-1}, breast milk 3.3 $\pm$ 2 mg l^{-1}, placental tissue 3.2-12.6 mg 100 ml^{-1}, amniotic fluid 6.2 $\pm$ 2.8 mg 100 ml^{-1} and umbilical cord blood 6.3 $\pm$ 3 mg 100 ml^{-1} (10).

Former smelter workers occupationally exposed to antimony had on average 12-times higher lung concentrations (315 μg kg^{-1}) than unexposed (26 μg kg^{-1}) (11). Inhalation of dust and fumes in humans causes nose and throat irritation, inflamation of the respiratory tract, pneumonitis, ulceration and perforation of the nasal septum. Headaches, dyspnoea, nausea, vomiting and diarrhoea have been reported in smelters (12).

Legislation

Limited under EEC Directive on Drinking Water Quality 80/778/EEC Antimony: maximum admissible concentration 10 μg l^{-1} (13).

Any other comments

Pollution by antimony, toxicity and health hazard reviewed (14).

Antimony content in human senile cataractous lenses was much lower that that in normal lenses (15).

Health effects assessment for antimony reported, suggesting tolerable exposure levels (16).

A review on the acute and chronic toxicological properties of antimony with primary focus in relation to exposure in an industrial environment where it is employed in metallurgical operations (17).

Physico-chemical properties, experimental toxicology and human health effects reviewed (18,19).

Avoid conditions in which nascent hydrogen will react with antimony to form toxic stibine.

References

1. Birge, W. J; et al. *Aquatic Toxicity Tests on Inorganic Elements Occuring in Oil Shale* 1980, EPA 600/9-80-022
2. Khangarot, B. S; et al. *Ecotoxicol. Environ. Saf.* 1989, **18**(2), 109-120
3. Bradley, F. *Ind. Med.* 1941, **10**(2), 15
4. Bailly, R; et al. *Br. J. Ind. Med.* 1991, **48**(2), 93-97
5. *USEPA Drinking Water Criteria Document for Antimony* 1988, III-16, EPA 68-03-3417
6. *Encyclopedia of Occupational Health and Safety* 1983, **1-2**, 177, International Labour Office, Geneva
7. Trifonova, N. F. *Vop. Klin. Eksp. Onkol.* 1972, **8**, 283-285, (*Chem. Abstr.* **80**, 34232u)
8. Huang, J; et al. *Zhongguo Yaolixue Yu Dulixue Zazhi* 1988, **2**(4), 288-292, (Ch.) (*Chem. Abstr.* 1989, **110**, 52551q)
9. Otoshi, T; et al. *Taiki Osen Gakkaishi* 1991, **26**(3), 176-183, (Jap.) (*Chem. Abstr.* 1991, **115**, 165424x)
10. *USEPA Health and Environmental Effects Profile for Antimony Oxides* 1985, 58, EPA 600/X-85-271
11. Friberg, L; et al. *Handbook of the Toxicology of Metals* 2nd ed., 1986, **1-2**, Elsevier Science, Amsterdam
12. Renes, L. E. A. M. A. *Arch. Ind. Hyg. Occup. Med.* 1953, **7**, 99
13. *EC Directive Relating to the Quality of Water Intended for Human Consumption* 1982, 80/778/EEC, Office for Official Publications of the European Communities, 2 rue Mercier, L-2985 Luxembourg
14. Murata, T. *Gekkan Haikibutsu* 1991, **17**(5), 209-212, (Jap.) (*Chem. Abstr.* 1991, **115**, 165324q)
15. Wu, X; et al. *Shandong Yike Daxue Xuebao* 1989, **27**(3), 59-64, (Ch.) (*Chem. Abstr.* 1990, **112** 74812u)
16. *USEPA Report* 1987, EPA 600/8- 88/018, PB88-179445/GAR
17. James, R. M; et al. *Trans. Am. Foundrymen's Soc.* 1987, **95**, 883-885
18. *ECETOC Technical Report No. 30(4)* 1991, European Chemical Industry Ecology and Toxicology Centre, B-1160 Brussels
19. Izmerov, N. F. *Scientific Reviews of Soviet Literature on Toxicity and Hazards of Chemicals* 1991, **71**, Eng. Trans., Richardson, M. L. (Ed.), UNEP/IRPTC, Geneva

A243   Antimonyl potassium tartrate hemihydrate

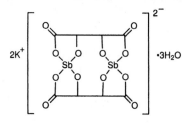

CAS Registry No. 28300-74-5

Synonyms antimonate(2-), bis[μ-[2,3- dihydroxybutanedioato(4-)- $O^1,O^2:O^3,O^4$]]di-, dipotassium, trihydrate; potassium antimonyl tartrate; antimonyl potassium tartrate; tartar emetic

Mol. Formula $C_8H_{10}K_2O_{15}Sb_2$ **Mol. Wt.** 667.86

Uses Mordant in textile and leather industry. Antischistomal drug. Formerly used as an emetic.

Physical properties
Specific gravity 2.607.

Solubility
Organic solvent: glycerol

Occupational exposure
US TLV (TWA) 0.5 mg m^{-3} (as Sb); **UK Long-term limit** 0.5 mg m^{-3} (as Sb); **UN No.** 1551; **HAZCHEM Code** 2X; **Conveyance classification** harmful substance; **Supply classification** harmful.

Risk phrases
Harmful by inhalation and if swallowed (R20/22)

Safety phrases
Do not breathe dust (S22)

Mammalian and avian toxicity

Acute data
LD$_{Lo}$ oral human 2 mg kg^{-1} (1).
LD$_{50}$ oral rat, rabbit 115 mg kg^{-1} (2,3).
LD$_{50}$ subcutaneous mouse 55 mg kg^{-1} (4).
LD$_{50}$ intravenous mouse 65 mg kg^{-1} (4).

Sub-acute data
Oral rat administration of antimony potassium tartrate in drinking water for 1000 days at 13.7 mg l^{-1} (average daily dose 1.07 mg kg^{-1}), 12.1 µg g^{-1} in the heart, 10.14 µg g^{-1} in kidney, 11.57 µg g^{-1} in liver, 17.67 µg g^{-1} in lung and 5.97 µg g^{-1} in spleen were detected (5).
Intraperitoneal F344 rats and B6C3F1 mice (90 day) 1.5, 3, 6, 12 and 24 mg kg^{-1} every other day. No clinical signs of toxicity nor gross or microscopic lesions in mice although elevated concentrations of antimony were detected in the liver and spleen of mice. Rats displayed dose-related mortality, body weight reduction and hepatoxicity. Hepatocellular degeneration and necrosis occurred in association with dose-related elevations in activities of the liver-specific serum enzymes sorbitol dehydrogenase and alanine aminotransferase (6).

Metabolism and pharmacokinetics
Following administration to rats, mice and monkeys, excretion was greater in faeces than urine indicating poor absorption from intestinal tract and no persistent accumulation of antimony in the body (7).
Intravenous administration of antimony potassium tartrate to monkey revealed >50% in liver, and detection in heart, kidney, thigh and thyroid. Maximum concentration in blood was reached at 8 hr after administration (8).
Rats were fed 8 mg kg^{-1} day^{-1} antimony potassium tartrate for 7.5 months (*ad libitum*); another group were fed for 6 months in doses increasing to 100 mg kg^{-1} day^{-1} antimony and maintained at that level for 6 months. Rabbits were fed at a dose of 8 mg kg^{-1} day^{-1} for 4 months. An average of 1 mg of antimony was found in the carcasses of exposed rats, regardless of the daily dose indicating that accumulation in animals is insignificant (9).

Legislation

Limited under EEC Directive on Drinking Water Quality 80/778/EEC. Antimony: maximum admissible concentration 10 μg l^{-1}. Potassium: guide level 10 mg l^{-1}, maximum admissible concentration 12 mg l^{-1} (10).

Any other comments

Carcinogenicity is currently being reviewed by the National Toxicology Program (11). Incompatible with acids and alkalis, salts of heavy metals, albumin, soap and tannins (12).

References

1. *Pesticide Chemicals Official Compendium* 1966, 1097, Assoc. of the American Pesticide Control Officials Inc., Topeka, KS
2. *Agric. Res. Serv.* 1966, **20**, 24
3. *Environ. Qual. Saf. Suppl.* 1975, **1**, 1
4. *Proceedings of the Society for Experimental Biology and Medicine* 1968, **129**, 284
5. *USEPA Health and Environmental Effects Profile for Antimony Oxides* 1985, 43, EPA 600/x- 85/271
6. Dieter, M. P; et al. *J. Toxicol. Environ. Health* 1991, **34**(1), 51-82
7. Browning, E. *Toxicity of Industrial Metals* 2nd ed., 1969, Appleton-Century-Crofts, New York
8. Abdel-Wahab, M. F; et al. *Egypt J. Bilharziasis* 1974, **1**(1), 101-106
9. *USEPA Drinking Water Criteria Document for Antimony* 1988, III-16, EPA 68-03-3417
10. *EC Directive Relating to the Quality of Water Intended for Human Consumption* 1982, 80/778/EEC, Office for Official Publications of the European Communities, 2 rue Mercier, L-2985 Luxembourg
11. *National Toxicology Program, Research and Testing Div.* 1992, NIEHS, Research Triangle Park, NC 27709
12. *Martindale: The Extra Pharmacopoeia* 29th ed., 1989, The Pharmaceutical Press, London

A244   Antimony pentachloride

SbCl₅

CAS Registry No. 7647-18-9
Synonyms antimonic chloride; antimony perchloride; antimony chloride; pentachloroantimony; antimony(V) chloride
Mol. Formula SbCl₅   **Mol. Wt.** 299.02
Uses Catalyst, chemical reagent.

Physical properties

M. Pt. 2.8°C; **B. Pt.** $_{22}$ 79°C; **Specific gravity** d_4^{20} 2.336; **Volatility** v.p. 1 mmHg at 22.7°C.

Solubility

Water: decomp. Organic solvent: chloroform, carbon disulfide, carbon tetrachloride

Occupational exposure

US TLV (TWA) 0.5 mg m^{-3} (as Sb); **UK Long-term limit** 0.5 mg m^{-3} (as Sb); **UN**

No. 1730 (liquid); 1731 (solution); **HAZCHEM Code** 4X (liquid); 2X (solution);
Conveyance classification corrosive substance; **Supply classification** corrosive.

Risk phrases
Causes burns – Irritating to respiratory system (R34, R37)

Safety phrases
In case of contact with eyes, rinse immediately with plenty of water and seek medical
advice (S26)

Mammalian and avian toxicity

Acute data
LD_{50} oral rat 1120 mg kg^{-1} (1).
LD_{50} oral guinea pig 900 mg kg^{-1} (2).

Sub-acute data
Inhalation effects tested in mice and rats. Affected mucous membranes of respiratory
tract and eyes (3).
Toxic effects in guinea pigs included weakness, drowsiness, adynamia and paresis of
the hind limbs (1).
Inhalation by rats of 15 µg l^{-1} for 2 hr daily for 4 months caused general loss of
weight and weight loss in the liver, kidneys and adrenal glands.
Histological changes were also found in the heart, liver, kidneys and thyroid due to
the irreversible accumulation of antimony (2).

Metabolism and pharmacokinetics
Distribution in rats after chronic poisoning by inhalation of antimony pentachloride
showed high antimony concentration in blood. Levels in liver, kidneys, spleen and
pancreas were similar (4).

Genotoxicity
Bacillus subtilis rec assay positive (5).

Legislation
Limited under EEC Directive on Drinking Water Quality 80/778/EEC. Antimony:
maximum admissible concentration 10 µg l^{-1}. Chlorides: guide level 25 mg l^{-1},
maximum admissible concentration 400 mg l^{-1} (6).

Any other comments
Physico-chemical properties, human health effects and experimental toxicology
reviewed (7).
Reacts explosively with phosphorium iodide at ambient temperature.

References
1. Arzamastsev, E. V. *Gig. Sanit.* 1964, **29**(12), 16-21
2. Chekunova, M. P; et al. *Gig. Tr. Prof. Zabol.* 1969, **13**(10), 25-29, (Russ.) (*Chem. Abstr.* **72**, 109212h)
3. Poteryaeva, G. E. *Gig. Tr. Prof. Zabol.* 1958, **2**(6), 22-25, (Russ.) (*Chem. Abstr.* **53**, 11664c)
4. Chekunova, M. P. *Gig. Tr. Prof. Zabol.* 1971, **15**(3), 31-34, (Russ.) (*Chem. Abstr.* **75**, 3443f)
5. Kanematsu, N; et al. *Mutat. Res.* 1980, **77**, 109-116

6. *EC Directive Relating to the Quality of Water Intended for Human Consumption* 1982, 80/778/EEC, Office for Official Publications of the European Communities, 2 rue Mercier, L-2985 Luxembourg

7. *ECETOC Technical Report No. 30(4)* 1991, European Chemical Industry Ecology and Toxicology Centre, B-1160 Brussels

A245   Antimony pentafluoride

SbF_5

CAS Registry No. 7783-70-2
Synonyms antimony(V) fluoride
Mol. Formula SbF_5 **Mol. Wt.** 216.74
Uses Chemical intermediate in the fluorination of organic compounds.

Physical properties

M. Pt. 7°C; **B. Pt.** 149.5°C; **Specific gravity** d_{23} 2.99;
Volatility v.p. 10 mmHg at 25°C; v. den. 2.2.

Occupational exposure

US TLV (TWA) 0.5 mg m^{-3} (as Sb); **UK Long-term limit** 0.5 mg m^{-3} (as Sb); **UN No.** 1732; **HAZCHEM Code** 4WE; **Conveyance classification** corrosive substance; toxic substance; **Supply classification** harmful.

Risk phrases
Harmful by inhalation and if swallowed (R20/22)

Safety phrases
Do not breathe dust (S22)

Mammalian and avian toxicity

Acute data
LD$_{50}$ inhalation mice 0.27 mg l^{-1} (1).

Sub-acute data
TD inhalation rats 0.015 mg l^{-1} for 2hr day^{-1} for 3.5 month, produced alterations in liver and thyroid function and changes in cardiac, renal and lung morphology (1).

Metabolism and pharmacokinetics
Distribution in rats after chronic exposure by inhalation of antimony pentafluoride showed high antimony concentration in blood. Levels in liver, kidneys, spleen and pancreas were similar. Antimony was retained for a long time and could be detected in organs examined months after the study was discontinued (2).

Legislation

Limited under EEC Directive on Drinking Water Quality 80/778/EEC. Antimony: maximum admissible concentration 10 µg l^{-1}. Fluorides: maximum admissible concentration 1500 µg l^{-1} at 8-12°C and 700 µg l^{-1} at 25- 30°C (3).

Any other comments

Soluble fluorides are rapidly absorbed from the gastrointestinal tract. Part is rapidly excreted via the urine, and the remainder is distributed to bone. Fluoride is a general protoplasmic poison, and poisoning can result from any soluble compound which dissociates fluoride ion. Effects include enzyme inhibition, hypocalcaemia, cardiovascular collapse and damage to the kidneys and brain (4,5).

Ignites on contact with phosphorus and reacts violently with water.

References

1. Chekunova, M. P; et al. *Gig. Sanit.* 1970, **35**(7), 25-28, (Russ.) (*Chem. Abstr.* **73**, 8590z)
2. Chekunova, M. P. *Gig. Tr. Prof. Zabol.* 1971, **15**(3), 31-33, (Russ.) (*Chem. Abstr.* **72**, 109212h)
3. *EC Directive Relating to the Quality of Water Intended for Human Consumption* 1982, 80/778/EEC, Office for Official Publications of the European Communities, 2 rue Mercier, L-2985 Luxembourg
4. Deichmann, W. B; et al. *Toxicology of Drugs and Chemicals* 1969, Academic Press, London
5. Gosselin, R. E; et al. *Clinical Toxicology of Commercial Products* 5th ed., 1984, Williams & Wilkins, Baltimore

A246   Antimony trichloride

SbCl₃

CAS Registry No. 10025-91-9

Synonyms antimonous chloride; antimony(III) chloride; trichlorostibine; stibine, trichloro-

Mol. Formula SbCl₃   **Mol. Wt.** 228.11

Uses Chemical reagent, catalyst, mordant. Used as a mordant for patent leather and in dyeing, for bronzing iron, colouring zinc black, in the manufacture of other antimony salts, as a catalyst and in organic syntheses.

Physical properties

M. Pt. 73.4°C; **B. Pt.** 283°C; **Specific gravity** d_4^{20} 3.14; **Volatility** v.p. 1 mmHg at 49.2°C.

Solubility

Organic solvent: ethanol, benzene, chloroform, acetone

Occupational exposure

US TLV (TWA) 0.5 mg m⁻³ (as Sb); **UK Long-term limit** 0.5 mg m⁻³ (as Sb); **UN No.** 1733; **HAZCHEM Code** 4WE; **Conveyance classification** corrosive substance; **Supply classification** corrosive.

Risk phrases

Causes burns – Irritating to respiratory system (R34, R37)

Safety phrases

In case of contact with eyes, rinse immediately with plenty of water and seek medical advice (S26)

Mammalian and avian toxicity

Acute data

LD_{50} oral rat 530 mg kg^{-1} (1).
LD_{50} oral guinea pig 570 mg kg^{-1} (2).
TC_{Lo} inhalation human 73 mg m^{-3} pulmonary and gastrointestinal effects (1).

Metabolism and pharmacokinetics

Exposure of rats by inhalation resulted in about 10% of the body burden of antimony found in blood two wk later. Intratracheal exposure in rabbits and dogs had <1% of the blood antimony concentrations found in rats (3).

Inhalation exposure of rats showed atypical, rapid loss from lung with somewhat longer biological $t_{1/2}$ of 100 days (4).

Part of intravenously administered antimony salts is absorbed by erythrocytes, and the rest is distributed to other tissues, predominantly liver, adrenals, spleen and thyroid (5).

Trivalent antimonials are excreted slowly by kidney because of low plasma concentration. Therefore, following single therapeutic dose, $\approx$10% is recovered in urine within 24 hr and only 30% within 1 wk (6).

Any other adverse effects

Chronic inhalation exposure may lead to olfactory disorders (7).
Inhalation causes irritation of the nose, throat and upper respiratory tract and may cause sore throat, dyspnoea, gastrointestinal symptoms, such as abdominal pain and loss of appetite (8,9).

Legislation

Limited under EEC Directive on Drinking Water Quality 80/778/EEC. Antimony: maximum admissible concentration 10 µg l^{-1}. Chlorides: guide level 25 mg l^{-1}, maximum admissible concentration 400 mg l^{-1} (10).

Any other comments

Physico-chemical properties, human health effects and experimental toxicology reviewed (11).

References

1. Sax, N. I; et al. *Dangerous Property of Industrial Materials* 7th ed., 1989, Van Nostrand Reinhold, New York
2. Arzamestsev, E. V. *Gig. Sanit.* 1964, **29**, 16-21
3. Friberg, L; et al. *Handbook of the Toxicology of Metals* 2nd ed., 1986, **1,2**, 31, Elsevier Science Publishers, Amsterdam
4. Clayton, G. D; et al. *Patty's Industrial Hygiene and Toxicology* 3rd ed., 1981-1982, **2A, 2B, 2C**, 1512, John Wiley Sons, New York
5. Venugopal, B; et al. *Metal Toxicity in Mammals* 1978, **2**, 214, Plenum Press, New York
6. Gilman, A. G; et al. *Goodman and Gilman's The Pharmacological Basis of Therapeutics* 6th ed., 1980, 1028, Macmillan Publishing Co., New York
7. *Chemical Safety Data Sheets* 1989, **2**, 50-52, RSC, London
8. Cordasco, E. M. *Angiology* 1974, **25**, 590

9. Taylor, P. J. *Br. J. Ind. Med.* 1966, **23**, 318
10. *EC Directive Relating to the Quality of Water Intended for Human Consumption* 1982, 80/778/EEC, Office for Official Publications of the European Communities, 2 rue Mercier, L-2985 Luxembourg
11. *ECETOC Technical Report No. 30(4)* 1991, European Chemical Industry Ecology and Toxicology Centre, B-1160 Brussels

A247   Antimony trifluoride

SbF₃

CAS Registry No. 7783-56-4
Synonyms antimonous fluoride; antimony(III) fluoride; trifluorostibine; stibine, trifluoro-
Mol. Formula SbF_3 **Mol. Wt.** 178.75
Uses Chemical reagent. Used in ceramics, dyeing, chlorofluoride manufacture and as a catalyst.

Physical properties
M. Pt. 292°C; **B. Pt.** 319°C (sublimes); **Specific gravity** d^{21} 4.379.

Occupational exposure
US TLV (TWA) 0.5 mg m^{-3} (as Sb); **UK Long-term limit** 0.5 mg m^{-3} (as Sb);
Supply classification toxic.

Risk phrases
Toxic by inhalation, in contact with skin and if swallowed (R23/24/25)

Safety phrases
Keep container tightly closed – In case of contact with eyes, rinse immediately with plenty of water and seek medical advice – If you feel unwell, seek medical advice (show label where possible) (S7, S26, S44)

Mammalian and avian toxicity

Acute data
LD$_{50}$ subcutaneous rats 23 mg kg^{-1} (1).

Sub-acute data
In rats, doses of 30 g kg^{-1} day^{-1} for 14 days increased levels of citric acid in the liver and kidneys (2).

Metabolism and pharmacokinetics
Part of intravenously administered trivalent antimony salts is absorbed by erythrocytes, and the rest is distributed to other tissues, predominantly liver, adrenals, spleen and thyroid (3).
Trivalent antimonials are excreted slowly by kidney because of low plasma concentration. Therefore following single therapeutic dose, ≈10% is recovered in urine within 24 hr and only 30% within 1 wk (4).

Any other adverse effects

In humans, ingestion of antimony fluoride may result in nausea, vomiting, diarrhoea, abdominal pain, weakness, dyspnoea, cyanosis, coma and convulsions. Weight loss and anaemia may also result from chronic exposure (5).

Legislation

Limited under EEC Directive on Drinking Water Quality 80/778/EEC. Antimony: maximum admissible concentration 10 μg l^{-1}. Fluoride: maximum admissible concentration 1500 μg l^{-1} at 8- 12°C, 700 μg l^{-1} at 25-30°C (6).

Any other comments

Fluorides are absorbed from gastrointestinal tract, lung and skin and 95-96% of body burden of fluoride is in bones and teeth (7,8).
Physico-chemical properties, human health effects and experimental toxicology reviewed (9).

References

1. Levina, E. N; et al. *Gig. Tr. Prof. Zabol.* 1964, **8**(7), 25-31, (Russ.) (*Chem. Abstr.* **61** 16690g)
2. Kurayuki, Y. *Osaka Shiritsu Daigaku Igaku Zasshi* 1959, **8**, 817-820, (Jap.) (*Chem. Abstr.* **57**, 14109d)
3. Venugopal, B; et al. *Metal Toxicity in Mammals* 1978, **2**, 214, Plenum Press, New York
4. Gilman, A. G; et al. *Goodman and Gilman's The Pharmacological Basis of Therapeutics* 6th ed., 1980, 1028, Macmillan Publishing Co., New York
5. *Chemical Safety Data Sheets* 1989, **2**, 53-55, RSC, London
6. *EC Directive Relating to the Quality of Water Intended for Human Consumption* 1982, 80/778/EEC, Office for Official Publications of the European Communities, 2 rue Mercier, L-2985 Luxembourg
7. Gilman, A. G; et al. *Goodman and Gilman's The Pharmacological Basis of Therapeutics* 7th ed., 1985, Macmillan Publishing Co., New York
8. Jones, L. M; et al. *Veterinary Pharmacology and Therapeutics* 4th ed., 1977, 1274, Iowa State University Press, Ames
9. *ECETOC Technical Report No. 30(4)* 1991, European Chemical Industry Ecology and Toxicology Centre, B-1160 Brussels

A248 Antimony trioxide

Sb_2O_3

CAS Registry No. 1309-64-4
Synonyms diantimony trioxide; flowers of antimony; antimony white; Exitelite; Valentinite; Weisspiessglanz
Mol. Formula Sb_2O_3 **Mol. Wt.** 291.50
Uses Manufacture of tartar emetic, as mordant in paints, pigments, enamels, glasses, in flame-proofing canvas.
Occurrence Senartmontite, valentinite.

Physical properties
M. Pt. 655°C; **B. Pt.** 1550°C; **Specific gravity** 5.2.

Occupational exposure
US TLV (TWA) 0.5 mg m^{-3} (as Sb); **UK Long-term limit** 0.5 mg m^{-3} (as Sb) under review; **UN No.** 1549.

Ecotoxicity
Fish toxicity
LC_{50} (96 hr) bluegill sunfish, fathead minnow 530-833 mg l^{-1} (1,2).

Environmental fate
Degradation studies
Pure cultural study using *Stibiobacter senarmontii* autotroph grown in a mineral medium containing antimony trioxide oxidised the chemical at 45-52 mg month^{-1} for cubic and 13-20 mg month^{-1} for rhombic forms. Little oxidation occurred in sterile medium (3).

Abiotic removal
Loss of antimony oxides from water by evaporation is very unlikely under normal conditions due to their very low concentration and low concentration of their hydrolysis products (3).

Mammalian and avian toxicity
Acute data
LD_{50} oral rat >34.6 μg kg^{-1} (4).
LD_{50} percutaneous rabbit >2 μg kg^{-1} (4).

Sub-acute data
Intragastric administration of 10 g kg^{-1} to rats caused no deaths suggesting the compounds' low toxicity. However, chronic toxic effects and effects on reproduction and other biochemical indexes via inhalation are reported (5).

Teratogenicity and reproductive effects
In rats exposed during pregnancy to 0.082-0.27 mg m^{-3} antimony trioxide, embryo mortality (both pre- and post-implantion) was greater than controls and embryo mass was below controls (6).

Metabolism and pharmacokinetics
Rats and rabbits fed 2% antimony trioxide in a casein diet; antimony was found from 6.7-88 μg g^{-1} of tissue, found in the thyroid, adrenal glands, spleen, liver, lung, heart and kidney (7).
After single oral doses of 200 mg in 5 ml of water to rats, 3% was eliminated in urine. At a concentration of 2% for 8 months in the diet, antimony excretion in the faeces was much greater than in urine (8).
Part of intravenously administered trivalent salts are absorbed by erythrocytes, and the rest are distributed to other tissues, predominantly liver, adrenals, spleen and thyroid (9).

Genotoxicity

Bacillus subtilis gene conversion and mitotic recombination positive (10).

Any other adverse effects to man

Women exposed to unspecified amounts of antimony trioxide in a metallurgical plant had antimony concentrations in blood 10-times greater than in controls; levels in urine, breast milk, placental tissue, amniotic fluid and umbilical cord blood were 3.3-126 mg l^{-1} (3).

Lung reticulomicronodular images among employees of an antimony oxide production plant revealed collagenous fibrosis with 600-3000 μg g^{-1} antimony in biopsies from 2 workers showing there is a risk of pneumoconiosis due to dust inhalation in antimony trioxide plants (11).

Women occupationally exposed to antimony trioxide experienced increased rate of spontaneous late abortions, premature birth, gynaecological problems and reduced weight gain of children (12).

Any other adverse effects

Occasional exposure of rats did not cause chronic pneumonitis, but chronic inhalation caused pathological changes in the lungs of rats, pneumonitis and fatty degeneration of the liver in guinea pigs, and lipoid pneumonias in rats and rabbits (13-16).

Legislation

Limited under EEC Directive on Drinking Water Quality 80/778/EEC. Antimony: maximum admissible concentration 10 μg l^{-1} (17).

Any other comments

Physico-chemical properties, human health effects, experimental toxicology, epidemiology, workplace experience and environmental effects reviewed (18).

References

1. Buccafusco, R. J. *Bull. Environ. Contam. Toxicol.* 1981, **26**(4), 446-452
2. Curtis, M. W; et al. *I. Hydrol.* 1981, **51**, 359-367
3. *USEPA Health and Environmental Effects Profile for Antimony Oxides* 1985, EPA 600/X-85/271
4. *Documentation of the Threshold Limit Values* 4th ed., 1980, 21, Amercian Conference of Governmental Industrial Hygienists, Inc., Cincinnati, OH
5. Grin, N. V; et al. *Gig. Sanit.* 1989, (4), 68-69, (Russ.) (*Chem. Abstr.* 1989, **111**, 2278f)
6. Grin, N. V; et al. *Gig. Sanit.* 1987, (10), 85-86, (Russ.) (*Chem. Abstr.*1987, **107**, 231008c)
7. Clayton, G. D; et al. *Patty's Industrial Hygiene and Toxicology* 1981-1982, **2A, 2B, 2C**, 1512, John Wiley Sons, New York
8. Gross, P; et al. *Arch. Ind. Health* 1955, **11**, 473-478
9. Venugopal, B; et al. *Metal Toxicity in Mammals* 1978, **2**, 214, Plenum Press, New York
10. *Mutat. Res.* 1980, **77**, 109
11. Le Bouffant, L; et al. *Inst. Natl. Sante Rech. Med., [Colloq.]* 1987, **155**, 447-455 (Fr.) (*Chem. Abstr.* 1988, **109**, 78918u)
12. Belyeava, A. P. *Gig. Tr. Prof. Zabol.* 1967, **11**, 32, (Russ.) (*Chem. Abstr.* **66**, 108037u)
13. Cooper, D. A; et al. *Am. J. Roentgenol. Radium Ther. Nucl. Med.* 1968, **103**, 495
14. Gudzovskii, G. A. *Sb. Nauch. Rab. Kirg. Nauch.-Issled. Inst. Tuberk.* 1967, **5**, 82-87
15. Dernehl, C. U; et al. *J. Ind. Hyg. Toxicol.* 1945, **27**, 256
16. Gross, P; et al. *Arch. Ind. Health* 1955, **11**, 479

17. *EC Directive Relating to the Quality of Water Intended for Human Consumption* 1982, 80/778/EEC, Office for Official Publications of the European Communities, 2 rue Mercier, L-2985 Luxembourg
18. *ECETOC Technical Report No. 30(4)* 1991, European Chemical Industry Ecology and Toxicology Centre, B-1160 Brussels

A249   Antimycin A

CAS Registry No. 1397-94-0
Synonyms antipiricullin; virosin; antibiotic 720A
Mol. Formula $C_{28}H_{40}N_2O_9$ **Mol. Wt.** 548.64
Uses Antibiotic.
Occurrence Produced by *Streptomyces spp.*

Physical properties
M. Pt. 149-150°C.

Solubility
Organic solvent: ethanol, acetone, diethyl ether, chloroform

Ecotoxicity

Fish toxicity
LC_{50} (96 hr) northern pike, rainbow trout $\leq 55 \ \mu g \ l^{-1}$ (1).
LC_{50} (96hr) guppy 1.14 $\mu g \ l^{-1}$ (2).

Effects to non-target species
LD_{50} oral pigeon, duck 2.0-2.9 mg kg^{-1} (3).

Mammalian and avian toxicity

Acute data
LD_{50} oral rat, guinea pig 1.8-28 mg kg^{-1} (4).
LD_{50} intraperitoneal rat 0.8 mg kg^{-1} (4).
LD_{50} subcutaneous mouse 21 mg kg^{-1} (5).
LD_{50} intravenous mouse 0.89 mg kg^{-1} (5).

References
1. Kemp, H. T; et al. *Water Quality Data Book* 1973, **5**
2. Gupta, P. K; et al. *Arch. Hydrobiol.* 1984, **101**(4), 601-607
3. Lewis, R. J; et al. *Registry of Toxic Effects of Chemical Substances* 1984, National Institute for Occupational Safety and Health No. 83-107-4
4. Reif, A. E; et al. *Cancer Res.* 1953, **13**, 49-57
5. Nakayama, K; et al. *J. Antibiol. Ser. A* 1956, **9**(2), 63-66

A250　Arachidonic acid

CAS Registry No. 506-32-1
Synonyms 5,8,11,14-eicosatetraenoic acid
Mol. Formula $C_{20}H_{32}O_2$　　　　　　　　　　　　Mol. Wt. 304.48
Occurrence Essential fatty acid.
Precursor in biosynthesis of prostaglandins, thromboxanes and leukotrienes.

Physical properties

M. Pt. –49°C; B. Pt. 169-171°C; Flash point >110°C; Specific gravity d^{20} 1.4824;
Partition coefficient log P_{ow} 6.98.

Solubility

Organic solvent: ethanol, diethyl ether, acetone

Mammalian and avian toxicity

Acute data

LD_{50} intravenous mouse 33 mg kg^{-1} (1).
LD_{Lo} intravenous mouse, rat, rabbit 1-100 mg kg^{-1} (2).

Metabolism and pharmacokinetics

The major metabolite found in the human placenta was 12-hydroxy-
5,8,10,14-eicosatetraenoic acid with 5- and 15-hydroxy-5,8,10,14-eicosatetraenoic
acids as minor metabolites (3).
Renal glomerular and cortical metabolism of endogenous arachidonic acid by
cytochrome P-450 epoxygenase yields 8,9-, 11,12- and 14,15- epoxyeicosatrienoic
acids (4).
The production of eicosanoids by alveolar macrophages from the lungs of rats
exposed to bituminous coal dust for 2 wk significantly altered arachidonic acid
metabolism in a manner that promoted an inflammatory response which may be an
important factor in the pathophysiology of coal workers' pneumoconiosis (5).
Metabolites of 15(S)-hydroxy- 5,8,11,13-eicosatetraenoic acid in human bronchi and
in tissue specimens from three asthmatic patients were identified (6).

Irritancy

Application of arachidonic acid at 1 mg ear^{-1} day^{-1} for 5 day wk^{-1} for 3 wk caused a
marked swelling of the ear which plateaued on day 10. This was accompanied by
cellular infiltration and epidermal hyperplasia (7).

Any other comments

Arachidonic acid metabolism and the role of its metabolites in blood vessel-platelet
interactions, cytoprotection and tumour metastasis reviewed (8).
Cytochrome P-450 metabolism reviewed (9).

References

1. Myers, A; et al. *J. Pharmacol. Exp. Ther.*, 1983, **224**, 369
2. *Thrombosis Research* 1976, **9**, 67
3. Saeed, S. A. *Ind. J. Physiol. Allied Sci.* 1990, **44**(4), 161-169
4. Katoh, T; et al. *Am. J. Physiol.* 1991, **261**(4, Part 2), F578-F586
5. Kuhn, D. C; et al. *J. Toxicol. Environ. Health* 1990, **29**(2), 157-168
6. Kumlin, M; et al. *Adv. Prostaglandin, Thromboxane, Leukotriene Res.* 1990, **21A**, 441-444
7. Bouclier, M; et al. *Agents Actions* 1989, **26**(1-2), 227-228
8. McGiff, J. C. *Prev. Med.* 1987, **16**(4), 503-509
9. McGiff, J. C. *Ann. Rev. Pharmacol. Toxicol.* 1991, **31**, 339-369

A251   Aramite

CAS Registry No. 140-57-8

Synonyms sulfurous acid 2-chloroethyl
2-[4-(1,1-dimethylethyl)phenoxyl]-1-methylethyl ester; sulfurous acid
2-(*p-tert*-butylphenoxy)-1-methylethyl 2-chloroethyl ester; Compound 88R; ENT
16519; Araton; Niagaramite

Mol. Formula $C_{15}H_{23}ClO_4S$ **Mol. Wt.** 334.87

Uses Acaricide.

Physical properties

M. Pt. $-37.3°C$; **B. Pt.** $_{0.1}175°C$; **Specific gravity** d_{20}^{20} 1.145.

Solubility

Organic solvent: miscible with ethanol, diethyl ether, acetone, benzene

Ecotoxicity

Invertebrate toxicity

EC_{50} (48 hr) *Daphnia pulex, Simocephalus serrulatus* 0.16-0.18 mg l^{-1} (1).

Bioaccumulation

Estimated bioconcentration factor is 2265 based on water solubility of 0.1 ppm,
which suggests bioconcentration in aquatic organisms is likely (2).

Environmental fate

Abiotic removal

Aramite does not absorb UV irradiation above 290 nm which suggests that it does not
directly photolyse in the environment (3).

Absorption

Estimated soil adsorption coefficient is 15500 which indicates soil immobility (4).

Mammalian and avian toxicity

Acute data
LD_{50} oral rat, guinea pig 3900 mg kg^{-1} (5).
LD_{Lo} intraperitoneal mouse 200 mg kg^{-1} (6).

Sub-acute data
LC_{50} (5 day in diet) Japanese quail >5000 ppm (7).

Carcinogenicity and long-term effects
No adequate evidence for carcinogenicity to humans, sufficient evidence for carcinogenicity to animals, IARC classification group 2B (8).
Experimental carcinogen in livers of rats and mice and of the liver and biliary tracts of dogs via ingestion and produces liver tumours in rats (9-12).

Any other adverse effects
Aramite is an aminal carcinogen currently in commerical use. The US epidemiological studies of its use and production volume have not been carried out because of simultaneous exposure to other chemicals (13).

Legislation
Limited under EC Directive on Drinking Water Quality 80/778/EEC. Insecticide: maximum admissible concentration 0.1 µg l^{-1} (14).

References

1. Sanders, H. O; et al. *Trans. Am. Fish Soc.* 1966, **95**, 165-169
2. Lyman, W. J; et al. *Handbook of Chemical Property Estimation Methods* 1982, 5-10, McGraw-Hill, New York
3. Gore, R. C; et al. *J. Assoc. Off. Anal. Chem.* 1971, **54**, 1040-1082
4. Swann, R. L; et al. *Res. Rev.* 1983, **85**, 16-28
5. *Pesticide Chemicals Official Compendium* 1986, 61, Association of American Pesticide Control Officials Inc.
6. *Toxicol. Appl. Pharmacol.* 1972, **23**, 288
7. Hill, E. F; et al. *Lethal Dietary Toxicities of Environmental Contaminants and Pesticides to Coturnix. Fish and Wildlife Technical Report* 1986, 27, US Dept of Interior Fish and Wildlife Service, Washington, DC
8. *IARC Monograph*, 1987, **Suppl. 7**, 57
9. Truhaut, R; et al. *C. R. HEBD. Seances Acad. Sci., Ser. D.* 1975, **281**(9), 599 (*Chem. Abstr.* **84**, 1081y)
10. Popper, H; et al. *Cancer* 1960, **13**, 780
11. *IARC Monograph* 1974, **5**, 39
12. Popper, H; et al. *Cancer* 1960, **13**, 1035
13. Karstadt, M; et al. *Teratogen., Carcinogen. Mutagen.* 1982, **2**, 151-167
14. *EC Directive Relating to the Quality of Water Intended for Human Consumption* 1982, 80/778/EEC, Office For Official Publications of the European Communities, 2 rue Mercier, L-2985, Luxembourg

A252   Aroclor 1016

CAS Registry No. 12674-11-2
Synonyms polychlorinated biphenyl; PCB; Aroclor 1016
Uses Electrical insulator, heat transfer medium, hydraulic fluids and lubricants.

Occupational exposure

UN No. 2315; **HAZCHEM Code** 4X; **Conveyance classification** other dangerous substance; **Supply classification** harmful.

Risk phrases
Danger of cumulative effects (R33)

Safety phrases
This material and its container must be disposed of in a safe way (S35)

Ecotoxicity

Fish toxicity
LC_{Lo} (28 day) flow through bioassay intermittent sheepshead minnow 32 µg l^{-1} (1).
LD_{Lo} (42 day) *Lagodon rhomboides* 32 µg l^{-1} tissue damage found in pancreas (2).
LC_{50} (96 hr) rainbow trout, Atlantic salmon, brown trout, brook trout, fingerling longnose sucker, channel catfish, bluegill sunfish, yellow perch 134-800 µg l^{-1} (3).
Toxicity of Aroclor 1016 to different age groups of sheepshead minnows was studied in intermittent-flow bioassays lasting 26 days. 32 and 100 µg l^{-1} killed newly hatched fry and juveniles and adults. Accumulation was in proportion to the chemicals' concentration in test water. Fry contained 2500- 8100 times the concentration added to the test water, adults 1700-15,000 times and juveniles 10,000-34,000 times. 77 µg l^{-1} of Arochlor 1016 in eggs from exposed adults did not effect survival of embryos and fry (1).

Invertebrate toxicity
LC_{50} (72 hr) *Hydra oligactis* 5 mg l^{-1} (4).

Bioaccumulation
Ambient concentration 0.1-10 µg l^{-1} sheepshead minnow fry bioconcentration factor 2500-8100, adults 4700-14,000 (1).
Bioaccumulation factor for pinfish exposed to 1 µg l^{-1} for 56 days was a 17,000 fold increase over the nominal test concentration (2).

Environmental fate

Degradation studies
Using BOD dilution water, settled domestic waste water inoculum and 5 and 10 ppm Aroclor 1016, after 28 days 48% degradation at 5 ppm and 13% degradation at 10 ppm occurred (5).

Absorption
Experimental soil adsorption coefficients of 52,100-171,000 suggest strong adsorption to sediments during a study of 3 ponds (6,7).

Mammalian and avian toxicity

Carcinogenicity and long-term effects
Limited evidence for carcinogenicity to humans, sufficient evidence for carcinogenicity to animals, IARC classification group 2A (8).

A NIOSH retrospective cohort study was conducted of 3588 men and women employed for ≥ 1 day between 1st January 1957 and 31st March 1977 manufacturing capacitors with known exposure to PCBs in the midwest US. Aroclor 1016 was used from 1970-1977 as a dielectric fluid. Results provided some evidence for an association between occupational PCB exposure and mortality from malignant melanoma. There was an increased incidence of brain cancer among workers who had more than twice the estimated cumulative PCB dose than the comparison group (9).

Metabolism and pharmacokinetics
TD oral rat 3.6-6.9 mg kg^{-1} (daily) long term feeding study showed PCB distributed throughout body, but mainly concentrated in adipose tissue. Steady-state concentration found in adipose tissue after 4 months. Only small levels of PCB found in urine. After exposure ceased, the elimination of the PCB was slow, with measurable amounts still present 5-6 months later. Pathological changes were found in liver cells (10).

Rats fed 100 ppm of Aroclor 1016 (6.6-3.5 mg kg^{-1} day^{-1}) revealed PCBs in plasma, kidneys, urine, brain, liver and adipose tissue at 0.5-10 months exposure. PCB tissue levels determined 2, 4, 5 and 6 months after exposure was discontinued were highest in adipose tissue; little residue was excreted in urine (11).

173 workers occupationally exposed to Aroclor 1016 revealed levels of serum lipid PCBs equal to the adipose fat PCB level (12).

Legislation
Limited under EEC Directive on Drinking Water Quality 80/778/EEC. Pesticides and related products: maximum admissible concentration 0.5 µg l^{-1} (13).

Included in Schedule 5 (Release Into Water: Prescribed Substances) of Statutory Instrument 1991, No. 472 (14).

Any other comments
Epidemiology, toxicity and waste disposal reviewed (15-17).

Long-term neurobehavioural effects in monkeys reviewed (18).

References

1. Hansen, D. J; et al. *Trans. Am. Fish. Soc.* 1975, **104**(3), 584-588
2. Hansen, D. J; et al. *Environ. Res.* 1974, **7**(3), 363-373
3. *Handbook of Acute Toxicity of Chemicals to Fish and Aquatic Invertebrates* 1980, 67, US Dept. of Interior, Fish and Wildlife Service, US Govt. Print. Off., Washington, DC
4. Adams, J. A; et al. *Arch. Environ. Contam. Toxicol.* 1984, **13**(4), 493-499
5. Tabak, H. H; et al. *J. Water Pollut. Control Fed.* 1981, **53**, 1503-1518
6. Sklarew, D. S; et al. *Rev. Environ. Contam. Toxicol.* 1987, **98**, 1-41
7. Leifer, A; et al. *Environmental Transport and Transformation of Polychlorinated Biphenyls* 1983, 4-5, 4-14, EPA 560/5-83-025, NTIS PB84-142579
8. *IARC Monograph* 1987, **Suppl 7**, 322
9. *Gov. Rep. Announce. Index US* 1991, **91**(20), 156,257, Health Hazard Evaluation Report HETA 89-116-2094, Westinghouse Electric Corp., Bloomington, IN
10. Burse, V. W; et al. *J. Water Pollut. Control Fed.* 1981, **53**(10), 1503-1518

11. Burse, V. W; et al. *Arch. Environ. Health* 1974, **29**(6), 301-307
12. Brown, J. F; et al. *Bull. Environ. Contam. Toxicol.* 1984, **33**(3), 277-280
13. *EC Directive Relating to the Quality of Water Intended for Human Consumption* 1982, 80/778/EEC, Office for Official Publications of the European Communities, 2 rue Mercier, L-2985 Luxembourg
14. *S. I. 1991, No. 472 The Environmental Protection (Prescribed Processes and Substances) Regulations* 1991, HMSO, London
15. MAFF *Polychlorinated biphenyl (PCB) residues in food and human tissues* Food surveillance Paper No. 13, 1983, HMSO, London
16. Rogers, C. G; et al. *Toxicology* 1983, **26**(2), 113-124
17. *Polychlorinated Biphenyls* 1988, **107**, International Registry of Potentially Toxic Chemicals, United Nations Environmental Programme, Geneva
18. Schantz, S. L; et al. *Environ. Toxicol. Chem.* 1991, **10**(6), 747-756

A253   Aroclor 1221

CAS Registry No. 11104-28-2
Synonyms Arochlor 1221; chlorodiphenyl (21% Cl)
Uses Electrical insulator, solvent, dispersants, lubricants.

Occupational exposure

UN No. 2315; **HAZCHEM Code** 4X; **Conveyance classification** other dangerous substance; **Supply classification** harmful.

Risk phrases
Danger of cumulative effects (R33)

Safety phrases
This material and its container must be disposed of in a safe way (S35)

Ecotoxicity

Fish toxicity
LC$_{50}$ (96 hr) constant flow harlequin fish 1.05 mg l^{-1} (1).
LC$_{50}$ (96 hr) cutthroat trout 1.2 mg l^{-1} (2).

Bioaccumulation
Experimental bioconcentration factors of 955-13,804 have been reported for the dominant polychlorinated biphenyl (PCB) congeners present in Arochlor 1221 (3).

Environmental fate

Degradation studies
Mixed culture of *Pseudomonas sp.* HV3 and *Nocardia sp.* degraded 40-87% of Aroclor 1221 in 6-21 days releasing chlorobenzoic acidmetabolites (4).
Using BOD dilution water, settled domestic waste water inoculum and 5 and 10 ppm Aroclor 1221, 100% was biodegraded during 28 days (5).

Abiotic removal
Processes such as hydrolysis and oxidation do not significantly degrade Aroclor 1221 in the aquatic environment (6,7).

Absorption

Estimated soil adsorption coefficients for the primary PCB congeners present in Aroclor 1221 were 10,965- 75,858, suggesting that Aroclor 1221 is adsorbed onto solids and is often immobilised in sediment, although eventual re- solution can occur (8).

Removal by biological adsorption in activated sludge was 96.3% at an aeration time of 24 hr (9).

Mammalian and avian toxicity

Acute data

LD_{50} oral rat 3980 mg kg^{-1} (10).
LC_{50} oral Japanese quail 12,000 ppm (11).
LD_{50} dermal rabbit 2000 mg kg^{-1} (10).

Carcinogenicity and long-term effects

Limited evidence for carcinogenicity to humans, sufficient evidence for carcinogenicity to animals, IARC classification group 2A (12).

A NIOSH retrospective cohort study was conducted of 3588 men and women employed for ≥1 day between 1st January 1957 and 31st March 1977 manufacturing capacitors with known exposure to PCBs in the midwest US. Results provided some evidence for an association between occupational PCB exposure and mortality from malignant melanoma. There was an increased incidence of brain cancer among workers who had more than twice the estimated cumulative PCB dose than the comparison group (13).

Legislation

Limited under EEC Directive on Drinking Water Quality 80/778/EEC. Pesticides and related products: maximum admissible concentration 0.5 µg l^{-1} (14).

Included in Schedule 5 (Release Into Water: Prescribed Substances) of Statutory Instrument 1991, No. 472 (15).

Any other comments

Toxicity and waste disposal reviewed (16,17).

Pyrolysis at low temperatures (200-600°C) can produce even more toxic materials such as polychlorinated dibenzofuran and 2,3,7,8-tetrachlorodibenzodioxin (dioxin).

References

1. Tooby, T. E; et al. *Chem. Ind. (London)* 1975, 523-526
2. *Handbook of Acute Toxicity of Chemicals to Fish and Aquatic Invertebrates* 1980, 67, US Dept. of Interior, Fish and Wildlife Service, US Govt. Printing Office, Washington, DC
3. Leifer, A; et al. *Environmental Transport and Transformation of Polychlorinated Biphenyls* 1983, 5-11, EPA 560/5-83-025, NTIS PB84-142579
4. Kilpi, S; et al. *FEMS Microbiol. Ecol.* 1988, **53**(1), 19-26
5. Tabak, H. H; et al. *J. Water Pollut. Contr. Fed.* 1981, **53**, 1503-1518
6. Mabey, W. R; et al. *Aquatic Fate Process Data for Organic Priority Pollutants* 1981, 115-128, EPA 440/4-81-014
7. Callahan, M. A; et al. *Water-Related Environmental Fate of 129 Priority Pollutants* 1979, EPA 440/4-79-029a
8. *Treatability Manual* 1981, **1**, 2-1, EPA 600/8-80-042
9. Aly, A. O; et al. *Egypt. J. Microbiol.* 1985, **20**(2), 277-283 (*Chem. Abstr.* **105**, 177810b)
10. Fishbein, L. *Ann. Rev. Pharmacol.* 1974, **14**, 139-156

11. *Lethal Dietary Toxicities of Environmental Pollutants to Birds* 1975, 9, US Dept. of the Interior, Fish and Wildlife Service, Bureau of Sports Fisheries and Wildlife, US Govt. Print. Off., Washington, DC
12. *IARC Monograph* 1987, **Suppl. 7**, 322
13. *Gov. Rep. Announce. Index US* 1991, **91**(20), 156,357, Health Hazard Evaluation Report HETA 89-116-2094, Westinghouse Electric Corp., Bloomington, IN
14. *EC Directive Relating to the Quality of Water Intended for Human Consumption* 1982, 80/778/EEC, Office for Official Publications of the European Communities, 2 rue Mercier, L-2985 Luxembourg
15. *S. I. 1991, No. 472 The Environmental Protection (Prescribed Processes and Substances) Regulations* 1991, HMSO, London
16. MAFF *Polychlorinated biphenyl (PCB) residues in food and human tissue* Food Surveillance Paper No. 13, 1983, HMSO, London
17. *Polychlorinated Biphenyls* 1988, **107**, International Registry for Potentially Toxic Chemicals, United Nations Environmental Programme, Geneva

A254   Aroclor 1232

CAS Registry No. 11141-16-5
Synonyms Arochlor 1232; chlorodiphenyl (32% Cl)
Uses Electrical insulator, heat transfer medium, hydraulic fluids and lubricants.

Occupational exposure

UN No. 2315; **HAZCHEM Code** 4X; **Conveyance classification** other dangerous substance; **Supply classification** harmful.

Risk phrases
Danger of cumulative effects (R33)

Safety phrases
This material and its container must be disposed of in a safe way (S35)

Ecotoxicity

Fish toxicity
LC_{50} (96 hr) constant flow harlequin fish 0.32 mg l^{-1} (1).
LC_{50} (96 hr) static assay cutthroat trout 2.5 mg l^{-1} (2).

Bioaccumulation
Bioconcentration factor for white sucker 5500 (3).

Environmental fate

Degradation studies
Using BOD dilution water, settled domestic wastewater inoculum and 5 and 10 ppm Aroclor 1232, after 28 days 100% was biodegraded (4).

Abiotic removal
Processes such as hydrolysis and oxidation do not significantly degrade Aroclor 1232 in the aquatic environment (5,6).

Absorption

Removal by biological adsorption in activated sludge, 94.2% at aeration time of 24 hr (7). Estimated soil adsorption coefficients of primary polychlorinated biphenyl congeners in Aroclor 1232 are 10,965-177,828 suggesting Arochlor 1232 is readily adsorbed onto soils and are often immobilized in sediment, although eventual re-solution can occur (8).

Mammalian and avian toxicity

Acute data
LD_{50} oral rat 4470 mg kg^{-1}.
LD_{50} dermal rabbit >1260 mg kg^{-1} (9).

Sub-acute data
LD_{50} oral Japanese quail, Northern bobwhite 5000-3002 mg kg^{-1} diet 5 day^{-1} on treated diet plus 3 days untreated (10).

Carcinogenicity and long-term effects
Limited evidence for carcinogenicity to humans, sufficient evidence for carcinogenicity to animals, IARC classification group 2A (11).
A NIOSH retrospective cohort study was conducted of 3588 men and women employed for ≥1 day between 1st January 1957 and 31st March 1977 manufacturing capacitors with known exposure to PCBs in the midwest US. Results provided some evidence for an association between occupational PCB exposure and mortality from malignant melanoma. There was an increased incidence of brain cancer among workers who had more than twice the estimated cumulative PCB dose than the comparison group (12).

Teratogenicity and reproductive effects
Diet containing 20 ppm fed to hens for 9 wk caused decreased hatchability and teratogenic effects in embryos.
TD hen 20 ppm diet (9 wk) teratogenic, reproductive effects (13).

Legislation

Limited under EEC Directive on Drinking Water Quality 80/778/EEC. Pesticides and related products: maximum admissible concentration 0.5 μg l^{-1} (14).
Included in Schedule 5 (Release Into Water: Prescribed Substances) of Statutory Instrument 1991, No. 472 (15).

Any other comments

Toxicity and waste disposal reviewed (16-18).
Incomplete combustion can lead to the production of polychlorinated dibenzofurans and 2,3,7,8-tetrachlorodibenzodioxin, which are more toxic.

References

1. Tooby, T. E; et al. *Chem. Ind. (London)* 1975, 523-526
2. Mayer, F. L; et al. *Arch. Environ. Contam. Toxicol.*1977, **5**, 501-511
3. *USEPA Ambient Water Quality Criteria: Polychlorinated Biphenyls* 1980, EPA 440/5-80-068
4. Tabak, H. H; et al. *J. Water Pollut. Contr. Fed.* 1981, **53**, 1503-1518
5. Mabey, W. R; et al. *Aquatic Fate Process Data for Organic Priority Pollutants* 1981, 115-128, EPA 440/4-81-014

6. Callahan, M. A; et al. *Water-Related Environmental Fate of 129 Priority Pollutants* 1979, EPA 440/4-79-029a
7. Aly, A. O; et al. *Egypt. J. Microbiol.* 1985, **20**(2), 277-283 (*Chem. Abstr.* **105**, 177810b)
8. *USEPA Treatability Manual* 1981, **1**, 3-1, EPA 600/8-80-042
9. Fishbein, L. *Ann. Rev. Pharmacol.* 1974, **14**, 139-156
10. *Polychlorinated Biphenyls Hazards to Fish, Wildlife and Invertebrates: A Synoptic Review Biol. Report No. (85)1.7* 1986, US Dept. of Interior/ Fish and Wildlife Service Contaminant Reviews
11. *IARC Monograph* 1987, **Suppl. 7**, 322
12. *Gov. Rep. Announce. Index US* 1991, **91**(20), 156,257, Health Hazard Evaluation Report HETA 89-116-2094, Westinghouse Electric Corp., Bloomington, IN
13. Cecil, H. C; et al. *Bull. Environ. Contam. Toxicol.* 1974, **11**(6), 489-495
14. *EC Directive Relating to the Quality of Water Intended for Human Consumption* 1982, 80/778/EEC, Office for Official Publications of the European Communities, 2 rue Mercier, L-2985 Luxembourg
15. *S. I. 1991, No. 472 The Environmental Protection (Prescribed Processes and Substances) Regulations* 1991, HMSO, London
16. Kimbrough, R. D. *Ann. Rev. Pharmacol. Toxicol.* 1987, **27**, 87-111
17. MAFF *Polychlorinated biphenyl (PCB) residues in food and human tissues* Food Surveillance Papers No. 13, 1983, HMSO, London
18. *Polychlorinated Biphenyls* 1988, **107**, International Registry of Potentially Toxic Chemicals, United Nations Environmental Programme, Geneva

A255   Aroclor 1242

CAS Registry No. 53469-21-9
Synonyms Arochlor 1242; chlorodiphenyl (42% Cl); PCB 1242
Uses In production of lubricants, plasticisers. Dielectric liquids (electrical insulators).

Physical properties

Specific gravity $d_{15.5}^{65}$ 1.41; **Volatility** v.p. 55 mmHg at 225°C.

Occupational exposure

US TLV (TWA) 1 mg m^{-3}; **UN No.** 2315; **HAZCHEM Code** 4X;
Conveyance classification other dangerous substance; **Supply classification** harmful.

Risk phrases
Danger of cumulative effects (R33)

Safety phrases
This material and its container must be disposed of in a safe way (S35)

Ecotoxicity

Fish toxicity
LC$_{50}$ (96 hr) constant flow harlequin fish 0.37 mg l^{-1} (1).
LC$_{50}$ (96 hr) cutthroat trout, yellow perch, scud ≤5420 µg l^{-1} (2).
LC$_{50}$ (5-15 day) rainbow trout, bluegill sunfish, channel catfish 54-125 µg l^{-1} (2).

Invertebrate toxicity
LC_{50} (96 hr) *Ischnura verticalis*, *Macromia sp.* 400-800 μg l^{-1} (2).

Bioaccumulation
Macoma balthica accumulated 60 ppm after 30 days of exposure (ambient concentration not reported) (3).
Bioconcentration factor fathead minnow 32,000-274,000 (4).

Effects to non-target species
LC_{50} oral ring-necked pheasant 2078 mg kg^{-1} diet (5 days on treated diet plus 3 days untreated) (2).
LD_{50} oral mallard duck 3182 mg kg^{-1} diet (50 days on treated diet plus 3 days untreated) (2).

Environmental fate

Anaerobic effects
Dechlorination was achieved by anaerobic microorganisms in lake sediments with the loss of *meta* plus *para* chlorines ranging from 15-85%. Maximum dechlorination rate was 0.3 μg-atoms of chlorine removed per g of sediment per week (5).

Degradation studies
Pseudomonas putida LB400 degraded 50% of Aroclor 1242 in soil containing the transformed contaminant at 525 ppm in 15 wk (6).
Using BOD dilution water, settled domestic waste water inoculum and 5 and 10 ppm Arochlor 1242, after 28 days 66% of 5 ppm was degraded, and 0% of 10 ppm was degraded (7).

Abiotic removal
90.9% removed by biological adsorption in activated sludge in aeration time of 24 hr (8).
Processes such as hydrolysis and oxidation do not significantly degrade Aroclor 1242 in the aquatic environment (9,10).
Evaporation rate is not rapid, but total loss from soil over time may be important because of the stability and persistence of Aroclor 1242 (11).

Absorption
Soil R_f values of 0.02-0.04, which are indicative of soil immobility, were measured for Aroclor 1252 in Ottawa sand, Catlin loam, Ara silty clay loam and Catlin silt loam using water and a landfill leachate (12).
Experimental soil adsorption coefficients of 2240- 150,000 reported (11).

Mammalian and avian toxicity

Acute data
LD_{50} oral rat 794-1269 mg kg^{-1}.
LD_{50} dermal rabbit 8650 mg kg^{-1} (13).

Carcinogenicity and long-term effects
Limited evidence for carcinogenicity to humans, sufficient evidence for carcinogencity to animals, IARC classification group 2A (14).
A NIOSH retrospective cohort study was conducted of 3588 men and women employed for ≥1 day between 1st January 1957 and 31st March 1977 manufacturing capacitors with known exposure to PCBs in the midwest US. Aroclor 1242 was used

---

as a dielectric fluid until 1970. Results provided some evidence for an association between occupational PCB exposure and mortality from malignant melanoma. There was an increased incidence of brain cancer among workers who had more than twice the estimated cumulative PCB dose than the comparison group (15).

Teratogenicity and reproductive effects
TD oral mink ≥5 ppm diet complete reproductive failure (16).
TD oral ferrets ≥20 ppm diet complete reproductive failure (16).

Metabolism and pharmacokinetics
PCBs (1016 and 1242) fed to rats over prolonged periods were metabolised into mono and dihydroxy derivatives of di-, tri-, and tetrachlorobiphenyl (17).
Rats fed 3.89-6.6 mg kg^{-1} daily showed traces of PCB throughout body, but mainly in adipose tissue.
Steady state reached in adipose tissue after 4 months.
Little of PCB metabolites seen in urine.
After administration ceased, elimination of PCB is slow, measurable traces found 5-6 months later.
Pathological changes found in liver cells (18).

Any other adverse effects to man
Workers exposed to 0.32-2.22 mg m^{-3} (air) complained of a burning sensation on face and hands, nausea and a persistant body odour.
Chloracne and an eczematous rash on arms and legs was observed (19).
Fasting blood samples from 173 capacitor workers occupationally exposed to Aroclor 1242 detected 12-392 ppb which were dependent on serum concentration of lipids, but not albumin (20).

Legislation
Limited under EEC Directive on Drinking Water Quality 80/778/EEC. Pesticides and related products: maximum admissible concentration 0.5 µg l^{-1} (21).
Included in Schedule 5 (Release Into Water: Prescribed Substances) of Statutory Instrument 1991, No. 472 (22).

Any other comments
Human health effects of PCB ingestion reviewed (14,23,24).
Human health effects, epidemiology, workplace experience and experimental toxicology reviewed (25).

References
1. Tooby, T. E; et al. *Chem. Ind. (London)* 1975, 523-526
2. *Polychlorinated Biphenyls Hazards to Fish, Wildlife and Invertebrates: A Synoptic Review Biol. Report No. (85)1.7* 1986, US Dept. of Interior/Fish and Wildlife Service Contaminant Reviews
3. Langston, W. J. *Mar. Biol. (Berlin)* 1978, **45**(3), 265-272, (Ger.) (*Chem. Abstr.* **88**, 165045y)
4. *Fed. Regist.* 1977, **42**, 6532
5. Quensen, J. F., III; et al. *Appl. Environ. Microbiol.* 1990, **56**(8), 2360-2369
6. Unterman, R; et al. *USEPA Report* 1987, EPA 600/9-87/018F, 259-264
7. Tabak, H. H; et al. *J. Water Pollut. Contr. Fed.* 1981, **53**, 1503-1518
8. Aly, A. O; et al. *Egypt. J. Microbiol.* 1985, **20**(2), 277-283 (*Chem. Abstr.* **105**, 177810b)

9. Mabey, W. R; et al. *Aquatic Fate Process Data for Organic Priority Pollutants* 1981, 115-128, EPA 440/4-82-014
10. Callahan, M. A; et al. *Water-Related Environmental Fate of 129 Priority Pollutants* 1979, EPA 440/4-79-029a
11. Sklarew, D. S; et al. *Rev. Environ. Contam. Toxicol.* 1987, **98**, 1-41
12. Leifer, A; et al. *Environmental Transport and Transformation of Polychlorinated Biphenyls* 1983, 4-11, EPA 560/5-83-025, NTIS PB84-142579
13. *USEPA Ambient Water Quality Criteria: Polychlorinated Biphenyls* 1980, EPA 440/5/80/068
14. *IARC Monograph* 1987, **Suppl. 7**, 322
15. *Gov. Rep. Announce. Index US* 1991, **91**(20), 156,257, Health Hazard Evaluation Report HETA 89-116-2094, Westinghouse Electric Corp., Bloomington, IN
16. Bleavins, M. R; et al. *Arch. Environ. Contam. Toxicol.* 1980, **9**(5), 627-635
17. Burse, V. W; et al. *Bull. Environ. Contam. Toxicol.* 1976, **15**(1), 122-128
18. Burse, V. W; et al. *Arch. Environ. Health* 1974, **29**(6), 301-307
19. Ouw, H. K; et al. *Arch. Environ. Health* 1976, **31**(4), 189-194
20. Brown, J. F., jun; et al. *Bull. Environ. Contam. Tox.* 1984, **33**(3), 277-280
21. *EC Directive Relating to the Quality of Water Intended for Human Consumption* 1982, 80/778/EEC, Office for Official Publications of the European Communities, 2 rue Mercier, L-2985 Luxembourg
22. *S. I. 1991, No. 472 The Environmental Protection (Prescribed Processes and Substances) Regulations* 1991, HMSO, London
23. Kimbrough, R. D., *Ann. Rev. Pharmacol. Toxicol.* 1987, **27**, 87-111
24. MAFF *Polychlorinated Biphenyl (PCB) Residues in Food and Human Tissues* Food Surveillance Paper No. 13, 1983, HMSO, London
25. *ECETOC Technical Report No. 30(4)* 1991, European Chemical Industry Ecology and Toxicology Centre, B-1160 Brussels

A256   Aroclor 1248

CAS Registry No. 12672-29-6
Synonyms Arochlor 1248; chlorodiphenyl (48% Cl)
Uses Electrical insulator, heat transfer medium, hydraulic fluids and lubricants.

Occupational exposure

UN No. 2315; **HAZCHEM Code** 4X; **Conveyance classification** other dangerous substance; **Supply classification** harmful.

Risk phrases
Danger of cumulative effects (R33)

Safety phrases
This material and its container must be disposed of in a safe way (S35)

Ecotoxicity

Fish toxicity
LC_{50} (30 days, calculated) flow-through bioassay fathead minnow larvae (<8 hr old) 4.7 µg l^{-1} (1).
LC_{50} (96 hr) cutthroat trout, rainbow trout, channel catfish, bluegill sunfish, yellow perch ≤5750 µg l^{-1} (2).

Invertebrate toxicity
EC_{50} (2 wk) *Daphnia magna* 2.6 µg l^{-1}.
LC_{50} (48 hr) pink shrimp 32 µg l^{-1} (3).

Bioaccumulation
Bioconcentration factors channel catfish 56,400, bluegill sunfish 52,000, scud 108,000, fathead minnow 120,000 (3).

Environmental fate

Anaerobic effects
Dechlorination was achieved by anaerobic microorganisms in lake sediments with the loss of *meta* plus *para* chlorines ranging from 15-85%. Maximum dechlorination rate was 0.3 µg-atoms of chlorine removed per g of sediment per wk (4).

Degradation studies
Seven agricultural or forest soils degraded Aroclor 1248 at least 70% in 14 days and >90% in 112 days (5).

Absorption
Removal by biological adsorption in activated sludge, 85.4% at aeration time of 24 hr. (6).

Mammalian and avian toxicity

Acute data
LD_{50} oral rat 11 g kg^{-1} (7).
TD_{Lo} oral rhesus monkey 1500 mg kg^{-1}, single dose, produced anorexia and lethargy, increasing with time (8).
LD_{50} dermal rabbit >794 mg kg^{-1} (7).
Pathological changes included liver damage, gastric hypertrophy, hyperplasia and ulceration of stomach lining (8).

Sub-acute data
LC_{50} (5 day) oral bobwhite quail, Japanese quail 1175-4844 ppm in diet (9).

Carcinogenicity and long-term effects
Limited evidence for carcinogenicity to humans, sufficient evidence for carcinogenicity to animals, IARC classification group 2A (10).
A NIOSH retrospective cohort study was conducted of 3588 men and women employed for ≥1 day between 1st January 1957 and 31st March 1977 manufacturing capacitors with known exposure to PCBs in the midwest US. Results provided some evidence for an association between occupational PCB exposure and mortality from malignant melanoma. There was an increased incidence of brain cancer among workers who had more than twice the estimated cumulative PCB dose than the comparison group (11).

Teratogenicity and reproductive effects
TD_{Lo} oral rhesus monkey 2.5-5.0 ppm (diet) during gestation and lactation, caused death in some young and pathological changes in the others, such as bone marrow and lymphoid tissue changes. Young also showed behavioural and learning difficulties. The PCBs were passed *in utero* and through breast feeding (12).

Metabolism and pharmacokinetics
Pretreatment of rats and plant tissues with Aroclor 1248 significantly increased microsomal protein concentrations and enzyme activities linked to cytochrome P450 (13). After 4 wk intraperitoneal injection of 1 g of Aroclor 1248 into rats, adipose tissue contained 338 µg tetrachlorinated biphenyl residues g^{-1} and there was a shift to penta- and hexachlorobiphenyls (up to 63%) (14).

Over 90% of a single oral dose (1.5 or 3 g kg^{-1}) of Aroclor 1248 was absorbed from the gastrointestinal tract of monkeys, the major route via biliary excretion. By the 14th day after administration, 5.6% of the original dose had been eliminated in the urine and faeces (15).

Legislation
Limited under EEC Directive on Drinking Water Quality 80/778/EEC. Pesticides and related products: maximum admissible concentration 0.5 µg l^{-1} (16).

Included in Schedule 5 (Release Into Water: Prescribed Substances) of Statutory Instrument 1991, No. 472 (17).

Any other comments
Toxicity and waste disposal methods reviewed (18-20).

Long-term neurobehavioural effects in monkeys reviewed (21).

Incomplete combustion can lead to the production of polychlorinated dibenzofurans and 2,3,7,8- tetrachlorodibenzodioxin which are more toxic.

References
1. Defoe, D. L; et al. *J. Fish. Res. Board Can.* 1978, **35**(7), 997-1002
2. *Handbook of Acute Toxicity of Chemicals to Fish and Aquatic Invertebrates* 1980, US Dept. of Interior, Fish and Wildlife Service, US Govt. Print. Off., Washington, DC
3. *USEPA Ambient Water Quality Criteria: Polychlorinated Biphenyls* 1980, EPA 440/5-80-068
4. Quensen, J. F., III; et al. *Appl. Environ. Microbiol.* 1990, **56**(8), 2360-2369
5. Hankin, L; et al. *Soil Sci.* 1984, **48**(6), 1581-1586
6. Aly, A. O; et al. *Egypt. J. Microbiol.* 1985, **20**(2), 277-283 (*Chem. Abstr.* **105**, 177810b)
7. Fishbein, L. *Ann. Rev. Pharmacol.* 1974, **14**, 139-156
8. Allen, J. R; et al. *Arch. Environ. Contam. Toxicol.* 1974, **2**(1), 86-95
9. *Lethal Dietary Toxicities of Environmental Pollutants to Birds. Special Report – Wildlife No. 191* 1975, 10, US Dept. of the Interior, Fish and Wildlife Service, Bureau of Sports Fisheries and Wildlife, US Govt. Print. Off., Washington, DC
10. *IARC Monograph* 1987, **Suppl 7**, 322
11. *Gov. Rep. Announce. Index US* 1991, **91**(20), 156,257, Health Hazard Evaluation Report HETA 89-116-2094, Westinghouse Electric Corp., Bloomington, IN
12. Garstens, L. A; et al. *Anim. Monit. Environ. Pollut. [Symp. Pathobiol. Environ. Pollut.: Anim. Models Wildl. Monit.] 1977* 1979, 233-240
13. Borlakoglu, J. T; et al. *Comp. Biochem. Physiol., C: Comp. Pharmacol. Toxicol.* 1989, **94C**(2), 613-617
14. Benthe, H. F; et al. *Arch. Toxicol.* 1973, **30**(3), 207-214
15. *IARC Monograph* 1978, **18**, 73
16. *EC Directive Relating to the Quality of Water Intended for Human Consumption* 1982, 80/778/EEC, Office for Official Publications of the European Communities, 2 rue Mercier, L-2985 Luxembourg
17. *S. I. 1991, No. 472 The Environmental Protection (Prescribed Processes and Substances) Regulations* 1991, HMSO, London

18. Kimbrough, R. D. *Ann. Rev. Pharmacol. Toxicol.* 1987, **27**, 87-111
19. MAFF *Polychlorinated biphenyl (PCB) residues in food and human tissues* Food Surveillance Paper No.13, 1983 HMSO, London
20. *Polychlorinated Biphenyls* 1988, **107**, International Registry of Potentially Toxic Chemicals, United Nations Environmental Programme, Geneva
21. Schantz, S. L; et al. *Environ. Toxicol. Chem.* 1991, **10**(6), 747-756

A257   Aroclor 1254

CAS Registry No. 11097-69-1
Synonyms polychlorinated biphenyl; Clophen; Fenclor; Kanechlor; Phyralene; Santotherm; Arochlor 1254; PCB 1254
Mol. Formula $C_{12}H_5Cl_5$ **Mol. Wt.** 326.44
Uses Dielectric liquids, transformers, capacitors, vacuum pumps, and gas turbines.

Physical properties

B. Pt. 365-390°C; **Partition coefficient** log P_{ow} 6.9 (extrapolated) (1).

Occupational exposure

US TLV (TWA) 0.5 mg m^{-3}; **UN No.** 2315; **HAZCHEM Code** 4X; **Conveyance classification** other dangerous substance; **Supply classification** harmful.

Risk phrases

Danger of cumulative effects (R33)

Safety phrases

This material and its container must be disposed of in a safe way (S35)

Ecotoxicity

Fish toxicity

Sublethal effects on Atlantic cod 1-50 µg g^{-1} over 5.5 months.
Abnormalities in testes, gills, livers (2).
LC_{50} (96 hr) crayfish, cutthroat trout, channel catfish, bluegill sunfish, yellow perch ≤42 mg l^{-1} (3).

Invertebrate toxicity

LC_{50} (96 hr) *Ischnura venticalis*, grass shrimp 200, 6.1-7.8 µg l^{-1}, respectively (3,4).
EC_{50} (14-21 day) *Daphnia magna* 24-1.3 mg l^{-1} (4).

Bioaccumulation

Small amounts accumulate in food chain. PCBs are fat soluble and stored in lipids of animals, resisting metabolic change and concentrating in animals high in the food chain (2).
Bioconcentration factors American oyster 85,000, rotifer 340,000, crayfish 5100, scud 27,000, *Daphnia sp.* 3800 (4,5).
Bioconcentration factors *Sphaerium striatinum* 1000-25,118, fathead minnow 31,622-316,227 (6).

Effects to non-target species
Pheasants fed either a single dose of 50 mg or 17 weekly doses of 12.5 or 50 mg showed up to 82% absorbed from the gastrointestinal tract and up to 50 mg kg[-1] in their eggs (7).

Environmental fate

Nitrification inhibition
PCBs at >10 μg l[-1] inhibited nitrification, principally ammonium oxidation (8).

Anaerobic effects
Dechlorination was achieved by anaerobic microorganisms in lake sediments with the loss of *meta* plus *para* chlorines ranging from 15-85%. Maximum dechlorination rate was 0.2 μg-atoms of chlorine removed per g of sediment per wek (9).

Degradation studies
Using BOD dilution water, settled domestic wastewater inoculum and 5 and 10 ppm Arochlor 1254, no biodegradation occurred during 28 days (10).
Carbon dioxide evolution of Arochlor 1254 from 3 New Mexico soils over 240 days incubation in either unamended soil or soil amended with sewage sludge ranged from 1-11% with greatest carbon dioxide evolution occurring in the presence of sewage sludge (11).

Abiotic removal
Only destroyed by incineration at >1100°C with long residence time (2).
Calculated $t_{1/2}$ 10.3 hr based on evaporative loss for a water depth of 1 m at 25°C (12).
208 mg of Arochlor 1254 in wastewater influent was diluted to a BOD of 200 ppm so that concentration was 1 ppm. After 17 hr wk treatment with a labscale biomass, 54% was recovered in effluent plus biomass and 30-39% was lost by evaporation (13).
Transformation processes such as hydrolysis and oxidation do not significantly degrade Aroclor 1254 in aquatic environment (14,15).
Vapour loss of Aroclor 1254 from 3 soils was 40- 50% over 2-4 months (16).
Evaporation was the major process by which Arochlor 1254 was lost from three untreated calcareous soils from New Mexico during 240 days; treating the soils with sewage sludge decreased the rate of evaporation (17).

Absorption
No effect on waste treatment efficiency nor toxicity to biomass although biomass accumulated 6.2 g kg[-1] Aroclor 1254 (18).
Soil R_f values which are indicative of soil immobility are 0.02-0.04 in Ottawa sand, Catlin loam, Ava silty clay loam and Catlin silt loam using a water and landfill leachate (19).
Soil adsorption coefficients range from 110,000-1,330,000 (20).

Mammalian and avian toxicity

Acute data
LD$_{50}$ oral rat 1300 mg kg[-1].
LD$_{50}$ intravenous rat 358 mg kg[-1] (21).
LD$_{50}$ intraperitoneal mouse 2840 mg kg[-1] (22).

Sub-acute data
LC$_{50}$ oral bobwhite quail, Japanese quail 604-2898 ppm in 5-day diet (23).

Intragastric ♂ Fischer 344 rat (5, 10 or 15 wk) 0-25 mg kg^{-1}. After 5, 10 and 15 wk urinary alkaline phosphatase and lactate dehydrogenase activities were elevated and the kidney-to-body weight ratios were elevated at the 10 and 25 mg kg^{-1} dose after 10 and 15 wk exposure indicating nephrotoxicity (24).

Carcinogenicity and long-term effects
Limited evidence for carcinogenicity to humans, sufficient evidence for carcinogenicity to animals, IARC classification group 2A (25).
National Toxicology Program tested ♀ and ♂ rat with Arochlor 1254 via feed (concentration unspecified), equivocal results (26).
A NIOSH retrospective cohort study was conducted of 3588 men and women employed for ≥1 day between 1st January 1957 and 31st March 1977 manufacturing capacitors with known exposure to PCBs in the midwest US. Results provided some evidence for an association between occupational PCB exposure and mortality from malignant melanoma. There was an increased incidence incidence of brain cancer among workers who had more than twice the estimated cumulative PCB dose than the comparison group (27).

Metabolism and pharmacokinetics
Sherman rats administered 10 or 50 mg kg^{-1} day^{-1} Aroclor 1254 on days 7-15 of pregnancy showed average PCB concentrations in foetuses taken by Caesarean section on day 20 of pregnancy as 0.63 and 1.38 mg kg^{-1}, respectively, compared with <0.12 mg kg^{-1} in controls (7).
Evidence is given of 2 cases of Aroclor 1254 transmission from transformer maintenance workers to their wives whose household exposure occurred through contaminated clothes (28).
Fasting blood samples from 173 capacitor workers occupationally exposed to Aroclor 1254 revealed 4-129 ppb in serum, dependent on serum lipids but not albumin (29).
In a medical surveillance programme for persons potentially exposed to PCBs from an electrical transformer fire, about 820 litres of dielectric fluid containing 65% Aroclor 1254 leaked from a transformer. 450 subjects were monitored 6-12 months after the fire, 147 firemen and other persons in the building for ≥25 hr were questioned about symptoms and examined for physical abnormalities. Mean serum PCB concentrations were positively correlated with exposure extent and liver enzyme and lipid concentrations. About 50% of the subjects reported skin lesions (30).

Genotoxicity
Chinese hamster ovary cells with Aroclor 1254- induced S9 caused chromosomal abberations at exceptionally high levels eg. up to 40 per 100 cells against a normal range of 0-10 (31).
Mouse lymphoma L5178Y tk+/tk- with metabolic activation induced DNA strand breaks (32).
Intraperitoneal common carp, tench, grasscarp (48 hr) increased dose-response in micronucleus frequency and a species response dependency (33).

Any other adverse effects to man
Excessive industrial exposure to PCBs often leads to symptoms including chloracne, brown pigmentation of the skin and nails, temporary visual disturbances, swelling of eyelids, eye discharge and some gastrointestinal symptoms with liver abnormalities and jaundice (34).

Any other adverse effects

Oral rat (4 day) 300 mg kg^{-1} day^{-1} caused weight loss and increased urinary nitrogen excretion as urea (35).

Oral Japanese quail 0.5 g kg^{-1} increased renal and hepatic concentrations of porphyrins, the activities of δ-aminolevulinate synthetase and uroporphyrinogen I synthetase, and faecal concentrations of δ-aminolevulinate and porphyrins, the main porphyrins being hepta-, octa-, penta-, hexa- and tetracarboxyporphyrin (36).

Aroclor 1254 fed to boars revealed PCBs in urine and faeces (37).

Aroclor 1254 pretreatment induced P450IA1 in rabbit bone marrow microsomes 11-fold (38).

Intraperitoneal injection (dose unspecified) of Aroclor 1254 induced hepatic cytochrome P450 after 5 days in rats but not pigeons (39).

Legislation

US Toxic Substances Control Act contains provision of discontinuance of use and eventual disposal (40-42).

US Regulations on Storage and Disposal of PCBs specify incineration as only acceptable method of disposal (43-45).

Limited under EEC Directive on Drinking Water Quality 80/778/EEC. Pesticides and related products: maximum admissible concentration 0.5 μg l^{-1} (46).

Included in Schedule 5 (Release into Water: Prescribed Substances) of Statutory Instrument 1991, No. 472 (47).

Any other comments

Human health effects, epidemiology, workplace experience, waste disposal and toxicology reviewed (48,49).

Once widely used for high stability but now concern over long term deleterious environmental and health effects.

References

1. Rapaport, R. A; et al. *Environ. Sci. Technol.* 1975, **9**, 13
2. Freeman, H. C; et al. *Sci. Total Environ.* 1982, **24**(1)
3. *Handbook of Acute Toxicity of Chemicals to Fish and Aquatic Invertebrates. Resource Publication No. 137* 1980, 68, US Dept. of Interior, Fish and Wildlife Service, US Govt. Print. Off., Washington, DC
4. *Polychlorinated Biphenyls Hazards to Fish, Wildlife and Invertebrates: A Synoptic Review Biol. Report No. (85) 1.7* 1986, US Dept. of Interior/ Fish and Wildlife Service Contaminant Review
5. Kenaga, E. E; et al. *Aquatic Toxicology* 1980, 78-115
6. Rice, C. P; et al. *Environ. Toxicol. Chem.* 1987, **6**(4), 259-274
7. *IARC Monograph* 1974, **7**, 279
8. Sayler, G. S; et al. *Appl. Environ. Microbiol.* 1982, **43**(4), 949
9. Quensen, J. F., III; et al. *Appl. Environ. Microbiol.* 1990, **56**(f), 2360-2369
10. Tabak, H. H; et al. *J. Water Pollut. Contr. Fed.* 1981, **53**, 1503-1518
11. Fairbanks, B. C; et al. *J. Environ. Qual.* 1987, **16**, 18-24
12. Mackay, D; et al. *Environ. Sci. Technol.* 1975, **9**, 13
13. *USEPA Drinking Water Criteria Document for Polychlorinated Biphenyls* 1987, N.29
14. Mabey, W. R; et al. *Aquatic Fate Process Data for Organic Priority Pollutants* 1981, 115-128, EPA 440/4-81-014
15. Callahan, M. A; et al. *Water-related Environmental Fate of 129 Priority Pollutants* 1979, EPA 440/4-79-029a

16. Pal, D; et al. *Res. Rev.* 1980, **74**, 45-98
17. Fairbanks, B. C; et al. *J. Environ. Qual.* 1987, **16**, 18-24
18. Viteus, T; et al. *J. Water Pollut. Control Fed.* 1985, **57**(9), 1
19. Leifer, A; et al. *Environmental Transport and Transformation of Polychlorinated Biphenyls* 1983, 4-11, EPA 560/5-83-025, NTIS PB84-142579
20. Sklarew, D. S; et al. *Rev. Environ. Contam. Toxicol.* 1987, **98**, 1-41
21. Linder, R. E; et al. *Food Cosmet. Toxicol.* 1974, **12**, 63
22. *Bull. Environ. Contam. and Toxicol.* 1972, **8**, 245
23. *Lethal Dietary Toxicities of Environmental Pollutants to Birds. Special Scientific Report – Wildlife No. 191* 1975, 10, US Dept. of the Interior, Fish and Wildlife Service, Bureau of Sports, Fisheries and Wildlife, US Govt. Print. Off., Washington, DC
24. Andrews, J. E. *Toxicology* 1989, **57**(1), 83-86
25. *IARC Monograph* 1987, **Suppl. 7**, 322
26. *National Toxicology Program, Research and Testing Div.* 1992, Report No. TR-038, NIEHS, Research Triangle Park, NC 27709
27. *Gov. Rep. Announce. Index US* 1991, **91**(20), 156,257, Health Hazard Evaluation Report HETA 89-116-2094, Westinghouse Electric Corporation, Bloomington, IN
28. Fischbein, A; et al. *Br. J. Ind. Med.* 1987, **44**(4), 284-286
29. Brown, J. F; et al. *Bull. Environ. Contam. Tox.* 1984, **33**(3), 277-280
30. Fitzgerald, E. F; et al. *Arch. Environ. Health* 1986, **41**(6), 368-376
31. Kirkland, D. J; et al. *Mutat. Res.* 1989, **214**(1), 115-122
32. Garberg, P; et al. *Mutat. Res.* 1988, **203**(3), 155-176
33. Al-Sabti, K. *Cytobios* 1986, **47**(190-191), 147-154
34. Heuper, W. C. *Lawyers' Medical Cyclopedia, Medicolegal Considerations of Occupational and Non-occupational Environmental Cancers*
35. Ebner, K. V; et al. *Fundam. Appl. Toxicol.* 1987, **8**(1), 89-96
36. Miranda, C. L; et al. *Biochem. Pharmacol.* 1986, **35**(20), 3637-3639
37. *Foreign Compound Metabolism in Mammals* 1975, **3**, 433, The Chemical Society, London
38. Schnier, G. G; et al. *J. Pharmacol. Exp. Ther.* 1989, **251**(2), 790-796
39. Borlakoglu, J. T; et al. *FEBS Lett.* 1989, **247**(2), 327-329
40. *Public Law* 1976, 94-469
41. *Fed. Regist.* 1978, **43**(34), 7150
42. *Fed. Regist.* 1978, **43**(110), 24802
43. *Interdepartmental Task Force on PCBs: Polychlorinated Biphenyls and the Environment* 1972, Com 72-10419, Washington, DC
44. *Final Report; PCBs in the US, Industrial Use and Environmental Distribution* 1976, Report to USEPA Task 1
45. *Proceedings of the National Conference on Polychlorinated Biphenyls* 1975, 19-21, EPA 560/10075-004, Chicago, IL
46. *EC Directive Relating to the Quality of Water Intended for Human Consumption* 1982, 80/778/EEC, Office for Official Publications of the European Communities, 2 rue Mercier, L-2985 Luxembourg
47. *S. I. 1991, No. 472, The Environmental Protection (Prescribed Processes and Substances) Regulations* 1991, HMSO, London
48. *ECETOC Technical Report No. 30(4)* 1991, European Chemical Industry Ecology and Toxicology Centre, B-1160 Brussels
49. *Polychlorinated Biphenyls* 1988, **107**, International Register of Potentially Toxic Chemicals, United National Environment Programme, Geneva

A258   Aroclor 1260

CAS Registry No. 11096-82-5
Synonyms polychlorinated biphenyl; Clophen; Fenclor; Kanechlor; Phyralene;
Santotherm; PCB 1260
Uses Dielectric liquids, thermostatic fluids, swelling agents for transmission seals,
additives or base for lubricants, oils, greases, plasticiser for cellulosics, vinyls,
chlorinated rubbers.

Physical properties

B. Pt. 385-420°C; **Specific gravity** 1.41 at 15.5°C; **Partition coefficient** log P_{ow} 7.7
(extrapolated) (1); **Volatility** v.p. 50 mmHg at 225°C.

Occupational exposure

UN No. 2315; **HAZCHEM Code** 4X; **Conveyance classification** other dangerous
substance; **Supply classification** harmful.

Risk phrases
Danger of cumulative effects (R33)

Safety phrases
This material and its container must be disposed of in a safe way (S35)

Ecotoxicity

Fish toxicity
LC_{50} (96 hr) cutthroat trout, yellow perch ≤61 mg l^{-1}.
LC_{50} (20 day) rainbow trout 21 μg l^{-1}.
LC_{50} (30 day) bluegill sunfish 150 μg l^{-1} (1).

Bioaccumulation
Bioconcentration factor for ♀ fathead minnow 270,000 (2).
Small amounts accumulate in food chain. PCBs are fat soluble and stored in lipids of
animals, resisting metabolic change and concentrating in animals high in the food
chain (3).

Effects to non-target species
LD_{50} oral mallard duck >2000 mg kg^{-1} (4).

Environmental fate

Nitrification inhibition
PCB's at >10 μg l^{-1} inhibited nitrification, principally ammonium oxidation (5).

Anaerobic effects
Dechlorination was achieved by anaerobic microorganisms in lake sediments with the
loss of meta plus para chlorines ranging from 15-85%. Maximum dechlorination rates
for Hudson River and Silver Lake organisms for Aroclor 1260 were 0.04 and 0.21
μg-atoms of chlorine removed per g of sediment wk^{-1}, respectively (6).

Degradation studies
Only destroyed by incineration at >1100°C with long residence time (3).

Results of a static flask screening procedure utilising BOD dilution water, settled domestic wastewater inoculum and 5 and 10 ppm of Aroclor 1260 found 0% biodegraded at the end of 28 days' incubation (7).

Abiotic removal
Biological $t_{1/2}$ in *Rhabsodargus holubi* during depuration was estimated at 50 days (8).
Calculated $t_{1/2}$ time based on evaporative loss for a water depth of 1 m at 25°C: 10.2 hr (3).
Abiotic transformation processes, such as hydrolysis and oxidation, do not significantly degrade Aroclor 1260 in the aquatic environment (9).

Mammalian and avian toxicity

Acute data
LD_{50} oral rat 1320 mg kg^{-1} (10).
LD_{50} dermal rabbit 10 g kg^{-1} (11).

Sub-acute data
LD_{50} oral northern bobwhite 747 mg kg^{-1} 5 days on treated diet plus 3 days untreated (1).
Excessive industrial exposure to PCBs often leads to symptoms including chloracne, brown pigmentation of the skin and nails, temporary visual disturbances, swelling of eyelids, eye discharge and some gastrointestinal symptoms with liver abnormalities and jaundice (12).
Hepatic changes, neoplastic modules and liver adenofibrosis in young rats in short time after exposure (13).
♂ Wistar rats (120 day) fed 50 and 100 ppm significantly increased the activity of liver succinate dehydrogenase. Lactate dehydrogenase activity increased at 50 ppm level and decreased at 100 ppm. Alanine and aspartate aminotransferases and alkaline and acid aspartate aminotransferases and alkaline and acid phosphatases showed remarkable decrease in activity. In both Aroclor 1260-fed groups, liver showed centrilobular hypertrophy, hepatocellular damage, hyperplasia, karyolysis and karyorrhexis. The kidney showed glomerulonephritis, degenerative changes in the proximal and distal tubules and increased cellularity of glomeruli whilst the thyroid showed degeneration of follicles, fibrosis of follicles and lymphocytic infiltration followed by thyroiditis (14).

Carcinogenicity and long-term effects
Limited evidence for carcinogenicity to humans, sufficient evidence for carcinogenicity to animals, IARC classification group 2A (15).
A NIOSH retrospective cohort study was conducted of 3588 men and women employed for ≥1 day between 1st January 1957 and 31st March 1977 manufacturing capacitors with known exposure to PCBs in the midwest US. Results provided some evidence for an association betwwen occupational PCB exposure and mortality from malignant melanoma. There was an increased incidence of brain cancer among workers who had more than twice the estimated cumulative PCB dose than the comparison group (16).

Any other adverse effects to man
Fasting blood samples obtained from 173 capacitor workers occupationally exposed

to Aroclor 1260 revealed 4-129 ppb in serum, concentration being dependent on serum lipids but not of albumin (17).

Legislation

US Toxic Substances Control Act contains provision of discontinuance of use and eventual disposal (18-20).

US Regulations 1978 on Storage and Disposal of PCBs specify incineration as only acceptable method of disposal (21-23).

Limited under EEC Directive on Drinking Water Quality 80/778/EEC. Pesticides and related products: maximum admissible concentration 0.5 μg l^{-1} (24).

Included in Schedule 5 (Release into Water: Prescribed Substances) of Statutory Instrument 1991, No. 472 (25).

Any other comments

Analysis of ready to eat foods collected in markets of 20 US cities as conducted by the US FDA found Aroclor 1260 in one of 360 food composites (positive in meat, fish and poultry category) in 1978 (26).

Toxicity and waste disposal methods reported (27).

Once widely used for high stability but now concern over long term deleterious environmental and health effects.

References

1. *Polychlorinated Biphenyls Hazards to Fish, Wildlife and Invertebrates: A Synoptic Review Biol. Report No. (85)1.7* 1986, US Dept. of Interior/Fish and Wildlife Service Contaminant Reviews
2. De Kock, A. C; et al. *Chemosphere* 1988, **17**(12), 2381-2390
3. *USEPA Ambient Water Quality Criteria: Polychlorinated Biphenyls* 1980, B-22, EPA 440/5-80-068
4. Mackay, D; et al. *Environ. Sci. Technol.* 1975, **9**, 13
5. *Handbook of Toxicity of Pesticides to Wildlife* 1970, US Dept. of the Interior, Fish and Wildlife Service, US Govt. Print. Off., Washington, DC
6. Sayler, G. S; et al. *Appl. Environ. Microbiol.* 1982, **43**(4), 949
7. Quensen, J. F., III; et al. *Appl. Environ. Microbiol.* 1990, **56**(8), 2360-2369
8. Tabak, H. H; et al. *J. Water Pollut. Contr. Fed.* 1981, **53**, 1503-1518
9. Mabey, W. R; et al. *Aquatic Fate Process Data for Organic Priority Pollutants* 1987, 115-128, EPA 440/4-82-014
10. Linder, R. E; et al. *Food Cosmet. Toxicol.* 1974, **12**, 63
11. *USEPA Ambient Water Quality Criteria: Polychlorinated Biphenyls* 1980, EPA 440/5-80-068
12. Heuper, W. C. *Lawyers' Medical Cyclopedia; Medicolegal Considerations of Occupational and Non-Occupational Environmental Cancers*
13. Rao, C. V; et al. *Cancer Lett.* 1988, **39**(1), 59
14. Rao, C. V; et al. *Indian J. Exp. Biol.* 1990, **28**(2), 149-151, 152-154
15. *IARC Monograph* 1987, **Suppl 7**, 322
16. *Gov. Rep. Announce. Index US* 1991, **91**(20), 156,257, Health Hazard Evaluation Report HETA 89-116-2094, Westinghouse Electric Corp., Bloomington, IN
17. Brown, J. F; et al. *Bull. Environ. Contam. Tox.* 1984, **33**(3), 277-280
18. *Public Law* 1976, 94-469
19. *Fed. Regist.* 1978, **43**(34), 7150
20. *Fed. Regist.* 1978, **43**(110), 24802
21. *Interdepartmental Task Force on PCBs; Polychlorinated Biphenyls and the Environment*, 1972, Com 72-10419, Washington, DC

22. *Final Report: PCBs in the US, Industrial Use and Environmental Distribution* 1976, Report to USEPA Task 1
23. *Proceedings of the National Conference on Polychlorinated Biphenyls* 1975, 19-21, EPA 56010075-004
24. *EC Directive Relating to the Quality of Water Intended for Human Consumption* 1982, 80/778/EEC, Office for Official Publications of the European Communities, 2 rue Mercier, L-2985 Luxembourg
25. *S. I. 1991, No. 472, The Environmental Protection (Prescribed Processes and Substances) Regulations* 1991, HMSO, London
26. Podrebarac, D. S. *J. Assoc. Off. Anal. Chem.* 1984, **33**(3), 277-280
27. *Polychlorinated Biphenyls* 1988, **107**, International Register of Potentially Toxic Chemicals, United Nations Environmental Programme, Geneva

A259 Arsenic

As

CAS Registry No. 7440-38-2
Synonyms arsenic-75; arsenic black; arsenicals; colloidal arsenic; grey arsenic; metallic arsenic
Mol. Formula As   **Mol. Wt.** 74.92
Uses Metallurgy, hardening copper, lead alloys, manufacture of glass.
Occurrence *The toxicity of arsenic is dependent on the ability of the organism to absorb it. Therefore toxicity data refers to bioavailable forms such as the ion in solution or particulate matter.* Found widely in nature.
Produced by roasting then reducing sulfide ore.

Physical properties

M. Pt. 814°C at 36 atm; **B. Pt.** sublimes at 612°C; **Specific gravity** black crystals 5.724 at 14°C, black amorphous 4.7.

Occupational exposure

US TLV (TWA) 0.2 mg m^{-3}; **UK Long-term limit** MEL 0.1 mg m^{-3} (as As);
UN No. 1558; **HAZCHEM Code** 2X; **Conveyance classification** toxic substance;
Supply classification toxic.

Risk phrases

Toxic by inhalation and if swallowed (R23/25)

Safety phrases

Keep locked up and out of reach of children – When using do not eat, drink or smoke – After contact with skin, wash immediately with plenty of water – If you feel unwell, seek medical advice (show label where possible) (S1/2, S20/21, S28, S44)

Ecotoxicity

Fish toxicity

EC_{50} (96 hr) fathead minnow 141-144 mg l^{-1} (1).
LC_{50} (96 hr) knifefish 31 mg l^{-1} (2).
Oral (24 wk) rainbow trout 0.52 mg kg^{-1} day^{-1} caused chronic inflammatory changes

in subepithelial tissues of the gall bladder wall in 71% of group (3).
LC$_{50}$ (96 hr) striped bass 30 mg l^{-1} (4).

Invertebrate toxicity
EC$_{50}$ (96 hr) *Daphnia magna* 4.3 mg l^{-1} (with food), 1.5 mg l^{-1} (without food) (5).
LC$_{50}$ (48 hr) *Aplexa hypnorum* 24.5 mg l^{-1} (6).

Mammalian and avian toxicity

Acute data
LD$_{Lo}$ subcutaneous rabbit 300 mg kg^{-1} (7).
LD$_{Lo}$ intraperitoneal guinea pig 10 mg kg^{-1} (8).

Carcinogenicity and long-term effects
Sufficient evidence for carcinogenicity to humans, sufficient evidence for
carcinogenicity to animals, IARC classification group 1 (9,10).
Intratracheal ♂ Syrian golden hamster total dose 3.75 mg arsenic once a week for 15
wk. Results concerning arsenic compounds inconclusive (11).

Teratogenicity and reproductive effects
LD$_{50}$ (14 day) chick embryo 9 μg egg/-1. The gross malformations observed were
reduced body size, micromelia, twisted neck, haemorrhage, everted viscera and
microphthalmia (12).
LD$_{Lo}$ oral pregnant mouse 120 mg kg^{-1} reduced foetal weight and survival. Some
gross and skeletal malformations reported. Effects of oral administration were less
than those after intraperitoneal injection of 40 mg kg^{-1} (13).

Metabolism and pharmacokinetics
Trivalent arsenic was metabolised by rat liver, kidney, lung slices to
monomethylarsenic acid and dimethylarsenic acid, the liver had the greatest
methylating capacity. GSH regulated trivalent arsenic metabolism, in contrast to
trivalent arsenic, pentavalent arsenic was not extensively taken up by the hepatocyte
and was poorly methylated (14).
Inhalation animals (species and dose unspecified) increased tissue levels of arsenic
during the first week or months but levels decrease if exposure is prolonged. In rats
after single exposure the biological t$_{1/2}$ is substantial due to the accumulation of
arsenic in blood (15).
The ability of the rat to sequester arsenic is greater than that of humans (16).
Normal blood arsenic values for individuals are 1-5 μg l^{-1} whole blood. Cigarette
smokers showed mean blood arsenic levels approximately 50% higher than
nonsmokers (17).

Genotoxicity
In vivo intraperitoneal ♂ Swiss Albino mice neither chromatid nor chromosome
aberrations were observed in bone marrow cells and spermatogonia (18).

Legislation
Limits under EC Directive on Drinking Water Quality 80/778/EEC.
Arsenic: maximum admissibe concentration 50 μg l^{-1} (19).

Any other comments
The aquatic biota, environmental fate, toxicology, neoplastic transformations,

carcinogenicity, human health effects, occupational hazards, experimental toxicology, environmental effects have been extensively reviewed (20-40).

References

1. Lima, A. R; et al. *Arch. Environ. Contam. Toxicol.* 1984, **13**(5), 595
2. Ghosh, A. R; et al. *Environ. Ecol.* 1990, **8**(2), 576-579
3. Cockell, K. A; et al. *Arch. Environ. Contam. Toxicol.* 1991, **21**(4), 518-527
4. Palacoski D; et al. *Trans. Am. Fish Soc.* 1985, **114**, 748-753
5. Lundi, J. *Science Food Agiculture* 1970, **21**, 242
6. Holcombe, G. W; et al. *Ecotoxicol. Environ. Saf.* 1983, **8**(2), 106-117
7. *Archivio di Science Biologiche* 1938, **24**, 442, Cappelli Editore, Bologna
8. *Comptes Rendus des Seances de la Societe de Biologie et ses Filiales* 1918, **81**, 164, Masson et Cie, Paris
9. *IARC Monograph* 1980, **23**, 39
10. *IARC Monograph* 1987, **Suppl. 7**, 100
11. Yamamoto, A; et al. *Int. J. Cancer* 1987, **40**(2), 220-223
12. Gilani, S. H; et al. *J. Toxicol. Environ. Health* 1990, **30**(1), 23-31
13. Thacker, G. T; et al. *Teratology* 1977, **15**, 30A-31A
14. Georis, B; et al. *Toxicology* 1990, **63**(1), 73-84
15. Friberg, L; et al. *Handbook of the Toxicology of Metals* 2nd ed., 1986, **2**, 55, Elsevier, Amsterdam
16. Ducoff, H. S; et al. *Proc. Soc. Exp. Biol.* 1948, **69**, 548
17. *USEPA Health Assessment Document: Inorganic Arsenic* 1984, 421, EPA 600/8-83-021
18. Poma, K. *Experentia* 1981, **37**(2), 129
19. *EC Directive Relating to the Quality of Water Intended for Human Consumption* 1982, 80/778/EEC, Office for Official Publications of the European Communities, 2 rue Mercier, L-2985 Luxembourg
20. Wahrendorf, J. et al. *Energy Res. Abstr.* 1990, **15**(20), 45631
21. Phillips, D. J. M. *Aquat. Toxicol.* 1990, **16**(3), 151-186
22. Trevors, J. T. *Hydrobiologia* 1989, **188-189**, 143-147
23. Dybing, E; et al. *Rev. Biochem. Toxicol.* 1989, **10**, 139-186
24. Hoelzel, G; et al. *NDZ. Neue Deliva-Z.* 1989, **40**(3), 122-124, (Ger.) (*Chem. Abstr.* **111**, 12145z)
25. Barrett, J. C; et al. *Biol. Trace Elem. Res.* 1988, **21**, 421-429
26. Maeda, S. *Kagaku Kogye* 1989, **40**(7), 618-623 (Jap.) (*Chem. Abstr.* **111**, 128297e)
27. Marcus, W. L; et al. *Adv. Mod. Environ. Toxicol.* 1988, **15**, 133-158
28. Grajeta, H. *Rocz. Panstw. Zakl. Hig.* 1987, **38**(4-5), 356-362, (Pol.) (*Chem. Abstr.* **110**, 49687n)
29. Cavanagh, J. B. *NATO ASI Ser., Ser. A* 1988, **100**, 177-202
30. Irgolic, K. J. *Appl. Organomet. Chem.* 1988, **2**(4), 303-307
31. Eisler, R. *Gov. Rep. Announce Index (US)* 1988, **88**(11), Abstr. No. 828339
32. Harper, M. *Br. J. Ind. Med.* 1988, **45**(9), 602-605
33. Kaise, T. *Kanagawoi-ken Eisei Kenkyusho Kenkyu Hokoku* 1987, (17), 1-7, (Jap.) (*Chem. Abstr.* **108**, 199520w)
34. Piscator, M. *Life Sci. Res. Rep.* 1986, **33**, 59-70
35. O'Neil, I. K; et al. *IARC Sci. Publ.* 1987, **71**(Environ. Carcinog.: Sel. Methods Anal. Vol. 8), 3-13
36. Pershagen, G. *Environ. Carcinogen.: Sel. Methods. Anal.* 1987, **8** 45-61
37. Stohrer, G. *Arch. Toxicol.* 1991, **65**(7), 525-531
38. *Dangerous Prop. Ind. Mater. Rep.* 1989, **9**(4), 2-19
39. *ECETOC Technical Report No. 30(4)* 1991, European Chemical Industry Ecology and Toxicology Centre, B-1160 Brussels
40. Izmerov, N. F; *Scientific Reviews of Soviet Literature on Toxicity & Hazards of Chemicals* 1991, **20**, Eng. Trans. Richardson, M. L. (Ed.), UNEP/IRPTC, Geneva

A260   Arsenic acid

$$HO-\overset{\displaystyle \overset{O}{\|}}{\underset{\displaystyle |}{As}}-OH$$
$$OH$$

CAS Registry No. 7778-39-4
Synonyms orthoarsenic acid
Mol. Formula AsH_3O_4 **Mol. Wt.** 141.94
Uses In the preparation of arsenate salts. Manufacture of insecticides.

Physical properties

M. Pt. 35.5°C; **Specific gravity** 2.0-2.5.

Solubility
Organic solvent: ethanol, glycerol

Occupational exposure

US TLV (TWA) 0.2 mg m^{-3} (as As); **UK Long-term limit** MEL 0.1 mg m^{-3} (as As);
UN No. 1553 (liquid); 1554 (solid); **HAZCHEM Code** 2X;
Conveyance classification toxic substance; **Supply classification** toxic.

Risk phrases
Toxic by inhalation and if swallowed (R23/25)

Safety phrases
Keep locked up and out of reach of children – When using do not eat, drink or smoke
– After contact with skin, wash immediately with plenty of water – If you feel unwell,
seek medical advice (show label where possible) (S1/2, S20/21, S28, S44)

Mammalian and avian toxicity

Acute data
LD_{50} oral mouse, rat, guinea pig, rabbit 57-470 mg kg^{-1} (1).
LD_{Lo} oral rabbit, dog, chicken 5-125 mg kg^{-1} (2).
LD_{50} intratracheal mouse 105 mg kg^{-1} (1).
LD_{50} intravenous rabbit 6 mg kg^{-1} (2).

Carcinogenicity and long-term effects
Sufficient evidence for carcinogenicity to humans, limited evidence for carcinogenicity to
animals, IARC classification group 1 for arsenic and arsenic compounds (3).
Two major studies have been carried out on the mortality of workers involved in the
production of arsenical insecticides where arsenic acid was an important starting
material. In both cases arsenic acid was produced *in situ* from arsenic trioxide. In
study 1 arsenic acid could not be confirmed as a contributory factor in the cancer
mortality of workers (4-6).
In study 2, the mortality of workers who retired between 1960-1972 was investigated.
Insufficient information was available to evaluate the reliability of reported results,
but the authors stated of 17 cancer deaths in men, 10 were respiratory tract and 3
lymphosarcomas (4,7,8).

Metabolism and pharmacokinetics

Analysis of urine from treated hamsters showed the presence of inorganic arsenic and dimethylarsinic acid (9).

Oral ♂, ♀ hamster (single dose) 0.01 μg arsenic as arsenic acid. The $t_{1/2}$ for 1st component (98% of dose) was 0.29 days, and the $t_{1/2}$ for 2nd component (2% of dose) 3-8 days (10).

Mice given radio-labelled arsenic acid and x-rayed for distribution revealed a high proportion of arsenic concentrated in the intestinal mucosa and the kidneys (11).

Any other adverse effects

Human leucocytes incubated with 1 mg l^{-1} cytotoxicity test positive (9).
Human fibroblasts incubated with 100 ppb cytotoxicity test positive (10).

Legislation

Limited under EC Directive on Drinking Water Quality 80/778/EEC. Arsenic: maximum admissible concentration 50 mg l^{-1} (12).

Any other comments

Toxicity reviewed under United Nations Environmental Programme (13).

References

1. Iskandarov, T. *Med. Zh. Uzbekistana* 1965, (3), 22-24, (Russ.) *(Chem. Abstr.* **63** 8948e)
2. Joachimoglu, G. *Biochem. Z.* 1915, **70**, 144-157, (Ger.) *(Chem. Abstr.* **9**, 2944)
3. *IARC Monograph* 1987, **Suppl 7**, 100
4. *Health Effects of Occupational Lead and Arsenic Exposure* 1976, 296-298/313-343, Carnow, B. W. (Ed.), Washington DC
5. Ott, G. M; et al. *Arch. Environ. Health.* 1974, **29**, 250-255
6. Sobel, W; et al. *Am. J. Ind. Med.* 1988, **13**, 263-270
7. Baetjer, A. M; et al. *Abstracts of the 18th International Congress on Occupational Health* 1975, Brighton, UK., 14-19 Sept
8. Mabuchi, K; et al. *Prev. Med.* 1980, **9**, 51-77
9. Nakamuro, K; et al. *Mutat. Res.* 1981, **88**, 73
10. Charbonneau, S. M; et al. *Toxicol. Lett* 1980, **5** (3-4), 175-182
11. Deak, S. T; et al. *J. Toxicol. Environ. Health* 1976, **1**, 981-984
12. *EC Directive Relating to the Quality of Water Intended for Human Consumption* 1982, 80/778/EEC, Office for Official Publications of the European Communities, 2 rue Mercier, L-2985 Luxembourg
13. *International Registry of Potentially Toxic Chemicals: Arsenic Compounds* 1991, **20**, United Nations Environmental Programme, Geneva

A261   Arsenic disulfide

As_2S_2

CAS Registry No. 12044-79-0
Synonyms arsenic sulfide; arsine, thioxo-; arsenic monosulfide; arsenic sulfide red; C.I. 77085
Mol. Formula As_2S_2 **Mol. Wt.** 213.97

Occurrence Arsenic ore. Realgar.

Physical properties
M. Pt. α 267°C, β 307°C; B. Pt. 565°C; Specific gravity α 3.506, β 3.254.

Occupational exposure
UK Long-term limit MEL 0.1 mg m^{-3} (as As); Supply classification toxic.

Risk phrases
Toxic by inhalation and if swallowed (R23/25)

Safety phrases
Keep locked up and out of reach of children – When using do not eat, drink or smoke – After contact with skin, wash immediately with plenty of water – If you feel unwell, seek medical advice (show label where possible) (S1/2, S20/21, S28, S44)

Mammalian and avian toxicity

Carcinogenicity and long-term effects
Sufficient evidence for carcinogenicity to humans, limited evidence ofor carcinogenicity to animals, IARC classification group 1 for arsenic and arsenic compounds. This evaluation applies to the group of chemicals as a whole and not necessarily to all individual chemicals within the group (1).

Legislation
Limited under EC Directive on Drinking Water Quality 80/778/EEC. Arsenic: maximum admissible concentration 50 mg l^{-1} (2).

References
1. *IARC Monograph* 1987, **Suppl. 7**, 100
2. *EC Directive Relating to the Quality of Water Intended for Human Consumption* 1982, 80/778/EEC, Office for Official Publications of the European Communities, 2 rue Mercier, L-2985 Luxembourg

A262   Arsenic trichloride

AsCl₃

CAS Registry No. 7784-34-1
Synonyms arsenic chloride; arsenic(III) chloride; arsenous chloride; arsenious chloride; arsenious trichloride; trichloroarsine
Mol. Formula AsCl$_3$ Mol. Wt. 181.28
Uses In specialised ceramic manufacture.

Physical properties
M. Pt. –16°C; B. Pt. 130°C; Specific gravity d^{20} 2.163; Volatility v.p. 10 mmHg at 23.5°C; v. den. 6.25..

Solubility
Organic solvent: diethyl ether, chloroform, carbon tetrachloride

Occupational exposure
US TLV (TWA) 0.2 mg m^{-3} (as As); **UK Long-term limit** MEL 0.1 mg m^{-3} (as As); **UN No.** 1560; **HAZCHEM Code** 2X; **Conveyance classification** toxic substance; **Supply classification** toxic.

Risk phrases
Toxic by inhalation and if swallowed (R23/25)

Safety phrases
Keep locked up and out of reach of children – When using do not eat, drink or smoke – After contact with skin, wash immediately with plenty of water – If you feel unwell, seek medical advice (show label where possible) (S1/2, S20/21, S28, S44)

Mammalian and avian toxicity

Acute data
LC_{Lo} (10 min) inhalation mouse 338 ppm (1).
LD_{Lo} intravenous dog 120 mg kg^{-1} (2).

Carcinogenicity and long-term effects
Sufficient evidence for carcinogenicity in humans, limited evidence for carcinogenicity in animals, IARC classification Group 1 for arsenic and arsenic compounds. This evaluation applies to the group of chemiclas as a whole and not necessarily to all individual chemicals within the group (3).

Genotoxicity
Bacillus subtilis rec+rec- positive (4).
Escherichia coli DNA damage repair test positive (5).

Legislation
Limited under EC Directive on Drinking Water Quality 80/778/EEC. Arsenic: maximum admissible concentration 50 mg l^{-1}. Chloride: guide level 25 mg l^{-1} (6).

Any other comments
Toxicity reviewed under the United Nations Environmental Programme (7). Explodes on contact with sodium, potassium and aluminium.

References
1. *Handbook Toxicol.* 1956, **1**, 324
2. Tinao, M. M; et al. *Farmcoterap. Actual.* 1948, **5**, 548-552, (*Chem. Abstr.* **43**, 5119i)
3. *IARC Monograph* 1987, **Suppl. 7**, 57
4. Kanematsu, N; et al. *Mutat. Res* 1980, **77**(2), 109-116
5. Yagi, T; et al. *Doshisha Daigaku Rikogaku Kenkyu Hokoku* 1977, **18**(2), 63-70, (Jap.) (*Chem. Abstr.* **89**, 54935q)
6. *EC Directive Relating to the Quality of Water Intended for Human Consumption* 1982, 80/778/EEC, Office for Official Publications of the EC, 2 rue Mercier, L-2985 Luxembourg
7. *International Register of Potentially Toxic Substances: Arsenic Compounds* 1991, **20**, United Nations Environmental Programme, Geneva

A263   Arsenic trioxide

As_2O_3

CAS Registry No. 1327-53-3

Synonyms arsenic oxide; arsenic(III) oxide; arsenic sesquioxide; arsenious acid; arsenious oxide; arsenious trioxide; arsenous acid; arsenous acid anhydride; arsenous anhydride; arsenous oxide; arsenous oxide anhydride; crude arsenic; diarsenic trioxide; white arsenic

Mol. Formula As_2O_3 **Mol. Wt.** 197.84

Uses In the manufacture of glass, Paris green and enamels. Used in weed killers. For preserving hides and in sheep dips. Used in rodenticides and insecticides.

Occurrence Occurs in nature as the mineral claudetite (As_2O_3).

Physical properties

M. Pt. claudetite (monoclinic crystal structure) 313°C, arsenolite (cubic crystal structure) 275°C; **B. Pt.** 465°C; **Specific gravity** claudetite 3.865, arsenolite 4.15 at 25°C.

Occupational exposure

UK Long-term limit MEL 0.1 mg m^{-3} (as As); **UN No.** 1561; **HAZCHEM Code** 2Z; **Conveyance classification** toxic substance; **Supply classification** very toxic.

Risk phrases

May cause cancer – Very toxic if swallowed – Causes burns (R45, R28, R34)

Safety phrases

Avoid exposure-obtain special instructions before use – In case of accident or if you feel unwell, seek medical advice immediately (show label where possible) (S53, S45)

Ecotoxicity

Fish toxicity

Rainbow trout (8 wk) 1-137 µg arsenic g^{-1} in diet, no observed effects, at high exposures 137-1477 µg arsenic g^{-1} in diet reduced growth, reduced feed behaviour have been reported (1).

LC_{50} (48 hr) *Channa punctatus* 14.7 mg (2).

Mammalian and avian toxicity

Acute data

LD_{50} oral rat 20 mg kg^{-1} (3).

LD_{Lo} intravenous rabbit 4 mg kg^{-1} arsenic as arsenic trioxide (4).

LD_{Lo} subcutaneous guinea pig 6 mg kg^{-1} (5).

Single subcutaneous, unspecified dose ♂ guinea pig showed 18.5 µg arsenic g^{-1} in liver. An increase in the content of pyruvate, citrate and malate, and a decrease in hydroxybutrate was found (6).

Sub-acute data

Subcutaneous ♂ guinea pig twice per day for 5 days (dose unspecified), had 3.3 µg arsenic g^{-1} in liver. Pyruvate concentration was decreased, no correlation was found between metabolic changes and arsenic burden of liver after single and repeated treatment of arsenic oxide (6).

Ulceration and perforation of the nasal septum is caused by airborne arsenic trioxide if proper precautions are not observed. However these injuries have not been associated with malignancy (7).

Highly toxic. Effects of excess exposure include dermatitis, acute or chronic poisoning (8).

Carcinogenicity and long-term effects

Intratracheal installation Syrian golden hamster (15 wk) 3.75 mg total dose, tumour incidence 3.3% although arsenic trioxide had no apparent carcinogenicity or tumorigenicity (9).

Sufficient evidence for carcinogenicity to humans, limited evidence for carcinogenicity to animals, IARC classification group 1 for arsenic and arsenic compounds (10).

Teratogenicity and reproductive effects

A woman who swallowed 30 ml of rat poison containing 1.32% elemental arsenic as the trioxide in the 30th wk of pregnancy gave birth 4 days later to a live baby weighing 1100 g, who died 11 hrs later (11).

Smelter workers exposed to arsenic compounds during pregnancy had an excess of low birth weight babies, increased frequency of spontaneous abortion and increased occurrence of multiple malformations (11).

Metabolism and pharmacokinetics

Oral rat (52 days) 215 ppm arsenic as arsenic trioxide, highest levels found in kidneys, liver and relatively lower levels in hair, brain, bone, muscle and skin (12).

Oral monkey (duration unspecified) 1 mg kg^{-1} arsenic as arsenic trioxidce resulted in approximately 80% absorption from gut, 75% of dose excreted within 14 days (11).

In vivo hamster (single dose) methylated into methylarsenic acid and dimethylarsenic acid. Inorganic arsenic accounted for the major portion of total arsenic deposited in organs and tissues. The dose was followed by excretion of an amount of arsenic equivalent to approximately 60% of the administered dose, 49% in urine, 11% in faeces (13).

Genotoxicity

The frequency of sister chromatid exchanges in human peripheral lymphocytes exposed to 2 $\mu g\ ml^{-1}$ was above that of controls (14).

Intraperitoneal ♂ mouse (12-48 hr) 0- 12 mg arsenic kg^{-1}, neither chromatid nor chromosome aberrations were observed in spermatogonia and bone marrow cells (15).

Any other adverse effects to man

In humans, symptoms of acute poisoning can occur from 30 min to several hours after ingestion, and include dryness and irritation of the mouth, difficulty swallowing, vomiting, abdominal pain, diarrhoea, dehydration, fall in blood pressure, cyanosis and collapse (16).

Any other adverse effects

The effect of trivalent arsenic oxide and pentavalent arsenic oxide on gluconeogenesis from various substrates in the liver and kidneys of rats was investigated. Decreased acetyl CoA, 3-hydroxybutyrate, and reduced glutathione was found in suspensions of isolated rat kidney tubes or hepatocytes incubated with trivalent arsenic oxide. About 10-times higher concentrations of pentavalent arsenic oxide were needed to induce a similar extent of inhibition of gluconeogenesis (17).

Legislation

Limited under EC Directive on Drinking Water Quality 80/778/EEC. Arsenic: maximum admissible concentration 50 mg l^{-1} (18).

Any other comments

Chemical and physical properties, uses, risks, mutagenicity and carcinogenicity, human health effects, epidemiology and experimental toxicology reviewed (19-21).

References

1. Cockell, K. A; et al. *Aquat. Toxicol.* 1988, **12**(1), 73-82
2. Burton, G. A; et al. *Bull. Environ. Contam. Toxicol.* 1987, **38**, 491
3. *AMA Arch. Ind. Health* 1958, **17**, 118
4. *Biochem. Zeitschrift* 1915, **70**, 144
5. *Heffter's Handbuch der Experimentelle Pharmakologie* 1927, **3**, 479
6. Reidl, F. X; et al. *Trace Elem. Anal. Chem. Med. Biol. Proe. Int. Workshop 5th* 1988, 581-586, Braetter, P., (Ed.) Berlin
7. Drinker, P; et al. *Industrial Dust* 1954, 2nd ed., McGraw Hill, New York, 1954
8. *Office of Toxic Substances: Summary Characteristics of Selected Chemicals of Near Terms Interest* 1976, 1-2, USEPA, Washington, DC
9. Ohyama, S; et al. *Appl. Organomet. Chem.* 1988, **2**(4), 333-337
10. *IARC Monograph* 1987, **Suppl. 7**, 57
11. *IARC Monograph* 1980, **23**, 39,91
12. *IARC Monograph* 1973, **2**, 59
13. Yamauchi, Y. Y. *Toxicol.* 1985, **34**(2), 113-121
14. Bassendowska-Karska, E; et al. *Bromotol. Chem. Toksykol.* 1986, **19**(4), 255-257, (Pol.) (*Chem. Abstr.* **107**, 2426v)
15. Poma, K; et al. *Experientia* 1981, **37**(2), 129-130
16. *Encyclopaedia of Occupational Health and Safety* 1971, International Labor Organisation, Geneva
17. Szinicz, L; et al. *Arch. Toxicol.* 1988, **61**(6), 444-449
18. *EC Directive Relating to the Quality of Water Intended for Human Consumption* 1982, 80/778/EEC, Office for Official Publications of the European Communities, 2 rue Mercier, L-2985 Luxembourg
19. *ECETOC Technical Report No. 30(4)* 1991, European Chemical Industry Ecology and Toxicology Centre, B-1160 Brussels
20. *Cah. Notes Doc.* 1989, **136**, 543-548, (Fr.) (*Chem. Abstr.* **112**, 61848k)
21. *International Registry of Potentially Toxic Chemicals* 1991, **20**, United Nations Environmental Programme, Geneva

A264   Arsenic trisulfide

As_2S_3

CAS Registry No. 1303-33-9

Synonyms arsenic sesquisulfide; arsenic sulfide; arsenious sulfide; arsenous sulfide; diarsenic trisulfide; Kings gold

Mol. Formula As_2S_3 **Mol. Wt.** 246.04

Uses Manufacture of specialist glasses, pyrotechnics, electronics.

Occurrence Ore: orpiment.

Physical properties

M. Pt. 300°C; **B. Pt.** 707°C; **Specific gravity** 3.43.

Solubility
Organic solvent: ethanol

Occupational exposure

UK Long-term limit MEL 0.1 mg m^{-3} (as As); **Supply classification** toxic.

Risk phrases
Toxic by inhalation and if swallowed (R23/25)

Safety phrases
Keep locked up and out of reach of children – When using do not eat, drink or smoke – After contact with skin, wash immediately with plenty of water – If you feel unwell, seek medical advice (show label where possible) (S1/2, S20/21, S28, S44)

Ecotoxicity

Fish toxicity
LC_{50} (96 hr) fathead minnow 135 mg l^{-1} (1).

Invertebrate toxicity
LC_{50} (96 hr) white shrimp 500 mg l^{-1} (2).

Mammalian and avian toxicity

Acute data
LD_{50} oral mouse, rat 185-255 mg kg^{-1}.
LD_{50} dermal rat 936 mg kg^{-1}.
LD_{50} intraperitoneal mouse, rat 86-215 mg kg^{-1} (3).
A single application of 0.05 LD_{50} to eye of rats caused severe keratoconjunctivitis leading to blindness (3).

Sub-acute data
Oral doses totalling 570 mg kg^{-1} in rats caused hypoglycaemia, blood abnormalities and pathological changes in the kidneys, liver and spleen in the first 2 wk of the study (3).

Carcinogenicity and long-term effects
Sufficient evidence for carcinogenicity in humans, limited evidence for carcinogenicity in animals, IARC classification group 1 for arsenic and arsenic compounds (4).
Intratracheal ♂ Syrian golden hamsters 3 mg kg^{-1} (as As) of As_2S_3 induced lung adenoma in 1 of the 28 animals tested (5).
Intratracheal installation (15 wk) ♂ Syrian golden hamster weekly administration of arsenic sulfide equivalent to a total dose of 3.75 mg arsenic induced 1 lung adenoma in 22 hamsters in the arsenic sulfide group compared with 1 lung adenosquamous carcinoma in 21 hamsters of the control group. The authors conclude the results were inconclusive (6).

Metabolism and pharmacokinetics
Intratracheal (species unspecified) retention of arsenic trisulfide in lung 10 times higher compared to arsenic trioxide (7).

Legislation

Limited under EEC Directive on Drinking Water Quality 80/778/EEC. Arsenic: maximum admissible concentration 50 mg l^{-1} (8).

Any other comments

Human health effects and experimental toxicology reviewed (9,10).

References

1. Curtis, M. W; et al. *J. Hydrol.* 1981, **51** 359-367
2. Curtis, M. W; et al. *Water Res.* 1979, **13**(2), 137-142
3. Davydova, V. I; et al. *Aktual. Vopr. Gig. Tr. Profpatol. Prom-sti. Sel'sk. Khoz* 1984, 43-49, (Russ.) (*Chem. Abstr.* **103**, 191133y)
4. *IARC Monograph* 1987, **Suppl. 7**, 57
5. Pershagen, G; et al. *Cancer Lett.* 1985, **27**(1), 99-104 (*Chem. Abstr.* **103** 49545r)
6. Yamamoto, A; et al. *Int. J. Cancer* 1987, **40**(2), 220-223
7. *Natl. Res. Council Drinking Water and Health* 1986, **6**, 279, National Academy Press, Washington, DC
8. *EC Directive Relating to the Quality of Water Intended for Human Consumption* 1982, 80/778/EEC, Office for Official Publications of the European Communities, 2 rue Mercier, L-2985 Luxembourg
9. *ECETOC Technical Report No. 30(4)* 1991, European Chemical Industry Ecology and Toxicology Centre, B-1160 Brussels
10. *International Register of Potentially Toxic Chemicals: Arsenic Compounds* 1991, **20**, United Nations Environment Programme, Geneva

A265   Arsine

AsH₃

CAS Registry No. 7784-42-1

Synonyms arsenic hydride; arsenic trihydride; arseniuretted hydrogen; arsenous hydride; hydrogen arsenide

Mol. Formula AsH₃   **Mol. Wt.** 77.95

Uses Doping agent for semiconductors. Organic synthesis. Military poison gas. Transistor manufacture.

Physical properties

M. Pt. –116.3°C (decomp. at 300°C); **B. Pt.** –62.5°C; **Specific gravity** 1.640 at –64.3°C (liquid); **Volatility** v. den. 2.7.

Solubility

Water: 20 ml 100 g^{-1}. Organic solvent: chloroform, benzene

Occupational exposure

US TLV (TWA) 0.05 ppm (0.16 mg m^{-3}); **UK Long-term limit** 0.05 ppm (0.2 mg m^{-3}); **UN No.** 2188; **Conveyance classification** toxic gas; flammable gas; **Supply classification** toxic.

Risk phrases
Toxic by inhalation and if swallowed (R23/25)

Safety phrases
Keep locked up and out of reach of children – When using do not eat, drink or smoke – After contact with skin, wash immediately with plenty of water – If you feel unwell, seek medical advice (show label where possible) (S1/2, S20/21, S28, S44)

Ecotoxicity

Fish toxicity
LC_{50} (96 hr) bluegill sunfish 1.1 µg l^{-1} (1).

Abiotic removal
Rapid hydrolysis in water to arsenic acids and hydrides (2).

Mammalian and avian toxicity

Acute data
LC_{Lo} (15 min) inhalation rat 300 mg m^{-3} (3).
LC_{Lo} (15 min) inhalation rabbit 500 mg m^{-3} (3).
No effects on haematopoietic system observed following single exposure to 0.5 ppm (10 times threshold limit value set by American Conference of Governmental Industrial Hygienists), repeated exposure to 0.025 ppm caused significant anaemia in rats (4).

Sub-acute data
Inhalation rat, mouse (28 day) 10 ppm produced 100% mortality within 4 days. Regenerative anaemia developed after prolonged exposure. Arsine-induced disturbances in haematopoietic system produce marked, specific changes in excretion of urinary proporphyrins (5).
Inhalation B6C3F₁ mouse, Fischer 344 rat (90 days exposure) 0.5- 5.0 ppm. No changes in body weight gain observed in mouse, decrease in body weight gain in ♂ rat exposed to 5.0 ppm for 28 days. Significant exposure-related increases in relative spleen weights occurred in both sexes of mouse and rat in all exposure groups. Decreased packed cell volumes, haematological profiles (in rats) and increased δ-aminolevulinic acid dehydratase activity in all species. Arsenic content measured in rat liver after 90 days increased in concentration. Histopathological changes include increased haemosiderosis, extramedullary haematopoiesis in spleen and intracanalicular bile stasis in liver (mice only). Additional bone marrow hyperplasia in rats observed (4).
Inhalation B6C3F₁ mouse, Fischer 344 rat (14 days subacute exposure), Fischer 344 rat, Syrian golden hamster (28 days subacute exposure) 0.5-5.0 ppm. No changes in body weight gain in either sex of mouse or hamster. Rat exposed to 5.0 ppm (28 days) incurred decrease in body weight gain. Significant exposure-related increases in relative spleen weight occurred in both sexes of mouse and rat in the 0.5 ppm (except 14-day ♀ rats), 2.5, 5.0 ppm exposure groups from all studies and in hamsters in 2.5, 5.0 ppm exposure groups. Decreased packed cell volumes and an increase in δ-aminolevulinic acid dehydratase activity observed in all species and haematological profiles in rats were also affected (4).

Carcinogenicity and long-term effects
Sufficient evidence for carcinogenicity in humans, limited evidence for

carcinogenicity in animals, IARC classification group 1 for arsine and arsine compounds (6).

Teratogenicity and reproductive effects

Rats and mice (6-15 days pregnancy) 0.025-2.5 ppm. In rats, maternal spleens enlarged in 2.5 ppm group and packed red cell volume, decreased in pregnant rats. Arsine did not adversely affect endpoints of developmental toxicity (7).

Any other adverse effects to man

Acute arsine poisoning in industrial metallurgy worker produced jaundice, vomiting and blood- stained urine (8). Renal failure, fever, nausea, vomiting, diarrhoea, abdominal pain, haemoglobinuria, intravascular haemolysis in sailors exposed to leaking cylinder (9).

Any other adverse effects

Inhalation of 250 ppm is instantly lethal. Exposure to 25-50 ppm for 30 min is lethal (species unspecified) (10).
Inhalation mouse (duration unspecified) 0.025-2.5 mg l^{-1}, about 60% is absorbed. In rabbits, highest concentrations were found in liver, lungs and kidneys (11).

Any other comments

Human health effects, epidemiology and experimental toxicology reviewed (12).

References

1. U.S. Dept. Interior, Fish and Wildlife Service *Handbook Acute Toxicity Chem. Fish Aquatic Invertebrates* 1980, No. 137, Washington, DC
2. Compton, J. A. F. *Military Chem. Biol. Agents* 1987, 102
3. *Farmakol. Toksikol. (Moscow)* 1967, **30**(2), 226
4. Blair, P. C; et al. *Fundam. Appl. Toxicol.* 1990, **14**(4), 776-787
5. Fowler, B. A; et al. *Hazard Assess. Control Technol. Semicond. Manuf. [Pap. Symp.]* 1989, 85-89
6. *IARC Monograph* 1987, **Suppl. 7**, 57
7. Morrissey, R. E; et al. *Fundam. Appl. Toxicol.* 1990, **15**(2), 350-356
8. Hockeny, A. G; et al. *Br. J. Ind. Med.* 1970, **27**(1), 56-60
9. Wilkinson, S. P; et al. *Br. Med. J.* 6 Sep 1975, 3(5983), 559-562
10. *Encyclopaedia of Occupational Health and Safety* 3rd ed., 1983, International Labour Organisation, Geneva
11. *IARC Monograph* 1980, **23**, 91
12. *ECETOC Technical Report No. 30(4)* 1991, European Chemical Industry Ecology and Toxicology Centre, B-1160 Brussels

A266   Asbestos

CAS Registry No. 1332-21-4
Synonyms Chrysotile (white) $Mg_3(Si_2O_5)(OH)_4$; Amosite (brown) $(Fe^{2+}Mg)_7Si_8O_{22}(OH)_2$; Crocidolite (blue) $Na_2Fe^{3+}_2(Fe^{2+}Mg)_3Si_8O_{22}(OH)_2$; Actinolite $Ca_2(MgFe^{2+})_5Si_8O_{22}(OH)_2$; Tremolite $Ca_2Mg_5Si_8O_{22}(OH)_2$; Anthophyllite $Mg_7Si_8O_{22}(OH)_2$

Uses Heat resistant insulators, cements, furnace and hot pipe coverings, inert filler medium, fireproof gloves, clothing, brake linings.

Occurrence In two large groups of rock-forming minerals, the serpentines and the amphiboles as fibrous mineral silicates. Extensive deposits in Russia, China and South Africa.

The main commercil varieties are chrysotile (a serpentine mineral), and crocidolite and amosite (amphiboles) (1).

Physical properties

M. Pt. Decomposes to pyroxenes and silica; Specific gravity chrysotile 2.55; anthophyllite 2.85-3.1; amosite 3.43; crocidolite 3.37; tremolite 2.9-3.2; actinolite 3.0- 3.2.

Occupational exposure

US TLV (TWA) 0.2 mg m^{-3} f cc^{-1} (intended change);amosite 0.5 fibre/cc; chrysotile and other forms 2 fibres/cc;crocidolite 0.2 fibre/cc; UN No. 2212 blue; 2590 white; Conveyance classification other dangerous substance.

Mammalian and avian toxicity

Sub-acute data

Pulmonary fibrosis reported in rats, monkeys, hamsters, rabbits and guinea pigs following inhalation of chrysotile and amphiboles. Exposure was in the order of 10 mg m^{-3} for 6-12 months. The ability of asbestos fibres to cross the gut wall is still debated; it seems likely that if it does occur it is very limited (1).

Inhalation of asbestos dust causes the following pathological changes: asbestosis, carcinoma, mesothelioma, pleural plaques (2).

Carcinogenicity and long-term effects

Bronchial carcinomas and pleural mesotheliomas developed in rats following inhalation exposure to chrysotile and amphibole asbestos. Shorter fibres are less fibrogenic and carcinogenic; Potts hypothesis being that maximum carcinogenicity is from fibres 20 μm long and 0.125 μm in diameter.

Many cohort studies since 1955 in industrial populations showed excess lung cancer risk due to asbestos exposure (1).

Sufficient evidence for carcinogenicity in humans and animals, IARC overall evaluation group 1 (3).

Metabolism and pharmacokinetics

Inhaled fibres deposit by sedimentation, diffusion, impaction and interception in airways of the respiratory system. Asbestos particles move through the epithelium to the lung interstitium where the fibres react with macrophages and fibroblasts. Two human studies gave evidence for the penetration and migration of asbestos. Amphibole asbestos has been detected in the urine of Minnesota residents who ingested drinking water contaminated with 5×10^7 fibres (4).

After intrapleural or subcutaneous inoculation (species unspecified), occasional asbestos fibres or bodies reported in other tissues, including pancreas, spleen and thyroid. There is no information on how fibres reach these sites (5).

Peritoneal mesothelioma in humans, excess cancer of the stomach, colon, rectum and cancers at other non- respiratory sites could result from the migration of fibres to and across the gastrointestinal mucosa, by transdiaphragmatic migration or lymphatic-haematogenous transport (6).

Oral rat (dose unspecified) for up to 1 yr. There was no evidence of asbestos retention within gut lumen, and no sign of cell penetration or damage to intestinal mucosa were observed (7).

Intraperitoneal Wistar rat (2 day intervals from day 10-14 of gestation) total dose 4-12 mg. Asbestos fibres were found to cross the placenta but the extent of this occurrence was highly variable (8).

Legislation

Federally regulated carcinogen (NIOSH) in USA.
Use in UK controlled by legislation (9).
Release into the air: prescribed substance, schedule 4, regulation 6(1) (10).

Any other comments

Human health effects, epidemiology and experimental toxicologyextensively reviewed (1,11).
Toxicity and hazards reviewed, to be updated in 1992-1993 (12).

References

1. *Enviromental Health Criteria 53: Asbestos* 1986, WHO/IPCS, Geneva
2. Michaels, L; et al. *Asbestos Vol. 1, Properties Applications and Hazards*, 1979, Wiley & Sons, New York
3. *IARC Monograph* 1987, **Suppl. 7**, 106
4. Seiler, H. G; et al. *Handbook on the Toxicity of Inorganic Compounds* 1988, 603-604, Marcel Dekker, Inc., New York
5. *IARC Monograph* 1973, **2**, 29
6. *USEPA Asbestos Health Assessment Update (Draft)* 1984, 74-76, EPA 600/8-84-003A
7. Bolton, R. E; et al. *Am. Occup. Hyg.* 1976, **19**(2), 121-128
8. Cunningham, H. M; et al. *Arch. Environ. Contam. Toxicol.* 1977, **6**, 507-513
9. *Control of Asbestos at Work Regulations* 1987, HMSO, London
10. *S. I. 472 Environmental Protection (Prescribed Processes and Substances) Regulations 1991* 1991, HMSO, London
11. *ECETOC Technical Report No. 30(4)* 1991, European Chemical Industry Ecology and Toxicology Centre, B-1160 Brussels
12. Izmerov, N. F. *Scientific Reviews of Soviet Literature on Toxicity & Hazards of Chemicals* 1991, **2**, Eng. Trans. Richardson, M. L. (Ed.), UNEP/IRPTC, Geneva

A267   Ascorbic acid

CAS Registry No. 50-81-7
Synonyms Vitamin C; *L*-3-ketothreohexuronic acid lactone; Cevitex; Ascorin; Proscorbin; Vitacin
Mol. Formula $C_6H_8O_6$ **Mol. Wt.** 176.13
Uses Vitamin supplement. Preservative and antioxidant in foodstuffs.

Occurrence Widely distributed in plant and animal kingdom. Good sources are citrus fruits and hip berries.

Physical properties

M. Pt. 193°C; **Specific gravity** 1.65.

Solubility
Water: 333 g l^{-1}. Organic solvent: ethanol

Mammalian and avian toxicity

Acute data
LD$_{50}$ intraperitoneal mouse, rat 2630- 3120 mg kg^{-1} (1,2).
LD$_{50}$ intravenous mouse 518 mg kg^{-1} (3).

Sub-acute data
Oral rabbit (10 days) 50 mg kg^{-1} day^{-1} induced immunostimulant activity by activating lymphocyte production by facilitating DNA and RNA formation (4).

Carcinogenicity and long-term effects
National Toxicology Program tested ♂, ♀ rats and mice with pharmaceutical grade and naturally occurring L- ascorbic acid in feed (concentration unspecified). Results negative (5).

Metabolism and pharmacokinetics
Absorbed ascorbic acid is ubiquitously distributed in all body tissues. Highest concentrations found in glandular tissue, the lowest in muscle and stored fat (6).

Genotoxicity
Genetic alteration in *Neurospora crassa* at 350 mg l^{-1} (7).
Chromosomal abnormalities in cultivated hamster cells at 300 mg l^{-1} (8).
Chinese hamster lung V79 assay (6 hr) with metabolic activation increased the incidence of cells with chromosomal aberration (9).

Any other adverse effects
Using the Ames *Salmonella*/microsome assay, the antimutagenic activities of vitamin C against solvent extracts of coal dust, diesel emission particles, airborne particles, fried beef and tobacco snuff were compared. Vitamin C inhibited <39% of the activity of the complex mixtures studied, but enhanced the mutagenicity of airborne particles 10). In vitro effect of L-ascorbic acid in the presence of a trace level of copper was investigated. L-Ascorbic acid exerted a strong killing effect on *Bacillus subtilis*, *Escherichia coli*, *Klebsiella pneumoniae*, *Proteus morganii*, *Proteus vulgaris*, *Pseudomonas phaseolicola*, *Salmonella typhimurium*, *Serratia marcescens* and *Staphylococcus auresu*. It had no effect on *Hansenula anomala*, *Pichia membranaefaciens*, *Rhodotorula rubra*, *Saccharomyces cerevisiae*, *Saccharomyces rouxii*, *Schizosaccharomyces octospolus* and *Sporobolomyces salmonicolor*. The results indicated that L-ascorbic acid exerts a killing effect on a number of bacteria and proaryotic cells, but not on yeasts and eukaryotic cells (11).

Any other comments
Metabolism and toxicology reviewed for fish, cancer prevention, toxicity, human health effects, metabolism and experimental toxicology reviewed (12-19).

References

1. Lipkan, G. N. *Farmakol. Toksikol.* 1971, **6**, 127, (Russ.) (*Chem. Abstr.* **76**, 136642n)
2. Lipkan, G. N. *Farmakol. Toksikol.* 1972, **7**, 128, (Russ.) (*Chem. Abstr.* **79**, 27340d)
3. *Res. Prog. Organ. Biol. Med. Chem.* 1970, **27**, 269
4. Gurer, Firdeus *Turk Hij. Deneysel Biyol. Derg.* 1988, **45**(2), 243-259, (Turk.) (*Chem. Abstr.* **111**, 146449k)
5. *National Toxicology Program, Research and Testing Division* 1992, Report No. 247, NIEHS, Research Triangle Park, NC 27709
6. Goodman, L.S; et al. *The Pharmacological Basis of Therapeutics* 5th ed., 1975, Macmillan, New York
7. Munkres, K. D. *Mech. Age. Develop.* 1979, **10**, 249
8. Stich, W; et al. *Food Cosmet. Toxicol.* 1980, **18**, 497
9. Kojima, H; et al. *Nippikyo Janaru* 1986, **9**(2), 97-101, (Jap.) (*Chem. Abstr.* **108**, 1926h)
10. Ong, T; et al. *Mutat. Res.* 1989, **222** (1), 19-25
11. Murata, A; et al. *Bitamin* 1990, **64**(12), 709-713, (Jap.) (*Chem. Abstr.* **115**, 68277u)
12. Sandnes, K. *Fiskeridiv. Skr., Ser. Ernaer.* 1991, **4**(1), 3-32
13. Florence, T. M. *Proc. Nutr. Soc. Res.* 1988, **15**, 88-93
14. Hanck, A. B. *Prog. Clin. Biol. Res.* 1988, **259**, 307-320
15. Glatthaar, B. E; et al. *Adv. Exp. Med. Biol.* 1986, **206**, 357-377
16. Moser, U; et al. *Food Sci. Technol.* 1991, **40** (Handbook Vitamin, 2nd ed.,) 195-232
17. Rivers, J. M. *Int. J. Vitam. Nutr. Res., Suppl.* 1989, **30**, 95-102
18. Stier, A. *Schriftenr, Bundesapothekerkammer Wiss. Forrbild., Gelbe Reihe* 1986, **14**, 255-271, (Ger.) (*Chem. Abstr.* **109**, 53582s)
19. *ECETOC Technical Report No. 30(4)* 1991, European Chemical Industry Ecology and Toxicology Centre, B-1160 Brussels

A268   Asphalt, fumes

CAS Registry No. 8052-42-4
Synonyms asphaltum; bitumen; Judgan pitch; mineral pitch; petroleum pitch; road tar
Uses Making roads, roofing, waterproofing and insulating materials. Also an ingredient of some paints and varnishes.
Occurrence Consitituent of crude petroleum. Occurs in natural deposits in pits or lakes resulting from the evaporation and oxidation of liquid petroleum. Separated from petroleum by refining.

Physical properties

B. Pt. 470°C; **Flash point** 204°C (closed cup); **Specific gravity** 0.95-1.1.

Solubility

Organic solvent: turpentine oil, petroleum, carbon disulfide

Occupational exposure

US TLV (TWA) 5 mg m^{-3}; **UK Long-term limit** 5 mg m^{-3}; **UK Short-term limit** 10 mg m^{-3}.

Mammalian and avian toxicity

Acute data
LD$_{50}$ intragastric rat 3-8 g kg^{-1} (1).

Carcinogenicity and long-term effects

Inadequate evidence for carcinogenicity in humans, limited to sufficient evidence for carcinogenicity in animals, IARC classification group 2B (for steam-refined and cracking-residue-; air-refined-; extracts of steam-refined and air-refined–bitumens) (2).

Any other comments

Human health effects and experimental toxicology reviewed (4).

A significant contribution to the polycyclic aromatic hydrocarbon content of street dust from material associated with asphalt (5).

Experiments were conducted to find out whether the use of petroleum-asphalt seal coating in ductile-iron pipe would contribute significant concentrations of polycyclic aromatic hydrocarbons in drinking water distribution systems. The results of the analyses were compared with the WHO recommendation for maximum allowable concentration of polycyclic aromatic hydrocarbons in drinking water of 200 ng l^{-1}. The highest concentration found in 3 experiments was 5 ng l^{-1} (6).

References

1. Mlynarczyk, W. *Bromatol. Chem. Toksykol.* 1984, **17**(2), 119 (*Chem. Abstr.* **102**, 1560u)
2. *IARC Monograph* 1985, **35**, 39
3. *Encyclopaedia of Occupational Health and Safety* 1971, **1**, 125, International Labor Organisation, Geneva
4. *ECETOC Technical Report No. 30(4)* 1991, European Chemical Industry Ecology and Toxicology Centre, B-1160 Brussels
5. Wakeham, S. G; et al. *Geochim. Cosmochim. Acta* 1980, **44**(3), 403-414
6. Miller, H. C; et al. *Am. Water Works Assoc. J.* 1982, **74**(3), 151-156

A269   Astatine

$$_{85}At^{200}-_{85}At^{219}$$

CAS Registry No. 7440-68-8

Mol. Formula At   **Mol. Wt.** 210.00

Uses Used in radiobiology, mainly At^{211}. Cancer diagnostics and treatment.

Occurrence Minute quantities of At^{215}, At^{218} and At^{219} exist, in nature, in equilibrium uranium and chorium isotopes.

Physical properties

M. Pt. 302°C (estimated); **B. Pt.** 337°C (estimated).

Mammalian and avian toxicity

Sub-acute data

Mice injected with a single dose containing 61 kBq At^{211} showed pathological changes in spleen, lymph nodes, bone marrow, gonads, thyroid, salivary glands, and stomach, when killed at 14 or 56 days (1).

Any other adverse effects

Is thought to concentrate in the thyroid, similarly to iodine (1-3).

References

1. Cobb, L. M; et al. *Hum. Toxicol.* 1988 **7**(6), 529-534
2. Cobb, L. M; et al. *Radiother. Oncol.* 1988, **13**(3), 203-209
3. Mitchell, J. S; et al. *Experientia* 1985, **41**(7), 925-928

A270   Asulam

CAS Registry No. 3337-71-1

Synonyms methyl 4-aminophenylsulfonylcarbamate; methyl 4-aminobenzene sulfonylcarbamate; methyl sulfanilylcarbamate; methyl ((4- aminophenyl)sulfonyl) carbamate; carbamic acid, [(4- aminophenyl)sulfonyl]-, methyl ester

Mol. Formula $C_8H_{10}N_2O_4S$ **Mol. Wt.** 230.24

Uses Systemic herbicide.

Physical properties

M. Pt. 143-144°C (decomp.); **Volatility** v.p. $<7.5 \times 10^{-6}$ mmHg at 20°C.

Solubility

Organic solvent: dimethyl formamide, acetone, methanol, ethanol

Ecotoxicity

Fish toxicity

LC_{50} (96 hr) rainbow trout, channel catfish, goldfish, bluegill sunfish >3000 mg l^{-1} (1).

Effects to non-target species

Non-toxic to bees.

LD_{50} oral mallard duck, pheasant, pigeon >4000 mg kg^{-1} (2).

Environmental fate

Degradation studies

In soil $t_{1/2}$ 6-14 days (2).

Mammalian and avian toxicity

Acute data

LD_{50} oral rat, mouse, rabbit, dog >4000 mg kg^{-1}.

LD_{50} percutaneous rat >1200 mg kg^{-1}.

LC_{50} (6 hr) inhalation rat >1.8 mg l^{-1} (2).

Sub-acute data

Oral rat (90 day) 400 mg kg^{-1} in diet no significant ill-effects observed (2).

Metabolism and pharmacokinetics

Asulam administered by mouth or intravenously to rats was excreted within 24 hr. Excretory products consisted mainly of unchanged asulam, plus minor amounts of *N*-acetylasulam and *N*-acetylsulfanilamide (3).

Legislation

Limited under EC Directive on Drinking Water Quality 80/778/EEC. Herbicide: maximum admissible concentration $0.5 \mu g \, l^{-1}$ (4).

References

1. Ingham, B; et al. *Bull. Environm. Contam. Toxicol.* 1975, **13**(2), 194-199
2. *The Agrochemicals Handbook* 3rd ed., 1991, RSC, London
3. Heijbroek, W. M. H; et al. *Xenobiotica* 1984, **14**(3), 235-247
4. *EC Directive Relating to the Quality of Water Intended for Human Consumption* 1982, 80/778/EEC, Office for Official Publications of the European Communities, 2 rue Mercier, L-2985 Luxembourg

A271 Atrazine

CAS Registry No. 1912-24-9

Synonyms 2-chloro-4-ethylamino-6-isopropylamino-1,3,5-triazine; 6-chloro-*N*-2-ethyl-*N*-4-isopropyl-1,3,5-triazine-2,4-diamine; 6-chloro-*N*-ethyl-*N'*-(1-methylethyl)-1,3,5-triazine-2,4-diamine; 2-chloro-4-ethylamino-6-isopropylamino-*s*-triazine; AAtrex; Aktikon; Atazinax; Atratol; Mebazine; Gesaprim; Primatol A

Mol. Formula $C_8H_{14}ClN_5$ **Mol. Wt.** 215.69

Uses A selective herbicide and plant growth regulator.

Occurrence Atrazine residues have been detected in numerous tapwater, groundwater and river samples.

Physical properties

M. Pt. 171-174°C; **Specific gravity** d_{20} 1.187; **Partition coefficient** log P_{ow} 2.34 (1); **Volatility** v.p. 3.0×10^{-7} mmHg at 20°C.

Solubility

Water: 70 mg l^{-1} at 22°C. Organic solvent: acetone, dimethyl sulfoxide, ethanol, chloroform, ethyl acetate, diethyl ether

Occupational exposure

US TLV (TWA) 5 mg m^{-3}; **UK Long-term limit** 10 mg m^{-3} (under review).

Ecotoxicity

Fish toxicity

LC_{50} (96 hr) rainbow trout, bluegill sunfish, carp, catfish, guppy 4.3-8.8 mg l^{-1} (1). Rainbow trout were exposed to 1.4-2.8 mg l^{-1} atrazine for 96 hr and 5-80 $\mu g \, l^{-1}$ for 28 days. Necrosis of endothelial cells and renal tissue were observed at high doses for both time periods (2).

Invertebrate toxicity

EC_{50} (48 hr) *Daphnia magna* >39 mg l^{-1} (3).
Chlamydomonas reinhardtii (24-48 hr) 21.6 μg l^{-1} caused change in cell number of algae species observed (4).
Cell multiplication inhibition test, *Pseudomonas putida* 10 mg l^{-1}, *Microcytes aeruginosa* 3 μg l^{-1} (5).

Bioaccumulation

Bioconcentration factor values for snails 2-15, algae 10-83, fish 3-10 (5).
Calculated bioconcentration factors of 0.3-2 have been reported for mottled sculpin, golden ide, fathead minnow, whitefish and catfish. These values indicate environmental bioaccumulation of atrazine is unlikely (6-11).

Effects to non-target species

Non-toxic to bees (1).
LD_{Lo} bullfrog, leopard frog 0.41 mg l^{-1} (12).

Environmental fate

Anaerobic effects

Under anaerobic conditions, 0.59% atrazine degraded to carbon dioxide (13).

Degradation studies

In soil, microbial degradation occurs with a $t_{1/2}$ of 6-10 wk (5).
Dealkylation is major mechanism of microbial degradation of atrazine, s-triazine is ring resistant to microbial degradation (14-16).
Moderately to highly mobile in soils ranging in texture from clay to gravelly sand, respectively, determined by soil thin layer chromatography. The metabolite hydroxyatrazine has low mobility in sandy loam and silty clay loam soils (17).
Soil metabolites include diethylatrazine, diisopropylatrazine, dealkylatrazine and hydroxyatrazine, with the exception of dealkylatrazine, $t_{1/2}$ 17-26 days (18).

Abiotic removal

Evaporation is not significant for the removal of atrazine from the environment (19).
Chemical hydrolysis to hydroxyatrazine was the principal pathway of detoxification in soil. Biological dealkylation without dehalogenation occurs simultaneously leading to the formation 2-chloro-4-amino-6-isopropyl-s-triazine (20).

Absorption

Atrazine binds strongly to soil and sediments, the process is reversible and dependent on factors such as temperature, moisture and pH (21, 22).

Mammalian and avian toxicity

Acute data

LD_{50} oral rat, mouse, rabbit 750-3080 mg kg^{-1} (1,12,20,23).
LC_{50} inhalation rat (4 hr) 5200 mg m^{-3} (24).
LD_{50} percutaneous rabbit, rat >3100-7500 mg kg^{-1} (1).

Sub-acute data

LC_{50} (8 day dietary) bobwhite quail, mallard duck 5760-19,650 mg kg^{-1} as 80% wettable powder formulation (1).
Six month feeding to rabbits, maize from treated fields (approx. 2.5 kg ha^{-1}) led to loss of appetite, general debility, depression and anaemia (25).

Ingestion rat (72 hr) 65% excreted in urine, 20% in faeces and 0.1% exhaled. Highest concentrations observed in muscle, tissue and fat (26).

Carcinogenicity and long-term effects
Inadequate evidence for carcinogenicity to humans, limited evidence for carcinogenicity to animals, IARC classification group 2B (27).
F344/LATI rats (126 wk) 0-750 ppm caused dose dependent depression of body weight. Increased incidence of mammary tumours, uterine carcinomas and leukaemias/lymphomas (28).
Sprague-Dawley rats fed 0- 50 mg kg^{-1} day^{-1} high doses caused decreased body weight, hyperplastic changes to mammary gland, bladder, prostate, myeloid tissue of bone marrow and transitional epithelium of the kidney. Mammary gland tumour observed (29).

Teratogenicity and reproductive effects
Pregnant rats (6-15 day gestation) 0-1000 mg kg^{-1} day^{-1} increased number of embryonic and foetal death, decreased foetal weight and retarded skeletal development at high doses. No teratogenic effects observed. Induced 23% maternal mortality at 1000 mg kg^{-1} day^{-1} (28).

Metabolism and pharmacokinetics
Rats administered a single oral dose of 0.53 mg by gavage excreted 20% in faeces and 65% in urine. At 72 hr 16% of the administered dose was retained in the liver, kidneys and lung (30).
Rats administered a single dose of 3000 mg kg^{-1} by gavage and sacrificed 24-48 hr later showed evidence of pulmonary oedema, cardiac dilation and microscopic haemorrhages in the liver and spleen, cerebral oedema and histochemical alterations in lungs, liver and brain (31).
A single 0.1 g dose was administered by gavage to 3-5 months old Pittman-Moore pigs, the major urinary products detected were the parent compound atrazine and its metabolite diethylatrazine (32).
In rats the major metabolic pathways for the detoxification of the compound were identified as dechlorination of the triazine ring and N-dealkylation. Secondary metabolic routes appeared to be oxidation of alkyl substituents (33). *In vitro* hepatic microsomal systems from rats, goats, sheep, pigs, rabbits were used to investigate metabolic action of atrazine. Phase I reactions were cytochrome P450 mediated and phase II products were reduced glutathione conjugates and monodealkylated products (34).
In rats (3 day) (unspecified route) 0.005-50 mg day^{-1}, major urinary metabolite was 2-chloro-4, 6-diamino-s-traizine (35).

Irritancy
Dermal rabbit (duration unspecified) 38 mg caused mild irritation, while 6.3 mg instilled into rabbit eye caused severe irritation (36).
A farm worker exposed to atrazine formulation was diagnosed with severe contact dermatitis. Clinical signs included red, swollen and blistered hands with haemorrhagic bullae between the fingers (37).
In a primary irritation study in rats 2800 mg kg^{-1} caused erythema but no systemic effects (30).

Genotoxicity
Salmonella typhimurium TA97, TA98, TA100 with and without metabolic activation negative (38).

Escherichia coli PQ37 with and without metabolic activation negative (39).

Legislation

Included in Schedule 1 of Statutory Instrument No. 1156, 1989, and Schedule 5 (Release Into Water: Prescribed Substances) of Statutory Instrument No. 472, 1991 (40-41). Water criteria include EEC Drinking Directive 80/778/EEC 0.1 µg l^{-1}, British Department of the Environment Guidelines for Drinking Water Quality 2.0 µg l^{-1} and USEPA Drinking Water Advisory 3.0 µg l^{-1} (42).

Any other comments

Mammalian and aquatic toxicity reviewed (43- 45).
Epidemiological, carcinogenicity and genotoxicity studies reported (31).
The British Government announced a ban on all non-farm uses of atrazine from August 1992. The ban includes aerial spraying of crops and the amount of atrazine used by farmers each year is to be restricted (46).

References

1. *The Agrochemicals Handbook* 3rd ed., 1991, RSC, London
2. Fischer-Scherl, T; et al. *Arch. Environ. Contam. Toxicol.* 1991, **20**(4), 454-461
3. Marchini, S; et al. *Ecotoxicol. Environ. Saf.* 1988, **16**, 148-157
4. Hersch, C. M; et al. *Bull. Environ. Contam. Toxicol.* 1987, **39**, 47
5. Verschueren, K. *Handbook of Environmental Data of Organic Chemicals* 2nd ed., 1983, 224, Van Nostrand Reinhold, New York
6. Lynch, T. R. et al. *Environ. Toxicol. Chem.* 1982, **1**, 179-192
7. Freitag, D; et al. *Chemosphere* 1985, 1589-1616
8. Veith, G. D; et al. *J. Fish Res. Board Can.* 1979, **36**, 1040-1048
9. Gunkel, G; et al. *Water Res.* 1980, **14** 1573-1584
10. Ellgehausen, H; et al. *Ecotoxicol. Environ. Safety* 1980, **4**, 134-157
11. Macek, K. J; et al. *Chronic Toxicity of Atrazine to Selected Aquatic Invertebrates and Fishes* 1976, EPA 600/3-76-047
12. Birge, W. J; et al. *Report* 1980, RR-121, W80-03438, OWRT-A-074-KY(2) (*Chem. Abstr.* **93, 162104j**)
13. Goswani, K. P; et al. *Environ. Sci. Technol.* 1971, **5**(5), 426-429
14. Kaufman, D. O; et al. *Res. Rev.* 1970, **32**, 235-265
15. *Health and Environmental Effects Profile for Atrazine* 1984, ECAO-CIN-PO98
16. Skipper, H. D; et al. *Weed Sci.* 1972, **20**, 344-347
17. Helling, C. S. *Proc. Soil Sci. Soc. Am.* 1971, **35**, 737-748
18. Winkelman, D. A; et al. *DECHEMA Biotechnol. Conf. 3* 1989, 49-52, Memphis State Univ.
19. Lyman, W. J; et al. *Handbook of Chemical Property Estimation Methods* 1982, **15**, 15-29
20. *Neurobehav. Toxicol. Teratol.* 1983, **5**, 503
21. Geller, A. *Arch. Environ. Contam. Toxicol.* 1980, **9**, 289-305
22. *Herbicide Handbook* 5th ed., 1983, 33, Weed Sci. of America, IL
23. Dalgaard-Mikklesen, S; et al. *Pharmacol. Rev.* 1962, **14**, 225
24. *Farm Chemicals Handbook* 1983, C3, Meister Publishing Co, Willoughby, OH
25. Salem, F. M. S; et al. *Vet. Med. J.* 1985, **33**(2), 239-250
26. *Doc. TLV Biol. Exp. Ind.* 5th ed., 1986, 44, Amer. Conf. Gov. Ind. Hyg., Cincinnati, OH
27. *IARC Monograph* 1991, **53**, 441-466
28. Pinter, A; et al. *Neoplasma* 1990, **37**(5), 533-544
29. *Drinking Water Health Advisory, Pesticides* 1989, 43-67, Lewis Publishers, Chelsea, MI
30. Pakke, J. E; et al. *J. Agric. Food Chem.* 1972, **20**, 602-607
31. Molner, V. *Rev. Med.* 1971, **17**, 271-274
32. Erickson, M. D; et al. *Agric. Food Chem.* 1979, **27**(4), 743-746

33. Hansworth, J. W. *Summary on Atrazine Toxicity Studies* 1988, EPA ABR-87048, 87087, 87115, 87116, 85104, AG-520, Ciba-Geigy
34. Adams, N. H; et al. *J. Agric. Food Chem.* 1990, **38**(6), 1411-1417
35. Bradway, D. E; et al. *J. Agric. Food Chem.* 1982, *30*, 244-247
36. *Ciba-Geigy Toxicology Data/Indexes* 1977, Ciba-Geigy Corp., New York
37. Schlicher, J. E; et al. *J. Iowa Med. Soc.* 1972, **62**, 419-420
38. Butler, M. A; et al. *Bull. Environ. Contam. Toxicol.* 1989, **43**(6), 797-804
39. Mersch-Sundermann, V; et al. *Zentralbl. Hyg. Umweltmed.* 1989, **189**(2), 135-146
40. *S. I. 1989 No. 1156. The Trade Effluent (Prescribed Processes and Substances) Regulations* 1989, HMSO, London
41. *S. I. 1991 No. 472 The Environmental Protection (Prescribed Processes and Substances) Regulations* 1991, HMSO, London
42. Richardson, M. L. Ed., *Chemistry, Agriculture and the Environment* 1991, 444, RSC, London
43. Izmerov, N. F. *Scientific Reviews of Soviet Literature Toxicity & Hazards of Chemicals* 1982, **18**, Eng. Trans., Richardson, M. L. (Ed.), UNEP/IRPTC, Geneva
44. *Dangerous Prop. Ind. Mater. Rep.* 1990, **10**(3), 28-41
45. Premazzi, G; et al. *Comm. Eur. Commun. Report* 1990, 1-100, EUR-12569
46. *The Guardian* 1992, 13th May

A272   Atropine

CAS Registry No. 51-55-8

Synonyms benzeneacetic acid, α-(hydroxymethyl)-8-methyl-8-azobicyclo[3.2.1]o ct-3-yl ester endo-(±)-; *DL*-hyoxyomene; *DL*-tropyl tropate

Mol. Formula $C_{17}H_{23}NO_3$ **Mol. Wt.** 289.38

Uses Antimuscarinic drug. Antidote in poisoning cases.

Occurrence Obtained from *Atropase belladonna, Datura stramonium L.* and other *Salmanaceae.*

Physical properties

M. Pt. 118-119°C.

Solubility
Organic solvent: benzene, diethyl ether, ethanol, chloroform

Occupational exposure

Supply classification toxic.

Risk phrases
Very toxic by inhalation and if swallowed (R26/28)

Safety phrases
Keep locked up – Avoid contact with eyes – In case of accident or if you feel unwell, seek medical advice immediately (show label where possible) (S1, S25, S45)

Mammalian and avian toxicity

Acute data

LD_{50} oral mouse 75 mg kg^{-1} (1).
LD_{50} intraperitoneal rat 280 mg kg^{-1} (2).
LD_{50} intravenous rabbit 50 mg kg^{-1} (1).
LD_{Lo} subcutaneous guinea pig 450 mg kg^{-1} (3).

Sub-acute data

Rat (90 day subchronic exposure) in feed, 0.5, 1.58, 5.0% jimson weed seed. The alkaloid content was 2.71 mg atropine and 0.66 mg scopolamine g^{-1} of seed. Decreased body weight gain, serum albumin and serum calcium, increased liver and testes weights (as percentage of body weight) serum alkaline phosphatase, and blood urea nitrogen. ♀ developed decreased serum total protein and cholesterol, and increased serum glutamic-pyruvic transaminase activity, red blood cell count, haemoglobin concentration and packed red cell volume. No histological lesions associated with ingestion of jimson weed seed at 5% (4).

Irritancy

Dermatitis from the use of eyedrops containing atropine has been reported (5).

Any other adverse effects

Symptoms of atropine poisoning are dryness of skin, mouth and throat, tachycardia, flushed skin and face, irritability and restlessness (6).

References

1. *Arch. Inter. Pharmacodyn. Therap.* 1938, **59**, 149
2. *J. Pharmacol. Exp. Therap.* 1952, **105**, 166
3. *Biochem. Zeitschrift* 1914, **66**, 389
4. Dugan, G. M; et al. *Food Chem. Toxicol.* 1989, **27**(8), 501-510
5. Van der Willigen, A. M. *Contact Dermatitis* 1987, **17**(1), 56-57
6. North, R. V; et al. *Opthalmol. Physiol. Opt.* 1987, **7**(2), 109-114

A273   Auramine (Base)

CAS Registry No. 492-80-8
Synonyms bis(*p*-dimethylaminophenyl)methyleneimine; 4,4'-carbonimidoylbis(*N,N*-dimethylbenzenamine); 4, 4'-dimethylaminobenzophenonimide; 4,4'-(imidocarbonyl)bis(*N,N*-dimethylaniline); tetramethyldiaminodiphenylacetimine; CI Solvent Yellow 34; Aniline, 4,4-(imidocarbonyl)bis*N,N*-Dimethyl-; Auramine O; 4,4-bisdimethylaminobenzophenoneimide; Apyonin; Yellow pyoctenin
Mol. Formula $C_{17}H_{21}N_3$ **Mol. Wt.** 267.38
Uses Dyestuff for paper, textiles and leather.

Physical properties

M. Pt. 136°C.

Solubility
Organic solvent: dimethylformamide, acetone, ethanol

Ecotoxicity

Bioaccumulation
Calculated bioconcentration factor 288 (1).

Environmental fate

Degradation studies
BOD_5 1.5% reduction of dissolved oxygen concentration using standard dilution techniques (2).

Abiotic removal
Calculated hydrolytic $t_{1/2}$ 65 days at pH5 increasing to 74 days at pH9. Michler's ketone has been detected as a product of hydrolysis (3).

Mammalian and avian toxicity

Acute data
LD_{50} intraperitoneal mouse 103 mg kg^{-1} (4).

Carcinogenicity and long-term effects
Sufficient evidence for carcinogenicity to humans, IARC classification group 1 for the manufacture of auramine.
Inadequate evidence for carcinogenicity to humans, sufficient evidence for carcinogenicity to animals, IARC classification group 2B for technical grade auramine (5).
Oral mice (1 yr) 0.1% in diet, total dose 728 mg administered as 14 mg wk^{-1} induced hepatomas, lymphomas, especially lymphosarcoma and reticulum-cell sarcoma and intestinal tumours (6).
Oral ♂ rat (87 wk) 0.1% in diet induced hepatomas. Accompanying cholangiomatous areas were seen in 30% of tumour bearing livers.
Subcutaneous rat (21 wk) 0.1 ml of 2.5% suspension in arachis oil 5 day wk^{-1} induced hepatomas, intestinal carcinoma and subcutaneous sarcoma at injection site (7).

Irritancy
Absorption through skin may result in dermatitis and burns (8).

Genotoxicity

Salmonella typhimurium TA1535, TA1537, TA1538, TA98, TA100 with and without metabolic activation positive (9).

Any other comments

Human health effects, epidemiology and experimental toxicology reviewed (10).

References

1. Saito, T; et al. *Frezenius Z. Anal. Chem.* 1984, **319**, 433-434
2. Ellington, J. J; et al. *USEPA Report* 1982, EPA600/3-88/02, NTIS PB88-23042
3. Lyman, W. J; et al. *Handbook of Chemical Property Estimation Methods* 1982, 4-9, McGraw-Hill, New York

4. Salamone, M. F. *Prog. Mutat. Res.* 1981, **1**, 682-685
5. *IARC Monograph* 1987, **Suppl. 7**, 57
6. Bonser, G. M; et al. *Br. J. Cancer* 1956, **10**, 653-667
7. Walpole, A. L. *Int. J. Cancer* 1963, **19**, 483
8. *Prog. Mutat. Res.* 1981, **1**, 626
9. Simmon, V. F; et al. *Prog. Mutat. Res.* 1981, **1**, 333-342
10. *ECETOC Technical Report No. 30(4)* 1991, European Chemical Industry Ecology and Toxicology Centre, B-1160 Brussels

A274   5-Azacytidine

CAS Registry No. 320-67-2
Synonyms 4-amino-1-β-d-ribofuranosyl-d-triazin-2(1*H*)-one;
4-amino-1-β-d-ribofuranosyl-1,3,5-triazin-2(1*H*)-one; azacytidine; azacitidine;
Ladakamycin
Mol. Formula $C_8H_{12}N_4O_5$ **Mol. Wt.** 244.21
Uses Antineoplastic agent.
Occurrence Obtained from *Streptoverticillium ladakanus*.

Physical properties
M. Pt. 228-230°C.

Solubility
Organic solvent: methanol

Ecotoxicity

Effects to non-target species
LD_{50} oral starling 100 mg kg^{-1} (1).

Mammalian and avian toxicity

Acute data
LD_{50} oral mouse 572 mg kg^{-1} (2).
LD_{50} intraperitoneal mouse 68 mg kg^{-1} (3).

Sub-acute data
Repeated doses of ≥0.55 mg kg^{-1} (5 days) produced bone marrow depression and liver degeneration in dogs (2).
Nausea, vomiting and bone marrow suppression found in patients treated with azacytidine (4).
LD_{50} (14 days) intravenous rhesus monkey 2.2 mg kg^{-1} (5).

Carcinogenicity and long-term effects
No adequate evidence for carcinogenicity to humans, limited evidence for carcinogenicity to animals, IARC classification group 3 (6,7).

National Toxicology Program tested ♂, ♀ rats and mice with 5-azacytidine via intraperitoneal injection. Results were positive in ♀ mice, but the study was inadequate to achieve valid results for other species (8).
Intraperitoneal ♂ rat (12 months) 10 mg kg^{-1} twice weekly induced multiple tumours sites included the testis, skin, bronchus. Acute leukaemia and malignant reticuloendotheliosis also occurred. No hepatic tumours were found unless a prior initiating dose of N-nitrosodiethylamine was given (9).

Teratogenicity and reproductive effects
Pregnant mouse (route unspecified) 1 mg kg^{-1} after 7.5-8.5 days of pregnancy development of malformed foetuses (10).

Metabolism and pharmacokinetics
It is phosphorylated to the mono-, di- and triphosphates.
It has been shown to be capable of incorporation into DNA and RNA, and is in inhibitor of uridine kinase and of orotidylic acid decarboxylase (7,11,12).

Genotoxicity
Salmonella typhimurium TA98, TA100 with and without metabolic activation positive (13).
C3H IOTI/2 mouse embryo fibroblasts positive chromosome aberrations in DNA methylation (14).
Human lymphoblast TK6 induced mutation at thymidine kinase (TK) locus and the hypoxanthine-guanine phosphoribosyltransferase (HGPRT) locus (13).
5-Azacytidine can reactivate genes on the inactive human X-chromosome. May act by causing demethylation of the DNA at specific sites (15).
Exposure to 5-azacytidine in Chinese hamster ovary cells increased the incidence of sister chromatid exchange (16).

Any other comments
Human health effects and experimental toxicology reviewed (17).

References
1. Schafer, E. W; et al. *Arch. Environ. Contam. Toxicol.* 1983, **12**(3), 355-382
2. Palm, P. E; et al. *Toxicol. Appl. Pharmacol.* 1971, **19**, 382
3. *Experientia* 1966, **22**, 53
4. *Martindale: The Extra Pharmacopoeia* 29th ed., 1989, The Pharmaceutical Press, London
5. Palm, P. E; et al. *US Nat. Tech. Inform. Serv.* PB Rep. 1972, PB-210642 (*Chem. Abstr.* **78**, 52629s)
6. *IARC Monograph* 1987, **Suppl 7**, 57
7. *IARC Monograph* 1981, **26**, 37
8. *National Toxicology Program, Research and Testing Div.* 1992, Report No. 42, NIEHS, Research Triangle Park, NC 27709
9. Carr, B. I; et al. *IARC Sci. Publ.* 1984, **56**(Models, Mech. Etiol. Tumour Promot.), 409-412
10. Takeuchi, I. K; et al. *J. Anat.* 1985, **140**(3), 403-412
11. Von Hoff, D. D; et al. *Clinical Brochure, 5-Azacytidine National Cancer Institute* 1975, NSC 102816
12. Van Hoff, D. D; et al. *Ann. Intern. Med.* 1976, **85**, 237
13. Call, K. M; et al. *Mutat. Res.* 1986, **160**(3), 249-257
14. Hsiao, W. L. W; et al. *J. Virol.* 1986, **57**(3), 1119-1126
15. Venolia, L; et al. *Proc. Natl. Acad. Sci. USA* 1982, **79**, 2352-2354
16. Perticone, P; et al. *Carcinogenesis (London)* 1987, **8**(8), 1059-1063
17. *ECETOC Technical Report No. 30(4)* 1991, European Chemical Industry Ecology and Toxicology Centre, B-1160 Brussels

A275 Azamethiphos

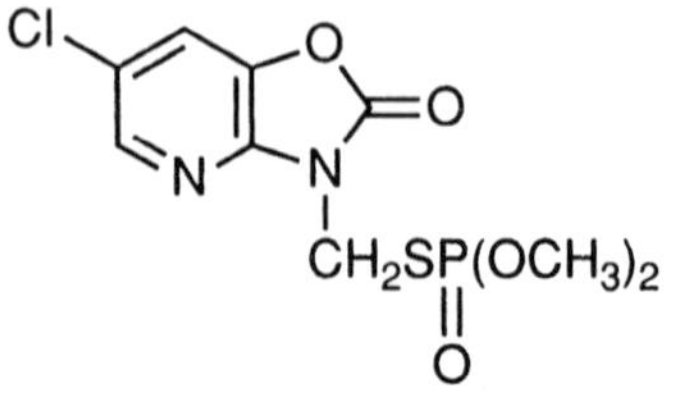

CAS Registry No. 35575-96-3

Synonyms *S*-6-chloro-2,3-dihydro-2-oxo-1,3-oxazolo[4,5-b]pyridin-3-ylmethyl *O,O*-dimethyl phosphorothioate; *S*-6-chloro-2-oxooxazolo[4,5-b]pyridin-3-ylmethyl *O,O*-dimethyl phosphorothioate; 6-chloro-3-dimethoxyphosphinoylthiomethyl-1,3-oxazolo[4,5-b]pyridin-2(3H)-one; *S*-[(6-chloro-2-oxooxazolo[4,5-b]pyridin-3(2H)-yl)methyl]*O,O*-dimethyl phosphorothioate

Mol. Formula $C_9H_{10}ClN_2O_5PS$ **Mol. Wt.** 324.68

Uses Insecticide and acaricide.

Physical properties

M. Pt. 89°C; **Specific gravity** 1.6 at 20°C; **Volatility** 3.69 x 10^{-7} mmHg at 20°C.

Solubility
Organic solvent: benzene, dichloromethane, methanol, octanol

Ecotoxicity

Fish toxicity
LC_{50} (96 hr) rainbow trout, bluegill sunfish, carp 0.2-6 mg l^{-1} (1).

Mammalian and avian toxicity

Acute data
LD_{50} oral rat 1180 mg kg^{-1} (2).
LD_{50} percutaneous rat >2150 mg kg^{-1} (1).

Legislation

Limited under EC Directive on Drinking Water Quality 80/778/EEC.
Organophosphorus insecticide: maximum admissible concentration 0.1 μg l^{-1} (3).

References

1. *The Agrochemicals Handbook* 2nd ed, 1987, RSC, London
2. Ruefenacht, K; et al. *Helv. Chim. Acta* 1976, **59**(5), 1593-1612
3. *EC Directive Relating to the Quality of Water Intended for Human Consumption* 1982, 80/778/EEC, Office for Official Publications of the European Communities, 2 rue Mercier, L-2985 Luxembourg

A276 Azaserine

$$HO_2CCH(NH_2)CH_2O_2CCHN_2$$

CAS Registry No. 115-02-6

Synonyms azaserin; diazoacetate (ester)-1-serine; *O*-diazoacetyl-1-serine; 1-serine diazoacetate; 2-amino-3-hydroxypropionic acid

Mol. Formula $C_5H_7N_3O_4$ **Mol. Wt.** 173.13

Uses Glutamine antagonist which inhibits purine biosynthesis.
Antifungal and antitumour agent.

Occurrence Present in cultures of *Streptomyces fragilis*.

Physical properties

M. Pt. 153-155°C (decomp.).

Solubility
Organic solvent: acetone, ethanol

Abiotic removal
Hydrolysis in aqueous solution at pHs of 3, 7 and 11 at 25°C corresponds to $t_{1/2}$ 2.1 hr, 111 days and 425 days, respectively. Photochemical reaction with atmospheric hydroxyl radicals corresponds to an atmospheric $t_{1/2}$ of 10 hr at an atmospheric concentration of $5 \times 10\ 5^{-1}$ hydroxyl radicals cm^{-3} (1).

Mammalian and avian toxicity

Acute data
LD_{50} oral rat, mouse 150-170 mg kg^{-1} (2).
LD_{50} intraperitoneal rat, mouse 70-100 mg kg^{-1} (2,3).
LD_{50} subcutaneous mouse 50 mg kg^{-1} (4).

Carcinogenicity and long-term effects
No adequate evidence for carcinogenicity to humans, sufficient evidence for carcinogenicity to animals, IARC classification group 2B (5,6).
Intraperitoneal single dose (7-wk old) ♂ rat 30 mg kg^{-1}. After 6-18 months all treated rats had acidophilic and basophilic foci and nodules present in pancreas. At 9 months, incidence of carcinoma *in situ* was 30%; by 18 months, 100% incidence of pancreatic cancers (58% *in situ*, 42% carcinoma) (7).

Genotoxicity

A 50% reduction in colony formation and unscheduled DNA synthesis was observed in rat pancreatic epithelial cells exposed to 52 mg l^{-1} without metabolic activation (8).
In vitro mouse L1210 leukaemia cells inhibition of N-formylglycineamidine ribotide synthetase and glucosamine-6-phosphate isomerase and large accumulations of N-formylglycineamide ribotide and its di- and tri-phosphate derivatives, which could interfere with the biosynthesis of nucleic acids, have been reported (9).
Chinese hamster ovary V-79 cells without metabolic activation positive (10).

Any other comments

Carcinogenicity, human health effects and experimental toxicology reviewed (11,12).

References

1. Atkinson, R. *Intern. J. Chem. Kin.* 1987, **19**, 799-828
2. Sternberg, S; et al. *Cancer* 1957, **10**, 889
3. Thiersch, J. B. *Proc. Soc. Exptl. Biol. Med.* 1957, **94**, 27-32
4. *CRC Handbook of Antibiotic Compounds* 1980, **4**(1), 432
5. *IARC Monograph* 1987, **Suppl. 7**, 57
6. *IARC Monograph* 1976, **10**, 73

7. Roebuck, B. D; et al. *Carcinogenesis (London)* 1987, **8**(12), 1831-1835
8. Shepherd, J; et al. *Exp. Mol. Pathol.* 1990, **53**(3), 203-210
9. Lyons, S. D; et al. *J. Biol. Chem.* 1990, **265**(19), 11377-11381
10. Schaeffer, B. K; et al. *Pancreas* 1987, **2**(5), 518-522
11. Longnecker, D. S. *Exp. Pancreatic Carcinog.* 1987, 117-130
12. *ECETOC Technical Report No. 30(4)* 1991, European Chemical Industry Ecology and Toxicology Centre, B-1160 Brussels

A277   Azathioprine

CAS Registry No. 446-86-6

Synonyms 6-[(1-methyl-4-nitro-1*H*-imidazol-5-yl)thio]-1*H*-purine; 6-(1^1-methyl-4^1-nitro-5^1-imidazolyl)-mercaptopurine; 6-(methyl-*p*-nitro-5-imidazolyl)-thiopurine; 6-[(1-methyl-4-nitroimidazol-5-yl)thio]purine

Mol. Formula $C_9H_7N_7O_2S$ **Mol. Wt.** 277.27

Uses Immunosuppressive drug. Antirheumatic.

Physical properties

M. Pt. 243-244°C (decomp.).

Solubility

Organic solvent: ethanol, chloroform, acetone, methanol.

Mammalian and avian toxicity

Acute data

LD_{50} oral mouse, rat 535-1389 mg kg^{-1} (1).
LD_{50} intraperitoneal rat 300 mg kg^{-1} (2).
LD_{50} subcutaneous mouse 350 mg kg^{-1} (3).

Sub-acute data

C57BL and C3H mice were examined at 1, 4 and 10 wk after sub-acute treatment (5 days) 500 mg kg^{-1} caused 1.2-3.4 % increase in sperm abnormalities (4).

Carcinogenicity and long-term effects

Sufficient evidence of carcinogenicity to humans, limited evidence of carcinogencity to animals, IARC classification group 1 (5,6).

Intraperitoneal ♂ and ♀ Sprague-Dawley (CD strain) rat and Swiss-Webster mouse 7.5-37 mg kg^{-1} dose^{-1} induced lymphosarcoma and lung tumours in both ♂ and ♀ mice and uterus tumours in ♀ mice. Skin, pituitary tumours and sarcomas tumours induced in both ♂ and ♀ rats, lymphosarcoma in ♂ rat and breast tumour and leukaemia in ♀ rat (7).

Metabolism and pharmacokinetics

In rats, presence of glutathione lead to formation of 1-methyl-4-nitro-5-(*S*-glutathionyl)- imidazole, then to 1-methyl-4-nitro-5-(*N*-acetyl- *S*-cysteinyl)

---

imidazole and 1-methyl-4-nitro-5- thioimidazole (8).

The immunosuppressive agent azathioprine is metabolised to the purine antagonist 6-mercaptopurine and to 5-substituted 1-methyl-4-nitro-5-thioimidazoles or aminoimidazoles (9).

Genotoxicity

Mutation in microorganisms *Klebsiella pneumoniae* 0.277 mg l^{-1} (10). *Salmonella typhimurium* TA100, TA1535 with and without metabolic activation positive. Induced chromosome aberrations but not sister chromatid exchanges in human lymphocytes *in vitro*. Induced dominant lethal mutations in mice, chromosome aberrations but not sister chromatid exhanges in Chinese hamster bone marrow cells, and induced micronuclei in mice, rabbits and hamsters *in vivo* (11).

Any other adverse effects

Isolated rat hepatocytes exposed to the immunosuppressant azathioprine showed a marked decrease in both oxygen uptake and cell viability (12).

Any other comments

Renal transplant patients were at high risk from non-Hodgkins lymphoma, squamous cell cancers, hepatobiliary carcinomas and mesenchipmal tumours (5).
Toxicity and genotoxicity reviewed (13,14).

References

1. *Drugs in Japan* 6th ed., 1982, **6**, 3
2. *J. Reprod. Fert.* 1962, **4**, 297
3. *J. Med. Chem.* 1975, **18**, 320
4. Wyrobek, A. J; et al. *Proc. Natl. Acad. Sci. USA* 1975, **72**(11), 4425-4429
5. *IARC Monograph* 1987, **Suppl. 7**, 119
6. *IARC Monograph* 1981, **26**, 47
7. Weisburger, E. K. *Cancer* 1977, **40**, 1935-1949
8. de Miranda, P; et al. *J. Pharmacol. Exp. Ther.* 1975, **95**, 50
9. Sauer, H; et al. *Arzneim.-Forsch* 1988, **38**(6), 820-824
10. Voogd, C. E; et al. *Mutat. Res.* 1979, **66**, 207
11. *IARC Monograph* 1987, **Suppl. 6**, 86-88
12. Uno, H; et al. *Josai Shika Daigaku Kiyo* 1986, **15**(2), 418-423, (Jap.) (*Chem. Abstr.* **106**, 168667d)
13. Chan, G. L. C; et al. *Pharmacotherapy (Carlisle, Mass.)* 1987, **7**(5), 165-177
14. Voogd, C. E. *Mutat. Res.* 1989, **221**(2), 133-152

A278   1-Azetidinecarbonyl chloride

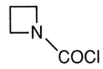

CAS Registry No. 75485-12-0
Mol. Formula C_4H_6ClNO **Mol. Wt.** 119.55
Uses Chemical intermediate (1,2).

References

1. Nuessleinck, L. (Schering A-G.) *German Patent-Ger. Offen. 2, 901,659 (Chem. Abstr.* **93**, P204439e)
2. Alekperov, P. K; et al. *Khim. Geterotsikl. Soedin* 1987(7), 912-914 (*Chem. Abstr.* **108** 55186u)

A279   Azinphos-ethyl

CAS Registry No. 2642-71-9

Synonyms *S*-(3,4-dihydro-4-oxobenzo-[*d*]- [1,2,3]-triazin-3-ylmethyl)*O,O*-diethyl phosphorodithioate; *O, O*-diethyl *S*-(4-oxobenzotriazino-3-methyl)phosphorodithioate; *O,O*- diethyl *S*-[(4-oxo-1,2,3-benzotriazin-3(4*H*)- yl)methyl]phosphorodithioic acid ester; ethyl guthion

Mol. Formula $C_{12}H_{16}N_3O_3PS_2$ **Mol. Wt.** 345.38

Uses Insecticide. Acaricide (1).

Physical properties

M. Pt. 53°C; **B. Pt.** 111°C; **Specific gravity** d_4^{20} 1.284; **Volatility** v.p. $<2.2 \times 10^7$ mmHg at 20°C.

Solubility
Water: 4-5 mg l^{-1}.

Occupational exposure

Risk phrases
Very toxic by inhalation, in contact with skin and if swallowed (R26/27/28)

Safety phrases
Keep locked up – Keep away from food, drink and animal feeding stuffs – In case of accident or if you feel unwell, seek medical advice immediately (show label where possible) (S1, S13, S45)

Ecotoxicity

Fish toxicity
LC$_{50}$ (96 hr) guppy, goldfish 0.01-0.1 mg l^{-1} (1).
LC$_{50}$ (96 hr) rainbow trout 0.019 mg l^{-1} (2).

Invertebrate toxicity
LC$_{50}$ (96 hr) *Penaeus monodon* 120 ppb. *Penaeus monodon* exposed to 1.5-150 ppb had 27-53% shell softening. Histopathological changes in gills and hepatopancreas included slight hyperplasia of the gill epithelium, delamination of the

hepatopancreatocytes and general necrosis and degeneration of these tissues (3).
LC_{50} (48 hr) *Daphnia pulex* 3.2 µg l^{-1} (4).

Effects to non-target species
Toxic to bees (1).

Mammalian and avian toxicity

Acute data
LC_{50} (duration unspecified) inhalation rat 390 mg m^{-3} (5).
LD_{50} dermal rat 250 mg kg^{-1} (7).
LD_{50} intraperitoneal rat 7500 µg kg^{-1} (6).

Sub-acute data
Oral rat (90 day) 2 mg kg^{-1} via feed no ill effects reported (1).

Metabolism and pharmacokinetics
In plants, metabolised to azinphos-ethyl oxon, benzazimide, and dimethyl benzazimide sulfide and disulfide. Oral mammal (species unspecified) >90% eliminated in faeces and urine in 2 days (1).

Genotoxicity
Salmonella typhimurium TA100 without metabolic activation positive (7).

Any other comments
Rapidly hydrolysed in alkaline conditions.

References
1. *The Agrochemicals Handbook* 3rd ed., 1991, RSC, London
2. Verschueren, K. *Handbook of Environmental Data of Organic Chemicals* 1983, 1310, Van Nostrand Reinhold, New York
3. Baticades, C. L; et al. *Aquaculture* 1991, **93**(1), 9-19
4. Sanders, H. O; et al. *Trans. Amer. Fish Soc.* 1966, **95**(2), 165-169
5. Klimmer, O. R. *Pflanzenschutz-und Schaedlingsbekaempfungsmittel: Abriss einer Toxikologie und Therapie von Vergiftungen* 1971, 2nd ed., 16
6. *Guide to the Chemicals Used in Crop Protection* 1973, **6**, 24, Information Canada, Ontario
7. Diril, N; et al. *Doga: Truk Muhendislik Cevre Bilimleri Derg.* 1990, **14**(2), 272-279, (Turk.) (*Chem. Abstr.* **113**, 147089w)

A280   Azinphos-methyl

CAS Registry No. 86-50-0
Synonyms *S*-3,4-dihydro-4-oxobenzo-[d]-1,2,3-triazin-3-ylmethyl; *O,O*-dimethyl phosphorodithioate; Bayer 17147; ENT 23233; R 1582; Cotnion-methyl; Gusathion
Mol. Formula $C_{10}H_{12}N_3O_3PS_2$ **Mol. Wt.** 317.33

Uses Insecticide. Acaracide.

Physical properties

M. Pt. 73-74°C; **Specific gravity** d_4^{20} 1.44.

Solubility
Water: 33 mg l^{-1} at 25°C. Organic solvent: methanol, ethanol, propylene glycol, xylene

Occupational exposure
US TLV (TWA) 0.2 mg m^{-3}; **UK Long-term limit** 0.2 mg m^{-3}; **UK Short-term limit** 0.6 mg m^{-3}.

Risk phrases
Very toxic by inhalation, in contact with skin and if swallowed – Irritating to eyes and skin (R26/27/28, R36/38)

Safety phrases
Keep locked up – Keep away from food, drink and animal feeding stuffs – In case of accident or if you feel unwell, seek medical advice immediately (show label where possible) (S1, S13, S45)

Ecotoxicity

Fish toxicity
LC_{50} (96 hr) channel catfish 3.3 mg l^{-1}.
LC_{50} (96 hr) rainbow trout 14 μg l^{-1} (1).
LC_{50} (96 hr) goldfish 4.3 mg l^{-1}.
LC_{50} (96 hr) bluegill sunfish 20 μg l^{-1} (2).
LC_{50} (96 hr) fathead minnow 65 μg l^{-1} (3).

Invertebrate toxicity
LC_{50} (30 days) *Pteronarcys dorsata* 5 μg l^{-1} (4).
LC_{50} (96 hr) *Aplex hypnorum* 3.7 mg l^{-1} (3).
LC_{50} (48 hr) American oyster eggs 620 ppb in a static laboratory bioassay.
LC_{50} (12 hr) hard clam larvae 860 ppb in a static laboratory bioassay.
LC_{50} (96 hr) *Gammarus lacustris* 0.15 μg l^{-1}.
LC_{50} (96 hr) *Gammarus fasciatus* 0.10 μg l^{-1} (4).

Effects to non-target species
LD_{50} oral mallard duck 13.6 mg kg^{-1}.
Toxic to bees (5).

Environmental fate

Degradation studies
After 44 and 197 days incubation of ^{14}C- labelled compound in a soil, about 50 and 93%, respectively, was degraded and after 222 days incubation, 18.6% of the radiolabel was recovered (6).
The main degradation products in soil and selected by soil microorganisms are benzazimide, thiomethylbenzazimide, bis-(benzozimidyl- methey)disulfide and anthranilic acid (7).

Abiotic removal

Calculated bioconcentration factor for azinphos-methyl is 72 (8).
Hydrolysis $t_{1/2}$ in water at pH 8.6 was 36.4, 27.9, 7.2 days at 6, 25, 40°C, respectively (9).

Mammalian and avian toxicity

Acute data

LD_{50} guinea pig 4-80 mg kg^{-1}.
LD_{50} oral rat 7 mg kg^{-1} (10).
LC_{50} (1 hr) inhalation rat 0.385 mg l^{-1}.
LD_{50} percutaneous rat 220 mg kg^{-1} (4,5).

Carcinogenicity and long-term effects

In 2 year feeding trials, rats receiving 2.5 mg kg^{-1} in the diet showed no ill- effects.
LC_{50} (5 day) bobwhite quail 540 mg kg^{-1} in diet (5).
Intraperitoneal mouse (4-64 hr) 2.08 mg kg^{-1} caused changes in liver tissue after 4 hr.
In liver parenchyma cells, granular endoplasmic reticulum increased in the liver,
whereas glycogen decreased. In the liver parenchyma cells, one of the ribosomes
dissociated from the endoplasmic reticulum and scattered in the cytoplasm (12).
National Toxicology Program tested ♂, ♀ rats and mice with azinphos-methyl via feed
(concentration unspecified). Results showed equivocal evidence in ♂ rat and negative
in mice and ♀ rat (13).

Metabolism and pharmacokinetics

In mammals following oral administration >95% is eliminated in the urine and faeces
within 2 days (5).
In vivo percutaneous absorption in humans is 16%. Oncculsion 56% dose absorbed
(11).

Genotoxicity

Saccharomyces pombe without metabolic activation positive (14).
Salmonella typhimurium TA98, TA100, TA1535, TA1537 with and without
metabolic activation weakly positive (15).
Failed to induce sister chromatid exchanges in cultured human lymphocytes (16).

Any other adverse effects

Intravenous human (dose unspecified), radioactivity equivalent to about 1.5% of
administered dose hr^{-1} recovered in urine during the first 12 hr. Recovery decreased
gradually but still slightly over 0.1% 96-120 hr after injection. Total recovery during
120 hr was 69.% of the dose following intravenous administration and 15.9%
following dermal application (17).

Legislation

Included in Schedule 5 (Release Into Water: Prescribed Substances) of Statutory
Instrument No. 472, 1991 (18).
Limited under EC Directive on Drinking Water Quality 80/778/EEC.
Organophosphorous insecticide: maximum admissible concentration 0.1 μg l^{-1} (19).

References

1. Macek, K. J; et al. *Trans Amer. Fish Soc.* 1970, **99**(1), 20-27
2. Katz, M. *Trans Amer. Fish Soc.* 1961, **90**(3), 264-268

3. Holcombe, G. W; et al. *Arch. Environ. Contam. Toxicol.* 1987, **16**, 697
4. Verschueren, K. *Handbook of Environmental Data of Organic Chemicals* 2nd ed., 1983, Van Nostrand Reinhold, New York
5. *The Agrochemicals Handbook* 3rd ed., 1991, RSC, London
6. Engelhardt, G; et al. *J. Agric. Food Chem.* 1984, **32**, 102-108
7. Engelhardt, G; et al. *Chemosphere* 1983, **12**, 955-960
8. Lyman, W. J; et al. *Handbook of Chemical Property Estimation Methods. Environmental Behaviour of Organic Compounds* 1982, McGraw-Hill, New York
9. Heuer, B; et al. *Bull. Environ. Contam. Toxicol.* 1974, **11**, 532-537
10. *J. Pharm. Pharmacol.* 1961, **13**, 435
11. Wester, R. C; et al. *J. Toxicol. Environ. Health* 1985, **16**(1), 25-38
12. Ozata, A; et al. *Doga: Diyol. Ser.* 1986, **10**(3), 447-451, (Turk.) (*Chem. Abstr.* **106**, 15164q)
13. *National Toxicology Program, Research and Testing Div.* 1992, Report No. 69, NIEHS, Research Triangle Park, NC 27709
14. Gilot-Del Halle, J; et al. *Mutat Res.* 1983, **117**, 139
15. Zeiger, E; et al. *Environ. Mutagen* 1987, 9(Suppl. 9), 1
16. Gomez-Arroyo, S; et al. *Contam. Ambient.* 1987, **3**(1), 63-70
17. Hayes, W. J, Jr. *Pesticides Studied in Man* 1982, 359, Williams & Wilkins, Baltimore
18. *S. I. No. 472 The Environmental Protection (Prescribed Processes and Substances) Regulations* 1991, HMSO, London
19. *EC Directive Relating to the Quality of Water Intended for Human Consumption* 1982, 80/778/EEC, Office for Official Publications of the European Communities, 2 rue Mercier, L-2985 Luxembourg

A281   Aziprotryne

CAS Registry No. 4658-28-0
Synonyms aziprotryn; 4-azido-N-(1-methylethyl)-6-methylthio-1,3,5-triazin-2-amine; 2-azido-4-(isopropylamino)-6-(methylthio)-s-triazine; 4-azido-4-isopropylamino-6-methylthio-1,3,5-triazine
Mol. Formula $C_7H_{11}N_7S$ **Mol. Wt.** 225.28
Uses Herbicide. Fungicide.

Physical properties

M. Pt. 94.5-95.5°C; **Specific gravity** 1.4 at 20°C; **Volatility** v.p. 2.67×10^{-5} mmHg at 20°C.

Solubility

Organic solvent: acetone, ethyl acetate, dichloromethane, benzene

Ecotoxicity

Fish toxicity

LC_{50} (96 hr) largemouth bass, bluegill sunfish >1 mg l^{-1} (1).

Mammalian and avian toxicity

Acute data

LD$_{50}$ (8 day) oral mallard duck, quail >4000 mg kg^{-1} in diet (2).
LD$_{50}$ oral rat, mouse 2970-3600 mg kg^{-1} (3).
LD$_{50}$ (6 hr) inhalation rat >208 mg m^{-3} (1).
LD$_{50}$ dermal rat >3000 mg kg^{-1} (4).
LD$_{50}$ intraperitoneal mouse 265 mg kg^{-1} (3).

Legislation

Limited under EC Directive on Drinking Water Quality 80/778/EEC. Herbicide: maximum admissible concentration 0.5 µg l^{-1} (5).

References

1. *The Pesticide Manual* 9th. ed., 1991, Worthing, C. R. (Ed.), British Crop Protection Council, Farnham
2. *The Agrochemicals Handbook* 3rd ed., 1991, RSC, London
3. Yamada, K; et al. *Eisei Kagaku* 1971, **17**(5), 310-326, (Jap.) (*Chem. Abstr.* **76**, 122607t)
4. Gabrkyan, S. G; et al. *Zh. Eksp. Klin. Med.* 1975, **15**(5), 32-36, (Russ.) (*Chem. Abstr.* **84**, 174819y)
5. *EC Directive Relating to the Quality of Water Intended for Human Consumption* 1982, 80/778/EEC, Office for Official Publications of the European Communities, 2 rue Mercier, L-2985 Luxembourg

A282   Aziridine

CAS Registry No. 151-56-4
Synonyms ethylenimine; aminoethylene; dihydroazirene; azacyclopropane; azirane; dimethyleneimine; ethylimine; ENT-50324; RCRA Waste No. P054; TL 337; dihydro-1*H*-azirine
Mol. Formula C$_2$H$_5$N   **Mol. Wt.** 43.07
Uses Chemical intermediate.

Physical properties

M. Pt. −71.5°C; **B. Pt.** 56.72°C; **Flash point** −11°C (closed cup); **Specific gravity** d$_4^{20}$ 0.832; **Volatility** v.p. 160 mmHg at 20°C.

Solubility

Water: miscible. Organic solvent: ethanol, benzene, acetone, ether

Occupational exposure

US TLV (TWA) 0.5 ppm (0.88 mg m^{-3}); **UK Long-term limit** 0.5 ppm (1 mg m^{-3}) under review; **UN No.** 1185; **Conveyance classification** toxic substance; flammable liquid; **Supply classification** highly flammable and toxic.

Risk phrases

Highly flammable – Very toxic by inhalation, in contact with skin and if swallowed – Possible risk of irreversible effects (R11, R26/27/28, R40)

Safety phrases

Keep container in a well ventilated place – Do not empty into drains – Wear suitable protective clothing – In case of accident or if you feel unwell, seek medical advice immediately (show label where possible) (S9, S29, S36, S45)

Mammalian and avian toxicity

Acute data

LD_{50} oral rat 15 mg kg^{-1} (1).
LC_{50} (2 hr) inhalation rat, mouse, rabbit 100-400 mg m^{-3} (2).
LD_{Lo} (8 hr) inhalation rat, guinea pig 25 ppm. (3).

Carcinogenicity and long-term effects

No adequate data for carcinogenicity to humans, limited evidence for carcinogenicity to animals, IARC classification group 3 (4).
Oral ♂ and ♀ mouse 4.64 mg kg^{-1} day^{-1}, subsequently 13 mg kg^{-1} diet 77-78 wk. The number of hepatomas and pulmonary tumours combined was significantly greater than that in controls (5).
Subcutaneous ♂ and ♀ albino rat twice weekly, total dose 20 mg kg^{-1} over 67 injections. Total number of tumours at injection site greater than that in controls (6).

Genotoxicity

Chromosomal aberrations in Syrian hamster cells induced by 43 mg l^{-1} (7).
Inhibits DNA transformation by *Bacillus subtilis* at 86 mg l^{-1} (8).
Chinese hamster ovary cells at 2 mg l^{-1} positive at five independent genetic loci (Emt^r, Drb^R, Oua^R, Mbg^r and Thg^r) (9).
Drosophilia melanogaster induction of sex-linked recessive lethals and ring-X loss in ♂ adult (10).

Any other adverse effects to man

Symptomatic effects, which appeared 3-7 hr after a 2 hr exposure to aziridine, included vomiting, irritation of eyes and respiratory tract (11).

Any other adverse effects

Inhalation rat 0.01 mg l^{-1} (4 hr for 1.5 months) caused catarrhal bronchitis, diminishing of lymphatic elements in lymph glands and degenerative changes in liver and kidney (12).

Any other comments

Human health effects and experimental toxicology reviewed (13-15).

References

1. Smyth, H. F; et al. *J. Ind. Hyg. Toxicol.* 1948, **30**, 63-68
2. Izermov, N. F; et al. *Toxicometric Parameters of Industrial Toxic Chemicals under Single Exposure* 1982, Centre of International Projects, GKNT, Moscow
3. Carpenter, C. P; et al. *J. Ind. Hyg. Toxicol.* 1948, **30**, 2-6
4. *IARC Monograph* 1987, **Suppl. 7**, 58
5. Innes, J. R. M; et al. *J. Nat. Cancer Inst.* 1969, **42**, 1101-1114

6. Walpole, A. L; et al. *Br. J. Pharmacol.* 1954, **9**, 306-323
7. Dubinin, N. P; et al. *Dokl. Akad. Nauk. SSSR* 1973, **210**(2), 464-467, (Russ.) (*Chem. Abstr.* **79**, 133304e)
8. Daciulyte, J. *Liet. TSR. Mokslu. Akad. Darb., Ser. C* 1974, (1), 11-22, (Russ.) (*Chem. Abstr.* **81**, 163879e)
9. Gupta, R. S; et al. *Mutat. Res.* 1982, **94**(2), 449-466
10. Zijlstra, J. A; et al. *Mutat. Res.* 1988, **201**(1), 27-38
11. Weightman, J; et al. *J. Am. Med. Assoc.* 1964, **189**(7), 543-545
12. Zaeva, G. N; et al. *Toksikol. Novykh. Prom. Khim. Veschchestv* 1966, **8**, 41-60
13. Santodonato, J; et al. *Report 1985, SRC-TR-84-740*
14. Verschaeve, L; et al. *Mutat. Res.* 1990, **238**(1), 39-55
15. *ECETOC Technical Report No. 30(4)* 1991, European Chemical Industry Ecology and Toxicology Centre, B-1160 Brussels

A283   Aziridine, 1,1α,1α-phosphinylidene tris-

CAS Registry No. 545-55-1
Synonyms 1,1',1''-phosphinylidynetrisaziridine; *tris*(1-aziridinyl)phosphine oxide; phosphoric acid triethylene imide; Aphoxide; TEPA; APO
Mol. Formula $C_6H_{12}N_3OP$　　　　　**Mol. Wt.** 173.16
Uses Used in dyeing, creaseproofing and flameproofing textiles. Stabiliser for polymers. In photographic emulsion hardening. Antineoplastic agent. Insect sterilant.

Physical properties
M. Pt. 41°C; **B. Pt.** 90-91°C at 23°C.

Solubility
Organic solvent: ethanol, diethyl ether, acetone

Mammalian and avian toxicity

Acute data
LD_{50} oral rat 37 mg kg^{-1} (1).
LD_{Lo} intraperitoneal mouse 156 μg kg^{-1} (2).
LD_{50} intravenous mouse 178 mg kg^{-1} (3).

Carcinogenicity and long-term effects
No adequate data for carcinogenicity to humans, inadequate evidence for carcinogenicity to animals, IARC classification group 3 (4).

Metabolism and pharmacokinetics
Excreted in urine as a thiophosphamide metabolite (5).
Intraperitoneal rat and mouse (dose unspecified) radioactivity was not localised selectively in any of the tissues examined. During first 24 hrs after treatment, 60-75% of dose excreted in urine, 2-5% in faeces in mouse with 80% of radioactivity

identified as inorganic phosphate. In rat, 80% of radioactivity in blood, associated with haemoglobin. During first 24 hr, 89-90% of radioactivity excreted in urine, however, 50-70% of urinary radioactivity was present as unchanged TEPA (6).
The sperm-rich fraction of boar semen was treated *in vitro* for 10 min with an equal volume of 1% solution. On average, 0.8% was taken up in the spermatozoa, 69% was associated with the heads and 31% with the tails and acrosomes (7).

Genotoxicity
Drosophila melanogaster wing spot test positive (8).

Any other comments
Human health effects and experimental toxicology reviewed (9).

References
1. *Bull. World Health Organisation* 1964, **31**, 737
2. *Toxicol. Appl. Pharmacol.* 1972, **23**, 288
3. Sax, N. I; et al. *Dangerous Properties of Industrial Materials* 7th ed., 1989, Van Nostrand Reinhold, New York
4. *IARC Monograph* 1987, Suppl. 7, 73
5. Chistyakov, V. V; et al. *Khim. Farm. Zh.* 1988, **22**(10), 1158-1162, (Russ.) (*Chem. Abstr.* **10**, 689213v)
6. *IARC Monograph* 1973, **9**, 75
7. Stokes, J. B; et al. *Agric. Res. Results* 1981, 1-7
8. Graf, U; et al. *Mutat. Res.* 1989, **222**(4), 359-373
9. *ECETOC Technical Report No. 30(4)* 1991, European Chemical Industry Ecology and Toxicology Centre, B-1160 Brussels

A284   Azobenzene

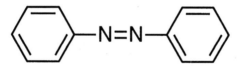

CAS Registry No. 103-33-3
Synonyms diphenyl diazine; azobisbenzene; azodibenzene; azodibenzeneazofume; benzenazobenzene; diazobenzene
Mol. Formula $C_{12}H_{10}N_2$ **Mol. Wt.** 182.23
Uses Acaricide. Chemical and dyestuff intermediate.

Physical properties

M. Pt. 68°C; **B. Pt.** 297°C; **Specific gravity** d_4^{20} 1.203; **Partition coefficient** log P_{ow} 3.82 (1); **Volatility** v.p. 1 mmHg at 103.5°C.

Solubility
Organic solvent: ethanol, diethyl ether, acetone, benzene

Occupational exposure

Supply classification harmful.

Risk phrases
Harmful by inhalation and if swallowed (R20/22)

Safety phrases
After contact with skin, wash immediately with plenty of water (S28)

Environmental fate

Nitrification inhibition
Nitrosomonas sp. no inhibition of ammonia oxidation at concentrations of 100 mg l^{-1} (2).

Mammalian and avian toxicity

Acute data
LD$_{50}$ oral rat 1000 mg kg^{-1} (3).
LD$_{50}$ intraperitoneal mouse 500 mg kg^{-1} (4).

Carcinogenicity and long-term effects
No adequate evidence for carcinogenicity to humans, limited evidence for carcinogenicity to animals, IARC classification group 3 (5).
National Toxicology Program tested ♂, ♀ rats and mice with azobenzene via feed. Results positive in rats, negative in mice (6).

Metabolism and pharmacokinetics
Rabbit (route unspecified) 500 mg kg^{-1}, 30% appeared in faeces, 23% excreted in urine (7).

Genotoxicity
Salmonella typhimurium TA100 with metabolic activation positive (8).

Any other comments
Human health effects and experimental toxicology reviewed (9,10).

References

1. Verschueren, K. *Handbook of Environmental Data on Organic Chemicals* 2nd ed., 1983, Van Nostrand Reinhold, New York
2. Hockenbury, M. R; et al. *J. Water Pollut. Contr. Fed.* 1977, **49**(5), 768-777
3. Sax, N. I; et al. *Dangerous Properties of Industrial Materials* 7th ed., 1989, Van Nostrand Reinhold, New York
4. *NTIS Report* AD 277-689, Nat. Tech. Info. Serv. Springfield, VA
5. *IARC Monograph* 1987, **Suppl. 7**, 58
6. *National Toxicology Program, Research and Testing Div.* 1992, Report No. 154, NIEHS, Research Triangle Park, NC 27709
7. *IARC Monograph* 1973, **878**
8. McCann, J; et al. *Proc. Nat. Acad. Sci. USA* 1975, **72**(12), 5135-5139
9. *ECETOC Technical Report No. 30(4)* 1991, European Chemical Industry Ecology and Toxicology Centre, B-1160 Brussels
10. *Dangerous Prop. Ind. Mater. Rep.* 1987, **7**(1), 38-47

A285 Azocyclotin

CAS Registry No. 41083-11-8

Synonyms tri(cyclohexyl)-1H-1,2,4-triazol-1-yltin; (1H-1,2,4-triazolyl-1-yl) tricyclohexyl stannane; (1H-1,2,4-triazolyl)tricylohexyl stannane; 1-(tricyclohexylstannyl)-1H-1,2,4-triazole

Mol. Formula $C_{20}H_{35}N_3Sn$ **Mol. Wt.** 436.21

Uses Acaricide.

Physical properties

M. Pt. 218.8°C; **Volatility** <3.76×10^{-6} mmHg.

Solubility

Water: < 1 mg l^{-1} at 20°C. Organic solvent: dichloromethane, isopropanol

Occupational exposure

US TLV (TWA) 0.1 mg m^{-3} (as Sn).

Ecotoxicity

Fish toxicity

LC_{50} (96 hr) carp, goldfish, rainbow trout 0.005-0.1 mg l^{-1} 25% wettable powder formulation (1).

Effects to non-target species

Non-toxic to bees (1).

Mammalian and avian toxicity

Acute data

LD_{50} oral rat, guinea pig, mouse 100-450 mg kg^{-1} (1).
LD_{50} oral chicken 250-370 mg kg^{-1} (1).
LD_{50} dermal rat 1000 mg kg^{-1} (2).

Sub-acute data

Oral rat (90 day) 15 mg kg^{-1} in diet, no adverse effects reported (1).

Legislation

Limited under EC Directive on Drinking Water Quality 80/778/EEC. Pesticide: maximum admissible concentration 0.1 µg l^{-1} (3).

References

1. *The Agrochemicals Handbook* 3rd ed., 1991, RSC, London
2. *Farm Chemicals Handbook* 1983, C182, Meister Publishing Co, Willoughby, OH
3. *EC Directive Relating to the Quality of Water Intended for Human Consumption* 1982, 80/778/EEC, European Commission, Office for Official Publications of the European Communities, 2 rue Mercier, L-2985 Luxembourg

A286   Azodicarbonamide

H₂NCON=NCONH₂

CAS Registry No. 123-77-3

Synonyms azoformamide; 1,1-azobisformamide; azobiscarbonamide; azodicarboxylic acid diamide; azodicarboxamide; diazenedicarboxamide

Mol. Formula $C_2H_4N_4O_2$ **Mol. Wt.** 116.08

Uses Blowing agent for foams, plastics and rubbers. Maturing agent for flour. Food additive No. E927 (UK).

Physical properties

M. Pt. 225°C (decomp.); **Specific gravity** d_{20}^{20} 1.65.

Solubility
Organic solvent: diethyl ether

Mammalian and avian toxicity

Sub-acute data

Inhalation rat, mouse (2 wk) 2-207 mg m^{-3} caused no exposure-related mortality or abnormal clinical signs in rats or mice during or after exposure. No lesions noted, on either gross or histological evaluation of rats or mice. Inhalation rat, mouse (13 wk) 50-204 mg m^{-3} no exposure-related mortality or abnormal clinical signs related were observed. No histopathological lesions reported in mice. In rats exposed to 50 mg m^{-3} lung weights increased and enlarged mediastinal and/or tracheobronchial lymph nodes were observed. No exposure- related lesions observed microscopically in rats exposed to 100 or 200 mg m^{-3} (1).

Any other adverse effects to man

Inhalation of azodicarbonamide dust at levels of 2-5 mg m^{-3} during its manufacture were found to cause sensitisation and asthma in workers. Symptoms ceased on removal from source of irritancy (2).

Any other comments

Human health effects and experimental toxicology reviewed (3).
British Rubber Manufacturers' Association recommends exposure should be reduced to <1 mg m^{-3} (4).

References

1. Medinsky, M. A; et al. *Fundam. Appl. Toxicol.* 1990, **15**(2), 308-319
2. Slovak, A. J. M. *Thorax* 1981, **36**(12), 906-909
3. *ECETOC Technical Report No. 30(4)* 1991, European Chemical Industry Ecology and Toxicology Centre, B-1160 Brussels
4. *Occup. Health* 1986, **38**(3), 72

A287   2,2'-Azobis(2-methylpropionitrile)

$$NCC(CH_3)_2N=NC(CH_3)_2CN$$

CAS Registry No. 78-67-1

Synonyms Propanenitrile, 2,2'azobis(2 methyl)-; 2,2'-azobisisobutyronitrile; α,α'-azobisisobutylonitrile; Aceto azib; Porofor 57; Genitron

Mol. Formula $C_8H_{12}N_4$ **Mol. Wt.** 164.21

Uses Initiator for free radical reactions, blowing agent for elastomers and plastics.

Physical properties

M. Pt. 107°C (decomp.).

Solubility
Organic solvent: methanol, ethanol

Mammalian and avian toxicity

Acute data
LD_{Lo} oral rat 670 mg kg^{-1} (1).
LD_{50} oral mouse 700 mg kg^{-1} (2).
LD_{50} intraperitoneal mouse 25 mg kg^{-1} (3).

Metabolism and pharmacokinetics
After oral administration to mice 700 mg kg^{-1} (duration unspecified) 2,2'-azobis(2-methylpropionitrile) formed hydrogen cyanide which was detected in the blood, liver and brain. It was not absorbed through the skin (2).

Any other comments

Explosive decomposition can occur.

References

1. Deichmann, W. B. *Toxicology of Drugs and Chemicals* 1969, Academic Press, London
2. Rusin, V. Y. *Trudy Nauch. Sessii Leningrad. Nauch.-Issledovatel. Inst. GigieNew York Truda i Profzabolev.* 1958, 247-251, (Russ.) (*Chem. Abstr.* **56**, 2682f)
3. *NTIS Report* AD 691-490, National Technical Information Service, Springfield, VA

A288   Azoprocarbazine

$$CH_3N=NCH_2-\langle\!\langle\ \rangle\!\rangle-CONHCH(CH_3)_2$$

CAS Registry No. 2235-59-8

Synonyms 4-[(methylazo)methyl]-*N*-(1-methylethyl)-benzamide; *N*-isopropyl-1alp-(2-methylazo)-*p*-toluamide

Mol. Formula $C_{12}N_{17}N_3O$ **Mol. Wt.** 440.27

Mol. Formula $C_{12}N_{17}N_3O$　　　　　　　　**Mol. Wt.** 440.27

Occurrence As the primary oxidative metabolite of the drug procarbazine (1).

Mammalian and avian toxicity

Metabolism and pharmacokinetics
Metabolised by rat liver microsomes to azoxy metabolites (1).

Any other comments
Found to be active as procarbazine in increasing the lifespan of mice implanted with leukaemia cells (2).

References
1. Cummings, S. W; et al. *Drug Metab. Dispos,* 1982, **10**(5), 459-464
2. Sluba, D. A; et al. *Cancer Chemother. Pharmacol.* 1983, **11**(2), 124-129

A289　Azothoate

CAS Registry No. 5834-96-8

Synonyms *O*-4-(4-chlorophenylazo)phenyl *O,O*-dimethylphosphorothioate; *O*-[4-(4-chlorophenyl)azo]phenyl]-*O, O*-dimethyl phosphorothioate; O-[*p*-[*p*-chlorophenyl)azo]phenyl]-*O,O*-dimethyl phosphorothioate; Slam C

Mol. Formula $C_{14}H_{14}ClN_2O_3PS$　　　　　　**Mol. Wt.** 356.77

Uses Insecticide, acaricide.

Physical properties
M. Pt. 76.5-78°C.

Solubility
Organic solvent: methanol

Occupational exposure

Risk phrases
Harmful by inhalation and if swallowed (R20/22)

Safety phrases
Keep out of reach of children – Keep away from food, drink and animal feeding stuffs (S2, S13)

Legislation
Limited under EC Directive on Drinking Water Quality 80/778/EEC. Organopohosphorus insecticide: maximum admissible concentration 0.1 µg l^{-1} (1).

References

1. *EC Directive Relating to the Quality of Water Intended for Human Consumption* 1982, 80/778/EEC, Office for Official Publications of the European Communities, 2 rue Mercier, L-2985 Luxembourg

A290   Azoxybenzene

CAS Registry No. 495-48-7

Synonyms diazene, diphenyl 1-oxide; azobenzene oxide; azoxybenzide; azoxydibenzene; diphenyldiazene 1-oxide

Mol. Formula $C_{12}H_{10}N_2O$ **Mol. Wt.** 198.23

Uses Acaricide. Chemical intermediate.

Physical properties

M. Pt. 36°C; **B. Pt.** decomp.; **Specific gravity** d_4^{26} 1.1590.

Solubility
Organic solvent: ethanol, diethyl ether, acetone

Occupational exposure

Supply classification harmful.

Risk phrases
Harmful by inhalation and if swallowed (R20/22)

Safety phrases
After contact with skin, wash immediately with plenty of water (S28)

Mammalian and avian toxicity

Acute data
LD_{50} oral rat, mouse 515-700 mg kg^{-1} (1,2).
LD_{50} dermal rabbit 1090 mg kg^{-1} (1).
LD_{Lo} intraperitoneal mouse 500 mg kg^{-1} (3).
LD_{Lo} subcutaneous rabbit 250 mg kg^{-1} (3).

Sub-acute data
Caused swelling of spleen and liver, icteritious skin, methaemoglobinemia, and atrophy of the testes and epididymis in rats and mice after oral administration of LD_{50} (2).

Carcinogenicity and long-term effects
No adequate data for carcinogenicity to humans, limited evidence for carcinogenicity to animals, IARC classification group 3 (4).

Irritancy

10 mg applied to rabbit skin for 24 hr caused mild irritancy, and 500 mg instilled into rabbit eye for 24 hr caused mild irritancy (1).

Genotoxicity

Escherichia coli without metabolic activation positive (5).

Any other comments

Oral administration to ♂ and ♀ rat, 0-100 mg kg^{-1} daily for ≤7 days caused time- and dose-dependent decrease in cytochrome P450 and in the activity of aminopyrine-*N*-demethylase and aniline hydroxylase in the hepatic microsomes of both sexes.

Retardation and body weight gain, increase in relative liver weight and increases in microsomal cytochrome b5 were reported (6).

Human health effects and experimental toxicology reviewed (7).

References

1. Smyth, H. F; et al. *Arch. Ind. Hyg. Occup. Med.* 1954, **10**, 61-68
2. Nakamura, E; et al. *Kyoritsu Yakka Daigaku Kenkyu Nempo 1976* 1977, **21**, 25-47, (Jap.) (*Chem. Abstr.* **87**, 16710p)
3. Sax, N. I; et al. *Dangerous Properties of Industrial Materials* 7th ed., 1989, Van Nostrand Reinhold, New York
4. *IARC Monograph* 1987, **Suppl. 7**, 58
5. Rosenkranz, H. S; et al. *Prog. Mutat. Res.* 1981, **1**, 210-218
6. Plass, R; et al. *Nahrung* 1988, **32**(10), 989-997, (Ger.), (*Chem. Abstr.* **110**, 226808w)
7. *ECETOC Technical Report No. 30(4)* 1991, European Chemical Industry Ecology and Toxicology Centre, B-1160 Brussels

B1 Bacitracin

CAS Registry No. 1405-87-4
Synonyms Altracin; Fortracin; Penitracin; Topitracin; Ayfivin; Zutracin
Mol. Formula $C_{66}H_{103}N_{17}O_{16}S$ **Mol. Wt.** 1422.73
Uses Antibiotic applied topically to treat infections of the skin, nose or eye. An animal feed supplement.
Occurrence Antimicrobial polypeptide produced by certain strains of *Bacillus licheniformis* and *Bacillus subtilis*.
Commercial bacitracin is a mixture of at least 9 bacitracins.

Physical properties

Solubility
Organic solvent: ethanol, methanol, cyclohexanol

Ecotoxicity

Effects to non-target species
LD$_{50}$ oral red-winged blackbird >100 mg kg^{-1}.
LD$_{50}$ oral Japanese quail >316 mg kg^{-1} (1).

Mammalian and avian toxicity

Acute data

LD_{50} oral guinea pig 2000 mg kg^{-1} (2).
LD_{50} intraperitoneal, intravenous rat, mouse 190-360 mg kg^{-1} (3).
LD_{50} subcutaneous mouse 1300 mg kg^{-1} (3).

Metabolism and pharmacokinetics

There is no appreciable absorption from the gastroinestinal tract. Following intramuscular injection ≈30% of a single injected dose is excreted in urine within 24 hr. Bacitracin readily diffuses into the pleural and ascitic fluids but little passes into the cerebrospinal fluid (4).

Following parenteral administration of large single doses, significant concentrations of the drug persisted in the blood stream for as long as 7-8 hr (5).

Any other comments

Physico-chemical properties, experimental toxicology and health effects reviewed (6,7).

References

1. Schafer, E. W. Jr. *Arch. Environ. Toxicol.* 1983, **12**, 355-382
2. Radowski, J. L; et al. *Antibiol. Chemother.* 1954, **4**, 304-307
3. Scudi, J. V; et al. *Proc. Soc. Exp. Biol. Med.* 1947, **64**, 503-506
4. *Martindale: The Extra Pharmacopoeia* 29th ed., 1989, The Pharmaceutical Press, London
5. Scudi, J. V; et al. *Proc. Soc. Exp. Biol. Med.* 1947, **65**, 9-13
6. Wolstenholme C. E. W; et al. *Ciba Foundation Symposium on Amino Acids and Peptides with Antimetabolic Activity* 1958, 226-246, Little, Brown, Boston
7. *Dangerous Prop. Ind. Mater. Rep.* 1988, **8**(4), 23-26

B2   Bandrowski's base

CAS Registry No. 20048-27-5
Synonyms
N,N''-(2,5-diamino-2,5-cyclohexadiene-1,4-diylidene)bis-1,4-benzenediamine
Mol. Formula $C_{18}H_{18}N_6$ **Mol. Wt.** 318.38
Occurrence Water samples from oxidation of *p*-phenyldiamine.

Genotoxicity

Salmonella typhimurium TA1538 without metabolic activation positive (1).
Salmonella typhimurium TA98, TA1538 with and without metabolic activation induced frame shift mutations (2).
Salmonella typhimurium TA98 without metabolic activation highly mutagenic on exposure to ultra violet light (3).

References

1. Ames, B. N; et al. *Proc. Nat. Acad. Sci. USA* 1975, **72**, 2423
2. Shah, M. J; et al. *Toxicol. Appl. Pharmacol.* 1979, **48**, A49
3. Niski, K; et al. *Mutat. Res.* 1982, **104**, 347

B3   Barbaloin

CAS Registry No. 1415-73-2
Synonyms (*R*)-10-β-D-glucopyranosyl-1,8-dihydroxy-3-(hydroxymethyl)-9(10*H*)-anthracenone
Mol. Formula $C_{21}H_{22}O_9$ **Mol. Wt.** 418.40
Uses Laxative.
Occurrence Various species of aloe.

Physical properties

M. Pt. 148-9°C.
Solubility
Organic solvent: pyridine, methanol

Mammalian and avian toxicity

Acute data
LD_{Lo} oral cat 500 mg kg^{-1} (1).
LD_{50} intravenous mouse 200 mg kg^{-1} (2).
LD_{Lo} subcutaneous rabbit 200 mg kg^{-1} (1).

Metabolism and pharmacokinetics
Under anaerobic conditions the C-glucosyl bond of barbaloin, a major purgative principle of aloe, was cleaved with human intestinal bacteria, yielding aloe-emodin anthrone and aloe-emodin bianthrone. The faecal flora of humans had the most potent transforming activity whereas those of rats and mice had less or no activity (3).

Any other comments

Has been replaced by safer purgatives (4,5).

References

1. *Abdernalden's Handbuch der Biologischen Arbeitsmethoden* 1935, **4**, 1298
2. *CRC Handbook of Antibiotic Compounds* 1982, **8**(2), 314
3. Hattori, M; et al. *Chem. Pharm. Bull.* 1988, **36**(11), 4462-4466
4. *The Pharmaceutical Codex* 11th ed., 1979, The Pharmaceutical Press, London
5. *Martindale: The Extra Pharmacopoeia* 29th ed., 1989, The Pharmaceutical Press, London

B4 Barban

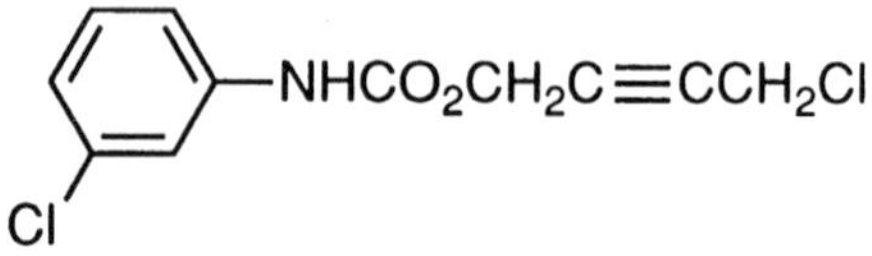

$$\langle\text{phenyl ring with Cl}\rangle-NHCO_2CH_2C\equiv CCH_2Cl$$

Cl

CAS Registry No. 101-27-9

Synonyms Barbamat; Chlorinat; 4-chlorobut-2-ynyl; 3-chlorocarbanilate; 4-chloro-2-butynyl(3-chlorophenyl)carbamate; (3-chlorophenyl)carbamic acid 4-chloro-2-butynyl ester; Barbane

Mol. Formula $C_{11}H_9Cl_2NO_2$ **Mol. Wt.** 258.11

Uses Post-emergence herbicide.

Physical properties

M. Pt. 75-76°C; **Flash point** 81°C; **Specific gravity** 1.403 at 25°C; **Volatility** v.p. 3.76×10^{-5} mmHg.

Solubility

Water: 11 mg l^{-1} at 25°C. Organic solvent: benzene, hexane, kerosene, ethylene dichloride

Occupational exposure

Supply classification harmful.

Risk phrases
Harmful by inhalation, in contact with skin and if swallowed (R20/21/22)

Safety phrases
Keep out of reach of children – Keep away from food, drink and animal feeding stuffs (S2, S13)

Ecotoxicity

Fish toxicity
LC_{50} (96 hr) rainbow trout, bluegill sunfish, goldfish, guppy 0.6-1.3 mg l^{-1} (1).
Effects to non-target species
LC_{50} (8 day) mallard duck >1000 mg kg^{-1}.
LC_{50} (8 day) bobwhite quail >1000 mg kg^{-1}.
Non-toxic to bees (1).

Environmental fate

Degradation studies
Degrades in soil to give 3-chloroaniline.
Residual activity in soil $\approx$2-3 months (1).
Absorption
Carbanilates resist leaching into the soil profile. Immobile in soil and activated by adsorption to soil organic matter (2).

Mammalian and avian toxicity

Acute data

LD_{50} oral mouse, rat, rabbit 322-600 mg kg^{-1} (3,4).

LD_{50} percutaneous rabbit, rat >2000 mg kg^{-1} (1).

Effects on blood cholinesterase activity were reported in rats following inhalation exposure to 80 mg m^{-3} for 4 hr (5).

Sub-acute data

Daily oral administration to guinea pigs and rabbits for 4-6 months caused fatty dystrophy of the liver and kidneys, haemosiderosis of the spleen and vascular hyperemia of the liver, brain, kidneys, spleen and gastric mucosa. Daily doses of 20-40 mg kg^{-1} for 4-6 months to rabbits caused a significant decrease in liver glycogen content (5).

Carcinogenicity and long-term effects

In 2 yr feeding trials the no-effect levels were oral rat 150 mg kg^{-1} in diet^{-1} and oral dog 5 mg kg^{-1} in diet^{-1} (6).

Chronic inhalation exposure caused irritation to the mucous membranes (5).

Metabolism and pharmacokinetics

Following oral administration to rats, chloroaniline, 2-amino-4-chlorophenol and 4-amino-2- chlorophenol were excreted free and in conjugated form. In addition to aniline and *m*-chloroaniline, hydroxycarbamate was found in blood and in all organs; the urine contained *p*-aminophenol (7).

Genotoxicity

Salmonella typhimurium TA100, TA1535 with metabolic activation positive (8).

Legislation

Limited under EEC Directive on Drinking Water Quality 80/778/EEC. Pesticides and related products: maximum admissible concentration 0.5 µg l^{-1} (9).

References

1. *The Agrochemicals Handbook* 1987, 2nd ed., RSC, London
2. Doull, J; et al. *Casarett and Doull's Toxicology* 2nd ed., 1980, MacMillan Publishing Co., New York
3. Dreisbach, R. H. *Handbook of Poisoning, Prevention, Diagnosis and Treatment* 1983, 11th ed., 141, Manuzen Asian Edition, Lange Medical Publications, California
4. *Pharmacol. Rev.* 1962, **14**, 225
5. *Pestline Material Safety Data Sheets for Pesticides and Related Chemicals* 1990, **2**, 1377, Occupational Health Services Inc., Van Nostrand Reinhold, New York
6. Worthing, C. R. *The Pesticide Manual* 1987, 8th ed., BCPC, Surrey
7. Menzie, C. M. *Metabolism of Pesticides, An Update* 1974, 44, US Dept. of the Interior, Fish and Wildlife Service, Special Scientific Report – Wildlife No. 184
8. De Lorenzo, F; et al. *Cancer Res.* 1978, **38**(11), 13-15
9. *EC Directive Relating to the Quality of Water Intended for Human Consumption* 1982, 80/778/EEC, Office for Official Publications of the European Communities, 2 rue Mercier, L-2985 Luxembourg

B5   Barium

Ba

CAS Registry No. 7440-39-3
Mol. Formula Ba   Mol. Wt. 137.34
Uses Used in the radio, ceramics, glass, electronics and computer industries. In the
manufacture of alloys and valves. Extinguisher for uranium and plutonium fires.
Occurrence The toxicity of barium is dependent on the ability of the organism to
absorb it. Therefore toxicity data refer to bioavailable forms, such as the ion in
solution or particulate matter.
A relatively abundant element found in soils, rocks and minerals, also in plant and
animal tissue. Some surface and sea waters also contain barium. Occurs in lead and
zinc ore deposits.
Emissions of barium into the air from mining, refining and processing barium ore can
occur during loading and unloading, stockpiling, materials handling and grinding and
refining the ore.
Fossil fuel combustion may also release barium into the air.
The detonation of nuclear devices in the atmosphere is a source of radioactive barium.

Physical properties

M. Pt. 725°C (metal); **B. Pt.** 1640°C; **Specific gravity** d^{20} 3.51; **Volatility** v.p. 10
mmHg at 1049°C.
Solubility
Organic solvent: ethanol (decomp.)

Occupational exposure

US TLV (TWA) 0.5 mg m^{-3}; **UK Long-term limit** 0.5 mg m^{-3}; **UN No.** 1400;
Conveyance classification substance which in contact with water emits flammable gas.

Ecotoxicity

Fish toxicity
LC_{50} (96 hr) sheepshead minnow >500 mg l^{-1} (1).
LC_{50} (96 hr) salmon 78 mg l^{-1} (2).

Invertebrate toxicity
LC_{50} (48 hr) *Daphnia magna* >530 mg l^{-1} (3).

Mammalian and avian toxicity

Acute data
Acute effects of barium ingestion in experimental animals include salivation, nausea,
diarrhoea, tachycardia, hypokalaemia, twitching, flaccid aralysis of sketetal muscle,
respiratory muscle paralysis and ventrilcular ibrilation (4).

Sub-acute data
Guinea pigs administered 12 mg kg^{-1} daily subcutaneously for 26 wk developed
myeloid hyperplasia of the spleen, liver and bone marrow (5).
Chronic barium feeding induced hypertension and associated cardiovascular
abnormalities (6).

Rats exposed to 100 ppm barium in drinking water for 16 months exhibited depressed rates of cardiac contraction and depressed excitability in the heart. Barium-induced increase in the blood pressure of rats was modest but a comparable mild hypertension in humans would have major health implications (6).

Metabolism and pharmacokinetics

Insoluble barium salts (e.g. barium sulphate) are not absorbed by the intestine and excreted in the faeces of both humans and animals (7).

In humans, a major part of this element is concentrated in the bones (nearly 91%), remainder being in soft tissues such as the brain, heart, kidney, spleen, pancreas and lungs. The skeletal metabolism of barium in humans is qualitatively similar to calcium although the incorporation of these 2 elements is quantitatively different. The $t_{1/2}$ of barium in bones is 50 days (8).

During the first 4 hr after administration, barium absorption from the nasal passages was approximately 61% compared with 11% gastric absorption (9).

Excretion of barium following intravenous injection in a 60 yr old man was 0.22-0.33 and 0.181% in saliva and seminal fluid, respectively after 3-6 hr. The percentage of the injected dose eliminated via the faeces and urine was 20% after 24 hr, 70% after 3 days and 85% after 10 days. (10).

Rats were fed a basal diet containing 25% Brazil nuts (containing 996 ppm barium) and a basal diet to which a concentration of barium as barium chloride was added at 249 ppm equivalent to that in the Brazil nut diet. Corresponding total skeletal deposition of barium was 72.8 and 91.3 ppm respectively. Excretion of barium in the faeces was 10-20- fold greater than that in urine (11).

Any other adverse effects to man

Negative results were obtained for *in vitro* human lymphocyte testing of barium by (phytohaemagglutinin) PHA-induced blastogenesis using [3H]thymidine incorporation (12).

A study of the association between age and sex-adjusted cardiovascular death rates and barium levels in drinking water in 16 Illinois communities was reported (13).

Four cases of pneumonoconiosis were reported in barium miners in Scotland, three of whom developed progressive massive fibrosis, from which two died, and one developed a nodular simple pneumonoconiosis after leaving the industry. There was a complete absence of barium in the lungs suggesting that much of the barium inhaled is not taken into pulmonary tissues, but remains in alveolar macrophages and is eventually removed by the mucociliary mechanism (14).

Toxic effects from doses as low as 0.2-0.5 mg kg^{-1} in adult humans include acute gastroenteritis, loss of deep reflexes with onset of muscular paralysis and progressive muscular paralysis (4).

Workers occupationally exposed to soluble barium compounds in welding fumes revealed 31-234 µg l^{-1} barium in urine after 3 hr and 20- 110 µg l^{-1} in 12 hr (4).

Any other adverse effects

Barium ion contracture tonically activates myocardium while preserving cellular integrity. Myocardium in barium ion contracture is metabolically stable for 30 min (species unspecified) (15).

Legislation

Limited under EEC directive on Drinking Water Quality 80/778-EEC. Barium: guide level 100 µg l^{-1}, maximum admissible concentration 1000 µg l^{-1} (16).

Any other comments

Barium has been reported to inhibit growth and cellular process in microorganisms. It has also been observed to affect the development of germinating bacterial spores (4).

Physico-chemical properties, human health effects, experimental toxicology, ecotoxicology, environmental effects, exposure levels and epidemology reviewed (4,17).

Industrial hygiene and toxicology during use of barium compunds discussed (18).

General information and properties, pharmacokinetics, health effects in humans and animals, quantification of toxicological effects and other criterial guidance and standards summarised (19).

Relatively high levels of barium were found in childrens' colouring materials and coloured magazines printed in Turkey. Health hazards from exposure were discussed (20).

Gastrointestinal absorption of barium relating to physiological differences between infant rats and infant humans reviewed (21).

Variability occurred in the barium concentration in bottled waters (7-660 µg l^{-1}) and groundwaters (7-1160 µg l^{-1}) while concentrations were homogenous in treated waters (13-140 µg l^{-1}). The median value for groundwater was higher than the maximum allowed concentration according to EEC Guidelines. Barium levels were not altered significantly during water transportation. Barium drinking water standards appear to have been developed in the absence of studies examining the human health effects associated with drinking water intake (22).

References

1. Heitmuller, P. T; et al. *Bull. Environ. Contam. Toxicol.* 1981, **27**, 596-604
2. Boutet, C; et al. *Crit. Rev. Soc. Biol. (Paris)* 1973, **167**, 1933
3. Le Blanc, G. A. *Bull. Environ. Contam. Toxicol.* 1980, **24**, 684
4. *Enrionmental Health Criteria 107: Barium* 1990, World Health Organisation, Geneva 5. Kolpakov, V. V. *Petol. Fiziol. Eksp. Ter.* 1971, **5**, 64
6. Perry, H. M. Jr; et al. *J. Toxicol. Environ. Health* 1989, **28**, 373
7. Clary, J. J; et al. *Toxicol. Appl. Pharmacol.* 1974, **29**, 139
8. Bauer, G. C. H; et al. *Acta Orth. Scand.* 1957, **26**, 241
9. Cuddihy, R. G; et al. *Health Phys.* 1973, **25**, 219-224
10. Harrison, G. E; et al. *Nature (London)* 1966, **209**, 526-527
11. Stoewsand, G. S; et al. *Nutr. Rep. Int.* 1988, **38**(2), 259-262
12. Borella, P; et al. *J. Trace Elem. Electrolytes Health Dis.* 1990, **4**(2), 87-95
13. Pendergrass, E. P; et al. *Arch. Ind. Health* 1953, **7**, 44
14. Seaton, A; et al. *Thorax* 1986, **41**(8), 591-595
15. Shibata, T; et al. *Am. J. Physiol.* 1990, **259**(5/2), H1566-H1574
16. *EC Directive Relating to the Quality of Water Intended for Human Consumption* 1982, 80/778/EEC, Office for Official Publications of the European Communities, 2 rue Mercier, L-2985 Luxembourg
17. *ECETOC Technical Report No. 30(4)* 1991, European Chemical Industry Ecology and Toxicology Centre, Brussels
18. Gerasimova, I. L; et al. *Aktal. Probl. Gigieny. Truda. v. Sovrem. Usloviyakh, M.* 1986, 45-50, (Russ.) (*Chem. Abstr.* 1988, **108**, 43181a)
19. *Environmental Protection Agency Report* 1987, PB87-205613

20. Orkun, T; et al. *Chim. Acta. Turc.* 1988, **16**(1), 83-89
21. Golden, R. J; et al. *Int. J. Environ. Stud.* 1989, **34**(3), 217-225
22. Lanciotti, E; et al. *Bull. Environ. Contam. Toxicol.* 1989, **43**(6), 833-837

B6   Barium acetate

$Ba(O_2CCH_3)_2$

CAS Registry No. 543-80-6
Synonyms acetic acid, barium salt; barium diacetate
Mol. Formula $C_4H_6BaO_4$ **Mol. Wt.** 255.43
Uses In fabric printing, lubricating oils. Catalyst in organic synthesis. Paint and varnish drier.

Physical properties

Specific gravity 2.468.
Solubility
Water: 588 g l^{-1} at 0°C. Organic solvent: ethanol

Occupational exposure

US TLV (TWA) 0.5 mg m^{-3}; **UK Long-term limit** 0.5 mg m^{-3};
Supply classification harmful.

Risk phrases

Harmful by inhalation and if swallowed (R20/22)

Safety phrases

After contact with skin, wash immediately with plenty of water (S28)

Ecotoxicity

Fish toxicity

Listed as negative in tests on trout, bluegill sunfish and goldfish. However the authors state that the high mineral content of the water used in these studies adds an additional source of error. Therefore the compounds listed as negative might be toxic in softer water supplies (1).

Mammalian and avian toxicity

Acute data

LD_{50} oral rat, rabbit 240-920 mg kg^{-1} (2,3).
LD_{50} intravenous mouse 23 mg kg^{-1} (4).
LD_{Lo} intravenous rabbit 12 mg kg^{-1} (3).
LD_{Lo} subcutaneous rabbit 96 mg kg^{-1} (3).

Carcinogenicity and long-term effects

Lifetime exposure to barium acetate 0 or 5 mg l^{-1} in drinking water of rat. After 540 days slight increased mortality in ♀ rat. After 150 days increased growth rate in exposed rats over controls (5).

Any other adverse effects to man

The symptoms of barium poisoning from soluble barium salts arise from stimulation of all forms of muscle and include vomiting, colic, diarrhoea, hypertension, convulsive tremors and muscular paralysis. Hypokalaemia is common. Death from cardiac or respiratory failure may occur within one to many hr (6).

Legislation

Limited under EEC Directive on Drinking Water Quality 80/778/EEC. Barium: guide level 100 μg l^{-1}, maximum admissible concentration 1000 μg l^{-1} (7).

Any other comments

Experimental toxicology, environmental fate and human health effects reviewed (8).

References

1. *The Toxicity of 3400 Chemicals to Fish* 1987, EPA 560/6-87-002 PB-200-275, Washington, DC
2. Sax, N. I; et al. *Dangerous Properties of Industrial Materials* 7th ed., 1989, Van Nostrand Reinhold, New York
3. *Environ. Qual. Saf. Suppl.* 1975, **1**, 1
4. Yed, S; et al. *Toxicol. Appl. Pharmacol.* 1972, **22**, 150
5. Schroeder, H. A; et al. *J. Nutr.* 1975, **105**, 421-427
6. *Martindale: The Extra Pharmacopoeia* 29th ed., 1987, The Pharmaceutical Press, London
7. *EC Directive Relating to the Quality of Water Intended for Human Consumption* 1982, 80/778/EEC, Office for Official Publications of the European Communities, 2 rue Mercier, L-2985 Luxembourg
8. *Environmental Health Criteria 107: Barium* 1990, World Health Organisation, Geneva

B7   Barium bromate

Ba(BrO3)2

CAS Registry No. 13967-90-3
Synonyms Bromic acid, barium salt
Mol. Formula BaBr$_2$O$_6$ **Mol. Wt.** 393.15
Uses Analytical reagent. Oxidation agent. Corrosion inhibitor in low carbon steel. Used to prepare rare earth bromates.

Physical properties

M. Pt. 260°C (decomp.); **Specific gravity** d^{18} 3.99.
Solubility
Water: 9.6 g l^{-1} at 30°C. Organic solvent: acetone

Occupational exposure

US TLV (TWA) 0.5 mg m^{-3}; **UK Long-term limit** 0.5 mg m^{-3}; **UN No.** 2719;
Conveyance classification oxidising substance; **Supply classification** harmful.

Risk phrases
Harmful by inhalation and if swallowed (R20/22)

Safety phrases

After contact with skin, wash immediately with plenty of water (S28)

Ecotoxicity

Bioaccumulation

Barium obtained from soluble salts in soil accumulates in some plants and algae (1,2).

Any other adverse effects to man

In humans symptoms of barium poisoning from soluble barium salts arise from stimulation of all forms of muscle and include vomiting, colic, diarrhoea, hypertension, convulsive tremors and muscular paralysis. Hypokalaemia is common. Death from cardiac or respiratory failure may occur within one to many hr (3).

Legislation

Limited under EEC Directive on Drinking Water Quality 80/778/EEC. Barium: guide level 100 μg l^{-1}, maximum admissible concentration 1000 μg l^{-1} (4).

Any other comments

Experimental toxicology, environmental fate and human health effects reviewed (5).

References

1. Robinson, W. O; et al. *US Dep. Agric. Tech. Bull.* 1950, **1013**, 1-36
2. Havlik, B. *J. Hyg. Epidemiol. Microbiol. Immunol.* 1980, **24**(4), 396-404
3. *Martindale: The Extra Pharmacopia* 29th ed., 1989, The Pharmaceutical Press, London
4. *EC Directive Relating to the Quality of Water Intended for Human Consumption* 1982, 80/778/EEC, Office for Official Publications of the European Communities, 2 rue Mercier, L-2985 Luxembourg
5. *Environmental Health Criteria 107: Barium* 1990, World Health Organisation, Geneva

B8   Barium bromide

BaBr$_2$

CAS Registry No. 10553-31-8
Mol. Formula BaBr$_2$ **Mol. Wt.** 297.16
Uses As an intermediate to prepare bromide compounds and phosphors. In photographic compounds.

Physical properties

M. Pt. 847°C; **B. Pt.** (decomp.); **Specific gravity** d^{24} 4.781.
Solubility
Water: 1041 g l^{-1} at 20°C. Organic solvent: methanol

Occupational exposure

US TLV (TWA) 0.5 mg m^{-3}; **UK Long-term limit** 0.5 mg m^{-3}; **Supply classification** harmful.

Risk phrases
Harmful by inhalation and if swallowed (R20/22)

Safety phrases
After contact with skin, wash immediately with plenty of water (S28)

Ecotoxicity

Bioaccumulation
Barium obtained from soluble salts in soil accumulates in some plants and algae (1,2).

Any other adverse effects to man

In humans symptoms of barium poisoning from soluble barium salts arise from stimulation of all forms of muscle and include vomiting, colic, diarrhoea, hypertension, convulsive tremors and muscular paralysis. Hypokalaemia is common. Death from cardiac or respiratory failure may occur within one to many hr (3).

Legislation

Limited under EEC Directive on Drinking Water Quality 80/778/EEC. Barium: guide level 100 μg l^{-1}, maximum admissible concentration 1000 μg l^{-1} (4).

Any other comments

Soluble barium salts are readily absorbed in mammals (5).
In general, the Ba^{2+} ion is toxic or inhibitory to cellular processes in bacteria, fungi, mosses and algae. Environmental fate, experimental toxicology and human health effects reviewed (6).

References

1. Robinson, W. O; et al. *US Dep. Agric. Tech. Bull.* 1950, **1013**, 1-36
2. Havlik, B; et al. *J. Hyg. Epidemiol. Microbiol. Immunol.* 1980, **24**(4), 396-404
3. *Martindale: The Extra Pharmacopoeia* 29th ed., 1989, The Pharmaceutical Press, London
4. *EC Directive Relating to the Quality of Water Intended for Human Consumption* 1982, 80/778/EEC, Office for Official Publications of the European Communities, 2 rue Mercier, L-2985 Luxembourg
5. McCauley, P. T; et al. *Advances in Modern Environ. Toxicology* 1985, Princeton Publ., NJ
6. *Environmental Health Criteria 107: Barium* 1990, World Health Organisation, Geneva

B9   Barium carbonate

BaCO3

CAS Registry No. 513-77-9
Synonyms Witherite; carbonic acid, barium salt (1:1); C.I. 77099; C.I. Pigment White 10
Mol. Formula BaCO$_3$ **Mol. Wt.** 197.35
Uses A rodenticide. Used in ceramics, paints, enamels, marble substitutes and rubber. In the manufacture of paper, barium salts, electrodes and optical glasses. An analytical reagent.
Occurrence In nature as the mineral Witherite.

Physical properties

M. Pt. 1300°C (decomp.).

Solubility
Water: 0.022 g l^{-1} at 20°C. Organic solvent: ethanol

Occupational exposure

Supply classification harmful.

Risk phrases
Harmful by inhalation and if swallowed (R20/22)

Safety phrases
After contact with skin, wash immediately with plenty of water (S28)

Mammalian and avian toxicity

Acute data
LD_{50} oral mouse, rat 200-420 mg kg^{-1}.
LD_{50} intraperitoneal mouse 50 mg kg^{-1} (1).
LD_{Lo} oral human 57 mg kg^{-1} (2).
TD_{Lo} oral human 11 mg kg^{-1} gastro- intestinal tract effects (3).
TD_{Lo} oral human 29 mg kg^{-1} peripheral and central nervous system effects (4).
Oral rabbit, mouse lethal dose 170-300 mg kg^{-1}, respectively (5).

Sub-acute data
♂ rats exposed to 1.15 and 5.2 mg m^{-3} 4 hr day^{-1} 6 days wk^{-1} for 4 months
experienced decreased weight gain, low blood sugar and haemoglobin as well as
leucocytosis and thrombopenia in the high-dose group (6).

Carcinogenicity and long-term effects
A benign pneumoconiosis (baritosis) may result from the inhalation of barium
carbonate dust (7).

Teratogenicity and reproductive effects
TC_{Lo} oral ♀ rat 26 mg kg^{-1} 29 days before conception and throughout pregnancy
caused increased mortality of offspring, low birth weights but no teratogenesis (8).

Metabolism and pharmacokinetics
Following intramuscular injection into hind leg of rats, aerosolised barium carbonate
dissolved rapidly leaving injection site within 3 days (9).

Irritancy
Can cause dermatitis in humans (10).

Any other adverse effects to man

In a reported case of an attempted suicide, the ingestion of 40 g barium carbonate
resulted in a 30% reduction in normal plasma potassium level and induced muscle
weakness, respiratory failure and complete paralysis. Normal muscular and renal
function was regained within 7 days (11).
Out of >100 people who had consumed sausages made with barium carbonate instead
of potato meal, 19 people were hospitalised with symptoms ranging from mild
vomiting and diarrhoea to partial paralysis. One patient died suddenly after
developing right facial paralysis and left hemiplegia (12,13).

Any other adverse effects

Poison to human by ingestion, systemic effects include stomach ulcers, muscle weakness, paresthesias and paralysis (14).

Legislation

Limited under EEC Directive on Drinking Water Quality 80/778/EEC. Barium: guide level 100 μg l^{-1}, maximum admissible concentration 1000 μg l^{-1} (15).

Any other comments

Incompatible with BrF$_3$; 2-furanpercarboxylic acid.

In vivo solubility of the carbonate salt was studied in rats after intramuscular injection. It was found the carbonate and chloride salts were equally soluble in the soft tissues and were absorbed from the injection site very rapidly (16).

Experimental toxicology, human health effects and environmental fate reviewed (6).

References

1. Izmerov, N. F; et al. *Toxicometric Parameters of Industrial Toxic Chemicals under Single Exposure* 1982, 23, Moscow
2. *Pesticide Chemicals Official Compendium* 1966, 95, Assoc. Am. Pesticide Contr. Official, KS
3. *Yakkyoku Pharmacy* 1980, **31**, 1247
4. *Israel J. Med. Sci.* 1967, **3**, 565
5. *USEPA Drinking Water Criteria Document for Barium* 1985, V-2, TR-540-60F
6. *Environmental Health Criteria 107: Barium* 1990, World Health Organisation, Geneva
7. Brenniman, G. R; et al. *Environ. Res.* 1979, **20**, 318
8. Tarasenko, N. Y; et al. *J. Hyg. Epidemiol. Microbiol. Immunol.* 1977, **21**, 361-373
9. Thomas, R. G; et al. *Amer. Ind. Hyg. Ass. J.* 1973, **34**(8), 350-359
10. *Dangerous Prop. Ind. Mater. Rep.* 7th ed., 1981, **1**(6)
11. Phelan, D. M; et al. *Br. Med. J.* 1984, **289**, 882
12. Lewi, Z; et al. *Lancet* 1964, **2**, 342-343
13. Diengott, D; et al. *Lancet* 1964, **2**, 343-344
14. Sax, N. I; et al. *Dangerous Properties of Industrial Materials* 7th ed., 1989, Van Nostrand Reinhold, New York
15. *EC Directive Relating to the Quality of Water Intended for Human Consumption* 1982, 80/778/EEC, Office for Official Publications of the European Communities, 2 rue Mercier, L-2985 Luxembourg
16. Thomas, R. G; et al. *Am. Ind. Hyg. Assoc. J.* 1973, **34**, 350-359

B10   Barium chlorate

Ba(ClO$_3$)$_2$

CAS Registry No. 13477-00-4

Synonyms chloric acid, barium salt

Mol. Formula BaCl$_2$O$_6$ **Mol. Wt.** 304.24

Uses In pyrotechnics. A textile mordant. In the manufacture of other chlorates and of explosives and matches.

Physical properties

M. Pt. 414°C; **B. Pt.** 250°C loses O_2; **Specific gravity** 3.18.
Solubility
Water: 274 g l^{-1} at 15°C. Organic solvent: ethylamine, ethanol, acetone

Occupational exposure

US TLV (TWA) 0.5 mg m^{-3}; **UK Long-term limit** 0.5 mg m^{-3}; **UN No.** 1445;
HAZCHEM Code 2YE; **Conveyance classification** oxidising substance; **Supply classification** oxidising and harmful.

Risk phrases
Explosive when mixed with combustible material – Harmful by inhalation and if swallowed (R9, R20/22)

Safety phrases
Keep away from food, drink and animal feeding stuffs – Take off immediately all contaminated clothing (S13, S27)

Ecotoxicity

Bioaccumulation
Barium obtained from soluble salts in soil accumlates in some plants and algae (1,2).

Mammalian and avian toxicity

Irritancy
An eye, skin and respiratory tract irritant (3).

Any other adverse effects to man

In humans symptoms of barium poisoning from soluble barium salts arise from stimulation of all forms of muscle and include vomiting, colic, diarrhoea, hypertension, convulsive tremors and muscular paralysis. Hypokalaemia is common. Death from cardiac or respiratory failure may occur within one to many hr (4).

Legislation

Limited under EEC Directive on Drinking Water Quality 80/778/EEC. Barium: guide level 100 μg l^{-1}, maximum admissible concentration 1000 μg l^{-1} (5).

Any other comments

In general the Ba^{2+} ion is toxic or inhibitory to cellular processes in bacteria, fungi, mosses and algae (3).
Principal toxic effects of chlorates are the production of methaemoglobin in the blood and destruction of red blood corpuscles.
Damage to heart muscle and kidney has been reported.
Soluble barium salts are readily absorbed in mammals (6).
Chlorates are absorbed by ingestion or by inhalation of dust (7).
Physico-chemical properties, environmental fate, experimental toxicology and human health effects reviewed (3,8).

References

1. Robinson, W. O; et al. *US Dep. Agric. Tech. Bull.* 1950, **1013**, 1-36
2. Havlik, B; et al. *J. Hyg. Epidemiol. Microbiol. Immunol.* 1980, **24**(4), 396-404

3. *Environmental Health Criteria 107: Barium* 1990, World Health Organisation, Geneva
4. *Martindale: The Extra Pharmacopoeia* 29th ed., 1989, The Pharmaceutical Press, London
5. *EC Directive Relating to the Quality of Water for Human Consumption* 1982, 80/778/EEC, Office for Official Publications of the European Communities, 2 rue Mercier, L-2985 Luxembourg
6. McCauley, P. T; et al. *Advances in Modern Environmental Toxicology* 1985, Princeton Publ., NJ
7. *Encyclopedia of Occupational Health and Safety* 1983, **1-2**, 458, International Labour Office, Geneva
8. *ECETOC Technical Report No 30(4)* 1991, European Chemical Industry Ecology and Toxicology Centre, B-1160 Brussels

B11   Barium chloride

BaCl$_2$

CAS Registry No. 10361-37-2
Synonyms barium dichloride; SBa 0108E; NCI-C61
Mol. Formula BaCl$_2$ **Mol. Wt.** 208.25
Uses In the manufacture of pigments, colour lakes, glass. A mordant for acid dyestuffs. Used in the manufacture of pesticides, lube oil additives, boiler compounds and aluminium refining. In leather tanning and finishing. In photographic paper and textiles. A flux in the manufacture of Mg metal. Formerly used as purgative in horses and in cattle.

Physical properties

M. Pt. 962°C; **B. Pt.** 1560°C; **Specific gravity** d^{24} 3.856.
Solubility
Water: 375 g l^{-1} at 26°C.

Occupational exposure

US TLV (TWA) 0.5 mg m^{-3}; **UK Long-term limit** 0.5 mg m^{-3}; **Supply classification** harmful.

Risk phrases
Harmful by inhalation and if swallowed (R20/22)

Safety phrases
After contact with skin, wash immediately with plenty of water (S28)

Mammalian and avian toxicity

Acute data
LD$_{50}$ oral guinea pig, rat 76-188 mg kg^{-1} (1,2).
LD$_{Lo}$ oral mouse, dog, rabbit 70- 170 mg kg^{-} (3-5).
LD$_{50}$ subcutaneous rat 178 mg kg^{-1} (3).
LD$_{Lo}$ subcutaneous guinea pig 55 mg kg^{-1} (4).
LD$_{50}$ intravenous mouse 8.2-19.2 mg kg^{-1} (6-8).
LD$_{50}$ intravenous cat 40 mg kg^{-1} (5).
LD$_{Lo}$ ingestion human 11.4 mg kg^{-1} (9).

Metabolism and pharmacokinetics

In rats nasal absorption of barium chloride was estimated at 60-80% after 4 hr and alveolar absorption may be of a similar magnitude. In hamsters receiving barium chloride by intragastric intubation, 11-32% of the dose was absorbed (10).

When barium was intramuscularly injected into the hind leg of rats, barium chloride dissolved rapidly, leaving injection site within 3 days (11).

Following intramuscular injection of barium chloride solution into five children and two adults, the concentration of barium in serum fell rapidly during the first 6 days and then more slowly. The amount detected on day 6 was <0.0002% of the initial dose (12).

After a single oral dose (1, 5, 25 or 125 mg kg^{-1}) to ♂ weanling rats (weight and strain unspecified) and sacrifice 0.5, 2, 4, 8, 16 and 24 hr after exposure, results indicated rapid absorption of barium from the gastrointestinal tract. Peak barium concentrations occurred 30 min post-administration in blood and soft tissues and 2 hr in the femur (13).

Pigmented and albino ♂ mice (25-30 g) and late gestation ♀ (40 g) were intravenously administered 63 μg kg^{-1} barium containing barium chloride and sacrificed. The calcified tissues, cartilage, kidney and melanin containing tissues of the ♂ and ♀ pigmented animals had highest barium activity at 20 minutes. The calcified and cartilage tissues retained activity even at the longest survival time, 32 days. After four days the only sites other than bone, cartilage and melanin containing tissue positive for barium were gastrointestinal contents, urinary tract and salivary gland (14).

Genotoxicity

Saccharomyces cerevisiae gene conversion and mitotic recombination positive (15). *Salmonella typhimurium* TA97, TA100, TA1535, TA1737, TA1538 with and without metabolic activation negative (16).

Any other adverse effects to man

In humans symptoms of barium poisoning from soluble barium salts arise from stimulation of all forms of muscle and include vomiting, colic, diarrhoea, hypertension, convulsion and muscular paralysis. Hypokalaemia is common. Death from cardiac or respiratory failure may occur within one to many hr (17).

A patient suffered acute renal failure associated with barium chloride poisoning (18).

In a 10 wk study barium chloride was administered in the drinking water of 11 healthy ♂ volunteers at concentratins of 0 mg l^{-1} (1-2 wk), 5 mg l^{-1} (3- 6 wk) and 10 mg l^{-1} (7-10 wk). The levels corresponded to levels found in drinking water of some communities in the US. Subjects ranged in age from 27-61 yr and had no previous history of diabetes, hypertension or cardiovascular disease of any kind. No change was reported in blood pressure, total cholesterol, triglycerides, high-density lipoprotein or low-density lipoprotein cholesterol levels. Serum potassium and glucose levels and urinary metanephrine levels were also unchanged. No significant arrhythmias were noted during the barium exposure period (19).

Any other adverse effects

Rats exposed to 250 ppm barium ion for 5 months were challenged with an arrhythmagenic dose of 1-norepinephrine administered intravenously as 5 mg kg^{-1}. No significant ECG changes occurred (20).

Oral rat (16 months) 1, 10, 100 mg l^{-1} caused depressed cardiac rates and excitabiity, and increased systolic pressure, while decreasing conductivity and conduction (21).

Infusion of barium chloride in conscious rabbits caused severe ventricular dysrhythmias (22).

TD$_{Lo}$ (1 day) intratesticular rat 16 mg kg^{-1} caused reproductive effects to testes, epididymis and spermduct (23).

Legislation

Limited under EEC Directive on Drinking Water Quality 80/778/EEC. Barium: guide level 100 μg l^{-1}, maximum admissible concentration 1000 μg l^{-1}. Chloride: guide level 25 mg l^{-1}, maximum admissible concentration 400 mg l^{-1} (24).

Any other comments

Experimental toxicology, human health effects and environmental fate reviewed (9).

References

1. *Food Res.* 1942, **7**, 313
2. Perry, H. M; et al. *Inorganics in Drinking Water and Cardiovascular Disease* 1985, **20**, 221-229, Calabrese (Ed.), Princeton Publ., NJ
3. *Drug Dosages in Laboratory Animals* 1973, 53
4. Kopp, S. J; et al. *Toxicol. Appl. Pharmacol.* 1985, **72**, 303-314
5. *Environ. Quality Safety Suppl.* 1975, **1**, 1
6. *The Merck Index* 10th ed., 1983, 140, Merck & Co. Inc., Rahway, NJ
7. Venugopal, B; et al. *Metal Toxicity in Mammals* 1978, 66, Plenum Press, New York
8. Syed, I. B; et al. *Toxicol. Appl. Pharmacol.* 1972, **22**, 150-152
9. *Environmental Health Criteria 107: Barium* 1990, World Health Organisation, Geneva
10. Friberg, L; et al. *Handbook of the Toxicology of Metals* 2nd ed., 1986, Elsevier Science Publishers, Amsterdam
11. Thomas, R. G; et al. *Amer. Ind. Hyg. Ass. J.* 1973, **34**(8), 350-359
12. Hayes, W. J. *Pesticides Studied in Man* 1982, 2, Williams & Wilkins, Baltimore
13. Clary, J. J; et al. *Toxicol. Appl. Pharmacol.* 1974, **29**, 139
14. Dencker, L; et al. *Acta. Radiol.* 1976, **15**(4), 273-287
15. *Mutat. Res.* 1986, **1**, 21
16. Monaco, M; et al. *Med. Lav.* 1990, **81**(1), 54-64, (Ital.) (*Chem. Abstr.* 1990, **113**, 53974d)
17. *Martindale: The Extra Pharmacopoeia* 29th ed., 1989, The Pharmaceutical Press, London
18. Wetherill, S. F; et al. *Ann. Intern. Med.* 1981, **95**, 187
19. Wones, R. G; et al. *Environ. Health. Perspect.* 1990, **85**, 1-13
20. McCauley, P. T; et al. *Adv. Mod. Environ. Toxicol.* 1985, 197-210
21. Kolpakov, V. V. *Petol. Fizrod. Elesp. Ter.* 1971, **5**, 64
22. Mathla, M. J; et al. *Arch. Toxicol., Suppl.* 1986, **9**, 205-208
23. *J. Reprod. Fertil.* 1964, **7**, 21
24. *EC Directive Relating to the Quality of Water Intended for Human Consumption* 1982, 80/778/EEC, Office for Official Publications of the European Communities, 2 rue Mercier, L-2985 Luxembourg

B12   Barium cyanide

Ba(CN)$_2$

CAS Registry No. 542-62-1
Synonyms barium dicyanide; RCRA Waste No. PO13

Mol. Formula BaC_2N_2 **Mol. Wt.** 189.38
Uses In metallurgy and electroplating processes.

Physical properties
Solubility
Water: 800 g l^{-1} at 14°C. Organic solvent: ethanol

Occupational exposure
US TLV (TWA) 0.5 mg m^{-3}; **UK Long-term limit** 0.5 mg m^{-3}; **UN No.** 1565;
Conveyance classification toxic substance; **Supply classification** harmful.

Risk phrases
Harmful by inhalation and if swallowed (R20/22)

Safety phrases
After contact with skin, wash immediately with plenty of water (S28)

Ecotoxicity

Fish toxicity
Threshold concentrations for fish 5 ppm Ba^{2+}.
Threshold concentration for fresh and sea water fish 0.02 ppm CN^{-1}.
LD_{100} (120 hr) trout 0.05 mg l^{-1}.
LD_{50} (24 hr) bluegill sunfish 0.18 mg l^{-1}.
LD_{50} (1.5 hr) carp 10 mg l^{-1} (1).

Invertebrate toxicity
Cyanides are moderate chronic poisons and may have accumulative effects. Marine
waters should not exceed concentrations of 1:20 LD_{50} (1).

Bioaccumulation
Bioconcentration factor goldfish 150. Bioconcentration factors in marine and fresh
water plants 500, fish 10, invertebrates 100. Barium was toxic to Euvasian water
milfoil at 40-100 ppm. Cyanide is the more toxic ion, it is toxic by all routes (1).
Ba^{2+} accumulates in the soil from the dissolution of soluble barium salts and can
accumulate in some plants (2).

Any other adverse effects to man
Workers such as electroplaters and picklers can develop a cyanide rash characterised
by itching, and muscular, papular and vesicular eruptions. Case studies and
epidemiological studies have reported symptoms of cyanide poisoning to include
headache, dizziness and thyroid enlargement (3).
A blood cyanide level of >1.0 µg ml^{-1} is lethal (species unspecified) (1).
In humans symptoms of barium poisoning from soluble barium salts arise from
stimulation of all forms of muscle and include vomiting, colic, diarrhoea,
hypertension, convulsive tremors and muscular paralysis. Hypokalaemia is common.
Death from cardiac or respiratory failure may occur within one to many hr (4).

Legislation
Limited under EEC Directive on Drinking Water Quality 80/778/EEC. Barium: guide
level 100 µg l^{-1}, maximum admissible concentration 1000 µg l^{-1}. Cyanides: maximum
admissible concentration 50 µg l^{-1} (5).

Any other comments

Soluble barium salts are readily absorbed in mammals (6).

The Ba^{2+} ion is toxic or inhibitory to cellular processes in bacteria, fungi, mosses and algae (7).

Cyanide ions are rapidly absorbed after oral or parental administration, from skin and mucosal surfaces and are extremely dangerous when inhaled (8,9).

Experimental toxicology, environmental fate and human health effects reviewed (7).

References

1. *Dangerous Prop. Ind. Mater. Rep.* 1983, **3**(4), 31-32
2. Robinson, W. O; et al. *US Dep. Agric. Tech. Bull.* 1950, **1013**, 1-36
3. *NIOSH* 1976, 77-108, Washington DC
4. *Martindale: The Extra Pharmacopoeia* 29th ed., 1989, The Pharmaceutical Press, London
5. *EC Directive Relating to the Quality of Water Intended for Human Consumption* 1982, 80/778/EEC, Office for Official Publications of the European Communities, 2 rue Mercier, L-2985 Luxembourg
6. McCauley, P. T; et al. *Advances in Modern Environ. Toxicol.* 1985, Princeton Publ., NJ
7. *Environmental Health Criteria: Barium* 1990, **107**, World Health Organisation, Geneva
8. Gilman, A.G; et al. *Goodman and Gilman's The Pharmacological Basis of Therapeutics* 7th ed., 1985, 1642, MacMillan Publishing Co., New York
9. Arena, J.M. *Poisoning: Toxicology-Symptoms Treatments* 3rd ed., 1974, 136, Charles C. Thomas, Springfield, IL

B13   Barium disulfide

BaS_2

CAS Registry No. 12230-99-8
Synonyms barium polysulfide
Mol. Formula BaS_2 **Mol. Wt.** 201.47

Physical properties

M. Pt. 925°C; **Specific gravity** 3.84 (calculated).

Occupational exposure

Supply classification irritant.

Risk phrases

Contact with acids liberates toxic gas – Irritating to eyes, respiratory system and skin (R31, R36/37/38)

Safety phrases

After contact with skin, wash immediately with plenty of water (S28)

Any other adverse effects to man

In humans symptoms of barium poisoning from soluble barium salts arise from stimulation of all forms of muscle and include vomiting, colic, diarrhoea, hypertension, convulsive tremors and muscular paralysis. Hypokalaemia is common. Death from cardiac or respiratory failure may occur within one to many hr (1).

Legislation

Limited under EEC Directive on Drinking Water Quality 80/778/EEC. Barium: guide level 100 μg l^{-1}, maximum admissible concentration 1000 μg l^{-1} (2).

Any other comments

Physico-chemical and spectra properties reviewed (3).

References

1. *Martindale: The Extra Pharmacopoeia* 29th ed., 1989, The Pharmaceutical Press, London
2. *EC Directive Relating to the Quality of Water Intended for Human Consumption* 1982, 80/778/EEC, Office for Official Publications of the European Communities, 2 rue Mercier, L-2985 Luxembourg
3. Mellor, J. W. *Comprehensive Treatise on Inorganic and Threshold Chemistry* 3, 752-753, Longman, Harlow

B14   Barium nitrate

Ba(NO3)2

CAS Registry No. 10022-31-8
Synonyms nitrobarite; barium dinitrate; nitric acid, barium salt
Mol. Formula BaN_2O_6 Mol. Wt. 261.35
Uses In pyrotechnics, for green signal lights. In vacuum tubes. In the manufacture of barium oxides. Incendary devices. Ceramic glazes. Rodenticide.

Physical properties

M. Pt. 592°C; B. Pt. decomp.; Specific gravity d^{23} 3.24.
Solubility
Water: 87 g l^{-1} at 20°C.

Occupational exposure

US TLV (TWA) 0.5 mg m^{-3}; UK Long-term limit 0.5 mg m^{-3}; UN No. 1446;
Conveyance classification oxidising substance; toxic substance;
Supply classification harmful.

Risk phrases

Harmful by inhalation and if swallowed (R20/22)

Safety phrases

After contact with skin, wash immediately with plenty of water (S28)

Ecotoxicity

Fish toxicity

Listed as negative in tests on trout, bluegill sunfish and goldfish. However the authors state that the high mineral content of the water used in these studies adds an additional source of error. Therefore the compounds listed as negative might be toxic in softer water supplies (1).

Bioaccumulation

Barium obtained from soluble salts accumulates in some plants and algae (2,3).

Mammalian and avian toxicity

Acute data

LD_{50} oral rat 355 mg kg^{-1} (4).
LD_{Lo} oral dog 800 mg kg^{-1} (5).
LD_{Lo} oral rabbit 150 mg kg^{-1} (5).
LD_{50} intravenous mouse 8 mg kg^{-1} (6).

Irritancy

Dermal rabbit (24 hr) 500 mg caused mild irritation and 100 mg instilled into rabbit
eye caused moderate irritation (4).

Any other adverse effects to man

In humans symptoms of barium poisoning from soluble barium salts arise from
stimulation of all forms of muscle and include vomiting, colic, diarrhoea,
hypertension, convulsive tremors and muscular paralysis. Hypokalaemia is common.
Death from cardiac or respiratory failure may occur within one to many hr (7).

Legislation

Limited under EEC Directive on Drinking Water Quality 80/778/EEC Barium: guide
level 100 μg l^{-1}, maximum admissible concentration 1000 μg l^{-1}. Nitrates: guide level
25 mg l^{-1}, maximum admissible concentration 50 mg l^{-1} (8).

Any other comments

In general the Ba^{2+} ion is toxic or inhibitory to cellular processes in bacteria, fungi,
mosses and algae (9).
Soluble barium salts are readily absorbed in mammals (10).
Experimental toxicology, environmental fate and human health effects reviewed (9).

References

1. *The Toxicity of 3400 Chemicals to Fish* 1987, EPA 560/6-87-002 PB 87-200-275,
   Washington, DC
2. Robinson, W. O; et al. *US Dep. Agric. Tech. Bull.* 1950, **1013**, 1-36
3. Havlik, B; et al. *J. Hyg. Epidiemol. Microbiol. Immunol.* 1980, **24**(4), 396-404
4. Marhold, J. *Sbornik Vysledku Toxixologickeho Vystreni Latek A Pripravku* 1972, 10, Prague
5. *Yakkyoka Pharmacy* 1980, **31**, 1247
6. *Toxicol. Appl. Pharmacol.* 1972, **22**, 150
7. *Martindale: The Extra Pharmacopoeia* 29th ed., 1989, The Pharmaceutical Press, London
8. *EC Directive Relating to the Quality of Water Intended for Human Consumption* 1982,
   80/778/EEC, Office for Official Publications of the European Communities, 2 rue Mercier,
   L-2985 Luxembourg
9. *Environmental Health Criteria 107: Barium* 1990, World Health Organisation, Geneva
10. McCauley, P. T; et al. *Advances in Modern Environmental Toxicology* 1985, Princeton
    Publ., NJ

B15   Barium oxide

BaO

CAS Registry No. 1304-28-5
Synonyms barium monoxide; barium protoxide; baryta; calcined baryta
Mol. Formula BaO   **Mol. Wt.** 153.34
Uses For drying solvents and gases. In the manufacture of lubricating oil detergents.
Chemical intermediate in the production of barium methoxide.

Physical properties

M. Pt. 1923°C; **B. Pt.** 2000°C; **Specific gravity** 5.7.
Solubility
Water: 34.8 g l^{-1} at 20°C. Organic solvent: methanol

Occupational exposure

US TLV (TWA) 0.5 mg m^{-3}; **UK Long-term limit** 0.5 mg m^{-3}; **UN No.** 1884;
HAZCHEM Code 2Z; **Conveyance classification** harmful substance;
Supply classification harmful.

Risk phrases
Harmful by inhalation and if swallowed (R20/22)

Safety phrases
After contact with skin, wash immediately with plenty of water (S28)

Ecotoxicity

Fish toxicity
Threshold concentration for toxicity to young sturgeon 50 mg l^{-1} (1).

Mammalian and avian toxicity

Carcinogenicity and long-term effects
The long-term effects of barium was investigated in Long-Evans rats and CD mice
given 0 or 5 mg barium $litre^{-1}$ (administered as soluble salts) in drinking water
throughout their lifetime. The incidence of tumours in treated animals was not
significanly different to that of controls. It was concluded that under these conditions
barium was not carcinogenic (2).

Irritancy
Dusts of barium oxide are potential dermal and nasal irritants (3).

Any other adverse effects to man

Heavy exposure to dust may cause baritosis (4).
In humans symptoms of barium poisoning from soluble barium salts arise from
stimulation of all forms of muscle and include vomiting, colic, diarrhoea,
hypertension, convulsive tremors and muscular paralysis. Hypokalaemia is common.
Death from cardiac or respiratory failure may occur within one to many hr (5).

Legislation

Limited under EEC Directive on Drinking Water Quality 80/778/EEC. Barium: guide level 100 µg l^{-1}, maximum admissible concentration 1000 µg l^{-1} (6).

Any other comments

Experimental toxicology, environmental fate and human health effects reviewed (7).

References

1. Movsumov, A. A; et al. *Dispersuye Sist. Bureu.* 1977, 117-119
2. Schroeder, H. A; *J. Nutr.* 1975, **105**(4), 421-427
3. *Documentation of TLV's and Biological Exposure Indices* 1986, American Conf. of Governmental Ind. Hyg. Cincinnati, OH
4. Browning, E. *Toxicology of Industrial Metals* 2nd ed., 1969, Butterworth, London
5. *Martindale: The Extra Pharmacopoeia* 29th ed., 1989, The Pharmaceutical Press, London
6. *EC Directive Relating to the Quality of Water Intended for Human Consumption* 1982, 80/778/EEC, Office for Official Publications of the European Communities, 2 rue Mercier, L-2985 Luxembourg
7. *Environmental Health Criteria 107: Barium* 1990, World Health Organisation, Geneva

B16   Barium pentasulfide

CAS Registry No. 12448-68-9
Synonyms barium polysulfides (BaS$_5$)
Mol. Formula BaS$_5$ **Mol. Wt.** 297.66

Occupational exposure

Supply classification irritant.

Risk phrases
Contact with acids liberates toxic gas – Irritating to eyes, respiratory system and skin (R31, R36/37/38)

Safety phrases
After contact with skin, wash immediately with plenty of water (S28)

Any other adverse effects to man

In humans symptoms of barium poisoning from soluble barium salts arise from stimulation of all forms of muscle and include vomiting, colic, diarrhoea, hypertension, convulsive tremors and muscular paralysis. Hypokalaemia is common. Death from cardiac or respiratory failure may occur within one to many hr (1).

Legislation

Limited under EEC Directive on Drinking Water Quality 80/778/EEC. Barium: guide level 100 µg l^{-1}, maximum admissible concentration 1000 µg l^{-1} (2).

Any other comments

Appears to exist only in solution. Decomposes to barium tetrasulfide, barium sulfite, hydrogen sulfide, sulfur and water (3).

References

1. *Martindale: The Extra Pharmacopoeia* 29th ed., 1989, The Pharmaceutical Press, London
2. *EC Directive Relating to the Quality of Water Intended for Human Consumption* 1982, 80/778/EEC, Office for Official Publications of the European Communities, 2 rue Mercier, L-2985 Luxembourg
3. Guitteau, L. *Compt. Rend.* 1916, **163**, 390-391

B17   Barium perchlorate

$Ba(ClO_4)_2$

CAS Registry No. 13465-95-7
Synonyms perchloric acid, barium salt
Mol. Formula $BaCl_2O_8$ **Mol. Wt.** 336.24
Uses Used in the determination of ribonuclease. Desiccant. Used in explosives and as an experimental rocket fuel.

Physical properties

M. Pt. 505°C; **Specific gravity** 3.20.
Solubility
Water: 1985 g l^{-1} at 25°C. Organic solvent: ethanol

Occupational exposure

US TLV (TWA) 0.5 mg m^{-3}; **UK Long-term limit** 0.5 mg m^{-3}; **UN No.** 1447; **HAZCHEM Code** 2W; **Conveyance classification** oxidising substance; toxic substance; **Supply classification** oxidising and harmful.

Risk phrases
Explosive when mixed with combustible material – Harmful by inhalation and if swallowed (R9, R20/22)

Safety phrases
Take off immediately all contaminated clothing (S27)

Any other adverse effects to man

In humans symptoms of barium poisoning from soluble barium salts arise from stimulation of all forms of muscle and include vomiting, colic, diarrhoea, hypertension, convulsive tremors and muscular paralysis. Hypokalaemia is common. Death from cardiac or respiratory failure may occur within one to many hr (1).

Legislation

Limited under EEC Directive on Drinking Water Quality 80/778/EEC. Barium: guide level 100 μg l^{-1}, maximum admissible concentration 1000 μg l^{-1} (2).

Any other comments

Physico-chemical properties, environmental fate, experimental toxicology and human health effects reviewed (3,4).

References

1. *Martindale: The Extra Pharmacopoeia* 29th ed., 1989, The Pharmaceutical Press, London
2. *EC Directive Relating to the Quality of Water Intended for Human Consumption* 1982, 80/778/EEC, Office for Official Publications of the European Communities, 2 rue Mercier, L-2985 Luxembourg
3. *ECETOC Technical Report No. 30(4)* 1991, European Chemical Industry Ecology and Toxicology Centre, B-1160 Brussels
4. *Environmental Health Criteria 107: Barium* 1990, World Health Organisation, Geneva

B18 Barium permanganate

Ba(MnO4)2

CAS Registry No. 7787-36-2
Synonyms permanganic acid, barium salt
Mol. Formula $BaMn_2O_8$ **Mol. Wt.** 375.21
Uses Dry cell depolariser. As a strong disinfectant.

Physical properties

M. Pt. 200°C (decomp.); **Specific gravity** 3.77.
Solubility
Water: 625 g l^{-1} at 11°C. Organic solvent: ethanol (decomp.)

Occupational exposure

US TLV (TWA) 0.5 mg m^{-3}; **UK Long-term limit** 0.5 mg m^{-3}; **UN No.** 1448;
Conveyance classification oxidising substance; toxic substance;
Supply classification harmful.

Risk phrases
Harmful by inhalation and if swallowed (R20/22)

Safety phrases
After contact with skin, wash immediately with plenty of water (S28)

Any other adverse effects to man

In humans symptoms of barium poisoning from soluble barium salts arise from stimulation of all forms of muscle and include vomiting, colic, diarrhoea, hypertension, convulsive tremors and muscular paralysis. Hypokalaemia is common. Death from cardiac or respiratory failure may occur within one to many hr (1).

Legislation

Limited under EEC Directive on Drinking Water Quality 80/778/EEC. Barium: guide level 100 μg l^{-1}, maximum admissible concentration 1000 μg l^{-1} (2).

Any other comments

Environmental fate, experimental toxicology and human health effects reviewed (3).

References

1. *Martindale: The Extra Pharmacopoeia* 29th ed., 1989, The Pharmaceutical Press, London
2. *EC Directive Relating to the Quality of Water Intended for Human Consumption* 1982, 80/778/EEC, Office for Official Publications of the European Communities, 2 rue Mercier, L-2985 Luxembourg
3. *Environmental Health Criteria 107: Barium* 1990, World Health Organisation, Geneva

B19   Barium peroxide

BaOO

CAS Registry No. 1304-29-6
Synonyms barium dioxide; barium superoxide
Mol. Formula BaO_2 **Mol. Wt.** 169.34
Uses For bleaching, in textile dyeing and printing. In hydrogen peroxide and oxygen manufacture. Glass decolouriser.

Physical properties

M. Pt. 450°C; **B. Pt.** 800°C (decomp.); **Specific gravity** 4.96.

Occupational exposure

UN No. 1449; **Conveyance classification** oxidising substance; toxic substance; **Supply classification** oxidising and harmful.

Risk phrases

Contact with combustible material may cause fire – Harmful by inhalation and if swallowed (R8, R20/22)

Safety phrases

Keep away from food, drink and animal feeding stuffs – Take off immediately all contaminated clothing (S13, S27)

Mammalian and avian toxicity

Acute data

LD_{50} subcutaneous mouse 50 mg kg^{-1} (1).

Legislation

Limited under EEC Directive on Drinking Water Quality 80/778/EEC. Barium: guide level 100 μg l^{-1}, maximum admissible concentration 1000 μg l^{-1} (2).

Any other comments

Physico-chemical properties, environmental fate, experimental toxicology and human health effects reviewed (3,4).

References

1. *Zhurnal Vsesoyunogo Khimicheskogo Obshchesta im D. I. Mendeleeva* 1974, **19** 186
2. *EC Directive Relating to the Quality of Water Intended for Human Consumption* 1982, 80/778/EEC, Office for Official Publications of the European Communities, 2 rue Mercier, L-2985 Luxembourg

3. *ECETOC Technical Report No. 30 (4)* 1991, European Chemical Industry Ecology and Toxicology Centre, B-1160 Brussels
4. *Environmental Health Criteria 107: Barium* 1990, World Health Organisation, Geneva

B20   Barium polysulfides (Ba₂S₃)

CAS Registry No. 53111-28-7
Synonyms dibarium trisulfide
Mol. Formula Ba_2S_3 **Mol. Wt.** 370.87

Physical properties

Specific gravity d_4^{25} 4.13.

Occupational exposure
Supply classification irritant.

Risk phrases
Contact with acids liberates toxic gas – Irritating to eyes, respiratory system and skin (R31, R36/37/38)

Safety phrases
After contact with skin, wash immediately with plenty of water (S28)

Any other adverse effects to man

In humans symptoms of barium poisoning from soluble barium salts arise from stimulation of all forms of muscle and include vomiting, colic, diarrhoea, hypertension, convulsive tremors and muscular paralysis. Hypokalaemia is common. Death from cardiac or respiratory failure may occur within one to many hr (1).

Legislation

Limited under EEC Directive on Drinking Water Quality 80/778/EEC. Barium: guide level 100 µg l⁻¹, maximum admissible concentration 1000 µg l⁻¹ (2).

References
1. *Martindale: The Extra Pharmacopoeia* 29th ed., 1989, The Pharmaceutical Press, London
2. *EC Directive Relating to the Quality of Water Intended for Human Consumption* 1982, 80/778/EEC, Office for Official Publications of the European Communities, 2 rue Mercier, L-2985 Luxembourg

B21   Barium polysulfides (Ba₄S₇)

CAS Registry No. 50864-67-0
Synonyms Solbar; tetrabarium heptasulfide; Polybarit
Mol. Formula Ba_4S_7 **Mol. Wt.** 773.81
Uses Insecticide. Fungicide.

Occupational exposure

Supply classification irritant.

Risk phrases

Contact with acids liberates toxic gas – Irritating to eyes, respiratory system and skin (R31, R36/37/38)

Safety phrases

After contact with skin, wash immediately with plenty of water (S28)

Mammalian and avian toxicity

Acute data

LD_{50} oral rat 375 mg kg^{-1} (1).
TD_{Lo} oral human 226 mg kg^{-1} (2).

Any other adverse effects to man

In humans symptoms of barium poisoning from soluble barium salts arise from stimulation of all forms of muscle and include vomiting, colic, diarrhoea, hypertension, convulsive tremors and muscular paralysis. Hypokalaemia is common. Death from cardiac or respiratory failure may occur within one to many hr (3,4).

Legislation

Limited under EEC Directive on Drinking Water Quality 80/778/EEC. Barium: guide level 100 $\mu g \, l^{-1}$, maximum admissible concentration 1000 $\mu g \, l^{-1}$ (5).

Any other comments

Oxidised in air to barium sulfite. Loses water of crystallization at 100°C-230°C to give hydrogen sulfide and barium sulfite (6).

References

1. *Arch. Ind. Med.* 1973, **132**, 891
2. *Farm Chemicals Handbook* 1991, **C36**, Meister Publishing, Willoughby, OH
3. *Environmental Health Criteria 107: Barium* 1990, World Health Organisation, Geneva
4. *Martindale: The Extra Pharmacopoeia* 29th ed., The Pharmaceutical Press, London
5. *EC Directive Relating to the Quality of Water Intended for Human Consumption* 1982, 80/778/EEC, Office for Official Publications of the European Communities, 2 rue Mercier, L-2985 Luxembourg
6. Mellor, J. W. *Comprehensive Treatise on Inorganic and Threshold Chemistry* **3**, 752-753, Longman, Harlow

B22   Barium sulfate

BaSO$_4$

CAS Registry No. 7727-43-7
Synonyms Barite; actybaryte; Artifical Heavy Spar; blanc fixe; Enamel White; sulfuric acid, barium salt (1:1)
Mol. Formula BaSO$_4$ **Mol. Wt.** 233.40

Uses In the manufacture of photographic paper and cellophane. A filler for rubber, lino, oil cloth. Pigment in white paint. Diagnostic aid (radiopaque medium).
Occurrence In nature as the mineral Barite.

Physical properties

M. Pt. 1580°C (decomp.); Specific gravity 4.5 at 15°C.
Solubility
Water: 0.0022 g l^{-1} at 50°C.

Occupational exposure

US TLV (TWA) 10 mg m^{-3}; UK Long-term limit 2 mg m^{-3} (respirable dust).

Ecotoxicity

Fish toxicity
Barium sulfate used to weight drilling fluid was not toxic to fish at concentrations up to 100,000 ppm in fresh or sea water. Therefore does not constitute the introduction of a toxic material into the environment (1).

Environmental fate

Anaerobic effects
In systems having pH ≤5 barium sulfate did not dissolve even in anaerobic conditions (2).

Mammalian and avian toxicity

Metabolism and pharmacokinetics
In ♀ beagle dog, barium sulfate was cleared from lung with $t_{1/2}$ 8-9 days indicating some solubility in body fluids, possibly in colloidal form (3).
After rat inhalation of barium sulfate 40 mg m^{-3} 5 hr $^{-1}$ daily for 2 months, lymphatic transport was slight. Skeletal concentration of barium was 0.8-1.5 mg day^{-1} and skeletal uptake decreased with advancing age (3).
In patients with a normal gastrointestinal tract barium sulfate is excreted unchanged within 24 hr after oral administration. After rectal administration the compound is usually excreted when the enema is expelled, however some may remain in the colon for several wk (4).

Sensitisation
Patients with history of atopy may be susceptible to allergic reactions if exposed to barium sulfate (5).

Any other adverse effects to man

A benign pneumoconiosis (baritosis) may result from the inhalation of barium sulfate dust (6- 8).
Baritosis was reported among workers exposed to finely ground barium sulfate in the US, Germany and Czechoslovakia, but no firm conclusions could be drawn (9).

Any other adverse effects

Intrauterine subcutaneous injection into rabbit fetuses caused acute inflammatory response in both vascular and cellular components (10).
Subpannicular injection into new born rabbits also produced an inflammatory response. Constipation may occur after oral or rectal barium sulfate administration.

Impaction, obstruction and appendicitis have occurred. Cramping and diarrhoea have also been reported. Intravasation has led to the formation of emboli, deaths have occurred. Perforation of the bowel has led to peritonitis, adhesions, granulomas and death. The use of barium sulfate for bronchography by aspiration into the lungs has led to pneumonitis or granuloma formation. Cardiac arrhythmias have occurred during the use of barium sulfate enemas (11).

In vitro mouse peritoneal macrophages exposed to barium sulfate for up to 144 hr showed marked cytoplasmic vacuolisation with only partial recovery (12).

Legislation

Limited under EEC Directive on Drinking Water Quality 80/778/EEC. Barium: guide level 100 μg l^{-1}, maximum admissible concentration 1000 μg l^{-1}. Sulfates: guide level 25 mg l^{-1}, maximum admissible concentration 250 mg l^{-1} (13).

Any other comments

Experimental toxicology, environmental fate, workplace experience and and human effects reviewed (14-16).

References

1. Grantham, C. K; et al. *Environ. Aspects Chem. Use Well Drill Oper. Conf. Proc.* 1975, 103
2. Jernelov, A. *Inst. Vatten. Luftvardsforsk Publ.* 1976, **B277**, 25
3. Friberg, L; et al. *Handbook of the Toxicology of Metals* 2nd ed., 1986, **1-2**, v2, Elsevier Science Publishers, Amsterdam
4. *American Hospital Formulary Service – Drug Information 88* 1988, 1317, American Society of Hospital Pharmacists, Bethesda, MD
5. Feczko, P. J; et al. *Am. J. Roetenology* 1989, **153**, 275
6. *Br. Med. J.* 1972, **2**, 5813
7. Browning, E. *Toxicity of Industrial Metals* 2nd ed., 1969, 61-66, Butterworth and Co, London
8. Brenniman, G. R; et al. *Environ. Res.* 1979, **20**, 318
9. Pendergrass, E. P; et al. *Arch. Ind. Health* 1953, **7**, 44
10. Low, W. C. A; et al. *J. Path.* 1977, **121**(3), 159
11. *Martindale: The Extra Pharmacopoeia* 29th ed., 1989, Pharmaceutical Press, London
12. Rae, T. *J. Biomed. Mater. Res.* 1977, **11**, 839-846
13. *EC Directive Relating to the Quality of Water Intended for Human Consumption* 1982, 80/778/EEC, Office for Official Publications of the European Communities, 2 rue Mercier, L-2985 Luxembourg
14. *ECETOC Technical Report 30(4)* 1991, European Chemical Industry Ecology and Toxicology Centre, B-1160 Brussels
15. *Environmental Health Criteria 107: Barium* 1990, World Health Organisation, Geneva
16. *BIBRA Toxicity Profile* 1987, British Industrial Biological Research Association, Carshalton, Surrey

B23   Barium tetrasulfide

CAS Registry No. 12248-67-8
Synonyms barium polysulfides (BaS$_4$)
Mol. Formula BaS$_4$ **Mol. Wt.** 265.60
Uses Pesticide. Insecticide.

Physical properties

Specific gravity 3.11.

Solubility
Water: 410 g l^{-1} at 15°C.

Occupational exposure

Supply classification irritant.

Risk phrases
Contact with acids liberates toxic gas – Irritating to eyes, respiratory system and skin (R31, R36/37/38)

Safety phrases
After contact with skin, wash immediately with plenty of water (S28)

Any other adverse effects to man

In humans symptoms of barium poisoning from soluble barium salts arise from stimulation of all forms of muscle and include vomiting, colic, diarrhoea, hypertension, convulsive tremors and muscular paralysis. Hypokalaemia is common. Death from cardiac or respiratory failure may occur within one to many hr (1).

Legislation

Limited under EEC Directive on Drinking Water Quality 80/778/EEC. Barium: guide level 100 µg l^{-1}, maximum admissible concentration 1000 µg l^{-1} (2).

Any other comments

Piezoelectric (3).
Decomposes >200- 300°C to sulfur, water, hydrogen sulfide and barium trisulfide.

References

1. *Martindale: The Extra Pharmacopoeia* 29th ed., 1989, The Pharmaceutical Press, London
2. *EC Directive Relating to the Quality of Water Intended for Human Consumption* 1982, 80/778/EEC, Office for Official Publications of the European Communities, 2 rue Mercier, L-2985 Luxembourg
3. Abrahams, S. C; et al. *Acta Crystallogr.* 1954, **7**, 423-429

B24   Barium trisulfide

CAS Registry No. 12231-01-5
Synonyms barium polysulfide (BaS$_3$)
Mol. Formula BaS$_3$ **Mol. Wt.** 233.53

Physical properties

M. Pt. 554-555°C; **Specific gravity** 3.94.

Occupational exposure

Supply classification irritant.

Risk phrases

Contact with acids liberates toxic gas – Irritating to eyes, respiratory system and skin (R31, R36/37/38)

Safety phrases

After contact with skin, wash immediately with plenty of water (S28)

Any other adverse effects to man

In humans symptoms of barium poisoning from soluble barium salts arise from stimulation of all forms of muscle and include vomiting, colic, diarrhoea, hypertension, convulsive tremors and muscular paralysis. Hypokalaemia is common. Death from cardiac or respiratory failure may occur within one to many hr (1).

Legislation

Limited under EEC Directive on Drinking Water Quality 80/778/EEC. Barium: guide level 100 μg l^{-1}, maximum admissible concentration 1000 μg l^{-1} (2).

Any other comments

Stable to 250°C.
Decomposes >400°C to barium disulfide (3,4). Oxidised in air to barium carbonate and barium sulfite.
Oxidised in aqueous solution to barium hexahydrate (5).
Evaporation of aqueous solution gives barium hexahydrate and barium tetrahydrate.

References

1. *Martindale: The Extra Pharmacopoeia* 29th ed., 1989, The Pharmaceutical Press, London
2. *EC Directive Relating to the Quality of Water Intended for Human Consumption* 1982, 80/778/EEC, Office for Official Publications of the European Communities, 2 rue Mercier, L-2985 Luxembourg
3. Philpot, E; et al. *C. R. Acad. Sci., Paris, Ser. C* 1968, **266**, 1499-1501
4. Von Schnering, H. G; et al. *Naturuinenschaften* 1974, **61**, 272
5. Mellor, J. W. *Comprehensive Treatise on Inorganic and Theoretical Chemistry* 3, 752-753, Longman, Harlow

B25   Beetroot Red

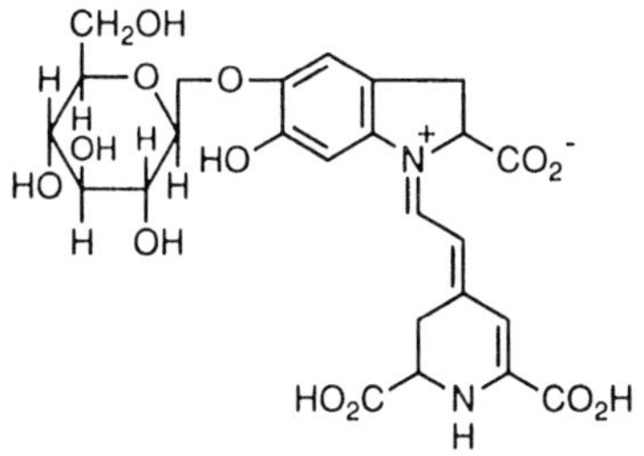

CAS Registry No. 57917-55-2

Synonyms C.I. Natural Red 33; E162; 2, 6-piperidinedicarboxylic acid 4-[2-[2-carboxy-5- (β-D-glucopyranosyloxy)-2,3-dihydro-6-hydroxy-1-*H*-indol-1-yl]ethylidene

Mol. Formula $C_{24}H_{26}N_2O_{13}$ **Mol. Wt.** 550.48
Uses Food dyestuff.
Occurrence Extract from red beetroot (*Beta vulgaris*) consists of red and yellow quaternary ammonium amino acid pigments of the betalamino class.

Genotoxicity
Salmonella typhimurium TA100 with and without metabolic activation positive.
In vitro Chinese hamster fibroblast cells negative (1).

Any other comments
Experimental toxicology and human health effects reviewed (2).
The chemistry, manufacture and safety, of beetroot red used as a food additive reviewed (3).
Nitrate levels in liquid and powdered beet red varied from 391-34,181 mg kg^{-1}.
Acceptable daily intake was not established for beetroot red as the compound was not considered hazardous to health (4).
See Betanin for structure.

References
1. Ishidate, M. J. R; et al. *Food Chem. Toxicol.* 1984, **22**(8), 623-636
2. *BIBRA Toxicity Profile* 1987, British Industrial Biological Research Assoc., Carshalton
3. Yoshizumi; et al. *Shokuhin Kogyu* 1975, **18**(20), 67-72, (Jap.) (*Chem. Abstr.* **84**, 57415c)
4. Lara, W. H; et al. *Rev. Inst. Adolto Lutz* 1984, **44**(2), 109-114

B26   Benalaxyl

CAS Registry No. 71626-11-4
Synonyms methyl-*N*-phenylacetyl-*N*-2,6-xylyl-DL-alaninate; methyl-*N*-(2, 6-dimethylphenyl)-*N*-(phenylacetyl)-DL-alaninate
Mol. Formula $C_{20}H_{23}NO_3$ **Mol. Wt.** 325.41
Uses Fungicide.

Physical properties
M. Pt. 78-80°C; **Flash point** 195°C (open cup); **Specific gravity** 1.27 at 25°C; **Volatility** v.p. 5.0×10^{-6} mmHg at 25°C.

Solubility
Water: 37 mg l^{-1} at 25°C. Organic solvent: ethanol, diethyl ether, acetone

Ecotoxicity

Fish toxicity
LC$_{50}$ (96 hr) rainbow trout, carp, guppy, goldfish 4-8 mg l^{-1} (1,2).

Effects to non-target species
LD$_{50}$ oral mallard duck >6000 mg kg^{-1} (2).

LD$_{50}$ oral Japanese quail >5000 mg kg^{-1} (2).
LD$_{50}$ (5 day) oral chicken 4600 mg kg^{-1} diet (2).
Non-toxic to bees (2).

Environmental fate

Degradation studies
t$_{1/2}$ in soil 20-100 days (2).

Mammalian and avian toxicity

Acute data
LD$_{50}$ oral rat 4200 mg kg^{-1} (1).
LD$_{50}$ oral mouse 6800 mg kg^{-1} (1).
LC$_{50}$ (4 hr) inhalation rat >10 mg l^{-1} (2).
LD$_{50}$ percutaneous rat >5000 mg kg^{-1} (1).
LD$_{50}$ intraperitoneal rat 1100 mg kg^{-1} (1).

Sub-acute data
No effect level for rats in 90 day feeding trials 1000 mg kg^{-1} diet.
No effect level for dogs in 1 yr feeding trials 200 mg kg^{-1} diet (2).

Metabolism and pharmacokinetics
Oral rat, rapidly metabolised and eliminated in urine 23% faeces 75% within 2 days (2).

Legislation

Limited under EEC Directive on Drinking Water Quality 80/778/EEC. Pesticides and related products: maximum admissible concentration 0.5 µg l^{-1} (3).

Any other comments

Incompatible with alkaline materials (2).

References

1. Bergamaschi, P; et al. *Proc. Brit. Crop. Prot. Conf. Pests Dis.* 1981, **19**
2. *The Agrochemicals Handbook* 3rd ed., 1991, RSC, London
3. *EC Directive Relating to the Quality of Water Intended for Human Consumption* 1982, 80/778/EEC, Office for Official Publications of the European Communities, 2 rue Mercier, L-2985 Luxembourg

B27   Benazolin

CAS Registry No. 3813-05-6
Synonyms benazoline; 4-chloro-2-oxybenzothiazolin-3-ylacetic acid; 4-chloro-2, 3-dihydro-2-oxobenzothiazol-3-ylacetic acid; 4-chloro-2-oxo-3(2*H*)-benzothiazoleacetic acid; 4-chloro-2-oxo-3-benzothiazolineacetic acid; Cresopur
Mol. Formula C$_9$H$_6$ClNO$_3$S   **Mol. Wt.** 243.67

Uses Herbicide.

Physical properties

M. Pt. 193°C (in acid); **Partition coefficient** log P_{ow} 2.48 (1); **Volatility** v.p. 0.003 mmHg at 20°C.

Solubility

Water: 600 mg l^{-1} at 20°C. Organic solvent: acetone, ethanol, carbon disulfide

Ecotoxicity

Fish toxicity

LC_{50} (24 hr) bluegill sunfish 204 mg l^{-1} (1).
LC_{50} (96 hr) trout 27 mg l^{-1} (1).
LC_{50} (24 hr) harlequin fish 360 ppm (2).

Mammalian and avian toxicity

Acute data

LD_{50} oral rat, mouse 3000-3200 mg kg^{-1} (1).
LD_{50} percutaneous rat >5000 mg kg^{-1} (1).
LD_{50} oral japanese quail >10 g kg^{-1} (1,2).

Sub-acute data

Oral rat (90 day) <1000 mg kg^{-1} in feed caused no adverse effects (1).

Legislation

Limited under EEC Directive on Drinking Water Quality 80/778/EEC. Pesticides and related products: maximum admissible concentration 0.5 μg l^{-1} (3).

Any other comments

The alkali-metal and diethanolamine salts are readily soluble in water (2).

References

1. *The Agrochemicals Handbook* 3rd ed., 1991, RSC, London
2. Alabaster, J. S. *Int. Pest. Contr.* 1969, **11**(2), 29
3. *EC Directive Relating to the Quality of Water Intended for Human Consumption* 1982, 80/778/EEC, Office for Official Publications of the European Communities, 2 rue Mercier, L-2985 Luxembourg

B28   Benfluralin

CAS Registry No. 1861-40-1
Synonyms Benfluraline; Balan; Benefin; *N*-butyl- *N*-ethyl-α,α,α- trifluoro-2, 6-dinitro-*p*-toluidine; *N*-butyl-*N*-ethyl-2,6-dinitro-4-(trifluoromethyl)benzenamine; *N*-butyl-*N*-ethyl-2,6-dinitro-4-trifluoromethylaniline

Mol. Formula $C_{13}H_{16}F_3N_3O_4$ **Mol. Wt.** 335.29

Uses Herbicide.

Physical properties

M. Pt. 65-66°C; **B. Pt.** $_{0.5}$ 121- 122°C; **Flash point** 25.6°C; **Volatility** v.p. 2.78×10^{-5} mmHg at 25°C.

Solubility

Water: <1 mg l^{-1} at 25°C. Organic solvent: acetone, ethanol, chloroform, dimethylformamide

Ecotoxicity

Fish toxicity

LC_{50} (96 hr) bluegill sunfish, goldfish, fathead minnow, harlequin fish 0.064-1.2 mg l^{-1} (1-3).

Invertebrate toxicity

LC_{50} (96 hr) freshwater shrimp 1.1 mg l^{-1} (4).

Effects to non-target species

LD_{50} oral mallard duck and chicken >2 g kg^{-1} (1).

Environmental fate

Degradation studies

Residual activity in soil 4-8 months (1).
Estimated degradation in soil at 30°C, $t_{1/2}$ 0.4-1.8 months (5).

Absorption

Strongly adsorbed on soil with negligible leaching, surface material was removed by photodecomposition (6).

Mammalian and avian toxicity

Acute data

LD_{50} oral rat, mouse, rabbit 2-10 g kg^{-1} (1,7).

Carcinogenicity and long-term effects

In 2 yr feeding trial with rats the no-effect level was 1 g kg^{-1} diet (1).

Metabolism and pharmacokinetics

Oral goat (5 day) after administration of an unspecified dose 90% and 10% were excreted in faeces and urine, respectively (8).

Irritancy

Skin and eye irritation reported in humans (9).

Legislation

Limited under EEC Directive on Drinking Water Quality 80/778/EEC. Pesticides and related products: maximum admissible concentration 0.5 µg l^{-1} (10).

Any other comments

Hazardous properties reviewed (11).
Decomposes in ultra-violet light.

References

1. *The Agrochemicals Handbook* 3rd ed., 1991, RSC, London
2. *Handbook of Acute Toxicity of Chemicals to Fish and Aquatic Invertebrates* 1980, 81, US Dept. Int. Fish and Wildlife Serv. No. 137
3. Tooby, T. E; et al. *Chem. Ind. (London)* June 21, 1975
4. Sanders, H. O. *J. Water Pollut. Control Fed.* 1970, **42**, 1544
5. Zimoahl, R. L. *Weed Sci.* 1977 **25**(3), 247
6. *Herbicide Handbook* 4th ed., 1979, 43, Weed Science Society of America, Champaign, IL
7. Worthing, C. R. (Ed.) *The Pesticide Manual* 9th ed., 1991, British Crop Protection Council, Farnham
8. Menzie, C. M. *Metabolism of Pesticides* 1974, 49, US Dept. Int. Fish and Wildlife Serv. No. 184
9. Gosselin, R. E. *Clinical Toxicology of Commercial Products* 5th ed., 1984, Williams & Wilkins, London
10. *EC Directive Relating to the Quality of Water Intended for Human Consumption* 1982, 80/778/EEC, Office for Official Publications of the European Communities, 2 rue Mercier, L-2985 Luxembourg
11. *Dangerous Prop. Ind. Mater. Rep.* 1990, **10**(5), 44-50

B29   Benfuracarb

CAS Registry No. 82560-54-1

Synonyms ethyl-*N*-[2,3-dihydro-2,2- dimethylbenzofuran-7-yloxycarbonyl(methyl) aminothio] -*N*-isopropyl-β-alaninate; 2,3-dihydro- 2,2-dimethyl-7-benzofuranyl, 2-methyl-4-(1-methylethyl)-7- oxo-8-oxa-3-thia-2,4-diazadecanoate; Benfuracarbe; Oncol; 2- methyl-4-(1-methylethyl)-7-oxo-8-oxa-3-thia-2,4- diazadecanoic acid 2,3 dihydro-2,2-dimethyl-7-benzofuranyl ester

Mol. Formula $C_{20}H_{30}N_2O_5S$ **Mol. Wt.** 410.54
Uses Insecticide.

Physical properties

M. Pt. 225°C (decomp.); **Specific gravity** 1.171; **Partition coefficient** log P_{ow} 4.30; **Volatility** v.p. 2×10^{-7} mmHg at 20°C.

Solubility
Water: 8 mg l^{-1} at 20°C. Organic solvent: benzene, dichloromethane, methanol, acetone, hexane, xylene, ethyl acetate

Ecotoxicity

Fish toxicity
LC$_{50}$ (48 hr) carp 0.65 mg l^{-1} (1).

Invertebrate toxicity
EC$_{50}$ (3 hr) *Daphnia magna* >10 mg l^{-1} (1).

Effects to non-target species
LD_{50} oral hen 92 mg kg^{-1} (1).

Environmental fate

Degradation studies
$t_{1/2}$ 2-3 days in soil. Metabolites include carbofuran and 3-hydroxycarbofuran (1).

Mammalian and avian toxicity

Acute data
LD_{50} oral rat, mouse 138-175 mg kg^{-1} (2).
LD_{50} oral dog 300 mg kg^{-1} (2).
LD_{50} percutaneous rat >2000 mg kg^{-1} (1).

Any other adverse effects
Cholinesterase inhibitor (3).

Legislation
Limited under EEC Directive on Drinking Water Quality 80/778/EEC. Pesticides and related products: maximum admissible concentration 0.1 µg l^{-1} (4).

Any other comments
The metabolism of ring 14C-labelled benfuracarb was examined in cotton, bean and corn. Initial degradation step is N-S bond cleavage giving rise to carbofuran as the first major metabolite. Carbofuran is further metabolised to 3-hydroxy carbofuran which is in turn conjugated. Carbofuran is significant in the early period after treatment while 3-hydroxycarbofuran is important during later stages (5).

References
1. *The Agrochemicals Handbook* 3rd ed., 1991, RSC, London
2. Goto, T; et al. *Proc. 10th Conf. Int. Cong. Plant Protect.* 1983, **1**, 360
3. Worthing, C. R. (Ed.) *The Pesticide Manual* 9th ed., 1991, British Crop Protection Council, Farnham
4. *EC Directive Relating to the Quality of Water Intended for Human Consumption* 1982, 80/778/EEC, Office for Official Publications of the European Communities, 2 rue Mercier, L-2985 Luxembourg
5. Tanaka, A; et al. *J. Agric. Food Chem.* 1985, **33**, 1049

B30   Benodanil

CAS Registry No. 15310-01-7
Synonyms 2-iodobenzanilide; 2-iodo-*N*-phenylbenzamide; calirus; BAS 3170F
Mol. Formula $C_{13}H_{10}INO$ Mol. Wt. 323.14

Uses Fungicide for rust diseases.

Physical properties
M. Pt. 137°C; **Volatility** v.p. $<7.52 \times 10^{-8}$ mmHg at 20°C.

Solubility
Water: 20 mg l^{-1} at 20°C. Organic solvent: acetone, ethanol, ethylacetate, chloroform

Ecotoxicity

Fish toxicity
LC_{50} (96 hr) trout 6.4 mg l^{-1} (1).

Effects to non-target species
Non-toxic to bees (1).

Environmental fate

Degradation studies
The $t_{1/2}$ in loamy and humous sandy soil 3-4 wk (1).

Mammalian and avian toxicity

Acute data
LD_{50} oral rat, guinea pig >6400 mg kg^{-1} (1).
LD_{50} percutaneous rats >2000 mg kg^{-1} (1).

Sub-acute data
No effect level for rats in 90 day feeding trials was 100 mg kg^{-1} diet (1).

Metabolism and pharmacokinetics
After oral administration to rats 80% is excreted in faeces and 16% in urine within 6 days. Principal metabolites are the 4-hydroxy derivative and its glucuronide and sulfate conjugates (1).

Reference

1. *The Agrochemicals Handbook* 3rd ed., 1991, RSC, London.

B31   Benomyl

CAS Registry No. 17804-35-2
Synonyms methyl 1-(butylcarbomyl)benzimidazol-2-ylcarbamate; methyl 1-[(butylamino)carbonyl]-1*H*-benzimidazol-2-ylcarbamate; methyl 1-(butylcarbamoyl)-2-benzimidazolecarbamate; Benlate; Benex; Fundazol

Mol. Formula $C_{14}H_{18}N_4O_3$ **Mol. Wt.** 290.32

Uses Fungicide. Acaricide. Used to increase the biological oxidation rate of sewage and fertilisers.

Physical properties

Solubility
Water: 2 mg l^{-1} at 25°C. Organic solvent: chloroform, dimethylformanide, acetone, xylene, ethanol

Occupational exposure
US TLV (TWA) 0.84 ppm (10 mg m^{-3}); **UK Long-term limit** 10 mg m^{-3}; **UK Short-term limit** 15 mg m^{-3}.

Ecotoxicity

Fish toxicity
LC_{50} (96 hr) rainbow trout 170 µg l^{-1} at 12°C – technical material (1).
LC_{50} (96 hr) fathead minnow 2200 µg l^{-1} at 22°C – technical material (1).
LC_{50} (96 hr) channel catfish 29 µg l^{-1} at 22°C – technical material (1).
LC_{50} (96 hr) fathead minnow, bluegill sunfish 1200-1900 µg l^{-1} at 22°C – wettable powder (1).
LC_{50} (96 hr) channel catfish 28 µg l^{-1} at 22°C – wettable powder (1).

Invertebrate toxicity
LC_{50} (96 hr) crayfish 1032 mg l^{-1} (2).

Effects to non-target species
LD_{50} oral starling >100 mg kg^{-1} (3).
LC_{50} 8 day dietary mallard duck, bobwhite quail >500 mg kg^{-1} (3).
Non-toxic to *Apis melliferra* (3).
Highly toxic to earthworms (4).

Environmental fate

Nitrification inhibition
Nitrobacter agilis effect of 0 ppm benomyl addition, 289.5 µg nitrite ml^{-1}, 10 ppm addition (513.2 days) and 100 ppm addition (440.5 days) 444.2 µg nitrite ml^{-1} (5).
Nitrosomonas sp. effect of 0 ppm benomyl addition 3.8 µg nitrite ml^{-1}, 10 ppm addition (2.1 days) 2.0 µg nitrite ml^{-1}, 100 ppm addition (2 days) 1.7 µg nitrite ml^{-1} (5).

Anaerobic effects
Laboratory and greenhouse experiments showed that benomyl and its two soil metabolites, methyl 2-benzimidazole carbamate and 2-aminobenzimidazole were immobile in soils (organic content ranged from 0.7-83.5%). No leaching or significant movement from the site of application (6).

Degradation studies
$t_{1/2}$ in soil 6-12 months (3).
Mixed bacterial cultures grew with benomyl as sole carbon source, but rates of breakdown to methyl benzimidazol-2-ylcarbamate were small and 2-aminobenzimidazole did not support growth (7).

Calculated bioconcentration factor 290 (8).

1000 ppm benomyl inhibited nitrification in a simulated oxidised surface of a flooded soil during a 30 day incubation period. Nitrification was not strongly inhibited at ≤100 ppm. Hydrolysis products detected were methyl-2-benzimidazolecarbamate and 2-aminobenzimidazole, both were less inhibitory of nitrification than benomyl (9).

Abiotic removal
In aqueous solutions under acidic conditions, benomyl is hydrolysed to methyl 2-benzimidazole carbamate and butyl isocyanate. In water, the conversion to benomyl to methylbenzimidazole-2-yl carbamate (carbendazim) is completed within 1 wk (10).

Mammalian and avian toxicity

Acute data
LD_{50} oral rat >9590 mg kg^{-1} (11).
LC_{50} (4 hr) inhalation rat >2 mg l^{-1} in air (3).
LD_{50} percutaneous rabbit >10,000 mg kg^{-1} (3).

Sub-acute data
Inhalation ♂, ♀ rat (90 day) ≤200 mg m^{-3} 6 hr day^{-1} 5 day wk^{-1} caused degeneration of the olfactory epithelium in all animals exposed to 200 mg m^{-3} (12).

Carcinogenicity and long-term effects
No effect level for rats in 2 yr feeding trials >2500 mg kg^{-1} diet and for dogs it was 500 mg kg^{-1} diet (3).

Teratogenicity and reproductive effects
Gavage pregnant Sprague-Dawley rats (7-21 day gestation) 31.2-125 mg kg^{-1}. Malformations increased in incidence and severity with increasing benomyl dosage and nearly doubled with a protein-deficient diet. Effects included foetal resorptions and late foetal death, anomalies included hydrocephalus, exencephaly and periventicular overgrowth. Common systemic malformations included cleft palate, micromelia, hydroureter and misshapen tails (13).
Oral gravid CD1 mice (5 day) 200 mg kg^{-1} in corn oil. Positive teratogenic effects included reduced litter size and viability (14).

Metabolism and pharmacokinetics
In animals, the butylcarbamoyl group is split off to give the relatively stable carbendazim, followed by slow degradation to non-toxic 2-aminobenzimidazole. Hydroxylation also occurs, and the principal metabolite 5- hydroxybenzimidazolecarbamate is converted to the O- and N- conjugates. In rats and dogs, following a single oral dose, 99% of benomyl and its metabolites are excreted in the urine and faeces within 72 hr. There is no accumulation of benomyl or its metabolites in animal tissue (3).

Sensitisation
Extreme sensitisation observed in guinea pig maximisation test (15).

Genotoxicity

Salmonella typhimurium TA98, TA100, TA1535, TA1537 with and without metabolic activation negative (16).
Salmonella typhimurium TA100, TA98, TA1535, TA1537, TA1535 and *Escherichia coli* WP2 hcr with and without metabolic activation negative (17).
In vitro human lymphocytes, 0.5-2.0 µg ml^{-1} caused decreases in the number of cells

undergoing third division, 0.25-4.0 µg ml^{-1} strongly increased the number of aneuploid cells and increased sister chromatid exchange (18).

Any other adverse effects

The effect of benomyl on the DNA turnover in various organs of the mouse was evaluated by measuring the incorpoation of [3H]thymidine 24 hr after administration of single oral doses of 377, 740 or 1480 mg kg^{-1}. Mice were sacrificed at 24 hr and inhibition of [3H]thymidine was observed in liver and kidney at the lowest dose while at the higher doses inhibition of [3H]thymidine occurred in the thymus, spleen and testis (19).

Any other comments

Experimental toxicology, genotoxicity, teratogenicity and carcinogenicity reviewed (17,20-22).
Toxicity and hazards reviewed (23).

References

1. US Dept. Interior, Fish and Wildlife Service *Handbook of Acute Toxicity of Chemicals to Fish and Aquatic Invertebrates. Resource Publication No. 137* 1980, **15**, US Government Printing Office, Washington, DC
2. Cheah, M. L; et al. *Prog. Fish. Cult.* 1980, **42**(3), 169
3. *The Agrochemicals Handbook* 3rd ed., 1991, RSC, London
4. Stringer, A; et al. *Pest. Sci.* 1973, **4**, 165
5. Vershueren, K. *Handbook of Environmental Data of Organic Chemicals* 1983, 223, 2nd ed., Van Nostrand Reinhold, New York
6. Rhodes, R. C; et al. *Bull. Environ. Contam. Toxicol.* 1974, **12**, 385-393
7. Fuchs, A; et al. *Antonie van Leeuwenhoek* 1978, **44**(3-4), 293-309
8. Kenaga, E. E. *Ecotox. Environ. Safety* 1980, **4**, 26-38
9. Ramakrishna, C; et al. *Bull. Environ. Contam. Toxicol.* 1979, **21**(3), 328-333
10. Peterson, C. A; et al. *J. Agric. Food. Chem.* 1969, **17**, 898-899
11. Schafer, E. W. *Toxicol. Appl. Pharmacol.* 1972, **21**, 315
12. Warheit, D. B; et al. *Fundam. Appl. Toxicol.* 1989, **12**(2), 333-345
13. Ellis, W. G; et al. *Teratog. Carcinog. Mutagen.* 1988, 8(6), 377-391
14. Chernoff, N. *J. Toxicol. Environ. Health* 1982, **10**, 541-550
15. Matsushita, T; et al. *Ind. Health* 1977, **15**(3-4), 141
16. Ficsor, G; et al. *Mutat. Res.* 1978, **51**(2), 151-164
17. Moriya, M; et al. *Mutat. Res.* 1983, **116**(3-4), 185-216
18. Georgieva, V; et al. *Environ. Mol. Mutagen.* 1990, **16**(1), 32-36
19. Hellman, B; et al. *Food Chem. Toxicol.* 1990, **28**(10), 701-706
20. Murakami, M. *Bull. Environ. Contam. Toxicol.* 1986, **37**(3), 326-329
21. *ECETOC Technical Report No. 30(4)* 1991, European Chemical Industry Ecology and Toxicology Centre, B-1160 Brussels
22. Biyashev, A. M. *Izat. Akad. Nauk. Kaz. SSR, Ser. Biol.* 1984, (6), 1-3, (Russ.) (*Chem. Abstr.* **102**, 107958v)
23. Izmerov, N. F. *Scientific Reviews of Soviet Literature on Toxicity & Hazards of Chemicals* 1992-1993, **112**, Eng. Trans, Richardson, M. L. (Ed.), UNEP/IRPTC, Geneva

B32 Benorylate

$$O_2CCH_3$$

(structure) —CO_2—(ring)—$NHCOCH_3$

CAS Registry No. 5003-48-5
Synonyms 4-acetamidophenylacetylsalicylate; 4- (acetamido)phenyl 2-acetoxybenzoate; 2-(acetyloxy)benzoic acid 4-(acetylamino)phenyl ester; Benoral; Salipran; Benortan
Mol. Formula $C_{17}H_{15}NO_5$ **Mol. Wt.** 313.31
Uses Analgesic. Anti-inflammatory and antipyretic.

Physical properties

M. Pt. 175-176°C.

Mammalian and avian toxicity

Acute data

LD_{50} oral mouse 2000 mg kg^{-1}.
LD_{50} intraperitoneal mouse, rat 1260-1830 mg kg^{-1} (1).
LD_{50} oral mouse 2700 mg kg^{-1} death due to respiratory arrest (2).

Sub-acute data

TD_{Lo} (8 day) ingestion woman 1280 mg kg^{-1} caused central nervous system and pulmonary effects (3).
Oral rat (4 day) 300 mg kg^{-1} every 6 hr caused no detrimental effect on stomach mucosa. A dose of 435 mg kg^{-1} showed better kidney tolerance in rat than equimolar dose of aspirin (2).

Metabolism and pharmacokinetics

Metabolised in humans to salicylic acid and paracetamol which are then excreted in the urine (4- 6).
Benorylate was hydrolysed *in vitro* to paracetamol and salicylate by human liver cytosol (7).
Following administration of 4 g to healthy volunteers benorylate was eliminated in the urine within 21 hr. The urinary metabolites detected were similar to those observed for aspirin and paracetamol (2).

Genotoxicity

Salmonella typhimurium TA98, TA100, TA1535, TA1537, TA1538 with and without metabolic activation negative (8).

Any other adverse effects to man

Human systemic effects by unspecified route include respiratory stimulation, dehydration and distorted perception (1).

Any other adverse effects

The humoral response in benorylate treated mice was 40-50% lower than the controls. Mitogen induced proliferation of spleen lymphocytes was also inhibited in treated mice, slight inhibition was observed where cells were activated by bacterial lipo-polysaccharide. When lymphocytes from drug treated animals were further cultured in the presence of the same drug a variable inhibition of mitogen induced proliferation was observed (9).

Any other comments
Review of benorylate as a antirheumatic drug (10).

References

1. Sax, N. I; et al. *Dangerous Properties of Industrial Materials* 7th ed., 1989, Van Nostrand Reinhold, New York
2. Rohrbach, P; et al. *Therapie* 1977, **32**(1), 89
3. *Br. Med. J.* 1984, **288**, 1344
4. Cailleux, P; et al. *Therapie* 1979, **34**(1), 81
5. Richardson, M. L; et al. *J. Pharm. Pharmacol.* 1985, **37**, 1-12
6. DeVries, J. *Biochem. Pharmacol.* 1981, **30**(5), 399
7. Williams, F. M; et al. *Br. J. Clin. Pharmacol.* 1989, **28**(6), 703-708
8. Jasiewicz, M. L; et al. *Mutat. Res.* 1987, **190**(2), 95-100
9. Bara Soain, I; et al. *Immunopharmacology* 1980, **2**(4), 293
10. Wright, V. *Scand. J. Rheumatol. Supplement* 1976, **13**(5)

B33   Benoxaprofen

CAS Registry No. 51234-28-7
Synonyms 2-(4-chlorophenyl)-α-methyl-5-benzoxazoleacetic acid; Opren; Oraflex; Uniprofen; Coxigon
Mol. Formula $C_{16}H_{12}ClNO_3$ **Mol. Wt.** 301.73
Uses Anti-inflammatory, analgesic and antipyretic drug.

Physical properties
M. Pt. 189-190°C.

Mammalian and avian toxicity

Acute data
LD_{50} oral mouse 800 mg kg^{-1} (1).
LD_{50} oral rat 118 mg kg^{-1} observed behavioural change in motor activity, ataxia, and aplastic anemia in blood (2).
LD_{50} intraperitoneal rat, mouse 129-398 mg kg^{-1} (2).
LD_{50} subcutaneous rat, mouse 121-482 mg kg^{-1} (2).

Metabolism and pharmacokinetics
Well absorbed after oral administration of 1- 10 mg kg^{-1} in dog, mouse, rat, rabbit,

rhesus monkey, and human. Only unchanged drug was detected in plasma, bound to plasma proteins, highest binding occurred in humans. Plasma elimination: human (33 hr); rat and mouse (28 and 24 hr respectively); all other species <13 hr. Oral dose of ^{14}C benoxaprofen 20 mg kg^{-1} to ♀ rats, tissue concentration was highest in liver, kidney, lungs, adrenals and ovaries (3).

Administration (dose, route unspecified) to rat induced the cytochrome P450 I and IV families and peroxisomal proliferation in the liver (4).

Irritancy
Skin disorders have been reported including photosensitivity reactions, erythema multiforme (the Stevens-Johnson syndrome) (5).

Sensitisation
Patients receiving 600 mg day^{-1} reported photosensitisation (6).

Genotoxicity
Human leukocyte 15 mg l^{-1} induced DNA damage (7).

Any other adverse effects
Onycholysis and other nail disorders, gastrointestinal disturbances including peptic ulceration and bleeding, blood disorders, liver disorders and heart failure have been reported (5,8).

Three hundred patients received 600 mg day^{-1} in the course of their treatment. Reported side effects included cutaneous, gastrointestinal and central nervous system disorders (6).

In vitro 3.75 μg ml^{-1} caused dose-related spontaneous activation of luminol- and lucigenin-enhanced chemiluminescence in human polymorphonuclear leukocytes (9).

Any other comments
Benoxaprofen's association with a variety of adverse reactions and fatalities in elderly people caused the drug to be withdrawn in 1982 (4).

Benoxaprofen is discussed in a retrospective study (10).

References
1. Evans; et al. *J. Med. Chem.* 1975, **18**, 53
2. *Pharmacol. Ther.* 1972, **1**
3. Chatfield, D. H; et al. *Xenobiotica* 1978, **8**, 133
4. Ayrton, A. D; et al. *Biochem. Pharmacol.* 1991, **42**(1), 109-115
5. *Martindale: The Extra Pharmacopoeia*, 29th ed., 1989, The Pharmaceutical Press, London
6. Halsey, J. P; et al. *Br. Med. J.* 1982, **284**, 1365
7. Schwalb, G; et al. *Cancer Res.* 1988, **48**, 3094
8. Gaudie, B. M; et al. *Lancet* 1982, **1**, 959
9. Van Reusburg, A. J; et al. *Agents Actions* 1991, **33**(3-4), 292-299
10. Lewis, D. F. V; et al. *Toxicology* 1990, **65**, 33-47

B34 Benquinox

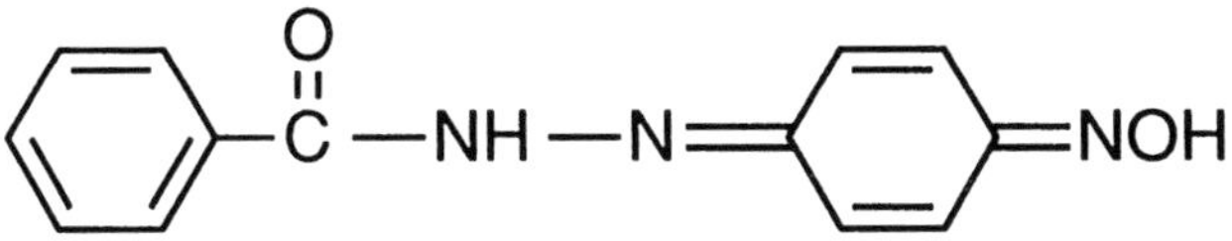

CAS Registry No. 495-73-8

Synonyms 1,4-Benzoquinone-N'-benzoylhydrazone oxime; 4-benzoyl-hydrazona-1,4-benzoquinone oxime; benzoic acid [4-hydroxyimino)-2,5-cyclohexadien-1-ylidene)hydrazide; Bayer 15080; Ceredon; Tillantox

Mol. Formula $C_{13}H_{11}N_3O_2$ Mol. Wt. 241.25

Uses Fungicide. Modifier for isoprene rubber.

Physical properties

M. Pt. 209-210°C.

Occupational exposure

Supply classification toxic.

Risk phrases

Toxic by inhalation, in contact with skin and if swallowed (R23/24/25)

Safety phrases

Keep out of reach of children – Keep away from food, drink and animal feeding stuffs – If you feel unwell, seek medical advice (show label where possible) (S2, S13, S44)

Mammalian and avian toxicity

Acute data

LD$_{50}$ oral rat, mouse 100 mg kg^{-1} (1,2).

Any other comments

Human health effects and experimental toxicology reviewed (3).

References

1. *Farm Chemicals Handbook* 1983, **48**, Meister Publishing Co., Willoughby, OH
2. *Guide to the Chemicals used in Crop Protection* 1973, **6**, 34, Canada
3. *ECETOC Technical Report No. 30(4)* 1991, European Chemical Industry Ecology and Toxicology Centre, B-1160 Brussels

B35 Bensulide

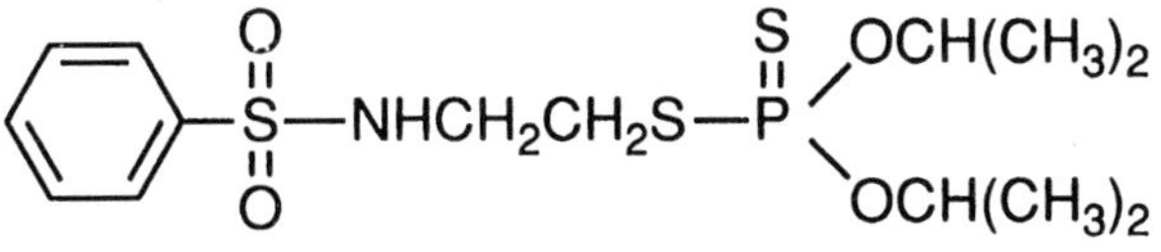

CAS Registry No. 741-58-2

Synonyms (*N*-(2-ethylthio)benzene sulfonamido-*S,O,O*-diisopropylphosphorodithioate; *S-O,O*diisopropylphosphrodithioate)of *N*-(2-mercaptoethyl)benzenesufonamide;

R4461; Betasan; Prefar; Exporsan; O,O-bis(1-methylethyl)S-[2-[(phenylsulfonyl) amino]ethyl]phosphorodithoate; phosphorodithioic acid, O,O-diisopropyl ester S-ester with N-(2-mercaptoethyl)benzenesulfonamide

Mol. Formula $C_{14}H_{24}NO_4PS_3$ **Mol. Wt.** 397.52

Uses Herbicide.

Physical properties

M. Pt. 34.4°C; **Partition coefficient** $\log P_{ow}$ 4.217 (1).; **Volatility** v.p. $<1 \times 10^{-6}$ mmHg at 25°C.

Solubility
Water: 25 mg l^{-1} at 20°C. Organic solvent: kerosene, misicible with acetone, ethanol, methyl isobutyl ketone, xylene

Occupational exposure

Supply classification harmful.

Risk phrases
Harmful by inhalation, in contact with skin and if swallowed (R20/21/22)

Safety phrases
Keep out of reach of children – Keep away from food, drink and animal feeding stuffs (S2, S13)

Ecotoxicity

Fish toxicity
LC_{50} (96 hr) rainbow trout 0.72 mg l^{-1}.
LC_{50} (96 hr) goldfish 1-2 mg l^{-1} (1).
LC_{50} (96 hr) channel catfish 0.379 mg l^{-1} (2).
LC_{50} (96 hr) bluegill sunfish 0.8 mg l^{-1} at 24°C (3).

Invertebrate toxicity
LC_{50} (96 hr) water shrimp 1.4 mg l^{-1} at 15°C (3).

Effects to non-target species
LD_{50} 0.0016 mg bee^{-1} (1).

Environmental fate

Degradation studies
Slowly degraded by microbial action. Residual activity in soil 4-6 months at 21-27°C (1).
Bensulide is degraded slowly in soil by microorganisms (4).

Mammalian and avian toxicity

Acute data
LD_{50} oral rat 270 mg kg^{-1} (5).
LD_{50} oral bobwhite quail 1390 mg kg^{-1} (1).
LD_{50} oral ♂ rat 1080 mg kg^{-1} (6).
LD_{50} oral ♂ albino rat 340 mg kg^{-1}.
LC_{50} (1 hr) inhalation rat >8 mg l^{-1} air (1).
LD_{50} dermal albino rat 3950 mg kg^{-1} (7).
LD_{50} percutaneous rabbit 3950 mg kg^{-1} (1).

Sub-acute data

In 90 day feeding study, well tolerated by rats and dogs given up to 250 ppm and 625 ppm in diet, respectively (7).

Genotoxicity

Salmonella typhimurium TA100, TA98, TA1535, TA1537, TA1538 with and without metabolic activation negative (8).
Escherichia coli WP2 hcr with and without metabolic activation negative (8).

References

1. *The Agrochemicals Handbook* 3rd ed., 1991, RSC, London
2. McCorkle, F. M; et al. *Bull. Environ. Contam. Toxicol.* 1977, **8**, 267
3. US Dept. Interior, Fish and Wildlife Service *Handbook of Acute Toxicity of Chemicals to Fish and Aquatic Invertebrates. Resource Publication No. 137* 1981, US Govt. Print. Off., Washington, DC
4. Weed Science Society of America *Herbicide Handbook* 4th ed., 1979 Champaign, IL
5. *Farm Chemical Handbook* 1991, Meister Publishing Co., Willoughby, OH
6. *Pesticide Dictionary* 1976, Meister Publishing Co., Willoughby, OH
7. Hubert, M. *The Pesticide Manual* 5th ed., 1968, British Crop Protection Council, Farnham
8. Moriya, M; et al. *Mutat. Res.* 1983, **116**(3-4), 185-216

B36   Bensultap

CAS Registry No. 17606-31-4

Synonyms *S,S'*-2-dimethylaminotrimethylene di(benzenethiosulfonate); *S,S'*-[2-(dimethylamino)-1,3- propanediyl benzenesulfonothionate; *S,S'*-[2-(dimethylamino)trimethylene]bis(benzenethiosulfonate); Bancol; Victenon; ZZ-Doricida; benzenesulfonic acid, thio- *S,S'*-(2-(dimethylamino)trimethylene)ester

Mol. Formula $C_{17}H_{21}NO_4S_4$ **Mol. Wt.** 431.62

Uses Insecticide.

Physical properties

M. Pt. 83-84°C; **B. Pt.** 150°C (decomp.).
Solubility
Organic solvent: ethanol, methanol, xylene, acetone, acetonitrile, *N,N*-dimethylformamide, chloroform

Ecotoxicity

Fish toxicity
LC_{50} (48 hr) carp, guppy, rainbow trout 15 mg l^{-1}, 17 mg l^{-1}, and 0.76 mg respectively (1).
LC_{50} (72 hr) carp, guppy, rainbow trout 8 mg l^{-1}, 16 mg l^{-1}, 0.76 mg l^{-1} respectively (1).

Effects to non-target species

LD$_{50}$ (8 day) oral pheasant 2730 mg kg^{-1} diet (1).
LD$_{50}$ (24 hr) topical application *Apis melliferra* 14.75 mg bee^{-1} (1).
LD$_{50}$ (24 hr) oral *Apis melliferra* 25.9 mg bee^{-1} (1).

Mammalian and avian toxicity

Acute data

LD$_{50}$ oral Japanese quail 60 mg kg^{-1} (2).
LD$_{50}$ oral ♂ rat 1110 mg kg^{-1} (3).
LD$_{50}$ oral ♂ mouse 415 mg kg^{-1} (4).
LD$_{50}$ percutaneous rabbit >2000 mg kg^{-1} (1).

Carcinogenicity and long-term effects

In 2 yr feeding trials the no-effect level for rat was 10 mg kg^{-1} day^{-1} and for mice 3.6 mg kg^{-1} day^{-1}. Reported to be non-oncogenic (1).

Any other comments

Incompatible with alkaline pesticides (1).

References

1. *The Agrochemicals Handbook* 3rd ed., 1991, RSC, London
2. Worthing, C. R. (Ed.) *The Pesticide Manual* 1987, 3rd ed., British Crop Protection Council, Farnham
3. *Farm Chemical Handbook* 1989, C32, Meister Publishing Co., Willoughby, OH
4. *Agric. Biol. Chem.* 1968, **32**, 678

B37   Bentazone

CAS Registry No. 25057-89-0
Synonyms 3-isopropyl-(1*H*)-benzo-2;1,3-thiadiazin-4-one-2,2-dioxide; 3-(1-methylethyl)-(1*H*)-2,1,3-benzothiadiazin-4(3*H*)-one-2,2-dioxide; 3-isopropyl-2,1,3-benzothiadiazin-4-one-2,2-dioxide; Bentazon; Bendioxide; Basagran; 1*H*-2,1,3-benzthiadiazin-4(3*H*)-one, 3-isopropyl-2,2-dioxide-
Mol. Formula $C_{10}H_{12}N_2O_3S$ **Mol. Wt.** 240.28
Uses Post-emergence herbicide.

Physical properties

M. Pt. 137-139°C; **Volatility** v.p. 3.459×10^{-6} mmHg at 20°C.

Solubility

Water: 500 mg l^{-1} at 20°C. Organic solvent: acetone, ethanol, ethyl acetate, diethyl ether, chloroform, benzene, cyclohexane

Occupational exposure

Supply classification harmful.

Risk phrases
Harmful by inhalation, in contact with skin and if swallowed (R20/21/22)

Safety phrases
Keep out of reach of children – Keep away from food, drink and animal feeding stuffs (S2, S13)

Ecotoxicity

Fish toxicity
LC_{50} (96 hr) rainbow trout 190 mg l^{-1} (1).
LC_{50} (96 hr) bluegill sunfish 616 mg l^{-1} (1).

Invertebrate toxicity
Chronic study *Daphnia magna* 0.25-1.0 mg l^{-1} affected post-embryonic growth period, which was delayed by 2.3-2.7 days compared to control (2).

Bioaccumulation
Calculated bioconcentration factor 19 (3).

Effects to non-target species
Non-toxic to bees (4).

Environmental fate

Degradation studies
Persistent in soil $t_{1/2}$ <2-14 wk. Rapidly metabolised in tolerant plants to extractable conjugates which are incorported into plant components. Potential to contaminate surface waters as result of its mobility in run off water and application to rice fields (5).

Mammalian and avian toxicity

Acute data
LD_{50} oral rat, mouse, rabbit, cat, quail 400-1100 mg kg^{-1} (6).
LD_{50} dermal rat, rabbit 2500-4000 mg kg^{-1} (6).
Acute inhalation studies included a study on saturated air containing bentazone volatiles at 1.2 mg l^{-1} at 20°C. Inhalation rat (8 hr) did not result in any mortality (7).

Carcinogenicity and long-term effects
In 2 yr feeding trials the no-effect level in rat was 350 mg kg^{-1} (1).
Oral ♂, ♀ rat (2 yr) 4000 ppm lesions observed included testicular Leydig cell adenomas, urinary bladder transitional cell papilloma and mesenchymal tumour in the thoracic cavity. The authors conclude there was no evidence of bentazone related tumour induction or of induced non-neoplastic histopathological changes. The no observed adverse effect level was 200 ppm (7).

Teratogenicity and reproductive effects
TD_{Lo} (6 days) pregnant oral rat 25 mg kg^{-1} caused both reduced fertility and foetotoxic effects including implantation mortality and specific developmental abnormalities to musculoskeletal system (8).
Administration of 0, 2.5, 7.5, 25 and 75 mg kg day^{-1} of bentazone at different times in gestation (6- 16 days) to beagle dogs caused prostatis in ♂ animals at highest dose and

1/3 ♂ and 2/3 ♀ animals died (8).

Oral pregnant ♀ albino rat (6-16 days gestation) 25-200 mg kg^{-1}. Foetus sizes below normal, absence of coccygeal vertebrae recorded in all groups. Metacarpal and metatarsal phlanges absent in all but one group (8).

In studies in rabbits no external malformations were noted in the 118 (control), 114 (low-dose), 129 (mid-dose) or 114 (high-dose) foetuses examined. No abnormalites of the abdominal or thoracic soft tissues, or of the skull skeletal structures were observed. A single foetus exhibited hydrocephaly at 150 mg kg^{-1} dose. In the absence of any dose effect relationship, this observation was not considered to be compound related. Skeletal variants were observed in a number of foetuses, but there was no evidence of a dose or compound relationship in their incidence. A no observed adverse effect level of 150 mg kg^{-1} was determined based on the total litter loss occuring in 1 female at 375 mg kg^{-1}. There was no evidence of teratogenicity at dose levels up to 375 mg kg^{-1} (7).

Metabolism and pharmacokinetics

Gavage ♂, ♀ rats received 0.8 mg ^{14}C-bentazone in 1 ml of 50% ethanol. Of the administered dose 91% was excreted in urine as parent compound within 24 hr, faeces contained 1% of administered dose (9,10).

Whole body autoradiography of rats indicated high levels of radioactivity in the stomach, liver, heart, kidneys after 1 hr of dosing with ^{14}C-bentazone. Radioactivity was not observed in brain or spinal cord (9).

Irritancy

3 White Vienna rabbits had 33 mg bentazone instilled into the conjunctival sac of one eye. Corneal opacity, iris congestion, conjunctival redness, chemosis and discharge were observed. Symptoms cleared by day 15 post-dosing (7).

Sensitisation

Sensitisation tests in guinea pigs using the Magnusson and Kligman Maximisation Test and the Open Epicutaneous Test indicated that bentazone has sensitising potential (7).

Genotoxicity

Salmonella typhimurium TA100, TA98, TA1535, TA1537, TA1538 and *Escherichia coli* WP2 hcr with and without metabolic activation negative (11).

In vitro Chinese hamster ovary HGPRT (hypoxanthine-guanine phosphoribosyl-transferase) point mutation negative. Chinese hamster ovary chromosomal aberrations negative. *In vivo* mouse bone marrow micronuclei negative. *In vivo* rat, mouse dominant lethal assay single 195 mg kg^{-1} (mouse) and 13 wk 20-180 ppm in diet (rat) negative (7).

Any other adverse effects

After oral dosing to rats, guinea pigs, rabbits, cats and dogs (dose unspecified), signs of toxicity included dyspnoea, apathy, staggering gait and, in cats and dogs, vomiting. Convulsions were also observed in cats (7).

Any other comments

Comprehensive report on toxicity of bentazone (7).

Resistant to hydrolysis in acid and alkali. Bentazone 1 ppm was stable to hydrolysis for up to 122 days in unbuffered water (pH 5, 7, 9) at 22°C (8).

References

1. *The Agrochemicals Handbook* 3rd ed., 1991 RSC, London
2. Shcherban, E. P; et al. *Gidrobiol. Zh.* 1986, **22**(5), 80-83, (Russ.) (*Chem. Abstr.* **106**, 97691u)
3. Kenaga, E. E. *Ecotox. Environ. Safety* 1980, **4**, 26-28
4. *Farm Chemicals Handbook* 1986, Meister Publishing, Willoughby, OH
5. *Drinking Water Health Advisory on Pesticides* 1989, 91-99, Lewis Publishers, Chelsea, MI
6. Spencer, E. Y. *Guide to the Chemicals Used in Crop Protection* 6th ed., 1976, 36, Research Institute, Agriculture Canada, Ottawa
7. *Pesticide Residues in Food 1991: Toxicology Evaluations* 1991, 25-54, World Health Organisation, Geneva
8. El-Mahdi, M. M; et al. *Arch. Exp. Veterinaermed.* 1988, **42**(2), 261-266
9. Chasseaud, L. F; et al. *Xenobiotica* 1972, **2**(3), 269-276
10. Gosselin, R. E; et al. *Clinical Toxicology of Commercial Products* 1984, 340, 5th ed., Williams & Wilkins, Baltimore
11. Moriya, M; et al. *Mutat. Res.* 1983, **116**(3-4), 185-216

B38   Bentonite

CAS Registry No. 1302-78-9
Synonyms Albagel premium USP4444; magbond; wilkinite; mineral soap; soap clay
Uses General purpose food additive (UK Additive No. 558). Pharmaceutical aid. Suspending and stabilising agent. Adsorbent or clarifying agent (1).
Occurrence In clay, predominantly montmorillonite.

Physical properties

Solubility
Water: Absorbs water to form insoluble gel (1).

Mammalian and avian toxicity

Acute data
LD_{50} intravenous rat 35 mg kg^{-1} (2).

Carcinogenicity and long-term effects
TD_{Lo} oral mouse 12000 g kg^{-1} administered over 28 wk caused equivocal tumorigenic effects (3).

Any other comments
Under consideration by the EEC for an E prefix food additive designation (4).
Human health effects and experimental toxicology reviewed (5,6).

References

1. *Martindale: The Extra Pharmacopoeia* 29th ed., 1989, The Pharmaceutical Press, London
2. *Bolletino della Societe Italiana di Biologa Sperimentale* 1968, **44**, 1685
3. *Ann. New York Acad. Sci.* 1954, **57**, 678
4. Hanssen, M. *E For Additives Completely Revised Edition* 1987, Thorsons, Vermont

5. *BIBRA Toxicity Profile* 1989, British Industrial Biological Research Association, Carshalton
6. *ECETOC Technical Report No. 30(4)* 1991, European Chemical Industry Ecology and Toxicology Centre, B-1160 Brussels

B39   Benzaldehyde

CAS Registry No. 100-52-7
Synonyms benzene aldehyde; artificial essential oil of almond; benzenecarbonal; benzene carbonal
Mol. Formula C_7H_6O **Mol. Wt.** 106.13
Uses In the manufacture of dyestuffs. Used in the perfume and flavouring industries. Reducing agent. Solvent.
Occurrence In kernels of bitter almonds.

Physical properties

M. Pt. −56.5°C; **B. Pt.** 179°C; **Flash point** 62°C; **Specific gravity** d_4^{15} 1.050;
Partition coefficient log P_{ow} 1.48; **Volatility** v.p. 1 mmHg at 26.2°C; v. den. 3.66.

Solubility
Organic solvent: miscible ethanol

Occupational exposure

Supply classification harmful.

Risk phrases
Harmful if swallowed (R22)

Safety phrases
Avoid contact with skin (S24)

Ecotoxicity

Fish toxicity
Exposure to 14.5 mg l^{-1} caused growth to stop early in *Phoxinus phoxinus* (1).
LC_{50} (96 hr) rainbow trout 11 mg l^{-1} (2).
LC_{50} (96 hr) bluegill sunfish 1.1 mg l^{-1} (2).

Invertebrate toxicity
Cell multiplication test *Microcystis aeruginosa* 20 mg l^{-1}, *Pseudomonas putida* 132 mg l^{-1}, *Entosiphon sulcatum* 0.29 mg l^{-1} (3,4).
EC_{50} (5, 15, 30 min) *Photobacterium phosphoreum* 5.32 mg l^{-1} Microtox test (5).

Effects to non-target species
LC_{50} (48 hr) *Ambystoma mexicanum* mexican axolotl (3-4 wk after hatching) 370 mg l^{-1} (6).
LC_{50} (48 hr) clawed toad 13-4 wk after hatching 190 mg l^{-1} (6).

Environmental fate

Degradation studies
99% degradation in adapted activated sludge (7).
ThOD: 2.4 g g^{-1}; DOC: 0.79 g g^{-1} (8).
BOD5 150% reduction in dissolved oxygen (9).
Confirmed biodegradable (10).

Absorption
Adsorbability 0.188 g g^{-1} carbon 94% reduction (11).

Mammalian and avian toxicity

Acute data
LD50 oral guinea pig 1000 mg kg^{-1} (12).
LDLo subcutaneous rat 5000 mg kg^{-1} (13).

Carcinogenicity and long-term effects
National Toxicology Program tested ♂, ♀ rat and mouse with benzaldehyde via oral gavage (dose unspecified). No evidence of carcinogenic activity in rats, some evidence of carcinogenic activity in mice (14).
Gavage ♂, ♀ F344/N rat and B6C3F1 mouse (2 yr) 0-400 mg kg^{-1}. No evidence of carcinogenic activity in rats, some evidence of carcinogenic activity in mice, indicated by increased incidences of squamous cell papillomas and hyperplasia of the forestomach (14).

Metabolism and pharmacokinetics
Metabolised by the liver, aromatic aldehydes, such as benzaldehyde, are oxidised to the corresponding acids (16).
Intraperitoneal rat, approximately 30% was excreted in urine as hippuric acid (17).

Genotoxicity
Bacillus subtilis M45, H17 liquid microsome rec-assay without metabolic activation DNA damaging potential negative (18).
In vitro Chinese hamster ovary cells with and without metabolic activation sister chromatid exchange positive (19).
Mouse lymphoma L5178Y cell forward mutation assay without metabolic activation positive (20).

Any other comments
Human health and effects experimental toxicology reviewed (21, 22).
Estimated acceptable daily intake from all food additive sources ≤5 mg kg^{-1} (23).

References
1. McKee, J. E; et al. *California State Water Control Board Report* 1963
2. Phipps, G. L; et al. *Environ. Pollut.* 1985, **38**, 141-157
3. Verschueren, K. *Handbook of Environmental Data on Organic Chemicals* 1983, Van Nostrand Reinhold, New York
4. Bringmann, G; et al. *Water Res.* 1980, **14**, 231
5. Kaiser, K. L. E; et al. *Water Pollut. Res. J. Canada* 1991, **26**(3), 361-431
6. Sloof, W; et al. *Bull. Environm. Contam. Toxicol.* 1980, **24**, 939-443
7. Pitter, P. *Water Res.* 1976, **10**, 231
8. Urano, K; et al. *J. Haz. Mat.* 1986, **13**, 147-159

9. US Coast Guard, Dept. Transport. *CHRIS – Hazardous Chemical Data* 1978, **2**, U.S. Govt. Print. Off., Washington
10. *MITI Report* 1984, Ministry of International Trade and Industry, Tokyo
11. Guisti, D. M; et al. *J. Water Pollut. Control Fed.* 1974, **46**(5), 947-965
12. *Food Cosmet. Toxicol.* 1964, **2**, 327
13. *Food Cosmet. Toxicol.* 1976, **14**, 693
14. *National Toxicology Program, Research and Testing Div.* 1991, Report No. TR-378, NIEHS, Research Triangle Park, NC 27709
15. Bishop, J. *Gov. Rep. Announce. Index (US)* 1990, **90**(20), Abstr. No. 052,283
16. Patty, F. A. (Ed.) *Industrial Hygiene and Toxicology* 2nd ed., 1965, **2**, Interscience Publishers, New York
17. Opdyke, D. L. J. (Ed.) *Monographs on Fragrance Raw Materials* 1979, 116, Pergamon Press, New York
18. Matsui, S; et al. *Water Sci. Technol.* 1989, **21**(8-9), 875-887
19. Galloway, S. M; et al. *Environ. Mol. Mutagen.* 1987, **10**(Suppl. 10), 1-175
20. McGregor, D. B; et al. *Environ. Mol. Mutagen.* 1991, **17**(3), 196-219
21. *ECETOC Technical Report No. 30(4)* 1991, European Chemical Industry Ecology and Toxicology Centre, B-1160 Brussels
22. *Dangerous Prop. Ind. Mater. Rep.* 1989, **9**(6), 61-70
23. *Tech. Rep. Ser. World Health Org. No. 383* 1968

B40   Benzamide

CONH$_2$

CAS Registry No. 55-21-0
Synonyms benzoic acid amide; benzoyl amide; phenylcarboxy amide
Mol. Formula C$_7$H$_7$NO   **Mol. Wt.** 121.14
Uses Intermediate in organic synthesis. Antiviral agent.

Physical properties

M. Pt. 128-129°C; **B. Pt.** 288°C; **Specific gravity** 1.341 at 4°C; **Partition coefficient** log P$_{ow}$ 0.64.

Solubility
Water: <1 mg ml^{-1} at 22°C. Organic solvent: acetone, dimethyl sulphoxide, ethanol

Ecotoxicity

Invertebrate toxicity
EC$_{50}$ (5, 15, 30 min) *Photobacterium phosphoreum* 59.3 mg l^{-1} Microtox test (1).

Environmental fate

Degradation studies
Biodegradable (2).

Mammalian and avian toxicity

Acute data
LD$_{50}$ oral mouse 1160 mg kg^{-1} (3).

Genotoxicity
Salmonella typhimurium TA97, TA98, TA100, TA1535, TA1537 with and without metabolic activation negative (4-6).
In vitro Chinese hamster ovary cells sister chromatid exchange positive (7).
Mouse lymphoma cell line L1210 sister chromatid exchange positive (8).

Any other adverse effects
Benzamide exhibited antitransforming action to human fibroblasts exposed to increasing ultraviolet radiation at low intracellular drug concentration (9).
Phorbol-ester induced promotion of initiated mouse skin keratinocytes to papillomas was largely prevented by the administration of benzamide, and other nicotinamide analogues (10).
Chronic administration oral mouse 200 mg kg^{-1} inhibits poly-ADP-ribose synthetase (11).

Any other comments
Central nervous system effects of benzamides are reviewed (12).

References

1. Kaiser, K. L. E; et al. *Water Pollut. Res. J. Canada* 1991, **26**(3), 361-431
2. *MITI Report* 1984, Ministry of International Trade and Industry, Tokyo
3. *Toxicol. Appl. Pharmacol.* 1971, **19**, 20, Academic Press, New York
4. Zeiger, E; et al. *Environ. Mol. Mugaten.* 1988, **11**(12), 1-158
5. Rossman, T. G; et al. *Mutat. Res.* 1991, **260**(4), 349-367
6. Zeiger, E; et al. *Environ. Mol. Mutagen.* 1988, **11**(Suppl. 2), 1-157
7. Morris, S. M; et al. *Mutat. Res.* 1984, **126**, 63-71
8. Lindahl-Kiesslung, K; et al. *Carcinogensis* 1987, **8**(9), 185-1188
9. Milo, G.E; et al. *Teratog. Carcinog. Mutagen.* 1989, **9**(3), 167-176
10. Ludwig, A; et al. *Cancer Res.* 1990, **50**(8), 2470-2475
11. Chieli, E; et al. *Mutat. Res.* 1987, **192**, 141-143
12. Boyer, P. *Sem. Hop.* 1983, **59**(12), 832-835, (Fr.) (*Chem. Abstr.* **98**, 209427n)

B41   Benz[*a*]anthracene

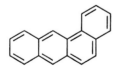

CAS Registry No. 56-55-3
Synonyms 1,2-benzanthracene; benzanthrene; naphthanthracene; tetraphene; 2,3-benzphenanthrene
Mol. Formula C$_{18}$H$_{12}$ **Mol. Wt.** 228.30
Occurrence In gasoline, bitumen, crude oil, wood preservative sludge, oil and waxes. Content in domestic effluent 0.191-0.319 ppb (1).

Physical properties

M. Pt. 157-159°C; **B. Pt.** 435°C.
Solubility
Water: 9-14 µg l^{-1}. Organic solvent: acetone, dimethylsulfoxide, ethanol, benzene, diethyl ether

Occupational exposure

Supply classification toxic.

Risk phrases
May cause cancer (R45)

Safety phrases
Avoid exposure-obtain special instructions before use – If you feel unwell, seek medical advice (show label where possible) (S53, S44)

Ecotoxicity

Fish toxicity
Skin and liver tumours observed in brown bullhead fish inhabiting the Black River, an industrialised tributary of Lake Erie. The river was particularly contaminated with benz[a]anthracene and benzo[a]pyrene (2).

Invertebrate toxicity
Eastern oyster (2 day) 4.1 µg g^{-1} $t_{1/2}$ depuration 9 days (3).

Bioaccumulation
Bioconcentration factor for *Daphnia sp.* 54 and oysters 20 (4-6).
Shrimp exposed to 2.8 ppb benz[a]anthracene showed rapid uptake during the first 6 hr. At 96 hr exposure shrimp exhibited continual accumulation. Accumulation in the tissues was in the order stomach, intestine, hepatopancreas, cephalothorax and abdomen. When transferred to seawater, shrimp appeared to depurate benz[a]anthracene rapidly to 805 after 7 days (7).

Environmental fate

Degradation studies
Soil contaminated with benz[a]anthracene showed losses of 22-88% in 400 days (8).
Degraded to CO_2 in estuarine water (9).
In seawater-sediment slurry, 1.4-1.8% degradation was observed per wk which indicated $t_{1/2}$ of 199-252 days (10,11).
$t_{1/2}$ of 290 days in sediment from an oil-contaminated stream, but $t_{1/2}$ 10-400 times longer in an uncontaminated stream (12).
Microbial degradation is faster in the upper layers of sediment (10).

Abiotic removal
Photolytic $t_{1/2}$ in early March is 5 hr. Based on seasonal variations, $t_{1/2}$ would be 2.9 and 7.8 hr in summer and winter, respectively (13).
Photolytic $t_{1/2}$ in an oligotrophic lake is 10 hr, in a stream 20 hr and eutrophic pond or lake 50 hr (14).

Absorption
Adsorption in estuarine waters of 3 µg l^{-1} recorded, 59% absorbed on particles after 3 hr (3).

Benzoanthracene is strongly adsorbed on sediments with a soil adsorption coefficient of $0.55 \times 10^6 - 1.87 \times 10^6$ in 3 sediments (14).

Mammalian and avian toxicity

Acute data
LD_{Lo} intravenous mouse 10 mg kg^{-1} (15).

Sub-acute data
LD_{50} (72 hr) chick embryo 79 µg kg^{-1} following injection into 7 day pre-incubated air sacs (16).

Carcinogenicity and long-term effects
No adequate evidence for carcinogenicity to humans, sufficient evidence of carcinogenicity to animals, IARC classification group 2A (17).
In a CASE-SAR analysis of polycyclic aromatic hydrocarbon carcinogenicity, benz[a]anthracene was classified as a suspect carcinogen with low to moderate potency (18).
Application to exposed mammary gland of ♀ Sprague-Dawley rats, inactive (19).
In ♂ mice administered 0.6 mg of benz[a]anthracene the hepatic tumour incidence was 79% (duration unspecified) (20).
♂ Syrian golden hamster buccal pouch surfaces painted for ≤20 wk with benzo[a]anthracene dissolved in paraffin oil developed carcinomas of the buccal pouch (21).

Metabolism and pharmacokinetics
In mammals benz[a]anthracene appears to be metabolized via 3,4 epoxide to 3,4 diol 3,4 diol 1,2 epoxide, or 8,9-dihydroxybenz[a]anthracene-10, 11 epoxide (22-24). Benzoanthracene was not detected in human liver, but was present in fatty tissues (25). Induces cytochrome P450 activity in rats. Benz[a]anthracene also has a self-inducing effect on its own metabolism as well as other polycyclic aromatic hydrocarbons e.g. pyrene and fluoroanthene (26).

Genotoxicity
Salmonella typhimurium TA100, TA98 with and without metabolic activation positive. Rat bone marrow cells limited ability to induce chromosome aberrations (27).
In vitro human peripheral blood lymphocytes benz[a]anthracene caused the formation of readily measurable levels of DNA adducts (28).
The frequency of sister chromatid exchanges in rat mammary epithelial cells was 0.4. Exposure of rat mammary epithelial cells to 5 µg ml^{-1} for 24 hr resulted in a 45-62% reduction in the ^{3}HTdR labelling index of rat cells; human mammary epithelial cells resulted in a 50-90% depression. In mammary epithelial cell mediated assays for rat and humans, frequencies of sister chromatid exchange per chromosome were 3 and 4.5, respectively (29).

Any other comments
Air pollution emissions, values for Europe using leaded and unleaded petrol 32.4 µg l^{-1} fuel burnt (30).
Metabolism, toxicity and carcinogenic potential of benz[a]anthracene reviewed (31,32).
Human health effects, experimental toxicology and environmental effects reviewed (33-35).

References

1. Borneff, G. *Arch. Hyg. Bakt.*, 1962, **146**, 183
2. Baumann, P. C; et al. *Trans. Am. Fish Soc.* 1987, **116**(1), 79-86
3. Lee, F. R; et al. *Environ. Sci. Tech.*, 1978, **12**(7), 832-838
4. Southworth, G. R; et al. *Water Res.* 1978, **12**, 973-977
5. Le, R. F. *Fate and Effects of Petroleum Hydrocarbons in Marine Organisms and Ecosystems* 1977, **6**, 60-70
6. Baughman, G. L; et al. *CRC Crit. Rev. Microbiol.* 1981, **8**, 205-228
7. Fox, F. R; et al. *Carcinogen. Polynucl. Aromat. Hydrocarbons Mar. Environ.* 1982, 336-349, EPA 600/9-82-013
8. Bulman, T. L; et al. *Oil Freshwater: Chem. Biol. Countermeans Tech. Proc. Symp. Oil Pollut. Freshwater*, 1987, 231 – 251, Pergamon, New York
9. Lee, F. R. *Fate of Petroleum Components in Estuarine Waters of the S.E. USA*, 1977, Oil Spill. Conference, US
10. Gardiner, W. S; et al. *Water Air Soil Pollut.* 1979, **11**, 339-347
11. Sims, R. C; et al. *Res. Rev.* 1983, **83**, 1-68
12. Herbes, S. E; et al. *Appl. Environ. Microbiol.* 1978, **35**, 306-316
13. Mill, T; et al. *Chemosphere* 1981, **10**, 1281-1290
14. Smith, J. H; et al. *Environmental Pathways of Selected Chemicals in Freshwater Systems* 1978, 39-63, EPA 600/7-78-074
15. *J. Natl. Cancer Inst.*, 1940, **1**, 225
16. Brunstroem, B; et al. *Arch. Toxicol.* 1991, **65**(6), 485-489
17. *IARC Monograph*, 1987, **suppl. 7**, 58
18. Richard, A. M; et al. *Mutat Res.*, 1990, **242**(4), 285-303
19. Cavalieri, E; et al. *J. Cancer Res. Clin. Oncol.*, 1988, **114**(1), 3-9
20. Wislocki, P. G; et al. *Carcinogensis*, 1986, **7**(8), 1317-1322
21. Solt, D. B; et al. *J. Oral Pathol.* 1987, **16**(6), 294-302
22. Casarett, L. J. (Ed.) *Toxicology, The Basic Science of Poisons* 1975, Macmillan, New York
23. Jerina, D. M. *Fed. Proc.*, 1978,**37**(16), 1383
24. Miller, E. C; et al. *Chemical Carcinogens* 1974, **16** Amer. Chem. Soc., Washington DC
25. Obana, H; et al. *Bull. Environ. Contam. Toxicol.* 1981, **27**(1), 23-27
26. Brandys, J; et al. *Folia Med. Cracov.* 1990, **31**(3), 157-163, (Pol.) (*Chem. Abstr.* **115**, 177144a)
27. Ito, Y; et al. *Mutat Res.*, 1988, **206**(1), 55-63
28. Gupta, R. C; et al. *Proc. Natl. Acad. Sci. USA*, 1988, **85**(10), 3513-3517
29. Mane, S. S; et al. *Environ. Mol Mutag.* 1990, **15**(2), 78-82
30. Candeli, A; et al. *A. Environ.* 1974, **8**, 693-705
31. Yang Shen, K. *Polycyclic Aromatic Hydrocarbon Carcinogens: Smict-Act. Relat.* 1988, **1**, 129-149
32. *IARC Monograph* 1983, **32**, 135-145
33. *Toxicological Profile for Benz[a]anthracene*, 1990, 1-81, Dept Health, Human Services, Atlanta, GA
34. *Gov. Rep. Announce. Index US* 1990, **90**(19), 048,363
35. *ECETOC Technical Report No. 30(4)* 1991, European Chemical Industry Ecology and Toxicology Centre, B-1160 Brussels

B42   Benzarone

CAS Registry No. 1477-19-6
Synonyms (2-ethyl-3-benzofuranyl)(4-hydroxyphenyl); methanone;
2-ethyl-3-benzofuranyl-p-hydroxyphenyl ketone;
2-ethyl-4'-hydroxy-3-benzoylbenzofuran; Benzaron; Fagivil; Fragivix
Mol. Formula $C_{17}H_{14}O_3$ Mol. Wt. 266.30
Uses Therapeutically for increased capillary fragility, thrombolytic agent.

Physical properties

M. Pt. 124.3°C.

Mammalian and avian toxicity

Acute data
LD_{50} intraperitoneal mouse 200 mg kg^{-1} (1).

Teratogenicity and reproductive effects
TD_{Lo} (9-14 days) oral ♀ pregnant rat 1020 mg kg^{-1} both maternal toxicity to ovaries,
fallopian tubes and abnormal musculoskeletal development in foetus (2).
TD_{Lo} (7-12 days) oral ♀ pregnant mouse specific development abnormalities
musculoskeletal system (2).

Metabolism and pharmacokinetics
In man 70% dose eliminated in urine, the metabolites present as conjugates (mainly
glucuronides). The 2 major aglycones were the hydroxylated metabolites
2-(1-hydroxyethyl)-3-(4- hydroxybenzoyl)benzofuran and 2-(1-hydroxyethyl)-3-(4-
hydroxybenzoyl)-hydroxybenzofuran, 26% and 8% respectively. Small amounts (2-3%
dose) of benzarone and metabolites 2-ethyl-3-(4-hydroxybenzoyl)- hydroxybenzofuran
and 2-(1,2-dihydroxyethyl)-3-(4-hydroxybenzoyl)benzofuran were detected. The
same conjugated components were present in human plasma and metabolites
2-(1-hydroxyethyl)-3-(4-hydroxybenzoyl)benzofuran and 2-ethyl-3-(4-hydroxybenzoyl)-
hydroxybenzofuran were the major components in the faeces. Metabolism was less
extensive in rat and dog, benzarone (free and conjugated) representing >70% of the
total material in bile, urine and faecal extracts from dogs and in bile from rats (3).
The metabolic fate of benzarone in rat 2 mg kg^{-1}, dog 0-5 mg kg^{-1}, and human 100
mg, all taken orally, was compared. Humans excreted 73% in urine, 19% in faeces,
dog and rat excreted >80% in faeces, mostly during first 48 hr (4).

References

1. *Arch. Int. Pharmacodynamie Therapie* 1965, **154**, 94
2. *Kiso to Rinsho. Clinical Report* 1969, **3**, 961
3. Hawkins, D. R. *Biotransformations* 1989, **1**, 352-353
4. Wood, S. G; et al. *Xenobiotica* 1987, **17**(7), 881-896

B43 Benzathine

$$CH_2NHCH_2CH_2NHCH_2$$

CAS Registry No. 140-28-3
Synonyms 1,2-bis(benzylamino)ethane; DBED; *N,N'*-dibenzylethylenediamine;
1,2-ethanediamine-*N, N*-bis(phenylmethyl)-
Mol. Formula $C_{16}H_{20}N_2$ **Mol. Wt.** 240.35
Uses Precursor for benzathine penicillin.

Physical properties

M. Pt. 26°C; **B. Pt.** $_{12}$ 212-213; **Specific gravity** d_4^{20} 1.024.

Mammalian and avian toxicity

Acute data
LD_{50} oral mouse 388 mg kg^{-1} (1).
LD_{50} intraperitoneal mouse 50 mg kg^{-1} (2).

References

1. *Cancer Chemother. Rep.* 1968, **52**, 579
2. *NTIS Report* AD 277-689, Nat. Tech. Inf. Ser., Springfield, VA

B44 Benzathine penicillin

$$CH_2NHCH_2CH_2NHCH_2$$

CAS Registry No. 1538-09-6
Synonyms benzathine benzylpenicillin; Benzethacil; Penoepon; penicillin benzatin;
penicillin g benzathine; 4- thia-1-azabicyclo[3.2.0]heptane-2-carboxylic acid,3,3-
dimethyl-7-oxo-6-[(phenylacetyl)amino]- [2*S*(2α6β]-, compound with
N,N'-bis(phenylmethyl)-1,2-ethanediamine (2:1)
Mol. Formula $C_{48}H_{56}N_6O_8S_2$ **Mol. Wt.** 909.14
Uses Antibacterial agent. Used in the treatment of syphilis.

Physical properties

M. Pt. 123-124°C.
Solubility
Water: 0.15 mg ml^{-1}. Organic solvent: benzene, acetone, formamide

---

Mammalian and avian toxicity

Acute data
LD_{50} oral mouse 2000 mg kg^{-1} (1).
LD_{50} intraperitoneal mouse 460 mg kg^{-1} (1).
Teratogenicity and reproductive effects
TD_{Lo} (4 day) ♀ rat 23 mg kg^{-1} caused post implantation mortality (2).
Metabolism and pharmacokinetics
In humans hydrolysed to benzylpenicillin. Major excretory route via urine (3).

Any other adverse effects

Administered to guinea pigs at 100,000-400,000 V kg^{-1} can alter erythrocyte membrane functions and may lead to haemolysis. Highest dose increased membrane cholesterol and decreased phospholid levels. All doses decreased Na, K ATPase activity (4).

Any other comments

Converts to benzylpenicillin and benzatine (5).
Toxicity reviewed (6).

References

1. *Drugs Jpn.* 1982, **6**, 774
2. *Bull Exp. Biol. Med.* 1976, **82**,1076
3. *Martindale: The Extra Pharmacopoeia* 29th ed., 1989, The Pharmaceutical Press, London
4. Turkozkan, N; et al. *Biyokim Derg.* 1988, **13**(2), 47-51, (Turk.) (*Chem. Abstr.* **110** 33371q)
5. Richardson, M. L. *J. Pharm. Pharmacol.* 1985, **37**, 1
6. Kreuzig, F. *Analytical Profiles of Drug Substances* 1982, **11**, 463-482, K. Florey (Ed.) New York

B45   Benzene

CAS Registry No. 71-43-2
Synonyms benzol; benzole; coal naphtha; mineral naphtha; phenylhydride; pyrobenzol; pyrobenzole
Mol. Formula C_6H_6 **Mol. Wt.** 78.11
Uses Solvent for fats, inks, oils, paints, plastics and rubber. Starting material in chemical manufacture of resins, plastics, nylon-66, polyamides and styrene. Used in the manufacture of detergents, explosives and pharmaceuticals (1).
Occurrence Originally produced by coal carbonisation, but now largely derived from petroleum or by cyclisation and aromatisation of paraffinic hydrocarbons.

Physical properties

M. Pt. 5.5°C; **B. Pt.** 80.1°C; **Flash point** –11°C; **Specific gravity** d_4^{20} 0.8786;
Partition coefficient log P_{ow} 2.15; **Volatility** v.p. 76 mmHg at 20°C.

Solubility
Water: 1780 mg l^{-1} at 20°C. Organic solvent: miscible acetone, ethanol, diethyl ether

Occupational exposure
US TLV (TWA) 10 ppm (32 mg m^{-3}); **UK Long-term limit MEL** 5 ppm (16 mg m^{-3}); **UN No.** 1114; **HAZCHEM Code** 3WE; **Conveyance classification** flammable liquid; **Supply classification** highly flammable and toxic.

Risk phrases
May cause cancer – Highly flammable – Toxic by inhalation, in contact with skin and if swallowed – Danger of serious damage to health by prolonged exposure (R45, R11, R23/24/25, R48)

Safety phrases
Avoid exposure-obtain special instructions before use – Keep away from sources of ignition – No Smoking – Do not empty into drains – If you feel unwell, seek medical advice (show label where possible) (S53, S16, S29, S44)

Ecotoxicity

Fish toxicity
LC$_{50}$ (24-96 hr) fathead minnow, bluegill sunfish, goldfish 36-22 mg l^{-1} (1).
LC$_{50}$ (96 hr) bass 6-11 ppm (1).

Invertebrate toxicity
LC$_{50}$ (96 hr) grass shrimp 20-27 ppm (1,2).
Cell multiplication inhibition test, *Pseudomonas putida* 92 mg l^{-1}, *Microcystis aeruginosa* >1400 mg l^{-1}, *Entosiphon sulcatum* >700 mg l^{-1} (1).
EC$_{50}$ (8 day) *Selenastrum capricornutum* 41 mg l^{-1} (3).

Bioaccumulation
Bioconcentration factor in pacific herring larvae, eel 3.5-3.9 (1).

Effects to non-target species
LC$_{50}$ (48 hr) Mexican axolotl (3-4 wk after hatching) 370 mg l^{-1} (4).

Environmental fate

Nitrification inhibition
Not inhibited at 500 mg l^{-1} (5).

Degradation studies
ThOD 3.07 g g^{-1}, COD 0.92 g g^{-1} (6).
BOD$_{10}$ 67% reduction of dissolved oxygen in acclimatised sludge (1).
Benzene is subject to rapid volatilisation in water and from soil surfaces, and is very mobile in soil. Biodegradation may occur in shallow aerobic ground water, but not under anaerobic conditions (7).
Confirmed biodegradable (8).

Mammalian and avian toxicity

Acute data
LD$_{50}$ oral rat, mouse 3400-4700 mg kg^{-1} (9,10).
Short-term acute exposure in humans may cause initial exhilaration, followed by

dizziness, headache, nausea, drowsiness and pulmonary irritation. 7500 ppm and above for approximately 30 min may produce narcosis and death (11).

Sub-acute data

Exposure of rats to 50 ppm benzene vapour for several wk, led to a reduction in red and white blood cells and platelets; exposure to concentrations >100 ppm produced leukopenia and aplasia (11).

♂ mice were fed 0-790 mg l^{-1} in drinking water for 28 days. Stimulation of the hypothalamic-pituitary-adrenocortical axis and increased circulatory levels of corticosterone were observed at high dose levels (12).

Carcinogenicity and long-term effects

Chronic exposure to benzene, in humans, at concentrations that produce changes in the blood may result in leukaemia, especially acute myelogenous leukaemia (11). There is clear evidence of carcinogenicity in mice and rats treated by gavage (103 wk) 100 and 200 mg l^{-1}. Tumours have been reported in various tissues including adrenals, lung, liver, ovary, oral cavity, stomach and skin (13).

Benzene administered by gavage produced ovarian atrophy, cysts, hyperplasia and neoplasia in mice (14).

Sufficient evidence for carcinogenicity to humans and animals, IARC classification group 1 (15).

Teratogenicity and reproductive effects

Teratogenicity has been reported at high concentrations in rats, but there is no evidence of foetal malformations at concentrations which produce no maternal toxicity. Women are considered hypersusceptible to benzene, particularly during pregnancy and breast feeding, however there are no reports of teratogenic effects or any increase in spontaneous abortion in women occupationally exposed to benzene (14,16).

Metabolism and pharmacokinetics

Benzene is partly eliminated unchanged in the breath and urine of humans. Oxidation occurs producing benzene epoxide, phenols and diphenols, including catechol, hydroquinone, benzoquinone and 1,2,4-benzenetriol, which are in turn conjugated in the liver and excreted in the urine (11,17).

Toxic amounts of benzene can readily be absorbed through the skin (11).

Irritancy

Dermal rabbit (24 hr) 15 mg caused mild irritation, and 2 mg instilled into rabbit eye caused severe irritation (18,19).

Genotoxicity

Salmonella typhimurium TA98 TA100, TA1535, TA97/TA1537 with and without metabolic activation negative (13,20).

In vivo rodent bone marrow autogenetic test positive induction of micronuclei and chromosomal aberrations (20).

Chromosomal aberrations in white blood cells and bone marrow in humans which could initiate leukaemia have been reported, but there is no evidence of aberrations at exposure levels of 25 ppm or less (11).

Inhalation ♂ mice 1 ppm induced chromosomal aberrations in spermatocytes and sister chromatid exchange in spermatogonia (21,22).

Any other adverse effects to man

A mortality study of 259 ♂ employees working in a benzene chemical plant between 1947 and 1960, found 4 deaths resulted from lymphoreticular cancers where 1 would have been expected. 3 deaths were due to leukaemia and 1 to multiple myeloma. The authors conclude occupational exposure to benzene causes leukaemia and is linked to multiple myeloma (23).

Haematological and immunochemical investigation of 270 workers with chronic exposure to benzene evidenced changes to lymphocyte nuclei and disorders of the humoral immune response (24).

A retrospective cohort study was carried out in 1982-1983 on 28,460 workers exposed to benzene in China. Mortality was significantly higher in exposed persons as was mortality from all malignant neoplasms (25).

Any other comments

Benzene exposure, experimental toxicology, epidemiology studies, human health and environmental effects have been extensively reviewed (26-43).

World Health Organisation guidelines on drinking water, provisional limit 10 μg l⁻¹.

References

1. Verschueren, K. *Handbook of Environmental Data on Organic Chemicals* 2nd ed., 1983, Van Nostrand Reinhold, New York
2. Bringman, G; et al *Nat. Res.* 1980, **14**, 231
3. Herman, D. C; et al. *Aquat. Toxicol.* 1990, **18**(2), 87-100
4. Slooff, W; et al. *Bull. Environ. Contam. Toxicol.* 1980, **24**, 439
5. Richardson, M. *Nitrification Treatment of Sewage* 1985, RSC, London
6. Urano, K. *J. Haz. Mat.* 1986, **13**, 147
7. Howard P. H. (Ed.) *Fate and Exposure Data for Organic Chemicals* 1990, **2**, 29-39
8. *MITI Report* 1984, Ministry of International Trade and Industry, Tokyo
9. *Raw Materials Handbook* 1974, **1**, 5
10. *Hyg. Sanit.* 1967, **32**, 349
11. *Chemical Data Safety Sheets* 1988, **1**, 14-18, RSC, London
12. Hsieh, G. C; et al. *Immunopharmacology* 1991, **21**(1), 23-31
13. Ashby, J. *Mutat. Res.* 1988, **204**, 17-115
14. Maronpot, R. R. *Environ. Health Perspect.* 1987, **73**, 125-130
15. *IARC Monograph* 1987, **Suppl. 7**, 120
16. Fishbein; et al. *IARC Environ. Carcinog.* 1988, 10
17. Lewis, J. G; et al. *Toxicol. Appl. Pharmacol.* 1988, **92**(2), 246-254
18. *Am. Ind. Hyg. Assoc. J.* 1962, **95**, 23
19. Marhold, J. V. *Sbornik Vysledku Toxixologickeho Vysetreni Latek A Pripravku* 1972, 23, Prague
20. Shelby, M. D. *Mutat. Res.* 1988, **204** 3-15
21. Au, W. W; et al. *Teratog. Carcinog. Mutagen.* 1990, **10**(2), 125-134
22. Zhu, Yufen; et al. *Weisheng Dulixue Zazhi* 1989, 3(4), 228-230, (Ch.) (*Chem. Abstr.*, **113**, 72845d)
23. Decoufle, P; et al. *Environ. Res.* 1983, **30**(1), 16-25
24. Chirco, V; et al. *Rev. Roum. Med. Interne.* 1981, **19**(4), 373-378
25. Yin, S. N; et al. *Environ. Health Perspect.* 1989, **82**, 207-213
26. Jin, C; et al. *Br. J. Ind. Med.* 1987, **44**(2), 124-128
27. Sandler, D. P; et al. *Amer. J. Epidemol.* 1987, **126**(6), 1017-1032
28. Schwartz, E. *Amer. J. Ind. Med.* 1987, **12**(1), 91-99
29. Damrau, J. Z. *Klin. Med.* 1990, **45**(23), 2063-2064, (Ger.) (*Chem. Abstr.*, **114**, 87806m)
30. Kipen, H. M; et al. *Adv. Mod. Environ. Toxicol.* 1989, **16**, 67-86

31. Aksoy, M. *Adv. Mod. Environ. Toxicol.* 1989, **16**, 87-98
32. Lamm, S. H. *Environ. Health Perspect.* 1989, **82**, 289-297
33. Sabourn, P. J; et al. *Adv. Mod. Environ. Toxicol.* 1989, **16**, 153-176
34. Spitzer, H. L. *Adv. Mod. Environ. Toxicol.* 1989, **16**, 141-152
35. Moszczynski, P; et al. *Med. Pr.* 1985, **36**(5), 316-324
36. Kalt, G. T. *CRC Crit. Rev. Toxicol.* 1987, **18**(2), 141-159
37. King, A. G; et al. *Mol. Pharmacol.* 1987, **32**(6), 807-812
38. *ECETOC Technical Report No. 30(4)* 1991, European Chemical Industry Ecology and Toxicology Centre, B-1160 Brussels
39. Goldstein, B. D. *Adv. Mod. Environ. Toxicol.* 1989, **16**, 55-65
40. *Gov. Rep. Announce Index US* 1989, **16**, 131-139
41. Bailer, A. J; et al. *Adv. Mod. Environ. Toxicol.* 1989, **16**, 131-139
42. Wallace, L. A. *Adv. Mod. Environ. Toxicol.* 1989, **16**, 113-130
43. Izmerov, N. F. *Scientific Reviews of Soviet Literature on Toxicity & Hazards of Chemicals* 1992-1993, **90**, Eng. Trans, Richardson, M. L. (Ed.), UNEP/IRPTC, Geneva

B46   Benzene, 1-chloromethyl-4-nitro-

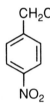

CAS Registry No. 100-14-1
Synonyms 4-nitrobenzyl chloride; α-chloro-4-nitrotoluene; α-chloro-*p*-nitrotoluene; *p*-nitrobenzyl chloride; 1-chloromethyl-4-nitrobenzene
Mol. Formula $C_7H_6ClNO_2$ **Mol. Wt.** 171.58

Physical properties
M. Pt. 70-73°C.

Occupational exposure
UN No. 2433; **HAZCHEM Code** 2X; **Conveyance classification** harmful substance.

Ecotoxicity

Invertebrate toxicity
EC_{50} (5, 15, 30 min) *Photobacterium phosphoreum* 2.6 mg l^{-1} Microtox test (1).

Mammalian and avian toxicity

Acute data
LC_{Lo} (4 hr) inhalation rat 280 mg m^{-3} (2).

Genotoxicity
Salmonella typhimurium TA100, TA98 without metabolic activation positive (3).
Salmonella typhimurium TA98, TA100, TA1535, TA1537 with and without metabolic activation positive (4).

Salmonella typhimurium TA100 without metabolic activation weakly positive (5).
In vitro Chinese hamster ovary cells sister chromatid exchange equivocal (6).

Any other comments

Genotoxicity, biological properties and human exposure limits reported (7).

References

1. Kaiser, K. L. E; et al. *Water Pollut. Res. J. Canada* **26**(3), 361-431
2. *Toxicologist* 1984, **4**, 66
3. Ball, J. C; et al. *Mutat. Res.* 1984, **138**(2-3), 145-151
4. Zeiger, E; et al. *Environ. Mutagen.* 1987, **9**(Suppl. 9), 1-109
5. Shimizu, M; et al. *Mutat. Res.* 1986, **170**(1-2), 11-22
6. Hemminki, K; et al. *J. Appl. Toxicol.* 1983, **3**(4), 203-207
7. Kawai, A; et al. *Sangyo Igaku* 1987, **29**(1), 34-54, (Jap.) (*Chem. Abstr.* **107**, 72503s)

B47   Benzene, 1,3-dichloro-5-nitro-

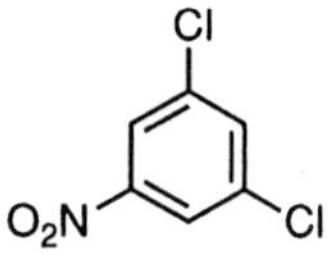

CAS Registry No. 618-62-2

Synonyms 3,5-dichloronitrobenzene; 1,3-dichloro-5-nitrobenzene

Mol. Formula $C_6H_3Cl_2NO_2$ **Mol. Wt.** 192.00

Uses Pesticide.

Physical properties

M. Pt. 64-65°C; **Specific gravity** d_4^{14} 1.69.

Solubility

Organic solvent: diethyl ether, ethanol

Ecotoxicity

Invertebrate toxicity

EC_{50} (5, 15, 30 min) *Photobacterium phosphoreum* 17.1 mg l^{-1} Microtox test (1).

Any other comments

Experimental toxicology reviewed (2).

References

1. Kaiser, K. L. E; et al. *Water Pollut. Res. J. Canada* 1991, **26**(3), 361-431
2. *ECETOC Technical Report No. 30(4)* 1991, European Chemical Industry Ecology and Toxicology Centre, B-1160 Brussels

B48   Benzene, 1,4-dichloro-2-nitro-

CAS Registry No. 89-61-2
Synonyms 2,5-dichloronitrobenzene; 1,4-dichloro-2-nitrobenzene
Mol. Formula $C_6H_3Cl_2NO_2$ **Mol. Wt.** 192.00
Uses Used in the manufacture of dyestuff intermediates. In the manufacture of
p-chloro-o-nitrophenol.

Physical properties
M. Pt. 54-57°C; **B. Pt.** 266-269°C; **Specific gravity** d^{22} 1.669.

Solubility
Organic solvent: chloroform, hot ethanol, benzene, diethyl ether

Occupational exposure

Risk phrases
Harmful if swallowed – Irritating to eyes – Irritating to skin (R22, R36, R38)

Ecotoxicity

Fish toxicity
Rainbow trout and threespine stickleback exposed to 10 mg l^{-1} suffered loss of
equilibrium in 3-4 hr and death in 6-8 hr. Sockeye salmon exposed to 10 mg l^{-1}
suffered loss of equilibrium in 2-3 hr and death in 3-4 hr (1).
Bluegill sunfish exposed to 5 ppm died within 2 hr and trout exposed
to 5 ppm died with 23 hr (2).

Invertebrate toxicity
EC_{50} (5, 15, 30 min) *Photobacterium phosphoreum* 8.78 mg l^{-1} Microtox test (3).

Environmental fate

Degradation studies
Biodegradation by activated sludge was >99% (4).

Mammalian and avian toxicity

Acute data
LD_{50} oral mouse, rat 1210-5089 mg kg^{-1} (5,6).

Metabolism and pharmacokinetics
Major metabolite was 2,5 dichloroaniline (7).

Irritancy
500 mg applied to rabbit skin for 24 hr caused mild irritation, while 100 mg instilled
into rabbit eye for 24 hr caused moderate irritation (4, 8).

Genotoxicity

Salmonella typhimurium TA97, TA98, TA1535, TA1537, TA1538 without metabolic activation positive (9).

Induced SOS reactions in *Salmonella typhimurium* and *Escherichia coli* containing umu-lac fused gene. Mutagenic effects in *Salmonella typhimurium* were enhanced by plasmids pKM$_{101}$ and pSK$_{1002}$ (10).

Any other adverse effects

Absorption into blood stream may lead to methaemoglobinaemia, onset can be delayed 2-4 hr (11).

Any other comments

Experimental toxicology reviewed (12).

References

1. MacPhee, C; et al. *Fish Toxicity Screening Data. Part 2. Lethal Effects of 2014 Chemicals upon Sockeye Salmon, Steelhead Trout and Threespine Stickleback* April 1989, EPA 560/6-89-001, PB89-156715
2. *The Toxicity of 3400 Chemicals to Fish* 1987, EPA 560/6-87-02, Washington, DC
3. Kaiser, K. L. E; et al. *Water Pollut. Res. J. Canada* 1991, **26**(3), 361-431
4. Jakobczyk, J; et al. *Chem. Tech. (Leipzig)*, 1984, **36**(3), 112
5. Shansi Institute of Fertiliser and Pesticide (Shansi, Peop. Rep. China) *Hua Hsueh Tung Pro* 1976, **5**, 271, (Ch.) (*Chem. Abstr.* **86**, 119937b)
6. *Sigma Aldrich Catalogue of Chemical Safety Data* 2nd ed., 1990, 1147, Sigma Aldrich, Milwaukee, WI
7. Hafsah, Z; et al. *Nippon Noyaku Gakkaishi* 1984, **9**(1), 117, (Jap.) (*Chem. Abstr.* **102**, 19073x)
8. *Prehled. Prumyslove. Toxicol. Org. Latky* 1986, 601
9. Shimizu, M; et al. *Mutat. Res.* 1983, **116**, 217
10. Jin, Z; et al. *Weisheng Dulixue Zazhi* 1991, **5**(1), 20-23
11. *Toxic and Hazardous Industrial Chemicals Safety Manual* 1975, The International Technical Information Institute
12. *ECETOC Technical Report No. 30(4)* 1991, European Chemical Industry Ecology and Toxicology Centre, B-1160 Brussels

B49   Benzene, 2,4-dichloro-1-nitro-

CAS Registry No. 611-06-3
Synonyms 2,4-dichloronitrobenzene; 1,3-dichloro-4-nitrobenzene
Mol. Formula $C_6H_3Cl_2NO_2$ **Mol. Wt.** 192.00
Uses Lubricating oil additive. In the manufacture of the pharmaceutical agent diazoxide.

Physical properties

M. Pt. 29-32°C; **B. Pt.** 258°C; **Flash point** 112°C; **Specific gravity** d_4^{78} 1.551.

Solubility

Organic solvent: diethyl ether, ethanol

Ecotoxicity

Invertebrate toxicity

EC_{50} (5, 15, 30 min) *Photobacterium phosphoreum* 3.19 mg l^{-1} Microtox test (1).

Environmental fate

Degradation studies

Less than 0.1% of applied 2,4- dichloronitrobenzene (concentration unspecified) was degraded in a 5 day biodegradation study using activated sludge from a municipal sewage treatment plant (2).

A 3 day exposure bioconcentration factor of 80 was determined for golden ide fish and a 1 day exposure bioconcentration factor of 150 was found for *Chlorella fusca* (2). Degradation was investigated in *Mucor javavicus* and other fungi species. Major metabolites were 2,4- dichloroaniline and chloro-2-nitrothioanisole (3).

Abiotic removal

Photodegradation induced by sunlight (4).

Photochemical reaction with atmospheric hydroxyl radicals has been estimated to be 1.237×10^{-13} cm^3 molecules sec^{-1} at 25°C which corresponds to an atmospheric $t_{1/2}$ of about 130 days at normal atmospheric concentrations (5).

Mammalian and avian toxicity

Metabolism and pharmacokinetics

Metabolised in rats via glutathione S-aryl-transferase to form the conjugated compound (6).

Genotoxicity

Salmonella typhimurium TA98, TA100, TA1535, TA1537, TA1538 without metabolic activation positive (7,8).

Any other adverse effects

Absorption into the body leads to the formation of methaemoglobin which in sufficient concentration causes cyanosis (9).

Any other comments

Human health effects, experimental toxicology, environmental effects and ecotoxicology reviewed (10,11).

2,4-Dichloronitrobenzene has been qualitatively detected in drinking water concentrates collected from Cincinnati, OH, 1978 and Seattle, WA, 1976 (12). Genotoxicity, biological properties and human exposure limits reported (13).

References

1. Kaiser, K. L. E; et al. *Water Pollut. Res. J. Canada* 1991, **26**(3), 361-431
2. Freitag, D; et al. *Chemosphere* 1985, **14**, 1589-1616

3. Hafsah, Z; et al. *Nippon Noyaku Gakkaishi* 1984, **9**(1), (Jap.) (*Chem. Abstr.* **102**, 19073x)
4. Simmons, M. S; et al. *Water Res.* 1986, **20**(7), 899
5. Atkinson, R. *Inter. J. Chem. Kinet.* 1987, **19**, 799-828
6. Casarett, L. J. (Ed.) *Toxicology, The Basic Science of Poisons* 3rd ed., 1986, Macmillan, New York
7. Haworth, S; et al. *Environ. Mutagen.* 1983, **5**(Suppl. 1), 3
8. Shimizu, M; et al. *Mutat. Res.* 1983, **116**(3-4), 217-238
9. *Sangyo Igaku* 1987, **29**, 34
10. Freitag, D; et al. *Chemosphere* 1988, **14**(10), 1989
11. *ECETOC Technical Report No. 30(4)* 1991, European Chemical Industry Ecology and Toxicology Centre, B-1160 Brussels
12. Lucas, S. V. *GC/MS Analysis of Organics in Drinking Concentrates and Advanced Waste Treatment Concentrates* 1984, **1**, 146, EPA 600/1-84-020a
13. Kawai, A; et al. *Sangyo Igaku* 1987, **29**(1), 34-54

B50   Benzene, 1,2-dimethyl-3-nitro-

CAS Registry No. 83-41-0
Synonyms 1,2-dimethyl-3-nitrobenzene; 2,3-dimethylnitrobenzene; 3-nitro-*o*-xylol
Mol. Formula $C_8H_9NO_2$ **Mol. Wt.** 151.17
Uses Chemical intermediate. Used in the manufacture of plasticisers.

Physical properties

M. Pt. 7-9°C; **B. Pt.** 245°C; **Flash point** 107°C; **Specific gravity** d^{15} 1.147.

Solubility
Organic solvent: ethanol

Ecotoxicity

Invertebrate toxicity
EC_{50} (15 min) *Photobacterium phosphoreum* 0.536 mg l^{-1} Microtox test (1).

Genotoxicity
Salmonella typhimurium TA98, TA100 with and without metabolic activation positive (2).

Any other comments
Human health effects and experimental toxicology reviewed (3).

References
1. Kaiser, K. L. E; et al. *Water Pollut. Res. J. Canada* 1991, **26**(3), 361-431
2. Maynard, A. T; et al. *Mol. Pharmacol.* 1986, **29**(6), 629
3. *ECETOC Technical Report No. 30(4)* 1991, European Chemical Industry Ecology and Toxicology Centre, B-1160 Brussels

B51   Benzene, 1,2-dimethyl-4-nitro-

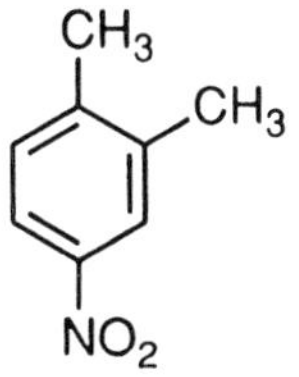

CAS Registry No. 99-51-4
Synonyms 4-nitro-o-xylene; 1,2-dimethyl-4-nitrobenzene
Mol. Formula $C_8H_9NO_2$ Mol. Wt. 151.17

Physical properties
M. Pt. 29-31°C; B. Pt. $_{20}$ 143°C.

Solubility
Organic solvent: ethanol

Ecotoxicity

Invertebrate toxicity
EC_{50} (15 min) *Photobacterium phosphoreum* 2.14 mg l^{-1} Microtox test (1).

Abiotic removal
Sunlight induced photodegradation (2).

Genotoxicity
Salmonella typhimurium TA98, TA100 with and without metabolic activation positive (3).

Any other comments
Human health effects, experimental toxicology reviewed (4).

References

1. Kaiser, K. L. E; et al. *Water Pollut. Res. J. Canada* 1991, **26**(3), 361-431
2. Simmons, M. S; et al. *Water Res.* 1986, **20**(7), 899
3. Maynard, A. T; et al. *Mol. Pharmacol.* 1986, **29**(6), 629
4. *ECETOC Technical Report No. 30(4)* 1991, European Chemical Industry Ecology and Toxicology Centre, B-1160 Brussels

B52   Benzene, 1,3-dimethyl-2-nitro-

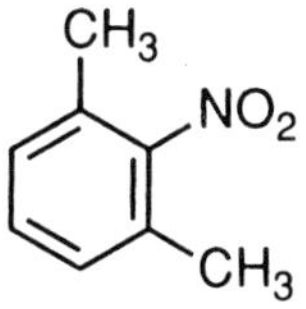

CAS Registry No. 81-20-9
Synonyms 1,3-dimethyl-2-nitrobenzene; 2-nitro-m-xylene; 2,6-dimethylnitrobenzene

Mol. Formula $C_8H_9NO_2$ **Mol. Wt.** 151.17

Physical properties

M. Pt. 14-16°C; **B. Pt.** $_{744}$ 225°C; **Specific gravity** 1.147.

Abiotic removal

Sunlight induced photodegradation (1).

Genotoxicity

Salmonella typhimurium TA98, TA100 with and without metabolic activation positive (2).

Any other comments

Human health effects and experimental toxicology reviewed (3).
Genotoxicity, biological properties and human exposure limits reported (4).

References

1. Simmons, M. S; et al. *Water Res.* 1986, **20**(7), 899
2. Maynard, A. T; et al. *Mol. Pharmacol.* 1986, **29**(6), 629
3. *ECETOC Technical Report No. 30(4)* 1991, European Chemical Industry Ecology and Toxicology Centre, B-1160 Brussels
4. Kawai, A; et al. *Sangyo Igaku* 1987, **29**(1), 34-54, (Jap.) (*Chem. Abstr.* **107**, 72503s)

B53   Benzene, 2,4-dimethyl-1-nitro-

CAS Registry No. 89-87-2
Synonyms 4-nitro-*m*-xylene; 2,4-dimethylnitrobenzene
Mol. Formula $C_8H_9NO_2$ **Mol. Wt.** 151.17

Physical properties

M. Pt. 2°C; **B. Pt.** 244°C.

Solubility

Organic solvent: diethyl ether, acetone

Abiotic removal

Sunlight induces photodegradation (1).

Mammalian and avian toxicity

Acute data

LD$_{50}$ intravenous mouse 45 mg kg^{-1} (2).

Any other comments

Human health effects and experimental toxicology reviewed (3).

All reasonable efforts have been taken to find information on isomers of this compound, but no relevant data are available.

References

1. Simmons, M. S; et al. *Water Res.* 1986, **20**(7), 899
2. *US Army Data NIOSH Exch. Chem. Rep.* No. NK 00236
3. *ECETOC Technical Report No. 30(4)* 1991, European Chemical Industry Ecology and Toxicology Centre, B-1160 Brussels

B54   Benzene, 1-nitro-2-trifluoromethyl-

CAS Registry No. 384-22-5
Synonyms 2-nitrobenzotrifluoride; 2-nitro-α,α,α-trifluorotoluene
Mol. Formula $C_7H_4F_3NO_2$ **Mol. Wt.** 191.11
Uses Organic synthesis.

Physical properties

M. Pt. 31-32°C; **B. Pt.** $_{20}$ 104-105°C; **Flash point** 95°C.

Solubility
Organic solvent: acetone, diethyl ether, benzene

Occupational exposure

UN No. 2306; **HAZCHEM Code** 2X; **Conveyance classification** toxic substance.

Abiotic removal
Sunlight induced photodegradation (1).

Genotoxicity

Bacillus subtillis recE4 DNA damage and repair test negative (2).
Salmonella typhimurium TA98, TA100, TA1535, TA1537, TA1538 with and without metabolic activation negative (2).
Saccharomyces cerevisiae 6117 with and without metabolic activation gene conversion and mitotic crossing over negative (2).

Any other comments

All reasonable efforts have been taken to find information on isomers of this compound, but no relevant data are available.

References

1. Simmons, M. S; et al. *Water Res.* 1986, **20**(7), 899
2. Mazza, G; et al. *Farmaco, Ed. Prat.* 1986, **41**(7), 215-225

B55   Benzene, 1-nitro-3-trifluoromethyl-

NO$_2$

CF$_3$

CAS Registry No. 98-46-4

Synonyms 3-nitrobenzotrifluoride; 3-nitro- α,α,α-trifluorotoluene; *m*-nitrotrifluorotoluene; *m*- (trifluoromethyl)nitrobenzene; 3-trifluoromethylnitrobenzene

Mol. Formula C$_7$H$_4$F$_3$NO$_2$ **Mol. Wt.** 191.11

Uses Bactericide.

Physical properties

M. Pt. –5°C; **B. Pt.** 200-205°C; **Flash point** 102°C (open cup); **Specific gravity** d$_{15.5}^{15.5}$ 1.437; **Volatility** v.p. 0.3 mmHg at 25°C.

Solubility
Organic solvent: diethyl ether, acetone, benzene

Occupational exposure

UN No. 2306; **HAZCHEM Code** 2X; **Conveyance classification** toxic substance.

Abiotic removal
Sunlight induced photodegradation (1).

Mammalian and avian toxicity

Acute data
LD$_{50}$ oral rat, mouse 520-610 mg kg^{-1} (2).
LC$_{50}$ (2 hr) inhalation mouse 800 mg m^{-3} (3).
LD$_{50}$ intraperitoneal mouse 100 mg kg^{-1} (4).

Genotoxicity

Salmonella typhimurium TA98, TA100 with and without metabolic activation negative (5).
Bacillus subtilis recE4 DNA damage and repair test negative (5).
Saccharomyces cerevisiae 6117 with and without metabolic activation gene conversion and mitotic crossing-over negative (5).
Drosophila melanogaster larvae administered in food increased the incidence of dominant lethal mutations among the offspring and increased the percentage of unfertilised eggs (6).

References

1.   Simmons, M. S; et al. *Water Res.* 1986, **20**(7), 899
2.   *Toxicology of New Industrial Chemical Sciences* 1968, **10**, 131
3.   Izmerov, N. F. *Toxicometric Parameters of Industrial Toxic Chemicals under Single Exposure* 1982, Moscow
4.   *NTIS Report* AD 277-689, Nat. Tech. Inf. Ser., Springfield, VA
5.   Mazza, G; et al. *Farmaco. Ed. Prat.* 1986, **41**(7), 215-225
6.   Ilichkina, A. G; et al. *Mol. Mekh. Genet. Protsessov.* 1976, 291, (Ital.) (*Chem. Abstr.* **86**, 157742g)

B56   Benzene, 1-nitro-4-trifluoromethyl-

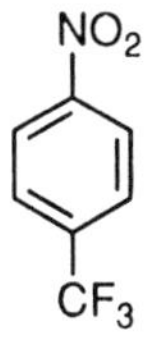

CAS Registry No. 402-54-0

Synonyms 4-nitrobenzotrifluoride; 4-nitro-α,α,α-trifluorotoluene; 4-(trifluoromethyl)nitrobenzene

Mol. Formula $C_7H_4F_3NO_2$ **Mol. Wt.** 191.11

Physical properties

M. Pt. 38-40°C; **B. Pt.** $_{10}$ 81-83°C.

Occupational exposure

UN No. 2306; **HAZCHEM Code** 2X; **Conveyance classification** toxic substance.

Ecotoxicity

Invertebrate toxicity

EC_{50} (5, 15, 30 min) *Photobacterium phosphoreum* 15.2 mg l^{-1} Microtox test (1).

Effects to non-target species

LD_{Lo} subcutaneous frog 870 mg kg^{-1} (2).

Environmental fate

Degradation studies

Sunlight induced photodegradation (3).

References

1. Kaiser, K. L. E; et al. *Water Pollut. Res. J. Canada* 1991, **26**(3), 361-431
2. *Kirk-Othmer Encyclopedia of Chemical Technology* 1982, **10**, 521, John Wiley & Sons, New York
3. Simmons, M. S; et al. *Water Res.* 1986, **20**(7), 899

B57   Benzene sulfohydrazide

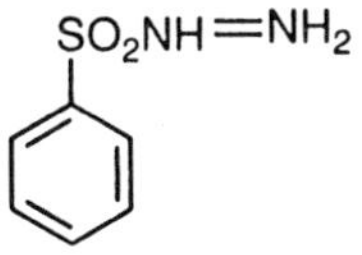

CAS Registry No. 80-17-1

Synonyms Celogen BSH; Genitron BSHF; phenyl sulfohydrazide; phenylsulfonyl hydrazide; Parofor BSH

Mol. Formula $C_6H_8N_2O_2S$ **Mol. Wt.** 172.21

Uses Gas generating agent for use in making foam rubber and foam plastics. Blowing agent in rubber industry.

Physical properties
M. Pt. 101-103°C; **Flash point** 110°C; **Specific gravity** 1.369.

Solubility
Organic solvent: diethyl ether

Occupational exposure
UN No. 2970; **Conveyance classification** flammable solid.

Mammalian and avian toxicity

Acute data
LD_{Lo} oral rat 50 mg kg^{-1} (1).

Sensitisation
Eczema reported in humans from contact with benzene sulfohydrazide (2).

Any other comments
ED_{50} 3 day chicken embryos 15-560 ^mug egg^{-1} positive embryo toxic effect (3). Human health effects and experimental toxicology reviewed (4).

References
1. Chernykh, V. P; et al. *Farmakol. Toxicol. (Moscow)* 1979, **42**(3), 285, (Russ.) (*Chem. Abstr.* **91**, 6830k)
2. English, J. S. C; et al. *Contact Dermatitis* 1986, **14**(3), 183
3. Kartionen, A; et al. *Arch. Contam. Toxicol.* 1982, **11**(6), 753-759
4. *ECETOC Technical Report No. 30(4)* 1991, European Chemical Industry Ecology and Toxicology Centre, B-1160 Brussels

B58   Benzene, 1,3,5-trinitro-

CAS Registry No. 99-35-4
Synonyms TNB; 1,3,5-trinitrobenzene; *s*-trinitrobenzene; trinitrobenzene, wet
Mol. Formula $C_6H_3N_3O_6$ **Mol. Wt.** 213.11
Uses Explosive, less sensitive to impact than TNT but more powerful.

Physical properties
M. Pt. 122°C; **B. Pt.** decomp.; **Specific gravity** d_4^{20} 1.760.

Solubility
Water: 3.5 mg kg^{-1}. Organic solvent: benzene, methanol, diethyl ether, carbon disulfide, ethanol, petroleum ether, acetone

Occupational exposure

UN No. 0214/1354; **Conveyance classification** flammable solid.

Ecotoxicity

Fish toxicity
Administration of 0.13-0.61 mg l^{-1} to bluegill sunfish caused significant changes in ventilatory depth respiration and body movement (1).
LC$_{50}$ (96 hr) fathead minnow 1.03 mg l^{-1} (2).

Invertebrate toxicity
EC$_{50}$ (48 hr) *Daphnia magna* 3 mg l^{-1} (2).

Bioaccumulation
Estimated bioconcentration factors 5 and 23 based on a log P$_{ow}$ of 1.18 and a water solubility of 340 mg l^{-1} at 20°C, respectively, which suggests accumulation is insignificant in aquatic organisms (3,4).

Environmental fate

Degradation studies
Microbial degradation was incomplete and unsustained in Tennessee river water. Nitro group reduction occurred in the presence of added nutrients and lab cultures of Tennessee river microorganisms (5).
At an initial concentration of 100 ppm 1,3,5-trinitrobenzene was found to be resistant to biodegradation when incubated 180 minutes in a phenol-adapted mixed culture of microorganisms obtained from garden soil, compost, river sediment and a petroleum refinery waste lagoon (6).

Abiotic removal
Chromophores absorb ultraviolet light at >290 nm which suggests potential hydrolysis when exposed to sunlight (7).
Estimated photochemical t$_{1/2}$ 35 yr (8).
Volatilisation from water is not expected to be an environmentally important fate process (3).

Absorption
Estimated soil adsorption coefficients of 104 and 178 suggest moderate to high mobility in soil and moderate to low adsorption of suspended solids and sediments in water (3,4,9).

Mammalian and avian toxicity

Acute data
LD$_{50}$ oral mouse, rat, guinea pig 280-730 mg kg^{-1} (10-12).
LD$_{50}$ intravenous mouse 32 mg kg^{-1} (13).

Irritancy
Skin irritant characterised by hyperaemia, oedema and haemorrhages in experimental animals (12).

Genotoxicity

Salmonella typhimurium TA1535, TA1537, TA1538, TA98, TA100 with and without metabolic activation positive (14).

Any other adverse effects

A single oral dose of 85 mg kg^{-1} to rat caused 50% methaemoglobinaemia (15).

Any other comments

General literature review of aromatic amino and nitro compounds (16).

Human health effects, experimental toxicology and environmental effects reviewed (17).

All reasonable efforts have been taken to find information on isomers of this compound, but no relevant data are available.

References

1. Van der Schalie, W. H. *Aquatic Toxicol. Hazard Assess.* 1988, **10**, 307-3152
2. Pearson, J. G; et al. *Aquatic Toxicology* 1979, 284-301, ASTM STP 667
3. Lyman, W. J; et al. *Handbook of Chemical Property Estimation Methods* 1982, McGraw-Hill, New York
4. Hansch, C; et al. *Medchem Project* 1985, 25, Pomona College, Claremont, CA
5. Mitchell, W. R; et al. *Report* 1982, Iss USAMBRDL- TR-8201, AD-A116651
6. Tabak, H. H; et al. *J. Bacteriol.* 1964, **87**, 910-919
7. Mill, T; et al. *Environmental Exposure from Chemicals* 1985, **1**, CRC Press, Boca Raton, FL
8. Atkinson, R. *Intern. J. Chem. Kinet.* 1987, **19**, 799-828
9. Swann, R. L; et al. *Res. Rev.* 1983, **85**, 17-28
10. Izmerov, N. F; et al. *Toxicometric Parameters of Industrial Toxic Chemicals under Single Exposure* 1982, 117, Moscow
11. *Gig. Sanit.* 1977, **42**(10), 12-77, (Russ)
12. Timatievskaya, L. A. *Toxicol. New Ind. Chem. Subst.* 1973, **13**, 138-133 (Russ.)
13. *US Army Armament Research and Development Command* NX00192
14. McGregor, D. B; et al. *Environ. Mol. Mutagen.* 1980, 2(4), 531-541
15. Senczuk, W; et al. *Bromatologica i Chemia Totsykologiczna* 1976, 9(3), 289-294 (Pol.)
16. Von-Oettingen, W. F. *Public Health Bull.* 1941, **271**, 1-228
17. *ECETOC Technical Report No.30(4)* 1991, European Chemical Industry Ecology and Toxicology Centre, B-1160 Brussels

B59   Benzenearsonic acid

CAS Registry No. 98-05-5
Synonyms phenyl arsenic acid; phenyl arsonic acid
Mol. Formula $C_6H_7AsO_3$ **Mol. Wt.** 202.04
Uses Reagent for tin. In the preparation of carbarsone (an antiamoebic agent in humans). Veterinary medicine. In the preparation of tryparsamide.

Physical properties

M. Pt. 160°C; **Specific gravity** 1.760.

Solubility
Water: 25 g l^{-1}.

Occupational exposure
US TLV (TWA) 0.2 mg m^{-3} (as As).

Ecotoxicity

Fish toxicity
Exposure to 5 ppm caused trout to die within 23 hr (1).
Invertebrate toxicity
EC$_{50}$ (30 min) *Photobacterium phosphoreum* 583 mg l^{-1} Microtox test (2).

Mammalian and avian toxicity

Acute data
LD$_{50}$ oral rat 50 mg kg^{-1} (3).
LD$_{50}$ intravenous rabbit 16 mg kg^{-1} (4).

Metabolism and pharmacokinetics
Major excretory route (species unspecified) urine (5).

References
1. *The Toxicity of 3400 Chemicals to Fish* 1987, EPA 560/6-87-002, Washington, DC
2. Kaiser, K. L. E; et al. *Water Pollut. Res. J. Canada* 1991, **26**(3), 361-431
3. Karel, L. *J. Pharm. Exp. Therapeutics* 1948, **93**, 287-293
4. Morgan, R. B; et al. *J. Pharm. Exp. Therapeutics* 1944, **80**, 93-113
5. Parke, D. V. *The Biochemistry of Foreign Compounds* 1968, 171, Pergamon Press, Oxford

B60   1,2-Benzenedicarboxylic acid, dimethyl ester

CAS Registry No. 131-11-3
Synonyms dimethyl phthalate; dimethyl 1,2-benzenedicarboxylate; dimethyl benzeneorthodicarboxylate; Avolin; methyl phthalate
Mol. Formula C$_{10}$H$_{10}$O$_4$ **Mol. Wt.** 194.19
Uses Manufacture of latex, cellulose and acetate film. Constituent of lacquers, plastics, rubber and coating agents. In the manufacture of safety glasses and moulding powders. Insect repellant. Perfumes.

Physical properties

M. Pt. 6°C; **B. Pt.** 284°C; **Flash point** 146°C; **Specific gravity** d$_{25}^{25}$ 1.19; **Volatility** v.p. 1 mmHg at 20°C.

Solubility
Water: 4 g l^{-1} at 20°C. Organic solvent: miscible with ethanol, acetone, diethyl ether, chloroform

Occupational exposure
US TLV (TWA) 5 mg m^{-3}; US TLV (STEL) 10 mg m^{-3}; UK Long-term limit 5 mg m^{-3}; UK Short-term limit 10 mg m^{-3}.

Ecotoxicity

Fish toxicity
LC_{50} (96 hr) bluegill sunfish, sheepshead minnow 50-58 mg l^{-1} (1,2).

Invertebrate toxicity
EC_{50} (96 hr) *Selenastrum capricornutum, Selenastrum costatum* 26.1-42.7 mg l^{-1} (3).
EC_{50} (96 hr) *Gymnodinium breve* 54-125 mg l^{-1} (4,5).
LC_{50} (96 hr) shrimp 62 mg l^{-1} (6).
EC_{50} (48 hr) *Daphnia magna* 33 mg l^{-1} (7).
EC_{50} (5, 15, 30 min) *Photobacterium phosphoreum* 18.1 mg l^{-1} Microtox test (8).

Bioaccumulation
Bioconcentration factors in brown shrimp, sheepshead minnow and bluegill sunfish are 4.7, 5.4 and 57 respectively. The former two demonstrating low concentration and the latter elevated because of metabolites were included with the parent compound in the results (9,10).

Environmental fate

Anaerobic effects
Complete degradation occurred in undiluted and 10% municipal sludge in 1 and 10 days, respectively, with 82% mineralisation occurring in the 10% sludge (11).

Degradation studies
$t_{1/2}$ 1.9 days after a 2.7-day lag using a soil/sewage inoculum. After 28 days, >99% had disappeared and 86% mineralisation had occurred (12).
Biodegradation is expected to be the principal loss process in lakes and ponds with an estimated $t_{1/2}$ of 13-17hr (13).
Bacillus sp. utilised 1,2- benzenedicarboxylic acid dimethyl ester as a sole source of carbon isolated from garden soil with monomethyl terephthalate and terephthalic acid as intermediates (14).

Abiotic removal
Photolytic $t_{1/2}$ in surface waters estimated as 145 days (15).
$t_{1/2}$ in pure water irradiated at >290 nm was 12.7 hr, reduced to 2.8 hr in the presence of nitrogen dioxide (16).
Hydrolytic $t_{1/2}$ estimated as 3.2 yr at 30°C under neutral conditions. At pH 9 estimated $t_{1/2}$ are 11.6 and 25 days at 30°C and 18°C, respectively (13,15).
Photochemical $t_{1/2}$ estimated as 23.8 hr (17).

Absorption
Estimated soil adsorption coefficients of 160 and 44 indicate low adsorption to soil and sediment (15,18).

Mammalian and avian toxicity

Acute data

LD_{50} oral mouse, rat, guinea pig 2400- 7200 mg kg^{-1} (19).
LD_{50} intraperitoneal mouse, rat 1580-3380 mg kg^{-1} (19).

Carcinogenicity and long-term effects

National Toxicology Program two year skin paint study in mice currently scheduled for peer review (20).

Teratogenicity and reproductive effects

On gestational days 6-15, dietary concentrations of 0-5% were administered to Sprague-Dawley rats providing an average intake of 0-3.6 g kg^{-1} day^{-1}. Results suggested the apparent toxic effects of high doses on body weight reflected the unpalatability of 1,2-benzene dicarboxylic acid dimethyl ester in feed. The no-observed-adverse-effect level for maternal toxicity was 1% and for developmental toxicity was 5% (21).
Doses of 0.5, 1 and 2 ml day^{-1} epicutaneously administered to rats did not induce any teratogenic effects (22).

Metabolism and pharmacokinetics

After oral administration to mice and rats, absorption was rapid and excretion was mostly in urine and faeces in 12 hr. Maximum concentration was attained at 20 min in blood and organs, with kidney having the highest concentration followed by liver and fatty tissues. In 24 hr, 90.2% and 4.3% was excreted in urine and faeces, respectively (23).
In vitro study of human and rat epidermal membranes found human skin to be less permeable (24).
Intraperitoneal rats 2 g kg^{-1}, urine collected over 24 hr was not mutagenic to *Salmonella typhimurium* TA100 and contained an equivalent of 1.96 mg ml^{-1} of the administered phthalate. More than 97% of the phthalic acid-containing derivatives present in the urine consisted of monomethyl phthalate (25).

Genotoxicity

Salmonella typhimurium TA100, TA1535 without metabolic activation weakly positive. Cytotoxic at high concentrations (26).

Any other adverse effects

Irritating to mucous membranes and eyes and can cause central nervous system depression when ingested (27).

Any other comments

Human health effects and experimental toxicology reviewed (28).

References

1. Buccafusio, R. J; et al. *Bull. Environ. Contam. Toxicol.* 1981, **26**, 446
2. Heitmuller, P. T; et al. *Bull. Environ. Contam. Toxicol.* 1981, **27**, 596-604
3. *In-Depth Studies on Health and Environmental Impacts of Selected Water Pollutants* 1978, EPA 68-01-4646, Washington, DC
4. Wilson, W. B; et al. *Bull. Environ. Contam. Toxicol.* 1978, **20**, 149
5. Linden, E; et al. *Chemosphere*, 1979, **11-12**, 843-851
6. Le Blanc, G. A. *Bull. Environ. Contam. Toxicol.* 1980, **24**, 684-691

7. Kaiser, K. L. E; et al. *Water Pollut. Res. J. Canada* 1991, **26**(3), 361-431
8. Wofford, H. W; et al. *Ecotox. Environ. Safety* 1981, **5**, 202-210
9. Barrows, M. E; et al. *Dyn. Exposure Hazard Assess.* 1980, 279-392, Ann. Arbor, MI
10. Johnson, L. D; et al. *J. Water Pollut. Control Fed.* 1983, **55**, 1441-1449
11. Sugatt, R. H; et al. *Appl. Environ. Microbiol.* 1984, **47**, 601-606
12. Shelton, D. R; et al. *Environ. Sci. Technol.* 1984, **18**, 93-97
13. Callahan, M. A; et al. *Water-Related Fate of 129 Pollutants* 1979, **2**, 94-1-94-28, EPA 440/4-79-029b, Washington, DC
14. Sivamurthy, K; et al. *J. Ferment. Bioeng.* 1989, **68**(5), 375-377
15. Wolfe, N. L; et al. *Chemosphere* 1980, **9**, 393-408
16. Kotzias, D; et al. *Naturwissenschaften* 1982, **69**, 444-445
17. *GEMS; Graphical Exposure Modeling System. Fate of Atmospheric Pollutants Data Base* 1986, Office of Toxic Substances, USEPA, Washington, DC
18. Kenaga, E. E. *Ecotox. Environ. Safety* 1980, **4**, 26-38
19. Autian, J. *Environ. Health Perspect.* 1973, 3
20. *National Toxicology Program, Research and Testing Div.* 1992, Report No. 429, NIEHS, Research Triangle Park, NC
21. *Gov. Rep. Announce. Index US* 1989, **89**(12), 933,227
22. Hansen, E; et al. *Pharmacol. Toxicol. (Copenhagen)* 1989, **64**(2), 237-238
23. Ioku, M; et al. *Osaka Furitsu Koshu Eisei Kenkyusho Kenkyu Hokoku, Yakuji Shido Hen* 1975, **9**, 5-10
24. Scott, R. C; et al. *Environ. Health Perspect.* 1987, **74**, 223-227
25. Kozambo, W. J; et al. *J. Toxicol. Environ. Health* 1991, **33**(1), 29-46
26. Agarwal, D. K; et al. *J. Toxicol. Environ. Health* 1985, **16**(1), 61-69
27. Gosselin, R. E; et al. *Clinical Toxicology of Commercial Products* 5th ed., 1984, 205, Williams & Wilkins, Baltimore, MD
28. *ECETOC Technical Report No. 30(4)* 1991, European Chemical Industry Ecology and Toxicology Centre, B-1160 Brussels

B61   1,2-Benzenediol

CAS Registry No. 120-80-9

Synonyms pyrocatechol; *o*-benzenediol; *o*-hydroquinone; *o*-hydroxyphenol; 2-hydroxyphenol; catechol

Mol. Formula $C_6H_6O_2$ **Mol. Wt.** 110.11

Uses Manufacture of polymerisation inhibitors and antioxidants. Photographic developing agent. Manufacture of pesticides and dyestuffs.

Occurrence Detected in onions, crude beet sugar, crude wood tar, coal and cigarette smoke.

Physical properties

M. Pt. 105°C; **B. Pt.** 246°C; **Flash point** 127°C; **Specific gravity** d_4^{15} 1.37; **Partition coefficient** log P_{ow} 0.88-1.01; **Volatility** v.p. 10 mmHg at 118°C; v. den. 3.8.

Solubility
Water: 451 g l^{-1} at 20°C. Organic solvent: ethanol, diethyl ether

Occupational exposure

US TLV (TWA) 5 ppm (23 mg m^{-3}); **UK Long-term limit** 5 ppm (20 mg m^{-3}); **Supply classification** harmful.

Risk phrases
Harmful in contact with skin and if swallowed – Irritating to eyes and skin (R21/22, R36/38)

Safety phrases
Do not breathe dust – In case of contact with eyes, rinse immediately with plenty of water and seek medical advice – Wear suitable gloves (S22, S26, S37)

Ecotoxicity

Fish toxicity
LC$_{50}$ (48 hr) goldfish 14 mg l^{-1} (1).
LC$_{50}$ (96 hr) fathead minnow, rainbow trout 3-9 mg l^{-1} (2).

Invertebrate toxicity
EC$_{50}$ (30 min) *Photobacterium phosphoreum* 29.6 mg l^{-1} mg l^{-1} Microtox test (3).

Bioaccumulation
Estimated bioconcentration factor of 3 based on a measured soil adsorption coefficient of 7.6 which suggests concentration in aquatic organisms is unlikely (4).

Environmental fate

Degradation studies
Adapted activated sludge (at 20°C) 96% COD, 55.5 mg COD g^{-1} dry inoculum hr^{-1} (1).
Aerobic degradation products include the pyruvate, succinate and acetaldehyde derivatives (1).

Abiotic removal
Autoxidation in aqueous environment t$_{1/2}$ 447 hr at 25°C and pH7 (1).
Nonionised 1,2-benzediol will not be subject to significant direct photolysis since it does not significantly absorb light above 290 nm in methanol (5).
Estimated atmospheric t$_{1/2}$ 14 hr (6).
Reaction of 1,2-benzenediol with nitrate radicals at night may be a significant removal process (7).

Absorption
Estimated soil adsorption coefficient of 72 suggests 1,2-benzenediol will be highly mobile in soil (8).

Mammalian and avian toxicity

Acute data
LD_{50} oral rat 260 mg kg^{-1} (9).
LD_{50} intraperitoneal mouse 175 mg kg^{-1} (10).
LD_{50} percutaneous rabbit 800 mg kg^{-1} (11).

Carcinogenicity and long-term effects
Inadequate evidence for carcinogenicity to humans, limited evidence for carcinogenicity to animals, IARC classification group 3 (12).
Oral ♂, ♀ rat (104 wk), ♂, ♀ mice (96 wk) 0.8% 1,2-benzenediol induced glandular stomach adenocarcinomas in 54% and 43% of ♂ and ♀ rats, respectively, but not in mice (13).

Metabolism and pharmacokinetics
Sulfate and glucuronide conjugation products detected in urine of chickens and dogs (14).
Metabolised to o-benzoquinone by human leucocytes (15).
Rapidly absorbed from gastrointestinal tract and through skin. The conjugated fraction hydrolyses in urine forming free compound which is oxidised to a dark substance (16).

Irritancy
Systemic effects from inhalation in man include sore throat, cough and dyspnoea (19).

Sensitisation
Allergen and can cause dermatitis in man (14).

Genotoxicity
In vitro V79 Chinese hamster cells induced sister chromatid exchange. Mutations to 6- thioguanine resistance induced (17).
Intraperitoneal ♂ mice 40 mg kg^{-1} induced micronuclei in bone marrow cells more readily than following oral administration (18).

Any other adverse effects
At <1 mg l^{-1} 1,2-benzenediol acts as a redox catalyst on the myeloperoxidase-Cl^--hydrogen peroxide antimicrobial/cytotoxic system of the human neutrophil. Higher concentrations did not induce haemolysis. Methaemoglobin formulation was stimulated in a concentration-dependent fashion (20).
♂ Rats administered 10-90 mg kg^{-1} by gastric intubation induced ≤19-fold increase in ornithine decarboxylase activity with a maximum after 8 hr and ≤8-fold increase in replicative DNA synthesis with a maximum after 24 hr in the pyloric mucosa of the stomach. At 37-90 mg kg^{-1} DNA single strand scission in the pyloric mucosa after 2 and 6 hr or unscheduled DNA synthesis after 2 and 12 hr did not occur (21).
Depression of central nervous system, hypertension, and methaemoglobinaemia have been reported (17,22).

Legislation
Prohibited for use in cosmetics in Czechoslovakia (23).
Maximum admissible concentrations in surface waters 0.1 mg l^{-1} (USSR) (23).

Any other comments
Human health effects, experimental toxicology, environmental effects, workplace experience and physico-chemical properties reviewed (24,25).

References

1. Vershueren, K. *Handbook of Environmental Data on Organic Chemicals* 2nd ed., 1983, Van Nostrand Reinhold, New York
2. De Graeve, G. M; et al. *Arch. Environ. Contam. Toxicol.* 1980, **9**, 543
3. Kaiser, K. L. E; et al. *Water Pollut. Res. J. Canada* 1991, **26**(3), 361-431
4. Lyman, W. J; et al. *Handbook of Chem. Property Estimation Methods Environ. Behaviour of Organic Compounds* 1982, 5-4-5-10, McGraw-Hill, New York
5. *Sadtler Standard UV Spectra No. 15924* 1970, Sadtler Research Lab., Philadelphia
6. Atkinson, R. *Intern. J. Chem. Kinet.* 1987, **19**, 799-828
7. Atkinson, R; et al. *Environ. Sci. Technol.* 1987, **21**, 1123-1126
8. Swann, R. L; et al. *Res. Rev.* 1983, **85**, 17-28
9. *Advances in Food Research* 1951, **3**, 197
10. Podlesnaya, A. I. *Bull. Exp. Biol. Med.* 1966, **61**, 291-294
11. Flickinger, C. W. *Am. Ind. Hyg. Assoc. J.* 1976, **37**, 596-606
12. Hirose, M; et al. *Jpn. J. Cancer Res.* 1990, **81**(3), 207-212
13. *IARC Monograph* 1987, **Suppl. 7**, 64
14. *IARC Monograph* 1977, **15**, 155
15. Hawkins, D. R. *Biotransformations* 1989, **2**, RSC, London
16. *Encyclopedia of Occupational Health and Safety* 1971, **1-2**, 1045, McGraw-Hill, New York
17. *Substances Hazardous to Health* 1990, Croner Publications, Surrey
18. Glatt, H; et al. *Environ. Health Perspect.* 1989, **82**, 81-89
19. Ciranni, R; et al. *Mutat. Res.* 1988, **209**(1-2), 23-28
20. Van Zyl, J. M; et al. *Toxicology* 1991, **68**(1), 37-49
21. Furihata, C; et al. *Jpn. J. Cancer Res.* 1989, **80**(11), 1052-1057
22. *Patty's Industrial Hygiene and Toxicology* 3rd ed., 1981, **2A**, Clayton, G. D; et al. (Eds.), John Wiley & Sons, New York
23. *Legal File* 1986, **2**, International Register of Potentially Toxic Chemicals, Geneva
24. *Dangerous Prop. Ind. Mater. Rep.* 1988, **8**(3), 87-94
25. *ECETOC Technical Report No. 30(4)* 1991, European Chemical Industry Ecology and Toxicology Centre, B-1160 Brussels

B62   1,3-Benzenediol

CAS Registry No. 108-46-3
Synonyms resorcinol; 1,3-dihydroxybenzene; *m*-hydroquinone; *m*-hydroxyphenol; 3-hydroxyphenol; resorcin
Mol. Formula $C_6H_6O_2$ **Mol. Wt.** 110.11
Uses Manufacture of pharmaceuticals, adhesives, synthetic resins and dyestuffs. Used in antiseptic preparations.
Occurrence Found in wood smoke and cigarette smoke. In effluents resulting from production of coal-tar chemicals (1).

Physical properties

M. Pt. 109-111°C; **B. Pt.** 276-280°C; **Flash point** 171°C;

Specific gravity d_4^{25} 1.28; **Partition coefficient** log P_{ow} 0.77-0.80; **Volatility** v.p. 53 mmHg at 190°C.

Solubility
Water: 1 g in 0.9 ml. Organic solvent: benzene, tetrachloromethane, ethanol, dietyl ether, acetone, acetic acid

Occupational exposure
US TLV (TWA) 10 ppm (45 mg m^{-3}); **US TLV (STEL)** 20 ppm (90 mg m^{-3}); **UK Long-term limit** 10 ppm (45 mg m^{-3}); **UK Short-term limit** 20 ppm (90 mg m^{-3}); **UN No.** 2876; **HAZCHEM Code** 2X; **Conveyance** classification harmful substance; **Supply classification** harmful.

Risk phrases
Harmful if swallowed – Irritating to eyes and skin (R22, R36/38)

Safety phrases
In case of contact with eyes, rinse immediately with plenty of water and seek medical advice (S26)

Ecotoxicity

Fish toxicity
LC_{50} (48,96 hr) fathead minnow 72.6-53.4 mg l^{-1} (2).
LC_{50} (48 hr) goldfish 57 mg l^{-1} (3).

Invertebrate toxicity
LD_0 (24 hr) *Daphnia magna* 0.8 mg l^{-1} (4).
LC_{50} (24-96 hr) grass shrimp 170-42 mg l^{-1} (4).
EC_{50} (5, 15, 30 min) *Photobacterium phosphoreum* 264 mg l^{-1} Microtox test (5).

Effects to non-target species
LC_{50} (48 hr) *Palaemontes pugio* 78 mg l^{-1} (2).
LD_{Lo} (48 hr) *Rana palustris* 270 mg kg^{-1} (6).

Environmental fate

Carbonaceous inhibition
Inhibition of glucose degradation in *Pseudomonas fluorescens* 200 mg l^{-1} (4).

Degradation studies
Adapted culture, 89% removal after 48 hr incubation (feed) 446 mg l^{-1}.
Decomposition by soil microflora in 8 days (4).
Adapted activated sludge, 90% COD, 57.5 mg COD g^{-1} dry inoculum hr^{-1} (4).
$t_{1/2}$ of 0.16-0.24 days in an aerobic screening test using activated sludge acclimated to cresols (7).
BOD_5 61% reduction of dissolved oxygen using a sewage inocular (8).
COD_5 90% in activated sludge system (9).
Biodegradable (10).

Abiotic removal
Autoxidation in aquatic environment at pH 9 $t_{1/2}$ 67 days at 25°C (4).
Does not significantly absorb ultraviolet light above 295 nm at low concentrations in

methanol or acidic aqueous solution. However, in dilute alkaline solutions 1,3-benzenediol does absorb ultraviolet light above 295 nm (11,12).
The pK$_a$ value of 9.15 suggests it will increasingly dissociate with an increase of pH. Direct photolysis in environmental waters will not occur under acidic conditions, but may potentially occur in alkaline waters (13).
Reacts with photochemically produced hydroxyl radicals atmospheric t$_{1/2}$ estimated as 1.9 hr (14).
Phenols are generally resistant to aqueous environmental hydrolysis, therefore 1,3-benzenediol is not expected to hydrolyse significantly in water (15).

Absorption
Based on a water solubility of 1230 g l^{-1} and a 10 g K$_{ow}$ of 0.8, the soil adsorption coefficient is estimated at 2-65 indicating high to very high soil mobility (16).

Mammalian and avian toxicity

Acute data
LD$_{50}$ oral rat, guinea pig 301-370 mg kg^{-1} (1,17).
LD$_{Lo}$ oral human 29 mg kg^{-1} (18).
LD$_{50}$ dermal rabbit 3360 mg kg^{-1} (19).
LD$_{50}$ subcutaneous rat 450 mg kg^{-1} (1).

Carcinogenicity and long-term effects
No adequate evidence for carcinogenicity to humans, insufficient evidence for carcinogenicity in experimental animals, IARC classification group 3 (20).
National Toxicology Program gavage rats and mice long term study. Post peer review technical report in progress, no evidence of carcinogenicity (21).

Metabolism and pharmacokinetics
Readily absorbed from gastrointestinal tract and in suitable solvent is readily absorbed through human skin. Excreted in urine, free and conjugated with hexuronic, sulfuric or other acids (22).
Absorption and metabolic fate of 2% 1, 3-benzenediol applied topically in aqueous alcohol solution was determined in 3 human subjects. After 2 weeks of application (twice daily) of 800 mg to about 30% of body surface, an average of 1.64% of the dosage was excreted in 24 hr urine specimens as the glucuronide or as the sulfate conjugate (23).
Oral administration of 112 mg kg^{-1} to 3 ♂ and 3 ♀ Fischer 344 rats, sacrificed after one day, revealed no evidence of specific organ accumulation of the compound. More than 90% of the total administered dose was recovered from the excreta in 24 hr, primarily in urine with 1-2% eliminated in the faeces and <0.1% as carbon dioxide. Cannulation of the common bile duct followed by intravenous injection of 11.2 mg kg^{-1} 1,3-benzenediol indicated that excretion in bile was rapid and procedes via enterohepatic circulation to be excreted in urine. Less than 50% of the parent compound was excreted in urine. Major metabolites included glucuronide conjugates (10-20%), and diconjugate with glucuronide and sulfate (5-10%) and diglucuronide conjugate (<2%). Repeated daily exposure of 225 mg kg^{-1} for 5 days did not alter the rate or relative metabolite ratio of excretion (24).

Irritancy
100 mg instilled into rabbit eye caused severe irritation (17).

Sensitisation

Application of 3-25% solutions of 1,3- benzenediol to skin has been shown to cause allergic contact dermatitis in man (1).

Genotoxicity

Salmonella typhimurium TA98, TA100, TA1535, TA1537 with and without metabolic activation negative (25).

Human lymphocytes treated with 1,3-benzenediol produced minor increases in the number of micronucleated cells (26).

Any other adverse effects to man

Prolonged exposure of rubber workers to ≤ 0.3 mg m^{-3} 1,3-benzenediol in air caused no reported adverse effects (27).

Human pathology reports from poisoning include siderosis of the spleen and kidney tubule damage (28).

Any other adverse effects

Ingestion of 8 g 1,3-benzenediol by a child resulted in hypothermia, hypotension, decreased respiration, tremors, icterus and haemoglobinaemia (1).

Absorption can cause methaemoglobinaemia, cyanosis, convulsions and death (29).

Legislation

Maximum admissible concentration (USSR) for surface waters 0.1 mg l^{-1} (30).

Maximum permitted concentration (USA) in food products 2.5 g kg^{-1} (food) (30).

Any other comments

Human health effects, experimental toxicology and exposure conditions reviewed (31).

References

1. *IARC Monograph* 1977, **15**, 155
2. Curtis, M. W; et al. *Water Res.* 1979, **13**, 137-141
3. McKee, J. W; et al. *The Resources Agency of California, State Water Quality Control Board* 1963
4. Vershueren, K. *Handbook of Environmental Data on Organic Chemicals* 2nd ed., 1983, Van Nostrand Reinhold, New York
5. Kaiser, K. L. E; et al. *Water Pollut. Res. J. Canada* 1991, **26**(3), 361-431
6. Heubner, W. *Arch. Exper. Pathol. Pharmakol.* 1913, **72**, 241-281
7. Urushigawa, Y; et al. *Kogai Shigen Kenkyusho Iho* 1984, **13**, 59-65
8. Heukelekian, H; et zl. *J. Water Pollut. Control Fed.* 1955, **29**, 1040-1053
9. Pitter, P. *Water Res.* 1976, **10**, 453-460
10. *MITI Report* 1984, Ministry of International Trade and Industry, Tokyo
11. Sadtler *2572 UV* 1961, Sadtler Res. Labs, Philadelphia, PA
12. Perbet, G; et al. *J. Chim. Phys. Phys.-Chim. Biol.* 1979, **76**, 89-96
13. Varagnat, J. *Kirk-Othmer Encycl. Chem. Technol.* 3rd ed., 1981, **13**, 40, Wiley-Interscience, New York
14. Atkinson, R. *J. Inter. Chem. Kinet.* 1987, **19**, 799-828
15. Lyman, W. J; et al. *Handbook of Chemical Property Estimation Methods* 1982, 7-4, McGraw-Hill, New York
16. Swann, R. L; et al. *Res. Rev.* 1983, **85**, 16-28
17. *Biofax Industrial Biological Testing Lab.* April 1970

18. Lewis, R. J; et al. *National Institute for Occupational Safety and Health Report* 1984, 83-107-4
19. Flickinger, C. W; et al. *Am. Ind. Hyg. Assoc. J.* 1976, **37**, 596-606
20. *IARC Monograph* 1987, **Suppl. 7**, 71
21. *National Toxicology Program, Research and Testing Div.* 1992, Report No. 403, NIEHS, Research Triangle Park, NC
22. Clayton, G. D; et al. *Patty's Industrial Hygiene and Toxicology* 1981-1982, **2A, 2B, 2C**, 2588
23. Yeung, D; et al. *Int. J. Dermatol.* 1983, **22**(5), 321-324
24. Kim, Y. C; et al. *Fundam. Appl. Toxicol.* 1987, **9**(3), 409-414
25. McCann, J; et al. *Proc. Nat. Acad. Sci. USA* 1975, **72**(12), 5135-5139
26. Robertson, M. L; et al. *Mutat. Res.* 1991, **249**(1), 201-209
27. Gamble, J. F; et al. *Am. Ind. Hyg. Assoc. J.* 1976, **37**, 499-505
28. *Toxic and Hazardous Industrial Chemical Safety Manual* 1975, The International Technical Information Institute
29. *The Merck Index* 11th ed., 1989, Merck & Co. Inc., Rahway, NJ
30. *Legal File* 1986, **2**, International Register of Potentially Toxic Chemicals, Geneva
31. *ECETOC Technical Report No. 30(4)* 1991, European Chemical Industry Ecology and Toxicology Centre, B-1160 Brussels

B63 1,4-Benzenediol

CAS Registry No. 123-31-9

Synonyms hydroquinone; quinol; *p*-benzenediol; *p*-hydroxyphenol; Tecquinol; Quinnone

Mol. Formula $C_6H_6O_2$ **Mol. Wt.** 110.11

Uses Photographic reducer and developer. Antioxidant. Depigmentation agent for the skin.

Physical properties

M. Pt. 170-171°C; **B. Pt.** 285-287°C; **Specific gravity** d_{15} 1.332.

Solubility

Water: 1 g in 14 ml water. Organic solvent: ethanol, diethyl ether

Occupational exposure

US TLV (TWA) 2 mg m^{-3}; **UK Long-term limit** 2 mg m^{-3}; **UK Short-term limit** 4 mg m^{-3}; **UN No.** 2662; **HAZCHEM Code** 2Z;

Conveyance classification harmful substance; **Supply classification** harmful.

Risk phrases

Harmful by inhalation and if swallowed (R20/22)

Safety phrases
Keep out of reach of children – Avoid contact with skin and eyes – Wear eye/face protection (S2, S24/25, S39)

Ecotoxicity

Fish toxicity
LC_{50} (96 hr) fathead minnow 0.1-0.18 mg l^{-1} (1).
In stickleback and steelhead trout (10 mg l^{-1}) death occurred in 1-2 and 0-1 hr, respectively (2).
Exposure to 5 ppm caused death of trout, bluegill sunfish and goldfish in 2-6 hr.
Exposure to 1 ppm caused death of trout in 7 hr (3).
LC_{50} (96 hr) rainbow trout 0.097 mg l^{-1} (4).

Invertebrate toxicity
EC_{50} (48 hr) *Daphnia magna* 0.05 mg l^{-1} (1).
EC_{50} (5,15,30 min) *Photobacterium phosphoreum* 0.0382 mg l^{-1} Microtox test (5).

Bioaccumulation
Estimated biocencentration factor for golden ide 40. Experimental 24 hr bioconcentration factor range in algae were 40-65. Data indicate that bioconcentration in fish and aquatic organisms is insignificant (6-8).

Environmental fate

Degradation studies
BOD_5 7.5%, 0.05 mg l^{-1} when inoculated with an activated sludge seed (6).
Pure culture oxidation produced 1,4-benzoquinone, 2- hydroxy-1,4-benzoquinone and β-ketoacidic acid (9).
BOD_{10} 0.48% reduction of dissolved oxygen concentration using a sewage seed inoculum (10).
COD (<120 hr) 54.2% reduction of dissolved oxygen concentration at 200 mg l^{-1} using a thickened adapted activated sludge under aerobic conditions (11).
$ThOD_5$ 53% reduction of dissolved oxygen concentration in a settled raw waste water seed inoculum (12).
Bacteria from soil and related environments raised on a wide variety of different phenolic compounds utilised 95% of 300 ppm 1,4-benzenediol within 1-2 days under aerobic conditions (13).
Complete degradation of 0.55-2.2 g l^{-1} using a mixed microbial seed under anaerobic conditions (14).
Biodegradable (15).

Abiotic removal
Photolysis in water by natural sunlight produced superoxide anion and ultimately hydrogen peroxide (16).
Photolysis in dilute aqueous solutions and in the presence of oxygen produced quinone through the intermediary of semiquinoic radicals (17,18).
$t_{1/2}$ 12 min for oxidation by alkylperoxy radicals present in sunlit waters in organic solvents (19).
Estimated photochemical $t_{1/2}$ 15.7 hr (20).

Absorption

Estimated soil adsorption coefficients of 9 and 50 suggest that 1,4-benzenediol will display high to very high mobility in soil (21,22).

Mammalian and avian toxicity

Acute data

LD$_{50}$ oral rat 320 mg kg^{-1} (23).
LD$_{Lo}$ oral human 29 mg kg^{-1} (24).
LD$_{50}$ intraperitoneal rat 170 mg kg^{-1} (25).
LD$_{50}$ intravenous rat 115 mg kg^{-1} (23).

Carcinogenicity and long-term effects

No adequate evidence for carcinogenicity to humans, limited evidence for carcinogenicity to animals, IARC classification group 3 (26).
National Toxicology Program tested rats and mice via gavage. No evidence of carcinogenicity in ♂ mice, some evidence for carcinogenicity (increased incidence of chemically related neoplasms, malignant, benign or combined) in ♂ rats and ♀ rats and ♀ mice (27).

Metabolism and pharmacokinetics

Absorbed from gastrointestinal tract and possibly through skin and is partially excreted in urine as 1,4-benzenediol and in conjugation with hexuronic, sulfuric and other acids (28).

Irritancy

A 2-5% solution applied to human skin caused mild to severe irritation (29).

Genotoxicity

Salmonella typhimurium TA98, TA100, TA1535, TA1537 with and without metabolic activation negative (30).
Human lymphocyte HL-60 cells treated with 1, 4-benzenediol produced DNA adducts at levels from 0.05-10 adducts per 10^7 nucleotides as a function of treatment time and concentration. Induced a 3-fold increase in micronucleated cells (31).

Any other adverse effects to man

Ingestion of 1 g by humans has caused tinnitus, nausea, vomiting, sense of suffocation, shortness of breath, cyanosis, convulsions, delirium and collapse. Death has followed ingestion of 5 g. Irritation of intestinal tract occurs with oral ingestion and dermatitis can result from skin contact. Staining and opacification of cornea occur in workers exposed for prolonged periods to concentrations of vapour not high enough for production of systemic effects (32).
Patchy pigmentation of the palm, forefinger and base of the neck was reported in a West Indian woman after using a cosmetic cream containing 1,4- benzenediol (33). Localised exogenous ochronosis of the face developed in a black woman who had used a proprietary bleaching cream containing 2% 1,4-benzenediol up to 6 times daily for about 30 months. Eighteen months after discontinuing use, there was clearing of the hyperpigmentation except for some residual changes in the periorbital areas (34).

Any other comments

Human health effects, experimental toxicology and environmental fate reviewed (35-37).

References

1. *Env. Eff. Photoproc. Chem.* 1974, **18**, 219
2. *Fish Toxicity Screening Data* 1989, EPA 560/6-89-001, Washington, DC
3. *The Toxicity of 3400 Chemicals to Fish* 1987, EPA 560/6-87-002, Washington, DC
4. De Graeve, G. M; et al. *Arch. Environ. Contam. Toxicol.* 1980, **9**, 543
5. Kaiser, K. L. E; et al. *Water Pollut. Res. J. Canada* 1991, **26**(3), 361-431
6. Freitag, D; et al. *Chemosphere* 1985, **14**, 1589-1616
7. Freitag, D; et al. *Ecotox. Environ. Saf.* 1982, **6**, 60-81
8. Geyer, H. J; et al. *Chemosphere* 1981, **10**, 1307-1313
9. Harbison, K. G; et al. *Environ. Toxicol. Chem.* 1982, **1**, 9-15
10. Heukelekian, H; et al. *J. Water Pollut. Control Fed.* 1955, **29**, 1040-1053
11. Pitter, P. *Water Res.* 1976, **10**, 231-235
12. Young, R. H. F; et al. *J. Water Pollut. Control Fed.* 1968, **40**, 354-368
13. Tabak, H. H; et al. *J. Bacteriol.* 1964, **87**, 910
14. Szewzyk, U; et al. *Appl. Microbiol. Technol.* 1989, **32**, 346-349
15. *MITI Report* 1984, Ministry of International Trade and Industry, Tokyo
16. Choudhry, G. G; et al. *Res. Rev.* 1985, **96**, 19-136
17. Perbet, G; et al. *J. Chimie Physique* 1979, **76**, 89-96
18. Tissot, A; et al. *Chemosphere* 1985, **14**, 1221-1230
19. Mill, T. *Environ. Toxicol. Chem.* 1982, **1**, 135-141
20. Atkinson, R; *Int. J. Chem. Kinet.* 1987, **19**, 799-828
21. Lyman, W. J; et al. *Handbook of Chemical Property Estimation Methods* 1982, McGraw-Hill, New York
22. Swann, R. L; et al. *Res. Rev.* 1983, **85**, 17-28
23. *Fed. Proc. Fed. Am. Soc. Expt. Biol.* 1949, **8**, 348
24. Deichmann, W. B; et al. *Toxicology of Drugs and Chemicals* 1969, 321, Academic Press, London
25. *J. Ind. Hyg. Toxicol.* 1949, **31**, 79
26. *IARC Monograph* 1977, **15**, 169
27. *National Toxicology Program Research Testing Division* 1992, Report No. TR-366, NIEHS, Research Triangle Park, NC
28. Florin, T; et al. *Toxicology* 1980, **18**, 210
29. *Archives of Dermatology* 1966, **93**, 589
30. McCann, J; et al. *Proc. Nat. Acad. Sci. USA* 1975, **72**(12), 5135-5139
31. Robertson, M. L; et al. *Mutat. Res.* 1991, **249**(1) 201-209
32. *The Merck Index* 11th ed., 1989, Merck & Co. Inc., Rahway, NJ
33. Ridley, C. M; et al. *Br. Med. J.* 1984, **288**, 1537
34. Cullison, D; et al. *J. Am. Acad. Derm.* 1983, **8**, 882
35. *Dangerous Prop. Ind. Mater. Rep.* 1988, **8**(1), 51-60
36. Devillers, J; et al. *Ecotoxiol. Environ. Saf.* 1990, **19**(3), 327-354
37. *ECETOC Technical Report No. 30(4)* 1991, European Chemical Industry Ecology and Toxicology Centre, B-1160 Brussels

B64 Benzenesulfonamide, *N*-butyl-

SO₂NH(CH₂)₃CH₃

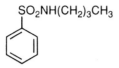

CAS Registry No. 3622-84-2
Synonyms Cetamoll BMB; Plastomoll BMB; BBSA
Mol. Formula $C_{10}H_{15}NO_2S$ **Mol. Wt.** 213.30
Uses Plasticizer for polyamides and other plastics.
Occurrence Contaminant in ground and surface water.

Physical properties
M. Pt. −35°C; **B. Pt.** 192.5°C; **Specific gravity** 1.147.

Mammalian and avian toxicity

Sub-acute data
Young adult New Zealand white rabbits inoculated repeatedly by intracisternal or intraperitoneal routes developed dose-dependent motor dysfunction characterised by limb splaying, hyperreflexia, hypertomia, gait impairment abnormal righting reflexes. Histopathological changes consisted of intramedullary thickening of the ventral horn axons, random neuroaxonal spheriods confined to brain stem nuclei and spinal motor neurons and swollen dendritic processes of spinal motor neurons (dose and duration unspecified) (1).

References

1. Strong, M.J; et al. *Acta Neuropathol.* 1991, **81**(3), 235-241

B65 Benzenesulfonyl chloride

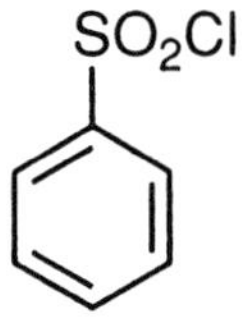

SO₂Cl

CAS Registry No. 98-09-9
Synonyms benzenesulfochloride; benzenesuphonchloride; benzenesulfonic (acid) chloride; phenylsulfonyl chloride
Mol. Formula $C_6H_5ClO_2S$ **Mol. Wt.** 176.62
Uses Intermediate for dyestuff manufacture. Accelerator in alkyl resin formation. Intermediate in manufacture of phenol and resorcinol.

Physical properties
M. Pt. 14.5°C; **B. Pt.** 246°C (decomp.); **Flash point** 110°C;

Specific gravity d_4^{23} 1.378.

Solubility
Organic solvent: ethanol, diethyl ether

Occupational exposure
UN No. 2225; HAZCHEM Code 2X; Conveyance classification corrosive substance.

Ecotoxicity

Fish toxicity
LC_{50} (48 hr) brown trout yearlings 3 mg l^{-1} static bioassay (1).
Trout, bluegill sunfish, yellow perch and goldfish exposed to 5 ppm died within 2-6 hr (2).

Bioaccumulation
Since benzenesulfonyl chloride rapidly hydrolyses in water, bioconcentration in aquatic organisms is not expected to be significant (3).

Environmental fate

Degradation studies
Since benzenesulfonyl chloride rapidly hydrolyses in water, biodegradation probably will not be an important process in the environment (3).

Abiotic removal
Hydrolytic $t_{1/2}$ 5.1 min at 21°C. Calculated hydrolytic $t_{1/2}$ 12.9 min at 10°C, producing benzenesulfonic acid and hydrogen chloride (3).
Atmospheric $t_{1/2}$ 7.9 days (4).
Based upon an estimated vapour pressure of 0.068 mmHg at 25°C, volatilisation from surfaces and near-surface dry soil may be a significant process (5).

Absorption
Adsorption to soil is not expected to be a significant removal process (3).

Mammalian and avian toxicity

Acute data
LD_{50} oral rat 1960 mg kg^{-1} (6).
LC_{50} (1 hr) inhalation rat 32 ppm (7).
LD_{50} intraperitoneal rat 76 mg kg^{-1} (6).

Legislation
Limited under EEC Directive on Drinking Water Quality 80/778/EEC. Chlorides: guide level 25 mg l^{-1}, maximum admissible concentration 400 mg l^{-1} (8).

References
1. Woodiwiss, F. S. *J. Water Pollut. Control Fed.* 1974, **73**, New York
2. *The Toxicity of 3400 Chemicals to Fish* 1987, EPA560/6-87-002, Washington, DC
3. Haughton, A.R; et al. *J. Chem. Soc. Perkin Trans. 2* 1975, **6**, 637-643
4. Atkinson R. *Intern. J. Chem. Kinetics* 1987, **1**, 98
5. Perry R.H; et al. *Perry's Chemical Handbook. Physical and Chemical Data* 6th ed., 1984, McGraw-Hill, New York
6. *Medycyna Pracy* 1969, **20**, 513

7. *Toxicol. Appl. Pharmacol.* 1977, **42**, 417
8. *EC Directive Relating to the Quality of Water Intended for Human Consumption* 1982, 80/778/EEC, Office for Official Publications of the European Communities, 2 rue Mercier, L-2985 Luxembourg

B66   1,2,4,5-Benzenetetracarboxylic 1,2:4,5-dianhydride

CAS Registry No. 89-32-7

Synonyms pyromellitic dianhydride; PMDA; 1,2,4,5-benzenetetracarboxylic anhydride; 1*H*,3*H*-benzo(1,2-c; 4, 5-c')difuran-1,3,5,7-tetrone

Mol. Formula $C_{10}H_2O_6$ **Mol. Wt.** 218.12

Uses Principal commercial use is as a raw material for polyimide resins. Also used as an intermediate in the production of film, enamels and varnishes.

Physical properties

M. Pt. 285-287°C; **B. Pt.** 390°C; **Specific gravity** d_4^{25} 1.68.

Occupational exposure

Supply classification irritant.

Risk phrases
Irritating to eyes, respiratory system and skin (R36/37/38)

Safety phrases
Avoid contact with eyes (S25)

Mammalian and avian toxicity

Acute data
LD_{50} oral guinea pig, rat, mouse 1595-2595 mg kg^{-1} (1,2).

Sub-acute data
Oral rat (6 month) no adverse effect level was determined to be 0.07 mg kg^{-1} day^{-1}. Doses >4.4 mg kg^{-1} affected the kidney, liver and reproductive tract (2).

Teratogenicity and reproductive effects
Oral rat reported to be non-embryotoxic at 230 mg kg^{-1} (≤0.1 LD_{50}) (2).

Sensitisation
Unspecified route rat (6 month) >4.4 mg kg^{-1} weak allergen (2).

Any other adverse effects to man

In humans, mucosal and skin irritant and allergen. Clinical symptoms include asthma, a flu-like syndrome and pulmonary haemorrhage (3).

Any other comments

Human health effects and experimental toxicology reviewed (4).

References

1. *Toksikol. Novykh Promyslennykh. Khim. Veshehestv.* 1975, **14**, 125
2. Kondratyuk, V. A; et al. *Gig. Sanit.* 1986, (11), 79-80, (Russ.) (*Chem. Abstr.* 1987, **106**, 97512m)
3. Venables, K. *Br. J. Med.* 1989, **46**(4), 222-232
4. *ECETOC Technical Report No. 30(4)* 1991, European Chemical Industry Ecology and Toxicology Centre, B-1160 Brussels

B67   Benzenethiol

SH

CAS Registry No. 108-98-5
Synonyms thiophenol; phenylmercaptan
Mol. Formula C_6H_6S **Mol. Wt.** 110.18
Uses In the manufacture of pharmaceuticals. Chemical intermediate. Mosquito larvicide.

Physical properties

M. Pt. −15°C; **B. Pt.** 168-169°C; **Flash point** 50°C; **Specific gravity** d_4^{25} 1.078; **Partition coefficient** log P_{ow} 2.52; **Volatility** v.p. 1 mmHg at 18.6°C.

Solubility
Organic solvent: miscible diethyl ether, benzene and carbon disulfide, soluble in ethanol

Occupational exposure

US TLV (TWA) 0.5 ppm (2 mg m^{-3}); **UK Long-term limit** 0.5 ppm (2 mg m^{-3}); **UN No.** 2337; **HAZCHEM Code** 3WE; **Conveyance classification** toxic substance.

Ecotoxicity

Fish toxicity
Trout exposed to 5 ppm benzenethiol died within 1 hr, bluegill sunfish died within 6 hr, yellow perch within 3 hr, and gold fish within 23 hr (1).

Invertebrate toxicity
EC_{50} (5,15,30 min) *Photobacterium phosphoreum* 0.875 mg l^{-1} Microtox test (2).

Bioaccumulation

Estimated bioconcentration factors 14-48 based on a log K_{ow} of 2.52 and water solubility of 836 mg l^{-1} suggest accumulation in aquatic organisms will not be significant (3-5).

Effects to non-target species

LD_{50} oral redwing blackbird, starling 24-32 mg kg^{-1} (6).

Environmental fate

Degradation studies

Theoretical BOD_6 30-42% in activated sludge inocula (7).
Benzenethiol was oxidised by isolated cells of *Thiobacillus thiooxidans* (8).

Abiotic removal

At 19.8 mg l^{-1} in cyclohexane solution, very slight absorption of ultraviolet light at >290 nm occurs which suggests benzenethiol is not susceptible to direct photolysis in air or water, or on soil surfaces (9).
Atmospheric $t_{1/2}$ 8.8 hrs (10).
A vapour pressure of 2 mmHg at 25°C suggests benzenethiol would volatilise rapidly from dry soil surfaces (11).

Absorption

Estimated soil adsorption coefficients of 108- 560 suggest moderate adsorption to suspended solids and sediments in water and moderate mobility in soil. However, benzenethiol is an acidic compound which should exist predominantly in its ionised form under neutral and alkaline conditions. Significance of adsorption is therefore uncertain since ionisation could cause more or less mobility than its estimated value indicates (12).

Mammalian and avian toxicity

Acute data

LD_{50} oral rat 46 mg kg^{-1} (13).
LC_{50} (4 hr) inhalation rat 149 mg m^{-3} (14).
LD_{50} dermal rat 300 mg kg^{-1} (13).
LD_{Lo} intraperitoneal mouse 25 mg kg^{-1} (15).

Sub-acute data

Subacute inhalation studies in mice (duration and concentration unspecified) showed systemic effects, including kidney damage, liver necrosis and lung haemorrhages, in exposed animals (14).

Metabolism and pharmacokinetics

Rapidly metabolised via thiomethane and sulfone. Further biotransformation occurred to yield diphenyl disulfide (species unspecified) (14).
Methaemoglobin formation increased as a result of oxidative stress following exposure of 27.5-55 mg l^{-1} benzenethiol in human erythrocytes (19).

Irritancy

108 mg instilled into rabbit eye caused severe irritation (16).

Genotoxicity

Salmonella typhimurium TA98, TA100, TA1535, TA1537 with and without metabolic activation negative (17,18).

Any other adverse effects

Glycolysis was inhibited following administration of 55 mg l^{-1} and at the lowest dose the erythrocyte response appears to represent a level of oxidative stress to which the cell is capable of adaptive metabolic response (19).

Inhalation may be fatal as a result of spasm. Systemic effects include inflammation and oedema of the larynx and bronchi and pneumonitis (16).

Legislation

Federal Republic of Germany regulates air emissions of benzenethiol according to class 1 organic compounds (20).

Any other comments

Human health effects and experimental and industrial toxicology reviewed (21,22).

References

1. *The Toxicity of 3400 Chemicals to Fish* 1987, EPA 560/6-87-002 PB-200-275, Washington, DC
2. Kaiser, K. L. E; et al. *Water Pollut. Res. J. Canada* 1991, **26**(3), 361-431
3. Hansch, C; et al. *Medichem. Project* 1985, Pomona College, Claremont, CA
4. Hine, J; et al. *J. Org. Chem.* 1975, **40**, 292-298
5. Lyman, W. J; et al. *Handbook of Chemical Property Estimation Methods* 1982, 5-6, McGraw-Hill, New York
6. Schafer, E. W. *Toxicol. Appl. Pharmacol.* 1972, **21**, 315
7. Lutin, P. A; et al. *Purdue Univ. Eng. Bull. Ext. Ser.* 1965, **118**, 131-145
8. Tano, T; et al. *Hakko Kyokaishi* 1968, **26**, 322-327
9. *Sadtler Standard UV Spectra No. 2919* 1961, Sadtler Research Lab., Philadelphia, PA
10. Atkinson, R. *Intern. J. Chem Kinet.* 1987, **19**, 799-8828
11. Perry, R.H; et al. *Perry's Chemical Engineer's Handbook* 50th ed., 1984, 3-61, MacGraw-Hill, New York
12. Swann, R.L; et al. *Res. Rev.* 1983, **85**, 17-28
13. *Am. Ind. Hyg. Assoc. J.* 1988, **19**, 171
14. *Patty's Industrial Hygiene and Toxicology* 3rd ed., 1981, **2B**, Clayton, G. D; et al. (Eds.), John Wiley & Sons, New York
15. *NTIS Report* AD 277-689, Nat. Tech. Inf. Serv., Springfield, VA
16. Amrolia, P; et al. *J. Appl. Toxicol.* 1989, **9**(2), 113-118
17. *The Sigma-Aldrich Library of Chemical Safety Data* 1985
18. *IARC Monograph* 1977, **15**, 155-175
19. Shatun, M. M; et al. *Mutat. Res.* 1980, **78**(3), 213-218
20. *Legal File* 1986, **1**, International Register of Potentially Toxic Chemicals, Geneva
21. Flickinger, C. W. *Am. Ind. Hyg. Assoc. J.* 1976, **37**(10), 596-606
22. *ECETOC Technical Report No. 30(4)* 1991, European Chemical Industry Ecology and Toxicology Centre, B-1160 Brussels

B68　Benzene-1,2,4-tricarboxylic acid, 1,2-anhydride

CAS Registry No. 552-30-7

Synonyms trimellitic anhydride; 5-isobenzofuran carboxylic acid
1,3-dihydro-1,3-dioxo-; 1,2,4-benzene tricarboxylic acid anhydride;
4-carboxyphthalic anhydride; NIC-C56633; TMA

Mol. Formula $C_9H_4O_5$ **Mol. Wt.** 192.13

Uses Preparation of resins, adhesives, polymers, dyes and printing inks.

Physical properties

M. Pt. 161-163.5°C; **B. Pt.** $_{14}$ 240-245°C; **Flash point** 227°C.

Solubility
Organic solvent: acetone, ethyl acetate, dimethyl formamide, carbon tetrachloride, xylene

Occupational exposure

US TLV (TWA) 0.005 ppm (0.039 mg m^{-3}); **UK Long-term limit** 0.04 mg m^{-3};
Supply classification irritant.

Risk phrases
Irritating to eyes, respiratory system and skin – May cause sensitisation by inhalation
(R36/37/38, R42)

Safety phrases
Do not breathe dust – After contact with skin, wash immediately with plenty of water
(S22, S28)

Mammalian and avian toxicity

Acute data
LD$_{50}$ oral rat 5600 mg kg^{-1} (1).

Sub-acute data
In a 90-day feeding study in rats and dogs, the no-effect level was 1,000 and 20,000
ppm, respectively (2).
Inhalation rats 0-300 μg m^{-3} for varying durations. After 10 exposures for 6 hrs day^{-1}
rats experienced external haemorrhagic lung foci, alveolar macrophage accumulation,
alveolar haemorrhage, pneumonitis and lung and mediastinal lymph node damage.
Rats exposed and rested 12 days were nearly recovered from these effects, however,
rats rested 12 days and subsequently challenged exhibited lesions similar to those
seen immediately following exposure (3).

Any other adverse effects to man

Three types of syndrome have been reported in workers exposed to fume or dust
during its manufacture. The first is a direct irritant response, with sneezing, occasional

nose bleed, cough, laboured breathing and, rarely, wheezing which abates after 8
hours and rarely lasts into the night. The second, ''TMA asthma'' is rhinitis/asthma
which develops after a sensitisation period of weeks or months. In the third, known as
late onset respiratory systemic syndrome (LRSS) or ''TMA flu'', a sensitisation
period of weeks or months is required and symptoms include wheezing, coughing,
malaise, chills, fever and muscle and joint pains and is associated with relatively high
levels of exposure. A fourth type of syndrome has been reported in workers exposed
to fumes during the coating of hot pipes with epoxy resin containing benzene-1,2,4-
tricarboxylic acid, 1,2-anhydride. It is severe and potentially fatal and effects include
dyspnoea, pulmonary haemorrhages and haemolytic anaemia (4-6).

Any other comments
Hazards and precautions to be taken when used in the paint and ink industries reported (7).
Human health effects and experimental toxicology reviewed (8).

References
1. *TLV's in Air, American Conference of Govermental Industrial Hygienists* 1980, **4**, 415
2. *Toxicity Review 8, part 1: Trimetallic anhydride* 1983, HMSO, London
3. Leach, C. L; et al. *Toxicol. Appl. Pharmacol.* 1987, **87**(1), 67-80
4. Zeiss, C. R; et al. *J. Allergy Clin. Immunol.* 1977, **60**, 96
5. Ahmed, D; et al. *Lancet* 1979, **2**, 328
6. AMOCO Chem. Corp. *Industrial Hygiene, Toxicology & Safety Data Sheet – Trimetallic anhydride* 1976, Med. & Health Serv. Dept.
7. Scott, I.C. *J. Oil Colour Chem. Assoc.* 1986, **69**(11), 310-313
8. *ECETOC Technical Report No.30(4)* 1991, European Chemical Industry Ecology and Toxicology Centre, B-1160 Brussels

B69   Benzethonium chloride

CAS Registry No. 121-54-0
Synonyms benzenemethanaminium *N,N*-dimethyl N-[2-
[2-[4-[1,1,3,3-tetramethylbutyl]phenoxy]ethoxy]ethyl] chloride; Hyamine 1622;
Phemeridesp; Quatrachlor; antiseptol; BZT
Mol. Formula $C_{27}H_{42}NO_2Cl$ **Mol. Wt.** 448.09
Uses Topical anti-infective agent. Antiseptic. Cationic detergent.

Physical properties
M. Pt. 164-166 °C.

Solubility
Organic solvent: ethanol, acetone, chloroform

Ecotoxicity

Fish toxicity
LC$_{50}$ (96 hr) fathead minnow, bluegill sunfish 1.4-1.6 mg l^{-1} (1).
LC$_{50}$ (96 hr) coho salmon 53 mg l^{-1} (1).

Environmental fate

Degradation studies
Nonbiodegradable (2).

Mammalian and avian toxicity

Acute data
LD$_{50}$ oral rat, mouse 338-368 mg kg^{-1} (3).
LD$_{50}$ intraperitoneal rat 119 mg kg^{-1} (4).

Carcinogenicity and long-term effects
National Toxicology Program currently carrying out a 2 yr skin painting study on rats and mice (5).

Irritancy
30 μg instilled in rabbit eye caused severe irritation (6).
Mild skin irritation at 5% or lower. Is not considered to be a sensitiser and is considered to be safe at 0.5% in cosmetics applied to the skin and at a maximum concentration of 0.02% in cosmetics used in the eye area (7).

Any other adverse effects to man

Ingestion may cause vomiting, collapse, convulsions and coma in humans (8).

Legislation

Limited under EEC Directive on Drinking Water Quality 80/778/EEC. Chlorides: guide level 25 mg l^{-1}, maximum admissible concentration 400 mg l^{-1} (9).

Any other comments

Incompatible with soap and anionic detergents (8).

References

1. Surber, E. W; et al. *Progr. Fish. Cult.* 1962, **24**(4), 164
2. *MITI Report* 1984, Ministry of International Trade and Industry, Tokyo
3. *Proc. Soc. Exptl. Biol. Med.* 1965, **120**, 511
4. Kirschstein, R. L. *Dev. Biol. Stand.* 1974, **24**, 203
5. *National Toxicology Program, Research and Testing Div.* 1992, Report No. TR-438, NIEHS, Research Triangle Park, NC
6. *Proc. Sci. Sect. Toilet. Goods Assoc.* 1953, **20**, 16
7. *J. Am. Coll. Toxicol.* 1985, **4**, 65
8. *The Merck Index* 11th ed., 1989, Merck & Co. Inc., Rahway, NJ
9. *EC Directive Relating to the Quality of Water Intended for Human Consumption* 1982, 80/778/EEC, Office for Offical Publications of the European Communities, 2 rue Mercier, L-2985 Luxembourg

B70 Benzidine

$$H_2N-\langle\bigcirc\rangle-\langle\bigcirc\rangle-NH_2$$

CAS Registry No. 92-87-5
Synonyms biphenyl-4,4'-enediamine; 4,4'-diaminobiphenyl; p,p'-bianiline;
[1,1'-biphenyl]-4,4'-diamine
Mol. Formula $C_{12}H_{12}N_2$ **Mol. Wt.** 184.24
Uses Intermediate for dyestuffs. Reagent for hydrogen peroxide.
Occurrence Released as emissions and in wastewater during its production and use
as an intermediate in the manufacture of direct azo dyestuffs.

Physical properties

M. Pt. 128°C; **B. Pt.** $_{740}$ 400°C; **Specific gravity** d_4^{20} 1.250; **Partition coefficient** log
P_{ow} 2.007.

Solubility
Water: 400 mg l^{-1} at 12°C. Organic solvent: ethanol, diethyl ether, dimethyl sulfoxide

Occupational exposure

UN No. 1885; **Conveyance classification** toxic substance; **Supply classification**
toxic.

Risk phrases
May cause cancer – Harmful if swallowed (R45, R22)

Safety phrases
Avoid exposure-obtain special instructions before use – If you feel unwell, seek
medical advice (show label where possible) (S53, S44)

Ecotoxicity

Fish toxicity
LC_{50} (96 hr) sheepshead minnow 64 ppm. 1 ppm caused proliferative liver lesions
and $\geq$50 ppm caused various anomalies in embryos in sheepshead minnows in a 1 wk
study period (1,2).
LC_{50} (24 hr) red killifish 16.5 mg l^{-1} (2).

Environmental fate

Degradation studies
Initial benzidine concentration of 20 mg l^{-1} in activated sludge from mixed industrial
and domestic treatment plants was depleted by 85-93% after 6 hr at 25°C. (3).
Benzidine in sludge applied to a sandy loam soil in a biological soil reactor, $t_{1/2}$
averaged 76 days (4).
Benzidine has a low decomposition rate in natural soils due to its strong adsorption to
clay soils and its toxicity to microorganisms at high concentrations (5,6).
Biodegradable (7).

Abiotic removal

Removed from the environment by photolysis. Benzidine is rapidly oxidised by ferric ions and by complexing fulvic acids and clay minerals in water. Removal by evaporation and hydrolysis are insignificant (8,9).

Mammalian and avian toxicity

Acute data

LD_{50} oral rat, mouse 214-309 mg kg^{-1} (10).
LD_{50} intraperitoneal mouse 110 mg kg^{-1} (11).

Carcinogenicity and long-term effects

0-160 ppm administered to BALB/c and C57BL mice in drinking water for 33 months produced treatment related lung tumours, reticulum-cell sarcomas and hepatocellular carcinomas. Other dose related effects included pigmentation of the spleen, hepatic cytological alterations, hyperplasia of the bile ducts, megakaryocytosis of the bone marrow, vacuolisation of the brain, adenoma of the Harderian gland, atrophy of the ovaries and angioma of the uterus (12).
Sufficient evidence of carcinogenicity to both humans and animals, IARC classification group 1 (13).
Oral administration of benzidine to mouse, hamster and dog induced liver and bladder carcinomas while administration to rat by oral, subcutaneous and intraperitoneal routes caused mammary and Zymbal gland neoplasms (14-18).

Metabolism and pharmacokinetics

Metabolites of benzidine include N, N'- diacetylbenzidine, N-hydroxy-N, N'-diacetylbenzidine, N-hydroxy-N-acetylbenzidine, N,N'-dodecylbenzidine, 3-hydroxybenzidine and 4-amino- 4-hydroxybiphenyl (13,19).
Intravenous dog 1 mg kg^{-1}, plasma $t_{1/2}$ of benzidine and metabolites 3 hr, $t_{1/2}$ measured for benzidine 30 minutes. A significant amount of binding observed in DNA from liver, kidney and bladder (20).

Genotoxicity

Salmonella typhimurium TA1538 with metabolic activation positive (21).
Was non-mutagenic, but induced aneuploidy, gene conversion and DNA damage in yeast. *In vitro*, rodent cells induced chromosomal aberrations, sister chromatid exchanges, unscheduled DNA synthesis and DNA strand breaks. *In vivo*, rodent cells induced micronuclei, sister chromatid exchange, DNA strand breaks and unscheduled DNA synthesis (22).
Induced transformation in Syrian hamster embryo and BALB/c 3T3 cells (13,23).
In vitro L51178Y mouse lymphoma cells with and without metabolic activation positive mutagenic response (24).
Chinese hamster V-79 cells *in vitro* with and without metabolic activation negative (25-27).

Any other adverse effects to man

A study of workers exposed to benzidine between 1945-1979 showed a significant excess of bladder cancer incidence (28).
Of 25 benzidine workers at a US plant, 13 developed bladder cancer, all cases had been exposed to benzidine for a minimum of 6 yr (29).

An investigation of 244 workers in a Japanese dyestuffs manufacturing plant revealed 9 cases of bladder cancer occurring between 1968-1981, all had been working in benzidine production (30).

Of 1601 workers in the chemical industry in China exposed to benzidine, 21 cases of bladder carcinoma were confirmed (31).

Any other adverse effects

50% of an unspecified concentration of benzidine in acetone solution was absorbed through the skin of rats in 24 hr (32).

Legislation

Use prohibited in UK, controlled in US (33-35).

Any other comments

Experimental toxicology, environmental and human health effects reviewed (36,37). Toxicity and hazards reviewed (38).

References

1. Martin, B. J. *USEPA Report* 1982, EPA 600/3-82-09 (*Government Rep. Announcement Index* 1983, **83**(6), 99)
2. Tonogai, Y; et al. *J. Toxicol. Sci.* 1982, **7**(3), 193-203
3. Baird, R; et al. *J. Water Pollut. Control Fed.* 1977, 1609-1615
4. Kincannon, D. F; et al. *Proc. Ind. Waste Conf.* 1985, **40**, 607-619
5. Freitag, D. *Chemosphere* 1985, **14**, 1589-1616
6. Graveel, J. G; et al. *J. Environ. Qual.* 1986, **15**, 53-59
7. *MITI Report* 1984, Ministry of International Trade and Industry, Tokyo
8. Callahan, M. A; et al. *Water Related Environmental Fate of 129 Priority Pollutants* 1979, **2**(102), 1-7, EPA 440/4-79-029b
9. Lu, P. Y; et al. *Arch. Environ. Contam. Toxicol.* 1977, **6**, 129-142
10. *NTIS Report* PB 214-270, Nat. Tech. Inf. Serv., Springfield, VA
11. *Prog. Mut. Res.* 1981, **1**, 682
12. Littlefield, N. A; et al. *J. Toxicol. Environ. Health* 1983, **12**(4-6), 671-685
13. *IARC Monograph* 1987, **Suppl. 7** 123-125
14. *IARC Monograph* 1982, **29**, 149-183, 391-398
15. Littlefield, N. A; et al. *Fundam. Appl. Toxicol.* 1984, **4**, 69-80
16. Littlefield, N. A; et al. *J. Toxicol. Environ. Health* 1985, **15**, 357-367
17. Nelson, C. J; et al. *Toxicol. Appl. Pharmacol.* 1982, **64**, 171-186
18. Vesselinovitch, S. D. *Biol. Res. Preg. Perinatol.* 1983, **4**, 22-25
19. Haley, T. J; *Drug Metab. Rev.* 1982, **13**, 473-483
20. Lakshmi, V. M; et al. *Carcinogenesis (London)* 1990, **11**(1),139-144
21. McCann, J; et al. *Proc. Natl. Acad. Sci. USA* 1975, **72**, 5135-5139
22. *IARC Monograph* 1987, **Suppl. 6**, 96-100
23. Harbach, P. R; et al. *Mutat. Res.* 1991, **252**(2), 149-155
24. Sarkar, F. H; et al. *Mutat. Res.* 1990, **242**(4), 319-328
25. Phillips, B. J; et al. *Mutagenesis* 1990, **5 Suppl.**, 15-20
26. O'Donovan, M. R. *Mutagenesis* 1990, **5 Suppl.**, 9-13
27. Fassina, G. et al. *J. Toxicol. Environ. Health* 1990, **29**(1), 109-130
28. Meigs, J. W; et al. *J. Natl. Cancer Inst.* 1986, **76**, 1-8
29. Horton, A. W; et al. *J. Natl. Cancer Inst.* 1977, **58**, 1225-1228
30. Yamaguchi, N; et al. *Am. J. Ind. Med.* 1982, **3**, 139-148
31. Sun, L. D; et al. *Chin. J. Surg.* 1980, **18**, 491-493
32. Shah, P. V; et al. *Bull. Environ. Contam. Toxicol.* 1983, **31**, 73-78

33. Carcinogenic Substances Regulations, 1968
34. *Fed. Regist.* 1975, **40**(May 28), 23072, 29 CFN 1910-1010
35. *Statutory Instruments* No. 879, 1967, Ministry of Labour, London
36. *ECETOC Technical Report No. 30(4)* 1991, European Chemical Industry Ecology and Toxicology Centre, B-1160 Brussels
37. *Cah. Notes. Doc.* 1990, **138**, 229-233 (*Chem. Abstr.*, **113**, 28471e)
38. Izmerov, N. F. *Scientific Reviews of Soviet Literature on Toxicity & Hazards of Chemicals* 1992-1993, **123**, Eng. Trans., Richardson, M. L. (Ed.), UNEP/IRPTC, Geneva

B71   Benzimidazole

CAS Registry No. 51-17-2
Synonyms benzoimidazole; BZI; NSC 759; 3-azindole; 1,3-diazaindene; *N,N'*-methyenyl-*o*-phenylenediamine
Mol. Formula $C_7H_6N_2$ **Mol. Wt.** 118.14
Occurrence Tobacco smoke constituent.

Physical properties
M. Pt. 170.5°C; **B. Pt.** >360°C.

Solubility
Organic solvent: boiling xylene, ethanol

Mammalian and avian toxicity

Acute data
LD_{Lo} oral rat 500 mg kg^{-1} (1).
LD_{50} intravenous mouse 280 mg kg_{-1} (2).
LD_{50} intraperitoneal mouse 445 mg kg^{-1} (3).

Any other adverse effects
Treatment of rats with 80 mg kg^{-1} day^{-1} for 30 days caused severe testicular atrophy and arrest of spermatogenesis. Plasma follice stimulating hormone was markedly increased by 40 and 80 mg kg^{-1} day^{-1} doses administered for 30 days and remained high for 90 days after discontinuation of the drug (4).

References
1. *Nat. Acad. Sci.* 1953, **5**, 22
2. Foye, W. O. (Ed.) *Principles of Med. Chem.* 1974, Lea & Febiger, Philadelphia, PA
3. *Pharm. Chem. J.* 1980, **14**, 130
4. Asamoah, K. A; et al. *Diabetes Res. Clin. Pract.* 1990, **9**(3), 273-278

B72 1,2-Benzisothiazolin-3(2*H*)-one 1,1-dioxide

CAS Registry No. 81-07-2

Synonyms saccharin; benzosulfimide; *o*-sulfobenzimide; Gluside;
1,2-dihydro-2-ketobenzisosulfonazole; 2,3-dihydro-3-oxobenzisosulfonazole;
o-sulfobenzoic acid imide

Mol. Formula $C_7H_5NO_3S$ **Mol. Wt.** 183.19

Uses Non-caloric sweetening agent. Food additive. Formulation for electroplating
bath brightness. Normally used as sodium (RN 128-44-9) or calcium (RN 6485-34-3)
salt.

Physical properties

M. Pt. 229°C; **Specific gravity** 0.828.

Solubility
Water: 3.4 g l^{-1} at 25°C. Organic solvent: ethanol, acetone, glycerol, alkali
carbonates

Ecotoxicity

Fish toxicity
LC_{50} (96 hr) fathead minnow 18.3 g l^{-1} (1).

Bioaccumulation
Bioconcentration factors of 1.58 and 2.1 based on a water solubility of 4000 mg l^{-1} at
25°C and log K_{ow} of 0.91 which suggests bioconcentration in aquatic organisms is
unlikely (2).

Environmental fate

Abiotic removal
Respective acid and alkaline hydrolysis products are o-sulfamoylbenzoic acid and
ammonium o- sulfolbenzoic acid (3).
Stable enough in aqueous solution for most normal food applications, especially
beverages, therefore suggesting hydrolysis in environmental media is unlikley (4).
Atmospheric $t_{1/2}$ 6.5 hr (4).

Absorption
Calculated solid adsorption coefficient from 45-75 indicate high mobility in soil. It
should not partition from the water column to organic matter contained in sediments
and suspended solids (5).

Mammalian and avian toxicity

Acute data
LD_{50} oral mouse, rat 17 g kg^{-1} (6).

Sub-acute data

Albino mice (1 yr) 0-1.5 g kg^{-1} day^{-1} in diet. A marked decrease in total erythrocyte count, haemoglobin content and packed cell volume, at 1 and 1.5 g kg^{-1} day^{-1} suggesting occurrence of anaemia. Differential leukocyte count revealed an increase in the number of polymorphonuclear neutrophils (7).

Carcinogenicity and long-term effects

No adequate evidence for carcinogenicity to humans, sufficient evidence for carcinogencity to animals, IARC classification group 2B (8).
Dose-related increase in bladder tumours in F_1 ♂ rats fed sodium saccharin at 5-7.5% in 2-generation studies. Increased incidence of bladder tumours in mice (8).

Teratogenicity and reproductive effects

Intraperitoneal, intragastric and oral ingestion mice 2000 mg kg^{-1} no teratogenic effects reported (9).

Metabolism and pharmacokinetics

Transplacental transfer following intravenous infusion to rhesus monkeys in late pregnancy was rapid, but slight. Clearance was slower from foetal than maternal blood and was distributed in all foetal tissues examined with limited biotransformation and rapid excretion (10).
In three volunteers, 85-92% of 1 g doses administered orally for 21 days was excreted unchanged in the urine within 24 hr; no metabolites were found. Within 48 hr, 92.3% of a 500 mg dose was excreted in the urine and 5.8% in the faeces (11).

Genotoxicity

Salmonella typhimurium TA98, TA100 with and without metabolic activation negative (12).
Failed to induce micronuclei in bone marrow cells of mice, caused DNA strand breaks in rat hepatocytes but no chromosome aberrations in Chinese hamster cells, did not induce sister chromatid exchange in cultured human lymphocytes (8).
In human RSa cells, 10-22.5 mg ml^{-1} induced a dose-dependent increase in the number of mutations to ouabain resistance which is the first reporting of mutagenic activity in human cells (13).

Any other adverse effects

1969 study of 10:1 mixture of cyclamates and saccharin found bladder tumours in rats which led to a ban on cyclamates in the US and a ban on saccharin in Canada. However, with normal ingestion there is little or no risk of bladder cancer. It should be noted that the equivalent to a daily intake by humans would be 500 g day^{-1} and that foods containing such doses would be inedible (1,14).

Any other comments

In 1976, 23% of all saccharin consumed in the US was added to non-food items. In particular, 10% was used in cosmetics such as toothpaste, mouthwash and lipstick, 7% in pharmaceuticals such as pill coatings, 2% in smokeless tobacco products such as chewing tobacco and snuff, 2% in electroplating, 1% in cattle feed and 1% in miscellaneous uses (15).
Toxicology and carcinogenicity reviewed (16- 18).
Human health effects and experimental toxicology reviewed (19).

References

1. Geiger, D. L; et al. *Acute Toxicities of Organic Chemicals to Fathead Minnows (Pimephales Promelas)* 1988, **4**, 141, University of Wisconsin-Superior, Superior, WI
2. Lyman, W. J; et al. *Handbook of Chemical Property Estimation Methods* 1982, 5-4, 5-10, McGraw-Hill, New York
3. Windholz, M; et al. *Merck Index* 10th ed., 1983, Rahway, NJ
4. Atkinson, R. *Intern. J. Chem. Kinet.* 1987, **19**, 799-828
5. Swann, R. L; et al. *Res. Rev.* 1983, **85**, 16-28
6. Renner, H. W. *Experientia* 1979, **35**, 1364
7. Prasad, O; et al. *Toxicol. Lett.* 1987, **36**(1), 81-88
8. *IARC Monograph* 1987, **Suppl. 7**, 334-339
9. Drophkin, R. H; et al. *Arch. Toxicol.* 1984, **56**(4), 283
10. The Chemical Society. *Foreign Compound Metabolism in Mammals* 1972, **2**, 150, The Chemical Society, London
11. *IARC Monograph* 1980, **22**, 151
12. Herbold, B. A. *Mutat. Res.* 1981, **90**(4), 365
13. Suzuki, H; et al. *Mutat. Res.* 1988, **209**(1-2), 13-16
14. FAO Nutrition Meetings Report Series, No. 44A, WHO/Food Add./68.33, 1967
15. *IARC. Saccharin* 1980, **22**, 111-171
16. Arnold, D. L. *Fundam. Appl. Toxicol.* 1984, **4**, 674
17. Renwick, A. G; et al. *Comments Toxicol.* 1989, **3**(4), 289-305
18. Arnold, D. L. *Mutat. Res.* 1989, **221**(2), 69-132
19. *ECETOC Technical Report No. 30(4)* 1991, European Chemical Industry Ecology and Toxicology Centre, B-1160 Brussels

B73   1,2-Benzisothiazolin-3-one

CAS Registry No. 2634-33-5
Synonyms 1,2-benzisothiazol-3(2*H*)-one; Proxel P. L.
Mol. Formula $C_7H_5NO_5$ **Mol. Wt.** 183.12
Uses Antibacterial agent. Slimicide.

Physical properties
M. Pt. 157-158°C.

Ecotoxicity
Invertebrate toxicity
LC_{50} (96 hr) shrimp 25 mg l^{-1} (1).

Mammalian and avian toxicity
Acute data
LD_{50} oral rat, mouse 1020-1150 mg kg^{-1} (2).

Sensitisation
Weak sensitiser in guinea pig maximisation test (3).

Any other comments
Allergic effects of paint preservatives containing 1,2-benziosthiazolin-3-one reviewed and recommendations are made not to use it as a preservative in water-based paints (4).

References
1. Linden, E; et al. *Chemosphere* 1979, **11-12**, 843-851
2. Betaccini, G; et al. *Pharm. Res. Comm.* 1971, **3**(4), 385
3. Anderson, K. E. *Food Chem. Toxicol.* 1984, **22**(8), 655
4. Hansen, M.K. *Faerg Lack Scand.* 1987, **33**(5), 81-83, (Dan.) (*Chem. Abstr.* 1988, **108**, 50660j)

B74   Benzocaine

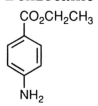

CO$_2$CH$_2$CH$_3$

NH$_2$

CAS Registry No. 94-09-7
Synonyms ethyl *p*-aminobenzoate; anaesthesin; americaine; *p*-aminobenzoic acid, ethyl ester
Mol. Formula C$_9$H$_{11}$NO$_2$ **Mol. Wt.** 165.19
Uses Topical anaesthetic. Suntan preparations.

Physical properties
M. Pt. 88-90°C.

Solubility
Organic solvent: ethanol, diethyl ether, chloroform

Ecotoxicity

Effects to non-target species
LD$_{50}$ oral starling 316 mg kg^{-1} (1).

Environmental fate

Nitrification inhibition
Nitrosomonas sp. exposure to 100 mg l^{-1} resulted in 20-50 % inhibition of ammonia oxidation (2).
Nitrosomonas sp. 10 mg l^{-1} 0% inhibition of ammonia oxidation (3).

Mammalian and avian toxicity

Acute data
LD$_{50}$ intraperitoneal mouse 216 mg kg^{-1} (4).

Irritancy

Dermal (24 hr) guinea pig 2% solution caused mild irritation (5).

Any other adverse effects

Methaemoglobinaemia has been reported following the use of benzocaine (6).

References

1. Schafer, E. W. *Arch. Environ. Toxicol.* 1983, **12**, 355
2. Richardson, M. L. (Ed.) *Nitrification Inhibition in the Treatment of Sewage* 1985, RSC, London
3. Hockenbury, M. R; et al. *J. Water Pollut. Control Fed.* 1977, **49**(5), 768-777
4. *J. Med. Chem.* 1974, **17**, 900
5. *J. Soc. Cosmet. Chem.* 1977, **28**, 357
6. *Martindale The Extra Pharmacopoeia* 29th ed., 1989, The Pharmaceutical Press, London

B75   Benzo[*b*]fluoranthene

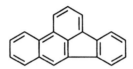

CAS Registry No. 205-99-2
Synonyms 3,4-benz[*e*]acephenanthrylene; 2,3-benzfluoranthene;
3,4-benzfluoranthene; benzo[*e*]fluoranthene; B(*b*)F
Mol. Formula $C_{20}H_{12}$ **Mol. Wt.** 252.32
Occurrence In crude and motor oil. From manmade pollution sources including cigarette smoke and gasoline exhaust (1,2).

Physical properties

M. Pt. 168°C.

Solubility
Organic solvent: benzene, acetone

Occupational exposure

Supply classification toxic.

Risk phrases
May cause cancer (R45)

Safety phrases
Avoid exposure-obtain special instructions before use -- If you feel unwell, seek medical advice (show label where possible) (S53, S44)

Ecotoxicity

Bioaccumulation
The high estimated log P_{ow} 6.124 suggests that it will bioconcentrate appreciably in aquatic organisms. The presence of microsomal oxidase in fish, however, suggests

that benzo[b]fluroanthene will not accumulate in fish due to the anticipated rapid metabolism (1,2).

Environmental fate

Degradation studies
Suject to co-metabolism, but as a sole carbon source is not expected to biodegrade (3).

Abiotic removal
Ozonolytic $t_{1/2}$ were 52.7, 10.8 and 2.9 hrs for ozone concentrations of 0.19, 0.70 and 2.28 ppm, respectively. Photolytic $t_{1/2}$ irradiated with 290-400 nm light in the absence of ozone was 8.7 hrs, and in the presence of ozone at above concentrations was 4.2, 3.6 and 1.9 hrs, respectively (4).
Photolysis in the presence of oxygen is likely to produce quinone (5).
Volatilisation $t_{1/2}$ in streams, rivers and lakes were 10, 14 and 5586 hrs, respectively. Volatilisation from soil is not expected to be significant (6).

Absorption
Estimated soil adsorption coefficient 358 suggests immobility in soil. Leaching to groundwater is therefore expected to be very slow and it is also expected to partition to sediments (6,7).

Mammalian and avian toxicity

Carcinogenicity and long-term effects
No adequate data for evidence of carcinogenicity to humans, sufficient evidence for carcinogenicity to animals, IARC classification group 2B (8).
Dermal ♀ CD-1 mice 0.25-1.0 µg l^{-1} per mouse 10 subdoses administered every other day. Induced a 100% incidence of tumour-bearing mice averaging 8.5 tumours per mouse at a total initiation dose of 0.25 µg l^{-1} (9).

Metabolism and pharmacokinetics
Highly soluble in adipose tissue and lipids (10).
Topical administration of benzo[b]fluoranthene (B*b*F) to mice caused death at various intervals. Metabolites detected in the epidermis were 4-, 5- and 6-hydroxy B*b*F. Sulfate and glucuronide conjugates of these hydroxy B*b*F were also detected. Minor metabolites included 12-hydroxy-B*b*F, B*b*F-1,2-diol and B*b*F-11,12-diol (11).

Any other adverse effects to man

A case-control study was undertaken in Montreal to investigate the possible associations between occupational exposures and cancer of oesophagus, stomach, colorectum, liver, pancreas, lung, prostrate, bladder, kidney, melanoma and lymphoid tissue. In total 3726 cancer patients were interviewed, between 1979 and 1985, to obtain lifetime job histories, which were translated into a history of occupational exposures to polycyclic aromatic hydrocarbons (PAHs). 75% of all subjects had some exposure to PAHs. At the levels of exposure experienced, the preliminary analysis reported here revealed no clear evidence of an increased risk of any type of cancer among exposed workers (12).

Any other comments
Human health effects, experimental toxicology and environmental effects reviewed (13,14).

References

1. *GEMS: Graphical Exposure Modeling System. Fate of Atmospheric Pollutants Data Base* 1986, Office of Toxic Substances, USEPA, Washington, DC
2. Santofonato, J; et al. *Health and Ecological Assessment of Polynuclear Aromatic Hydrocarbons* 1981, Pathotox. Publ. Park, Forest South, IL
3. Sims, R. C; et al. *Res. Rev.* 1983, **88**, 1-68
4. Lane, D. A; et al. *Adv. Environ. Sci. Technol.* 1977, **8**, 137-154
5. Graedel, T. E. *Chemical Compounds in the Atmosphere* 1978, 130, Academic Press, New York
6. Lyman, W. J; et al. *Handbook of Chemical Property Estimation Methods* 1982, McGraw-Hill, New York
7. Kenagu, E. E. *Ecotox. Env. Saf.* 1980, **4**, 26-38
8. *IARC Monograph* 1987, **Suppl. 7**, 58
9. Weyand, E. H; et al. *Cancer Lett. (Shannon, Irel.)* 1990, **52**(3), 229-233
10. Sittig, M. *Handbook of Toxic and Hazardous Chemicals and Carcinogens* 2nd ed., 1985, 741, Noyes Data Corporation, Park Ridge, NJ
11. Geddie, J. E; et al. *Carcinogenesis (London)* 1987, **8** (11), 1579-1584
12. Krewski, D; et al. *Environ. Sci. Res.* 1990, **39**, 343-352
13. *ECETOC Technical Report No. 30(4)* 1991, European Chemical Industry Ecology and Toxicology Centre, B-1160 Brussels
14. *Gov. Rep. Announce. Index US* 1990, **90**(19), 048,363

B76   Benzo[*ghi*]fluoranthene

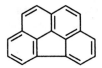

CAS Registry No. 203-12-3

Synonyms benzo[*mno*]fluoranthene; 2,13-benzofluoranthene; 7,10-benzofluoranthene

Mol. Formula $C_{18}H_{10}$ **Mol. Wt.** 226.28

Occurrence In exhaust condensate of gasoline engines, wood preservative sludge and water sediment (1-3).

Physical properties

M. Pt. 147-149°C.

Solubility
Organic solvent: petroleum ether

Mammalian and avian toxicity

Carcinogenicity and long-term effects
No adequate evidence for carcinogenicity to humans, inadequate evidence for carcinogenicity to animals, IARC classification group 3 (4).

Any other adverse effects to man

A case-control study was undertaken in Montreal to investigate the possible associations between occupational exposures and cancers of oesophagus, stomach,

---

colorectum, liver, pancreas, lung, prostrate, bladder, kidney, melanoma and lymphoid tissue. In total, 3726 cancer patients were interviewed, between 1979 and 1985, to obtain detailed lifetime job histories, which were translated into a history of occupational exposures to polycyclic aromatic hydrocarbons (PAHs). 75% of all subjects had some exposure to PAHs. At the levels of exposure experienced, the preliminary analysis reported here revealed no clear evidence of an increased risk of any type of cancer among exposed workers (5).

Any other comments
Human health effects and experimental toxicology reviewed (6).

References
1. Grimmer, G; et al. *Zbl. Bakt. Hyg. I. Abt. Orig. B*. 1977, **164**, 218
2. Lao, R. C; et al. *J. Chromatog*. 1975, **112**, 681
3. Brown, R. A; et al. *Mar. Pollut. Bull*. 1978, **9**, 162
4. *IARC Monograph* 1987, **Suppl. 7**, 58
5. Krewski, D; et al. *Environ. Sci. Res*. 1990, **39**, 343-352
6. *ECETOC Technical Report No 30(4)* 1991, European Chemical Industry Ecology and Toxicology Centre, B-1160 Brussels

B77   Benzo[*j*]fluoranthene

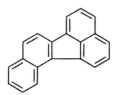

CAS Registry No. 205-82-3
Synonyms 10,11-benzofluoranthene; 7,8-benzofluoranthene; benz[*j*]fluoranthene; benzo(1)fluoranthene; dibenzo[*a,j,k*]fluorene
Mol. Formula $C_{20}H_{12}$ **Mol. Wt.** 252.32
Occurrence From manmade pollution sources including gasoline exhausts, cigarette smoke, soot and sewage sludge effluent (1-4).

Physical properties
M. Pt. 165.5°C.

Solubility
Organic solvent: ethanol

Occupational exposure
Supply classification toxic.

Risk phrases
May cause cancer (R45)

Safety phrases
Avoid exposure-obtain special instructions before use – If you feel unwell, seek medical advice (show label where possible) (S53, S44)

Ecotoxicity

Bioaccumulation
Calculated bioconcentration factor was 55.1- 83.1 which suggests potential accumulation in aquatic systems (1).

Environmental fate

Degradation studies
After 1280 days in soil treated with oil sludge, 79% of the original benzo[j]fluoranthene was recovered (1).

Abiotic removal
May potentially undergo photolysis in sunlit media since absorbs ultraviolet light at >290 nm. Atmospheric $t_{1/2}$ was 7 hr (2,3).
Volatilisation $t_{1/2}$ estimated as >400 yr (4).

Absorption
Based on water solubility of 6.76 µg l^{-1} and an estimated log K_{ow} of 6.12, the calculated soil adsorption coefficient of 51,000- 68,000 suggests strong adsorption to soils and sediments (5).

Mammalian and avian toxicity

Carcinogenicity and long-term effects
No adequate evidence for carcinogenicity to humans, sufficient evidence for carcinogenicity to animals, IARC classification group 2B (6).
Dermal mouse 0.1 and 0.5% benzo[j]fluoranthene solutions applied twice weekly for life. At 9 months all animals had died. Incidence of skin papillomas was 70-95% and skin carcinomas 75-100% in low and high dose groups, respectively (7).

Any other adverse effects to man

A case-control study was undertaken in Montreal to investigate the possible associations between occupational exposures and cancers of oesophagus, stomach, colorectum, liver, pancreas, lung, prostrate, bladder, kidney, melanoma and lymphoid tissue. In total 3726 cancer patients were interviewed, between 1979 and 1985, to obtain detailed lifetime job histories, which were translated into a history of occupational exposures to polycyclic aromatic hydrocarbons (PAHs). 75% of all subjects had some exposure to PAHs. At the levels of exposure experienced, the preliminary analysis reported here revealed no clear evidence of an increased risk of any type of cancer among exposed workers (8).

Any other comments
Human health effects and experiment toxicology reviewed (9).

References

1. Lyman, W. J; et al. *Handbook of Chemical Property Estimation Methods* 1982, McGraw-Hill, New York
2. Atkinson, R. *Intern. J. Chem. Kinet.* 1987, **19**, 799-828
3. *IARC Monograph* 1972, **32**, 155-161
4. *USEPA; EXAMS II Computer Simulation* 1987
5. Swann, R. L; et al. *Res. Rev.* 1983, **85**, 16-28
6. *IARC Monograph* 1987, **Suppl. 7**, 58

7. Wynder, E. L; et al. *Cancer* 1959, **12**, 1194
8. Krewski, D; et al. *Environ. Sci. Res.* 1990, **39**, 343-352
9. *ECETOC Technical Report No. 30(4)* European Chemical Industry Ecology and Toxicology Centre, B-1160 Brussels

B78   Benzo[*k*]fluoranthene

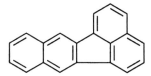

CAS Registry No. 207-08-9
Synonyms 8,9-benzofluoranthene; 2,3,1'8'-binaphthylene; 11,12-benzofluoranthene; dibenzo[*bjk*]fluorene
Mol. Formula $C_{20}H_{12}$ **Mol. Wt.** 252.32
Occurrence From manmade pollution sources including gasoline exhausts and sewage sludge (1-5).
Occurs as a pollutant in tapwater and groundwater (6).

Physical properties
M. Pt. 217°C; **B. Pt.** 480°C.

Solubility
Organic solvent: benzene, ethanol, acetic acid

Occupational exposure
Supply classification toxic.

Risk phrases
May cause cancer (R45)

Safety phrases
Avoid exposure-obtain special instructions before use – If you feel unwell, seek medical advice (show label where possible) (S53, S44)

Ecotoxicity

Bioaccumulation
From calculated log K_{ow} of 6.84, the estimated bioconcentration factor for fish is 144. However no accumulation is likely due to the presence of microsomal mixed function oxidases which enables it to be metabolised (7,8).
Short-necked clam cultured in artificial seawater at 21-25°C for 10 days revealed a decrease in benzo[*k*]fluoranthene of ≈20% on day 8. When clams were placed in a basket and kept in harbour water, only a small increase in polycyclic aromatic hydrocarbons were found within a short period of time. After 1 month, a 2.5-9-fold increase was observed (9).

Environmental fate

Degradation studies
When soil treated with seven applications of oil sludge containing polynucleated

aromatic hydrocarbons over a two yr period, was monitored for an additional 18 months, the benzo[*k*]fluoranthene residue in the soil decreased by 57%. In a static biodegradability test employing a domestic wastewater inoculum, 50-70% of benzo[*k*]fluoranthene was degraded in four successive weekly subcultures (10).

Abiotic removal
Demonstrates considerable atmospheric stability. Pollution resulting from emissions can be found far from source (11).
Atmospheric $t_{1/2}$ 14 hr in sunlight without ozone, $t_{1/2}$ 3-9 hr with ozone and $t_{1/2}$ 35 hr with ozone in the dark (11).
Very reactive with chlorine and ozone in solution which suggests that the levels of these chemicals will be lowered considerably in treated drinking water (12,13).
Generally resistant to hydrolysis (7).
Under ultraviolet irradiation $t_{1/2}$ was 111 min (14).

Absorption
Volatilistion from soil is expected to be low due to its low vapour pressure (15).
Estimated soil adsorption coefficient of 678 suggests strong adsorption to soils and sediments will occur (16).
Leaching may occur from soils with low organic content or high porosity (sand), or from sites that have been exposed to spills or chemical wastes containing benzo[*k*]fluoranthene (17).

Mammalian and avian toxicity

Sub-acute data
LD_{50} (72 hr) chick embryo 14 μg kg^{-1} following injection into 7 day pre-incubated air sacs (18).

Carcinogenicity and long-term effects
No adequate evidence for carcinogenicity to humans, sufficient evidence for carcinogenicity to animals, IARC classification group 2B (19).

Metabolism and pharmacokinetics
Readily absorbed from the gastrointestinal tract and lung (species unspecified) (17).
Major metabolites of benzo[*k*]fluoranthene (B*k*F) formed *in vitro* formed by incubation with rat liver S9 metabolic fraction were 8,9-dihydro-8,9-dihydro-B*k*F, the 2,3-quinone of B*k*F and 3-, 8- and 9-hydroxy-B*k*F (20).

Any other adverse effects to man

A case-control study was undertaken in Montreal to investigate the possible associations between occupational exposures and cancers of the oesphagus, stomach, colorectum, liver, pancreas, lung, prostrate, bladder, kidney, melanoma and lymphoid tissue. In total 3726 cancer patients were interviewed, between 1979 and 1985, to obtain detailed lifetime job histories, which were translated into a history of occupational exposure to polycyclic hydrocarbons (PAHs). 75% of all subjects had some occupational exposure to PAHs. At the levels of exposure experienced, the preliminary analysis reported here revealed no clear evidence of an increased risk of any type of cancer among exposed workers (21).

Any other comments

Human health effects, experimental toxicology and environmental effects reviewed (22).

References

1. Grimmer, G; et al. *Zbl. Bakt. Hyg. I. Abt. Orig. B.* 1977, **164**, 218
2. Candeli, A; et al. *Atm. Environ.* 1974, **8**, 693
3. Borneff, J; et al. *Arch. Hyg. Bakt.* 1962, **146**, 183
4. Borneff, J; et al. *Arch. Hyg. Bakt.* 1963, **146**, 572
5. Borneff, J; et al. *Arch. Hyg. Bakt.* 1965, **149**, 226
6. Borneff, J; et al. *Arch. Hyg. Bakt.* 1969, **153**, (3), 220
7. Lyman, W. J; et al. *Handbook of Chemical Property Estimation Methods. Environmental Behaviour of Organic Chemicals* 1982, 7.1-7.41, McGraw-Hill, New York
8. Santodonato, J; et al. *Health and Ecological Assessment of Polynuclear Aromatic Hydrocarbons* 1981, 160-176, Pathotox. Publishers Inc., Park Forest South, IL
9. Obano, H; et al. *Osaka-Furitsu Koshu Eisei Kenkyusho Kenkyu Hokoku, Shokuhin Eisei Hen* 1981, **12**, 91-94
10. Tabak, H. H; et al. *94th Ann. Mtg. Assoc. Off. Anal. Chem.* 1981, 267
11. Bjorseth, A; et al. *Handbook of Polycyclic Aromatic Hydrocarbons* 1983, 507-524, Marcel Dekker Inc., New York
12. Redding, S. B; et al. *The Environmental Fate of Selected Polynuclear Aromatic Hydrocarbons* 1976, 131 USEPA-560/5-75-009
13. Sorrell, R. K; et al. *Environ. Inter.* 1980, **4**, 245-254
14. Paalme, L; et al. *Vopr. Onkol.* 1983, **29**(7), 74-80
15. Callahan, M. A; et al. *Water Related Environmental Data of 129 Priority Pollutants* 1979, 97.1-97.23, USEPA-440/4-79-029b
16. Sims, R. C; et al. *Res. Rev.* 1983, **88**, 1-68
17. *USEPA; Health and Environmental Effects Profile for Benzo[k]fluoranthene* 1987, ECAO-CIN-P229
18. Brunstroem, B; et al. *Arch. Toxicol.* 1991, **65**(6), 485-498
19. *IARC Monograph* 1987, **Suppl. 7**, 58
20. Weyand, E. H; et al. *Carcinogenesis (London)* 1988, **9**(7), 1277-1281
21. Krewski, D; et al. *Environ. Sci. Res.* 1990, **39**, 343-352
22. *ECETOC Technical Report No. 30(4)* 1991, European Chemical Industry Ecology and Toxicology Centre, B-1160 Brussels

B79 Benzo[*a*]fluorene

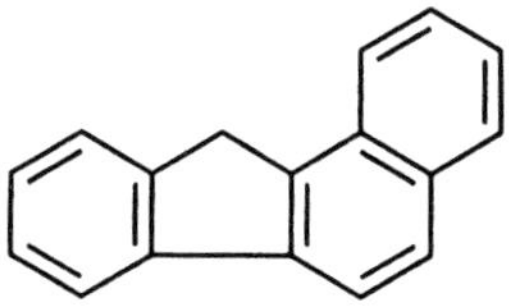

CAS Registry No. 30777-18-5
Synonyms 1,2-benzofluorene; 11*H*-benzo[*a*]fluorene; chrysofluorene; α-naphthofluorene
Mol. Formula $C_{17}H_{12}$ **Mol. Wt.** 216.29
Occurrence From manmade pollution sources including bitumen, coal tar, gasoline and cigarette smoke (1).

Physical properties

M. Pt. 189-190°C; **B. Pt.** 413°C.

Solubility

Organic solvent: diethyl ether, chloroform, hot benzene

Mammalian and avian toxicity

Carcinogenicity and long-term effects

No adequate evidence for carcinogenicity to humans, inadequate evidence for carcinogenicity to animals, IARC classification group 3 (2).

Any other adverse effects to man

A case-control study was undertaken in Montreal to investigate the possible associations between occupational exposures and cancers of oesophagus, stomach, colorectum, liver, pancreas, lung, prostrate, bladder, kidney, melanoma and lymphoid tissue. In total 3726 cancer patients were interviewed, between 1979 and 1985, to obtain detailed lifetime job histories, which were translated into a history of occupational exposure to polycyclic aromatic hydrocarbons (PAHs). 75% of all subjects had some occupational exposure to PAHs. At levels of exposure experienced, the preliminary analysis reported here revealed no clear evidence of an increased risk of any type of cancer among exposed workers (3).

References

1. Grimmer, G; et al. *Zbl. Bakt. Hyg. I. Abt. Orig. B.* 1977, **164**, 218
2. *IARC Monograph* 1987, **Suppl. 7**, 58
3. Krewski, D; et al. *Environ. Sci. Res.* 1990, **39**, 343-352

B80   Benzofuran

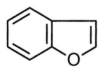

CAS Registry No. 271-89-6

Synonyms benzo(b)furan; 2,3-benzofuran; coumarone; benzofurfuran; NCI-C56166; 1-oxindene

Mol. Formula C_8H_6O **Mol. Wt.** 118.14

Uses Manufacture of coumarone and indene resins.

Physical properties

M. Pt. <-18°C; **B. Pt.** 170-173°C; **Specific gravity** $d_4^{22.7}$ 1.0913; **Partition coefficient** $\log P_{ow}$ 2.67.

Solubility

Organic solvent: miscible with benzene, ethanol, diethyl ether, petroleum ether, acetone

---

Mammalian and avian toxicity

Acute data

LD$_{50}$ intraperitoneal mouse 500 mg kg^{-1} (1).

Carcinogenicity and long-term effects

National Toxicology Program tested ♂ and ♀ rats and mice with benzofuran via gavage (dose unspecified). Results showed no evidence of carcinogenic activity in ♂ rat, some evidence of carcinogenic activity in ♀ rat and clear evidence of carcinogenic activity in both ♂ and ♀ mice (2).

Any other comments

Human health effects, experimental toxicology and environmental effects reviewed (3).

References

1. Fatome, M; et al. *Eur. J. Med. Chem.* 1977, **12**, 383
2. *National Toxicology Program, Research and Testing Div.* 1992, Report No. TR-370, NIEHS, Research Triangle Park, NC
3. *ECETOC Technical Report No. 30(4)* 1991, European Chemical Industry Ecology and Toxicology Centre, B-1160 Brussels

B81　Benzoguanamine

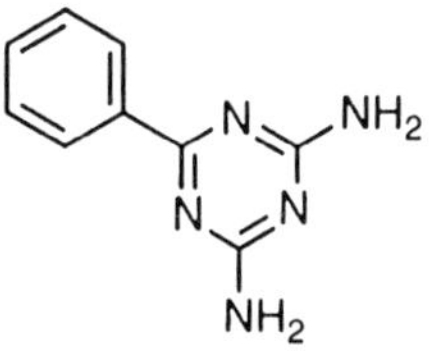

CAS Registry No. 91-76-9

Synonyms benzochinamide; benzquinamide; BZQ; quantril; P2647

Mol. Formula C$_{22}$H$_{32}$N$_2$O$_5$　　　　　　　　　**Mol. Wt.** 404.51

Uses In the manufacture of thermosetting resins, pesticides, pharmaceuticals and dyestuffs.

Physical properties

M. Pt. 227°C; **Specific gravity** d$_4^{25}$ 1.40.

Solubility

Organic solvent: ethanol, diethyl ether

Occupational exposure

Supply classification harmful.

Safety phrases

Do not breathe dust (S22)

Ecotoxicity

Effects to non-target species
LD_{50} oral redwing blackbird 100 mg kg^{-1} (1).

Mammalian and avian toxicity

Acute data
LD_{50} oral rat 1050 mg kg^{-1} (2).
LD_{50} intraperitoneal mouse 320 mg kg^{-1} (3).

References

1. Schafer, E. W; et al. *Toxicol. Appl. Pharmacol.* 1972, **21**, 315
2. Usdon, *Psychotropic Drugs and Related Compounds* 2nd ed., 1972
3. *Dissertationes Pharmaceuticae* 1965, **17**, 145

B82   Benzoic acid

CAS Registry No. 65-85-0
Synonyms benzenecarboxylic acid; phenylformic acid; dracylic acid; Retordex; E210
Mol. Formula $C_7H_6O_2$ **Mol. Wt.** 122.12
Uses Permitted foodstuff preservative in fats and fruit juices. In the manufacture of dyestuffs. Antifungal agent.
Occurrence Component of berries, fruits and vegetables.

Physical properties

M. Pt. 122.4°C; **B. Pt.** 249°C; **Flash point** 121-131°C; **Specific gravity** 1.27;
Partition coefficient log P_{ow} 1.87 at 20°C; **Volatility** v.p. 1 mmHg at 96° (sublimes).

Solubility
Water: 2.9 g l^{-1} at 20°C. Organic solvent: ethanol, diethyl ether, acetone, chloroform, carbon tetrachloride, benzene, carbon disulfide

Ecotoxicity

Fish toxicity
LC_{50} (96 hr) mosquito fish 180 mg l^{-1} (1).

Invertebrate toxicity
Cell multiplication inhibition test, *Pseudomonas putida* 480 mg l^{-1}, *Microcystis aeruginosa* 55 mg l^{-1}, *Scenedesmus quadricauda* 1630 mg l^{-1}, *Entosiphon sulcatum* 218 mg l^{-1}, *Uronema parduczi* 31 mg l^{-1} (2-4).

Bioaccumulation
Bioconcentration factor in golden ide and *Chlorella fusca* <10 (5).

Bioconcentration factors of mosquito fish, algae, mosquito larvae are 21, 100 and 138, respectively, and of *Daphnia* and snail are 1800 and 2800, respectively (6).

Environmental fate

Degradation studies
99% removal within 24 hr using adapted activated sludge (7).

ThOD 1.97 g g^{-1}; COD 0.69 g g^{-1} (8).

BOD_5 at 20°C 1.34-1.4 using normal sewage seed and 1.36 using acclimated sewage seed. BOD_{10} at 20°C 1.4 using normal sewage seed. BOD_{20} 1.45 at 20°C (9).

The $t_{1/2}$ for mineralisation using a Captina silt loam inoculum was 4.5 hr after a 30 min lag (10).

$t_{1/2}$ of 0.85 and 3.6 days in a polluted river and reservoir, respectively (11).

$t_{1/2}$ of 0.22 days for mineralisation in eutrophic water in a range of 32 ng l^{-1} to 50 µg l^{-1}. From 63-83% was lost in 6 hr and >94% in 58 hr (12).

Under anaerobic conditions, 91% of benzoic acid was converted to methane and carbon dioxide in 18 days including an 8 day lag period (13).

86-93% conversion to methane and carbon dioxide occurred in 14 days with a sewage sludge inoculum (14).

Benzoic acid is biodegradable under aerobic conditions by bacteria present in crude municipal wastewater at ≤200 g m^{-3} (15).

$t_{1/2}$ for carboxyl and ring-labelled benzoic acid was 3.9 and 7.2 hr, respectively (16).

Confirmed biodegradable (17).

Abiotic removal
The pk_a of 4.2 suggests almost exclusive existance in the dissociated form at environmental pHs (18).

In the vapour phase, estimated $t_{1/2}$ with photochemically produced hydroxyl radicals 2 days (19).

Absorption
Benzoic acid did not adsorb appreciably to two different sandy soils, a clay subsoil and montmorillonite clay (20,21).

Mammalian and avian toxicity

Acute data
LD_{Lo} oral human 500 mg kg^{-1} (22).

LD_{50} oral rat 1700 mg kg^{-1} (22).

LD_{Lo} subcutaneous rabbit 2000 mg kg^{-1} (23).

Metabolism and pharmacokinetics
Following intravenous administration to weaning Yorkshire sows in physiological saline (200 µg, 10µCi) of benzoic acid and collection of excrement over 6 days revealed >80% in urine. In ethanol (200 µg, 10 µCi), topical doses applied at a surface concentration of 40 µg cm^{-2} penetrated at 25.7%; faecal clearance of radiolabel, expressed as a percentage of total excretion was greater after topical administration (24).

Human clearance time 9-15 hr (25).

When administered orally, benzoic acid is rapidly absorbed from the gastrointestinal tract, conjugated with glycine in the liver to form hippuric acid which is rapidly excreted in the urine within 12 hr (up to 97% within the first 4 hr). When taken in large doses, some benzoic acid may be excreted as benzoylglucuronic acid (26). Ruminants excrete much larger quantities of aromatic acids, such as benzoic acid, in their urine than do nonruminants, particularly when fed a high-roughage diet (27). In 3 hr, biliary excretion as percentage dose excreted is 1.2 in rat, 1.7 in guinea pig, 0.7 in rabbit, 0.8 in dog, 1.2 in cat and 0.5 in hen (28).

Irritancy
Dermal rabbit (24 hr) 500 mg caused mild irritation and 100 mg instilled into rabbit eye caused severe irritation (29).

Sensitisation
Asthmatics or urticaria sufferers may be sensistive to benzoic acid (25).

Genotoxicity
Salmonella typhimurium TA98, TA100 with and without metabolic activation negative (30).

Any other adverse effects
Neonates with fatal toxic syndrome attributable to benzyl alcohol have been reported to have elevated urinary concentrations of its metabolite benzoic acid. The accumulation of benzoic acid in blood, possibly due to the liver's diminished metabolic capacity in premature infants, might be responsible for the metabolic acidosis observed (31,32).

Any other comments
Reacts with the preservative sodium hydrogen sulfite (*E222*) (25).
Found in groundwater in Australia underlying an area where acid wastes from a manufacturing process of a chemical company was stored in unlined ponds. Since the chemical was only found in the aquifier down stream from the believed source of pollution and not close to this source, it was either formed by bacterial action or came from another source (33).
Hazardous properties, human health effects and experimental toxicology reviewed (34,35).

References
1. Jones, H. R. *Environmental Control in the Organic and Petrochemical Industries* 1971, Noyes Data Corporation, Park Ridge, NJ
2. Bringman, G; et al. *Z. Wasser/Abwasser, Forsch.)* 1980, **1**, 26
3. Bringmann, G; et al. *Water Res.* 1980, **14**, 231-241
4. Bringmann, G; et al. *Gwf-Wasser-Abwasser* 1976, **117**(9)
5. Freitag, D; et al. *Chemosphere* 1985, **14**, 1589-1616
6. Lu, P. Y; et al. *Environ. Health Perspect.* 1975, **10**, 269-284
7. Alexander, M; et al. *J. Agr. Food Chem.* 1966, **14**, 410-413
8. Urano, K; et al. *J. Haz. Mater.* 1986, **13**, 147-159
9. Verschueren, K. *Handbook of Environmental Data of Organic Chemicals* 2nd ed., 1983, 258, Van Nostrand Reinhold, New York
10. Dao, T. H; et al. *Soil Sci.* 1987, **143**, 66-72
11. Banerjee, S; et al. *Environ. Sci. Technol.* 1984, **18**, 416-422
12. Subba-Rao, R. V; et al. *Appl. Environ. Microbiol.* 1982, **43**, 1139-1150
13. Healy, J. B., Jr; et al. *Appl. Environ. Microbiol.* 1979, **38**, 84-89
14. Nottingham, P. M; et al. 1969, **98**, 1170-1172
15. Jaroszynski, T; et al. *Environ. Prot. Eng.* 1979, **5**(4), 375

16. Ward, T. E. *Environ. Toxicol. Chem.* 1985, **4**, 727-737
17. *MITI Report* 1984, Ministry of International Trade and Industry, Tokyo
18. Serjeant, E. P; et al. *Ionization Constants of Organic Acids in Aqueous Solution* 1979, 989, Pergamon Press, New York
19. *GEMS; Graphical Exposure Modeling System. Fate of Atmospheric Pollutants Data Base* 1987, Office of Toxic Substances, USEPA, Washington, DC
20. Loekke, H. *Water Air Soil Pollut.* 1984 , **22**, 373-387
21. Bailey, G. W; et al. *Soil Sci. Soc. of Amer. Proc.* 1968, **32**, 222-234
22. Patty, F. A. (Ed.) *Industrial Hygiene and Toxicology* 1967, **2**, Intersciencs Publishers, New York
23. *Handbook of Toxicology* 1959, **5**, 23
24. Carver, M. P; et al. *Fundam. Appl. Toxicol.* 1989, **13**(4), 714-722
25. Hanissen, M. *E for Additives* 1987, Thorsons Publ., Wellingborough
26. *Martindale The Extra Pharmacopoeia* 28th ed., 1982, 1283, The Pharmaceutical Press, London
27. Cheeke, P. R; et al. *Natural Toxicants in Feeds and Poisonous Plants* 1985, 335, AVI Publishing Co. Inc., Westport, CT
28. La Du, B. N; et al. *Fundamentals of Drug Metabolism and Disposition* 1971, 142, Williams & Wilkins, Baltimore, MD
29. *BIOFAX Industrial Bio-Test Laboratories, Inc.* 1973
30. McCann, J; et al. *Proc. Nat. Acad. Sci., USA* 1975, **72**, 5185
31. Gershanik, J; et al. *New Engl. J. Med.* 1982, **307**, 1384
32. Brown, W. J; et al. *Lancet* 1982, **1**, 1250
33. Stepan, S; et al. *Austral. Water Resources Council Conf. Ser.* 1981, **1**, 415-424
34. *Dangerous Prop. Ind. Mater. Rep.* 1989, **9**(6). 11-29
35. *ECETOC Technical Report No. 30(4)* 1991, European Chemical Industry Ecology and Toxicology Centre, B-1160 Brussels

B83   Benzoic acid, ethyl ester

$CO_2CH_2CH_3$

CAS Registry No. 93-89-0
Synonyms ethyl benzoate; benzoic ether; essence of Niobe
Mol. Formula $C_9H_{10}O_2$ **Mol. Wt.** 150.18
Uses Perfumery.

Physical properties

M. Pt. –34.6°C; **B. Pt.** 213.4°C; **Specific gravity** d_4^{25} 1.050.

Solubility
Organic solvent: miscible in petroleum ether, diethyl ether, chloroform, ethanol

Mammalian and avian toxicity

Acute data
LD_{50} oral rat, rabbit 2100-2630mg kg^{-1} (1).

Irritancy
Dermal (24 hr) rabbit 10 mg caused mild irritation (2).

References
1. Graham, B. E; et al. *J. Pharmacol. Exp. Therap.* 1945, **84**, 358
2. Smyth, H. F; et al. *Archiv. Ind. Hyg. Occup. Med.* 1954, **10**, 61

B84   Benzoic acid, methyl ester

CO$_2$CH$_3$

CAS Registry No. 93-58-3
Synonyms methyl benzoate; niobe oil; oil of niobe
Mol. Formula C$_8$H$_8$O$_2$ **Mol. Wt.** 136.15
Uses Perfume manufacture. Flavourings. Pesticides.
Occurrence In oils of cloves, ylang ylang and tuberose.

Physical properties
M. Pt. −12.5°C; **B. Pt.** 199.6°C; **Flash point** 82.8°C (closed cup); **Specific gravity** d$_4^{20}$ 1.0888.

Solubility
Organic solvent: miscible with ethanol, diethyl ether, methanol

Occupational exposure
UN No. 2938; **HAZCHEM Code** 3⊠.

Ecotoxicity

Fish toxicity
Rainbow trout exposed to 5 mg l^{-1} experienced loss of equilibrium in 14-16 hr (1).

Invertebrate toxicity
EC$_{50}$ (30 min) *Photobacterium phosphoreum* 4.61 mg l^{-1} Microtox test (2).

Environmental fate

Degradation studies
Nonbiodegradable/qualified (3).

Mammalian and avian toxicity

Acute data
LD$_{50}$ oral rat, mouse 1350-3330 mg kg^{-1} (4).

Irritancy
Dermal (24 hr) rabbit 10 mg caused mild irritation (5).

Any other comments

Methyl benzoate was slow to penetrate the skin of rats (6).
Human health effects and experimental toxicology reviewed (7).

References

1. *Fish Toxicity Screening Data. Part 1. Lethal effects of 964 chemicals upon Steelhead Trout and Bridgelip Sucker* 1989, EPA 560/6-89-001, Washington, DC
2. Kaiser, K. L. E; et al. *Water Pollut. Res. J. Canada* 1991, **26**(3), 361-431
3. *MITI Report* 1984, Ministry of International Trade and Industry, Tokyo
4. Jenner, P. M; et al. *Food Cosmet. Toxicol.* 1964, **2**, 327
5. Smyth, A. F; et al. *Arch. Ind. Hyg. Occup. Med.* 1954, **10**, 61
6. Opdyke, D. L. J. (Ed.) *Monographs on Fragrance Raw Materials* 1979, 537, Pergamon Press, New York
7. *ECETOC Technical Report No. 30(4)* 1991, European Chemical Industry Ecology and Toxicology Centre, B-1160 Brussels

B85   Benzoin

CAS Registry No. 119-53-9

Synonyms benzoylphenylcarbinol; bitter almond oil camphor; α-hydroxybenzyl phenyl ketone; α-hybroxy-α-phenylacetophenone; 2-hydroxy-2-phenylacetophenone; NCI-CS011

Mol. Formula $C_{14}H_{12}O_2$ **Mol. Wt.** 212.25

Uses Organic synthesis.

Physical properties

M. Pt. 133-135°C; **B. Pt.** $_{768}$ 344°C.

Solubility

Organic solvent: acetone, boiling ethanol

Environmental fate

Degradation studies

Confirmed biodegradable (1).

Mammalian and avian toxicity

Acute data

LD_{50} oral rat 10 g kg^{-1} (2).
LD_{50} dermal rabbit 8870 mg kg^{-1} (2).

Carcinogenicity and long-term effects

National Toxicolgy Program tested rats and mice via feed. Negative results obtained in both species (3).

Genotoxicity

Salmonella typhimurium TA97, TA98, TA100, TA1535 without metabolic activation positive (4).

Any other comments

Human health effects, genotoxicity and experimental toxicology reviewed (5-7).

References

1. *MITI Report* 1984, Ministry of International Trade and Industry, Tokyo
2. Mortelmans, K. *Environ. Mutagen.* 1986, **8**(Suppl. 7), 1-119
3. *National Toxicology Program, Research and Testing Div.* 1992, Report No. TR-204, NIEHS, Research Triangle Park, NC
4. Zeiger, E; et al. *Prog. Mutat. Res.* 1985, **5**, 187-199
5. Brady, A. L; et al. *Mutat. Res.* 1989, **224**(3), 391-403
6. Ashby, J; et al *Mutat. Res.* 1989, **224**(3), 321-324
7. *ECETOC Technical Report No. 30(4)* 1991, European Chemical Industry Ecology and Toxicology Centre, B-1160 Brussels

B86   Benzo[*b*]naphtho[1,2-*d*]thiophene

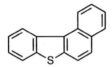

CAS Registry No. 205-43-6
Synonyms 3,4-benzo-9-thiafluorene; 7-thia-7(H)-benzo[*c*]fluorene; naphtho[2,1-*b*]thianaphthene
Mol. Formula $C_{16}H_{10}S$ **Mol. Wt.** 234.32
Occurrence Combustion product from fossil fuels, particularly from diesel engines. Found in metal working oils and machine oils (1).

Physical properties

M. Pt. 103-4°C.

Genotoxicity

Salmonella typhimurium TA98, TA100, TA1535, TA1538 with metabolic activation positive (2).

Any other adverse effects to man

A case-control study was undertaken in Montreal to investigate the possible associations between occupational exposures and cancers of oesophagus, stomach, colorectum, liver, pancreas, lung, prostrate, bladder, kidney, melanoma and lymphoid tissue. In total 3726 cancer patients were interviewed between 1979 and 1985, to obtain detailed lifetime job histories, which were translated into a history of occupational exposures to polycyclic aromatic hydrocarbons (PAHs). 75% of all subjects had some occupational exposure to PAHs. At levels of exposure experienced, the preliminary analysis reported here revealed no clear evidence of an increased risk of any type of cancer among exposed workers (3).

References

1. Eyres, A. R. *Inst. Pet., (Tech. paper)* 1981, 13, P81-002 (*Chem. Abstr.* **95**, 10295w)
2. Petroy, R. A; et al. *Mutat. Res.* 1983, **117**(1-2), 31
3. Krewski, D; et al. *Environ. Sci. Res.* 1990, **39**, 343-352

B87   Benzo[*b*]naphtho[2,1-*d*]thiophene

CAS Registry No. 239-35-0
Synonyms naphtho[1,2-*b*]thianaphthene
Mol. Formula $C_{16}H_{10}S$ **Mol. Wt.** 234.32
Occurrence Combustion product from fossil fuels, especially diesel engines. Major impurity in materials related to coal tar fractions 2-3 wt%.

Physical properties

M. Pt. 185-186°C.

Mammalian and avian toxicity

Carcinogenicity and long-term effects
Using a beeswax/trioctanoin mixture as vehicle, 6, 3 and 1 mg were injected into the lungs of 35 ♀ rabbits per group. A dose-dependent tumour incidence was found, but an increasing incidence of neoplasms was not seen between the median and high dose. Estimated ED_{10} value is 1.65 mg (1).

Any other adverse effects to man
A case-control study was undertaken in Montreal to investigate the possible associations between occupational exposures and cancers of oesophagus, stomach, colorectum, liver, pancreas, lung, prostrate, bladder, kidney, melanoma and lymphoid tissue. In total 3726 cancer patients were interviewed between 1979 and 1985, to obtain detailed lifetime job histories, which were translated into a history of occupational exposures to polycyclic aromatic hydrocarbons (PAHs). 75% of all subjects had some occupational exposure to PAHs. At levels of exposure experienced, the preliminary analysis reported here revealed no clear evidence of an increased risk of any type of cancer among exposed workers (2).

References

1. Wenzel-Hartung, R; et al. *Exp. Pathol.* 1990, **40**(4), 221-227
2. Krewski, D; et al. *Environ. Sci. Res.* 1990, **39**, 343-352

B88 Benzo[*b*]naphtho[2,3-*d*]thiophene

CAS Registry No. 243-46-9
Synonyms naphtho[3,2-*b*]thianaphthene
Mol. Formula C$_{16}$H$_{10}$S   **Mol. Wt.** 234.32
Occurrence Combustion product from fossil fuels especially from diesel engines. Found in metal working oils, engine sump oils, quenching oils.

Physical properties
M. Pt. 158-9°C.

Occupational exposure

Safety phrases
Keep away from sources of ignition – No Smoking – In case of contact with eyes, rinse immediately with plenty of water and seek medical advice – Do not empty into drains (S16, S26, S29)

Any other adverse effects to man

A case-control study was undertaken in Montreal to investigate the possible associations between occupational exposures and cancers of oesophagus, stomach, colorectum, liver, pancreas, lung, prostrate, bladder, kidney, melanoma and lymphoid tissue. In total 3726 cancer patients were interviewed between 1979 and 1985, to obtain detailed lifetime job histories, which were translated into a history of occupational exposures to polycyclic aromatic hydrocarbons (PAHs). 75% of all subjects had some occupational exposure to PAHs. At levels of exposure experienced, the preliminary analysis reported here revealed no clear evidence of an increased risk of any type of cancer among exposed workers (1).

References
1. Krewski, D; et al. *Environ. Sci. Res.* 1990, **39**, 343-352

B89 Benzonitrile

CAS Registry No. 100-47-0
Synonyms benzenenitrile; cyanobenzene; benzoic acid nitrile; phenyl cyanide
Mol. Formula C$_7$H$_5$N   **Mol. Wt.** 103.12

Uses Solvent.

Physical properties

M. Pt. −13°C; **B. Pt.** 191°C; **Specific gravity** d_{15}^{15} 1.01;
Volatility log P_{ow} 1.56.

Solubility
Organic solvent: ethanol, diethyl ether, acetone, benzene

Occupational exposure

UN No. 2224; **HAZCHEM Code** 3X; **Conveyance classification** toxic substance;
Supply classification harmful.

Risk phrases
Harmful in contact with skin and if swallowed (R21/22)

Safety phrases
Do not breathe vapour (S23)

Ecotoxicity

Fish toxicity
LC_{50} (96 hr) fathead minnow 78-135 mg l^{-1} range dependent on pH (1).
LC_{50} (96 hr) bluegill sunfish, guppy 78-400 mg l^{-1} (1,2).
Trout and bluegill sunfish exposed to 5 ppm died in 22 hr (3).

Invertebrate toxicity
Cell multiplication inhibition test *Pseudomonas putida* 11 mg l^{-1}, *Microcystis aeruginosa* 3.4 mg l^{-1}, *Scenedesmus quadricauda* 75 mg l^{-1}, *Entosiphon sulcatum* 30 mg l^{-1}, *Uronema parduczi* 119 mg l^{-1} (4- 6).
EC_{50} (5-30 min) *Photobacterium phosphoreum* 11.6 mg l^{-1} Microtox test (7).

Environmental fate

Degradation studies
Confirmed biodegradable (8).

Mammalian and avian toxicity

Acute data
LD_{50} oral mouse 970 mg kg^{-1} (9).
LD_{50} intraperitoneal rabbit 1250 mg kg^{-1} (10).

Irritancy
Dermal rabbit (24 hr) 500 mg caused moderate irritation (11).

Any other comments

Physico-chemical properties, human health effects and experimental toxicology
reviewed (12).

References

1. Jones, H. R. *Environmental Control in the Organic and Petrochemical Industries* 1971, Noyes Data Corp., Park Ridge, NJ
2. Henderson, C; et al. *Purdue Univ. Eng. Bull. Ext. Ser.* 1960, **106**, 120

3. Hollis, E. H; et al. *The Toxicity of 1085 Chemicals to Fish* 1954, US Fish and Wildlife Service, Kearneysville
4. Bringmann, G; et al. *Z. Wasser/Abwasser Forsch.* 1980, **1**, 26
5. Bringmann, G; et al. *Water Res.* 1980, **14**, 231
6. Bringmann, G; et al. *GWF.-Wasser/Abwasser* 1976, **117**, 9
7. Kaiser, K. L. E; et al. *Water Pollut. Res. J. Canada* 1991, **26**(3), 361-431
8. *MITI Report* 1984, Ministry of International Trade and Industry, Tokyo
9. *Nippon Eiseig Aki Zasshi. Jap. J. Hyg.* 1984, **39**, 423
10. *Food Cosmet. Toxicol.* 1979, **17**, 723
11. *Food Cosmet. Toxicol.* 1979, **17**, 695
12. *ECETOC Technical Report No. 30(4)* 1991, European Chemical Industry Ecology and Toxicology Centre, B-1160 Brussels

B90   Benzo[*ghi*]perylene

CAS Registry No. 191-24-2
Synonyms 1,12-benzperylene; 1,12-benzoperylene
Mol. Formula $C_{22}H_{12}$ **Mol. Wt.** 276.34
Occurrence Coal tar. Gasoline engine exhausts. Contaminant in tap water, groundwater and sediment. Occurs in domestic effluent (1-4).

Physical properties
M. Pt. 278-280°C; **Partition coefficient** log P_{ow} 6.90.

Solubility
Water: 0.26 µg l^{-1} at 25°C.

Ecotoxicity

Fish toxicity
LC_{50} (96 hr) fathead minnow 78-135 mg l^{-1}, dependent on pH (1).
LC_{50} (96 hr) bluegill sunfish, guppy 400 mg l^{-1} in soft water (1,2).

Invertebrate toxicity
EC_{50} (48 hr) *Daphnia magna* 0.2 µg l^{-1} (5).

Bioaccumulation
Based on a water solubility of 0.26 µg l^{-1} at 25°C and an estimated log K_{ow} of 6.58, calculated bioconcentration factors were 85-110 which indicates potential accumulation in aquatic systems (6,7).

Environmental fate

Degradation studies
Bacillus megaterium metabolises benzo[*ghi*]perylene, however, data are insufficient

to predict the significance of soil biodegradation (8).

If released to the atmosphere, may be susceptible to slow biodegradation under aerobic conditions. It is not expected to leach or volatilise from soils (8).

Within 7 days, an aerobic aqueous screening test inoculated with sewage showed a 60% loss of an initial 1 ppm benzo[*ghi*]perylene concentration (9).

After 240 days, in an unacclimated agricultural sandy loam soil incubated at 10 and 20°C, 81 and 76%, respectively, of an initial concentration of 9.96 μg g^{-1} remained. Corresponding estimated $t_{1/2}$ were 650 and 600 days (10).

After 1280 days, 78.3% of an inital 3.1 μg g^{-1} remained in a soil treated with oil sludge at 17.0 μg g^{-1} (11).

Abiotic removal
Photolytic $t_{1/2}$ on silica gel, alumina, fly ash and carbon black were 7, 22, 29 and >1000 hr, respectively (12).

Atmospheric $t_{1/2}$ using photochemically produced hydroxyl radicals was estimated to be 2 hr (13).

Volatilisation $t_{1/2}$ from a model river 1 m deep, flow rate 1 m sec^{-1} with a windspeed of 3 m sec^{-1} has been estimated to be 38 days (6).

Absorption
Estimated soil adsorption coefficients of >1 million and the widespread detection of benzo[*ghi*]perylene in various USA sediments indicate that adsorption to suspended particulate matter and sediments is an important environmental process (8).

Estimated soil adsorption coefficients of 90,000-400,000 indicate high immobility in soil (14).

Mammalian and avian toxicity

Carcinogenicity and long-term effects
No adequate evidence for carcinogenicity to humans, inadequate evidence for carcinogenicity to animals, IARC classification group 3 (15).

Dermal mouse (1 yr) 0.05 or 0.1% solution applied three times a week induced a low tumour incidence (16).

Metabolism and pharmacokinetics
Absorbed readily from the gastrointestinal tract and lung, is highly lipid soluble and can pass across epithelial membranes (8).

Any other adverse effects to man
A case-control study was undertaken in Montreal to investigate the possible associations between occupational exposures and cancers of oesophagus, stomach, colorectum, liver, pancreas, lung, prostrate, bladder, kidney, melanoma and lymphoid tissue. In total 3726 cancer patients were interviewed between 1979 and 1985, to obtain detailed lifetime job histories, which were translated into a history of occupational exposures to polycyclic aromatic hydrocarbons (PAHs). 75% of all subjects had some occupational exposure to PAHs. At levels of exposure experienced, the preliminary analysis reported here revealed no clear evidence of an increased risk of any type of cancer among exposed workers (17).

Any other comments
Human health effects, experimental toxicology and environmental effects reviewed (18).

References

1. Candeli, A; et al. *ATM. Environ.* 1974, **8**, 693
2. Borneff, J; et al. *Arch. Hyg.* 1969, **153**, (3), 220-229
3. Edward, J. D; et al. *Bull. Environ. Contam. Toxicol.* 1979, **22**, 653-659
4. Borneff, J; et al. *Arch. Hyg. Bakt.* 1962, **146**, 183-197
5. Newsted, J. L; et al. *Environ. Toxicol. Chem.* 1987, **6**, 445
6. Lyman, W. J; et al. *Handbook of Chemical Property Estimation Methods* 1982, McGraw-Hill, New York
7. Mackay, D; et al. *J. Chem. Eng. Data* 1977, **22**, 399-402
8. *USEPA Health and Environmental Effects Profile for Benzo[ghi]perylene* 1987, EPA 600/X-87/395
9. Fochtman, E. G. *Biodegradation and Carbon Adsorption of Carcinogenic and Hazardous Organic Compounds* 1981, 38, EPA 600/S2-81-032
10. Coover, M. P; et al. *Haz. Waste Haz. Mat.* 1987, **4**, 69-82
11. Bossert, I. D; et al. *Appl. Environ. Microbiol.* 1984, **47**, 763-767
12. Behymer, T. D; et al. *Environ. Sci. Technol.* 1985, **19**, 1004-1006
13. Atkinson, R. *Intern. J. Chem. Kinet.* **19**, 799-828
14. Swann, R. L; et al. *Res. Rev.* 1983, **85**, 16-28
15. *IARC Monograph* 1987, **Suppl. 7**, 58
16. Hoffman, D; et al. *Z. Krebforsch* 1966, **68**, 137-149
17. Krewski, D; et al. *Environ. Sci. Res.* 1990, **39**, 343-352
18. *ECETOC Technical Report No. 30(4)* 1991, European Chemical Industry Ecology and Toxicology Centre, B-1160 Brussels

B91   Benzophenone

CAS Registry No. 119-61-9
Synonyms diphenyl ketone; benzoylbenzene; diphenylmethanone
Mol. Formula $C_{13}H_{10}O$ **Mol. Wt.** 182.22
Uses Used in the manufacture of antihistamines, hypnotics and insecticides. Used as a fixative for heavy perfumes, especially when used in soaps.
Occurrence In Baltic Sea shale tar.

Physical properties

M. Pt. 48.5°C; **B. Pt.** 305°C; **Specific gravity** d_4^{18} 1.1108.

Solubility
Organic solvent: ethanol, diethyl ether, benzene, acetone, chloroform

Ecotoxicity

Fish toxicity
LC_{50} (96 hr) fathead minnow 15.3 mg l^{-1} (1).

Invertebrate toxicity

EC_{50} (30 min) *Photobacterium phosphoreum* 8.92 mg l^{-1} Microtox test (2).

Bioaccumulation

Non-accumulative or low accumulative (3).

Environmental fate

Degradation studies

Biodegradable (3).

Mammalian and avian toxicity

Acute data

LD_{50} oral mouse 2900 mg kg^{-1} (4).
LD_{50} intraperitoneal mouse 727 mg kg^{-1} (4).

Carcinogenicity and long-term effects

Selected for general toxicology study by National Toxicology Program (5).

Genotoxicity

Escherichia coli polA without metabolic activation negative (6).
Salmonella typhimurium TA98, TA100 with and without metabolic activation negative (7).

Any other comments

Human health effects and experimental toxicology reviewed (8).

References

1. Veith, G. D; et al. *Am. Soc. Test. Mater.* 1983, 90, ASTM STP 803, Philadelphia, PA
2. Kaiser, K. L. E; et al. *Water Pollut. Res. J. Canada* 1991, **26**(3), 361-431
3. *MITI Report* 1984, Ministry of International Trade and Industry, Tokyo
4. Caprino, L; et al. *Journal Europeen de Toxicologie* 1976, **9**, 99
5. *National Toxicology Program, Research and Testing Div.* 1992, NIEHS, Research Triangle Park, NC
6. *EPA, TSCA, Chemical Inventory* 1986
7. Mortelmans, K; et al. *Environ. Mutagen.* 1986, **8**, 1-119
8. *ECETOC Technical Report No. 30(4)* 1991, European Chemical Industry Ecology and Toxicology Centre, B-1160 Brussels

B92   Benzo[*a*]pyrene

CAS Registry No. 50-32-8
Synonyms benzo[*d,e,f*]chrysene; 3,4-benzopyrene; 6,7-benzopyrene; 3,4-benzpyrene; benz[*a*]pyrene

Mol. Formula $C_{20}H_{12}$ **Mol. Wt.** 252.32

Occurrence Occurs in cigarette smoke and from the combustion of fuels (1-3).

Physical properties

M. Pt. 179°C; **B. Pt.** 495°C; **Partition coefficient** log P_{ow} 6.35.

Solubility
Organic solvent: benzene, toluene, xylene

Occupational exposure

Supply classification toxic.

Risk phrases
May cause cancer – May cause heritable genetic damage – May cause birth defects (R45, R46, R47)

Safety phrases
Avoid exposure-obtain special instructions before use – If you feel unwell, seek medical advice (show label where possible) (S53, S44)

Ecotoxicity

Fish toxicity
LD_{50} (96 hr) *Neanthes arenaceodentata* in seawater at 22°C <1 ppm (in static bioassay) (4).

Invertebrate toxicity
EC_{50} (96 hr) *Daphnia pulex* 0.05 mg l^{-1} (5).

Bioaccumulation
Administered to northern pike via the diet or water at levels found in moderately polluted water and fish were observed for 10 hr to 21 days after initial exposure. Uptake was through gastrointestinal system and the gills, metabolised in the liver and excreted in urine and bile. Benzo[a]pyrene and its metabolites were heterogeneously distributed in the kidneys, little accumulation in adipose tissue. Since recovery in different organs after 8.5 days of exposure was in the form of metabolites, metabolism is thought to play an important role in the bioaccumulation and disposition in fish (6). Bioconcentration factors in oysters 3000, rainbow trout 920, bluegill sunfish 2657, *Daphnia magna* 1000, *Daphnia pulex* 13,000 (7-12).

Environmental fate

Degradation studies
50-80% degradation occurred in soil inocculated with N5 and N13 bacterial strains. After 8 days the highest degradation occurred in soil inocculated with pretreated bacteria (13,14).
Degradation of 56-67% was observed in nutrient/benzo[a]pyrene solutions seeded with acclimated sewage and incubated for 4-7 day periods (15).
Soil biodegradation temperature dependent, $t_{1/2}$ 2-694 days from <15->25°C (16).

Abiotic removal
Absorbs light at >290 nm and may undergo direct photolysis (17).
Calculated photolytic $t_{1/2}$ is 0.54 hr in surface waters (18).
Estimated $t_{1/2}$ with photochemically produced hydroxyl radicals is 21 hr (19).

Reported estimated theoretical maximum $t_{1/2}$ for volatilisation from a model river 1 m deep, flowing at 1 m sec^{-1} with a wind velocity of 4 m sec^{-1} was 18 days. It is concluded that vaporisation will be significant (20).

Absorption
Adsorption of benzo[a]pyrene on kaolinite clay inhibited photolysis and absorption on fly ash showed 17% degradation when exposed to sunlight for 3-4 hr (16,21). Soil adsorption coefficient, binding to dissolved organic carbon in 3 natural waters 18,000-52,000 and to Aldrich humates 890,000 (22).

Mammalian and avian toxicity

Acute data
LD_{Lo} intraperitoneal mouse 500 mg kg^{-1} (23).
LD_{50} subcutaneous rat 50 mg kg^{-1} (24).

Carcinogenicity and long-term effects
No adequate evidence for carcinogenicity to humans, sufficient evidence for carcinogenicity to animals, IARC classification group 2A (25).
Dermal ♀ mouse (9 wk) 8, 16, 32, 64 μg wk^{-1} caused initial linear increase in DNA adducts with dose in the epidermis, but the increase was much less steep above 32 μg wk^{-1}. This did not correlate with the sharp rise in tumour response above 32 μg wk^{-1}. There was little epidermal hyperplasia but an associated dose-dependent increase in incidence of pyknotic and dark cells at 64 μg wk^{-1} indicated that benzo[a]pyrene produced extensive cytotoxicity and cell death with regenerative proliferation. Giant keratinocytes occurred in all dose groups. Tumours were initially papillomas and required an average of 8 wk to convert to carcinomas (26).

Metabolism and pharmacokinetics
A single oral or topical dose was administered to ♂ SENCAR mice. After 6-48 hr, high concentrations were found in skin following topical applications but very little material reached the organs after oral doses. The internal organs contained more material after oral administration (27).
Urine samples of pitch-coke plant workers showed increased levels of benzo[a]pyrene both before and after work shifts, which indicated slow excretion (28).
Intraperitoneal administration in rats showed maximum uptake of benzo[a]pyrene after 1 hr in aortic media and liver, levels in kidney and urine continued to increase. Smooth arterial muscle cells in culture incubated with tritium-benzo[a]pyrene took up the radioactivity. This was inversely related to extracellular lipid content and ^{3}H release was increased by proteins (29).
Readily absorbed from the intestinal tract, localised primarily in body fat and fatty tissues. Single intravenous injection rat benzo[a]pyrene detected in blood and liver, $t_{1/2}$ is <5 min and 10 min, respectively (30).
♂ rats, with cannulated bile duct, received intravenous injections of radiolabelled benzo[a]pyrene non-covalently bound to lipoproteins in equimolar amounts. Cumulative biliary excretions were 39.6, 24.6 and 21.2% for very-low-, low- and high-density lipoproteins complexed to benzo[a]pyrene, respectively. 60-80% of injected benzo[a]pyrene, and 50-60% of injected benzo[a]pyrene metabolites were not excreted immediately in control or induced animals (31).

Genotoxicity

Escherichia coli PQ37 with metabolic activation positive (32).
Salmonella typhimurium TA98, TA100, TA1535, TA1538 with metabolic activation positive (33).

Any other adverse effects to man

A case-control study was undertaken in Montreal to investigate the possible associations between occupational exposures and cancers of oesophagus, stomach, colorectum, liver, pancreas, lung, prostrate, bladder, kidney, melanoma and lymphoid tissue. In total 3726 cancer patients were interviewed, between 1979 and 1985, to obtain detailed lifetime job histories, which were translated into a history of occupational exposures to polycyclic aromatic hydrocarbons (PAHs). 75% of all subjects had some occupational exposure to PAHs. In this study benzo[*a*]pyrene is the only PAH for which adequate data are available. However the authors conclude on the basis of their available data and at the levels of exposure experienced, the preliminary analysis reported revealed no clear evidence of an increased risk of any type of cancer among exposed workers (34).

Any other comments

Risk assessment review (35).
Comparative genotoxicites with pyrene reviewed (36).
Physico-chemical properties, human health effects, experimental toxicology, environmental effects, ecotoxicology, epidemiology and workplace experience reviewed (37).

References

1. *Identification of Organic Compounds in Effluents from Industrial Sources* 1975, EPA 560/3-75-002
2. Wynder, *Tobacco and Tobacco Smoke* 1967, Academic Press, New York
3. Candeli, A; et al. *Atm. Environ.* 1974, **8**, 693
4. Rossi, S. S; et al. *Mar. Pollut. Bull.* 1978, **9**, 220
5. Govers, H; et al. *Chemosphere* 1984, **13**, 227
6. Balk, L; et al. *Toxicol. Appl. Pharmacol.* 1984, **74**(3), 430-449
7. Verschueren, K. *Handbook of Environmental Data on Organic Chemicals* 2nd ed., 1983, 27, Van Nostrand Rheinhold, New York
8. Spehar, R. L; et al. *J. Water Pollut. Control Fed.* 1980, **52**, 1703-1774
9. McCarthy, J. F; et al. *Environ. Sci. Technol.* 1985, **19**, 1072-1076
10. Leversee, G. J; et al. *Can. J. Fish. Aquat. Sci.* 1983, **40**, 63-69
11. McCarthy, J. F. *Arch. Environ. Contam. Toxicol.* 1983, **12**, 559-568
12. Biddinger, G. R; et al. *Res. Rev.* 1984, **91**, 103-145
13. Poglazova, M. N; et al. *Life Science* 1967, **6**, 1053-1063
14. Gibson, D. T. *Microbial Degradation of Polycyclic Aromatic Hydrocarbons* 1976, Univ. Texas at Austin, TX
15. Fochtman, E. G. *Biodegradation and Carbon Adsorption of Carcinogenic and Hazardous Organic Compounds* 1981, EPA 600/S2-81-032
16. Sims, R. C; et al. *Res. Rev.* 1983, **88**, 1-68
17. Callahan, M. A; et al. *Water-Related Environmental Fate of 129 Priority Pollutants* 1979, **2**, 98.1-98.27, EPA 440/4-79-029b
18. Herbes, S. E; et al. *The Scientific Basis of Toxicity Assessment* 1980, 113-128, Elsevier/North Holland Biomedical Press
19. *GEMS: Graphical Exposure Modeling System. Fate of Atmospheric Pollutants Data Base* 1986, Office of Toxic Substances, USEPA, Washington, DC

20. Southworth, G. R; et al. *Bull. Environ. Contam. Toxicol.* 1979, **21**, 507-514
21. Korfmacher, W. A; et al. *Environ. Sci. Technol.* 1980, **14**, 1094-1099
22. Landrum, P. F; et al. *Environ. Sci. Technol.* 1984, **18**, 187-192
23. Epstein, S. S; et al. *Toxicol. Appl. Pharmacol.* 1972, **23**, 288
24. *Zeitschrift fur Krebsforschung* 1967, **697**, 103
25. *IARC Monograph* 1987, **Suppl. 7**, 58
26. Albert, R. E; et al. *Carcinogenesis (London)* 1991, **12**(7), 1273-1280
27. Carlson, G. P; et al. *Environ. Health Perspect.* 1986, **68**, 53-60
28. Bieniek, G; et al. *Med. Pr.* 1989, **40**(3), 183-191, (Pol.) (*Chem. Abstr.* 1990, **113**, 196891k)
29. Pessah-Rasmussen, H; et al. *Pharmacol. Toxicol. (Copenhagen)* 1989, **65**(1), 78-80
30. *National Research Council. Drinking Water & Health* 1981, **4**, 257, National Academy Press, Washington, DC
31. Shu, H. P; et al. *Cancer Res.* 1983, **43**(2), 485-490
32. Hass, B. S; et al. *Mutat. Res.* 1979, **60**, 395
33. De Flora, S; et al. *Carcinogenesis* 1984, **5**, 505
34. Krewski, D; et al. *Environ. Sci. Res.* 1990, **39**, 343-352
35. Collins, J. F; et al. *Regul. Toxicol. Pharmacol.* 1991, **13**(2), 170-184
36. Forster, R. *Eval. Short-Term Tests Carcinog.* 1988, **1**, 1.45-1.56
37. *ECETOC Technical Report No. 30(4)* 1991, European Chemical Industry Ecology and Toxicology Centre, B-1160 Brussels

B93   2-Benzoquinone

CAS Registry No. 583-63-1
Synonyms 3,5-cyclohexadiene-1,2-dione; 1,2-benzoquinone; *o*-benzoquinone
Mol. Formula $C_6H_4O_2$ Mol. Wt. 108.10
Uses Oxidising agent.

Physical properties
M. Pt. 67-70°C.

Genotoxicity
Salmonella typhimurium TA100 without metabolic activation positive (1).

References
1. Nazar, M. A; et al. *Mutat. Res.* 1981, **89** (1), 45-55

B94   4-Benzoquinone

CAS Registry No. 106-51-4

Synonyms quinone; 2,5-cyclohexadiene-1,4-dione; 1,4-benzoquinone; 1,4-cyclohexadienedione; USAF P-220; NCI-C55845; *p*-benzoquinone

Mol. Formula $C_6H_4O_2$ **Mol. Wt.** 108.10

Uses Manufacture of dyestuffs and hydroquinone. Used in photography and the tannery industry. Oxidising agent.

Occurrence Occurs naturally in a variety of arthropods (1).

Physical properties

M. Pt. 113-115°C; **B. Pt.** sublimes; **Flash point** 38.93°C; **Specific gravity** 1.318 at 20°C; **Partition coefficient** log P_{ow} 0.20; **Volatility** v.p. 0.09 mmHg at 20°C; v. den. 3.7.

Solubility

Water: 10 g l^{-1} at 25°C. Organic solvent: ethanol, diethyl ether

Occupational exposure

US TLV (TWA) 0.1 ppm (0.4 mg m^{-3}); **US TLV (STEL)** 0.3 ppm (1.2 mg m^{-3}); **UK Long-term limit** 0.1 ppm (0.4 mg m^{-3}); **UK Short-term** limit 0.3 ppm (1.2 mg m^{-3}); **UN No.** 2587; **HAZCHEM Code** 2Z; **Conveyance classification** toxic substance; **Supply classification** toxic.

Risk phrases

Toxic by inhalation and if swallowed – Irritating to eyes, respiratory system and skin (R23/25, R36/37/38)

Safety phrases

In case of contact with eyes, rinse immediately with plenty of water and seek medical advice – After contact with skin, wash immediately with plenty of water (S26, S28)

Ecotoxicity

Fish toxicity

Exposure to 1 mg l^{-1} caused death in 1-2 hr in rainbow trout and steelhead trout, and in 3-4 hr in bridgelip sucker (2).

LC_{50} (96 hr) fathead minnow, rainbow trout 0.045-0.125 mg l^{-1} (3).

Invertebrate toxicity

Cell multiplication inhibition test, *Escherichia coli* 55 mg l^{-1}, *Scenedemus quadricauda* 6 mg l^{-1} (4).

Inhibited Calvin cycle in *Chlorella pyrenoidosa* (5).

EC_{50} (5-30 min) *Photobacterium phosphoreum* 2.09 mg l^{-1} Microtox test (6).

Bioaccumulation

Based an a log K_{ow} of 0.20, estimated bioconcentration factor is 0.84 (7).

Environmental fate

Nitrification inhibition
Inhibiton of photosynthesis of freshwater *Selenastrum capricornutum* 0.1 mg l^{-1} 37% ^{14}C fixation, 1 mg l^{-1} 17% ^{14}C fixation, 10 mg l^{-1} 7-13% ^{14}C fixation, 100 mg l^{-1} 1% ^{14}C fixation (8).

Degradation studies
Concentration of 0.2 mg l^{-1} inhibited degradation of glucose by *Pseudomonas fluorescens* and 55 mg l^{-1} inhibited degradation of glucose by *Escherichia coli* (9).
Fourteen strains of phenol-utilising bacteria from soil did not grow visibly when using 4-benzoquinone as a carbon source in an aqueous mineral salt media incubation period 5 days (10).

Abiotic removal
Estimated $t_{1/2}$ is 110 min with photochemically produced hydroxyl radicals in the atmosphere (11).
Combined atmospheric $t_{1/2}$ due to reaction with both hydroxyl radicals and ozone is 33 min (11).
Estimated $t_{1/2}$ for the vapour-phase reaction with ozone in the atmosphere is 48 min (11).
Photolysis of 4-benzoquinone in aqueous solution using natural sunlight and artifical light in excess of 290 nm has yielded unidentifiable polar products (12).
Based on its vapour pressure of 0.09 mmHg at 20°C and ability to sublime, volatilisation of solid particles from soil surfaces may be a significant transfer mechanism (13).

Absorption
Estimated soil adsorption coefficient of 30 indicates high mobility in soil (7).

Mammalian and avian toxicity

Acute data
LD$_{50}$ oral rat 130 mg kg^{-1}.
LD$_{50}$ intravenous rat 25 mg kg^{-1} (14).
LD$_{50}$ intraperitoneal mouse 8.5 mg kg^{-1} (15).

Carcinogenicity and long-term effects
No adequate evidence for carcinogenicity to humans, inadequate evidence for carcinogenicity to animals, IARC classification group 3 (16).

Metabolism and pharmacokinetics
Readily absorbed from gastrointestinal tract and subcutaneous tissues, partially excreted unchanged in urine, but bulk is eliminated in conjugation with hexuronic, sulfuric and other acids (17).

Irritancy
Exposure to 0.1 ppm caused irritation to human eye (duration unspecified) (18).

Genotoxicity
Salmonella typhimurium TA100 with metabolic activation positive (19).
Treatment of HL-60 cells produced DNA adducts (0.05-10 adducts per 10^7 nucleotides) (20).
In vitro human lymphocytes resulted in 2- and 5-fold increases in micronuclei (21).

Any other adverse effects

Reported to cause dermatitis in humans. Inhibitor of sulfhydryl enzymes (22).
Readily absorbed from the gastroenteric tract and subcutaneous tissues, but no
systemic poisonings reported in humans (23).

Poisoning (species unspecified) caused convulsions, breathing difficulties and death
due to medullary paralysis. Lung damage from excretion into alveoli and effects on
haemoglobin also played a role in death from asphyxia (24).

Any other comments

Physico-chemical properties, human health effects and experimental toxicology
reviewed (25).

References

1. Harrison, V. J; et al. *J. Chromat.* 1967, **31**, 258
2. *Fish Toxicity Screening Data. Part 1. Lethal Effects of 964 Chemicals upon Steelhead Trout and Bridgelip Sucker* 1989, EPA560/6-89-001, Washington, DC
3. Degrave, G. M; et al. *Arch. Environ. Contam. Toxicol.* 1980, **9**, 557
4. Heinck, F; et al. *Les Eaux Residuaires Industrielles* 1970
5. Pristavn, N. *Proc. 3rd Int. Cong. Photosynthesis* 1974, 1541
6. Kaiser, K. L. E; et al. *Water Pollut. Res. J. Canada* 1991, **26**(3), 361-431
7. Lyman, W. J; et al. *Handbook of Chemical Property Estimation Methods. Environmental Behaviour of Organic Compounds* 1982, McGraw-Hill, New York
8. Giddings, J. M. *Bull. Environ. Contam. Toxicol.* 1979, **23**, 360
9. Bringmann, G; et al. *GWF-Wasser/Abwasser*, 1960, **81**, 337
10. Kramer, N; et al. *Arch. Biochem. Biophys.* 1950, **26**, 401-405
11. *GEMS; Graphical Exposure Modeling System. Fate of Atmospheric Pollutants Data Base* 1985, Office of Toxic Substances, USEPA, Washington, DC
12. Ware, G. W; et al. *Arch. Environ. Contam. Toxicol.* 1980, **9**, 135-146
13. *IARC Monograph* 1977, **15**, 255
14. Woodward, G; et al. *Fed. Proc.* 1949, **8**, 348
15. *Biochemical Pharmacology* 1963, **12**, 885
16. *IARC Monograph* 1987, **Suppl. 7**, 71
17. Clayton, G. D; et al. *Patty's Industrial Hygiene and Toxicology* 3rd ed., 1981-1982, **2A, 2B, 2C**, 2593, John Wiley Sons, New York
18. Henschler, D. *Toxikologysch-Arbeitshedizinische Begrundung von Mak-Werte* Verlag Chemie
19. Koike, N; et al. *Sangyo Igaku* 1988, **30**(6), 475-480
20. Levay, G; et al. *Carcinogenesis (London)* 1991, **12**(7), 1181-1186
21. Yager, J. W; et al. *Cancer Res.* 1990, **50**(2), 393-399
22. Brandt, H. J. *Phenol. Abwasser/Abwasserphenol. eine Monographlische Studie* 1958, Wiss Abhalt Akademie Verlag, Berlin
23. Gosselin, R. E; et al. *Clinical Toxicology of Commerical Products* 5th ed., 1984, Williams & Wilkins, Baltimore, MD
24. Liu, S. *Biochem. Z.* 1928, **195**, 248
25. *ECETOC Technical Report No. 30(4)* 1991, European Chemical Industry Ecology and Toxicology Centre, B-1160 Brussels

B95   4-Benzoquinone monophenylhydrazone

CAS Registry No. 1689-82-3

Synonyms 4-(phenylozo)phenol; *p*-benzeneazophenol; C.I. Solvent Yellow 7; 4-hydroxyazobenzene

Mol. Formula $C_{12}H_{10}N_2O$ Mol. Wt. 198.23

Uses Dyestuff.

Physical properties

M. Pt. 155-157°C; B. Pt. $_{20}$ 220-230°C.

Solubility
Organic solvent: diethyl ether

Ecotoxicity

Fish toxicity
LC_{50} (96 hr) fathead minnow 1.1 mg l^{-1} (1).

Invertebrate toxicity
EC_{50} (30 min) *Photobacterium phosphoreum* 0.927 mg l^{-1} Microtox test (2).

Mammalian and avian toxicity

Acute data
LD_{50} intraperitoneal mouse 75 mg kg^{-1} (3).

Sub-acute data
Oral Sprague-Dawley rats (9 month) $\leq$ 53 mg kg^{-1} no adverse effects reported (4).

Carcinogenicity and long-term effects
No adequate evidence for carcinogenicity to humans, inadequate evidence for carcinogenicity to animals, IARC classification group 3 (5).

Metabolism and pharmacokinetics
In mice metabolised to 2-acetamidophenol, 4- acetamidophenol, conjugated 4-aminophenol, conjugated 2- aminophenol. Doses of 500 and 700 mg, which did not produce toxic effects, were completely absorbed from the diet of rabbits and were excreted mainly in the urine as glucuronide (6).

Any other comments
Experimental toxicology and human health effects reviewed (7).

References

1. Holcombe, G. W; et al. *Environ. Pollut. (Series A)* 1984, **35**, 367-381
2. Kaiser, K. L. E; et al. *Water Pollut. Res. J. Canada* 1991, **26**(3), 361-431
3. *NTIS Report* AD 691-490, Nat. Tech. Inf. Ser., Springfield, VA

4. Miller, J. A; et al. *J. Exper. Med.* 1948, **87**, 139
5. *IARC Monograph* 1987, **Suppl. 7**, 64
6. Smith, J. N; et al. *Biochem. J.* 1951, **48**, 546
7. *ECETOC Technical Report No. 30(4)* 1991, European Chemical Industry Ecology and Toxicology Centre, B-1160 Brussels

B96   Benzothiazole

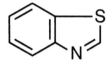

CAS Registry No. 95-16-9
Synonyms benzosulfonazole; 1-thia-3-azaindene; 0-2857; USAF EK-4812
Mol. Formula $C_7H_5N_1S$ **Mol. Wt.** 135.19
Uses In organic synthesis.
Occurrence In flowers and fruit.

Physical properties

M. Pt. 2°C; **B. Pt.** 233-235°C; **Specific gravity** d_4^{20} 1.246.

Solubility
Organic solvent: ethanol, carbon disulfide

Ecotoxicity

Invertebrate toxicity
EC_{50} (5-30 min) *Photobacterium phosphoreum* 1.81-1.87 mg l⁻ Microtox test (1).

Effects to non-target species
LD_{50} oral redwing blackbird 96 mg kg^{-1} (2).

Environmental fate

Degradation studies
Benzothiazole is oxidised by activated sludge. The sulfur and nitrogen moieties are converted to sulfate and ammonium ions (3).

Mammalian and avian toxicity

Acute data
LD_{50} oral mouse 900 mg kg^{-1} (4).
LD_{50} intravenous mouse 95 mg kg^{-1} (5).
LD_{50} intraperitoneal mouse 100 mg kg^{-1} (6).

Any other comments

Provides fungal resistance to cedar, red pine and beech trees (7).

References
1. Kaiser, K. L. E; et al. *Water Pollut. Res. J. Canada* 1991, **26**(3), 361-431

2. Schafer, E. W; et al. *Arch. Environ. Contam. Toxicol.* 1983, **12**(3), 335
3. Mainprize, J. *J. Appl. Bacteriol.* 1976, **40**(3), 285
4. *Drug Chem. Toxicol.* 1980, **3**, 249
5. Domino, E. F; et al. *J. Pharmacol. Expt. Ther.* 1952, **105**, 486
6. *NTIS Report* AD 277-689, Nat. Tech. Inf. Ser., Springfield, VA
7. Nishimoto, K; et al. *Jpn. Kokai Tokkyo Koho J. P.* 1961, **41**, 501, [86 41,501] (*Chem. Abstr.* **104**, 226568u)

B97   Benzo[*b*]thiophene

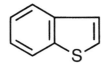

CAS Registry No. 95-15-8
Synonyms benzothiofuran; 1-benzothiophene; thionaphthene; thionaphthalene
Mol. Formula C_8H_6S **Mol. Wt.** 134.20
Uses Manufacture of pharmaceuticals, thio-indigo.
Occurrence In lignite tar.

Physical properties
M. Pt. 29-32°C; **B. Pt.** 221-222°C; **Partition coefficient** log P_{ow} 3.12.

Solubility
Organic solvent: ethanol, acetone, diethyl ether, benzene

Ecotoxicity

Fish toxicity
Artemia salina egg hatchability 0% at 500 ppm (1).

Any other comments
Inhibition of photosynthesis of a freshwater, nonaxenic unialgal culture of *Selenastrum capricornutum* at 1% saturation, 89% [14]C fixation reported, at 10% saturation ^{14}C fixation was 82% [14]C fixation and at 100% saturation 4% [14]C fixation occurred (2).

References
1. Kuwabara, K; et al. *Seitai Kaguku*, 1984, **7**(2), 33, (Jap.) (*Chem. Abstr.* **104**, 63489c)
2. Giddings, J. M. *Bull. Environ. Contam. Toxicol.* 1979, **23**, 360

B98   1,2,3-Benzotriazole

CAS Registry No. 95-14-7

Synonyms 1*H*-benzotriazole; 1,2-aminozophenylene; azimidobenzene; benzene azimide; NCI-C03521; U-6233

Mol. Formula $C_6H_5N_3$ **Mol. Wt.** 119.13

Uses Anticorrosive in metal working.

Physical properties

M. Pt. 95-97°C; **B. Pt.** $_{15}$ 205°C.

Solubility

Water: 20 g l^{-1}. Organic solvent: ethanol, benzene, toluene, chloroform

Environmental fate

Degradation studies

1,2,3-Benzotriazole was biologically stable as a source of organic carbon for activated sludge inoculum (1).

Mammalian and avian toxicity

Acute data

LD_{50} oral rat 500 mg kg^{-1} (2).

LD_{50} intravenous mouse 240 mg kg^{-1} (3).

LD_{50} intraperitoneal mouse 860 mg kg^{-1} (4).

Carcinogenicity and long-term effects

National Toxicology Program tested rats and mice via feed. Equivocal results in ♂ and ♀ rats and ♀ mice, negative results in ♂ mice (5).

Genotoxicity

Escherichia coli without metabolic activation positive (6).

Any other comments

Human health effects and experimental toxicology reviewed (7).

References

1. Pitter, P; et al. *Sb. Vys. Sk. Chem- Technol. Praze* 1978, **F22**, 93
2. *National Academy of Sciences, National Research Council* 1953, **5**, 22
3. *J. Pharmacol. Exp. Ther.* 1952, **105**, 486
4. Belen'Kaya, N. M; et al. *Pharm. Chem. J.* 1978, **12**, 1277
5. *National Toxicology Program, Research and Testing Div.* 1992, Report No. Tr-088, NIEHS, Research Park, NC
6. McGregor, D. B; et al. *Environ. Mutagen.* 1985, **7**(Suppl. 5), 1
7. *ECETOC Technical Report No. 30(4)* 1991, European Chemical Industry Ecology and Toxicology Centre, B-1160 Brussels

B99   Benzotrichloride

CCl₃

CCl$_3$

CAS Registry No. 98-07-7
Synonyms α,α,α-trichlorotoluene; (trichloromethyl)benzene; phenylchloroform;
benzenyltrichloride; benzoic trichloride; benzyl trichloride
Mol. Formula $C_7H_5Cl_3$ **Mol. Wt.** 195.48
Uses In dyestuff chemistry.

Physical properties

M. Pt. −3 to −5°C; **B. Pt.** $_{15}$ 90-91°C; **Specific gravity** d_4^{20} 1.3756.

Solubility
Organic solvent: ethanol, benzene, diethyl ether

Occupational exposure

UN No. 2226; **HAZCHEM Code** 2X; **Conveyance classification** corrosive
substance; **Supply classification** harmful.

Risk phrases
Harmful by inhalation (R20)

Safety phrases
Avoid contact with skin and eyes (S24/25)

Ecotoxicity

Fish toxicity
LC_{50} (48 hr) golden orfe 4140 mg l^{-1} (1).

Invertebrate toxicity
Cell multiplication inhibition test, *Pseudomonas putida* >100 mg l^{-1}, *Entosiphon
sulcatum* 56 mg l^{-1}, *Uronema parduczi* >80 mg l^{-1} (2,3).
EC_{50} (48 hr) *Daphnia magna* >100 mg l^{-1} (4).
EC_{50} (5-30 min) *Photobacterium phosphoreum* 17.8 mg l^{-1} Microtox test (5).

Bioaccumulation
Based on a log K_{ow} of 2.92, bioconcetration factor is estimated to be 98. Due to rapid
hydrolysis in water, accumulation in aqautic organisms is not expected to occur (6).

Abiotic removal
Hydrolysed in the presence of moisture to form benzoic acid and hydrochloric acid (7).
At 25°C, hydrolytic $t_{1/2}$ is 11 sec at 5.1°C $t_{1/2}$ is 3 min (8,9).
Virtually no adsorption occurs above 290 nm, therefore direct photolysis in the
environment is not expected to occur (10).
Reacts with photochemically produced hydroxyl radicals in the atmosphere, estimated
$t_{1/2}$ 2 day (11).
Due to rapid hydrolysis in water, volatilisation is not expected to be important.

Evaporation from dry surfaces may be expected due to its fuming in air and estimated vapour pressure of 0.23 mmHg at 20°C (12,7).

Mammalian and avian toxicity

Acute data
LD_{50} oral rat 6000 mg kg^{-1} (13).

Sub-acute data
LD_{50} (14 day) oral mouse 770 mg kg^{-1} (14).
LD_{50} (14 day) dermal rabbit 4 g kg^{-1} (14).

Carcinogenicity and long-term effects
Inadequate evidence for carcinogenicity to humans, sufficient evidence for carcinogenicity in animals, IARC classification group 2B (15,16).

Irritancy
Benzotrichloride has been reported to be highly irritating to skin and mucous membranes (17).
Dermal rabbit (24 hr) 10 mg caused severe irritation and 50 µg instilled into rabbit eye caused severe irritation (18).

Genotoxicity
Salmonella typhimurium TA98, TA100, TA1535 and *Escherichia coli* WP2 hcr with and without metabolic activation positive (17).
Bacillus subtilis differential toxicity assay without metabolic activation positive (17, 19).

Any other adverse effects to man
The standard mortality ratios for all causes, i.e. all cancers and all non-cancers in 1961-1984 were 138, 163 and 129, respectively, for a group of 951 workers exposed to benzotrichloride and chlorinated toluenes in England and Wales. Statistically significant excesses were found for lung cancer and for Hodgkin's disease (standardised mortality ratios 180 and 741, respectively). However, 95% confidence intervals were large and the authors conclude that risks did not approach statistical significance, but that chlorinated toluenes should continue to be considered potential carcinogens until further research is carried out (20).

Any other comments
Physico-chemical properties, human health effects, experimental toxicology, workplace experience, epidemiology and environmental effects reviewed (21).

References
1. Juhki, I; et al. *Z. Wasser/Abwasser Forsch.* 1978, **11**, 161-164
2. Bringmann, G; et al. *Water Res.* 1980, **14**, 231
3. Bringmann, G; et al. *Z. Wasser/Abwasser Forsch* 1980, **1**, 26
4. Bringmann, V. G; et al. *Z. Wasser/Abwasser Forsch.* 1977, **10**(5), 161-166
5. Kaiser, K. L. E; et al. *Water Pollut. Res. J. Canada* 1991, **26**(3), 361-431
6. Lyman, W. J; et al. *Handbook of Chemical Property Estimated Methods* 1982, 5.4, McGraw-Hill, New York
7. *The Merck Index* 11th ed., 1989, Merck Co. Inc., Rahway, NJ
8. Mabey, W; et al. *J. Phys. Chem. Ref. Data* 1978, **7**, 383

9. Laughton, P. M; et al. *Can. J. Chem.* 1959, **37**, 1491
10. *Sadtler Standard UV Spectra No. 380* 1966, Sadtler Research Lab., Philadelphia, PA
11. *GEMS: Graphical Exposure Mgdeling System. Fate of Atmospheric Pollutants Data Base*, 1986, Office of Toxic Substances, USEPA, Washington, DC
12. Weber, R. C; et al. *Vapour Pressure Distribution of Selected Organic Chemicals* 1981, EPA 600/2-81-021
13. Smyth, H. F; et al. *Arch. Ind. Hyg. Occup. Med.* 1951, **4**, 119
14. *USEPA Chemical Hazard Information Profile: Benzotrichloride* 1982, 7, Washington, DC
15. *IARC Monograph* 1982, **29**, 49, 65, 73
16. *IARC Monograph* 1987, **Suppl. 7**, 148
17. Yasuo, K; et al. *Mutat. Res.* 1978, **58**, 143
18. *AMA Arch. Ind. Hyg. Occup. Med.* 1951, **4**, 119
19. *IARC Monograph* 1987, **Suppl. 6**, 101
20. Sorahan, T; et al. *Br. J. Ind. Med.* 1989, **46**(6), 425-427
21. *ECETOC Technical Report No. 30(4)* 1991, European Chemical Industry Ecology and Toxicology Centre, B-1160 Brussels

B100   Benzotrifluoride

CF₃

CAS Registry No. 98-08-8
Synonyms α,α,α-trifluorotoluene; (trifluoromethyl)benzene; phenylfluoroform; benzenyl fluoride; benzylidyne fluoride; ω-trifluorotoluene
Mol. Formula $C_7H_5F_3$ **Mol. Wt.** 146.11
Uses In dyestuff chemistry. Manufacture of substituted benzotrifluorides containing an ethylenic group. Used in dielectric fluids such as transformer oils.

Physical properties

M. Pt. –29.05°C; **B. Pt.** 104°C; **Flash point** 12°C;
Specific gravity d^{20} 1.1886; **Volatility** v.p. 4.0×10^{-5} mmHg; v. den. 5.0.
Solubility
Organic solvent: acetone, benzene

Occupational exposure

UN No. 2338; **HAZCHEM Code** 2YE; **Conveyance classification** flammable liquid; **Supply classification** highly flammable.

Risk phrases
Highly flammable (R11)

Safety phrases
Keep away from sources of ignition – No Smoking – Do not breathe vapour (S16, S23)

Ecotoxicity

Invertebrate toxicity
EC_{50} (5-30 min) *Photobacterium phosphoreum* 32.0 mg l^{-1} Microtox test (1).

Effects to non-target species
LD_{Lo} subcutaneous frog 870 mg kg^{-1} (2).

Mammalian and avian toxicity

Acute data
LD_{50} oral mouse, rat 10-15 g kg^{-1} (3).
LC_{50} (4 hr) inhalation rat 71 g m^{-3} (4.).
LC_{50} (2 hr) inhalation mouse 92 g m^{-3} (4.).
LD_{50} intraperitoneal mouse 100 mg kg^{-1} (4).

Any other comments

Physico-chemical properties, human health effects and experimental toxicology reviewed (5).

References

1. Kaiser, K. L. E; et al. *Water Pollut. Res. J. Canada* 1991, **26**(3), 361-431
2. *Naunym-Schmiedekery's Arch. f. Exp. Path. u. Pharm.* 1928, **130**, 250
3. *Toksikologia Novykh Pronyshlennylch Klinmicheskikh Veshchestr (Toxicology of New Industrial Chemical Sciences)* 1968, **10**, 131
4. *NTIS Report* AD 277-689, Nat. Tech. Inf. Ser., Springfield, VA
5. *ECETOC Technical Report No. 30(4)* 1991, European Chemical Industry Ecology and Toxicology Centre, B-1160 Brussels

B101 Benzoximate

CAS Registry No. 29104-30-1
Synonyms Benzomate; Citrazon; Azomate; Artaban; Acarmate; ethyl*O*-benzol-3-chloro-2,6-dimethoxybenzohydroximate
Mol. Formula $C_{18}H_{18}ClNO_5$ **Mol. Wt.** 363.80
Uses A non-systemic acaricide.

Physical properties

M. Pt. 73°C.

Solubility
Organic solvent: benzene, dimethyl formamide, xylene, *n*-hexane

Ecotoxicity

Fish toxicity
LC_{50} (48 hr) carp 1.75 mg l^{-1} (1).

Mammalian and avian toxicity

Acute data
LD_{50} oral rat >10 g kg^{-1} (1-3).
LD_{50} percutaneous rat >15 g kg^{-1} (1).
LD_{50} intraperitoneal rat, mouse 4220-4260 mg kg^{-1} (1-3).

Carcinogenicity and long-term effects
In 2 yr feeding trials, rats receiving up to 400 mg kg^{-1} diet
showed no ill effects (1).

References

1. *The Agrochemicals Handbook* 2nd ed., 1987, RSC, London
2. World Health Organisation, 1986-1987, VBC/86.1, 19
3. *Special Publication of the Entomological Society of America* 1978, **1**, 23

B102   Benzoyl chloride

CAS Registry No. 98-88-4
Synonyms benzenecarbonyl chloride; benzoic acid chloride; α-chlorobenzaldehyde
Mol. Formula C_7H_5ClO **Mol. Wt.** 140.57
Uses Used in acylation reactions. Manufacture of benzoyl peroxide. Production of
dyestuff intermediates, drugs and agrochemicals. Inorganic analysis for preparation of
benzoyl derivatives for identification purposes.

Physical properties

M. Pt. −1.0°C; **B. Pt.** 197.2°C; **Flash point** 88°C;
Specific gravity d_4^{25} 1.207; **Volatility** v.p. 3.8×10^{-7} mmHg at 20°C 0.5; v. den. 4.9.

Solubility
Organic solvent: miscible with diethyl ether, benzene, carbon disulfide, dimethyl
sulfoxide

Occupational exposure

UN No. 1736; **HAZCHEM Code** 2X; **Conveyance classification** corrosive
substance; **Supply classification** corrosive.

Risk phrases
Causes burns (R34)

Safety phrases
In case of contact with eyes, rinse immediately with plenty of water and seek medical advice (S26)

Ecotoxicity

Fish toxicity
LC_{50} (24-96 hr) fathead minnow 43-35 mg l^{-1} (1).

Invertebrate toxicity
LC_{50} (96 hr) *Palomonetes pugio* 180 mg l^{-1} (2).
EC_{50} (5-30 min) *Photobacterium phosphoreum* 12.2 mg l^{-1} Microtox test (3).

Bioaccumulation
Due to the rapid rate of hydrolysis, bioconcentration is not expected to be an important factor (4).

Environmental fate

Degradation studies
Hydrolysis is so rapid that biodegradation is not expected to be an important factor in the fate of benzoyl chloride (4).

Abiotic removal
Using photochemically produced hydroxyl radicals, estimated atmospheric $t_{1/2}$ is 2-10 days (4).
Hydrolytic $t_{1/2}$ 16 secondary products detected benzoic acid and hydrochloric acid (5,6).

Absorption
Due to the rapid rate of hydrolysis, benzoyl chloride is not expected to persist in the soil (4).

Mammalian and avian toxicity

Acute data
LD_{50} oral rat 2460 mg kg^{-1} (7).
LC_{50} (2 hr) inhalation rat 1870 mg m^{-3} (8).
TC_{Lo} (1 min) inhalation human 2 ppm, affects sense organs. Causes olfactory effects and damage to lungs, thorax and respiration (8).
TD_{Lo} dermal mouse 9200 mg kg^{1} (9).
LD_{50} dermal rabbit 790 mg kg^{-1} (7).

Carcinogenicity and long-term effects
Inadequate evidence for carcinogenicity to humans, inadequate evidence for carcinogenicity to animals, IARC classification group 3 (10).
Dermal mouse (50 wk intermittently) caused equivocal tumourigenic effects to lungs, thorax or respiration system (9,11).

Genotoxicity

Salmonella typhimurium TA100 with and without metabolic activation negative.
Escherichia coli WP2 hcr with and without metabolic activation positive (12).

Any other adverse effects to man

Six cases of respiratory cancer reported among workers in 2 small factories where benzoyl chloride and its chlorinated precursors were produced. Epidemiologic data provided limited evidence that employment in the production of chlorinated toluenes presented carcinogenic risk to man (13).

Legislation

Limited under EC Directive on Drinking Water 80/778/EEC. Organochlorines: guide level 1 μg l^{-1}; Haloform concentrations must be as low as possible (14).

Any other comments

Physico-chemical properties, human health effects, experimental toxicology and workplace experience reviewed (15).
Corrosive. Lachrymatory. Decomposes violently with water.

References

1. Verschueren, K. *Handbook of Environmental Data of Organic Chemicals* 2nd ed., 1983, 282, Van Nostrand Reinhold, New York
2. Curtis, M. W; et al. *Water Res.* 1979, **13**, 137
3. Kaiser, K. L. E; et al. *Water Pollut. Res. J. Canada* 1991, **26**(3), 361-431
4. *GEMS: Graphical Exposure Modeling System. Fate of Atmospheric Pollutants (FAP)* 1985, Office of Toxic Substances, USEPA, Washington, DC
5. Mabey, W; et al. *J. Phys. Chem. Ref. Data* 1978, **77**, 383-415
6. Morrison, R. T; et al. *Organic Chemistry* 3rd ed., 1973, 665
7. *USEPA Chemical Hazard Information Profile: Benzoyl Chloride* 1982, Washington, DC
8. *Handbook of Organic Industrial Solvents* 1961, **2**, 31
9. Yoshimura, H; et al. *Carcinogenicity (Chem. Abstr.* **106**, 80137v)
10. *IARC Monograph* 1987, **Suppl. 7**, 58
11. Gann, *Japanese Journal of Cancer Res.* 1981, **72**, 655
12. Yasou, K; et al. *Mutat. Res.* 1978, **58**(2-3), 143-150
13. *IARC Monograph* 1982, **29**, 49
14. *EC Directive Relating to the Quality of Water Intended for Human Consumption* 1982, 80/778/EEC, Office for Official Publications of the European Communities, 2 rue Mercier, L-2985 Luxembourg
15. *ECETOC Technical Report No. 30(4)* 1991, European Chemical Industry Ecology and Toxicology Centre, B-1160 Brussels

B103   Benzoyl peroxide

CAS Registry No. 94-36-0
Synonyms dibenzoyl peroxide; benzoyl superoxide; acetoxyl; benoxyl; debroxide; panoxyl
Mol. Formula $C_{14}H_{10}O_4$ **Mol. Wt.** 242.23

Uses Bleaching agent for flour, fats, oils, waxes. Drying agent for unsaturated oils. Pharmaceutical and cosmetic ingredient. Rubber vulcaniser without sulfur. Production of embossed vinyl flooring. Burnout agent for acetate yarns, radical reactions. Polymerisation catalyst.

Physical properties
M. Pt. 104-106°C (decomp.).

Solubility
Water: <1 g l^{-1}. Organic solvent: ethanol, diethyl ether, acetone, benzene, chloroform

Occupational exposure
US TLV (TWA) 5 mg m^{-3}; **UK Long-term limit** 5 mg m^{-3}; **UN No.** 2085 (pure or >52% with inert solid); 2087 (≤72% as a paste);2088 (>77% but <95% with water); 2089 (from 30% to maximum 52% with inert solid); 2090 (≤ 77% with water); **Supply classification** explosive and irritant.

Risk phrases
Extreme risk of explosion by shock, friction, fire or other sources of ignition – Irritating to eyes, respiratory system and skin (R3, R36/37/38)

Safety phrases
Keep container tightly closed, in a cool well ventilated place – Keep away from incompatible materials (consult manufacturer) – Take off immediately all contaminated clothing – Avoid shock and friction – Wear suitable gloves and eye/face protection (S3/7/9, S14, S27, S34, S37/39)

Mammalian and avian toxicity

Acute data
LD$_{50}$ oral rat 7710 mg kg^{-1} (1).
LD$_{Lo}$ intraperitoneal mouse 250 mg kg^{-1} (2).

Carcinogenicity and long-term effects
Inadequate evidence for carcinogenicity to humans, inadequate evidence for carcinogenicity to aminals, IARC classification group 3 (3).

Metabolism and pharmacokinetics
Approximately 50% of a dose was absorbed following topical application to the forearm of primates (4).
Percutaneous penetration and metabolism on human skin *in vivo* and *in vitro* on 5 patients with leg ulcers revealed absorption by the skin *in vitro* with conversion to benzoic acid principally in the dermis (5).

Irritancy
500 mg instilled into rabbit eye caused mild irritation within 24 hr (1).

Sensitisation
Some patients were sensitised after repeated applications for acne therapy (6,7).
A case of purpuric contact dermatitis has been reported (8).
A number of cases of skin reactions to formulations used in the treatment of acne have been described (9).
Contact reactions are not linked to susceptibility to skin irritants in atopic dermatitis patients, but tend to increase with time (10).

Any other adverse effects to man

Applications may produce an initial stinging effect (11).

Any other adverse effects

Applications may produce an initial stinging effect (11).

Any other comments

Hazards and legislation in France, recommendations for its storage and handling, and medical treatment in case of accidents reviewed (12).

Physico- chemical properties, human health effects, experimental toxicology, epidemiology, workplace experience and exposure reviewed (13,14).

References

1. Marhold, J. V. *Sbornik Vysledku Toxixologickeho Vysetreni Latik a Pripravtu* 1972, 52, Prague
2. *Yakk. Yoku* 1980, **31**, 855
3. *IARC Monograph* 1987, **Suppl. 7**, 58
4. *Am. Med. Assoc., AMA Dept. of Drugs, AMA Drug Evaluations* 5th ed., 1982, 1070, Am. Med. Assoc., Chicago, IL
5. Morsches, B; et al. *Arzneimittelforsch* 1982, **32**(3), 298-300
6. Poole, R. L; et al. *Arch. Dermatol.* 1970, **102**, 635
7. Eaglestein, H. *Arch. Dermatol.* 1968, **97**, 527
8. Van Joost, T; et al. *J. Am. Acad. Dermatol.* 1990, **22**, 359
9. *NIOSH Criteria for a Recommended Standard* US Dept. Health, Education & Welfare, June 1977
10. Lisi, P; et al. *Dermatologica* 1985, **171**, 1
11. *Martindale The Extra Pharmacopoeia* 29th ed., 1989, 916, The Pharmaceutical Press, London
12. *Dangerous Prop. Ind. Mater. Rep.* 1989, **9**(3), 59-67
13. *Cah. Notes Doc.* 1988, **130**, 173-176, (Fr.) (*Chem. Abstr.* **108**, 136871r)
14. *ECETOC Technical Report No. 30(4)* 1991, European Chemical Industry Ecology and Toxicology Centre, B-1160 Brussels

B104   Benzoylprop-ethyl

CAS Registry No. 22212-55-1

Synonyms *L*-alanine, *N*-benzoyl-*N*-(3,4-dichlorophenyl)ethyl ester; *N*-benzoyl-*N*-(3,4-dichlorophenyl)-*L*-alanine ethyl ester; propionic acid, 2-(*N*-benzoyl-*N*-(3,4-dichlorophenyl)amino ethyl ester; ethyl *N*-benzoyl-*N*-(3,4-dichlorophenyl)-2-aminopropionate; Enaven; Suffix

Mol. Formula $C_{18}H_{17}Cl_2NO_3$ **Mol. Wt.** 366.25

Uses A selective systemic herbicide.

Physical properties

M. Pt. 71°C (rhombic crystals);55°C (needles); **Volatility** v.p. 3.5×10^{-8} mmHg at 20°C.

Solubility
Water: 20 mg l^{-1} at 25°C. Organic solvent: ethanol, acetone

Ecotoxicity

Fish toxicity
LC_{50} (96 hr) rainbow trout, harlequin fish 2.2-5 mg l^{-1} (1).

Effects to non-target species
LD_{50} oral hen >1000 mg kg^{-1} (1).
Bees non-toxic (1).

Abiotic removal
Photochemically and hydrolytically stable range pH 3-6 (1).

Mammalian and avian toxicity

Acute data
LD_{50} oral rat, guinea pig, rabbit 1000- 1550 mg kg^{-1} (2,3).
LD_{50} oral mouse 716 mg kg^{-1} (4).
LD_{50} percutaneous rat >1000 mg kg^{-1}(3).

Sub-acute data
In 90-day feeding trials, rats receiving up to 1000 mg kg^{-1} diet and dogs 300 mg kg^{-1} diet showed no ill effects (1).

Any other comments

In plants, metabolism involves hydrolysis of the ester group to give benzoyl moiety which ultimately converts to a biologically inactive conjugate (5).

References

1. *The Agrochemicals Handbook* 2nd ed., 1987, RSC, London
2. *Guide to the Chemicals used in Crop Protection* 1973, **6**, 256
3. *Pesticide Index* 1976, **5**, 24
4. *Worksubstanzen der Pflanzenschutz und Schadlingsbekanpfungsmittel* 1971-1976, Werner Perkow, Berlin
5. Crayford, J. V; et al. *Pestic. Sci.* 1976, **1**, 559

B105   Benzphetamine

CAS Registry No. 156-08-1
Synonyms *N*,α-dimethy-*N*-(phenylmethyl)benzenethanamine;
N-benzyl-*N*,α-dimethylphenethylamine
Mol. Formula $C_{17}H_{21}N$ **Mol. Wt.** 239.36

Uses Anorexic used in obesity treatment.

Physical properties

B. Pt. $_{0.02}$ 127°C.

Solubility
Organic solvent: ethanol, methanol, chloroform, diethyl ether, acetone, benzene

Mammalian and avian toxicity

Acute data
LD_{50} intraperitoneal mouse 32 mg kg^{-1} (1).

Metabolism and pharmacokinetics
Methylbenzylamine derived from benzphetamine, could in the nitrosating environment of the gastrointestinal tract yield the carcinogenic compound methylbenzylnitrosamine (3).

In vitro metabolism studied using rat liver microsomal preparation. Five metabolites were isolated and identified as benzylamphetamine, 1-(p-hydroxyphenyl)-2-(N-methyl-N-benzylamino)propane, 1-(p- hydroxyphenyl)-2-(N-benzylamino)propane, methamphetamine and amphetamine. Their formation was catalysed by the microsomal mixed function oxidase system. *In vitro* metabolism was mediated by three different types of cytochrome P450 enzymes (4).

In vivo studies in rats showed the major metabolite formed by aromatic hydroxylation and N-demethylation, was 1- (p-hydroxyphenyl)-2-(N-benzylamino)propane. Of the 8 other metabolites obtained, one was methamphetamine. After administration, 40% of the dose was recovered as urinary metabolites during 3 days (5).

Nine metabolites and traces of the unchanged drug were excreted in the urine of human volunteers. The major metabolite was 1-(p- hydroxyphenyl)-2-(N-benzylamino) propane. Minor metabolites included methamphetamine, amphetamine and their hydroxylated products. Identified urinary metabolites excreted during 3 days following administation accounted for 30-44% of the dose (6).

Legislation

Controlled substance (stimulant) listed in the US Code of Federal Regulations (7).

Any other comments

Excessive use may lead to tolerance and physical dependence (8).

References

1. *J. Pharmacol.* 1986, **17**, 37
2. *Current Ther. Res.* 1960, **2**, 33
3. Akintonwa, D. A. A. *J. Theor. Biol.* 1986, **20**, 303
4. Inoue, T; et al. *Xenobiotica* 1983, **13**(4), 241-249
5. Niwaguchi, T; et al. *Xenobiotica* 1982, **12**(10), 617-625
6. Inoue, T; et al. *Xenobiotica* 1986, **16**(7), 691-698
7. *US Code of Federal Regulations* 1985, Title 21, Pat 1308.13
8. *Martindale The Extra Pharmcopoeia* 29th ed., 1989, The Pharmaceutical Press, London

B106 Benzthiazuron

CAS Registry No. 1929-88-0
Synonyms 1-(2-benzothiazoyl)-3-methylurea; BSI; ISO
Mol. Formula $C_9H_9N_3OS$ **Mol. Wt.** 207.26
Uses Herbicide.

Physical properties
M. Pt. 287°C (decomp.).

Solubility
Water: 12 mg l^{-1} at 20°C. Organic solvent: acetone, xylene

Occupational exposure
Supply classification harmful.

Risk phrases
Harmful by inhalation, in contact with skin and if swallowed (R20/21/22)

Safety phrases
Keep out of reach of children – Keep away from food, drink and animal feeding stuffs
(S2, S13)

Ecotoxicity
Fish toxicity
LC_{50} (96 hr) harlequin fish 400 mg l^{-1} (1).

Environmental fate
Degradation studies
$t_{1/2}$ in soil is 10-12 wk (2).

Mammalian and avian toxicity
Acute data
LD_{50} oral rat 1280 mg kg^{-1} (2).

Sub-acute data
60-day feeding trials rats at 130 mg kg^{-1} daily dose no adverse effects reported (2).

Genotoxicity
Environmental Protection Agency Genotox Program 1988, *Saccharomyces cerevisiae*
gene conversion negative (3).

Legislation
Limited under EC Directive on Drinking Water Quality 80/778/EEC. Pesticides and
related products: maximum admissible concentration 0.5 µg l^{-1} (4).

References

1. Tooby, T. E; et al. *Chem. and Ind.* 1975, **21**, 523-525
2. Worthing, C. R. (Ed.) *The Pesticide Manual* 1991, 9th ed., British Crop Protection Council, Farnham
3. *Chem. Pestic.* 1971, **277**, 71
4. *EC Directive Relating to the Quality of Water Intended for Human Consumption* 1982, 80/778/EEC, Office for Official Publications of the European Communities, 2 rue Mercier, L-2985 Luxembourg

B107   Benzyl acetate

CH$_2$O$_2$CCH$_3$

CAS Registry No. 140-11-4
Synonyms acetic acid, phenylmethyl ester; acetic acid, benzyl ester; benzyl ethanoate; α-acetoxytoluene
Mol. Formula C$_9$H$_{10}$O$_2$ **Mol. Wt.** 150.18
Uses In perfumery. In the manufacture of lacquers, polishes, printing ink and varnish removers. Solvent for cellulose acetate and cellulose nitrate.
Occurrence In jasmine, gardenia and other essential oils (1,2).

Physical properties

M. Pt. −51.5°C; **B. Pt.** 213°C; **Flash point** 102°C;
Specific gravity d$_4^{25}$ 1.05; **Volatility** v.p. 1.0×10^{-6} mmHg at 45°C; v. den. 5.2.

Solubility

Water: <1 mg ml^{-1} at 25°C. Organic solvent: dimethyl sulfoxide, ethanol, acetone

Ecotoxicity

Invertebrate toxicity

EC$_{50}$ (30 min) *Photobacterium phosphoreum* 4.54 mg l^{-1} Microtox test (3).

Environmental fate

Degradation studies

Readily biodegradable (4).

Mammalian and avian toxicity

Acute data

LD$_{50}$ oral mouse 830 mg kg^{-1} (5).
LD$_{50}$ oral rat, rabbit, guinea pig 2200-2490 mg kg^{-1} (6,7).
LC$_{Lo}$ (22 hr) inhalation mouse 1300 mg m^{-3} (7).

Sub-acute data

Oral B6C3F1 mice (13 wk) 3130 – 50,000 ppm in diet decreased absolute weight (increased relative weight) of reproductive organs (testis, epididymis and cauda epididymis). There was no effect on sperm motility, density or normality (11).

Carcinogenicity and long-term effects

No adequate data for carcinogenicity to humans, limited evidence for carcinogenicity to animals, IARC classification group 3 (8).

National Toxicology Program evaluation of benzyl acetate in rats by gavage. Equivocal evidence of carcinogenic activity in ♂ rats (target organs not specified), no evidence of carcinogenicity in ♀ rats (9).

Gavage F344 rats 0, 250 and 500 mg kg^{-1} and B6C3F1 mice 0, 500 and 1000 mg kg^{-1}. Dosed once daily, 5 days wk^{-1} for 103 wk. No evidence of carcinogenicity in ♀ rats. Increased incidence of acinar cell adenoma of the exocrine pancreas in ♂ rats. Increased incidence of hepatocellular neoplasms and squamous cell neoplasms of the forestomach in ♂ and ♀ rats (10).

Metabolism and pharmacokinetics

The major urinary metabolite in rats and mice was hippuric acid, other metabolites detected in minor quantities included benzoyl glucuronide, benzoic acid and benzylmercapturic acid (12).

Changes in minor routes of metabolism and excretion occur with age, but formation of hippuric acid from benzyl acetate is unaffected by aging (13).

Genotoxicity

Salmonella typhimurium TA1535, TA1537, TA97, TA98 and TA100 with and without metabolic activation negative (14).

Mouse lymphoma L5178Y cell forward mutation assay without metabolic activation negative, with metabolic activation positive (15).

Chinese hamster ovary cells with and without metabolic activation did not produce chromosome aberrations or sister chromatid exchange (16).

Any other adverse effects to man

In humans if ingested can cause gastrointestinal irritation with vomiting and diarrhoea. Skin, eye and respiratory tract irritant (17).

Any other comments

Experimental toxicology and human health effects reviewed (18).

References

1. Archlander, S. *Perfume and Flower Materials of Natural Origin* 1960, 313, Elizabeth, NJ
2. *Dictionary of Organic Compounds* 4th ed., Chapman & Hall, London, **1**, 178
3. Kaiser, K. L. E; et al. *Water Pollut. Res. J. Canada* 1991, **26**(3), 361-431
4. *JETOC No.5* 1991, Japan Chemical Industry Ecology-Toxicology and Information Center, Tokyo
5. *Gig. Sanit.* 1985, **50**, 17
6. *Food Cosmet. Toxicol.* 1964, **2**, 327
7. *J. Pharmacol. Exp. Ther.* 1984, **358**, 45
8. *IARC Monograph* 1987, **Suppl.7**, 58
9. *National Toxicology Program Research and Testing Division* 1992, Report No-TR-250, NIEHS, Research Triangle Park, NC 27709
10. Abdo, K. M; et al. *Toxicology* 1985, **37**, 159
11. Morrissey, R. E; et al. *Fundam. Appl. Toxicol* 1988, **11**, 343-358

12. Chidgey, M. A. J; et al. *Food Chem.Toxicol.* 1987, **257**, 521
13. McMahon, T. F; et al. *Drug Metab. Dispos.* 1989, **17**(5), 506-512
14. Mortelmans, K; et al. *Environ. Mutagen.* 1986, 8(Suppl. 7), 1-119
15. Caspary, W. J; et al. *Mutat. Res.* 1988, **196** 61-81
16. Gallaway, S. M; et al. *Environ. Mol. Mutagen.* 1987, **10**(Suppl.10), 1-175
17. *The Merck Index* 11th ed., 1989, Merck & Co. Inc., Rahway, NJ
18. *ECETOC Technical Report No. 30(4)* 1991, European Chemical Industry Ecology and Toxicoogy Centre, B-1160 Brussels

B108   Benzyl alcohol

CH$_2$OH

CAS Registry No. 100-51-6
Synonyms benzenemethanol; α-hydroxytoluene; phenylcarbinol; phenylmethanol; 2-hydroxytoluene
Mol. Formula C$_7$H$_8$O   **Mol. Wt.** 108.14
Uses Manufacture of other benzyl compounds. In perfumery and flavourings. Pharmaceutical aid. Bacteriostat. Solvent for inks. Surfactant.
Occurrence Constituent of jasmine, hyacinth, ylang ylang oils. Identified in wastewater effluent from photographic processing industry, paper mills, secondary effluent from wastewater treatment plants, and wastewater of a petrochemical plant (1-4).
Also found in a test waste incinerator and as a volatile flavour component of baked potatoes, cheese and roast nuts (5-8).

Physical properties

M. Pt. −15°C; **B. Pt.** 205°C; **Flash point** 100°C (closed cup); **Specific gravity** d$_4^{20}$ 1.045; **Partition coefficient** log P$_{ow}$ 1.10; **Volatility** v.p. 1 mmHg at 58°C; v. den. 3.72.

Solubility
Water: 35 g l^{-1} at 20°C. Organic solvent: ethanol, diethyl ether, chloroform, acetone, benzene, aromatic hydrocarbons

Occupational exposure

Supply classification harmful.

Risk phrases
Harmful by inhalation and if swallowed (R20/22)

Safety phrases
In case of contact with eyes, rinse immediately with plenty of water and seek medical advice (S26)

Ecotoxicity

Fish toxicity

LC_{50} (48-96 hr) fathead minnow 770-460 mg l^{-1}, static bioassay at 18-22°C (9).
LC_{50} (96 hr) bluegill sunfish 10 ppm, static bioassay in fresh water at 23°C with mild aeration after 24 hr (9).
LC_{50} (96 hr) tidewater silverside fish 15 ppm, static bioassay in synthetic seawater at 23°C with mild aeration after 24 hr (9).
Trout, bluegill sunfish and goldfish exposed to 5 ppm benzyl alcohol died within 2-6 hr (10).

Invertebrate toxicity

EC_{50} (5, 15, 30 min) *Photobacterium phosphoreum* 71.4 mg l^{-1} Microtox test (11).
EC_{50}(48 hr) *Daphnia magna* 400 mg l^{-1}, EC_{100} (48 hr) *Daphnia magna* 500 mg l^{-1} (12).
EC_{10} (48 hr) *Pseudomonas putida* 658 mg l^{-1} (12).
EC_{50} (24 hr) *Haematococcus pluvalis* 2600 mg l^{-1} (12).
Cell multiplication inhibition test, *Pseudomonas fluorescens* 350 mg l^{-1} (13).

Bioaccumulation

Calculated bioconcentration factor 4 (14).

Effects to non-target species

LD_{50} oral starling >100 mg kg^{-1} (15).

Environmental fate

Anaerobic effects

Initial concentration of 50 ppm benzyl alcohol underwent >75% degradation to carbon dioxide and methane within 8 wk using municipal sewage sludge inocula under anaerobic conditions (16).
Using sediment from anoxic salt marsh, 1080 mg l^{-1} benzyl alcohol underwent degradation to carbon dioxide and methane after a 2 month incubation period (17).

Degradation studies

ThOD 2.52 g g^{-1} DOC 0.78 g g^{-1} (18).
BOD_5 70% degradation under aerobic conditions with an acclimated mixed microbial culture (19).
At an initial concentration of 500 ppm ThOD (12 hr) 52%, 42% and 43% using settled sewage sludge acclimated to phenol, benzoic acid and catechol respectively under aerobic conditions (20).

Abiotic removal

Exposure of benzyl alcohol to sunlight for 4 hr in natural water did not produce any detectable oxidising species (detection limit 160 μg l^{-1}) showing photochemically induced oxidation did not occur within 4 hr (21).

Mammalian and avian toxicity

Acute data

LD_{50} oral rat 1230 mg kg^{-1} (22).
LC_{50} (2 hr) inhalation rat 200-300 ppm (23).
LD_{Lo} intravenous dog 50 mg kg^{-1} (24).
LD_{50} intraperitoneal mouse 650 mg kg^{-1} (25).

Carcinogenicity and long-term effects
National Toxicology Program evaluation of benzyl alcohol in rats and mice by gavage. No evidence of carcinogenicity (26).

Teratogenicity and reproductive effects
TD_{Lo} oral mouse (6-13 day pregnant) 6 g kg^{-1} (total dose) reduced birth weight and maternal weight gain (27).
No evidence of developmental toxicity (28).

Metabolism and pharmacokinetics
Oxidised to benzoic acid and excreted as hippuric acid (29).
Rabbits given 1 g benzyl alcohol subcutaneously eliminated 300-400 mg hippuric acid within 24 hr. Rabbits given 0.4 g kg^{-1} orally eliminated 65.7% as hippuric acid in the urine (30).

Irritancy
16 mg applied to human skin caused mild irritation within 24 hr (31).

Sensitisation
Human hypersensitivity to benzyl alcohol reported (32).

Genotoxicity
Salmonella typhimurium TA1535, TA1537, TA98, TA100 with and without metabolic activation negative (33).
Chinese hamster ovary cells chromosome aberration without metabolic activation negative, with metabolic activation positive; sister chromatid exchange with or without metabolic activation weakly positive (34).
Mouse lymphoma L5178Y cell forward mutation assay with or without metabolic activation equivocal responses (35).
Did not induce micronuclei in mice after single intraperitoneal injection (36).

Any other adverse effects
In humans prolonged exposure may cause lung damage, gastrointestinal disturbances and narcotic effects (37).
Can cause death in neonates. Systemic effects include central nervous system depression, respiratory distress and renal failure (38).
Aseptic meningitis observed following intrathecal administration of radio pharmaceuticals containing benzyl alcohol in preservative (39).

Any other comments
Gasoline exhaust content >0.1 to 47 ppm (40).
Toxicology reviewed (41,42).
Hygroscopic.

References
1. Dagon, T. J. *J. Water Pollut. Control Fed.* 1973, **45**, 2123-2135
2. Keith, L. H. *Environ. Sci. Technol.* 1976, **10**, 555-564
3. Ellis, D. D; et al. *Arch. Environ. Contam. Toxicol.* 1982, **11**, 373-382
4. Keith, L. H. *Sci. Total. Environ.* 1974, **20**, 1153-1159
5. James, R. H; et al. *J. Proc. APCA 77th annual meeting* 1984, paper 84-18.5
6. Coleman, E. C; et al. *J. Agric. Food Chem.* 1981, **29**, 42-48
7. Dumont, J. P; et al. *J. Agric. Food. Chem.* 1978, **26**, 364-367

8. Kinlin, T. E; et al. *J. Agric. Food. Chem.* 1972, **20**, 1021-1028
9. Verschueren, K. *Handbook of Environmental Data of Organic Chemicals* 2nd ed., 1983, Van Nostrand Reinhold, New York
10. *Toxicity of 3400 Chemicals to Fish* 1087, EPA560/6-87-002, Washington, DC
11. Kaiser, K. L. E; et al. *Water Pollut. Res. J. Canada* 1991, **26** (3),361-431
12. Bringmann, G; et al. *Wasser/Abwasser Forsch.* 1982, **15**(1-5), 1-6
13. Schafer, E. W; et al. *Arch. Environ. Contam. Toxicol.* 1983, **12**, 355
14. Urano, K; et al. *J. Haz. Mater.* 1986, **13**, 147
15. Hansch, C; el al.*Medchem. Project Issue No. 26*, 1985, Pomona College, Claremont, CA
16. Shelton, D. R; et al. *Appl. Environ. Microbiol.* 1984, **47**, 850-857
17. Balba, M. T. M; et al. *Biochem. Soc. Trans.* 1981, **9**, 230-231
18. McCloskey, S. E; et al. *J. Pharm. Sci.* 1986, **75**, 702
19. Baben, L; et al. *J. Ind. Microbiol.* 1987, **2**, 107-115
20. McKinney, R. E; et al. *Sew. Indust. Wastes* 1956, **28**, 547-557
21. Draper, W. M; et al. *Arch. Environ. Contam. Toxicol.* 1983, **12**, 121-126
22. Ouvel, W. A. *J. Water Pollut. Control Fed.* 1975, **47**(1)
23. *Toxicol. Appl. Pharmacol.* 1977, **18**, 60
24. Hardin, B. D; et al. *Teratogen. Carcinogen. Mutagen.* 1987, **7**, 29
25. *Food Cosmet. Toxicol.* 1964, **2**, 327, Pergammon Press, Oxford
26. *National Toxicology Program Research and Testing Division* 1992, Report No. TR-343, NIEHS, Research Triangle Park, NC27709
27. NIOSH *Experimental Toxicol. Branch Biomed. Behav. Sci.* 1986, *200-84-2753*, 1-59
28. Bignami, M. *Mutat. Res.* 1977, **46**, 395
29. *J. Pharmacol. Exp. Ther.* 1945, **84**, 358
30. *Patty's Industrial Hygiene and Toxicology* 3rd ed., 1981-82, Clayton, G. D. (Ed.), John Wiley, New York
31. Occupational Health Services Inc *Pestline* 1991, **1**, 117, Van Nostrand Reinhold, New York
32. Stimunes, E. *Arch. Derm.* 1984, **120**, 2000
33. Mortelmans, K; et al. *Environ. Mol. Mutagen.*1986, **7**(Suppl. 7), 1-119
34. Anderson, B. E; et al. *Environ. Mol. Mutagen.* 1990, **16**(Suppl. 18), 55-137
35. McGregor, D. B. *Environ. Mol. Mutagen.* 1988, **12**, 85-154
36. Hayashi, M; et al. *Food Chem. Toxicol.* 1988, **26**(6), 487-500
37. Grant, J. A; et al. *New Engl. J. Med.* 1982, **306**, 108
38. Anderson, C. W; et al. *Am. J. Obstet. Gynecol.* 1984, **148**, 344
39. Delano, F. H. *Toxicol. Appl. Pharmacol.* 1973, **25**(2), 153-156
40. Seizinger, E. D; et al. *J. Air. Pollut. Control. Assoc.* 1972, **22**(1)
41. *BIBRA Toxcity Profile* 1989, British Industrial Biological Research Association, Carshalton
42. *Cah. Notes Doc.* 1990, **140**, 665-668 (Fr.) (*Chem. Abst.* **114**, 233938z)

B109   Benzylamine

CAS Registry No. 100-46-9
Synonyms aminotoluene; benzenemethanamine; Moringine; phenylmethylamine
Mol. Formula C_7H_9N **Mol. Wt.** 107.16
Uses Chemical intermediate for dyestuffs, pharmaceuticals and polymers.

Physical properties

M. Pt. 10°C; **B. Pt.** 184-185°C; **Flash point** 60°C (closed cup); **Specific gravity** d_4^{19} 0.983; **Partition coefficient** log P_{ow} 1.09.

Solubility
Water: miscible. Organic solvent: ethanol, diethyl ether

Occupational exposure

Supply classification corrosive.

Risk phrases
Causes burns (R34)

Safety phrases
In case of contact with eyes, rinse immediately with plenty of water and seek medical advice (S26)

Ecotoxicity

Fish toxicity
Steelhead trout exposed to 6 mg l^{-1} died within 6-8 hr (1).

Invertebrate toxicity
Toxic dose *Daphnia magna* 60 mg l^{-1}. Toxic dose *Scenedesmus quadricauda* 6 mg l^{-1} (2).
EC_{50} (30 min) *Photobacterium phosphoreum* 17 mg l^{-1} Microtox test (3).

Environmental fate

Nitrification inhibition
Nitrosomonas sp. 100 mg l^{-1} inhibited ammonia oxidation by 26%, 50 mg l^{-1} inhibited 10%, 10 mg l^{-1} inhibited 0% (4).

Degradation studies
Pseudomonas fluorescens 400 mg l^{-1} inhibited glucose degradation. *Escherichia coli* >1000 mg l^{-1} inhibited glucose degradation (5).
Confirmed biodegradable (6).

Genotoxicity
Mouse lymphoma L5178Y cell forward mutation assay without metabolic activation negative (7).

Any other adverse effects to man
Metabolic inactivation of benzylamine in myometrium from women at term, with normal labour and with uterine inertia was studied. Benzylamine deamination was highest in uterine inertia. Possible pathogenic implications of the selective changes in the myometrial activity associated with uterine inertia are discussed (8).

Any other comments
General literature review of toxicity of aromatic amino and nitro compounds (9).
Corrosive. Lachrymatory. Incompatible with *N*-chlorosuccinimide.

References

1. *Fish Toxicity Screening Data. Part 2. Lethal Effects of 2014 Chemicals upon Sockeye Salmon, Steelhead Trout and Threespine Stickleback* 1989, EPA 560/6-89-001, PB89-156715, Washington, DC
2. Meinck, F; et al. *Les Eaux Residuaires Industrielles* 1970
3. Kaiser, K. L. E; et al. *Water Pollut. Res. J. Canada* 1991, **26** (3), 361-431
4. Hockenbury, M. R; et al. *J. Water Pollut. Control Fed.* 1977, **49** (5), 768-777
5. Bringmann, G; et al. *GWF- Wasser/Abwasser* 1960, **81**, 337
6. *MITI Report* 1984, Ministry of International Trade and Industry, Tokyo
7. McGregor, D. B; et al. *Environ. Mol. Mutagen.* 1988, **11** (4), 523-544
8. Matesha, E. I; et al. *Akush Ginekol. (Moscow)* 1988, **2**, 48-50, (Russ) (*Chem. Abstr*, **109**, 86737z)
9. Von Oettingen, W. F. *Public Health Bull.* 1941, **271**, 1-228

B110   Benzyl benzoate

CAS Registry No. 120-51-4
Synonyms Ascabin; benzoic acid benzyl ester; benzoic acid phenylmethyl ester; Benylate; benzylbenzenecarboxylate; Venzonate
Mol. Formula $C_{14}H_{12}O_2$ **Mol. Wt.** 212.25
Uses Scabicide, pediculicide. Component in dyestuffs, perfume fixatives. Solvent for cellulose acetate, nitrocellulose and artificial musk. Camphor substitute in celluloid and plastic pyroxylin compounds. Flavour for confectionary and chewing gum.
Occurrence Natural substance in the volatile bark oils of cinnamon (1).

Physical properties

M. Pt. 18-20°C; **B. Pt.** 323°C; **Flash point** 147°C;
Specific gravity d_4^{25} 1.118; **Partition coefficient** log P_{ow} 3.97;
Volatility v.p. 1.3 mmHg at 44°C; v. den 7.31.

Solubility
Organic solvent: miscible with ethanol, chloroform, diethyl ether

Occupational exposure
Supply classification harmful.

Risk phrases
Harmful if swallowed (R22)

Safety phrases
Avoid contact with eyes (S25)

Mammalian and avian toxicity

Acute data

LD_{50} oral guinea pig, mouse, rabbit, rat 1000-1700 mg kg^{-1} (3,4).
LD_{50} dermal rabbit 4000 mg kg^{-1} (4).

Metabolism and pharmacokinetics

Hydrolyses to benzyl alcohol and benzoic acid (6).

Irritancy

Irritant to eyes and mucous membranes and it may be an irritant to skin (7).

Sensitisation

Hypersensitivity reactions have been reported (7).
The dyestuffs residue associated with benzyl benzoate was detected in cloth samples.
The skin test of these cloth samples revealed irritating and sensitising effects (8).

Any other adverse effects to man

Blood pressure lowering capabilities (9).

Any other adverse effects

In experimental animals ingestion causes progressive incoordination, excitation, convulsions and death (10).
Expectorant, respiratory sedative and mild local anaesthetic (11).
Diuretic (12).

Any other comments

Toxicity and sensitisation effects of benzyl benzoate reviewed (13-15).
Physico-chemical properties, human health effects and experimental toxicology reviewed (16).

References

1. Nolen, G. A; et al. *Toxicol. Appl. Pharmacol.* 1975, **31**, 430
2. Beck, J. E. *Lancet* 1978, **I**, 8061, 444
3. Draize; et al. *J. Pharmacol. Exp. Ther.* 1948, **93**, 26
4. *Food Cosmet. Toxicol.* 1973, **11**, 1015
5. *J. Pharmacol. Exp. Ther.* 1945, **84**, 358
6. Gruber, C. M. *J. Lab. Clin. Med.* 1923, **9**, 15-33, 92-112
7. *Martindale The Extra Pharmacopoeia* 29th ed., 1989, The Pharmaceutical Press, London
8. Aoyama, M; et al. *Igaku to Seibutsugaku* 1988, **116**(2), 85-88, (Jap.) (*Chem. Abstr.* **108**, 217407w)
9. Lee, P. T; et al. *Res. Comm. Chem. Path. Pharmacol.* 1979, **23**(3), 597-609
10. Verrett, M. J; et al. *Ann. New York Acad. Sci.* 1969, **160**, 334
11. Occupational Health Services *Pestline* 1991, 118-119, Van Nostrand Reinhold, New York
12. Macht, D. I. *J. Pharmacol. Exp. Ther.* 1914, **13**, 509-511
13. Gruber, C. M. *J. Lab. Clin. Med.* 1924, **10**, 284-294
14. *BIBRA Toxicity Profile* 1989, British Industrial Biological Research Association, Surrey
15. Lehman, A. J. *Med. Bull.* 1956, **16**(3), 243-246 (reprint of Assoc. Food Drug Office US Quarterly Bull) 1955, **19**(3)
16. Opdyke, D. L. J. *Food Cosmet. Toxicol.* 1973, **11**, 1011-1081
17. *ECETOC Technical Report No. 30(4)* 1991, European Chemical Industry Ecology and Toxicology Centre, B-1160 Brussels

B111   Benzyl bromide

$$CH_2Br$$

CAS Registry No. 100-39-0
Synonyms (bromomethyl)benzene; bromophenylmethane; ω-bromotoluene;
α-bromotoluene; benzene, bromomethyl-
Mol. Formula C_7H_7Br **Mol. Wt.** 171.04
Uses In tear gas. Foaming and frothing agent. Intermediate in organic synthesis.
Filling for fire system sprinkler heads.

Physical properties

M. Pt. –4.0°C; **B. Pt.** 198°C; **Flash point** 86°C; **Specific gravity** d_0^{22} 1.438; **Partition
coefficient** log P_{ow} 2.92;
Volatility v. den. 5.8.

Solubility
Organic solvent: ethanol, diethyl ether

Occupational exposure

UN No. 1737; **HAZCHEM Code** 2X; **Conveyance classification** toxic substance;
Supply classification irritant.

Risk phrases
Irritating to eyes, respiratory system and skin (R36/37/38)

Safety phrases
Wear eye/face protection (S39)

Environmental fate

Abiotic removal
Estimated hydrolytic $t_{1/2}$ 79 min (1).
At high concentration, 297 mg l^{-1} benzyl bromide in methanol, absorbs ultraviolet
light >290 nm which suggests direct photolysis is unlikely (2).
Photochemical reaction with atmospheric hydroxyl radicals estimated $t_{1/2}$ 6-7 days (3).
Due to its high vapour pressure, volatilisation is expected to be fairly rapid from dry
soil surfaces (4).

Absorption
Estimated soil adsorption coefficients of 154-923 suggests benzyl bromides would
not be susceptible to significant leaching in soil and that adsorption to suspended
solids and sediments in water would be negligible (5).

Any other adverse effects
Narcotic (6).

Any other comments

Physico-chemical properties, human health effects and experimental toxicology reviewed (9).
Corrosive. Lachrymatory.

References

1. Mabey, W; et al. *J. Phys. Chem. Ref. Data* 1978, **7**, 383-415
2. *Sadtler Standard UV Spectra No. 660* 1966, Sadtler Research Lab., Philadelphia, PA
3. Atkinson, R. *Inter. J. Chem. Kinet.* 1987, **19**, 799-828
4. Hine, J; et al. *J. Org. Chem. 1975*, **40**, 292-298
5. Swann, R. L; et al. *Res. Rev.* 1983, **85**, 17-28
6. *The Merck Index* 1989, 11th ed., Merck & Co. Inc., Rahway, NJ
7. *ECETOC Technical Report No. 30(4)* 1991, European Chemical Industry and Toxicology Centre, B-1160 Brussels

B112   Benzyl bromoacetate

$CH_2CO_2CH_2Br$

CAS Registry No. 5437-45-6
Synonyms acetic acid, bromo-, benzyl ester; acetic acid, bromo-, phenylmethyl ester
Mol. Formula $C_9H_9BrO_2$ **Mol. Wt.** 229.08
Uses Plastics additive. Antimicrobial agent. Active ingredient in pesticides.

Physical properties

B. Pt. $_{22}$ 166-170°C; **Flash point** >110°C; **Specific gravity** 1.446.

Any other comments

Benzyl bromoacetate may be used under the US Federal, Food, Drug and Cosmetic Act as an antimicrobial preservative (1).

References

1. *Food and Drug Admin. Fed. Regist.* 1973, **38**(64), 8594, Washington, DC

B113   Benzyl butyl phthalate

$-CO_2CH_2-$
$CO_2(CH_2)_3CH_3$

CAS Registry No. 85-68-7
Synonyms BBP; 1,2-benzenedicarboxylic acid, butyl phenylmethyl ester; butyl benzyl phthalate; phthalic acid, benzyl butyl ester; Saniticiser 160

Mol. Formula $C_{19}H_{20}O_4$ **Mol. Wt.** 312.37

Uses Plasticiser for synthetic resins, chiefly polyvinylchloride and cellulose resins.

Occurrence In effluent from industrial and sewage plants (1,2).

Physical properties

M. Pt. –35°C; **B. Pt.** 370°C; **Flash point** 199°C (closed cup);

Specific gravity d_{25}^{25} 1.100; **Partition coefficient** log P_{ow} 4.91 (3); **Volatility** v.p. 8.6 $\times 10^{-6}$ mmHg at 20°C; v. den. 10.8.

Solubility

Water: 1.2 mg l^{-1}.

Occupational exposure

UK Long-term limit 5 mg m^{-3}.

Ecotoxicity

Fish toxicity

LC$_{50}$ (96 hr) bluegill sunfish, fathead minnow, sheepshead minnow, rainbow trout 1.7-5.3 mg l^{-1} (4).

LC$_{50}$ (24 hr, 96 hr) bluegill sunfish 62 mg l^{-1} and 43 mg l^{-1} (5).

Invertebrate toxicity

EC$_{50}$ (96 hr) *Selenastrum capricornutum, Skeletonema costatum* 110-170 μg l^{-1} (6).

Bioaccumulation

Bluegill sunfish (21 day) 9.73 mg l^{-1} bioconcentration factor 663 (7).

Uptake efficiency 42.2 % in English sole gills exposed to 20-250 mg l^{-1} for 3 hr (8).

Environmental fate

Anaerobic effects

Degraded >90% of 20 mg ml^{-1} in 1 wk under anaerobic conditions with 10% sludge (9).

Undiluted anaerobic sludge, neutralised 20 μg ml^{-1} in >7 days. In sludge diluted to 10% in an anaerobic salt medium, 76-103% of the phthalate ester carbon expected as methane was found as methane with 30 days. Degradation pathway apparently butyl benzyl phthalate forming monobutyl phthalate forming phthalic acid (9).

Degradation studies

Benzylbutyl phthalate is readily degraded in water and sediment, t$_{1/2}$ ≤2 days (10).

93% primary degradation rate and 5 mg-cycle addition rate in a semi-continuous activated sludge test (11).

Abiotic removal

Insoluble in water, tends to partition to soil, sediment and biota in aqueous environments (4).

Mammalian and avian toxicity

Acute data

LD$_{50}$ oral rat 2330 mg kg^{-1} (12).

LD$_{50}$ oral mouse 4170 mg kg^{-1} (12).

LD$_{50}$ intraperitoneal mouse 3160 mg kg^{-1} (13).

Sub-acute data

14 day dietary study in adult ♂ Fischer 344 rats at 0, 0.625, 1.25 and 5.0% benzyl butyl phthalate, 2.5% and 5% reduced total body, thymus, testis, epididymis, prostrate and seminal vesicle weights, and plasma tesosterone. Dose dependent atrophy of testis, prostrate and seminal vesicles at 2.5% and 5%. Atrophy of thymus and epididymis at 5% (14). Fed in diet of ♂ F334 rats at 25,000 ppm for 10 wk benzyl butyl phthalate caused aspermia (15).

Carcinogenicity and long-term effects

National Toxicology Program evaluation of benzyl butyl phthalate in feed. Positive ♀ rats, negative ♂ rats and ♂, ♀ mice (16).
Fed in diet at 0, 6000 and 12,000 ppm to rats and mice for 102-106 wk. Benzyl butyl phthalate did not induce tumours in ♂ or ♀ mice. Significantly increased incidence of myelomonocytic leukaemia in ♀ rats (17).

Teratogenicity and reproductive effects

Fed in diet of ♀ Sprague-Dawley rats on day 6 of pregnancy at 0, 0.5, 1.25, and 2%. No observed adverse effects level 0.5% for maternal and development toxicity. Significant maternal toxicity and minimal developmental toxicity at 1.25%. Significant maternal and developmental toxicity at 2%. Increased incidence of malformations reported (18).
Oral ♀ rats (0-20 day gestation) 0, 0.25, 0.5, 1.0 or 2%. No significant effect on preimplantation loss. Complete resorption of all embryos at 2%. No observable effect levels were 0.5% and 1% for maternal and embryo-foetal toxicity respectively. Teratogenic effects, cleft palate and fusion of sternebrae detected in foetuses of high dose animals (19).

Metabolism and pharmacokinetics

Male Fischer-344 rats were dosed with ^{14}C- labelled butyl benzyl phthalate (BPP) at 2, 20, 200, or 2000 mg kg orally or 20 mg kg^{-1} intravenously to detect the effects of dose on rates and routes of excretion. In 24 hr, 61-74% of the dose was excreted in the urine and 13-19% in the faeces at 2-200 mg kg^{-1}. At 2000 mg kg^{-1}, 16% of the ^{14}C was excreted in the urine and 57% in the faeces. Urinary ^{14}C was composed of monophthalate glucuronides derivatives (10-42% of the dose) and monophthalate glucuronides (2-21% of the dose). At 4 hr after intravenous administration of BBP (20 mg kg^{-1}), 53-58% of the dose was excreted in the bile of anaesthetised rats. BBP was not found in the bile, but monobutyl glucuronide and monobenzyl phthalate glucuronide (26 and 13% of the dose, respectively) and trace amts of free monoesters (2% of the dose). The half-lives of BBP, monophthalates, and total ^{14}C in blood (20 mg kg^{-1} intravenous) were 10 min, 5.9 hr, and 6.3 hr, respectively (20).

Irritancy

Rabbit skin (intact and abraded) 0.5 ml benzyl butyl phthalate held in continuous contact for 24 hr no irritation. Fifteen daily applications of undiluted benzyl butyl phthalate (dose unspecified) over 3 wk in 200 humans no primary irritation or sensitisation. 0.1 ml undiluted benzyl butyl phthalate instilled into rabbits' eyes caused slight irritation, subsiding within 48 hr (21).

Genotoxicity

Mouse lymphoma L5178Y cell forward mutation assay with or without metabolic activation negative (22).

Chinese hamster ovary cells without metabolic activation sister chromatid exchange equivocal (23).

Any other adverse effects to man

Exposure of dialysis patients to benzylbutyl phthalate can cause sodium wastage and polyuria, defect is resistant to arginine vasopression (24).

Any other adverse effects

In a 21-day feeding study, in F344 rats 25,000 ppm induced a moderate increase in peroxisomes in the liver (21).

Any other comments

Published research on toxicity and environmental effects of phthalic acid ester, since 1978, has been reviewed (25).

Potential occupational hazards, experimental toxicology and carcinogenicity reviewed (12,21,26,27).

Extensive studies have been carried out under a variety of laboratory conditions, including activated sludge, static flask, anaerobic microrganisms in river, lake and seawater (28-37).

References

1. Hites, R. A. *Natl. Conf. Munic. Sludge Manage 8th* 1979, 107-119
2. Staples, C. A; et al. *Environ. Tox. Chem.* 1985, **4**, 131-142
3. Howard, P. H. *Fate and Exposure Data for Organic Chemicals* 1990, **1**, 107-113, Lewis Publishers, Chelsea, MI
4. Gledhill, W. E; et al. *Environ. Sci. Technol.* 1980, **14**(3), 301-305
5. Buccafusco, R. J; et al. *Bull. Environ. Contam. Toxicol.* 1981, **26**, 446-452
6. *Studies on Health and Environmental Impacts of Selected Water Pollutants* 1978, EPA 68-01-4646
7. Barrow, M. E; et al. *Dynamics, Exposure, Hazard Assessment of Toxic Chemicals* 379-392, Ann Arbor Sci. Publ., MI
8. Boese, B. L. *Can. J. Fish. Aquat. Sci.* 1984, **41**, 1713-1718
9. Shelton, D. R; et al. *Environ. Sci. Technol.* 1984, **18** (2), 93-97
10. Adams, W. J; et al. *ASTM Spec. Tech. Publ.* 1988, **11**, 19-40
11. Saeger, V. W; et al. *Tech. Pap. Reg. Tech. Conf. Soc. Plast. Eng. Palisades Sect.* 1973, March 20-22, 105-113
12. *IARC Monograph* 1982, **29**, 193-201
13. *Environ. Health Perspect.* 1973, **4**, 3
14. Kluwe, W. M; et al. *Environ. Health Perspect.* 1982, **45**, 129-133
15. *Toxicologist* 1984, **4**, 136
16. *National Toxicoogy Program Research Testing Division* 1992, Report No. TR-213, NIEHS, Research Triangle Park, NC
17. Haseman, J. K; et al. *Environ. Mol. Mutagen.* 1990, **16** (Suppl.18), 15-31
18. *NTIS Report* 1989 NTP-89-246 (Order No. PB90-115346), Nat. Tech. Inf. Ser., Springfield, VA
19. Ema, M; et al. *J. Appl. Toxicol.* 1992, **12**(3), 179-183
20. Eigenberg, P. A; et al. *J. Toxicol. Environ. Health* 1986, **17**(4), 445-456
21. Hammond, B. G; et al. *Toxicol. Ind. Health* 1987, **3**(2), 79-98
22. Myhr, B; et al. *Environ. Mol. Mutagen.* 1991, **18**(1), 51-83
23. Galloway, S. M; et al. *Environ. Mol. Mutagen.* 1987, **B10**(Suppl. 10), 1-175
24. Sabatini, S; et al. *J. Pharmacol. Exp. Ther.* 1989, **250**(3), 910-914
25. Bogyo, D. A; et al. *Centre for Chemical Hazard Assessment* 980, 1-31, Report No. PE-4 O-LS-SRC TR80-507, Life Met. Sci. Div. Syracuse, New York

26.   Calley, D; et al. *J. Pharmaceut. Sci.* 1966, **55**(2), 158-162
27.   Wilbown, J; et al. *Environ. Health Perspect.* 1982, **145**, 127-128
28.   Gledhill, W. E; et al; *Environ. Sci. Technol.* 1980, **14**, 301-5
29.   Patterson, J. W; et al. *Chem. Eng. Prog.* 1981, **77**, 48-55
30.   Petrasek, A. C; et al. *J. Water Pollut. Control Fed.* 1983, **55**, 1286-1296
31.   O'Grady, D. P; et al. *Appl. Env. Microbiol.* 1985, **49**, 443-445
32.   Saeger, V. W; et al. *Biodegradation of Phthalate Esters Tech. Pap. Reg. Conf. Soc. Plast. Eng. Palisades Sect.* 1973, 105-113
33.   Saeger, V. W; et al. *Appl. Env. Microbiol.* 1976, **31**, 29-34
34.   Sugatt, R. H; et al. *Appl. Env. Microbiol.* 1984, **47**, 601-606
35.   Horowitz, A; et al. *Dev. Ind. Microbiol.* 1982, **23**, 435-144
36.   Shelton, D.R; et al. *Appl. Env. Microbiol.* 1984, **47**, 850-857
37.   Taylor, B. F; et al. *Appl. Env. Microbiol.* 1981, **42**, 590-595

B114   Benzyl chloride

CAS Registry No. 100-44-7
Synonyms (chloromethyl)benzene; α-chlorotoluene; tolyl chloride
Mol. Formula C_7H_7Cl **Mol. Wt.** 126.59
Uses A dye intermediate. Pharmaceutical precursor. Manufacture of perfumes, synthetic tannins and artificial resins.

Physical properties

M. Pt. –43-48°C; **B. Pt.** 177-181°C; **Flash point** 73°C; **Specific gravity** d_{20}^{20} 1.100; **Partition coefficient** log P_{ow} 2.30 (1); **Volatility** v.p. 1 mmHg at 22°C.

Solubility
Water: 493mg l^{-1} 20°C. Organic solvent: acetone, ethanol, dimethyl sulfoxide

Occupational exposure

US TLV (TWA) 1 ppm (5.2 mg m^{-3}); **UK Long-term limit** 1 ppm (5 mg m^{-3}); **UN No.** 1738; **HAZCHEM Code** 2W; **Conveyance classification** toxic substance; **Supply classification** irritant.

Risk phrases
Irritating to eyes, respiratory system and skin (R36/37/38)

Safety phrases
Wear eye/face protection (S39)

Ecotoxicity

Fish toxicity
LC$_{50}$ (96 hr) fathead minnow, trout, carp 6-17 mg l^{-1}. Exposed fish suffered paralysis (2).

Invertebrate toxicity
LC_{50} (24-96 hr) white shrimp 7-4 mg l^{-1} (3).
EC_{50} (5-30 min) *Photobacterium phosphoreum* 2.97 mg l^{-1} Microtox test (4).
Cell multiplication inhibition test, *Pseudomonas putida* 4.8 mg l^{-1}, *Scenedesmus quadricauda* 50 mg l^{-1}, *Entosiphon sulcatum* 25 mg l^{-1} (5).

Bioaccumulation
Based on a log P_{ow} 2.30 and water solubility 493 ppm at 20°C, estimated bioconcentration factor range 16-33 (6-8).

Environmental fate

Degradation studies
Readily biodegradable under the Japanese MITI test (9,10).
During a 2-day incubation period using raw sewage and raw sewage acclimated to non-chlorinated compounds, biodegradation was significant with the formation of dechlorinated products (11).
Confirmed biodegradable (12).

Abiotic removal
Aqueous hydrolysis products are benzyl alcohol and hydrogen chloride (13).
Hydrolytic $t_{1/2}$ 14 hr-19 day independent of pH up to pH 13.0 (14).
Atmospheric residence time due to vapour-phase reaction with hydroxyl radicals estimated at 3 days (15).

Absorption
Soil sorption coefficient estimated at 123-482 which indicates medium to high mobility in soil (16).

Mammalian and avian toxicity

Acute data
LD_{50} oral mouse, rat 1150-1660 mg kg^{-1} (17,18).
LC_{50} (2 hr) inhalation rat 150 ppm (19).

Carcinogenicity and long-term effects
Maximum tolerated dose by gavage 3 doses wk^{-1} for 104 wk, mice and rats was 100 mg kg^{-1} and 30 mg kg^{-1} respectively. Exposure induced papillomas and carcinomas of forestomach, aveolar and bronchiolar adenomas, carcinomas of lung, hemangiosarcomas, liver carcinomas and thyroid cell adenomas (20).
Application to skin mice (560 day) caused 15% incidence of skin cancer (21).

Teratogenicity and reproductive effects
Oral ♀ Sprague-Dawley rat (6-15 day gestation) 100 mg kg^{-1} positive foetotoxic effect, but reported to be non-teratogenic (22).

Metabolism and pharmacokinetics
Absorbed through the lung and gut (23).
After a single oral dose of 25 mg kg^{-1} to adult ♂ and ♀ rats elimination was predominantly in the urine. Excretion was faster in ♀ rats and slightly lower tissue concentration was maintained with the exception of blood and kidneys. Recovery was 90% in urine and faeces of ♀ rats at 24 hr compared with 80% in ♂ rats (24).

Genotoxicity

Salmonella typhimurium TA100 without metabolic activation weakly positive (25).
Salmonella typhimurium TA98, TA100, TA1535, TA1537, TA97 with and without metabolic activation weakly positive (26).
Salmonella typhimurium TA100 and *Escherichia coli* WP2 with and without metabolic activation negative (27).
Saccharomyces cerevisiae induced mitotic gene conversion (28).
Mouse lymphoma L5178V tk+/tk- without metabolic activation positive (29).

Any other adverse effects

Large doses cause central nervous system depression (30).
Intradermal injection mice caused depigmentation of hair (31).

Legislation

Limits under EC Directive on Drinking Water Quality 80/778/EEC. Other organochlorides: guide level 1 μg l^{-1}. Haloform concentrations must be as low as possible (32).

Any other comments

Toxicological and skin corrosion effects and safety handling practices discussed (33,34).
Applications, physical-chemical properties, genotoxicity, carcinogenicity, pathology, fire risk and toxicity of benzyl chloride reviewed (35-37).
Physico-chemical properties, human health effects, experimental effects and exposure reviewed (38).

References

1. Howard, P. H. *Handbook of Environmental Fate Exposure Data for Organic Chemicals* 1990, 78 Lewis Publishers, Chelsea, MI
2. Meinck, F; et al. *Les Eaux Residuraires Industrieles* 1970
3. Curtis, M. W; et al. *Water Res.* 1979, **13** ,137-141
4. Kaiser, K. L. E; et al. *Water Pollut. Res. J. Canada* 1991, **26**(3), 361-431
5. Bringmann, G; et al. *Water Res.* 1980, **14**, 231-241
6. Hansch, C; et al. *Medchem Project* 1981, Pomona College, Claremont, CA
7. Ohnishi, R; et al. *Bull. Chem. Soc. Japan* 1971, **41**, 2647-2649
8. Lyman, W. J; et al. *Handbook of Chemical Property Estimation Methods* 1982, 5.4, 5.10, McGraw-Hill, New York
9. Kitano, M, *Biodegradation Bioaccumulation Test on Chemical Substances* 1978, OECD, Toyko TSU-NO 3
10. Sasaki, S. *Aquatic Pollutants Transformations and Biological Effects* 1978, 283-298, Pergamon Press, New York
11. Jacobson, S. N; et al. *Appl. Environ. Microbiol.* 1981, **42**, 1062-1066
12. *MITI Report* 1984, Ministry of International Trade and Industry, Tokyo
13. Alvery, W. J; et al. *J. Chem. Soc. Chem. Comm.* 1972, 425-426
14. Tanabe, K; et al. *Hokkaido Daigaku* 1962, **10**, 173-182
15. Atkinson, R; et al. *Int. J. Chem. Kinet.* 1982, **14**, 13-18
16. Swann, R. L; et al. *Residue Rev.* 1983, **85**, 17-28
17. Izmerov, N. F. *Toxicometric Parameters of Industrial Toxic Chemicals Under Single Exposure* 1982, Moscow
18. Vernot, E. H; et al. *Toxicol. Appl. Pharm.* 1977, **42**, 417-423
19. *IARC Monograph* 1976, **11**, 217
20. Lijinsky, W. *J. Natl. Cancer Inst.* 1986, **76**(6), 1231-1236
21. Fukuda, K. *Gann Monograph* 1981, **72**(5), 655-664

22. Skowronski, G; et al. *J. Toxicol Environ. Health* 1986, **17**(1), 51-56
23. *IARC Monograph* 1982, **29**, 57
24. Bunner, B. L; et al. *J. Toxicol. Environ. Health* 1982, **10**(4-5), 837-846
25. Varley, R. B; *Mutat Res.* 1982, **100**(1-4), 45-47
26. Zeiger, E; et al. *Environ. Mol. Mutagen.* 1987, **9** (Suppl. 9), 1-110
27. Yasou, K; et al. *Mutat. Res.* 1978, **58**(2-3), 143-150
28. Parry, J. M; *Mutat Res.* 1982, **100** (1-4), 145-151
29. McGregor, D. B; et al. *Environ. Mol. Mutagen.* 1988, **11**(1), 91-118
30. *The Merck Index*, 11th Ed. 1989, Merck & Co. Inc., Rahway, NJ
31. AW, T. C., *IRCS Med. Sci. Lib. Compend.* 1981 **9**(1), 29-30
32. *EC Directive Relating to the Quality of Water Intended for Human Consumption* 1982, 80/778/EEC, Office for Official Publications of the European Communities, 2 rue Mercier, L-2985 Luxembourg
33. Vernot, E. M; et al. *Toxicol. Appl. Pharmacol.* 1977, **42**(2), 417-423
34. Cohen, J. L; et al. *Dhew NIOSH P. US* 1978, 78-160, SRI Instit., Menlo Park, CA
35. *IARC Monograph* 1987, **Suppl. 6**, 105-109
36. *IARC Monograph* 1982, **29**, 49-71
37. Farhi, M. et al. *French Natl. Res. Safety Inst. Fiche Toxicol. No. 90*, Cahiers de notes documentairs – securite et hygiene du travail No. 64, 759-64-71, 361-364
38. *ECETOC Technical Report No. 30(4)* 1991, European Chemical Industry Ecology and Toxicology Centre, B-1160 Brussels

B115   2-Benzyl-4-chlorophenol

CAS Registry No. 120-32-1
Synonyms 4-chloro-α-phenyl-*o*-cresol; 2-benzyl-4-chlorophenol; 5-chloro-2-hydroxydiphenylmethane; chlorophene; Bio-clave; *o*-benzyl-*p*-chlorophenol
Mol. Formula $C_{13}H_{11}ClO$ **Mol. Wt.** 218.69
Uses Germicide. Hard surface cleaner. Used in disinfectant solutions and soaps.

Physical properties

M. Pt. 49°C; **B. Pt.** 175°C at 5 mmHg; **Specific gravity** d_{25}^{55} 1.2.

Solubility
Organic solvent: dimethyl sulfoxide, ethanol, acetone

Mammalian and avian toxicity

Acute data
LD_{50} oral rat 1700 mg kg^{-1} (1).
LD_{50} oral mouse 65 mg kg^{-1} (2).

Sub-acute data
Oral rats, mice (12 exposures) 0-1000 mg kg^{-1} resulted in dose-related rectal dilation and nephrosis (3).

Oral gavage in corn oil rats 0-480 mg kg/-1 mice 0-1000 mg kg^{-1} (13 wk) 5 days wk^{-1} caused urogenital staining in rats and rough/oily haircoats in mice. Main target organ is the kidney (3).

Irritancy
Formulations containing 10% or more 2-benzyl-4-chlorophenol are primary skin irritants (4).

Genotoxicity
Salmonella typhimurium TA1535, TA1537, TA98, TA100 with or without metabolic activation (preincubation modification) negative (5).
Mouse lymphoma cell L5178Y forward mutation assay without metabolic activation positive. Human lymphoblast TK6 cells without metabolic activation positive. Chinese hamster ovary cells with or without metabolic activation chromosome aberration, sister chromatid exchange negative (6).

References
1. *J. Pharm. Sci.* 1974, **63**, 1068
2. *Pharmacol. Toxicol.* 1959, **22**, 270
3. Birnbaum, L. S; et al. *Fundam. Appl. Toxicol.* 1986, **7**(4), 615-625
4. Gosselin, R. E. *Clinical Toxicol. Commercial Products* 5th ed., 1984, **2**(193), 520
5. Mortelmans, K; et al. *Environ. Mutagen.* 1986, **8**(Suppl. 7), 1-119
6. Caspary, W. J; et al. *Mutat. Res* 1988, **196**(1) 61-81

B116   Benzyl cyanide

CH$_2$CN

CAS Registry No. 140-29-4
Synonyms benzenacetonitrile; ω-cyanotoluene; phenylacetonitrile; α-tolunitrile
Mol. Formula C$_8$H$_7$N   **Mol. Wt.** 117.15
Uses Manufacture of rubber. Chemical intermediate.
Occurrence In garden cress and other plants.

Physical properties
M. Pt. –24°C; **B. Pt.** 233-234°C; **Flash point** 101°C (closed cup); **Specific gravity** d$_{15}^{15}$ 1.021; **Volatility** v.p. 0.1 mmHg at 20°C.

Solubility
Organic solvent: ethanol, diethyl ether

Occupational exposure
UN No. 2470; **HAZCHEM Code** 3X; **Conveyance classification** harmful substance.

Ecotoxicity

Effects to non-target species
EC$_{50}$ (30 min) *Photobacterium phosphoreum* 1.35 mg l^{-1} Microtox test (1).

LD$_{Lo}$ subcutaneous frog 1500 mg kg^{-1} (2).

Environmental fate

Degradation studies
Microbial degradation occurs in water, initial concentration 500 mg l^{-1} at 20°C, 84% degraded within 24 hr (3).

Mammalian and avian toxicity

Acute data
LD$_{50}$ oral mouse 45 mg kg^{-1} (4).
LC$_{50}$ (2 hr) inhalation mouse 100 mg m^{-3} (5).
LD$_{50}$ dermal rabbit 270 mg kg^{-1} (6).
LD$_{Lo}$ intraperitoneal rat 25 mg kg^{-1} (7).

Sub-acute data
LD$_{Lo}$ ♂ and ♀ rats 0.2-0.3 g kg^{-1} by gavage. Sub lethal oral doses were nephrotoxic (5). Subcutaneous injection to rabbits (21 day) 0.01-0.5 ml day^{-1} produced thyroid hyperplasia (8).

Metabolism and pharmacokinetics
Metabolised to cyanide by rat liver and nasal microsomes (9,10).

Irritancy
Dermal rabbit (24 hr) 500 mg caused mild irritation (5).

Genotoxicity
Salmonella typhimurium TA97, TA98, TA100, TA1535, TA1537 with and without metabolic activation negative (11).
Exposure to concentrations between 15 ppm and 180 ppm induced mitotic aneuploidy in yeast (strain unspecified) (12).

Any other adverse effects
Reported to cause damage to liver, kidney, heart, spleen, lungs, brain and cerebral tissue respiration in experimental animals (species unspecified) (13).

References

1. Kaiser, K. L. E; et al. *Water Pollut. Res. J. Canada* 1991, **26**(3), 361-431
2. *Arch. Int. Pharmacodynamic Therapie* 1899, **55**, 161
3. Worne, H. E. *Activity of Mutant Microorganisms in the Biological Treatment of Industrial Wastes* Tijdschrift van het BECEWA, Liege
4. *Arch. Toxicol.* 1984, **55**, 47
5. Guest, A; et al. *Toxicol. Lett.* 1982, **10**(2-3), 265-272
6. *Food Chem. Toxicol.* 1982, **20**, 803
7. Izmerov, N. F; et al. *Toxicometric Parameters of Industrial Toxic Chemicals Under Single Exposure* 1982, Moscow
8. Dahl, A. R; et al. *Xenobiotica* 1989, **19**(11), 1201-1205
9. Dahl, A. R; et al. *Annual Report No. LMF-120 of the Inhalation Toxicology Research Institute* 1986-1987, 397-399
10. Zeiger, E; et al. *Environ. Mol. Mutagen.* 1988, **11**(Suppl. 12), 1-58
11. Zimmermann, F. K. *Mutat. Res.* 1985, **150**(1-2), 203-210
12. Manni, D; et al. *Proc. Soc. Exp. Biol. Med.* 1932, **29**, 772-775
13. Galbin, G. P; et al. *Hyg. Sanit.* 1967, **32**(8), 176-181

B117   Benzylidene chloride

CHCl$_2$

CAS Registry No. 98-87-3
Synonyms (Dichloromethyl) benzene; benzyl dichloride; α, α-dichlorotoluene; benzal chloride; benzylene chloride
Mol. Formula C$_7$H$_6$Cl$_2$ **Mol. Wt.** 161.03
Uses In manufacture of benzaldehyde and cinnamic acid.

Physical properties

M. Pt. –16°C; **B. Pt.** 205°C; **Flash point** 85°C; **Specific gravity** 1.26.

Solubility
Organic solvent: ethanol, diethyl ether

Occupational exposure

UN No. 1886; **HAZCHEM Code** 2X; **Conveyance classification** toxic substance; **Supply classification** irritant.

Risk phrases
Irritating to eyes, respiratory system and skin (R36/37/38)

Safety phrases
Wear eye/face protection (S39)

Ecotoxicity

Invertebrate toxicity
EC$_{50}$ (30 min) *Photobacterium phosphoreum* 5.85 mg l^{-1} Mictrotox test (1).

Bioaccumulation
Calculated bioconcentration factor 164 (2).

Abiotic removal
Readily biodegrades in water (3).
Hydrolyses to benzaldehyde under both acid and alkaline conditions (4).

Mammalian and avian toxicity

Acute data
LD$_{50}$ oral rat 3250 mg kg^{-1} (5).
LC$_{50}$ (2 hr) inhalation rat, mouse 30-60 ppm (6).

Carcinogenicity and long-term effects
Evidence for carcinogenicity to humans inadequate, evidence for carcinogenicity to animals limited, IARC classification group 2B (7).
Dermal mice (560 day) dose unspecified induced 15% incidence of skin cancers (8).

Genotoxicity

Salmonella typhimurium TA100 with and without metabolic activation positive.
Escherichia coli WP2 with and without metabolic activation positive (9).

Any other comments

Experimental toxicology and human health effects reviewed (10).
Lachrymatory.

References

1. Kaiser, K. L. E; et al. *Water Pollut. Res. J. Canada* 1991, **26** (3), 361-431
2. Lyman, W. J; et al. *Handbook of Chemical Property Estimation Methods. Environmental Behaviour of Organic Compounds* 1982, McGraw-Hill, New York
3. Steinhauser, K. G; et al. *Vom. Wasser* 1986, **67**, 147-154
4. *Kirk-Othmer Encyclopedia of Chemical Technology* 3rd ed., 1979, **5**, 828
5. *NTIS Report* PB214-270, Nat. Tech. Inf. Ser., Springfield, VA
6. *IARC Monograph* 1982, **29**, 65
7. *IARC Monograph* 1987, **Suppl. 7**, 147
8. Fukuda, K; et al. *Gann* 1981, **72**(5), 655-664
9. Yasuo, K; et al. *Mutat. Res.* 1978, **58**(2-3), 143-150
10. *ECETOC Technical Report No. 30(4)* 1991, European Chemical Industry Ecology and Toxicology Centre, B-1160 Brussels

B118   Benzyl isoeugenol

CAS Registry No. 120-11-6
Synonyms benzyl isoeugenol ether; benzyl 2-methoxy-4-proprenylphenyl ether
Mol. Formula $C_{17}H_{18}O_2$ **Mol. Wt.** 254.33
Uses Intermediate in perfume manufacture.

Physical properties

M. Pt. 58-60°C.

Mammalian and avian toxicity

Acute data
LD_{50} oral rat 4900 kg^{-1} (1).

Sub-acute data
Gavage rat (28 day) 240 mg kg^{-1} caused significant decreases in body weight, blood glucose, blood urea and liver weights. No dose related histopathological changes were observed in any organs. The no effect level was 60 mg kg^{-1} day^{-1} (2).

Irritancy
Dermal rat (24 hr) 500 mg caused irritation (1).

Sensitisation

A 5% concentration of benzyl isoeugenol in petrolatum applied to the skin of 25 volunteers produced no sensitisation reactions (1).

Genotoxicity

Salmonella typhimurium TA98 with and without metabolic activation reduced mutation frequency 30-80% (3).

Any other comments

Human health effects and experimental toxicology reviewed (4,5).

References

1. Opdyke, D. L. J. *Food Cosmet. Toxicol.* 1973, **1** 1011-1081
2. Boe, M; et al. *Drug Chem. Toxicol.* 1989, **12**(2) 165-171
3. Yazava, I; et al. *J. P. Kokai Tokkyo Koho J. P.* 63, 167, 791, [88, 167, 791] 186, 4, (*Chem Abstr.* **110**, 52817f)
4. *Bibra Toxicity Profile*, 1987, British Industrial Biological Research Association, Carlshalton
5. *ECETOC Technical Report No. 30(4)* 1991, European Chemical Industry Ecology and Toxicology Centre, B-1160 Brussels

B119   Benzylpenicillin

CAS Registry No. 61-33-6

Synonyms Penicillin G; 3,3-dimethyl-7-oxo-6-phenyl-acetamido-4-thia-1-azabicyclo[3,2,0] heptane- 2-carboxylic acid; benzylpenicillinic acid; 6-phenylacetamidopenicillinic acid

Mol. Formula $C_{16}H_{18}N_2O_4S$ **Mol. Wt.** 334.40

Uses Antibacterial.

Occurrence Produced by growing certain strains of *Penicillium notatum* (1).

Physical properties

Solubility

Organic solvent: methanol, ethanol, diethyl ether, ethyl acetate, benzene, acetone

Mammalian and avian toxicity

Acute data

LD_{50} oral hamster 24 mg kg^{-1} (2).
LD_{50} intracerebal mouse 7500 mg kg^{-1} (3).
LD_{50} intracerebal rabbit 653-1118 mg kg^{-1} (3).
LD_{50} intraspinal dog 4940 mg kg^{-1} (3).
LD_{50} intravenous mouse 329 mg kg^{-1} (4).

Teratogenicity and reproductive effects

Post-implantation rat embryo culture system negative (5).

Metabolism and pharmacokinetics

Degradation products include penillic, penicillenic and penicillic acids (1).

Widely distributed in the body; apparent volume of distribution is in 50% total body water. >90% in blood is in plasma; <10% in erthythrocyre; 65% reversibly bound to plasma albumin (6).

Significant amounts appear in liver, bile, kidney, semen, lymph and intestine, but does not readily enter cerebrospinal fluid when meninges normal (5).

Oral doses of 500 mg potassium penicillin to humans. Urinary concentrations 2 and 4 hr after dosing 600 μg ml^{-1} and 300 μg ml^{-1} respectively. Poor placental transfer due to low lipid solubility and low ionisation constant of penicillin G (7).

Sensitisation

Allergic reactions include exfoliative dermatitis, interstitial nephritis and vasculitis (1). Sodium and potassium salts cause allergic sensitisation reactions in some patients due to penicillin metabolites. Desensitsation has been attempted where appropriate (8). Sensitised patients may also react to cephalosporins (1).

Genotoxicity

Bacillus subtilis: H17/M45 agar incorporation test (AT) and spot test (ST) negative; HLL3 g/HJ15 AT equivocal, ST negative (9).

Escherichia coli: AB1157/JC5547 AT negative, ST positive; AB1157/JC2921 AT negative, ST positive; AB1157/JC2926 AT negative, ST positive; AB1157/JC5519 AT negative (9).

Any other adverse effects to man

TD_{Lo} child 15000 units kg^{-1} change in cochlear structure or function. Convulsions or effect on seizure threshold, lungs, thorax, respiration (dyspnoea) (10).

Administration of benzylpenicillin to a hypersensitive patient may result in anaphylactic shock with collapse and death occuring within minutes. Angiodema or bronchospasm may also occur. Generalised sensitivity reaction with urticaria, fever, joint pains and eosinophilia can develop within a few hours to several weeks after starting treatment with penicillin. Haemolytic anaemia and leucopenia have been reported usually following high intravenous doses of benzylpenicillin. Prolongation of bleeding time and defective platelet function has also been observed. Convulsions and other signs of toxicity to the central nervous system may occur with very high doses of benzylpenicillin, particularly when administered intravenously or to patients with renal failure. Encephalopathy may also follow intrathecal administration. Disturbances of blood electrolytes may follow the administration of large doses of the potassium and sodium salts of benzylpenicillin. Patients with syphilis may experience a Jarisch-Herxheimer reaction, symptoms are fever, chills, headache and reactions at lesion sites shortly after treatment with penicillin. Benzylpenicillin may produce diarrhoea, nausea and heartburn following oral administration. A sore mouth or tongue or a black hairy tongue have occasionally been reported (1).

Any other comments

Benzylpenicillin is commonly used to describe either benzylpenicillin potassium or benzylpenicillin sodium as these are the forms in which benzylpenicillin is used. Incompatible with metal ions, some acid and alkaline drugs and some rubber products (1).

References

1. *Martindale The Extra Pharmacopoeia* 29th ed., 1989, Pharmaceutical Press, London
2. *Toxicol. Appl. Pharmacol.* 1969, **14**, 510
3. *J. Lab. Clin. Med.* 1949, **34**, 126
4. *Biochem. Pharmacol.* 1967, **16**, 1365
5. Cicurel, L; et al. *Xenobiotica* 1988, **18**(6), 617-624
6. Goodman, L. S; et al. *The Pharmacological Basis of Therapeutics* 5th ed., 1975, Macmillan, New York
7. *Foreign Compound Metabolism in Mammals* 1975, The Chemical Society, London
8. Sullivan, T. J; et al. *J. Allergy Clin. Immunol.* 1982, **62**, 275
9. Suter, W; et al. *Mutat. Res.* 1982, **97**, 1-18
10. *Antimicrob. Agents Chemother.* 1980, **17**, 572

B120   Benzyl Violet 4B

CAS Registry No. 1694-09-3

Synonyms Benzenemethanaminium,
N-[4-[[4-(dimethylamino)phenyl][4-[ethyl[(3-sulfophenyl)methyl]amino]phenyl]
methylene]-2,5-cyclohexadien-1-ylidene]-*N*-ethyl-3-sulfo-, hydroxide, inner salt,
sodium salt; C.I. Acid Violet 49; A.F. Violet No 1; Benzyl Violet 3B; Food Violet 2;
Wool Violet 4BN

Mol. Formula $C_{39}H_{40}N_3O_6S_2Na$ **Mol. Wt.** 733.89

Uses In dyestuffs for wool, silk, nylon, leather and paper. Used as biological and wood stain.

Occupational exposure

Supply classification harmful.

Risk phrases
Possible risk of irreversible effects (R40)

Safety phrases
Wear protective clothing and gloves (S36/37)

Mammalian and avian toxicity

Carcinogenicity and long-term effects
No adequate evidence for carcinogenicity to humans, sufficient evidence for
carcinogenicity in animals, IARC classification group 2B (1).
Oral or subcutaneous administration to ♀ rats (unspecified dose) induced mammary
and squamous carcinomas and local fibrosarcomas following subcutaneous injection
in ♂ and ♀ rats (2).

Metabolism and pharmacokinetics

Oral administration to rats and dogs, less than 5% absorbed, mainly excreted in faeces (2).

Levels of colour in liver, kidney, abdominal muscle and blood serum ranged from 1 to 3 $\mu g\ g^{-1}$ tissue in rats fed 5% in the diet (3).

Genotoxicity

Salmonella typhimurium TA1535 without metabolic activation positive (4).

Any other comments

Withdrawn as food colourant, drug and cosmetics additive US 1973, Europe 1978 (2). Toxicity reviewed (5).

References

1. *IARC Monograph* 1987, **Suppl. 7**, 56, 58
2. *IARC Monograph* 1978, **16**, 153-162
3. Minegishi, K; et al. *Toxicology* 1977, **7**(3), 367-383
4. *Mutat. Res.* 1981, **89**, 21
5. *Ann. Rev. Pharmacol.* 1974, **14**, 127

B121 Beryllium

Be

CAS Registry No. 7440-41-7

Synonyms beryllium-9; glucinum; glucinium

Mol. Formula Be   **Mol. Wt.** 9.01

Uses Used in alloys, applications in the electrical, nuclear and aerospace industries.

Occurrence The toxicity of beryllium is dependent on the ability of the organism to absorb it. Therefore toxicity data refers to bioavailable forms, such as the ion in solution or particulate matter.

In nature, beryllium occurs in the minerals beryl, phenacite, bertrandite, bromellite and chrysoberyl.

Physical properties

M. Pt. 1278°C; **B. Pt.** $_5$ 2970°C; **Specific gravity** d^{20} 1.85; **Volatility** v.p. 7.6 mmHg at 1810°C.

Occupational exposure

US TLV (TWA) 0.002 mg m^{-3}; **UK Long-term limit** 0.002 mg m^{-3}; **UN No.** 1567;
Conveyance classification toxic substance; flammable solid;
Supply classification toxic.

Risk phrases

Very toxic by inhalation and in contact with skin – Irritating to respiratory system – Danger of very serious irreversible effects (R26/27, R37, R39)

Safety phrases
In case of contact with eyes, rinse immediately with plenty of water and seek medical advice – After contact with skin, wash immediately with plenty of water – In case of accident or if you feel unwell, seek medical advice immediately (show label where possible) (S26, S28, S45)

Ecotoxicity

Fish toxicity
Acute toxicity range to fish (24-96 hr) (species unspecified) in fresh water 87-0.87 µg l^{-1} (1).
LC_{50} (96 hr) fathead minnow 150 µg l^{-1} (2).

Invertebrate toxicity
EC_{50} (48 hr) *Daphnia magna* 1.88 mg l^{-1} (3).

Bioaccumulation
Bioconcentration factor for marine and freshwater fish, invertebrates and plants is 100 (1).

Mammalian and avian toxicity

Acute data
TC_{Lo} inhalation human 300 mg m^{-3} pulmonary effects include bronchitis, pneumonitis and oedema (4).
Single exposure (50 min) inhalation rat 800 µg m^{-3} beryllium resulted in severe, acute, chemical pneumonitis progressing to chronic-active fibrosing pneumonitis. The authors conclude that induced lung lesions in rats appears to be due to direct chemical toxicity and foreign-body-type reactions in contrast to the immunological mediated granulomatous lung disease in humans (5).
LD_{50} intravenous rat 496 mg kg^{-1} (6).

Carcinogenicity and long-term effects
Limited evidence for carcinogenicity to humans, sufficient evidence for carcinogenicity to animals, IARC classification 2A (7,8).
TD_{Lo} intratracheal rat 13 mg kg^{-1} can cause bronchogenic carcinoma (9).
TD_{Lo} intravenous rabbit 20 mg kg^{-1} muscular skeletal tumours (10).

Metabolism and pharmacokinetics
Beryllium crosses the placenta to a small extent (11).
Unabsorbed beryllium is excreted in the faeces of rats and other animals, 96% of a single dietary dose is excreted within 24 hr, while only 1% of absorbed beryllium is excreted in urine and faeces in 24 hr (12).
$t_{1/2}$ in total human body 180 days (1).
Beryllium is stored in bones, with transient retention by liver, kidney and lungs (13).

Irritancy
Beryllium dermatitis, granulomatous ulcerations of the skin and conjunctivitis reported (14).

Sensitisation
Cutaneous hypersensitivity to beryllium observed in guinea pigs may be significant for humans exposed to beryllium (15,16).

Genotoxicity

Escherichia coli, HeLa cells and Ehrlich ascites tumour cells, DNA cell binding assays positive (17).

Chinese hamster ovary and rat lung epithelial cells, 20 hr exposure, positive cytotoxic effects (18).

Oral rat (6 month) in drinking water, caused cytotoxicity at the toxic dose level and induced chromosomal aberrations, but was negtive in dominant lethal assays (19).

Any other adverse effects to man

Concentrations of beryllium were determined in human breast milk implications for childhealth are highlighted (20).

Any other adverse effects

Ionic beryllium has been shown to inhibit a number of enzymes *in vitro* including phosphatases, phosphoglutamase, hexokinase, deoxythymidine kinase, lactate dehydrogenase and amylase (21).

Any other comments

Exposure levels, ecotoxicology, experimental toxicology, human health and environmental effects have been extensively reviewed (1,8,22-30).

Incompatible with halocarbons causes flash or spark on impact. Reacts violently with trichloroethylene. Moderate fire and explosion hazard.

References

1. *Dangerous Prop. Ind. Mater. Rep.* 1981, **1**(3), 36-38, *Beryllium*
2. Schwitzgebel, K; et al. *Trace Element Discharge from Coal fired Power Plants* 1975, **2**, 146
3. Khangarot, B. S; et al. *Ecotoxicol. Environ. Saf.* 1989, **18**(2), 109-120
4. *Arch. Environ. Health* 1964, **9**, 473
5. Haley, P. J; et al. *Fund Appl. Toxicol.* 1990, **15**(4), 767-778
6. *Lab. Invest.* 1966, **15**, 176
7. *IARC Monograph* 1980, **23**, 143
8. *IARC Monograph* 1987, **Suppl. 7**, 58
9. *Environ. Res.* 1980, **21**, 63
10. *Lancet* 1950, **1**, 463
11. Puzanova, L. *Folia Morphol.* 1980, **28**(4), 354, Prague
12. Venugopal, B; et al. *Metal Toxicity in Mammals* 2nd ed., 1978, Plenum Press, New York
13. Browning, E. *Toxicity of Industrial Metals* 2nd ed., 1969, Appleton & Lange, East Norwalk, CT
14. Curtis, G. H. *Arch. Dermatol. Syphilol.* 1951, **64**, 47
15. Reeves, A. L; et al. *Trans. New York Acad. Sci.* 1974, **36**(1), 78
16. Boman, A; et al. *Contact Dermatitis* 1979, **5**(5), 332
17. Groth, D. H; et al. *Mutat. Res.* 1981, **89**, 95
18. Finch, G. L; et al. *In Vitro Toxicol.* 1988, **2**(4), 287-297
19. Nikiforova, V. Y; et al. *Tsitol. Genet.* 1989, **23**(4), 27-30, (Russ.) (Chem. Abstr., **111**, 148616m)
20. Durrant, S. F. *J. Micornutr. Anal.* 1989, **5**(2), 111-126
21. Reeves, A. L. *Handbook on Toxicol. Metals: Beryllium* 1979, 329-343
22. Aller, A. J. *Trace Elem. Electrolytes Health Dis.* 1990, **4**(1), 1-6
23. Wagoner, J. K. *Environ. Res.* 1980, **21**, 15-34
24. Kjellstrom, T; et al. *Arbete Och Halsa* 1984, **2**
25. *Gov. Rep. Announce Index US* EPA 600/8-84/026F 1988, **88**(12), 830682

26. *Gov. Rep. Announce Index US* 1989, **89**(11), 930605
27. O'Neil, I. K; et al. *IARC Sci. Publ.* 1986, **8**, 3-13
28. Skilleter, D. N. *Adv. Mod. Environ. Toxicol.* 1987, **11**, 61-86
29. *ECETOC Technical Report No. 30(4)* 1991, European Chemical Industry Ecology and Toxicology Centre, B-1160 Brussels
30. Izmerov, N. F. *Scientific Reviews of Soviet Literature on Toxicity & Hazards of Chemicals* 1992-1993, **63**, Eng. Trans, Richardson, M. L. (Ed.), UNEP/IRPTC, Geneva

B122   Beryllium chloride

$BeCl_2$

CAS Registry No. 7787-47-5
Synonyms beryllium dichloride
Mol. Formula $BeCl_2$ **Mol. Wt.** 79.92
Uses Manufacture of beryllium catalysts.

Physical properties

M. Pt. 399°C; **B. Pt.** 482°C; **Specific gravity** d_4^{25} 1.899;
Volatility v.p. 1 mmHg at 291°C.

Solubility
Organic solvent: ethanol, diethyl ether, pyridine

Occupational exposure

US TLV (TWA) 0.002 mg m^{-3}; **UK Long-term limit** 0.002 mg m^{-3}; **UN No.** 1566; **Conveyance classification** toxic substance; **Supply classification** toxic.

Risk phrases
Very toxic by inhalation and in contact with skin – Irritating to respiratory system – Danger of very serious irreversible effects (R26/27, R37, R39)

Safety phrases
In case of contact with eyes, rinse immediately with plenty of water and seek medical advice – After contact with skin, wash immediately with plenty of water – In case of accident or if you feel unwell, seek medical advice immediately (show label where possible) (S26, S28, S45)

Ecotoxicity

Fish toxicity
LC$_{50}$ (96 hr) fathead minnow 150 µg l^{-1} (soft water), 20,000 µg l^{-1} (hard water) static bioassay (1).
LC$_{50}$ (96 hr) bluegill sunfish 1300 µg l^{-1} (soft water) 12,000 µg l^{-1} (hard water) static bioassay (2).

Invertebrate toxicity
EC$_{50}$ (48 hr) *Daphnia magna* 2500 µg l^{-1} (2).

Mammalian and avian toxicity

Acute data
LD_{50} oral rat 86 mg kg^{-1}.
LD_{50} intraperitoneal rat 4.4 mg kg^{-1}.
LD_{50} intramuscular mouse 12 mg kg^{-1}.
LD_{50} intraperitoneal guinea pig 50 mg kg^{-1} (3).

Carcinogenicity and long-term effects
Limited evidence for carcinogenicity to humans, sufficient evidence for carcinogenicity to animals, IARC classification group 2A, for beryllium and beryllium compounds (4,5).
Inhalation exposure of rats to 0.2 or 0.4 mg m^{-3} for 1 hr daily 5 days a wk for 4 months. Adenocarcinomas and trabecular adenomas found 18 months after termination of exposure (6).

Teratogenicity and reproductive effects
A single injection of 3-300 µg into chicken embryos killed most of the embryos and caused severe damage to those surviving. Embryos treated with 0.03-0.3 µg survived but showed malformations, defects included cardiac malformations, malpositions and caudal regression (7).
Intravenous ICR mice administered before fertilisation and on days 7-14 of pregnancy. Beryllium permeates placenta with difficulty. Part of administered dose circulated in blood long enough to penetrate foetuses (8).

Metabolism and pharmacokinetics
Intraperitoneal injection into rats at 1.2 mg kg^{-1} every other day for 3 months accumulated in liver and spleen > kidney > heart > lung (9).

Sensitisation
0.5% challenge caused skin sensitisation in guinea pigs at 24 hr and 48 hr (10).

Genotoxicity
Salmonella typhimurium TA98, TA100 with and without metabolic activation negative. Mild positive effect in the rec. assay. Induced sister chromatid exchanges in V79 Chinese hamster cells (11).
In vitro domestic pig peripheral lymphocytes and primary kidney cells mitotic delay and chromosome aberrations positive (12,13).

Any other adverse effects
Respiratory diseases from inhalation of soluble beryllium compounds include rhinitis, pharyngitis, tracheobronhitis and pneumonitis (14).

Legislation
Limited under EC Directive on Drinking Water Quality 80/778/EEC. Chlorides : guide level 25 mg l^{-1}, maximum admissible concentration 400 mg l^{-1} (15).

Any other comments
Human health effects, experimental toxicology and environmental effects reviewed (16,17).
Beryllium has been measured in 59 samples of surface water from 15 US/Canadian rivers. Highest concentration was less than 0.22 µg l^{-1}. Beryllium was found in 85%

of samples from 15 major river basins in the USA at concentrations of 0.01-1.22 μg l⁻¹. Beryllium has been reported to occur in US drinking water at 0.1-0.7 μg l⁻¹ with a mean of 0.013 μg l⁻¹ total beryllium (18).

References

1. Tarzwell, C. M; et al. *Toxicity of Less Common Metals to Fish from Industrial Wastes* 1960, **5**, 12
2. *USEPA Ambient Water Quality Criteria Doc: Beryllium* 1980, Washington, DC
3. Luckey, T. D. *Metal Toxicity in Mammals* 1977, **2**, 43, Plenum Press, New York
4. *IARC Monograph* 1980, **23**, 143
5. *IARC Monograph* 1987, **Suppl. 7**, 58
6. Litvinov, N. N; et al. *Gig. Tr. Prof. Zabol.* 1975, **7**, 34-37
7. Puzanova, L. *Sb. Lek.* 1978, **80**(4), 105, Czech
8. Bencko, V; et al. *J. Hyg. Epidemiol. Microbiol. Immunol. (Prague)* 1979, **23**(4), 361
9. Shima, S; et al. *J. Sci. Labour* 1982, **58**(12), 635-644
10. Boman, A; et al. *Contact Dermatitis* 1979, **5**(5), 332-333
11. Kuroda, K; et al. *Mutat. Res.* 1991, **264**, 163-170
12. Luke, M. Z; et al. *Biochem. Biophys. Res. Commun.* 1975, **62**, 497-501
13. Talluri, M. V; et al. *Caryologia* 1967, **20**, 355-367
14. Reeves, A. L; et al. *Handbook on Toxicology of Metals – Beryllium* 1979, 329, Elsevier, Amsterdam
15. *EC Directive Relating to the Quality of Water Intended for Human Consumption* 1984, 80/778/EEC, European Communities, 2 rue Mercier, L-2985 Luxembourg
16. *ECETOC Technical Report No. 30(4)* 1991, European Chemical Industry Ecology and Toxicology Centre, B-1160 Brussels
17. *Dangerous Prop. Ind. Mater. Rep.* 1988, **8**(6), 17-23
18. *National Research Council Drinking Water and Health* Report 1, 1977, National Academy Press, Washington, DC

B123   Beryllium fluoride

BeF_2

CAS Registry No. 7787-49-7
Synonyms beryllium difluoride
Mol. Formula BeF_2 **Mol. Wt.** 47.01
Uses Commercial production of beryllium metals and glass. Used in nuclear reactors.

Physical properties

M. Pt. 555°C; **B. Pt.** 1160°C; **Specific gravity** d^{25} 1.986.

Solubility
Water: miscible. Organic solvent: diethyl ether, ethanol

Occupational exposure

US TLV (TWA) 0.002 mg m⁻³; **UK Long-term limit** 0.002 mg m⁻³; **UN No.** 1566;
Conveyance classification toxic substance; **Supply**
classification toxic.

Risk phrases
Very toxic by inhalation and in contact with skin – Irritating to respiratory system – Danger of very serious irreversible effects (R26/27, R37, R39)

Safety phrases
In case of contact with eyes, rinse immediately with plenty of water and seek medical advice – After contact with skin, wash immediately with plenty of water – In case of accident or if you feel unwell, seek medical advice immediately (show label where possible) (S26, S28, S45)

Ecotoxicity

Fish toxicity
LC_{50} (96 hr) fathead minnow 150 µg l^{-1} (soft water); 20,000 µg l^{-1} (hard water) static bioassay (1).
LD_{50} (96 hr) bluegill sunfish 1300 µg l^{-1} (soft water); 12,000 µg l^{-1} (hard water) static bioassay (1).

Invertebrate toxicity
EC_{50} (48 hr) *Daphnia magna* 2500 µg l^{-1} (2).

Mammalian and avian toxicity

Acute data
LD_{50} oral mouse, rat 98-100 mg kg^{-1} (3).
LD_{50} subcutaneous mouse 20 mg kg^{-1} (3).
LD_{50} intraperitoneal hamster 21 mg kg^{-1} (3).

Carcinogenicity and long-term effects
Limited evidence for carcinogenicity to humans, sufficient evidence for carcinogenicity to animals, IARC classification group 2A, for beryllium and its compounds (4).
Inhalation (1 hr) or intratracheal (single instillation) ♂ rats neoplasms in the lung observed after 16 months (5).
Inhalation rat (4 month) 0.2 mg m^{-3} 1 hr daily, 5 × wk, induced the development of large foci of catharal pneumonia, atelectasis and fields of emphysema. Twelve months after termination, chronic interstital pneumonia, interspersed with emphysema, occurred. In a number of cases hyperplasia with polymorphism of rapidly reproducing cells was observed. Flat-cell cancer infiltrating surrounding tissue, and adenocarcinomas were observed in several cases. Eighteen months after termination of the inhalation, trabecular adenoma and adenocarcinomas with infiltrating growth and hepatic metastic foci were found (6).

Metabolism and pharmacokinetics
Humans exposed to 3 µg Be m^{-3} inhaled air (duration unspecified), excretion mainly in urine. Urinary excretion is prolonged with detection of beryllium in urine for 10 yr after exposure (7).
Accumulation, distribution and excretion of inhaled aerosol beryllium fluoride in rats depends on age of rat. Beryllium clearance from nose, oral cavity, and trachea was slower, retention in stomach and small intestine longer in 1 wk old than adult rats (8).

Sensitisation
Hartley strain II and III guinea pigs were sensitised to beryllium fluoride by painting a 20% solution of beryllium fluoride in detergent each day for 3 days on the left ear. Fourteen days after starting sensitisation treatment, they were skin painted with 1%

beryllium fluoride in 1% Triton X 100 on the flank to determine sensitivity. Only strain II animals could be sensitised to beryllium fluoride (9).

Any other adverse effects

Inhalation monkey (7-16 day) 27 γ ft^3 6 hr day^{-1} beryllium fluoride (corresponds to 5.2 γ ft^3 beryllium) caused severe pulmonary reactions and changes to liver, kidney, adrenals, pancreas, thyroid and spleen (10).

Symptoms of acute inhalation (species unspecified) include rhinitis, pharyngitis, tracheobronchitis and pneumonitis (11).

Legislation

Limited under EC Directive on Drinking Water Quality 80/778/EEC. Fluorides: maximum admissible concentration 1500 mg l^{-1} (8-12°C); 700 mg l^{-1} (25-30°C) (12).

Any other comments

Human health effects, experimental toxicology and environmental effects reviewed (13,14).

References

1. Tarzwell, C. M; et al. *Toxicity of Less Common Metals to Fish from Industrial Wastes* 1960, **5**, 12
2. *USEPA Ambient Water Quality Criteria Document: Beryllium* 1980, Washington, DC
3. Vacher, J; et al. *Toxicol. Appl. Pharmacol.* 1973, **24**, 497
4. *IARC Monograph* 1987, **Suppl. 7**, 58
5. *IARC Monograph* 1980, **23**, 143
6. Litvinov, N. N; et al. *Gig. Tr. Prof. Zabol.* 1975, **7**, 34-37, (Russ.) (*Chem. Abstr.* **83**, 173624y)
7. Venugopal, B; et al. *Metal Toxicity in Mammals* 1978, Plenum, New York
8. Bugryshev, P. F; et al. *Gig. Tr. Prof. Zabol.* 1984, **6**, 52-53
9. Turk, J. L; et al. *Int. Arch. Allergy. Appl. Immunol.* 1969, **36**(1-2), 75-81
10. Schepers, G. W. H. *Ind. Med. Surg.* 1964, **33**, 1-16
11. Reeves, A. L. *Handbook on Toxicology of Metals* 1979, 329, Amsterdam
12. *EC Directive Relating to the Quality of Water Intended for Human Consumption* 1982, 80/778/EEC, Office for Official Publications of the European Communities, 2 rue Mercier, L-2985 Luxembourg
13. *Cah. Notes Doc. Toxicol. Data Sheets No 92, Beryllium* 1989, **136**, 549-554, (Fr.) (*Chem. Abstr.* **111**, 238722x)
14. *ECETOC Technical Report No. 30*(4) 1991, European Chemical Industry Ecology and Toxicology Centre, B-1160 Brussels

B124   Beryllium nitrate

Be(NO3)2

CAS Registry No. 13597-99-4
Mol. Formula BeN$_2$O$_6$ **Mol. Wt.** 133.02
Uses Used to stiffen mantles in gas and acetylene lamps.

Physical properties

M. Pt. ≈60°C; **B. Pt.** 100-200°C (decomp.).

Solubility
Water: miscible. Organic solvent: ethanol

Occupational exposure

US TLV (TWA) 0.002 mg m^{-3}; **UK Long-term limit** 0.002 mg m^{-3}; **UN No.** 2464;
Conveyance classification oxidising substance; toxic substance;
Supply classification toxic.

Risk phrases
Very toxic by inhalation and in contact with skin – Irritating to respiratory system –
Danger of very serious irreversible effects (R26/27, R37, R39)

Safety phrases
In case of contact with eyes, rinse immediately with plenty of water and seek medical
advice – After contact with skin, wash immediately with plenty of water – In case of
accident or if you feel unwell, seek medical advice immediately (show label where
possible) (S26, S28, S45)

Ecotoxicity

Fish toxicity
LC$_{50}$ (96 hr) fathead minnow 150 μg l^{-1} (soft water) 20,000 μg l^{-1} (hard water) static
bioassay (1).
LC$_{50}$ (96 hr) bluegill sunfish 1300 μg l^{-1} (soft water) 12,000 μg l^{-1} (hard water) static
bioassay (1).

Invertebrate toxicity
EC$_{50}$ (48 hr) *Daphnia magna* 2500 μg l^{-1} (2).

Bioaccumulation
Bioconcentration of 100-fold can occur under constant exposure. Not significant in
spill conditions (3).

Effects to non-target species
LD$_{Lo}$ subcutaneous frog 1041 mg kg^{-1} (4).

Mammalian and avian toxicity

Acute data
LD$_{50}$ intraperitoneal, intravenous mouse 0.5-3.0 mg kg^{-1} (5,6).
LD$_{Lo}$ intraperitoneal guinea pig 100 mg kg^{-1} (5).

Carcinogenicity and long-term effects
Limited evidence for carcinogenicity to humans, sufficient evidence for
carcinogenicity to animals, IARC classification group 2A, for beryllium and its
compounds (7).

Metabolism and pharmacokinetics
Intravenous rat (dose unspecified) circulating beryllium carried to all tissues. Analysis
2.5 hr after administration gave measurable levels in most organs. Organ distribution
was dose-dependent, favouring skeleton for smaller doses and liver for larger doses
(8).

Genotoxicity

Salmonella typhimurium TA100 and TA98 with and without metabolic activation negative. Mild positive effect in rec. assay. Induced sister chromatid exchanges in V79 Chinese hamster cells (9).

Any other adverse effects

Intratesticular injections (7 days) caused decrease of testis weight from 651 mg in control to 580 mg in treated rat. Partial necrosis occurred in 2 days and total necrosis occurred within 7 days (10).

Legislation

Limited under EC Directive on Drinking Water Quality 80/778/EEC. Nitrates : guide level 25 mg l^{-1}, maximum admissible concentration 50 mg l^{-1} (11).

Any other comments

Beryllium has been measured in 59 samples of surface water from 15 US/Canadian rivers. Highest concentration was less than 0.22 μg l^{-1}. Beryllium was found in 85% of samples from 15 major river basins of the conterminous USA at concentrations of 0.01-1.22 μg l^{-1}. Beryllium has been reported to occur in US drinking water at 0.1-0.7 μg l^{-1} with a mean of 0.013 μg l^{-1} total beryllium (12).

Human health effects, experimental toxicology and environmental effects reviewed (13,14).

References

1. Tarzwell, C. M; et al. *Toxicity of Less Common Metals to Fish from Industrial Wastes* 1960, **5**, 12
2. *Ambient Water Quality Criteria Doc: Beryllium* 1980, EPA 400/5-80-024
3. *CHRIS – Hazardous Chemical Data* 1984-1985, US Coastguard, US Govt. Print. Off., Washington, DC
4. Richter, V. *Beitrage zur Pharmakologie der Berylliums* Dissertation Universitat Wurzburg 1930
5. *Environ. Qual. Saf. Suppl.* 1975, **1**, 1, Academic Press, London
6. *Current Science* 1986, **55**, 899
7. *IARC Monograph* 1987, **Suppl. 7**, 58
8. Frieberg, L; et al. (Ed.) *Handbook of the Toxicology of Metab.* 2nd ed., 1986, Elsevier, Amsterdam
9. Kuroda, K; et al. *Mutat. Res.* 1991, **264**, 163-170
10. Kamboj, V. P; et al. *J. Reprod. Fert.* 1964, **7**, 21-28
11. *EC Directive Relating to the Quality of Water Intended for Human Consumption* 1982, 80/778/EEC, Office for Official Publications of the European Communities, 2 rue Mercier, L-2985 Luxembourg
12. *National Research Council Drinking Water and Health* **1**, 1977, National Academy Press, Washington, DC
13. *Dangerous Prop. Ind. Mater. Rep.* 1989, **9**(5), 29-37
14. *ECETOC Technical Report No. 30(4)* 1991, European Chemical Industry Ecology and Toxicology Centre, B-1160 Brussels

B125 Beryllium sulfate

BeSO₄

CAS Registry No. 13510-49-1
Synonyms sulfuric acid, beryllium salt
Mol. Formula BeO_4S **Mol. Wt.** 105.07
Uses In x-ray media.

Physical properties
M. Pt. 550-600°C (decomp.); **Specific gravity** d^{25} 2.443.

Occupational exposure
US TLV (TWA) 0.002 mg m^{-3}; **UK Long-term limit** 0.002 mg m^{-3}; **UN No.** 1566;
Conveyance classification toxic substance; **Supply**
classification toxic.

Risk phrases
Very toxic by inhalation and in contact with skin – Irritating to respiratory system –
Danger of very serious irreversible effects (R26/27, R37, R39)

Safety phrases
In case of contact with eyes, rinse immediately with plenty of water and seek medical
advice – After contact with skin, wash immediately with plenty of water – In case of
accident or if you feel unwell, seek medical advice immediately (show label where
possible) (S26, S28, S45)

Mammalian and avian toxicity

Acute data
LD$_{50}$ oral mouse, rat 80 mg kg^{-1} (1).
LD$_{50}$ subcutaneous rat 1.5 mg kg^{-1} (1).
LD$_{50}$ intravenous, intraperitoneal rat 7-18 mg kg^{-1} (1).
LD$_{50}$ intravenous monkey 0.6 mg kg^{-1} (1).

Sub-acute data
Intravenous rats, rabbits, 0.5 or 0.75 mg Be^{-1} kg^{-1} (injected as sulfate solution) caused
death within 72 hr. Symptomatic effects included low blood sugar levels and necrotic
liver lesions (2).

Carcinogenicity and long-term effects
Limited evidence for carcinogenicity to humans, sufficient evidence for
carcinogenicity to animals IARC classification group 2A, beryllium and its
compounds (3,4).
TC$_{Lo}$ (26 wk) inhalation rat 432 μ g m^{-3} induced tumours of lungs, thorax or
respiratory system (5).
TD$_{Lo}$ (2 wk) intratracheal rat 17 mg kg^{-1} intermittent doses induced tumours of lungs,
thorax or respiratory system (6).

Metabolism and pharmacokinetics
Oral rat (dose unspecified) in drinking water, most of the beryllium precipated as
phosphate in the gut lumen and was excreted in the faeces. Ultimate site of

accumulation of beryllium was in the skeleton (7).

Intravenous rat (dose unspecified) circulating beryllium carried to all tissues. Analysis 2.5 hr after administration gave measurable levels in most organs. Organ distribution was dose-dependent, favouring skeleton for smaller doses and liver for larger doses (8).

Beryllium sulfate forms beryllium phosphate in plasma, uptake by liver cells in rats (16).

Irritancy
Beryllium dermatitis, granulomatous ulcerations and conjunctivitis reported (9).

Sensitisation
Depressed lymphocyte stimulation in sensitised animals demonstrated delayed skin reactivity and macrophage migration inhibition (9).

Animals immunised with beryllium sulfate developed skin reactivity as well as antigen-specific alveolar macrophage migration inhibition (10).

Genotoxicity
Salmonella typhimurium TA1530, TA1538, TA1535 and *Saccharomyces cerevisiae* D3 host mediated assay using Swiss-Webster mice negative (11).

Bacillus subtilis H17, M45 without metabolic activation weekly positive (12).

Mouse embryo cell line C3H/10T1/2 equivocal mutagen, transformation assay negative (13).

Syrian hamster embryo cells exposure to 59 μg l^{-1} induced morphological transformation (14).

Any other adverse effects
In vitro pulmonary alveolar macrophages of dogs exposed to beryllium sulfate cytotoxic (15).

On nose-only inhalation rat (1 hr) to beryllium sulfate aerosol caused granulomatosis and a high prevalence of pneumonitis (16).

A single exposure inhalation rat (21 day) 3.3 or 7.0 μg Be l^{-1} (administered as beryllium sulfate) caused lung injury enhanced by lactate dehydrogenuse and alkaline phosphotase activity (17).

Rats exposed to 66 γ ft^3 6 hr day^{-1} beryllium sulfate (corresponds to 5.6 γ ft^3 beryllium) caused malaise and apathy, anorexia and dyspnea (18).

Legislation
Limited under EC Directive on Drinking Water Quality 80/778/EEC. Sulfates: guide level 25 mg l^{-1}, maximum admissible concentration 250 mg l^{-1} (19).

Any other comments
Beryllium has been measured in 59 samples of surface water from 15 US and Canadian rivers. Highest concentration was less than 0.22 μg l^{-1}. Beryllium was found in 85% of samples from 15 major river basins of the conterminous US at concentrations of 0.01-1.22 μg l^{-1}. Beryllium has been reported to occur in US drinking water at 0.1-0.7 μg l^{-1} with a mean of 0.013 μg l^{-1} total beryllium (20).

Experimental toxicology, human health and environmental effects reviewed (21,22).

Insoluble in cold water, converts to tetrahydrate in hot water (1).

References

1. Luckey, T. D. *Metal Toxicity in Mammals* 1977, **2**, 43, Plenum Press, New York
2. Aldridge, W. N. *Br. J. Exp. Path.* 1950, **31**, 473
3. *IARC Monograph* 1980, **23**, 143
4. *IARC Monograph* 1987, **Suppl. 7**, 58
5. *Prog. Exp. Tumor. Res.* 1961, **2**, 203
6. *Arch. Ind. Health* 1959, **19**, 19
7. *IARC Monograph* 1972, **1**, 24
8. Frieberg, L; et al. (Ed.) *Handbook of the Toxicology of Metals* 2nd ed., 1986, Elsevier, Amsterdam
9. Bice, D; et al. *J. Allergy Clin. Immunol.* 1977, **59**(6), 425-436
10. Curtis, G. H. *Arch. Dermatol. Syphitol.* 1951, **64**, 470
11. Simmon, V. F; et al. *J. Natl. Cancer. Inst.* 1979, **62**, 911-918
12. Kanematsu, N; et al. *Mutat. Res.* 1980, **77**, 109
13. Dunkel, V. C; et al. *Environ. Mol. Mutagen.* 1988, **12**(1), 21-31
14. Diapolo, J. A. *Cancer Res.* 1979, **39**, 1008
15. Finch, G. L. *Toxicol. Lett.* 1988, **41**(2), 97-105
16. Reeves, A. L; et al. *Trans. N.Y. Acad. Sci.* 1976, **36**(1), 78-93
17. Sendelbach, L. E. *Toxicol. Appl. Pharmacol.* 1987, **90**(2), 322-329
18. Shepers, G. W. H. *Ind. Med. Surg.* 1964, **33**, 1-16
19. *EC Directive Relating to the Quality of Water Intended for Human Consumption* 1982, 80/778/EEC, Office for Official Publications of the European Communities, 2 rue Mercier, L-2985 Luxembourg
20. *National Research Council Drinking Water & Health* 1977, **1**, National Academy Press, Washington, DC
21. *ECETOC Technical Reoprt No. 30(4)* 1991, European Chemical Industry Ecology Toxicology Centre, B-1160 Brussels
22. *Cah. Notes Doc.* 1989, **136**, 549-554, (Fr.) (*Chem. Abstr.* **111**, R238722x)

B126 Betanin

CAS Registry No. 7659-95-2

Synonyms 5-β-*D*-glucoside; E162; phytolaccanin; 2,6-pyridinedicarboxylic acid, 4-[2-[2-carboxy-5-(β-*D*-glucopyranosyloxy)-2,3-dihydro-6-hydroxy-1-*H*-indol-1-yl]ethenyl]-2,3-dihydro-, [5-(R,R)]-

Mol. Formula $C_{24}H_{26}N_2O_{13}$ **Mol. Wt.** 550.48

Uses Natural dyestuff. Taxonomically important.

Occurrence Obtained from red beet extracts, *Beta vulgaris* (1).

Mammalian and avian toxicity

Carcinogenicity and long-term effects

50 mg kg^{-1} pure or degraded betanin administered to partially hepatectomised

Sprague-Dawley rats pretreated with phenobarbital for 6 months showed no cancer initiating activity (2).

Any other comments

Red Beetroot contains both red and yellow pigments of the class betaines. Used in food to replace delisted FD&C Reds 2 and 4 (1).
Principal colouring compound is β-D-glucopyranoside of betanidine. Colour is unstable in many food processing conditions (most stable at pH 4.0-5.0). May contain sodium nitrate E251 up to 25 mg kg^{-1} so may need to be eliminated from the diets of babies and young children (3).

References

1. *Kirk-Othmer Encyclopedia of Chemical Technology* 3rd ed., 1979, **8**, John Wiley & Sons, New York
2. Goldsworthy, T. L; et al. *J. Toxicol. Environ. Health* 1985, **16**(3-4), 389-402
3. Hanssen, M. *The New E for Additives* 1987, Thorsons Publishers, VT

B127   Bifenox

CAS Registry No. 42576-02-3
Synonyms methyl 5-(2,4-dichlorophenoxy)-2-nitrobenzoate;
5-(2,4-dichlorophenoxy)-2-nitrobenzoic acid methyl ester;
2,4-dichlorophenyl-3-(methoxycarbonyl)-4-nitrophenyl ether
Mol. Formula $C_{14}H_9Cl_2NO_5$ **Mol. Wt.** 342.14
Uses Pre-emergence herbicide.

Physical properties

M. Pt. 84-86°C; **Partition coefficient** log P_{ow} 4.5; **Volatility** v.p. 2.4×10^{-6} mmHg at 30°C.

Solubility

Water: 0.35 mg l^{-1} at 25°C. Organic solvent: acetone, chlorobenzene, xylene, ethanol

Ecotoxicity

Fish toxicity

LC$_{50}$ (96 hr) bluegill sunfish, rainbow trout 0.64-0.87 mg l^{-1} (1,2).

Effects to non-target species

LD$_{50}$ (8 day) dietary duck, pheasant >5000 mg kg^{-1} (1,2).

Environmental fate

Degradation studies

In soil t$_{1/2}$ 7-14 days. Residual activity 6-8 weeks (1,2).

Mammalian and avian toxicity

Acute data
LD_{50} oral rat >6400 mg kg^{-1} (1).
LD_{50} oral mouse 4556 mg kg^{-1} (1).
LD_{50} dermal rabbit >20,000 mg kg^{-1} (1,2).

Carcinogenicity and long-term effects
In 2 yr feeding trials the no effect level for rats and dogs was 600 mg kg^{-1} and for mice the no effect level was 50 mg kg^{-1} (1,2).
Oral ♂, ♀ B6C3F1 mice (18 month) dose unspecified induced liver adenomas and carcinomas in ♂ mice only (3).

Teratogenicity and reproductive effects
In rats Bifenox caused low incidence of bloody tears but did not decrease survival to term or to weaning in rats or mice. Did not reduce Harderian gland weight in mice. The authors concluded Bifenox is non-teratogenic (4).

Any other comments
Structurally related to the probable human carcinogen Acifluorfen (3).

References
1. *The Agrochemicals Handbook* 3rd ed., 1991, RSC, London
2. *The Pesticide Manual* 8th ed., 1987, British Crop Protection Council, Farnham
3. Quest, J. A; et al. *Regul. Toxicol. Pharmcol.* 1989, **10**(2), 149-159
4. Francis, B. M. *J. Environ. Sci. Health, Part B* 1986, **B21**(4), 308-317

B128   Bifenthrin

CAS Registry No. 82657-04-3
Synonyms
2-methylbiphenyl-3-ylmethyl(**Z**)-(1RS,3RS)-3-(2-chloro-3,3,3-trifluoroprop-1-enyl)-2, 2-dimethylcyclopropanecarboxylate; [1α,3α(**Z**)]-(±)-(2-methyl[1,1'-biphenyl]-3-yl)methyl 3-(2-chloro-3,3,3-trifluoro-1-propenyl)-2,2-dimethylcyclopropanecarboxylate; Brigade; Talstar; FMC 54800; Biphenthrin
Mol. Formula $C_{23}H_{22}ClF_3O_2$ **Mol. Wt.** 422.88
Uses Contact insecticide and acaricide. Synthetic pyrethroid.

Physical properties
M. Pt. 68-70.6°C; **Flash point** 165°C (open cup); **Specific gravity** d^{25} 1.21; **Volatility** v.p. 1.81×10^{-7} mmHg at 25°C.

Solubility
Water: 0.1 mg l^{-1}. Organic solvent: acetone, chloroform, dichloromethane, diethyl ether, toluene

Ecotoxicity

Fish toxicity
LC$_{50}$ (96 hr) rainbow trout, bluegill sunfish 150-350 mg l^{-1} (1).

Invertebrate toxicity
EC$_{50}$ (48 hr) *Daphnia magna* 160 mg l^{-1} (1).

Effects to non-target species
LD$_{50}$ oral mallard duck >4450 mg kg^{-1} (1).
LD$_{50}$ (8 day) bobwhite quail 4450 mg kg^{-1} in diet (1).

Mammalian and avian toxicity

Acute data
LD$_{50}$ oral rat 54 mg kg^{-1} (1).
LD$_{50}$ dermal rabbit >2000 mg kg^{-1} (1).

Teratogenicity and reproductive effects
Non-teratogenic in rats given <2 mg kg^{-1}, day^{-1}. Non-teratogenic in rabbits given 8 mg kg^{-1}, day^{-1} (1).

References
1. *The Agrochemicals Handbook* 3rd ed., 1991, RSC, London.

B129   Binapacryl

CAS Registry No. 485-31-4
Synonyms 2-butenoic acid, 3-methyl-2-(1-methylpropyl)-4, 6-dinitrophenyl ester; 2-*sec*-butyl-4,6-dinitrophenyl 3-methylcrotonate; 2-(1-methylpropyl)-4, 6-dinitrophenyl-3-methyl-2-butenoate; Dinoseb-methacrylate; crotonic acid, 3-methyl-2-*sec*-butyl-4,6-dinitrophenyl ester; Endosan
Mol. Formula C$_{15}$H$_{18}$N$_2$O$_6$ **Mol. Wt.** 322.32
Uses Acaricide. Fungicide.

Physical properties
M. Pt. 66-67°C; **Specific gravity** d^{20} 1.2; **Volatility** v.p. 1×10^{-4} mmHg at 60°C.

Solubility
Water: 1 mg l^{-1} at 20°C. Organic solvent: acetone, xylene, dichloromethane, isophorone, toluene, ethyl acetate, ethanol, methanol

Occupational exposure

Supply classification toxic.

Risk phrases
Toxic by inhalation, in contact with skin and if swallowed (R23/24/25)

Safety phrases
Keep out of reach of children – Keep away from food, drink and animal feeding stuffs – If you feel unwell, seek medical advice (show label where possible) (S2, S13, S44)

Ecotoxicity

Fish toxicity
LC_{50} (96 hr) channel catfish, bluegill sunfish, rainbow trout 15-50 µg l^{-1} as the technical material (1).

Invertebrate toxicity
Asellus brevicaudus (96 hr) 29 µg l^{-1} at 16°C as the technical material (1).

Effects to non-target species
Non-toxic to bees (2).

Environmental fate

Degradation studies
Residual activity in soil 15-25 days. Degraded in the environment to the amine and carboxylic acid (2).

Mammalian and avian toxicity

Acute data
LD_{50} oral rat, guinea pig, rabbit, dog 150-640 mg kg^{-1} (2).
LD_{50} oral mice 1600-3200 mg kg^{-1} (2).
LD_{50} percutaneous rat 750 mg kg^{-1} (in acetone) (2).

Carcinogenicity and long-term effects
Rats administered 500 mg kg^{-1} in diet for 2 yr and dogs receiving 50 mg kg^{-1} in diet for 2 yr showed no ill effects (2).

Metabolism and pharmacokinetics
In mammals (species unspecified) after oral administration binapacryl was eliminated as the glucuronic acid conjugate (2).
Cytochrome P450, lipoperoxidase and xanthine oxidase in liver and blood glutathione levels were altered by an unspecified concentration of binapacryl (3).
After a single dose of binapacryl, 17% was excreted in the urine of rats and rabbits within 48 hr. 0.12% could still be detected in the urine of rats after 10 days (4).

Genotoxicity

Salmonella typhimurium TA100 without metabolic activation positive (5).

Any other comments

Listed as compound of little commercial interest, although it is noted that the pesticide may still be widely used elsewhere (6).
Physico-chemical properties, human health effects and experimental toxicology are reviewed (7).

References

1. *Handbook of Acute Toxicity of Chemicals to Fish and Aquatic Invertebrates* 1980, No. 137, US Dept. Int. Fish & Wildlife
2. *The Agrochemicals Handbook* 2nd ed., 1987, RSC, London
3. Popovic, M; et al. *Arh. Hig. Rada. Toksikol.* 1989, **40**(3), 277-283
4. Hayes, W. J., Jr. *Pesticides Studied in Man* 1982, 470, Williams & Wilkins, London
5. *Mutat. Res.* 1983, **116**, 185
6. Worthing, C. R. (Ed.) *The Pesticide Manual* 9th ed., 1991, British Crop Protection Council, Farnham
7. *ECETOC Technical Report No. 30(4)* 1991, European Chemical Industry Ecology and Toxicology Centre, B-1160 Brussels

B130   1,1'-Biphenyl

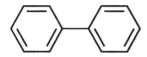

CAS Registry No. 92-52-4
Synonyms diphenyl; phenylbenzene; Lemonene; Xenene
Mol. Formula $C_{12}H_{10}$ **Mol. Wt.** 154.21
Uses In organic synthesis. Heat transfer medium. Formerly fumigant for oranges during shipment.

Physical properties

M. Pt. 69-71°C; **B. Pt.** 254-255°C; **Flash point** 113°C;
Specific gravity 1.041; **Partition coefficient** log P_{ow} 3.98.

Solubility
Water: 17.5 mg l^{-1}. Organic solvent: ethanol, diethyl ether

Occupational exposure

US TLV (TWA) 0.2 ppm (1.5 mg m^{-3}); UK Long-term limit 0.2 ppm (1.5 mg m^{-3}); UK Short-term limit 0.6 ppm (4 mg m^{-3}).

Ecotoxicity

Invertebrate toxicity
EC_{50} (24,48 hr) *Daphnia magna* 1.3-0.36 mg l^{-1} (1).
Paracentrotus lividus and *Sphaerechinus granularis* (sea urchins) ≥1.5 mg l^{-1} caused developmental defects and mitotic abnormalities, following exposure of embryos or by pretreatment of sperm or eggs (2).

Environmental fate

Degradation studies
After 24 hr incubation with normal sewage sludge 0% degradation, 135 hr incubation 79% degradation. Incubation with acclimated sewage sludge, 24 hr 87% degradation and 135 hr 100% degradation had occurred (3).

Pseudomonas sp. and gram negative isolate degrade 1,1'-biphenyl to yeild a variety of products. Metabolites include 2,3-dihydroxy-2,3-dihydroxybiphenyl; 2,3-dihydroxybiphenyl; 2-hydroxy-6-oxo-6-phenylhexa-2,4-dienoate; 2-hydroxy-3-phenyl-6-oxohexa-2;4-dienoate; benzoic acid, 2-oxopenta-4-enoate and phenylpyruvic acid (4,5).

Abiotic removal
Calculated $t_{1/2}$ in water at 25°C and 1 m depth 7.5 hr, based on an evaporation rate of 0.092 m hr^{-1} (6).

Mammalian and avian toxicity

Acute data
LD$_{50}$ oral rat, rabbit 2400-3280 mg kg^{-1} (7,8).

Carcinogenicity and long-term effects
Oral mice (2 yr) 0.25% in feed induced tumours of mammary glands, lungs, lymphocytic tissue, liver, stomach, skin, ovary, uterus, pituitary gland (9).

Genotoxicity
Salmonella typhimurium TA97, TA98, TA100 with and without metabloic activation negative (10).
Saccharomyces cerevisiae D7 with and without metabolic activation positive (2).

Any other adverse effects
Central nervous system depression, paralysis, convulsions have been observed in rats (8).

Any other comments
Human health effects, experimental toxicology and environmental effects reviewed (11).
Aerobic and anaerobic microbial degradation reviewed (12).

References
1. Gersich, F. M; et al. *Bull. Environ. Contam. Toxicol.* 1989, **43** (3), 355-362
2. Pagano, G; et al. *Teratog. Carcinog. Mutagen.* 1983, **3**(4), 377-393
3. Gaffney, P. E. *J. Water Pollut. Control Fed.* 1976, **48**(12), 2731-2737
4. Lunt, D; et al. *Biochem. J.* 1970, **118**, 54
5. Catelani, D; et al. *Biochem. J.* 1973, **134**, 1063
6. MacKay, D; et al. *Environ. Sci. Tech.* 1975, **9**(13), 1178-1180
7. *J. Nagoya City Univ. Med. Assoc.* 1977, **28**, 983
8. Deichmann; et al. *J. Ind. Hyg. Toxicol.* 1947, **29**, 1
9. Imai, S; et al. *Nara Igaku Zasshi* 1983, **34**(5), 512-522, (Jap.) (*Chem. Abstr.* **100**, 187082b)
10. Brams, A; et al. *Toxicol. Lett.* 1987, **38**, 123-133
11. *ECETOC Technical Report No. 30(4)* 1991, European Chemical Industry Ecology and Toxicology Centre, B-1160 Brussels
12. Arvin, E; et al. *Int. Conf. Physiochem. Biol. Detox. Hazard. Wastes 1988* 1989, **2**, 828-847

B131   Biphenylmethane

CAS Registry No. 101-81-5
Synonyms 1,1-methylenebisbenzene; diphenylmethane; Ditan; benzylbenzene
Mol. Formula $C_{13}H_{12}$ **Mol. Wt.** 168.24
Occurrence Wastewater from 4000 industrial and publicly owned treatment works
surveyed. Biphenylmethane was identified in discharges from the timber, paint and
ink, printing and publishing, coal mining, organics and plastics, synthetics, rubber
processing, pesticide manufacturing, pharmaceuticals, explosives, electronics, oil and
gas, and organic chemicals industries. Also in publicly owned treatment plants.
Highest levels of 29,554 ppm were from the paint and ink industry (1).

Physical properties
M. Pt. 22-24°C; **B. Pt.** 264°C; **Flash point** 110°C;
Specific gravity 1.006.

Solubility
Water: 3.0 mg l^{-1} at 24°C. Organic solvent: ethanol, diethyl ether, chloroform,
hexane, benzene

Ecotoxicity

Invertebrate toxicity
Cell multiplication inhibition test, *Uronema parduczi* 2.2 mg l^{-1} (2).

Bioaccumulation
Calculated bioconcentration factor 825 (3).

Environmental fate

Degradation studies
Degraded by microorganisms isolated from sewage, water and soil (4).
Extensively degraded in 40 days when seeded with a soil inoculum, following a 30
day lag (5).
Three metabolic pathways identified to 1,1,1′, 1′-tetraphenyldimethyl ether,
phenylacetic acid and benzhydrol (4).

Abiotic removal
Biphenylmethane has an absorption band extending beyond 290 nm, hence direct
photolysis likely. Irradiation as a thin film with simulated sunlight causes
photooxidation to diphenylmethyl hydroperoxide, diphenylmethanol and
benzophenone (6).

Mammalian and avian toxicity

Acute data
LD_{Lo} oral rat 5000 mg kg^{-1} (7).

References

1. Shackelford, W. M; et al. *Anal. Chim. Acta* 1983, **146**, 15-27
2. Bringmann, G; et al. *Z. Wasser/Abwasser Forsch.* 1980, **1**, 26-31
3. Lyman, W. J; et al. *Handbook of Chemical Property Estimation Methods* 1982, McGraw-Hill, New York
4. Subba-Rao, R. V; et al. *J. Agric. Food Chem.* 1977, **33**, 101-108
5. Subba-Rao, R. V; et al. *J. Agric. Food Chem.* 1977, **33**, 327-329
6. Aksnes, G; et al. *Chemosphere* 1983, **12**, 385-396
7. *Arch. Ind. Health* 1959, **19**, 403

B132   2,2'-Bis(acryloyloxymethyl)butyl acrylate

$(H_2C{=}CHCO_2CH_2)_3CCH_2CH_3$

CAS Registry No. 15625-89-5

Synonyms trimethylolpropane triacrylate; TMPTA; 2-ethyl-2-(hydroxymethyl)-1,3-propanediol triacrylate; MFA (multifunctional acrylate); MFM (multifunctional monomer); 2-ethyl-2-[[(1-oxo-2-propenyl)oxy]methyl]-1,3-propanediyl propenoate

Mol. Formula $C_{15}H_{20}O_6$ **Mol. Wt.** 296.32

Uses In radiation curing of urethanes, epoxy resins, and polyether resins. Manufacture of synthetic lubricants, coatings and ultra violet-cured inks.

Physical properties

M. Pt. <0°C; **B. Pt.** >315.5°C; **Flash point** >110°C (closed cup); **Specific gravity** d_4^{25} 1.10; **Volatility** v.p. <0.01 mmHg at 20°C.

Occupational exposure

Supply classification irritant.

Risk phrases

Irritating to eyes and skin – May cause sensitisation by skin contact (R36/38, R43)

Safety phrases

Wear eye/face protection (S39)

Mammalian and avian toxicity

Acute data

LD_{50} oral mouse, rat 5170-5190 mg kg^{-1} (1,2).
LD_{50} (24 hr) dermal rabbit 7000 mg kg^{-1} (1).
Inhalation rat (6 hr) exposure to air saturated with bis(cryloyloxymethyl)butylacrylate at 60°C caused no deaths (2).

Sub-acute data

Dermal rabbits (2 wk) 500 mg kg^{-1} day^{-1} 5 day wk^{-1} caused skin corrosion (2).

Carcinogenicity and long-term effects

Oral rats, mice 250-1000 ppm and 1250-2500 ppm, respectively in food, induced hepatocellular carcinomas and hemangiosarcomas (3).

Positive correlation between hepatocarcinogenic effects in rats and binding of metabolites to liver DNA (4).

Selected for general toxicology study by National Toxicology Program (5).

Irritancy

Single application of 1-10% solution causes mild to moderate skin irritation. Single application of 0.1% solution, no reaction (6-8).

Sensitisation

Dermal guinea pig undiluted, no sensitisation in 10 animals (2).

Dermal guinea pig 1-5% solution, some sensitivity. Strong correlation between lymph node alterations and sensitisation potential (6).

Dermal guinea pigs caused no sensitisation on its own, but sensitised 6 out of 6 animals previously sensitised by other acrylates (7).

2,2'-Bis(acryloyloxymethyl)butyl acrylate caused contact hypersensitivity reactions in guinea pigs immunised with it in Freund's Complete Adjuvant. Cross reactivity to pentaerythriol triacrylate, methyl acrylate, methyl vinyl ketone, and 4-vinyl pyridine was reported (8).

Humans exposed to aerosols have developed dermatitis (3).

Genotoxicity

Salmonella typhimurium TA1535 with metabolic activation positive. Mouse lymphoma L5178Y tk+/tk- assay without metabolic activation positive, with metabolic activation negative (9).

Induced mutations, aberrations and micronuclei in L5178Y mouse lymphoma cells without metabolic activation (10).

Any other comments

Human health effects and experimental toxicology reviewed (11).

References

1. Carpenter, C. P; et al. *Toxicol. Appl. Pharmacol.* 1974, **28**, 313-319
2. *Am. Ind. Hyg. Assoc. J.* 1981, **42**(11), B53-B54
3. Emmet, E. A; et al. *JOM, J. Occup. Med.* 1977, **19**(2), 113-115
4. Bjoerkner, B. *Contact Dermatitis* 1984, **11**(4), 236-246
5. *National Toxicology Program Research and Testing Div.* 1992, Management Status Report, NIEHS, Research Triangle Park, NC
6. Bull, J. E; et al. *J. Invest. Dermatol.* 1985, **85**(5), 403-406
7. Parker, D; et al. *Contact Dermatitis* 1983, **9**(1), 55-60
8. Parker, D; et al. *Contact Dermatitis* 1985, **12**(3), 146-154
9. Cameron, T. P; et al. *Environ. Mol. Mutagen.* 1991, **17**(4), 264-271
10. Dearfield, K. L; et al. *Mutagenesis* 1989, **4**(5), 381-393
11. *ECETOC Technical Report No. 30(4)* 1991, European Chemical Industry & Toxicology Centre, B-1160 Brussels

B133   1,4-Bis(bromoacetoxy)-2-butene

$$BrCH_2CO_2CH_2CH=CHCH_2CO_2CH_2Br$$

CAS Registry No. 20679-58-7
Synonyms acetic acid, bromo-, 2-butene-1,4-diyl ester; Fennosan F-50; Grace GR 861; Slimicide V-10
Mol. Formula $C_8H_{10}Br_2O_4$ **Mol. Wt.** 329.98
Uses Controls slime in the water systems of paper mills.

Physical properties

M. Pt. (E) form 56-58°C; **B. Pt.** (Z) form $_{0.005}$ 135-136°C.

Ecotoxicity

Fish toxicity
LC_{50} (96 hr) bleak 520 μg l^{-1} (1).

Invertebrate toxicity
EC_{50} (24 hr) *Chlorella kessleri* 0.1-3.0 mg l^{-1} but more effective against *Monoraphidium griffithii* and *Selenastrum capricornutum* in lower water levels (2).
LC_{50} (96 hr) *Nitocra spinipes* 0.24 mg l^{-1} (1).

Environmental fate

Degradation studies
Biodegradation in fresh water $t_{1/2}$ 6-10 hr and in sea water $t_{1/2}$ ≈24 hr (3).

Mammalian and avian toxicity

Acute data
LD_{50} oral mouse 125 mg kg^{-1} (4,5).

Genotoxicity

Salmonella typhimurium TA98, TA100 with and without metabolic activation negative (4).

References

1.   Linden, E; et al. *Chemosphere* 1979, **11**(12), 843-851
2.   Vuorinen, P. *Finn. Fish Res.* 1982, **4**, 62
3.   Varpu, E. *Pop. Puu* 1982, **64**, 129 (*Chem. Abstr.* **97**, 2200s)
4.   Rasonen, L; et al. *Bull. Environ. Contam. Toxicol.* 1977, **18**, 565 (*Chem. Abstr.* **88**, 184173k)
5.   Linder, E; et al. *Chemosphere* 1978, **8**, 843

B134   2,2-Bis(bromomethyl)-1,3-propanediol

$$C[(CH_2Br)_2](CH_2OH)_2$$

CAS Registry No. 3296-90-0
Synonyms dibromoneopentyl glycol; dibromopentaerythritol; pentaerythritol dibromide; DBNPG

Mol. Formula $C_5H_{10}Br_2O_2$ Mol. Wt. 261.95

Uses In organic synthesis, principal use is in unsaturated polyester resins where it is used to replace part of the regular glycol to yield a resin with a desirable bromine content.

Physical properties

Solubility

Water: ≤1 gl^{-1} at 19°C. Organic solvent: dimethyl sulfoxide, ethanol

Mammalian and avian toxicity

Acute data

LD$_{50}$ oral rat 3460 mg kg^{-1} (1).

Sub-acute data

Mouse gavage (13 wk) 25-400 mg kg^{-1} reduced absolute weights of testis, epididymis and cauda epididymis. Rat gavage 50-800 mg kg^{-1} reduced absolute weight of epididymis, increased relative weight of testis and increased sperm density (2).

Carcinogenicity and long-term effects

Gavage rat, mouse (13 wk) 0-800 mg kg^{-1} and 0-400 mg kg^{-1} day wk^{-1}, respectively, in corn oil. Tumours of kidney and urinary bladder evidenced in high dose animals (2).

National Toxicology Program 2 yr feed study on mice and rats in progress (3).

Teratogenicity and reproductive effects

♀ Swiss CD-1 mice (98 day) 0.1, 0.2 and 0.4% in feed for 7 days. Impaired fertility reported, absence of an effect on reproductive organ weights and oestrual cyclicity (4). Subsequent decreased number of litters and pups per litter in breeding pairs of CD-1 mice (♀ more affected than ♂ in crossover breeding studies). Effects on ♀ reproduction in subsequent continuous breeding studies in mice reported (5).

Metabolism and pharmacokinetics

Rats fed 100 mg kg^{-1} day^{-1} had increased tissue bromide content (1).

Genotoxicity

Salmonella typhimurium TA98, TA100, TA1535, TA1537 with and without metabolic activation negative (6).

In vitro Chinese hamster ovary cells with metabolic activation chromosome aberration positive (7,8).

References

1. Keyes, D. G; et al. *J. Combust. Toxicol.* 1980, **7**, 77
2. Elwell, M. R; et al. *Fundam. Appl. Toxicol.* 1989, **12**(3), 480-490
3. *National Toxicology Program Research & Testing Div.* 1992, Management Status Report, NIEHS, Research Triangle Park, NC
4. Treinen, K. A; et al. *Fundam. Appl. Toxicol.* 1989, **13**(2), 245-255
4. Morrissey, R. E; et al. *Fundam. Appl. Toxicol.* 1988, **11**, 343-358
6. Mortelmans, K; et al. *Environ. Mol. Mutagen.* 1986, **8**(Suppl. 7), 1-119
7. Galloway, S. M. *Environ. Mol. Mutagen.* 1987, **10**(Suppl. 10), 1-175
8. Gulati, D. K; et al. *Environ. Health Research Test, US Gov. Rep Announce Index* 1986, **86**(12)

B135   Bis(4-bromophenyl) ether

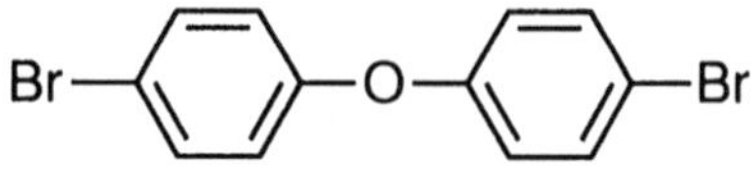

Br—⟨◯⟩—O—⟨◯⟩—Br

CAS Registry No. 2050-47-7
Synonyms 4,4'-dibromodiphenyl ether; 1,1'-oxybis(4-bromobenzene);
di-4-bromophenyl ether
Mol. Formula $C_{12}H_8Br_2O$ **Mol. Wt.** 328.01

Physical properties
M. Pt. 58.5°C; **B. Pt.** 338-340°C.

Mammalian and avian toxicity

Acute data
LD_{50} intraperitoneal mouse 125 mg kg^{-1} (1).

Sub-acute data
Oral ♂ rat (14 day) 33 mg kg^{-1} increased liver weight and increased activity of
cytochrome C reductase and cytochrome P450 content (2).

Any other comments
Human health effects, experimental toxicology and environmental effects reviewed (3).

References
1. *NTIS Report* AD 277-689, Nat. Tech. Inf. Ser., Springfield VA
2. Carlson, G. P. *Toxicol. Lett.* 1980, **5**(1), 19-25
3. *ECETOC Technical Report No.30(4)* 1991, European Chemical Industry Ecology and
   Toxicology Centre, B-1160 Brussels

B136   1,1-Bis(*tert*-butyldioxy)-3,3,5-trimethyl-cyclohexane

H$_3$C⟍   ⟋OOC(CH$_3$)$_3$
               ⟍OOC(CH$_3$)$_3$
H$_3$C⟋ ⟍CH$_3$

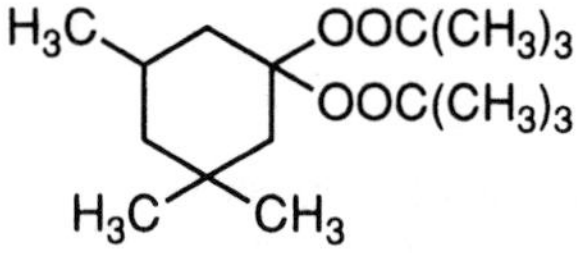

CAS Registry No. 6731-36-8
Synonyms 1,1-bis(*tert*-butylperoxy)-3,3,5-trimethylcyclohexane; 3,3,
5-trimethylcyclohexylidene bis(1,1-dimethylethyl peroxide); DIGIF;
3,3,5-trimethylcyclohexylidene bis(*tert*-butyl peroxide)
Mol. Formula $C_{17}H_{34}O_4$ **Mol. Wt.** 302.46
Uses Catalyst in cross-linking rubber manufacture. Vulcanisation agent.

Mammalian and avian toxicity

Irritancy
Causes local skin irritation, affects eye mucosa and upper respiratory tract. No apparent cumulative action (1).

Any other comments
IC_{50} *Plasmodium falciparum* clone D-G 50 mg l^{-1} antimalarial activity positive. Mice treated with *Plasmodium berghei* at concentrations of 40, 160 and 640 mg kg^{-1} respectively, survived 0.9, 0.3 and 0.3 days longer than the control group. The compound is considered active if mice survive 6.2 days longer than controls (2). Explosion hazards reviewed (3).

References
1. *Kauch. Rezina* 1985, (12), 28-29, (*Chem. Abstr.* **104**, 63708y)
2. Vennerstrom, J. L; et al. *Drug Des. Delivery* 1988, **4**(1), 45-54
3. Matsunaga, T; et al. *Anzen Kogaku* 1984, **23**(2), 82-87, (*Chem. Abstr.* **101**, 113378b)

B137　Bis(4-chlorobenzoyl) peroxide

CAS Registry No. 94-17-7
Synonyms 4-chlorobenzoyl peroxide; *p*-chlorobenzoyl peroxide; di-(4-chlorobenzoyl) peroxide
Mol. Formula $C_{14}H_8Cl_2O_4$　　　　**Mol. Wt.** 311.12
Uses Bleaching agent. Polymerisation catalyst. Vulcanising agent.

Physical properties
M. Pt. 137-138°C.

Occupational exposure
UN No. 2113 (maximum 75% with water);2114 (maximum 52% as paste);2115 (maximum concentration 52% in solution); **Supply classification** explosive and irritant.

Risk phrases
Extreme risk of explosion by shock, friction, fire or other sources of ignition – Irritating to eyes, respiratory system and skin (R3, R36/37/38)

Safety phrases
Keep container tightly closed, in a cool well ventilated place – Keep away from organic materials – Take off immediately all contaminated clothing – Avoid shock and friction – Wear suitable gloves and eye/face protection (S3/7/9, S14, S27, S34, S37/39)

Mammalian and avian toxicity

Acute data

LD_{Lo} intraperitoneal mouse 500 mg kg^{-1} (1).

Any other comments

Human health effects and experimental toxicology reviewed (2).
Strong oxidising agent. Explosion hazard over 38°C. Will ignite on contact with organic materials.

References

1. Summary Tables of Biological Tests Natl. Res. Council Chem. Biol. Coord. Centre (Natl. Acad. Sci. Library), Washington, DC
2. *ECETOC Technical Report No. 30(4)* 1991, European Chemical Industry Ecology and Toxicology Centre, B-1160 Brussels

B138   Bis(2-chloroethoxy)methane

(ClCH2CH2O)2CH2

CAS Registry No. 111-91-1
Synonyms bis(β-chloroethyl)formal; dichloroethyl formal; formaldehyde bis(β-chloroethyl)acetal; 1,1'-methylenebis(oxy)bis(2-chloroethane)
Mol. Formula $C_5H_{10}Cl_2O_2$ **Mol. Wt.** 173.04
Uses Solvent. Intermediate for polysulfide rubber.
Occurrence Contaminant in industrial effluent (1,2).

Physical properties

M. Pt. −32.8°C; **B. Pt.** 217.5°C; **Flash point** 110°C (open cup); **Specific gravity** d_{20}^{20} 1.2339; **Volatility** v. den. 5.9.

Ecotoxicity

Bioaccumulation
Calculated bioconcentration factor 0.84 − 2.2 (3).

Environmental fate

Degradation studies
No biodegradation occurred using settled domestic wastewater inoculum under aerobic conditions through 3 successive subcultures (4).

Abiotic removal
Estimated hydrolytic $t_{1/2}$ 6 month-2 yr independent of pH (5).
Estimated $t_{1/2}$ for atmospheric reaction with photochemically produced hydroxyl radicals 10 hr (6).
Direct degradation is not as important a fate process as bis(2-chloroetnoxy)methane does not have chromophores that absorb visible or near ultraviolet radiation (5).
Based on estimated $\log_{ow}$ (K_{ow}) of 0.75 and estimated water solubility of 1.2×10^5 mg l^{-1} at 25°C, K_{oc} values are 7-61, suggesting high to very high mobility in soil (7).

Mammalian and avian toxicity

Acute data

LD_{50} oral rat 65 mg kg^{-1} (8).

LC_{50} (4 hr) inhalation rat 62 ppm (9).

LD_{50} dermal guinea pig 170 mg kg^{-1} (8).

Irritancy

Dermal rabbit (24 hr) 10 mg caused irritation and 500 mg administered into rabbit eye caused irritation (10).

Any other comments

Human health effects and experimental toxicology reviewed (11, 12).

Decomposed by mineral acids.

References

1. Francis, A. J; et al. *Nuc. Tech.* 1980, **50**, 158-163
2. Staples, C. A; et al. *Environ. Toxicol. Chem.* 1985, **4**, 131-142
3. Lyman, W. J; et al. *Handbook of Chemical Property Estimation Methods* 1982, McGraw-Hill, New York
4. Tabak, H. H; et al. *Proc. Symp. Assoc. Off. Anal. Chem.* 1981, 94th Ann. Mtg., Washington, DC
5. Callahan, M. A; et al. *Water Related Fate of 129 Priority Pollutants* 1979, **2**, USEPA-440/4-79-029B
6. Atkinson, R. *Int. J. Chem. Kinetics* 1987, **19**, 799-828
7. Swann, R. L; et al. *Res. Rev.* 1983, **85**, 17-28
8. *J. Ind. Hyg. Toxicol.* 1948, **30**(6), 3
9. *J. Ind. Hyg. Toxicol.* 1949, **31**, 343
10. *Am. J. Ophthalmology* 1946, **29**, 1363
11. *Dangerous Prop. Ind. Mater. Report* 1987, **7**(4), 39-42
12. *ECETOC Technical Report No. 30*(4) 1991, European Chemical Industry Ecology and Toxicology Centre, B-1160 Brussels

B139   5-Bis(2-chloroethyl)aminouracil

CAS Registry No. 66-75-1

Synonyms aminouracil mustard; 5-(bis(2-chloroethyl)amino)-2,4(1H,3H)pyrimidinedione; 5-N, N-bis(2-chloroethyl)-aminouracil; 2,6-dihydroxy-5-bis(2-chloroethyl)aminopyramidine; 2,4(1H3H)pyramidinedione, 5-bis(2-chloroethylamino); uracil mustard

Mol. Formula $C_8H_{11}Cl_2N_3O_2$ **Mol. Wt.** 252.10

Uses Antineoplastic agent. Used in treatment of chronic lymphocytic leukaemia and malignant lymphomas. Occasionally used to treat mycosis fungoides, polycythaemia, thrombocytosis. Adjunct in treatment of carcinoma of ovary and lung.

Physical properties

M. Pt. 206°C (decomp.).

Solubility
Water: <1 g l^{-1} at 20°C. Organic solvent: dimethyl sulfoxide, ethanol, acetone

Ecotoxicity

Bioaccumulation
Calculated bioconcentration factor 0.783 (1).

Mammalian and avian toxicity

Acute data
LD$_{50}$ oral rat 7.5 mg kg^{-1} (2).
LD$_{50}$ intraperitoneal rat 1250 µg kg^{-1} (3).
LD$_{Lo}$ intraperitoneal mouse 3 mg kg^{-1} (4).

Carcinogenicity and long-term effects
Inadequate evidence of carcinogenity to humans, sufficient evidence of carcinogenity to animals, IARC classification group 2B (5).
Intravenous mice (24 wk) 40, 20 and 8 mg kg^{-1} all developed lung adenomas and adencarcinomas (6).

Teratogenicity and reproductive effects
TD$_{Lo}$ (21 day) intraperitoneal pregnant rat 0.3-0.6 mg kg^{-1} reported malformations in surviving offspring including exencephaly, retarded and club appendages and deformed paws and tails (7).

Metabolism and pharmacokinetics
Oral administration of 2 mg kg^{-1} or intravenous injection of 1 mg kg^{-1} in dogs. No evidence of the drug detected at 2 hr. Less than 1% dose recovered unchanged in urine (8).

Genotoxicity

Salmonella typhimurium TA1535 without metabolic activation positive (9).

Any other adverse effects

Symptoms of poisoning include nausea, vomiting, diarrhoea, bone marrow depression (7).

Any other comments

Chemical and physical properties, human health effects, carcinogenicity and experimental toxicology reviewed (10-12).

References

1. Lyman, W. J; et al. *Handbook of Chemical Estimation Methods* 1982, McGraw-Hill, New York
2. Sax, N. I; et al. *Dangerous Properties of Industrial Materials* 1989, 7th ed., Van Nostrand Reinhold, New York
3. *Advances in Teratology* 1968, **3**, 181
4. *Toxicol. Appl. Pharmacol.* 1972, **23**, 288
5. *IARC Monograph* 1987, **Suppl. 7**, 370-371

6. Stoner, G. D; et al. *Cancer Res.* 1973, **33**, 3069-3085
7. Chaube, S; et al. *Cancer Chem. Ther. Rep.* 1967, **51**, 363-376
8. Gilman, A. G; et al. (Ed.) *Goodman and Gilman's The Pharmacological Basis of Therapeutics* 7th ed., 1985, Macmillan, New York
9. *J. Nat. Cancer Inst.* 1979, **62**, 893
10. *Dangerous Prop. Ind. Mater. Report* 1987, **7**(4), 43-45
11. *IARC Monograph* 1974, **9**, 235-241
12. *ECETOC Technical Report No. 30(4)* 1991, European Chemical Industry Ecology and Toxicology Centre, B-1160 Brussels

B140   Bis(2-chloroethyl) ether

$ClCH_2CH_2OCH_2CH_2Cl$

CAS Registry No. 111-44-4

Synonyms bis(β-chloroethyl) ether; 1-chloro-2-(β-chloroethoxy)ethane; 2, 2'-dichlorethyl ether; 1,1'-oxybis(2-chloro)ethane; Chlorex; *sym*-dichloroethyl ether

Mol. Formula $C_4H_8Cl_2O$ **Mol. Wt.** 143.01

Uses Soil fumigant. Used to make polysulfide polymers. Solvent for resins, rubbers, cellulox posters in paints industry. Extracting agent in petroleum industry.

Occurrence Found in water samples probably as an artifact in chemical analysis procedures (1).

Bis (2-chloroethyl) ether residues found in treated wastewater effluent of organic chemicals manufacturing, plastics industry, paint and ink formulation and synthetic rubber plants (2,3).

Physical properties

M. Pt. −50°C; **B. Pt.** 178.5°C; **Flash point** 55°C (closed cup); **Specific gravity** d_{20}^{20} 1.222; **Partition coefficient** log P_{ow} 1.29 (2); **Volatility** v.p. 0.7 mmHg at 20°C; v. den. 4.93..

Solubility

Water: 10.2 g l^{-1} at 20°C. Organic solvent: miscible with aromatics

Occupational exposure

US TLV (TWA) 5 ppm (29 mg m^{-3}); **US TLV (STEL)** 10 ppm (58 mg m^{-3}); **UN No.** 1916; **HAZCHEM Code** 2W; **Conveyance classification** toxic substance; **Supply classification** flammable and toxic.

Risk phrases

Flammable – Very toxic by inhalation, in contact with skin and if swallowed – Possible risk of irreversible effects (R10, R26/27/28, R40)

Safety phrases

Keep container tightly closed and in a well ventilated place – Take off immediately all contaminated clothing – In case of insufficient ventilation, wear suitable respiratory equipment – In case of accident or if you feel unwell, seek medical advice immediately (show label where possible) (S7/9, S27, S38, S45)

Ecotoxicity

Fish toxicity
LC_{50} (96 hr) bluegill sunfish 600 mg l^{-1} (4).

Invertebrate toxicity
EC_{50} (48 hr) *Daphnia magna* 238-240 mg l^{-1} static assay (5,6).

Bioaccumulation
Bioconcentration factor of 11 observed in bluegill sunfish 14 day exposure (7).

Environmental fate

Nitrification inhibition
Does not inhibit anaerobic digestion, laboratory scale at 100 mg l^{-1} (8).

Degradation studies
Concentrations of $\geq$300 mg l^{-1} significantly inhibited overall biodegradation during treatment of wastewater from an organic chemical manufacturing plant (9).

Abiotic removal
Estimated hydrolysis $t_{1/2}$ 20 yr at 25°C (9).
Abiotic estimated atmospheric $t_{1/2}$ 13.44 hr for reaction of bis(2-chloroethyl) ether with hydroxy radicals (10).

Mammalian and avian toxicity

Acute data
LD_{50} oral rat, mouse 75-112 mg kg^{-1} (11).
LC_{50} (4 hr) inhalation rat 330 mg m^{-3} (11).
LC_{50} (2 hr) inhalation mouse 650 mg m^{-3} (11).
LD_{50} dermal guinea pig 300 mg kg^{-1} (12).

Carcinogenicity and long-term effects
No adequate evidence for carcinogenicity to humans, limited evidence for carcinogenicity to animals, IARC classification group 3 (13).

Metabolism and pharmacokinetics
Rapidly distributed throughout body tissues, with the kidney and lung taking up greatest amounts (species unspecified) (14).
After injection of 5 mg into rats the metabolites thiodiglycolic acid and hydroxyethyl mercapturic acid were detected in 24 hr urine samples (15).
Oral ♂ rat 40 mg kg^{-1} dose ^{14}C-labelled bis (2-chloroethyl) ether. Excreted in urine and respired air for 48 hr. Expired $^{14}CO_2$ accounted for 11.5% of dose, urinary ^{14}C 64.7% and faeces 2.4% $t_{1/2}$ for elimination 12 hr. Inhalation exposure ♂ rat (8 hr) 10, 50, 100 and 500 ppm. Metabolites hydroxymethyl mercapturic acid and thiodiglycolic acid determined in 24 hr urine samples (15).

Irritancy
Dermal rabbit (24 hr) 10 mg caused irritation (12).
20 mg instilled into rabbit eye caused irritation (16).

Genotoxicity
Salmonella typhimurium TA1535, TA1537, TA97, TA98, TA100 with and without metabolic activation positive (17).

Any other adverse effects

Narcotic at high concentrations. In humans, vapour is highly irritant to eyes, nose and respiratory passages (14).

Rapidly absorbed through skin although skin itself showed no marked irritation (18).

Any other comments

Discontinued by Union Carbide for use as a soil fumigant (19).

Toxicity of bis(2-chloroethyl)ether reviewed (20).

Adverse health implications of bis(2-chloroethyl)ether and other compounds reported (21).

Human health effects and experimental toxicology reviewed (22,23).

References

1. Richardson, M. L. *Compendium of Toxicological Ecological Data on Chemicals found by GC-MS in Water Samples* 1985, Thames Water Authority
2. Howard, P. H. *Fate and Exposure Data for Organic Chemicals* 1989, **1**, 84-89, Lewis Publishers, Chelsea, MI
3. Durkin, P.R; et al. *Investigation of Selected Potential Environmental Contaminants: Haloethers, final report* 1975, EPA 560/2-75-006
4. Buccafusco, R. J; et al. *Bull. Environ. Contam. Toxicol.* 1981, **26**, 446
5. Le Blanc, G. A. *Bull. Environ. Contam. Toxicol.* 1980, **24**, 684-691
6. *USEPA: Ambient Water Quality Criteria: Chloro-alkyl ethers* 1980, Washington, DC
7. Mabey, W. R; et al. *Aquatic Fate Process Data for Organic Priority Pollution* 1981, EPA 440/4-81-014, Washington, DC
8. Richardson, M. *Nitrification Inhibition the Treatment of Sewage* 1985, RSC, London, Thames Water Reading
9. Davis, E. M; et al. *Water Sci. Technol.* **21**(12), 1833-1836
10. *GEMS: Graphical Exposure Modelling System: Fate of Atmospheric Pollutants Database* 1986, USEPA Office of Toxic Substances, Washington, DC
11. Izmerov, N. F; et al. *Toxicometric Parameters of Industrial Toxic Chemicals under Single Exposure* 1982, 45, Moscow
12. *J. Ind. Hyg. Toxicol.* 1948, **30**, 63
13. *IARC Monograph* 1987, **(Suppl. 7)**, 58
14. Browning, E. *Toxicity Metabolism of Industrial Solvents* 1965, **7**, 513-516, Elsevier, Amsterdam
15. Norpoth, K; et al. *J. Cancer Res. Clin. Oncol.* 1986, **112**(2), 125-130
16. *Amer. J. Ophthalmol.* 1946, **29**, 1363
17. Mortelmans, K; et al. *Environ. Mutagen.* 1986, 8(Suppl. 7), 1-119
18. Allen, H. *Chem. Prod.* 1986, **19**, 482
19. *Farm Chemicals Handbook* 1991, 103, Meister Publishing, Willoughby, OH
20. *Dangerous Prop. Ind. Mater. Report* 1987, **7**(4), 62-67
21. *Gov. Rep. Announc. Index* 1988, **88**(13) 833,876g, EPA 600/8-88/023
22. *ECETOC Technical Report No. 30(4)* 1991, European Chemical Industry Ecology and Toxicology Centre, B1160 Brussels
23. *NTIS Report* 1989, ATSDR/TP-89/02 (Order No. P1390-168683), Nat. Tech. Inf. Ser., Springfield, VA

B141   1,3-Bis(chloroethyl)-1-nitrosourea

$$Cl(CH_2)_2N(NO)CONH(CH_2)_2Cl$$

CAS Registry No. 154-93-8

Synonyms 1,3-bis(β-chloroethyl)-1-nitrosourea; carmustine; Nitrumon; BCNU; urea, N,N'-bis(2-chloroethyl)-N-nitroso-; Carmubris

Mol. Formula $C_5H_9Cl_2N_3O_2$ **Mol. Wt.** 214.05

Uses Anticancer drug.

Physical properties

M. Pt. 27-32°C.

Solubility
Water: 4 g l^{-1} at 18°C. Organic solvent: dimethyl sulfoxide, diethyl ether, ethanol

Mammalian and avian toxicity

Acute data
LD_{50} oral dog, mouse 5-19 mg kg^{-1} (1,2).
LD_{Lo} intraperitoneal, intravenous rat, monkey 10-40 mg kg^{-1} (3-6).

Sub-acute data
LD_{Lo} (52 wk intermittently) intravenous child 78 mg kg^{-1} (7).
Treatment with 1,3-bis(chloroethyl)-1-nitrosourea can cause acute interstitial pneumonitis in children with rhabdomyosarcoma or brain tumour (dose unspecified). A review of lung damage is included (8).

Carcinogenicity and long-term effects
Limited evidence for carcinogenity to humans, sufficient evidence for carcinogenicity in experimental animals, IARC classification group 2A (9).
Intravenous rat (2 yr) dose unspecified induced malignant lung tumours (10).
Skin painting mouse (23 wk) dose unspecified induced papilloma (11).

Teratogenicity and reproductive effects
Intravenous rat 0.25-1.5 mg kg^{-1} prior to breeding and during gestation decreased implantations and 1-4 mg kg^{-1} during organogenesis was teratogenic. Intravenous rabbit 0.5-4 mg kg^{-1} day^{-1} on days 6-18 of gestation caused weight loss, abortion, maximum dose increased mortality but was not teratogenic (12).

Metabolism and pharmacokinetics
1,3-bis(chloroethyl)-1-nitrosourea is readily absorbed from the gastrointestinal tract. It is rapidly metabolised, $t_{1/2}$ <15 min, excreted in urine and exhaled as carbon dioxide (13). Sensitive to oxidation and hydrolysis, forms alkylating and carbamoylating intermediates, $t_{1/2}$ 1 hr at neutral pH (14).

Genotoxicity

Salmonella typhimurium TA1535 with and without metabollic activation positive (15).
Escherichia coli multitest negative. Recombin-agenic and SOS inducing (16).
Sacchromyces cerevisiae D5 mitotic crossing over positive (17).
Saccharomyces cerevisiae diploid homozygous ra 18 enhanced mutagenic and recombination potential (18).

Human lymphocytes (1 hr) induced sister chromatid exchange and chromosomal aberrations (19).

C6B3F1 mice single and repeated dose caused proliferation of bone marrow stem cells. DNA crosslinking and myelotoxicity (20).

Direct acting, bifunctional alkylating agent which induced chromosomal abberations, micronuclei and sister chromatid exchanges in cells of mice treated *in vivo*. *In vitro* human cells, rodents, bacteria caused DNA damage and mutation. Induced gene conversion in yeast and sex linked recessive lethal mutations in *Drosophila melanogaster* (21).

Any other adverse effects to man

Two cases of non-lymphoblastic leukaemia reported among 1621 patients treated with the drug for brain tumours. Considered leukaemogenic in humans (22,23).

Any other adverse effects

Rats given a single intraperitoneal injection 20 mg kg^{-1} and sacrificed at day 14 showed decreased cytochrome P450, ethylmorphine N-demethylase activity and hepatic delta aminolevulinic acid synthetase activity. Produced cholestasis which preceded its effect on microsomal mixed function oxygenase activity (24).

Single injection ♂ mice (unspecified dose) killed differential spermatogenesis and caused some stem cell killing (25).

In humans, systemic effects include nausea, vomiting, diarrhoea, dyspnoea, flushing of skin, oesophagitis, cytotoxic effect in liver, kidneys and central nervous system, leukopenia and thrombocytopenia (26).

Delayed and cumulative bone marrow depression has been reported in humans. A potentially fatal pulmonary toxin, serious risk occurs at cumulative doses of 1.2-1.5 g m^{-2} body $surface^{-1}$ (27).

Any other comments

The pharmacokinetics of 1,3-bis(chloroethyl)-1-nitrosourea reviewed (28).

Antiviral, antibacterial and antifungal activity reported (29-31).

Can cross the blood/brain barrier (32).

Human health effects and experimental toxicology reviewed (33,34).

Very soluble in lipids. Store between 2-8°C, above 27°C the drug liquefies and decomposes.

References

1. *Toxicol. Appl. Pharmacol.* 1972, **21**, 405
2. *Adv. Cancer Res.* 1972, **16**, 273
3. Linden, C. J. *In Vivo* 1989, 3(4), 259-262
4. *Oncology* 1980, **37**, 177
5. *Develop. Toxicol. Environ. Science* 1980, **8**, 273
6. *J. Surg. Oncol.* 1965, **2**, 202
7. *Cancer* 1978, **42**, 74
8. Smelhaus, V; et al. *Prak. Lek.* 1989, **69**(5), 176-178 (Czech.) (*Chem. Abstr.* **110**, 225130g)
9. *IARC Monograph* 1987, **Suppl. 7**, 59
10. Eisenbrand, G; et al. *Dev. Toxicol. Environ. Sci.* 1980, **8**, 273-278
11. Zackheim, H. S; et al. *Experientia* 1980, **36** (10), 1211-1212
12. Thompson, D. J; et al. *Toxicol. Appl. Pharmacol.* 1974, **30**(3), 422-439
13. *Martindale The Extra Pharmacopoeia* 29th ed., 1989, The Pharmaceutical Press, London
14. Schein, P. S; et al. *Fund. Cancer Chemother. Antibiot. Chemother* 1978, **23**, 64-75

15. Stolzenbach, J. C; et al. *Cancer Chemother. Pharmacol.* 1990, **25**(4), 227-235
16. Stahl, W; et al. *Mutat. Res.* 1988, **206**(4), 459-465
17. Quinto, I; et al. *Mutat. Res.* 1987, **181**(2), 235-242
18. Ferguson, L. R; et al. *Mutat. Res.* 1988, **204**(2), 239-249
19. Ferguson, L. R. *Mutat. Res.* 1990, **241**(4), 369-377
20. Wiencke, J. K; et al. *Cancer Res.* 1985, **45**(10), 4798-4803
21. Berger, M. R; et al. *J. Can. Res. Clin. Onc.* 1985, **110**(3), 185-190
22. Meistrich, M. L; et al. *Cancer Res.* 1982, **42**, 122-131
23. Green, M. H; et al. *New Eng. J. Med.* 1985, **313**, 579
24. Lichtman, S. M. *Ann. Intern. Med.* 1985, **103**, 964
25. *IARC Monograph* 1987, **Suppl. 6**, 60
26. *Compendium of Safety Data Sheets for Research and Industrial Chemicals* 1987, 2208-2209, VCH, Deerfield Beach
27. Weiss, R. B; et al. *Cancer Treat. Rev.* 1981, **8**, 111
28. Balis, F. M; et al. *Clin. Pharmacokin.* 1983, **8**, 202
29. Sidwell, R. W; et al. *Appl. Microbiol.* 1965, **13**, 579-589
30. Pittillo, R. F; et al. *Cancer Res.* 1964, **24**, 1222-1228
31. Hunt, D. E; et al. *Antimicrobial Agents for Chemotherapy* 1965, **5**, 710-716
32. Goodman, L. S. et al. *The Pharmacological Basis of Therapeutics* 7th ed., 1985, Macmillan, New York
33. *IARC Monograph* 1981, **26**, 79-95
34. *ECETOC Technical Report No.30(4)* 1991, European Chemical Industry Ecology and Toxicology Centre, B-1160 Brussels

B142   Bis-1,4-(chloromethoxy)-*p*-xylene

CH$_2$OCH$_2$Cl

CH$_2$OCH$_2$Cl

CAS Registry No. 56894-91-8
Synonyms 1,4-bis(chloromethoxymethyl)benzene; benzene, 1,4-bis[(chloromethoxy)methyl]-; terephthalyl alcohol bis(chloromethyl) ether
Mol. Formula C$_{10}$H$_{12}$Cl$_2$O$_2$ **Mol. Wt.** 235.11
Uses In the preparation of ion-exchange resins.

Mammalian and avian toxicity

Carcinogenicity and long-term effects
No adequate data for evidence of carcinogenicity to humans, limited evidence for carcinogenicity to animals, IARC classification group 3 (1).

Any other comments
Experimental toxicology, human health and environmental effects are reviewed (2,3).

References
1. *IARC Monograph* 1987, **Suppl. 7**, 58
2. *IARC Monograph* 1977, **15**, 37-39
3. *ECETOC Technical Report No. 30(4)* 1991, European Chemical Industry Ecology and Toxicology Centre, B-1160 Brussels

B143   Bis(2-chloro-1-methylethyl) ether

CICH2CH(CH3)OCH(CH3)CH2Cl

CAS Registry No. 108-60-1

Synonyms 2-chloro-1-methylethyl ether; 2,2′-dichlorodiisopropyl ether; Nemamort; propane, 2,2′-oxybis(1-chloro-); 2,2′-oxybis(1-chloropropane)

Mol. Formula $C_6H_{12}Cl_2O$ **Mol. Wt.** 171.07

Uses Nematocide. Extractant. Ingredient in paint and varnish removers. Solvent for fats, waxes and greases. Ingredient in potting and cleaning solutions. A component in textile processing.

Occurrence Occurs as a contaminant in surface drinking water and ground water in Europe and USA, concentration range 0.11-19 µg l^{-1} (1-3).

Occurs in industrial effluent from propylene glycol production (4-7).

Physical properties

M. Pt. –97 – –102°C; **B. Pt.** 187.8°C; **Flash point** 85°C (open cup); **Specific gravity** d_4^{20} 1.103; **Partition coefficient** log P_{ow} 1.76 (8); **Volatility** v.p. 0.10 mmHg at 20°C; v. den. 6.0.

Solubility

Water: 1.7 g l^{-1}. Organic solvent: dimethyl sulfoxide, ethanol

Ecotoxicity

Bioaccumulation

Weighted average bioconcentration factor in edible part of freshwater and estuarine aquatic organisms eaten by Americans calculated as 2.47 (1).

Environmental fate

Degradation studies

Static culture flask biodegradation at 5 mg l^{-1} yeast extract the original culture achieved 85% biodegradation in 7 days and subcultures 100% in 7 days (2).

No biodegradation after 5 days at 20°C incubated with Ohio river water at initial concentration 33 mg l^{-1} (5).

Abiotic removal

Estimated degradation $t_{1/2}$ 59 day and 3 day in Rhine basin and Rhine river respectively (9).

Direct photolysis would not be expected in surface waters as bis(2-chloro-1-methylethyl) ether has no chromophores to absorb visible or near ultraviolet radiation (10).

Calculated atmospheric $t_{1/2}$ 30 hr at atmospheric concentration of 5×10^5 hydroxyl radicals cm^{-3} (11).

Estimated K_{oc} 73 based on measured water solubility of 1700 ppm at 20°C, suggesting high mobility in soil and hence potential leaching (12,13).

Volatilisation $t_{1/2}$ estimated at 6 days from model environmental pond (14).

Mammalian and avian toxicity

Acute data

LD50 oral rat 240 mg kg^{-1} (15).

LC_{Lo} (5 hr) inhalation rat 700 ppm (15).
LD_{50} dermal rabbit 3000 mg kg^{-1} (16).

Carcinogenicity and long-term effects

No adeqate evidence for carcinogenicity to humans, limited evidence for carinogenicity in animals, IARC classification group 3 (17).

No evidence of carcinogenicity inhalation ♀ rats. Positive evidence of carcinogenicity in ♀ and ♂ mice following gavage dosing. Tumours in liver or lung (18).

National Toxicology Program evaluation of bis(2-chloro-1-methylethyl)ether by gavage, rat negative, mouse positive (19).

Metabolism and pharmacokinetics

Oral rat 90 mg kg^{-1} radioactive ^{14}C urinary excretion was 48% within 48 hr or intraperitoneal rat 30 mg kg^{-1} radioactive ^{14}C 50% elimination 19 hr (20).

Urinary metabolites included 1-chloropropan-2-ol, propylene oxide; 2-(2-chloro-1-methylethoxy)propanoic acid and N-acetyl-S-(2-hydroxypropyl)cystenic (20,21).

Irritancy

Dermal rabbit (24 hr) 50 mg caused mild irritation and 500 mg instilled into rabbit eye caused mild irritation (22).

Genotoxicity

Salmonella typhimurium TA98 with metabolic activation positive (23).

Mouse lymphoma L5178Y without metabolic activation negative (24,25).

Chinese hamster ovary cell lines with metabolic activation induced sister chromatid exchange and chromosome aberrations (26).

Any other adverse effects

Inhalation rat (duration unspecified) 350 ppm 4 hr day^{-1} caused respiratory distress and affected liver, kidney and spleen (27).).

Any other comments

Can be used as an agent for preventing the growth of sulfate reducing bacteria (28).

Toxicity of bis(2-chloro-1-methylethyl)ether has been reviewed (29).

Human health effects and experimental toxicology reviewed (30).

References

1. *Ambient Water Quality Criteria Document: Chloroalkyl ethers* 1980, EPA 440/5-80-030
2. *Chemical Hazard Information Profile: bis(2-chloro-1-methylethyl) ether* 1983, USEPA
3. Pret, G. J; et al. *J. Am. Water Works Assoc.* 1980, **72**, 400-404
4. Staples, C. A; et al. *Environ. Toxicol. Chem.* 1985, **41**, 31-42
5. Kleopfer, R. D; et al. *Environ. Sci. Technol.* 1985, **6**, 1036-1037
6. Fishbein, L. *Sci. Total Environ.* 1979, **11**, 223-257
7. Hanser, T.R; et al. *Environ. Monit. Assess.* 1982, **2**, 249-272
8. *IARC Monograph* 1986, **41**, 149-160
9. Callahan, M.A; et al. *Water-related Environmental Fate of 129 Priority Pollutants* 1979, Volume 1, USEPA, Washington DC
10. Zoeteman, B. J. C; et al. *Chemosphere* 1980, **9**, 231-249
11. Atkinson, R. *Int. J. Chem. Kinetics* 1987, **19**, 799-828
12. Yalkowsky, S. H; et al. *Arizona Database of Aquaeous Solubility* 1987
13. Swann, R. L; et al. *Res. Rev.* 1983, **85**, 17-28

14. *EXAMS II computer simulation* 1987, USEPA
15. *Br. J. Ind. Med.* 1970, **27**, 1
16. *Arch. Ind. Hyg. Occup. Med.* 1951, **4**, 119
17. Ashby, J; et al. *Mutat. Res.* 1988, **204**, 17-115
18. *IARC Monograph* 1987, **Suppl. 7**, 59
19. *National Txoicology Program Research and Testing Div.* 1992, Report Nos. TR-191, TR-239, NIEMS, Research Triangle Park, NC
20. Lingg, R. O; et al. *Arch. Environ. Contam. Toxicol.* 1982, **11**, 173-183
21. Smith, C. C. *Ann. New York. Acad. Sci.* 1977, **298**, 111-123
22. *Prehled Prumyslove Toxikol. Org. Latky* 1986, 543
23. Mortelmans, K; et al. *Environ. Mutagen.* 1986, **8**(Suppl. 7), 1-119
24. McGregor, D. B. *Environ. Mol. Mutagen.* 1988, **12**(1), 85-154
25. Tenant, R. W. *Science* 1987, **236**, 933-941
26. Galloway, S. M. *Environ. Mol. Mutagen. Suppl. 10* 1987, **10**, 1-175
27. Gage, J. C. *Br. J. Ind. Med.* 1970, **27**, 1-18
28. Andreson, R. K. *Otkrytiya Izobret* 1987, **25**, 66
29. Kirwin, C. J; et al. *Patty's Industrial Hygiene and Toxicology* 3rd ed., **2A**, 2505, 2519-2520
30. *ECETOC Technical Report no. 30(4)* 1991, European Chemical Industry Ecology and Toxicology Centre, B-1160 Brussels

B144   Bis(4-chlorophenyl)acetic acid

CAS Registry No. 83-05-6
Synonyms bis(*p*-chorophenyl)acetic acid; dichlorodiphenylacetic acid;
p,p'-dichlorodiphenylacetic acid; benzeneacetic acid, 4-chloro-α-(4-chlorophenyl)
Mol. Formula $C_{14}H_{10}Cl_2O_2$ Mol. Wt. 281.14
Occurrence Metabolite of DDT.

Mammalian and avian toxicity

Acute data
LD_{50} oral mouse 590 mg kg^{-1} (1).

Metabolism and pharmacokinetics
Compound is a major urinary metabolite of DDT. In 11 human volunteers with no
known exposure to DDT the levels of bis(*p*-chlorophenyl)acetic acid ranged from
0.025-0.120 μg ml^{-1} of urine (2).

Genotoxicity
Drosophila melanogaster sex chromosome loss and non-disjunction positive (3).

Any other comments
Human health effects and experimental toxicology reviewed (4).

References

1. *Archives Internationales Pharmacodynamie Therapie* 1946, **73**, 128
2. Banerjee, B. D. *Bull. Environ. Contam. Toxicol.* 1987, **38**(5), 798-804
3. *Mutat. Res.* 1972, **16**, 157
4. *ECETOC Technical Report No. 30(4)* 1991, European Chemical Industry Ecology and Toxicology Centre, B-1160 Brussels

B145   2,2-Bis(4-chlorophenyl)ethanol

CAS Registry No. 2642-82-2
Synonyms benzeneethanol, 4-chloro-β-(4-chlorophenyl)-;
2,2-bis(*p*-chlorophenyl)ethanol; 2, 2-bis(4-chlorophenyl)-1-hydroxyethane; DDOH; DDOM
Mol. Formula $C_{14}H_{12}Cl_2O$ **Mol. Wt.** 267.16
Occurrence Metabolite of DDT (1).

Physical properties

M. Pt. 100-102°C.

Mammalian and avian toxicity

Metabolism and pharmacokinetics
The compound is a metabolite of DDT and is detoxified in rat kidney (2,3).

Genotoxicity

Salmonella typhimurium TA100, TA98 with and without metabolic activation negative (4).
Drosophila melanogaster genotoxicity research into sex linked lethal mutations inconclusive (5).

Any other comments

Details of DDT metabolism and effects on the environment reported (6,7).
Human health effects and experimental toxicology reviewed (8).

References

1. *Science* 1990, **168**, 582
2. Fawcett, S. C; et al. *Xenobiotica* 1987, **17**(5), 525-538
3. Datta, P. R. *Pest. Symp. Collect Pap. Inter. Amer. Conf. Toxicol. Occup. Med 6th 7th 1968-1970* 1970, 41-45
4. Planche, G; et al. *Chem. Biol. Interact.* 1979, **25**(2-3), 157-175
5. *Registry of Toxic Effects Chemical Substances* 1990, **21**, 16352
6. Duszeln, J. N. *Pesticide Contamination and Pesticide Control in Developing Countries: Costa Rica, Central America* Richardson, M. L. (Ed.) *Chemistry, Agriculture and the Environment* 1991, 410-428, RSC, London

7. Manno, M. *Toxicology and Risk Assessement of Pesticides* Richardson, M. L. (Ed.) *Chemistry Agriculture and the Environment* 1991, 466-490, RSC, London
8. *ECETOC Technical Report No.30(4)* 1991, European Chemical Industry Ecology and Toxicology Centre, B-1160 Brussels

B146   Bis(2,4-dihydroxyphenyl)methanone

CAS Registry No. 131-55-5
Synonyms benzophenone-2,2',4,4'-tetrahydroxybenzophenone; THBP; Uvinol D-50
Mol. Formula $C_{13}H_{19}O_5$ **Mol. Wt.** 255.29
Uses Ultra violet absorbing agent used in sunscreens and cosmetics.

Physical properties
M. Pt. 196-198°C.

Mammalian and avian toxicity

Acute data
LD_{50} oral rat 1220 mg kg^{-1} (1).

Irritancy
100 mg instilled into rabbit eye for 24 hr caused mild irritation (1).

Genotoxicity
Salmonella typhimurium TA1537 with metabolic activation positive (1,2).
Mouse lymphocyte sister chromatid exchange positive (3).

References
1. *Food Chem. Toxicol.* 1982, **20**, 427
2. Popkin, D. J; et al. *Mutat. Res.* 1989, **224**(4), 453-464
3. *J. Amer. Coll. Toxicol.* 1983, **2**(5), 35

B147   4,4'-Bis(dimethylamino)benzophenone

CAS Registry No. 90-94-8
Synonyms bis-[4-(dimethylamino)phenyl]methanone; Michler's ketone; di-*p*-dimethylaminophenyl ketone; *p, p'*-bis(*N*,*N*-dimethylamino)benzophenone; NCI-C02006; tetramethyldiaminobenzophenone

Mol. Formula $C_{17}H_{20}N_2O$ **Mol. Wt.** 268.36

Uses Manufacture of dyestuffs and photo sensitiser.

Occurrence Contaminant in drinking water (1).

Physical properties

M. Pt. 172°C; **B. Pt.** >360°C (decomp.).

Solubility

Water: 0.4 g l^{-1} 20-25°C. Organic solvent: ethanol, quinoline, pyridine

Ecotoxicity

Bioaccumulation

Non accumulative or low accumulative (2).

Effects to non-target species

LD_{50} oral redwing blackbird 100 mg kg^{-1} (3).

Mammalian and avian toxicity

Carcinogenicity and long-term effects

Oral rat, mouse (2 yr) 250-1000 ppm and 1250-2500 ppm, respectively, induced hepatocellular carcinomas and hemangiosarcomas (4).

Positive correlation between rats fed 4,4'-bis(dimethylamino)benzophenone and hepatocarcinogenic effects in food in rats and binding of metabolites to liver DNA and RNA (5).

National Toxicology Program evaluation of 4, 4-bis(dimethylamine)benzene rat and mouse in feed positive (6).

Genotoxicity

Salmonella typhimurium TA100, TA98, TA1535, TA1537, TA1538 with and without metabolic activation negative (7).

Escherichia coli PQ37 SOS chromotest negative (8).

In vitro rat hepatocyte DNA repair test positive (6).

In vitro Chinese hamster fibroblasts chromosome damage positive (9).

Any other comments

Human health effects and experimental toxicology reviewed (10).

References

1. Noordsij, A; et al. *Sci. Total Environ.* 1985, **47**, 273-292
2. *MITI Report* 1984, Ministry of International Trade and Industry, Tokyo
3. Schafer, E. W; et al. *Arch. Environ. Contam. Toxicol.* 1983, **12**(3), 355-382
4. *Government Report* 1979, 1737 *Govt. Rep. Announce. Index (US)* 1979, **79**(25), 105 (*Chem. Abstr.* **92**, 123109z)
5. Scribner, J. D; et al. *Cancer Lett. (Shannon, Irel.)* 1980, **9**(2), 117-121
6. *National Toxicology Program Research and Testing Div.* 1992, Report No. TR-181, NIEHS, Research Triangle Park, NC
7. Dunkel, V. C; et al. *Environ. Mutagen.* 1985, **7**(Suppl. 5), 1-248
8. Von der Hude, W; et al. *Mutat. Res.* 1988, **203**, 81-94
9. Lafi, A; et al. *Mutagenesis* 1986, **1**(1), 17-20
10. *ECETOC Technical Report No 30(4)* 1991, European Chemical Industry Ecology and Toxicology Centre, B-1160 Brussels

B148   1,3-Bis(2,3-epoxypropoxy)benzene

CAS Registry No. 101-90-6

Synonyms Araldite ERE 1359; *m*-bis(2,3-epoxypropoxy)benzene;
1,3-diglycidyloxybenzene; resorcinol diglycidyl ether; oxirane,
2,2′-[1,3-phenylenebis(oxymethylene)]bis-

Mol. Formula $C_{12}H_{14}O_4$ **Mol. Wt.** 222.24

Uses Limited application in the aerospace industry. A liquid epoxy resin and diluent in manufacture of other epoxy resins.

Physical properties

B. Pt. $_{0.88}$ 172°C; **Flash point** 176°C (open cup); **Specific gravity** d^{25} 1.21.

Occupational exposure

Supply classification toxic.

Risk phrases
Toxic by inhalation, in contact with skin and if swallowed – Possible risk of irreversible effects – May cause sensitisation by skin contact (R23/24/25, R40, R43)

Safety phrases
Do not breathe vapour – Avoid contact with skin – If you feel unwell, seek medical advice (show label where possible) (S23, S24, S44)

Mammalian and avian toxicity

Acute data
LD_{50} oral mouse, rabbit, rat 980-2570 mg kg^{-1} (1).
LD_{50} intraperitoneal rat, mouse 178-243 mg kg^{-1} (1).

Carcinogenicity and long-term effects
No adequate evidence for carcinogenicity to humans, sufficient evidence for carcinogenicity to animals, IARC classification group 2B (2).
Gavage ♂ ♀ rat (103 wk) 81% (in corn oil) 5 hr wk^{-1} corresponding to 25-50 mg kg^{-1} or 50-100 mg kg^{-1}. High mortality observed in 50 mg kg^{-1} group. Hyperkeratosis and hyperplasia of forestomach, squamous cell papillomas, squamous cell carcinomas in forestomach of ♂ and ♀ mice and rats (3).
National Toxicology Program evaluation of 1,3-bis(2,3-epoxypropoxy)benzene in rats and mice by gavage positive (4).

Irritancy
Dermal rabbit (25 hr) 500 mg hr caused moderate irritation (5).
In humans severe burns in contact with skin (6).

Genotoxicity

Induced chromosome aberrations and sister chromatid exchanges in Chinese hamster cells with or without metabolic activation (7).

Mouse lymphoma cell L5178Y tk+/tk- forward mutation assay without metabolic activation positive (8).

Drosophila melanogaster sex-linked recessive lethal induction by feeding and reciprocal translocation induction (9).

Drosophila melanogaster sex chromosome loss, nondisjunction and translocation (10).

Did not induce micronuclei in bone marrow of mice *in vivo* (11).

Any other comments

Human health effects and experimental toxicology reviewed (12).

References

1. Hine, C. H; et al. *Arch. Ind. Health* 1958, **17**, 129-144
2. *IARC Monograph* 1987, **(Suppl. 7)**, 58
3. *Natl. Toxicol. Program Tech. Rep. Ser.* 1986, **257**, 222
4. *National Toxicology Program Research and Testing Div.* 1992, Report No. 257, NIEHS, Research Triangle Park, NC
5. Goodman, L. S; et al. *The Pharmacological Basis of Therapeutics* 2nd ed., 1955, Macmillan, New York
6. *IARC Monograph* 1985, **36**, 183-188
7. Gulati, D. K; et al. *Environ. Mol. Mutagen.* 1989, **13**(2), 133-193
8. McGregor, D. B; et al. *Environ. Mol. Mutagen.* 1988, **12**, 85-154
9. Valencia, R; et al. *Environ. Mutagen.* 1985, **7**, 325-348
10. Canter, D. A; et al. *Mutat. Res.* 1986, **172**(2), 105-138
11. Seiler, J. P. *Mutat. Res.* 1984, **135**, 159
12. *ECETOC Technical Report No. 30(4)* 1991, European Chemical Industry Ecology and Toxicology Centre, B-1160 Brussels

B149   Bis[4-(2,3-epoxypropoxy)phenyl]propane

CAS Registry No. 1675-54-3

Synonyms bisphenol A diglycidyl ether; DGEBPA; 2,2′-[(1-methylethylidene)bis(4,1-phenyleneoxymethylene)]bisoxirane]; diphenylolpropane glycidyl ether; 4,4′-isopropylidenediphenol diglycidyl ether; epoxide A

Mol. Formula $C_{21}H_{24}O_4$ **Mol. Wt.** 340.42

Uses Manufacture of adhesives and epoxy resins. Photopolymers as a co-polymer and cross-linking agent. Plasticiser for vinyl polymers.

Occurrence Found in drinking waters from pipes sealed with its resins and polymers (1,2).

Physical properties

M. Pt. 43°C.

Occupational exposure

Supply classification irritant.

Risk phrases
Irritating to eyes and skin – May cause sensitisation by skin contact (R36/38, R43)

Safety phrases
After contact with skin, wash immediately with plenty of water – Wear suitable gloves and eye/face protection (S28, S37/39)

Mammalian and avian toxicity

Acute data
LD$_{50}$ oral rat 11 g kg^{-1} (3).

Carcinogenicity and long-term effects
No adequate evidence for carcinogenicity to humans, limited evidence for carcinogenicity in experimental animals, IARC classification group 3 (4).
Total dose (2 yr, intermittent) dermal mouse 16.4 g kg^{-1} induced tumours of lymph system and kidney adenomas (5).
0.2 ml of a 1 or 10% (wt./vol.) solution applied in acetone to skin of ♂ and ♀ mice twice weekly for 1 yr. Mild irritation, very low incidence of benign and malignant skin tumours. Increased incidence of lymphoreticular/haematopoietic tumours in ♀ mice (6).

Teratogenicity and reproductive effects
Applied daily to skin of rabbits for 6 hr day^{-1} at 0, 30, 100 or 300 mg kg^{-1} day^{-1} on days 6 to 18 pregnancy. 300 and 100 mg kg^{-1} doses were maternally toxic, but there was no embryo/foetal toxicity or teratogenicity (7).

Metabolism and pharmacokinetics
Dermal mouse (3 day): 56 mg kg^{-1} 20% eliminated in faeces, 3% in urine, 66% recovered from application site. (3 day) oral ♂ mice: 55 mg kg^{-1} 80% in faeces, 11% in urine (8).

Sensitisation
Conclusive allergen guinea pig, sensitises 80-100% of animals challenged (9).
Skin humans positive reaction in 3 out of 12 men tested (1% w/w) causes rashes of eyelids, face, forearms, hands (10).

Genotoxicity

Salmonella typhimurium TA100 with and without metabolic activation positive (11,12).
Escherichia coli with and without metabolic activation positive (13).

Any other comments

Human health effects and experimental toxicology reviewed (14).

References

1. Watts, C. D; et al. *Comm. Eur. Communities, [Rep] EUR* 1984, EUR 8518, Anal. Org. Micropollut. Water, 120-131 (*Chem. Abstr.* **101**, 177042v)
2. Crathorne, B; et al. *Environ. Sci. Technol.* 1984, **18**, 797-802
3. *Union Carbide Data Sheet* 1967, (4), 21, Union Carbide Corp., New York
4. Zakova, N; et al. *Food Chem. Toxicol.* 1985, **23**, 1081-1089

5. Peristianis, G. C; et al. *Food Chem. Toxicol.* 1988, **26**(7), 611-624
6. *IARC Monograph* 1989, **47**, 237-261
7. Breslin, W. J; et al. *Fundam. Appl. Toxicol.* 1988, **10**(4), 736-743
8. Climie, I. J. G; et al. *Xenobiotica* 1981, **11**(6), 391-424
9. Thorgeirrson, A; et al. *Acta Derm.-Venereol.* 1978, **58**(1), 17-21
10. Burrows, D; et al. *Contact Dermatitis* 1984, **11**(2), 80-82
11. Wade, M. J; et al. *Mutat. Res.* 1979, **66**(14), 367-371
12. Oerstavik, D; et al. *Biomaterials* 1985, **6**(2), 129-132
13. Hemminki, H; et al. *Arch. Toxicol.* 1980, **46**, 277-285
14. *ECETOC Technical Report No. 30(4)* 1991, European Chemical Industry Ecology and Toxicology Centre, B-1160 Brussels

B150   *N,N*-Bis(2,3-epoxypropyl)aniline

CAS Registry No. 2095-06-9
Synonyms *N,N*-diglycidylaniline; bis(2,3-epoxypropyl)aniline; bis(expoxypropyl)phenylamine; *N*-(oxiranylmethyl)-*N*-phenyl-oxiranemethanamine
Mol. Formula $C_{12}H_{15}NO_2$ **Mol. Wt.** 205.26
Uses Intermediate in polymer synthesis.

Mammalian and avian toxicity

Acute data
LD_{50} oral rat 1620 mg kg^{-1} (1).
LD_{50} dermal rabbit 3560 mg kg^{-1} (1).

Irritancy
500 mg applied to rabbit skin and 500 mg instilled into rabbit eye for 24 hr caused severe irritation (2).

References

1. *Am. Ind. Hyg. Assoc. J.* 1969, **30**, 470
2. Sax, N. I; et al. *Dangerous Properties of Industrial Materials* 7th ed., 1989, Van Nostrand Reinhold, New York

B151   Bis(2-ethylhexyl) phosphate

$$[CH_3(CH_2)_3CH(CH_2CH_3)CH_2O]_2P(=O)OH$$

CAS Registry No. 298-07-7
Synonyms bis(2-ethylhexyl) hydrogen phosphate; bis(2-ethylhexyl)orthophosphoric acid; di(2-ethylhexyl)phosphate; Dehpa extractant; 2-ethyl-1-hexanol hydrogen phosphate

Mol. Formula $C_{16}H_{35}O_4P$ **Mol. Wt.** 322.43

Uses Cation extracting agent. Heavy metal extraction.

Physical properties

M. Pt. –60°C; **Flash point** 171°C (open cup); **Specific gravity** d_4^{20} 0.965.

Mammalian and avian toxicity

Acute data

LD_{50} oral rat 4940 mg kg^{-1} (1).
LD_{50} dermal rabbit 1250 mg kg^{-1} (1).
LD_{50} intraperitoneal rat 50 mg kg^{-1} (2).
LD_{Lo} intraperitoneal mouse 63 mg kg^{-1} (3).

Sub-acute data

Exposure of rats to 1% or 3% bis(2-ethylhexyl)phosphate in the diet for 5 days results in 2-3 fold induction of liver cytosolic epoxide hydrolase activity and microsomal cytochrome P450 content (4).

Irritancy

Dermal rabbit (24 hr) 5 mg caused mild irritation (5).
250 mg instilled into eye rabbit for 24 hr caused severe irritation (5).

Any other comments

Human health effects and experimental toxicology reviewed (6).
Ingredient in barrier cream for protecting hands from sensitising metals (7).

References

1. *Union Carbide Data Sheet* 1972, Union Carbide Corp., New York
2. *Hydrometallurgy* 1978, **3**, 201
3. *Summary Tables of Biological Tests* Nat. Res. Cri. Chem.-Biol. Coor. Center 1957, **9**, 132, Washington, DC
4. Lundgren, B. *Xenobiotica* 1987, **17**(5), 585-593
5. Marhold, A. *Prehled Prumyslove Toxicol. Org. Latky* 1986, 1130
6. *ECETOC Technical Report No. 30(4)* 1991, European Chemical Industry Ecology and Toxicology Centre, B-1160 Brussels
7. Yurtov, E. V; et al. *Orkrytiya Izobret* 1989, **24**, 40

B152   *N,N*-Bis(hydroxyethyl)-coco-amides

CAS Registry No. 68603-42-9

Synonyms amides, coco, *N,N*-bis(hydroxyethyl); Clindrol Superamide 100 CG; coconut oil diethanolamine; Amidet B112

Uses Mild surfactant used in hand gels, hand washing liquids, shampoos and some dishwashing liquids.

Ecotoxicity

Bioaccumulation
Slow 6% degradation occurred in 24 hr (1).

Mammalian and avian toxicity

Acute data
LD$_{50}$ oral rat ≈2000-3000 mg kg^{-1} (2).

Sensitisation
A 27 yr old mine worker reported acute eczema. Patch-testing with 0.5% petrolatum coconut diethanolamide gave positive result (3).
Reported dermatitis, positive sensitisation results in tests on a woman (4).

Genotoxicity

Salmonella typhimurium TA98, TA100 with and without metabolic activation negative. No morphological transformation of cryopreserved primary hamster embryo cells in culture (2).

Any other comments

The Toxicology Design Committee has approved pre-chronic testing of coconut diethanolamide by skin painting in rats and mice (5).
Toxicity has been evaluated (6).
Human health effects and experimental toxicology reviewed (7).

References

1. Maloney, G. W; et al. *J. Water Pollut. Control Fed.* 1969, R18-37
2. Inove, K; et al. *Food Cosmet. Toxicol.*1980, **18**, 289
3. Hindson, C; et al. *Contact Dermatitis* 1983, **9**(2), 168
4. Nurse, D. S. *Contact Dermatitis* 1980, **6**(7),502
5. Toxicology Information Program *Tox-Tips* 1987, **130**, 27, Bethesda, MD
6. *BIBRA Toxicity Profile* 1990, British Industrial Biological Research Association, Carlshalton
7. *ECETOC Technical Report No. 30(4)* 1991, European Chemical Industry Ecology and Toxicology Centre, B-1160 Brussels

B153 *N,N*-Bis(2-hydroxyethyl)dodecanamide

CH$_3$(CH$_2$)$_{10}$C(O)N[(CH$_2$)$_2$OH]$_2$

CAS Registry No. 120-40-1
Synonyms lauroyl diethanolamide; lauric acid diethanilamide;
N,N-bis(hydroxyethyl)lauramide; *N, N*-bis(β-hydroxyethyl)lauramide
Mol. Formula C$_{16}$H$_{33}$NO$_3$ **Mol. Wt.** 287.45
Uses In pharmaceuticals and cosmetics. Detergents and cleansers. Acaracide and mite repellent.

Physical properties

Solubility
Water: <1 mg ml^{-1} at 24°C. *Organic solvent:* ethanol, acetone, dimethyl sulfoxide

---

Environmental fate

Degradation studies
Slow 6% biodegrades in 24 hr (1).

Mammalian and avian toxicity

Acute data
LD_{50} oral rat 2700 mg kg^{-1} (2).

Carcinogenicity and long-term effects
National Toxicology Program prechronic study in progress (3).

Genotoxicity
Chinese hamster ovary cells with and without metabolic activation sister chromatid exchanges positive and chromosome aberrations negative (4).

Any other adverse effects
The Toxicology Design Committee has approved prechronic testing of lauric acid diethanolamine condensate, a detergent, in rats and mice by skin painting (5). Toxicity has been evaluated (6).

References

1. *J. Soc. Cosmet. Chem.* 1962, **13**, 469
2. Cosmetic, Toiletry and Fragrance Assoc. *J. Am. Coll. Toxicol.* 1986, **5**(5), 415-454 (*Chem. Abstr.* **106**, 72674c)
3. *National Toxicology Program Research and Testing Div.* 1992, Management Status Report, NIEHS, Research Triangle Park, NC
4. Loveday, K. S; et al. *Environ. Mol. Mutagen.* 1990, **16**(4), 272-303
5. National Toxicology Program *Tox-Tips* 1987, **30**, 31, Bethesda, MD
6. *BIBRA Toxicity Profile* 1990, British Industrial Research Association, Carlshalton

B154   Bis(2-hydroxy-4-methoxyphenyl)methanone

CAS Registry No. 131-54-4

Synonyms methanone, bis(2-hydroxy-4-methoxyphenyl)-; 2,2'-dihydroxy-4,4'-dimethoxybenzophenone; benzophenone 6; Univul D49; Cyasorb UV12

Mol. Formula $C_{15}H_{14}O_5$ **Mol. Wt.** 274.28

Uses Ultra violet protecting agent, used in sunscreen and cosmetics.

Physical properties

M. Pt. 139-140°C.

Solubility
Organic solvent: ethanol, toluene

Mammalian and avian toxicity
Sensitisation
Can cause contact sensitisation (1,2).

References

1. Pariser, R. J. *Contact Dermatitis* 1977, **3**, 172
2. Thompson, G; et al. *Arch. Derm.* 1977, **113**, 1252

B155 2,2-Bis[(methacryloyloxy)methyl]butyl methacrylate

$(H_2C=CCH_3CO_2CH_2)_3CCH_2CH_3$

CAS Registry No. 3290-92-4
Synonyms TPT-MA; 2-ethyl-2-[[(2-methyl-1-oxo-2-propenyl)oxy]methyl]-1,3-propanediyl; 2-methyl-2-propenoate-2-ethyl-2-hydroxymethyl-1,3-propanediol trimethacrylate; trimethylol trimethacrylate
Mol. Formula $C_{18}H_{26}O_6$ **Mol. Wt.** 338.40
Uses Manufacture of acrylic polymers, adhesives and cleaning agents. Cross-linking agent, vulcanisation agent and plasticiser. Polymer used in dental composites. Photoimaging materials.

Physical properties

M. Pt. −14°C; **B. Pt.** $_1$ 155°C; **Flash point** >112°C (closed cup); **Specific gravity** d_4^{20} 1.02; **Partition coefficient** log P_{ow} 3.53.

Mammalian and avian toxicity

Acute data
LD_{50} intraperitoneal mouse 2889 mg kg^{-1} (1).

Teratogenicity and reproductive effects
ED_{50} (intraovary) 3 day-old white leghorn chick embryos 2.64 mg egg^{-1} caused a maximum of 30% malformed embryos. LD_{50} 4.4 mg egg^{-1} caused early death (3-5 days after hatching) (2,3).

Sensitisation
Did not induce contact sensitivity on guinea-pig skin (4).

Genotoxicity
Salmonella typhimurium TA1535, with metabolic activation positive. Mouse lymphoma L5178Y without metabolic activation induced mutations, chromosome aberrations and micronuclei (5).
Mouse lymphoma L5178Y tk+/tk- assay with or without metabolic activation negative (6).

Any other comments

Human health effects and experimental toxicology reviewed (7).

References

1. Lawrence, W. H; et al. *J. Dental Res.* 1972, **51**(2), 526-535
2. Korhonen, A; et al. *Scand. J. Work, Environ. Health* 1983, **9**(2), 115
3. Korhonen, A. *Acta Pharmacol. Toxicol.* 1983, **52**(2), 95-99
4. Parker, D; et al. *Contact Dermatitis* 1983, **9**(1), 55-60
5. Dearfield, K. L; et al. *Mutagenesis* 1989, **4**(5), 381-393
6. Cameron, T. P; et al. *Environ. Mol. Mutagen.* 1991, **17**(4), 264-271
7. *ECETOC Technical Report No. 30(4)* 1991, European Chemical Industry Ecology and Toxicology Centre, B-1160 Brussels

B156   Bis(methoxythiocarbonyl)disulfide

CH₃OC(S)SSC(S)OCH₃

CAS Registry No. 1468-37-7

Synonyms bis(methylxanthogen)disulfide; *O,O*-dimethyl dithiobis(thioformate); dimethyl dixanthogen; dimethyl thioperoxydicarbonate; dimethyl xanthic disulfide; TRI-PE

Mol. Formula $C_4H_6O_2S_4$ **Mol. Wt.** 214.35

Uses Reagent for oxidation of thiols to disulfides. Herbicide. Insecticide.

Physical properties

M. Pt. 22.5-23°C; **B. Pt.** 122°C (decomp.); **Specific gravity** d_4^{20} 1.3886.

Solubility
Water: 25 mg l^{-1}.

Occupational exposure

Supply classification harmful.

Risk phrases
Harmful by inhalation, in contact with skin and if swallowed (R20/21/22)

Safety phrases
Keep out of reach of children – Keep away from food, drink and animal feeding stuffs (S2, S13)

Mammalian and avian toxicity

Acute data
LD_{50} oral rat 240 mg kg^{-1} (1).

Any other comments

Best stored at –20°C (2).
Its use as a herbicide and insecticide has been superceded (3).

References

1. Ben-Dyke, R; et al. *World Rev. Pest Control* 1970, **9**, 119
2. Barany, G; et al. *J. Org. Chem.* 1983, **48**(24), 4750-4761
3. Feekes, F. H; *Mededel. Landbouwhogesthool Opzoehingssta Staat Gent* 1962, **27**(3), 1289-1307, (Ger.) (*Chem. Abstr.*, **63**, 13957d)

B157   Bismuth telluride

Bi_2Te_3

CAS Registry No. 1304-82-1
Synonyms bismuth sesquitelluride; tellurobismuthite; bismuth tritelluride; dibismuth tritelluride
Mol. Formula Bi_2Te_3 **Mol. Wt.** 800.76
Uses In electronics as semiconductor and for thermoelectric cooling. Power generation applications.
Occurrence As ingot, single crystals.

Physical properties

M. Pt. 585°C; **Specific gravity** d^{20} 7.7.

Occupational exposure

US TLV (TWA) 10 mg m^{-3}; **UK Long-term limit** 10 mg m^{-3}; **UK Short-term limit** 20 mg m^{-3}.

Any other adverse effects

Symptoms of poisoning include vomiting, reduced appetite, insomnia, unconsciousness, liver and kidney damage (1).

References

1. *Chemical Safety Data Sheets* 1989, **2**, 110-112, RSC, London

B158   Bismuth subgallate

CAS Registry No. 99-26-3
Synonyms bismuth gallate, basic; gallic acid bismuth basic salt; Dermatol; BSG; bismuth oxygallate; 2,7-dihydroxy-1,3,2-benzodioxabismole-5-carboxylic acid
Mol. Formula $C_7H_5BiO_6$ **Mol. Wt.** 394.09
Uses Astringent. Antacid. Suppository in the treatment of haemorrhoids. Used as a treatment for gastrointestinal disorders.

Physical properties

M. Pt. 100°C (decomp.).

Ecotoxicity

Bioaccumulation

Predicted concentration for the River Lee, UK, 0.15 µg l^{-1} (1).

Any other adverse effects

Can reduce clotting time of whole blood (2).
Systemic effects include ulcerative stomatitis, anorexia, headache, kidney tubule damage and mild jaundice (3).

Any other comments

Major active ingredient of Bungast (4).
Has an inhibitory effect on *Camplyobacter pylori* (5).
Combined with gallic acid is used as a dusting powder in dermatology (6).
Thermolabile compound (7).

References

1. Richardson, M. L; et al. *J. Pharm. Pharmacol.* 1985, **37**, 1-12
2. Thorisdottir, H; et al. *J. Lab. Clin. Med.* 1988, **112**(4), 481-486
3. Gosselin, R. E; et al. *Clin. Toxicol. Commer. Prod.* 1984, 5th ed., **11**, 134
4. *The Merck Index* 1989, 11th ed., Merck & Co. Inc., Rahway, NJ
5. Vogt, K; et al. *Zentral. Bakteriol.* 1989, **271**(3), 304-310
6. *Kirk-Othmer Encyclopedia of Chemical Technology* 1982, **18**, 677, John Wiley & Sons, New York
7. Astrakhanova, M. M; et al. *Farmatstya (Moscow)* 1988, **37**(4), 76-79, (Russ.) (*Chem. Abstr.* **109**, 115848u)

B159   *trans*-1,2-bis(propylsulfonyl)ethylene

$$CH_3(CH_2)_2S(O)_2CH=CHS(O)_2(CH_2)_2CH_3$$

CAS Registry No. 1113-14-0

Synonyms Chemagro B-1843; Vancide PA; propane, 1,1-[1,2-ethenediylbis(sulfonyl)]bis-(E)-; *trans*-1, 2-bis-(*n*-propylsulfonyl)ethylene; B 1843; C 272; CHE 1843

Mol. Formula $C_8H_{16}O_4S_2$ **Mol. Wt.** 240.34

Uses Fungicide.

Ecotoxicity

Effects to non-target species

LC$_{50}$ (8 day) Japanese quail (14 day old) >5000 ppm in food, based on active ingredient (1).

Mammalian and avian toxicity

Acute data
LD_{50} oral rat 200 mg kg^{-1} (2).
LD_{50} intraperitoneal rat 11,500 mg kg^{-1} (3).
LD_{Lo} intraperitoneal guinea pig 11,500 mg kg^{-1} (3).

Any other comments
All reasonable efforts have been taken to find information on isomers of this compound, but no relevant data are available.

References

1. Hill, E. F; et al; *US Fish Wildlife Serv. Spec. Sci. Rep. Wildlife* 1975, 191, 1-61
2. *Farm Chemicals Handbook* 1983, C49, Meister Publishing, Willoughby, OH
3. Deichmann, W. B; et al. *Toxicology of Drugs and Chemicals* 1969, 161-162, Academic Press, London

B160   Bis(2,3,3,3-tetrachloropropyl) ether

(Cl3CCHClCH2)2O

CAS Registry No. 127-90-2
Synonyms octachlorodipropyl ether; S 421; 1,1'-oxybis[2,3,3,3-tetrachloropropane]; Ent 25,456; Monsanto CP-16226
Mol. Formula $C_6H_6Cl_8O$ **Mol. Wt.** 377.74
Uses Lubricant additive. Insect repellent. Synergist for DDT, carbamates, pyrethrin and organophosphorus insecticides.

Physical properties
B. Pt. 296-298°C; **Flash point** 177°C; **Specific gravity** 1.65.

Ecotoxicity

Fish toxicity
LC_{50} (96 hr) guppy 1.7 mg l^{-1} (1).

Mammalian and avian toxicity

Acute data
LD_{50} oral rabbit, rat 2500-3630 mg kg^{-1} (3,4).

Sub-acute data
Rats fed 400-2500 mg kg^{-1} in diet for 13 wk showed increase in absolute weight of liver and kidney and thyroid (σ), and hypophysis ($\female$).
ED oral rat 10-60 mg kg^{-1} via feed increased kidney weight (5).
ED oral rat 3200-7812 mg kg^{-1} disturbed circulatory system (6).
ED (28 day) oral rat 30-3000 mg kg^{-1} liver and kidney degeneration (6).

Metabolism and pharmacokinetics
Oral rat (24 hr) 37.8 mg kg^{-1} formed 0.014-0.018% trichloroacrylic acid in urine (7).

Any other adverse effects to man

Found as a residue in human milk (0.5 ppb in women) ($\approx$14 ppb fat-milk basis) or 1.5 ppb in 9712 women ($\approx$32 ppb fat-milk basis) (1).

Any other comments

Toxicity reviewed (8).

References

1. *The Agrochemicals Handbook* 3rd ed., 1991, RSC, London
2. *MITI Report* 1984, Ministry of International Trade and Industry, Tokyo
3. *Agriculture Research Service* 1966, **20**, 15, USDA Information Memorandum, Bettsville
4. Kenaga, E. E. *Bull. Entomol. Soc. Am.* 1966, **12**, 161-217
5. Kita, K; et al. *Organika* 1980, 175-183, (Pol.) (*Chem. Abstr.* **95**, 92005t)
6. Kita, K; et al. *Pestcycydy (Warsaw)* 1980, (4), 7-12, (Pol.) (*Chem. Abstr.* **96**, 63792n)
7. Marsden, P. J; et al. *J. Agric. Food Chem.* 1982, **30**(4), 627-631
8. Fishbein, L; et al. *Ind. Med. Surg.* 1968, **37**(11), 848-863

B161   Bis(2,4,6-trinitrophenyl)amine

CAS Registry No. 131-73-7

Synonyms 2,4,6,2′,4′,6′-hexanitrodiphenylamine; hexyl; 2,4,6-trinitro-*N*-(2,4,6-trinitrophenyl)benzenamine; dipicrylamine; benzenamine, 2,4,6-trinitro-*N*-2,4,6-trinitrophenyl; C.I. 10360; Aurantia

Mol. Formula $C_{12}H_5N_7O_{12}$ **Mol. Wt.** 439.21

Uses A booster explosive. Analysis for potassium.

Physical properties

M. Pt. 245-246°C (decomp.).

Occupational exposure

Supply classification explosive and toxic.

Risk phrases

Risk of explosion by shock, friction, fire or other sources of ignition – Very toxic by inhalation, in contact with skin and if swallowed – Danger of cumulative effects (R2, R26/27/28, R33)

Safety phrases

This material and its container must be disposed of in a safe way – Wear suitable protective clothing – If you feel unwell, seek medical advice (show label where possible) (S35, S36, S44)

Genotoxicity

Salmonella typhimurium TA100, TA98, TA1537 and TA1538 without metabolic activation positive (1).

Any other comments

The effects of bis(2,4,6-trinitrophenyl)amine on the lipid membrane transport of ionised drugs was studied. Highly lipophilic counterions accumulate in the lipid membrane and act as a carrier for ionised drugs (2).

A powerful and violent explosive superior to TNT. Soluble in alkali and warm acetic or nitric acid. Explodes on shock or exposure to heat.

Human health effects and experimental toxicology reviewed (3).

References

1. Whong, W.-Z; et al. *Mutat. Res.* 1984, **136**, 209-215
2. Neubert, R; et al. *Pharmazie* 1984, **39**(6), 401-403, (Ger.) (*Chem. Abstr.* **10**, 183402d)
3. *ECETOC Technical Report No. 30(4)* 1991, European Chemical Industry Ecology and Toxicology Centre, B-1160 Brussels

B162   Bitertanol

$(CH_3)_3CCH(OH)CHO$—(biphenyl)

CAS Registry No. 55179-31-2

Synonyms Baycor; (1*RS*,2*RS*:1*RS*,2*SR*)-1-biphenyl-4-yloxy)-3,3-dimethyl-1-(1*H*-1,2,4-triazol-1-yl)butan-2-ol [ratio of race moles (1*RS*,2*RS* (B) and (1*RS*,2*SR*) (A) ca. 20:80]; β-[(1,1′-biphenyl)-4-yloxy]-α-(1-1-dimethylethyl)-1*H*-1,2,4-triazole-1-ethanol; Sibutol

Mol. Formula $C_{20}H_{23}N_3O_2$ **Mol. Wt.** 337.43

Uses Fungicide. Steroid demethylation inhibitor.

Physical properties

M. Pt. 136.7°C (diastereomer A); 145.2°C (diastereomer B); 118°C (eutectic of A and B); **Volatility** v.p. 2.857×10^{-5} mmHg at 100°C (diastereomer A); v.p. 2.406×10^{-5} mmHg at 100°C (diastereomer B).

Solubility

Water: 2.9 mg l^{-1} (A), 1.6 mg l^{-1} (B), 5.0 mg l^{-1} (A+B). Organic solvent: isopropanol, toluene, methylene chloride, cyclohexanone

Ecotoxicity

Fish toxicity

LC_{50} (96 hr) rainbow trout 2.2-2.7 mg l^{-1} (1).
LC_{50} (48 hr) carp 2.5 mg l^{-1} (2).

Effects to non-target species
Non-toxic to bees (1).

Mammalian and avian toxicity

Acute data
LD_{50} oral rat >5000 mg kg^{-1} (technical grade) (1).
LC_{50} (4 hr) inhalation rat >0.55 mg l^{-1} air (aerosol) (2).
LC_{50} (4 hr) inhalation rat >1.2 mg l^{-1} air (dust) (2).
LD_{50} dermal rat >5000 mg kg^{-1} (technical grade) (1).

Carcinogenicity and long-term effects
In 2 yr feeding trials the no-effect level for rats was 100 mg kg^{-1} in diet (2).

Any other adverse effects to man
A study of 55 people working with bitertanol was undertaken. From the results no conclusions could be drawn regarding the effects on workers' health (3).

Any other comments
Calculated allowable daily intake for humans 0.01 mg kg^{-1} (2).

References
1. *Farm Chemicals Handbook* 1991, C38, Meister Publishing, Willoughby, OH
2. Worthing, C. R. (Ed.) *The Pesticide Manual* 9th ed., 1991, British Crop Protection Council, Farnham
3. Nehez, M; et al. *Regul. Toxicol. Pharmacol.* 1988, **8**, 37-44

B163   Bitoscanate

CAS Registry No. 4044-65-9
Synonyms benzene, 1,4-diisothiocyanato-; Biscomate; isothiocyanic acid, *p*-phenylene ester; 1,4,-phenylenediisothiocyanic acid; phenylene thiocyanate; WM 842
Mol. Formula $C_8H_4N_2S_2$ **Mol. Wt.** 192.26
Uses Antihelmintic used in the treatment of hookworm.

Physical properties
M. Pt. 132°C.

Ecotoxicity

Invertebrate toxicity
EC_{50} (30 min) *Photobacterium phosphoreum* 0.0179 mg l^{-1} Microtox test (1).

Mammalian and avian toxicity

Acute data

LD_{50} oral rat 21 mg kg^{-1}(2).

LD_{50} intraperitoneal mouse 21 mg kg^{-1}(3).

TD_{Lo} oral human 3 mg kg^{-1} reported central nervous system, gastrointestinal tract effects (2).

Metabolism and pharmacokinetics

After oral administration to dogs, renal elimination and fall-off of blood concentrations were biphasic. Similar results were obtained in humans except that urinary excretion was more delayed and more was excreted renally. Excretion was mostly complete in 5 days from both species, but excretion of the remainder was protracted. After 3 wk, the dogs had excreted 80% in the faeces and 12% in urine, and after 5 wk, humans had excreted 55% and 28%, respectively by these routes; 3% still circulated in blood. Foetuses contained 3% of the dose, 16 hr after administration to a pregnant dog (4).

Bitoscanate is partly absorbed from the gastrointestinal tract and slowly excreted (5).

Any other adverse effects

Reacts with albumin and red blood cells (3).

Ingestion caused hallucinations and nausea. Can cause gastrointestinal disturbance and dizziness and central nervous system (5).

Any other comments

5 mg ml^{-1} in ethylene glycol showed inhibition against western equine encepholomyelitis viruses (6).

References

1. Kaiser, K. L. E; et al. *Water Pollut. Res. J. Canada* 1991, **26**(3), 361-431
2. *J. Trop. Med. Hyg.* 1969, **72**, 252-253
3. *Fortschritte der Arzneimittelforschung* 1973, **17**, 108
4. *Foreign Compound Metabolism in Mammals* 1972, **2**, 65, The Chemical Society, London
5. *Martindale: The Extra Pharmacopoeia* 29th ed. 1989, Press London Pharmaceutical
6. Rada, B. Et al *Acta. Virol.* 1971, **15**, 329-332

B164   Bordeaux mixture

$$CuSO_4.3Cu(OH)_2.3CaSO_4$$

CAS Registry No. 8011-63-0

Synonyms basic cupric sulfate; tribasic copper mixture; Comac; Bordo mixture

Mol. Formula $H_2Ca_3Cu_4S_2O_{10}$ **Mol. Wt.** 600.54

Uses Foliar fungicide. Insect repellent.

Occupational exposure

UN No. 2775; **Conveyance classification** toxic substance.

Mammalian and avian toxicity

Acute data
LD_{50} oral rat >4000 mg kg^{-1} as wettable powder (1).

Any other adverse effects
Humans are exposed to high copper sulfate content, 392-555 g kg^{-1}, by repeated administration of the Bordeaux mixture. Systemic effects include gastroenteritis, haemolysis, jaundice, albuminuria, haemoglobinurea and delayed sudden increase in the blood bilirubin, both free and conjugated with glucoronic acid. Uremia indicated injury of the glomerular apparatus in kidney. Haemolytic crises were associated with hypercupremia. Liver copper content increased 8-10-fold (2).

Legislation
Limited under EC Directive on Drinking Water Quality 80/778/EEC. Pesticides and related products: maximum admissible concentration 0.5 μg l^{-1} (3).

Any other comments
Dissolves in ammonium hydroxide to form a cuprammonium complex. Incompatible with alkali-sensitive pesticides such as organophosphorus compounds and carbamates and strongly alkaline pesticides such as lime sulfur (1).

References
1. *The Agrochemicals Handbook* 3rd ed., 1991, RSC, London
2. Bauer, M. *Vet. Arh.* 1975, **45**(9-10), 257-267, (Croat) (*Chem. Abstr.* **84**, 160298r)
3. *EC Directive Relating to the Quality of Water Intended for Human Consumption* 1982, 80/778/EEC, Office for Official Publications of the European Communities, 2 rue Mercier, L-2985 Luxembourg

B165   Borneol

CAS Registry No. 507-70-0
Synonyms bicyclo[2.2.1]heptan-2-ol, 1,7,7- trimethylendo-; endo-2-bornanol; bornyl alcohol; camphol; endo-2-camphanol; endo-2-hydroxybornane
Mol. Formula $C_{10}H_{18}O$ **Mol. Wt.** 154.25
Uses In textile industry. Solvent and bactericide in soap manufacture. Flotation agent. Preservative in paints. Perfumes, flavourings and medicinals.

Physical properties
M. Pt. 206-208°C; **B. Pt.** 212°C (sublimes); **Flash point** 65°C (closed cup); **Specific gravity** d_4^{25} 1.011; **Volatility** v. den. 5.31.

Solubility

Organic solvent: ethanol, diethyl ether, petroleum ether, benzene, toluene, acetone, decalin

Occupational exposure

UN No. 1312; HAZCHEM Code 1☒; Conveyance classification flammable solid.

Environmental fate

Degradation studies

Adapted activated sludge used product as sole carbon source, 90.3% COD removal at 8.9 mg COD^{-1} dry $inoculum^{-1}$ hr^{-1} (1).
Biodegradable (2).

Mammalian and avian toxicity

Acute data

LD_{50} oral rat, mouse, rabbit 500-2000 mg kg^{-1} (3-5).

Any other adverse effects

In humans systemic effects include nausea, vomiting, mental confusion, dizziness and convulsions (6).

References

1. Pitter, P. *Water Res.* 1976, **10**, 231
2. *MITI Report* 1984, Ministry of International Trade and Industry, Tokyo
3. *Shika Gakuho* 1975, **75**, 934
4. *Fr. Demand Pat. Doc. No. 2448856*
5. *Arch. Exp. Path. Pharmacol.* 1883, **17**, 363
6. *The Merck Index* 11th ed., 1989, 204, Merck & Co Inc., Rahway, NJ

B166   Boron oxide

B_2O_3

CAS Registry No. 1303-86-2
Synonyms boric anhydridge; boron sesquioxide; diboron trioxide
Mol. Formula B_2O_3 **Mol. Wt.** 69.62
Uses Herbicide. Metallurgy. Analysis for silicates and blowpipe analysis.

Physical properties

M. Pt. 450°C; **B. Pt.** 1860°C; **Specific gravity** 2.460.

Solubility

Organic solvent: ethanol

Occupational exposure

US TLV (TWA) 10 mg m^{-3}; UK Long-term limit 10 mg m^{-3}; UK Short-term limit 20 mg m^{-3}.

Mammalian and avian toxicity

Acute data

LD_{50} oral mouse 3136 mg kg^{-1} (1).
LD_{50} intraperitoneal mouse 1868 mg kg^{-1} (1).

Irritancy

Irritant to eye and skin (2).

Any other comments

Toxicity and hazards reviewed (3).
Hygroscopic. Mixed with CaO and fused into $CaCl_2$ the mixture incandesces.

References

1. Izmerov, N. F; et al. *Toxicometric Parameters of Industrial Toxic Chemicals under Single Exposure* 1982, 27, Moscow
2. Wilding, J. L. *Am. Ind. Hyg. Assoc. J.* 1959, **20**, 284
3. Izmerov, N. F. *Scientific Reviews of Soviet Literature on Toxicity & Hazards of Chemicals* 1992-1993, **116**, Engl. Trans., Richardson, M. L. (Ed.), UNEP/IRPTC, Geneva

B167   Boron tribromide

BBr_3

CAS Registry No. 10294-33-4
Synonyms boron bromide; tribromoborane; tribromoboron; Trona
Mol. Formula BBr_3 **Mol. Wt.** 250.54
Uses In the manufacture of diborane and of ultra high purity boron. In the manufacture of anhydrous metal bromides. Catalyst. Doping agent for semiconductors.

Physical properties

M. Pt. –46°C; **B. Pt.** 90°C; **Specific gravity** 2.650 at 0°C; **Volatility** v.p. 40 mmHg at 14°C.

Solubility

Water: (decomp.). Organic solvent: ethanol (decomp.)

Occupational exposure

US TLV (TWA) Ceiling limit 1 ppm (10 mg m^{-3}); **UK Short-term limit** 1 ppm (10 mg m^{-3}); **UN No.** 2692; **HAZCHEM Code** 4WE; **Conveyance classification** corrosive substance; **Supply classification** toxic.

Risk phrases

Reacts violently with water – Very toxic by inhalation and if swallowed – Causes severe burns (R14, R26/28, R35)

Safety phrases

Keep container in a well ventilated place – In case of contact with eyes, rinse immediately with plenty of water and seek medical advice – After contact with skin, wash immediately with plenty of water – Wear suitable protective clothing – In case of accident or if you feel unwell, seek medical advice immediately (show label where possible) (S9, S26, S28, S36, S45)

Environmental fate

Absorption
Some boron is adsorbed by iron and aluminium hydroxy compounds and clay minerals, maximum sorption in the range pH 7-9 (1).

Mammalian and avian toxicity

Irritancy
Corrosive to tissues of mucous membranes, upper respiratory tract, eyes and skin (2).

Any other adverse effects

In humans systemic effects include vomiting and abdominal pain. Burns on contact and can be absorbed through the skin. Exposure to vapour causes pain, redness, watering, blurred vision in eyes and coughing and shortness of breath. Can cause pulmonary oedema, severe vomiting and diarrhoea (3).

Legislation

Limited under EC Directive on Drinking Water Quality 80/778/EEC. Boron: guide level 1000 μg l^{-1}, maximum admissible concentration 2000 μg l^{-1} (4).

Any other comments

Toxicity and hazards reviewed (5).
Physico-chemical properties, human health effects, experimental toxicology and workplace experience reviewed (6).
Mixtures are shock sensitive and liable to explode on impact (7).
Corrosive. Light and moisture sensitive. Incompatible with sodium and potassium. Hydrolyses to boric acid and hydrogen bromide.

References

1. Brown, K. W; et al. *Hazardous Waste Land Treatment* 1983, 211, Butterworth Publishers, Boston, MA
2. Lenga, R. E. *Sigma Aldrich Library of Chemical Safety Data* 1985, Sigma Aldrich, Milwaukee, WI
3. Croner *Substances Hazardous to Health* 1990, Croner Publ., London
4. *EC Directive Relating to the Quality of Water Intended for Human Consumption* 1982, 80/778/EEC, Office for Official Publications of the European Communities, 2 rue Mercier, L-2985, Luxembourg
5. Izmerov, N. F. *Scientific Reviews of Soviet Literature of Toxicity and Hazards of Chemicals* 1992-1993, **116**, Eng. Trans., Richardson, M. L. (Ed.), UNEP/IRPTC, Geneva
6. *ECETOC Technical Report No.30(4)* 1991, European Chemical Industry Ecology and Toxicology Centre, B-1160 Brussels
7. Mellor, J. W. *Comprehensive Treatise on Inorganic and Theoretical Chemistry, Supplement 3* 1937, **2**, 1571

B168   Boron trichloride

BCl₃

CAS Registry No. 10294-34-5
Synonyms boron chloride; trichloroborane; trichloroboron
Mol. Formula BCl₃   Mol. Wt. 117.17
Uses Manufacture and purification of boron. Catalyst for organic reactions.
Semiconductors. Bonding of iron and steels. Purification of metal alloys to remove
nitrides and carbides.

Physical properties

M. Pt. –107°C; B. Pt. 12.5°C; Specific gravity d^{15} 1.35; Volatility v.p. 1128 mmHg
at 20°C.

Solubility
Water: (decomp.). Organic solvent: ethanol (decomp.)

Occupational exposure

UN No. 1741; Conveyance classification toxic gas; corrosive substance;
Supply classification toxic.

Risk phrases
Reacts violently with water – Very toxic by inhalation and if swallowed – Causes
burns (R14, R26/28, R34)

Safety phrases
Keep container in a well ventilated place – In case of contact with eyes, rinse
immediately with plenty of water and seek medical advice – After contact with skin,
wash immediately with plenty of water – Wear suitable protective clothing – In case
of accident or if you feel unwell, seek medical advice immediately (show label where
possible) (S9, S26, S28, S36, S45)

Environmental fate

Absorption
Some boron is adsorbed by iron and aluminium hydroxy compounds and clay
minerals, maximum sorption ranging from pH 7-9 (1).

Mammalian and avian toxicity

Acute data
LC_{Lo} (7 hr) inhalation mouse, rat 20 ppm (2).
LC_{50} (1 hr) inhalation ♂ rat 2541 ppm (3).

Irritancy
Exposure can cause dermatitis in humans (4).

Any other adverse effects
Destructive to tissues of the mucous membranes and upper respiratory tract, eyes and
skin. In humans, inhalation can be fatal as a result of spasm, inflammation, and

oedema of larynx and bronchi, chemical pneumonitis and pulmonary oedema. Symptoms of exposure may include burning sensation, coughing, laryngitis, shortness of breath, wheezing, headache, nausea and vomiting. Can cause nervous system disturbance (5).

Legislation

Limited under EC Directive on Drinking Water Quality 80/778/EEC. Boron: guide level 1000 μg l^{-1}, maximum admissible concentration 2000 μg l^{-1}. Chlorides: guide level 25 mg l^{-1}, maximum admissible concentration 400 mg l^{-1} (6).

Any other comments

Reacts with hydrogen at 1200°C (7).
Hydrolyses to hydrochloric and boric acids (8).
Toxicity and hazards reviewed (9).
Physico-chemical properties, human health effects and experimental toxicology reviewed (5,10).
Nonflammable gas. Corrosive.

References

1. Brown, K. W; et al. *Hazardous Waste Land Treatment* 1983, 211, Butterworth Publishers, Boston, MA
2. Stokinger, H. E; et al. *Pharmacology and Toxicology of Uranium Compounds* 1953, **4**, McGraw Hill, New York
3. Vernot, E. H; et al. *Toxicol. Appl. Pharm.* 1977, **42**(2), 417-423
4. Adams, R. M. *Boron, Metallo Boron Compounds and Boranes* 1964, John Wiley & Sons, New York
5. Lenga, R. E. *The Sigma Aldrich Library of Chemical Safety* 1985, Sigma Aldrich, Milwaukee
6. *EC Directive Relating to the Quality of Water Intended for Human Consumption* 1982, 80/778/EEC, Office for Official Publications of the European Communities, 2 rue Mercier, L-2985, Luxembourg
7. Hawley, G. G. *The Condensed Chemical Dictionary* 10th ed., 1981, Van Nostrand Reinhold, New York
8. Hunter, D. *Diseases of Occupation* 1987, Hodder Stoughton, London
9. Izmerov, N. F. *Scientific Reviews of Soviet Literature on Toxicity & Hazards of Chemicals* 1992-1993, **116**, Eng. Trans., Richardson, M. L. (Ed.), UNEP/IRPTC, Geneva
10. *ECETOC Technical Report No.30(4)* 1991, European Chemical Industry Ecology and Toxicology Centre, B-1160 Brussels

B169   Boron trifluoride

$\dot{B}F_3$

CAS Registry No. 7637-07-2
Synonyms boron fluoride; trifluoroborane; trifluoroboron; ANCA 1040
Mol. Formula BF$_3$ **Mol. Wt.** 67.81
Uses Initiation and polymerisation catalyst. Catalyst in organic synthesis.

Physical properties
M. Pt. –127°C; **B. Pt.** –100°C; **Specific gravity** 2.99; **Volatility** v.p. >1 mmHg at 20°C.

Solubility
Water: 3320 g l^{-1} at 0°C.

Occupational exposure
US TLV (TWA) Ceiling limit 1 ppm (2.8 mg m^{-3}); **UK Short-term limit** 1 ppm (3 mg m^{-3}); **UN No.** 1008; **Conveyance classification** toxic gas; **Supply classification** toxic.

Risk phrases
Reacts violently with water – Very toxic by inhalation – Causes severe burns (R14, R26, R35)

Safety phrases
Keep container in a well ventilated place – In case of contact with eyes, rinse immediately with plenty of water and seek medical advice – After contact with skin, wash immediately with plenty of water – Wear suitable protective clothing – In case of accident or if you feel unwell, seek medical advice immediately (show label where possible) (S9, S26, S28, S36, S45)

Environmental fate

Absorption
Some boron is adsorbed by iron and aluminium hydroxy compounds and clay minerals, maximum range pH 7-9 (1).

Mammalian and avian toxicity

Acute data
LC$_{50}$ (4 hr) inhalation rat 1180 mg m^{-3} (2).
LC$_{50}$ (2 hr) inhalation mouse 3460 mg m^{-3}.
LC$_{50}$ (4 hr) inhalation guinea pig 109 mg m^{-3} (3).

Metabolism and pharmacokinetics
During inhalation exposure of rats for up to 6 months at 12.8 ppm, 3-4 ppm and 1.5 ppm, the average fluorine contents of teeth and bone were elevated at all exposure levels but not in soft tissues, lung, liver and blood (4).
Fluoride exchanged with hydroxy groups of hydroxyapatite in bone, to form fluorohydroxyapatite. Unretained fluoride was excreted rapidly in urine (5).

Irritancy
Inhalation rat 17 mg m^{-3} showed irritation of mucous membranes and eyes (6).

Any other adverse effects to man
A study of 78 workers exposed for 10-15 yr to boron trifluoride in the USSR showed workers suffered from dryness and bleeding of nasal mucosa, bleeding gums, dry and scaly skin, and pain in joints. Exposure levels were not reported (7).
In the USA 13 workers exposed to boron trifluoride concentration range 0.1-1.8 ppm showed reduced pulmonary function (8).

Any other adverse effects

Kidney necrosis, rales, lachrymation, reversible depression of serum total protein and globulin, and increased urinary, serum and bone fluoride reported in rats exposed to 17 mg m^{-3}, 6 hr day^{-1}, 5 day wk^{-1} for 13 wk (6).

Repeated exposure of rats, rabbits, and guinea pigs to concentrations of 6.4, 3.85 and 1.5 ppm of boron trifluoride caused respiratory irritation. The highest exposures caused respiratory failure in guinea pigs (6).

Legislation

Limited under EC Directive on Drinking Water Quality 80/778/EEC. Boron: guide level 1000 μg l^{-1}, maximum admissible concentration 2000 μg l^{-1}. Fluoride: maximum admissible concentration 1500 μg l^{-1} at 8-12°C, 700 μg l^{-1} at 25-30°C (9).

Any other comments

Toxicity and hazards reviewed (6).

Physico- chemical properties, experimental toxicology, human health effects, epidemiology and workplace experience reviewed (10).

Incompatible with alkali metals, alkaline earth metals (except magnesium), alkyl nitrates and lime (CaO), hydrofluoric and hydrofluoroboric acids.

References

1. Brown, K. W; et al. *Hazardous Waste Land Treatment* 1983, 211, Butterworth Publishers, Boston, MA
2. Izmerov, N. F; et al. *Toxicometric Parameters of Industrial Toxic Chemicals under Single Exposure* 1982, 27, Moscow
3. Kasparov, A. A; et al. *Farmakologiya i Toksikologiya* 1972, **35**, 369, Moscow
4. Clayton, G. D; et al. *Patty's Industrial Hygiene and Toxicology* 3rd ed., 1981-1982, **2A, 2B, 2C**, 2998, John Wiley & Sons, New York
5. *USEPA Office of Drinking Water Criteria Document: Fluoride* 1985, III-19
6. Rusch, G. M; et al. *Toxicol. Appl. Pharmacol.* 1986, **83**(1), 69-78
7. Kirii, V. G. *Zh. Otd. Vyp. Farmakol. Khimioter. Sredstva Toksikol.* 1967, **54**(ii), 990
8. Torkelson, T. R; et al. *Am. Ind. Hyg. Assoc. J.* 1961, **22**, 263-270
9. *EC Directive Relating to the Quality of Water Intended for Human Consumption* 1982, 80/778/EEC, Office for Official Publications of the European Communities, 2 rue Mercier, L-2985 Luxembourg
10. *ECETOC Technical Report No.30(4)* 1991, European Chemical Industry Ecology and Toxicology Centre, B-1160 Brussels

B170   Boron trifluoride diethyl etherate

(CH3CH2)2O.BF3

CAS Registry No. 109-63-7

Synonyms boron trifluoride etherate; boron trifluoride diethyl ether

Mol. Formula $C_4H_{10}BF_3O$ **Mol. Wt.** 141.93

Uses In copolymer initiation reactions.

Physical properties

M. Pt. –60.4°C; **B. Pt.** 125.7°C; **Flash point** 64°C (open cup); **Specific gravity** 1.154.

Occupational exposure

UN No. 2604; **HAZCHEM Code** 4WE; **Conveyance classification** corrosive substance.

Mammalian and avian toxicity

Metabolism and pharmacokinetics

Borates are rapidly absorbed from mucous membranes and abraded skin, but not from intact or unbroken skin. Borate excretion occurs mainly through kidneys, ≈50% is excreted in first 12 hr and remainder is eliminated over 5-7 days (1).

Irritancy

Contact with skin causes burning and redness. Exposure to vapour caused pain, redness, watering and blurred vision in eyes (2).

Any other adverse effects

Systemic effects include vomiting, abdominal pain and diarrhoea (2).
Destructive to tissues of mucous membranes and upper respiratory tract, eyes and skin. Inhalation can be fatal due to spasm, inflammation and oedema of the larnyx and bronchi, chemical pneumonitis and pulmonary oedema, symptoms of exposure may include burning sensation, coughing, shortness of breath, wheezing, laryngitis, headache, nausea and vomiting (3).

Legislation

Limited under EC Directive on Drinking Water Quality 80/778/EEC. Boron: guide level 1000 μg l^{-1}, maximum admissible concentration 2000 μg l^{-1}. Fluoride: maximum admissible concentration 1500 μg l^{-1} at 8-12°C, 700 μg l^{-1} at 25-30°C (4).

Any other comments

Toxicity and hazards reviewed (5).
Corrosive. Moisture sensitive. Incompatible with oxidizing materials. Can form peroxides. Fire hazard.

References

1. Gosselin, R. E; et al. *Clinical Toxicology of Commercial Products* 5th ed., 1984, III-67, Williams & Wilkins, Baltimore
2. *Substances Hazardous to Health* 1990, Croner Publ., London
3. Lenga, R. E. *The Sigma Aldrich Library of Chemical Safety Data* 1985, Sigma Aldrich, Milwaukee
4. *EC Directive Relating to the Quality of Water Intended for Human Consumption* 1982, 80/778/EEC, Office for Official Publications of the European Communities, 2 rue Mercier, L-2985 Luxembourg
5. Izmerov, N. F. *Scientific Reviews of Soviet Literature on Toxicity & Hazards of Chemicals* 1992-1993, **116**, Eng. Trans, Richardson, M. L. (Ed.), UNEP/IRPTC, Geneva

B171   Boron trifluoride dimethyl etherate

$(H_3C)_2CO.BF_3$

CAS Registry No. 353-42-4
Mol. Formula $C_2H_6BF_3O$ **Mol. Wt.** 113.88
Uses Used to produce ^{10}B isotope.

Physical properties

M. Pt. −15°C; **B. Pt.** p26-127°C; **Flash point** 35°C; **Specific gravity** 1.239.

Occupational exposure

UN No. 2965; **HAZCHEM Code** 4WE; **Conveyance classification** substance which in contact with water emits flammable gas.

Mammalian and avian toxicity

Acute data
LC_{Lo} (4 hr) inhalation guinea pig 50 ppm (1).

Any other adverse effects

Destructive to tissues of mucous membranes and upper respiratory tract, eyes and skin. Inhalation may be fatal as a result of spasm, inflammation and oedema of larynx and bronchi, chemical pneumonitis and pulmonary oedema. Symptoms of exposure include burning sensation, coughing, wheezing, laryngitis, shortness of breath, headache, nausea and vomiting (2).

Legislation

Limited under EC Directive on Drinking Water Quality 80/778/EEC. Fluoride: maximum admissible concentration 1500 µg l^{-1} at 8-12°C, 700 µg l^{-1} at 25-30°C (4).

Any other comments

Toxicity and hazards reviewed (3).
Flammable liquid. Corrosive. Reacts violently with water. Incompatible with acid, bases, alcohols and alkali metals. Decomposes to produce toxic fumes of carbon monoxide, carbon dioxide and hydrogen fluoride.

References

1. Adams, R. M. *Boron, Metallo-Boron Compounds and Boranes* 1964, John Wiley & Sons, New York
2. Lenga, R. E. *The Sigma Aldrich Library of Chemical Safety Data Sheets* 1985, Sigma Aldrich, Milwaukee)
3. *EC Directive Relating to the Quality of Water Intended for Human Consumption* 1982, 80/778/EEC, Office for Official Publications of the European Communities, 2 rue Mercier, L-2985 Luxembourg
4. Izmerov, N. F. *Scientific Reviews of Soviet Literature on Toxicity & Hazards of Chemicals* 1992-1993, **116**, Eng. Trans, Richardson, M. L. (Ed.), UNEP/IRPTC, Geneva

B172   Brodifacoum

CAS Registry No. 56073-10-0
Synonyms Talon rodenticide; 3-(3-(4'-bromobiphenyl-4-yl)-1,2,3,4-
tetrahydronaphth-1-yl)-4- hydroxycoumarin; Havo; 3-[3-(4'-bromo
[1,1'-biphenyl]-4-yl)-1,2,3,4-tetrahydro-1-naphthalenyl]-
4-hydroxy-2H-1-benzopyran- 2-one
Mol. Formula $C_{31}H_{23}BrO_3$ **Mol. Wt.** 523.44
Uses Anticoagulant rodenticide which inhibits prothrombin synthesis.

Physical properties
M. Pt. 228-230°C; **Volatility** v.p. <9.776×10^{-7} mmHg at 25°C.

Solubility
Water: <10 mg l^{-1} at 20°C. Organic solvent: chloroform, benzene, acetone, ethanol

Ecotoxicity

Fish toxicity
LC_{50} (96 hr) rainbow trout, bluegill sunfish 91-165 μg l^{-1} (1).
LC_{50} (3-7 days) carp, Crucian carp, tench 0.1->1 mg l^{-1} (2).

Effects to non-target species
LD_{50} oral mallard duck 5.0 mg kg^{-1} (1).

Environmental fate

Degradation studies
Degraded in soils, pH 5.5-pH 8, under aerobic and flooded conditions (1).

Mammalian and avian toxicity

Acute data
LD_{50} oral rat, rabbit mouse 160-400 μg kg^{-1} (3,4).
LD_{50} oral gerbil 1 mg kg^{-1} (5).
LD_{50} dermal rat 50 μg kg^{-1} (6).
LD_{50} percutaneous rat 50 mg kg^{-1} (technical grade compound) (1).
Administration of the LD_{50} to rats increased bleeding time, coagulation time, white
blood cells and prothrombin. Decreases were recorded in red blood cell count,
haemoglobin content and haemocrit values (7).
In one day feeding trials 0.005% brodifacoum caused 100% mortality in rats (8).

Irritancy
Dust non-irritating in Vermac tests (9).

Any other adverse effects

1 μg brodifacoum caused complete inhibition of vitamin K formation in warfarin-resistant rats (10).

In rabbits treated with brodifacoum no direct effect was observed on the clearance of vitamin K from either plasma or liver (11).

Any other comments

Baits containing >100 mg kg^{-1} are hazardous to humans and availability should be restricted (12).

A review with references on the rodenticide brodifacoum (13).

Water is a suitable vehicle for brodifacoum (14).

References

1. Worthing, C. R. (Ed.) *The Pesticide Manual* 9th ed., 1991, British Crop Protection Council, Farnham
2. Wohlgemuth, E. *Agrochemie (Bratislava)* 1988, **28**(4), 126-128, (Czech.) (*Chem. Abstr.* **109**, 87766v)
3. *J. Am. Med. Assoc.* 1984, **252**, 3005
4. *Farm Chemicals Handbook* 1991, C227, Meister Publishing, Willoughby, OH
5. *New Zealand J. Exp. Agric.* 1981, **9**, 23, 147
6. *Malaysian Agric. J.* 1979, **52**(4), 1
7. Abdel-Raheem, K; et al. *Proc. Zool. Soc., A. R. Egypt* 1986, (10), 9-20
8. Saxena, Y. S; et al. *Natl. Acad. Sci. Lett.* 1987, **10**(10), 361-363
9. *J. Hyg.* 1981, **87**, 179
10. Trivedi, L. S. *Arch. Biochem. Biophys.* 1988, **264**(1), 67-73
11. Winn, M. J. *Br. J. Pharmacol.* 1988, **94**(9), 1077-1084
12. *Tech. Rep. Ser. World Health Org.* 1985, Vector Biology and Control **No 720**
13. Kitahara, E. *Shokubutsu Boeki* 1988, **42**(3), 144-147, (Jap.) (*Chem. Abstr.* 109, 50141m)
14. Sheikher, C. *Proc. Indian Natl. Sci. Acad. Part B* 1986, **52**(6), 341-345

B173   Bromacil

CAS Registry No. 314-40-9

Synonyms 5-bromo-3-*sec*-butyl-6-methyluracil; 5-bromo-6-methyl-3-(1-methylpropyl)-2,4(1H,3H)-pyrimidinedione;Cyanogen; Hyvar

Mol. Formula $C_9H_{13}BrN_2O_2$ **Mol. Wt.** 261.13

Uses Herbicide.

Physical properties

M. Pt. 158-159°C; **Specific gravity** 1.55; **Volatility** v.p. 2.0 x 10^{-5} mmHg at 25°C.

Solubility

Water: 815 mg l^{-1} at 25°C. Organic solvent: ethanol, acetone, acetonitrile, xylene

Ecotoxicity

Fish toxicity
LC_{50} (48 hr) bluegill sunfish, rainbow trout 71-75 mg l^{-1}.
LC_{50} (48 hr) carp 164 mg l^{-1} (1).
Fathead minnows 30 days old were exposed to bromacil. LC_{50} values were 185, 183, 182 and 167 mg l^{-1} at 25, 48, 96 and 168 hr, respectively. Eggs, newly hatched fry and juvenile fish were continuously exposed to lower concentrations for 64 days. Growth was significantly reduced at 1 mg l^{-1} (2).

Invertebrate toxicity
EC_{50} (5 min) *Photobacterium phosphoreum* 6.71 mg l^{-1} Microtox test (3).

Effects to non-target species
LC_{50} (8 day) mallard duck >10,000 mg kg^{-1} in diet.
Non-toxic to bees (1).

Environmental fate

Degradation studies
Residual activity in soil ≈7 months (1).
Of 55 fungal and 73 bacterial cultures isolated from soil, only 4 fungi were capable of degrading 5-bromo-3-*sec*-butyl-6-methyluracil; one culture was identified as *Penicillium paraherquei* Abe (4).

Abiotic removal
Photolytic and hydrolytic removal of bromacil occurs via the formation of intermediates 5-bromo-6-methyluracil and 6-methyluracil (5).
Loss from soil due to volatilisation is negligible (6).

Absorption
Weakly absorbed by soils and was uniformly distributed in the soil, but did exhibit slight retardation in the heavier soils. After several cycles of wetting and drying, it was completely leached from around the emitter in soils and was concentrated at the outer edges of the wetted zone (7).

Mammalian and avian toxicity

Acute data
LD_{50} oral mouse, rat 3040-5200 mg kg^{-1} (1,8).
LC_{50} (4 hr) inhalation rat >4.8 mg l^{-1}.
LD_{50} percutaneous rabbit >5000 mg kg^{-1}.
LC_{50} (8 day) bobwhite quail >10,000 mg kg^{-1} in diet (1).

Carcinogenicity and long-term effects
In a 2 yr feeding study the no effect level in rats and dogs was 250 mg kg^{-1} (1).

Teratogenicity and reproductive effects
TC_{Lo} (2 hr) inhalation (7-14 day pregnant) rat 165 mg m^{-3}, no prenatal or teratogenic effects observed (9).
Not teratogenic in New Zealand white rabbits when fed in the diet at 0, 50 or 250 ppm on day 8-16 of pregnancy. Dietary level of 250 ppm had no adverse effects on reproduction or lactation in a 3 generation 6 litter study (10).

Metabolism and pharmacokinetics

In humans, bromacil is excreted unchanged in the urine or as the 5-bromo-3-*sec*-butyl-6-hydroxymethyluracil metabolite primarily as the glucuronide and/or sulfonate conjugate (9).

Concentration in cow feed of 5 and 30 ppm, secretion of intact compound in milk reached 0.019 and 0.13 ppm, respectively. Absent in urine and faeces (4).

Irritancy

50% aqueous suspension applied to intact and abraded skin of albino ♂ guinea pigs. Reaction after 24 hr indicated mild irritation in young animals and slightly greater irritation in older animals. 10 mg (as dry powder) sprinkled onto eyes of ♂ rabbits caused mild conjunctivitis with no corneal injury (10).

Sensitisation

50% aqueous suspension applied to abraded skin of albino ♂ guinea pigs, 3 times per wk for 3 wk. Animals were reexposed after 2 and 3 wk rest periods. No evidence of sensitisation (10).

Genotoxicity

Salmonella typhimurium TA100, TA1535, TA1537, TA1538 with and without metabolic activation negative (9).
Saccharomyces cerevisiae D3, D7 with and without metabolic activation negative (11).
Drosophila melanogaster sex chromosome loss and nondisjunction negative (12).

Any other adverse effects

Rat hepatocytes high toxicity displayed (dose unspecified) but lower toxicity in rat bone marrow cell cultures (13).

Legislation

Limited under EC Directive on Drinking Water Quality 80/778/EEC. Pescticides and related products: maximum admissible concentration 0.5 µg l^{-1} (14).

References

1. *The Agrochemicals Handbook* 3rd ed., 1991, RSC, London
2. Call, D. J; et al. *Arch. Environ. Contam. Toxicol.* 1987, **16**(5), 607-613
3. Kaiser, K. L. E; et al. *Water Poll. Res. J. Canada* 1991, **26**(3), 361-431
4. Menzie, C. M. *Metabolism of Pesticides, An Update. US Dept of the Interior, Fish, Wildlife Service, Special Scientific Report* 1974, 387, US Govt. Printing Office, Washington, DC
5. Moilanen, K. W; et al. *Arch. Environ. Contam. Toxicol.* 1974, **2**, 3
6. *Herbicide Handbook* 4th ed., 1979, 69, Weed Science Society of America, Champaign, IL
7. Gersch, Z; et al. *Soil Sci. Soc. Am. J.* 1983, **47**(3), 478-483
8. *J. Environ. Sci. Health* 1980, **15**, 867
9. *Drinking Water Health Advisory on Pesticides* 1989, 101-116, Lewis Publishers, Chelsea, MI
10. Sherman, K; et al. *Toxicol. Appl. Pharmacol.* 1975, **34**, 189-196
11. Riccio, E. G; et al. *Environ. Mol. Mutagen* 1981, **3**, 327
12. Gopalan, H. N. B; et al. *Genetics* 1981, **97**, 544
13. Parent, D; et al. *Meded. Fac. Landbouwwet; Rijksuniv Gent* 19909, **55**(3b), 1369-1375, (Fr.) (*Chem. Abstr.* 1991, **114**, 180014m)
14. *EC Directive Relating to the Quality of Water Intended for Human Consumption* 1982, 80/778/EEC, Office for Official Publications of the European Communities, 2 rue Mercier, L-2985 Luxembourg

B174 Bromadiolone

CAS Registry No. 28772-56-7

Synonyms Bromadialone; Bromone; 3-(α- (*p*-(*p*-bromophenyl)-β-hydroxyphenethyl)benzyl)-4-hydroxycoumarin; 3-(3- (4′-bromo-(1,1′-biphenyl)-4-yl)3-hydroxy-1-phenylpropyl)-4-hydroxy-2*H*-1- benzopyran-2-one; Canadien 2000

Mol. Formula $C_{30}H_{23}BrO_4$ **Mol. Wt.** 527.42

Uses An anticoagulant rodenticide which inhibits prothrombin synthesis.

Physical properties

M. Pt. 200-210°C.

Solubility

Water: 19 mg l^{-1} at 20°C. Organic solvent: acetone, dimethylformamide, ethanol

Ecotoxicity

Fish toxicity
LC_{50} (96 hr) rainbow trout 1.4 mg l^{-1} (1).

Effects to non-target species
Non-toxic to bees (1).

Mammalian and avian toxicity

Acute data
LD_{50} oral rabbit, rat, mouse 1125-1750 μg kg^{-1} (1,2).
LD_{50} oral ♀ rat 0.59 mg kg^{-1} (3).
LD_{50} percutaneous rabbit 2100 μg kg^{-1} (4).
In 1 day feeding trials 0.1% bromodiolone solution caused 100% mortality in rats (3).

Metabolism and pharmacokinetics
Oral rat 0.8 and 3 mg kg^{-1} elimination only from liver 30% of administered dose is excreted in bile as a glucuronide conjugate during first 8 hr after dosing (5).

Any other comments

Baits containing >100 mg kg^{-1} are hazardous to humans and their availability should be restricted (6).

Bittrex, the taste deterrent for humans and non-target animals, increased the baits palatibility to rats and mice (7).

References

1. *Farm Chemicals Handbook* 1991, C50, Meister Publishing, Willoughby, OH
2. *Phytiatrie-Phytopharmacie* 1976, **25**, 69
3. Lam, Y. M. *MARDI Res. Bull.* 1985, **13**(3), 303-308
4. *The Agrochemicals Handbook* 2nd ed., 1987, RSC, London
5. Nahas, K. *Pharmacol. Res. Comm.* 1987, **19**(11), 767-775

6. *Tech. Rep. Ser. Wld. Health Org.* 1985, Vector Biology and Control **No 720**
7. *Res. Discl.* 1988, **287**, 130

B175   Bromethalin

CAS Registry No. 63333-35-7
Synonyms *N*-methyl-2,4-dinitro-*N*-(2,4,6-
tribromophenyl)-6-)(trifluoromethyl)benzenamine; α,α,α, -trifluoro-
N-methyl-4,6-dinitro-*N*-(2,4, 6-tribromophenyl)- *o*-toluidine; Ven
Mol. Formula $C_{14}H_7Br_3F_3N_3O_4$ **Mol. Wt.** 577.95
Uses Rodenticide effective against warfarin resistant mice and rats.

Physical properties
M. Pt. 150-151°C; **Volatility** v.p. 0.776×10^{-8} mmHg at 25°C.

Solubility
Water: <0.01 mg l^{-1}. Organic solvent: chloroform, dichloromethane, methanol

Mammalian and avian toxicity

Acute data
LD_{50} oral rat, mouse 2-5 mg kg^{-1} (technical grade in propane-1,2-diol).
LC_{50} (1 hr) inhalation rat 0.024 mg l^{-1} (in air) (1).
LD_{50} percutaneous ♂ rabbit 1000 mg kg^{-1} (2).
Doses in excess of the LD_{50} 2 mg kg^{-1} in rats caused death within 8-12 hr and was
preceded by 1-3 episodes of clonic convulsions with death usually due to respiratory
arrest. Guinea pigs could tolerate ≥1000 mg kg^{-1} without effect (3).

Sub-acute data
In 90 day feeding trials the no effect level for rats and dogs was 0.025 mg kg^{-1} daily (4).

Metabolism and pharmacokinetics
Metabolised to the desmethyl analogue in rats (5).

Any other adverse effects
Multiple low doses of sublethal intoxications yields hind leg weakness and loss of
tactile sensation in rodents. Spongy degeneration of the white matter (intramycleric
odema) occurred in the brain and spinal cord of these animals. No inflammation or
cellular destruction of neuronal tissue was noted (3).
In rats a potent uncoupler of oxidative phosphorylation (5).

Any other comments
The toxicity, mechanism of action and rodenticidal efficiency of bromethalin is
reviewed (6).

Mode of action, toxicity, clinical effects and treatment of efficacy in rats, dogs and cats reviewed (7).

References

1. Dreikorn, B. A; et al. *Proc. Brit. Crop. Prot. Conf: – Pests Dis.* 1979, 491
2. *The Agrochemicals Handbook* 2nd ed., 1987, RSC, London
3. Van Lier, R. B. L; et al. *Fundam. Appl. Toxicol.* 1988, **11**(4), 664-672
4. Worthing, C. R. (Ed.) *The Pesticide Manual* 9th ed., 1991, British Crop Protection Council, Farnham
5. Van Lier *Fundam. Appl. Toxicol.* 1988, **11**(4), 664-672, RBL
6. Spaulding, S. R. *Monogr. Br. Crop. Prot. Counc.* 1987, **37**, 137-147
7. Dorman, D. C. *Diss. Abstr. Int. B* 1990, **51**(4), 1688

B176   Bromine

Br_2

CAS Registry No. 7726-95-6

Mol. Formula Br_2 **Mol. Wt.** 159.82

Uses Water disinfectant. Bleaching fibres and silk. Manufacture of medicinal bromine compounds and dyestuffs. Manufacture of ethylene dibromide (anti-knock gasoline). Used in form of adduct with a quaternary ammonium compound in the treatment of plantar warts (1).

Occurrence Occurs in igneous rock and in seawater.

Physical properties

M. Pt. –7.25°C; **B. Pt.** 58.73°C; **Volatility** v.p. 175 mmHg at 21°C; v. den. 55.

Solubility

Organic solvent: ethanol, diethyl ether, chloroform, tetrachloromethane, carbon disulfide

Occupational exposure

US TLV (TWA) 0.1 ppm (0.66 mg m^{-3}); **US TLV (STEL)** 0.3 ppm (2 mg m^{-3}); **UK Long-term limit** 0.1 ppm (0.7 mg m^{-3}); **UK Short-term limit** 0.3 ppm (2 mg m^{-3}); **UN No.** 1744; **HAZCHEM Code** 2XE; **Conveyance classification** corrosive substance; toxic substance; **Supply classification** corrosive.

Risk phrases

Very toxic by inhalation – Causes severe burns (R26, R35)

Safety phrases

Keep container tightly closed and in a well ventilated place – In case of contact with eyes, rinse immediately with plenty of water and seek medical advice (S7/9, S26)

Mammalian and avian toxicity

Acute data

LD_{Lo} oral human 14 mg kg^{-1} (2).

LC_{50} (9 min) inhalation mouse 750 ppm (3).

LC$_{Lo}$ (6.5 hr) inhalation rabbit 180 ppm (4).

Metabolism and pharmacokinetics
Bromine vapours enter body by respiratory system, skin and digestive system, deposited in tissues as bromides (5).

Irritancy
Destructive and painful burns to skin and eyes from contact with liquid or vapour (6).

Any other adverse effects to man
Chronic bronchitis was the most common disorder of workers in the manufacture of bromine and bromine-containing materials in the USSR. Pathological changes in the respiratory tract, contact and allergic dermatitis and arteral hypertension were observed and related to exposure levels (7).

Any other adverse effects
Severe exposure may result in pulmonary oedema (6).
Ingestion may cause severe gastroenteritis and death (1).
Inhalation of vapours may cause chemical pneumonitis (6).

Any other comments
Levels of bromide in excess of 50 mg kg^{-1} have been detected in herbs imported into Switzerland, although plants were not treated with any bromide containing pesticide (8).
Physico-chemical properties, human health effects, experimental toxicology, epidemiology, workplace experience and environmental effects reviewed (9).
Corrosive. Explosion risk.

References
1. *Martindale: The Extra Phamacopoeia* 29th ed., 1988, The Pharmaceutical Press, London
2. Deichmann, W. B; et al. *Toxicology of Drugs and Chemicals* 1969, 645, Academic Press, London
3. *Am. Ind. Hyg. Assoc. Journal* 1978, **39**, 129
4. Saunders, W. B. *Handbook of Toxicology* 1956, Volume I-V, **1**, 324
5. *Encyclopedia of Occupational Health and Safety* 1983, **I-II**, 327, International Labour Office, Geneva
6. Gosselin, R. E; et al. *Clinical Toxicology of Commerce Products* 5th ed., 1984, 2(99), 25
7. Krasnyuk, E. P; et al. *Gig. Sanit.* 1987, (7), 81-82, (Russ.) *(Chem. Abstr.,)* 1987, **107**, 182619s
8. Corvi, C; et al. *Mitt. Geb. Lebensmittelunters. Hyg.* 1989, **80**(2), 215-222, (Fr) *(Chem. Abstr.* **111**, 152284u)
9. *ECETOC Technical Report No.30(4)* 1991, European Chemical Industry Ecology and Toxicology Centre, B-1160 Brussels

B177   Bromine chloride

BrCl

CAS Registry No. 13863-41-7
Synonyms bromine monochloride; bromochloride
Mol. Formula BrCl   Mol. Wt. 115.36
Uses In organic addition and substitution reactions. Disinfectant in wastewater
treatment.

Physical properties

M. Pt. −66°C; B. Pt. 10°C (decomp.).

Solubility
Organic solvent: carbon disulfide, diethyl ether

Ecotoxicity

Fish toxicity
Brominated compounds were detected in fathead minnows from bromine/chloride
disinfected sewage, concentrations of bromine 5-200 ng g^{-1} for various compounds
(1).
Fathead minnows and lake trout with >2 hr exposure to chlorobrominated effluent
were capable of tolerating high levels of total residual bromine chloride for longer
periods than for fish not previously exposed to the wastewater disinfectant (2).
LC_{50} (96 hr) Atlantic menhaden, Atlantic silverside 0.21-0.23 mg l^{-1} (3).

Invertebrate toxicity
LC_{50} (24,48,96 hr) grass shrimp 1.1, 0.8, 0.6 mg l^{-1}, respectively.
LC_{50} (48-96 hr) blue crab 1.2, 0.8 mg l^{-1} (4).
LC_{50} (48 hr) oyster, copepods 0.10-0.21 mg l^{-1}.
LC_{50} (96 hr) shrimp 0.70 mg l^{-1} (5).

Environmental fate

Degradation studies
In water, interacts with ammonia to form bromoamines which are less toxic to fish
than chloramines formed when chlorine is used (6).

Legislation
Limited under EC Directive on Drinking Water Quality 80/778/EEC. Chlorides:
guide level 25 mg l^{-1}, maximum admissible concentration 400 mg l^{-1} (7).

Any other comments
Readily hydrolyses to HOBr over wide pH range (4).
Oxidising agent.

References

1. Jolley, R. L; et al. *Water Chlorination, Environmental Impact and Health Effects* 1978, **2**, 175-192
2. DeGraeve, G. M. *J. Water Pollut. Control Fed.* 1977, **49**(10), 2172-2178
3. Roberts, M. H; et al. *Marine Environ. Res.* 1978, **1**(1), 19-30

4. Barton, D. T; et al. *Bull. Environ. Toxicol.* 1978, **19**(2), 131-138
5. Roberts, M. H; et al. *Mar. Environ. Research* 1978, **1**(1), 19-30
6. Mills, J. F. *Disinfectants: Waterwaste treatment* 1975, 113
7. *EC Directive Relating to the Quality of Water Intended for Human Consumption* 1982, 80/778/EEC, Office for Official Publications of the European Communities, 2 rue Mercier, L-2985 Luxembourg

B178   Bromine pentafluoride

BrF$_5$

CAS Registry No. 7789-30-2
Mol. Formula BrF$_5$ **Mol. Wt.** 174.90
Uses Fluorinating agent in organic synthesis. Formation of uranium fluorides for isotope enrichment and for fuel element reprocessing.

Physical properties

M. Pt. –61.3°C; **B. Pt.** 40.5°C; **Specific gravity** d^{25} 2.466; **Volatility** v. den. 6.05.

Occupational exposure

US TLV (TWA) 0.1 ppm (0.72 mg m^{-3}); **UK Long-term limit** 0.1 ppm (0.7 mg m^{-3}); **UK Short-term limit** 0.3 ppm (2 mg m^{-3}); **UN No.** 1745;
HAZCHEM Code 4WE; **Conveyance classification** oxidising substance; toxic substance; corrosive substance.

Legislation

Limited under EC Directive on Drinking Water Quality 80/778/EEC. Fluoride: maximum admissible concentration 1500 µg l^{-1} at 8-12°C, 700 µg l^{-1} at 25-30°C (1).

Any other comments

Fire risk. Corrosive. Very reactive, therefore must be handled in resistant materials like nickel, Monel metal or Teflon plastics. Reacts with every known element except inert gases, nitrogen and oxygen (2).
Human health effects, experimental toxicology and environmental effects reviewed (3).

References

1. *EC Directive Relating to the Quality of Water Intended for Human Consumption* 1982, 80/778/EEC, Office for Official Publications of the European Communities, 2 rue Mercier, L-2985 Luxembourg
2. *Kirk-Othmer Encyclopedia of Chemical Technology* 1978, **4**, 246, John Wiley & Sons, New York
3. *ECETOC Technical Report No.30(4)* 1991, European Chemical Industry Ecology and Toxicology Centre, B-1160 Brussels

B179   Bromine trifluoride

BrF₃

CAS Registry No. 7787-71-5
Mol. Formula BrF₃   **Mol. Wt.** 136.90
Uses As fluorinating agent in organic synthesis. Electrolyte solvent. Rocket propellant.

Physical properties

M. Pt. 9°C (decomp.); **B. Pt.** 125°C; **Specific gravity** d^{25} 2.803.

Occupational exposure

UN No. 1746; **HAZCHEM Code** 4WE; **Conveyance classification** oxidising substance; toxic substance; corrosive substance; **Supply classification** toxic.

Risk phrases
Reacts violently with water – Very toxic by inhalation – Causes severe burns (R14, R26, R35)

Safety phrases
Keep container in a well ventilated place – In case of contact with eyes, rinse immediately with plenty of water and seek medical advice – Wear suitable protective clothing – In case of accident or if you feel unwell, seek medical advice immediately (show label where possible) (S9, S26, S36, S45)

Any other adverse effects

Corrosive, irritating to skin, eyes, mucous membranes and respiratory tract (1,2).

Legislation

Limited under EC Directive on Drinking Water Quality 80/778/EEC. Fluoride: maximum admissible concentration 1500 µg l⁻¹ at 8-12°C, 700 µg l⁻¹ at 25-30°C (3).

Any other comments

Very reactive, must be handled in resistant materials such as nickel, Monel metal, Teflon plastics (4).
Numerous violent explosive reactions can occur with organic and inorganic materials.

References

1. *The Merck Index* 1989, 11th ed., Merck & Co. Inc., Rahway, NJ
2. Hawley, G. G; et al. *The Condensed Chemical Dictionary* 1981, 10th ed., Van Nostrand Reinhold, New York
3. *EC Directive Relating to the Quality of Water Intended for Human Consumption* 1982, 80/778/EEC, Office for Official Publications of the European Communities, 2 rue Mercier, L-2985 Luxembourg
4. *Kirk-Othmer Encyclopedia of Chemical Technology* 1978, **4**, 246, John Wiley & Sons, New York

B180   Bromoacetic acid

BrCH2CO2H

CAS Registry No. 79-08-3
Synonyms bromoethanoic acid; α-bromoethanoic acid; monobromoacetic acid
Mol. Formula $C_2H_3BrO_2$ **Mol. Wt.** 138.95

Physical properties

M. Pt. 49-51°C; **B. Pt.** 208°C; **Flash point** >110°C; **Specific gravity** d^{20} 1.93.

Solubility
Organic solvent: ethanol

Occupational exposure

UN No. 1938; **HAZCHEM Code** 2R; **Conveyance classification** corrosive substance; **Supply classification** toxic.

Risk phrases
Toxic by inhalation, in contact with skin and if swallowed – Causes severe burns (R23/24/25, R35)

Safety phrases
Wear suitable protective clothing, gloves and eye/face protection – If you feel unwell, seek medical advice (show label where possible) (S36/37/39, S44)

Mammalian and avian toxicity

Acute data
LD_{50} oral mouse 100 mg kg^{-1} (as 5% solution) (1).
LD_{50} intraperitoneal rat, mouse 50-66 mg kg^{-1} (2,3).
LD_{Lo} intravenous mouse 45 mg kg^{-1} (4).

Carcinogenicity and long-term effects
Inhalation rat (30 month) total exposure 114 g m^{-3} caused irritation effects to respiratory system (3).

Genotoxicity

In vitro mouse L-1210 leukocytes induced DNA damage (5).

Any other adverse effects

Corrosive to the skin, although the effects are not necessarily immediate, blisters may not appear for 12 hr or more. Application of a 10% aqueous solution to rabbit skin caused death within 16 hr (6).

Any other comments

Physico-chemical properties, human health effects and experimental toxicology reviewed (7).

References

1. Morrison, J. I. *J. Pharm. Exp. Ther.* 1946, **86**, 336
2. *J. Natl. Cancer Inst.* 1963, **31**, 297

3. *Russ. Pharm. Toxicol.* 1978, **41**, 113
4. *Naunyn-Schmadebeng's Arch. f. Exp. Path. u. Pharmak.* 1931, **160**, 551
5. Stratton, C. E; et al. *Biochem. Pharm.* 1981, **30**(12), 1497-1500
6. *Chemical Safety Data Sheets* 1991, **4a**, 74-76, RSC, London
7. *ECETOC Technical Report No.30(4)* 1991, European Chemical Industry Ecology and Toxicology Centre, B-1160 Brussels

B181   Bromoacetonitrile

BrCH2CN

CAS Registry No. 590-17-0
Synonyms Bromomethyl cyanide
Mol. Formula C_2H_2BrN **Mol. Wt.** 119.95
Occurrence Contaminant in potable water supplies from action of chlorine on humic material (1).

Physical properties

B. Pt. $_{24}$ 60-2°C; **Flash point** 110°C; **Specific gravity** 1.722.

Genotoxicity

Salmonella typhimurium TA98, TA100 with metabolic activation negative (2).

Any other comments

Corrosive. Lachrymatory.

References

1. Bull, R. J; et al. *Environ. Sci. Technol.* 1982, **16**(10), 554A
2. Mortelmans, K; et al. *Environ. Mutagen.* 1986, **8**(Suppl 7), 1-119

B182   4-Bromoaniline

CAS Registry No. 106-40-1
Synonyms *p*-bromoaniline; 4-bromobenzenamine; *p*-bromophenylamine
Mol. Formula C_6H_6BrN **Mol. Wt.** 172.03
Uses Synthesis of azo dyestuffs and dihydroquinazolines.

Physical properties

M. Pt. 62-64°C; **Specific gravity** $d_4^{99.6}$ 1.497.

Solubility
Organic solvent: ethanol, diethyl ether

Ecotoxicity

Bioaccumulation
Calculated bioconcentration factor 4.4 (1).

Environmental fate

Degradation studies
Purified enzymes of the soil fungus *Geotrichum candidum* biotransformed
4-bromoaniline to 4,4'-dibromoazobenzene (2).
A strain of *Moraxella sp.* used 4-bromoaniline as sole source of carbon and nitrogen (3).

Abiotic removal
Significantly absorbs UV light above 290 nm in alcohol solution indicating a
potential for direct photolysis in the environment (4).
Estimated $t_{1/2}$ with sunlight-produced hydroxyl radicals in a typical ambient
atmosphere 2 days (5).

Absorption
Soil adsorption studies using four silt loam soils and a 2 hr adsorption period
measured a soil adsorption coefficient of 7 (6).
Undergoes rapid and reversible covalent bonding with humic materials in aqueous
solution (7).

Mammalian and avian toxicity

Acute data
LD_{50} oral mouse, rat 289-456 mg kg^{-1} (8,9).
LD_{50} intraperitoneal mouse 248 mg kg^{-1} (9).

Genotoxicity
Salmonella typhimurium TA97, TA98, TA100, TA1535, TA1537 with and without
metabolic activation negative (10).
In vitro rat liver cells induced unscheduled DNA synthesis (11).

Any other comments
Toxicity and hazards reviewed (12).
All reasonable efforts have been taken to find information on isomers of this
compound, but no relevant data are available.

References

1. Lyman, W. J; et al. *Handbook of Chemical Property Estimation Methods* 1982, 5.4, McGraw-Hill, New York
2. Bordeleau, L. M; et al. *Can. J. Microbiol.* 1972, **18**, 1873-1882
3. Zeyer, J; et al. *Appl. Environ. Microbiol.* 1985, **50**, 447-453
4. *Sadtler; 1829 UV* S.P. Sadtler & Sons, Philadelphia, PA
5. *GEMS: Graphical Exposure Modeling System. Fate of Atmospheric Pollutants* 1987

6. Briggs, G. G. *J. Agric. Food Chem.* 1981, **29**, 1050-1059
7. Parris, G. E. *Environ. Sci. Technol.* 1980, **14**, 1099-1106
8. *Czechoslovak Hyg.* 1978, **23**, 168
9. *Gig. Sanit.* 1979, **44**(12), 19
10. Zeiger, E; et al. *Environ. Mol. Mutagen.* 1988, **11**(Suppl. 12), 1-158
11. *Environ. Mutagen.* 1981, **3**, 11
12. Izmerov, N. F. *Scientific Reviews of Soviet Literature on Toxicity & Hazards of Chemicals* 1992-1993, **60**, Eng. Trans., Richardson, M. L. (Ed.), UNEP/IRPTC, Geneva

B183　Bromobenzene

CAS Registry No. 108-86-1
Synonyms phenyl bromide
Mol. Formula C_6H_5Br **Mol. Wt.** 157.02
Uses Intermediate in organic synthesis. Solvent. Motor oil and fuel additive.

Physical properties

M. Pt. –31°C; B. Pt. 156°C; Flash point 51°C; Specific gravity d_4^{20} 1.4952;
Partition coefficient log P_{ow} 2.99; Volatility v.p. 3.3 mmHg at 20°C; v. den 5.4.

Solubility
Water: 500 mg l^{-1} at 20°C. Organic solvent: miscible with chloroform, benzene, ethanol, petroleum spirit, diethyl ether

Occupational exposure

UN No. 2514; HAZCHEM Code 2**Y**; Conveyance classification flammable liquid; Supply classification flammable and toxic.

Risk phrases
Flammable – Irritating to skin (R10, R38)

Ecotoxicity

Invertebrate toxicity
EC_{50} (30 min) *Photobacterium phosphoreum* 9.46 mg l^{-1} Microtox test (1).

Mammalian and avian toxicity

Acute data
LD_{50} oral guinea pig, rat 1700-3300 mg kg^{-1} (2).
Inhalation mice, rats, rabbit (4 hr) 250- 3400 ppm. 48 hr after termination of exposure

revealed injury to Clara cells and adjacent epithelium in mouse bronchioli at a concentration of 250 and 1000 ppm and to Clara cells of rat bronchi and bronchioli (1000 ppm) and rabbit bronchi (2500 ppm and 3400 ppm). Kidney toxicity was observed in mice (20% showed tubular necrosis and elevated concentration of plasma area) and rats (all had elevated plasma concentrations of creatinine) exposed to 1000 ppm (3).

LD_{50} intraperitoneal rat 3880 mg kg^{-1} (4).

Mice given a single intraperitoneal dose of ≥754 mg l^{-1} revealed degeneration and necrosis of the glands of Bowman and degenerative changes in the olfactory epithelium. Focal degeneration and necrosis were found in the lateral nasal glands and cyst-like dilation of acini in the lateral nasal glands (5).

Metabolism and pharmacokinetics
Absorbed through lungs, gastrointestinal tract and intact skin. Excreted as catechol derivatives both free and conjugated with sulfate or mercapturic acid (6).

4 hr after intraperitoneal administration to rats, bromobenzene was found in adipose tissue ≥300-fold than in other tissues. 85% was excreted in urine in 24 hr (7).

May be metabolised to an epoxide and then excreted in bile, reabsorbed through enterohepatic circulation and metabolised in several steps to S-P-bromophenyl mercapturic acid which is then excreted in urine (8).

Intragastric administration mice (unspecified dose) caused liver necrosis, increased lipid peroxidation, decreased protein thiols, GSH content and calcium uptake (9).

Genotoxicity
Salmonella typhimurium TA1535/pSK1002 *umu* test negative (10).
Escherichia coli polA+/A- DNA modifying activity weakly positive (11).
Micronucleus test intraperitoneal mouse (24 hr) 125 mg kg^{-1} positive (12).
Increased formation of micronucleated polychromatic erythrocytes in bone marrow of mice after intraperitoneal injection of up to 70% of LD_{50} (13).

Any other comments
Physico-chemical properties, human health effects and experimental toxicology reviewed (14,15).

References
1. Kaiser, K. L. E; et al. *Water Pollut. Res. J. Canada* 1991, **26**(3), 361-431
2. Izmerov, N. F; et al. *Toxicometric Parameters of Industrial Toxic Chemicals under Single Exposure* 1982, 28, Moscow
3. *Toxicol. Appl. Pharmacol.* 1986, **83**, 108
4. Dahl, J. E; et al. *Arch. Toxicol.* 1990, **64**(5), 370-376
5. Brittebo, E. B; et al. *Arch. Toxicol.* 1990, **64**(1), 54-60
6. Gosselin, R. E; et al. *Clinical Toxicology of Commercial Products* 4th ed., 1976, II-114, Williams & Wilkins, Baltimore
7. *Foreign Compound Metabolism in Mammals* 1972, **2**, 158, The Chemical Society, London
8. *National Research Council. Drinking Water & Health* 1977, **1**, 693, National Academy Press, Washington, DC
9. Casini, A; et al. *Basic Life Sci.* 1988, **44** (O_2 Radicals Biol. Med.), 773-776
10. McCann, J; et al. *Proc. Natl. Acad. Sci. (USA)* 1975, **72**, 5135-5139
11. *J. Natl. Cancer Inst.* 1979, **62**, 873
12. *Mutagenesis* 1987, **2**, 111

13. Mohtashamipur, E; et al. *Mutagenesis* 1987, **2**(2), 111-113
14. *ECETOC Technical Report No.30(4)* 1991, European Chemical Industry Ecology and Toxicology Centre, B-1160 Brussels
15. Lau, S. S; et al. *Life Sci.* 1988, **42**(13), 1259-1269

B184   α-Bromobenzyl cyanide

CHBrCN

CAS Registry No. 5798-79-8
Synonyms α-bromophenylacetonitrile; α-bromo-α-tolunitrile; Camite; α-bromobenzylnitrile
Mol. Formula C_8H_6BrN **Mol. Wt.** 196.05
Uses Chemical weapon.

Physical properties

M. Pt. 29°C; **B. Pt.** 242° (decomp.); **Specific gravity** d_4^{29} 1.539; **Volatility** v.p. 0.012 mmHg at 20°C; v. den. 6.8.

Solubility
Organic solvent: ethanol, diethyl ether, chloroform, acetone

Occupational exposure

UN No. 1694; **HAZCHEM Code** 2XE; **Conveyance classification** toxic substance.

Mammalian and avian toxicity

Acute data
LD_{Lo} oral rat 100 mg kg^{-1} (1).
LC_{50} (duration unspecified) inhalation human 3500 mg m^{-3} (2).

Any other comments
Strong lachrymator (3).

References
1. *Natl. Acad. Sci.* 1953, **5**, 32
2. *Science Journal* 1967, **4**, 33
3. *The Merck Index* 11th ed., 1989, Merck & Co. Inc., Rahway, NJ

B185   2-Bromo-2-(bromomethyl)pentanedinitrile

$BrCH_2C(Br)(CN)CH_2CH_2CN$

CAS Registry No. 35691-65-7
Synonyms 1,2-dibromo-2,4-dicyanobutane; Tektamer 38
Mol. Formula $C_6H_6Br_2N_2$ **Mol. Wt.** 265.95
Uses Biocide in liquid soaps and cosmetic formulations. Paper coatings for food products. Adhesives. Latex paints.

Physical properties

Solubility
Organic solvent: acetone, chlorofrom, ethyl acetate, benzene, methanol, ethanol, diethyl ether

Ecotoxicity

Fish toxicity
LC_{50} (96 hr) bluegill sunfish, rainbow trout 9-12 mg l^{-1} (1).

Environmental fate

Degradation studies
Using ^{14}C compound in activated sludge, no significant impact on the microbial population in a 10 day period. Complete degradation occurred in 24 hr at exposure of ≤99 ppm (2).

Mammalian and avian toxicity

Acute data
LD_{50} oral rat 1800 mg kg^{-1} (1).

Any other comments

Effective antimicrobial compared to other cosmetic preservatives. Activity decreased with heat treatment (3).
Evaluation of fungitoxicity *in vitro* to 26 species of phytopathogenic fungi (1).
Can cause plasmid-mediated bacterial resistance to *Pseudomous sp.* (2).

References
1.   Martinelli, J. A; et al. *Summa Phytopathol.* 1984, **10**, 273
2.   Candal, F. J; et al. *Int. Biodetior.* 1984, **20**, 221
3.   Diehl, K. H. *Seifen, Oele, Fette, Wachse* 1985, **111**, 222, (Ger.) (*Chem. Abstr.* **103**, 165914j)

B186   1-Bromobutane

$CH_3(CH_2)_3Br$

CAS Registry No. 109-65-9
Synonyms butyl bromide; butane, 1-bromo-
Mol. Formula C_4H_9Br **Mol. Wt.** 137.03

Uses In preparation of drugs.

Physical properties
M. Pt. −112.4°C; **B. Pt.** 101.4°C; **Flash point** 18.3°C (open cup); **Specific gravity** d_4^{20} 1.258; **Partition coefficient** log P_{ow} 2.75; **Volatility** v. den. 4.72.

Solubility
Organic solvent: ethanol, diethyl ether, acetone, chloroform

Occupational exposure
UN No. 1126; **HAZCHEM Code** 2▨E; **Conveyance classification** flammable liquid.

Mammalian and avian toxicity

Acute data
LC_{50} (30 min) inhalation rat 237 g m^3 (1).
LD_{50} intraperitoneal rat, mouse 4450-6680 mg kg^{-1} (2).

Carcinogenicity and long-term effects
The effect on lung tumour frequency in laboratory mice at low doses was investigated. The toxicity of butyl bromide was too great to allow testing at dosages used for other halides (3).

Any other comments
Review of toxic effects in laboratory animals (4).
Toxicity of brominated hydrocarbons given (5).
Flammable.

References
1. *Fiziol. Akt. Vershestra* 1975, **7**, 35
2. Fischer, G. W; et al. *J. Prakt. Chem.* 1978, **320**(1), 133
3. Poirer, L. A; et al. *Cancer Res.* 1975, **35**(6), 1411-1415
4. Chenoweth, M. B; et al. *Ann. Rev. Pharmacol.* 1962, **2**, 363-398
5. Rabotnikova, L. V; et al. *Izuch. Biol. Deistviya Nov. Productov Organ. Sinteza i Prirod. Soedin Perm* 1981, 105-107, (Russ) (*Chem. Abstr.* **97**, 105112u)

B187   2-Bromobutane

CH3CH2CHBrCH3

CAS Registry No. 78-76-2
Synonyms *sec*-butyl bromide; methylethylbromomethane
Mol. Formula C_4H_9Br **Mol. Wt.** 137.03

Physical properties
M. Pt. −112°C; **B. Pt.** 91°C; **Flash point** 21°C; **Specific gravity** d_4^{25} 1.2530.

Solubility
Organic solvent: ethanol, diethyl ether

Occupational exposure
UN No. 2339; HAZCHEM Code 2▤E; Conveyance classification flammable liquid.

Mammalian and avian toxicity
Carcinogenicity and long-term effects
Intraperitoneal injection of 3000 mg kg^{-1} (total dose given in 24 injections over 24 wk) to 8 wk old mice caused a slight but significant increase in lung tumours (1).

Any other comments
Narcotic (2).
All reasonable efforts have been taken to find information on isomers of this compound, but no relevant data are available.

References
1. Poirer, L. A; et al. *Cancer Res.* 1975, **35**, 1411-1415
2. *Merck Index* 1989, 11th ed., Rahway, NJ

B188   Bromobutide

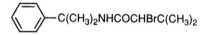

CAS Registry No. 74712-19-9
Synonyms Butanamide, 2-bromo-3,3-dimethyl-*N*-(1-methyl-1-phenylethyl)-; 2-bromo-*N*-(α, α-dimethylbenzyl)-3,3-dimethylbutyramide
Mol. Formula C$_{15}$H$_{22}$BrNO   **Mol. Wt.** 312.26
Uses Selective herbicide.

Physical properties
M. Pt. 180°C.

Solubility
Water: 3.54 mg l^{-1}.

Ecotoxicity

Fish toxicity
LC$_{50}$ (48 hr) carp >10 mg l^{-1} (1,2).

Mammalian and avian toxicity

Acute data
LD$_{50}$ oral rat >5000 mg kg^{-1}.
LD$_{50}$ percutaneous rat >5000 mg kg^{-1} (1,2).

References
1. *The Agrochemicals Handbook* 3rd ed., 1991, RSC, London
2. Worthing, C. R. (Ed.) *The Pesticide Manual* 9th ed., 1991, British Crop Protection Council, Farnham

B189　Bromochloroacetonitrile

BrCHCl(CN)

CAS Registry No. 83463-62-1
Mol. Formula C$_2$HBrClN　　　　　　　　Mol. Wt. 154.40
Occurrence Contaminant in water samples, caused by interaction of chlorine on humic substances (1,2).

Abiotic removal
Hydrolytic t$_{1/2}$ 55 hr at pH8.3 (3).

Mammalian and avian toxicity

Teratogenicity and reproductive effects
In vivo teratology screening rat (route/dose unspecified) reduced post natal growth (4).

Legislation
Limited under EC Directive on Drinking Water Quality 80/778/EEC. Organochlorine compounds: guide level 1 μg l^{-1}. Haloform concentrations must be as low as possible (5).

Any other comments
Human health effects and experimental toxicology reviewed (6,7).

References

1. Trehy, M. L; et al. *Advances in the Identification and Analysis of Organic Pollutants in Water* 1981, Keith, L. H. (Ed.), **2**, Ann Arbor Science
2. Zoeteman, B. L. J; et al. *Mutagenic Activity Associated with Products of Drinking Water Disinfection by Chlorine, Chlorine Dioxide, Ozone and UV irradiation* Int. Symp. on Health Eff. of Drinking Water Disinfection and Disinfection By Products (21-24 April 1981, Cincinnati)
3. Bieber, T. I. *Water Chlorination Environmental Impact and Health Effects* 1982, **4**(1), (Ch5) 85, Jolley, R. L. et al. (Ed.), Ann Arbor Science
4. Smith, M. K; et al. *Toxicology* 1987, **46**(1), 83-93
5. *EC Directive Relating to the Quality of Water Intended for Human Consumption* 1982, 80/778/EEC, Office for Official Publications of the European Communities, 2 rue Mercier, L-2985 Luxembourg
6. Bull, R. J. *Environ. Sci. Technol.* 1982, **46**(10), 554A
7. *ECETOC Technical Report No.30(4)* 1991, European Chemical Industry Ecology and Toxicology Centre, B-1160 Brussels

B190　3-Bromo-1-chloro-5,5-dimethylhydantoin

CAS Registry No. 126-06-7
Synonyms 3-bromo-1-chloro-5,5-dimethyl-2,4-imidazolidinedione

---

Mol. Formula $C_7H_8BrClO_2$　　　　　　　　　　**Mol. Wt.** 239.50

Uses Disinfecting agent for swimming pool and water cooling systems.

Physical properties

M. Pt. 160-164°C.

Ecotoxicity

Fish toxicity

LC_{50} (96 hr) adult fathead minnow 0.46- 0.57 mg l^{-1} and juvenile fathead minnow 0.28-0.41 mg l^{-1} (1).

LC_{50} (96 hr) bluegill sunfish 0.56- 0.71 mg l^{-1} (1).

LC_{50} (96 hr) rainbow trout 0.87 mg l^{-1}.

LC_{50} (24 hr) sheepshead minnow 20 mg l^{-1} (2).

Invertebrate toxicity

EC_{50} (48 hr) *Daphnia magna* 0.47 mg l^{-1}.

LC_{50} (48 hr) grass shrimp 13 mg l^{-1}.

LC_{50} (48 hr) American oyster >640 mg l^{-1} (2).

Environmental fate

Degradation studies

During a ^{14}C biodegradation study with activated sludge it was observed that dehalogenation to 5,5-dimethylhydantoin (RN 77-77-4) occurred, which in turn decreased to <1 ppm in 3 days and by day 19, 94% of the ^{14}C has been recovered as carbon dioxide (2).

Mammalian and avian toxicity

Acute data

LC_{50} oral rat 200 mg kg^{-1} (2).

Sub-acute data

Three 5 month old ♀ rats receiving 10 and 60 mg kg^{-1} day^{-1} (duration and route of exposure unspecified) showed no gross pathological changes, no significant changes to haemoglobin, erythrocytes, leucocytes, no internal disturbances and no treatment related lesions on autopsy to heart, lungs, gastrointestinal tract and kidneys (2).

Teratogenicity and reproductive effects

Oral ♀ rat 5, 7.5 and 10 mg kg^{-1} uterine examination values were unaffected, however embryo lethality was noted in the high dose groups. No malformations or increased developmental variants were observed in the range 500-4500 mg kg^{-1} (2).

Metabolism and pharmacokinetics

Using ^{14}C in rats at 20 and 100 mg kg^{-1} doses, an average of 91% was found in the urine, with 88% elimination during the first 24 hr. No measurable ^{14}C was observed in tissues from the 20 mg kg^{-1} dose, but some ^{14}C was found in kidney and bone of rats receiving the higher dose (2).

Genotoxicity

Salmonella typhimurium TA98, TA100 with and without metabolic activation negative.

Saccharomyces cerevisiae with and without metabolic activation negative (2).

Legislation

Limited under EC Directive on Drinking Water Quality 80/778/EEC. Organochlorine compounds: guide level 1 µg l^{-1}. Haloform concentrations must be as low as possible (3).

Any other comments

Maximum permissible concentration of dimethylhydantoin in water reservoir 1 mg l^{-1} (4).

References

1. Wilde, E. W; et al. *Biol. Environ. Contam. Toxicol.* 1983, **31**, 309
2. *Material Safety Data Sheet* 1986, Great Lakes Chemical Inc.
3. *EC Directive Relating to the Quality of Water Intended for Human Consumption* 1982, 80/778/EEC, Office for Official Publications of the European Communities, 2 rue Mercier, L-2985, Luxembourg
4. Korelev, A. A; et al. *Gig. Sanit.* 1982, (2), 76, (Russ.) (*Chem. Abstr.* **97**, 508625)

B191   1-Bromo-2-chloroethane

BrCH2CH2Cl

CAS Registry No. 107-04-0
Synonyms ethylene chlorobromide; *syn*-chlorobromoethane
Mol. Formula C$_2$H$_4$BrCl   **Mol. Wt.** 143.42

Physical properties

M. Pt. –16.6°C; **B. Pt.** 106-107°C; **Specific gravity** d$_4^{20}$ 1.7392.

Ecotoxicity

Bioaccumulation
Calculated bioconcentration factor 9.6 suggesting accumulation in aqueous organisms is unlikely (1).

Environmental fate

Abiotic removal
Estimated photochemical t$_{1/2}$ using hydroxyl radicals of 49.4 days (2). Volatilisation t$_{1/2}$ is 4.7 hr from a model river 1 m deep with a 1 m second^{-1} current and a 3 m sec^{-1} wind speed (1).

Absorption
Using the water solubility of 6.83 g l^{-1} at 30°C, estimated soil adsorption coefficient is 34 which suggests adsorption to soil and sediment is unlikely (1).

Mammalian and avian toxicity

Acute data
LD$_{50}$ oral rat 64 mg kg^{-1} (3).

Carcinogenicity and long-term effects

Selected for General Toxicology Study by the National Toxicology Program through Special Initiatives (4).

Genotoxicity

Salmonella typhimurium TA100 with metabolic activation positive (5).
In vitro Chinese hamster ovary cells sister chromatid exchange positive (6).

Legislation

Limited under EC Directive on Drinking Water Quality 80/778/EEC. Organochlorine compounds: guide level 1 μg l^{-1}. Haloform concentrations must be as low as possible (7).

Any other comments

Human health effects, experimental toxicology and environmental effects reviewed (8). All reasonable efforts have been taken to find information on isomers of this compound, but no relevant data are available.

References

1. Lyman, W. J; et al. *Handbook of Chemical Property Estimation Methods* 1982, McGraw-Hill, New York
2. Atkinson, R. *Int. J. Chem. Kinet.* 1987, **19**, 799-828
3. Frear, E. H. (ed.), *Pesticide Index* 1969, **4**, 73
4. *National Toxicology Program, Research & Testing Div.* 1992, NIEHS, Research Triangle Park, NC
5. Bladeren, van P. J; et al. *Carcinogenesis* 1981, **2**, 499-505
6. *Mutat. Res.* 1981, **90**, 183
7. *EC Directive Relating to the Quality of Water Intended for Human Consumption* 1982, 80/778/EEC, Office for Official Publications of the European Communities, 2 rue Mercier, L-2985 Luxembourg
8. *ECETOC Technical Report No.30(4)* 1991, European Chemical Industry Ecology and Toxicology Centre, B-1160 Brussels

B192   Bromochloromethane

BrCH2Cl

CAS Registry No. 74-97-5

Synonyms chlorobromomethane; methylene chlorobromide; Fluorocarbon 1011; Halon 1011

Mol. Formula CH_2BrCl **Mol. Wt.** 129.39

Uses Solvent. Component in fire extinguishers. Diesel fuel additive. Nail varnish remover.

Occurrence Contaminant of water samples in most industrial countries.

Physical properties

M. Pt. −88°C; **B. Pt.** 68°C; **Specific gravity** d_4^{20} 1.9344.

Solubility
Organic solvent: ethanol, diethyl ether, acetone, benzene

Occupational exposure
US TLV (TWA) 200 ppm (1060 mg m^{-3}); UK Long- term limit 200 ppm (1050 mg m^{-3}); UK Short-term limit 250 ppm (1300 mg m^{-3}); UN No. 1887; HAZCHEM Code 2Z; Conveyance classification flammable liquid.

Ecotoxicity

Bioaccumulation
Based on the water solubility of 16.7 g l^{-1} at 25°C and log P_{ow} 1.41, bioconcentration factors have been calculated as 3 and 7, respectively, which suggests accumulation in fish and aquatic organisms will not occur to any significant extent (1,2).

Environmental fate

Degradation studies
In a screening test, bromochloromethane at 5 or 10 mg l^{-1} underwent 100% degradation within 7 days using a settled domestic wastewater inoculum under aerobic conditions. Complete degradation followed with 3 successive subcultures (3,4).
Reported to undergo microbial degradation under anoxic conditions when cultured with soil bacteria (5).

Abiotic removal
Does not absorb ultraviolet light at >290 nm which suggests direct photochemical degradation in the atmosphere or water is unlikely (6).
Estimated photochemical t$_{1/2}$ with hydroxyl radicals 160 days suggesting it will not be a significant removal process (7).
Hydrolysis in environmental waters is not expected to be a significant method of removal estimated t$_{1/2}$ is 44 yr (8).
Volatilisation t$_{1/2}$ ≈1 hr, significant removal process from either moist or dry soil (9).

Absorption
Soil adsorption coefficients of 21 and 139 have been calculated based on water solubility and log P_{ow} respectively indicating high to very high mobility in soil (10).

Mammalian and avian toxicity

Acute data
LD$_{50}$ oral mouse, rat 4300-5000 mg kg^{-1} (11,12).
LC$_{50}$ (7 hr) inhalation mouse 3000 ppm (13).

Metabolism and pharmacokinetics
Dogs exposed to 1000 ppm bromochloromethane in air 7 hr day^{-1} 5 day wk^{-1} detected inorganic bromide in blood serum and urine. During the 3rd wk, serum inorganic bromide had increased from 5-10 mg 100 ml^{-1} to >200 mg. By 13th and 14th wk, concentration was >300 mg 100 ml^{-1} of inorganic bromide in blood (14).

Genotoxicity
Salmonella typhimurium TA98, TA100, TA1535, TA1537, TA1538 without metabolic activation negative (15).

Any other adverse effects

Affects mycardial energy metabolism. Anaesthetised dogs 45 min exposure to
0.3-1.0% in oxygen resulted in elevation of venus blood PO_2 and O_2 content (16,17).
In rats metabolic products included carboxyhaemoglobin (18).

Any other comments

Human health effects, epidemiology, workplace experience, experimental toxicology
and environmental effects reviewed (19).
Moderate narcotic action.

References

1. Tewari, Y. B; et al. *J. Chem. Eng. Data* 1982, **27**, 451-454
2. Wasik, S. P; et al. *Octanol Water Partition Coefficients and Aqueous Solubilities of Organic Compounds* 1981, 66, NBSIR81-2406, US Dept. Comm. Natl. Bur. Std., Washington, DC
3. Tabac, H. H; et al. *J. Water Pollut. Control Fed.* 1981, **53**, 1503-1518
4. Tabak, H. H; et al. *Test Protocols for Environmental Fate and Movement of Toxicants Proc. Sym. Assoc. Off. Anal. Chem.* 1981, 267-328
5. Kobayashi, H; et al. *Environ. Sci. Tech.* 1982, **16**, no A-183A
6. Cadwell, P; et al. *Trans. Faraday Soc.* 1966, **62**, 631-641
7. Atkinson, R; et al. *Chem. Rev.* 1985, **85**, 69-201
8. Mill, T; et al. *Validation of Estimation Techniques for Predicting Environmental Transformation of Chemicals* 1982, USEPA 68-01-6269, Washington, DC
9. Lyman, W. J; et al. *Handbook of Chemical Property Estimation Methods* 1982, 15.1-15.29, McGraw-Hill, New York
10. Swann, R. L; et al. *Res. Rev.* 1983, **85**, 17-28
11. Deichmann, W. B; et al. *Toxicology of Drugs and Chemicals* 1969, Academic Press, London
12. *J. Ind. Hyg. Tox*, 1947, **29** 382
13. *Documentation of the Threshold Limit Values and Biological Exposure Indices* 5th ed., 1986, 125, American Conference of Governmental Industrial Hygienists, Cincinnati, OH
14. Clayton, G. D; et al. *Patty's Industrial Hygiene and Toxicology* 3rd ed., 1981-1982, **2A, 2B,2C**, 3457, John Wiley & Sons, New York
15. Simmon, V. F; et al. *Der. Toxicol. Environ. Sec.* 1977, **2**, 249
16. Terpolinni, R. N; et al. *Toxicology* 5th ed., 1986, 125, American Conference of Governmental Industrial Hygienists, Cincinnati, OH
17. Van Stee, E. W; et al. *Toxicol. Appl. Pharmacol.* 1975, **34**, 62
18. Kubic, V; et al. *Drug Metab. Dispos.* 1974, **2**, 53
19. *ECETOC Technical Report No. 30(4)* 1991, European Chemical Industry Ecology & Toxicology Centre, B-1160 Brussels

B193   1-Bromo-3-chloropropane

$Br(CH_2)_3Cl$

CAS Registry No. 109-70-6

Synonyms 3-bromopropyl chloride; 1,3-CHBP; ω-chlorobromopropane; trimethylene bromochloride

Mol. Formula C_3H_6BrCl **Mol. Wt.** 157.44

Physical properties

M. Pt. −59°C; **B. Pt.** 144-145°C; **Specific gravity** d_4^{20} 1.5969.

Solubility

Organic solvent: ethanol, diethyl ether, chloroform

Ecotoxicity

Fish toxicity
LC_{50} (24 hr) goldfish 75 mg l^{-1} (1).

Mammalian and avian toxicity

Acute data
LD_{50} oral rat, mouse 930-1290 mg kg^{-1} (2,3).
LC_{Lo} (2 hr) inhalation mouse 7270 mg kg^{-1} (3).

Any other comments

All reasonable efforts have been taken to find information on isomers of this compound, but no relevant data are available.

References

1. Bridie, A.L; et al. *Water Res.* 1979, **13**, 623
2. *Tox. New Ind. Chem. Sci, Moscow* 1971, **12**, 93
3. Izmerov, N. F; et al. *Toxicometric Parameters of Industrial Toxic Chemicals under Single Exposure* 1982, 35, Moscow

B194   Bromodichloromethane

CHBrCl₂

CAS Registry No. 75-27-4
Synonyms dichlorobromomethane
Mol. Formula $CHBrCl_2$ **Mol. Wt.** 163.83
Uses Fire extinguisher fluid. Solvent. Intermediate in organic synthesis.
Occurrence From chlorination of water. Contaminant in drinking, surface, groundwater and seawater (1-4).
In effluent from industrial wastewater discharges and wastewater treatment facilities (5,6).

Physical properties

M. Pt. −55°C; **B. Pt.** 90°C; **Specific gravity** 1.97; **Volatility** v.p. 50 mmHg at 20°C.

Environmental fate

Nitrification inhibition
>50% degradation in bacterial cultures after 8 wk under anaerobic conditions. No degradation in sterile controls (7).

Anaerobic effects

Total degradation within 2 wk in anaerobic tests using mixed methanogenic bacterial cultures from sewage effluents. No degradation in aerobic tests in sterile or seeded conditions (8).

IC_{50} 1.6 mg l^{-1} at 35°C, anaerobic toxicity assay with methanogenic bacteria (9).

Degradation studies

Degradation 28 day incubation bromodichloromethane in static flask screening test removed 51-59% (10).

Abiotic removal

Aqueous hydrolysis at 25°C and pH 7, estimated $t_{1/2}$ 137 yr (11).

Water infiltration study in the Rhine river, the Netherlands, detected high soil mobility (12).

50% of applied amount volatilised from soil columns in laboratory studies of transport and fate mechanisms (13).

Mammalian and avian toxicity

Acute data

LD_{50} oral rat 450 mg kg^{-1}(14).

Carcinogenicity and long-term effects

Inadequate evidence for carcinogenicity in humans, sufficient evidence in animals, IARC classification group 2B (15).

National Toxicology Program evaluation of bromodichloromethane in rats and mice by gavage, positive evidence of carcinogenicity (16).

Teratogenicity and reproductive effects

Gavage Sprague-Dawley rats (6-15 day gestation) 0, 50, 100, 200 mg kg^{-1} decreased maternal weight gain and kidney weight at 200 mg kg^{-1}. No increase in incidence of resorptions, litter size, foetal weight, external or visceral malformations. Increased incidence of sternum aberrations at 100 or 200 mg kg^{-1} (17).

Metabolism and pharmacokinetics

Oral ♂ rats single doses of 1 or 100 mg kg^{-1} and 10 day repeat doses of 10 or 100 mg kg^{-1} radiolabelled bromodichloromethane. 80-90% metabolised within 24 hr. Urinary and faecal elimination 4- 5% and 1-3% of dose, respectively. Persistence in tissues was 3-4% 24 hr after single dose, mainly in liver (18).

Oral rat single dose 20 mg kg^{-1} ^{14}C-bromodichloromethane was cleared rapidly. 32% recovered from gut and carcass after 3 hr and 41% after 6 hr. Most recovered from stomach. Fat contained more than other tissue. <1% eliminated via urine (19).

Genotoxicity

Salmonella typhimurium TA100, TA1535, TA1537, TA98 with and without metabolic activation negative (20).

Salmonella typhimurium TA100 with metabolic activation positive, provided the bacteria are exposed in closed container to the vapour (21).

In vitro Chinese hamster ovary cells with metabolic activation chromosome aberrations positive (22).

In vitro human lymphocyte sister chromatid exchange positive (23).

In vivo mice without metabolic activation micronucleus test negative (24).

Any other adverse effects

Oral rat evidence of fatty liver infiltration and haemorrhaging in the adrenals, lungs and brain (14).

Any other comments

Human health effects and experimental toxicology reviewed (25, 26).
Mean levels of bromodichloromethane of 7-22 ng g^{-1} (dry wt.) detected in marine algae (27).

References

1. Westrick, J. J; et al. *J. Am. Water Works Assoc.* 1984, **76**, 52-59
2. Trussel, A. R; et al. *Water Chlorination Environ. Impact Health Effects* 1980, **3**, 39-53
3. Staples, C. A; et al. *Environ. Toxicol. Chem.* 1985, **4**, 131-142
4. Class, T; et al. *Chemosphere* 1986, **15**, 429-436
5. Perry, D. L; et al. *Identification of Organic Compounds in Industrial Effluent Discharges* 1979, EPA -600/4-79-016, p. 42-43
6. Dunorant, V. S; et al. *J. Water Pollut. Control Fed.* 1986, 886-895
7. Bouwer, E. J; et al. *Appl. Environ. Microbiol.* 1983, **45**, 1295-1299
8. Bouwer, E. J; et al. *Environ. Sci. Technol.* 1981, **15**, 596-599
9. Blum, D. J. W; et al. *J. Water Pollut. Contr. Fed.* 1991, **63**(3), 198-207
10. Tabak, H. H; et al. *J. Water Pollut. Control. Fed.* 1981, **53**, 1503-1518
11. Mabey, W; et al. *J. Phys. Chem. Ref. Data* 1978, **7**, 383-415
12. Kaczmr, S. W; et al. *Environ. Toxicol. Chem.* 1984, **3**, 31-35
13. Wilson, J. T; et al. *J. Environ. Qual.* 1981, **10**, 501-506
14. Bowman, F. J; et al. *Toxicol. Appl. Pharmacol.* 1978, **44**, 213
16. *National Toxicology Program Research & Testing Div.* 1992, Report No. TR-321, NIEHS, Research Triangle Park, NC27709
15. *IARC Monograph* 1991, **52**, 179-212
17. Ruddick, J. A; et al. *J. Environ. Sci. Res.* 1983, **B18**, 333-349
18. Matthews, J. M; et al. *J. Toxicol. Environ. Health* 1990, **30**(1), 15-22
19. *National Research Council, Drinking Water and Health* 1980, Vol. 3, National Academy Press, Washington, DC
20. Mortelmans, K; et al. *Environ. Mutagen.* 1986, **8**(Suppl. 7), 1-119
21. McGregor, D. B; et al. *Environ. Mol. Mutagen.* 1988, **11**(1), 91-118
22. Ishidate, M. Jr. *Data Book of Chromosomal Aberration Tests in vitro* 1987, Life-Sci. Info. Center, Tokyo
23. *Environ. Res.* 1983, **32**, 72-79
24. Hayashi, M; et al. *Food Chem. Toxicol.* 1988, **26**, 487-500
25. *NTIS Report* 1989, ATSDR/TP-89/04, National Technical Information Service, Springfield, VA
26. *ECETOC Technical Report No. 30(4)* 1991, European Chemical Industry Ecology & Toxicology Centre, B-1160 Brussels
27. Gschwend, P. M; et al. *Science* 1985, **227**, 1033-1035 1990, **30**(1), 15-22

B195   4-Bromodiphenyl ether

CAS Registry No. 101-55-3

Synonyms 4-bromophenyl phenyl ether; 1-bromo-4-phenoxybenzene

Mol. Formula $C_{12}H_9BrO$ **Mol. Wt.** 249.11

Occurrence Contaminant in water samples. Occurs in effluent from sewage sludge (1,2).

Physical properties

M. Pt. 18°C; **B. Pt.** 305°C; **Flash point** >110°C; **Specific gravity** 1.423.

Ecotoxicity

Invertebrate toxicity
EC_{50} (48 hr) *Daphnia magna* 0.36 mg l^{-1} (3).

Environmental fate

Nitrification inhibition
Does not inhibit anaerobic digestion in laboratory trials at 100 mg l^{-1} (4).

Degradation studies
Biodegradeable (5).
Metabolised in soil by *Pseudomonas sp.* to phenol and 2-phenoxymuconic acid derivatives (6).

Any other adverse effects

Central nervous system depression and behavioural changes. The liver and immune system were sensitive target areas ♂ more sensitive than ♀ (7).

Any other comments

Toxicity reviewed (8).
Removal from drinking water discussed (9).
Human health effects, experimental toxicology and environemntal effects reviewed (10).
All reasonable efforts have been taken to find information on isomers of this compound, but no relevant data are available.

References

1. Lao, R. C; et al. *Pergamon Ser. Environ. Sci.* 1982, **7**, (Anal. Tech. Environ. Chem. 2), 107
2. Belve, H; et al. *Water Management Res.* 1989, **7**, 43
3. LeBlanc, G. A. *Bull. Environ. Contam. Toxicol.* 1980, **24**, 684-691
4. Richardson, M. L. *Nitrification Inhibition in the Treatment of Sewage* 1985, RSC, London
5. Tabak, H. H; et al. *J. Water Pollut. Control Fed.* 1981, **53**, 1503
6. Takase, I; et al. *Agric. Biol. Chem.* 1986, **50**, 681
7. Borzelleca, J. F. *Proc. Am. Water Works Assoc. Water Qual. Technol. Conf.* 1982, 225
8. *Dang. Prop. Ind. Mater. Rep.* 1986, **6**, 43
9. Van Dyke, K; et al. *Am. Lab.* 1986, **18**, 118
10. *ECETOC Technical Report No. 30(4)* 1991, European Chemical Industry Ecology and Toxicology Centre, B-1160 Brussels

B196　Bromoethane

CH₃CH₂Br

CAS Registry No. 74-96-4
Synonyms ethyl bromide; bromic ether; Halon 2001; hydrobromic ether
Mol. Formula C_2H_5Br　　　　　　　　　　　**Mol. Wt.** 108.97
Uses Ethylating agent in organic synthesis. Refrigerant and extraction solvent.
Investigated as a possible substitute for chlorofluorocarbons in compression heat
pumps.

Physical properties

M. Pt. −114°C; **B. Pt.** 37-40°C; **Flash point** −23°C; **Specific gravity** d_4^{20} 1.460;
Partition coefficient log P_{ow} 1.61; **Volatility** v.p. 400 mm Hg at 21°C; v. den. 3.75.

Solubility
Water: 10.67 g l⁻¹ at 0°C. Organic solvent: miscible with ethanol, diethyl ether,
chloroform

Occupational exposure

US TLV (TWA) 200 ppm (891 mg m⁻³); **US TLV (STEL)** 250 ppm (1110 mg m⁻³);
UK Long-term limit 200 ppm (891 mg m⁻³); **UK Short-term limit** 250 ppm (1110
mg m⁻³); **UN No.** 1891; **HAZCHEM Code** 3YE; **Conveyance classification** toxic
substance; **Supply classification** harmful.

Risk phrases
Harmful by inhalation, in contact with skin and if swallowed (R20/21/22)

Safety phrases
After contact with skin, wash immediately with plenty of water (S28)

Environmental fate

Degradation studies
Biodegraded by *Acinetobacter sp.* strain GJ70 (1).

Mammalian and avian toxicity

Acute data
LD₅₀ oral rat 1350 mg kg⁻¹ (2).
LC₅₀ (1 hr) inhalation mouse 16,230 ppm (3).
LC₅₀ (1 hr) inhalation mouse 16,230 ppm (4).
LD₅₀ intraperitoneal mouse, rat 1750-2850 mg kg⁻¹ (5).

Sub-acute data
All ♂ and ♀ Fischer 344/N rats and B6C3F1 mice exposed by inhalation (2 wk) to
4000 ppm and 2000 ppm for 6 hr day⁻¹ 5 day wk⁻¹ died. Signs of toxicity were
prostration, dyspnoea, lachrymation, haemorrhage and congestion in the respiratory
tract (4).

Carcinogenicity and long-term effects

No adequate evidence for carcinogenicity to humans, limited evidence for carcinogenicity to animals, IARC classification group 3 (6).

Rats and mice inhalation (0, 100, 200 or 400 ppm) 6 hr day^{-1} 5 day wk^{-1} for 103 or 104 wk. Induced pheochromocytomas of the adrenal gland, neoplasms of brain and lung in ♂ F344/N rats; equivocal evidence in ♀ rats indicated by marginally increased incidences of brain and lung neoplasms. Equivocal evidence in ♂ B6C3F1 mice based on marginally increased lung neoplasm evidence; neoplasms of uterus in ♀ mice (4).

Teratogenicity and reproductive effects

In a 14 wk inhalation study with B6C3F1 mice and Fischer 344 rats exposed to 100-1600 ppm for 6 hr day^{-1} 5 day wk^{-1}, severe testicular atrophy was observed in all rats but not in mice at 1600 ppm. Four of ten ♂ rats in the 1600 ppm group died. In ♀ mice the size and number of corpora lutea in the ovary decreased at 1600 ppm and at 800 ppm (4).

Genotoxicity

Salmonella typhimurium TA100, TA1535 with and without metabolic activation positive (7).

Mutations were not induced in *Drosphila melanogaster* and chromosomal aberrations were not induced in cultured mammalian cells (6).

Increased incidence of sister chromatid exchange in Chinese hamster ovary cells (8).

Any other adverse effects

Reported to be narcotic at high concentrations (9).

Any other comments

Toxicity reviewed (10).

Physico-chemical properties, human health effects, experimental toxicology, epidemiology and workplace experience reviewed (11).

References

1. Janssen, D. B; et al. *Appl. Environ. Microbiol.* 1987, **53**, 561
2. *Toxicometric Parameters of Industrial Toxic Chemicals Under Single Exposure* 1982, 65
3. *Aerospace Med. Res. Lab. Rep.* TR-72-62/72
4. Roycroft, J. *Report* 1989, NTP-TR-363, NIH/PUB-90-2818
5. *J. Praktische Chemie* 1978, **320**, 133
6. *IARC Monograph* 1990, **52**, 299-314
7. Barber, E. D; et al. *Mutat. Res.* 1981, **90**, 31-48
8. Hatch, G; et al. *Environ. Mutagen.* 1983, **5**, 442
9. *The Merck Index* 11th ed., 1989, Merck & Co. Inc., Rahway, NJ
10. *Health Effects Assessment of Bromoethane* 1987, EPA 600/8-88/022, PB-179478
11. *ECETOC Technical Report No.30(4)* 1991, European Chemical Industry Ecology and Toxicology Centre, B-1160 Brussels

B197   Bromofenoxim

CAS Registry No. 13181-17-4

Synonyms Benzaldehyde, 3,5-dibromo-4-hydroxy-, O-(2,4-dinitrophenyl)oxime; bromophenoxim

Mol. Formula $C_{13}H_7Br_2N_3O_6$ **Mol. Wt.** 461.04

Uses Herbicide.

Physical properties

M. Pt. 196-197°C; **Partition coefficient** log P_{ow} 3.2; **Volatility** v.p. 9.8×10^{-8} mmHg at 20°C.

Solubility

Water: 0.1 mg l^{-1}. Organic solvent: hexane, isopropanol, benzene

Occupational exposure

Supply classification harmful.

Risk phrases

Harmful by inhalation and if swallowed (R20/22)

Safety phrases

Keep out of reach of children – Keep away from food, drink and animal feeding stuffs (S2, S13)

Ecotoxicity

Fish toxicity

LC_{50} (96 hr) goldfish, rainbow trout, catfish 0.056-0.24 mg l^{-1} (1).

Effects to non-target species

Non-toxic to birds and bees (1).

Environmental fate

Degradation studies

No persistance in soil (1).

Abiotic removal

At 70°C, 50% hydrolysis in 41.4 hr at pH 1, in 9.6 hr at pH 5, and 0.76 hr at pH 9 (1).

Mammalian and avian toxicity

Acute data

LD_{50} oral dog, rat 1000-1200 mg kg^{-1} (1).
LC_{50} (6 hr) inhalation rat >0.24 mg l^{-1} (2).
LC_{50} percutaneous rat >3000 mg kg^{-1} (1).

Sub-acute data

In a 90 day feeding trial, no-effect level in rats was 300 mg kg^{-1} diet and in dogs 100 mg kg^{-1} (1).

Metabolism and pharmacokinetics

In plants metabolised to 3,5-dibromo-4-hydrobenzoic acid (1).
Following oral administration to rats metabolised to dinitrophenol and hydroxdibromobenzoynitrile (1).

Legislation

Limited under EC Directive on Drinking Water Quality 80/778/EEC. Pesticides and related products: maximum admissible concentration 0.5 µg l^{-1} (3).

References

1. *The Agrochemicals Handbook* 3rd ed., 1991, RSC, London
2. Worthing, C. R. (Ed.) *The Pesticide Manual* 9th ed., 1991, British Crop Protection Council, Farnham
3. *EC Directive Relating to the Quality of Water Intended for Human Consumption* 1982, 80/778/EEC, Office for Official Publications of the European Communities, 2 rue Mercier, L-2985 Luxembourg

B198   Bromoform

HCBr$_3$

CAS Registry No. 75-25-2
Synonyms tribromomethane
Mol. Formula CHBr$_3$ **Mol. Wt.** 252.75
Uses Solvent. Sedative. Antitussive.
Occurrence Contaminant in water samples, concentration range <0.8-92 µg l^{-1} (1).

Physical properties

M. Pt. 8.3°C; **B. Pt.** 150-151°C; **Specific gravity** d$_4^{20}$ 2.894; **Volatility** v.p. 5.6 mmHg at 25°C; v. den. 8.7.

Solubility
Water: 0.8 g l^{-1}. Organic solvent: ethanol, benzene, chloroform, diethyl ether, acetone

Occupational exposure

US TLV (TWA) 0.5 ppm (5 mg m^{-3}); **UK Long-term limit** 0.5 ppm (5.2 mg m^{-3}); **UN No.** 2515; **HAZCHEM Code** 2X; **Conveyance classification** harmful substance; **Supply classification** toxic.

Safety phrases
After contact with skin, wash immediately with plenty of water – If you feel unwell, seek medical advice (show label where possible) (S28, S44)

Ecotoxicity

Fish toxicity
LC$_{50}$ (96 hr) bluegill sunfish 29 mg l^{-1} static bioassay.
LC$_{50}$ (96 hr) sheepshead minnow 17 mg l^{-1} static bioassay (1).
LC$_{50}$ (96 hr) bluegill sunfish 29 mg l^{-1} (2).

Invertebrate toxicity
EC$_{50}$ (48 hr) *Daphnia magna* 46 mg l^{-1} (4).
LD$_{50}$ (48 hr) eastern oyster larvae 1 mg l^{-1} static bioassay (3).
LC$_{50}$ (96 hr) mysid shrimp 24 mg l^{-1} (1).

Mammalian and avian toxicity

Acute data
LD$_{50}$ oral rat, mouse 1150-1400 mg kg^{-1} (5,6).
LD$_{50}$ subcutaneous mouse 1820 mg kg^{-1} (7).

Sub-acute data
Intravenous guinea pig (10 day) 100-200 mg kg^{-1} day^{-1} resulted in pathological changes in liver and kidney (8).

Carcinogenicity and long-term effects
Inadequate evidence for carcinogenicity to humans, limited evidence for carcinogenicity, IARC classification group 3 (9).
National Toxicology Program evaluation of bromoform in rats and mice by gavage, clear evidence in ♀ rats, no evidence in ♂ or ♀ mice, some evidence in ♂ rats (10).

Teratogenicity and reproductive effects
Gavage Swiss CD-1 mice (18 wk) 0, 50, 100, 200 mg kg^{-1} continuous breeding protocol. No adverse effects on fertility. Reduced body and kidney weight, increased liver weight at 200 mg kg^{-1} and 100 mg kg^{-1}. Reduced neonatal survival at 200 mg kg^{-1} (11).
Gavage Sprague-Dawley rats (6-15 day gestation) 0, 50, 100, 200 mg kg^{-1} no maternal toxicity, no increased incidence of resorptions, litter size or foetal weight observed. Dose related increase in incidence of skeletal variations (12).

Metabolism and pharmacokinetics
Rectal or inhalation administration to rabbits. Biotransformed in the liver to inorganic bromides, which were later found in tissues and urine (13).
Rectal anaesthesia with bromoform, 0.3-1.2% recovered in urine as sodium bromide (14).
Intragastric administration to Sprague-Dawley rats and B6C3F1 mice. Total radioactivity of bladder, brain, kidneys, liver, lungs, pancreas and thymus 3-6% of total dose in rats and 5-14% in mice. Urine contained <5% at 8 hr after administration, and <10% of total radiolabel at 36-48 hr (15).

Irritancy
Absorbed through rabbit skin causing moderate irritation, lethargy and weight loss. Moderately irritating to eyes (16).

Genotoxicity

Salmonella typhimurium TA98, TA100 with and without metabolic activation positive when tested in a desiccator and in vapour phase (not using Agar) (17,18). *Salmonella typimurium* TA97, TA98 without metabolic activation positive, with metabolic activation negative (19).

In vitro Chinese hamster ovary cells without metabolic activation sister chromatid exchange and chromosome aberrations positive, with metabolic activation negative (20).

In vitro human lymphocytes without metabolic activation sister chromatid exchange positive (21).

In vivo mouse bone marrow cells sister chromatid exchange positive, chromosome aberrations negative.

In vivo B6C3F mice micronucleus test positve (20).

In vitro mouse lymphoma L5178Y tk+/- cell forward mutation assay without metabolic activation positive, with metabolic activation negative (22).

Sister chromatid exchange (marine fish) did not increase the rate of dividing leucocytes (23).

Any other adverse effects to man

Epidemiology studies indicated higher oesophagus and stomach cancer mortality where there are high levels of all haloalkanes in drinking water (24).

Legislation

US Drinking Water recommendation 2 μg l^{-1} (25).

Any other comments

Human health effects and experimental toxicology reviewed (26,27).

References

1. *Ambient Water Quality Criteria Doc.: Halomethanes* 1980, EPA 440/5-80-051
2. Buccafusco, R. J; et al. *Bull. Environ. Contam. Toxicol.* 1981, **26**, 446
3. Heitmuller, P. T; et al. *Environ. Contam. Toxicol.* 1981, **27**, 596-604
4. Verschueren, K. *Handbook of Environmental Data of Organic Chemicals* 2nd ed., 1983, Van Nostrand Reinhold, New York
5. LeBlanc, G. A. *Bull. Environ. Contam. Toxicol.* 1980, **24**, 684-691
6. *Toxicol. Appl. Pharmacol.* 1980, **52**, 351
7. *Toxicol. Appl. Pharmacol.* 1978, **4**, 213
8. Kulob, S. D; et al. *Toxicol. Appl. Pharmacol.* 1962, **4**, 354
9. Dykan, V. A. *Nauchn. Tr. Ukr. Nauchn. Issled. Inst, Gig. Tr. Profzabol.* 1962, **29**, 82
10. *IARC Monograph* 1991, **52**, 213-242
11. *National Toxicology Program Research & Testing Div.* 1992, Report No. Tr-350, NIEHS, Research Triangle Park, NC27709
12. Gulati, D. K; et al. *Bromoform: Reproductive Fertility Assessment* 1989, NIEHS, Research Triangle Park, NC
13. Ruddick, J. A; et al. *J. Environ. Sci. Health* 1983, **B118**, 333-349
14. Lucas, G. H. W. *J. Pharm. Exp. Therm.* 1929, **34**, 223
15. *Drinking Water and Health* 1977, **1**, National Academy Press, Washington, DC
16. Mink, F. L; et al. *Bull. Environ. Contam. Toxicol.* 1986, **37**(5), 752-758
17. *Chemical Safety Data Sheets Vol. 1 Solvents* 1988, RSC, London
18. Simmon, V. F; et al. *Prog. Genet. Toxicol.* 1977, **2**, 249
19. Zeiger, E. *Environ. Mol. Mutagen.* 1990, **16**(Suppl. 18), 32-54

20. Mersch-Sunderman, V. *Zbl. Bakt. Hyg.B.* 1989, **187**, 230-243
21. Pereiri, M. A. *Environ. Health Perspect.* 1982, **46**, 151
22. Morimoto, K; et al. *Environ. Res.* 1983, **32**, 72-79
23. Myhr, B; et al. *Environ. Mol. Mutagen.* 1990, **16**(Suppl. 18), 138-167
24. Maddock, M. B; et al. *Water Chlornation Environ. Impact Health Eff.* 1980, **3**, 848
25. Kool. H. J; et al. *Sci. Total Environ.* 1981, **18**, 135
26. Theiss, J. C; et al. *Cancer Res.* **27**, 2717
27. *ECETOC Technical Report No. 30 (4)* 1991, European Chemical Industry Ecology & Toxicology Centre, B-1160 Brussels

B199 Bromomethane

CH3Br

CAS Registry No. 74-83-9
Synonyms methyl bromide; Embafume; Dowfume; Halon 1001; Bromo-*O*-Gas
Mol. Formula CH$_3$Br   **Mol. Wt.** 94.94
Uses Ionisation chambers. Degreasing wool. Extracts oils for nuts, seeds and flowers. Insect fumigant.
Occurrence In seawater collected off Dorset, UK, in 1975 ranging from 1.5-3.9 μg l^{-1} (1).
Man-made and natural sources contribute to the concentration of bromomethene in ambient air (2).
Residues have been reported in some foodstuffs after fumigation with bromomethane (3).

Physical properties

M. Pt. –94°C; **B. Pt.** 4°C; **Specific gravity** d$_4^0$ 1.730.

Solubility
Water: 17.5 g l^{-1} at 20°C, 748 mmHg. Organic solvent: ethanol, chloroform, diethyl ether, carbon disulphide, benzene

Occupational exposure

US TLV (TWA) 5 ppm (19 mg m^{-3}); **UK Long-term limit** 5 ppm (20 mg m^{-3}); **UK Short-term limit** 15 ppm (60 mg m^{-3}); **UN No.** 1062; **HAZCHEM Code** 2XE; **Conveyance classification** toxic gas; **Supply classification** toxic.

Risk phrases
Very toxic by inhalation (R26)

Safety phrases
Keep locked up and out of reach of children – Keep container tightly closed and in a well ventilated place – Avoid contact with skin and eyes (S1/2, S7/9, S24/25)

Ecotoxicity

Fish toxicity
LC$_{50}$ (96 hr) inland silverside, bluegill sunfish 11-12 ppm (4).

Invertebrate toxicity

LD_{50} (24 hr) *Coleoptera* 4.505 mg l^{-1} (5).

Effects to non-target species

Non-toxic to bees (6).

Abiotic removal

Primarily removed from the atmosphere by reaction with hydroxyl radicals. Residence time in urban atmosphere was estimated to be 289 days, with a daily rate of loss (12 sunlit hr) of 0.4% (1).

Mammalian and avian toxicity

Acute data

LC_{100} (6 hr) inhalation rat 0.63 mg l^{-1} (6).
LC_{50} (2 hr) inhalation mouse 1540 mg m^{-3} (7).
LC_{Lo} (2 hr) inhalation child 1 mg m^{-3} (8).

Carcinogenicity and long-term effects

Inadequate evidence for carcinogenicity to humans, limited evidence for carcinogenicity to animals, IARC classification group 3 (9).

National Toxicology Program technical reports in progress in long-term inhalation study using mice; no evidence of carcinogenic activity is demonstrated (10).

♂ and ♀ Wistar rats inhalation (29 month) 0, 3, 30 or 90 ppm 6 hr day^{-1}, 5 day wk^{-1}, 5 day wk^{-1} and 10 rats sex^{-1} group^{-1} were killed after 13, 52 and 104 wk. Mortality was increased by wk 114 in 90 ppm and body weights were lower. Increased incidences of degenerative and hyperplastic changes of the nasal olfactory epithelium were observed in all groups. Exposure to 90 ppm induced lesions in the heart and hyperkeratosis in the oesophagus and forestomach (11).

Metabolism and pharmacokinetics

Readily absorbed through lungs (12).

Serum bromide levels achieved in serious cases of bromomethane poisoning are considerably lower than those required for poisoning by inorganic bromides, suggested as being due to greater lipid solubility of bromomethane and hence greater penetration into the brain (13).

Following absorption in rat blood levels of residual non volatile bromide increased. Bromomethane rapidly distributed to various tissues and broken down to inorganic bromide. Storage, only as bromides, occurred mainly in lipid-rich tissues (3). Elimination was initially rapid, largely through the lung as bromomethane and the remainder was eliminated in urine as bromomethane/bromide. In rats fed bromomethane-fumigated diets with residual bromide levels, higher tissue bromide levels were found in their eyes, lung, blood, spleen and testes, while the lowest tissue levels were in the fat, skeletal muscle, bone and liver (3).

♂ Fischer-344 rats were exposed (nose only) to a vapour concentration of 9 ppm at 25°C for 6 hr and urine, faeces, expired air and tissues were collected for up to 65 hr after exposure. Carbon dioxide was the major route of excretion (47% of total absorbed). Distribution of the compound was in lung, adrenal, kidney, liver and nasal turbinates in decreasing order of concentration (14).

Irritancy

Unintentional exposure of the skin of 6 persons to 40 g m^{-3} for 40 min led to redness and blistering. Plasma bromide levels were highest immediately following exposure averaging 9 ± 1.4 mg l^{-1} and decreased to average 6.8 ± 2.3 mg l^{-1} 12 hr after exposure (15).

Thirty adult ♂ Long-Evans rats (2 wk) 200 ppm 4 hr day^{-1}, 4 day wk^{-1} extensive damage to olfactory epithelium and impaired function on first day of exposure; even with continuous exposure, function was essentially normal after 4 days of exposure. Repair of the epithelium was in progress thereafter (16).

Genotoxicity

Mutagenic to plants and bacteria. *In vitro* human lymphocytes induced sister chromatid exchange. *In vitro* mouse lymphoma cells positive mutagen. Inhalation rat induced micronuclei in bone marrow and peripheral blood cells. DNA methylation of liver and spleen observed in mouse. *Drosphila melanogaster* induced sex-linked recessive lethal mutations (17).

Any other comments

Physico-chemical properties, human health effects, experimental toxicology, environmental effects, ecotoxicology, exposure levels, epidemiology and workplace experience reviewed (18).

Hazardous properties reviewed (19).

Industrial poisoning, its diagnosis and therapy reviewed (20).

References

1. *IARC Monograph* 1986, **41**, 192
2. Dennis, N. M; et al. *J. Econ. Entomol.* 1972, **65**, 1753
3. *USEPA Ambient Water Quality Criteria: Haloethenes* 1980, EPA 440/5-80-051
4. Gaynor, W. D; et al. *J. Haz. Mat.* 1975, **1**
5. *USDA/Forest Service; Pesticide Background Statements. Fungicides and Fumigants* 1986, **2**, MB/C-28
6. *The Agrochemicals Handbook* 3rd ed., 1991, RSC, London
7. Izmerov, N. F; et al. *Toxicometric Parameters of Industrial Toxic Chemicals under Single Exposure* 1982, Moscow
8. *Jap. J. Legal Med.* 1969, **23**, 241
9. *IARC Monograph* 1987, **Suppl. 7**, 243
10. *National Toxicology Program, Research and Testing Div.* 1992, NIEHS, Research Triangle Park, NC
11. Reuzel, P. G. J; et al. *Food Chem. Toxicol.* 1991, **29**(1), 31-39
12. *Patty's Industrial Hygiene and Toxicology* 3rd ed., 1981-1982, **2A, 2B, 2C**, 3444, Clayton, G. D. (Ed.), John Wiley & Sons, New York
13. Doull, J; et al. *Casarett and Doull's Toxicology* 3rd ed., 1986, 567, Macmillan Co. Inc., New York
14. Bond, J. A; et al. *Toxicol. Appl. Pharmacol.* 1985, **78**(2), 259-267
15. Zwaveling, J. H; et al. *Hum. Toxicol.* 1987, **6**(6), 491-496
16. Hastings, L; et al. *Chem. Senses* 1991, **16**(1), 43-55
17. *IARC Monograph* 1987, **Suppl. 6**, 386
18. *ECETOC Technical Report No. 30(4)* 1991, European Chemical Industry Ecology and Toxicology Centre, B-1160 Brussels
19. *Cah. Notes Doc.* 1987, **127**, 293-297, (Fr.) (*Chem. Abstr.* 1987, **107**, 120210t)
20. Inoue, N; et al. *Sangyo Igaky Janaruy* 1987, **10**(4), 38-43, (Jap.) (*Chem. Abstr.* 1987, **107**, 182391m)

B200   1-Bromo-3-methylbutane

$(CH_3)_2CHCH_2CH_2Br$

CAS Registry No. 107-82-4
Synonyms isoamyl bromide; isopentyl bromide; 3-methylbutyl bromide
Mol. Formula $C_5H_{11}Br$ **Mol. Wt.** 151.05
Uses Organic synthesis.

Physical properties

M. Pt. –112°C; **B. Pt.** 120-121°C; **Flash point** 32°C; **Specific gravity** d_4^{15} 1.210.

Solubility

Organic solvent: miscible with ethanol, diethyl ether

Mammalian and avian toxicity

Acute data

LD_{50} intraperitoneal rat 6150 mg kg^{-1}.
LD_{50} intraperitoneal mouse 13,750 mg kg^{-1} (1).

Any other comments

All reasonable efforts have been taken to find information on isomers of this compound, but no relevant data are available.

References

1.   Izmerov, N. F; et al. *Toxicometric Parameters of Industrial Toxic Chemicals under Single Exposure* 1982, 76, Moscow

B201   2-Bromo-2-nitro-1,3-propanediol

$HOCH_2CBr(NO_2)CH_2OH$

CAS Registry No. 52-51-7
Synonyms Bronopol; β-bromo-β-nitrotrimethylene glycol; Bronosal
Mol. Formula $C_3H_6BrNO_4$ **Mol. Wt.** 199.99
Uses Disinfectant. Bacteriostat. Bactericide especially effective against *Pseudomonas aeruginosa*.

Physical properties

M. Pt. 130-133°C; **Volatility** v.p. 1.3×10^{-5} mmHg at 20°C.

Solubility

Water: 250 g l^{-1} at 22°C. Organic solvent: ethanol, isopropanol, acetone, ethyl acetate

Mammalian and avian toxicity

Acute data

LD_{50} oral rat 180-400 mg kg^{-1}.
LD_{50} oral mouse 250-500 mg kg^{-1}.

LC$_{50}$ (6 hr) inhalation rat 5 mg l^{-1}.
LD$_{50}$ percutaneous rat >1600 mg kg^{-1} (1).

Sub-acute data
In 72 day feeding trials, rats receiving up to 1000 mg kg^{-1} diet showed no ill-effects (1).

Irritancy
Contact dermatitis from 2-bromo-2-nitro-1,3-propanediol (milk preservative) identified in a milk recorder with hand eczema (2).
A study of 149 eczematous patients determined that 0.25% in soft yellow paraffin caused mild irritation. No evidence of sensitisation or cross sensitisation (3).
Acute allergic contact dermatitis reported in 7 patients using Eucerin cream preserved with 2- bromo-2-nitro-1,3-propanediol. Patients patch tested positive to 2-bromo-2-nitro-1,3-propanediol (4).

Legislation
Limited under EC Directive on Drinking Water Quality 80/778/EEC. Pesticides and related products: maximum admissible concentration 0.5 mg l^{-1} (5).

Any other comments
Human health effects and experimental toxicology reviewed (6).

References
1. *The Agrochemicals Handbook* 1987, 2nd ed., RSC, London
2. Grattan, E. H; et al. *Br. J. Derm.* 1985, **113**(Suppl. 29), 43
3. Croshaw, B. *J. Soc. Cosmet. Chem.* 1977, **28**, 3
4. Storrs, F. J; et al. *J. Am. Acad. Derm.* 1983, **8**, 157
5. *EC Directive Relating to the Quality of Water Intended for Human Consumption* 1982, 80/778/EEC, Office for Official Publications of the European Communities, 2 rue Mercier, L-2985 Luxembourg
6. *ECETOC Technical Report No. 30(4)* 1991, European Chemical Industry Ecology and Toxicology Centre, B-1160 Brussels

B202   β-Bromo-β-nitrostyrene

CH=CBr(NO$_2$)

CAS Registry No. 7166-19-0
Synonyms 2-Bromo-2-nitroethenylbenzene; BNS-10%; Slimetrol RX-39
Mol. Formula C$_8$H$_6$BrNO$_2$ **Mol. Wt.** 228.05
Uses Biocide in paper making.

Ecotoxicity

Fish toxicity
LC$_{50}$ (96 hr) bluegill sunfish, fathead minnow, rainbow trout 0.32-1.2 mg l^{-1} (1).

Environmental fate

Degradation studies
Biodegrades to form bromonitromethane (1).

Abiotic removal
Usually treated with sodium sulfide for detoxification (1).

Mammalian and avian toxicity

Acute data
LD$_{50}$ oral rat 966 mg kg^{-1} (1).

Sub-acute data
In a 30 day oral study, rat 110 mg kg^{-1} caused no ill-effects and 300 mg kg^{-1} was of minimal toxicity (1).

Carcinogenicity and long-term effects
National Toxicology Program prechronic study in progress in rats and mice via gavage (2).

References

1. *Material Safety Data Sheets* 1981, Betz Chemical Company
2. *National Toxicology Program, Research and Testing Div.* 1992, NIEHS, Research Triangle Park, NC

B203   Bromophos

CAS Registry No. 2104-96-3
Synonyms phosphorothioic acid, O-(4-bromo-2,5-dichlorophenyl)-, O,O-dimethyl ester; bromophos-methyl; Brofene
Mol. Formula $C_8H_8BrCl_2O_3PS$ **Mol. Wt.** 366.00
Uses Insecticide.

Physical properties

M. Pt. 53-54°C; **B. Pt.** $_{0.01}$ 140- 142°C; **Volatility** v.p. 1.2×10^{-4} mmHg at 20°C.

Solubility
Water: 40 mg l^{-1}. Organic solvent: acetone, dichloromethane, xylene, methanol

Ecotoxicity

Fish toxicity
LC$_{50}$ (96 hr) guppy, rainbow trout 0.05-0.5 mg l^{-1} (1).

Invertebrate toxicity
EC$_{50}$ (48 hr) *Daphnia pulex* 0.0064 mg l^{-1} (2).

Effects to non-target species
Toxic to bees (1).

Mammalian and avian toxicity

Acute data
LD_{50} oral mouse, rat 3300-8000 mg kg^{-1} (1).
LD_{50} oral guinea pig 1500 mg kg^{-1} (1).
LD_{50} percutaneous rabbit 2190 mg kg^{-1} (1).

Carcinogenicity and long-term effects
In 2 yr feeding trials the no-effect level in rats was 0.63 mg kg^{-1} day^{-1} and in dogs 1.5 mg kg^{-1} day^{-1} (1).

Metabolism and pharmacokinetics
Following oral administration in mammals, >90% excreted in urine in 24 hr. The principal metabolite is 4-bromo-2,5-dichlorophenol (1).

Legislation
Limited under EC Directive on Drinking Water Quality 80/778/EEC. Pesticides and related products: maximum admissible concentration 0.1 mg l^{-1} (3).

Any other comments
Toxicity and hazards reviewed (4,5).

References
1. *The Agrochemicals Handbook* 3rd ed., 1991, RSC, London
2. Frear, D. E. H; et al. *Int. J. Econ. Entomol.* 1967, **60**, 1228-1238
3. *EC Directive Relating to the Quality of Water Intended for Human Consumption* 1982, 80/778/EEC, Office for Official Publications of the European Communities, 2 rue Mercier, L-2985 Luxembourg
4. Izmerov, N. F. *Scientific Reviews of Soviet Literature on Toxicity & Hazards of Chemicals* 1992-1993, **96**, Eng. Trans., Richardson, M. L. (Ed.), UNEP/IRPTC, Geneva
5. Pachaly, P. *Dtsch. Apoth. Ztg.* 1989, **129**(27), 1447, (Ger.) (*Chem. Abstr.* **111** 148686j)

B204   Bromophos-ethyl

CAS Registry No. 4824-78-6
Synonyms phosphorothioic acid, *O*-(4-bromo-2, 5-dichlorophenyl)-, *O,O*-diethyl ester; bromofos-ethyl; Filoriol; Nexagan
Mol. Formula $C_{10}H_{12}BrCl_2O_3PS$ **Mol. Wt.** 394.06
Uses Insecticide.

Physical properties

B. Pt. $_{0.001}$ 122-123°C; **Specific gravity** d^{20} 1.52; **Volatility** v.p. 4.5×10^{-5} mmHg at 30°C.

Solubility

Water: 2 mg l^{-1}. Organic solvent: miscible with ethanol, benzene, acetone

Occupational exposure

Supply classification toxic.

Risk phrases

Toxic by inhalation, in contact with skin and if swallowed (R23/24/25)

Safety phrases

Keep out of reach of children – Keep away from food, drink and animal feeding stuffs – If you feel unwell, seek medical advice (show label where possible) (S2, S13, S44)

Ecotoxicity

Fish toxicity

LC_{50} (96 hr) rainbow trout, guppy 0.14-0.4 mg l^{-1} (1).

Effects to non-target species

LD_{50} oral quail 200 mg kg^{-1}.
Toxic to bees (1).

Mammalian and avian toxicity

Acute data

LD_{50} oral rat, mouse 52-210 mg kg^{-1}.
LD_{50} percutaneous\rabbit 1366 mg kg^{-1} (1).

Carcinogenicity and long-term effects

In 2 yr feeding trials no effect level rats 78 mg kg^{-1} day^{-1} and dogs 0.26 mg kg^{-1} day^{-1} (1).

Metabolism and pharmacokinetics

Following oral administration in mammals, 85-90% excreted in urine and faeces in 4 days. The principal metabolite is 4-bromo-2,5-dichlorophenol (1).

Any other adverse effects

Cholinesterase inhibitor (1).

Legislation

Limited under EC Directive on Drinking Water Quality 80/778/EEC. Pesticides and related products: maximum admissible concentration 0.1 mg l^{-1} (2).

References

1. Sax, N. I; et al. *Dangerous Properties of Industrial Materials* 7th ed., 1989, Van Nostrand Reinhold, New York
2. *EC Directive Relating to the Quality of Water Intended for Human Consumption* 1982, 80/778/EEC, Office for Official Publications of the European Communities, 2 rue Mercier, L-2985 Luxembourg

B205   1-Bromopropane

CH₃CH₂CH₂Br

CAS Registry No. 106-94-5
Synonyms propyl bromide
Mol. Formula C_3H_7Br Mol. Wt. 123.00
Uses Organic synthesis.

Physical properties

M. Pt. –110°C; B. Pt. 71°C; Flash point 25°C; Specific gravity 1.354.

Solubility
Water: 0.4 g l⁻¹. Organic solvent: ethanol

Occupational exposure

UN No. 2344; Conveyance classification flammable liquid; Supply classification highly toxic and flammable.

Risk phrases
Highly flammable – Very toxic by inhalation, in contact with skin and if swallowed (R11, R26/27/28)

Safety phrases
Keep container tightly closed and in a well ventilated place – Do not empty into drains – In case of accident or if you feel unwell, seek medical advice immediately (show label where possible) (S7/9, S29, S45)

Mammalian and avian toxicity

Acute data
LD_{Lo} oral rat 4000 mg kg⁻¹ (1).
LC_{50} (30 min) inhalation rat 253 g m⁻³ (2).
LD_{50} intraperitoneal mouse, rat 2530-2950 mg kg⁻¹ (3).

Metabolism and pharmacokinetics
Parent compound and metabolities excreted in urine as mercapturic acid in rats (4).

Genotoxicity
Salmonella typhimurium TA1535 with and without metabolic activation negative (5).

Any other adverse effects
Narcotic (3).

Any other comments
Human health effects and experimental toxicology reviewed (6).

References

1. *Mutat. Res.* 1982, **101**, 321
2. *Fiziologiekeshi Aktivnye Veshihestva (Physcologically Active Substances)* 1975, **7**, 35
3. Izmerov, N. F; et al. *Toxicometric Parameters of Industrial Toxic Substances under Single Exposure* 1982, 102, Moscow

4. Jones, A. R; et al. *Xenobiotica* 1980, **9**(12), 763-772
5. Barker, E. V; et al. *Environ. Sci. Rev.* 1982, **25**, 3-18
6. *ECETOC Technical Report No. 30 (4)* 1991, European Chemical Industry Ecology and Toxicology Centre, B-1160 Brussels

B206   2-Bromopropane

CH3CHBrCH3

CAS Registry No. 75-26-3
Synonyms isopropyl bromide
Mol. Formula C_3H_7Br **Mol. Wt.** 123.00
Uses Intermediate in organic synthesis.

Physical properties
M. Pt. –89°C; **B. Pt.** 59°C; **Flash point** 19°C; **Specific gravity** 1.310.

Solubility
Organic solvent: ethanol, benzene, chloroform, diethyl ether

Occupational exposure
UN No. 2344; **HAZCHEM Code** 2ME; **Conveyance classification** flammable liquid.

Genotoxicity
Salmonella typhimurium TA1535 with and without metabolic activation negative (1).

Any other comments
Modelling risk assessment for nursing infants exposed to volatile organics including 2-bromopropane via mother's occupational inhalation exposure is discussed (2).

References
1. Barker, E. V; et al. *Environ. Sci. Rev.* 1982, **257**, 3-18
2. Shelly, M. L; et al. *Appl. Ind. Hyg.* 1989, **4**(1), 21-26

B207   Bromopropylate

CAS Registry No. 18181-80-1
Synonyms 4-bromo-α-(4-bromophenyl)- α-hydroxybenzeneacetic acid 1-methylethyl ester; 4,4'-dibromobenzelic acid isopropyl ester; 4,4'-dibromobenzilate; phenisobromolate; Acanal; Neoron
Mol. Formula $C_{17}H_{16}Br_2O_3$ **Mol. Wt.** 428.13

Uses Acaricide.

Physical properties
M. Pt. 77°C; **Volatility** v.p. 5.1×10^{-8} mmHg at 20°C.

Solubility
Water: <5 mg l^{-1} at 20°C.

Ecotoxicity

Fish toxicity
LC_{50} (96 hr) rainbow trout, bluegill sunfish, carp 0.35-2.4 mg l^{-1} (1).

Effects to non-target species
LD_{50} oral Japanese quail >200 mg kg^{-1}.
LC_{50} (8 day) oral duck 600 mg kg^{-1} in diet.
LC_{50} (8 day) oral Japanese quail 1000 mg kg^{-1} in diet.
Not toxic to bees (1).

Mammalian and avian toxicity

Acute data
LD_{50} oral rat >5000 mg kg^{-1}.
LD_{50} percutaneous rat >1000 mg kg^{-1} (1).

Carcinogenicity and long-term effects
In 2 yr feeding trials the no-effect level for rats was 500 mg kg^{-1} (1).

Irritancy
Dermal rabbit (duration unspecified) 600 μg moderate irritation effects, while 600 μg instilled into rabbit eye caused mild irritation (2).

Legislation
Limited under EC Directive on Drinking Water Quality 80/778/EEC. Pesticides and related products: maximum admissible concentration 0.1 μg l^{-1} (3).

Any other comments
Admissible daily intake human 8 μg kg^{-1} (total intake from all sources) (4).

References
1. *The Agrochemicals Handbook* 3rd ed., 1991, RSC, London
2. *Ciba-Geigy Toxicology Data* 1977
3. *EC Directive Relating to the Quality of Water Intended for Human Consumption* 1982, 80/778/EEC, Office for Official Publications of the European Communities, 2 rue Mercier, L-2985 Luxembourg
4. Worthing, C. R. (Ed.) *The Pesticide Manual* 1991, 9th ed., British Crop Protection Council, Farnham

B208 Bromoxynil

CAS Registry No. 1689-84-5
Synonyms 3,5-dibromo-4-hydroxybenzonitrile; 3,5-dibromo-4-hydroxyphenyl cyanide; 2,6-dibromo-4-cyanophenol; Broxynil; Brominil; Buctril
Mol. Formula $C_7H_3Br_2NO$ **Mol. Wt.** 276.93
Uses Herbicide.

Physical properties

M. Pt. 194-195°C; **Volatility** v.p. 7.6×10^{-6} mmHg at 20°C.

Solubility
Water: 130 mg l^{-1} at 25°C. Organic solvent: methanol, ethanol, acetone, cyclohexanone, tetrahydrofuran

Occupational exposure

Supply classification toxic.

Risk phrases
Toxic by inhalation, in contact with skin and if swallowed (R23/24/25)

Safety phrases
Keep out of reach of children – Keep away from food, drink and animal feeding stuffs – If you feel unwell, seek medical advice (show label where possible) (S2, S13, S44)

Ecotoxicity

Fish toxicity
LD_{50} (48 hr) harlequin fish 5.0 mg l^{-1} (potassium salt) (1).
LD_{50} (48 hr) rainbow trout 0.15 mg l^{-1} (1).

Environmental fate

Nitrification inhibition
Inhibits nitrification in soil at 50 ppm (2).

Degradation studies
In soil $t_{1/2} \approx 10$ days. Degraded by hydrolysis and debromination to less toxic substances such as hydroxybenzoic acid. In plants, the ester and nitrile groups are hydrolysed, and debromination also occurs (1).
Flexibacterium BR4 rapidly degraded bromoxynil. After 5 wk only 5% remained. Benzamide and benzoic acid metabolites were identified (3).

Mammalian and avian toxicity

Acute data
LD_{50} oral mouse, rat 110-190 mg kg^{-1} (1).
LD_{50} percutaneous rat >2000 mg kg^{-1} (1).

Sub-acute data
In 90 day feeding trials, the no-effect level for rats was 16.6 mg kg^{-1} (K salt) (1).

Teratogenicity and reproductive effects
Gavage pregnant rats and mice 15 and 96.4 mg kg^{-1} day^{-1} respectively on days 6-15 of gestation. Frequency of supernumeracy ribs were determined in foetuses at term and in offspring on postnatal days 6, 20 and 40. In rats supernumeracy ribs occurred in 62% of treated foetuses and in mice 45%. In mice the elevated incidence persisted through day 40 (42.3%), but no significant difference was observed in rats (4).

Metabolism and pharmacokinetics
Studies on fate of bromoxynil in cows have shown that no residual bromoxynil was present in either milk or faeces. Nine days after feeding, <20% was excreted in urine as parent compound (5).

Any other comments
Physico-chemical properties, human health effects and experimental toxicology reviewed (6).

References
1. *The Agrochemicals Handbook* 3rd ed., 1991, RSC, London
2. Parr, J. F; *Pesti. Soil Water* 1974, 321-340
3. Menzie, C. M. *Metabolism of Pesticides, Update II. US Dept. of the Interior, Fish Wildlife Service, Special Scientific Report – Wildlife No. 212* 1978, 52, US Govt. Print. Off., Washington, DC
4. Chernoff, N.; et al. *Fundam. Appl. Toxicol.* 1991, **17**(3), 448-453
5. Kearney, P. C; et al. *Herbicides: Chemistry, Degradation and Mode of Action* 2nd ed., 1975, **1-2**, 584, Marcel Dekker Inc., New York
6. *ECETOC Technical Report No. 30(4)* 1991, European Chemical Industry Ecology and Toxicology Centre, B-1160 Brussels

B209   Buminafos

CAS Registry No. 51249-05-9
Synonyms dibutyl[1-(butylamino)cyclohexyl]phosphorate; 1-butylaminocyclohexanephosphoric acid, butyl ester
Mol. Formula C$_{18}$H$_{38}$NO$_3$P   **Mol. Wt.** 347.48
Uses Herbicide.

Physical properties
M. Pt. –25°C; **B. Pt.** $_1$ 95-99°C; **Specific gravity** d^{20} 0.969; **Volatility** v.p. 7.5 × 10^{-4} mmHg at 20°C.

Solubility
Water: 170 mg l^{-1}. Organic solvent: acetone, methanol, xylene

Ecotoxicity

Fish toxicity
LC$_{50}$ (96 hr) guppy 7 mg l^{-1} (1).

Environmental fate

Degradation studies
t$_{1/2}$ in soil 8-11 days (1).

Abiotic removal
Hydrolysis 50% at pH 6 in 3 days (1).
In aqueous medium 50% hydrolysis occurs in 13 days at pH 6, 20 hr at pH 8 and 2.75 hr at pH 11 (1).

Mammalian and avian toxicity

Acute data
LD$_{50}$ oral mouse, rat 3475-7000 mg kg^{-1} (1,2).
LD$_{50}$ percutaneous rat 12-15 g kg^{-1} (1).
LD$_{50}$ percutaneous rabbit 5000-8000 mg kg^{-1} (1).

Sub-acute data
In 130-day feeding trials, the no-effect level for rats was 140 mg kg^{-1} day^{-1} (1).

Metabolism and pharmacokinetics
Oral administration (species unspecified), rapidly metabolised by dealkylation at the oxygen and nitrogen atoms, followed by deamination (1).

Legislation

Limited under EC Directive on Drinking Water Quality 80/778/EEC. Pesticides and related products: maximum admissible concentration 0.5 mg l^{-1} (3).

References

1. *The Agrochemicals Handbook* 2nd ed., 1987, RSC, London
2. *Environ. Qual. Saf.* 1975, **3**, 686
3. *EC Directive Relating to the Qualtiy of Water Intended for Human Consumption* 1982, 80/778/EEC, Office for Official Publications of the European Communities, 2 rue Mercier, L-2985 Luxembourg

B210   Bupirimate

CAS Registry No. 41483-43-6
Synonyms sulfamic acid, dimethyl-,5-butyl-2- (ethylamino)-6-methyl-4-pyrimidinyl ester; 5-butyl-2- ethylamino-6-methylpyrimidin-4-yldimethyl sulfamate; Nimrod
Mol. Formula C$_{13}$H$_{24}$N$_4$O$_3$S   **Mol. Wt.** 316.43
Uses Fungicide.

Physical properties

M. Pt. 50-51°C; **Volatility** v.p. 4.1×10^{-7} mmHg at 20°C.

Solubility
Water: 22 mg l^{-1} at 20°C. Organic solvent: acetone, ethanol, diethyl ether

Ecotoxicity

Fish toxicity
LC_{50} (96 hr) rainbow trout 1.7 mg l^{-1} (1).

Effects to non-target species
LD_{50} oral pigeons >2700 mg kg^{-1} (1).
No effect oral bees 0.20 mg bee^{-1} (2).

Abiotic removal
The $t_{1/2}$ in soil was ≈ 6-7 wk. Major degradation product is ethirimol (1).

Mammalian and avian toxicity

Acute data
LD_{50} oral rat, mouse, rabbit, guinea pig >5000 mg kg^{-1} (1,3).
LD_{50} percutaneous rat >500 mg kg^{-1} (1).

Carcinogenicity and long-term effects
In 2 yr feeding trials, the no-effect level for rats was 100 mg kg^{-1} diet (1).

Metabolism and pharmacokinetics
Oral administration (species unspecified), 68% of dose eliminated in urine within 24 hr. 77% eliminated in urine and 21% in faeces within 10 days (1).

Legislation

Limited under EC Directive on Drinking Water Quality 80/778/EEC. Pesticides and related products: maximum admissible concentration 0.5 µg l^{-1} (4).

References

1. *The Agrochemicals Handbook* 3rd ed., 1991, RSC, London
2. Worthing, C. R. (Ed.) *The Pesticides Manual* 9th ed., 1991, British Crop Protection Council, Farnham
3. *Werksubstanzen der Pflanzenschutz und Scandlingsbekumpfungsmittel* 1971-1976
4. *EC Directive Relating to the Quality of Water Intended for Human Consumption* 1982, 80/778/EEC, Office for Official Publications of the European Communities, 2 rue Mercier, L-2985 Luxembourg

B211 Buprofezin

CAS Registry No. 69327-76-0
Synonyms 4*H*-1,3,5-thiadiazin-4-one, 2-[(1, 1-dimethylethyl)imino]tetrahydro-3-

(1-methylethyl)-5-phenyl-; 2-*tert*-butylimino-3-isopropyl-5-phenyl-
3,4,5,6-tetrahydro-2*H*-1,3,5-thiadiazin-4-one; Applaud
Mol. Formula $C_{16}H_{23}N_3OS$ **Mol. Wt.** 305.45
Uses Insecticide. Acaricide.

Physical properties
M. Pt. 104.5-105.5°C; **Volatility** v.p. 9.3×10^{-6} at 23°C.

Solubility
Water: 0.9 mg l^{-1} at 25°C. Organic solvent: chloroform, toluene, acetone, ethanol,
n-hexane

Ecotoxicity

Fish toxicity
LC_{50} (48 hr) carp 2.7 mg l^{-1} (1).

Invertebrate toxicity
EC_{50} (3 hr) *Daphnia magna* 50.6 mg l^{-1} (2).

Effects to non-target species
No direct effects on honey bees at 2000 mg l^{-1} (2).

Mammalian and avian toxicity

Acute data
LD_{50} oral ♂, ♀ rats 2200-2350 mg kg^{-1} (1).
LD_{50} oral mouse >10,000 mg kg^{-1} (1).
LC_{50} (4 hr) inhalation rat >4.6 mg l^{-1} (3).
LD_{50} dermal rat >5000 mg kg^{-1} (1).

Sub-acute data
Oral Sprague-Dawley rats (90 day) 0, 40, 200, 1000 or 5000 ppm. No effects on
mortality nor clinical appearance were observed. Body weight gains were suppressed
in ♂ and ♀ at 5000 ppm and in ♀ at 1000 ppm. Haematocrit, haemoglobin red cell
count, serum glucose and triglyceride were decreased in high dose animals, while
levels of total cholesterol and phospholipids were increased. Liver and thyroid were
identified as target organs (3).
Oral beagle dog (13 wk) 0, 2, 10, 50 or 300 mg kg^{-1} day^{-1}. No mortalities observed
throughout study. Slight ataxia and abdominal distension observed in high dose
animals. At necropsy no treatment-related lesions were noted. The no-observed
adverse effect level determined as 10 mg kg^{-1} day^{-1} based on changes in liver (3).

Carcinogenicity and long-term effects
Oral mice (2 yr) 0, 20, 200, 2000 or 5000 ppm in diet. High dose animals exhibited
retarded growth, decreased specific gravity of urine, reduced levels of protein in the
urine, elevation in platelet and lymphocyte count, increased absolute and relative liver
weight and an increased incidence of hepatocellular swelling (centrilobular and
diffuse) and hepatocellular hyperplasia were seen in both sexes. The incidence of
hepatoceullar adenoma was increased in high-dose ♀, but the combined incidence of
hepatocellular adenomas and carcinomas was not significant. However, the overall
incidence of lung adenoma and carcinoma in high-dose ♂ animals was significantly
higher than controls. The no-observed adverse effect level for ♂ animals was

determined to be 1.82 mg kg^{-1} buprofezin day^{-1} (3).
Oral rat (2 yr) 0, 5, 20, 200, 2000 ppm in feed. No effect of treatment on clinical observations was observed, survival in all groups was >40%. In high-dose ♀ increased incidence of cystitis, chronic nephrosis and interstitial oedema in the heart observed (3).

Teratogenicity and reproductive effects

Oral rat (13 wk) 0, 10, 100 or 1000 ppm in diet. F_0, F_1 and F_2 animals were studied. Litter data revealed decrease in survival for F_0 pups during 0-4 day lactation from high dose animals. Lower mean live pup weight observed in all dose groups. The authors conclude at doses ≤1000 ppm buprofezin has no influence on reproductive performance (3).

Gavage Sprague-Dawley rat (6-15 day gestation) 0, 50, 200 or 800 mg kg^{-1} day^{-1}. Maternal toxicity in high-dose animals evidenced by reduced food intake, decreased body weight, loose faeces, urogenital staining, lethargy, hunched posture, thin appearance and piloerection. At 800 mg kg^{-1} day^{-1} 4 ♀ showed total resorption and increased early post-implantation loss, reduced litter size and foetal weight were recorded. Foetuses in highest dose group showed significant increased incidence of subcutaneous oedema and signs of slight foetal immaturity including reduced mean foetal weight. The no-observed adverse effect level was 50 mg kg^{-1} day^{-1} for maternal toxicity and ≤166-188 mg kg^{-1} day^{-1} for embryotoxicity (3).

Metabolism and pharmacokinetics

♂ Sprague-Dawley rats 10 or 100 ^{14}C buprofezin in olive oil. Rats were sacrificed 2, 5, 9, 24 or 96 hr after dosing. Following administration of 10 mg kg^{-1}, the highest radioactivity was detected in the urinary bladder 9 hr after dosing. Maximum dose detected in liver 5 hr after dosing. Radioactivity was also detected in adipose tissue, kidney, adrenal gland, pancreas and blood. Following administration of 100 mg kg^{-1} highest radioactivity was detected in adipose tissue after 9 hr. Distribution pattern to other organs and tissues was similar to lower dose animals. The $t_{1/2}$ in all tissues and organs examined was 3.5-15 hr (between 9 and 24 hr) after dosing and 15-72 hr (between 24 and 96 hr) (3).

Oral rat single 10 mg kg^{-1} dose excreted in urine, faeces and bile within 24 hr. Main route of metabolism was via hydroxylation of phenyl ring and oxidation of sulfur. Hydroxylation of phenyl ring gave 4-hydroxy; 3,4-dihydroxy and 3-hydroxy-4-metoxy buprofezin and some of these were conjugated with glucuronic acid or sulfate. The oxidation products of sulfur were thought to form isopropyl phenyl urea through cleavage of the thiadiazin ring. 12% of dose excreted into faeces as parent compounds. In urine and bile only more polar metabolites were detected (3).

Oral rat single dose 10 or 100 mg kg^{-1} polar metabolites in faeces and urine included 1-(4-hydroxyphenyl)-3- isopropylurea 4-aminophenol and 4-acetamido phenol. The sulfuric acid conjugate of 4-acetamidophenol was the major metabolite in urine, accounting for 3.9% of the dose (3).

Genotoxicity

Salmonella typhimurium TA98, TA100, TA1535, TA1537, TA1538 with and without metabolic activation negative. *Escherichia coli* WP2 uvrA with and without metabolic activation negative. *In vitro* mouse lymphoma L5178Y tk+/tk- with and without metabolic activation negative. *In vivo* mouse bone marrow micronucleus test negative (3).

Any other adverse effects to man

Medical surveillance of workers who routinely handled buprofezin in a factory in Japan has been undertaken. The survey revealed no effects which could be attributed to exposure to buprofezin (3).

Legislation

Limited under EC Directive on Drinking Water Quality 80/778/EEC. Pesticides and related products: maximum admissible concentration 0.1 μg l^{-1} (4).

Any other comments

The toxicity of buprofezin has been extensively reviewed (3).

References

1. *The Agrochemicals Handbook* 3rd ed., 1991, RSC, London
2. Worthing, C. R. (Ed.) *The Pesticide Manual* 9th ed., 1991, British Crop Protection Council, Farnham
3. *Pesticide Residues in Food 1991: Toxicology Evaluations* 1991, 75-95, World Health Organisation, Geneva
4. *EC Directive Relating to the Quality of Water Intended for Human Consumption* 1982, 80/778/EEC, Office for Official Publications of the European Communities, 2 rue Mercier, L-2985 Luxembourg

B212   Butachlor

CAS Registry No. 23184-66-9

Synonyms acetamide, *N*-(butoxymethyl)-2-chloro-*N*-(2,6-diethylphenyl)-*N*-butoxymethyl-2-chloro-2′, 6′-diethylacetanilide; Lambart; Butanex; Pillarsete; Butanox

Mol. Formula $C_{17}H_{26}ClNO_2$ **Mol. Wt.** 311.86

Uses Herbicide.

Physical properties

M. Pt. <5°C; **B. Pt.** $_{0.5}$ 156°C; **Specific gravity** d^{25} 1.070; **Volatility** v.p. 4.2×10^{-6} mmHg at 25°C.

Solubility

Water: 20 g mg l^{-1} at 20°C. Organic solvent: diethyl ether, acetone, benzene, ethanol, ethyl acetate, *n*-hexane

Ecotoxicity

Fish toxicity

LC$_{50}$ (96 hr) carp, bluegill sunfish, rainbow trout 0.32-0.52 mg l^{-1} (1).

Invertebrate toxicity
EC_{50} (48 hr) *Daphnia magna* 2.4 mg l^{-1} (2).

Effects to non-target species
LD_{50} oral mallard duck, bobwhite quail >10,000 mg kg^{-1} (1).
LD_{50} (8 day) oral bobwhite quail, mallard ducks >10,000 mg kg^{-1} diet (1).

Mammalian and avian toxicity

Acute data
LD_{50} oral rat 2000 mg kg^{-1} (1).
LD_{50} dermal rabbit 4080 mg kg^{-1} (2).

Sub-acute data
One year no-effect level for dogs 5 mg kg^{-1} daily (2).

Carcinogenicity and long-term effects
In 2 yr feeding trials rats and dogs receiving 1000 mg kg^{-1} diet exhibited no ill effects (1).

Genotoxicity
Chinese hamster ovary cells induced chromosome aberrations with and without metabolic activation (3).

Legislation
Limited under EC Directive on Drinking Water Quality 80/778/EEC Pesticides and related products: maximum admissible concentration 0.5 μg l^{-1} (4).

References

1. *The Agrochemicals Handbook* 3rd ed., 1991, RSC, London
2. *Farm Chemicals Handbook* 1983, **39**, Meister Publishing, Willoughby, OH
3. Lin, M. F; et al. *Mutat. Res.* 1987, **188**(3), 241-250
4. *EC Directive Relating to the Quality of Water Intended for Human Consumption* 1982, 80/778/EEC, Office for Official Publications of the European Communities, 2 rue Mercier, L-2985 Luxembourg

B213   1,3-Butadiene

$H_2C=CHCH=CH_2$

CAS Registry No. 106-99-0
Synonyms buta-1,3-diene; bivinyl; divinyl; vinylethylene; pyrrolylene
Mol. Formula C_4H_6 **Mol. Wt.** 54.09
Uses In the manufacture of polymers and synthetic rubbers.
Occurrence Detected, but not quantified in drinking water (1).
Fugitive emission from petrochemical processes (2).

Physical properties

M. Pt. −109°C; **B. Pt.** −4.5°C; **Flash point** −70°C;
Specific gravity d^{20} 0.6211 (liquified); **Volatility** v.p. 2.5 mmHg at 20°C; v. den. 1.9.

Occupational exposure

US TLV (TWA) 10 ppm (22 mg m^{-3}); **UK Long-term limit** MEL 10 ppm (22 mg m^{-3}); **UN No.** 1010; **HAZCHEM Code** 2WE; **Conveyance classification** flammable gas; **Supply classification** highly flammable.

Risk phrases

Extremely flammable liquefied gas (R13)

Safety phrases

Keep container in a well ventilated place – Keep away from sources of ignition – No Smoking – Take precautionary measures against static discharges (S9, S16, S33)

Ecotoxicity

Fish toxicity

LC$_{50}$ (24 hr) pinperch 71.5 mg l^{-1} (3).

Bioaccumulation

Calculated bioconcentration factor of 19.1 indicates that environmental accumulation is unlikely (4).

Environmental fate

Degradation studies

Nocardia sp. 249 degraded butadiene by catabolic mechanisms using butadiene as sole carbon energy source (5).
Oxidised by methylotrophic bacteria and utilised by a *Nocardia sp.* as sole carbon source (6).

Abiotic removal

Estimated t$_{1/2}$ due to reactions with hydroxyl radicals and ozone is 4.9 hr (4).
Stable reaction products of photo-oxidation are acetaldehyde and acrolein. Due to its low boiling point (-4.5°C), it would be expected to evaporate rapidly from soils (7-9).

Absorption

Estimated adsorption coefficients in soils and sediments are 72-228 indicating that appreciable absorption is unlikely (4).

Mammalian and avian toxicity

Acute data

LD$_{50}$ oral rat, mouse 3.2-5.5 g kg^{-1} (10).
LC$_{50}$ (23 min) inhalation rabbit 250,000 ppm (in air) (11).
LC$_{50}$ (2 hr) inhalation mouse 270,000 mg m^{-3} (12).
LC$_{50}$ (4 hr) inhalation rat 285,000 mg m^{-3} (12).

Carcinogenicity and long-term effects

Inadequate evidence for carcingenicity to humans, limited evidence for carcinogenicity to animals, IARC classification group 2B (13).
Carcinogenic to ♂ and ♀ rats producing haemangiosarcomas of the heart, malignant

lymphomas, alveolar/bronchiolar adenomas and carcinomas, papillomas and carcinomas of the stomach, heptacellular adenomas and carcinomas, mammary gland carcinomas and ovarian granulomas (14).

National Toxicology Program post peer review technical reports in progress on long-term mice inhalation study. Clear evidence of carcinogenic activity (15).

♂ and ♀ B6C3F1 mice (65 wk) exposed to 62.5-625 ppm. Anaemia occurred at >62.5 ppm, testicular atrophy was induced at 625 ppm and ovarian atrophy was observed ≥20 ppm. During the first 50 wk, lymphocytic lymphom was the major cause of death at 625 ppm. Heart, forestomach, lung, Harderian gland, mammary gland, ovary and liver neoplasms were observed in mice that died between 40 and 65 wk (16).

Teratogenicity and reproductive effects

Pregnant Sprague-Dawley rats and Swiss (CD-1) mice (0, 40, 200 or 1000 ppm) 6 hr day^{-1} on days 6-15 of gestation and killed on day 18 (mice) or 20 (rats). In rats, maternal toxicity in the form of reduced extra gestational weight gain was observed in animals dosed with 1000 ppm. No evidence of developmental toxicity recorded. In mice, the foetuses were more susceptible than the dam; maternal toxicity was observed at 200 and 1000 ppm and mean body weights of ♀ foetuses were reduced (17).

Metabolism and pharmacokinetics

Rats exposed for 2 hr to an airborne concentration of 130,000 ppm revealed highest accumulation in the perirenal fat (152 mg%). Lower concentrations (36-51 mg%) were found in the liver, brain, spleen and kidney. After 90 min exposure to 130,000 ppm for 1 hr the tissue concentrations were minimal (18).

Intraperitoneal administration to ♂ B6C3F1 mice, most of the dose was exhaled unchanged, carbon dioxide was the next largest pool, lesser amounts in urine and faeces; little remained in the carcass 65 hr later (19).

♂ Sprague Dawley rats and B6C3F1 mice exposed by inhalation (nose only) for 3.4 hr to 1220 and 121 μg l^{-1} in air, respectively. 1,3-Butadiene was distributed in lung, trachea, nasal turbinates, small and large intestine, liver, kidneys, bladder and pancreas within 1 hr after exposure in both species. Reported $t_{1/2}$ of 2-10 hr (20).

In mice, 1,3-butadiene is metabolised to 1,2-epoxybut-3-ene at <1000 ppm and 2000 ppm at twice the metabolic rate of rats (21).

Irritancy

Inhalation human (1 min) 100,000 ppm caused irritation to respiratory system (22). Conjunctivitis was reported in mice exposed to 90,000-140,000 ppm and rabbits exposed to 150,000-250,000 ppm. No eye injury was observed in rabbits exposed to 6700 ppm for 7.5 hr day^{-1} 6 day wk^{-1} for 8 month. It has been suggested that the conjuntivitis is due to the presence of the dimer 4-vinyl-1-cyclohexane (23).

Genotoxicity

Salmonella typhimurium TA1530 with metabolic activation positive (24). Human lymphocytes (2 hr) induced sister chromatid exchange with and without metabolic activation (25).

Potent *in vivo* genotoxin but weak genotoxin *in vitro*: *Salmonella typhimurium* TA1535 with metabolic activation weakly positive; did not induce sister chromatid exchanges in human whole blood lymphoctyes with or without metabolic activation; induced sister chromatid exchanges and micronuclei in bone marrow cells of mice but not rats after inhalation exposure to 10-10,000 ppm, 6 hr day^{-1} for 2 days; failed to induce unscheduled DNA synthesis in rat or mouse hepatocytes *in vivo* (26).

Induced micronuclei and sister chromatid exchanges in bone-marrow of mice but not rats *in vivo* (24).

In vivo mouse bone marrow induced micronuclei and sister chromatid exchange (27).

Any other adverse effects to man

Several studies have shown elevated standardised mortality ratios for cancers in various organs in rubber industry workers. The results could however, be complicated by exposure to other chemicals (28).

Any other comments

Toxicity and hazards reviewed (29).

1,3- Butadiene rubber-based plastic containers can contaminate foodstuffs (30).

Physico-chemical properties, human health effects, experimental toxicology, epidemiology, workplace experience, exposure and environmental effects reviewed (31).

Metabolism, genotoxicity, carcinogenicity and developmental toxicity in the oil refining and petrochemical industry reviewed (32).

Future directions in toxicology studies reviewed (33).

Cytogenic evaluation of *in vivo* genotoxic and cytotoxic activity using rodent somatic cells reviewed (34).

Worldwide regulatory activity for occupational exposure with reference to control technology and economics reviewed (35).

Carcinogenicity in susceptible mouse strains and its risk assessment in humans reviewed (36).

Toxicity studies, its metabolic fate in different animal species and human epidemiology data reviewed (37).

All reasonable efforts have been taken to find information on isomers of this compound, but no relevant data are available.

References

1. *USEPA: CHIP (Draft) 1,3-Butadiene* 1981, 20
2. Hughes, T. W; et al. *Chem. Eng. Prog.* 1979, **75**, 35-39
3. Jones, H. R. *Environmental Control in the Organic and Petrochemical Industries* 1971, Noyes Data Corp, Park Ridge, NJ
4. Lyman, W. J. *Handbook of Chemical Property Estimation Methods. Environmental Behaviour of Organic Compounds* 1982, McGraw-Hill, New York
5. Wilkinson, R. J; et al. *The Microbial Utilization of Butadiene* 1976, Shell Research Ltd, Sittingbourne Research Centre
6. Hou, C. T; et al. *Appl. Environ. Microbiol.* 1979, **38**(1), 127-134
7. Atkinson, R. *Atmos. Environ.* 1983, **24A**, 1-41
8. Atkinson, R; et al. *Chem. Rev.* 1984, **84**, 437-470
9. Verschueren, K. *Handbook of Environmental Chemicals* 2nd ed., 1983, Van Nostrand Reinhold, New York
10. *USEPA Health Assessment Document: 1, 3-Butadiene* 1985, EPA 600/8-85-004A
11. Carpenter; et al. *J. Ind. Hyg. Toxicol.* 1944, **26**, 69
12. *USEPA Chemical Hazard Information Profile: 1,3-Butadiene* 1981, 10
13. *IARC Monograph* 1987, **Suppl. 7**, 136
14. *IARC Monograph* 1986, **39**, 155-179
15. *National Toxicology Program, Research and Testing Div.* 1992, NIEHS, Research Triangle Park, NC

16. Melnick, R. L; et al. *Environ. Health Perspect.* 1990, **86**, 27-36
17. Morrissey, R. E; et al. *Environ. Health Perspect.* 1990, **86**, 79-84
18. Shugger, B. *Arch. Environ. Health* 1969, **18**, 878-882
19. *USEPA Ambient Water Quality Criteria: Halomethanes* 1980, EPA 440/5-80-051
20. Bond, J. A; et al. *Am. Ind. Hyg. Assoc. J.* 1987, **48**(10), 867-872
21. Laib, R. J; et al. *Ann. N.Y. Acad. Sci.* 1988, **534**, (Living Chem. World), 663-670
22. Verschueren, K. *Handbook of Environmental Data on Organic Chemistry* 2nd ed., 1983, Van Nostrand Reinhold, New York
23. *Toxicity Review 11: 1,3-Butadiene and Related Compounds* 1985, HSE, London
24. de Meester, L. F; et al. *Biochem. Biophys. Res. Commun.* 1978, **80**, 298
25. Sasiadek, M; et al. *Mutat. Res.* 1991, **261**(2), 117-121
26. Arce, G. T; et al. *Environ. Health Perspect.* 1990, **86**, 75-78
27. *IARC Monograph* 1987, **Suppl. 6**, 126-128
28. *IARC Monograph* 1982, **28**, 183-230
29. Izmerov, N. F. *Scientific Reviews of Soviet Literature on Toxicity & Hazards of Chemicals* 1992-1993, **127**, Eng. Trans., Richardson, M. L. (Ed.), UNEP/IRPTC, Geneva
30. McNeal, T. P; et al. *J. Assoc. Off. Anal. Chem.* 1987, **70**, 18-21
31. *ECETOC Technical Report No. 30(4)* 1991, European Chemical Industry Ecology and Toxicology Centre, B-1160 Brussels
32. Mehlman, M. A; et al. *Toxicol. Ind. Health* 1991, 7(3), 207-220
33. Bird, M. G. *Environ. Health Perspect.* 1990, **86**, 99-102
34. Tice, R. R. *Cell Biol. Toxicol.* 1988, **4**(4), 475-486
35. McGraw, J. L. *Annu. Meet. Proc. – Int. Inst. Synth. Rubber Prod.* 1988, **29**, 39-55
36. Hinderer, R. K. *Annu. Meet. Proc. – Int. Inst. Synth. Rubber Prod.* 1988, **29**, 32-38
37. Loeser, E. *Annu. Meet. Proc. – Int. Inst. Synth. Rubber Prod.* 1988, **29**, 13

B214   2,3-Butadione

CH3COCOCH3

CAS Registry No. 431-03-8
Synonyms diacetyl; biacetyl; 2,3-diketobutane; dimethyl diketone; dimethylglyoxal; 2,3-butanedione
Mol. Formula $C_4H_6O_2$ **Mol. Wt.** 86.09
Uses In essential oils and butter. Aroma agent especially for coffee.

Physical properties
B. Pt. 88°C; **Flash point** 26°C; **Specific gravity** 0.981.

Environmental fate

Degradation studies
2,3-Butadione has been identified as an intermediate in microbial oxidation. Since 2-butanol is biodegradable using river water or sewage inoculums with extensive mineralisation, it can be predicted that 2,3- butadione would be biodegradable (1).

Abiotic removal
Absorbs visible radiation <460 nm and is subject to loss by photolysis as well as reaction with reactive atmospheric species. Photolytic $t_{1/2}$ is 0.7 hr compared with $t_{1/2}$ estimated for reaction with hydroxyl radicals of 621 hr (2).

Mammalian and avian toxicity

Acute data
LD$_{50}$ oral guinea pig, rat 990-1580 mg kg^{-1} (3).
LD$_{50}$ intraperitoneal rat 400 mg kg^{-1} (4).

Irritancy
Dermal rat (24 hr) 500 mg caused moderate irritation (5).

Genotoxicity
Salmonella typhimurium TA98, TA100 with and without metabolic activation positive (6).
In vitro human embryo test equivocal (7).

Any other comments
Human health effects and experimental toxicology reviewed (8).
All reasonable efforts have been taken to find information on isomers of this compound, but no relevant data are available.

References

1. Lijmbach, G. W. M; et al. *Antoine von Leeuwenhoek* 1973, **39**, 415-423
2. Plum, C. N; et al. *Environ. Sci. Technol.* 1983, **17**, 479-484
3. Jenner, P. M; et al. *Food. Cosmet. Toxicol.* 1964, **2**, 237
4. *Food Cosmet. Toxicol.* 1969, **7**, 571
5. Opdyke, D. L. J. *Food Cosmet. Toxicol.* 1979, **17**(Suppl.), 695
6. *Mutat. Res.* 1979, **67**, 367
7. *Bull. Exp. Biol. Med.* 1972, **74**, 828
8. *ECETOC Technical Report No. 30(4)* 1991, European Chemical Industry Ecology and Toxicology Centre, B 1160 Brussels

B215   Butamifos

CAS Registry No. 36335-67-8
Synonyms Phosphoromidothiroic acid, (1-methylpropyl), *O*-ethyl-*O*-(5-methyl-2-nitrophenyl)ester; *O*-ethyl, *O*-6-nitro-*m*-tolyl sec-butylphosphoramidothioate; Cremart; Tufler
Mol. Formula C$_{13}$H$_{21}$N$_2$O$_4$PS   **Mol. Wt.** 332.36
Uses Herbicide.

Physical properties
Specific gravity d^{25} 1.188; **Volatility** v.p. 5.5×10^{-4} mmHg at 27°C.

Solubility
Water: 5 mg l^{-1}. Organic solvent: acetone, methanol, xylene

Ecotoxicity

Fish toxicity

LC_{50} (48 hr) carp 2.4 mg l^{-1} (1).

Mammalian and avian toxicity

Acute data

LD_{50} oral ♂, ♀ rat 1070, 845 mg kg^{-1} respectively (1).
LD_{50} percutaneous rat >5000 mg kg^{-1} (1).

Legislation

Limited under EC Directive on Drinking Water Quality 80/778/EEC. Pesticides and related products: maximum admissible concentration 0.5 mg l^{-1} (2).

References

1. *The Agrochemicals Handbook* 3rd ed., 1991, RSC, London
2. *EC Directive Relating to the Quality of Water Intended for Human Consumption* 1982, 80/778/EEC, Office for Official Publications of the Europeon Communities, 2 rue Mercier, L-2985 Luxembourg

B216   Butanal

CH3(CH2)2CHO

CAS Registry No. 123-72-8
Synonyms butyraldehyde; butylaldehyde; butyric aldehyde
Mol. Formula C_4H_8O **Mol. Wt.** 72.11
Uses Rubber accelerators. Synthetic resins. Solvents. Plasticisers.

Physical properties

M. Pt. –99°C; **Flash point** –6.6°C; **Specific gravity** d_4^{20} 0.8016; **Partition coefficient** log P_{ow} 1.18.

Solubility
Organic solvent: ethanol, diethyl ether, ethyl acetate, acetone, toluene

Occupational exposure

UN No. 1129; **HAZCHEM Code** 3WE; **Conveyance classification** flammable liquid; **Supply classification** highly flammable.

Risk phrases
Highly flammable (R11)

Safety phrases
Keep container in a well ventilated place – Do not empty into drains – Take precautionary measures against static discharges (S9, S29, S33)

Ecotoxicity

Fish toxicity
LC$_{50}$ (96 hr) fathead minnow 25.8 mg l^{-1} (1).

Invertebrate toxicity
EC$_{50}$ (5 min) *Photobacterium phosphoreum* 16.5 mg l^{-1} Microtox test (2).
Cell multiplication inhibition test, *Pseudomonas putida* 100 mg l^{-1}, *Scenedesmus quadricauda* 83 mg l^{-1} *Entosiphon sulcatum* 4.2 mg l^{-1}, *Uronema parduczi* 98 mg l^{-1} (3).

Environmental fate

Degradation studies
Activated sludge, 22.8% ThOD in 24 hr (4).
Biodegradable (2).

Abiotic removal
85% removal by air stripping in 8 hr (5).

Mammalian and avian toxicity

Acute data
LD$_{50}$ oral rat 5890 mg kg^{-1} (6).
LC$_{50}$ (30 min) inhalation rat 60,000 ppm (7).

Carcinogenicity and long-term effects
Butanal is currently being investigated by the US National Toxicology Program (8).

Irritancy
Dermal rabbit (24 hr) 500 mg caused severe irritation and 20 mg instilled into rabbit eye for 24 hr caused moderate irritation (9).

Genotoxicity

Salmonella typhimurium TA97, TA98, TA100, TA1535, TA1537 with and without metabolic activation negative (10).
V79 Chinese hamster lung cells induced a dose-dependent increase in 6-thioguanine- and ouabain- resistant mutants (11).
Cell degeneration and polyploidy during spermatogenesis, chromosome aberrations and altered sperm morphology were reported in mice (12).

Any other comments

Physico-chemical properties, human health effects and experimental toxicology reviewed (13).

References

1. Curtis, M. W; et al. *J. Hydrol.* 1981, **51**, 359
2. Kaiser, K. L. E; et al. *Water Pollut. Res. J. Canada* 1991, **26**(3), 361-431
3. Bringmann, G; et al. *Water Res.* 1980, **14**, 231-241
4. Gerhold, R. M; et al. *J. Water Pollut. Contr. Fed.* 1966, **38**(4), 562
5. Meinck, F; et al. *Les Eaux Residuaires Industrielles* 1970
6. Smyth; et al. *Arch. Ind. Hyg. Occup. Med.* 1951, **4**, 119
7. Verschueren, K. *Handbook of Environmental Data on Organic Chemicals* 2nd ed., 1983, Van Nostrand Reinhold, New York
8. Toxicology Information Program *Tox-Tips* 1987, **130-25**, Bethseda, MD

9. Marhold, J. V. *Sbornik Vystedku Toxixologickeho Vysetieni Latek A Pripravku* 1972, 40, Prague

10. Mortelmans, K; et al. *Environ. Mol. Mutagen.* 1986, **8**(7), 1-119

11. Brambilla, G; et al. *Mutagenesis* 1989, **4**(4), 277-279

12. *Mutat. Res.* 1977, **39**, 317

13. *ECETOC Technical Report No. 30(4)* 1991, European Chemical Industry Ecology and Toxicology Centre, B-1160 Brussels

B217   Butane

CH3CH2CH2CH3

CAS Registry No. 106-97-8

Synonyms *n*-Butane

Mol. Formula C_4H_{10} **Mol. Wt.** 58.12

Uses Fuel. Aerosol propellant.

Occurrence Natural gas. Contaminant in drinking, surface and ground water. Released into air from waste incinerators and landfill sites. Product of gasoline combustion (1-7).

Physical properties

M. Pt. $-135°C$; **B. Pt.** $-0.5°C$; **Flash point** $-138°C$; **Specific gravity** d_4^{20} 0.5788; **Partition coefficient** log P_{ow} 2.89; **Volatility** v. den. 2.0.

Solubility
Water: 61 μg g^{-1} at 20°C. Organic solvent: ethanol, diethyl ether

Occupational exposure

US TLV (TWA) 800 ppm (1900 mg m^{-3}); **UK Long- term limit** 600 ppm (1430 mg m^{-3}); **UK Short-term limit** 750 ppm (1780 mg m^{-3}); **UN No.** 1011; **HAZCHEM Code** 2WE; **Conveyance classification** flammable gas; **Supply classification** highly flammable.

Risk phrases
Extremely flammable liquefied gas (R13)

Safety phrases
Keep container in a well ventilated place – Keep away from sources of ignition – No Smoking – Take precautionary measures against static discharges (S9, S16, S33)

Ecotoxicity

Bioaccumulation
Calculated bioconcentration factor of 1.9 indicates that environmental accumulation is unlikely (8).

Environmental fate

Degradation studies
With oxygen, butane supports the growth of *Neurospora crassa* as well as the

germination of *N. ascrospores* and growth of *Escherichia coli* B and Sd4, thus rendering butane potentially biodegradable (9).

Incubation with natural flora in the groundwater in presence of the other components of high octane gasoline (100 μg l^{-1}) biodegradation 0% after 192 hr at 13°C (initial concentration 0.63 μl l^{-1}) (10,11).

Within 24 hr, butane was oxidised to 2-butanone and 2-butanol by cell suspensions of over 20 methyltrophic organisms isolated from lake water and soil samples (12). Degradation by microorganisms occurs via the beta-oxidation pathway (13). Nonbiodegradable/qualified (14).

Abiotic removal

A model for the reaction of butane and nitrous oxides in air determined products of the photooxidation as 2-butyl nitrate, butyraldehyde, 1-butyl nitrate, methyl nitrate, peroxyacetyl nitrate, propene oxide, propionaldehyde, formaldehyde and acetaldehyde (15).

Estimated lifetime under photochemical smog conditions in south east England 15 hr (10).

Absorption

Calculated soil adsorption coefficient range from 450-900 which indicates a medium to low soil mobility (16,17).

Mammalian and avian toxicity

Acute data

LC_{50} (4 hr) inhalation rat, mouse 660-680 g m^3 (18).

Metabolism and pharmacokinetics

Inhalation (4 hr) rats and mice exposed to lethal concentration (28-29%) revealed highest concentrations in perinephric fat (2086 ppm), then brain (750 ppm), spleen (522 ppm), liver (492 ppm) and kidney (441 ppm) (19).

Irritancy

Direct contact of eye and skin with liquified butane may cause burns or frostbite. Repeated exposure may cause dermatitis (20).

Genotoxicity

Salmonella typhimurium TA98, TA100, TA1535, TA1537, TA1538 with and without metabolic activation negative (21).

Any other adverse effects to man

An autopsy of a 25 yr old man who had been abusing lighter refill gas for 10 yr revealed butane in blood, gastric content, brain, lung, heart, liver, kidney, spleen and pancreas. Bronchi and alveolar abnormalities were present. It was concluded that death was caused by asphyxia due to respiratory obstruction (22).

Any other comments

Physico-chemical properties, human health effects, experimental toxicology, epidemiology, workplace experience and exposure reviewed (23).

References

1. Abrams, E. F; et al. *Identification of Organic Compounds in Effluents from Industrial Sources* 1975, USEPA-560/3-75-002

2. Kool, H. J; et al. *Crit. Rev. Env. Control* 1982, **12**, 307-357
3. Kopfler, F. C; et al. *Adv. Environ. Sci. Technol.* 1977, **8**, 419-433
4. McFall, J. A; et al. *Chemosphere* 1985, **14**, 1253-1265
5. Stump, F. D; et al. *Atmos. Environ.* 1989, **23**, 307-320
6. Zweidinger, R. B; et al. *Environ. Sci. Tech.* 1988, **22**, 956-962
7. Altwicker, E. R; et al. *Atmos. Environ.* 1978, **12**, 1289-1296
8. Lyman, W. J; et al. *Handbook of Chemical Property Estimation Methods* 1982, 5.4, 5.10, McGraw-Hill, New York
9. Clayton, G. D; et al. *Patty's Ind. Hyg. and Toxicol.* 3rd ed., 1981-1982, **2A, 2B, 2C** (Toxicology), 3183, John Wiley & Sons, New York
10. Verschueren, K. *Handbook of Environmental Data of Organic Chemicals* 2nd ed., 1983, Van Nostrand Reinhold Co., New York
11. Jamison, V. W; et al. *Biodegradation of High Octane Gasoline* Proc. 3rd Int. Biograd. Symp., 1976, Applied Science Publishers, New York
12. Patel, R. N; et al. *Appl. Anviron. Microbiol.* 1980, **39**, 720-733
13. Parr, J. F; et al. *Land Treatment of Hazardous Wastes* 1983, 327, Noyes Data Corporation, Park Ridge, NJ
14. *MITI Report* 1984, Ministry of International Trade and Industry, Tokyo
15. Carter, W. P. L; et al. *Int. J. Chem. Kinet.* 1979, **11**(1), 45-102
16. Lyman, W. J; et al. *Handbook of Chemical Property Estimation Methods* 1982, 4-9, McGraw-Hill, New York
17. Swann, R. L; et al. *Res. Rev.* 1983, **85**, 16-28
18. *Farmakologiya i Toksikologiya (Moscow)* 1967, **30**, 102
19. *Ethyl Browning's Toxicity and Metabolism of Industrial Solvents* 2nd ed., 1987, **1**, 169, Elsevier, Amsterdam
20. Lenga, R. E. *The Sigma Aldrich Library of Chemical Safety Data* 2nd ed., 1988, Sigma Aldrich, Milwaukee, WI
21. Kawata, F; et al. *Hochudoku* 1990, **8**(2), 68-69, (Jap.) (*Chem. Abstr.* 1991, 115, 86903z)
22. *In Vitro Microbiological Mutagenicity Studies of Philips Petroleum Company Hydrocarbons Propellants and Aerosols* 1977, Stanford Research Institute, CA
23. *ECETOC Technical Report No. 30(4)* 1991, European Chemical Industry Ecology and Toxicology Centre, B-1160 Brussels

B218   1,3-Butanediamine

CH3CH(NH2)CH2CH2NH2

CAS Registry No. 590-88-5
Synonyms 1,3-diaminobutane
Mol. Formula $C_4H_{12}N_2$ **Mol. Wt.** 88.15
Uses Intermediate in organic synthesis.

Physical properties

B. Pt. 142-150°C; **Flash point** 52°C (open cup); **Volatility** v. den. 3.0.

Solubility
Organic solvent: ethanol

Mammalian and avian toxicity

Acute data
LD_{50} oral rat 1350 mg kg^{-1} (1).
LD_{50} dermal rat 430 mg kg^{-1} (1).

Irritancy
Dermal rabbit (24 hr) 10 mg and 250 µg instilled into rabbit eye caused severe irritation (1).

Any other comments
All reasonable efforts have been taken to find information on isomers of this compound, but no relevant data are available.

References
1. *AMA Arch. Ind. Hyg. Occup. Med.* 1951, **4**, 119

B219   1,4-Butanediamine

$H_2N(CH_2)_4NH_2$

CAS Registry No. 110-60-1
Synonyms 1,4-diaminobutane; putrescine; tetramethylenediamine
Mol. Formula $C_4H_{12}N_2$ **Mol. Wt.** 88.15
Uses Organic synthesis. Biochemical research.
Occurrence Product of decomposition of animal matter and sewage sludge.

Physical properties
M. Pt. 27-28°C; **B. Pt.** 158-160°C; **Flash point** 51°C; **Specific gravity** d^{25} 0.877.

Environmental fate

Anaerobic effects
Biodegradable (1).

Mammalian and avian toxicity

Acute data
LD_{Lo} oral mouse, rabbit 1600 mg kg^{-1} (2,3).
LD_{Lo} intravenous rabbit 80 mg kg^{-1} (3).
LD_{Lo} rectal rabbit 400 mg kg^{-1} (3).
LD_{Lo} subcutaneous rabbit, rat 200-360 mg kg^{-1} (4).

Teratogenicity and reproductive effects
Intraperitoneal injection of 78.5 mg kg^{-1}, four times at 3 hr intervals, on days 10-14 pregnancy. Reduced foetal weight but no gross malformations, maternal or foetal death reported (5).

Metabolism and pharmacokinetics
Deaminated by rat liver monoamine oxidase A and B (6).

Genotoxicity

Stimulates chromatin transcription (7).

HeLa cells cytotoxicity 170 mg l^{-1} (8).

Mouse liver unscheduled DNA synthesis at 170 mg l^{-1} (9).

Mouse ascites tumour DNA inhibition 880 mg l^{-1} (9).

Mouse liver DNA inhibition 1700 mg l^{-1} (9).

References

1. Schenk, B. L. *Anaerobic Digestion (Proc. Int. Symp.)* 1988, 459-464, Pergamon Press, Oxford
2. *Arch. Env. Contam. Toxicol.* 1985, **14**, 111
3. *Comp. Residue* 1920, **83**, 481
4. *Z. Exp. Pathol. Ther.* 1915, **17**, 59
5. Manen, C. A; et al. *Teratology* 1983, **28**, 237-242
6. Yu, P. H. *J. Pharm. Pharmacol.* 1989, **41**, 205-281
7. Pierce, D. A; et al. *Biochemistry* 1978, **17**, 102
8. *J. Cellular Physiology* 1971, **78**, 217
9. *Acta Medica Okayama* 1979, **33**, 149

B220   1,4-Butanediol diglycidyl ether

CAS Registry No. 2425-79-8

Synonyms 1,4-bis(2,3 epoxypropoxy) butane; 2,2'-[1, 4-butanediylbis(oxymethylene)bisoxirane]

Mol. Formula $C_{10}H_{18}O_4$ **Mol. Wt.** 202.25

Uses Binding and transfer agent for fibres. Crosslinking agent for epoxy resins. Used in the manufacture of adhesives and protective coatings.

Physical properties

B. Pt. 260°C; **Flash point** >112°C; **Specific gravity** d_4^{20} 1.100.

Occupational exposure

Supply classification toxic.

Risk phrases

Harmful by inhalation and in contact with skin – Irritating to eyes and skin – May cause sensitisation by skin contact (R20/21, R36/38, R43)

Safety phrases

In case of contact with eyes, rinse immediately with plenty of water and seek medical advice – After contact with skin, wash immediately with plenty of water – Wear suitable gloves and eye/face protection (S26, S28, S37/39)

Mammalian and avian toxicity

Acute data
LD$_{50}$ oral rat 2980 mg kg^{-1}(1).
LD$_{50}$ dermal rat 1130 mg kg^{-1} (1).

Sensitisation
Skin sensitiser guinea pigs, concentrations ≥0.5% cause cross-reaction in 95% of animals tested (2).
Guinea pig skin sensitivity maximisation test positive (3).
Sensitiser to skin patch tests for allergens in humans (4).

Genotoxicity
Salmonella typhimurium TA1535, TA98, TA100 with and without metabolic activation positive (5).
Escherichia coli PQ37 SOS chromotest with and without metabolic activation positive (6).

Any other comments
Human health effects and experimental toxicology reviewed (7).

References
1. Cornish, H. H; et al. *Arch. Environ. Health* 1959, **20**, 390-398
2. Clemmensen, S. *Drug. Chem. Toxicol.* 1984, **7**(6), 527-540
3. Thorgeirsson, A. *Acta Derm. Venereol.* 1978, **58**(11), 329-331
4. Jolanki, R; et al. *Contact Dermatitis* 1987, **16**, 87-92
5. Carter, D. A; et al. *Mutat. Res.* 1986, **172**, 105-138
6. Van der Hude, W; et al. *Mutat. Res.* 1990, **231**(2), 205-218
7. *ECETOC Technical Report No. 30(4)* 1991, European Chemical Industry Ecology & Toxicology Centre, B 1160 Brussels

B221   1,2-Butanediol

CH$_3$CH$_2$CH(OH)CH$_2$OH

CAS Registry No. 584-03-2
Synonyms 1,2-butylene glycol
Mol. Formula C$_4$H$_{10}$O$_2$ **Mol. Wt.** 90.12
Uses Polymerisation agent. Antimicrobiol. Pharmaceutical preparations.

Physical properties

M. Pt. 48-55°C; **B. Pt.** 193.5-195°C; **Flash point** 90°C; **Specific gravity** d$_4^{20}$ 1.0024; **Volatility** v. den. 3.1.

Solubility
Organic solvent: ethanol, acetone

Environmental fate

Degradation studies
Nonbiodegradable/qualified (1).

Mammalian and avian toxicity

Acute data
LD_{50} oral mouse 3720 mg kg^{-1} (2).
LD_{50} oral rat 16 g kg^{-1} (3).

Any other comments
Toxicity reviewed (4).
1,4-butanediol isomer is toxic to fish (5).
Human health effects and experimental toxicology reviewed (6).

References

1. *MITI Report* 1984, Ministry of International Trade and Industry, Tokyo
2. *Toxicol. Appl. Pharmacol.* 1979, **49**, 385
3. *Prehled Prumyslove Toxikol Org. Latky* 1986, 207
4. *BIBRA Toxicity Profile* 1989, British Industrial Biological Research Association, Farnham
5. *The Toxicity of 3400 Chemicals to Fish* 1987, EPA 560/6-89-001, Washington, DC
6. *ECETOC Technical Report No. 30(4)* 1991, European Chemical Industry Ecology and Toxicology Centre, B-1160 Brussels

B222   1,3-Butanediol

CH₃CH₂OHCH₂CH₂OH

CAS Registry No. 107-88-0
Synonyms 1,3-butylene glycol; 1,3-dihydroxybutane; β-butylene glycol; methylmethylene glycol
Mol. Formula $C_4H_{10}O_2$ **Mol. Wt.** 90.12
Uses Intermediate in the manufacture of polyester. Plasticiser. Humectant for cellophane, tobacco. In the preparation of some cosmetic and pharmaceuticals.

Physical properties

B. Pt. 203-204°C; **Flash point** 121°C; **Specific gravity** d_{20}^{20}; **Volatility** v.p. 0.6 mmHg at 20°C; v. den. 3.1.

Solubility
Organic solvent: acetone, methyl ethyl ketone, ethanol, dibutyl phthalate, castor oil

Environmental fate

Degradation studies
Nonbiodegradable/qualified (1).

Mammalian and avian toxicity

Acute data
LD_{50} oral guinea pig, mouse, rat 11-23 g kg^{-1} (2-4).
LD_{50} subcutaneous mouse, rat 16-20 g kg^{-1} (5).

Carcinogenicity and long-term effects
In 2 yr study test, 1-10% exposure to rats via food, and 0.5-3% exposure to dogs via food caused no discernable toxic efects (6).

Teratogenicity and reproductive effects
Gavage ♀ rat during organogenesis 0, 4236 and 7060 mg kg^{-1} day^{-1}. Maternal sedation observed at 4236 and 7060 mg kg^{-1}. Food consumption and maternal body weights were unaffected. A dose dependent decrease in offspring birth weights was observed (7).

Irritancy
Dermal rabbit, guinea pig, miniature pig, human (48 hr) (dose unspecified) open and closed patch tests, no irritation in any species except humans and only in closed patch test (8).

Any other adverse effects
Systemic effects include coughing, headache, pharyngitis, dizziness, nausea, and dysponea. Gastrointestinal irritation and diarrhoea may occur after exposure to high concentrations. Toxicity characterised by central nervous system depression (9).

Any other comments
The 1,4 butanediol isomer is toxic to fish (9).
Experimental toxicology and human health effects reviewed (9,11-13).
Physiological effects, chronic and acute toxicity and metabolism reviewed (14).

References
1. *MITI Report* 1984, Ministry of International Trade and Industry, Tokyo
2. Smyth, H. F; et al. *Arch. Ind. Hyg. Occup. Med.* 1951, **4**, 119
3. *J. Am. Pharmacol. Assoc.* 1956, **45**, 669
4. *J. Ind. Hyg. Toxicol.* 1941, **23**, 259
5. *Raw Material Data Handbook, Organic Solvents* 1974, **1**, 14
6. L. Scala, R. A; et al. *Toxicol. Appl. Pharmacol.* 1967, **10**, 160-164
7. Mankes, R. F; et al. *J. Am. Coll. Toxicol.* 1986, **5**(4), 189-196
8. Motoyoshi, K. et al. *Cosmet. Toiletries* 1984, **99**(10), 83-91
9. *Information Profiles on Potential Occupational Hazards* 1982, 1-208, Centre for Chemical Hazard Assessment, Syracuse, NY
10. *The Toxicity of 3400 Chemicals to Fish* 1987, EPA560/6-87-002 PB-200-275, Washington, DC
11. *J. Am. Coll. Toxicol.* 1985, **4**(5), 223
12. *BIBRA Toxicity Profile* 1990, British Industrial Biological Research Association, Carshalton
13. *ECETOC Technical Report No. 30(4)* 1991, European Chemical Industry Ecology and Toxicology Centre, B-1160 Brussels
14. Smyth, R. A; et al. *Am. Chem Soc. Monograph* **114**, 300-327

B223   1,4-Butanediol

HO(CH2)4OH

CAS Registry No. 110-63-4
Synonyms 1,4-dihydroxybutane; tetramethylene glycol; 1, 4-butylene glycol;
1,4-tetramethylene glycol
Mol. Formula $C_4H_{10}O_2$ **Mol. Wt.** 90.12
Uses Solvent. Chemical intermediate. Wood preservative.

Physical properties
M. Pt. 16°C; **B. Pt.** 230°C; **Flash point** >110°C; **Specific gravity** 1.017.

Ecotoxicity

Effects to non-target species
LC_{50} (duration unspecified) *Salientia sp.* tadpole <10,000 mg l^{-1} (1).

Environmental fate

Degradation studies
98.7% COD, 40 mg COD g dry inoculum^{-1} hr^{-1} with substance as sole carbon source (2).
Biodegradable (3).

Mammalian and avian toxicity

Acute data
LD_{50} oral guinea pig, rat, mouse, rabbit 1200-2500 mg kg^{-1} (4).
LD_{50} intraperitoneal rat 1370 mg kg^{-1} (5).
LD_{Lo} intraperitoneal mouse 500 mg kg^{-1} (6).

Carcinogenicity and long-term effects
National Toxicology Program has approved 1,4-butanediol for toxicology study (7).

Any other comments
Experimental toxicology and human health effects reviewed (8).

References
1. Nishuichi, Y. *Suisan Zoshoku* 1984, **32**, 115-119, (Jap.) (*Chem. Abstr.* **105**, 147635f)
2. Pitter, P. *Water Res.* 1976, **10**, 231
3. *MITI Report* 1984, Ministry of International Trade and Industry, Tokyo
4. *Gig. Sanit.* 1968, **33**, 41
5. *Toxicol. Appl. Biochem.* 1973, **25**, 461
6. *Summary Tables of Biological Tests* National Research Council Chemical Biological
   Coordination Centre 1951, **3**, 363
7. *National Toxicology Program Research and Testing Division* 1992, NIEHS, Research
   Triangle Park, NC 27709
8. *ECETOC Technical Report No. 30(4)* 1991, European Chemical Industry Ecology and
   Toxicology Centre, B-1160 Brussels

B224   2,3-Butanediol

CH3CH(OH)CH(OH)CH3

CAS Registry No. 513-85-9
Synonyms 2,3-butylene glycol; 2,3-dihydroxybutane; dimethylene glycol
Mol. Formula $C_4H_{10}O_2$ **Mol. Wt.** 90.12
Uses Solvent. Intermediate in organic synthesis.

Physical properties

M. Pt. 25°C; **B. Pt.** 183-184°C; **Flash point** 85°C; **Specific gravity** 0.995.

Environmental fate

Degradation studies
Nonbiodegradable/qualified (1).

Mammalian and avian toxicity

Acute data
LD_{50} oral mouse 5460 mg kg^{-1} (2).

Any other adverse effects

In vitro 10-day Albino Wistar rat embryo 2 day incubation, effects on embryonic protein, DNA, somite development, gross morphology and viability negative (2).

References

1.   *MITI Report* 1984, Ministry of International Trade and Industry, Tokyo
2.   *Toxicol. Appl. Pharmacol.* 1979, **49**, 385
3.   Priscott, P. K. *Biochem. Pharmacol.* 1985, **34**(4), 529-532

B225   1,4-Butanediol dimethanesulfonate

CH3SO2O(CH2)4OSO2CH3

CAS Registry No. 55-98-1
Synonyms Busulfan; Myleran; 1,4-bis(methanesulfonoxy)butane; bisulfane; 1,4-dimesyloxybutane; Sulfabutin; Myeloleukon
Mol. Formula $C_6H_{14}O_6S_2$ **Mol. Wt.** 246.30
Uses Treatment for chronic myeloid leukemia.

Physical properties

M. Pt. 116-117°C.

Solubility

Organic solvent: ethanol, acetone

Ecotoxicity

Effects to non-target species
LD_{50} oral redwing blackbird 56 mg kg^{-1} (1).

Mammalian and avian toxicity

Acute data
LD_{50} oral rat 1860 μg kg^{-1} (2).
LD_{50} intraperitoneal rat 18 mg kg^{-1} (3).
LD_{Lo} intravenous dog, monkey 8 mg kg^{-1} (4).

Carcinogenicity and long-term effects
Sufficient evidence for carcinogenicity to humans, limited evidence for carcinogenicity to animals, IARC classification group 1 (5).
Leukaemia patients who had been treated with Myleran developed cytological abnormalities and some developed carcinomas, effects were not dose related, although the cases were confirmed to those patients who had received no radiation and no other cytotoxic agent (6,7).

Teratogenicity and reproductive effects
Extreme intra-uterine arrest of growth, congenital anomalies of the eyes, palate, thyroid and ovaries, and disseminated cytomegaly reported in an infant whose mother received chemotherapy (4-6 mg day^{-1}) during pregnancy. The infant died at 10 wk (8).
Atypical cervical cytology reported after proloned therapy (9).
In mice, non-teratogenic low doses (5 or 10 mg kg^{-1}) given orally on day 13 of pregnancy reduced testis and ovary weight, and reduced fertility and reproductive performance of offspring (10).

Metabolism and pharmacokinetics
Largely excreted in the urine as sulfur containing metabolites (11).
It is readily absorbed from the gastrointestinal tract and rapidly disappears from the blood (11,12).

Genotoxicity
Salmonella typhimuriom TA100, TA1535 with and without metabolic activation positive (13).
Salmonella typhimurium TA98 with metabolic activation positive; without metabolic activation negative (14,15).
Human peripheral blood lymphocytes increased frequencies of sister chromatid exchange and chromosomal aberrations (5,14).
Drosophila melanogaster induced sex-linked recessive lethal mutations (14).
In vivo mouse induced dominant lethal mutations and increased frequency of chromosomal aberrations and micronuclei in bone-marrow cells (16).
Induced DNA damage but not mutation; covalent binding to DNA, RNA and protein in mice treated *in vivo* (17).

Any other adverse effects
Side effects at high dosage include leukopenia, thrombocytopenia and haemorrhagic symptoms, bone marrow depression which may not become apparent until several months after the start of treatment (11).

Any other comments
Experimental toxicology and human health effects reviewed (18).

References

1. *Arch. Environ. Contam. Toxicol.* 1983, **12**, 355
2. *Arzneimittel-Forschung* 1970, **20**, 1461
3. *Biochem. Pharmacol.* 1958, **1**, 39
4. *Cancer Chemotherapy Reports* 1965, **2**, 203
5. *IARC Monograph* 1987, **Suppl. 7**, 137-139
6. *IARC Monograph* 1974, **4**, 247-252
7. Stott, H; et al. *Br. Med. J.* 1977, (ii), 1513-1517
8. Diamond, I; et al. *Pediatrics* 1960, **25**, 85-90
9. Gureli, N; et al. *Obstet. Gynecol.* 1963, **21**(4), 466-470
10. Ooshima, Y. T; et al. *Teratology* 1984, **30**(1), 14A
11. *Martindale The Extra Pharamacopaeia* 29th ed., 1989, The Pharmaceutical Press, London
12. Feet, P. W; et al. *J. Pharm. Sci.* 1973, **62**, 1007
13. Mortelmans, K; et al. *Environ. Mol. Mutagen.* 1986, **8**(Suppl. 7), 1
14. *IARC Monograph* 1987, **Suppl. 6**, 129
15. Pak, K; et al. *Urol. Res.* 1979, **7**, 119
16. Bruce, W. R; et al. *Can. J. Genet. Cytol.* 1979, **21**, 319
17. Seino, Y; et al. *Cancer Res.* 1978, **38**, 2148
18. *ECETOC Technical Report No. 30(4)* 1991, European Chemical Industry Ecology and Toxicology Centre, B-1160 Brussels

B226   1,4-Butanesultone

CAS Registry No. 1633-83-6

Synonyms 4-hydroxy-1-butanesulfonic acid; δ-sultone; 1,4-butylene sulfone; δ-valerosultone; 1, 2-oxathiane-2,2-dioxide

Mol. Formula $C_4H_8O_3S$ **Mol. Wt.** 136.17

Uses Alkylating agent.

Occurrence Contaminant in water samples (India) (1).

Physical properties

M. Pt. 12.5-14.5°C; **B. Pt.** $_4$ 134-136°C; **Flash point** >110°C; **Specific gravity** 1.331.

Mammalian and avian toxicity

Acute data

LD_{50} oral rat 500 mg kg^{-1} (1).
LD_{50} subcutaneous rat 350 mg kg^{-1} (1).
LD_{50} intravenous rat 270 mg kg^{-1} (1).
LD_{50} intraperitoneal mouse 138 mg kg^{-1} (2).

Carcinogenicity and long-term effects

Subcutaneous rat (26 wk) 10, 15 or 30 mg kg^{-1} weak incidence of sarcomas at injection site (2).

Subcutaneous ICR/Ha Swiss mice (42 wk) 1680 mg kg^{-1} intermittently induced sarcomas at injection site (3).

Genotoxicity

Salmonella typhimurium TA98, TA100 with and without metabolic activation positive. *Escherichia coli* DNA repair positive (4).
Saccharomyces ceresvisiae D3 gene conversion and mitotic recombination positive (5).
Salmonella typhimurium TA1530, TA1535, TA1538 with and without metabolic activation positive (5).
Hamster kidney and hamster embryo oncogenic transformation positive (6).

Any other comments

Experimental toxicology and human health effects reviewed (7).

References

1. Kaul, B. L; et al. *Sci Cult.* 1972, **38**, 284
2. Druckrey, Z. *Krebsforschung* 1970, **75**(1), 69-84
3. Van Duuren, B. L; et al. *J. Natl. Cancer Inst.* 1974, **53**(3), 695-700
4. Rosenkranz, H. S; et al. *J. Natl. Cancer Inst.* 1979, **62**(4), 873-892
5. Simmon, V. F. *J. Natl. Cancer Inst.* 1979, **62**(4), 893-918
6. Pienta, R. J; et al. *Prev. Detect. Cancer Proc. Int. Symp.* 1976, **1**(2), 1993-2004
7. *ECETOC Technical Report No. 30(4)* 1991, European Chemical Industry Ecology and Toxicology Centre, B-1160 Brussels

B227   Butanoate

$$H_3C(CH_2)_2CO_2CH(CCl_3)P(O)(OCH_3)_2$$

CAS Registry No. 126-22-7
Synonyms butanoic acid, 2,2,2-trichloro- dimethoxyphosphinyl)ethyl ether; butilchlorofos; dimethyl(2,2,2-trichloro-1-hydroxyethyl- (phosphonate)butyrate; Tribufon
Mol. Formula $C_8H_{14}Cl_3O_5P$ **Mol. Wt.** 327.53
Uses Insecticide.

Physical properties

B. Pt. $_{0.5}$ 129°C.

Ecotoxicity

Effects to non-target species
LD_{50} oral chicken 210 mg kg^{-1} (1).

Environmental fate

Degradation studies
Eutrophic carp ponds were treated with 400-1000 µg butanoate. $t_{1/2}$ 46-108 hr (1).

Mammalian and avian toxicity

Acute data
LD_{50} oral guinea pig, mouse, rat 760-1100 mg kg^{-1} (2,3).
LD_{50} dermal dog 3080 mg kg^{-1} (2).

Metabolism and pharmacokinetics
Intraperitoneal ♂ mouse 200 and 400 mg kg^{-1} labelled with ^{14}C. Radioactivity detected in lung, kidney, testes and liver. $t_{1/2}$ liver 2 hr (4).

Genotoxicity
In vivo intraperitoneal mouse 200 and 400 mg kg^{-1} DNA methylation positive (4).

Legislation
Limited under EC Directive on Drinking Water Quality 80/778/EEC Pesticide: maximum admissible concentration 0.1 µg l^{-1} (5).

Any other comments
Toxicity reviewed (6).

References
1. Grahl, K; et al. *Acta Hydrochem. Hydrobiol.* 1981, **9**(2), 147-161
2. *Z. fuer Immunitactsforschung, Experimentelle und Klinische Immunologie* 1979, **25**, 512
3. *Agricultural Research Service* 1966, **20**(6), USDA Information Memorandum
4. Dedek, W; et al. *Pesticide Biochem. Physiology* 1976, **6**(2), 101-110
5. *EC Directive Relating to the Quality of Water Intended for Human Consumption* 1982, 80/778/EEC, Office for Official Publications of the European Communities, 2 rue Mercier, L-2985, Luxembourg
6. Knowles, C. O; *J. Agric. Food Chem.* 1966, **14**, 566

B228   Butanoic acid

$$CH_3(CH_2)_2CO_2H$$

CAS Registry No. 107-92-6
Synonyms butyric acid; ethylacetic acid; 1-propanecarboxylic acid; propylformic acid
Mol. Formula $C_4H_8O_2$ **Mol. Wt.** 88.11
Uses Manufacture of esters. Artificial flavouring ingredient in liqueurs, soda water syrups and confectionery. Varnishes. Decalcifier of hides.
Occurrence Ester content in butter 4.5%. Contaminant in surface and groundwater (1,2). Detected in effluent discharges from sewage treatment and landfill sites (1,3).

Physical properties

M. Pt. –7.9°C; **B. Pt.** 163.5°C; **Flash point** 77°C (closed cup); **Specific gravity** d_{20}^{20} 0.9590; **Partition coefficient** log P_{ow} 0.79; **Volatility** v.p. 0.433 mmHg at 20°C; v. den. 3.04.

Solubility
Water: ≥100 mg ml^{-1} at 19°C. Organic solvent: acetone, dimethyl sulfoxide, ethanol

Occupational exposure

UN No. 2820; **HAZCHEM Code 2R; Conveyance classification** corrosive substance; **Supply classification** corrosive.

Risk phrases
Causes burns (R34)

Safety phrases
In case of contact with eyes, rinse immediately with plenty of water and seek medical advice – Wear suitable protective clothing (S26, S36)

Ecotoxicity

Fish toxicity
LC_{50} (24 hr) bluegill sunfish 200 mg l^{-1} (4).

Invertebrate toxicity
EC_{50} (48 hr) *Daphnia magna* 61 mg l^{-1} (4).
Cell multiplication inhibition test *Pseudomonas putida* 875 mg l^{-1}, *Microcystis aeruginosa* 318 mg l^{-1}, *Scenedesmus quadricauda* 2600 mg l^{-1}, *Entosiphon sulcutam* 26 mg l^{-1} (5,6).

Environmental fate

Anaerobic effects
Methanogenic microbes raised on acetate completely removed butanoic acid after a 3 day lag period at a rate of 284 mg l^{-1} day^{-1}, initial concentration not provided (7).

Degradation studies
BOD_5 initial concentration 5 ppm 76% reduction in dissolved oxygen in fresh water and 72% reduction in sea water (8).
ThOD (6, 12, 18 and 24 hr) 17-27% with activated sludge seed at an initial concentration of 500 ppm (9).
Screening study using sewage seed theoretical BOD_5 72-78% reduction in dissolved oxygen content BOD_{20} 92-99% (10).
BOD_5 0.34 standard dilution; 0.90 standard dilution sewage sludge, 1.16 standard dilution acclimated sewage sludge (11).
Biodegradable (12).
ThOD (5hr) 72% with activated sludge, initial concentration 100 mg l^{-1} (13).

Absorption
Activated carbon adsorbability 0.119 g kg^{-1} carbon; 60% reduction; influent 1000 mg l^{-1}, effluent 405 mg l^{-1} (14).
Adsorbs to kalonite or montmorillonite clay. After (48 hr) at 22°C 14-20% was adsorbed, after (144 hr) increased to 24-31% (15).

Mammalian and avian toxicity

Acute data
LD_{50} oral rat 2940 mg kg^{-1} (16).
LD_{Lo} oral mouse 500 mg kg^{-1} (17).
LD_{50} dermal rabbit 530 mg kg^{-1} (18).

LD_{50} intraperitoneal mouse 3180 mg kg^{-1} (19).
LD_{50} intravenous mouse 800 mg kg^{-1} (20).

Metabolism and pharmacokinetics

Butanoic acid is a normal substrate of the mammalian fatty acid metabolic pathway. Butyric acid is also produced as a metabolic product by colonic bacteria (21). Intraperitoneal injection mice butanoic acid or arginine salts to investigate possible antitumour therapies. Rapid appearance in blood, longest retention in liver, $t_{1/2}$ <5 min. In humans rapid initial excretion $t_{1/2}$ 30 sec followed by a slower second elimination phase $t_{1/2}$ 13 min (22).

Irritancy

Dermal rabbit (duration unspecified) 500 mg caused moderate irritation (18). Dermal rabbit (24 hr) 10 mg caused severe irritation and 250 μg instilled into rabbit eye caused severe irritation (23).

Genotoxicity

Low concentrations stopped reversibly the proliferation of chick embryo fibroblasts and human HeLa cells by inhibiting DNA synthesis. Extensive acetylation of histones was also observed (24).
Human lymphocytes P3HR- inhibited DNA synthesis and cell growth (25).

Any other adverse effects

Intrarectal butanoic acid 1-12% induced consistent and reproducible colitis in mice (21).

Any other comments

The gastrointestinal tract is desensitised from food allergies by oral ingestion of a compound containing butanoic acid or its salts (26).
There is evidence that sodium butyrate inhibits tumour colony formation in adapted cell lines (27).
Experimental toxicology and human health effects reviewed (28,29).
All reasonable efforts have been taken to find information on isomers of this compound, but no relevant data are available.

References

1. Murtauch, J. J; et al. *J. Water Pollut. Control Fed.* 1965, **37**, 410-415
2. Goerlilz, D. F; et al. *ASTM Spec. Tech. Publ.* 1979, **686**, 256-257
3. Dunlop, W. J; et al. *Org. Pollut. Contrib. to Ground Water from Landfill* 1976, 96-110, USEPA-600/9-76-004
4. Dowden, B. F; et al. *J. Water Pollut. Control Fed.* 1965, **37**(9), 1310
5. Bringmann, G; et al. *Gwtf-Wasser/Abwasser* 1976, **117**(9), 119
6. Bringmann, G; et al. *Water Res.* 1980, **14**, 231-241
7. Schafer, E. W; et al. *Arch. Environ. Contam. Toxicol.* 1983, **12**, 355-382
8. Chou, W. L; et al. *Biotech. Bioeng. Symp.* 1979, **8**, 391-414
9. Takemoto, S; et al. *Suishitsu Odaku Kenkyu* 1981, **4**, 80-90
10. Malaney, G. W; et al. *J. Water Pollut. Control Fed.* 1969, **41**, 18-33
11. Gaffney, G. W; et al. *J. Water Pollut. Control Fed.* 1961, **33**, 1169-1183
12. Verschueren, K. *Handbook of Environmental Data on Organic Chemicals* 2nd ed., 1983, Van Nostrand Reinhold, New York
13. *MITI Report* 1984, Ministry of International Trade and Industry, Tokyo
14. Urano, K; et al. *J. Haz. Mat.* 1986, **13**, 135-159

15. Guisti, D. M; et al. *J. Water Pollut. Control Fed.* 1974, **46**(5), 947-965
16. Hemphill, L; et al. *Proc. 18th Ind. Waste Contr.* 1964, **18**, 204-217
17. *Arch. Ind. Hyg. Occup. Med.* 1951, **4**, 119
18. *Toxicol. New Ind. Chem. Sci.* 1962, **4**, 19
19. *Union Carbide Data Sheet* 1968 (10th Apr), Union Carbide Corp., New York
20. *J. Pharm. Pharmacol.* 1969, **21**, 85
21. *Acta Pharmacol. Toxicol.* 1961, **18**, 141
22. McCafferty, D. M; et al. *Int. J. Tissue React.* 1989, **11**(4), 165-168
23. Daniel, P; et al. *Clin. Chem. Acta* 1989, **181**(3), 255-263
24. *Arch. Ind. Hyg. Occup. Med.* 1954, **10**, 61
25. Neesby, T. E. *US Patent, US 4721716 Cl. 514-251 A61K31/19* 1988
26. Hagopisn, H. K; et al. *Cell* 1977, **12**, 855-860
27. Scirenji, T; et al. *Hematol. Oncol.* 1984, **2**(4), 381-389
28. Flatlow, E; et al. *Cancer Invest.* 1989, **7**(5), 423-436
29. *ECETOC Technical Report No. 30(4)* 1991, European Industry Ecology and Toxicology Centre, B-1160 Brussels

B229   Butanoic acid, 3-methyl-

(CH3)2CHCH2CO2H

CAS Registry No. 503-74-2
Synonyms isovalerianic acid; isopropylacetic acid; delphinic acid; isovaleric acid
Mol. Formula $C_5H_{10}O_2$ **Mol. Wt.** 102.13
Uses Used in flavours and perfumes. Intermediate in organic synthesis.
Occurrence Natural substance in hop oil, tobacco and valeriana.

Physical properties

M. Pt. –37°C; **B. Pt.** 175-177°C; **Specific gravity** d_4^{20} 0.931; **Partition coefficient** log P_{ow} 0.93 (calculated).

Solubility
Organic solvent: ethanol, chloroform, diethyl ether

Ecotoxicity

Bioaccumulation
Enters surface waters in the effluents from sewage works; typical concentrations range from 0.8 to 574 µg l^{-1} in the effluent, but because the logP_{ow} for the acid is only 0.93 it is improbable that aquatic fauna would be exposed to appreciable concentrations (1).

Mammalian and avian toxicity

Acute data
LD$_{50}$ oral rat 2000 mg kg^{-1} (2).
LD$_{50}$ dermal rabbit 310 mg kg^{-1} (2).
LD$_{50}$ intravenous mouse 1120 mg kg^{-1} (3).

Irritancy

Dermal rabbit (24 hr) 500 mg caused moderate irritation and 940 µg instilled into rabbit eye caused mild irritation (2,4).

Any other comments

Experimental toxicology and human health effects reviewed (5).
All reasonable efforts have been taken to find information on isomers of this compound, but no relevant data are available.

References

1. Murtaugh, J. J; et al. *J. Water Pollut. Control Fed.* 1965, **37**, 410-415
2. *Union Carbide Safety Data Sheets* 31 Jan 1972, Union Carbine Corp., NY
3. Ore, L; et al. *Acta Pharmacol. Toxicol.* 1961, **18**, 141
4. Opdyke, D. C. J. *Food Cosmet. Toxicol.* 1979, **17**, 841
5. *ECETOC Technical Report No. 30(4)* 1991, European Chemical Industry Ecology and Toxicology Centre, B-1160 Brussels

B230   Butanoic acid, butyl ester

$CH_3(CH_2)_2CO_2(CH_2)_3CH_3$

CAS Registry No. 109-21-7
Synonyms *n*-butyl-*n*-butyrate
Mol. Formula $C_8H_{16}O_2$ **Mol. Wt.** 144.22
Uses Flavouring.

Physical properties

B. Pt. 164-165°C; **Flash point** 49°C; **Specific gravity** 0.871.

Solubility
Organic solvent: ethanol, diethyl ether

Ecotoxicity

Fish toxicity
LC_{50} (96 hr) fathead minnow 11.6 mg l^{-1} (1).

Mammalian and avian toxicity

Acute data
LD_{50} oral rabbit 9520 mg kg^{-1} (2).
LD_{50} intraperitoneal rat 2300 mg kg^{-1} (3).
LD_{50} intraperitoneal mouse 8900 mg kg^{-1} (3).

Irritancy
Dermal rabbit (24 hr) 500 mg caused moderate irritation (3).

Any other comments

Undergoes hydrolysis to *n*-butanol and butanoic acid.
Experimental toxicology and human health effects reviewed (4).

References

1. Curtis, M. W; et al. *J. Hydrol.* 1981, **51**, 359
2. *Ind. Med. Surgery* 1972, **41**, 31
3. Opdyke, D. C. J. *Food Cosmet. Toxicol.* 1979, **17**, 521
4. *ECETOC Technical Report No. 30(4)* 1991, European Chemical Industry Ecology and Toxicology Centre, B-1160 Brussels

B231   1-Butanol

CH3(CH2)3OH

CAS Registry No. 71-36-3
Synonyms *n*-butyl alcohol; butryic alcohol; propylcarbinol; propylmethanol;
1-hydroxybutane; butylhydroxide
Mol. Formula $C_4H_{10}O$ **Mol. Wt.** 74.12
Uses Solvent for fats, waxes, resins, gums and varnishes. Solvent in urea
formaldehydes foams. Used in the manufacture of butyl acetate, butyl acrylate,
detergents, rayon, and lacquers. Dilutent for brake fluids. Extractant in
pharmaceutical synthesis of antibiotics, vitamins and hormones.
Occurrence Detected in industrial effluents, surface water (1,2).

Physical properties

M. Pt. −88.9°C; **B. Pt.** 117.5°C; **Flash point** 29°C (open cup); **Specific gravity** d_4^{20}
0.810; **Partition coefficient** log P_{ow} 0.84; **Volatility** v.p 7.02 mmHg at 25°C; v. den.
2.55 at 25°C.

Solubility

Water: 77 g l^{-1}. Organic solvent: acetone, benzene, diethyl ether, ethanol

Occupational exposure

UK Short-term limit 50 ppm (150 mg m^{-3}); **UN No.** 1120; **HAZCHEM Code** 3ME;
Conveyance classification flammable liquid; **Supply classification** flammable and
harmful.

Risk phrases

Flammable – Harmful by inhalation (R10, R20)

Safety phrases

Keep away from sources of ignition – No Smoking (S16)

Ecotoxicity

Fish toxicity
LD_{100} (24 hr) creek chub 1400 mg l^{-1} in Detroit River water (3).
LC_{50} (96 hr) fathead minnow 1910 mg l^{-1} 18-22°C in fresh water (4).

EC_{50} (96 hr) fathead minnow 1510 mg l^{-1} (33 day old) water hardness 47.7 mg l^{-1} calcium carbonate, temperature 24.7°C, pH 7.64, dissolved oxygen 6.3 mg l^{-1} (5).

Invertebrate toxicity

Cell multiplication inhibition test, *Pseudomonas putida* 650 mg l^{-1}, *Scenedesmus quadricauda* 875 mg l^{-1}, *Entosiphon sulcatum* 55 mg l^{-1} (6).
EC_{50} (30 min) *Photobacterium phosphoreum* 2187 mg l^{-1} Microtox test (7).
EC_{50} (48 hr) *Daphnia magna* 1980 mg l^{-1} (8).

Effects to non-target species

LD_{50} oral redwing blackbird 2500 mg kg^{-1} (9).

Environmental fate

Nitrification inhibition

Inhibition of NH_3 oxidation, pure culture, 50% at 8200 mg l^{-1} (10).

Degradation studies

Volatilisation from river water, $t_{1/2}$ estimated at 3-9 hr (11).
Biodegradable (12).
38 process wastewaters and 37 organic substances identified in the effluent from a petrochemical complex were subjected to the activated sludge degradability test, sludge was acclimated to the wastewater and organic substances. Water in the test container was sampled during aeration at 0 hr and 24 hr. After 1 day of acclimation, 100 mg l^{-1} 1-butanol resulted in chemical oxygen demand of 82% and 93% total organic carbon (13).

Abiotic removal

Sunlit urban atmosphere, $t_{1/2}$ estimated at 5 hr (14).
Poses an indirect hazard for the aquatic environment because it is readily biodegradable which may lead to oxygen depletion (15).

Absorption

Activated carbon absorbs 0.0107 g g^{-1} carbon, 53.4% reduction; influent 1000 mg l^{-1}, effluent 466 mg l^{-1} (16).

Mammalian and avian toxicity

Acute data

LD_{50} oral rat 4360 mg kg^{-1} (17).
LD_{50} oral rabbit 4250 mg kg^{-1} (18).
LC_{50} (4 hr) inhalation rat 8000 ppm (18).
LD_{50} intraperitoneal rat 1122 mg kg^{-1} (19).
LD_{50} intravenous rat 310 mg kg^{-1} (19).

Sub-acute data

Inhalation mouse (7 hr) 1650 ppm, no adverse effects reported (20).

Teratogenicity and reproductive effects

Inhalation rat (7 day) 0-8000 ppm, teratology assessment during gestation 1-19 days. Highest concentration produced maternal toxicity. Slight increase in skeletal malformations observed at 8000 ppm (21).

Metabolism and pharmacokinetics

Metabolised via butyraldehyde to butanoic acid which is eliminated mainly as carbon dioxide (22).

Volunteers exposed to concentrations of 100 and 200 ppm for 2 hr developed blood concentrations that were below 1 mg l^{-1}, whether at rest or during exercise. Exposure to an air concentration of 50 ppm for 2 hr resulted in blood levels < 0.08 mg l^{-1} (23).

Single oral dose rats (24 hr) 83% coverted to carbon dioxide, 4% excreted in the urine and 13% was retained in tissues (23).

Irritancy

Dermal rabbit (24 hr) 500 mg caused moderate irritation and 750 μg instilled in rabbit eye caused severe irritation (24).

Genotoxicity

Salmonella tymphimurium TA98, TA100, TA1535, TA1537 with metabolic activation negative (26).

In vitro Chinese hamster ovary cells sister chromatid exchange negative (27).

In vitro micronuclei assay V79 cell line (1 hr) chromosomal aberrations negative (28).

Any other adverse effects

Alcoholic intoxicant and narcotic. Can cause central nervous system depression, with headache, dizziness and drowsiness (14).

Following exposure degeneration of liver and kidneys and effects to lungs and blood has been reported (29).

Any other comments

Experimental toxicology and human health effects reviewed (16,30-32).
Biological studies of human exposure to organic solvents reported (33).

References

1. Yasuhara, A; et al. *Environ. Sci. Technol.* 1981, **15**, 570-573
2. Keith, L. H. *Sci. Total Environ.* 1974, **3**, 87-102
3. Gillette, L. A; et al. *Sewage. Ind. Wastes* 1957, **29**(6), 695-711
4. Mattson, V. R; et al. *Acute Toxicity of Selected Organic Compounds to Fathead Minnows* 1976, EPA 600/3-76-097
5. Verschuren, K. *Handbook of Environmental Data on Organic Chemicals* 2nd ed., 1983, Van Nostrand Reinhold, New York
6. Bringmann, G; et al. *Water Res.* 1980, **14**, 231-241
7. Kaiser, K. L. E; *Water Pollut. Res. J. Canada* 1991, **26**(3), 361-431
8. Kuhn, R; et al. *Water Res.* 1989, **23**(4), 495-499
9. Schafer, W; et al. *Arch. Environ. Contam. Toxicol.* 1983, **12**(3), 355-382
10. Richardson, M. L. *Nitrification Inhibition in the Treatment of Sewage* 1985, RSC, London
11. Lyman, W. T; et al. *Handbook of Chemical Property Estimation Methods Environmental Behaviour of Organic Compounds* 1982, McGraw-Hill, New York
12. *MITI Report* 1984, Ministry of International Trade and Industry, Tokyo
13. Matsui, S; et al. *Water Sci. Technol.* 1989, **20**(10), 201-210
14. Dilling, W. L; et al. *Environ. Sci. Tech.* 1976, **10**, 351-356
15. *Environmental Health Criteria No. 65: Butanols* 1987, World Health Organisation, Geneva
16. Guisti, D. M; et al. *J. Water Pollut. Control Fed.* 1974, **46**(5), 947-965
17. Smyth; et al. *Arch. Ind. Hyg. Occup. Med.* 1951, **4**, 119
18. *Raw Materials Data Handbook – Organic Solvents* 1974, **1**, Nat. Assoc. Print. Int. Res. Inst.

19. *Environ. Health Perspect.* 1985, **61**, 321
20. Patty, F. A. (Ed.) *Industrial Hygiene and Toxicology* 1967, **2**, Interscience Publishers, New York
21. Nelson, B. K; et al. *Fund. Appl. Toxicol.* 1989, **12**(3), 469-479
22. *Chemical Safety Data Sheets* 1989, **1**, RSC, London
23. Baselt, R. C. *Biological Monitoring Methods for Industrial Chemicals* 2nd ed., 1988, 51, PSG Publishing, Littleton, MA
24. Marhold, J. V. *Sbornik Vysledku Toxicologickeho Vysetreni Latek A Pripravku* 1972, Prague
25. Fregert, S; et al. *Acta Derm.-Venerol.* 1969, **49**, (5), 493-497
26. McCann, J; et al. *Proc. Natl. Acad. Sci. US* 1975, **72**, 5735-5739
27. Ube, G. *Mutat. Res.* 1977, **55**, 211-213
28. Lasne, C; et al. *Mutat. Res.* 1984, **130**(4), 273-282
29. *Martindale: The Extra Pharmacopoeia* 1989, 29th ed., The Pharmaceutical Press, London
30. Pellizzarri, E. D; et al. *Bull. Environ. Contam. Toxicol.* 1982, **28**, 322-328
31. Scientific Basis for Swedish Occupational Standards III *Arbete och Hals* 1982, **24**, 1-125
32. *ECETOC Technical Report No. 30(4)* 1991, European Chemical Industry Ecology and Toxicology Centre, B-1160 Brussels
33. Milling Pederson, L. *Pharmacol. Toxicol. (Copenhagen)* 1987, **61**(Suppl. 3), 1-55

B232   2-Butanol

CH3CH2CH(OH)CH3

CAS Registry No. 78-92-2

Synonyms *sec*-butylalcohol; butylene hydrate; 2-hydroxybutane; methyl ethyl carbinol

Mol. Formula $C_4H_{10}O_2$ **Mol. Wt.** 90.12

Uses Used in flavours and perfumes. Dyestuff synthesis. Paint removers and industrial cleaners.

Occurrence Contaminant in drinking and surface waters. Discharged from industrial effluent. Industries include mechanical products, petroleum refining and paint and inks (1-3).

Found as a volatile component in a diverse array of foodstuffs (4).

Physical properties

M. Pt. –89 to –108°C; **B. Pt.** 99.5°C; **Flash point** –114.7°C; **Specific gravity** d_4^{20} 0.808; **Volatility** v.p. 12 mmHg at 20°C.

Solubility
Water: 125 g l^{-1}. Organic solvent: ethanol, diethyl ether, acetone, benzene

Occupational exposure

UK Long-term limit 100 ppm (300 mg m^{-3}); **UK Short-term limit** 150 ppm (450 mg m^{-3}); **UN No.** 1120; **HAZCHEM Code** 3ⱲE; **Conveyance classification** flammable liquid; **Supply classification** flammable and harmful.

Risk phrases
Flammable – Harmful by inhalation (R10, R20)

Safety phrases
Keep away from sources of ignition – No Smoking (S16)

Ecotoxicity

Fish toxicity
LC_{50} (24 hr) goldfish 4300 mg l^{-1} (5).

Invertebrate toxicity
Cell multiplication inhibition, *Pseudomonas putida* 500 mg l^{-1}, *Scenedesmus quadricauda* 95 mg l^{-1}, *Entosiphon sulcatum* 1280 mg l^{-1}, *Microcystis aeruginosa* 312 mg l^{-1} (6,7).

Environmental fate

Anaerobic effects
100% degradation obtained after 14 days lag by acetate-acclimated cultures (8).
Long-term study using anaerobic upflow filters and acetate enriched cultures, 93% utilisation rate obtained after 52 days of operation (9).

Degradation studies
Adapted activated sludge with 2-butanol as sole carbon source: 98.5% COD, 55 mg COD g^{-1} dry inoculum^{-1} hr^{-1} (10).
Biodegradable (11).
$ThOD_5$ sewage seed or activated sludge 82-98% (12-14).

Mammalian and avian toxicity

Acute data
LD_{50} oral rat 6480 mg kg^{-1} (15).
LC_{Lo} (4 hr) inhalation rat 16,000 ppm (15).
LD_{50} intraperitoneal guinea pig, hamster, rat 1060-1200 mg kg^{-1} (16).
LD_{50} intravenous rat 138 mg kg^{-1} (16).
LD_{50} intraperitoneal rabbit 277 mg kg^{-1} (16).

Irritancy
16 mg instilled into rabbit eye caused irritation (17).

Genotoxicity
Saccharomyces cerevisiae cytotoxic at concentrations of 750 mg tube^{-1} (18).
In vivo rat bone marrow chromosomal aberrations and polyploidy positive (19).

Any other comments
Experimental toxicology and human health effects reviewed (20).

References
1. Lucas, S. V. *Anal. Org. Drink Water Conc. Adv. Treat. Conc.* 1984, **1**, USEPA-600/1-84-020a
2. *Invent. Chem. Sub. Ident. Gt. Lake Ecosyst.* 1983, Gt. Lakes Water Quality Board, Windsor, Canada
3. Shackleford, W. M; et al. *Anal. Chem. Acta* 1983, **146**, 15-27

4. Coleman, E. C; et al. *J. Agric. Food Chem.* 1981, **29**, 42-48
5. *Shell Chemie* 1 Jan 1975, Gravenhage
6. Bringmann, G; et al. *Water Res.* 1980, **14**, 231-241
7. Bringmann, G; et al. *Gwf.-Wasser Abwasser* 1976, **117**(9)
8. Yonezawa, Y; et al. *Chemosphere* 1979, **8**, 139-142
9. Pitter, P. *Water Res.* 1976, **10**, 231-235
10. Chow, W. L; et al. *Bioeng. Symp.* 1979, **8**, 391-414
11. *MITI Report* 1984, Ministry of International Trade and Industry, Tokyo
12. Bridie, A. L; et al. *Water Res.* 1979, **13**, 627-630
13. Wagner, R. *Vom Wagner* 1974, **42**, 271-305
14. Piffer, P. *Water Res.* 1976, **10**, 231-235
15. Smyth, H. F; et al. *Arch. Ind. Hyg. Occup. Med.* 1954, **10**, 61
16. *Env. Health Perspect.* 1985, **61**, 321
17. AMA *Arch. Ind. Hyg. Occup. Med.* 1954, **10**, 61
18. *Hereditas* 1947, **33**, 457
19. Barilyak, I. R; et al. *Tsitol. Genet.* 1988, **22**(2), 49-52, (Russ.) (*Chem. Abstr.* **109** 68620b)
20. *ECETOC Technical Report No. 30(4)* 1991, European Chemical Industry Ecology and Toxicology Centre, B-1160 Brussels

B233   *tert*-Butanol

$(CH_3)_3COH$

CAS Registry No. 75-65-0
Synonyms *tert*-butyl alcohol; *tert*-butyl hydroxide; 1,1-dimethylethanol; 2-methylpropan-2-ol; trimethylcarbinol
Mol. Formula $C_4H_{10}O$ **Mol. Wt.** 74.12
Uses Denaturant for ethanol. Manufacture of flotation agents. Agent in flavourings and perfumes. Used as a solvent in paint removers. Octane booster in gasoline.
Occurrence Contaminant in drinking water (1).

Physical properties

M. Pt. 25.3°C; **B. Pt.** 82.8°C; **Flash point** 10°C (closed cup); **Specific gravity** d_4^{20} 0.7887; **Partition coefficient** log P_{ow} 0.35; **Volatility** v.p. 40 mmHg at 24.5°C; v. den. 2.55.

Solubility
Water: miscible. Organic solvent: ethanol, acetone, diethyl ether, benzene

Occupational exposure
US TLV (TWA) 100 ppm (300 mg m^{-3}); **US TLV (STEL)** 150 ppm (455 mg m^{-3}); **UK Long-term limit** 100 ppm (300 mg m^{-3}); **UK Short-term limit** 150 ppm (450 mg m^{-3}); **UN No.** 1120; **HAZCHEM Code** 3ME;
Conveyance classification flammable liquid; **Supply classification** highly flammable and harmful.

Risk phrases
Highly flammable – Harmful by inhalation (R11, R20)

Safety phrases
Keep container in a well ventilated place – Keep away from sources of ignition – No
Smoking (S9, S16)

Ecotoxicity

Fish toxicity
LC_{50} (7 day) guppy 3500 ppm (2).
LD_{100} (24 hr) creek chub 6000 mg l^{-1} Detroit river water (3).

Invertebrate toxicity
Chlorella pyrenoidosa toxic effects at 24 g l^{-1} (4).

Bioaccumulation
Non-accumulative or low accumulative (5).

Effects to non-target species
LD_{Lo} parenteral frog 12 g kg^{-1} peripheral nerve and sensation effects (6).

Environmental fate

Nitrification inhibition
Nitrification inhibition limit concentration 39 g l^{-1} (7).

Degradation studies
Degradation rates for subsurface soils with *tert*-butyl alcohol were in the range
0.1-0.3 mg l^{-1} day^{-1} (8).
Degradation in anoxic groundwater systems was enhanced by presence of nitrate at
pH ≥7 (9).

Absorption
Adsorbability 0.059 g g^{-1} carbon; 29.5% reduction in influent sludge (10).

Mammalian and avian toxicity

Acute data
LD_{50} oral rabbit, rat 3500 mg kg^{-1} (11,12).
LD_{50} intravenous, intraperitoneal mouse 930-1530 mg kg^{-1} (13,14).

Teratogenicity and reproductive effects
Oral mouse (6-20 day gestation) 0.5, 0.75 and 1% wt. $vol.^{-1}$ produced development
delay in postparturition physiological and psychomotor performance scores.
Significant postnatal maternal nutritional and behavioural factors affecting lactation
or nesting behaviour were also evident (15).
TC_{Lo} (1-19 day gestation) inhalation rat 2000 ppm 7 hr day^{-1} observed foetotoxicity.
TC_{Lo} (1-19 day pregnant) inhalation rat 3500 ppm 7 hr day^{-1} specific developmental
abnormalities in musculoskeletal system (16).
Oral CBA/J and C57BL/6J mice (6-18 day gestation) 0.8 g kg^{-1} induced a significant
increase in resorptions per litter. No significant abnormalities reported (17).

Metabolism and pharmacokinetics
Oral rat single dose 2 g kg^{-1} rapid adsorption into blood, <1% excreted via urine.
Intraperitoneal rat 0.84 g kg^{-1}, blood $t_{1/2}$ 13 hr (18).

Irritancy

Exposure via inhalation or contact (species unspecified). Conjunctivitis and dermatitis reported (17).

Genotoxicity

Saccharomyces cerevisiae cytoxic at concentrations of 3.7 µg tube^{-1} (19).
In vitro mouse lymphona L5178Y tk+/tk- without metabolic activation equivocal (20).
L5178Y tk+/tk- mouse lymphoma cell forward mutation assay with and without metabolic activation negative (21).

Any other adverse effects to man

Central nervous system depressant, can cause liver and kidney damage. Ingestion effects include headache, vomiting, fatigue, ataxia and unconsciousness. Aspiration may cause respiratory failure and death (16).

Any other comments

Crystalline form hygroscopic. Physical properties, fire and health hazards, toxicology and safe handling recommendations reviewed (22-24).
Experimental toxicology and human health effects reviewed (25).
Crystalline form hygroscopic.

References

1. Coleman, W. E; et al. *Anol. Id. Organ. Sub. Water* 1976, 305-327, Keith L. (Ed.), Ann Arbor Sci
2. Konemann, W. H. *Quantitative Structure-Activity Relationships for Kinetic and Toxicity of Aquatic Pollutants and Their Substances in Fish* 1979, Utrecht
3. Gillette, L. A; et al. *Sewage Ind. Wastes* 1952, **24**, 1397-1401
4. Jones, H. *Environmental Control in the Organic and Petrochemical Industries* 1971, Noyes Data Corp, Park Ridge, NJ
5. *MITI Report* 1984, Ministry of International Trade and Industry, Tokyo
6. *Arch. Int. Pharmacodyn. Ther.* 1935, **50**, 296
7. Richardson M. L. (Ed). *Nitrification Inhibition in the Treatment of Sewage* 1985, RSC, London
8. Hickman, G. T; *Environ. Sci. Technol.* 1989, **23**(5), 535-532
9. Wilson, W. G; et al. *Proc. Ind. Waste Conf.* 42nd ed., 1987, 197-205
10. Gunster, D. M; et al. *J. Water Pollut. Control Fed.* 1974, **46**(5), 947-965
11. *Science* 1952, **116**, 663
12. *Ind. Med. Surg.* 1972, **41**, 31
13. *Arch. Int. Pharmacol. Ther.* 1962, **135**, 330
14. *Shell. Chem. Co.* 1962, -, 2
15. Daniel, M. A. *J. Pharmacol. Exp. Ther.* 1982, **222**, 294
16. Nelson, B. K; et al. *Fund. Appl. Toxicol.* 1989, **12**, 469-479
17. Faulkner, T. P; et al. *Life Sci.* 1989, **45**(21), 1989-1995
18. *Drinking Water & Health* 1977, **1**, 697, NRC, National Academy Press, Washington DC
19. Occupational Health Services *Pestline* 1991, **2**, Van Nostrand Reinhold, New York
20. *Hereditas* 1947, **33**, 457, Toernqvist Book Dealers, Sweden
21. McGregor, D. B; et al. *Environ. Mol. Mutagen.* 1988, **11**(1), 91-118
22. Saha, A. K; et al. *Environ. Ecol.* 1987, **5**(2), 321-324
23. *Ind. Hyg. Bull.* 1959, 57-97, Ind. Hyg. Dep. Shell Chemical, New York
24. Von Oettingen, W. F. *Publ. Health Bull.* 1943, **281**
25. *J. Am. Coll. Toxicol.* 1989, **8**(4), 627-641

B234   2-Butanone

CH₃CH₂COCH₃

CAS Registry No. 78-93-3
Synonyms methyl ethyl ketone; butan-2-one; ethyl methyl ketone; MEK; methyl acetone
Mol. Formula C_4H_8O **Mol. Wt.** 72.11
Uses Solvent. Paint stripper and cleaning fluid. Manufacture of cements and adhesives. Intermediate in inorganic synthesis. Extraction solvent in food processing.
Occurrence Contaminant from vehicle exhaust and water samples from sewage and leachate from PVC pipe cements (1).
Natural component of some foods (2).

Physical properties

M. Pt. –87°C; **B. Pt.** 80°C; **Flash point** –1°C; **Specific gravity** d_4^{20} 0.8054; **Partition coefficient** log P_{ow} 0.26; **Volatility** v.p. 71.2 mmHg at 20°C;v. den. 2.42.

Solubility
Organic solvent: acetone, benzene, diethyl ether, ethanol

Occupational exposure

US TLV (TWA) 200 ppm (590 mg m^{-3}); **US TLV (STEL)** 300 ppm (885 mg m^{-3}); **UK Long-term limit** 200 ppm (590 mg m^{-3}); **UK Short-term limit** 300 ppm (885 mg m^{-3}); **UN No.** 1193; **HAZCHEM Code** 2☒E; **Conveyance classification** flammable liquid; **Supply classification** highly flammable and harmful.

Risk phrases
Highly flammable – Irritating to eyes and respiratory system (R11, R36/37)

Safety phrases
Keep container in a well ventilated place – Keep away from sources of ignition – No Smoking – Avoid contact with eyes – Take precautionary measures against static discharges (S9, S16, S25, S33)

Ecotoxicity

Fish toxicity
LC$_{50}$ (24-96 hr) mosquito fish, goldfish, fathead minnow 5600-3220 mg l^{-1} (3-5).

Invertebrate toxicity
Cell multiplication inhibition test, *Pseudomonas putida* 1150 mg l^{-1}, *Entosiphon sulcatum* 190 mg l^{-1}, *Microcystis aeruginosa* 10 mg l^{-1}, *Scendesmus quadricauda* 4300 mg l^{-1} (6,7).
EC$_{50}$ (30 min) *Photobacterium phosphoreum* 3373 mg l^{-1} Microtox test (8).
EC$_{50}$ (48 hr) *Daphnia magna* >520 mg l^{-1} (9).

Environmental fate

Anaerobic effects
Degraded in anaerobic systems, but time required for acclimating degrading organisms was 1 wk (10).

Degradation studies

Readily degraded within 5-10 day in aerobic systems using activated sludge, sewage seed or inoculum from polluted surface water (11-15).

BOD_5 1.52-1.92 at 20°C, standard dilution technique and normal sewage seed (16).

Readily oxidised by microorganisms in activated sludge following selection and adaption, 80% removed within 24 hr. Metabolism in unacclimated sludges is slow (17).

Removed from soil by evaporation (18).

Abiotic removal

Reacts photochemically in the atmosphere to give acetaldehyde, $t_{1/2}$ 2 day (19).

Mammalian and avian toxicity

Acute data

LD_{50} oral rat 2740 mg kg^{-1} (20).

LC_{50} (2 hr) inhalation mouse 40 g m^{-3} (21).

LD_{50} dermal rabbit 13 g kg^{-1} (22).

LD_{Lo} intraperitoneal guinea pig 2000 mg kg^{-1} (23).

Teratogenicity and reproductive effects

Inhalation rat (6-15 day gestation) 0, 400, 1000 or 3000 ppm 7 hr day^{-1} maternal toxicity (decreased weight gain and increased water consumption). Animals exposed to 3000 ppm suffered slight foetal toxicity and skeletal aberrations. The authors conclude results do not indicate embryotoxic or teratogenic response in rats (24).

Inhalation rat (6-15 day gestation) 1000 or 3000 ppm 7 hr day^{-1} was embryotoxic, foetoxocity and potentially teratogenic. Effects reported include retardation of foetal development and low incidence of acaudia, imperforate anus and brachyngothia (25).

Pregnant rat (23 hr day^{-1}) inhalation 500-1500 ppm induced a concentration dependent increase of intrauterine mortatility, reduced body growth and a delay in the maturation of cerebellar cortex. Concentration dependent embryotoxic and foetotoxic effects but no teratogenic effects reported (26).

Swiss CD-1 mice (7 hr day) inhalation 0-3000 ppm on days 6-15 gestation, the 3000 ppm concentration group had reduced mean foetal body weight, malformation observed included cleft-palate, fused ribs, missing vertebrae, syndactyly and misaligned sternebrae (27).

Metabolism and pharmacokinetics

Readily absorbed through intact skin. Excretion occurs via the lung (28).

Measurable quantities (2-13 µg l^{-1}) were detected in expired air of adult humans 3 min after dermal exposure to 100 ml of 2-butanone applied to 90 cm^{-2} of skin (29). In workers exposed to 0.3 mg l^{-1} 2-butanone, lung uptake averaged 1.05 mg min^{-1}, blood concentration averaged 2.6 mg l^{-1}. 2-butanone did not persist in tissues. Urinary excretion averaged 487 mg l^{-1} (30).

Irritancy

Dermal rabbit (24 hr) 400 mg caused mild irritation and 80 mg instilled into rabbit eye caused irritation (22).

Genotoxicity

Salmonella typhimurium TA98, TA100 with and without metabolic activation, assays for mitotic gene conversion and chromosomal aberrations negative (22,31).

Saccharomyces cerevisiae D61.M strongly induced aneuploidy but not recombination and point mutation (32).

Any other comments

Pharmacokinetics, experimental toxicology, exposure and human health effects reviewed (33-36).

References

1. Richardson, M. L. *Compendium of Toxicological Ecotoxicological Data on Chemicals formed by GC-MS in Water Samples* 1985, Thames Water Authority, London
2. Lande, S. S; et al. *Investigation of Selected Potential Environmental Contaminants: Ketonic Solvents* 1976, 102, USEPA 560/2-76-003
3. *Shell Chemie – Shell Industrie Chemicalien* 1 Jan 1975, Gravenhage
4. Jones, H. R. *Environmental Control in the Organic and Petrochemical Industries* 1971, Noyes Data Corp., Park Ridge, NJ
5. Brooke, L. T; et al. *Acute Toxicities of Organic Chemicals to Fathead Minnows* 1984, 100, Lake Superior Centre for Environmental Studies, Univ. of Wisconsin
6. Bringmann, G; et al. *Water Res.* 1980, **14**, 231-241
7. Bringmann, G; et al. *Gwf-Wasser/Abwasser* 1976, **117**(9)
8. Kaiser, K. L. E. *Water Pollut. Res. J. Canada* 1991, **26**(3), 3661-431
9. LeBlanc, G. A. *Bull. Environ. Contam. Toxicol.* 1980, **24**, 684-691
10. Chou, W. L; et al. *Biotech. Bioeng. Symp.* 1979, **8**, 391-414
11. Dojlido, J. R. *Investigation of Biodgradability and Toxicity of Organic Compounds* 1979, USEPA 600/2-79-163
12. Price, K. S; et al. *J. Water Pollut. Control Fed.* 1974, **46**, 63-77
13. Bridie, A. L; et al. *Water Res.* 1979, **13**, 627-630
14. Dose, M; et al. *Trib. Cebedeau* 1975, **28**, 3-11
15. Heukelekian, H; et al. *J. Water Pollut. Control Fed.* 1955, **29**, 1040-1053
16. Veschueren, K. *Handbook of Experimental Data of Organic Chemicals* 2nd ed., 1983, Van Nostrand Reinhold, New York
17. *Methyl Ethyl Ketone III: Exposure Aspects* 1981, USEPA 68-01-6147
18. Mill, T; et al. *Laboratory Protocols for Evaluating the Fate of Organic Chemicals in Air and Water* 1982, 255, USEPA600/3-82-022
19. Cox, R. A; et al. *Environ. Sci. Technol.* 1981, **15**, 587-592
20. *Toxicol. App. Pharmacol.* 1971, **19**, 699
21. Izmerov, N. F; et al. *Toxicometric Parameters of Industrial Toxic Chemicals under Single Exposure* 1982, 83, Moscow
22. *Chemical Safety Data Sheets* 1988, **1**, RSC, London
23. *Food Cosmet. Toxicol.* 1977, **15**, 611
24. Deacon, M. M; et al. *Toxicol. Appl. Pharmacol.* 1981, **59**, 620
25. Schwetz, B. A; et al. *Toxicol. Appl. Pharmacol.* 1974, **28**(3), 227-232
26. Stoltenburg-Didinger, G; et al. *Neurotoxicol. Teratol.* 1990, **12**(6), 585-589
27. Schwelz, B. A; et al. *Fundam. Appl. Toxicol.* 1991, **16**(4) 742-748
28. Brooks, T. M; et al. *Mutagenesis* 1988, **3**(3), 227-232
29. Wurster, J. *J. Pharm. Sci.* 1965, **54**, 554
31. Perbellini, L; et al. *Intern. Arch. Occup. Environ. Health* 1984, **54**(1), 73-81
31. Brooks, T. M; et al. *Mutagenesis* 1988, **3**(3), 227-232
32. *Mutat. Res.* 1985, **149**, 339
33. *ECETOC Technical Report No. 30(4)* 1991, European Chemical Industry Ecology and Toxicology Centre, B-1160 Brussels
34. *Dangerous Prop. Ind. Mater. Rep.* 1990, **10**(3), 53-65
35. *Gov. Rep. Announce. Index US* 1990, **90**(7), 014356, EPA 600/8-89/093
36. Kessler, W; et al. *Biol. Mased Methods Cancer Risk ASSESS.* 1989, 123-139, NATO ASI Ser., Germany

B235   2-Butanone peroxide

$CH_3COCH(CH_3)OOCH(CH_3)COCH_3$

CAS Registry No. 1338-23-4
Synonyms methyl ethyl ketone peroxide; Lupersol; MEK peroxide; methyl ethyl ketone hydroperoxide; Quickset extra; Thermacure; Sprayset MEKP
Mol. Formula $C_8H_{16}O_8$ **Mol. Wt.** 240.21
Uses Cross-linking agent for thermosetting resins.
Occurrence Contaminant in air samples collected in the vicinity of a fibreglassing plant (1).

Physical properties

B. Pt. 118°C with (decomp.); **Flash point** 82°C; **Specific gravity** 1.170.

Occupational exposure

US TLV (TWA) Ceiling limit 0.2 ppm (1.5 mg m^{-3}); **UK Short-term limit** 0.2 ppm (1.5 mg m^{-3}); **UN No.** 3068 (≤40% in diisobutyl-nylonate, ≤8.2% available oxygen); 2563 (≤50% in solution with >10% available oxygen); 2550 (maximum concentration 50% with ≤10% available oxygen); 2127 (maximum 60%).

Mammalian and avian toxicity

Acute data
LD$_{50}$ oral mouse, rat 470-484 mg kg^{-1} (2,3).
LC$_{50}$ (4 hr) inhalation mouse, rat 170- 200 ppm (2).
LD$_{50}$ intraperitoneal rat 65 mg kg^{-1} (2).

Irritancy
Dermal rabbit (duration unspecified) 500 mg and 3 mg instilled into rabbit eye. Results for observed irritation equivocal (2).

Genotoxicity
Salmonella typhimurium TA98, TA100, TA1535, TA1537 with and without metabolic activation negative (4).

Any other comments

Usually supplied as a 50% solution in dimethyl phthalate (5).
Strong oxidising agent and contact with organic materials can create a fire hazard.
Experimental toxicology and human health effects reviewed (6).

References

1. *Res. Needs Methyl Ethyl Ketone Peroxide* 1989, USEPA-ECAO 68-C8-0004/122, Cincinnati, OH
2. *Am. Ind. Hyg. Ass. J.* 1958, **19**, 205
3. *J. Am. Med. Ass.* 1957, **165**, 201
4. Mortelmans, K; et al. *Environ. Mutagen.* 1986, **8**(Suppl. 7), 1-119
5. Lenga, R. E. *Sigma-Aldrich Library of Chemical Safety Data* 2nd ed., 1988, Sigma Aldrich, Milwaukee, WI
6. ECETOC Technical Report No. 30(4) 1991, European Chemical Industry Ecology and Toxicology Centre, B-1160 Brussels

B236   1-Butene

$$CH_3CH_2CH=CH_2$$

CAS Registry No. 25167-67-3
Synonyms butylene; α-butylene; butylene-1; ethylethylene
Mol. Formula C_4H_8 **Mol. Wt.** 56.11
Uses Butenes are weak anaesthetics. Used in petroleum and chemical industry for polymer synthesis.
Occurrence In diesel engine exhaust gas and refinery gases.

Physical properties

M. Pt. -185.35°C; **B. Pt.** -6.3°C; **Flash point** -80°C (closed cup); **Specific gravity** d_4^{20} 0.5951; **Volatility** v.p. 3480 mmHg at 21°C; v. den. 1.94.

Solubility

Organic solvent: diethyl ether, ethanol, benzene

Occupational exposure

UN No. 1012; **HAZCHEM Code** 2WE; **Conveyance classification** flammable gas; **Supply classification** highly flammable.

Risk phrases

Extremely flammable liquefied gas (R13)

Safety phrases

Keep container in a well ventilated place – Keep away from sources of ignition – No Smoking – Take precautionary measures against static discharges (S9, S16, S33)

Mammalian and avian toxicity

Acute data

Exposure to 20% concentration in mice resulted in respiratory failure within 2 hr and mice exposed to 40% concentration died within 10 min (1).

Metabolism and pharmacokinetics

Metabolised slowly to 1-hydroxy metabolite (1).

Genotoxicity

Butene (isomer unspecified) in the vapour phase, tested with *Salmonella typhimurium* TA97, TA98, TA100 with and without metabolic activation negative (2).

Any other adverse effects to man

Six workers assigned to removal of paraffinic oil residue including butylene had systemic effects of narcosis including generalised convulsions, nystagmus, finger tremors and gastritis (3).

Any other adverse effects

♂ Swiss Webster mice (5 min) 0.4 to 18 ppm exposure to photochemical oxidant mixture of 1-butene at 0, 1, 2, 3 and 4 hr of reaction period. Severe irritation of upper respiratory tract reported (4).

Narcotic, asphyxiant at high concentrations can cause respiratory failure (5).

Any other comments

Toxicology and anaesthetic potency of butylene discussed (5).
Working conditions and health status of workers involved in butylene production are reviewed (6,7).
Liquid form can cause burns and frostbite (1).
Fire hazard. Explodes in mixtures with oxygen.

References

1. Browning, E. *Toxicity and Metabolism of Industrial Solvents* 2nd ed. 1987, **1**, 362-367, Elsevier, Amsterdam
2. Hughes, T. J; et al. *Gov. Rep. Announc. Index* 1984, **84**(10), 65
3. Cali, V; et al. *Folia Medica* 1954, **37**(10), 827-836, (Ital.) *English translation available*
4. Kane, L. E. *Arch. Env. Health* 1978, **33**(5), 224-250
5. Riggs, L. K. *Proc. Soc. Exp. Biol. Med.* 1925, **22**, 269-270
6. Heuss, J. M. *Environ. Sci. Technol.* 1968, **2**(12), 1109-1116
7. Petko, L. E; et al. *Gig. Truka i Professional'nye Zabol.* 1966, **10**(11), 52-54

B237   *cis*-2-Butene

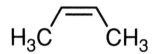

CAS Registry No. 590-18-1
Synonyms butene-2; β-butylene; butylene-2; dimethylethylene; pseudobutylene
Mol. Formula C_4H_8 **Mol. Wt.** 56.11
Uses In the production of gasolines, butadiene and other chemicals.
Occurrence Low environmental contaminant which occurs in diesel exhaust. Recovered from refining gases or by petroleum cracking.

Physical properties

M. Pt. –139.3°C; **B. Pt.** 3.7°C; **Flash point** –12°C; **Specific gravity** d_4^{20} 0.6213;
Partition coefficient log P_{ow} 2.33; **Volatility** v.p. 760 mmHg at 37°C; v. den. 1.9.

Solubility
Organic solvent: ethanol, diethyl ether, benzene

Mammalian and avian toxicity

Acute data
Exposure of mice to 13% concentration (duration unspecified) resulted in deep narcosis, exposure to 19% concentration was fatal (1).

Genotoxicity

Salmonella typhimurium TA97, TA98, TA100 butene (isomer unspecified) in vapour phase with and without metabolic activation negative (2).

Any other adverse effects

Inhalation ♂ Swiss Webster mice (5 min) 0.4-18 ppm at 0, 1, 2, 3 and 4 hr of reaction with nitrogen dioxide caused severe irritation to upper respiratory tract (3). Cardiac sensitiser (4).

Any other comments

Eye irritation of a range of hydrocarbons including *cis*-2-butene reviewed (5). Butenes are weak anaesthetics and asphyxiants at high concentrations. 2- Butene is more narcotic than the corresponding 1-isomer.

References

1. Van Oettingen, W. R. *Toxicity and Potential Dangers of Aliphatic and Aromatic Hydrocarbons* 1940, **256S**, Public Health Bulletin
2. Kane, L. E; et al. *Arch. Environ. Health* 1978, **33**(5), 244-250
3. Hughes, T. J; et al. *Gov. Rep. Announc. Index* 1984, **84**(10), 65
4. Krantz, J. C; et al. *J. Pharmacol. Exp. Ther.* 1948, **94**, 315
5. Heuss, J. M. *Environ. Sci. Technology* 1968, **2**(12), 1109-1116

B238   *trans*-2-Butene

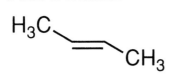

CAS Registry No. 624-64-6
Synonyms β-butylene; *sym*-dimethylethylene
Mol. Formula C_4H_8 **Mol. Wt.** 56.11
Uses In the production of gasolines. Intermediate in the synthesis of butadiene and other chemicals.
Occurrence In exhaust of diesel engines. Low environmental contaminant.

Physical properties

M. Pt. −105.5°C; **B. Pt.** 0.9°C; **Flash point** −6°C; **Specific gravity** d_4^{20} 0.6042; **Partition coefficient** log P_{ow} 2.31; **Volatility** v.p. 760 mmHg at 0.9 °C; v. den. 1.90.

Solubility

Organic solvent: ethanol, diethyl ether, benzene

Environmental fate

Degradation studies

Nocardia sp. are capable of growth using *trans*-2-butene as carbon source. Degradation occurs via crotonic acid (1).

Mammalian and avian toxicity

Acute data
Mice exposed to 13% concentrations (duration unspecified) suffered deep nacrosis, while 19% was fatal (2).

Metabolism and pharmacokinetics
In vitro conversion of simple prochiral and chiral alkenes into oxiranes by liver microsomes of treated rats, mice and humans (3).

Irritancy
Low mucous membrane irritant (2).

Genotoxicity

Salmonella typhimurium TA97, TA98, TA100 butene (isomer unspecified) in the vapour phase with and without metabolic activation negative (4).

Any other adverse effects

Cardiac sensitiser (5)

Any other comments

Eye irritation of hydrocarbons including *trans*-2-butene reviewed (6).
The more highly reactive trans-2-butene occurs at a much lower frequency in atmosphere than other comparable hydrocarbons (7).
Butenes are weak anaesthetics 2-butene is an asphyxiant and narcotic at high concentrations.

References

1. Van Ginkel, C. G; et al. *J. Gen. Microbiol.* 1987, **133**(7), 1713-1720
2. Von Oettingen, W. R. *Toxicity and Potential Dangers of Aliphatic and Aromatic Hydrocarbons* 1940, Public Health Bulletin No. 255
3. Wistruba, D; et al. *Chirality* 1989, **1**(2), 127-136
4. Hughes, T. J; et al. *Gov. Rep Announc. Index* 1984, **84**(10), 65
5. Krantz, J. C; et al. *J. Pharmacol. Exp. Ther.* 1948, **94**, 315
6. Heuss, J. M. *Environ. Sci Technology* 1968, **2**(12), 1109-1116
7. *Patty's Industrial Hygiene Toxicology* 3rd ed., 1981-1982, **2A, 2B, 2C**, 3302, Clayton, G. D. (Ed.), J. Wiley, New York

B239   (Z)-Butenedioic acid

CAS Registry No. 110-16-7
Synonyms maleic acid; toxilic acid; *cis*-butenedioic acid; *cis*-1,2-ethylenedicarboxylic acid; maleinic acid; malenic acid
Mol. Formula $C_4H_4O_4$ **Mol. Wt.** 116.07

Uses Manufacture of artificial resins. Retards rancidity of fats and oils. Used in the dyeing of wool, cotton, silk. In manufacture of plasticisers and polymers.
Occurrence Low level air pollutant (1).

Physical properties

M. Pt. 130.5°C; **B. Pt.** 135°C (decomp.); **Specific gravity** d_{20}^{20} 1.590; **Partition coefficient** log P_{ow} –0.79.

Solubility
Water: 788 g l^{-1} at 25°C. Organic solvent: acetone, benzene, methanol, ethanol, diethyl ether, chloroform, carbon tetrachloride

Ecotoxicity

Fish toxicity
LC$_{100}$ (24 hr) carp 130 mg l^{-1} (2).
LC$_{50}$ (24 hr) ide 106 mg l^{-1} (3).
LC$_{50}$ (24-48 hr) mosquito fish 240 ppm (4).
LC$_{50}$ (96 hr) fathead minnow 5 mg l^{-1} (5).

Invertebrate toxicity
EC$_{100}$ (24 hr) *Daphnia magna* 200 mg l^{-1} (3).

Bioaccumulation
Chlorella fenea var. *vacuolata* exposed to 50 μg l^{-1}, bioconcentration factor ranged 10-14 (6-8).
Ide melonate (3 day) exposure to median concentration of 31 μg l^{-1} at 20-25°C bioconcentration factor <10 (7,8).

Environmental fate

Carbonaceous inhibition
EC$_{10}$ (16-18 hr) *Pseudomonas putida* 1190 mg l^{-1} (3).

Degradation studies
ThOD (6, 12, 24 hr) 2.7, 4.5 and 2.7% respectively, using wastewater bench activated sludge (9).
Incubation (5 day, 23°C) 1 g l^{-1}, activated sludge and 50 μg l^{-1} maleic acid, volatilisation 26.9%; mineralisation 26.3% (CO_2); 41% sludge metabolite and non-extractible residues 96.8% (7,8).
ED$_{50}$ (4 hr) *Haematococcus pluvialis* flotow (80×10^6 cells l^{-1}) 125 mg l^{-1} (3).

Mammalian and avian toxicity

Acute data
LD$_{50}$ oral rat 708 mg kg^{-1} (10).
LD$_{50}$ oral mouse 2400 mg kg^{-1} (10).
LD$_{50}$ dermal rabbit 1560 mg kg^{-1} (11).

Metabolism and pharmacokinetics
♂ Wistar albino rats (7 day) 1 mg kg^{-1} unspecified route, 17.3% excreted via faeces and 7.7% via urine. Percentage of dose retained liver 0.45%, lungs 0.04% and carcass 6.7% (7,8).

Irritancy
Dermal rabbit (24 hr) 500 mg caused mild irritation and 100 mg instilled into rabbit eye caused severe irritation (11).

Any other adverse effects
Unspecified route rat 200 mg kg^{-1} hr^{-1} caused decrease in renal water clearance. Kidney nephron clearance controls 7.16%, drugged rats 4.03% (12,13).
Unspecified route rat (24 hr) 200 or 400 mg kg^{-1} caused proximal tubular necrosis in 40-90% of kidney (14).

Any other comments
Experimental toxicology and human health effects reviewed (15).

References

1. Kaplan, I. R; et al. *Environ. Sci. Technol.* 1987, **21**, 105-110
2. Nishiuchi, Y. *Suisan Zoshuku* 1975, **23**(3), 132-134, (Jap.) (*Chem. Abstr.* **85**, 117592r)
3. Knie, J; et al. *Dtsch. Gawaesserkd. Mitl.* 1983, **27**(3), 77-79, (Ger.) (*Chem. Abstr.* **99**, 170838p)
4. *Hazard. Chem. Data* 1984-5, **2**, Coastguard Dept of Transport, Washington, DC
5. Kemp, H. T; et al. *Water Poll. Control Ser. 09/73* 1973, **5**, USEPA, Washington, DC
6. Geyer, H; et al. *Chemosphere* 1981, **10**(11-12), 1307-1313
7. Freitag, D; et al. *Ecotoxicol. Environ. Sci.* 1982, **6**(1), 60-81
8. Freitag, D; et al. *Chemosphere* 1985, **14**(10), 1589-1616
9. Malaney, G. W; et al. *J. Water Pollut. Control Fed.* 1969, **41**(2), R18-R33
10. Burns, R. L; *J. Cell. Physiol.* 1976, **88**, 307-316
11. Industrial Bio Test Labs, Inc., Data Sheet 7-4, 1970
12. Boyland, E. *Biochem. J.* 1940, **34**, 1196
13. Brewer, E. D; et al *Am. J. Physiol.* 1983, **245**(3), F339-F344
14. Verani, R. R; et al. *Lab. Invest.* 1982, **46**, 79-88
15. *ECETOC Technical Report No. 30(4)* 1991, European Chemical Industry Ecology and Toxicology Centre, B-1160 Brussels

B240   *trans*-2-Butenoic acid

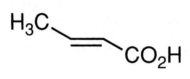

CAS Registry No. 107-93-7;003724-65-0
Synonyms Crotonic acid; 3-methylacrylic acid; α-butenoic acid; croton oil
Mol. Formula $C_4H_6O_2$ **Mol. Wt.** 86.09
Uses Manufacture of copolymers with vinyl acetate. Used in lacquers and paper sizing. Manufacture of softening agents for synthetic rubber. Medicinal chemistry.

Physical properties

M. Pt. 72°C; **B. Pt.** 185°C; **Flash point** 87.8°C (open cup); **Specific gravity** d_4^{15} 1.018; **Volatility** v.p. 0.19 mmHg at 20°C; v. den. 2.97.

Solubility

Water: 54.6 g l^{-1} at 20˚C. Organic solvent: ethanol, diethyl ether, acetone

Occupational exposure

UN No. 2823; HAZCHEM Code 2X; Conveyance classification corrosive substance.

Mammalian and avian toxicity

Acute data

LD$_{50}$ oral rat, mouse 1000-4800 mg kg^{-1} (1,2).
LD$_{50}$ dermal rabbit 600 mg kg^{-1} (3).
LD$_{50}$ intraperitoneal guinea pig, rat 60-100 mg kg^{-1} (4).
LD$_{50}$ subcutaneous mouse 3590 mg kg^{-1} (5).

Irritancy

Dermal rabbit (24 hr) 10 mg caused irritation (1).
Dermal mouse 0.5% or 20% croton oil, examination 1-24 hr and 48-96 hr periods showed marked swelling and focal epidermal ulceration of the ears (6).

Genotoxicity

Salmonella typhimurium TA1535, TA1537, TA1538, TA98, TA100 with and without metabolic activation negative (7).

Any other comments

Corrosive. Explosion risk.

References

1. *J. Ind. Hyg. Toxicol.* 1944, **26**, 269
2. *Biochem. J.* 1940, **34**, 1196
3. *Prehled Prumyslove Toxikol. Org. Latky* 1986, 309
4. Deichmann, W. B; et al. *Toxicology of Drugs and Chemicals* 1969, 190, Academic Press, London
5. *J. Pharm. Pharmacol.* 1969, **21**, 85
6. Kolde, G; et al. *J. Investig. Dermatol.* 1987, **89**(1), 19-23
7. Lijinsky, W; et al. *Teratogen. Carcinogen. Mutagen.* 1980, **1**(3), 259-267

B241   2-Buten-1-ol

$$H_3CCH=CHCH_2OH$$

CAS Registry No. 6117-91-5
Synonyms but-2-en-1-ol; crotyl alcohol; crotonyl alcohol; 1-hydroxy-2-butene; 3-methylallyl alcohol; Δ^2-1-butenol
Mol. Formula C_4H_8O Mol. Wt. 72.11
Uses Catalyst for polymerisation of ketones (1).
Shortening dough-mixing time (2).
Occurrence Natural occurrence in Victoria plums *Prunus domesticus*, lemon juice, colza oil and wood-alcohol oil (3-5).

Pollutant from gasoline exhaust (0.1-3.6 ppm).

Physical properties

M. Pt. < –30°C; **B. Pt.** 118-122°C; **Flash point** 33-37°C; **Specific gravity** d_4^{20} 0.8532; **Partition coefficient** log P_{ow} 0.54; (6).

Solubility
Water: 166 g l^{-1} at 20°C. Organic solvent: miscible ethanol

Abiotic removal
Wastewater treatment by reverse osmosis, 40°C, 600 psi. 18.3% rejection rate from a 0.72 g l^{-1} aqueous solution (7).

Mammalian and avian toxicity

Acute data
LD_{50} oral rat 930 mg kg^{-1} (8).
LC_{Lo} (4 hr) inhalation rat 2000 ppm (8).
LD_{50} dermal rabbit 1270 mg kg^{-1} (8).

Irritancy
Causes irritation, redness, blistering to skin and severe irritation, redness, corneal burns to eyes (8).

Genotoxicity

Salmonella typhimurium TA100 with and without metabolic activation positive (9,10).

Any other adverse effects

Ingestion can cause nausea, vomiting, symptoms of drunkenness. Inhalation systemic effects cough, tight chest and throat irritation. Exposure to high concentrations is destructive to mucous membranes, upper respiratory tract, skin, and eyes. Causes burning sensation, cough, wheeze, laryngitis, dyspnoea, nausea, vomiting and unconsciousness (11).

Any other comments

Metabolised by *Saccharomyces cerevisiae* 30% starved yeast in 6.8 g l^{-1} phosphoric acid, monopotassium salt solution consumed 1392 µg O_2 min^{-1} (12).

References

1. Kametani, Y; *Jpn. Kokai.* 76 96,899, 25 Aug 1976, (Jap.) (*Chem. Abstr.* **85**, 178941a)
2. Conn, J. F; et al. *US Pat.* 3,556,804, 19 Jan 1971
3. Ismail, H. M; et al. *J. Sci. Food Agric.* 1981, **32**(6), 613-619
4. Andre, E; et al. *Compt. Rend.* 1950, **231**, 872-874
5. Ruibin, S. I; et al. *J. Appl. Chem. USSR* 1933, **6**, 311-319
6. McCreery, M. J; et al. *Neuropharmacol.* 1978, **17**(7), 451-461
7. Duvel, W. A; et al. *J. Water Pollut. Control Fed.* 1975, **47**(1), 57-65
8. Smyth, H. F; et al. *Am. Ind. Hyg. Assoc. J.* 1962, **23**, 95-107
9. Lijinsky, W; et al. *Teratogen. Carcinogen. Mutagen.* 1980, **1**(3), 259-267
10. Lutz, D; et al. *Mutat. Res.* 1982, **93**(2), 305-315
11. Huhtanen, C. N; et al. *J. Food Prot.* 1985, **48**(7), 570-573
12. Maitra, P. K; et al. *Arch. Biocehm. Biophys.* 1967, **121**(1), 117-128

B242   *trans*-2-Buten-1-ol

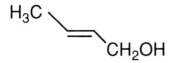

CAS Registry No. 504-61-0
Synonyms *trans*-but-2-en-1-ol; *E*-but-2-en-1-ol; *trans*-crotyl alcohol
Mol. Formula C_4H_8O **Mol. Wt.** 72.11
Uses Manufacture of epoxy resins. Electrochromic display solvents.
Occurrence Natural substance in cabbage *Brassica oleraceae* leaf (1,2).
In raspberry flavour (3).

Physical properties

B. Pt. 120-122°C; **Flash point** d_4^{20} 0.8454.

Mammalian and avian toxicity

Metabolism and pharmacokinetics
Subcutaneous rat 54 mg kg^{-1}, after 24 hr urine metabolites (via crotonaldehyde)
included 3-hydroxy-1-methylpropylmercapturic acid and 2-carboxy-1-
methylethylmercapturic acid (4).

References

1. Greenhalgh, J. R; et al. *New Phytol.* 1976, **77**(2), 391-398
2. Mitchell, N. D; et al. *New Phytol.* 1978, **81**(1), 189-200
3. Winter, M; et al. *Helv. Chim. Acta* 1962, **45**, 2212-2218
4. Gray, J. M; et al. *Xenobiotica* 1971, **1**(1), 55-67

B243   3-Buten-1-ol

CH2=CHCH2CH2OH

CAS Registry No. 627-27-0
Synonyms but-3-en-1-ol; allylcarbinol; Δ^3-1-butenol
Mol. Formula C_4H_8O **Mol. Wt.** 72.11
Uses Ethylene polymerisation catalyst.
Occurrence Occurs in *Aspalathus linearis* Rooibos tea (1).
Beef fat distillates (2).
Colza oil (3).

Physical properties

B. Pt. 112-114°C; **Flash point** 32°C; **Specific gravity** d_4^{20} 0.8424.

Solubility
Water: miscible. Organic solvent: miscible diethyl ether

Abiotic removal

Wastewater treatment, reverse osmosis, 40°C, 600 psi. From a 0.72 g l⁻¹ aqueous
solution, 28.3% of solute is rejected (4).

Any other adverse effects

IC$_{50}$ *Lactuia sativa* germination, 30°C 0.111 g l⁻¹ (5).
Induction of ethylene response in *Tabacum sp.* (6,7).

References

1. Habu, T; et al. *J. Agric. Food Chem.* 1985, **33**(2), 249-254
2. Ohnishi, S; et al. *J. Agric. Food Chem.* 1984, **32**(5), 987-992
3. Andre, E; et al. *Compt. Rend.* 1950, **231**, 872-874
4. Duvel, W. A; et al. *J. Water Pollut. Control. Fed.* 1975, **47**(1), 57-65
5. Reynolds, T; *Ann. Bot. (London)* 1977, **41**(173), 637-648
6. Sisler, E. C; et al. *Tobacco Sci.* 1978, **22**(99-105), 102-105
7. *Tobacco* 1980, (17), 60-63

B244   3-Buten-2-ol

CH₃C(OH)CH=CH₂

CAS Registry No. 598-32-3
Synonyms methyl vinyl carbinol; 1-methylpropenol; but-3-en-2-ol;
3-hydroxy-1-butene; 1-methylallyl alcohol; Δ³-2-butenol
Mol. Formula C$_4$H$_8$O   **Mol. Wt.** 72.11
Uses Vulcanising agent. Stabiliser for methylchloroform.
Occurrence Possible constituent of mushroom aroma [(±)isomer] (1).
Wood-spirit oil (2).

Physical properties

M. Pt. <-80°C (-100°C); **B. Pt.** 96-98°C; **Flash point** 16°C; **Specific gravity** d$_4^{20}$ 0.8318.

Mammalian and avian toxicity

Irritancy

Humans exposed to 50 ppm for 15 min suffered eye irritation. Exposure for 8 hr to 50
ppm caused irritation to both eyes and nose (3).

Genotoxicity

Salmonella typhimurium TA100 with and without metabolic activation weakly
positive (4).

References

1. Ney, K. H; et al. *Gordian* 1978, **78**(5), 144, 146, (Ger.) (*Chem. Abstr.* **89**, 174437a)
2. Pringsheim, H; et al. *Ber. Dtsch. Chem. Ges.* 1924, **57B**, 1561-1566, (Ger.) (*Chem. Abstr.* **19**, 2409)
3. Silverman, L; et al. *J. Ind. Hyg. Toxicol.* 1946, **28**, 262-266
4. Lutz, D; et al. *Mutat. Res.* 1982, **93**(2), 305-315

B245   Buthiobate

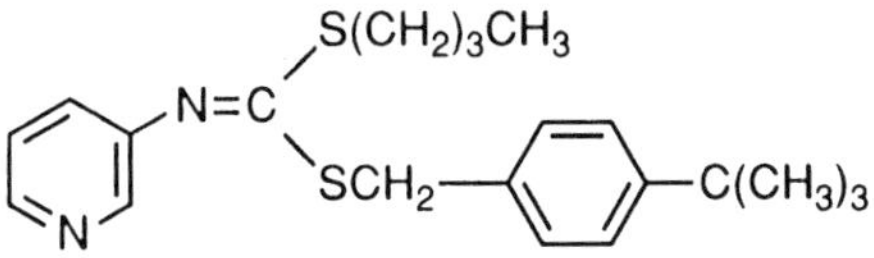

CAS Registry No. 51308-54-4

Synonyms butyl, 4-*tert*-butylbenzyl-*N*-(3- pyridyl)dithiocarbonim idate; butyl[4-(1,1-dimethylethyl)phenyl]methyl-3-pyridinylcarbonimi dodithioate; carbonimidodithioic acid, 3-pyridinyl-, butyl[4-(1,1-dimethylethyl)phenyl]methyl ester; Denmert

Mol. Formula $C_{21}H_{28}N_2S_2$ Mol. Wt. 372.60

Uses Fungicide.

Physical properties

M. Pt. 31-33°C; Specific gravity d_{25}^{25} 1.0865; Volatility v.p. 4.52×10^{-7} mmHg at 20°C.

Solubility

Water: 1 mg l^{-1} at 25°C. Organic solvent: methanol, xylene

Ecotoxicity

Fish toxicity

LC_{50} (48 hr) carp 6.4 mg l^{-1} (1).

Effects to non-target species

LD_{50} oral mallard duck >10,000 mg kg^{-1} (1).

Mammalian and avian toxicity

Acute data

LD_{50} oral rat 2700 mg kg^{-1} (2).

LD_{50} percutaneous rat >5000 mg kg^{-1} (1).

Metabolism and pharmacokinetics

Degradation by rat liver enzymes systems yielded *S*-*n*-butyl-*S'*-*p*-(1,1-dimethyl-2-hydroxyethyl)benzyl-*N*-3- pyridyldithiocarbonimidate and 2-(3'-pyridylimino)-4-carboxylthiazolidine. It is suggested the intermediates could decompose to 3-aminopyridine (3).

Genotoxicity

Allium cepa (4, 24 hr) 4, 20, 100 and 500 ppm inhibited mitosis and induced chromosomal aberrations (4).

Legislation

Limited under EC Directive on Drinking Water Quality 80/778/EEC. Fungicide: maximum admissible concentration 0.5 μg l^{-1} (5).

References

1. *The Agrochemicals Handbook* 2nd ed., 1987, RSC, London
2. *The Merck Index* 11th ed., 1989, Merck & Co. Inc., Rahway, NJ

3. Ohkawa, H; et al. *Agric. Biol. Chem.* 1976, **40**(6), 1175-1182
4. Badr, A. *Cytologia* 1988, **53**(4), 635-640
5. *EC Directive Relating to the Quality of Water Intended for Human Consumption* 1982, 80/778/EEC, Office for Official Publications of the European Communities, 2 rue Mercier, L-2985 Luxembourg

B246   Butobarbitone

CAS Registry No. 77-28-1

Synonyms 5-butyl-5-ethylbarbituric acid; 5-butyl-5- ethyl-2,4,6(1*H*,3*H*,5*H*)pyrimidinetrione; 5- ethyl-5-*N*-butylbarbituric acid; Butethal; Neonal; 2,4,6-(1*H*,3*H*,5*H*)-pyrimidinetrione, 5-butyl- 5-methyl

Mol. Formula $C_{10}H_{16}N_2O_3$ **Mol. Wt.** 212.25

Uses Central nervous system depressant. Sedative. Hypnotic.

Physical properties

M. Pt. 124-127°C.

Solubility

Organic solvent: ethanol, diethyl ether, chloroform

Mammalian and avian toxicity

Acute data

LD_{Lo} oral rat, mouse 100 mg kg^{-1} (1,2).

LD_{Lo} intraperitoneal rat, rabbit 115-135 mg kg^{-1} (3).

LD_{Lo} subcutaneous rabbit, rat 100- 190 mg kg^{-1} (4,5).

LD^{Lo} intravenous rabbit 90 mg kg^{-1} (6).

LD_{50} intraperitoneal mouse 320 mg kg^{-1} (7).

TD_{Lo} oral woman 166 mg kg^{-1} central nervous system effects observed (8).

Teratogenicity and reproductive effects

TD_{Lo} (7-19 day gestation) subcutaneous rat 1300 mg kg^{-1} total dose decreased rat litter size and foetal weight. No foetal malformations reported (9).

Metabolism and pharmacokinetics

Inactivated in liver by hydroxylation, small amounts excreted in urine as unchanged drug. Reported $t_{1/2}$ 40-55 hr (10).

Following single dose in human volunteers, butobarbitone was metabolised to 4'-hydroxybutobarbitone and 4'-hydroxypentobarbitone (11).

Any other adverse effects to man

A woman who took 6 g butobarbitone over 3 days had vertical gaze paralysis suggestive of brain lesions (12).

Any other adverse effects

Frog heart *Rana pipiens* exposed to 4.25 mg l^{-1} recorded mild depressive effects (13). TD_{Lo} intramuscular ♂ rat 28 mg kg^{-1} altered testis and caput and cauda epididymis metabolism. The changes were transient and reversible by administration of ascorbic acid (14).

Legislation

Controlled substance in the US (15).

Any other comments

Adult hypnotic dose orally 100-200 mg (16).
Toxicity reviewed (17).

References

1. *Arch. Int. Pharmacol. Ther.* 1953, **92**, 305
2. *J. Pharmacol. Exp. Ther.* 1932, **44**, 337
3. *J. Pharmacol. Exp. Ther.* 1932, **44**, 325
4. *J. Pharmacol. Exp. Ther.* 1925, **26**, 371
5. *J. Am. Chem. Soc.* 1923, **45**, 243
6. *J. Am. Chem. Soc.* 1935, **57**, 1961
7. *J. Pharmacol. Exp. Ther.* 1947, **89**, 356
8. *Br. Med. J.* 1955, **1**, 1238
9. Champakamalini, A. V; et al. *Experientia* 1977, **33**(4), 499-500
10. Gilbert, J. N. T; et al. *J. Pharm. Pharmacol.* 1974, **26**(Suppl.), 1-16
11. Al Shariff, M. A; et al. *Xenobiotica* 1983, **13**(3), 179-183
12. Edis, R. H. *Br. Med. J.* 1977, **1**, 144
13. Gruber, C. M; et al. *J. Pharmacol. Exp. Ther.* 1946, **87**(2), 104-108
14. Chinoy, N. J. *Indian J. Exp. Biol.* 1978, **16**(3), 316-322
15. *US Code Federal Regulations* 1987, **21**, 329.1, 1308.13
16. Gosselin, R. E. *Clin. Toxicol of Comm. Products* 5th ed., 1984, **II**, 354
17. Tatum, A. L. *Physiological Reviews* 1939, **19**(23), 472-502

B247   Butocarboxim

CH3SCH(CH3C(CH3)=NOCONHCH3

CAS Registry No. 34681-10-2
Synonyms 3-(methylthio)-*O*- (methylamino)carbonyl)oxime-2 –butanone; Afilene; 3-(methylthio)-2-butanone-*O*- [(methylamino)carbonyl]oxime; 3-(methylthio)butanone-*O*-methylcarbamoyloxime; 2-butanone, 3-(methylthio)- *O*-[(methylamino)carbonyl]oxime
Mol. Formula $C_7H_{14}N_2O_2S$ **Mol. Wt.** 190.27
Uses Systemic insecticide.

Physical properties

M. Pt. 32-37°C; **Specific gravity** 1.12 at 20°C; **Volatility** v.p. 7.97×10^{-5} at 20°C.

Solubility

Water: 35 g l^{-1} at 20°C.

Ecotoxicity

Fish toxicity
LC_{50} (48 hr) rainbow trout, golden orfe 35-55 mg l^{-1} (1).

Effects to non-target species
Toxic to bees (1).
LC_{50} (48 hr) Japanese quail 1180 mg kg^{-1} in diet (1).

Environmental fate

Degradation studies
Degrades in soil by losing the methylamine group, and sulfur oxidises to sulfoxide and sulfone, $t_{1/2}$ soil 3-5 days for unchanged compound 42 days for metabolites (1).

Mammalian and avian toxicity

Acute data
LD_{50} oral rat 153 mg kg^{-1} (2).
LD_{50} oral rabbit 275 mg kg^{-1} (3).
LD_{50} percutaneous rabbit 360 mg kg^{-1} (4).

Sub-acute data
In 90 day feeding trials, no effect level for dogs was 100 mg kg^{-1} (1).

Carcinogenicity and long-term effects
In 2 yr feeding trials the no effect level for rats was 100 mg kg^{-1} (1).

Metabolism and pharmacokinetics
In animals metabolised to butoxycarboxim and excreted in the urine (1).

Genotoxicity

Administration of 100 mg kg^{-1} (24 hr) to rat gave positive chromosome aberrations, 10 mg kg^{-1} dose negative (5).

Any other adverse effects

Cholinesterase inhibitor (6).

Legislation

Limited under EC Directive on Drinking Water Quality 80/778/EEC. Pesticide: maximum admissible concentration 0.1μg l^{-1} (7).

References

1. *The Agrochemicals Handbook* 3rd ed., 1991, RSC, London
2. *Worksubstanzen der Pflanzenschutz und Schadlingsbekampfungsmittel* 1971-1976
3. Worthing, C. R. (Ed.) *The Pesticide Manual* 9th ed., 1991, British Crop Protection Council, Farnham
4. *Farm Chemicals Handbook* 1983, C39, Meister Publishing, Willoughby, OH
5. Braginskii, M. L; et al. *Vopr. Eskp. Med. Genet.* 1982, 7-12, (Russ.) (*Chem. Abstr.* **101**, 145557a)
6. Occup. Health Services Inc. *Pestline* 1991, **2**, 1772-1774
7. *EC Directive Relating to the Quality of Water Intended for Human Consumption* 1982, 80/778/EEC, Office for Official Publications of the European Communities, 2 rue Mercier, L-2985 Luxembourg

B248   Butoxycarboxim

$$CH_3S(O)_2CH(CH_3)C(CH_3)=NOCONHCH_3$$

CAS Registry No. 34681-23-7

Synonyms 3-(sulfonyl)-*O*- ((methylamino)carbonyl)oxime-2-butanone;
2-methylsulfonyl- *O*-(*N*-methyl-carbamoyl)-buta-non-(3)-oxime; 3-
mesylbutanone-*O*-methylcarbamoyloxime; 3- (methylsulfonyl)-2-butanone-*O*-
[(methylamino)carbonyl]oxime; 2-butanone, 3-(methylsulfonyl)-
O-[(methylamino)carbonyl]oxime

Mol. Formula $C_7H_{14}N_2O_4S$ **Mol. Wt.** 222.26

Uses Systemic insecticide. Acaracide.

Physical properties

M. Pt. 85-89°C; **Volatility** v.p. 2×10^{-6} mmHg at 20°C.

Solubility
Water: 209 g l^{-1} at 20°C. Organic solvent: chloroform, acetone, isopropanol, toluene,
heptane

Ecotoxicity

Fish toxicity
LC_{50} (96 hr) rainbow trout 170 mg l^{-1} (1).

Effects to non-target species
LD_{50} oral hen 367 mg kg^{-1} (1).
Non-toxic to bees (1).

Mammalian and avian toxicity

Acute data
LD_{50} oral rabbit, rat 275-458 mg kg^{-1} (2,3).
LD_{50} subcutaneous rat 288 mg kg^{-1} (3).

Sub-acute data
In 90 day feeding trials, no effect level for rats 300 mg kg^{-1} (1).

Genotoxicity
Salmonella tymphimurium TA100, TA98, TA1535, TA1537 and TA1538 with and
without metabolic activation negative. *Escherichia coli* WP2 hcr with and without
metabolic activation negative (4).

Any other adverse effects
Cholinesterase inhibitor (1).

Legislation
Limited under EC Directive on Drinking Water Quality 80/778/EEC. Pesticide:
maximum admissible concentration 0.1 μg l^{-1} (5).

References

1. *The Agrochemicals Handbook* 3rd ed., 1991, RSC, London

2. *Farm Chemicals Handbook* 1983, C40, Meister Publishing Co., Willoughby, OH
3. *Wirksubstanzen der Pfanzenschutz und Schadlingsbekampfungsnittel* 1971-1976
4. Moriya, T; et al. *Mutat. Res.* 1983, **116**, 185-216
5. *EC Directive Relating to the Quality of Water Intended for Human Consumption* 1982, 80/778/EEC, Office for Official Publications of the European Communities, 2 rue Mercier, L-2985 Luxembourg

B249   2-Butoxyethanol phosphate

$[CH_3(CH_2)_3OCH_2CH_2O]_3PO$

CAS Registry No. 78-51-3
Synonyms KP 140; tri(2-butoxyethanol phosphate); tributyl cellosolve phosphate; TBEP; tributoxyethyl phosphate
Mol. Formula $C_{18}H_{39}O_7P$ **Mol. Wt.** 398.48
Uses Plasticiser.
Occurrence Contaminant in tap water in Japan 0 to 58.5 ppt detected over the period of 1 yr (1).

Physical properties

M. Pt. –70°C; **B. Pt.** 215-228°C at 4 mmHg; **Flash point** 224°C; **Specific gravity** d_{20}^{20} 1.02; **Volatility** v.p. 0.03 mmHg at 150°C; v. den. 13.8.

Mammalian and avian toxicity

Acute data
LD_{50} oral guinea pig, rat 3000 mg kg^{-1} (2,3).
LD_{50} intravenous mouse 180 mg kg^{-1} (4).

Sub-acute data
Gavage Sprague Dawley rats (18 wk) 0.25-0.5 ml kg day^{-1} 5 day wk^{-1} caused neurotoxicity (5-7).

Irritancy
Dermal rabbit (24 hr) 500 mg caused mild irritation and 500 mg instilled into rabbit eye caused mild irritation (8).

Any other adverse effects

Detected in human adipose tissue, 25-483 μg^{-1} in Canada (9).
Interacts with β adrenergic transport proteins, specific and non-specific binding sites and β adrenergic receptors. May alter catecholamine sensitive adenylate-cyclase activity (10).

Any other comments

Fish and shellfish captured in Okayama Prejecture Japan contained <0.005-0.019 μ g^{-1} (11).
Experimental toxicology and human health effects reviewed (12).

References

1. Adachi, K; et al. *Hyogo-ken Eisei Kenkyusho Kenkyu Hokoku* 1984, **19**(1-6), (Jap.) (*Chem. Abstr.* **103**, 11048t)
2. *Raw Mater. Data Handbook* 1975, **2**, 93
3. LeFaux, R. *Prac. Toxicol. Plastics* 1968, 336
4. *US Army Armament Res. Develop. Command*
5. Laham, S; et al. *J. Appl. Toxicol.* 1984, **4**(1), 42-48
6. Laham, S; et al. *Am. Ind. Hyg. Assoc.* 1985, **8**, 442-448
7. Laham, S; et al. *Chemosphere* 1984, **13**(7), 801-812
8. *Prehled Prumyslove Toxikol Org Latky* 1986, 1142
9. Le Bel, G. L; et al. *Bull. Environ. Contam. Toxicol.* 1989, **43**(2), 225-230
10. Sager, G; et al. *Biochem. Phamacol.* 1989, **38**(15), 2551-2557
11. Kenomochi, V; et al. *Okayama-Ken Kanko Hoken Senta Nenpo* 1981, 167-175, (Jap.) (*Chem. Abstr.* **99**, 68978z)
12. *ECETOC Technical Report No. 30(4)* 1991, European Chemical Industry Ecology and Toxicology Centre, B-1160 Brussels

B250   2-Butoxyethanol

$CH_3(CH_2)_3O(CH_2)_2OH$

CAS Registry No. 111-76-2
Synonyms ethylene glycol monobutyl ether; butyl cellosolve; glycol monobutyl ether; butyglycol; 2-butoxy-1-ethanol
Mol. Formula $C_6H_{14}O_2$ **Mol. Wt.** 118.18
Uses Solvent for nitrocellulose, resins, grease, oil and albumin. Dry cleaning solvent.
Occurrence Contaminant in drinking water, surface water and groundwater (1-3).

Physical properties

M. Pt. –70°C; **B. Pt.** 168-170°C; **Flash point** 62°C (closed cup); **Specific gravity** d_4^{20} 0.9030; **Volatility** v.p. 0.6 mmHg at 20°C; v. den. 4.07.

Solubility

Water: ≥100 mg l^{-1} at 22°C. Organic solvent: acetone, DMSO, ethanol

Occupational exposure

US TLV (TWA) 25 ppm (121 mg m^{-3}); **UK Long- term limit** MEL 25 ppm (120 mg m^{-3}); **UN No.** 2369; **HAZCHEM Code** 2R; **Conveyance classification** harmful substance; **Supply classification** harmful.

Risk phrases

Harmful by inhalation, in contact with skin and if swallowed – Irritating to respiratory system (R20/21/22, R37)

Safety phrases

Avoid contact with skin and eyes (S24/25)

Ecotoxicity

Fish toxicity

LC_{50} (24 hr) goldfish 1650 mg l^{-1} (4).

LC_{50} (7 day) guppy 983 ppm (4).

LC_{50} (96 hr) bluegill sunfish 1490 mg l^{-1} static bioassay in fresh water at 23°C, mild aeration applied after 24 hr (5).

LC_{50} (96 hr) Atlantic silverside 1250 mg l^{-1} static bioassay in synthetic seawater at 23°C , mild aeration applied after 24 hr (5).

Invertebrate toxicity

LC_{50} (48 hr) brown shrimp 600-1000 mg l^{-1} (6).

LC_{50} (96 hr) *Artemia sp.* 100 mg l^{-1} (6).

Cell multiplication inhibition test, *Pseudomonas putida* 700 mg l^{-1}, *Scenedesmus quadricauda* 900 mg l^{-1}, *Entosiphon sulcatum* 91 mg l^{-1}, *Uronema parduczi Chatton-Lwoff* 465 mg l^{-1}, *Microcystis aeruginosa* 35 mg l^{-1} (7-9).

Environmental fate

Absorption

Wastewater treatment: activated sludge adsorbabilty 0.112 g g^{-1} carbon; 55.9% reduction; influent 1000 mg l^{-1}; effluent 441 mg l^{-1} (10).

Mammalian and avian toxicity

Acute data

LD_{50} oral rat 1480 mg kg^{-1} (11).

LD_{50} oral rabbit 300 mg kg^{-1} (12).

LC_{50} (4 hr) inhalation rat 450 ppm (13).

LD_{50} dermal rabbit 490 mg kg^{-1} (14).

LD_{50} intravenous rat 340 mg kg^{-1} (15).

LD_{50} intraperitoneal rat 220 mg kg^{-1} (16).

TC_{Lo} (8 hr) inhalation human 195 ppm gastrointestinal tract effects (15).

The lethal oral dose in humans is $\approx$1.4 ml kg^{-1}, equivalent to $\approx$100 ml for a 70 kg person (17).

Sub-acute data

Skin penetration was investigated in 12 exposure experiments with 5 men. Presence of 2-butoxyethanol was detected in the blood and urine confirming its entry into the systemic circulation in man *in vivo* during dermal exposure. Epicutaneous administration of 5-100% of 2-butoxyethanol for 2 hr compared to administration of the undiluted compounds showed that water facilitates the absorption of the chemical (18).

Teratogenicity and reproductive effects

TC_{Lo} (6-15 day gestation) inhalation rat 100 or 200 ppm 6 hr day^{-1} caused maternal toxicity (decreased body weight and weight gain; decreased organ weight, decreased food and water consumption and anaemia) embryotoxicity (increased resorptions and decreased implantation rate) and foetoxicity reductions in skeletal ossification. No increase in foetal malformations. Inhalation rabbit (6-18 day gestation) 200 ppm 6 hr day^{-1} caused maternal toxicity and embryotoxicity. No treatment related foetotoxicity or foetal malformations observed (19).

Development of F-344 rat was not uniquely sensitive at oral dose 30-100 mg kg^{-1}

day^{-1}, however reduced prenatal viability was noted at 200 mg kg^{-1} day^{-1}.
2-Butoxyethanol was administered on days 9-11 of gestation (20).

Metabolism and pharmacokinetics

^{14}C 2-Butoxyethanol was administered for 24 hr to ♂ F344/N rats in drinking water, absorbed doses were 12 to 171 mg kg^{-1}. Elimination of radioactivity was monitored for 72 hr, 50-60% was eliminated in the urine as butoxyacetic acid, 8-10% as carbon dioxide and approximately 10% was excreted in the urine as ethylene glycol (21,22).
Gavage ♂ rats (unspecified dose) rapidly absorbed, 48 hr after administration was detected in the forestomach, liver, kidney, spleen and the glandular stomach. Major routes of elimination were via the urine and as carbon dioxide (23).
Following dermal exposure (unspecified dose) in 5 human volunteers, 2-butoxyethanol was detected in the blood and butoxyacetic acid in the urine.
Calculated dermal uptake rates ranged from 0.8-11 μg min^{-1} cm^{-2}.
The authors state that persons exposing large portions of their skin to butoxyethanol are at risk of absorbing acutely toxic doses (24).

Irritancy
Dermal rabbit (duration unspecified) 500 mg caused mild irritation (25).
Human reported irritation to eye, upper respiratory tract, dyspnoea and dermatitis (26).

Genotoxicity

Not mutagenic at HGPRT locus of Chinese hamster ovary cells *in vitro* with or without metabolic activation. Equivocal results for unscheduled DNA synthesis in rat primary hepatocyte test *in vitro* (27,28).

Any other adverse effects to man

A study of 9365 individuals employed in 2 US leather tanneries between 1940-1982 was undertaken. Mortality from all causes was lower than expected (29).

Any other adverse effects

Inhalation humans systemic effects include nausea, vomiting, headache, and eye injury. 2-Butoxyethanol is nephrotoxic, a haemolytic agent, and can cause central nervous system depression and liver and kidney damage (26).
Toxicity is associated with changes to blood and secondary effects on liver, kidney and spleen. Narcotic at high concentrations (15).
The haemolytic effect of 2-butoxyethanol can be attributed primarily to its metabolite butoxyacetic acid (30,31).

Any other comments

Experimental toxicology and human health effects reviewed (32).

References

1. Lucas, S. V. *Anal. Org. Drink Water Conc. Adv. Treat. Conc.* 1984, **1**, 397, USEPA-600/1-84-020A
2. Yasuhara, A; et al. *Environ. Sci. Technol.* 1981, **15**, 570-573
3. Stonebreaker, R. D; et al. *Control Haz. Mater. Spill* 1980, 1-10, Proc. Natl. Conf., Louisville, KY
4. Verschueren, K. *Handbook of Environmental Data on Organic Chemicals* 2nd ed., 1983, Van Nostrand Reinhold, NY
5. *Dangerous Prop. Ind. Mater. Rep.* 1984, **4**(2) 58-61

6. Blackmann, R. A. A. *Marine Pollut. Bull.* 1974, **5**(8), 116-118
7. Bringmann, G; et al. *Water Res.* 1980, **14**, 231-241
8. Bringmann, G; et al. Z. *Wasser/Abwasser Forsch* 1980, **1**(1), 26-31
9. Bringmann, G; et al. *Gwf Wasser/Abwasser* 1976, **117**,(9)
10. Ford, D. L; et al. *J. Water Pollut. Control Fed.* 1974, **46**(5), 947-965
11. *J. Ind. Hyg. Toxicol.* 1949, **23**, 259
12. *Yakkyoku* (Pharmacy), 1981, **32**, 1241, Tokyo, Japan
13. *Toxicol. Appl. Pharmacol.* 1983, **68**, 405
14. *Raw Material Data Handbook, Organic Solvents* 1974, **1**, 50, Lehigh Univ., Bethelehem, PA
15. *Archives of Industrial Health* 1956, **14**, 114, Heldreft Publ., Washington, DC
16. Izmerov, N. F; et al. *Toxicometric Parameters of Industrial Toxic Chemicals under Single Exposure* 1982, 67, Moscow
17. Doull, J; et al. *Casarett and Doull Toxicology* 3rd ed., 1986, 654, Macmillan & Co., New York
18. Johanson, G; et al. *Scand. J. Work Environ. Health* 1988, **14**(2), 95-100, 101-109
19. Tyl, R. W; et al; *Environ. Health Perspect.* 1984, **57**, 47-68
20. Sleet, R. B. *US Gov. Rep. Announce Index* 1989, **89**(13)
21. Jonsson, A. K. *Acta Pharmacol. Toxicol.* 1978, **42**, 354-356
22. Medinsky, M. A. *Toxicol. Appl. Pharmacol.* 1990, **102**(3), 443-455
23. Ghanayem, B. I; et al. *Drug Metab. Dispos.* 1987, **15**(4), 478-484
24. Johanson, G; et al. *Scand. J. Work Environ. Health* 1988, **14**(2), 101-109
25. *Union Carbide Data Sheet* Union Carbide Corp., CT
26. Occupational Health Services *Pestline* 1991, **1**, 156
27. *Chemical Safety Data Sheet* 1988, **1**, 152, RSC, London
28. McGregor, D. B. *Environ. Health Perspect.* 1984, **57**, 97-103
29. Stern, F. B; et al. *Scand. J. Work, Environ. Health* 1987, **13**(2), 108-117
30. Ghanayem, B. I; et al. *Biochem. Pharmacol.* 1989, **38**(10), 1679-1684
31. Bartnik, F. G. *Fundam. Appl. Toxicol.* 1987, **8**(1), 59-70
32. *ECETOC Technical Report No. 30(4)* 1991, European Chemical Industry Ecology and Toxicology Centre, B-1160 Brussels

B251 2-(2-Butoxyethoxy)ethanol

$$HO(CH_2)_2O(CH_2)_2O(CH_2)_3CH_3$$

CAS Registry No. 112-34-5
Synonyms diethylene glycol monobutyl ether; butoxydiglycol; butylcarbitol; Butyl Carbitol; butyl digol
Mol. Formula $C_8H_{18}O_3$ **Mol. Wt.** 162.23
Uses Industrial solvent. Mosquito repellent.
Occurrence Contaminant in drinking and surface water. Discharged in industrial effluent, industries include paint and ink, print and publishing, foundries and electronics (1- 3).

Physical properties

M. Pt. −68.1°C; **B. Pt.** 230.6°C; **Flash point** 66.7°C; **Specific gravity** d_4^{20} 0.9553; **Volatility** v.p. 0.02 mmHg at 20°C; v. den. 5.58.

Solubility

Water: miscible. Organic solvent: miscible

Ecotoxicity

Fish toxicity

LC_{50} (96 hr) inland silverside 2000 ppm static bioassay in synthetic seawater at 23°C, mild aeration applied after 24 hr (4).
LC_{50} (96 hr) bluegill sunfish 1300 ppm static bioassay in fresh water at 23°C mild aeration applied after 24 hr (4).
LC_{50} (24 hr) goldfish 2700 mg l^{-1} (5).

Mammalian and avian toxicity

Acute data

LD_{50} oral guinea pig, rat 2000-5660 mg kg^{-1} (6,7).
LD_{50} dermal rabbit 4120 mg kg^{-1} (8).
LD_{50} intraperitoneal mouse 850 mg kg^{-1} (9).
Lethal single oral dose to humans of diethylene glycols estimated as 1 ml kg^{-1} (10).

Irritancy

5 mg instilled into rabbit eye caused severe irritation (11).
Dermal rabbit 0, 100, 300, or 1000 mg kg^{-1} day^{-1} mild skin irritation observed (12).
A 0.1 ml solution of diethylene glycol monobutyl ether instilled into rabbit eye caused moderately severe conjunctivitis, with mild blepharitis and mild diffuse keratitis. Symptoms subsided within 14 days of cessation of exposure (13).

Genotoxicity

In vitro mouse lymphoma L5178Y without metabolic activation positive, with metabolic activation negative (14).

Any other adverse effects

Oral ♂ rat (2 month) 0, 250, 500 or 1000 mg kg day^{-1}. Oral ♀ rat (14 day) 0, 250, 500 or 1000 my kg day^{-1} no adverse effect on fertility in either sex (12,15).
LD_{50} intraperitoneal mice 850 mg kg^{-1} systemic effects included pulmonary congestion, atelectasis and oedema. Toxic reaction occurred in spleen and lymph tissue. Congestion of viscera, marked renal tubular damage (16).
Butyl carbitol can be absorbed through human skin, but only in toxic amounts if exposure is prolonged and continuous (17).

Any other comments

Toxic effects of glycol ethers reviewed (18).
Experimental toxicology and human health effects reviewed (19).
Flammable.

References

1. Perry, D. L; et al. *Id. Org. Comp. Ind. Effl. Disch.* 1979, 230, USEPA-600/4-79-016
2. Yasuhara, A; et al. *Environ. Sci. Technol.* 1981, **15**, 570-573
3. Lucas, S. V. *Anal. Org. Drink. Water Conc. Adv. Treat. Conc.* 1984, **1**, 397, USEPA-600/1-84-020A
4. Dawson, G. W. *J. Haz. Mat.* 1975/77, **1**, 303-318

5. Verschueren, K. *Handbook of Environmental Data of Organic Chemicals* 2nd ed., 1983, 524, Van Nostrand Reinhold, New York
6. *Dow Chemical Co. Rep.* 1941, MSD
7. *J. Ind. Hyg. Toxicol.* 1941, **23**, 259
8. *Union Carbide Data Sheet* 1966 (31st Jan)
9. *Fed. Proc. Amer. Soc. Exp. Biol.* 1947, **6**, 342
10. *Casarett & Doull Toxicology* 3rd ed., 1986, 565, Doull J; et al. (Eds.) MacMillan Co., New York
11. *Amer. J. Opthal.* 1946, **29**, 1363
12. Nolen, G. A; et al. *Fundam. Appl. Toxicol.* 1985, **5**, 1137-1143
13. Ballantyne, B. *J. Toxicol. Cutaneous. Ocul. Toxicol.* 1984, **3**(1), 7-15
14. Thompson, E. D. *Environ. Health Perspect.* 1984, **57**, 105-112
15. Ema, M; et al. *Drug Chemical Toxicology* 1988, **11**(2), 97-111
16. Karel, L; et al. *J. Pharmacol. Exp. Ther.* 1947, **90**, 338-347
17. Browning, E. *Toxicity Metab. Ind. Solvent* 1965, 635, Elsevier, New York
18. Hardin, B. D; et al. *Environ. Health Perspect.* 1984, **57**, 273-275
19. *ECETOC Technical Report No. 30(4)* 1991, European Chemical Industry Ecology and Toxicology Centre; B-1160 Brussels

B252   2-(2-Butoxyethoxy)ethanol acetate

$$CH_3CO_2[(CH_2)_2O](CH_2)_3CH_3$$

CAS Registry No. 124-17-4

Synonyms butyl carbitol acetate; diethylene glycol butyl ether acetate; DGBA; diglycol monobutyl ether acetate

Mol. Formula $C_{10}H_{20}O_4$ **Mol. Wt.** 204.27

Uses Insect repellant synergist. Used in plastics and cosmetics. Solvent.

Physical properties

M. Pt. $-32.2°C$; **B. Pt.** 247°C; **Flash point** 115.6°C (open cup); **Specific gravity** d_{20}^{20} 0.981; **Volatility** v.p. 0.01 mmHg at 20°C.

Solubility
Water: 6.5 wt at 20°C.

Ecotoxicity

Effects to non-target species
LD_{50} oral chicken 5000 mg kg^{-1} (1).

Mammalian and avian toxicity

Acute data
LD_{50} oral rat, mouse 6500-6600 mg kg^{-1} (2,3).
LD_{50} oral guinea pig, rabbit 2340-2600 mg kg^{-1} (3,4).

Metabolism and pharmacokinetics
In vitro studies on ♂ Sprague Dawley rats (0-14 min) 0-1 g l^{-1} hydrolysed to diethylene glycol monobutyl ether by rat blood (5).

In vivo gavage ♂ Sprague-Dawley rats 200 or 2000 mg kg^{-1} ^{14}C2-(2-butoxyethoxy) ethanol acetate. Urine, faeces and expired air were collected for 72 hr. After 24 hr, 82% excreted in urine, 2-3% in faeces, 5% as expired air. Major metabolite detected 2-(2- butoxyethoxy)acetic acid (5).

Irritancy

Dermal rabbit (duration unspecified) 500 mg caused mild irritation (6).
500 mg instilled into rabbit eye caused irritation (7).

Any other comments

Experimental toxicology and human health effects reviewed (8).
Designated unsafe for military use in US (9).

References

1. *J. Pharmacol. Exp. Ther.* 1948, **93**, 26
2. Frear, E. H; et al. *Pesticide Index* 1976, **5**, 32
3. *J. Pharmacol. Exp. Ther.* 1944, **82**, 377
4. *J. Pharmacol. Exp. Ther.* 1941, **23**, 259
5. Deisenger, P. G; et al. *Xenobiotica* 1989, **19**(9), 981-989
6. *Union Carbide Data Sheet* 1971 (29th Dec), Union Carbide Corp., NY
7. *Am. J. Ophthalmol.* 1946, **29**, 1363
8. *ECETOC Technical Report No. 30(4)* 1991, European Chemical Industry Ecology and Toxicology Centre, B-1160 Brussels
9. Draize, J. H. *J. Pharmacol. Exp. Ther.* 1948, **93**, 26-39

B253 Butoxypropanol

CH3(CH2)3O(CH2)3OH

CAS Registry No. 29387-86-8
Synonyms propylene glycol monobutyl ether; propylene glycol butoxy ether; Propasol B; Dowanol PnB
Mol. Formula $C_7H_{16}O_2$ **Mol. Wt.** 132.20
Uses Antifreeze agent. Foam stabiliser. Detergent base.

Occupational exposure

Supply classification irritant.

Risk phrases
Irritating to eyes, respiratory system and skin (R36/37/38)

Safety phrases
Avoid contact with eyes (S25)

Mammalian and avian toxicity

Acute data
LD$_{50}$ oral rat 1900 mg kg^{-1} (1).

Any other comments

Physical, chemical properties, experimental toxicology and human health effects reviewed (2).

References

1. *Patty's Industrial Hygiene and Toxicology* 3rd rev. ed., 1981, John Wiley & Sons, New York
2. *ECETOC Technical Report No. 30 (4)* 1991, European Chemical Industry Ecology and Toxicology Centre, B-1160 Brussels

B254   1-Butoxy-2-propanol

$CH_3(CH_2)_3OCH_2CH(OH)CH_3$

CAS Registry No. 5131-66-8
Synonyms 1,2-propylene glycol 1-monobutyl ether; 2-hydroxy-3-butoxypropane
Mol. Formula $C_7H_6O_2$ **Mol. Wt.** 122.12
Uses Used in plasticisers, detergents. Solvent for coatings and surfactants. In water desalination. Termite attractant. Solvent for nitrocellulose and acetylcellulose.

Physical properties

M. Pt. -100°C; **B. Pt.** 168°C; **Flash point** 59°C (closed cup); **Specific gravity** d_{20}^{20} 0.8789; **Volatility** v.p. <0.978 mmHg at 20°C.

Solubility
Water: 64 g l^{-1} at 20°C.

Occupational exposure

Supply classification irritant.

Risk phrases
Irritating to eyes and skin (R36/38)

Mammalian and avian toxicity

Acute data
LD_{50} oral rat 2200 mg kg^{-1} (1).
LD_{50} dermal rabbit 3100 mg kg^{-1} (2).

Any other comments

All reasonable efforts have been taken to find information on isomers of this compound, but no relevant data are available.

References

1. Patty, F. A. (Ed.) *Industrial Hygiene and Toxicology* 2nd ed., 1963, Interscience Publishers, New York
2. *Raw Materials Data Handbook: Organic Solvents* 1974, **1**, Nat. Assoc. Print. Ink Res. Inst., Lehigh Univ., Bethlehem, PA

B255   Butralin

$$CH_3CH_2-\overset{\overset{\displaystyle CH_3}{|}}{CH}-NH\text{—}\langle\text{ring with }NO_2,\ O_2N,\ C(CH_3)_3\rangle$$

CAS Registry No. 33629-47-9

Synonyms benzenamine,4-(1,1-dimethylethyl)-*N*-(1-methylpropyl)-2,6-dinitro-; *N*-*sec*-butyl-4-*tert*-butyl-2,6-dinitroaniline; dibutalin; 4-(1,1-dimethylethyl)-*N*-(1-methylpropyl)-2,6-dinitrobenzamine

Mol. Formula $C_{14}H_{21}N_3O_4$ **Mol. Wt.** 295.34

Uses Pre-emergence herbicide. Plant growth regulator.

Physical properties

M. Pt. 60-61°C; **B. Pt.** $_{0.5}$ 134-136°C; **Flash point** 36°C; **Volatility** v.p. 1.278×10^{-5} mmHg at 25°C.

Solubility

Water: 1 mg l^{-1} at 25°C. Organic solvent: acetone, benzene, butanone, carbon tetrachloride, methanol, xylene

Ecotoxicity

Fish toxicity

LC_{50} (48 hr) rainbow trout, bluegill sunfish 3.4-4.2 mg l^{-1} (1).

Effects to non-target species

LC_{50} (8 day) Japanese quail, mallard duck >10 mg kg^{-1} in diet (1).

Environmental fate

Degradation studies

Microbial degradation occurs to evolve carbon dioxide (1).
Major metabolite isolated from soil 4-*tert*-butyl-2,6-dinitroaniline. Microbial studies using *Aspergillus furnigatus* and *Fusanium oxysporum* major metabolite isolated was an oxygenated analogue of butralin, 3-(4-*tert*-butyl 2,6-dinitroaniline)-2-butanol (2).

Mammalian and avian toxicity

Acute data

LD_{50} oral rat 12 g kg^{-1} (1).
LD_{50} dermal rabbit 10 g kg^{-1} (1).

Legislation

Limited under EC Directive on Drinking Water Quality 80/778/EEC. Herbicide: maximum admissible concentration 0.5 μg l^{-1} (3).

Any other comments

Non-corrosive to metals but permeates or can distort plastic and rubbers (1).

References

1. *The Agrochemicals Handbook* 3rd ed., 1991, RSC, London
2. Kearney, P. C; et al. *J. Agric. Prod. Chem.* 1974, **22**, 856-859
3. *EC Directive Relating to the Quality of Water Intended for Human Consumption* 1982, 80/778/EEC, Office for Official Publications of the European Communities, 2 rue Mercier, L-2985 Luxembourg

B256   *tert*-Butyl acetate

$CH_3CO_2C(CH_3)_3$

CAS Registry No. 540-88-5
Synonyms acetic acid, *tert*-butyl ester; acetic acid, 1,1-dimethylethyl ester
Mol. Formula $C_6H_{12}O_2$ **Mol. Wt.** 116.16
Uses Gasoline additive. Used in the synthesis of *tert*-butyl esters, *N*-protected amino acids.
Occurrence Pollutant in air samples, source emissions from printing ink, paints and varnishes and automobile industry (1).

Physical properties

B. Pt. 97°C; **Flash point** 15°C (closed cup); **Specific gravity** d_4^{20} 0.8665.

Solubility
Organic solvent: ethanol, diethyl ether

Occupational exposure

US TLV (TWA) 200 ppm (950 mg m^{-3}); **UK Long- term limit** 200 ppm (950 mg m^{-3}); **UK Short-term limit** 250 ppm (1190 mg m^{-3}); **UN No.** 1123; **HAZCHEM Code** 3ME; **Conveyance classification** flammable liquid; **Supply classification** Highly flammable.

Risk phrases
Highly flammable (R11)

Safety phrases
Keep away from sources of ignition – No Smoking – Do not breathe vapour – Do not empty into drains – Take precautionary measures against static discharges (S16, S23, S29, S33)

Ecotoxicity

Invertebrate toxicity
Cell multiplication inhibition test *Pseudomonas putida* 78 mg l^{-1}, *Scenedesmus quadricauda* 3700 mg l^{-1}, *Entosiphon sulcatum* 970 mg l^{-1} (2).

Abiotic removal
Reacts with photochemically produced hydroxyl radicals in the atmosphere. Assuming atmospheric hydroxyl radical concentration of 8×10^5 molecules cm^{-3}, estimated $t_{1/2}$ is 26 days (3).

Any other comments

Experimental toxicology and human health effects reviewed (4).

References

1. Veulemans, H; et al. *Am. Ind. Hyg. Assoc.* 1987, **48**, 67-76
2. Bringmann, G; et al. *Water Res.* 1980, **14**, 231-241
3. *GEMS: Graphical Exposure Modeling System Fate of Atmosphere Pollutants Data Base* 1986, Office of Toxic Substances USEPA, Washington, DC
4. *ECETOC Technical Report No. 30(4)* 1991, European Chemical Industry Ecology and Toxicology Centre, B-1160 Brussels

B257   Butyl acetoacetate

$CH_3COCH_2CO_2(CH_2)_3CH_3$

CAS Registry No. 591-60-6
Synonyms acetoacetic acid butyl ester; 3-oxo-butanoic acid butyl ester
Mol. Formula $C_8H_{14}O_3$ **Mol. Wt.** 158.20
Uses Intermediate in organic synthesis. Manufacture of metal derivatives. Dyestuffs. Pharmaceuticals and flavourings.

Physical properties

B. Pt. 214°C; **Flash point** 85°C; **Specific gravity** 0.96; **Volatility** v.p. 0.19 mmHg at 20°C; v. den. 5.55.

Solubility

Organic solvent: ethanol, diethyl ether

Mammalian and avian toxicity

Acute data
LD_{50} oral rat 11 g kg^{-1} (1).

Irritancy
Dermal rabbit (24 hr) 500 mg caused mild irritation and 500 mg instilled into rabbit eye caused mild irritation (2).

Any other comments

Flammable.

References

1. *AMA Arch. Ind. Hyg. Occup. Med.* 1954, **10**, 61
2. *Prehled Prumysluve Toxicol. Org. Latky* 1986

B258   Butyl acid phosphate

$$CH_3(CH_2)_3H_2PO_4$$

CAS Registry No. 12788-93-1
Synonyms acid butyl phosphate; butyl phosphoric acid; phosphoric acid, butyl ester
Mol. Formula $C_4H_{11}PO_4$ **Mol. Wt.** 154.10
Uses Antifoaming agent for drilling muds. Catalyst. Flame retardant.
Occurrence Contaminant in natural and drinking water supplies from River Po at
Turin, Ferrara and Como in Northern Italy (1).

Physical properties

Flash point 110°C (open cup); **Specific gravity** d_{40}^{25} 1.120-1.125; **Partition
coefficient** log P_{ow} 0.28.

Solubility
Organic solvent: ethanol, acetone, toluene

Occupational exposure

UN No. 1718; **HAZCHEM Code** 2X; **Conveyance classification** corrosive
substance.

Ecotoxicity

Invertebrate toxicity
Cell multiplication inhibition test, *Pseudomonas putida* >100 mg l^{-1} *Microcystis
aeruginosa* 4.1 mg l^{-1} (1).

References

1.   Bringmann, G; et al. *Gwf. Wasser Abwasser.* 1976, **117**(9), 119

B259   Butylamine

$$CH_3(CH_2)_3NH_2$$

CAS Registry No. 109-73-9
Synonyms *n*-butylamine; 1-aminobutane; mono-*n*-butylamine; Norvalamine;
1-butanamine
Mol. Formula $C_4H_{11}N$ **Mol. Wt.** 73.14
Uses Chemical intermediate for organic synthesis. Used in the manufacture of
pharmaceuticals, dyestuffs, rubber chemicals, emulsifying agents, insecticides and
synthetic tanning agents.
Occurrence Contaminant in River Elbe in 1 of 2 sites tested. Concentration
determined 1.5 ppb (1).
Contaminant in surface water and advanced water treatment concentrates (1,2).

Physical properties

M. Pt. −50°C; **B. Pt.** 78°C; **Flash point** −12.2°C; **Specific gravity** d_4^{20} 0.741;
Partition coefficient log P_{ow} 0.86; **Volatility** v.p. 72 mmHg at 20°C; v. den. 2.52.

Solubility
Water: miscible. Organic solvent: miscible diethyl ether, ethanol

Occupational exposure

US TLV (TWA) 5 ppm (15 mg m^{-3}); **UK Short-term limit** 5 ppm (15 mg m^{-3}); **UN No.** 1125; **HAZCHEM Code** 2WE; **Conveyance classification** flammable liquid; **Supply classification** highly flammable and irritant.

Risk phrases
Highly flammable − Irritating to eyes, respiratory system and skin (R11, R36/37/38)

Safety phrases
Keep away from sources of ignition − No Smoking − In case of contact with eyes, rinse immediately with plenty of water and seek medical advice − Do not empty into drains (S16, S26, S29)

Ecotoxicity

Fish toxicity
LC_{50} (24 hr) creek chub 30-70 mg l^{-1} (3).
LC_{50} (96 hr) inland silverside, bluegill sunfish 24-32 mg l^{-1} aeration after 24 hr, static bioassay pH 7.6-pH 7.9 temperature range 20-23°C (4).

Invertebrate toxicity
Cell multiplication inhibition test, *Pseudomonas putida* 800 mg l^{-1}, *Entosiphon sulcatum* 9 mg l^{-1}, *Scenedesmus quadricauda* 0.53 mg l^{-1} (5).
LC_{50} (24 hr) brine shrimp 30- 70 mg l^{-1} (6).
EC_{50} (30 min) *Photobacterium phosphoreum* 18.3 g l^{-1} Microtox test (7).

Environmental fate

Degradation studies
Aerobacter sp. degraded 200 mg l^{-1} at 30°C parent strain degraded 100% in 22 hr and mutant strain 100% in 7 hr (8).
COD_2 >90% degradation using the Hoechst Bahl Method (9).
BOD_{12} 67% reduction in dissolved oxygen activated sludge (10).
BOD_6 50% reduction in dissolved oxygen using aniline acclimated activated sludge (11).
BOD (5, 10, 15 and 50 day) incubation settled sewage seed range 26-52% reduction in dissolved oxygen (12).

Abiotic removal
Evaporation is expected to be major removal process, $t_{1/2}$ of 2 days predicted (13).

Absorption
Activated carbon adsorbability 0.103 g g^{-1} carbon, 52% reduction; influent 1000 mg l^{-1}, effluent 480 mg l^{-1} (14).

Mammalian and avian toxicity

Acute data

LD_{50} oral rat, guinea pig 360-430 mg kg^{-1} (15,16).

LC_{50} (2 hr) inhalation mouse 800 mg m^{-3} (16).

LC_{Lo} (4 hr) inhalation rat 4000 ppm (17).

LD_{50} intraperitoneal mouse 629 mg kg^{-1} (18).

LD_{50} intravenous mouse 198 mg kg^{-1} (18).

LD_{50} dermal guinea pig, rabbit 370-850 mg kg^{-1} (19,20).

Irritancy

Dermal rabbit (24 hr) 500 mg caused severe irritation (20).

In humans, potent skin, eye, mucous membrane irritant. Direct skin contact causes severe primary irritation and blistering (21).

Genotoxicity

Salmonella typhimurium TA98, TA100, TA1535, TA1537 with and without metabolic activation negative (22).

Any other adverse effects

Inhalation of *n*-butylamine depressed the respiratory rate in normal and tracheally cannulated mice at concentrations of 121 and 300 ppm respectively. Sensory and pulmonary irritation reported (23).

Systemic effects include sedation, ataxia, nasal discharge, gasping, salivation and convulsions. Gavage rats 100-600 mg kg^{-1}, pathological examination showed pulmonary oedema (24).

Any other comments

Experimental toxicology and human health effects reviewed (25-27).

References

1. Newrath, G. B; et al. *Food Cosmet. Toxicol.* 1977, **15**, 275-282
2. Lucas, S. V. *Anal. Org. Drink. Water Conc. Adv. Waste Treatment* 19984, **2**, 397, USEPA 600/1-84-020B
3. McKee, J. E; *Water Quality Criteria* 1963, State Water Purity Board Res. Agency, CA
4. Dawson, G. W; et al. *J. Haz. Mat.* 1975-1977, **1**, 303-318
5. Bringmann, G; et al. *Water Res.* 1980, **14**, 231-241
6. Gillette, L. A; et al. *Sewage Ind. Works* 1975, **24**, 1397
7. Kaiser, K. L. E; et al. *Water Pollut. Res. J. Canada* 1991, **26**(3), 361-431
8. Howard, P. H. *Fate and Exposure Data for Organic Chemicals – Solvents* 1990, **2**, 66-69, Lewis Publishers, Chelsea, MI
9. Zahn, R; et al. *Z. Wasser Abwasser Forsch* 1980, **13**(1), 1-7
10. Yoshimura, K; et al. *J. Am. Oil Chem. Soc.* 1980, **57**, 238-241
11. Malaney, G. W. *J. Water Pollut. Control Fed.* 1960, **32**, 1300
12. Ettinger, M. B. *Ind. Eng. Chem.* 1956, **48**, 256-259
13. Verschueren, K. *Handbook of Environmental Data on Organic Chemicals* 2nd ed.,1983, 308, Van Nostrand Reinhold, New York
14. Guisti, D. M; et al. *J. Water Pollut. Control Fed.* 1974, **46**(5), 947-965
15. *Toxicol. Appl. Pharmacol.* 1982, **63**, 150
16. Izmerov, N. F; et al. *Toxicometric Parameters of Industrial Toxic Chemicals under Single Exposure* 1982, 28, Moscow
17. *J. Ind. Hyg. Toxicol.* 1949, **31**, 343

18. *J. Pharmacol. Exp. Ther.* 1946, **82**, 28
19. *Union Carbide Data Sheet* 1965, Union Carbide Corporation, New York
20. *J. Ind. Hyg. Toxicol.* 1944, **26**, 269
21. *The Merck Index* 11th ed., 1989, 237, Merck & Co. Inc., Rahway, NJ
22. Zeiger, E; et al. *Environ. Mol. Mutagen.* 1987, 9(Suppl. 9), 1-109
23. Neilson, G.D. *Pharmacol. Toxicol.* 1988, **63**(4), 293-304
24. Cheever, K. L; et al. *Toxicol. Appl. Pharmacol.* 1982, **63**(1), 150-152
25. *Dangerous Prop. Ind. Mater. Rep.* 1986, **6**(2), 45-48
26. Browning, E. *Br. J. Ind. Med.* 1959, **16**, 23-39
27. *ECETOC Technical Report No. 30(4)* 1991, European Chemical Industry Ecology and Toxicology Centre, B-1160 Brussels

B260   *sec*-Butylamine

CH3CH2CH(NH2)CH3

CAS Registry No. 13952-84-6
Synonyms 2-aminobutane; Butafume; 2-butanamine; Deccotane; 1-methylpropylamine; Tutane
Mol. Formula $C_4H_{11}N$ **Mol. Wt.** 73.14
Uses Fungicide.

Physical properties

M. Pt. −104°C; **B. Pt.** 63°C; **Flash point** −9.4°C (closed cup); **Specific gravity** d_{20}^{20} 0.724; **Volatility** v. den. 2.52.

Solubility
Water: ≥ 10 mg ml^{-1}. Organic solvent: ethanol, diethyl ether, acetone

Occupational exposure

UN No. 1125; **Supply classification** highly flammable and irritant.

Risk phrases
Highly flammable – Irritating to eyes, respiratory system and skin (R11, R36/37/38)

Safety phrases
Keep away from food, drink and animal feeding stuffs – Keep away from sources of ignition – No Smoking – Do not empty into drains (S13, S16, S29)

Ecotoxicity

Fish toxicity
LC$_{50}$ (24 hr) creek chub 20-60 mg l^{-1} (1).

Effects to non-target species
LD$_{50}$ oral redwing blackbird >96.0 mg kg^{-1} (2).

Environmental fate

Nitrification inhibition
Nitrosomonas sp. no inhibition of ammonia oxidation at 100 mg l^{-1} (3).

Degradation studies

Readily biodegradable (4).

Mammalian and avian toxicity

Acute data

LD_{50} oral rat, dog 150-225 mg kg^{-1} (5,6).
LD_{50} dermal rabbit 2500 mg kg^{-1} (6).

Carcinogenicity and long-term effects

In 2 yr study rats and dogs receiving 2500 mg kg^{-1} in diet suffered no ill-effects (7).

Genotoxicity

Salmonella typhimurium TA98, TA100, TA1535, TA1537 with and without metabolic activation negative (8).

Any other adverse effects

Systemic effects in rats which were given 100- 600 mg kg^{-1} by gavage included sedation, ataxia, nasal discharge, gasping, salivation convulsions and death at highest doses. Pathological examination showed pulmonary oedema (5).

Legislation

Limited under EC Directive Drinking Water Quality 80/778/EEC. Fungicide: maximum admissible concentration 0.5 µg l^{-1} (9).

Any other comments

Experimental toxicology and human health effects reviewed (10).

References

1. McKee, J. E. *Water Quality Criteria* 1963, State Water Purity Board, Res. Agency, CA
2. Schafer, E. W. *Arch. Environ. Contam. Toxicol.* 1982, **12**(3), 355-382
3. Hockenbury, M. R. *J. Water Pollut. Control Fed.* 1977, **49**, 768-777
4. Painter, H. A; et al. *Comm. Eur. Commun. Rep. EUR 9962* 1985, 105, Water Res. Cent., Herts
5. Cheever, K. L; et al. *Toxicol. Appl. Pharmacol.* 1982, **63**(1), 150-152
6. *Pesticide Index* 1976, **5**, 33
7. Worthing, C. R. (Ed.) *The Pesticide Manual* 9th ed., 1991, British Crop Protection Council, Farnham
8. Zeiger, E; et al. *Environ. Mol. Mutagen.* 1987, 9(Suppl 9), 1-109
9. *EC Directive Relating to the Quality of Water Intended for Human Consumption* 1982, 80/778/EEC, Office for Official Publications of the European Communities, 2 rue Mercier, L-2985 Luxembourg
10. *ECETOC Technical Report No. 30(4)* 1991, European Chemical Industry Ecology and Toxicology Centre, B-1160 Brussels

B261   *tert*-Butylamine

$(CH_3)_3CNH_2$

CAS Registry No. 75-64-9
Synonyms 2-aminoisobutane; 2-amino-2-methylpropane; trimethylaminomethane;
1,1-dimethylethylamine
Mol. Formula $C_4H_{11}N$ **Mol. Wt. 73.14**
Uses Intermediate in organic synthesis.
Occurrence Detected in leachate from a municipal refuse waste disposal site in the
Netherlands, at a concentration of 41 ppm (1).

Physical properties

M. Pt. $-72°C$; **B. Pt.** 44-46°C; **Specific gravity** d_4^{20} 0.69.

Solubility
Organic solvent: ethanol

Ecotoxicity

Fish toxicity
LC_{50} (96 hr) rainbow trout 28 mg l^{-1} (2).
EC_{50} (96 hr) *Selenastrum capricornutum* 16 mg l^{-1} (3).

Invertebrate toxicity
EC_{50} (24 hr) *Daphnia magna* 136 mg l^{-1} (3).

Environmental fate

Degradation studies
No biodegradation observed over a 12-day incubation period using Sapromat
respiration assays with river mud bacteria inocula, treatment plant sludge inocula and
adapted bacteria inocula. Initial concentration 10-100 ppm (2).

Mammalian and avian toxicity

Acute data
LD_{50} oral rat 80 mg kg^{-1} (4).
LD_{50} oral mouse 900 mg kg^{-1} (5).

Any other comments

The relationship between biotransformation of aromatic nitrogenous compounds in
rabbit liver microsomal preparations and carcinogenicity is discussed (6).
Experimental toxicology and human health effects reviewed (7).

References

1. Harmsen, J. *Water Res.* 1983, **17**, 699
2. Calamari, D; et al. *Chemosphere* 1980, **9**, 753
3. Calamari, D; et al. *Est. Haz. Aminies on Aquatic Life* 1982b, Report EUR 7549 EN/FR
4. *Toxicol. Appl. Pharmacol.* 1982, **63**, 150
5. Hann, W; et al. *Water Quality Characteristics of Hazardous Materials* 1974, **4**, Texas A &
   M University

6.   Stier, A; et al. *Xenobiotica* 1980, **10**(7-8), 661-673
7.   *ECETOC Technical Report No. 30(4)* 1991, European Chemical Industry Ecology and Toxicology Centre, B-1160 Brussels

B262   *N*-Butylaniline

CAS Registry No. 1126-78-9
Synonyms Benzenamine-*N*-butyl; *N*-(*n*-butyl)aniline; *N*-butylbenzenamine; 4-(phenylamino)butane
Mol. Formula $C_{10}H_{15}N$ **Mol. Wt.** 149.24
Uses Intermediate in organic synthesis. In dyestuff manufacture.

Physical properties

M. Pt. −14.4°C; **B. Pt.** 241.6°C; **Flash point** 107°C; **Specific gravity** d_4^{20} 0.932; **Partition coefficient** log P_{ow} 3.10.

Solubility
Organic solvent: ethanol, diethyl ether

Occupational exposure

UN No. 2738; **HAZCHEM Code** 3X; **Conveyance classification** toxic substance.

Mammalian and avian toxicity

Acute data
LD_{50} oral rat 1620 mg kg^{-1} (1).
LD_{50} dermal rabbit 5990 mg kg^{-1} (1).

Irritancy
Dermal rabbit (24 hr) 10 mg caused severe irritation (1).

References
1.   *AMA Arch. Ind. Hyg. Occup. Med.* 1954, **10**, 61

B263   Butylate

$$CH_3CH_2SC(O)N[CH_2CH(CH_3)_2]_2$$

CAS Registry No. 2008-41-5
Synonyms carbamothioic acid, bis(2-methylpropyl)-, *S*-ethyl ester; carbamic acid, diisobutylthio-, *S*-ethyl ester; diisobutylthiocarbamic acid, *S*-ethyl ester; Diisocarb;

S-ethyl bis(2-methylpropyl)carbamothioate; ethyl *N,N*-diisobutylthiocarbamate; Sutan

Mol. Formula $C_{11}H_{23}NOS$ **Mol. Wt. 217.38**
Uses Herbicide.

Physical properties

M. Pt. $_{21}$ 137.5-138°C; **Specific gravity** 0.9417; **Partition coefficient** log P_{ow} 4.146 (1).;
Volatility v.p. 1.3×10^{-3} mmHg at 25°C; v.den. 0.9402 at 25°C.

Solubility
Water: 45 mg l^{-1} at 25°C. Organic solvent: acetone, ethanol, kerosene, methyl, isobutyl ketone, xylene

Ecotoxicity

Fish toxicity
LC_{50} (96 hr) rainbow trout, bluegill sunfish 4.2-6.9 mg l^{-1} (1).
Invertebrate toxicity
LC_{50} (96 hr) scud 11 mg l^{-1} (2).
EC_{50} (30 min) *Photobacterium phosphoreum* 18.1 mg l^{-1} Microtox test (3).

Environmental fate

Degradation studies
Degrades in soil to ethylmercaptan, carbon dioxide and diisobutylamine. Residual activity 4 months (1).
Biodegradation $t_{1/2}$ in soil 1-3 wk (4).

Abiotic removal
Removed by vaporisation when applied to surface of wet soils without incorporation. Little loss occurs after application to dry soil surfaces (4).
In water exposed to sunlight 99% loss within 48 hr (5).

Mammalian and avian toxicity

Acute data
LD_{50} oral guinea pig 1660-4000 mg kg^{-1} (1,3).
LD_{50} percutaneous rabbit >5000 mg kg^{-1} (1).

Sub-acute data
In 90 day feeding study in rats 32 mg kg^{-1} day^{-1} caused no adverse effects (1).

Metabolism and pharmacokinetics
A major metabolic pathway in rats, accounting for 27-45% of the administered dose was thiocarbamate → thiocarbamate sulfoxide → *S*-(*N,N*-dialkylcarbamoyl)glutathione → *S*-(*N,N*-dialkylcarbamoyl) cysteine → *S*-(N,N-dialkylcarbamoyl)mercapturic acid,*S*-(*N,N*- dialkylcarbamoyl)mercaptoacetic acid and *N*[(*S*-*N,N*-dialkylcarbomoyl) mercaptoacetyl]glycine (6).

Genotoxicity
In vivo mouse bone marrow cells increased frequency of chromosomal aberrations (7).

Legislation

Limited under EC Directive on Drinking Water Quality 80/778/EEC. Herbicide: maximum admissible concentration 0.5 µg l^{-1} (8).

Any other comments

Physico-chemical properties, metabolic and environmental fate, genotoxicity and toxicology reviewed (9,10).

References

1. *The Agrochemicals Handbook*, 3rd ed., 1991, RSC, London
2. *Handbook of Acute Toxicity of Chemicals to Fish and Aquatic Invertebrates* 1980, **137**, 81, Dept. Fish & Wildlife, Washington, DC
3. Kaiser, K. L. E; et al. *Water Pollut. Res. J. Canada* 1991, **26**(3), 361-431
4. *Herbicide Handbook* 4th ed., 1979, 89, Weed Sci. Soc. Am., Champaign IL
5. *World Rev. Pest. Control* 1970, **9**, 119
6. Hubbell, J. P; et al. *J. Agric. Food Chem.* 1977, **25**, 404-413
7. Pilinskaya, M. A; et al. *Tsitol. Genet.* 1980, **14**(6), 41-47
8. *EC Directive Relating to the Quality of Water Intended for Human Consumption* 1982, 80/778/EEC, Office for Official Publications of the European Communities, 2 rue Mercier, L-2985 Luxembourg
9. WHO *Thiocarbamate pesticides: A general introduction* 1988, 1-49
10. Casido, J. E; et al. *Prelude Biochem. Physiology* 1975, **5**(1), 1-11

B264   Butylated hydroxyanisole

CAS Registry No. 25013-16-5
Synonyms BHA; phenol, (1,1-dimethylethyl)-4-methoxy-; phenol, *tert*-butyl-4-methoxy-; *tert*-butyl-4-hydroxyanisole; Antracine 12; Embanox
Mol. Formula $C_{11}H_{16}O_2$ **Mol. Wt.** 180.25
Uses Antioxidant in the polymer and food industries. Used in cosmetics and essential oils. Antimicrobial properties.

Physical properties

M. Pt. 48-55°C; **B. Pt.** $_{733}$ 264-270°C.

Solubility

Organic solvent: ethanol, propylene glycol, arachis oil, chloroform, diethyl ether

Ecotoxicity

Fish toxicity

Exposure of trout to 0.03-0.3% for 8 wk had no effect on hepatic tumour incidence (1).

Mammalian and avian toxicity

Acute data
LD_{50} oral rabbit, rat 2100-2200 mg kg^{-1} (2,3).
LD_{50} intraperitoneal rat 881 mg kg^{-1} (4).

Sub-acute data
Groups of 10 (5 wk old) ♂, ♀ Wistar rats were fed a diet containing 0-2% BHA *ad libitium* for 2 wk, BrdU (bromodeoxyuridine) was injected and incorporated into the DNA of cells during DNA synthesis. Cell kinetic parameters were measured and results showed the oesophagus, glandular stomach, small intestine, large bowel and forestomach were possible target tissues for the proliferation enhancing effects of BHA (5).

Carcinogenicity and long-term effects
Inadequate evidence for carcinogenicity to humans, sufficient evidence for carcinogenicity to animals, IARC classification group 2B (6).
Administration in diet (80 wk) 0-2% BHA to Japanese house musk shrews which have no forestomach. All high dose animals died of gastrointestinal bleeding within 8 wk of commencement of treatment. Other surviving animals exhibited adenomatous hyperplasia of the lung (7).
Lakeview (LVG) Syrian golden hamsters were fed diets containing 2% BHA for 30 wk, no papilloma or severe hyperplasia was found in any of the experimental or control groups. A significant increase in hamsters with mild hyperkeratous and mild hyperplasia was observed. 90% of Miski hamsters developed moderate to severe hyperplasia and 80% had papilloma in the forestomach (8).
In 2 yr feeding study on F344 rats 2% BHA induced carcinomas, epithelial downgrowths and cellular proliferation. Superficial hyperplasias, inflammatory lesions and many papillomas regressed when treatment was stopped after 12 months (9).
Fed in the diet of rats, hamsters and mice, 2% BHA caused stomach papillomas in 91.5%, 95% and 14.3% respectively. Squamous cell carcinomas occurred at lower rates (10).
Although humans do not have squamous epithelium in the stomach, possible tumour induction in squamous cell epithelium of the oesophagus is unlikely at food additive levels of use (11).

Teratogenicity and reproductive effects
Fed to rats in diet at 0, 0.125, 0.25 or 5.0% from prior to conception to 90 days old. No changes in maternal weight, reproductive performance, mortality, offspring growth post-weaning or brain weight. Marginal increase in mortality up to 30 days at 0.25%. Delayed startle development at 0.5 and 0.25% (12).
IC_{50} rat embryonic culture 50 mg l^{-1} inhibited production of differential foci in limb bud cells and 84 mg l^{-1} inhibited differentiation in mid-brain cells. Human embryonic platal mesenchyinal cell growth assay with microbial activation inhibited cell growth (13).

Metabolism and pharmacokinetics
Absorbed from gastrointestinal tract and excreted in urine as glucuronide and sulfate metabolites (14).
Following a single oral dose 0.5 mg kg^{-1} to humans, BHA recovery in urine and

faeces was 95%. BHA was excreted mainly as conjugated BHA in the urine and conjugated *tert*-butyl hydroquinone in the faeces. No free BHA was found in urine or faeces. In rats BHA is *O*-dimethylated to *tert*-butylhydroquinone (15).

Following oral administration in rat, rabbit and human BHA is absorbed and rapidly excreted with little evidence of long-term tissue storage (16).

Absorbed from gut by passive diffusion. No evidence of tissue storage in rats fed 0.12% in diet for 21 months (17).

Oral dog up to 100 mg kg^{-1} day^{-1} for 1 yr. No storage observed in fat, liver or brain (18).

27 to 77% excreted in urine by humans, mostly in 24 hr. No dealkylation or hydroxylation (19).

Sensitisation

Three positive reactions observed from a total 112 patients with eczematous dermatitis patch tested with 2% butylated hydroxyanisole (20).

Genotoxicity

Salmonella typhimurium TA97, TA102, TA104, TA100 with and without metabolic activation negative. Doses >100 μg plate^{-1} exhibited toxic effects (21,22).

Saccharomyces cerevisiae D6, D7, with metabolic activation induced epigenic action, gene conversion and reverse mutation (23).

Chinese hamster fibroblast cell line *in vitro* with metabolic activation only weakly induced chromosomal abberations (21).

BHA induced chromosome aberrations in Chinese hamster ovary cells with metabolic activation (24).

BHA did not induce micronuclei in mice after single intraperitoneal injection (25).

BHA stimulates superoxide formation in rat liver microsomes up to 10-fold with metabolic activation via the BHA metabolites *tert*- butyhydroquinone and *tert*-butylquinone. No oxygen-activating properties can be attributed to BHA itself (26).

Newt larvae negative results in micronucleus test using erythrocytes (27).

Any other adverse effects to man

Oral ♂ human 0.5 mg kg^{-1} BHA for 10 consecutive days had no effect on clinical, biochemical parameters and Phase I and Phase II biotransformation capacity (28).

An outbreak of toxic methaemoglobinaemia in a paediatric ward was attributed to the preservative in an infant feed formula (29).

BHA inhibits respiratory control by stimulating state 4 respiration thus acting as a membrane uncoupler (30).

Any other comments

Acceptable daily intake of 300 μg kg^{-1} body weight established (31).

The toxicity, physical and metabolic properties, genotoxicity and antitumour activity of BHA reviewed (11,16,32-39).

The toxicology of food additives evaluated (40).

Human health effects and experimental toxicology reviewed (41).

References

1. Goeger, D. E; et al. *Carcinogenesis* 1988, **9**(10), 1793-1800
2. Lehman, A. J; et al. *Adv. Food Res.* 1951, **3**, 197
3. *J. Am. Oil Chem. Soc.* 1977, **54**, 239
4. *Toxicol. Lett.* 1985, **27**, 15
5. Verhagen, H; et al. *Carcinogenesis* 1990, **11**(9), 1461-1468

6. *IARC Monograph* 1987, **Suppl. 7**, 59
7. Amo Hirayuki; et al. *Carcinogenesis* 1990, **11**(1), 151-154
8. Lam, L. K. T; *Carcinogenesis* 1988, **9**(9), 1611-1616
9. Nera, E. A; et al. *Toxicology* 1988, **53**(2-3), 251-268
10. Ito, N; et al. *Shokuhin Eisei Kenkyu* 1990, **40**(3), 7-18, (Jap.) (*Chem. Abstr.* **113**, 96278x)
11. Grice, H. C. *Food Chem. Toxicol.* 1988, **26**(8), 717-723
12. Vorhees, C. V; et al. *Neurobehavioural Toxicol. Teratol.* 1981, **3**, 321-392
13. Tsuchiya, T; et al. *Toxicol. In Vitro* 1988, **2**(4), 291-296
14. El-Rashidy, R; et al. *Biopharm. Drug Disposit.* 1983, **4**, 389
15. Verhagen, H. et al. *Toxicology* 1989, **27**(3), 151-158‾
16. Conneng, D. M; et al. *Food Chem. Toxicol.* 1986, **24**(10-11), 1145-1148
17. Clayton, G. D; et al. (Ed.) *Patty's Industrial Hygiene and Toxicology* 3rd ed., 1981-1982, John Wiley Sons, New York
18. Parke, D. V. *The Biochemistry of Foreign Compounds* 1968, Pergamon, Oxford
19. Furia, T. E. (Ed.) *CRC Handbook of Food Additives* 2nd ed., 1972, Chemical Rubber Co., Cleveland, OH
20. Roed-Petersen, J; et al. *Br. J. Dermatol.* 1976, **94**, 233
21. Matsuoka, A. *Mutat. Res.* 1990, **241**(2), 125-132
22. Hageman, G. J; et al. *Mutat. Res.* 1988, **203**(3-4), 207-211
23. Nouaim, R; et al. *Sci. Aliments* 1988, **8**(4), 431-445
24. Phillips, B. J; et al. *Mutat. Res.* 1989, **214**(1), 105-114
25. Hayashi, M; et al. *Food Chem. Toxicol.* 1988, **26**(6), 487-500
26. Kahl, R; et al. *Toxicology* 1989, **59**(2), 179-194
27. Fernander, M; et al. *Mutagenesis* 1989, **4**(1), 17-26
28. Verhagen, H; et al. *Toxicology* 1989, **8**(6), 451-459
29. Nitzan, M; et al. *Clin. Toxicol.* 1979, **15**, 273
30. Thompson, D; et al. *Biochem. Pharmacol.* 1988, **37**(11), 2201-2207
31. FAO/WHO Expert Committee on Food Additives *Tech. Rep. Ser. Wld. Hlth. Org.* 1987, No. 751
32. *IARC Monograph* 1986, **40**, 123-159, 444
33. Horman, G. *Int. J. Biochem.* 1988, **20**(7), 639-651
34. Wattenberg, L. W; *Food Chem.Toxicol.* 1986, **24**(10-11), 1099-1102
35. Williams, G. M; *Food Chem. Toxicol.* 1986, **24**(10-11), 1163-1166
36. *Cancer Res.* 1986, **46**, 165
37. *Jap. J. Cancer Res.* 1983, **74**, 459
38. *Jap. J. Cancer Res.* 1982, **73**, 332
39. *Carcinogenesis* 1983, **4**, 895
40. WHO *Evaluation of Certain Food Additives and Contaminants* 1983, 1-47, Geneva
41. *ECETOC Technical Report No.* **30**(4) 1991, European Chemical Industry Ecology and Toxicology Centre, B-1160 Brussels

B265   Butylated hydroxytoluene

CAS Registry No. 128-37-0
Synonyms BHT; 2,6-bis(1,1-dimethylethyl)-4- methylphenol; *p*-cresol, 2,6-di-*tert*-butyl-; antioxidant 4; E321; Improval; Tropanol

---

Mol. Formula $Cl_5H_{24}O$ **Mol. Wt.** 217.46

Uses Antioxidant used in food, petroleum products, synthetic rubbers, plastics animal and vegetable oils and soaps. Antiskinning agent in paints and inks.

Physical properties

M. Pt. 70°C; **B. Pt.** 265°C; **Flash point** 127°C; **Specific gravity** d_4^{20} 1.048; **Volatility** v. den. 7.6.

Solubility
Water: <1 mg ml^{-1}. Organic solvent: acetone, benzene, dimethyl sulfoxide, ethanol, isopropanol, methanol, toluene

Ecotoxicity

Fish toxicity
Non-toxic to goldfish in saturated solution 0.4 mgl^{-1} (1).

Mammalian and avian toxicity

Acute data
TD$_{Lo}$ oral human 80 mg kg^{-1} (2).
LD$_{Lo}$ oral rabbit 2100 mg kg^{-1} (3).
LD$_{50}$ intraperitoneal, intravenous mouse 138-180 mg kg^{-1} (4,5).

Carcinogenicity and long-term effects
No adequate evidence for carcinogenicity to humans, limited evidence for carcinogenicity to animals, IARC classification group 3 (6).
Oral F344 rats (76 wk) 100-6000 ppm in food, no increase in neoplasms at any site (7).

Metabolism and pharmacokinetics
After a single oral dose administered to mice, 80%-90% was excreted in the urine within 120 hr (8).
Cytochrome P450 catalyses oxidation of BHT to form 3 metabolites: quinone methide, hydroxy-tert-butyl analogue of BHT and hydroxy-quinone methide (9).

Irritancy
Dermal human (48 hr) 500 mg caused mild irritation (10).
Dermal rabbit (48 hr) 500 mg caused moderate irritation, and 100 mg instilled into rabbit eye caused moderate irritation (10,11).

Sensitisation
Contact sensitivity and allergic skin reactions have been reported (12).
Of 112 patients with eczematous dermatitis patch tested with 2% BHT, 3 had positive reactions (13).

Genotoxicity

Salmonella tymphimurium TA1535, TA1537, TA98, TA100 with or without metabolic activation negative (14,15).
Mouse lymphoma L5178Ytk+/tk- with metabolic activation cell forward mutagen assay positive (16).
Chinese hamster ovary cells with or without metabolic activation induction of chromosome aberrations and sister chromatid exchange negative (17).

Any other adverse effects

BHT in diet of doses of 0.875, 1.75 and 2.5% to mice significant differences in litter size, pup weight and litter weight reported (18).

Developmental toxicity of doses of 0.25, or 0.5%, little effect on adult behaviour and no special toxicity to the central nervous system reported (19).

A woman who ingested 4g BHT experienced epigastic cramping, nausea, vomiting and generalised weakness, dizziness, confusion and brief loss of conciousness (4).

Cutaneous, urticarial, disseminated eruption associated with chewing gum containing BHT reported (20).

BHT was tested for guinea pig, rat and human erythrocyte haemolysis. BHT induced complete lysis (21).

Adverse effects to kidneys, lungs, heart and blood reported in experimental animals (22).

Induced lung injury in mice (23).

Accumulates in human adipose tissue (24).

An outbreak of toxic methaemoglobinaemia in a paediatric ward was attributed to the antioxidant BHA, BHT and propyl gallate used as preservatives in a soybean infant food formula (25).

Legislation

The German Advisory Board on Existing Chemicals has compiled data on the environmental fate, ecotoxicity and toxicity of BHT (26).

Any other comments

The antioxidant mechanism of BHT, metabolism and pulmonary toxicity, antiviral properties, carcinogenic and anticarcinogenic effects reviewed (27-31).

References

1. *Shell Ind. Chemicalien Gids*, 1975, Shell Chemie, Gravenshagen
2. Shlian, D. M; et al. *N. Engl. J. Med.* 1986, **314**, 648
3. *AMA Arch. Ind. Health*, 1955, **11**, 93
4. *Toxicol. Appl. Pharmacol.* 1981, **61**, 475
5. *J. Med. Chem.*, 1980, **23**, 1350
6. *IARC Monograph* 1987, **Suppl 7**, 59
7. Williams, G. M; et al. *Food Chem. Toxicol.* 1990, **28**(12), 799-806
8. Daugherty, J. P; et al. *Res. Commun. Subst. Abuse* 1980, **1**(1), 99
9. Bolton, J. L; et al. *Drug Metab. Dispos.* 1991, **19**(2), 467-472
10. *AMA Arch. Ind. Hyg. Occup. Med.*, 1952, **5**, 311
11. *Sbornik Vysledku Toxicologickeho Vysetreni Latek A Pripravko*, 1972, 57
12. *Martindale The Extra Pharmacopoeia*, 29th ed., 1989, Pharmaceutical Press London
13. Roed-Peterson, J; et al. *Br. J. Derm.*, 1976, **94**, 233
14. Mortelmans, K; et al. *Environ. Mol. Mutagen.*, 1986, **8**(Suppl. 7), 1-119
15. Yoshida, Y. *Mutat. Res.* 1990, **242**(3), 209-217
16. McGregor, D. B; et al. *Environ. Mol. Mutagen*, 1988, **11**, 91-118
17. Galloway, S. M; et al. *Environ. Mol. Mutagen*, 1987, **10**(Suppl. 10), 1-175
18. Tanaka, T; et al. *Kenkyu-Tokyo-Koritsu Eisei Kenkyusho* 1989, **40**, 337-340, (Jap.) (*Chem. Abstr.*, **112**, 173749k)
19. Holder, G. M; et al. *J. Pharm. Pharmacol.* 1970, **22**, 375
20. Moneret Vautnn, D. A; et al. *Lancet*, 1986, **1**, 617
21. Sgaragli, G. P; et al. *Boll. Soc. Ital. Biol. Sper.*, 1975 **5**(22), 1712-1715, (Ital.) (*Chem Abstr.*, **85** 45020c)

22. *Bibra Toxicity Profiles*, 1989, British Industrial Biological Research Association, Carshalton
23. Kehrer, J. P; et al. *Toxicol. Lett.* 1990, **52**(1), 55-61
24. Conacher, H.B.S; et al. *Food Chem. Toxicol.* 1986, **24**(10-11), 1159-1162
25. Nitzan, M; et al. *Clin. Toxicol.* 1979, **15**, 273
26. Haltrich, W. G. *Vom Wasser* 1989, **73**, 11-24, (Ger.) (*Chem. Abstr.*, **112**, 124566q)
27. Kleinjans, J; et al. *Voeding* 1987, **48**(5), 153-158, (Neth.) (*Chem. Abstr.*, **107**, 95348j)
28. Witschi, H; et al. *Pharmacol. Toxicol.* 1989, **42**(1), 89-113
29. *Food Additives Series*, 1972, No. 3, WHO, Geneva
30. Tanaka, T. *Kenkyu Nenpo Tokyo-Toritsu Eisei Kenkyushu*, 1989, **40**, 337-340, (Jap.) (*Chem. Abstr.*, **112**, 173749k)
31. Richards, J. T; et al. *Antivirial Res.* 1985, **5**, 281

B266   Butylbenzene

$(CH_2)_3CH_3$

CAS Registry No. 104-51-8
Synonyms 1-phenylbutane
Mol. Formula $C_{10}H_{14}$ **Mol. Wt.** 134.22
Uses Intermediate in chemical synthesis. Solvent. Manufacture of pesticides.

Physical properties

M. Pt. –88°C; **B. Pt.** 183°C; **Flash point** 71°C; **Specific gravity** d_4^{20} 0.86; **Volatility** v.p. 1 mmHg at 22.7°C.

Solubility
Organic solvent: miscible ethanol, diethyl ether, benzene

Occupational exposure

UN No. 2709; **HAZCHEM Code** 3☒; **Conveyance classification** flammable liquid.

Mammalian and avian toxicity

Acute data
LD_{Lo} rat 5000 mg kg^{-1} (1).

Irritancy
Eye irritant (2).

References

1. *Arch. Ind. Health* 1959, **19**, 403
2. Yeung, L. K. K; et al. *Atm. Environ.* 1973, **7**, 551

B267   *tert*-Butylbenzene

C(CH₃)₃

CAS Registry No. 98-06-6
Synonyms 2-methyl-2-phenylpropane; (1,1-dimethylethyl)benzene;
trimethylphenylmethane; pseudobutyl benzene
Mol. Formula $C_{10}H_{14}$ **Mol. Wt.** 134.22
Uses Solvent. Intermediate in organic synthesis.

Physical properties

M. Pt. −58°C; **B. Pt.** 168°C; **Flash point** 60°C; **Specific gravity** d_4^{20} 0.86; **Partition coefficient** log P_{ow} 4.11; **Volatility** v.p. 1.5 mmHg at 20°C; v. den. 4.62.

Occupational exposure

UN No. 2709; **HAZCHEM Code** 3⊠; **Conveyance classification** flammable liquid.

Environmental fate

Degradation studies
Nonbiodegradable/qualified (1).

Absorption
Mediterranean red sandy clay soil samples with different moisture content (0, 0.8, 4 and 12% wt./wt.) were contaminated with vapour or liquid mixtures containing *tert*-butylbenzene. Adsorption on soil was 50 and 15 µg g⁻¹ at 7°C; 120 and 47 µg g⁻¹ at 17°C; 210 and 60 µg g⁻¹ at 27°C and 333 and 100 µg g⁻¹ at 34°C for oven dried and air dried samples, respectively. Volatisation was rapid 92-99% in 2-6 hr period (2).

Mammalian and avian toxicity

Acute data
LD_{Lo} oral rat 5000 mg kg⁻¹ (3).

Irritancy
Eye irritant (4).

References

1. *MITI Report* 1984, Ministry of International Trade and Industry, Tokyo
2. Acher, A. J; et al. *J. Contam. Hydrol.* 1989, **4**(4), 333-345
3. *Arch. Ind. Health* 1959, **19**, 403
4. Bach, R. W; et al. *Staub.-Reinhalt. Luft* 1973, **3**

B268   4-*tert*-Butylbenzoic acid

CO$_2$H

C(CH$_3$)$_3$

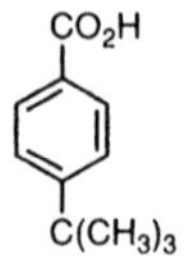

CAS Registry No. 98-73-7

Synonyms benzoic acid, 4-(1,1-dimethylethyl); benzoic acid, *p-tert*-butyl; *p-tert*-butylbenzoic acid; TBBA; *N,N'*,- terephthalidenebis-*tert*-(4-*tert*-butylaniline)

Mol. Formula C$_{11}$H$_{14}$O$_2$ **Mol. Wt.** 178.23

Uses Crosslinking agent with epoxy resins. Dispersing agent in pigments.

Physical properties

M. Pt. 166.3°C; **Specific gravity** d$_4^{20}$ 1.142.

Solubility
Organic solvent: benzene, ethanol

Ecotoxicity

Fish toxicity
LC$_{50}$ (96 hr) goldfish 33 mg l^{-1} (1).

Mammalian and avian toxicity

Acute data
LD$_{50}$ oral rat 735 mg kg^{-1} (2).

Sub-acute data
Dermal rat (13 wk) 70-140 mg kg^{-1} 5 day wk^{-1} caused changes to liver, testis, epididymis and kidneys (3).

Any other adverse effects

Dermal application or dust inhalation rat (7 day) dose unspecified cause reduction in testis weights, sperm counts and lactate dehydrogenase-Y enzyme levels. A reduction of absence of spermatogenic cell types was observed. Recovery following cessation of exposure would be anticipated (4).

Any other comments

No association reported between ♂ human occupational exposure to *4-tert*-butyl benzoic acid and sperm count (5).

Experimental toxicology, human health and reproductive effects, and occupational exposure reviewed (6,7).

All reasonable efforts have been taken to find information on isomers of this compound, but no relevant data are available.

References

1. *Shell Industrie Chemicalien Gids* 1975, Shell Chemie, Gravenshagen
2. *Food Cosmet. Toxicol.* 1965, **3**, 289
3. Cagen, S. Z; et al. *J. Am. Coll. Toxicol.* 1989, **8**(5), 1027-1038

4. Lu, C.C.J. *Am. Coll. Toxicol.* 1987, **6**(2), 233-243
5. Rosenberg, M. J; et al. *J. Occup. Med.* 1987, **29**(7), 584-591
6. Whortun, M. D; et al. *Scand. J. Work, Environ. Health* 1981, **7**(3), 204-213
7. *ECETOC Technical Report No. 30(4)* 1991, European Chemical Industry Ecology and Toxicology Centre, B-1160 Brussels

B269   Butyl chloride

$$CH_3(CH_2)_3Cl$$

CAS Registry No. 109-69-3
Synonyms butane, 1-chloro; *n*-butylchloride; 1-chlorobutane;
n-propylcarbinylchloride
Mol. Formula C_4H_9Cl **Mol. Wt.** 92.57
Uses Butylating agent in organic synthesis. Veterinary medicine as an anthelmintic.

Physical properties

M. Pt. –123.1°C; **B. Pt.** 77-78°C; **Flash point** –9.4°C (closed cup); **Specific gravity** d_{20}^{20} 0.8875; **Partition coefficient** log P_{ow} 2.64; **Volatility** v.p. 80.1 mmHg at 20°C; v. den. 3.21.

Solubility
Water: 660 mg l^{-1} at 12°C. Organic solvent: miscible ethanol, diethyl ether

Occupational exposure

UN No. 1127; **HAZCHEM Code** 3◙E; **Conveyance classification** flammable liquid; **Supply classification** highly flammable.

Risk phrases
Highly flammable (R11)

Safety phrases
Keep container in a well ventilated place – Keep away from sources of ignition – No Smoking – Do not empty into drains (S9, S16, S29)

Ecotoxicity

Fish toxicity
LC_{50} (7 day) guppy 97 ppm (1).

Invertebrate toxicity
EC_{50} (30 min) *Photobacterium phosphoreum* 735 mg l^{-1} Microtox test (2).

Environmental fate

Degradation studies
Activated sludge (24 hr) 2.6% ThoD (3).

Mammalian and avian toxicity

Acute data
LD_{50} oral rat 2200-2670 mg kg^{-1} (4,5).
LD_{50} oral mouse, guinea pig 5600-8000 mg kg^{-1} (5,6).
LC_{Lo} (4 hr) inhalation rat 8000 ppm (4).
LD_{Lo} dermal rabbit 20 g kg^{-1} (7).

Sub-acute data
Oral rat (6 month) 2 mg kg^{-1} day^{-1} raised the level of inorganic phosphate in blood and altered the activity of blood alkaline phosphatase and succinate dehydrogenase activity (6).

Carcinogenicity and long-term effects
No increase in the incidence of lung tumour observed in ♂, ♀ rats and mice (dose and duration unspecified) (8,9).
National Toxicology Program gavage rats, mice (2 yr) no evidence of carcinogenicity (10).

Teratogenicity and reproductive effects
Embryotoxic and teratogenic effects noted at the sublethal dose of 733 mg kg^{-1} (6).

Irritancy
Dermal rabbit (24 hr) 10 mg caused mild irritation and 500 µg instilled into rabbit eye caused mild irritation (4).

Genotoxicity

Salmonella typhimurium TA98, TA100, TA1535, TA1537, TA97 with and without metabolic activation negative (11).
DNA polymerase deficient *Escherichia coli* negative (12).

Any other comments

USSR maximum permissible concentration in open water 4 µg l^{-1} (5).
Experimental toxicology and human health effects reviewed (13).

References

1. Konemann, W. H. *Quantitative Structure-Activity Relationships for Kinetics and Toxicity of Aquatic Pollutants and their Mixtures in Fish* 1979, Univ. Utrecht
2. Kaiser, K. L. E; et al. *Water Pollut. Rep. J. Canada* 1991, **26**(3), 361-431
3. Gerhold, R. M; et al. *J. Water Pollut. Control Fed.* 1966, **38**(4), 562
4. *Arch. Ind. Hyg. Occup. Med.* 1954, **10**, 61
5. *Med. Prof.* 1979, **7**, 105
6. Rudner, M. I; et al. *Gig. Sanit.* 1979, **43**(3), 11-15
7. Deichmann, W. B; et al. *Toxicology of Drugs and Chemicals* 969, 745, Academic Press, London
8. Poirier, L. A; et al. *Cancer Res.* 1975, **35**, 1411-1415
9. *Prehled Prumyslove Toxikol. Org Latky* 1986, 100
10. *National Toxiology Program Research & Testing Div.* 1992, Report No. TR-312, NIEHS, Research Triangle Park, NC
11. Zeiger, E; et al. *Environ. Mutagen.* 1987, **9**(Suppl. 9), 1-110
12. Fluck, E. R; et al. *Chem. Biol. Interact.* 1976, **15**, 219
13. *ECETOC Technical Report No. 30(4)* 1991, European Chemical Industry Ecology and Toxicology Centre, B-1160 Brussels

B270   *n*-Butyl chloroformate

ClCO₂(CH₂)₃CH₃

CAS Registry No. 592-34-7
Synonyms butoxycarbonyl chloride; butylchlorocarbonate; carbonochloridic acid,
butyl ester
Mol. Formula $C_5H_9ClO_2$ **Mol. Wt.** 136.58
Uses Chemical synthesis of mixed or symetrical carbonates. Crosslinking agent for
cellulose textiles.

Physical properties

B. Pt. 142°C; **Flash point** 25°C; **Specific gravity** 1.074.

Occupational exposure

UK Long-term limit 1 ppm (5.6 mg m^{-3}); **UN No.** 2743; **HAZCHEM Code** 3W;
Conveyance classification toxic substance; **Supply classification** flammable and
toxic.

Risk phrases
Flammable – Toxic by inhalation – Causes burns (R10, R23, R34)

Safety phrases
In case of contact with eyes, rinse immediately with plenty of water and seek medical
advice – Wear suitable protective clothing – If you feel unwell, seek medical advice
(show label where possible) (S26, S36, S44)

Any other adverse effects

Systemic effects include burning sensation, coughing, wheezing, laryngitis, shortness
of breath, headache, nausea and vomiting. Inhalation, ingestion or absorption through
skin may be fatal as a result of spasm, inflammation and oedema of the bronchi and
larynx, chemical pneumonitis and oedema (1).
Vapours of lower chloroformates caused pneumonia (sometimes fatal) in animal
studies (2).

Any other comments

Butyl chloroformate is flammable and forms explosive mixtures in air. Incompatible
with strong oxidising agents, strong reducing agents and amines. May decompose on
exposure to moist air or water. Heat sensitive (1).
Experimental toxicology and human health effects reviewed (3).

References

1. Lenga, R. E. *The Sigma-Aldrich Library of Chemical Safety Data* 2nd ed., 1988, Sigma
    Aldrich, Milwaukee, WI
2. *Kirk-Othmer Encyclopedia of Chemical Technology* 1979-1983, **4**, 764-765, John Wiley &
    Sons, New York
3. *ECETOC Technical Report No. 30(4)* 1991, European Chemical Industry Ecology and
    Toxicology Centre, B-1160 Brussels

B271 *tert*-Butyl chromate

(CH3)3COCr(O)2OC(CH3)3

CAS Registry No. 1189-85-1

Synonyms *tert*-butylchromate (VI); chromic acid (H_2CrO_4), bis(1,1-dimethylethyl) ester; chromic acid, di-*tert*-butyl ester

Mol. Formula $C_8H_{18}CrO_4$ **Mol. Wt.** 230.23

Uses Oxidising agent. Catalyst for alkene polymerisation. Corrosion inhibitor.

Physical properties

M. Pt. –5-0°C.

Any other comments

Toxicity of chromium compounds reviewed (1).
Experimental toxicology and human health effects are reviewed (2).

References

1. *IARC Monograph* 1990, **49**, 49-256
2. *ECETOC Technical Report No. 30(4)* 1991, European Chemical Industry Ecology and Toxicology Centre, B-1160 Brussels

B272 *2-tert*-Butyl-1,4-dimethoxybenzene

CAS Registry No. 21112-37-8

Synonyms benzene, 2-*tert*-butyl-1,4-dimethoxy-; benzene, 2-(1,1-dimethylethyl)-1,4-dimethoxy-; mono-*tert*-butylhydroquinone dimethyl ether; Compound 77 B

Mol. Formula $C_{12}H_{18}O_2$ **Mol. Wt.** 194.28

Uses Antioxidant in polymer synthesis.

Physical properties

B. Pt. 117-118°C at 12 mmHg; **Specific gravity** d_4^{20} 1.007.

Any other adverse effects

Ability to induce glutathione S-transferase (EC 2.5.1.18) and NAD(P)H-quinone reductase (EC 1.6.99.2) in the cytosol of liver, mucosa of small intestine and the forestomach of female mice was tested. 2-*tert*-Butyl-1,4-dimethoxybenzene was fed to mice (gavage) in 5 daily doses of 4.8 mg l^{-1}. Increased enzyme activity observed in liver and intestine mucosa. No induction of enzyme in forestomach (1).

References

1. Prochaska, H. J; et al. *Biochem. Pharmacol.* 1985, **34**(21), 3909-3914

B273   1,3-Butylene glycol diacrylate

$$H_2C=CHCO_2CH(CH_3)(CH_2)_2O_2CCH=CH_2$$

CAS Registry No. 19485-03-1

Synonyms acrylic acid 1-methyltrimethylene ester; 1,3-butanediol diacrylate; 1,3-butylene diacrylate; 1-methyltrimethylene diacrylate; 2-propenoic acid, 1-methyl-1,3-propanediyl ester

Mol. Formula $C_{10}H_{14}O_4$ **Mol. Wt.** 198.22

Uses Polymer crosslinking agent. Coating agents.

Occupational exposure

Supply classification corrosive.

Risk phrases
Harmful in contact with skin – Causes burns – May cause sensitisation by skin contact (R21, R34, R43)

Safety phrases
In case of contact with eyes, rinse immediately with plenty of water and seek medical advice – Wear suitable protective clothing, gloves and eye/face protection (S26, S36/37/39)

Mammalian and avian toxicity

Acute data
LD_{50} oral rat 3540 mg kg^{-1} (1).
LD_{50} dermal rabbit 450 mg kg^{-1} (1).

Any other comments

Experimental toxicology and human health effects reviewed (2).

References

1. *Toxicol. Appl. Pharmacol.* 1974, **28**, 313
2. *ECETOC Technical Report No. 30(4)* 1991, European Chemical Industry Ecology and Toxicology Centre, B-1160 Brussels

B274   1,4-Butylene glycol diacrylate

$$H_2C=CHCO_2(CH_2)_4O_2CCH=CH_2$$

CAS Registry No. 1070-70-8

Synonyms acrylic acid, tetramethylene ester; 1,4-butanediol diacrylate; butylene diacrylate; 2-propenoic acid, 1,4-butanediyl ester; tetramethylene diacrylate

Mol. Formula $C_{10}H_{14}O_4$ **Mol. Wt.** 198.22

Uses Crosslinking agent for polymers.

Occupational exposure

Supply classification corrosive.

Risk phrases
Harmful in contact with skin – Causes burns – May cause sensitisation by skin contact (R21, R34, R43)

Safety phrases
In case of contact with eyes, rinse immediately with plenty of water and seek medical advice – Wear suitable protective clothing, gloves and eye/face protection (S26, S36/37/39)

Mammalian and avian toxicity

Sensitisation
Guinea pig maximisation test moderate to strong sensitiser. Cross reactivity is possible (1).

Genotoxicity

Salmonella typhimurium TA1535, TA1537, TA1538, TA98, TA100 with and without metabolic activation negative (2).

Any other adverse effects to man

In a study of 20 electron beam welding workers exposed to 1,4-butyleneglycol diacrylate showed delayed contact irritancy. Within 12-14 hr skin lesions developed with aching or itching. Healing was straightforward when contact removed (3). Data on allergic contact dermatitis from acrylates and 4 patients sensitised during routine patch testing are reported (4).

Any other comments

Sensitisation potential of diacrylates and dimethylacrylates reviewed (5,6).

References

1. Bjorkner, B. *Contact Dermatitis* 1984, **11**(4), 236-246
2. Waegernaekers, T. H. J. M; et al. *Mutat. Res.* 1984, **137**(2-3), 95-102
3. Malten, K. E. *Contact Dermatitis* 1979, **5**(3), 178-184
4. Kanerva, L. *Contact Dermatitis* 1988, **18**(1), 10-15
5. Malten, K. E. *Occup. Ind. Dermatol.* 1982, 301-314, Book Medical Publ., Chicago, IL
6. Roberts, D. *Contact Dermatitis* 1987, **17**(5), 281-289

B275   Butyl 2,3-epoxypropyl ether

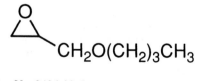

CAS Registry No. 2426-08-6
Synonyms butoxymethyloxirane; 1-butoxy-2,3-epoxypropane; 1-(2,3-epoxypropoxy)butane; glycidyl butyl ether; 2,3-epoxypropyl butyl ether; BGE

Mol. Formula $C_7H_{14}O_2$ **Mol. Wt.** 130.19

Uses Vicoscity-reducing agent for epoxy resins. Plasticiser. Lubricant antioxidant. In adhesives and cross-linking agents.

Physical properties

B. Pt. 164-166°C; **Specific gravity** d_4^{25} 0.91; **Volatility** v.p. 3.2 mmHg at 25°C.

Occupational exposure

US TLV (TWA) 25 ppm (133 mg m^{-3}); **UK Long-term limit** 25 ppm (135 mg m^{-3}); **Supply classification** harmful.

Risk phrases
Harmful by inhalation – May cause sensitisation by skin contact (R20, R43)

Safety phrases
Avoid contact with skin and eyes (S24/25)

Mammalian and avian toxicity

Acute data
LD_{50} oral gavage mouse, rat 1500-2300 mg kg^{-1} (1).
LC_{50} (4 hr) inhalation mouse >3500 ppm (2).
LD_{50} dermal rabbit 2520-4930 mg kg^{-1} (3).
LD_{50} intraperitoneal mouse, rat 700-1400 mg kg^{-1} (1).

Sub-acute data
Inhalation ♂ rats (35 hr) 300 ppm atrophied testes in 5 out of 10 animals. Inhalation ♂ rats (35 hr) 75 ppm patchy atrophy of testes in 1 out of 10 animals (4).

Teratogenicity and reproductive effects
♂ mice (8-10 wk) dermal application of 0, 0.375, 0.75 or 1.5 g kg^{-1}, 3 times wk^{-1} for 8 wk, then mated to unexposed ♀. No significant dose-related change in pregnancy rates of number of implants, but foetal deaths significantly higher in high dose group (5).

Metabolism and pharmacokinetics
Oral ♂ rats 20 mg kg^{-1} 87% eliminated via urine within 24 hr increased to 91% after 96 hr (6).
Oral ♂ New Zealand White rabbits 20 mg kg^{-1} 78% eliminated via urine within 24 hr, increased to 80% after 96 hr (7).

Sensitisation
Contact sensitiser to human skin (8).
10% of workers tested developed reaction to challenge by patch test (9,10).

Genotoxicity

Salmonella typhimurium TA100, TA1535 with and without metabolic activation positive (11-14).
Escherichia coli WP2 *uvrA* 48 hr incubation at 37°C induced DNA damage (12).
In vitro human lymphocytes induced DNA damage (15).
In vivo mouse bone marrow increased micronuclei (14).

Any other adverse effects

The exposure of two men to a spillage of 3.5 litres of butyl 2,3-epoxypropyl ether for

1.5 and 4 hr respectively produced irritation of gastrointestinal tract resulting in anorexia and vomiting. The severity of the symptoms were related to the length of time exposed. Other effects noted were persistent and severe headache and mild respiratory irritation (16,17).

Any other comments

Operation and handling recommendations reported (18,19).

References

1. Hine, C. H; et al. *AMA Arch. Ind. Health* 1956, **14**, 250-264
2. *Threshold Limit Values Biological Exposure Indices* 5th ed., 1986, 81, American Conference Governmental Industrial Hygienists, Cincinnatti, OH
3. Weil, C. S; et al. *Am. Ind. Hyg. Assoc. J.* 1963, **24**, 305-324
4. Lane, J. M; et al. *Vet. Hum. Toxicol.* 1980, **22**, 94-101
5. Whorton, E. B. Jr; et al. *Mutat. Res.* 1983, **124**, 224-233
6. Eadsforth, C. V. *Drug. Metab. Dispos.* 1985, **13**(2), 263-264
7. Eadsforth, C. V; et al. *Xenobiotica* 1985, **15**(7), 579-589
8. Kligman, A. M. *J. Invest. Dermatol.* 1966, **47**, 393-409
9. Fregert, S; et al. *Acta Allergol.* 1964, **19**, 269-299
10. Jolanki, R; et al. *Contact Dermatitis* 1987, **16**, 87-92
11. Wade, M. J; et al. *Mutat. Res.* 1979, **66**, 367-371
12. Hemminki, K; et al. *Arch. Toxicol.* 1980, **46**, 277-285
13. Thompson, E. D; et al. *Mutat. Res.* 1981, **90**(3), 213-231
14. Connor, T.H; et al. *Environ. Mutagen.* 1980, **2**, 284
15. Connor, T.H; et al. *Environ. Mutagen.* 1980, **2**(4), 521-530
16. Wallace, E. *J. Soc. Occup. Med.* 1979, **29**(4), 142-143
17. *NTIS Report* 1975, PB-267220, Nat. Tech. Inf. Serv., Springfield, VA
18. *IARC Monograph* 1989, **47**, 237-261
19. *ECETOC Technical Report No. 30(4)* 1991, European Chemical Industry Ecology and Toxicology Centre, B-1160 Brussels

B276   Butyl ethyl ketone

CH3(CH2)3COCH2CH3

CAS Registry No. 106-35-4

Synonyms *n*-butyl ethyl ketone; ethyl butyl ketone; 3-heptanone; heptan-3-one

Mol. Formula $C_7H_{14}O$ **Mol. Wt.** 114.19

Uses Used in solvent mixtures for air dried and baked finishes. Polyvinyl and nitrocellulose resins.

Physical properties

M. Pt. −36.7°C; **B. Pt.** 148°C; **Flash point** 46.1°C (open cup); **Specific gravity** d_{20}^{20} 0.8198; **Volatility** v.p 1.4 mmHg at 25°C; v. den. 3.93.

Solubility

Water: 14.3 g l^{-1} at 20°C. Organic solvent: ethanol

Occupational exposure

US TLV (TWA) 50 ppm (234 mg m^{-3}); **UK Long-term limit** 50 ppm (230 mg m^{-3}); **UK Short-term limit** 75 ppm (345 mg m^{-3}); **Supply classification** flammable and harmful.

Risk phrases

Flammable – Harmful by inhalation – Very toxic by inhalation (R10, R20, R26)

Safety phrases

Avoid contact with skin (S24)

Mammalian and avian toxicity

Acute data

LD$_{50}$ oral rat 2760 mg kg^{-1} (1).
LC$_{Lo}$ (4 hr) inhalation rat 2000 ppm (1).

Sub-acute data

Inhalation rat (24 wk) 700 ppm 72 hr wk^{-1}, no clinical signs of systemic toxicity or neurotoxicity (2).

Metabolism and pharmacokinetics

Metabolites in rats are 6-hydroxy-3-heptanone and 2,5-heptanedione (2).

Irritancy

Dermal rabbit (24 hr) 500 mg caused mild irritation (3).
Dermal rabbit (24 hr) 500 mg caused moderate irritation and 100 mg instilled into rabbit eye caused mild irritation (4).

Sensitisation

Non-sensitiser (5).

Any other comments

Health and safety standards discussed (6).
Toxicity of 3-heptanone including irritation, sensitisation, metabolism in invertebrates and microorganisms reviewed (4).
Experimental toxicology and human health effects reviewed (7).

References

1. *J. Ind. Hyg. Toxicol.* 1949, **31**, 60, 343
2. Katz, G. V. *Toxicol. Appl. Pharmacol.* 1980, **52**(1), 153-158
3. *Union Carbide Data Sheet* 1969 (12 Mar), Union Carbide Corp., New York
4. Opdyke, D. L. J. *Food Cosmet. Toxicol.* 1978, **16**(Suppl. 1), 731-732
5. Sharp, D. W. *Toxicology* 1978, **9**, 261-271
6. Clack, G. *Job Saf. Health* 1975, **3**(6), 5-10
7. *ECETOC Technical Report No. 30(4)* 1991, European Chemical Industry Ecology and Toxicology Centre, B-1160 Brussels

B277 *tert*-Butyl glycidyl ether

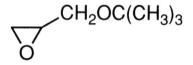

$CH_2OC(CH_3)_3$

CAS Registry No. 7665-72-7
Synonyms glycidyl-*tert*-butyl-ether; ((1,1-dimethylethoxy)methyl)-oxirane; T-BGE
Mol. Formula $C_7H_{14}O_2$ **Mol. Wt.** 130.19
Uses In epoxy resins for applications in protective coatings, reinforced plastics, bonding materials and adhesives.

Physical properties
B. Pt. 152°C; **Flash point** 43°C; **Specific gravity** 0.917.

Solubility
Water: 10-50 g l^{-1} at 23°C. Organic solvent: acetone, dimethyl sulfoxide, ethanol

Mammalian and avian toxicity

Acute data
LD$_{50}$ oral rat 2000 mg kg^{-1} (1).

Genotoxicity
Salmonella tymphimurium TA100, TA1535 with and without metabolic activation positive (2).
Escherichia coli without metabolic activation positive (3).
In vitro human peripheral blood lymphocytes induced unscheduled DNA damage (4).

References
1. *Gig. Tr. Prof. Zabol.* 1982, **26**(1) 43
2. Canter, D. A; et al. *Mutat. Res.* 1986, **172**, 105-138
3. *Environ. Mutagen.* 1980, **2**, 284
4. Frost, A. F; et al. *Mutat. Res.* 1982, **102**(2), 193-200

B278 Butylglycol acetate

$CH_3(CH_2)_3O(CH_2)_2O_2CH_3$

CAS Registry No. 112-07-2
Synonyms 2-butoxyethanol acetate; 2-butoxyethyl acetate; butyl cellusolve acetate; ethanol, 2-butoxy-acetate; glycolmonobutylether-acetate
Mol. Formula $C_8H_{16}O_3$ **Mol. Wt.** 160.21
Uses High boiling solvent for nitrocellulose lacquers. Epoxy resins.
Occurrence Air pollutant. Industrial sources include painting and inks and automotive industry (1,2).

Physical properties

M. Pt. −63.5°C; **B. Pt.** 192.3°C; **Flash point** 87.8°C; **Specific gravity** d_{20}^{20} 0.9424.

Solubility
Water: 11 g l^{-1}. Organic solvent: hydrocarbons

Occupational exposure

Supply classification harmful.

Risk phrases
Harmful by inhalation and in contact with skin (R20/21)

Safety phrases
Avoid contact with skin (S24)

Environmental fate

Degradation studies
Degradation >90% using Zahr-Wellen screening method, measured rate 12% day^{-1}, no observable lag period required (3).

Abiotic removal
Reacts in the atmosphere with photochemically produced hydroxyl radicals. At hydroxyl radical concentration of 5×10^5 hcm^3 estimated $t_{1/2}$ 18 hr (4).
The high water solubility of butyl glycol acetate predicts that the compound will be very mobile in soil (5).

Mammalian and avian toxicity

Acute data
LD$_{50}$ oral rat, mouse 2400-3200 mg kg^{-1} (6,7).
LD$_{50}$ dermal rabbit 1500 mg kg^{-1} (6).
Inhalation rat, rabbit (4 hr) 400 ppm non-toxic to either species (8).

Irritancy
Dermal rabbit (24 hr) 500 mg caused mild irritation and 500 mg instilled into rabbit eye caused mild irritation (9).

Any other adverse effects

Inhalation exposure of rabbits to butyl glycol acetate for one month caused haemoglobinuria and haematuria (10).

Any other comments

Experimental toxicology and human health effects reviewed (11).

References

1. Sexton, K; et al. *Environ. Sci. Technol.* 1980, **14**, 329-332
2. Lehman, E; et al. *Safe Health Asp. Org. Solv.* 1986, 31-41, Riihimaki, V. (Ed.), Proc. Int. Course Safety Asp. Org. Solv. Espoo, Finland
3. Zahn, R; et al. *Z. Wasser Abwasser Forsch* 1980, **13**, 1-7
4. Atkinson, R. *J. Inter. Chem. Kinet.* 1987, **19**, 799-828
5. Swann, R. L; et al. *Res. Rev.* 1983, **85**, 23
6. *Toxicol. Appl. Pharmacol.* 1979, **51**, 117
7. *Kodak Co. Reports* 1971 (21st May), Kodak Co., New York

8.   *Union Carbide Data Sheet* 1966, Union Carbide Corp., New York
9.   *Prehled Prumyslove Toxikol. Org. Latky* 1986, 713
10. Truhaut, R; et al. *Toxicol. Appl. Pharmacol.* 1971, **51**(1), 117-127
11. *ECETOC Technical Report No. 30(4)* 1991, European Chemical Industry Ecology and Toxicology Centre, B-1160 Brussels

B279   *tert*-Butyl hydroperoxide

(CH3)3COOH

CAS Registry No. 75-91-2
Synonyms 1,1-dimethylethyl hydroperoxide; 2-hydroproperoxy-2-methylpropane; Perbutyl H
Mol. Formula $C_4H_{10}O_2$ **Mol. Wt.** 90.12
Uses Polymerisation, oxidation and sulfonation catalyst. Reagent for selective oxygenation of olefins and acetylenes. Used in bleaching and deodorising.

Physical properties

M. Pt. –35°C; **B. Pt.** $_{20}$ 35°C; **Flash point** 26.7°C; **Specific gravity** d_4^{25} 0.880; **Volatility** v. den. 2.07.

Solubility
Organic solvent: ethanol, diethyl ether, chloroform

Mammalian and avian toxicity

Acute data
LD_{50} oral rat, mouse 400-700 mg kg^{-1} (1,2).
LD_{50} (4 hr) inhalation mouse, rat 350-500 ppm (1).
LD_{50} dermal rat 790 mg kg^{-1} (2).
LD_{50} intraperitoneal rat 90 mg kg^{-1} (1).

Irritancy
Dermal rabbit (24 hr) 500 mg caused moderate irritation and 100 mg instilled into rabbit eye caused severe irritation (3).
150 mg instilled into rabbit eye for 1 min then rinsed with water caused severe irritation (4).
Severe irritant and corrosive material to intact and abraded skin (5).

Genotoxicity
In vitro chinese hamster V79 cells chromosomal aberrations positive (6).

Any other adverse effects
tert-Butyl hydroperoxide is a substrate for glutathione peroxidase (7).
Effect of *tert*-butyl hydroperoxide on metabolism of liver and hepatoma reported.
Induces peroxidation of lipid membranes leading to cell damage (8).
Escherichia coli exposed to *tert*-butyl-hydroperoxide led to progressive and irreversible impairment of respiratory function (9).

The effects of *tert*-butyl hydroperoxide on the activity of antioxidant enzymes were investigated in cultured Chinese hamster V79 cells, incubation of cells with *tert*-butyl hydroperoxide for 1 hr significantly increased the activity of Cu-Zn superoxide dismutase up to a level 1.4 hr times of control cells (10).

Human erythrocytes were treated with *tert*-butyl hydroperoxide (10 min) 45 mg l^{-1} inhibited basal calcium and magnesium ATPase activity by 40% and calmodulin-stimulated activity by 54% (11).

Any other comments

Industrial toxicity of organic peroxides reported (5).

Metabolised by human carcinoma skin keratinocytes to free radicals. Study provided first direct evidence that human carcinoma skin cells can generate free radicals from organic hydroperoxides. The authors consider this metabolic capacity to be an important determinant of human cancer risk from hydroperoxides (12).

Dermatological and biological effects of organic peroxides reviewed (13,14).

Experimental toxicology and human health effects reviewed (15).

References

1. *Am. Ind. Hyg. Assoc.* 1958, **19**, 205
2. *Society of Plastics Industry Bulletin* 1/75-19B
3. *Sbornik Vysledku Toxicologickeho Vysetreni Latek A. Pripravku* 1972, -, 39, Prague
4. *Zentralblatt fuer Arbeitsmedizin und Arbeitsschutz* 1958, **19**, 205
5. *Dept. Toxicol. Product Safety* 1982, Arco Chem. Co., PA
6. Takahemi, O. *Mutat. Res.* 1989, **213**(2), 243-248
7. Oenfelt, A. *Mutat. Res.* 1987, **182**(3), 155-172
8. Borrello, S; et al. *Ann. N. Y. Acad. Sci.* 1988, **551**, 144-146
9. Rodriguez, L. C. *Biochem. Biophys. Acta* 1990, **1015**(3), 510-516
10. Ochi, R. *Toxicology* 1990, **61**(3), 229-239
11. Moore, R. B. *Arch. Biochem. Biophys.* 1989, **273**(2), 527-534
12. Mohammad, A; et al. *Carcinogenesis* 1989, **10**(8), 1499-1503
13. Kappus, D. P; et al. *Adv. Biosci.* 1989, **76**, 13-19
14. *Industrial Toxicology and Dermatology in the Production and Processing of Plastics* 1964, 211-217, Elsevier, Amsterdam
15. *ECETOC Technical Report No. 30(4)* 1991, European Chemical Industry Ecology and Toxicology Centre, B-1160 Brussels

B280   *tert*-Butylhydroquinone

CAS Registry No. 1948-33-0

Synonyms 1,4-benzenediol, 2-(1,1-dimethylethyl)-; hydroquinone, *tert*-butyl; mono-tertiarybutylhydroquinone; MTBHQ; TBHQ

Mol. Formula $C_{10}H_{14}O_2$ **Mol. Wt.** 166.22

Uses An antioxidant used in foods.

Physical properties
M. Pt. 127-129°C.

Mammalian and avian toxicity
Acute data
LD$_{50}$ oral rat, mouse 700-1000 mg kg^{-1} (1,2).
LD$_{50}$ intraperitoneal mouse, rat 150-300 mg kg^{-1} (1,3).

Sub-acute data
In diet (unspecified species) containing 2% *tert*-butylhydroquinone caused mild hyperplasia of the forestomach with increased hyperplasia of basal cells (4).

Carcinogenicity and long-term effects
National Toxicology Program (2 yr) study rats, mice via food currently in progress (5). Administration in diet to ♂ Syrian golden hamsters, *tert*-butylhydroquinone was inactive in causing forestomach hyperplasia and lesions (6).

Teratogenicity and reproductive effects
Concentrations of 0.125 to 0.5% in diet of pregnant rats produced no teratogenic effects (7).

Genotoxicity
Salmonella typhimurium TA92, TA102, TA104, TA100 with and without metabolic activation negative. Doses >100 μg plate^{-1} were cytotoxic (8).
In vivo mouse bone marrow cells sister chromatid exchange positive (9).

Any other comments
Chemical and physical properties, experimental toxicology and human health effects reviewed (10-13).
May have antimicrobial as well as antioxidant properties (14).
Acceptable daily intake of 200 μg established (15).

References
1. *J. Am. Oil Chem. Soc.* 1975, **52**, 53
2. *Kodak Co. Reports* 1971 (21st May), Kodak Co., New York
3. *Drug Chem. Toxicol.* 1984, **7**, 335
4. Altmann, H. J; et al. *Arch. Toxicol.* 1985, **Suppl. 8**, 114-116
5. *National Toxicology Program Res. and Test.Div.* 1992, NIEHS, Research Triangle Park, NC
6. Hirose, M; et al. *Carcinogenesis* 1986, **7**(8), 1285-1289
7. Krasawage, W. J. *Teratology* 1977, **16**(1), 31-34
8. Hageman, G. J; et al. *Mutat. Res.* 1988, **208**(3-4), 207-211
9. *Environ. Mol. Mutagen.* 1989, **13**, 234-237
10. *BIBRA Toxicity Profile* 1989, British Industrial Biological Research Association, Carshalton
11. *J. Am. Coll. Toxicol.* 1986, **5**(5), 329-351
12. Van Esch, G. J. *Food Chem. Toxicol.* 1986, **24**(10-11), 1063-1065
13. *ECETOC Technical Report No 30 (4)* 1991, European Chemical Industry Ecology and Toxicology Centre, B-1160 Brussels
14. Zeelu, J. J; et al. *Cosmet. Toilet.* 1982, **97**, 61
15. FAO, WHO Expert Committee on Food Additives *Tech Rep. Ser. Wld. Hlth. Org.* 1987, No. 751

B281   Butyl 4-hydroxybenzoate

$CO_2(CH_2)_3CH_3$

OH

CAS Registry No. 94-26-8

Synonyms benzoic acid, 4-hydroxy-, butyl ester; Butoben; Butyl paraben; butyl *p*-hydroxybenzoate; Butyl parasept; Butyl tegosept; *p*-hydroxybenzoic acid, butyl ester; Teposept B

Mol. Formula $C_{11}H_{14}O_3$ **Mol. Wt.** 194.23

Uses Antifungal pharmaceutical aid. Preservative in foods, creams, lotions, ointments and other cosmetics, drugs and dentifrices.

Physical properties

M. Pt. 68-69°C.

Solubility
Organic solvent: acetone, ethanol, diethyl ether, chloroform, propylene glycol

Mammalian and avian toxicity

Acute data
LD_{50} intraperitoneal mouse 230 mg kg^{-1} (1).

Sub-acute data
Oral mice (6 wk) 1.25% butyl-*p*- hydroxybenzoate marked atrophy of lymphoid tissue in organs including the spleen, thymus, lymph nodes, and multifocal degeneration and necrosis in liver parenchyma (2).

Carcinogenicity and long-term effects
Oral mice (2 yr) 0.15, 0.3, 0.6% induced tumours of haematopoietic system and lungs. However there was no significant difference from controls. The authors concluded that there was no evidence of tumourigenic effects at doses up to 0.6% (2).

Irritancy
Dermal guinea pig (48 hr) 5% butyl hydroxybenzoate caused mild irritation (1).

Any other adverse effects

In vitro incubation of butylhydroxy benzoate (30 mins) 1 g l^{-1} caused potent spermicidal activity against human spermatozoa. All spermatozoa were immobolised with no revival after 30 mins (3).

Any other comments

Experimental toxicology and human health effects reviewed (4,5).

References

1.   *J. Soc. Cosmet. Chem.* 1977, **28**, 357
2.   Inai, K; et al. *Food Chem. Toxicol.* 1985, **23**(6), 575-578
3.   Song, B; et al. *Contraception* 1989, **39**(3), 331-335

4. *BIBRA Toxicity Profile* 1989, British Industrial Biological Research Association, Carshalton
5. *ECETOC Technical Report No 30 (4)* 1991, European Chemical Industry Ecology & Toxicology Centre, B-1160 Brussels

B282   *N-n*-butylimidazole

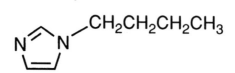

CAS Registry No. 4316-42-1
Synonyms 1-butylimidazole; *N*-butylimidazole; imidazole,1-butyl; 1*H*-imidazole, 1-butyl
Mol. Formula $C_7H_{12}N_2$ **Mol. Wt.** 124.19

Physical properties
B. Pt. 110°C at 11 mmHg.

Occupational exposure
UN No. 2690; **HAZCHEM Code** 2X; **Conveyance classification** toxic substance.

Ecotoxicity

Fish toxicity
LC_{50} (24 hr) goldfish 30 mg l^{-1} (1).

Any other adverse effects
Inhibits platelet aggregation induced by ADP or collagen (2).
Thromboxane synthetase inhibitor (3).

References
1. Bridie, A. L; et al *Water Res.* 1979, **13** 623-630
2. Avirans, M; et al. *Br. J. Clin. Pharmacol.* 1985, **19** (6) 715-719
3. Crockard, A; et al. *J. Cereb. Blood Flow Metals* 1982, **2** (1), 67-62

B283   Butyl isocyanate

$$CH_3(CH_2)_3NCO$$

CAS Registry No. 111-36-4
Synonyms BIC; butane, 1-isocyanato; 1-isocyanatobutane; isocyanic acid, butyl ester
Mol. Formula C_5H_9NO **Mol. Wt.** 99.13
Uses An active site specific reagent for yeast ethanol dehydrogenase. Intermediate in manufacture of pesticides, herbicides and pharmaceuticals.

Physical properties

B. Pt. 115°C; **Flash point** 26°C; **Specific gravity** d_4^{20} 0.880.

Occupational exposure

UN No. 2485.

Mammalian and avian toxicity

Acute data
LD_{50} oral mouse, guinea pig, rat 150-600 mg kg^{-1} (1).
LD_{50} intravenous mouse 1 mg kg^{-1} (2).

Genotoxicity

Pretreatment of H/r30R cells with butyl isocyanate did not alter x-ray induced mutagenesis (3).

Any other adverse effects

Eighteen persons exposed to *n*-butyl isocyanate developed acute respiratory system distress (4).

Any other comments

Growth of Ehrlich ascites tumours in Swiss ICR/HA mice was inhibited by butyl isocyanate (5).
The toxicity of isocyanates reviewed (6).

References

1. *Labour Hyg. Occup. Dis.* 1976, **20**(3), 53
2. *US Army Armament Res. Dev. Command* (NX 05701)
3. Kawazoe, Y. *Gann* 1982, **73**(2), 255-238
4. Engelhard, H. *Zeitschrift fur die Gesamte Hyg.* 1976, **22**(4), 235-238
5. Moos, G. E. *Cancer Res.* 1971, **31**(7), 937-941
6. Woolrich, P. F. *Technical Bull.* 1973, **106**, 1-19

B284   Butyl isovalerate

$$(CH_3)_2CHCH_2CO_2(CH_2)_3CH_3$$

CAS Registry No. 109-19-3
Synonyms butanoic acid, 3-methyl-, butyl ester; *n*-butyl isopentanoate; butyl 3-methylbutyrate; isovaleric acid, butyl ester; 3-methylbutanoic acid, butyl ester
Mol. Formula $C_9H_{18}O_2$ **Mol. Wt.** 158.24
Uses Fragances. Artificial flavourings.
Occurrence Natural substance in *Valerianella locusta*, in the oil from leaves of *Eriostermon coxii* and *Phebaluim dentatum* and in bananas and peas (1,2).

Physical properties

B. Pt. 150°C; **Specific gravity** 0.87; **Volatility** v. den. 5.45.

Mammalian and avian toxicity

Acute data
LD_{50} oral rat, rabbit 5000-8200 mg kg^{-1} (1,2).

Metabolism and pharmacokinetics
A solution 1 g in acetone fed through to the rumen of cows was detected unchanged in milk 2 hr after administration (3).

Irritancy
Dermal rabbit (24 hr) 500 mg caused mild irritation (2).
Humans closed patch tested with 1% in petrolatum showed no irritation effects 48 hr after application dose (2).

Any other adverse effects
Narcotic dose in rabbits 6000 mg kg^{-1} caused stupor and loss of voluntary movement in 50% of animals tested (4).

References

1. Goetz-Schmidt, E. M; et al. *Phytochemistry* 1988, **27**(3), 845-848
2. Opdyke, D. J. L. *Food Cosmet. Toxicol.* 1980, **18**, 659-660
3. Honkanen, E; et al. *Acta. Chem. Scand.* 1964, **18**, 612
4. Munch, J. C; et al. *Ind. Med. Surg.* 1972, **41**(4), 31-33

B285   Butyl mercaptan

CH3(CH2)3SH

CAS Registry No. 109-79-5
Synonyms 1-butanethiol; *n*-butanethiol; *n*-butyl thioalcohol; thiobutyl alcohol
Mol. Formula $C_4H_{10}S$ **Mol. Wt.** 90.19
Uses Solvent. Intermediate in organic synthesis. Guided missile propellant and oxidiser. Odour agent.
Occurrence Residues detected in the environment after commercial spraying with the defoliant DEF (1).

Physical properties

M. Pt. −116°C; **B. Pt.** 98°C; **Flash point** 12°C (closed cup); **Specific gravity** d_4^{25} 0.8365; **Partition coefficient** log P_{ow} 2.28; **Volatility** v. den. 3.1.

Solubility
Water: 590 mg l^{-1} at 22°C. Organic solvent: ethanol, diethyl ether

Occupational exposure

US TLV (TWA) 0.5 ppm (1.8 mg m^{-3}); **UN No.** 2347; **HAZCHEM Code** 3WE; **Conveyance classification** flammable liquid.

Ecotoxicity

Fish toxicity
LC_{50} (24-48 hr) bluegill sunfish 7.4-5.5 mg l^{-1} (2).

Exposure of catfish to 500 mg l^{-1} butyl mercaptan increased percentage of methaemoglobin to 16.5% of total haemoglobin (3).

Effects to non-target species
Oral hen >80 mg kg^{-1} severe cholinergic effects (4).

Environmental fate

Degradation studies
Alcaligenes faecolis, a microorganism in activated sludge flora, oxidised butyl mercaptan (5).

Abiotic removal
Reacts with oxygen atoms and hydrogen radicals in the troposphere, although the latter is more important, estimated $t_{1/2}$ 38 hr (6-8).

Mammalian and avian toxicity

Acute data
LC$_{50}$ oral rat 1500 mg kg^{-1} (9).
LC$_{50}$ (4 hr) inhalation mouse, rat 2500-4020 ppm (9).
LC$_{50}$ (30 min) inhalation dog 770 ppm (10).
LD$_{50}$ intraperitoneal rat 399 mg kg^{-1} (9).

Teratogenicity and reproductive effects
TC$_{Lo}$ inhalation mouse (6-16 day gestation) 10, 68 or 152 ppm 6 hr day^{-1} embryotoxic, increased post implantation loss and increased early resorption. Non significant increases in cleft palate. Maternal lethality $\geq$68 ppm (11).

Irritancy
83 mg instilled into rabbit eye caused irritation (4).

Any other adverse effects

Seven workers suffered acute poisoning from butylmercaptan in a laboratory producing acrylic resins. Chemical and toxicological properties and industrial applications are discussed (12).

Any other comments

Potential occupational hazards reviewed (13,14).
Experimental toxicology and human health effects reviewed (15).

References

1. Herman, B. W; et al. *Anal. Chem.* 1981, **53**, 1077-1082
2. Meinck, F; et al. *Les Eaux Residuaires Industrielles* 1970
3. Mather-Mihaich, E; et al. *Comp. Biochem. Physiol. C: Comp. Pharmacol. Toxicol.* 1986, **85c**(2), 427-432
4. Abov-Donia, M. B; et al. *Toxicol. Appl. Pharmacol.* 1980, **53** (3), 439-457
5. Marion, C. V; et al. *J. Water Pollut. Control Fed.* 1963, **35**, 1269-1284
6. Graedel, T. E; *Chemical Compounds in the Atmosphere* 1978, 306-322, Academic Press, New York
7. Sladge, I. R; et al. *Int. J. Chem. Kinet.* 1976, **8**, 451-457
8. *GEMS: Graphic Exposure Modeling System, Fate of Atmospheric Pollutants Data Base* 1984, Office of Toxic Substances, USEPA, Washington, DC
9. *Am. Ind. Hyg. Assoc. J* 1958, **19**, 171

10. *Prehled. Prumy. Toxicol. Org. Latky* 1986, 982
11. Thomas, W. C; et al. *Fundam. Appl. Toxicol.* 1987, **8**, 170-178
12. Gobbato, F; et al. *Folio Medica.* 1968, **51**(5), 329-341
13. NIOSH *Info. Profile Potent. Occup. Hazard* 1977, **210-77-0120**, 169-176
14. Krop, S. *Chem. Corp. Med. Lab.* 1954, **43**, 1-13
15. *ECETOC Technical Report No 30 (4)* 1991, European Chemical Industry Ecology and Toxicology Centre, B-1160 Brussels

B286   Butyl nitrate

$CH_3(CH_2)_3NO_3$

CAS Registry No. 928-45-0
Synonyms nitric acid, butyl ester
Mol. Formula $C_4H_9NO_3$ **Mol. Wt.** 119.12

Physical properties
B. Pt. 135.5°C; **Flash point** 36°C; **Specific gravity** 1.0228 at 30°C.

Solubility
Organic solvent: ethanol, diethyl ether

Genotoxicity
Escherichia coli T4B (24-48 hr) low mutagenic activity (1).

Any other comments
Explosion risk.

References
1. Kononova, S. D. *Teor. Khim. Mutageneza Mater. Vses, Soveshch. 4th* 1969, 34-36, (Russ.) (*Chem. Abstr.* **76**, 95128d)

B287   *tert*-Butyl peracetate

$(CH_3)_3COOCOCH_3$

CAS Registry No. 107-71-1
Synonyms acetyl *tert*-butyl peroxide; *tert*-butyl peroxyacetate; ethaneperoxoic acid-1, 1-dimethyl ester; peroxyacetic acid, *tert*-butyl ester
Mol. Formula $C_6H_{12}O_3$ **Mol. Wt.** 132.16
Uses Polymerisation initiator for vinyl monomers. Manufacture of polyethlene and polystyrene.

Physical properties
Flash point <26.7°C; **Specific gravity** 0.923; **Volatility** v.p. 50 mmHg at 26°C.

Occupational exposure
UN No. 2095 (maximum concn. 76% in solution);2096 (maximum concn. 52% in solution).

Mammalian and avian toxicity

Acute data

LD$_{50}$ oral mouse, rat 630-675 mg kg^{-1} (1).
LC$_{50}$ (4 hr) inhalation rat 8200 mg m^{-3} (1).
LC$_{50}$ (2 hr) inhalation mouse 6000 mg m^{-3} (1).

Any other adverse effects

Inhalation ♂ rats (4 month) 0.001 mg l^{-1} 4 hr day^{-1} had decreased numbers of spermatogonia but motility and fertility were not impaired (2).

Any other comments

Safety guidelines for organic peroxides discussed (3,4).
Experimental toxicology and human health effects reviewed (5).

References

1. Izmerov, N. F; et al. *Toxicometric Parameters of Industrial Toxic Chemicals under Single Exposure* 1982, 30, Moscow
2. Sanotskii, I. V; et al. *Tokisikol Nov. Prom Khinn. Veshchester* 1968, **10**, 44-55, (Russ.) (*Chem. Abstr.* **71**, 89634w)
3. Martin, J. J; *Ind. Eng. Chem.* 1960, **52**(4), 65A-68A
4. Walrud, D. H; et al. *Plast. Compd.* 1979, **2**(1), 52-54
5. *ECETOC Technical Report No. 30(4)* 1991, European Chemical Industry Ecology and Toxicology Centre, B-1160 Brussels

B288   *tert*-Butyl peroxybenzoate

$CO_2OC(CH_3)_3$

CAS Registry No. 614-45-9
Synonyms benzenecarboperoxoic acid, 1,1-dimethylethyl ester; benzoyl *tert*-butylperoxide; *tert*-butyl perbenzoate; Esperox 10; Novox; peroxybenzoic acid, *tert*-butyl ester; Trigonox C
Mol. Formula $C_{11}H_{14}O_3$ **Mol. Wt.** 194.23
Uses Polymerisation indicator for polyethylene, polystyrene, polyacrylates and polyesters. Intermediate in organic synthesis.

Physical properties

M. Pt. 8°C; **B. Pt.** 112°C (decomp.); **Flash point** 66°C (closed cup); **Specific gravity** d_{25}^{25} 1.04; **Volatility** v.p. 0.33 mmHg at 50°C.

Solubility

Organic solvent: ethanol, diethyl ether, esters, ketones

Mammalian and avian toxicity

Acute data

LD$_{50}$ oral rat, mouse 900-1000 mg kg^{-1} (1).

Carcinogenicity and long-term effects

Dermal application of *tert*-butyl peroxybenzoate to mice treated with an initiator and promoter increased rate of conversion of benign skin papillomas to carcinomas (2). Application of retinoic acid inhibits the process of malignant conversion induced by free radical generating compounds (3).

Irritancy

Dermal rabbit (24 hr) 500 mg caused mild irritation and 500 mg instilled into rabbit eye for 24 hr caused mild irritation (4).

100 mg instilled into rabbit eye for (1 min) and then rinsed with water caused mild irritation (5).

Genotoxicity

Salmonella typhimurium TA1535, TA1537, TA98, TA100 with and without metabolic activation positive (6).

Any other adverse effects

Metabolised by human carcinoma skin keratinocytes to free radicals. Study provides first direct evidence that human carcinoma skin cells can generate free radicals from organic hydroperoxides. The authors consider this metabolic capacity to be an important determinant of human cancer risk of hydroperoxides (7).

Any other comments

Safety and hazard classification tests for organic peroxides documented (8).

References

1. Izmerov, N. F; et al. *Toxicometric Parameters of Industrial Toxic Chemicals under Single Exposure* 1982, 30, Moscow
2. Mohammad, A; et al. *Carcinogenesis (London)* 1989, **10**(10), 1841-1845
3. Mohammad, A; et al. *Carcinogenesis (London)* 1991, **12**(12), 2325-2329
4. *Sbornik Vysledku Toxixologickeho Vysetreni Latek A Pripravku* 1972, -, 52, Prague
5. *Zentralblatt fuer Arbeitmedizin und Arbeitsschutz* 1958, **8**, 25
6. Mortelmans, K; et al. *Environ. Mol. Mutagen.* 1986, **8**(Suppl. 7), 1-119
7. Mohammad, A; et al. *Carcinogenesis (London)* 1989, **10**(8), 1499-1503
8. Noller, D. C; et al. *Ind. Eng. Chem.* 1964, **56**(12), 18-27

B289 *tert*-Butyl peroxyisobutyrate

(CH$_3$)$_3$COOC(O)CH(CH$_3$)$_3$

CAS Registry No. 109-13-7

Synonyms *tert*-butyl perisobutyrate; Esperox 24M; Lupersol 8; peroxyisobutyric acid, *tert*-butyl ester; propaneperoxoic acid, 2-methyl-1,1-dimethyl ester

Mol. Formula C$_8$H$_{16}$O$_3$ **Mol. Wt.** 160.21

Uses Polymerisation catalyst.

Physical properties
Flash point <26.6°C.

Occupational exposure
UN No. 2142 (>52% but ≤77% in solution); 2562 (≤52% in solution).

Any other comments
Hazard classification system discussed (1).
Guideline for safe storage and handling is given (2).

References

1. Noller, D. C; et al. *Ind. Eng. Chem.* 1964, **56**(1,2), 18-27
2. Donaldson, C. R. *Safety High Pressure Polyethylene Plants* 1973, 31-34, American Institute of Chemical Engineering, New York

B290   *tert*-Butyl peroxypivalate

(CH₃)₃CCO₂OC(CH₃)₃

CAS Registry No. 927-07-1
Synonyms Propaneperoxoic acid, 2,2-dimethyl-, 1, 1-dimethylethyl ester; Esperox 31M; Trigonoz 25-C75; peroxypivalic acid, *tert*-butyl ester; *tert*-butylperpivalate; *tert*-butyltrimethylperoxyacetate
Mol. Formula $C_9H_{18}O_3$ **Mol. Wt.** 174.24
Uses Polymerisation initiator. Catalyst.

Physical properties
M. Pt. 70°C (decomp.); **Flash point** 68-71°C (open cup); **Specific gravity** d_{25}^{25} 0.854.

Mammalian and avian toxicity

Acute data
LD_{50} oral rat 4300 μg kg^{-1} (1).

Any other comments
Rapid decomposition occurs at 21°C.

References

1. Society of Plastics *SPI Bull.* 1985, 1/75-19B

B291   *tert*-Butylperoxy-3,5,5-trimethylhexanoate

(H₃C)₃CCH₂CH(CH₃)CH₂CO₂C(CH₃)₃

CAS Registry No. 13122-18-4
Synonyms *tert*-butylperoxy-3,5,5-trimethylhexanoate; *tert*-butylperisononanoate;

1,1-dimethylethyl-3,5,5-trimethylhexaneperoxoate; *tert*-butyl-3,5,
5-trimethylperoxyhexanoate
Mol. Formula $C_{13}H_{26}O_3$ **Mol. Wt.** 230.35
Uses Polymerisation catalyst. Light stabilisers for ultra violet-transparent polymers.
Source of free radicals in polymer manufacture.

Physical properties

Solubility
Organic solvent: benzene

Occupational exposure

UN No. 2104.

Any other adverse effects

Systemic effects mild exposure (via inhalation and ingestion) dyspnoea, coughing,
sore throat, vomiting and abdominal pain. Severe exposure causes dyspnoea, lung
congestion vomiting at blood and unconsciousness (1).

References

1. Nakagawa, S. *Anzen Kogaku (J. Jpn. Soc. Saf. Engineering)* 1979, **18**(1), 14-21

B292   2-*sec*-Butylphenol

CAS Registry No. 89-72-5
Synonyms 2-*sec*-butylphenol; 2-(1-methylpropyl)phenol; phenol, *o-sec*-butyl;
phenol, 2-(1-methylpropyl)
Mol. Formula $C_{10}H_{14}O$ **Mol. Wt.** 150.22
Uses Chemical intermediate in the preparation of resins, plasticisers and surface
active agents.

Physical properties

M. Pt. 16°C; **B. Pt.** $_{25}$ mmHg 226-228°C; **Flash point** 107.2°C; **Specific gravity** d_{25}^{25}
0.981.

Solubility
Water: <1 mg ml^{-1} at 20°C. Organic solvent: acetone, dimethyl sulfoxide, ethanol

Occupational exposure

US TLV (TWA) 5 ppm or 30 mg m^{-3}; **UK Long-term limit** 5 ppm (30 mg m^{-3}); **UN
No.** 2228; **HAZCHEM Code** 3Ⓩ; **Conveyance classification** harmful substance.

Ecotoxicity

Fish toxicity
Lethal threshold Atlantic salmon 0.15-1.0 mg l^{-1} (1).

Invertebrate toxicity
Lethal threshold brown shrimp 0.15-2.7 mg l^{-1} (1).

Mammalian and avian toxicity

Acute data
LD$_{50}$ oral rat 2700 mg kg^{-1} (2).
LD$_{50}$ dermal guinea pig 600 mg kg^{-1} (3).
LD$_{50}$ intraperitoneal, intravenous mouse 60-63 mg kg^{-1} (4,5).

Irritancy
Dermal rabbit (24 hr) 500 mg caused severe irritation and 50 μg instilled into rabbit eye caused severe irritation (2).

Genotoxicity

Salmonella typhimurium TA1535, TA1537, TA97, TA98, TA100 with and without metabolic activation negative (5).

Any other comments

Experimental toxicology and human health effects reviewed (6).

References

1. McLeese, D. W; et al. *Chemosphere* 1981, **10**(7), 723-730
2. *Sbornik Vysledku Toxicologickehu Vystroni Latek A Pripravku* 1972, -, 55
3. *Doc. Threshold Limit Values of Substances in Workroom Air* 1980, **4**, 58
4. *J. Med. Chem.* 1980, **23**, 1350
5. Mortelmans, K; et al. *Environ. Mol. Mutagen.* 1986, **8** (Suppl.7), 1-119
6. *ECETOC Technical Report No 30(4)* 1991, European Chemical Industry Ecology and Toxicology Centre, B-1160, Brussels

B293   4-*tert*-Butylphenol

CAS Registry No. 98-54-4
Synonyms butylphen; 4-(1,1-dimethylethyl)phenol; 1- hydroxy-4-*tert*-butylbenzene; phenol, 4-(1,1- dimethylethyl)-
Mol. Formula $C_{10}H_{14}O$ **Mol. Wt.** 150.22
Uses Ingredient in de-emulsifiers for oil field use. In motor oil additives. Intermediate in the manufacture of varnish and lacquer resins. Soap antioxidant.

Physical properties

M. Pt. 98°C; **B. Pt.** 238°C; **Specific gravity** d_4^{114} 0.9081; **Partition coefficient** log P_{ow} 3.65; **Volatility** v.p. 1 mmHg at 70°C; v. den. 5.1.

Solubility
Water: 700 mg l^{-1}. Organic solvent: ethanol, diethyl ether

Occupational exposure

UN No. 2229; **HAZCHEM Code** 3☒; **Conveyance classification** harmful substance.

Ecotoxicity

Fish toxicity
Trout, bluegill sunfish and goldfish exposed to 5 ppm died within 2-6 hr (1).
LC_{50} (96 hr) fathead minnow 5.1 mg l^{-1} (2).
LC_{50} (96 hr) juvenile Atlantic salmon 0.74 mgl⁻ (3).

Invertebrate toxicity
EC_{50} (24 hr) *Daphnia magna* 4.2 mg l^{-1} (4).
EC_{50} (5 min) *Photobacterium phosphoreum* 0.21 mgl^{-1} Microtox test (5).

Bioaccumulation
Non-accumulative or low accumulative (3).

Environmental fate

Degradation studies
Nonbiodegradable (3).

Mammalian and avian toxicity

Acute data
LD_{50} oral rat 2950 mg kg^{-1} (6).
LC_{Lo} (4 hr) inhalation rat 5600 mg m^{-3} (7).
LD_{50} dermal rabbit 2288 mg kg^{-1} (6).
LD_{50} intraperitoneal mouse 78 mg kg^{-1} (8).

Irritancy
Dermal ♂, ♀ rabbit (24 hr) 16 g kg^{-1} caused local toxicity and irritation at site of application, but no mortality. Dermal application (4 hr) 0.5 g produced a range of effects from no reaction to necrosis (9).
Dermal rabbit (4 hr) 500 mg caused mild irritation (10).
50 µg instilled into rabbit eye for 24 hr caused severe irritation (11).

Sensitisation
Skin sensitiser (12).

Genotoxicity

Salmonella typhimurium TA98, TA100, TA1535, TA1537 with and without metabolic activation negative (13).

Any other adverse effects

Probable mechanism of action of alkyl phenols is a toxic effect on functional melanocytes. Exposure can cause depigmentation, hepatosplenomegaly and thyroid enlargement. May cause occupational leukoderma (7).

Inhalation rats (6 hr) exposure to respirable dust aerosol 5.6 mg l^{-1} or saturated vapour 6 mg l^{-1}. The vapour had no effect on body weight. Caused no adverse clinical signs, necropsy or mortality. The dust caused mucosal irritation and respiratory distress up to 7 days post exposure (9).

Any other comments

Occupational exposure, experimental toxicology and human health effects reviewed (14,15).

All reasonable efforts have been taken to find information on isomers of this compound, but no relevant data are available.

References

1. *Toxicity of 3400 Chemicals to Fish* 1987, EPA 560/6-87-002, Washington, DC
2. Holcombe, G. W; et al. *Environ. Pollut. Series A* 1984, **35**, 367-381
3. McLeese, D. W; et al. *Chemosphere* 1981, **10**, 723
4. Kuhn, R; et al. *Water Res.* 1989, **23**(4), 495-499
5. Kaiser, K. L. E; et al. *Water Pollut. Res. J. Canada* 1991, **26**(3), 361-431
6. *Am. Ind. Hyg. Assoc. J.* 1969, **30**, 470
7. Gellin, G. A. *Current Concepts in Cutaneous Toxicity* 1980, 213-220
8. *J. Med. Chem.* 1975, **18**, 868
9. Klonne, D. R; et al. *Drug. Chem. Toxicol.* 1988, **11**(1), 43-54
10. *Food Cosmet. Toxicol.* 1974, **12**, 807
11. *Prehled Prumyslove Toxikol Org Latky* 1986, 224
12. Occupational Health Services *Pestline* 1991, **2**, 747-749
13. Zeiger, E; et al. *Environ. Mol. Mutagen.* 1987, **9**(Suppl. 9), 1-110
14. Kosako, M; et al. *Occup. Environ. Health* 1989, **61**(7), 451-455
15. *ECETOC Technical Report No. 30(4)* 1991, European Chemical Industry Ecology and Toxicology Centre, B-1160 Brussels

B294   Butyl propionate

$CH_3CH_2CO_2(CH_2)_3CH_3$

CAS Registry No. 590-01-2

Synonyms butylpropanoate; propanoic acid butylester; *n*-butylpropionate

Mol. Formula $C_7H_{14}O_2$ Mol. Wt. 130.19

Uses Solvent for nitro cellulose. Retarder in lacquer thinner. Ingredient of perfumes and flavours.

Physical properties

M. Pt. −89.6°C; B. Pt. 145.4°C; Flash point 32.2°C; Specific gravity d_4^{20} 0.875; Volatility v. den. 4.49.

Solubility

Organic solvent: ethanol, diethyl ether

UN No. 1914; HAZCHEM Code 3Ⓨ; Conveyance classification flammable liquid;
Supply classification flammable.

Risk phrases
Flammable (R10)

Mammalian and avian toxicity

Acute data
LD$_{50}$ oral rat 5000 mg kg^{-1} (1).

Irritancy
Dermal rabbit (24 hr) 500 mg caused moderate irritation (1).

Any other comments
Experimental toxicology and human health effects reviewed (2,3).

References
1. *Food Cosmet. Toxicol.* 1980, **18**, 649
2. *BIBRA Toxicity Profile* 1990, British Industrial Biological Research Association, Carshalton
3. *ECETOC Technical Report No 30(4)* 1991, European Chemical Industry Ecology and
   Toxicology Centre, B-1160 Brussels

B295   4-*tert*-Butyltoluene

CAS Registry No. 98-51-1
Synonyms *p*-methyl-*tert*-butylbenzene; 1-methyl-4-*tert*-butylbenzene;
p-*tert*-butyltoluene; 8-methylparacymene; PTBT; benzene,
1,-(1,1-dimethylethyl)-4-methyl
Mol. Formula C$_{11}$H$_{16}$ Mol. Wt. 148.25
Uses Solvent in preparation of resins. Oil additive. Perfume component. Intermediate
in organic synthesis.

Physical properties
M. Pt. −62.5°C; B. Pt. 192.8°C; Flash point 68.3°C (closed cup); Specific gravity
d$_{20}^{20}$ 0.857; Volatility v.p. 0.65 mmHg at 25°C; v. den. 4.6.

Solubility
Organic solvent: ethanol, diethyl ether

Occupational exposure
US TLV (TWA) 10 ppm (61 mg m^{-3}); US TLV (STEL) 20 ppm (121 mg m^{-3}); UN
No. 2667; HAZCHEM Code 3Ⓩ; Conveyance classification harmful substance.

Ecotoxicity

Fish toxicity

LC_{50} (24 hr) goldfish 3 mg l^{-1} (1).

Mammalian and avian toxicity

Acute data

LD_{50} oral rabbit, rat 1500-2000 mg kg^{-1} (2).
LC_{50} (4 hr) inhalation mouse 248 ppm (2).
LD_{50} dermal rat 14-28 mg kg^{-1} (3).
TC_{Lo} (3 min) inhalation human 10 ppm gastrointestinal tract effects (2).
TC_{Lo} (5 min) inhalation human 20 ppm eye, gastrointestinal tract effects (3).

Sub-acute data

Inhalation rat (25 day) 50 ppm 7 hr day^{-1} no adverse effects reported (4).

Metabolism and pharmacokinetics

Metabolites include p-tert-butylbenzoic acid; 2-(p-carboxyphenyl)-2-methylpropan- 1-ol; p-tert-butylbenzoylglycine and 2-methyl-2- p-tolyl-propan-1-ol in rats exposed by inhalation. Accumulates in mesenteric fat, liver, kidney and the brain, rapidly eliminated via urine (5).

Irritancy

Dermal rabbit (24 hr) 500 mg caused mild irritation (2).
Exposure human (5 min) 8 ppm caused moderate eye irritation. Nasal mucosal and throat irritation observed at concentrations of 10 ppm and 60 ppm respectively (2).

Genotoxicity

Salmonella tymphimurium TA98, TA100, TA1537 with and without metabolic activation negative (6,7).
Escherichia coli, Saccharomyces cerevisiae with and without metabolic activation mitotic gene conversion and structural chromosome damage negative (7).

Any other adverse effects

Inhalation human systemic effects include: nausea, vomiting, conjunctival irritation, effects on sense of taste. Inhalation of vapours caused irritation to lungs and depression of central nervous system. Prolonged exposure may result in damage to liver and kidneys (3).

Any other comments

Hazards associated with p-tert-butyltoluene reviewed (8).
Experimental toxicology and human health effects reviewed (9).

References

1. Verscheuren, K. *Handbook of Environmental Data on Organic Chemicals* 2nd ed., 1983, Van Nostrand Reinhold, New York
2. Hine, C. H; et al. *AMA Arch. Ind. Hyg. Occup. Med.* 1954, **9**, 227
3. *Toxicol. Biochem. Aromat. Hydrocarb.* 1960, 156
4. Patty, F. A. (Ed.) *Industrial Hygiene and Toxicology* 1967, **2**, Interscience Publishers, New York
5. Ungar, H; et al. *Arch. Pathol.* 1955, **60**, 139-149
6. Zeiger, E; et al. *Environ. Mol. Mutagen.* 1987, **9**(Suppl. 9), 1-109

7. Dean, B. J; et al. *Mutat. Res.* 1985, **153**, 57-77
8. Fielden, M. *Chemical Hazard Information Profile* 1982, USEPA, Washington, DC
9. *ECETOC Technical Report No. 30(4)* 1991, European Chemical Industry Ecology and Toxicology Centre, B-1160 Brussels

B296   5-*tert*-butyl-2,4,6-trinitro-*m*-xylene

CAS Registry No. 81-15-2

Synonyms benzene, 1-(1,1-dimethylethyl)-3,5-dimethyl- 2,4, 6-trinitro; 1-*tert*-butyl-3,5-dimethyl-2,4,6- trinitrobenzene; Musk xylene; Musk xyldl; 2,4,6,-trinitro-1,3-dimethyl-5-*tert*-butyl-benzene

Mol. Formula $C_{12}H_{15}N_3O_6$ **Mol. Wt.** 297.27

Uses Ingredient in hand lotions, soaps and perfume.

Mammalian and avian toxicity

Acute data
LD_{50} oral rat >10 g kg^{-1} (1).
LD_{50} dermal rabbit >15 g kg^{-1} (1).

Sub-acute data
Dermal rat (90 day) 240 mg kg day^{-1} caused some organ weight changes but no associated histopathological changes in any tissue. No effect level was ♂ 75 mg kg^{-1} and ♀ 24 mg kg^{-1} (2).

Irritancy
Dermal human (48 hr) 5 mg caused mild irritation (3).

Sensitisation
Study using Colworth guinea pig photoallergy test showed that Musk xylene could cause a weak photoallergic reaction in humans (4).

References

1. Yamagishi, T; et al. *Bull. Environ. Contam. Toxicol.* 1981, **26**(5), 656-662
2. Ford, R. A; et al. *Food Chem. Toxicol.* 1990, **28**(1), 55-61
3. Opdyke, D. L. J. *Food Cosmet. Toxicol.* 1975, **13**, 881
4. Lovell, W. W; et al. *Int. J. Cosmet. Sci.* 1988, **10**(6), 271-279

B297   Butyl vinyl ether

$CH_3(CH_2)_3OCH=CH_2$

CAS Registry No. 111-34-2
Synonyms vinylbutylether; butoxyethene; 1-(ethenyloxy)butane; vinyl-*n*-butylether
Mol. Formula $C_6H_{12}O$ Mol. Wt. 100.16
Uses Intermediate in organic synthesis. Copolymerisation agent.
Occurrence Detected as a contaminant in air and water samples (1-3).

Physical properties

M. Pt. –92°C; B. Pt. 94.2°C; Flash point –9°C; Specific gravity d_{20}^{20} 0.7803;
Volatility v. den. 3.45.

Solubility
Organic solvent: ethanol, diethyl ether, acetone, benzene

Occupational exposure

UN No. 2352; HAZCHEM Code 3☒E; Conveyance classification flammable liquid.
Abiotic removal
Reacts with photochemically produced hydroxyl radicals and ozone in the
atmosphere, $t_{1/2}$ estimated 9 hr, at an atmosphere concentration of 5×10^5 hydroxyl
radicals cm^3 and 7×10^{11} ozone molecules cm^3. Direct photolysis is not expected to
be an important removal process since aliphalic ethers do not absorb light at
wavelength >290 nm (4).
Susceptible to hydrolysis in environmental waters especially at acidic pH. $t_{1/2}$ 9 hr at
pH5 and 10 yr at pH9 (5).

Mammalian and avian toxicity

Acute data
LD_{50} oral rat 10 g kg^{-1} (6).
LC_{50} (2 hr) inhalation mouse 62 g m^{-3} (7).
LC_{Lo} (4 hr) inhalation rat 8000 ppm (8).
LD_{100} (4 hr) inhalation rat 16,000 ppm (8).
LD_{50} dermal rabbit 4240 mg kg^{-1} (9).

Irritancy
Dermal rabbit (duration unspecified) 500 mg caused mild irritation (10).

Genotoxicity

Salmonella typhimurium TA100, TA100 with metabolic activation weakly positive (11).

Any other comments

Toxicity and physical properties reviewed (12).
Spontaneous explosion risk.

References

1.   Lucas, S. V. *Anal. Org. Drink. Water Conc. Adv. Treatment Conc.* 1984, **2**, 41
     USEPA-600/1-84-020B

2. *Invent. Chem. Subst. Id. Gt. Lakes Ecosyst.* 1983, 68, Gt. Lakes Water Quality Board

2. *Invent. Chem. Subst. Id. Gt. Lakes Ecosyst.* 1983, 68, Gt. Lakes Water Quality Board
3. Pellizzari, E. D; et al. *Devel. Anal. Tech. Meas. Ambient. Atmos. Carcinog. Vap.* 1975, 75, 115 USEPA-600/2-75-076
4. Calvert, J. G. *Photochemistry* 1966, 441-442, J. Wiley, New York
5. Saloman, P; et al. *Acta Chem. Scand.* 1966, **20**, 1790-1801
6. *Union Carbide Data Sheet* 1973, (28 Jun), Union Carbide Corp., New York
7. Izmerov, N. F; et al *Toxicol. Param. Ind. Toxic. Chem. Under Single Exposure* 1982, -, 119
8. *Patty's Ind. Hyg. Toxicol.* 3rd ed., 1981-1982, **2A, 2B, 2C**, 2503, J. Wiley, New York
9. *AMA Arch. Ind. Hyg. Occup. Med.* 1954, **10**, 61
10. *Union Carbide Data Sheet* 1972(28 Jun), Union Carbide Corp., New York
11. Sone, T; et al. *J. Pharmacobio-Dyn* 1989, **12**(6), 345-351
12. *EPA Chemical Profiles* 1985, USEPA, Washington, DC

B298   1,4-Butynediol

HOCH2C≡CCH2OH

CAS Registry No. 110-65-6
Synonyms bis(hydroxymethyl)acetylene; 2-butynediol; 2-butyne-1,4-diol
Mol. Formula $C_4H_6O_2$ **Mol. Wt.** 86.09
Uses In the manufacture of polyurethanes and synthetic rubber. Used in synthesis of the blood substitute polyvinylpyrrolidone.

Physical properties

M. Pt. 57.5°C; **B. Pt.** 238°C; **Flash point** 152°C; **Partition coefficient** log P_{ow} −1.83 (1).

Solubility
Organic solvent: acetone, ethanol

Occupational exposure

UN No. 2716; **Conveyance classification** harmful substance; **Supply classification** toxic.

Risk phrases
Toxic if swallowed – Causes burns (R25, R34)

Safety phrases
Do not breathe dust – Wear suitable protective clothing – If you feel unwell, seek medical advice (show label where possible) (S22, S36, S44)

Ecotoxicity

Fish toxicity
Exposure of steelhead trout to 6 mg l^{-1} caused loss of equilibrium in 4-6 hr and death within 6-8 hr (1).
LC_{50} (96 hr) fathead minnow 53 mg l^{-3} flow-through bioassay with measured concentrations 25°C, 6.8 mg l^{-1} dissolved oxygen content, hardness 46.5 mg l^{-1} calcium carbonate, pH 7.7. No loss of equilibrium prior to death (2).

Effects to non-target species
LD_{50} oral redwing blackbird 75 mg kg^{-1} (3).

Environmental fate

Degradation studies
Fusanum merismoides BII can utilise 2-butyne-1,4-diol as the sole carbon source with production of mannitol, 2,4,6 triketosuburic acid, 2,4,6,8-tetraketosebaric acid and phthalic acid using the enzyme 2-alkyne-1-ol dehydrogenase (4).

Mammalian and avian toxicity

Acute data
LD_{50} oral rat, rabbit, guinea pig 100-150 mg kg^{-1} (5).
LC_{Lo} (2 hr) inhalation rat 150 mg m^{-3} (6).

Sub-acute data
Oral rat (14 day) 1, 10 or 100 mg kg^{-1}. Toxic effects at 100 mg kg^{-1} included severe body weight changes, increased liver weight and serum cholesterol in both sexes. Increased serum calcium, and decreased red cell count, haemoglobin content and haemocrit values reported in ♀ at 100 mg kg^{-1} (7).

Any other comments

Toxicity reviewed (8).
The effect of 1,4- butynediol on animals reported (9).
Experimental toxicology and human health effects reviewed (10).

References

1. *Fish Toxicity Screening Data* 1989, EPA 560/6-89-001, US Dept. Wildlife and Fisheries
2. Geiger, D. L; et al. *Acute Toxicities Organic Chemicals to Fathead Minnows* 1988, **4**, 55, Univ. Wisconsin-Superior, WI
3. Schafer, E. W; et al. *Arch. Environ. Contam. Toxicol.* 1983, **12**, 355
4. Miyoshi, T; et al. *Biochem. Biophys. Acta* 1974 **358**, 231-239
5. *Hyg. Sanit.* 1968, **33**, 41
6. Izmerov, N. F; et al. *Toxicometric Parameters of Industrial Toxic Chemicals under Single Exposure* 1982, 30, Moscow
7. Komsta, E; et al. *Bull. Environ. Contam. Toxicol.* 1989, **43**(1), 87-94
8. *Information Profile Potential Occup. Hazard – Glycols* 1986, 1-208, Centre for Chemical Hazard Assessment, Syracuse, New York
9. Kutepov, E. N. *Hyg. Sanit.* 1968, **33**, 41-47
10. *ECETOC Technical Report No. 30(4)* 1991, European Chemical Industry Ecology and Toxicology Centre, B-1160 Brussels

B299   Butyraldoxime

$$CH_3(CH_2)_2CH=NOH$$

CAS Registry No. 110-69-0
Synonyms butanal oxime; butyraldehyde, oxime; *n*-butyraldehyde oxime; *N*-butyraldoxime

Mol. Formula C_4H_9NO **Mol. Wt.** 87.12

Uses Anti-skinning agent.

Physical properties

M. Pt. −29.5°C; **B. Pt.** 152°C; **Flash point** 57.8°C (closed cup); **Specific gravity** d_4^{20} 0.923; **Volatility** v. den. 3.01.

Solubility
Organic solvent: acetone, benzene, diethyl ether, ethanol

Occupational exposure

UN No. 2840; **Supply classification** toxic.

Risk phrases
Harmful if swallowed – Toxic in contact with skin – Irritating to eyes (R22, R24, R36)

Safety phrases
Do not breathe vapour – Wear suitable protective clothing – If you feel unwell, seek medical advice (show label where possible) (S23, S36, S44)

Mammalian and avian toxicity

Acute data
LD_{Lo} oral rabbit 100 mg kg^{-1} (1).
LD_{50} intraperitoneal mouse 200 mg kg^{-1} (2).

Genotoxicity

Salmonella tymphimurium TA98, TA100, TA1537, TA1538 with metabolic activation negative (3).
Mouse lymphoma L51787 tk+/tk- with and without metabolic activation negative (3).

Any other adverse effects

Exposure to butyraldoxime after consumption of alcohol can cause, flushing of face, red non-itching blotches, redness of eyes, drowsiness, shortness of breath and heart palpitations (4).

Any other comments

Experimental toxicology and human health effects reviewed (5).

References

1. *RTECS No. 26452* 1991, **31**, Reg. of Toxic Eff. of Chem. Subst., NIOSH, Dept. of Physical Sciences, Cincinnati, OH
2. *NTIS Report* AD 277-689, Natl. Tech. Inf. Serv., Springfield, VA
3. Rogero-Back, A. M; et al. *Mutat. Res.* 1988, **204**, 149-156
4. Lewis, W; et al. *Arch. Ind. Health* 1956, **13**, 76-79, 628-631
5. *ECETOC Technical Report No. 30(4)* 1991, European Chemical Industry Ecology and Toxicology Centre, B-1160 Brussels

B300 Butyric anhydride

$$CH_3(CH_2)_2 \qquad (CH_2)_2CH_3$$

CAS Registry No. 106-31-0

Synonyms butanoic acid anhydride; butanoic anhydride; butyric acid anhydride; *n*-butyric acid anhydride; butyryl oxide

Mol. Formula $C_8H_{14}O_3$ **Mol. Wt.** 158.20

Uses Vulcanisation retarder. Intermediate in organic synthesis.

Physical properties

M. Pt. –75°C; **B. Pt.** 199.4-201.4°C; **Flash point** 88°C (open cup); **Specific gravity** d_4^{20} 0.9668.

Solubility

Organic solvent: ethanol (decomp.), diethyl ether

Occupational exposure

UN No. 2739; **HAZCHEM Code** 2R; **Conveyance classification** corrosive substance; **Supply classification** corrosive.

Mammalian and avian toxicity

Acute data

LD_{50} oral mouse 2000 mg kg^{-1} (1).
LD_{50} oral rat 8790 mg kg^{-1} (2).

Any other comments

Toxicity reviewed (3).

References

1. Izmerov, N.F; et al. *Toxicometric Parameters of Industrial Toxic Chemicals Under Single Exposure* 1982, 31, Moscow
2. *Prehled Prumyslove Toxikol. Org. Latky* 1986, 321
3. *Information Profiles Potential Occup. Hazards: Organic Anhydrides* 1982, No 210-79-0030, 1-126, Centre for Chemical Hazard Assessment, Syracuse, New York

B301 β-Butyrolactone

CAS Registry No. 3068-88-0

Synonyms 3-hydroxybutanoic acid β-lactone; β-hydroxybutyric acid lactone; 3-hydroxybutyric acid lactone; 4-methyl-2-oxethanone; β-methylpropiolactone;

2-oxetanone, 4-methyl-
Mol. Formula $C_4H_6O_2$ **Mol. Wt.** 86.09
Uses Production of β-oxybutynl-*para*-phenelidine.

Physical properties

B. Pt. $_{10}$ 54-56°C; **Specific gravity** d_{20}^{20} 1.0555.

Mammalian and avian toxicity

Acute data
LD_{50} oral rat 17 g kg^{-1} (1).

Carcinogenicity and long-term effects
No adequate evidence for carcinogenicity to humans, sufficient evidence for
carcinogenicity to animals, IARC classification group 2B (2).
Subcutaneous rat, mouse (30 wk) 0.1-10 mg wk^{-1} and oral rat, mouse 5-100 mg wk^{-1}
via food. High dose 100 mg wk^{-1} in food induced gastric tumours in rats (3).
Subcutaneous rat (78 wk intermittent) 38 kg^{-1} (maximum tolerated dose) induced
sarcomas at injection site (4).

Irritancy
Dermal rabbit (duration unspecified) 500 mg caused moderate irritation (5).

Genotoxicity

Salmonella tymphimurium TA1535, pSIC1002, SOS-inducing activity detected by
Umu gene expression, suppressed by human urine (6).
Mammalian lymphocyte 860mg l^{-1} DNA damage (7).
Investigation of reactivity and adducts formed by lactones. β- Butyrolactone alkylated
guanosine, RNA and DNA. Chemical reactivity correlated with carcinogenic potency (8).

Any other adverse effects

The reactivity of β-butyrolactone with guanosine, RNA, DNA and
4-(p-nitrobenzyl)pyridine was studied. The rate of alkylation by the lactones was
guanosine >RNA, denatured DNA >double stranded DNA (8).

Any other comments

Toxicity reviewed (9).
Experimental toxicology and human health effects reviewed (10).

References

1. *Am. Ind. Hyg. Assoc. J.* 1969, **30**, 470
2. *IARC Monograph* 1987, **Suppl. 7**, 59
3. Van Duuren, B. L; et al. *J. Natl. Cancer Inst.* 1966, **37**, 825-838
4. Van Duuren, B. L; et al. *J. Natl. Cancer Inst.* 1967, **39**, 1213-1216
5. *Union Carbide Data Sheet* 1966, (20 Jan), Union Carbide Corp., New York
6. Nakamura, S; et al. *Chem. Express.* 1989, **4**(12), 817-820
7. Melzer, M. S. *Biochem. Biophys. Acta* 1967, **142**(2), 538-541
8. Hemminki, K. *Chem. Biol. Interact.* 1981, **34**, 323-331
9. *IARC Monograph* 1976, **11**, 225-229
10. *ECETOC Technical Report No. 30(4)* 1991, European Chemical Industry Ecology and
    Toxicology Centre, B-1160 Brussels

B302 γ-Butyrolactone

CAS Registry No. 96-48-0

Synonyms butyric acid lactone; γ-hydroxybutyrolactone; α-butyrolactone; dihydro-2-(3*H*)-furanone; 1,4-butanolide; 2,(3*H*)-furanone dihydro

Mol. Formula $C_4H_6O_2$ **Mol. Wt.** 86.09

Uses As a constituent of paint removers, textile aids and drilling oils. Intermediate in synthesis of polyvinylpyrrolidene. Solvent.

Occurrence Has been detected in a commercial natural liquid wood smoke preparation and in tobacco smoke condensates (1,2).

Physical properties

M. Pt. –44°C; **B. Pt.** 206°C; **Flash point** 98.3°C (open cup); **Specific gravity** d_4^{25} 1.124; **Volatility** v. den. 3.0.

Solubility
Water: miscible. Organic solvent: methanol, ethanol, acetone, diethyl ether, benzene

Ecotoxicity

Fish toxicity
Listed as negative in tests on trout, bluegill sunfish and goldfish. However the authors state that the high mineral content of the water used in the studies adds an additional source of error. Therefore the compound listed as negative might be toxic in softer water supplies (3).

Mammalian and avian toxicity

Acute data
LD_{50} oral rat 1260-1800 mg kg^{-1} (4,5).
LD_{50} intraperitoneal rat, mouse 1000-1100 mg kg^{-1} (6).
LD_{Lo} intravenous rabbit 500 mg kg^{-1} (6).

Carcinogenicity and long-term effects
No adequate evidence for carcinogenicity to humans, insufficient evidence for carcinogenicity to animals, IARC classification group 3 (7).

National Toxicology Program study, gavage rats, mice (2 yr) dose unspecified no evidence of carcinogenicity in either ♂ or ♀ rats, equivocal evidence in ♂ mice (8). Dermal Swiss-Millerton ♂ mice 100 mg 3 × wk^{-1} (in 10% benzene solution) 2 tumours and 1 cancer reported from 30 mice treated. The authors conclude γ-butyrolactone has a low order of activity and that strained lactone rings may favour carcinogenicity (9).

Teratogenicity and reproductive effects
Gavage Sprague-Dawley rats (days 6-15 gestation) 0, 10, 50, 125, 250 or 500 mg kg^{-1}.

Necropsy showed lung oedema, hyperaemia and emphysema. Foetal weights significantly increased in high dosage groups (10).

Metabolism and pharmacokinetics
Intravenous rat (dose unspecified) metabolised to γ-hydroxybutric acid (11).
86% of unspecified dose inhaled by rats was excreted as carbon dioxide within 18 hr (12).

Sensitisation
Negative results were obtained in guinea pig skin sensitisation tests using doses of 5.6 g kg^{-1} (13).

Genotoxicity
Bacillus subtilis H17, M45 with metabolic activation positve (14).
No chromosome damage in rat liver RL$_1$ cell line (15).
BHK-21 cell transformation test with metabolic activation positive (16).

Any other adverse effects
Metabolite γ-hydroxybutric acid can cause depression of central nervous system (10).

Any other comments
Butyrolactone protected chicken embryos from digitoxin toxicity (17).
The sodium salt has been used as an anaesthetic (18).
When administered orally or intravenously can cause drowsiness (19).
A review with references examined the analgesic action of γ- butyrolactone in humans and laboratory animals (20).
Experimental toxicology and human health effects reviewed (21).
Hygroscopic.

References

1. Fiddler, W; et al. *J. Agric. Food Chem.* 1970, **18**, 310-312
2. Neurath, G; et al. *Beitr. Tabakforsch.* 1971, **6**, 12-20, (Ger)
3. *The Toxicity of 3400 Chemicals to Fish* 1987, EPA560/6-87-002 PB 87-200-275, Washington, DC
4. Izmerov, N. F; et al. *Toxicometric Parameters of Industrial Toxic Chemicals under Single Exposure* 1982, 31, Moscow
5. Hampel, H; et al. *Arch. Int. Pharmacodyn. Ther.* 1968, **171**(2), 306
6. *Archiv. Immunol. Therap. Exp.* 1965, **13**, 70
7. *IARC Monograph* 1987, **Suppl. 7**, 59
8. *National Toxicology Program Research & Testing Division* 1992, Report No. TR-406, NIEHS, Research Triangle Park, NC
9. Duuren Van, B. L; et al. *J. Natl. Cancer Inst.* 1963, **31**, 41-55
10. Kronevi, T; et al. *Pharmacol. Toxicol.* 1988, **62**(1), 57-58
11. Roth, R. H; et al. *Biochem. Pharmacol.* 1965, **14**, 177-178
12. *IARC Monograph* 1976, **11**, 231
13. *Patty's Indust. Hygiene Toxicol.* 2nd ed., 1963, 1824-1825
14. Kada, T. *Prog. Mutat. Res.* 1981, **1**, 175-182
15. Dean, B. J. *Prog. Mutat. Res.* 1981, **1**, 570-579
16. Styles, J. A. *Prog. Mutat. Res.* 1981, **1**, 638-646
17. Cosmides, G. J. *J. Pharmacol. Exp. Ther.* 1956, **118**(3), 286-295
18. Walkenstein, S. S; et al. *Biochem. Biophys Acta* 1964, **86**, 640-642
19. *Kirk-Othmer Encyclopedia of Chemical Technology* 1981, **13**, 97, John Wiley & Sons, New York

20. Klosa, J. *Pharmazie* 1988, **43**(7), 516-517, (Ger.) (*Chem. Abstr.* **109**, 162754v)
21. *ECETOC Technical Report No. 30(4)* 1991, European Chemical Industry Ecology Toxicology Centre, B-1160 Brussels

B303   Butyronitrile

CH3(CH2)2CN

CAS Registry No. 109-74-0
Synonyms butanenitrile; *n*-butanenitrile; butyric acid nitrile; butyronitrile; 1-cyanopropane; propyl cyanide
Mol. Formula C_4H_7N **Mol. Wt.** 69.11
Uses Polymer synthesis. Intermediate in organic synthesis.

Physical properties

M. Pt. –112.6°C; **B. Pt.** 117°C; **Flash point** 26.1°C (open cup); **Specific gravity** d_4^{15} 0.7954; **Partition coefficient** log P_{ow} 0.60; **Volatility** v.p. 10 mm Hg at 15°C; v. den. 2.4.

Solubility
Organic solvent: ethanol, diethyl ether, benzene, dimethylformamide

Occupational exposure

UN No. 2411; **HAZCHEM Code** 3WE; **Conveyance classification** flammable liquid; **Supply classification** flammable and toxic.

Risk phrases
Flammable – Toxic by inhalation, in contact with skin and if swallowed (R10, R23/24/25)

Safety phrases
If you feel unwell, seek medical advice (show label where possible) (S44)

Ecotoxicity

Effects to non-target species
Highly toxic to birds (1).

Mammalian and avian toxicity

Acute data
LD_{50} oral mouse, rat 30-140 mg kg^{-1} (2,3).
LC_{Lo} (4 hr) inhalation rat 1000 ppm (2).
LC_{50} (1 hr) inhalation mouse 250 ppm (4).
LD_{50} dermal rabbit 500 mg kg^{-1} (3).
LD_{Lo} dermal guinea pig 100 mg kg^{-1} (5).
LD_{Lo} intravenous rabbit 980 mg kg^{-1} (6).
LD_{50} intraperitoneal mouse 38 mg kg^{-1} (7).

Metabolism and pharmacokinetics
Inhaled organonitriles metabolised to cyanide in the nasal cavity (species unspecified) (8).

Intraperitoneal rat (duration unspecified) 150 mg kg^{-1}. Accumulation in liver, stomach, intestine, kidney and testis. Elimination via urine accounted for very small amounts and was slow (9).

Irritancy

Dermal rabbit (duration unspecified) 395 mg caused mild irritation (10).
500 mg instilled into rabbit eye for 24 hr caused mild irritation (11).

Any other adverse effects

Systemic effects include bronchial tightness, gastric and respiratory distress, hypotension, conjunctivitis, chest pain, skin discolouration, tachypnea, dizziness, vomiting, convulsions, coma, cyanosis, retching, thyroid reaction and duodenal ulcers (12).

Any other comments

Toxicity of butyronitrile reviewed (13-15).
Experimental toxicology and human health efects reviewed (16).

References

1. Schafer, E. W; et al. *Arch. Environ. Contam. Toxicol.* 1983, **12**, 355-382
2. *Am. Ind. Hyg. Assoc. J.* 1962, **23**, 95
3. *Jpn. J. Hyg.* 1984, **39**, 423
4. Willhite, C. C; et al. *Clin. Toxicol.* 1981, **18**, 991
5. *Kodak Co. Rep.* 21st May 1971, Kodak Co., New York
6. *Comptes Rendus Hetbdomactaires des Seances, Acad. Sci.* 1911, **153**, 895
7. *Toxicol. Appl. Pharmacol.* 1981, **59**, 589
8. Dahl, A. R; et al. *Ann. Rep. Toxicol. Res. Institut.* 1986-1987, Report No. LMF-120, 397-399
9. Haguenoer; et al. *Bull. Soc. Pharm. Lille Iss.* 1974, **4**, 161
10. *Union Carbide Data Sheet* 1960 (May 17th), Union Carbide Corp., New York
11. *Prehled Prumyslove Toxikol. Org. Latky* 1986, 900
12. NIOSH *Criteria for A Recommended Standard Occupational Exposure to Nitriles* 1978, No 78-212, 1-163, US Dept. of Health
13. Johannsen, F. R; et al. *Fundam. Appl. Toxicol.* 1986, **7**(4), 690-697
14. Ahmed, A. E; et al. *Toxicol. Lett.* 1982, **12**(2-3), 157-163
15. Gage, J. C. *Br. J. Ind. Mech.* 1970, **27**, 1-18
16. *ECETOC Technical Report No. 30(4)* European Chemical Industry Ecolgy and Toxicology Centre, B-1160 Brussels

B304   Butyryl chloride

CH$_3$(CH$_2$)$_2$COCl

CAS Registry No. 141-75-3
Synonyms butanoyl chloride; butyric acid chloride; *n*-butyryl chloride
Mol. Formula C$_4$H$_7$ClO   **Mol. Wt.** 106.55
Uses Acylating agent. Polymerisation catalyst. Intermediate in organic synthesis.

Physical properties

M. Pt. −89°C; **B. Pt.** 101-102°C; **Flash point** 18°C (closed cup); **Specific gravity** d_{20}^{20} 1.028; **Volatility** v. den. 3.67.

Solubility
Water: (decomp.). Organic solvent: ethanol (decomp.), diethyl ether

Occupational exposure

UN No. 2353; **HAZCHEM Code** 2PE; **Conveyance classification** flammable liquid; **Supply classification** highly flammable and corrosive.

Risk phrases
Highly flammable – Causes burns (R11, R34)

Safety phrases
Keep away from sources of ignition – No Smoking – Do not breathe vapour – In case of contact with eyes, rinse immediately with plenty of water and seek medical advice – Wear suitable protective clothing (S16, S23, S26, S36)

Any other comments

Experimental toxicology and human health effects reviewed (1,2).

References

1. Smyth, H. F; et al. *Am. Ind. Hyg. Assoc. J.* 1962, **23**, 95
2. *ECETOC Technical Report No. 30(4)* 1991, European Chemical Industry and Toxicology Centre, B-1160 Brussels

GLOSSARY OF NAMES FOR EXPERIMENTAL ORGANISMS

Latin – Common

Abramis brama	bream
Acanthurus spp.	surgeon fish
Acartia sp.	oar-footed crustacean
Acartia tonsa	oar-footed crustacean
Accipiter nisus	sparrow hawk
Accipiter cooperi	Cooper's hawk
Acinetobacter sp.	bacteria
Acroneuria	stonefly
Acroneuria pacifica	stonefly
Acroynchia savaera	shrub
Aedes sp.	mosquito fly
Aerobacter sp.	
Aequipecten gibbus	calico scallop
Aequipecten (Pecten) irradians	bay scallop
Aeromonas hydrophila	bacteria
Agelaius tricolor	tricolored blackbird
Agelaius phoeniceus	redwing blackbird
Alburnus alburnus	bleak
Alcaligenes denitrificans	bacteria
Alcaligenes faecalis	bacteria
Ambystoma mexicanum	Mexican axolotl
Ameiurus melas	black bullhead
Amycolata autoptopica	bacteria
Anabaena sp.	blue-green algae
Anagasta kuehnie	Mediterranean flour moth
Anarchichas lupus	wolf-fish
Anas acuta	common pintail
Anas platyrhynchos	mallard
Anas discors	blue-winged teal
Anchoa mitchilly	bay anchovy
Anguilla anguilla	eel
Anguilla japonica	Japanese eel
Anguilla rostrata	American eel
Anguilla vulgaris	eel
Ankistrodesmus falcatus	green algae
Annelida spp.	segmented worms
Anodonta anatina	mollusc

Anodonta cygnea	freshwater clam
Anolis carolinensis	anole
Anopheles spp.	mosquito fly
Anser anser	greylag goose
Anthopleura elegantissima	sea anemone
Aphelocoma coerulescens	scrub jay
Apis fabae	bean aphid
Apis melliferra	honey bee
Apis spp.	honey bee
Aplexa	snail
Aplexa hypnorum	snail
Aposemus sylvaticus	field mouse
Aquila chrysaetos	golden eagle
Aratinga pertinax	brown-throated conure
Aratinga canicularis	orange-fronted conure
Arbacia punctulata	sea urchin
Ardea cinerea	heron
Ardeola spp.	egret
Artemia salina	brine shrimp
Artemisis dracunculus	tarragon
Arthrobacter sp.	bacteria
Aschelminthes spp.	round worms
Asellus	sowbug
Asellus aquaticus	sowbug
Asellus brevicaudus	sowbug
Aspergillus flavus	fungi
Aspergillus fumigatus	fungi
Aspergillus nidulans	fungi
Aspergillus parasiticus	fungi
Atropa belladonna	belladonna
Aulosira fertilissima	cyanobacteria
Bacillus licheniformis	bacteria
Bacillus megaterium	bacteria
Bacillus subtilis	bacteria
Baetis	mayfly
Balanus spp.	barnacle
Barbus conchonius	red barb
Barbus stigma	barb
Barbus ticto ticto	barb
Beta vulgaris	red beetroot
Biston betularia	peppered moth
Blarina brevicauda	short-tailed shrew
Bombus spp.	bumble bee
Bombycilla cedrorum	cedar waxwing
Brachydanio rerio	zebra fish
Brachidontes recurvis	mussel
Branta canadensis	Canada goose
Brassica nigra	black mustard seed

Brevoortia tyrannus	Atlantic menhaden
Brevoortia patronus	Gulf menhaden
Brevibacterium sp.	bacteria
Bufo americanus	American toad
Bufo fowleri	Fowler's toad
Bufo bufo	common toad
Buteo swainsoni	Swainson's hawk
Cadra cautella	almond moth
Calamospiza melanocory	lark bunting
Calanoida	oar-footed crustacean
Callinectes sapidus	blue crab
Cancer magister	Dungeness crab
Caranx spp.	pompano, jack cravally
Carassius auratus	goldfish
Carassius carassius	Crucian carp
Carcinus maenas	green crab
Carpodacus cassinii	cassins finch
Carpodacus mexicanus	house finch
Cassidix major	boat-tailed grackle
Catostomus catostomus	longnose sucker
Catostomus commersoni	white sucker
Catostomus spp.	suckers
Cemiscus nitilus	roach
Ceriodaphnia dubia	water flea
Chandrus crispus	sea moss
Channa striatus	snakehead fish
Chanos chanos	milkfish
Chaoborus	phantom midges
Chelon labrosus	grey mullet
Chelydra serpentina	snapping turtle
Chilomonas paramecium	protozoa, flagellate
Chironomus	midge
Chironomus plumosus	midge
Chironomus riparius	midge
Chlamydonas sp.	green algae
Chlamydomonas reinhardti	green algae
Chlamydomonas variabilis	green algae
Chlorella sp.	green algae
Chlorella fusca	green algae
Chlorella kesslerii	green algae
Chlorella pyrenoidosa	green algae
Chloristoneura occidentalis	budworm
Chlorococcum sp.	green algae
Chromobacterium sp.	bacteria
Chthamalus spp.	barnacles
Circus cyaneus	northern harrier, marsh hawk
Claassenia sabulosa	stonefly

Cladocera	water flea
Clarias batrachus	walking catfish
Clostridium difficile	bacteria
Clupea harengus	herring
Clupea harengus harengus	Altantic herring
Clupea sprattus	sprat
Clupea pallasi	Pacific herring
Coleoptera	beetle
Colinus virginianus	common bobwhite, bobwhite quail
Colisa fasciata	giant gourami
Colpidium	protozoa
Columba livia	common pigeon, rock dove
Columbina talpacoti	ruddy ground dove
Columbina passerina	common or ground dove
Conus spp.	cone shells
Coregonus albula	vendace
Coregonus lavaretus	rowan, lavaret
Corvus corax	northern raven
Corvus brachyrhynchos	American crow
Coturnix coturnix	Japanese quail
Coturnix coturnix japonica	
Crangon crangon	brown shrimp
Crangon septemspinosa	sand shrimp
Crangon spp.	snapping shrimp
Crangonyx pseudogracilis	crustacean
Crassostrea viringica	eastern oyster
Crassostrea gigas	Pacific oyster
Cryptotis parva	least shrew
Ctenopharyngodon	carp
Ctenopharyngodon idella	white amur grass carp
Culex	mosquito fly
Culex pipiens	mosquito fly
Culex theileri	mosquito fly
Cyanocitta cristata	blue jay
Cyanocorax yncas	green jay
Cymatogaster aggregata	shiner perch
Cypridopsis vidua	seed shrimp
Cyprinodon variegatus	sheepshead minnow
Cyprinus carpio	carp
Cytophaga	bacteria
Daphnia cucullata	water flea
Daphnia laevis	water flea
Daphnia longispina	water flea
Daphnia/Daphnia magna/Daphnia pulex	water flea
Dasypus novemcinctus	armadillo
Datura stramonium	jimsonweed

Delphinus delphis	common dolphin
Desulfovibrio vulgaris	bacteria
Dichapetalum cymosum	woody plant
Didelphis virginiana	opossum
Dipodomys sp.	kangaroo rat
Drosophila melanogaster	fruit fly
Dunaliella sp.	green algae
Dunaliella euchlora	green algae
Echinometra spp.	sea urchin
Eisenia sp.	kelp
Eisenia foetida	kelp
Elaphe sp.	rat snake
Engraulis engrasicolus	European anchovy
Engraulis mordax	northern anchovy
Entosiphon	protozoa
Entosiphon sulcatum	protozoa, flagellate
Ephemerella	mayfly
Eptesicus capensis	bat
Eptesicus fuscus	big brown bat
Eremophila alpestris	horned lark
Escherichia coli	bacteria
Esox lucius	northern pike
Euplectes orix	red bishop
Eutamias sp.	western chipmunk
Euthynnus pelamis	skipjack tuna
Falco peregrinus	peregrine falcon
Falco sparverius	American kestrel
Felis catus	cat
Fundulus	killifish
Fundulus heteroclitus	mummichog
Fusarium merismoides	ascomycetes
Fusarium oxysporum	ascomycetes
Gadus merlangus	whiting
Gadus morhua	Atlantic cod
Gadus pollachius	pollack
Gadus virens	coalfish
Gallus gallus	domestic chicken
Gambusia affinis	mosquito fish
Gammarus	scud
Gammarus fasciatus	scud
Gammarus pacustris	scud
Gammarus pseudolimnaeus	scud
Gammarus pulex	scud
Gasterosteus aculeatus	three spined stickleback
Glossina	tsetse fly
Glycera dibranchiata	bloodworm

Gobio gobio	gudgeon
Gobius minutus	goby
Gracilaria verucosa	red seaweed
Gracilaria foliifera	red seaweed
Grus canadensis	sandhill crane
Gryllus pennsylvanicus	field cricket
Gymnodinium breve	unicellar biflagellate algae
Haematococcus pluralis	algae
Haliotis spp.	abalone
Hansenula anomala	bacteria
Helcioniscus argentatus	saltwater limpet
Helcioniscus exaratus	saltwater limpet
Helix sp.	land snail
Hemigrapsus spp.	shore crabs
Heteropneustes fossilis	airsac catfish
Hexagenia	mayfly
Hippoglossus hippoglossus	halibut
Hirundinidae	swallows
Homarus americanus	northern lobster
Hyalella	shrimp
Hyalella azteca	shrimp
Hydra oligactis	protozoa
Hyla versicolor	treefrog
Hyla crucifer	spring peeper
Hypophthalmichthys molitrix	silver carp
Hypophthalmichthys nobilis	bighead
Ictalurus melas	black bullhead
Ictalurus nebulosus	brown bullhead
Ictalurus punctatus	channel catfish
Ictalurus natalis	yellow bullhead
Ictalurus nebulosus	brown bullhead
Ictalurus bebulosus	American catfish
Ictalurus	catfish
Ictiobus cyprinellus	bigmouth buffalo
Ictiobus	suckers
Ischnura verticalis	damselfly
Isoperla	stonefly
Jordanella floridae	American flagfish
Klebsiella pneumoniae	bacterium
Kuhlia sandvicensis	mountain bass
Labeo bicolor	red tailed black
Lagodon	common sunfish
Lagodon rhomboides	marine pin perch
Laminaria digibacta	Atlantic kelp

Laminaria agardhii	Atlantic kelp
Larus delawarensis	ring-billed gull
Lasiurus borealis	red bat
Lebistes reticulatus	guppy
Lepomis cyanellus	green sunfish
Lepomis macrochirus	bluegill sunfish
Lepomis megalotis	longear sunfish
Lepomis microlophus	redear sunfish
Lepomis gibbosus	pumpkinseed
Lepomis humilis	common sunfish
Leporidae	hare
Leptocottus armatus	staghorn sculpin
Leptotila verreauxi	white-fronted dove
Lepus europaeus	European hare
Lestes congener	damselfly
Leuciscus idus	ide
Leuciscus leuciscus	dace
Limanda limanda	common dab
Lophortyx californica	California quail
Lota lota	burbot
Lucania parva	rainwater killifish
Lumbricus terrestris	common earthworm
Lutianus campechanus	red snapper
Lytechinus spp.	sea urchin
Macaca fascicularis	macaque
Macaca mulatta	rhesus monkey
Macoma balthica	mollusc
Macrobrachium lamerrei	shrimp
Macrocystis pyrifera	kelp
Macromia	dragonfly
Macropodus cupanus	spike-tailed paradisefish
Macropodus opercularis	paradisefish
Mallotus villosus	capelin
Meleagris gallopavo	wild turkey
Meleagris	turkey
Melicope leptococca	citrus tree from New Caledonia
Melopsittacus undulatus	budgerigar
Menidia audens	Mississippi silverside
Menidia menidia	Atlantic silverside
Menidia beryllina	inland silverside
Menidia peninsulae	tidewater silverside
Mercenaria mercenaria	hard clam
Metapenaeus monoceros	shrimp
Microcystis aeruginosa	blue-green algae
Microcystis sp.	blue-green algae
Micromesistius poutassou	blue whiting
Micropogon undulatus	croaker
Micropterus dolomieui	smallmouth bass

Micropterus salmoides	largemouth bass
Microtus arvalis	common vole
Microtus longicaudus	vole
Microtus montanus	vole
Microtus pennsylvanicus	meadow vole
Mollienesia latipinna	sailfin molley
Molothrus ater	brown-headed cowbird
Molothrus bonariensis	shiny cowbird
Monoraphidium	green algae
Monoraphidium griffithii	diatom
Morone labra	sea bass
Morone saxatilis	striped bass
Mugil cephalus	striped mullet
Mulloidichthys spp.	goatfish
Mus musculus	house mouse
Musca domestica	housefly
Mustela vison	mink
Mya arenaria	soft shell clam
Mycobacterium tuberculosis	bacteria
Myiopsitta monachu	monk parakeet
Myocastor coypus	nutria
Myotis grisescens	gray bat
Myotis lucifugus	little brown bat
Mysidopsis	shrimp
Mysidopsis bahia	mysid shrimp
Mystus vittatus	striped catfish
Mytilus californianus	California sea mussel
Mytilus edulis	bay mussel
Myzus persicae	peach-potato aphid
Neanthes arenaceodentata	ragworm
Neotoma cinerea	pack rat
Nereis diversicolor	sandworm
Nereis vexillosa	sandworm
Nereis virens	sandworm
Neurospora crassa	mould
Nitrobacter agilis	nitrogen-fixing bacteria
Nitrosococcus oceanus	nitrogen-fixing bacteria
Nitrosomonas europaea	nitrogen-fixing bacteria
Nitocra	shrimp
Nitocra spinipes	shrimp
Nitrosomonas sp.	nitrogen-fixing bacteria
Nostoc linckia	starfish
Nostoc muscorum	blue-green algae
Notopterus notopterus	knifefish
Nycticeius schlieffeni	bat
Oncorhynchus gorbuscha	pink salmon
Oncorhynchus keta	chum salmon

Oncorhynchus masou	masu salmon
Oncorhynchus nerka	sockeye salmon, blueback salmon
Oncorhynchus nerka kennerlyi	kokanee
Oncorhynchus tschawytscha	king salmon
Oncorhynchus kisutch	coho salmon
Ophicephalus punctatus	snakehead fish
Orconectes nais	crayfish
Ortalis vetula	plain chachalaca
Oryctolagus cuniculus	domestic New Zealand white rabbit
Oryzias latipes	rice fish
Oscillatoria	blue-green algae
Ostracoda	shrimp
Otospermophilus beecheyi	California ground squirrel
Ovalipes ocellatus	calico crab
Ovis sp.	sheep
Palaemonon macrodactylus	glass shrimp
Palaemontes pugio	crustacean
Panuliris argus	spiny lobster
Panulirus japonicus	Pacific lobster
Panuliris pencillatus	lobster
Paracentrotus lividus	sea urchin
Paralichthys lethostigma	southern flounder
Paralichthys dentatus	summer flounder
Paralithodes camchatica	king crab
Paraphrys vetulus	English sole
Passer luteus	golden sparrow
Passer domesticus	house sparrow
Patella spp.	limpets
Penaeus monodon	shrimp
Penaeus setiferus	white shrimp
Penaeus duorarum	pink shrimp
Penaeus aztecus	brown shrimp
Pendalus spp.	shrimp
Penicillium notatum	bacteria
Perca fluviatilis	perch
Perca flavescens	yellow perch
Periplaneta americana	common cockroach
Peromyscus gossypinus	cotton mouse
Peromyscus leucopus	white-footed mouse
Peromyscus maniculatus	deer mouse
Peromyscus polionotus	old-field mouse
Petromyzon marinus	sea lamprey
Phaeodactylium tricornutum	diatom
Phanerochaete chrysoporium	fungi
Phasianus colchicus	ring-necked pheasant
Phebalium dentatum	mollusc

Photobacterium phosphoreum	bacteria
Phoxinus phoxinus	minnow
Phrynosoma cornutum	horned lizard
Pica nuttalli	yellow-billed magpie
Pica pica	black-billed magpie
Pichia membranefaciens	bacteria
Pimephales promelas	fathead minnow
Plasmodium berghei	protozoa
Plasmodium falciparum	bacteria
Platichthys stellatus	starry flounder
Platyhelminthes	flatworms
Pleuronectes platessa	plaice
Ploceus taeniopterus	northern masked weaver
Ploceus cucullatus	village weaver
Podophthalmus vigil	crab
Poecilia latipinna	sailfin molley
Poecilia reticulata	guppy
Pollachius virens	saithe, pollack
Polyodon spathula	paddlefish
Pomoxis annularis	white crappie
Pomoxis nigromaculatus	black crappie
Porphyra spp.	red algae
Porthetria dispar	gypsy moth
Portunus sanquinolentus	crab
Procambarus clarki	crayfish
Proteus morganii	bacteria
Proteus vulgaris	bacteria
Protococcus	green algae
Prototheca staminea	little neck clam
Psetta maxima, Scophthalmus maximus	turbot
Pseudomonas cepacial	bacteria
Pseudomonas fluorescens	bacteria
Pseudomonas minutus	copepod
Pdeusomonas phaseolicola	bacteria
Pseduromonas putida	bacteria
Pseudopleuronectes americanus	winter flounder
Pteronarcella	stonefly
Pteronarcella badia	stonefly
Pteronarcys	stonefly
Pteronarcys californica	stonefly
Pteronarcys dorsata	stonefly
Puntius conchonius	rosy barb
Puntius sophore	barb
Puntius ticto ticto	barb
Pygosteus pungitius	stickleback (12-pined)
Pylodictis olivaris	flathead catfish
Quelea quelea	red-billed quelea
Quiscalus quiscula	common grackle

Rana catesbeiana	bullfrog
Rana clamitans	green frog
Rana esculenta	edible frog
Rana palustris	pickerel frog
Rana clamitans	green frog
Rana temporaria	common frog
Rana sp.	frog
Rana sylvatica	woodfrog
Rana pipiens	leopard frog
Ranina serrata	crab
Rasbora daniconius	slender rasbora
Rasbora heteromorpha	harlequin fish
Rattus norwegicus	old world rat
Rhabdosargus holubi	
Rhodotorula rubra	bacteria
Roccus saxatilis	striped bass
Saccharomyces nounii	yeast
Saccobranchus fossilis	water flea
Saccharomyces cerevisiae	yeast
Sargassum fluitans	brown algae
Saimiri sciureus	squirrel monkey
Salmo trutta	brown trout
Salmo salar	Atlantic salmon
Salmo gairdneri	rainbow trout
Salmo clarki	cutthroat trout
Salmonella typhimurium	bacteria
Salmo trutta lacustris	brown trout
Salmo trutta m. trutta	sea trout
Salvelinus alpinus	arctic char
Salvelinus fontinalis	American char or brook trout
Salvelinus namaycush	lake trout
Sardina pilchardus	European pilchard
Sardinops caerula	Pacific sardine
Sardinops melanosticta	Japanese pilchard
Sardinops sagax	Chilean pilchard
Sarotheredon aureus	cichlid
Sarotheredon galilaeus	cichlid
Sarotherodon mossambicus	cichlid
Scardafella inca	inca dove
Scardafella squammata	scaly dove
Scenedesmus obliquus	green algae
Scenedesmes pannonicus	green algae
Scenedesmus quadricauda	green algae
Scenedesmus sp.	green algae
Schizosaccharomyces octopolus	yeast
Scomber japonicus	Pacific mackerel
Scomber scombrus	Atlantic mackerel
Scophthalmus maximus	European turbot

Scophthalmus rhombus	turbot
Selenastrum	green algae
Selenastrum capricornutum	green algae
Selenastrum costatum	green algae
Semotilus atromaculatus	creek chub
Senecio jacobaea	ragwort
Seriola lalandei	yellowtail
Serratia marcescens	bacteria
Shigella	bacteria
Sigmodon bispidus	cotton rat
Siliqua patula	razor clam
Silurus glanis	sheatfish
Simocephalus	water flea
Simocephalus serrulatus	water flea
Siphonaria normalis	ribbed limpet
Skeletonema costatum	algae
Solea solea	common sole, Dover sole
Sorex araneus	common shrew
Sorex cinereus	shrew sp.
Sorex vagrans	shrew sp.
Sphaerium stratinum	fingernail clam
Sphaerechinus granularis	sea urchin
Spisula solidissima	surf clam
Spiza americana	dickcissel
Sporophila minuta	ruddy-breasted seedeater
Sporobolomyces salmonicolor	yeast
Squalus cephalus	chub
Staphylococcus aureus	bacteria
Stenotomus chrysopsi	scup
Stibiobacter senarmontii	bacteria
Stizostedion lucioperca	pikeperch, sander
Stizostedion vitreum	walleye
Stolephorus purpureus	anchovy
Stomoxys calcitrans	stable fly
Streptomyces antiboticus	bacteria
Streptomyces aureus	bacteria
Streptomyces chrysomallus	bacteria
Streptomyces parvullus	bacteria
Streptomyces peucetius	bacteria
Streptomyces sanguis	bacteria
Streptomyvces violaceoruber	bacteria
Stretoverticillium ladakanus	bacteria
Strix aluco	tawny owl
Strongylocentrotus purpuratus	purple sea urchin
Sturnus vulgaris	European starling
Sylvilagus sp.	rabbit
Sus scrofa	wild boar
Sylvilagus floridanus	cottontail rabbit
Synechoccoccus sp.	algae

Talpa europaea	mole
Tamias striatus	eastern chipmunk
Tangavious aeneus	red-eyed cowbird or bronzed cowbird
Tenebrio sp.	mealworm
Terrapene sp.	box turtle
Tetrahymena	protozoa, ciliata
Petrahymena pyriformis	ciliata
Thamnophis sirtalis	garter snake
Theragra chalcogramma	Alaska pollack
Thiobacillus thiooxidans	bacteria
Thunnus albacares	yellowfin tuna
Thymallus thymallus	grayling
Tilapia mossambica	cichlid
Tilapia rendalli	cichlid
Tinca tinca	tench
Tivela stultorum	pismo clam
Toxostoma curvirostre	curve-billed thrasher
Toxostoma rufum	brown thrasher
Trachinotus carolinus	pompano
Trachurus capensis	Cape horse mackerel
Trachurus murphyi	Chilean jack mackerel
Tribolium spp.	flour beetles
Trichiurus lepturus	largehead hairtail
Trichogaster trichopterus	three spot gourami
Turdus migratorius	American robin
Turdus ericetorum	song thrush
Tyria jacobaeae	cinnabar moth
Tyto alba	barn owl
Ulva spp.	sea lettuce
Umbra pygmaea	mud minnow
Unio pictorium	mollusc
Uronema parduczi	protozoa, ciliata
Valerianella locusta	herb
Vicia faba	broad bean
Volatia jacarina	blue-black grassquit
Volsella demissa	Atlantic ribbed mussel
Westiellopsis prolifica	cyanobacteria
Xanthocephalus xanthocephalus	yellow-headed blackbird
Xenopus laevis	clawed frog
Xeroderma pigmentosum	human pigment cell line
Zebauda auriculata	eared dove
Zenaida macvouva	mourning dove
Zenaida asiatica	white-winged dove

Zonotrichia leucophrys	white-crowned sparrow
Zonotrichia atricapilla	golden-crowned sparrow

Common – Latin

abalone	*Haliotis sp.*
airsac catfish	*Heteropneustes fossilis*
Alaska pollack	*Theragra chalcogramma*
algae	*Haematococcus pluralis*
algae	*Lenastrum costatum*
algae	*Skeletonema costatum*
algae	*Synechoccoccus sp.*
almond moth	*Cadra cautella*
American catfish	*Ictalurus nebulosus*
American char	*Salvelinus fontinalis*
American crow	*Corvus brachyrhynchos*
American eel	*Anguilla rostrata*
American flagfish	*Jordanella floridae*
American kestrel	*Falco sparverius*
American robin	*Turdus migratorius*
American toad	*Bufo americanus*
anchovy	*Stolephorus purpureus*
anole	*Anolis carolinensis*
Arctic char	*Salvelinus alpinus*
armadillo	*Dasypus novemcinctus*
ascomycetes	*Fasarium merismoides*
ascomycetes	*Fusarium oxysporum*
Atlantic cod	*Gadus morhua*
Atlantic herring	*Clupea harengus harengus*
Atlantic kelp	*Laminaria digibacta, L. agardhii*
Atlantic mackerel	*Scomber scombrus*
Atlantic menhaden	*Brevoortia tyrannus*
Atlantic ribbed mussel	*Volsella demissa*
Atlantic salmon	*Salmo salar*
Atlantic silverside	*Menidia menidia*
bacteria	*Acinetobacter sp.*
bacteria	*Aeromonas hydrophila*
bacteria	*Alcaligenes denitrificans*
bacteria	*Alcaligenes faecalis*
bacteria	*Amycolata autoptophica*
bacteria	*Arthrobacter sp.*
bacteria	*Bacillus licheniformis*
bacteria	*Bacillus megaterium*
bacteria	*Brevibacterium sp.*
bacteria	*Chromobacterium sp.*
bacteria	*Clostridium difficile*
bacteria	*Cytophaga*
bacteria	*Desulfovibrio vulgaris*
bacteria	*Hansenula anomala*

bacteria	*Mycobacterium tuberculosis*
bacteria	*Penicillium notatum*
bacteria	*Photobacterium phosphoreum*
bacteria	*Pichia membranaefaciens*
bacteria	*Proteus morganii*
bacteria	*Proteus vulgaris*
bacteria	*Pseudomonas cepacia*
bacteria	*Pseudomonas fluorescens*
bacteria	*Pseudomonas phaseo*
bacteria	*Rhodococcus*
bacteria	*Rhodotarula rubra*
bacteria	*Serratia marcescens*
bacteria	*Shigella*
bacteria	*Stibiobacter senarmontii*
bacteria	*Streptomyces antiobotius*
bacteria	*Streptomyces chrysomallus*
bacteria	*Streptomyces ladakanus*
bacteria	*Streptomyces parvullus*
bacteria	*Streptomyces peucetius*
bacteria	*Streptococcus aureus*
bacteria	*Streptococcus sanguis*
bacteria	*Streptomyces violaceoruber*
bacteria	*Thiobacillus thiooxidans*
barb	*Puntius conchonius*
barb	*Puntius sophore*
barb	*Puntius ticto ticto*
barn owl	*Tyto alba*
barnacle	*Balanus sp.*
barnacles	*Chthamalus sp.*
bass	*Roccus saxatilis, Morone*
bat	*Eptesicus capensis*
	Nycticeius schlieffeni
bay anchovy	*Anchoa mitchilli*
bay mussel	*Mytilus edulis*
bay scallop	*Aequipecten (Pecten) irradians*
bean aphid	*Apis fabae*
beetle	*Coleoptera*
belladonna	*Atropa belladonna*
big brown bat	*Eptesicus fuscus*
bighead	*Hypophthalmichthys nobilis*
bigmouth buffalo	*Ictiobus cyprinellus*
black-billed magpie	*Pica pica*
black bullhead	*Ameiurus melas*
	Ictalurus melas
black crappie	*Pomoxis nigromaculatus*
bleak	*Alburnus alburnus*
bloodworm	*Glycera dibranchiata*
blue crab	*Callinectes sapidus*
blue-green algae	*Anabaena sp.*

blue jay	*Cyanocitta cristata*
blue whiting	*Micromesistius poutassou*
blue-black grassquit	*Volatia jacarina*
blue-green algae	*Microcystis aeruginosa*
	Microcystis sp.
	Nostoc muscorum
	Oscillatoria
blue-winged teal	*Anas discors*
blueback salmon	*Oncorhynchus nerka*
bluegill sunfish	*Lepomis macrochirus*
boat-tailed grackle	*Cassidix major*
bobwhite quail	*Colinus virginianus*
box turtle	*Terrapene sp.*
bream	*Abramis brama*
brine shrimp	*Artemia salina*
broad bean	*Vicia faba*
bronzed cowbird	*Tangavious aeneus*
brook trout	*Salvelinus fontinalis*
brown bullhead	*Ictalurus nebulosus*
brown shrimp	*Crangon crangon*
	Penaeus aztecus
brown thrasher	*Toxostoma rufum*
brown trout	*Salmo trutta*
brown trout, non-migratory	*Salmo trutta m. fario*
brown trout	*Salmo trutta m. lacustris*
brown-headed cowbird	*Molothrus ater*
brown-throated conure	*Aratinga pertinax*
budgerigar	*Melopsittacus undulatus*
budworm	*Choristoneura occidentalis*
bullfrog	*Rana catesbeiana*
bumble bee	*Bombus sp.*
burbot	*Lota lota*
calico crab	*Ovalipes ocellatus*
calico scallop	*Aequipecten gibbus*
California ground squirrel	*Otospermophilus beecheyi*
California quail	*Loportyx californica*
California sea mussel	*Mytilus californianus*
Canada goose	*Branta canadensis*
Cape horse mackerel	*Trachurus capensis*
capelin	*Mallotus villosus*
carp	*Ctenopharyngodon*
	Cyprinus carpio
cassins finch	*Carpodacus cassinii*
cat	*Felis catus*
catfish	*Ictalurus*
cedar waxwing	*Bombycilla cedrorum*
channel catfish	*Ameiurus nebulosus*
	Ictalurus punctatus

Chilean jack mackerel	*Trachurus murphyi*
Chilean pilchard	*Sardinops sagax*
chub	*Squalius cephalus*
chum salmon	*Oncorhynchus keta*
cichlid	*Sarotheredon aureus*
	Sarotheredon galilaeus
	Sarotherodon mossambicus
	Tilapia rendalli
cinnabar moth	*Tyria jacobaeae*
citrus tree	*Melicope leptococca*
clawed frog	*Xenopus laevis*
coalfish	*Pollachius virens*
cod	*Gadus morhua*
coho salmon	*Oncorhynchus kisutch*
common bobwhite	*Colinus virginianus*
common cockroach	*Periplaneta americana*
common dab	*Limanda limanda*
common dolphin	*Delphinus delphis*
common earthworm	*Lumbricus terrestris*
common frog	*Rana temporaria*
common grackle	*Quiscalus quiscula*
common or ground dove	*Columbina passerina*
common pigeon	*Columba livia*
common pintail	*Anas acuta*
common shrew	*Sorex araneus*
common sole	*Solea solea*
common sunfish	*Lepomis humilis*
common sunfish	*Lagodon*
common toad	*Bufo bufo*
common vole	*Microtus arvalis*
cone shells	*Conus sp.*
Cooper's hawk	*Accipiter cooperi*
copepod	*Pseudocalanus minutus*
cotton mouse	*Peromyscus gossypinus*
cotton rat	*Sigmodon hispidus*
cottontail rabbit	*Sylvilagus floridanus*
crab	*Podophthalmus vigil*
	Portunus sanquinolentus
	Ranina serrata
crayfish	*Orconectes nais*
	Procambarus clarki
creek chub	*Semolitus atromaculatus*
croaker	*Micropogon undulatus*
Crucian carp	*Carassius carassius*
crustacean	*Crangonyx pseudogracilis*
crustacean	*Palaemonetes pugio*
curve-billed thrasher	*Toxostoma curvirostre*
cutthroat trout	*Salmo clarki*
cyanobacteria	*Aulosira fertilissima*

cyanobacteria	*Westiellopsis prolifica*
dace	*Leuciscus leuciscus*
damselfly	*Ischnura verticalis*
	Lestes congener
deer mouse	*Peromyscus maniculatus*
diatom	*Monoraphidium graffithii*
	Phaeodactylium tricornutum
dickcissel	*Spiza americana*
domestic New Zealand white rabbit	*Oryctolagus cuniculus*
domestic chicken	*Gallus gallus*
dragonfly	*Macromia*
Dungeness crab	*Cancer magister*
eared dove	*Zebauda auriculata*
eastern chipmunk	*Tamias striatus*
eastern oyster	*Crassostrea virginica*
edible frog	*Rana esculenta*
eel	*Anguilla anguilla*
	Anguilla vulgaris
egret	*Ardeola sp.*
English sole	*Paraphrys vetulus*
European anchovy	*Engraulis engrasicolus*
European hare	*Lepus europaeus*
European pilchard	*Sardina pilchardus*
European starling	*Sturnus vulgaris*
European turbot	*Scophthalmus maximus, Psetta maxima*
fathead minnow	*Pimephales promelas*
field cricket	*Gryllus pennsylvanicus*
field mouse	*Aposemus sylvaticus*
flathead catfish	*Pylodictis olivaris*
flatworms	*Platyhelminthes*
flour beetles	*Tribolium sp.*
Fowler's toad	*Bufo fowleri*
freshwater clam	*Anodonta cygnea*
frog	*Rana sp.*
fruit fly	*Drosophila melanogaster*
fungi	*Aspergillus flavus*
fungi	*Aspergillus fumigatus*
fungi	*Aspergillus nidulans*
fungi	*Aspergillus parasiticus*
fungi	*Phanerorochaete chrysoporium*
garden pea	*Pisum sativum*
garter snake	*Thamophis sirtalis*
giant gourami	*Colisa fasciata*
glass shrimp	*Palaemonon macrodactylus*
goatfish	*Mulloidichthys sp.*

goby	*Gobius minutus*
golden eagle	*Aquila chrysaetos*
golden sparrow	*Passer luteus*
golden-crowned sparrow	*Zonotrichia atricapilla*
goldfish	*Carassius auratus*
gray bat	*Myotis grisescens*
grayling	*Thymallus thymallus*
green algae	*Ankistrodesmus falcatus*
green algae	*Chlamydonas sp.*
green algae	*Chlamydmonas reinhardti*
green algae	*Chlamydomonas variabilis*
green algae	*Chlorella fusca*
green algae	*Chlorella kesslerii*
green algae	*Chlorella pyrenoidosa*
green algae	*Chlorella sp.*
green algae	*Chlorococcum sp.*
green algae	*Dunaliella euchlora*
green algae	*Dunaliella sp.*
green algae	*Monoraphidium*
green algae	*Protococcus*
green algae	*Scenedesmus pannonicus*
green algae	*Scenedesmus quadricauda*
green algae	*Scenedesmus sp.*
green algae	*Selenastrum*
green algae	*Selenastrum capricornutum*
green algae	*Selenastrum costatum*
green crab	*Carcinus maenas*
green frog	*Rana clamitans*
green jay	*Cyanocorax yncas*
green sunfish	*Lepomis cyanellus*
grey mullet	*Chelon labrosus*
greylag goose	*Anser anser*
gudgeon	*Gobio gobio*
Gulf menhaden	*Brevoortia patronus*
guppy	*Lebistes reticulatus*
	Poecilia reticulata
gypsy moth	*Porthetria dispar*
halibut	*Hippoglossus hippoglossus*
hard clam	*Mercenaria mercenaria*
hare	*Leporidae*
harlequin fish	*Rasbora heteromorpha*
heron	*Ardea cinerea*
herring	*Clupea harengus*
honey bee	*Apis melliferra*
	Apis sp.
horned lark	*Eremophila alpestris*
horned lizard	*Phrynosoma cornutum*
house finch	*Carpodacus mexicanus*

house mouse	*Mus musculus*
house sparrow	*Passer domesticus*
housefly	*Musca domestica*
ide	*Leuciscus idus*
inca dove	*Scardafella inca*
inland silverside	*Menidia beryllina*
Japanese eel	*Anguilla japonica*
Japanese pilchard	*Sardinops melanosticta*
Japanese quail	*Coturnix coturnix*
jimsonweed	*Datura stramonium*
kangaroo rat	*Dipodomys sp.*
kelp	*Eisenia sp.*
kelp	*Eisenia foetida*
kelp	*Macrocystis pyrifera*
killifish	*Fundulus*
king crab	*Paralithodes camchatica*
king salmon	*Oncorhynchus tschawytscha*
knifefish	*Notopterus notopterus*
kokanee	*Oncorhynchus nerka kennerlyi*
lake trout	*Salvelinus namaycush*
land snail	*Helix sp.*
largehead hairtail	*Trichiurus lepturus*
largemouth bass	*Micropterus salmoides*
lark bunting	*Calamospiza melanocory*
latipinna	
least shrew	*Cryptotis parva*
leopard frog	*Rana pipiens*
limpets	*Patella sp.*
little brown bat	*Myotis lucifugus*
little neck clam	*Prototheca staminea*
lobster	*Panulirus pencillatus*
longear sunfish	*Lepomis megalotis*
longnose sucker	*Catostomus catostomus*
macaque	*Macaca fascicularis*
mallard	*Anas platyrhynchos*
marine pin perch	*Lagodon rhomboides*
marsh hawk	
masu salmon	*Oncorhynchus masou*
mayfly	*Baetis*
	Ephemerella
	Hexagenia
meadow vole	*Microtus pennsylvanicus*
mealworm	*Tenebrio sp.*
Mediterranean flour moth	*Anagasta kuehniella*

Mexican axolotl	*Ambystoma mexicanum*
midge	*Chironomus*
	Chironomus plumosus
	Chironomus riparius
milkfish	*Chanos chanos*
mink	*Mustela vison*
minnow	*Phoxinus phoxinus*
Mississippi silverside	*Menidia audens*
mole	*Talpa europaea*
monk parakeet	*Myiopsitta monacha*
mollusc	*Anodonta anatina*
mollusc	*Macoma balthica*
mollusc	*Phebalium dentalum*
mosquito fish	*Gambusia affinis*
mosquito fly	*Aedes sp.*
	Anopheles sp.
	Culex
	Culex pipiens
	Culex theileri
mould	*Neanthes arenaceodentata*
mould	*Neurospora crassa*
mountain bass	*Kuhlia sandvicensis*
mourning dove	*Zenaida macvouva*
mud minnow	*Umbra pygmaea*
mummichog	*Fundulus heteroclitus*
mussel	*Brachidontes recurvus*
mysid shrimp	*Mysidopsis bahia*
nitrogen-fixing bacteria	*Nitrobacter agilis*
nitrogen-fixing bacteria	*Nitrosomonas europaea*
nitrogen-fixing bacteria	*Nitrosocuccus oceanus*
northern anchovy	*Engraulis mordax*
northern harrier	*Circus cyaneus*
northern lobster	*Homarus americanus*
northern masked weaver	*Ploceus taeniopterus*
northern pike	*Esox lucius*
northern raven	*Corvus corax*
nutria	*Myocastor coypus*
oar-footed crustacean	*Acartia sp.*
	Acartia tonsa
	Calanoida
old world rat	*Rattus norwegicus*
old-field mouse	*Peromyscus polionotus*
opossum	*Didelphis virginiana*
orange-fronted conure	*Aratinga canicularis*
Pacific herring	*Clupea pallasi*
Pacific lobster	*Panuliris japonicus*

Pacific mackerel	*Scomber japonicus*
Pacific oyster	*Crassostrea gigas*
Pacific sardine	*Sardinops caerula*
pack rat	*Neotoma cinerea*
paddlefish	*Polyodon spathula*
paradisefish	*Macropodus opercularis*
peach-potato aphid	*Myzus persicae*
peppered moth	*Biston betularis*
perch	*Perca fluviatilis*
peregrine falcon	*Falco peregrinus*
phantom midges	*Chaoborus*
pickerel frog	*Rana palustris*
pikeperch, sander	*Stizostedion lucioperca*
pink salmon	*Oncorhynchus gorbuscha*
pink shrimp	*Penaeus duorarum*
pismo clam	*Tivela stultorum*
plaice	*Pleuronectes platessa*
plain chachalaca	*Ortalis vetula*
pollack	*Pollachius pollachius*
pompano	*Trachinotus carolinus*
pompano, jack cravally	*Caranx sp.*
powan, lavaret	*Coregonus lavaretus*
protozoa	*Colpidium*
protozoa	*Entosiphon*
protozoa	*Hydra oligaltis*
protozoa, ciliata	*Tetrahymena*
protozoa, ciliata	*Uronema parduczi*
protozoa, flagellate	*Chilomonas paramecium*
protozoa, flagellate	*Entosiphon sulcatum*
pumpkinseed	*Lepomis gibbosus*
purple sea urchin	*Strongylocentrotus purpuratus*
rabbit	*Sylvilagus sp.*
ragworm	*Neanthes arenaceodentata*
ragwort	*Senecio jacobaea*
rainbow trout	*Salmo gairdneri*
rainwater killifish	*Lucania parva*
rat snake	*Elaphe sp.*
razor clam	*Siliqua patula*
red algae	*Porphyra sp.*
red bat	*Lasiurus borealis*
red beetroot	*Beta vulgaris*
red bishop	*Euplectes orix*
red seaweed	*Gracilaria verucosa, G. foliifera*
red snapper	*Lutianus campechanus*
red tailed black	*Labeo bicolor*
red-billed quelea	*Quelea quelea*
red-eyed cowbird or	*Tangavious aeneus*
red-winged blackbird	*Agelaieus phoeniceus*

redear sunfish	*Lepomis microlophus*
rhesus monkey	*Macaca mulatta*
ribbed limpet	*Siphonaria normalis*
rice fish	*Oryzias latipes*
ring-billed gull	*Larus delawarensis*
ring-necked pheasant	*Phasianus colchicus*
roach	*Cemiscus nitilus*
rock dove	*Columba livia*
rosy barb, redbarb	*Puntius conchonius, Barbus conchonius*
round worms	*Aschelminthes sp.*
ruddy ground dove	*Columbina talpacoti*
ruddy-breasted seedeater	*Sporophila minuta*
sailfin molley	*Poecilia latipinna, Mollienesia*
saithe, pollack	*Pollachius virens*
saltwater limpet	*Helcioniscus argentatus*
	Helcioniscus exaratus
sand shrimp	*Crangon septemspinosa*
sandhill crane	*Grus canadensis*
sandworm	*Nereis diversicolor*
	Nereis virens, Nereis vexillosa
scaly dove	*Scardafella squammata*
scrub jay	*Aphelocoma coerulescens*
scud	*Gammarus*
	Gammarus fasciatus
	Gammarus lacustris
	Gammarus pseudolimnaeus
	Gammarus pulex
scup	*Stenotomus chrysops*
sea anemone	*Anthopleura elegantissima*
sea bass	*Morone labrax*
sea lamprey	*Petromyzon marinus*
sea lettuce	*Ulva sp.*
sea moss	*Chandrus crispus*
sea trout	*Salmo trutta m. trutta*
sea urchin	*Arbacia puntulata*
	Echinometra sp.
	Lytechinus sp.
	Paracentrotus lividus
segmented worms	*Annelida sp.*
sheatfish	*Silurus glanis*
sheep	*Ovis sp.*
sheepshead minnow	*Cyprinodon variegatus*
shiner perch	*Cymatogaster aggregata*
shiny cowbird	*Molothrus bonariensis*
shore crabs	*Hemigrapsus sp.*
short-tailed shrew	*Blarina brevicauda*
shrew	*Sorex cinereus*
	Sorex vagrans

shrimp	*Hyalella*
	Hyalella aztera
	Macrobrachium lamarrei
	Metapenaeus monoceros
	Mysidopsis
	Nitocra
	Nitocra spinipes
	Ostracoda
	Penaeus monodon
	Pendalus sp.
silver carp	*Hypophthalmichthys molitrix*
skipjack tuna	*Euthynnus pelamis*
slender rasbora	*Rasbora daniconius*
smallmouth bass	*Micropterus dolomieui*
snail	*Aplexa*
	Aplexa hypnorum
snakehead fish	*Channa striatus*
	Ophiocephalus punctatus
snapping shrimp	*Crangon sp.*
snapping turtle	*Chelydra serpentina*
sockeye salmon	*Oncorhynchus nerka*
soft shell clam	*Mya arenaria*
song thrush	*Turdus ericetorum*
southern flounder	*Paralichthys lethostigma*
sowbug	*Asellus*
	Asellus brevicaudus
sparrowhawk	*Accipiter nisus*
spike-tailed paradisefish	*Macropodus cupanus*
spiny lobster	*Panuliris argus*
sprat	*Clupea sprattus*
spring peeper	*Hyla crucifer*
squirrel monkey	*Saimiri sciureus*
stable fly	*Stomoxys calcitrans*
staghorn sculpin	*Leptocottus armatus*
starfish	*Nostoc linckia*
starry flounder	*Platichthys stellatus*
stickleback, three-spined	*Gasterosteus aculeatus*
stickleback, twelve-spined	*Pygosteus pungitius*
stonefly	*Acroneuria*
	Acroneuria pacifica
	Claassenia sabulosa
	Isoperla
	Pteronarcys dorsata
	Pteronarcella
	Pteronarcella badia
	Pteronarcys
	Pteronarcys californica
striped bass	*Morone saxatilis*
	Roccus saxatilis

striped catfish	*Mystus vittatus*
striped mullet	*Mugil cephalus*
suckers	*Catostomus sp.*
	Ictiobus
summer flounder	*Paralichthys dentatus*
surf clam	*Spirula solidissima*
surgeon fish	*Acanthurus sp.*
Swainson's hawk	*Buteo swainsoni*
swallows	*Hirundinidae*
taragon	*Artemisia dracunculus*
tawny owl	*Strix aluco*
tench	*Tinca tinca*
three spot gourami	*Trichogaster trichopterus*
tidewater silverside	*Menidia peninsulae*
treefrog	*Hyla versicolor*
tricolored blackbird	*Agelaeus tricolor*
tsetse fly	*Glossina*
turbot	*Pseta maxima*
	Scophthalmus rhombus
turkey	*Meleagris*
unicellular biflagellate algae	*Gymnodinium breve*
vendace	*Coregonus albula*
village weaver	*Ploceus cucullatus*
vole	*Microtus longicaudus*
	Microtus montanus
walking catfish	*Clarias batrachus*
walleye	*Stizostedion vitreum*
water flea	*Ceriodaphnia dubia*
	Cladocera
	Daphnia cucullata
	Daphnia laevis
	Daphnia longispina
	Daphnia sp.
	Saccobranchus fossilis
	Simocephalus
	Simocephalus serrulatus
watermint	*Mentha aquatica*
western chipmunk	*Eutamias sp.*
white amur grass carp	*Ctenopharyngodon idella*
white crappie	*Pomoxis annularis*
white shrimp	*Penaeus setiferus*
white sucker	*Catostomus commersoni*
white-crowned sparrow	*Zonotrichia leucophrys*
white-footed mouse	*Peromyscus leucopus*
white-fronted dove	*Leptotila verreauxi*

white-winged dove	*Zenaida asiatica*
whiting	*Gadus merlangus*
wild boar	*Sus scrofa*
wild turkey	*Meleagris gallopavo*
winter flounder	*Pseudopleuronectes americanus*
wolf-fish	*Anarchichas lupus*
woodfrog	*Rana sylvatica*
yeast	*Saccharomyces cerevisiae*
yeast	*Saccharomyces nounii*
yeast	*Schizosaccharomyces octopolus*
yeast	*Sporobolomyces salmonicolor*
yellow bullhead	*Ictalurus natalis*
yellow perch	*Perca flavescens*
yellow-billed magpie	*Pica nuttalli*
yellow-headed blackbird	*Xanthocephalus xanthocephalus*
yellowfin tuna	*Thunnus albacares*
yellowtail	*Seriola lalandei*
zebra fish	*Brachydanio rerio*

ABBREVIATIONS

ADI	acceptable daily intake
ATP	adenosine triphosphate
BAN	British Approved Name
BOD	Biological Oxygen Demand
Bq	Becquerel
BrdU	bromodeoxyuridine
BSI	British Standards Institute
c	centi
CAS	Chemical Abstracts Service
Ci	Curie
COD	Chemical Oxygen Demand
d_4^{20}	density (temperature stated where known)
DNA	deoxyribonucleic acid
decomp	decomposition
^{o}C	degree centigrade
DMSO	dimethyl sulfoxide
DO	dissolved oxygen
EC	effective concentration
ECETOC	European Chemical Industry Ecology & Toxicology Centre
ED	effective dose
Ed	editor
ed	edition
EEC	European Economic Community
ELISA	Enzyme-linked immunosorbent assay
EPA	Environmental Protection Agency (US)
et al.	and others (authors)
FAO	Food and Agriculture Organisation
GC-MS	gas chromatography-mass spectrometry
g	gram
GSH	reduced glutathione
ha	hectare
HGPRT	hypoxanthine-guanine phosphorisobutyltransferase
HMSO	Her Majesty's Stationery Organisation
HPLC	high performance liquid chromatography
hr	hour
HSC	Health & Safety Commission
HSE	Health & Safety Executive
5-HT	5-hydroxytryptamine

Hz	Hertz
IARC	International Agency for Research on Cancer
ILO	International Labour Office
IR	infrared
ISO	International Standards Organisation
i.u.	International Unit
JETOC	Japan Chemical Industry Ecology & Toxicology Information Center
J	Joule
IUPAC	International Union of Pure & Applied Chemistry
k	kilo
K_a	dissociation constant, acids
kg	kilogram
K_{ow}	dissociation constant, octanol: water
LC_{50}	lethal concentration – 50
LD_{50}	lethal dose – 50
Ltd	limited
l	litre
$\log P_{ow}$	$\log_{10}$ of partition of coefficient -octanol:water
u	micro
m	milli/metre
MetHb	methaemoglobin
mm	millimetre
mmHg	millimetres mercury
min	minute
ml	millilitre
M	Molar
N	Newton
n	normal
N	Normality
NAD^+	Nicotinamide – adenine dinucleotide (oxidised form)
NADH	Nicotinamide – adenine dinucleotide (reduced form)
NAPQI	*N*-acetyl-*p*-benzoquinone imine
n	nano
OECD	Organisation for Economic Cooperation and Development
PAH	polycyclic aromatic hydrocarbon
p	pico
pKa	log Ka
ppb	parts per billion (10^9)
ppm	parts per million (10^6)
ppt	parts per thousand (10^3)
pH	$\log_{10}$ hydrogen ion concentration
psi	pound force – square inch
RAD	Radiation absorbed dose 1 Ergs/g
RD_{50}	50% decrease in respiratory rate
RN	Registry Number
RNA	ribonucleic acid
RSC	The Royal Society of Chemistry
sec	second
$t_{1/2}$	half life

TCA	tricarboxylic acid
ThOD	Theoretical Oxygen Demand
TLV	Threshold Limit Value
TOC	Total Organic Carbon
TWA	Time-Weighted Average
UK	United Kingdom
UNEP	United Nations Environment Programme
USA	United States of America
UV	ultraviolet
v. den.	vapour density
vol	volume
v.p.	vapour pressure
v/v	volume/volume
w/v	weight/volume
W	Watt
wk	week
WHO	World Health Organization
wt	weight
w/w	weight/weight
yr	year

INDEX OF NAMES AND
SYNONYMS

Acetyl chloride A33
acetyl enheptin A39
acetyl ether A21
Acetyl iodide A35
Acetyl ketene A36
acetyl oxide A21
acetylaminobenzene A11
2-Acetylaminofluorene A31
N-acetyl-2-aminofluorene A31
N-acetyl-p-aminophenol A10
2-acetylamino-5-nitrothiazole A39
acetylaniline A11
acetylbenzene A28
1-(p-acetylbenzenesulfonyl)-3-cyclohexylurea
   A23
acetylen A34
Acetylene A34
2-(acetyloxy)benzoic acid
   4-(acetylamino)phenyl ester B32
2-acetyloxybenzoic acid A37
acetylphosphoramidothioic acid
   O,S-dimethyl ester A3
Acetylsalicylic acid A37
acetylthiourea A9
N-acetylthiourea A9
1-acetyl-2-thiourea A9
acid ammonium carbonate A182
acid butyl phosphate B258
acidum acetylsalicylicum A37
Acifluorfen A38
Acinitrazole A39
Aclofen A40
Aclonifen A40
aconitane A41
aconitin cristallisat A41
Aconitine A41
Acridine A42
acroleic acid A47
Acrolein A43
Acrolein dimer A44
acromycine A45
acronine A45
Acronycine A45
acrylaldehyde A43
Acrylamide A46
Acrylic acid A47
acrylic acid chloride A48
acrylic acid, 1-methyltrimethylene ester B273
acrylic acid, tetramethylene ester B274
acrylic aldehyde A43
acrylic amide A46
Acrylic chloride A48
Acrylonitrile A49
acrylyl chloride A48

Actedron A210
Actinolite Ca$_2$(MgFe^{2+})$_5$Si$_8$O$_{22}$(OH)$_2$ A266
Actinomycin C A50
Actinomycin C1 A51
Actinomycin D A51
activated aluminium oxide A109
actybaryte B22
Adamantane A52
Adipic acid A53
adipic acid dinitrile A54
Adiponitrile A54
1-(+)-adrenaline A55
Adrenaline-D A55
Adrenaline-L A56
Adriablastina A57
Adriamycin A57
AFBI A59
Afilene B247
Aflatoxicol A58
aflatoxin B A59
Aflatoxin B$_1$ A59
Aflatoxin B$_2$ A60
Aflatoxin G$_1$ A61
Aflatoxin G$_2$ A62
Aflatoxin M$_1$ A63
aflatoxin R$_o$ A58
Agar A64
agar-agar A64
Aktikon A271
Alachlor A65
alane A106
L-alanine, N-benzoyl-N-(3,4-
   dichlorophenyl)ethyl ester B104
Albagel premium USP4444 B38
Aldicarb A66
Aldol A67
aldoxime A7
Aldoxycarb A68
Aldrin A69
Alkyl(C$_{14}$-C$_{16}$)dimethylbenzyl-
   ammonium chloride A70
N-Alkyl(C$_8$-C$_{18}$)dimethyl-3,4-
   dichlorobenzylammonium chloride A72
N-Alkyl(C$_8$-C$_{18}$)dimethylbenzyl-
   ammonium chloride A71
N-Alkyl-1,3-propanediamine A73
Allethrin A74
allethrin-D A74
Allidochlor A75
Allopurinol A76
Alloxydim-sodium A77
alloxydimedon sodium A77
Allyl acetate A78
Allyl alcohol A79

allyl aldehyde A43
Allyl bromide A82
allyl butanoate A83
Allyl butyrate A83
Allyl caprylate A84
Allyl chloride A85
allyl chlorocarbonate A86
Allyl chloroformate A86
Allyl cinerin I A74
Allyl cinnamate A87
allyl cyclohexanepropionate A88
allyl β-cyclohexylpropionate A88
Allyl cyclohexylpropionate A88
allyl 2,3-epoxypropyl ether A91
Allyl ethyl ether A89
Allyl formate A90
Allyl glycidyl ether A91
allyl hexahydrophenyl propionate A88
Allyl iodide A92
Allyl isopropylacetamide A93
allyl isorhodanide A94
allyl isosulfocyanate A94
Allyl isothiocyanate A94
Allyl isovalerate A95
allyl isovalerianate A95
allyl octanoate A84
allyl octylate A84
Allyl phenoxyacetate A96
Allyl phenylacetate A97
allyl 3-phenylpropenoate A87
Allyl propyl disulfide A98
allyl sulfocarbamide A99
allyl thiocarbamide A99
allyl thiocarbonimide A94
allyl α-toluate A97
Allylamine A80
p-allylanisole A81
4-Allylanisole A81
allylcarbinol B243
allyl-3-cyclohexyl propionate A88
3-allylcyclohexyl propionate A88
1-(allyloxy)-2,3-epoxypropane A91
allyl-β-phenylacrylate A87
Allylthiourea A99
1-allyl-2-thiourea A99
Allyltrichlorosilane A100
Aloe Emodin A101
Altracin B1
alum A111
alum flour A111
alumina A109
α-alumina A109
β-alumina A109
γ-alumina A109

Aluminium A102
Aluminium bromide A103
Aluminium carbide A104
Aluminium chloride A105
aluminium fibre A102
aluminium flake A102
Aluminium hydride A106
Aluminium isopropoxide A114
Aluminium isopropylate A114
Aluminium lithium hydride A107
Aluminium nitrate nonahydrate A108
Aluminium oxide A109
Aluminium phosphide A110
Aluminium potassium sulfate A111
aluminium powder A102
Aluminium resinate A112
aluminium sesquioxide A109
Aluminium sulfate A113
aluminium trichloride A105
aluminium trihydride A106
α-aluminium trihydride A106
Aluminium triisopropoxide A114
aluminum A102
AM-FOC A178
AMBEN A126
Ambush A66
Amchlor A186
Amcide A202
americaine B74
Ametrex A115
Ametryn A115
amide C2 A8
amides, coco, N,N-bis(hydroxyethyl) B152
Amidet B112 B152
Amidithion A116
Amido-F acid A149
Amilfenol A223
aminitrozole A39
m-aminoanisole A235
o-aminoanisole A234
p-aminoanisole A236
2-aminoanthracenamine A117
β-aminoanthracene A117
2-Aminoanthracene A117
2-amino-9,10-anthracenedione A118
β-aminoanthraquinone A118
2-Aminoanthraquinone A118
4-Aminoantipyrene A119
4-Aminoazobenzene A120
aminobenzene A228
1-aminobenzene-3-sulfonic acid A122
2-Aminobenzenesulfonic acid A121
3-Aminobenzenesulfonic acid A122
4-Aminobenzenesulfonic acid A123

o-aminophenylsulfonic acid A121
p-aminophenylsulfonic acid A123
Aminophylline A161
1-amino-2-propanol A164
1-Aminopropan-2-ol A164
1-aminopropane A162
Aminopropane A162
2-Aminopropane A163
3-aminopropene A80
p-aminopropiophenone A165
4'-Aminopropiophenone A165
β-aminopropylbenzene A210
3-Aminopropyldiethylamine A166
3-Aminopropyldimethylamine A167
3-aminopropylene A80
aminopteridine A168
Aminopterin A168
amino-2-pyridine A169
amino-3-pyridine A170
amino-4-pyridine A171
m-aminopyridine A170
o-aminopyridine A169
p-aminopyridine A171
α-aminopyridine A169
γ-aminopyridine A171
2-Aminopyridine A169
3-Aminopyridine A170
4-Aminopyridine A171
aminopyrine A156
Aminosin A99
1-amino-4-sulfonaphthalene A145
N-(aminothioxomethyl)acetamide A9
aminotoluene B109
aminotriazole A177
3-amino-s-triazole A177
3-amino-1H-1,2,4-triazole A177
1-amino-2,4,5-trimethylbenzene A231
11-Aminoundecanoic acid A172
aminouracil mustard B139
1,2-aminozophenylene B98
Amiton A173
amiton hydrogen oxalate A174
Amiton oxalate A174
Amitraz A175
Amitriptyline A176
Amitrole A177
amizol A177
Ammate A202
Ammonia A178
ammonia gas A178
ammonia water A194
Ammonium acetate A179
ammonium acid arsenate A180
ammonium acid difluoride A193

ammonium amidosulfonate A202
ammonium amidosulfonate A202
ammonium aminoformate A184
Ammonium arsenate A180
Ammonium benzoate A181
Ammonium bicarbonate A182
Ammonium bichromate(VI) A189
ammonium bifluoride A193
Ammonium bisulfite A183
ammonium borofluoride A191
Ammonium carbamate A184
ammonium carbaminate A184
ammonium carbazoate A200
Ammonium carbonate A185
Ammonium chloride A186
Ammonium chloroplatinate A187
Ammonium chromate A188
Ammonium chromate(VI) A188
Ammonium dichromate A189
Ammonium fluoride A190
Ammonium fluoroborate A191
Ammonium fluorosilicate A192
ammonium fluosilicate A192
ammonium hexachloroplatinate(IV) A187
ammonium hexafluorosilicate A192
ammonium hydrogen carbonate A182
Ammonium hydrogen difluoride A193
ammonium hydrogen fluoride A193
Ammonium hydroxide A194
ammonium hyposulfite A208
Ammonium metavanadate A195
ammonium monosulfide A204
ammonium muriate A186
Ammonium nitrate A196
Ammonium oxalate A197
Ammonium perfluorooctanoate A198
ammonium peroxydisulfate A199
Ammonium persulfate A199
Ammonium picrate A200
ammonium picronitrate A200
ammonium platinic chloride A187
Ammonium polysulfide A201
ammonium rhodanate A207
ammonium rhodanide A207
ammonium saltpeter A196
ammonium silicofluoride A192
Ammonium sulfamate A202
ammonium sulfamidate A202
Ammonium sulfate A203
Ammonium sulfide A204
Ammonium sulfite A205
ammonium sulfocyanate A207
ammonium sulfocyanide A207
Ammonium tartrate A206

ammonium tetrafluoroborate A191
Ammonium thiocyanate A207
Ammonium thiosulfate A208
ammonium threonate A206
ammonium trisulfide A201
ammonium vanadate A195
amobarbitol A222
Amosite (brown) $(Fe^{2+}Mg)_7Si_8O_{22}(OH)_2$ A266
Amoxil Histocillin A209
Amoxycillin A209
Amphetamine A210
Ampicillin A211
ampyrone A119
AMS A202
n-amyl acetate A212
Amyl acetate A212
tert-Amyl acetate A213
n-amyl alcohol A214
Amyl alcohol A214
m-amyl butyrate A215
Amyl butyrate A215
Amyl chloride A216
Amyl mercaptan A219
Amyl nitrate A220
Amyl propionate A224
amylacetic ester A212
Amyl-3-cresol A217
amyl-m-cresol A217
6-amyl-m-cresol A217
1-Amyl-1-nitrosourea A221
N-amyl-N-nitrosourea A221
Amylene A218
Amylobarbitone A222
Amylol A214
4-tert-Amylphenol A223
p-tert-Amylphenol A223
Amyltrichlorosilane A225
anaesthesin B74
ANCA 1040 B169
Ancymidol A226
anhydride of ammonium carbonate A184
anhydrous aluminium sulfate A113
anhydrous ammonia A178
Anilazine A227
Aniline A228
Aniline hydrochloride A229
aniline oil A228
aniline salt A229
aniline yellow A120
Aniline, 2,4,5-trimethyl- A231
Aniline, 2,4,6-trimethyl- A232
Aniline, 3,5-dichloro- A230

Aniline, 4,4-(imidocarbonyl)bis-
    N,N-dimethyl- A273
aniline-2-sulfonic acid A121
aniline-4-sulfonic acid A123
o-anilinesulfonic acid A121
aniline-m-sulfonic acid A122
anilinium chloride A229
anilino-o-sulfonic acid A121
Anilofos A233
m-anisidine A235
o-anisidine A234
p-anisidine A236
2-Anisidine A234
3-Anisidine A235
4-Anisidine A236
Anisole A237
p-anisoyl chloride A238
Anisoyl chloride A238
p-anisyl chloride A238
m-anisylamine A235
p-anisylamine A236
Anthophyllite $Mg_7Si_8O_{22}(OH)_2$ A266
Anthracene A239
9,10-anthracenedione A241
9,10-anthracenedione,
    1,8-dihydroxy-3-(hydroxymethyl)-
    A101
9,10-anthracenedione, 2-amino- A118
2-anthracylamine A117
Anthraflavic acid A240
anthraflavin A240
2-anthramine A117
o-anthranilic acid A124
9,10-anthraquinone A241
Anthraquinone A241
2-anthrylamine A117
antibiotic 720A A249
antifebrin A11
antimonate(2-),
    bis[µ-[2,3-dihydroxybutanedioato(4-)-
    $O^1,O^2:O^3,O^4$]]di-, dipotassium,
    trihydrate A2
antimonic chloride A244
antimonous chloride A246
antimonous fluoride A247
Antimony A242
antimony black A242
antimony chloride A244
antimony(III) chloride A246
antimony(V) chloride A244
antimony(III) fluoride A247
antimony(V) fluoride A245
Antimony pentachloride A244
Antimony pentafluoride A245

antimony perchloride A244
antimony regulus A242
Antimony trichloride A246
Antimony trifluoride A247
Antimony trioxide A248
antimony white A248
antimonyl potassium tartrate A243
Antimonyl potassium tartrate
  hemihydrate A243
Antimycin A A249
antioxidant 4 B265
antipiricullin A249
antiseptol B69
Antracin A239
Antracine 12 B264
ANU A221
Anyvim A228
AP-S A201
Aphoxide A283
APO A283
Applaud B211
Apyonin A273
aqua ammonia A194
Arachidonic acid A250
Araldite ERE 1359 B148
Aramite A251
Araton A251
Arochlor 1221 A253
Arochlor 1232 A254
Arochlor 1242 A255
Arochlor 1248 A256
Arochlor 1254 A257
Aroclor 1016 A252
Aroclor 1221 A253
Aroclor 1232 A254
Aroclor 1242 A255
Aroclor 1248 A256
Aroclor 1254 A257
Aroclor 1260 A258
Arsenic A259
Arsenic acid A260
arsenic black A259
arsenic chloride A262
arsenic(III) chloride A262
Arsenic disulfide A261
arsenic hydride A265
arsenic monosulfide A261
arsenic oxide A263
arsenic(III) oxide A263
arsenic sesquioxide A263
arsenic sesquisulfide A264
arsenic sulfide A264
arsenic sulfide red A261
Arsenic trichloride A262

arsenic trihydride A265
Arsenic trioxide A263
Arsenic trisulfide A264
arsenic-75 A259
arsenicals A259
arsenious acid A263
arsenious chloride A262
arsenious oxide A263
arsenious sulfide A264
arsenious trichloride A262
arsenious trioxide A263
arseniuretted hydrogen A265
arsenous acid A263
arsenous acid anhydride A263
arsenous anhydride A263
arsenous chloride A262
arsenous hydride A265
arsenous oxide A263
arsenous oxide anhydride A263
arsenous sulfide A264
Arsine A265
arsine, thioxo- A261
Artaban B101
Artifical Heavy Spar B22
artificial essential oil of almond B39
Asbestos A266
Ascabin B110
Ascorbic acid A267
Ascorin A267
Asphalt, fumes A268
asphaltum A268
aspirin A37
Astatine A269
Asulam A270
ATA A175
Atazinax A271
Atratol A271
Atrazine A271
Atropine A272
Auramine (Base) A273
Auramine O A273
Aurantia B161
Avolin B60
Ayfivin B1
10-azaanthracene A42
azacitidine A274
azacyclopropane A282
azacytidine A274
5-Azacytidine A274
Azamethiphos A275
azaserin A276
Azaserine A276
Azathioprine A277
1-Azetidinecarbonyl chloride A278

2-azido-4-(isopropylamino)-6-
(methylthio)-s-triazine A281
4-azido-4-isopropylamino-6-
methylthio-1,3,5-triazine A281
4-azido-*N*-(1-methylethyl)-6-
methylthio-1,3,5-triazin-2-amine A281
azimidobenzene B98
3-azindole B71
Azinphos-ethyl A279
Azinphos-methyl A280
aziprotryn A281
Aziprotryne A281
azirane A282
Aziridine A282
Aziridine, 1,1α,1α-phosphinylidene,
tris- A283
Azobenzene A284
azobenzene oxide A290
2,2′-Azobis(2-methylpropionitrile) A287
azobisbenzene A284
azobiscarbonamide A286
1,1-azobisformamide A286
α,α′-azobisisobutylonitrile A287
2,2′-azobisisobutyronitrile A287
Azocyclotin A285
azodibenzene A284
azodibenzeneazofume A284
Azodicarbonamide A286
azodicarboxamide A286
azodicarboxylic acid diamide A286
azoformamide A286
Azoic Red 36 A142
azol A159
Azomate B101
Azoprocarbazine A288
Azothoate A289
Azoxybenzene A290
azoxybenzide A290
azoxydibenzene A290

B 1843 B159
B(*b*)F B75
Bacitracin B1
Balan B28
Bancol B36
Bandrowski's base B2
Barbaloin B3
Barbamat B4
barbamic A222
Barban B4
Barbane B4
Barite B22
Barium B5
Barium acetate B6

Barium bromate B7
Barium bromide B8
Barium carbonate B9
Barium chlorate B10
Barium chloride B11
Barium cyanide B12
barium diacetate B6
Barium dichloride B11
barium dicyanide B12
barium dinitrate B14
barium dioxide B19
Barium disulfide B13
barium monoxide B15
Barium nitrate B14
Barium oxide B15
Barium pentasulfide B16
Barium perchlorate B17
Barium permanganate B18
Barium peroxide B19
barium polysulfide B13
barium polysulfide (BaS$_3$) B24
barium polysulfides (BaS$_4$) B23
barium polysulfides (BaS$_5$) B16
Barium polysulfides (Ba$_2$S$_3$) B20
Barium polysulfides (Ba$_4$S$_7$) B21
barium protoxide B15
Barium sulfate B22
barium superoxide B19
Barium tetrasulfide B23
Barium trisulfide B24
baryta B15
BAS 3170F B30
BAS 9021 A77
Basagran B37
basic cupric sulfate B164
Baycor B162
Bayer 15080 B34
Bayer 17147 A280
Bayer's acid A149
BBP B113
BBSA B64
BCNU B141
Beetroot Red B25
Benalaxyl B26
Benaldehyde, 3,5-dibromo-4-hydroxy,
O-(2,4-dinitrophenyl)oxime B197
Benazolin B27
benazoline B27
Bendioxide B37
Benefin B28
Benex B31
Benfluralin B28
Benfluraline B28
Benfuracarb B29

Benfuracarbe B29
Bengal isinglass A64
Benlate B31
Benodanil B30
Benomyl B31
Benoral B32
Benortan B32
Benorylate B32
Benoxaprofen B33
benoxyl B103
Benquinox B34
Bensulide B35
Bensultap B36
Bentazon B37
Bentazone B37
Bentonite B38
Benylate B110
benzal chloride B117
Benzaldehyde B39
benzalkonium chloride A71
Benzamide B40
1,2-benzanthracene B41
benzanthrene B41
Benzaron B42
Benzarone B42
Benzathine B43
benzathine benzylpenicillin B44
Benzathine penicillin B44
benzenacetonitrile B116
benzenamine A228
benzenamine hydrochloride A229
benzenamine,
    2,4,6-trinitro-*N*-2,4,6-trinitrophenyl B161
benzenamine, 4-methoxy- A236
benzenamine,4-(1,1-dimethylethyl)-
    N-(1-methylpropyl)-2,6-dinitro- B255
Benzenamine-*N*-butyl B262
benzenazobenzene A284
Benzene B45
benzene aldehyde B39
benzene azimide B98
benzene carbonal B39
Benzene sulfohydrazide B57
benzene, 1,4-bis[(chloromethoxy)
    methyl]- B142
benzene, bromomethyl- B111
benzene, 2-*tert*-butyl-1,4-dimethoxy- B272
Benzene, 1-chloromethyl-4-nitro- B46
Benzene, 1,3-dichloro-5-nitro- B47
Benzene, 1,4-dichloro-2-nitro- B48
Benzene, 2,4-dichloro-1-nitro- B49
benzene, 1,4-diisothiocyanato- B163
benzene, 1-(1,1-dimethylethyl)-3,
    5-dimethyl-2,4,6-trinitro- B296

benzene, 2-(1,1-dimethylethyl)-
    1,4-dimethoxy- B272
benzene, 1-(1,1-dimethylethyl)-4-
    methyl- B295
Benzene, 1,2-dimethyl-3-nitro- B50
Benzene, 1,2-dimethyl-4-nitro- B51
Benzene, 1,3-dimethyl-2-nitro- B52
Benzene, 2,4-dimethyl-1-nitro- B53
Benzene, 1-nitro-2-trifluoromethyl- B54
Benzene, 1-nitro-3-trifluoromethyl- B55
Benzene, 1-nitro-4-trifluoromethyl- B56
Benzene, 1,3,5-trinitro- B58
benzeneacetic acid,
    α-(hydroxymethyl)-8-methyl-8-
    azobicyclo[3.2.1]oct-3-yl ester
    endo-(±)- A272
benzeneacetic acid, 2-propenyl ester A97
benzeneacetic acid,
    4-chloro-α-(4-chlorophenyl) B144
benzeneamine, 4-(phenylazo)- A120
Benzenearsonic acid B59
p-benzeneazophenol B95
benzenecarbonal B39
benzenecarbonyl chloride B102
benzenecarboperoxoic acid,
    1,1-dimethylethyl ester B288
benzenecarboxylic acid B82
1,2-benzenedicarboxylic acid, butyl
    phenylmethyl ester B113
1,2-Benzenedicarboxylic acid,
    dimethyl ester B60
o-benzenediol B61
p-benzenediol B63
1,2-Benzenediol B61
1,3-Benzenediol B62
1,4-Benzenediol B63
1,4-benzenediol, 2-(1,1-dimethylethyl)-
    B280
benzeneethanol, 4-chloro-β-
    (4-chlorophenyl)- B145
benzenemethanamine B109
benzenemethanaminium,
    N,N-dimethyl-*N*-[2-[2-[4-
    [1,1,3,3-tetramethylbutyl]phenoxy]
    ethoxy]ethyl]-, chloride B69
benzenemethanol B108
benzenenitrile B89
benzenesulfochloride B65
benzenesulfonamide,
    4-acetyl-*N*-[(cyclohexylamino)carbonyl]
    A23
Benzenesulfonamide, *N*-butyl- B64
benzenesulfonic (acid) chloride B65
benzenesulfonic acid, 2-amino A121

Bis(2,3,3,3-tetrachloropropyl) ether B160
m-bis(2,3-epoxypropoxy)benzene B148
1,3-Bis(2,3-epoxypropoxy)benzene B148
bis(2,3-epoxypropyl)aniline B150
N,N-Bis(2,3-epoxypropyl)aniline B150
Bis(2,4,6-trinitrophenyl)amine B161
Bis(2,4-dihydroxyphenyl)methanone B146
Bis(2-chloro-1-methylethyl) ether B143
Bis(2-chloroethoxy)methane B138
Bis(2-chloroethyl) ether B140
5-*N,N*-bis(2-chloroethyl)-aminouracil B139
5-(bis(2-chloroethyl)amino)-
   2,4(1*H*,3*H*)pyrimidinedione B139
5-Bis(2-chloroethyl)aminouracil B139
bis(2-ethylhexyl) hydrogen phosphate B151
Bis(2-ethylhexyl) phosphate B151
bis(2-ethylhexyl)orthophosphoric acid B151
Bis(2-hydroxy-4-methoxyphenyl)
   methanone B154
N,N-Bis(2-hydroxyethyl)dodecanamide B153
Bis(4-bromophenyl) ether B135
Bis(4-chlorobenzoyl) peroxide B137
2,2-bis(4-chlorophenyl)-1-hydroxyethane
   B145
Bis(4-chlorophenyl)acetic acid B144
2,2-Bis(4-chlorophenyl)ethanol B145
p,p'-bis(*N,N*-dimethylamino)benzophenone
   B147
2,2-bis(*p*-chlorophenyl)ethanol B145
bis(*p*-chorophenyl)acetic acid B144
bis(*p*-dimethylaminophenyl)methyleneimine
   A273
1,1-Bis(*tert*-butyldioxy)-3,3,5-
   trimethylcyclohexane B136
1,1-bis(*tert*-butylperoxy)-3,3,5-
   trimethylcyclohexane B136
2,2'-Bis(acryloyloxymethyl)butyl acrylate
   B132
1,2-bis(benzylamino)ethane B43
1,4-Bis(bromoacetoxy)-2-butene B133
2,2-Bis(bromomethyl)-1,3-propanediol B134
1,3-Bis(chloroethyl)-1-nitrosourea B141
1,4-bis(chloromethoxymethyl)benzene B142
4,4'-Bis(dimethylamino)benzophenone B147
bis(expoxypropyl)phenylamine B150
N,N-bis(hydroxyethyl)-coco-amides B152
N,N-bis(hydroxyethyl)lauramide B153
bis(hydroxymethyl)acetylene B298
1,4-bis(methanesulfonoxy)butane B225
Bis(methoxythiocarbonyl) disulfide B156
bis(methylxanthogen) disulfide B156
trans-1,2-bis(propylsulfonyl)ethylene B159
trans-1,2-bis(*n*-propylsulfonyl)ethylene B159
Bis-1,4-(chloromethoxy)-*p*-xylene B142

bis-[4-(dimethylamino)phenyl]methanone
   B147
bis(β-chloroethyl) ether B140
1,3-bis(β-chloroethyl)-1-nitrosourea B141
bis(β-chloroethyl)formal B138
Biscomate B163
4,4-bisdimethylaminobenzophenoneimide
   A273
N,N-bis(β-hydroxyethyl)lauramide B153
bismuth gallate, basic B158
bismuth oxygallate B158
bismuth sesquitelluride B157
Bismuth subgallate B158
Bismuth telluride B157
bismuth tritelluride B157
bisphenol A diglycidyl ether B149
bisulfane B225
2,2-Bis[(methacryloyloxy)methyl]butyl
   methacrylate B155
Bis[4-(2,3-epoxypropoxy)phenyl]propane
   B149
Bitertanol B162
Bitoscanate B163
bitter almond oil camphor B85
bitumen A268
bivinyl B213
blanc fixe B22
Blazer A38
Blue Oil A228
BNS-10% B202
Boktrysan A227
Bordeaux mixture B164
Bordo mixture B164
boric anhydride B166
Borneol B165
bornyl alcohol B165
boron bromide B167
boron chloride B168
boron fluoride B169
Boron oxide B166
boron sesquioxide B166
Boron tribromide B167
Boron trichloride B168
Boron trifluoride B169
boron trifluoride diethyl ether B170
Boron trifluoride diethyl etherate B170
Boron trifluoride dimethyl etherate B171
boron trifluoride etherate B170
Brigade B128
Britacil A211
BRL-2333 A209
Brodifacoum B172
Broenners acid A148
Brofene B203

buta-1,3-diene B213
Butachlor B212
1,3-Butadiene B213
2,3-Butadione B214
Butafume B260
Butamifos B215
butan-2-one B234
Butanal B216
butanal oxime B299
Butanamide,
    2-bromo-3,3-dimethyl-*N*-(1-methyl-1-
    phenylethyl)- B188
1-butanamine B259
2-butanamine B260
Butane B217
n-Butane B217
butane, 1-bromo- B186
butane, 1-chloro- B269
butane, 1-isocyanato- B283
1,3-Butanediamine B218
1,4-Butanediamine B219
1,4-butanedicarboxylic acid A53
butanedioic acid, 2,3-dihydroxy-,
    diammonium salt A206
1,2-Butanediol B221
1,3-Butanediol B222
1,4-Butanediol B223
2,3-Butanediol B224
1,3-butanediol diacrylate B273
1,4-butanediol diacrylate B274
1,4-Butanediol diglycidyl ether B220
1,4-Butanediol dimethanesulfonate B225
2,3-butanedione B214
2,2′-[1,4-butanediylbis(oxymethylene)
    bisoxirane] B220
butanenitrile B303
n-butanenitrile B303
1,4-Butanesultone B226
1-butanethiol B285
n-butanethiol B285
Butanex B212
Butanoate B227
Butanoic acid B228
butanoic acid anhydride B300
butanoic acid,
    2,2,2-trichlorodimethoxyphosphinyl)ethyl
    ether B227
Butanoic acid, 3-methyl- B229
butanoic acid, 3-methyl-, butyl ester B284
Butanoic acid, butyl ester B230
butanoic acid, pentyl ester A215
butanoic anhydride B300
1-Butanol B231
2-Butanol B232

tert-Butanol B233
1,4-butanolide B302
2-Butanone B234
2-Butanone peroxide B235
2-butanone,
    3-(methylsulfonyl)-*O*-[(methylamino)
    carbonyl]oxime B248
2-butanone, 3-(methylthio)-*O*-[(methylamino)
    carbonyl]oxime B247
Butanox B212
butanoyl chloride B304
2-Buten-1-ol B241
3-Buten-1-ol B243
trans-2-Buten-1-ol B242
3-Buten-2-ol B244
1-Butene B236
cis-2-Butene B237
trans-2-Butene B238
butene-2 B237
cis-butenedioic acid B239
(Z)-Butenedioic acid B239
α-butenoic acid B240
trans-2-Butenoic acid B240
2-butenoic acid,
    3-methyl-2-(1-methylpropyl)-4,
    6-dinitrophenyl ester B129
Δ^2-1-butenol B241
Δ^3-1-butenol B243
Δ^3-2-butenol B244
3-buteno-β-lactone A36
Butethal B246
Buthiobate B245
butilchlorofos B227
Butobarbitone B246
Butoben B281
Butocarboxim B247
2-butoxy-1-ethanol B250
1-butoxy-2,3-epoxypropane B275
1-Butoxy-2-propanol B254
butoxycarbonyl chloride B270
Butoxycarboxim B248
butoxydiglycol B251
2-Butoxyethanol B250
2-butoxyethanol acetate B278
2-Butoxyethanol phosphate B249
butoxyethene B297
2-(2-Butoxyethoxy)ethanol B251
2-(2-Butoxyethoxy)ethanol acetate B252
2-butoxyethyl acetate B278
butoxymethyloxirane B275
Butoxypropanol B253
Butralin B255
butryic alcohol B231
butyglycol B250

6-[(1-methyl-4-nitroimidazol-5-yl)thio]purine
A277
6-(1¹-methyl-4¹-nitro-5¹-imidazolyl)-
mercaptopurine A277
1-methyl-5*H*-pyrido[4,3-*b*]indol-3-amine
A143
N-methyl-*N*'-2,4-xylyl-*N*-(*N*-2,4-
xylylformimidoyl)formamidine A175
methyl-*N*-(2,6-dimethylphenyl)-*N*-
(phenylacetyl)- DL-alaninate B26
methyl-*N*-phenylacetyl-*N*-2,6-
xylyl-DL-alaninate B26
5-methyl-*o*-anisidine A142
6-(methyl-*p*-nitro-5-imidazolyl)-thiopurine
A277
p-methyl-*tert*-butylbenzene B295
3-methylacrylic acid B240
1-methylallyl alcohol B244
3-methylallyl alcohol B241
methylamine A141
O-[(methylamino)carbonyl]oxime A66
methylaminoethanolcatechol A56
4-[(methylazo)methyl]-*N*-(1-methylethyl)-
benzamide A288
(±)-α-methylbenzene ethanamine A210
3-methylbutanoic acid, 2-propenyl ester A95
3-methylbutanoic acid, allyl ester A95
3-methylbutanoic acid, butyl ester B284
3-methylbutyl bromide B200
methylcarbamic acid,
4-dimethylamino-*m*-tolyl ester A130
O-(methylcarbamoyl)oxime A66
methylcarbonitrile A27
methylene chlorobromide B192
1,1'-methylenebis(oxy)bis(2-chloroethane)
B138
1,1-methylenebisbenzene B131
1-methylethyl acetate A16
1-methylethyl ethanoate A16
3-(1-methylethyl)-(1*H*)-2,1,3-
benzothiadiazin-4(3*H*)-one-2,2-dioxide
B37
2-(1-methylethyl)-4-pentenamide A93
1-methylethylamine A163
methylethylbromomethane B187
2,2'-[(1-methylethylidene)bis(4,
1-phenyleneoxymethylene)]bisoxirane]
B149
2-(1-methylethylidene)hydrazine
carbothioamide A26
N,*N*'-[(methylimino)dimethylidyne]
di-2,4-xylidine A175
2-methyllactonitrile A25
methylmethylene glycol B222

8-methylparacymene B295
dl-α-methylphenethylamine A210
2-methylpropan-2-ol B233
1-methylpropenol B244
β-methylpropiolactone B301
2-(1-methylpropyl)-4,6-dinitrophenyl-
3-methyl-2-butenoate B129
2-(1-methylpropyl)phenol B292
1-methylpropylamine B260
3-(methylsulfonyl)-2-butanone-*O*-
[(methylamino)carbonyl]oxime B248
2-methylsulfonyl-*O*-(*N*-methyl-carbamoyl)-
butanone-3-oxime B248
3-(methylthio)-2-butanone-
O-[(methylamino)carbonyl]oxime B247
3-(methylthio)-*O*-(methylamino)
carbonyl)oxime-2 -butanone B247
3-(methylthio)butanone-*O*-methyl-
carbamoyloxime B247
1-methyltrimethylene diacrylate B273
Metramac A173
MFA (multifunctional acrylate) B132
MFM (multifunctional monomer) B132
Michler's ketone B147
mineral naphtha B45
mineral pitch A268
mineral soap B38
MIPA A164
Mitac A175
mono-*n*-butylamine B259
mono-*n*-propylamine A162
mono-*tert*-butylhydroquinone dimethyl ether
B272
mono-*tert*-butylhydroquinone B280
monoallylamine A80
monoammonium sulfamate A202
monobromoacetic acid B180
monofluoroacetic acid A15
monoisopropanolamine A164
monomethylamine A141
Monsanto CP-16226 B160
Moringine B109
MTBHQ B280
Musk xyldl B296
Musk xylene B296
mustard oil A94
Myeloleukon B225
Myleran B225

naphthanthracene B41
1,4-naphthionic acid A145
α-naphthofluorene B79
naphtho[1,2-*b*]thianaphthene B87
naphtho[2,1-*b*]thianaphthene B86

Parofor BSH B57
PCB A252
PCB 1242 A255
PCB 1254 A257
PCB 1260 A258
Penbritin A211
penicillin benzatin B44
Penicillin G B119
penicillin g benzathine B44
Penicline A211
Penitracin B1
Penoepon B44
pentachloroantimony A244
pentadecafluorooctanoic acid, ammonium
    salt A198
pentaerythritol dibromide B134
2,4-pentanedione A29
1-pentanethiol A219
1-pentanol A214
1-pentanol acetate A212
Pentaphen A223
Pentasol A214
pentyl acetate A212
1-pentyl acetate A212
tert-pentyl acetate A213
pentyl alcohol A214
pentyl butyrate A215
pentyl chloride A216
n-pentyl nitrate A220
n-pentyl propanoate A224
n-pentyl propionate A224
pentyl silicon trichloride A225
6-*m*-pentyl-*m*-cresol A217
N-pentyl-*N*-nitrosourea A221
pentylmercaptan A219
4-*tert*-pentylphenol A223
p-*tert*-pentylphenol A223
pentyltrichlorosilane A225
penymal A222
Perbutyl H B279
perchloric acid, barium salt B17
periethylenenaphthalene A1
permanganic acid, barium salt B18
peroxyacetic acid, *tert*-butyl ester B287
peroxybenzoic acid, *tert*-butyl ester B288
peroxyisobutyric acid, *tert*-butyl ester B289
peroxypivalic acid, *tert*-butyl ester B290
petroleum pitch A268
Phemeridesp B69
Phenedrine A210
phenisobromolate B207
phenol, (1,1-dimethylethyl)-4-methoxy-
    B264
phenol, 2-(1-methylpropyl) B292

phenol, 4-(1,1-dimethylethyl)- B293
phenol, *o-sec*-butyl B292
phenol, *tert*-butyl-4-methoxy- B264
3*H*-phenoxazine, actinomycin C deriv. A50
3*H*-phenoxazine, actinomycin D deriv. A51
phenyl bromide B183
phenyl cyanide B89
phenyl methyl ether A237
phenyl methyl ketone A28
phenyl sulfohydrazide B57
(4-phenyl)acetanilide A30
1-phenyl-2-aminopropane A210
N-phenylacetamide A11
6-phenylacetamidopenicillinic acid B119
p-phenylacetanilide A30
phenylacetic acid, allyl ester A97
phenylacetonitrile B116
phenylamine A228
4-(phenylamino)butane B262
2-phenylaniline A127
3-phenylaniline A128
m-phenylaniline A128
o-phenylaniline A127
p-phenylaniline A129
phenylarsenic acid B59
phenylarsonic acid B59
p-(phenylazo)aniline A120
phenylbenzene B130
1-phenylbutane B266
phenylcarbinol B108
phenylcarboxy amide B40
phenylchloroform B99
phenylene thiocyanate B163
1,4,-phenylenediisothiocyanic acid B163
1-phenylethanone A28
phenylfluoroform B100
phenylformic acid B82
(6*R*)-6-(α-*D*-phenylglycylamino)penicillanic
    acid A211
phenylhydride B45
(phenylisopropyl)amine A210
phenylmercaptan B67
phenylmethanol B108
phenylmethylamine B109
4-(phenylozo)phenol B95
phenylsulfonyl chloride B65
phenylsulfonyl hydrazide B57
1,1′,1″-phosphinylidynetrisaziridine A283
phosphoric acid triethylene imide A283
phosphoric acid, butyl ester B258
phosphorodithioic acid, *O,O*-diisopropylester,
    S ester with
    N-(2-mercaptoethyl)benzenesulfonamide
    B35

phosphoromidothioic acid, (1-methylpropyl)-,
    O-ethyl-O-(5-methyl-2-nitrophenyl) ester
    B215
phosphorothioic acid,
    S-[2-(diethylamino)ethyl]-, O,O-diethyl
    ester, oxalate A174
phosphorothioic acid, ester, O-(4-bromo-2,
    5-dichlorophenyl) O,O-diethyl ester B204
Phostoxin A110
phthalic acid, benzyl butyl ester B113
Phyralene A257
Phyralene A258
phytolaccanin B126
picramic acid A137
picrate of ammonia A200
Pillarsete B212
Piria's acid A145
Plastomoll BMB B64
platinic ammonium chloride A187
Pleocide A39
PMDA B66
Polybarit B21
polychlorinated biphenyl A257
polychlorinated biphenyl A258
polychlorinated biphenyl A252
Porofor 57 A287
potassium alum A111
potassium antimonyl tartrate A243
Prefar B35
primary amyl acetate A212
Primatol A A271
prop-2-enal A43
1-propanamine A162
2-propanamine A163
1-propanamine, 3-(10,11-dihydro-5H-dibenzo-
    (a,d)cyclohepten-5-ylidene,
    N,N-dimethyl- A176
propane, 1,1-[1,2-ethenediylbis
    (sulfonyl)]bis-(E)- B159
propane, 2,2'-oxybis(1-chloro-) B143
1-propanecarboxylic acid B228
propanenitrile, 2-hydroxy-2-methyl- A25
propanenitrile, -2,2'azobis(2 methyl) A287
propaneperoxoic acid, 2,2-dimethyl-,
    1,1-dimethylethyl ester B290
propaneperoxoic acid, 2-methyl-1,1-dimethyl
    ester B289
propanoic acid butylester B294
2-propanone A24
Propasol B B253
2-propen-1-amine A80
2-propen-1-ol A79
2-propenal A43
propenamide A46

2-propenamide A46
2-propenamine A80
propene acid A47
1-propene, 3-chloro- A85
2-propenenitrile A49
2-propenoic acid A47
2-propenoic acid, 1,4-butanediyl ester B274
2-propenoic acid, 1-methyl-1,3-propanediyl
    ester B273
2-propenoyl chloride A48
2-propenyl chloride A85
propenyl cinnamate A87
2-propenyl isothiocyanate A94
2-propenyl isovalerate A95
3-propenyl methanoate A90
2-propenyl phenylacetate A97
2-propenyl propyl disulfide A98
2-propenyl thiourea A99
2-propenylamine A80
p-propenylanisole A81
((2-propenyloxy)methyl)oxirane A91
propionic acid,
    2-(N-benzoyl-N-(3,4-dichlorophenyl)
    aminoethyl ester B104
n-propyl acetate A19
propyl bromide B205
propyl cyanide B303
n-propyl methanoate A19
2-propylamine A163
n-propylamine A162
sec-propylamine A163
propylcarbinol B231
n-propylcarbinylchloride B269
1,2-propylene glycol 1-monobutyl ether B254
propylene glycol butoxy ether B253
propylene glycol monobutyl ether B253
propylformic acid B228
propylmethanol B231
Proscorbin A267
Proxel P. L. B73
pseudobutyl benzene B267
pseudobutylene B237
pseudocumidine A231
PTBT B295
putrescine B219
Pynamin A74
2,4(1H3H)pyramidinedione,
    5-bis(2-chloroethylamino) B139
pyran aldehyde A44
2H-pyrancarboxaldehyde, 3,4-dihydro- A44
3H-pyrazol-3-one, 4-amino-1,2-dihydro-
    1,5-dimethyl-2-phenyl- A119
1H-pyrazolo[3,4-d]pyrimidin-4-ol A76
Pyresyn A74

INDEX OF CAS REGISTRY
NUMBERS

86-50-0 Azinphos-methyl A280
88-05-1 Aniline, 2,4,6-trimethyl- A232
88-21-1 2-Aminobenzenesulfonic
         acid A121
89-32-7 1,2,4,5-Benzenetetracarboxylic
         1,2:4,5-dianhydride B66
89-61-2 Benzene, 1,4-dichloro-
         2-nitro- B48
89-72-5 2-sec-Butylphenol B292
89-87-2 Benzene, 2,4-dimethyl-
         1-nitro- B53
90-04-0 2-Anisidine A234
90-41-5 2-Aminobiphenyl A127
90-94-8 4,4'-Bis(dimethylamino)
         benzophenone B147
91-76-9 Benzoguanamine B81
92-52-4 1,1'-Biphenyl B130
92-67-1 4-Aminobiphenyl A129
92-87-5 Benzidine B70
93-00-5 6-Aminonaphthalene-
         2-sulfonic acid A148
93-05-0 4-Amino-N,N-diethylaniline
         A133
93-58-3 Benzoic acid, methyl ester B84
93-71-0 Allidochlor A75
93-89-0 Benzoic acid, ethyl ester B83
94-09-7 Benzocaine B74
94-17-7 Bis(4-chlorobenzoyl)
         peroxide B137
94-26-8 Butyl 4-hydroxybenzoate B281
94-36-0 Benzoyl peroxide B103
95-14-7 1,2,3-Benzotriazole B98
95-15-8 Benzo[b]thiophene B97
95-16-9 Benzothiazole B96
95-55-6 2-Aminophenol A157
95-85-2 2-Amino-4-chlorophenol A131
96-48-0 γ-Butyrolactone B302
96-91-3 2-Amino-4,6-dinitrophenol A137
98-05-5 Benzenearsonic acid B59
98-06-6 tert-Butylbenzene B267
98-07-7 Benzotrichloride B99
98-08-8 Benzotrifluoride B100
98-09-9 Benzenesulfonyl chloride B65
98-46-4 Benzene, 1-nitro-3-
         trifluoromethyl- B55
98-51-1 4-tert-Butyltoluene B295
98-54-4 4-tert-Butylphenol B293
98-73-7 4-tert-Butylbenzoic acid B268
98-86-2 Acetophenone A28
98-87-3 Benzylidene chloride B117
98-88-4 Benzoyl chloride B102
99-05-8 3-Aminobenzoic acid A125
99-26-3 Bismuth subgallate B158
99-35-4 Benzene, 1,3,5-trinitro- B58

99-51-4 Benzene, 1,2-dimethyl-
         4-nitro- B51
99-57-0 2-Amino-4-nitrophenol A152
100-07-2 Anisoyl chloride A238
100-14-1 Benzene, 1-chloromethyl-
          4-nitro- B46
100-39-0 Benzyl bromide B111
100-44-7 Benzyl chloride B114
100-46-9 Benzylamine B109
100-47-0 Benzonitrile B89
100-51-6 Benzyl alcohol B108
100-52-7 Benzaldehyde B39
100-66-3 Anisole A237
100-73-2 Acrolein dimer A44
101-05-3 Anilazine A227
101-27-9 Barban B4
101-55-3 4-Bromodiphenyl ether B195
101-81-5 Biphenylmethane B131
101-90-6 1,3-Bis(2,3-epoxypropoxy)
          benzene B148
103-33-3 Azobenzene A284
103-84-4 Acetanilide A11
103-90-2 Acetaminophen A10
104-46-1 4-Allylanisole A81
104-51-8 Butylbenzene B266
104-78-9 3-Aminopropyldiethylamine A166
104-94-9 4-Anisidine A236
105-57-7 Acetal A4
106-31-0 Butyric anhydride B300
106-35-4 Butyl ethyl ketone B276
106-40-1 4-Bromoaniline B182
106-51-4 4-Benzoquinone B94
106-92-3 Allyl glycidyl ether A91
106-94-5 1-Bromopropane B205
106-95-6 Allyl bromide A82
106-97-8 Butane B217
106-99-0 1,3-Butadiene B213
107-02-8 Acrolein A43
107-04-0 1-Bromo-2-chloroethane B191
107-05-1 Allyl chloride A85
107-10-8 Aminopropane A162
107-11-9 Allylamine A80
107-13-1 Acrylonitrile A49
107-18-6 Allyl alcohol A79
107-29-9 Acetaldoxime A7
107-37-9 Allyltrichlorosilane A100
107-71-1 tert-Butyl peracetate B287
107-72-2 Amyl trichlorosilane A225
107-82-4 1-Bromo-3-methylbutane B200
107-88-0 1,3-Butanediol B222
107-89-1 Aldol A67
107-92-6 Butanoic acid B228
107-93-7 trans-2-Butenoic acid B240
108-21-4 Acetic acid, isopropyl ester A16

108-24-7   Acetic anhydride A21
108-46-3   1,3-Benzenediol B62
108-60-1   Bis(2-chloro-1-methylethyl)
           ether B143
108-86-1   Bromobenzene B183
108-98-5   Benzenethiol B67
109-13-7   *tert*-Butyl peroxyisobutyrate
           B289
109-19-3   Butyl isovalerate B284
109-21-7   Butanoic acid, butyl ester B230
109-55-7   3-Aminopropyldimethylamine
           A167
109-57-9   Allyl thiourea A99
109-60-4   Acetic acid, propyl ester A19
109-63-7   Boron trifluoride diethyl etherate
           B170
109-65-9   1-Bromobutane B186
109-69-3   Butyl chloride B269
109-70-6   1-Bromo-3-chloropropane B193
109-73-9   Butylamine B259
109-74-0   Butyronitrile B303
109-79-5   Butyl mercaptan B285
110-16-7   (Z)-Butenedioic acid B239
110-60-1   1,4-Butanediamine B219
110-63-4   1,4-Butanediol B223
110-65-6   1,4-Butynediol B298
110-66-7   Amyl mercaptan A219
110-69-0   Butyraldoxime B299
111-34-2   Butyl vinyl ether B297
111-36-4   Butyl isocyanate B283
111-44-4   Bis(2-chloroethyl) ether B140
111-69-3   Adiponitrile A54
111-76-2   2-Butoxyethanol B250
111-91-1   Bis(2-chloroethoxy)
           methane B138
112-07-2   Butylglycol acetate B278
112-34-5   2-(2-Butoxyethoxy)ethanol B251
115-02-6   Azaserine A276
116-06-3   Aldicarb A66
117-79-3   2-Aminoanthraquinone A118
118-92-3   2-Aminobenzoic acid A124
119-28-8   8-Aminonaphthalene-
           2-sulfonic acid A150
119-34-6   4-Amino-2-nitrophenol A154
119-53-9   Benzoin B85
119-61-9   Benzophenone B91
119-79-9   5-Aminonaphthalene-
           2-sulfonic acid A147
120-11-6   Benzyl isoeugenol B118
120-12-7   Anthracene A239
120-32-1   2-Benzyl-4-chlorophenol B115
120-40-1   *N,N*-Bis(2-hydroxyethyl)
           dodecanamide B153
120-51-4   Benzyl benzoate B110

120-71-8   1-Amino-2-methoxy-5-methyl-
           benzene A142
120-80-9   1,2-Benzenediol B61
121-47-1   3-Aminobenzenesulfonic
           acid A122
121-54-0   Benzethonium chloride B69
121-57-3   4-Aminobenzenesulfonic
           acid A123
121-66-4   2-Amino-5-nitrothiazole A155
121-88-0   2-Amino-5-nitrophenol A153
123-30-8   4-Aminophenol A159
123-31-9   1,4-Benzenediol B63
123-54-6   Acetyl acetone A29
123-72-8   Butanal B216
123-77-3   Azodicarbonamide A286
123-86-4   Acetic acid, butyl ester A13
124-04-9   Adipic acid A53
124-17-4   2-(2-Butoxyethoxy)ethanol
           acetate B252
126-06-7   3-Bromo-1-chloro-5,5-dimethyl-
           hydantoin B190
126-22-7   Butanoate B227
127-90-2   Bis(2,3,3,3-tetrachloropropyl)
           ether B160
128-37-0   Butylated hydroxytoluene B265
130-17-6   2-(4-Aminophenyl)-6-methyl-
           7-benzothiazolylsulfonic acid
           A160
131-11-3   1,2-Benzenedicarboxylic acid,
           dimethyl ester B60
131-54-4   Bis(2-hydroxy-4-methoxyphenyl)
           methanone B154
131-55-5   Bis(2,4-dihydroxyphenyl)
           methanone B146
131-73-7   Bis(2,4,6-trinitrophenyl)amine
           B161
131-74-8   Ammonium picrate A200
132-32-1   3-Amino-9-ethylcarbazole A139
137-17-7   Aniline, 2,4,5-trimethyl- A231
140-11-4   Benzyl acetate B107
140-28-3   Benzathine B43
140-29-4   Benzyl cyanide B116
140-31-8   *N*-Aminoethylpiperazine A140
140-40-9   Acinitrazole A39
140-57-8   Aramite A251
141-75-3   Butyryl chloride B304
141-78-6   Acetic acid, ethyl ester A14
142-04-1   Aniline hydrochloride A229
144-49-0   Acetic acid, fluoro- A15
150-05-0   Adrenaline-D A55
150-13-0   4-Aminobenzoic acid A126
151-56-4   Aziridine A282
154-93-8   1,3-Bis(chloroethyl)-1-
           nitrosourea B141

156-08-1	Benzphetamine B105	513-85-9	2,3-Butanediol B224
191-24-2	Benzo[*ghi*]perylene B90	536-90-3	3-Anisidine A235
203-12-3	Benzo[*ghi*]fluoranthene B76	540-18-1	Amyl butyrate A215
205-43-6	Benzo[*b*]naphtho	540-88-5	*tert*-Butyl acetate B256
	[1,2-*d*]thiophene B86	542-62-1	Barium cyanide B12
205-82-3	Benzo[*j*]fluoranthene B77	543-59-9	Amyl chloride A216
205-99-2	Benzo[*b*]fluoranthene B75	543-80-6	Barium acetate B6
207-08-9	Benzo[*k*]fluoranthene B78	545-55-1	Aziridine, 1,1α,1α-phos-
208-96-8	Acenaphthylene A2		phinylidene, tris- A283
239-35-0	Benzo[*b*]naphtho	552-30-7	Benzene-1,2,4-tricarboxylic
	[2,1-*d*]thiophene B87		acid, 1,2-anhydride B68
243-46-9	Benzo[*b*]naphtho	555-31-7	Aluminium triisopropoxide A114
	[2,3-*d*]thiophene B88	556-56-9	Allyl iodide A92
260-94-6	Acridine A42	557-31-3	Allyl ethyl ether A89
271-89-6	Benzofuran B80	563-68-8	Acetic acid, thallium(I) salt A20
281-23-2	Adamantane A52	583-63-1	2-Benzoquinone B93
298-07-7	Bis(2-ethylhexyl) phosphate	584-03-2	1,2-Butanediol B221
	B151	584-79-2	Allethrin A74
299-78-5	Allyl isopropylacetamide A93	590-01-2	Butyl propionate B294
300-62-9	Amphetamine A210	590-17-0	Bromoacetonitrile B181
301-04-2	Acetic acid, lead salt A17	590-18-1	*cis*-2-Butene B237
302-27-2	Aconitine A41	590-88-5	1,3-Butanediamine B218
309-00-2	Aldrin A69	591-08-2	Acetamide, *N*-(aminothio-
314-40-9	Bromacil B173		oxomethyl) A9
315-30-0	Allopurinol A76	591-27-5	3-Aminophenol A158
317-34-0	Aminophylline A161	591-60-6	Butyl acetoacetate B257
320-67-2	5-Azacytidine A274	591-87-7	Allyl acetate A78
353-42-4	Boron trifluoride dimethyl	592-34-7	*n*-Butyl chloroformate B270
	etherate B171	598-32-3	3-Buten-2-ol B244
384-22-5	Benzene, 1-nitro-	611-06-3	Benzene, 2,4-dichloro-
	2-trifluoromethyl- B54		1-nitro- B49
402-54-0	Benzene, 1-nitro-	613-13-8	2-Aminoanthracene A117
	4-trifluoromethyl- B56	614-45-9	*tert*-Butyl peroxybenzoate B288
431-03-8	2,3-Butadione B214	618-62-2	Benzene, 1,3-dichloro-
446-86-6	Azathioprine A277		5-nitro- B47
462-08-8	3-Aminopyridine A170	624-54-4	Amyl propionate A224
481-72-1	Aloe Emodin A101	624-64-6	*trans*-2-Butene B238
485-31-4	Binapacryl B129	625-16-1	*tert*-Amyl acetate A213
492-80-8	Auramine (Base) A273	626-43-7	Aniline, 3,5-dichloro- A230
494-44-0	7-Aminonaphthalene-	627-27-0	3-Buten-1-ol B243
	2-sulfonic acid A149	628-63-7	Amyl acetate A212
495-48-7	Azoxybenzene A290	631-61-8	Ammonium acetate A179
495-73-8	Benquinox B34	674-82-8	Acetyl ketene A36
503-74-2	Butanoic acid, 3-methyl- B229	712-68-5	2-Amino-5-(5-nitro-2-furyl)-
504-24-5	4-Aminopyridine A171		1,3,4-thiadiazole A151
504-29-0	2-Aminopyridine A169	741-58-2	Bensulide B35
504-61-0	*trans*-2-Buten-1-ol B242	814-68-6	Acrylic chloride A48
506-32-1	Arachidonic acid A250	834-12-8	Ametryn A115
506-87-6	Ammonium carbonate A185	919-76-9	Amidithion A116
506-96-7	Acetyl bromide A32	927-07-1	*tert*-Butyl peroxypivalate B290
507-02-8	Acetyl iodide A35	928-45-0	Butyl nitrate B286
507-70-0	Borneol B165	929-06-6	2-(2-Aminoethoxy)ethanol A138
513-35-9	Amylene A218	968-81-0	Acetohexamide A23
513-77-9	Barium carbonate B9	1002-16-0	Amyl nitrate A220

7429-90-5	Aluminium A102	10294-33-4	Boron tribromide B167
7440-36-0	Antimony A242	10294-34-5	Boron trichloride B168
7440-38-2	Arsenic A259	10361-37-2	Barium chloride B11
7440-39-3	Barium B5	10553-31-8	Barium bromide B8
7440-41-7	Beryllium B121	10589-74-9	1-Amyl-1-nitrosourea A221
7440-68-8	Astatine A269	11096-82-5	Aroclor 1260 A258
7446-70-0	Aluminium chloride A105	11097-69-1	Aroclor 1254 A257
7493-74-5	Allyl phenoxyacetate A96	11104-28-2	Aroclor 1221 A253
7637-07-2	Boron trifluoride B169	11141-16-5	Aroclor 1232 A254
7647-18-9	Antimony pentachloride A244	12044-79-0	Arsenic disulfide A261
7659-95-2	Betanin B126	12125-01-8	Ammonium fluoride A190
7664-41-7	Ammonia A178	12125-02-9	Ammonium chloride A186
7665-72-7	tert-Butyl glycidyl ether B277	12135-76-1	Ammonium sulfide A204
7726-95-6	Bromine B176	12230-99-8	Barium disulfide B13
7727-15-3	Aluminium bromide A103	12231-01-5	Barium trisulfide B24
7727-43-7	Barium sulfate B22	12248-67-8	Barium tetrasulfide B23
7727-54-0	Ammonium persulfate A199	12448-68-9	Barium pentasulfide B16
7773-06-0	Ammonium sulfamate A202	12672-29-6	Aroclor 1248 A256
7778-39-4	Arsenic acid A260	12674-11-2	Aroclor 1016 A252
7783-18-8	Ammonium thiosulfate A208	12771-68-5	Ancymidol A226
7783-20-2	Ammonium sulfate A203	12788-93-1	Butyl acid phosphate B258
7783-56-4	Antimony trifluoride A247	13122-18-4	tert-Butylperoxy-
7783-70-2	Antimony pentafluoride A245		3,5,5-trimethylhexanoate B291
7784-21-6	Aluminium hydride A106	13181-17-4	Bromofenoxim B197
7784-27-2	Aluminium nitrate nonahydrate	13465-95-7	Barium perchlorate B17
	A108	13477-00-4	Barium chlorate B10
7784-34-1	Arsenic trichloride A262	13510-49-1	Beryllium sulfate B125
7784-42-1	Arsine A265	13597-99-4	Beryllium nitrate B124
7784-44-3	Ammonium arsenate A180	13826-83-0	Ammonium fluoroborate A191
7787-36-2	Barium permanganate B18	13863-41-7	Bromine chloride B177
7787-47-5	Beryllium chloride B122	13952-84-6	sec-Butylamine B260
7787-49-7	Beryllium fluoride B123	13967-90-3	Barium bromate B7
7787-71-5	Bromine trifluoride B179	15310-01-7	Benodanil B30
7788-98-9	Ammonium chromate A188	15625-89-5	2,2'-Bis(acryloyloxymethyl)
7789-09-5	Ammonium dichromate A189		butyl acrylate B132
7789-30-2	Bromine pentafluoride B178	15972-60-8	Alachlor A65
7803-55-6	Ammonium metavanadate A195	16568-02-8	Acetaldehyde formylmethyl-
8001-54-5	N-Alkyl(C8-C18)dimethyl-		hydrazone A6
	benzylammonium chloride A71	16853-85-3	Aluminium lithium hydride A107
8011-63-0	Bordeaux mixture B164	16919-19-0	Ammonium fluorosilicate A192
8023-53-8	N-Alkyl(C8-C18)dimethyl-	16919-58-7	Ammonium chloroplatinate A187
	3,4-dichlorobenzylammonium	17606-31-4	Bensultap B36
	chloride A72	17804-35-2	Benomyl B31
8052-16-2	Actinomycin C A50	18181-80-1	Bromopropylate B207
8052-42-4	Asphalt, fumes A268	19485-03-1	1,3-Butylene glycol
9002-18-0	Agar A64		diacrylate B273
9080-17-5	Ammonium polysulfide A201	20048-27-5	Bandrowski's base B2
10022-31-8	Barium nitrate B14	20679-58-7	1,4-Bis(bromoacetoxy)-
10025-91-9	Antimony trichloride A246		2-butene B133
10043-01-3	Aluminium sulfate A113	20859-73-8	Aluminium phosphide A110
10043-67-1	Aluminium potassium sulfate	21112-37-8	2-tert-Butyl-1,4-dimeth
	A111		oxybenzene B272
10192-30-0	Ammonium bisulfite A183	22212-55-1	Benzoylprop-ethyl B104
10196-04-0	Ammonium sulfite A205	23184-66-9	Butachlor B212

23214-92-8	Adriamycin A57
25013-16-5	Butylated hydroxyanisole B264
25057-89-0	Bentazone B37
25167-67-3	1-Butene B236
26787-78-0	Amoxycillin A209
28300-74-5	Antimonyl potassium tartrate hemihydrate A243
28772-56-7	Bromadiolone B174
29104-30-1	Benzoximate B101
29387-86-8	Butoxypropanol B253
29611-03-8	Aflatoxicol A58
30560-19-1	Acephate A3
30777-18-5	Benzo[a]fluorene B79
33089-61-1	Amitraz A175
33629-47-9	Butralin B255
34256-82-1	Acetochlor A22
34681-10-2	Butocarboxim B247
34681-23-7	Butoxycarboxim B248
35575-96-3	Azamethiphos A275
35691-65-7	2-Bromo-2-(bromomethyl) pentanedinitrile B185
36335-67-8	Butamifos B215
41083-11-8	Azocyclotin A285
41483-43-6	Bupirimate B210
42576-02-3	Bifenox B127
50594-66-6	Acifluorfen A38
50864-67-0	Barium polysulfides (Ba$_4$S$_7$) B21
51234-28-7	Benoxaprofen B33
51249-05-9	Buminafos B209
51308-54-4	Buthiobate B245
53111-28-7	Barium polysulfides (Ba$_2$S$_3$) B20
53469-21-9	Aroclor 1242 A255
55179-31-2	Bitertanol B162
55635-13-7	Alloxydim-sodium A77
56073-10-0	Brodifacoum B172
56894-91-8	Bis-1,4-(chloromethoxy)-p-xylene B142
57917-55-2	Beetroot Red B25
61789-65-9	Aluminium resinate A112
61791-55-7	N-Alkyl-1,3-propanediamine A73
62450-06-0	3-Amino-1,4-dimethyl-5H-pyrido[4,3-b]indole A136
62450-07-1	3-Amino-1-methyl-5H-pyrido[4,3-b]indole A143
63333-35-7	Bromethalin B175
63449-41-2	Alkyl(C$_{14}$-C$_{16}$)dimethyl-benzylammonium chloride A70
64249-01-0	Anilofos A233
68603-42-9	N,N-bis(hydroxyethyl)-coco-amides B152
69327-76-0	Buprofezin B211
71626-11-4	Benalaxyl B26
74070-46-5	Aclonifen A40
74712-19-9	Bromobutide B188
75485-12-0	1-Azetidinecarbonyl chloride A278
77094-11-2	2-Amino-3,4-dimethylimidazo[4,5-f]quinoline A134
77500-04-0	2-Amino-3,8-dimethyl-imidazo[4,5-f]quinoxaline A135
82560-54-1	Benfuracarb B29
82657-04-3	Bifenthrin B128
83463-62-1	Bromochloroacetonitrile B189

INDEX OF MOLECULAR FORMULAE

C₂H₅Br	Bromoethane B196	C₄H₇ClO	Butyryl chloride B304
C₂H₅N	Aziridine A282	C₄H₇N	Butyronitrile B303
C₂H₅NO	Acetaldoxime A7	C₄H₇NO	Acetone cyanohydrin A25
C₂H₅NO	Acetamide A8	C₄H₈	trans-2-Butene B238
C₂H₅NO₄	Ammonium oxalate A197	C₄H₈	1-Butene B236
C₂H₆BF₃O	Boron trifluoride dimethyl etherate B171	C₄H₈	cis-2-Butene B237
		C₄H₈Cl₂O	Bis(2-chloroethyl) ether B140
C₂H₇NO₂	Ammonium acetate A179	C₄H₈N₂O	Acetaldehyde formylmethylhydrazone A6
C₂HBrClN	Bromochloroacetonitrile B189		
		C₄H₈N₂S	Allyl thiourea A99
C₃H₃ClO	Acrylic chloride A48	C₄H₈O	Butanal B216
C₃H₃N	Acrylonitrile A49	C₄H₈O	3-Buten-1-ol B243
C₃H₄O	Acrolein A43	C₄H₈O	2-Buten-1-ol B241
C₃H₄O₂	Acrylic acid A47	C₄H₈O	2-Butanone B234
C₃H₅Br	Allyl bromide A82	C₄H₈O	3-Buten-2-ol B244
C₃H₅Cl	Allyl chloride A85	C₄H₈O	trans-2-Buten-1-ol B242
C₃H₅Cl₃Si	Allyltrichlorosilane A100	C₄H₈O₂	Aldol A67
C₃H₅I	Allyl iodide A92	C₄H₈O₂	Acetic acid, ethyl ester A14
C₃H₅NO	Acrylamide A46	C₄H₈O₂	Butanoic acid B228
C₃H₆BrCl	1-Bromo-3-chloropropane B193	C₄H₈O₃S	1,4-Butanesultone B226
		C₄H₉Br	1-Bromobutane B186
C₃H₆BrNO₄	2-Bromo-2-nitro-1,3-propanediol B201	C₄H₉Br	2-Bromobutane B187
		C₄H₉Cl	Butyl chloride B269
C₃H₆N₂OS	Acetamide, N-(aminothiooxomethyl) A9	C₄H₉N₃S	Acetone thiosemicarbazide A26
		C₄H₉NO	Butyraldoxime B299
C₃H₆O	Acetone A24	C₄H₉NO₃	Butyl nitrate B286
C₃H₆O	Allyl alcohol A79	C₄H₁₀	Butane B217
C₃H₆O₂	Acetic acid, methyl ester A18	C₄H₁₀BF₃O	Boron trifluoride diethyl etherate B170
C₃H₇Br	1-Bromopropane B205		
C₃H₇Br	2-Bromopropane B206	C₄H₁₀NO₃PS	Acephate A3
C₃H₇N	Allylamine A80	C₄H₁₀O	1-Butanol B231
C₃H₉N	Aminopropane A162	C₄H₁₀O	tert-Butanol B233
C₃H₉N	2-Aminopropane A163	C₄H₁₀O₂	1,2-Butanediol B221
C₃H₉NO	1-Aminopropan-2-ol A164	C₄H₁₀O₂	1,3-Butanediol B222
C₄H₄N₃O₂S	2-Amino-5-nitrothiazole A155	C₄H₁₀O₂	1,4-Butanediol B223
		C₄H₁₀O₂	2,3-Butanediol B224
		C₄H₁₀O₂	2-Butanol B232
C₄H₄O₂	Acetyl ketene A36	C₄H₁₀O₂	tert-Butyl hydroperoxide B279
C₄H₄O₄	(Z)-Butenedioic acid B239		
C₄H₅ClO₂	Allyl chloroformate A86	C₄H₁₀S	Butyl mercaptan B285
C₄H₅NS	Allyl isothiocyanate A94	C₄H₁₁N	Butylamine B259
C₄H₆	1,3-Butadiene B213	C₄H₁₁N	sec-Butylamine B260
C₄H₆BaO₄	Barium acetate B6	C₄H₁₁N	tert-Butylamine B261
C₄H₆ClNO	1-Azetidinecarbonyl chloride A278	C₄H₁₁NO₂	2-(2-Aminoethoxy)ethanol A138
C₄H₆O₂	Allyl formate A90	C₄H₁₁PO₄	Butyl acid phosphate B258
C₄H₆O₂	2,3-Butadione B214	C₄H₁₂N₂	1,3-Butanediamine B218
C₄H₆O₂	trans-2-Butenoic acid B240	C₄H₁₂N₂	1,4-Butanediamine B219
C₄H₆O₂	1,4-Butynediol B298	C₄H₁₂N₂O₆	Ammonium tartrate A206
C₄H₆O₂	β-Butyrolactone B301	C₅H₄N₄O	Allopurinol A76
C₄H₆O₂	γ-Butyrolactone B302	C₅H₅N₃O₃S	Acinitrazole A39
C₄H₆O₂S₄	Bis(methoxythiocarbonyl) disulfide B156	C₅H₆N₂	2-Aminopyridine A169
C₄H₆O₃	Acetic anhydride A21		
C₄H₆O₄Pb	Acetic acid, lead salt A17		

C_7H_5N Benzonitrile B89
$C_7H_5N_1S$ Benzothiazole B96
$C_7H_5NO_3S$ 1,2-Benzisothiazolin-3(2*H*)-one 1,1-dioxide B72
$C_7H_5NO_5$ 1,2-Benzisothiazolin-3-one B73
$C_7H_6ClNO_2$ Benzene, 1-chloromethyl-4-nitro- B46
$C_7H_6Cl_2$ Benzylidene chloride B117
$C_7H_6N_2$ Benzimidazole B71
C_7H_6O Benzaldehyde B39
$C_7H_6O_2$ 1-Butoxy-2-propanol B254
$C_7H_6O_2$ Benzoic acid B82
C_7H_7Br Benzyl bromide B111
C_7H_7Cl Benzyl chloride B114
C_7H_7NO Benzamide B40
$C_7H_7NO_2$ 2-Aminobenzoic acid A124
$C_7H_7NO_2$ 3-Aminobenzoic acid A125
$C_7H_7NO_2$ 4-Aminobenzoic acid A126
$C_7H_8BrClO_2$ 3-Bromo-1-chloro-5,5-dimethylhydantoin B190
C_7H_8O Anisole A237
C_7H_8O Benzyl alcohol B108
C_7H_9N Benzylamine B109
C_7H_9NO 2-Anisidine A234
C_7H_9NO 3-Anisidine A235
C_7H_9NO 4-Anisidine A236
$C_7H_9NO_2$ Ammonium benzoate A181
$C_7H_{11}N_7S$ Aziprotryne A281
$C_7H_{12}N_2$ *N-n*-Butylimidazole B282
$C_7H_{12}O_2$ Allyl butyrate A83
$C_7H_{14}N_2O_2S$ Aldicarb A66
$C_7H_{14}N_2O_2S$ Butocarboxim B247
$C_7H_{14}N_2O_4S$ Butoxycarboxim B248
$C_7H_{14}N_2O_4S$ Aldoxycarb A68
$C_7H_{14}O$ Butyl ethyl ketone B276
$C_7H_{14}O_2$ Amyl acetate A212
$C_7H_{14}O_2$ *tert*-Amyl acetate A213
$C_7H_{14}O_2$ Butyl 2,3-epoxypropyl ether B275
$C_7H_{14}O_2$ *tert*-Butyl glycidyl ether B277
$C_7H_{14}O_2$ Butyl propionate B294
$C_7H_{16}NO_4PS_2$ Amidithion A116
$C_7H_{16}O_2$ Butoxypropanol B253
$C_7H_{18}N_2$ 3-Aminopropyldiethylamine A166
$C_8H_4F_{15}O_2N$ Ammonium perfluorooctanoate A198
$C_8H_4N_2S_2$ Bitoscanate B163
C_8H_6BrN α-Bromobenzylcyanide B184
$C_8H_6BrNO_2$ β-Bromo-β-nitrostyrene B202
C_8H_6O Benzofuran B80
C_8H_6S Benzo[*b*]thiophene B97
$C_8H_7ClO_2$ Anisoyl chloride A238
C_8H_7N Benzyl cyanide B116
$C_8H_8BrCl_2O_3PS$ Bromophos B203
C_8H_8O Acetophenone A28
$C_8H_8O_2$ Benzoic acid, methyl ester B84
C_8H_9NO Acetanilide A11
$C_8H_9NO_2$ Acetaminophen A10
$C_8H_9NO_2$ Benzene, 1,2-dimethyl-3-nitro- B50
$C_8H_9NO_2$ Benzene, 1,2-dimethyl-4-nitro- B51
$C_8H_9NO_2$ Benzene, 1,3-dimethyl-2-nitro- B52
$C_8H_9NO_2$ Benzene, 2,4-dimethyl-1-nitro- B53
$C_8H_{10}Br_2O_4$ 1,4-Bis(bromoacetoxy)-2-butene B133
$C_8H_{10}K_2O_{15}Sb_2$ Antimonyl potassium tartrate hemihydrate A243
$C_8H_{10}N_2O_4S$ Asulam A270
$C_8H_{11}Cl_2N_3O_2$ 5-Bis(2-chloroethyl) aminouracil B139
$C_8H_{11}NO$ 1-Amino-2-methoxy-5-methylbenzene A142
$C_8H_{12}ClNO$ Allidochlor A75
$C_8H_{12}N_4$ 2,2'-Azobis(2-methylpropionitrile) A287
$C_8H_{12}N_4O_5$ 5-Azacytidine A274
$C_8H_{14}ClN_5$ Atrazine A271
$C_8H_{14}Cl_3O_5P$ Butanoate B227
$C_8H_{14}O_2$ Allyl isovalerate A95
$C_8H_{14}O_3$ Butyl acetoacetate B257
$C_8H_{14}O_3$ Butyric anhydride B300
$C_8H_{15}NO$ Allyl isopropylacetamide A93
$C_8H_{16}O_2$ Butanoic acid, butyl ester B230
$C_8H_{16}O_2$ Amyl propionate A224
$C_8H_{16}O_3$ Butylglycol acetate B278
$C_8H_{16}O_3$ *tert*-Butyl peroxyisobutyrate B289
$C_8H_{16}O_4S_2$ *trans*-1,2-bis(propylsulfonyl)ethylene B159
$C_8H_{16}O_8$ 2-Butanone peroxide B235
$C_8H_{18}CrO_4$ *tert*-Butyl chromate B271
$C_8H_{18}O_3$ 2-(2-Butoxyethoxy)ethanol B251
$C_9H_4O_5$ Benzene-1,2,4-tricarboxylic acid, 1,2-anhydride B68
$C_9H_5Cl_3N_4$ Anilazine A227
$C_9H_6ClNO_3S$ Benazolin B27
$C_9H_7N_7O_2S$ Azathioprine A277

$C_{12}H_{11}N$	2-Aminobiphenyl A127	$C_{14}H_9NO_2$	2-Aminoanthraquinone A118
$C_{12}H_{11}N$	3-Aminobiphenyl A128	$C_{14}H_{10}$	Anthracene A239
$C_{12}H_{11}N$	4-Aminobiphenyl A129	$C_{14}H_{10}Cl_2O_2$	Bis(4-chlorophenyl)acetic
$C_{12}H_{11}N_3$	3-Amino-1-methyl-5H-pyrido		acid B144
	[4,3-b]indole A143	$C_{14}H_{10}O_4$	Benzoyl peroxide B103
$C_{12}H_{11}N_3$	4-Aminoazobenzene A120	$C_{14}H_{11}N$	2-Aminoanthracene A117
$C_{12}H_{11}N_5$	2-Amino-3,8-dimethylimidazo	$C_{14}H_{12}Cl_2O$	2,2-Bis(4-chlorophenyl)
	[4,5-f]quinoxaline A135		ethanol B145
$C_{12}H_{12}N_2$	Benzidine B70	$C_{14}H_{12}N_2O_3S_2$	2-(4-Aminophenyl)-
$C_{12}H_{12}N_4$	2-Amino-3,4-dimethylimidazo		6-methyl-7-benzothazolyl-
	[4,5-f]quinoline A134		sulfonic acid A160
$C_{12}H_{12}O_2$	Allyl cinnamate A87	$C_{14}H_{12}O_2$	Benzoin B85
$C_{12}H_{14}O_4$	1,3-Bis(2,3-epoxypropoxy)	$C_{14}H_{12}O_2$	Benzyl benzoate B110
	benzene B148	$C_{14}H_{13}NO$	4-Acetyl aminobiphenyl A30
$C_{12}H_{15}N_3O_6$	5-tert-Butyl-2,4,6-trinitro-m-	$C_{14}H_{14}ClN_2O_3PS$	Azothoate A289
	xylene B296	$C_{14}H_{14}N_2$	3-Amino-9-ethylcarbazole
$C_{12}H_{15}NO_2$	N,N-Bis(2,3-epoxypropyl)		A139
	aniline B150	$C_{14}H_{18}N_4O_3$	Benomyl B31
$C_{12}H_{16}N_3O_3PS_2$	Azinphos-ethyl A279	$C_{14}H_{20}ClNO_2$	Acetochlor A22
$C_{12}H_{17}N_3O$	Azoprocarbazine A288	$C_{14}H_{20}ClNO_2$	Alachlor A65
$C_{12}H_{18}O$	Amyl-3-cresol A217	$C_{14}H_{21}N_3O_4$	Butralin B255
$C_{12}H_{18}O_2$	2-tert-Butyl-1,4-	$C_{14}H_{24}NO_4PS_3$	Bensulide B35
	dimethoxy-benzene B272	$C_{15}H_8O_4$	Anthraflavic acid A240
$C_{12}H_{20}O_2$	Allyl cyclohexylpropionate	$C_{15}H_{10}O_5$	Aloe Emodin A101
	A88	$C_{15}H_{13}NO$	2-Acetylaminofluorene A31
$C_{12}H_{28}NO_7PS$	Amiton oxalate A174	$C_{15}H_{14}O_5$	Bis(2-hydroxy-4-methoxy-
$C_{13}H_7Br_2N_3O_6$	Bromofenoxim B197		phenyl)methanone B154
$C_{13}H_9N$	Acridine A42	$C_{15}H_{16}N_2O_2$	Ancymidol A226
$C_{13}H_{10}INO$	Benodanil B30	$C_{15}H_{18}N_2O_6$	Binapacryl B129
$C_{13}H_{10}O$	Benzophenone B91	$C_{15}H_{20}N_2O_4S$	Acetohexamide A23
$C_{13}H_{11}ClO$	2-Benzyl-4-chlorophenol	$C_{15}H_{20}O_6$	2,2'-Bis(acryloyloxymethyl)
	B115		butyl acrylate B132
$C_{13}H_{11}N_3O_2$	Benquinox B34	$C_{15}H_{22}BrNO$	Bromobutide B188
$C_{13}H_{12}$	Biphenylmethane B131	$C_{15}H_{23}ClO_4S$	Aramite A251
$C_{13}H_{13}N_3$	3-Amino-1,4-dimethyl-	$C_{15}H_{24}O$	Butylated hydroxytoluene
	5H-pyrido(4,3-b)indole A136		B265
$C_{13}H_{16}F_3N_3O_4$	Benfluralin B28	$C_{16}H_{10}S$	Benzo[b]naphtho
$C_{13}H_{17}N_3O$	4-Aminophenazone A156		[1,2-d]thiophene B86
$C_{13}H_{19}ClNO_3PS_2$	Anilofos A233	$C_{16}H_{10}S$	Benzo[b]naphtho
$C_{13}H_{19}O_5$	Bis(2,4-dihydroxyphenyl)		[2,1-d]thiophene B87
	methanone B146	$C_{16}H_{10}S$	Benzo[b]naphtho
$C_{13}H_{21}N_2O_4PS$	Butamifos B215		[2,3-d]thiophene B88
$C_{13}H_{24}N_4O_3S$	Bupirimate B210	$C_{16}H_{12}ClNO_3$	Benoxaprofen B33
$C_{13}H_{26}O_3$	tert-Butylperoxy-	$C_{16}H_{18}N_2O_4S$	Benzylpenicillin B119
	3,5,5-trimethylhexanoate	$C_{16}H_{19}N_3O_4S$	Ampicillin A211
	B291	$C_{16}H_{19}N_3O_5S$	Amoxycillin A209
$C_{14}H_7Br_2NO_2$	1-Amino-2,4-	$C_{16}H_{20}N_2$	Benzathine B43
	dibromoanthraquinone A132	$C_{16}H_{23}N_3OS$	Buprofezin B211
$C_{14}H_7Br_3F_3N_3O_4$	Bromethalin B175	$C_{16}H_{33}NO_3$	N,N-Bis(2-hydroxyethyl)
$C_{14}H_7ClF_3NO_5$	Acifluorfen A38		dodecanamide B153
$C_{14}H_8Cl_2O_4$	Bis(4-chlorobenzoyl)	$C_{16}H_{35}O_4P$	Bis(2-ethylhexyl) phosphate
	peroxide B137		B151
$C_{14}H_8O_2$	Anthraquinone A241	$C_{17}H_{12}$	Benzo[a]fluorene B79
$C_{14}H_9Cl_2NO_5$	Bifenox B127	$C_{17}H_{12}O_6$	Aflatoxin B1 A59